PENNAK'S FRESHWATER INVERTEBRATES OF THE UNITED STATES

Fourth Edition

PORIFERA TO CRUSTACEA

Douglas Grant Smith

John Wiley & Sons, Inc.

New York • Chichester • Weinheim • Brisbane • Singapore • Toronto

To the memory of

C. Juday

4th edition to the memory of my friends and mentors

David J. Klingener

Herbert E. Pottswald

This book is printed on acid-free paper. ♾

Published simultaneously in Canada.

Library of Congress Cataloging-in-Publication Data:

Smith, Douglas G. (Douglas Grant)
Pennak's freshwater invertebrates of the United States : porifera to crustacea / Douglas G. Smith.— 4th ed.
p. cm.
Includes bibligraphical references.
ISBN 0-471-35837-1 (cloth : alk. paper)
1. Freshwater invertebrates—United States. I. Title.

QL365.176'0973—cd21

00-043928

Printed in the United States of America.

10 9 8 7 6 5 4 3 2 1

PREFACE

This edition will mark the first time in almost fifty years that Robert Pennak has not written, edited, or revised his *Fresh-water Invertebrates of the United States.* Since the original 1953 volume and two subsequent editions, Pennak's treatise has provided instructors, students, and researchers alike an invaluable source of information on an enormous breadth of aquatic invertebrate diversity and biology.

With some humility I begin the preparation of the 4th edition of this book. My first course of action is to reconsider those groups of organisms to be included in the new edition. By the third edition, Dr. Pennak had realized that continuing to include insects in his manual had become an overwhelming and impractical task. The emergence of a separate and comprehensive literature on aquatic insects certainly alleviated the need to include the aquatic insects in his last edition. Advancements in the knowledge of many of the noninsect groups included in the third edition has once again required a serious review of the range of taxa a single volume can cover. Thus, my deliberation has led to the decision to drop from the book the Protoctista or Protista. Referred to in previous editions and elsewhere as the Protozoa, the immensely diverse and unique Kingdom Protoctista comprises some 30 currently recognized phyla of uni- and pluricellular eukaryotes lacking a tissue level of construction and a blastula stage during early development. All the remaining invertebrate animal phyla covered in the third edition have been retained in this edition.

All of the chapters have been retained in their original style. Beyond updating the chapters, certain sections and many of the keys have been completely revised. The sections on preparation of material in several chapters as well as many keys and illustrations have been modified after my 1995 manual (DGS) *Keys to the Freshwater Macroinvertebrates of Massachusetts*, second edition, Sunderland, Massachusetts.

I thank several people, many experts in their discipline, for providing various levels of review, revision, or support of either the previous edition or the current effort. Some of these people made notable contributions to a particular chapter and are cited here as well in the text. They are Jacek J. Kisielewski (Gastrotrichia), Giovanni Pilato (Tardigrada), Reyes Peña Santiago (Nematoda), and Eileen H. Jokinen (Gastropoda). To these people I extend my sincerest thanks. Others reviewed chapters, from the previous or the revised edition, bringing to my attention necessary changes and corrections. These people are Richard D. Campbell (Cnidaria), Mark J. Wetzel (Oligochaeta), Malcom F. Vidrine (Hydrachnida), Robert Hershler (Gastropoda), Kirsten Work (cladoceran Branchiopoda), and Koen Martens (Ostracoda). My thanks to you all.

Lastly, I would like to thank many friends and colleagues for their support and assistance. Dave Strayer provided me the opportunity to carry out this effort. Sally Klingener prepared the index. Ed and Libby Klekowski, Sean Werle, and Andrew Wier have graciously provided assistance and material since the beginning of the project.

In the end, responsibility for the final product rests with me and any or all errors are from my hand.

Douglas G. Smith

Amherst, Massachusetts
June 2001

PREFACE TO THE THIRD EDITION

Although only about eleven years have elapsed since the appearance of the second edition of this book, it seems to me that it is now time for this third edition, especially in view of the rapid journal publication rate of new material. During this interval I have examined more than 5000 new references pertaining to freshwater invertebrates of the United States, and have accordingly added, corrected, and revised both text and keys, wherever necessary. Major revisions have been necessary for such chapters as the Introductory Essay, Porifera, Annelida, Gastropoda, and Copepoda, but only minor revisions have been required for such taxa as the Protozoa, Turbellaria, Nematoda, Nematomorpha, Ostracoda, Pelecypoda, and Hydracarina. From the standpoint of nomenclature and taxonomy, some of these are "unpopular" taxa of genuine interest to few researchers.

With the advent of biochemical taxonomy and scanning electron microscopy, vast new fields of invertebrate systematics are being brought to attention. More specifically, ultra-details of structure are being discovered that appear to have taxonomic significance, resulting in increasing "splitting" of taxa into new subcategories. For routine identifications, however, it is mandatory that we rely on structural details that can be seen with an ordinary light microscope. The horrendous expense involved in using the SEM rules it out for such purposes. Ordinarily, biochemical techniques are also not practical.

Two editorial changes have been incorporated into this edition. First, I have numbered all figures separately for each chapter. Thus the first figure in each chapter is "Fig. 1." We think this device will simplify, avoid confusion, and speed the use of keys. Second, we are using only standard abbreviations of state names in the keys; this should likewise save space and time.

The decision to omit the aquatic insects from this edition was made only after long soul-searching, but it has the approval of John Wiley & Sons and also has the approval of many aquatic entomologists with whom I have had discussions.

I have always felt that the first two editions of this book have given unjustifiably short shrift to the insects. We should have included, for example, much more material on anatomy, ecology, behavior, and, above all, keys that go to the species level. Using today's level of insect information, however, we would end up with a 1500-page volume if insects were to be included in this third edition! The new information, "splitting," and taxonomic revisions for the Trichoptera, Ephemeroptera, Diptera, Plecoptera, and Coleoptera, for example, are reaching fantastic proportions.

And if I were to keep the insect chapters and expand them properly, this volume would not appear before 1994—an unconscionable delay which I cannot justify in my present functional retirement.

Fortunately, our aquatic entomologists are now producing excellent mongraphs, manuals, and text-references that will soon fill our needs.

To add a final bit of realism, I have done my stint, and it is now time for me to spend more of my days trout fishing.

ROBERT W. PENNAK

Boulder, Colorado
January 1989

PREFACE TO THE SECOND EDITION

About 25 years have elapsed since the publication of the first edition of this book. During this long interval I have done my best to gather the pertinent references and literature that would be necessary for the production of this second edition. This period has been characterized by a greatly increased interest in the biology of our fresh-water invertebrates. As a result, the germane literature since 1952 amounts to more than 5000 titles in technical journals. Colleagues and professional acquaintances everywhere have been generous in sending me reprints of their works, and I am greatly in their debt for such thoughtfulness. I am grateful to Dr. John Bushnell for his comments and patience in helping me with anatomical and nomenclatural problems in the Ectoprocta. Dr. Perry Holt generously supplied the key to the genera of Branchiobdellida.

Several of the shortest chapters have required only few changes, but most chapters have had a major overhaul. Nearly all keys have been rewritten and made longer in accordance with maturing systematics. A few of the old figures have been deleted, especially where I could find or make more suitable illustrations. Many new figures have been added, especially to clarify difficult key couplets. Lists of references at the end of each chapter have been revised and enlarged; about half of the citations are new.

When we refer to fresh-water invertebrates of the United States, we are including complete data only for the 48 contiguous states. In general, however, this volume covers essentially all the Alaskan fauna. The Hawaiian fresh-water fauna is impoverished and poorly known.

The Ronald Press Co. And I have been surprised and continuously gratified by the acceptance of the first edition by such a very broad spectrum of users. We hope this second edition will receive similar acceptance.

My greatest personal reward derived from producing the two editions of this book is the host of professional acquaintances I have made, not only in the United States but also in many foreign countries.

ROBERT W. PENNAK

Boulder, Colorado
March 1978

PREFACE TO THE FIRST EDITION

No comprehensive work on American fresh-water invertebrates has appeared since 1918, and in the meantime much new information has steadily accumulated in a wide variety of biological periodicals. To me, therefore, the need for the present volume is obvious. I hope it will prove useful to biologists and zoologists generally, and to aquatic zoologists, limnologists, fish biologists, and entomologists particularly. It will also serve as a text and reference pitched toward the level of the college senior and beginning graduate student.

Original material resulting from my own field and laboratory research forms only a small fraction of this book. Rather, my role as an author has been one of arrangement, organization, and selection from a very large mass of published information. Indeed, my most perplexing problem has been how *much* material on each taxonomic group to include within the covers of a single volume. General policies, however, have been clear from the outset; I have emphasized natural history, ecology, and taxonomy; I have minimized details in the sections on anatomy and physiology; I have tried to be complete, accurate, concise, and consistent from one chapter to another. During the preparation of this volume all of the chapters have been continually revised and kept up to date in accordance with new material appearing in the literature.

For the most part, only free-living, fresh-water invertebrates that occur in the United States are included. Cestodes, trematodes, parasitic nematodes, etc. are all omitted; their inclusion would have necessitated a much more extensive work and additional years of preparation. Furthermore, most aquatic biologists seldom need to refer to parasitic groups as contrasted with free-living groups.

It will be noted that some chapters contain keys to species, while others contain keys only as far as genera. In general, the chapters in the former category deal with small or comparatively stable taxonomic groups that are fairly well known and in which few new species are being described from time to time in the United States. The chapters in the latter category, however, are all concerned with larger taxonomic groups, as well as those that are less well known in the United States and in which new forms are being described so abundantly that any key to species is quickly out of date. It is assumed that users of this manual are familiar with the fundamentals of taxonomy and the construction of keys. Incidentally, each half of a key couplet consists of but a single sentence; changes in subject within a sentence are indicated by semicolons.

The list of references at the end of each chapter consists only of especially significant and comprehensive works; these include older classical papers as well as recent contributions. Except where sections are especially valuable, I have not listed the larger standard reference works such as Hyman's *The Invertebrates*, Schulze's *Biologie der Tiere Deutschlands*, Bronn's *Tierreich*, Kükenthal and Krumbach's *Handbuch der Zoologie*, Dahl's *Die Tierwelt Deutschlands*, Grassé's *Traité de Zoologie*, and *Faune de France*.

Almost all of the chapters (with or without figures) were sent to at least one specialist for criticisms, corrections, and suggestions. I am deeply grateful to all these specialists, who were most encouraging, generous, and cooperative. Without their help and assurance I would have had much less confidence in my efforts. Following is a list of these specialists, with the sections examined by each.

John L. Brooks (Cladocera); Royal B. Brunson (Gastrotricha); C. F. Byers (Odonata); Fenner A. Chace, Jr. (Crustacea Introduction, Mysidacea); B. G. Chitwood (Nematoda); W. R. Coe (Nemertea); R. E. Coker (Copepoda); Ralph W. Dexter (Eubranchiopoda); W. T. Edmondson (Rotatoria); T. H. Frison [deceased] (Plecoptera); R. E. Gregg (Insecta Introduction); H. H. Hobbs, Jr. (Decapoda); C. C. Hoff (Ostracoda); Leslie Hubricht (Amphipoda, Isopoda); H. B. Hungerford (Hemiptera); Libbie H. Hyman (Coelenterata, Turbellaria); F. P. Ide (Ephemeroptera); O. A. Johannsen (Diptera); M. W. de Laubenfels (Porifera); H. B. Leech (Coleoptera); James E. Lynch (Eubranchiopoda); J. G. Mackin (Isopoda); Ruth Marshall (Hydracarina); H. B. Mills (Collembola); J. Percy Moore (Hirudinea); L. E. Noland (Protozoa); Mary D. Rogick (Bryozoa); H. H. Ross (Plecoptera, Trichoptera); Henry van der Schalie (Gastropoda, Pelecypoda); Waldo L. Schmitt (Crustacea Introduction, Mysidacea); L. H. Townsend (Megaloptera, Neuroptera); H. C. Yeatman (Copepoda).

The introductory chapter was criticized by A. S. Pearse and by several of my graduate students. Other people have helped me in one way or another. A group of eight protozoologists, for example, selected the 300 most appropriate and common genera of Protozoa to be included in the Protozoa key. These men are: William Balamuth, Gordon H. Ball, A. M. Elliot, Harold Kirby [deceased]; J. B. Lackey, L. E. Noland, K. L. Osterud, and D. H. Wenrich.

Obviously, a book of this sort is bound to have errors and ambiguities, and certainly such faults are clearly my own responsibility. I hope readers will bring them to my attention.

In a sense, this book constitutes a plea and encouragement for more work on our rich but poorly known fresh-water invertebrate fauna. I hope these chapters will kindle sparks and raise questions in the minds of our students and beginning investigators in zoology, for it is among these young people that we shall find our authorities and specialists of tomorrow.

ROBERT W. PENNAK

Boulder, Colorado
1952

CONTENTS

1

INTRODUCTION*

It is estimated that the freshwater invertebrate fauna of the United States consists of over 11,000 described species. Ten years ago the figure was 10,000 species. Probably no aquatic biologist is of the opinion that this fauna is well known; certainly the total is bound to increase markedly in coming years, especially when the aquatic fauna of the western half of the country is more thoroughly studied. Although a few groups are fairly well known and reasonably stabilized, such as the Rotifera, Ectoprocta, and Pelecypoda, the majority of American freshwater invertebrate groups are in the process of taxonomic refinement. In expanding taxa composed mostly of noncosmopolitan species we are lagging behind European freshwater taxonomy by about 10 to 20 years. All taxonomic questions aside, we are seriously lacking in our understanding of geographic distribution patterns, physiology, natural history, and ecology of freshwater invertebrates in the United States. In a few groups, however, some remarkable progress has been made since the publication of the first edition of this book.

I (RWP) shall not recite long lists of taxa inhabiting freshwaters, comparing them with those found in saltwaters, but remember that only aquatic spiders, mites, insects, pulmonate gastropods, and perhaps the rotifers, cladocerans, and phyllopodus branchiopods are not derived directly from the sea. Rather I (RWP) am interested in considering the morphological and physiological adaptations that have appeared on a grand scale within the freshwater kingdom. In other words, what are the major evolutionary problems that have been solved as prerequisites to an existence within the rigorous freshwater environment, and how do they operate? Some of these items are obvious; others are more subtle and obscure. Furthermore, I (RWP) am restricting my remarks to the "true" freshwater invertebrate fauna, on the one hand, and the "true" marine invertebrate fauna, on the other hand, even though the brackish environment is the site where freshwater adaptations begin and where saltwater adaptations may be lost. A discussion of these comparative and complicated marine–brackish–freshwater faunal relationships and ecological transitions must await another effort. Some such material has already been brought together in several classical works (Green 1968; Newell 1970; Remane and Schlieper 1971). Incidentally, these authors emphasize the fact that the number of "true" brackish species, that is, those that complete their entire life

*Many paragraphs in this chapter are reproduced in either a verbatim or modified form from Pennak (1985), The freshwater invertebrate fauna: problems and solutions for evolutionary success. *Am. Zool.* 25:671–687. This material is used with the approval of the American Society of Zoologists.

cycles in brackish water, is insignificantly small compared with the numbers of true marine and true freshwater species.

Most of my remarks are centered around north temperate and midlatitude conditions as found in the United States. The reference list is selected and intentionally modest.

MAGNITUDE OF THE FRESHWATER ENVIRONMENT

Compared with the massive marine environment, the freshwater environment of the world is almost trivial. According to Todd (1970), the waters of the world are distributed as shown in Table 1.1. The oceans, in relation to lotic and lentic habitats together, constitute an unusual ratio of 97.3:0.0091. This means that the mass of saltwater is more than 10,000 times the mass of inhabitable freshwater. In terms of surface area, inland waters cover less than 2% of Earth's surface (Wetzel 1983), compared with the saltwater environment, which covers about 71%. Nevertheless, this small fraction of Earth's water has been thoroughly colonized by freshwater invertebrates.

This brings us to another point. Excluding the atmosphere, saltwaters constitute our finest example of a physically continuous environment. That is, any (small) part of the seas is theoretically accessible to any other part. There are no formidable dry land barriers—only other ecological barriers, such as temperature, distance, food conditions, substrate, and competition, which, I (RWP) submit, can often be at least temporarily tolerated.

The freshwaters of the world, on the other hand, are hopelessly fragmented into an enormous array of isolated habitats. To be sure, a large river system or a large lake is a considerable habitat mass, though not without internal ecological barriers, but consider the millions of reservoirs, small lakes, and ponds of the world, ranging down to such restricted habitats as a stock tank completely isolated on our short grass plains, 10 km from its nearest neighboring stock tank or water hole, and dependent on a windmill and well for its source of water. This, of course, is an extreme example of the environmental situation to which a segment of the freshwater fauna must be adapted. The small pond as an isolated microcosm is a neglected concept that was first emphasized a century ago in a classical paper by Forbes (1887).

THE CHEMICAL DIVERGENCE

The most obvious marine–freshwater distinction lies, of course, in the salt content of these two regimes. The great bulk of marine habitat mass ranges between only 3.30 and 3.70% salt, of which 80% consists of sodium chloride. The vast majority of "true" marine species, therefore, are restricted to a chemically monotonous environment (Fig. 1.1), and they may be said to constitute the "chloride fauna."

Most freshwater species, on the other

Table 1.1. Waters of the World

Water item	Volume (km^3)	Percentage of the world's total water
Lakes, ponds, and reservoirs	125,000	0.009
Average instantaneous volume of rivers	1,250	0.0001
Saline inland waters	104,000	0.008
Soil moisture and vadose water	67,000	0.005
Groundwater	8,350,000	0.61
Icecaps and glaciers	29,200,000	2.14
Atmosphere	13,000	0.001
World oceans	1,320,000,000	97.3

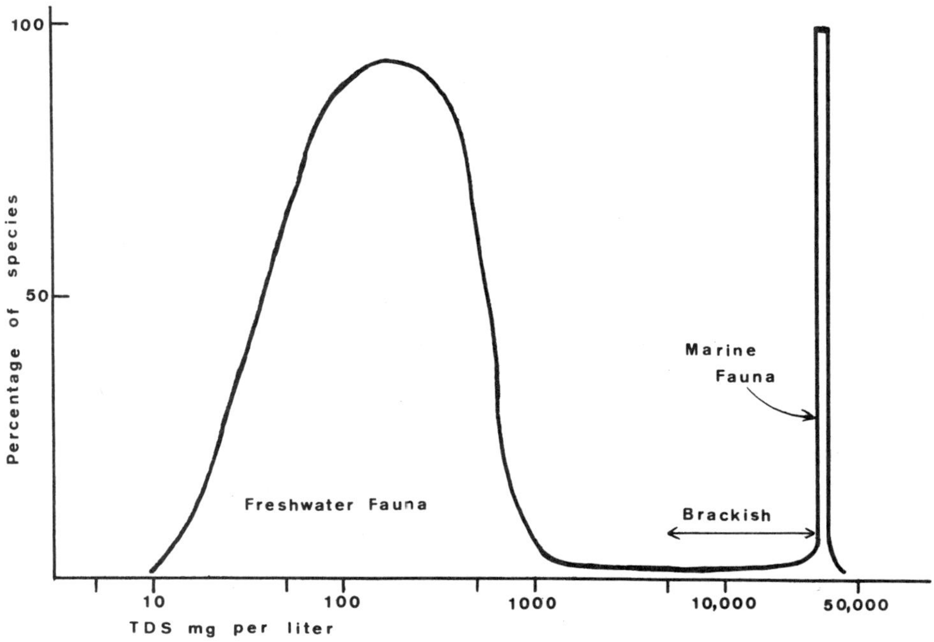

Fig. 1.1. Salt content and environmental preferences of the aquatic faunas. This figure shows only natural occurrence of the faunas; it does not show limits of tolerance.

hand, tolerate and occur naturally over a surprising range of total dissolved solids (TDS). From a chemical standpoint, and unlike the oceans, there is no such thing as an "average" freshwater habitat (Table 1.2). Each lotic or lentic situation has its own chemistry, but customarily the higher the altitude and the nearer a body of water is to the headwaters of its flowage system, the lower the TDS. Nevertheless, the dominant and most distinctive ions are the carbonate–bicarbonate complex, which usually accounts for more than 55% of the TDS. We may thus refer to freshwaters generally and collectively as "carbonate" waters, and to the fauna as the "carbonate fauna." It must be ac-

Table 1.2. A Comparison of the Total Dissolved Salt Content of Freshwaters, Brackish Waters, and Seawater. Maximum and minimum values are merely approximations. The most abundant salts in brackish waters are chlorides, carbonates, or bicarbonates, depending on the degree of brackishness. In freshwaters the most abundant salts are usually carbonates and bicarbonates. In alkali and saline lakes they are usually carbonates, bicarbonates, chlorides, and sulfates in varying proportions.

	Percentage concentration		Concentration in mg/L	
	Minimum	Maximum	Minimum	Maximum
Seawater	3.5		35,000	
Brackish waters	0.05	3.2	500	32,000
Freshwaters	0.001	0.05	10	500
Alkali and saline lakes	0.05	25.0	500	250,000

knowledged, however, that a significant minority of freshwaters are dominated by sulfates, rather than by the carbonate complex. Sodium chloride is present only in small quantities in most freshwaters, seldom exceeding 5% of the TDS.

Aside from the carbonate and sodium chloride situation, I (RWP) have long felt that another major ionic and biological distinction between the two types of aquatic environments is centered around potassium as a key element. The average potassium content of seawater is about 380 mg/L, while freshwaters usually contain less than 10 mg/L, and often only traces. Large quantities of potassium salts in freshwaters are known to be toxic to many invertebrates, and, conversely, marine groups making the evolutionary passage into freshwaters must have made an extreme physiological potassium adjustment.

The freshwater fauna is clearly distinguished by its occurrence in waters that, while dilute, have a wide range of dissolved salts, and thus it is, relatively speaking, a euryhaline fauna (Fig. 1.1). The true marine fauna, on the other hand, cannot reproduce and complete life cycles over wide ranges of salinity (Dorgelo 1976).

In general, we may arbitrarily set 1000 mg/L as the usual upper limit of TDS for the freshwater fauna and 10 mg/L as the lower limit. Within this wide range are permanent populations of many common species that tolerate, for example, a 10-fold range of 50 to 500 mg/L.

Running waters are especially notable for their annual variations in dissolved materials, depending on the local geochemistry and seasonal changes in discharge. Generally, an annual variation of ±50% in dissolved load is typical (Pennak 1977), and in exceptional cases ±80%. Most stream invertebrates endure such variations without difficulty.

Seawater seldom exceeds a range of pH 7.8 to 8.3 in surface samples, but the "normal" biological range for most freshwaters is about pH 4.4 to 8.6. A single habitat, during the course of a year, may vary as much as two full pH units, depending on the vagaries of photosynthesis, light, current, respiratory processes, biota, circulation, and so on. The extensive literature on this subject shows that many common species adjust to such variations with no difficulty.

ORIGINS OF THE FRESHWATER FAUNA

The generalization that most major freshwater invertebrate groups originated from marine ancestors is firmly established. Only a few groups, such as insects, mites, and pulmonate snails, are presumed to have clearly originated from terrestrial habitats. The freshwater fauna is therefore appropriately termed an immigrant fauna.

The fundamental problem that must be solved before any marine animal can make its way into freshwater involves major physiological readjustments. Body fluids of most marine invertebrates are roughly isotonic with seawater; that is, the internal dissolved salt concentrations are similar to or slightly higher than the 3.5% average salt concentration of seawater. It is further true that the majority of marine invertebrates cannot endure much dilution of seawater. Freshwaters commonly contain about 0.01 as much salts as the ocean, and the internal fluids of freshwater invertebrates usually contain 0.03 to 0.40 as much salts as the ocean. From an osmotic standpoint, water therefore tends to pass into the hypertonic tissues of freshwater animals. Consequently, any successful freshwater animal must have developed physiological mechanisms for maintaining a proper salt and water balance against this strong gradient.

It is difficult to imagine the appearance of such mechanisms de novo, and it is assumed that the transitions from marine to freshwater environments were not sudden and rapid processes, but rather series of slow evolutionary processes occurring by way of psammolittoral and phreatic waters, littoral zones, marshes, swamps, and river estuaries where there are transition zones between salt and fresh water.

Bear in mind, however, that a river estuary is not a constant environment. Salinities, currents, tides, food, temperatures, and other ecological factors vary widely from time to time, and these conditions are by no means favorable to the evolution and gradual development of forms suited to freshwater. Pearse (1950) has stated the problem well: An estuary has been called the doorway by which marine forms have populated freshwater. This statement is perhaps in part true, but an estuarine doorway is not wide open and easily passed through. There are many difficulties to be surmounted. Many animals struggle long ages to get through and fail. Only a few attain freshwater by this route.

Ideal conditions for the invasion of freshwater are afforded by such places as the Baltic Sea, where a large area is involved and where there is a permanent and very gradual transition from the seawater to freshwater. Yet the Baltic has not a single endemic brackish-water species of metazoan that has evolved since the most recent glaciation, although there are a very few subspecies or physiological varieties that may be endemic.

The generalization that marine invertebrates are isotonic with seawater is actually a slight exaggeration. Even pelagic, deep-sea, and the most primitive marine species maintain a dynamic steady state by which a difference in concentration of several ions commonly obtains across external membranes and which must be maintained through physiological regulatory processes. The internal concentrations of magnesium and sulfate ions are often much lower than those in seawater. Potassium concentrations may be considerably higher or lower than in seawater, and calcium and chloride are also variable. Many marine invertebrates have the remarkable ability to concentrate astonishing amounts of elements that are present in the seawater only as traces. Examples are vanadium, iodine, strontium, and bromine.

A wide range in type and degree of osmoregulatory control is found among littoral and estuarine invertebrates. Many species have limited ability to regulate the relative amounts of internal salt and water; their membranes are easily permeable so that the body fluids become more or less isotonic in diluted seawater, and death occurs rather promptly. A few stenohaline species can endure some dilution of seawater by regulating body volume and taking up more water into the tissues. A few other species can regulate their internal osmotic concentration only to a limited degree and venture into slightly brackish water; these species have a slight ability to remain hypertonic in diluted seawater. Many "typical" brackish-water species are euryhaline and frequently invade and persist in habitats containing 30 to 100% freshwater. Most of these invertebrates maintain a more or less hypertonic internal salt concentration, regardless of the degree of brackishness of their surroundings. They include representatives of many groups, especially arthropods, mollusks, and various kinds of worms. The active absorption of salts is an important mechanism, and perhaps in some species there is actually physiological control of the amounts of water absorbed. Marine and brackish species transferred directly to freshwater usually live a few hours at most, but when the transition is made very slowly, over days or weeks, they may live for weeks or months in freshwater.

Figure 1.2 is a diagram showing the relative composition of the aquatic fauna in relation to salinity. It is one of the most striking paradoxes of freshwater, brackish, and marine faunas that animals thrive well in either an environment that is very low in salts or one that is high in salts, but environments of intermediate salinities (brackish) have a poor fauna. From the standpoint of total number of all types of species present, it should be noted that the minimum appears at a salinity of 7‰ (7 parts per thousand), which is well toward the freshwater end of the diagram in Fig. 1.2. This situation results from the fact that the number of freshwater species drops very rapidly with a slight rise in salinity, whereas the number of marine species drops less rapidly with a decrease in salinity.

This diagram also poses a second paradox. It shows that specific brackish-water species, which occur chiefly or exclusively in such environments, are most abundant at salinities of 7 to 10‰ (ca400mg/L) or about the place where the freshwater forms decrease abruptly and where the total number of all species is smallest. This condition is not to

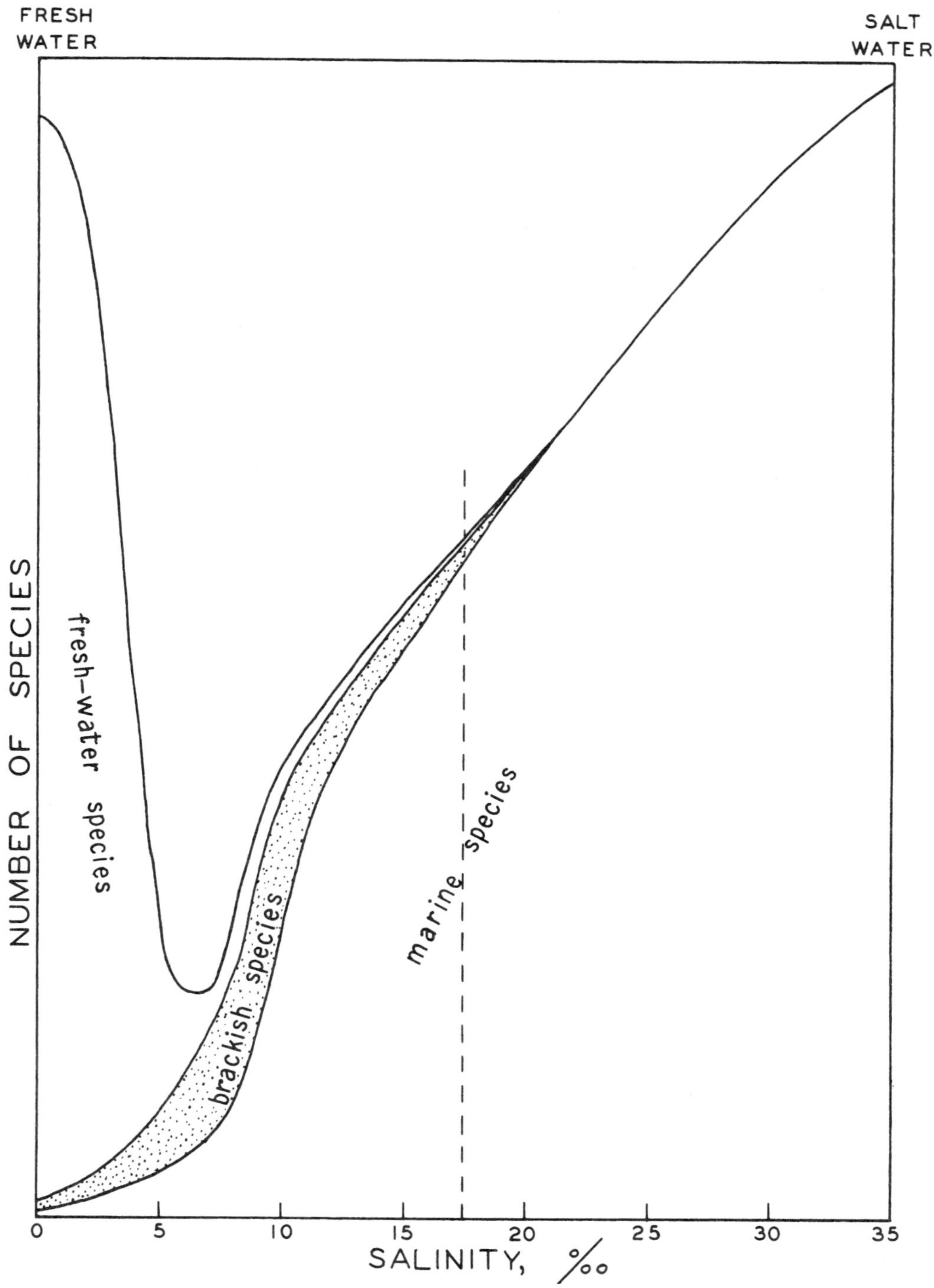

Fig. 1.2. The composition of the aquatic fauna in relation to salinity of the environment. (Modified from Remane and Schlieper 1971.)

be expected since most brackish-water species have obviously been derived from marine relatives and not from freshwater relatives. It would be more logical to expect the maximum number of brackish species much farther toward the right in the diagram.

The invasion of freshwaters is a continuing process, and some American species are undoubtedly new arrivals or are in the process of becoming adapted to freshwaters. The colonial cnidarian *Cordylophora* and a few species in each of the following groups are all typical examples: polychaetes, grapsoid crabs, shrimps, isopods, clams, and snails. A comparable list may be cited for Europe, where a few species are known to have invaded freshwaters within historical times. The mitten crab of the Old World has taken up permanent residence in rivers and returns to the sea only to breed. With few exceptions, the new freshwater forms are thought to be physiological varieties of their close marine relatives.

Several species of European crustaceans are marine in northern Europe and both brackish and fresh in central and southern Europe. Certain tropical areas, especially around the Bay of Bengal, Indonesia, the Malay Archipelago, Madagascar, and tropical America, are relatively rich in species that have only recently made their way into freshwaters from the sea.

In many man-made situations, clusters of species have been able to adapt to decreasing salinity conditions. At Lake Merced, California, for example, the channel to the Pacific Ocean was closed between 1869 and 1894. Subsequently, it slowly converted to freshwater, and a few species of marine ancestry were able to adapt. As of 1958, Miller reported the hydroid *Cordylophora*, the polychaete *Neanthes* (*Nereis*), the mysid *Neomysis*, the isopod *Exosphaeroma*, and the amphipod *Corophium*. *Cordylophora* and *Neomysis* are typical brackish species; the others are definitely marine. A similar pattern has been reported in Ljessel Lake, formerly a marine bay (Zuiderzee), in The Netherlands. In these instances and others, the transition to a freshwater physiology has been rather rapid. It remains to be seen if the species affected fully adapt to freshwater.

It is presumed that constant temperatures and heavy rainfalls facilitate freshwater invasions via estuaries. Alternative theories emphasize low temperature, chiefly because lower temperatures minimize the physiological effects of the various ions, or habitat specialization and reduced range in upper estuaries as important reasons for entry into freshwater. However, many considerations beyond osmotic adjustments need to be addressed in migrating to freshwater. Larval biology often needs to be altered extensively to accommodate the demands of freshwater existence. Typically, successful transformation involves the suppression of the extensive and multistaged larval biology characteristic of marine ancestors.

The invasion of phreatic estuarine water and hence to surface freshwater has been postulated several times. One proposed example (Pennak 1963) involves micrometazoans colonizing freshwaters from the sea via the psammolittoral and phreatic habitats. As shown in Fig. 1.3, micrometazoans could possibly move from intertidal and subtidal zones into the marine psammolittoral and (1) hence into the phreatic groundwater and thereafter inland to distant ponds or lakes via their psammolittoral, or (2) into the psammolittoral of an estuary, and hence progressively "upstream" in the river psammolittoral or deeper phreatic zone to inland freshwater localities. Porous substrates below typical stream beds have long been known to have well-developed phreatic currents. The phreatic brackish interstitial zone of mixing at the edge of the sea is relatively thin, but it is also relatively constant in position, and species in this habitat are in a much less hazardous situation than those species that must swim or crawl about on the substrate of an estuarine area.

FRESHWATER EMIGRANTS TO THE SEA

Once established in freshwater, few invertebrates have returned to brackish water and fewer still to undiluted seawater. Little is known about the major barriers to such migrations, but presumably they are physiological, and perhaps chiefly osmoregulatory.

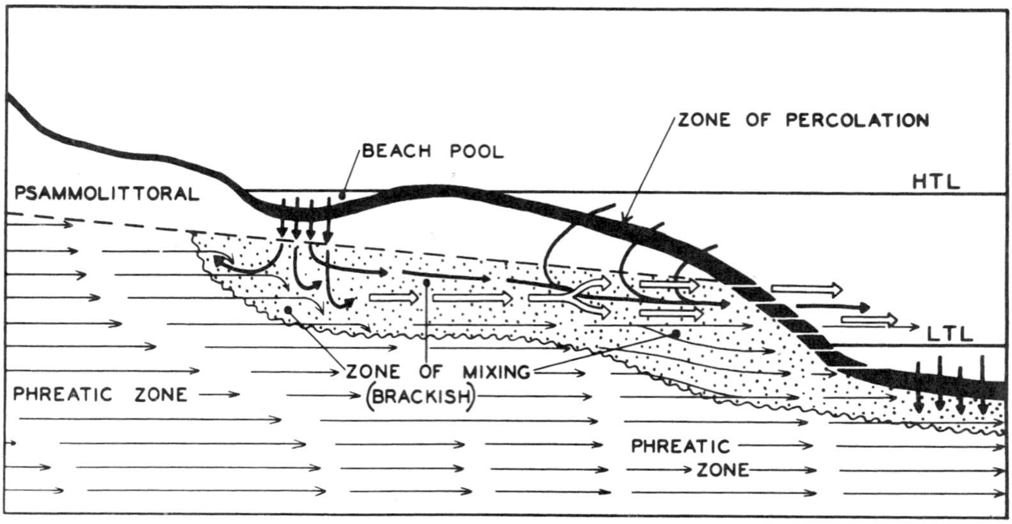

Fig. 1.3. Diagrammatic section of a marine shoreline in an area where the fresh phreatic waters are seeping into the saltwater. HTL, high tide level; LTL, low tide level; brackish zone of mixing shown by stippling; thick, open, and thin arrows show relative amounts of salt, brackish, and fresh waters, respectively. (From Pennak 1963, after Delamare Deboutteville 1960.)

Potassium salts, which are abundant in the sea, are toxic to freshwater invertebrates, but this effect may be partly neutralized by magnesium compounds. An impervious exoskeleton is advantageous for migrations into salt waters.

Only a few rotifers, mollusks, hemipterans, beetle larvae and adults, dipteran larvae and pupae, and several caddis larvae have successfully invaded estuaries, brackish waters, and the intertidal zone. Diptera larvae have occasionally been dredged from 10 to 15 fathoms off the coast of England. In Samoa there is a peculiar dipteran that is submarine for its entire life cycle. A few marine water striders occur regularly on the surface film of the seas far away from land. Several genera of the snail family Hydrobiidae have become fully adapted to estuarine environments.

Biologically, insects are considered a most successful group. They exhibit endless adaptations, occur in enormous numbers, and are found in almost every type of habitat. Yet they have not been successful in colonizing the seas. With the exception of some of the groups just noted, a salt content in excess of 2.5‰ (1.4 g/L) appears to be toxic. However, brackish ponds with salt contents ranging up to 1.0‰ (0.5 g/L) are quickly colonized by "freshwater" insects.

Occasionally, typical freshwater mollusks are reported from brackish estuaries. A few species of *Physa* and *Elimia* can endure up to 50% seawater, *Amnicola* and *Helisoma* up to 30%, and certain Lymnaeidae up to 25%.

MAJOR DISTINCTIONS BETWEEN MARINE AND FRESHWATER INVERTEBRATES

Osmoregulation. Continuing the discussion of the foregoing sections, one of the basic physiological distinctions centers around the occurrence of more specialized osmoregulatory mechanisms in freshwater organisms. Since their internal fluids are hypertonic to the greatly dilute surrounding water, there is a continuous and rapid inflow of water through all permeable surfaces of the body. Such surfaces may be epithelia, cuticle, chitin, gills, and other structures. The relative amounts of water absorbed through these various types of surfaces differ greatly from one taxon to another. In general, chitin and thick cuticle are relatively impermeable.

Contractile vacuoles, flame bulb systems, nephridia, and different types of glandular structures have the function of forming a highly dilute urine, which is excreted to the outside in considerable quantities. This urine is greatly hypotonic to the body fluids. By such means excessive water is prevented from accumulating in the body, and the relative concentrations of internal salts and water remain essentially constant. In a general way, freshwater invertebrates are specialized for retaining their low ionic concentrations, even during long periods of inactivity. Figure 1.4 shows these relationships in a diagrammatic fashion.

Marine invertebrates usually have excretory organs that are basically more or less similar to those of their freshwater relatives, but the urine they excrete is approximately isotonic with seawater and their body fluids. The fundamental problem that must be solved when a euryhaline animal invades freshwater is the physiological conversion of the osmoregulatory process to one that secretes dilute urine from hypertonic body fluids. Other things being equal, freshwater invertebrates use more oxygen than their close marine relatives, probably because of the greater energy needed to maintain the proper internal osmotic pressure.

It is significant that contractile vacuoles are absent from most marine protoctistans. Freshwater isopods, amphipods, and decapods have larger excretory (osmoregulatory) glands than their marine relatives, and presumably this indicates greater activity and greater physiological and anatomical complexity.

Although the urine excreted by freshwater invertebrates is highly dilute, it does contain traces of body salts. In most animals this loss is negligible and is apparently made up from food and by the slow selective ab-

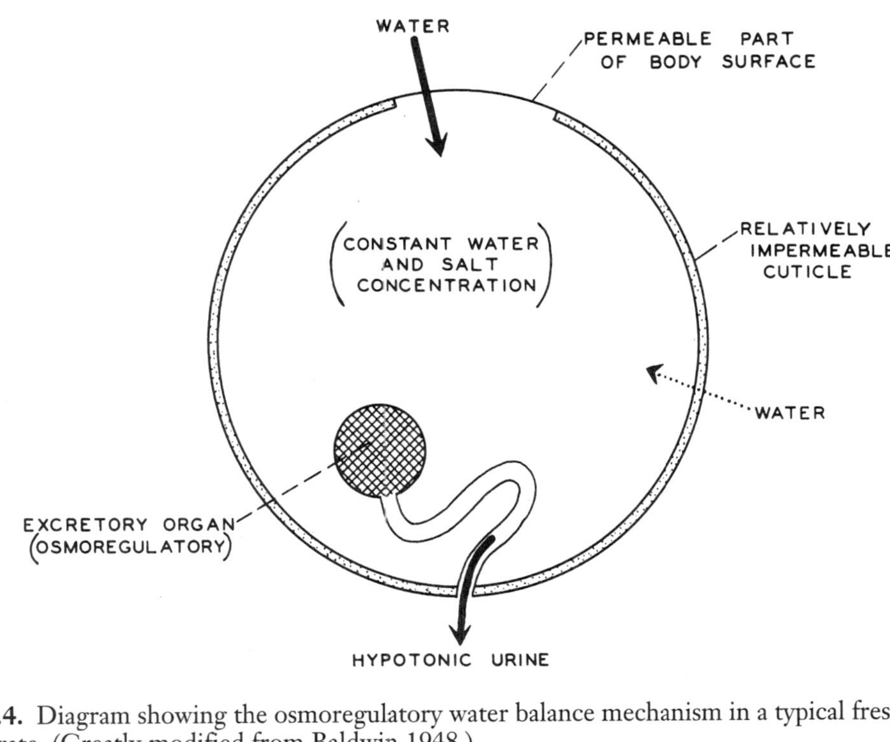

Fig. 1.4. Diagram showing the osmoregulatory water balance mechanism in a typical freshwater invertebrate. (Greatly modified from Baldwin 1948.)

sorption of ions through body surfaces. In a few groups of freshwater invertebrates, however, it has been shown that there is active absorption of ions by special parts of the body, particularly in snails, mussels, leeches, dragonfly nymphs, crayfish, and some Diptera larvae. At the same time, the ability to retain body salts is sometimes quite striking. Many insects and crustaceans may be placed in distilled water and kept there for weeks at a time without any apparent harm.

Although freshwater invertebrates are generally considered stenohaline organisms, this generalization is true only in the gross sense. "Freshwater" is highly variable in total salt content and in the relative quantities of different ions present, but many common species in almost all orders of freshwater invertebrates may be found in waters whose total salt content ranges from 0.002 to 0.04%. This is actually a 20-fold difference in dissolved salts! With the exception of a few species adapted to highly saline inland waters, however, the osmoregulatory mechanisms of most freshwater invertebrates quickly break down when they are placed in solutions containing in excess of 25% seawater (0.9% salt).

One of the most striking osmotic conditions may be found among invertebrates in our high mountain tarns, where there are usually populations of cladocerans, copepods, protoctistans, rotifers, nematodes, oligochaetes, many immature insects, and even fingernail clams (Sphaeriidae). Here the total dissolved solids may be only 10 mg/L, distributed as follows:

Dissolved organics	4 mg/L
Carbonate–bicarbonate complex	3 mg/L
Silicon complex	2 mg/L
All other salts	1 mg/L

Such water is more dilute than what passes for the usual "distilled water" in our laboratories. Within 1 mg of "all other salts" are the chlorides, sulfates, nitrates, phosphates, magnesium, and iron, plus a host of other essential ions—all present in trace quantities in such dilute mountain water. Evolutionary perfection of osmotic systems and integuments gives these animals the ability (1) to absorb from such water, in various ways, all (essential) ions on a basis that is seemingly atom by atom, and (2) to retain these absorbed internal ions as well as those obtained through ingested food. Indeed, this is illustrated by the fact that many freshwater arthropods will live in distilled water for weeks at a time.

A few marine species are specialized for absorbing and concentrating such exotic elements as zinc, vanadium, copper, and iodine (Galtsoff 1934; Vinogradov 1953), but I (RWP) know of no comparable examples of American freshwater invertebrates that selectively concentrate any unusual ions. Instead, inhabitants of exceptionally soft waters must take up essentially all ions against an unfavorable osmotic gradient. Macan (1963) points out that the fauna of the most dilute freshwaters tends to be ultimately of terrestrial origin. This has not been my experience, as shown for extremely soft high-altitude waters.

Eggs. No one seems to have collected extensive data to show conclusively whether there is a broad difference between fecundity in marine invertebrates and their freshwater counterparts, but most of us are familiar with a few examples. Thus, in a few freshwater hydroids such as *Hydra*, reproduction is restricted to one or a few fertilized eggs at a time. Its relative *Cordylophora* produces twice or more as many embryos in brackish as in fresh waters. Making allowances for the size difference, lobsters in berry carry proportionately more eggs than freshwater crayfish. Freshwater flatworms produce few eggs, and the same is often true, comparatively, for freshwater Mysidacea, Amphipoda, and Gastropoda. Within the shrimp genus *Palaemonetes*, brackish-water species produce more eggs than freshwater species. Needham (1930) and Barrington (1979) point out that the common oyster releases 1,800,000 eggs at once and that oviparous oysters may produce 100,000,000 eggs per spawning. The freshwater *Anodonta cygnea*, however, produces only 15,000 to 2,000,000 (rarely) eggs. As a whole, freshwater mussels produce several thousand to 3,000,000 embryos, depending on the taxon. The marine gastropod *Buccinum* may deposit 12,000 eggs and *Nucella* about 245 capsules containing 400 to 600 eggs each, while freshwater snails lay only a trivial 10 to hundreds at a time.

There has been some speculation about this fundamental difference in reproductive potential between the chloride fauna and the carbonate fauna. One such speculation argues that trace element nutrients are more difficult to accumulate in necessary quantities in freshwater invertebrates, and as compensation their eggs must be larger (to store such materials for early growth), and hence the total number of eggs is much smaller. This seems a weak explanation, especially in view of the ability of freshwater forms to live and reproduce in extremely dilute waters. Furthermore, most freshwater microcrustaceans are notable exceptions in having small eggs. And Thorson (1946) presents abundant evidence to show that eggs of marine invertebrates range widely in their yolk content, and in their small to large size. Careful experimental work on euryokous species might be revealing.

Egg-carrying habits may be related to this problem. In freshwater species the eggs are often carried by the parent until hatching and release of the young. This situation occurs notably in certain rotifers and most cladocerans, mussels, copepods, and malacostracans, as well as in other taxa. The egg-carrying habit is also true for some marine species, but it is by no means as common as it is in freshwaters. If eggs must be carried until the hatchling is sufficiently grown to assume an independent existence, we may reason that there is less room on the female body for the clutch of eggs; hence, the small number of (larger) eggs.

Another explanation for fewer (larger) eggs is that the freshwater organism must enclose with fully developed osmoregulatory capabilities, that is, must be at a more advanced stage to cope with the highly dilute surroundings. In marine forms the production of (lightweight) planktonic eggs is common, while freshwater eggs are heavy and remain or sink to the bottom.

Distinctive larval stages. Unlike most of their marine relatives, freshwater invertebrates produce special larval stages in only a few major taxa, as shown in Table 1.3. To this table could be added the phyllopodous Branchiopoda exception; this taxon is restricted to freshwaters and inland saline waters and does have nauplius stages. Barrington (1979) states that "the disadvantage of freshwater larval stages are so clear," but does not actually discuss the problem. I (RWP) believe that there is probably a variety of evolutionary reasons, operating through a very long geological time frame, to explain the paucity of special larval stages in freshwater invertebrates. It is, however, truly unusual that the freshwater Copepoda and the phyllopodous Branchiopoda, eminently successful groups, have retained the basic nauplius stages in their life history. The Copepoda, furthermore, is a remarkably euryokous group, very closely tied to its saltwater relatives.

Undoubtedly the greater abundance of yolky stored food in freshwater eggs is correlated with the general lack of special larval stages. In marine groups where there is little yolk the newly hatched larvae must begin foraging promptly.

One theory advanced to explain the infrequency of freshwater invasion by marine species is based on the vulnerability of marine larvae. Although free swimming, they are unable to make their way upstream against river currents. Even if the adults were to breed in an estuary or river, the unattached larvae would be swept downstream. It is also thought that the larval metabolism and hypertonic tissues are unable to withstand the severe and variable ecological conditions in freshwaters.

Nine major taxa are well represented in both fresh and salt waters, but special larval stages are present in neither environment, the young being hatched in an advanced stage of development; these are Gastrotricha, Rotifera, Nematoda, Tardigrada, Oligochaeta, Hirudinea, Mysidacea, Isopoda, and Amphipoda.

Phylogenetic differences. It is quite true that the seas are phylogenetically much richer than freshwaters, and there are many

Table 1.3. Comparison of Special Larval Stages in Those Major Taxa Occurring in Both Marine and Freshwater Environments

Taxon	Motile marine larval stages	Motile freshwater larval stages
Demospongiae	Coeloblasrula, stereogastrula	"Flagellated embryo" in some species
Hydrozoa	Medusa, planula, frustule (creeping), actinula	Frustule (creeping) in a few species, "planula"
Turbellaria	Müller's larva, Gotte's larva	
Nemertea	pilidium	
Entoprocta	"Ciliated larva" (trochophore)	Frustule, ciliated larva (rare)
Ectoprocta	Cystid, cyphonautes, etc.	"Ciliated larva" (rare), ancestrula
Copepoda	Nauplius stages, copepodid stages	Nauplius stages, copepodid stages
Ostracoda	Atypical nauplius	Atypical nauplius
Decapoda	Nauplius, protozoea, zoea, mysis, phyllosoma, etc.	
Gastropoda	Trochophore, veliger	
Pelecypoda	Trochophore, veliger	Veliger, glochidium (parasitic)

major taxonomic categories that are confined to salt water or are predominantly marine (lists A and B of Table 1.4). Only one class, the Phylactolaemata, and several subgroups of other classes, the Branchiopoda, the Syncarida, the several species of Thermosbaenacea, and one species of Spelaeogriphacea (known only from southern Africa) are exclusively freshwater. The Branchiopoda are unusual because some species are found in salt-poor waters (e.g., tundra ponds), whereas others are restricted to highly saline inland waters. The Rotifera, Nematomorpha, Hirudinea, Hexapoda, and Hydracarina have so few marine representatives, however, that these might just as well be considered freshwater groups. With few exceptions, those taxonomic categories that are well represented in both environments are much richer in species in marine waters than in freshwaters. This is especially true of the Porifera, Hydrozoa, Ectoprocta, Pelecypoda, Gastropoda, Malacostraca, and Copepoda. Because colonization of freshwaters probably began earlier than Devonian times, it is significant that many more freshwater groups have not made their appearance.

Body size. It is difficult to rationalize the fundamental difference in size (body weight) of marine as compared with freshwater invertebrates. In general, the temperate marine macrofauna consists of taxa whose average size is larger than that of their close freshwater relatives. It would seem, however, to be reasonable to expect this situation. Freshwater forms would have a metabolic and osmotic disadvantage in being larger since area (body surface) increases as the square, while body volume increases as the cube, thereby conferring a theoretical osmoregulatory disadvantage upon (large) freshwater forms in the dilute milieu. We suspect, however, that the answer might lie in the fact that food in freshwaters is less available and less abundant than in salt waters, and that this limiting factor has taken precedence over osmoregulatory and other physiological problems. Additionally, because of the variable nature of freshwater, invertebrate inhabitants tend to be short lived. At any rate, the evolutionary processes have appeared to screen out those large species having the potential of making the transition into freshwaters. In this respect, Brandt (1897) showed that the size of Baltic Sea mol-

Table 1.4. A CLASSIFICATION OF THE AQUATIC INVERTEBRATE ANIMAL KINGDOM ACCORDING TO ENVIRONMENTS. Taxa are limited to phyla and classes. Adult parasites are excluded.

A. Exclusively marine and brackish	B. Predominately marine, low in freshwater species	E. Predominately freshwater, marine species few or rare
Amphineura	Acoela	Branchiopoda
Anthozoa	Gymnolaemata	Hexapoda
Brachiopoda	Hydrozoa	Hirudinea
Cephalocarida	Entoprocta	Hydrachnida
Cephalopoda	Nemertea	Nematomorpha
Chaetognatha	Polychaeta	Oligochaeta
Cirripedia	Porifera	Rotifera
Ctenophora		
Cumacea	C. Predominately marine, many freshwater species	F. Exclusively freshwater
Echinodermata		Phylactolaemata
Echiuroida	Copepoda	
Gnathostomulida	Gastropoda	
Kinorhyncha	Malacostraca	
Leptostraca	Ostracoda	
Loricifera	Pelecypoda	
Monoplacophora		
Phoronida	D. Equal in fresh and salt water	
Placozoa	Gastrotricha	
Pogonophora	Nematoda	
Priapulida	Tardigrada	
Pycnogonida	"Turbellaria"	
Remipedia		
Scaphopoda		
Scyphozoa		
Sipunculida		
Stenolaemata		
Xiphosura		

lusks is correlated directly with salinity. In freshwaters only two taxa have bodies of any size: mussels and crayfish.

Anabiotic strategies. It is in the production of anabiotic devices that freshwater invertebrates excel, and we are here using the term "anabiosis" (Greek: *recovery of life*) in the broad sense to include all special resistant eggs, cysts, and other small resting stages that are produced to tide the animal over periods of desiccation, extreme cold, heat, anaerobic situations, lack of food, and other adverse conditions (Figs. 1.5 and 1.6). It seems logical to suppose that such freshwater mechanisms have arisen in response to variable and unpredictable environmental situations usually not found in the more stable and predictable marine surroundings. The fact that such devices are not present in marine species but are present in freshwater species in the same ancient taxon is indicative of the probable early appearance of these devices in the transitory primordial

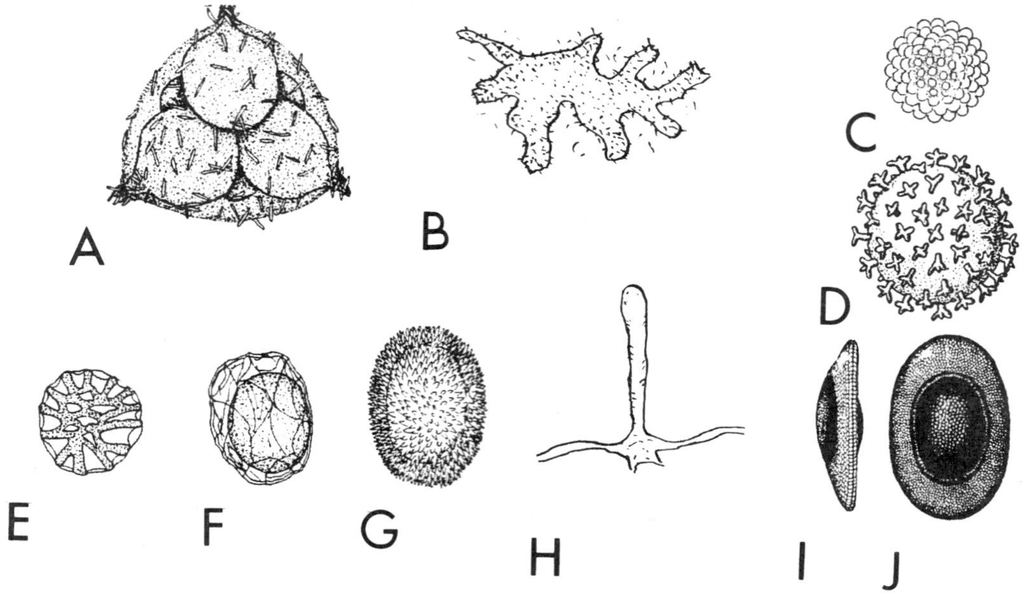

Fig. 1.5. Examples of freshwater invertebrate anabiotic devices. Various magnifications. A, cluster of sponge gemmules, *Eunapius fragilis;* B, reduction body of sponge; C and D, rotifer resting eggs; E, resting egg of clam shrimp, *Eulimnadia;* F and G, tardigrade resting eggs; H, hibernaculum of ectoproct *Paludicella;* I and J, ectoproct statoblasts, *Hyalinella punctata.* (C and D from Ramazotti 1972.)

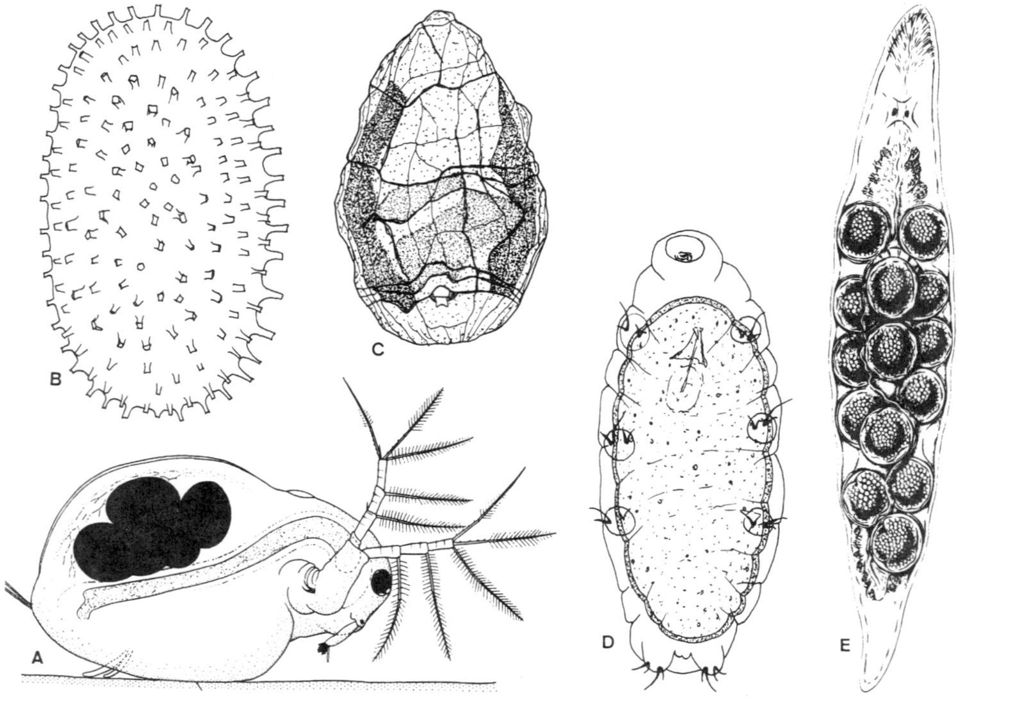

Fig. 1.6. Examples of freshwater invertebrate anabiotic devices. A, cladoceran, *Streblocerus,* carrying five ephippial eggs, ×25; B, opsiblastic egg of gastrotrich, *Chaetonotus,* ×120; C, desiccated rotifer, *Philodina,* ×490; D, cyst of tardigrade, *Hypsibius,* ×80; E, turbellarian, *Mesostoma,* carrying thick-shelled anabiotic eggs, ×50. (A from Fryer 1974; E from Riser and Morse 1974, *Biology of the Turbellaria,* with permission of McGraw-Hill Book. Co., New York.)

freshwater environment. Table 1.5 is an attempt to summarize the anabiotic devices for taxa. that occur in both the chloride and carbonate environments.

It is difficult to find accurate information as to the supposed existence of resistant devices among marine invertebrates, and the data in the middle column of Table 1.5 therefore are subject to minor modifications. In general, the third column in Table 1.5 shows the striking tendency for evolutionary development of thick-walled anabiotic stages, even though the associated instigating environmental conditions may be quite different from one time or one taxon to another. Indeed, this is a classical display of parallel adaptive evolution on a massive scale. Note the sharp break between ostracods and isopods in Table 1.5. It clearly separates the microscopic anabiotic groups from the macroscopic nonanabiotic groups. In addition to the material in this table, there are a few taxa essentially restricted to freshwaters (and originating in freshwaters?) that have also developed special anabiotic devices, including the highly resistant resting eggs of the phyllopodous Branchiopoda and the ephippial eggs of Cladocera (ignoring the very few marine Cladocera). Many phyllopodous Branchiopoda are macroscopic, and in this respect they are an exception to the size versus anabiotic stage principle just mentioned. Most, however, are restricted to astatic waters. And many aquatic insects, being largely restricted to freshwaters, have a diapause or diapauselike period in the larval, pupal, or nymphal stages that is resistant to unfavorable environmental conditions (see Mansingh 1971). Anabiotic stages of many freshwater taxa actually require a period of cold, freezing, drying, or anaerobiosis before they will hatch.

Many years ago I (RWP) acquired a quantity of dried mud that had just been collect-

Table 1.5. Relative Occurrence of Anabiotic Stages in Marine and Freshwater Invertebrates. Only those taxa occurring commonly in both environments are listed.

Freshwater and marine taxon	Anabiotic devices in marine species	Anabiotic devices in at least some freshwater species
Porifera	"Gemmulelike" structures rare	Gemmules, "reduction bodies"
Cnidaria		"Thecated embryos"
Turbellaria		Cocoons, fragmentation cysts, "winter" eggs"
Gastrotricha		Opsiblastic eggs
Rotifera		Resting eggs, winter eggs, anabiotic stage of adult
Nematoda		Resting eggs, desiccated individuals
Tardigrada		Various stages described as: tun, cyst stage, anabiotic stage, cryptobiotic stage
Nemertea		Cysts
Ectoprocta		Statoblasts (at least six types), hibernacula
Entoprocta		Hibernacula
Oligochaeta		Cysts
Copepoda		"Resting eggs" in a few spp., diapause cocoon stage
Ostracoda		Resting eggs, advanced instars able to withstand cold and drying in diapause
Malacostraca		
Pelecypoda		
Gastropoda		

ed from the bottom of a small pond in the Galapagos islands. Annually thereafter for 10 years I (RWP) put some of the mud in filtered pond water and periodically examined this microcosm for active invertebrates. Within a few days to weeks I (RWP) always found active excysted protoctistans, rotifers, tardigrades, nematodes, and microcrustaceans. Any student may duplicate these observations with dried pond mud. Dried intertidal marine mud, however, produces no such comparable populations when given similar treatment.

Also, I (RWP) have many unpublished observations on the anabiotic effectiveness of pond invertebrates at high altitudes where the entire water mass and as much as 0.5 in of the subtrate are annually frozen solid for 6 months or more. Each summer, beginning with ice melt, the normal population of pond invertebrates is restored.

Among many groups with resistant eggs or resting stages, there has been a remarkable convergence in construction for example, the gemmule of a sponge and the egg cyst of an anostracan branchiopod (Fig. 1.7). Obviously, a reinforced double-walled capsule with an inner framework has become a very successful plan for protecting the germinative cells or egg within. In the laboratory, the fantastic tolerances of cyst stages for drying, anaerobiosis, and temperature conditions far exceeding normal ranges are legendary (Von Brand 1946; Ashwood-Smith and Farrant 1980). Rotifer, protoctistan , crustacean, and tardigrade anabiotic stages are especially resistant, and a surprising fraction of anabiotic stages recover and become active after such drastic treatment.

Low-temperature tolerances. Aside from the resistant inactive anabiotic stages, the active stages of many freshwater invertebrates also withstand and recover from freezing and thawing of the substrates of lakeshores and streamsides, as well as in the substrate of ponds that freeze solidly for several weeks to as much as 9 months (Sernov 1927; Kenk 1949; Scholander et al. 1953; Neldner and Pennak 1955; Dougherty and Harris 1963; Danell 1981; Olsson 1981). In reality, ice and frozen substrates typically contain anabiotic stages in addition to immature and adult stages, depending on the particular species. In marine habitats, on the other hand, subtidal species are unable to recover from freezing, although some intertidal forms, especially mollusks, can withstand being frozen in ice for days or weeks (Kanwisher 1955).

Estivation. Estivation is a phenomenon closely associated with the production of anabiotic stages, but usually involves an inac-

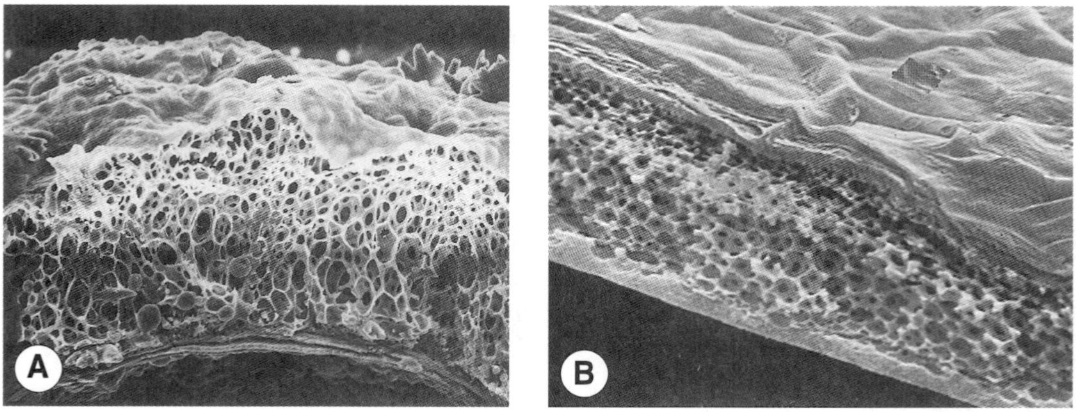

Fig. 1.7. Cyst walls of a (A) gemmule of a freshwater sponge and (B) cyst of an anostraca. (A from Smith, 1994, see Chap. 2, B from Rosowski et al. 1997, see Chap. 17, with permission.)

tive state of the whole mature animal, rather than an inactive state of an anabiotic device produced by that animal. We also prefer to think of estivation as applying only to the desiccation conditions of heat and dryness. In marine habitats temporary estivation is restricted to a relatively few species inhabiting rocky shores near the high-tide line, such as barnacles and mollusks.

Freshwater estivation is most clearly shown for populations of temporary pools and ponds; see especially the extensive study by Wiggins et al. (1980). These authors designate estivation forms as "overwintering residents," including juveniles and adults of such taxa as oligochaetes and leeches that secrete a protective coat of mucus, snails that secrete a mucoid epiphragm, a few copepods in the diapause stage, and a few immature and adult insects. These are species that actually survive true dry conditions, brought on before cold weather begins. In addition, there is always a large population of species that overwinter in crevices, in the water of crayfish burrows, in damp vegetable debris, and in damp pondside soil. These are not true examples of estivation, however, in the sense that we are using the term here.

Anaerobiosis. There seems to be no pervasive and consistent distinction between freshwater invertebrates and their marine relatives with respect to an ability to withstand anaerobic conditions. In both environments there are many related and unrelated taxa that have developed the ability to live for a few hours to a few months in the absence of oxygen or where oxygen is present in concentrations below 10% saturation. It is probable, nevertheless, that the freshwater fauna has more frequently developed the physiological mechanisms for withstanding anaerobic conditions for long periods than is the case for marine invertebrates. Examples include the fauna of sewage treatment waters, astatic waters, and the benthic and plankton faunas of eutrophic lakes and ponds whose bottom waters are anaerobic or near-anaerobic for months in both summer and winter. Both marine and freshwater environments, however, are commonly and continuously anaerobic or near-anaerobic below the surface of organic bottom muds. Perhaps our best marine example involves the subtidal thiobiotic microfauna.

Neuston. The surface film of salt water is virtually uninhabited by small metazoans, but the numerous freshwater organisms associated with the surface film either constantly or intermittently constitute a special community, the neuston. Many insects (hemipterans, collembolans), and even a few spiders run about on the upper face of the surface film, and some cladocerans and Protoctista are regularly attached to the underface of the film. Visitors to the film for varying lengths of time include insect larvae and pupae, adult insects (hemipterans, beetles), planarians, hydras, a few ostracods, and some snails. The surface film is visited chiefly for food or to make brief contact with air in order to renew the supply of gaseous oxygen held on the surface of the body or in the tracheal system.

The silt and current problem. Most running waters (except for spring brooks) exhibit periodic silting. Heavy loads of suspended silt may be present for only a few hours after a heavy shower or they may be present continuously for 2 or 3 months during spring and summer runoff. Suspended loads commonly reach levels of 1.0 g/L or more, and the water transparency approaches zero. Fine organic and inorganic particles have both abrasive and suffocative action, and while spates always have some decimating effects on lotic faunas, the stream bottom metazoans, especially insects, have remarkable powers of survival and resistance, so that "normal" population levels may soon be restored (Hynes 1970; Pennak 1977). Indeed, freshwater pelecypods are dependent on the continuous (but moderate) existence of a load of silt as a food source. Continuously clear streams seldom have appreciable populations of clams and mussels.

Stream species exhibit another cluster of adaptations that are minimally developed in marine invertebrates found in intertidal habitats. Bottom insects, for example, are often greatly flattened and can thus hide in narrow rubble crevices. They also remain sufficiently active to dislodge silt that accumulates on the body. Special devices to combat silting and swift currents include streamlining suckers, friction pads, hooks, sticky secretions, and protective ballast cases (Hynes 1970). To be sure, the intertidal zone may be temporarily silted by waves, tides, and storms, but, except for major storms and hurricanes, silt periods are brief and have not generally brought about major evolutionary adaptations. Morphological adaptation to turbulence in marine habitats seems to be mostly in the nature of crustose or flattened growth forms, for example, a few sponges, tunicates, ectoprocts, limpets, and sea anemones. Of additional importance, however, is the relatively small size of stream invertebrates as compared with marine invertebrates of the intertidal zone, and this may be a protective adaptation.

Another important distinction is the fact that the current in a swift stream microhabitat is constant and relatively unidirectional in the force exerted upon a metazoan, while intertidal turbulence is intermittent and multidirectional (Denny 1983). Possibly this distinction is responsible for the wide divergence in adaptations in streams as compared with tidal zone turbulence.

EVOLUTIONARY SHORTCOMINGS IN FRESHWATERS

Thus far we have been concerned chiefly with adaptations that have come about as the result of migrations of marine taxa ancestors into freshwaters (except freshwater insects, which originated from terrestrial ancestors), where the new adaptations have permitted the incipient fauna to become established. In so doing, however, it appears that some features of the marine ancestors have become lost, and, for the most part, are not to be found in freshwaters. Three of these features are briefly discussed in the following paragraphs.

Bioluminescence. We need give only passing mention of the sharp separation of freshwater and saltwater faunas with respect to bioluminescence. It occurs commonly in many marine taxa, and a wide variety of functions has been ascribed to the phenomenon (Harvey 1952). In striking contrast, it is rare in freshwater invertebrates. The proper combinations of anatomical adaptations, physiological mechanisms, bacteria, and essential enzyme systems have been lost in the environmental transition or have not yet generally evolved in the shallower carbonate waters.

In some forms bioluminescence is produced by bacteria on the body surface. Bacteria-caused bioluminescence has been observed in the amphipod *Hyalella azteca*. In other forms the phenomenon is produced by luciferin–luciferase mechanisms in the animal's tissues. A bioluminescent shrimp is known from Japan, and in New Zealand streams there is a small limpet that secretes light-producing mucus. It is *Latia neritoides*, first discovered in 1890. A single, rare, light-producing enchytraeid annelid is known from the nearshore sandy bottoms of Russian streams.

Morphological embellishments. Many marine species are notable for an abundance of external growths on the body surface, such as excessive setation, spines, palps, fleshy protuberances, and respiratory devices. Such embellishments occur generally but are most common among Crustacea, Polychaeta, and Mollusca. Some devices are presumed to retard sinking in plankton species; others are sensory; a few are feeding devices; and perhaps most frequently they are assumed to be respiratory. It seems, however, that excessive respiratory surfaces, such as a rich array of palps and antennae in polychaetes, are an unnecessary waste of energy, and that most marine forms should be able to get along quite well with more modest respiratory surface areas, even where the species is a burrower. There seems to be no correlation between the total expanded body surface and the usual oxygen content of the habitat of marine invertebrates.

In making the evolutionary adjustments to become established in freshwaters, the great majority of freshwater invertebrates have either lost or never had an imposing array of external protuberances. Exceptions (Fig. 1.8)

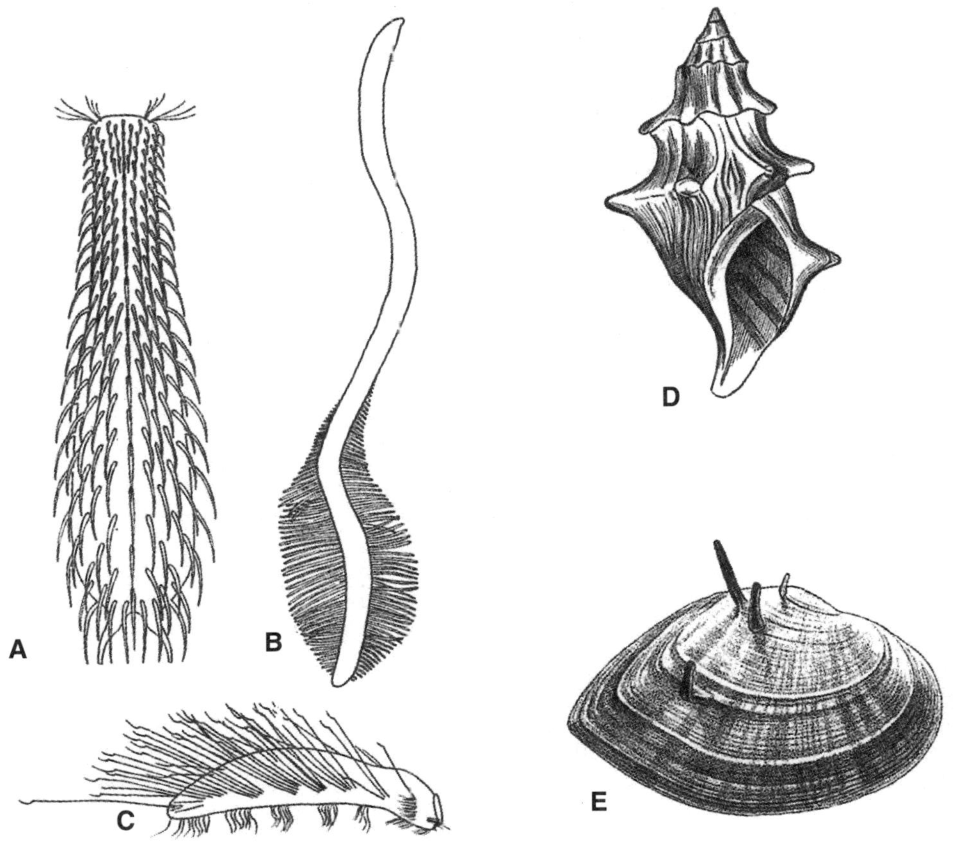

Fig. 1.8. Examples of some of the few freshwater invertebrates having morphological embellishments. Various magnifications. A, gastrotrich, *Chaetonotus;* B, oligochaete, *Branchiura sowerbyi;* C, gastrotrich, *Dasydytes;* D, shell of gastropod *Io fluvialis;* E, shell of pelecypod *Elliptio spinosa.* (D from Tryon 1875, *Smithson. Misc. Coll.* 253:1–435, E from Burch, 1973, *Biota Freshwater Ecosystems, E.P.A. Identification Manual* No. 11:1–176.)

include a few species of freshwater gastrotrichs that have excessive spination; an oligochaete *Branchiura sowerbyi* with excessive "gills"; a few mollusks with spinose shells, and several species of polychaetes that breed in freshwaters, that are probably recent invaders, and that have retained their excess parapodia, palps, and cirri. In the United States these polychaetes include one species each of *Ficopomatus, Lycastoides, Manayunkia, Nereis,* and *Troglochaetus.* Entoprocts and ectoprocts in both environments have equally developed tentacular crowns but these are chiefly food-gathering devices.

The only other freshwater taxa having modest to significant morphological embellishments are (1) oligochaetes with proboscides, elongate setae, and epidermal outgrowths, and (2) the phyllopodous Branchiopoda (fairy shrimps, tadpole shrimps, and clam shrimps), all of which possess 10 to 71 pairs of complicated flat respiratory appendages on the segments. The phyllopodous Branchiopoda are presumed to have originated in freshwaters, so neither group has immediate marine relatives, and thus the modifications are freshwater acquisitions.

We are therefore left with the great majority of freshwater invertebrates, none of which appear to have wasted significant growth energetics on overgrowths or external structures that have no obvious or critical functions. Indeed, it is remarkable that freshwater invertebrates have not developed a whole series of special respiratory embellishments, especially since so many species are exposed to anaerobic or near-anaerobic conditions at some time during the life cycle. A modest complement of gills and other respiratory structures seems to be adequate.

Coloration and color patterns. One of the most pervasive differences between marine and freshwater invertebrates is the general deficiency of bright colors and contrasting patterns among the latter. Some freshwater groups are very colorful: Freshwater sponges are often bright green owing to their intracellular algal populations. Leeches show an array of pigment patterns, as do the shells of most freshwater mussels and pleurocerid and viviparid snails from clean water. Some high-altitude copepods are bright scarlet or bluish. A few gammarids have an orange coloration. Some water mites are bright red, blue, yellow, or green. Red copepods are said to be protected against short-wave radiation (Hairston 1976); red water mites are said to be aposematic (first noted by Elton 1922); and the red in some chironomid larvae is produced by hemoglobin. In addition, most small and translucent metazoans may have transient colors. Distinct and contrasting patterns in coloration are also evident in planarians, which have a finely dotted coloration pattern, and many water mites have elaborate patterns produced by pigments in the body wall and by the varied contents of the visceral organs. Probably the most dramatic color in freshwater invertebrates is seen in crayfish. The section of plates in this book shows many examples of the development of color in freshwater invertebrates.

Marine invertebrates, on the other hand, are often brilliantly colored and patterned. One need look no further than postcards of tropical reefs or any of a myriad of televised specials on marine life. Variable, or chromatophore, coloration is almost restricted to marine forms, where it sometimes reaches astounding levels, especially in certain mollusks and crustaceans. I (RWP) know of very few American freshwater species that can make (rapid) changes in coloration. One example is *Palaemonetes* where chromatophores can color-adapt to the background in less than 24 hours, but this is a far cry from what we see in so many marine species.

No one has suggested the fundamental evolutionary reasons for the notable coloration disparity between marine and freshwater species. It must stem from the differing results of an evolutionary complex of taxon age, genetics, behavior, food, excretory mechanisms, and so on. Indeed, the whole matter of colors and color patterns among aquatic forms is a fundamental problem that has scarcely been touched.

The cellulase problem. The sporadic occurrence of cellulose digestion in the freshwater invertebrate assemblage is a continuing problem. Usually the criterion is the presence of cellulase in the digestive tract, elaborated either by the cells of the gut wall or by the normal bacterial flora of the gut. The "strongest" cellulase has been demonstrated (Monk 1976) for a few *Gammarus* and some snails and limpets. "Weak" cellulase has been found in a few aquatic insects, especially beetles and caddis larvae. Unfortunately, however, there is no correlation between the occurrence of cellulase and the abundance of cellulose in the diet. In view of the fact that there are so very many taxa that ingest large quantities of cellulose (e.g., leaves and other organic detritus), we must conclude that nutrients are derived chiefly from the smaller bulk of noncellular organic matter associated and ingested along with the true cellulose organic matter.

Euryoky of freshwater invertebrates. If estuarine habitats are omitted, even the most casual comparison of marine and freshwater environments emphasizes the much greater variations in ecological factors in the latter. This generalization holds true from one comparable freshwater habitat to another as well as from time to time within the same habitat. Seasonal and vertical temperature gradients in lakes and the greater variations in quiet shallows of ponds are much more pronounced than in saltwater. Currents may vary seasonally from imperceptible to torrential in the same stream, with accompanying variations in suspended silt. Stream invertebrates, in general, are capable of withstanding the severity of violent spring runoffs, intermittent cloudburst waters, and even short-term industrial pollution incidents produced by ions that are normally considered highly toxic. Other chemical conditions, including dissolved salts, dissolved organic matter, dissolved oxygen, and pH are highly variable and cannot be compared with the more monotonous conditions in typical marine habitats. Drying up of the habitat is a frequent and important hazard.

In general, therefore, our freshwater invertebrates consist of a group of euryokous species that are physiologically, morphologically, and behavioristically adapted for maintaining themselves under trying circumstances that would be disastrous or impossible for the great majority of marine forms.

ATYPICAL FRESHWATER HABITATS

As a whole, freshwater habitats are greater in variety (and often in complexity) than those of the marine environment, especially when the former include the complete series of running water habitats and many small, peculiar types of bodies of water. If one is interested in collecting a wide variety of freshwater invertebrates, the most favorable locations are small, weedy lakes and ponds. Such habitats afford large numbers of niches occupied by a correspondingly large variety of animals, especially mollusks, aquatic insects, crustaceans, and many representatives of the worms and wormlike groups. Food and population relationships are intricate, and presumably this complexity is the chief reason why so little comprehensive ecological work has been done on ponds and small lakes.

At the other extreme are highly specialized bodies of water in which one or more ecological conditions are so severe that the fauna is restricted to a small number of species. Frequently, however, these few species attain great population densities, perhaps because of the lack of interspecific competition.

There is little advantage to be gained by outlining the many divisions and subdivisions of freshwater communities. Several such outlines may be found in ecology texts. It may be worthwhile, nevertheless, to characterize some of the more peculiar habitats that represent extremes in ecological conditions. As compared with the usual sort of freshwater habitat, these peculiar habitats are featured by less variable ecological conditions.

Alkali and salt lakes. Alkali and salt lakes and ponds are common in many areas of the western states, especially where there are closed drainage basins, semiarid conditions, and soils containing large quantities of soluble salts. Depending on surrounding soil chemistry, these waters may contain large quantities of chlorides, sulfates, or carbonates, especially those of sodium, potassium, and calcium. The expressions "salt" and "alkali" lakes are by no means sharply defined and are often interchangeable, but, in general, where the dissolved salts are of such composition and concentration as to produce persistent pH readings in excess of 8.8, a body of water is referred to, in common usage, as an "alkali lake."

At low concentrations, up to about 1 to 3% of dissolved salts, the fauna is still varied in species but much more restricted than in nonalkaline waters. Above this concentration species diminish rapidly in numbers, and the metazoan plankton becomes characterized by a few copepods and highly alkaline rotifers, particularly some species of *Brachionus* and *Hexarthra*, which may be extremely abundant. The insect fauna becomes highly restricted and consists mostly of Chironomidae, occasionally one species of *Aedes*, and a few beetles.

Great Salt Lake, which contains about 25% dissolved salt (chiefly sodium chloride), has been studied relatively carefully and has an extremely limited fauna. Protists include many ciliates, flagellates, and amoeboid forms; metazoans consist of only two species of *Ephydra* (the brine fly) and

Artemia fransiscana (the brine shrimp). *Artemia* populations, sometimes attain an average density of 50 individuals per liter. In contrast to the situation in most normal lake beaches, no interstitial micrometazoan fauna has been found in the sandy beaches of Great Salt Lake.

Polluted waters. Waters grossly polluted with organic matter, especially domestic or "city" sewage, likewise have a highly restricted fauna, one capable of thriving in very low concentrations of oxygen and high concentrations of dissolved and particulate organic matter. The chief metazoans are a few species of tubificid oligochaetes, red chironomid dipteran larvae, psychodid larvae, and the rat-tailed maggot larva, *Eristalis*. Sometimes the bottom of polluted rivers is literally covered with a waving, writhing mass of tubificids. This habitat also supports large populations of ciliates and colorless flagellates.

At a certain distance below the entry of domestic pollutants a river enters a "zone of recovery" in which the dissolved oxygen progressively increases and the polluting materials are oxidized or otherwise sufficiently removed from the water, especially as the result of bacterial action, so that the pollution fauna is slowly replaced by a "normal" river fauna. Depending on the intensity of pollution, size of the river, and so on, the zone of recovery may begin anywhere between one quarter-mile and 100 miles downstream from the source of pollution.

Biologists are prone to overlook the fact that the zone of recovery is frequently the most productive part of a flowage. It is here that the periphyton is most dense. Dense metazoan populations include such forms as *Physa* (gastropod), isopods, planarians, leeches, amphipods, and insects (dipterans, caddis larvae, and mayfly nymphs).

In addition to organic pollution, the situation has been further severely aggravated by industrial toxicants, continuous silting, and more and more irrigation system water withdrawals. The history lesson of Lake Erie demonstrates the inherent recovery potential of our freshwater fauna, but many of our waters have been continuously fouled for such long periods, especially by industry, that they are beyond recovery and are simply biological deserts. For this reason those species indicated as having highly restricted ranges in this volume should not be taken at face value, especially in areas where the tradition of lake and stream despoliation has had a long and continuing history.

Aquatic biologists have been searching many years for the ideal common species or group of "indicator" species that would be (1) present unequivocally in the absence of pollution and (2) absent unequivocally in the presence of pollution. Unfortunately, this search appears to be a lost cause chiefly because of the euryokous nature of most species and because of individual differences within the same species. We are therefore forced to rely on facies or clusters of clean-water species and polluted-water species, differing in composition from one area to another, but nevertheless serving as rough indicators of relative purity or pollution.

Astatic ponds. All freshwater lakes and ponds are in truth temporary. The degree of impermanence is, of course, variable, ranging from seasons of the year to thousands of years. Generally, however, lakes and ponds that are persistent from year to year are treated as permanent water bodies. Temporary ponds are those that are dry at some point of the year or become dry every few to several years with regularity. This seemingly arbitrary classification is actually supported by biological data. The fauna of temporary ponds varies in many important respects from that of permanent ponds.

The division of temporary ponds, those that dry annually and those drying at least every few years, follows the arrangement of Hartland-Rowe (1972) who defined the former as seasonally astatic and the latter as perennially astatic. Seasonally astatic ponds are more easily conceptualized because of their annual nature. In North America, these types of temporary ponds are separated generally into those that are presently in spring, thus termed vernal, and those present from fall into spring, referred to as autumnal. The North American summer is the period when both vernal and autumnal ponds are dry.

Perennially astatic ponds can not be so easily categorized. Some can be filled for several years before completely drying. Nevertheless, the biotic communities include most of the higher taxa found in seasonally astatic ponds even though there is usually an increase in diversity of taxa and some of the species are different.

Both seasonally and perennially astatic ponds and their biological communities have several important features in common. First they are isolated, with no permanent streams supplying water. Filling is entirely the result of rainfall, rising water table, or flooding. Another feature defining these water bodies is the complete absence of permanent predators, particularly fish, but including certain other large invertebrate predators such as hirudinid leeches and omnivorous crayfish. Finally, the aquatic animal community that inhabits astatic ponds consists of species whose adaptive strategies have been controlled more by the dry period than the wet period. The requirements of surviving sustained dryness has evoked a wide variety of life history patterns among the inhabitants of the temporary pond habitat. The variations are nearly as diverse as the taxa themselves.

A common thread running through many of the species inhabiting tempoary ponds is rapid development and maturation and a reproductive biology involving precocious young or resting (anabiotic) stages of sorts that enable survival of the population during dry periods. Several of these anabiotic stages are shown in Fig. 1.5.

Among the more conspicuous invertebrate groups that exist in temporary ponds are the phyllopodous Branchiopoda. The "phyllopods" include the clam shrimp, fairy shrimp, and tadpole shrimp. These wide-ranging and common inhabitants of astatic ponds have evolved an egg that resists drying and even requires aerial exposure or severe atmospheric conditions such as freezing before hatching can occur. Less noticeable components of astatic ponds are flatworms of the Platyhelminthes that produce protective cocoons containing developing juveniles. The cocoons ensure survival of the juveniles through dry periods. Many "microturbellarians" have adjusted egg laying and egg physiology to brief and long periods of immersion, allowing them to exist in a variety of temporary to nearly permanent habitats (Heitkamp 1988). One species of flatworm typically found in temporary ponds has assured its survival beyond reproductive adaptations by undergoing fragmentation. The adult worm fragments the posterior region into several pieces that become encysted and remain dormant until favorable conditions return. Refilling of the pond stimulates regeneration of the encysted fragments, each giving rise to a new worm.

Mollusks are moderately represented in astatic ponds. Several species of air-breathing pulmonate snails and certain sphaeriid clams have developed different physiological mechanisms to survive dry periods, including capabilities to retain moisture and reduce metabolism. Insects have evolved larval stages that utilize the rich resources and protection of temporary ponds before maturing and exiting prior to desiccation of the habitat. Certain crayfish of the family Cambaridae have found a unique method of existence in temporary ponds. They have become accomplished burrowers, tunneling down to the water table where survival is possible during dry phases of the pond.

Temporary pond inhabitants are by no means limited to the several examples discussed above. Other groups, including the Rotifera, cladoceran and copepod Crustacea, Nematomorpha, and Tardigrada, all with anabiotic resting stages or other methods of persisting while open water is absent, are found living in temporary ponds. Many of these organisms appear at different successional stages of the pond and their presence is controlled somewhat by the availability of resources. Studies have shown (e.g., Lake et al. 1989) that early occupants comprise shredders and grazers with high turnover rates. Resources "mobilized" by the filling of the basin with water provide nutrition for the emerging consumers. Filter feeders and algae eaters tend to become prominent as suspended materials and phytoplankton become available. Finally, predators, mostly insect larvae, enter the system but do not become a significant part of the fauna until near the end of the pond's existence. Typically, as the pond begins to dry, predators come to outnumber prey species.

Unfortunately, astatic ponds, especially the most ephemeral seasonal ponds, are often lost before they are ever discovered. Their nature of being dry most of the warm part of the year renders them "invisible" and prone to alteration or destruction by human activity. In several parts of the country, the rate at which astatic ponds are being filled in is alarming and only the actions of observant biologists, naturalists, and concerned citizens can slow the pace of destruction by mapping and cataloging such important habitats.

Subterranean waters. Subterranean streams, cave ponds, and artesian wells sometimes contain endemic, primitive, or highly specialized turbellarians, hydrobiid snails, amphipods, isopods, decapods, and copepods. Recent discoveries have revealed freshwater sponges (*Eunapius*), bivalves of the genus *Congeria*, and the cnidarian *Cordylophora* living in freshwater caves. The Ozarks and similar limestone areas in the United States are particularly rich in underground aquatic habitats. Most subterranean species (Fig. 1.9) are colorless, translucent, or whitish; eyes are nonfunctional or absent; tactile structures are often well developed and include long antennae and abundant long setae. The subterranean environment is completely monotonous, with constant ecological conditions. Basic food materials consist of fungi and small bits of organic debris washed underground. Populations are generally small, owing to the greatly limited food supply.

Endemic subterranean species are thought to have been derived from ancestral surface species that underwent evolutionary change and specialization concomitantly with the gradual submergence of their surface drainage habitats. Compared with Europe, little is known of the subterranean microcrustacean fauna of the United States, but it is clear we do have a rich fauna. Recently, the

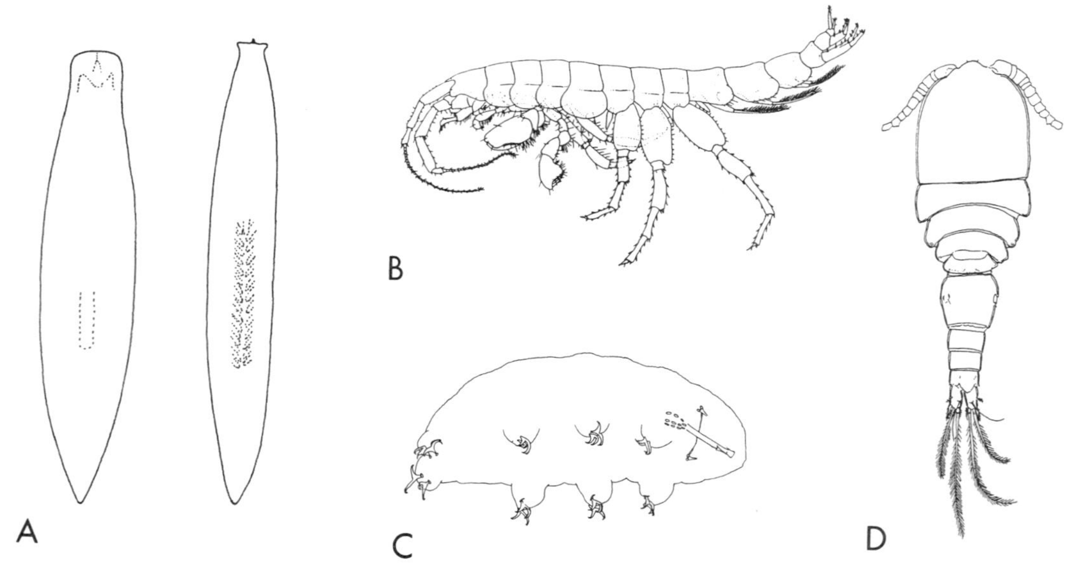

Fig. 1.9. Examples of aquatic invertebrates from subterranean and phreatic habitats. Various magnifications. A, *Sphalloplana* spp., blind planarians from caves; B, *Stygobromus tenuis*, blind amphipod from a cave; C, *Isohypsibius* sp., phreatic tardigrade; D, *Rheocyclops carolinianus*, phreatic copepod. (A from Kenk 1977, *Smithson. Contrib. Zool.* 246:1–38; C from Pilato 1996, *N. Z. J. Zool.* 23:67–71, with permission of the Royal Society of New Zealand; D from Reid et al., 1999, *J. Crustacean Biol.* 19:384–396.)

subterranean fauna of several states has been extensively reviewed (Peck and Lewis 1977; Peck and Christiansen 1990; Franz et al. 1994; Peck 1998). It should be added that accessible subterranean streams and ponds usually also contain common species that are regular members of aquatic communities on the surface of the ground.

Phreatic and psammolittoral waters. These waters are ecologically closely related to subterranean waters and at times the two systems overlap in hydrology and fauna. However, there is a distinction, primarily in the size of the organisms inhabiting each zone and the relationship of each to surface waters. Pennak (1988) has presented some criteria for hyporheic and psammolittoral habitats. Hydrologically, hyporheic and psammolittoral waters are those in close contact with surface water, lotic or lentic, respectively. Pennak (1988) differentiates hyporheic and psammolittoral regions from surface water by the distance beneath or away from standing water and the nature of the materials holding water. Depth of these habitats can range as much as 50 cm beneath the surface but are usually much less (Pennak 1988). The damp sandy beaches of lakes, the upper few centimeters of mud at the pond bottom, or the sand and gravel shoreline of a stream all constitute types of hyporheic and psammolittoral habitats (Fig. 1.10). The common factor uniting all these habitats is the extreme small size of the inhabitable space, or interstitial space, in

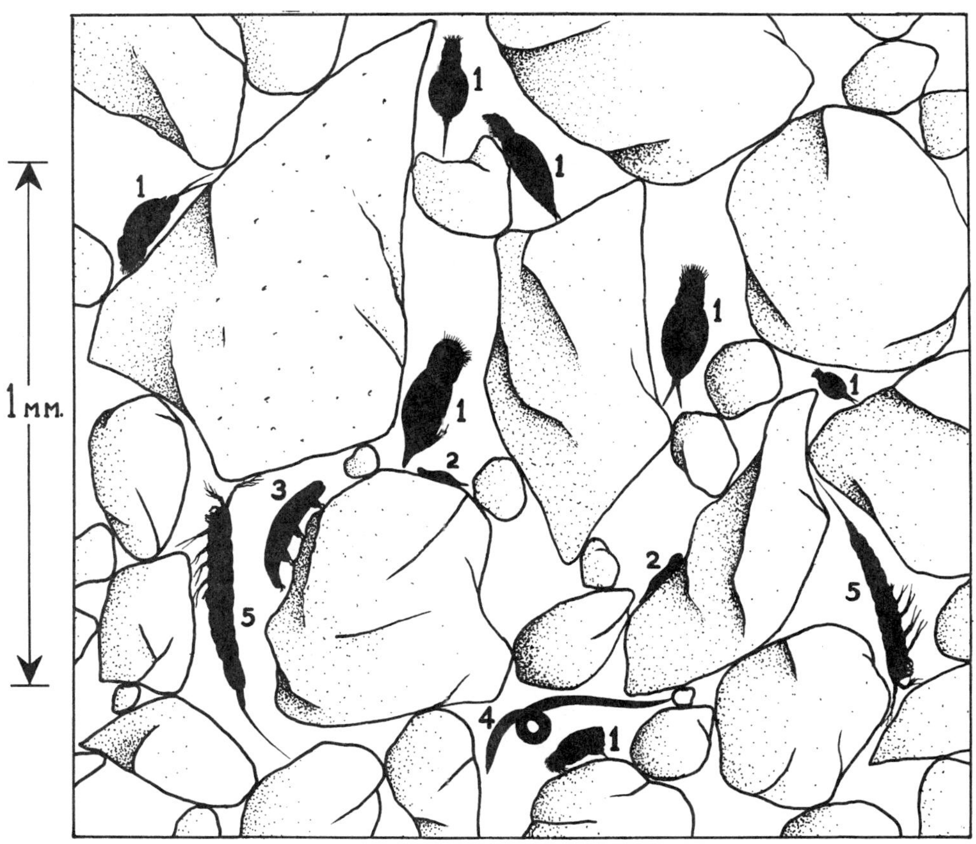

Fig. 1.10. Optical section of a small portion of a sandy beach, 15 cm deep and 150 cm from the water's edge, showing relative sizes of sand grains and some of the common micrometazoans. 1, rotifers; 2, gastrotrichs; 3, tardigrade; 4, nematode; 5, harpacticoid copepod. (From Pennak 1939.)

which organisms can live. As a consequence, organisms living in psammolittoral or phreatic environments are necessarily small, usually less than 2 mm in size (Pennak 1988) and mobile. Animals with a sessile lifestyle have not been reported in freshwater hyporheic and psammolittoral habitats.

Small size is a necessary adaptation to life between particles and an important adaptation to this lifestyle has been to decrease the diameter of the body (Giere 1993). Elongation is a consequence of the need to move beneath grains of sand so animals that can undergo substantial size reduction in company with a high length-to-diameter ratio have become successful in groundwater habitats. Added to size and dimension considerations is mobility and the ability to anchor to particles (Giere 1993). Thus rotifers, tardigrades (Fig. 1.9), nematodes, gastrotrichs, mites, annelida, copepods (Fig. 1.9), micromalacostracan crustaceans and "microturbellarians" are well represented in these habitats.

Because of the extreme small size of the inhabitants of phreatic and psammolittoral waters, endemism is very high. Nevertheless, some characteristic groundwater animals are widely distributed. A possible answer for the wide distribution of some species is the observation of some studies that some organisms confined to hyporheic and psammolittoral habitats are apparently unaffected by variations in certain chemical properties and resource abundance (Strayer 1994), but additional study has suggested otherwise (Strayer et al. 1997). It does seem that uncontaminated groundwater environments with a food source, adequate oxygen, and appropriate sediment characteristics should support phreatic invertebrate inhabitants.

Although neglected in the past (Pennak 1988), hyporheic and psammolittoral environments in North America have enjoyed active study in recent years. With increased research it is becoming apparent that groundwater habitats contain a rich and as yet hardly understood fauna.

Cold springs, spring brooks. Typical cold spring communities are restricted to a small area at and just below the point where the water issues from the ground. As soon as it flows any distance as a spring brook it becomes warmer, loses it characteristic growth of watercress, and contains a different animal community. Depending on local conditions, the typical spring and spring brook fauna zone usually extends from 10 to 200 inches below the source. Cold springs have an average temperature of about 8°C in temperate areas, and the associated animals are mostly cold stenotherms. The list of characteristic animals is relatively small, including planarians, amphipods, and some of the following: beetles, caddis larvae, leeches, isopods, sphaeriid clams, hydrobiid snails, and insect larvae. Frequently, two or more of these species are numerically dominant. If the spring emerges from appropriate underground water courses in limestone areas it occasionally brings up typical subterranean species. The ecology of springs and spring brooks has been sadly neglected in the United States.

Hot springs. In addition to their high temperatures (about 35°C), hot springs are often characterized by large quantities of dissolved salts, Sometimes the water is also high in dissolved carbon dioxide, hydrogen sulfide, or sulfur dioxide. Thus, it is often difficult to distinguish between the limiting effects of temperature and the limiting effects of other ecological factors. In general, however, 25 to 30°C seems to be the dividing line between a rich fauna at lower temperatures and a poor fauna at higher temperatures. Unlike cold springs, there do not seem to be any typical genera or species in hot springs, nor are hot spring animals ever abundant. Beetles are among the most characteristic metazoan inhabitants, many species having been reported from a wide variety of hot springs in the western states. Insects (Diptera larvae, especially Chironomidae) are also common inhabitants. Sometimes a few Hemiptera and Odonata may be collected. Most such records of insects are from springs whose temperatures range from 37 to 50°C. Amphipods, ostracods, and Hydrachnida are reported from springs with temperatures between 32 and 51°C. Micrometazoans are often present in exceptionally hot water, a few nematodes and rotifers having been taken as high as 60°C. Since the thermal death point of most freshwater invertebrates is between 30 and 40°C, many hot spring animals must have developed considerable thermal acclimation.

DISPERSAL AND BARRIERS

The successful establishment of the ancient marine ancestors of freshwater animals in swamps and near estuaries was actually only the beginning of the story. From such footholds they must have developed a higher degree of euryoky as they expanded to occupy new inland habitats.

Migration up and downstream throughout a single drainage system is comparatively easy for many freshwater invertebrates over a long period of time. Superficially, it would appear that they are "landlocked" in separate drainage systems, with little migration from one to another. In the long-time viewpoint, however, land barriers between different lakes and rivers are not effective for most groups within the confines of a single continental land mass. The ability of almost all microscopic species to form thick-shelled eggs, cysts, or resting stages that withstand partial drying or desiccation is a primary factor in ease of dispersal. Such small resting bodies are easily picked up and wind blown overland for long distances. If they drop into a favorable aquatic habitat with the appropriate combination of ecological conditions, they again become active and may produce a new population nucleus.

The literature contains scattered but authentic records of live fish being transported overland by storms, tornadoes, and other strong winds, and there is no reason to doubt that plankton and bottom invertebrates are transported in the same manner. Much of the "dust" of dust storms is organic debris, and it is not improbable that a small fraction consists of cysts and eggs. Ducks, shore birds, and aquatic insects are now known to be important for increasing the ranges of many groups, and the literature contains many accounts of such transport. Both resting and active stages are transported in the detritus and mud that frequently sticks to the appendages. Snails are frequently found on migratory birds, and there is one authentic record of a 2.5-oz. live mussel being transported overland by a duck! Many microzoan cysts and resting eggs are carried to new habitats via the intestinal tract of birds. Adult aquatic insects capable of flight easily migrate overland, often in what appears to be a completely random fashion, and come to occupy suitable new areas. Sometimes such flights are greatly aided by winds. Anyone who follows the biological development of a new, man-made reservoir is bound to be impressed with the promptness with which the reservoir becomes colonized with aquatic animals. Stock tanks and stock ponds, newly constructed in grazing lands where the nearest water is 10 km away, quickly acquire a varied invertebrate population. The literature of anabiotic transport by wind and vagile animals consists chiefly of fragmentary and widely scattered papers; a few of the more significant are Niethammer (1953), Lansbury (1955), Fernando (1958), Loffler (1963), Maguire (1963), Malone (1965), Proctor and Malone (1965), Dundee et al. (1967), Stewart et al. (1970), Carroll and Viglierchio (1981).

It is not easy to explain the faunal composition of small spring brooks on the basis of anabiotic structures and overland transport. Such habitats are abundant and widely scattered over the United States, but are effectively isolated from each other by a matter of a few to many kilometers. The community makeup is readily predictable. A dense growth of watercress (*Nasturtium officinale*) is almost invariably present, the substrate is usually gravel, temperature is constant and cool, and the whole community extends from the emergence point of water from the ground to the mouth of the brook, or to the point where watercress is no longer present. Few spring-inhabiting species have anabiotic stages that would facilitate geographic distribution. Adult spring brook insects can, of course, easily migrate from one site to another, but most of the other spring brook species are presumably transported on the bodies of flying insects, waterbirds, and shorebirds. Since spring brooks are first-order streams, it would be difficult for the various species to migrate actively and extensively downstream and upstream throughout large flowages, and through unfavorable habitats in so doing. We have seen many small spring brooks on steep, rocky canyon walls; active migrations into such sites would be impossible, except for adult insects.

It is tempting to theorize that spring brooks are stable habitats, and that their corresponding faunal community therefore is composed largely of species that do not produce anabiotic or diapause stages. While it is now convenient to explain the colonization of aquatic areas by disseminules distributed especially by insects and birds, one wonders about the zoogeographic spread of freshwater invertebrates in the long-time sense. How, for example, was it brought about before the evolutionary appearance of insects, birds, and other terrestrial vertebrates in Earth's history? Surely a freshwater fauna existed before birds and insects. Were winds, flooding, and changing drainage patterns then the chief means of distributing disseminules?

Some highly euryokous and vagile species, mostly microscopic, have spread so effectively that they are essentially cosmopolitan and occur the world over in suitable habitats. The list includes a great many Protoctista and Rotatoria, several sponges, *Paracyclops fimbriatus*, perhaps *Plumatella repens*, a few oligochaetes, some Tardigrada, and several Cladocera, such as *Daphnia longispina* and *Chydorus sphaericus*. Many more species could be added to this list.

The freshwater invertebrate faunas of North America and Europe, and to some extent northern Asia, have many of the species in the following groups in common: Protoctista, Rotatoria, Tardigrada, Entoprocta, Ectoprocta, Oligochaeta, and Cladocera. American gastrotrichs and nematodes are so poorly known that their European affinities are not well established. In all the other major taxa with low vagility, however, the great majority of species that occur in the United States are endemic and are not found in Europe.

The zoogeographic distribution of freshwater invertebrates has been rather well investigated in Europe, and zoogeographic regions have been suggested for several taxonomic groups, especially some of the orders of crustaceans and insects and the mollusks. Numerous distribution patterns have been correlated with the effects of the advances and retreats of glaciers on the European continent. In the United States, however, comprehensive studies of freshwater invertebrate zoogeography are largely lacking, with the notable exceptions of the mollusks, decapods, and perhaps a few other groups.

Europe, Asia, and Africa have certain very ancient lakes where primitive species have been preserved and where geographic isolation has been effective for such a long time that independent speciation has resulted in species "flocks" comprising many unique and endemic species, especially among Porifera, Turbellaria, Oligochaeta, Gastropoda, and crustaceans. Some of the better known lakes of this type are Baikal in Asia, Tanganyika and Nyasa in Africa, Ochrid in Jugoslavia, and the lakes of the Malili River system in the central Celebes. No such ancient lakes occur in the United States.

THE EPIDEMIC OF EXOTICS

One of the most obvious and significant changes occurring in the aquatic fauna of the United States and North America in general is the rapid and unabated spread of exotic species. Originating both within North America and elsewhere, species have been moved about the continent with such success that some exotic species have or are becoming the dominant representatives of certain higher taxa wherever they occur. First identified as a potential ecological problem by Elton (1958) and Laycock (1966), the effect of exotic organisms on native faunae was neglected for years, rarely discussed or considered in conservation science (Temple 1990). It has only recently become an important issue among ecologists and conservation biologists. In the aquatic environment, it took two biofouling mollusks, the asian clam (*Corbicula fluminea*) and later the zebra mussel (*Dreissena polymorpha*), and millions of dollars of restoration and refit to stimulate the scientific community to begin seriously investigating the effects of exotic species on native species and their habitats. In the last ten years, much has been written on the subject and comprehensive works are now being published on the biology of invasive species (e.g., Claudi and Leach 1999). A partial list (Table 1.6) of exotic freshwater species believed introduced into the United States shows the breadth of taxa that become established. The list is dominated by mollusks, as these invertebrates have been the most intensively researched (e.g., Robinson 1999; Strayer 1999). A few are shown in Fig. 1.11 and Plate 16. The list includes parasitic species. Many species whose exotic status is debatable have been left off the list. Most of the species on the list have been introduced into the Great Lakes or the Florida region.

When introductions of aquatic organisms into the United States from elsewhere actually commenced is unclear. It is likely that they began occurring as soon as Europeans settled North America, bringing with them casks of drinking water containing the resting stages (anabiotic stages, see above) of any number of species picked up along the coast of Europe and Africa. Later, immigrants, especially from Asia, came to the continent, bringing with them exotic species, such as snails and plants, to be consumed during the trip. With increased shipping and immigration the rate of introductions increased. Canal systems were subsequently developed, linking the coast with inland bodies of freshwater, permitting ships to penetrate deeper into the continent to discharge their alien cargo into large, resource rich freshwaters.

One well-documented vehicle of invasion is the ballast of oceangoing vessels. The loading of fresh ballast water in foreign ports has been identified as a major cause of introduction of exotic species into North America. Representatives of 16 higher (order and above) taxa have been recorded living or dead in the ballast of oceanic vessels (Locke et al. 1993). A recent and now infamous result of foriegn ships pumping ballast contaminated with exotic species is the establishment of the pelecypod mollusk family Dreissenidae into the Great Lakes and beyond. The dreissenid populations in North America—at least two species are now present—presumably originated in the Caspian or Black Sea region (Ludyanskiy 1993). The Dreissenidae are generally considered to be recent immigrants to freshwater and retain much of their marine habitats, including the production of a planktonic veliger larva. After settling and following metamorphosis, the young mussel produces a holdfast or byssus produced by glands in the foot. The byssus cements the animal to the substrate. Within months, the animal matures and begins to reproduce. Although many species in the family retain a estuarine physiology, several species have become fully adapted to freshwater and have spread throughout the major river systems of Europe.

As the story goes, veligers of dreissenid mussels living in basins most likely entered the ballast of ships embarking for the United States. Obviously, the ballast of some ships had conditions enabling survival, maturation, and the production of new veligers while enroute. Upon arriving in the Great Lakes, with canals and streams linking to nearly every major drainage system in the eastern United States, the vector ship, or ships, simply pumped its veliger-laden ballast. This particular event and its consequences have been documented in almost every form of media. The problems posed by the transport of exotic species in ship ballast has been discussed at length, most notably by Carlton (1989).

Aquatic exotic species are not limited to those originating outside of North America. Many species native to parts of the United States have been either deliberately or accidentally introduced into other parts of the country. A particular group of aquatic invertebrates that has been deliberately moved about is the Decapoda, a crustacean order including crayfish and shrimp. Authorities have purposefully introduced exotic crayfish species to new areas to enhance the food supply for sport fish. One species of crayfish, *Orconectes rusticus*, has been included in

Table 1.6. Some Exotic Freshwater Noninsect Invertebrate Species Known or Suspected to Be Established in the United States.

Species	Date of entry	Origin
Porifera		
Radiospongilla cerebellata	Before 1970	Southern Asia
Cnidaria		
Polypodium hydriforme	Ca. 1960	Russia
Craspedacusta sowerbii	Before 1900	China
Platyhelminthes		
Rectocephala exotica	Before 1950	?
Schmidtea polychroa	Ca. 1960	Europe
Nemertea		
Apatronemertes albimaculosa	Ca. 1980	?
Annelida		
Allonais paraguayensis	Before 1990	South America
Eiseniella tetraedra	Before 1900	Europe
Psammoryctides barbatus	Before 1970	Europe
Mollucsa		
Bellamya japonica	Ca. 1900	Japan
Bellamya chinensis	Ca. 1900	China
Bithynia tentaculata	Before 1900	Europe
Corbicula fluminea	Before 1950	China
Dreissena polymorpha	Ca. 1980	Eastern Europe
Dreissena bugensis	Ca. 1980	Eastern Europe
Marissa cornuaurietus	Ca. 1960	South America
Melanoides tuberculata	Before 1970	Southeast Asia
Pisidium amnicum	Before 1900	Europe
Pomacea bridgesi	Ca. 1960	Brazil
Potamopyrgus antipodum	Ca. 1985	New Zealand
Radix auricularia	Before 1900	Europe
Sphaerium corneum	Before 1900	Europe
Tarebia granifera	Before 1970	Southeast Asia
Valvata piscinalis	Before 1900	Europe
Viviparus viviparus	Before 1980	Europe
Ectoprocta		
Plumatella vaihiriae	Before 1940	Polynesia?
Lophopodella carteri	Before 1930	Asia
Crustacea		
Asellus hilgendorfii	Before 1995	Asia
Argulus japonica	Ca. 1960	Asia
Bythotrephes cederstroemi	Ca. 1984	Europe
Cercopagis pengoi	Before 1998	Caspian region
Cyclops strenuus	After 1959	Europe
Daphnia lumholtzi	Before 1990	Africa
Echinogammarus sp.	Ca. 1990	Caspian region
Eubosmina coregoni	Before 1984	Europe
Exopalaemon modestus	After 1980	China
Macrobrachium olfersi	Before 1950	Mexico
Macrobrachium rosenbergii	Before 1980	Malaysia
Megacyclops viridis	After 1959	Europe
Nitocra hibernica	After 1959	Europe
Platychirograpsus spectabilis	Ca. 1930	Mexico

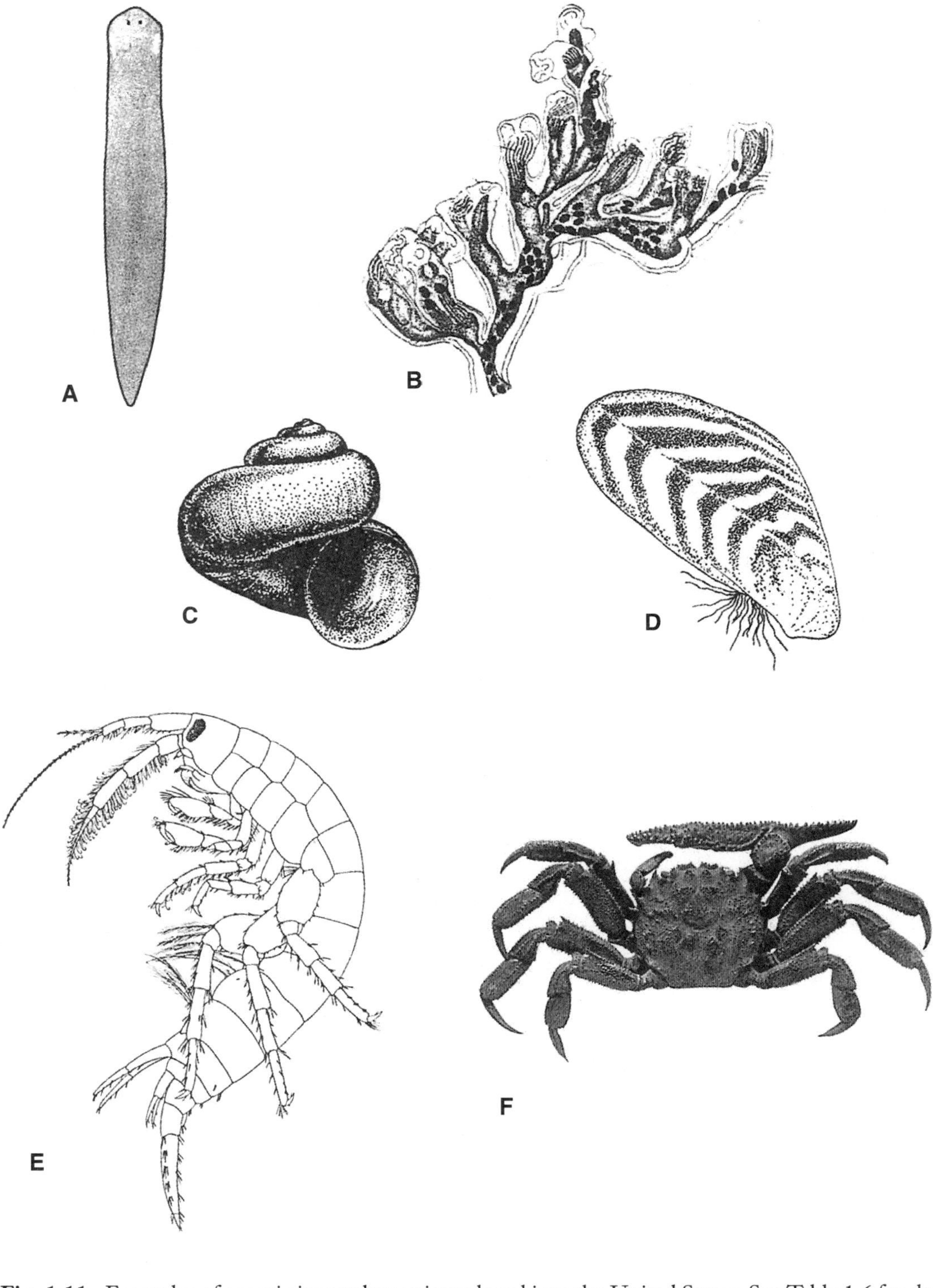

Fig. 1.11. Examples of aquatic invertebrates introduced into the United States. See Table 1.6 for details. Various magnifications. A, *Schmidtea polychroa;* B, *Plumatella vaihiriae;* C, *Valvata piscinalis;* D, *Dreissena polymorpha;* E, *Echinogammarus* sp.; F, *Platychirograpsus spectibilis.* (A from Kenk 1977, *Smithson. Contrib. Zool.* 246:1–38; B from Rogick and Brown 1942, *Ann. N. Y. Acad. Sci.* 43:123–144; C from Macan 1977, *Freshwater Biological Assoc. Sci. Publ. 13;* E from Barnard and Barnard 1983, *Freshwater Amphipoda of the World I: Evolutionary Patterns,* Hayfield Associates, Mt. Vernon, Virginia; F from Rathbun 1914, *Proc. U. S. Nat. Mus.* 47:117–129.)

such practice and now occurs in many regions outside of its natural range. Unfortunately, this species has turned out to be the top consumer, not the fish it was supposed to nourish. Several studies have revealed that *O. rusticus* hybridizes with native species and can eventually exclude native crayfish and can exhaust entire littoral habitats, rendering them useless for other species, including young fish. Although states have outlawed this species it continues to spread. It is now known in at least eight states outside of its natural range.

The impacts of other less-heralded introductions involving more cryptic species such as the jellyfish *Craspedacusta sowerbii* and the turbellarian *Schmidtea polychroa*, remain unknown. All we can hope is that these smaller less-conspicuous intruders have little affect on aquatic communities, because they are virtually established in the United States and occur in nearly every state.

The success of many exotic aquatic species will forever alter the makeup of many freshwater habitats. In many places native species, unable to quickly adapt or adjust to their new neighbors, will decline. Eradication of exotic species seems unlikely since the methods to eliminate unwanted aquatic pests are expensive, labor intensive, and invasive, often negatively affecting the native species as much as the exotic. Aquatic scientists will have to become familiar with these new members of the fauna and consider them seriously in any efforts dealing with aquatic ecology or management.

FOOD WEBS

During the past two generations the literature of aquatic biology has been replete with "productivity" and "trophic level" studies. It is unfortunate that most of these studies are misleading because they (1) are seasonal rather than year round, (2) have been conducted on complex (rather than simple) ecosystems, and (3) are based on only a fraction of the species making up a particular ecosystem. As a consequence, the generalizations postulated in many of these investigations are open to doubt and modification. Actually, a much more fruitful approach of population dynamics and interrelationships involves habitats characterized by few species and few niches. Examples are alkali and saline ponds, relatively barren alpine lakes, spring brooks, very small ponds, warm springs, and small mountain streams. Using such bodies of water, it should be relatively easy to determine food habits of individual species and to make quantitative estimates of populations. Productivity indexes, food cycles, energy transfers from one trophic level to another, species interactions, and seasonal population fluctuations are all important ecological concepts that should be derived from such studies on a year-round basis. It is important to note that the measurement of associated physical and chemical factors in small streams, lakes, and ponds is simpler than in more complex types of aquatic habitats and in terrestrial habitats.

Nevertheless, even in the simplest aquatic habitats the determination of pyramids of numbers and food chains (or, better, "food webs") is an involved problem, and only a very few thorough and inclusive studies have been attempted. The great majority of trophic investigations are concerned with only a few energy or biomass transpositions, the remainder being only roughly estimated. In any such study the basic role played by detritus and bacterial action should not be minimized. In a schematic fashion Fig. 1.11 shows the fundamental food interrelationships for any aquatic habitat, regardless of size and complexity. Only recently have aquatic ecologists come to realize the overwhelming role of detritus in the food web. The significance of the various segments of this diagram vary enormously from one habitat to another. Some habitats, for example, have few or no fishes; other habitats have no rooted aquatics or a poor substrate fauna.

The solid lines in Fig. 1.12 indicate consumption or utilization of food. Broken lines indicate death and disintegration, or defecation by animals. The dotted line indicates disintegration of detritus as the result of certain bacterial activities. The circle arrows originating and ending at "zooplankton," "fish," and "substrate fauna" are meant to

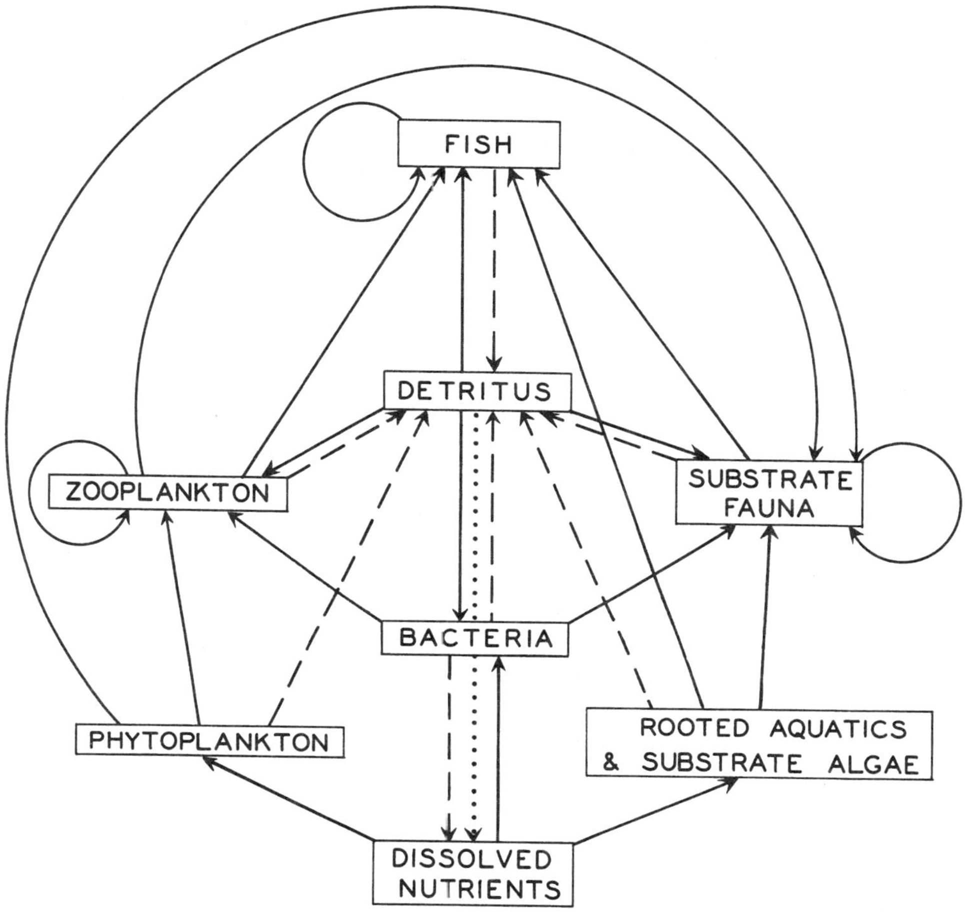

Fig. 1.12. Basic features of aquatic food webs. See text for explanation.

show that certain species in each of these categories are carnivorous and feed on other individuals within the same category. Actually each such category may represent two, three, or more trophic levels. "Substrate fauna" includes all bottom invertebrates as well as the protoctistans, micrometazoans, and macrometazoans associated with the surface of rooted aquatic plants and other objects. By inference, "bacteria" also includes molds and their activities.

For simplicity, certain minor aspects of this food web have been omitted. For example, the contribution of excretory materials to the main reservoir of dissolved nutrients is not shown. By inference, "fish" should include other vertebrates, such as frogs, salamanders, and turtles.

Not shown, but nevertheless intimately concerned with the continuing balance within the web, are the nitrogen, sulfur, and phosphorus cycles. These are fundamentally similar to the same cycles operating in terrestrial habitats.

COLLECTING ETHICS

A well-documented crisis affecting ecosystems of the entire world is the decline of many species of plants and animals. In some cases the observed decline is astonishing, with entire species disappearing in as little as a human lifetime. The loss of both density, in terms of standing population size, and diversity at the species and even genus level is acutely apparent in freshwater environments.

The rate of decline among freshwater species has led many states to develop lists of species in various categories of special concern and to impose regulations controlling activities that involve these species. A 20-year-old compilation (Berger and Neuner 1979) of the regulations existing for each state, where data existed, shows that by 1979, nearly every state of the contiguous United States required permits for the collecting of zoological and plant specimens. Although protection was infrequently extended to invertebrates at the time, most states have since included invertebrate species, especially those occurring in freshwater, within the scope of protection.

Most collecting permits stipulate the range of taxa to be sampled and the size of the samples allowed to be collected. Permits typically regulate the higher profile and more endangered groups of freshwater invertebrate species, more precisely the mollusks and crustaceans. The less familiar and smaller groups are generally regulated, but not as strictly. Nevertheless, many states with high concentrations of endemic species and especially habitat-restricted forms, such as those adapted to cave environments, have become very aggressive in limiting access to and collecting of listed species.

Scientists may find these restictions an inconvenience, but should understand that these policies represent an attempt to monitor activity affecting species as much as an effort to prevent overcollecting. Whereas it is essential to gather data in any study of freshwater invertebrate communities, which obligates the collecting of specimens, it is becoming more important for investigators to weigh their research needs against the ability of the dwindling population to replace the removal of specimens.

COMMENT

By way of a brief summary, we must accept the proposition that incipient colonizers of freshwaters must have first acquired, in varying degrees, whole clusters of special devices and mechanisms in order to establish successful and permanent populations in the harsh and variable freshwater environment. At the same time, certain specializations for the marine existence must have been lost. Our understanding of evolutionary and genetic processes does not permit the de novo establishment of freshwater taxa. Rather we think of these groups as only very rarely and precariously being successful in the new environment. We visualize the ages of freshwater taxa as being quite different from each other. Some are probably "new"; others are "ancient." But, in my (RWP) own estimation, and regardless of age, the presiding requisite for life in freshwaters is that of euryoky.

A number of freshwater invertebrate rarities and ecological curiosities in the United States are associated in various ways with the topics we have enumerated. Some involve "escapes," that is, certain uncommon freshwater invertebrates with close marine ancestry that have made the marine (brackish?) to freshwater transition accidentally and contrary to the situation among their closest relatives.

Undoubtedly, many who read the introduction will disagree with some of the generalizations and suggestions. If so, I (RWP) shall be pleased in knowing that I (RWP) have been responsible for such attention being accorded to the freshwater invertebrate biota. I (RWP) hope such controversy will generate future field and laboratory research.

REFERENCES

Adolph, E. F. 1925. Some physiological distinctions between freshwater and marine organisms. *Biol. Bull.* 48:327–335.

Anger, K. 1995. The conquest of freshwater and land by marine crabs: adaptations in life-history patterns and larval bioenergetics. *J. Exp. Marine Biol. Ecol.* 193:119–145.

Ashwood-Smith, M. J., and J. Farrant. 1980. *Low Temperature Preservation in Biology and Medicine.* 323 pp. Pittman Medical, Kent, UK.

Ax, P., and R. Ax. 1970. Das Verteilungsprinzip des subterranen Psammon am Cübergang Meer-Susswässer. *Mikrofauna des Meeresbodens* 1: 1–51.

Baldwin, E. 1948. *An Introduction to Comparative Biochemistry.* 164 pp. Cambridge Univ. Press, Cambridge, UK.

Bardachj, E., et al. 1972. *Aquaculture.* 868 pp. Wiley-Interscience, New York.

Barnes, T. C., and T. L. Jahn. 1934. Properties of water of biological interest. *Q. Rev. Biol.* 9: 292–341.

Barrington, E. J. W. 1979. *Invertebrate Structure and Function,* 2d ed. 765 pp. Halsted, New York.

Beadle, L. C. 1943. Osmotic regulation and the faunas of inland waters. *Biol. Rev.* 18:172–183.

——. 1957. Osmotic and ionic regulation in aquatic animals. *Annu. Rev. Physiol.* 19:329–358.

——. 1969. Osmotic regulation and the adaptation of freshwater animals to inland saline waters. *Proc. Int. Assoc. Theor. Appl. Limnol.* 17: 421–429.

Berger, T. J., and A. M. Neuner. 1979. *Directory of State Protected Species: A Reference to Species Controlled by Non-game Regulations.* Association of Systematics Collections, Lawrence, KS.

Bick, H., et al. 1972. Das Zooplankton der Binnengewdsser. *Die Binnengewässer* 26(l):1–294.

Bogatov, V. V., et al. 1980. The luminescence of freshwater Oligochaeta. *Hydrobiol. J.* 13:42–47.

Bond, R. M. 1933. A contribution to the study of the natural food cycle in aquatic environments. *Bull. Bingham Oceanogr. Coll.* 4:1–89.

Bowman, T. E., and F. Phillips. 1984. Bioluminescence in the freshwater amphipod *Hyalella azteca* caused by pathogenic bacteria. *Proc. Biol. Soc. Wash.* 97:526–528.

Brandt, K. 1897. Die Fauna der Ostsee, in besondere die der Kieler Bucht. *Verh. Deutsch Zool. Gesellschafit* (1897):10–34.

Brooks, J. L. 1950. Speciation in ancient lakes. *Q. Rev. Biol.* 25:30–60, 131–176.

Brues, C. T. 1932. Further studies on the fauna of North American hot springs. *Proc. Am. Acad. Arts Sci.* 67:185–303.

Cairns, J., and K. L. Dickson (eds.). 1973. Biological methods for the assessment of water quality. *ASTM Special Tech. Publ.* 528:1–256.

Carlton, J. T. 1989. Man's role in changing the face of the ocean: biological invasions and implications for conservation of near-shore environments. *Conservation Biol.* 3:265–273.

Carpenter, K. E. 1928. *Life in Inland Waters.* Sidgwick & Jackson, London.

Carroll, J. J., and D. R. Viglierchio. 1981. On the transport of nematodes by the wind. *J. Nematol.* 13:476–482.

Claudi, R., and J. Leach (eds.). 1999. *Non-indigenous Freshwater Organisms. Biology and Impact.* 464 pp. Lewis, Boca Raton, FL.

Coker, R. E. 1954. *Streams Lakes Ponds.* 327 pp. Univ. North Carolina Press, Chapel Hill.

Coull, B. B., and J. Grant. 1981. Encystment discovered in a marine copepod. *Science* 212: 342–343.

Danell, K. 1981. Overwintering of invertebrates in a shallow northern Swedish lake. *Int. Rev. gesamten Hydrobiol.* 66:837–845.

Delamare Deboutteville, C. 1960. *Biologie des Faux Souterraines Littorales et Continentales.* Hermann, Paris.

Denny, M. W. 1983. A simple device for recording the maximum force exerted on intertidal organisms. *Limnol. Oceanogr.* 28:1269–1274.

Dorgelo, J. 1976. Salt tolerance in Crustacea and the influence of temperature upon it. *Biol. Rev.* 51:255–290.

Dougherty, E. C., and L. G. Harris. 1963. Antarctic micrometazoa: freshwater species in the McMurdo Sound area. *Science* 140:497–498.

Dundee, D. S., et al. 1967. Snails on migratory birds. *Nautilus* 80:89–91.

Edmondson, W. T., and G. G. Winberg. 1971. Secondary productivity in fresh waters. *I.B.P. Handbook* 17:1–358.

Elton, C. 1922. On the colours of water-mites. *Proc. Zool. Soc. London* 1922:1231–1239.

Elton, C. S. 1958. *The Ecology of Invasions by Animals and Plants.* 181 pp. Chapman & Hall, London.

Fernando, C. H. 1958. The colonization of small freshwater habitats by aquatic insects, 1. general discussion, methods and colonization in the aquatic Coleoptera. *Ceylon J. Sci. Biol. Sci.* 1: 117–154.

Forbes, S. A. 1887. The lake as a microcosm. *Bull. Peoria Sci. Assoc.* 1887:77–87.

Foster, N. 1972. Freshwater polychaetes (Annelida) of North America. *Biota of Freshwater Ecosystems Identification Manual,* No. 4:1–15.

Fox, H. M., and B. G. Simmons. 1933. Metabolic rates of aquatic arthropods from different habitats. *J. Exp. Biol.* 10:67–74.

Franz, R., et al. 1994. Review of biologically significant caves and their faunas in Florida and south Georgia. *Brimleyana* 20:1–109.

Frey, D. G. (ed.). 1962. *Limnology in North America.* 734 pp. Univ. Wisconsin Press, Madison.

Fryer, G. 1974. Evolution and adaptive radiation

in the Macrothricidae (Crustacea: Cladocera): a study in comparative functional morphology and ecology. *Philos. Trans. R. Soc. London Ser.* 269:142–272.

Galtsoff, P. S. 1934. The biochemistry of the invertebrates of the sea. *Ecol. Monogr.* 4:481–490.

Giere, O. 1993. *Meiobenthology—The Microscopic Fauna in Aquatic Sediments.* 328 pp. Springer, Berlin.

Golterman, H. L. 1969. Methods for chemical analysis of fresh waters. *I.B.P. Handbook* 8:1–166.

Green, J. 1968. *The Biology of Estuarine Animals.* 401 pp. University of Washington, Seattle.

Grimaldi de Zio, S., et al. 1982. *Neostygarctos acanthophorus* n. gen., n. sp., nuovo Tardigrado marino del Mediterraneo. *Cah. Biol. Mar.* 23:319–323.

Hairston, N. G. 1976. Photoprotection by carotenoid pigments in the copepod *Diaptomus nevadensis. Proc. Natl. Acad. Sci. USA* 73:971–974.

Hardy, D. 1956. *The Open Sea.* William Collins, London.

Harnisch, O. 1951. Hydrophysiologie der Tiere. *Die Binnengewässer* 19:1–299.

Hart, C. W., and S. L. H. Fuller (eds). 1974. *Pollution Ecology of Freshwater Invertebrates.* 389 pp. Academic, New York.

Hartland-Rowe, R. 1972. The limnology of temporary waters and the ecology of Euphyllopoda. In: *Essays in Hydrobiology*, pp. 15–31, edited by R. B. Clark and R. Wootton. University of Exeter Press, Exeter, UK.

Harvey, E. N. 1952. *Bio-luminescence.* 649 pp. Academic, New York.

Heitkamp, U. 1988. Life cycles of microturbellarians of pools and their strategies of adaptation to their habitats. In: *Free-living and Symbiotic Plathelminthes*, pp. 449–456, edited by P. Ax, U. Ehlers, and Sopott-Ehlers. *Progress in Zoology*, 36. Gustav Fischer, Stuttgart.

Higgins, R. P., and H. Thiel, eds. 1988. *Introduction to the Study of Meiofauna.* 488 pp. Smithsonian Institution Press, Washington, DC.

Hora, G. L. 1930. Ecology, economics, and evolution of the torrential fauna. *Philos. Trans. R. Soc. London* 218B:171–282.

Hubalt, E. 1927. Contribution a l'etude des invertebres torrenticoles. *Bull. Biol. Fr. Belg., Suppl.* 9:1–390.

Husmann, S. 1967. Klassifizierung mariner, brackiger und limnischer Grundwasserbiotope. *Helgol. Wiss. Meeresunters.* 16:271–278.

——. 1975. *First International Symposium on Groundwater Ecology.* 232 pp. Swets Publ. Service, Amsterdam.

Hutchinson, G. E. 1957. *A Treatise on Limnology*, Vol. I: *Geography, Physics, and Chemistry.* Wiley, New York.

——. 1967. Vol. II: *Introduction to Lake Biology and the Limnoplankton.* 1115 pp. Wiley, New York.

——. 1975. Vol. III: *Limnological Botany.* 660 pp. Wiley, New York.

Hynes, H. B. N. 1960. *The Biology of Polluted Waters.* 199 pp. Liverpool Univ. Press, Liverpool, UK.

——. 1970. *The Ecology of Running Waters.* 555 pp. Univ. Toronto Press, Toronto.

Illies, J. (ed.). 1978. *Limnofauna Europaea.* 532 pp. Gustav Fischer, Stuttgart.

Jacobs, W. 1935. Das Schweben der Wasserorganismen. *Erg. Biol.* 11:131–218.

Johannes, R. E., and M. Satomi. 1967. Measuring organic matter retained by aquatic invertebrates. *J. Fish. Res. Board Can.* 24:2467–2471.

Kajak, Z., and A. Hillbricht-Ilkowska (eds.). 1972. *Productivity Problems of Freshwater.* 918 pp. P.W.N. Polish Scientific, Warsaw, Poland.

Kanwisher, J. W. 1955. Freezing in intertidal animals. *Biol. Bull.* 109:56–63.

Kenk, R. 1949. The animal life of temporary and permanent ponds in southern Michigan. *Misc. Publ. Mus. Zool. Univ. Mich.* 71:1–66.

Krogh, A. 1939. *Osmotic Regulation in Aquatic Animals.* 242 pp. Cambridge University Press, Cambridge, UK.

——. 1941. *The Comparative Physiology of Respiratory Mechanisms.* 172 pp. Univ. Pennsylvania Press, Philadelphia.

Lake, P. S., et al. 1989. The phenology of a temporary pond in western Victoria, Australia, with special reference to invertebrate succession. *Arch. Hydrobiol.* 115:171–202.

Lansbury, I. 1955. Some notes on invertebrates other than Insecta found attached to waterbugs (Hemiptera–Heteroptera). *Entomologist* 88:139–140.

Laycock, G. 1966. *The Alien Animals.* 240 pp. Natural History, New York.

Lee, C. E. 1999. Rapid and repeated invasions of fresh water by the copepod *Eurytemoura affinis. Evolution* 53:1423–1434.

Lee, C. E., and M. A. Bell. 1999. Causes and consequences of recent freshwater invasions by saltwater animals. *TREE* 14:284–288.

Locke, A., et al. 1993. Ballast warer exchange as a means of controlling dispersal of freshwater organisms by Ships. *Can. J. Aquatic Sci.* 50: 2086–2093.

Loffler, H. 1963. Vogelzug und Crustaceen Verbreitung. *Verh. Deut. Zool. Gesell. München* 1963:311–316.

Lucas, C. E. 1947. The ecological effects of external metabolites. *Biol. Rev.* 22:270–295.

Ludyanskiy, M. L. 1993. Recent introductions of *Dreissena* and other forms to North America—the Caspian Sea/Black Sea connection. In: *Zebra Mussels—Biology, Impacts, and Control,* pp. 699–704, edited by T. F. Nalepa and D. W. Schloesser. Lewis, Boca Raton, FL.

Macan, T. T. 1961. Factors that limit the range of freshwater animals. *Biol. Rev.* 36:151–198.

——. 1963. *Freshwater Ecology.* 338 pp. Longmans, London.

——. 1975. *Life in Lakes and Rivers,* 3d ed. 320 pp. Collins New Naturalist, London.

Mackenthun, K. M. 1973. *Toward a Cleaner Aquatic Environment.* 273 pp. U.S.E.P.A. Office of Air and Water Programs, Washington, DC.

Maguire, B. 1963. The passive dispersal of small aquatic organisms and their colonization of isolated bodies of water. *Ecol. Monogr.* 33: 161–175.

Malone, C. R. 1965. Killdeer (*Charadrius vociferus* Linnaeus) as a means of dispersal for aquatic gastropods. *Ecology* 46:551–552.

Mansingh, A. 1971. Physiological classification of dormancy in insects. *Can. Entomol.* 103: 983–1009.

Melchiorri-Santolini, U., and J. W. Hopton (eds.). 1972. Detritus and its role in aquatic ecosystems. *Mem. Ist. Ital. Idrobiol.* 29 (Suppl.): 1–540.

Miller, R. C. 1958. The relict fauna of Lake Merced, San Francisco. *J. Mar. Res.* 17:375–382.

Miner, D. 1950. *Field Book of Seashore Life.* 888 pp. Putnam, New York.

Monk, D. C. 1976. The distribution of cellulase in freshwater invertebrates of different feeding habits. *Freshwater Biol.* 6:471–475.

Morton, B., et al. 1998. Biology and anatomy of the 'living fossil' *Congeria kusceri* (Bivalvia: Dreissenidae) from subterranean rivers and caves in the Dinaric karst of the former Yugoslavia. *J. Zool. London* 254:147–174.

Naumann, E. 1931. *Limnologische: Terminologie.* Handbuch der biol. Arbeitsmeth., Sect. 9, Part 8, Nos. 1–5: 1–776.

Needham, J. 1930. On the penetration of marine organisms into freshwater. *Biol. Zentralbl.* 50: 504–509.

Neldner, K. H., and R. W. Pennak. 1955. Seasonal faunal variations in a Colorado alpine pond. *Am. Midl. Nat.* 53:419–430.

Newell, R. C. 1970. *Biology of Intertidal Animals.* 555 pp. American Elsevier, New York.

Nichol, E. A. T. 1935. The ecology of a salt marsh. *J. Mar. Biol. Assoc. U. K.* 20:203–261.

Nielsen, A. 1950. The torrential invertebrate fauna. *Oikos* 2:176–196.

Niethammer, G. 1953. Zurn Transport von Susswassertierchen durch Vogel. *Zool. Anz.* 151: 41–42.

Olsson, T. I. 1981. Overwintering of benthic macroinvertebrates in ice and frozen sediment in a north Swedish river. *Holarctic Ecol.* 4: 161–166.

Palmer, A. R. 1960. Miocene copepods from the Mojave Desert, California *J. Paleontol.* 34: 447–452.

Park, T., et al. 1946. Dynamics of production in aquatic populations. *Ecol. Monogr.* 16:311–391.

Pax, F. 1951. Die Grenzen tierischen Lebens in mitteleuropäischen Thermen. *Zool. Anz.* 147: 275–284.

Pearse, A. S. 1928. On the ability of certain marine invertebrates to live in diluted sea water. *Biol. Bull.* 54:405–409.

——. 1932. Animals in brackish water ponds and pools at Dry Tortugas. *Publ. Carnegie Inst. Wash.* 435:125–142.

——. 1950. *The Emigrations of Animals from the Sea.* 210 pp. Dryden, New York.

Peck, S. B. 1998. A summary of diversity and distribution of the obligate cave-inhabiting faunas of the United States and Canada. *J. Cave Karst Stud.* 60:18–26.

Peck, S. B., and Christiansen, K. 1990. Evolution and zoogeography of the invertebrate cave faunas of the Driftless Area of the upper Mississippi River valley of Iowa, Minnesota, Wisconsin, and Illinois, U.S.A. *Can. J. Zool.* 68:73–88.

Peck, S. B., and J. J. Lewis. 1977. Zoogeography and evolution of the subterranean invertebrate faunas of Illinois and southeastern Missouri. *NSS Bull.* 40:39–63.

Pennak, R. W. 1939. The microscopic fauna of the sandy beaches. *Publ. Am. Assoc. Adv. Sci.* 10:94–106.

——. 1963. Ecological affinities and origins of freeliving acelomate freshwater invertebrates.

In: *The Lower Metazoa*, pp. 435–551, edited by E. C. Dougherty., Univ. California Press, Berkeley.

——. 1977. Trophic variables in Rocky Mountain trout streams. *Arch. Hydrobiol.* 80:253–285.

——. 1988. Ecology of the freshwater meiofauna. In: *Introduction to the Study of Meiofauna*, pp. 39–60, edited by R. P. Higgins and H. Thiel. Smithsonian Institution Press, Washington, DC.

Prescott, G. W. 1969. *How to Know the Aquatic Plants.* 171 pp. Brown, Dubuque, IA.

——. 1970. *How to Know the Freshwater Algae.* 348 pp. Brown, Dubuque, IA.

Proctor, V. W. 1964. Viability of crustacean eggs recovered from ducks. *Ecology* 45:656–658.

Proctor, V. W., and C. R. Malone. 1965. Further evidence of the passive dispersal of small aquatic organisms via the intestinal tract of birds. *Ecology* 46:728–729.

Proctor, V. W., et al. 1967. Dispersal of aquatic organisms: viability of disseminules recovered from the intestinal tract of captive killdeer. *Ecology* 48:672–676.

Ramazotti, G. 1972. Il phylum Tardigrada. *Mem. Ist. Ital. Idrobiol. Dott. Marco de Marchi.* 28: 1–732.

Remane, A., and C. Schlieper. 1971. Biology of brackish water. *Die Binnengewässer* 25:1–372.

Riser, N., and M. P. Morse, eds. 1974. *Biology of the Turbellaria.* 530 pp. McGraw–Hill, New York.

Robinson, D. G. 1999. Alien invasions: the effects of the global economy on non-marine gastropod introductions into the United States. *Malacologia* 41:413–438.

Russell-Hunter, W. D. 1970. *Aquatic Productivity.* 306 pp. Macmillan, New York.

Scholander, P. F., et al. 1953. Studies on the physiology of frozen plants and animals in the Arctic. *J. Cell. Comp. Physiol.* 42 (Suppl. 1):1–56.

Schulz, E. 1935. *Actinarctus doryphorus* nov. gen. nov. spec., ein Markwürdiger Tardigrad aus der Nordsee. *Zool. Anz.* 111:285–288.

Schwoerbel, J. 1967. Das hyporheische interstitial als Grenzbiotop zwischen oberirdischem und subterranem Ökosystem und seine Bedeutung für die Primar-Evolution von Kleinsthöhlenbewornern. *Arch. Hydrobiol. Suppl.* 33: 1–62.

Segerstrale, S. G. 1957. Baltic Sea: Treatise on marine ecology and paleoecology. *Mem. Geol. Soc. Am.* 67:751–800.

Sernov, S. A. 1927. Über die Überwinterung der Wasserorganismen im Eise und in der gefroren Erde nach dem Material von N. Boldyreva, P. P. Scharmina, und J. D. Schmeleva. *Atti Congr. Int. Limn. Teor. Appl.* 4:555–563.

Sket, B., and M. Veliknja. 1986. Troglobitic freshwater sponges (Porifera, Spongillidae) found in Yugoslavia. *Stygologia* 2:254–266.

Smith, D. G. 1981. Evidence for hybridization among two crayfish species (Decapoda: Cambaridae: *Orconectes*) in a Massachusetts stream. *Am. Midl. Nat.* 105:405–407.

Steuer, H. 1910. *Planktonkunde.* 723 pp. Teubner, Leipzig.

Stewart, K. W., et al. 1970. Dispersal of algae, protoctistans, and fungi by aquatic Hemiptera, Trichoptera, and other aquatic insects. *Ann. Entomol. Soc. Am.* 63:139–144.

Strayer, D. L. 1994. Limits to biological distributions in groundwater. In: *Groundwater Ecology*, pp. 287–310, edited by J. Gilbert, D. Danielopol, and J. A. Stanford. Academic, New York.

——. 1999. Effects of alien species on freshwater mollusks in North America. *J. N. Am. Benthol. Soc.* 18:74–98.

Strayer, D. L., et al. 1997. Oxygen, organic matter, and sediment granulometry as controls on hyporheic animal communities. *Arch. Hydrobiol.* 140:131–144.

Streever, W. J. 1992. First record of the colonial cnidarian *Cordylophora lacustris* within a flooded cave system. *NSS Bull.* 54:77–78.

Taylor, C. A., and M. Redmer. 1996. Dispersal of the crayfish *Orconectes rusticus* in Illinois, with notes on species displacement and habitat preference. *J. Crustacean Biol.* 16:547–551.

Temple, S. A. 1990. The nasty necessity: eradicating exotics. *Conservation Biol.* 4:113–115.

Thienemann, A. 1950. Verbreitungsgeschichte der Süsswassertierwelt Europas. *Die Binnengewässer* 18:1–809.

Thorson, G. 1946. Reproduction and larval development of Danish marine bottom invertebrates. *Medd. Komm. Danmarks Fish. og Havunders. Ser. Plankton* 4:1523.

Valkanov, A. 1968. Das Neuston. *Limnologica* 6: 381–404.

Todd, D. K. (ed.). 1970. *The Water Encyclopedia.* 559 pp. Water Information Center, Port Washington, NY.

Vinogradov, A. P. 1953. The elementary chemical composition of marine organisms. *Mem. Sears Found. Mar. Res.* No. 11.

Von Brand, T. 1946. Anaerobiosis in invertebrates. Normandy, MO.

Ward, H. B., and G. C. Whipple (edited by W. T. Edmondson). 1959. *Freshwater Biology.* 1248 pp. Wiley, New York.

Warren, C. F. 1971. *Biology and Water Pollution.* 434 pp. Saunders, Philadelphia.

Wesenberg-Lund, C. 1939. *Biologie der Susswassertiere.* 817 pp. Vienna, Austria.

Wetzel, R. G. 1983. *Limnology. 2nd Edition.* 767 pp. Saunders College Pub., Fort Worth.

Whitton, B. A. 1975. *River Ecology.* 725 pp. Univ. of California Press, Berkeley.

Wiggins, G. B., et al. 1980. Evolutionary and ecological strategies of animals in annual temporary pools. *Arch. Hydrobiol. Suppl.* 58:97–206.

Williams, D. D. 1983. The natural history of a Nearctic temporary pond in Ontario with remarks on continental variation in such habitats. *Int. Rev. gesamten Hydrobiol.* 68:239–253.

2

PORIFERA

(Sponges)

The great majority of the 4500 species of sponges are marine, only a single family, the Spongillidae, is represented in U.S. freshwaters. This family consists of about 170 species, of which about 30 have been reported from the United States.

Freshwater sponges are common in clean ponds, lakes, streams, and rivers. Because they are sessile and because of their inconspicuous green, brown, gray, or yellowish coloration, however, they are frequently unnoticed or sometimes confused with algae. Indeed, until their morphology and physiology were first partially understood (beginning in 1825), sponges were often considered as plants. Syngamic reproduction has been known only since 1856. A suitable substrate for the matlike sponge growth may be provided by almost any stable submerged object, including rocks (Fig. 2.1), pebbles, aquatic vegetation (Fig. 2.2), logs, branches (Fig. 2.3), and twigs. Sponges may be found on the upper surfaces, sides, or lower surfaces.

The Phylum Porifera may be divided into two major groups, which are distinguished on the basis of their skeletal and cellular structure. The Hexactinellida are the glass sponges, entirely marine; their skeleton is an openwork structure of siliceous triaxon spicules or some modification of triaxon. Their cellular structure is completely syncitial. The Cellularia have cells with distinct membranes. The group includes the Class Calcarea, which consists of small marine sponges having a skeleton composed of calcium carbonate spicules, which may be one, three, or four rayed. In the Class Demospongea the skeleton may consist of horny fibers (spongin), siliceous spicules (not triaxon), or both. Commercial sponges, for example, in the Order Keratosa have a skeleton consisting entirely of spongin, spicules being absent. In four other orders there is no spongin, the skeleton being composed entirely of tetraxon spicules. In the three remaining orders, however, the skeleton consists of both monaxon spicules and variable amounts of spongin. One of these orders, the Haplosclerida, contains the Spongillidae, one of few freshwater sponge families. In addition to the skeletal materials already mentioned, it should be emphasized that the noncellular matrix, or mesohyl, of all sponges also functions in support and cohesion of the living cells.

General characteristics. The sponge body is simply constructed. There are no organs or true tissues. The closest to a tissue in a sponge is the outer dermal membrane (Fig. 2.4) composed of layers of specialized cells called pinacocytes. In some sponges the surface is patrolled by amoebocyte cells presumably involved with preventing foreign material from settling on the surface. The external surface bears openings of two general sizes. The very abundant microscopic ostia are merely more or less circular open-

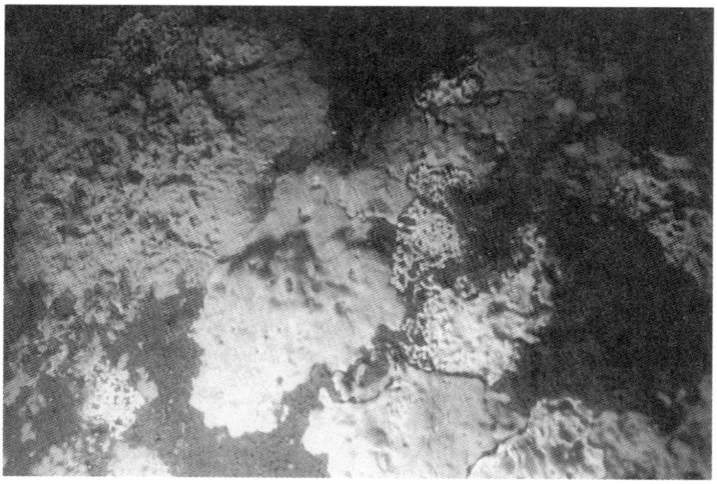

Fig. 2.1. *Ephydatia muelleri* growing on outcrop at 30 m, Connecticut River. (Photo © Ed Klekowski, with permission.)

ings in the dermal membrane. The ostia are surrounded by pinacocytes and penetrate the various cell layers of the dermal zone. The few oscular openings (Fig. 2.4) are scattered over the sponge surface. Usually they are 0.5 to 2.0 mm in diameter but may be as wide as 5 to 10 mm. An osculum is typically located at the distal end of a small, chimneylike projection. By means of a carmine suspension it can be easily shown that minute currents of water pass into the sponge through the ostia and out through the oscula.

Internally, the sponge is a maze of interconnected and ramifying spaces, channels, and chambers (Figs. 2.5, 2.6). In typical freshwater species the ostia open into an extensive subdermal cavity from which an abundance of minute narrow incurrent canals lead more deeply into the sponge. At their inner ends these incurrent canals open into small, more or less spherical cavities, the flagellated chambers. At the opposite side each flagellated chamber empties into another minute canal, the excurrent canal, which, along with many other excurrent canals, empties into a much larger irregular central cavity. Each central cavity opens to the outside through an osculum. The opening of the incurrent canal into the flagellated chamber is called the prosopyle; the larger opening from the chamber into the excurrent canal is called the apopyle. The entire arrangement of canals and chambers in the Spongillidae is the rhagon type.

In contrast to the situation in some marine sponges, the flagellated chambers of the Spongillidae are minute and lined with only a small number of collar cells, or choanocytes. Each choanocyte has a distal collar of microvilli surrounding a single flagellum. Except for the flagellated chambers, all internal chambers and canals of a sponge are lined with a flat layer variously characterized as a syncitium, semisyncitium, or layer of colloidal material with contained cells.

The space between layers of the dermis is filled with a gelatinous mesohyl containing an abundance of several kinds of totipotent archeocytes that have a wide variety of functions. Some secrete the skeleton; some function in nutrition, storage, transport, and excretion; others function in reproduction. Many archeocytes are motile and move about slowly in the mesohyl.

The size of a single freshwater sponge is enormously variable, depending on the species, age, and various ecological conditions. Some species, for example, even when mature may consist of a thin, slimy, delicate mat having a surface area of only a few square centimeters. Other species may cover an

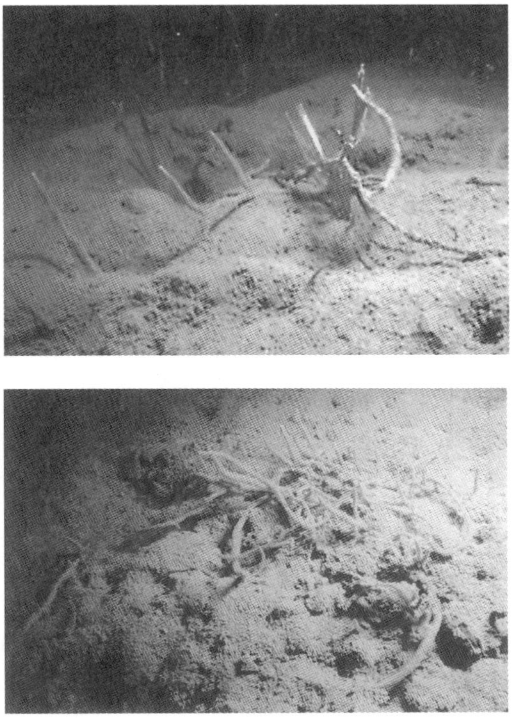

Fig. 2.2. *Corvomeyenia everetti* growing freely and on vegetation, 5 m. (Photos © Ed Klekowski, with permission.)

area of as much as 40 m^2 under highly favorable conditions. The thickness of a sponge growth is likewise variable; sometimes it may be only 1 or 2 mm thick, but under other circumstances it may be as much as 4 cm thick. The growth form may be typically encrusting and matlike (Fig. 2.1), or there may be numerous protuberances, fingerlike outgrowths (Fig. 2.3), branches (Fig. 2.2), or an extremely irregular lobed growth form.

Furthermore, few of our American species are limited to one type of growth form, so that field identification by this characteristic is usually not feasible. *Eunapius fragilis*, for example, may be found as a fine-textured uniform crust; or it may be more robust, with an irregular surface and loose texture; or it may consist of luxuriant branching lobes. *Spongilla lacustris* (Fig. 2.3) has every variation, from a simple encrustation to tufts of long fingerlike processes, and *Ephydatia muelleri* (Fig. 2.1) has a similar range. *Corvomeyenia everetti* (Fig. 2.2) produces long, slender, and delicate branches. In general, individuals of a particular species are usually more lobed and luxuriant in standing waters than in running waters.

Fig. 2.3. *Spongilla lacustris* growing on a branch of a sunken tree, 10 m. (Photo © Ed Klekowski, with permission.)

Sponges are often collected that are as much as 30 to 60 mm thick. However, when such growths are examined carefully it is usually found that the living portion of the mass forms only an outermost stratum 4 to 20 mm thick, whereas all the underlying material is composed of the dead remains of the previous one or more years' growth.

Coloration of sponges may be produced by several factors. Most species found growing on the upper sides of objects are some shade of green (Plate 1A), owing to con-

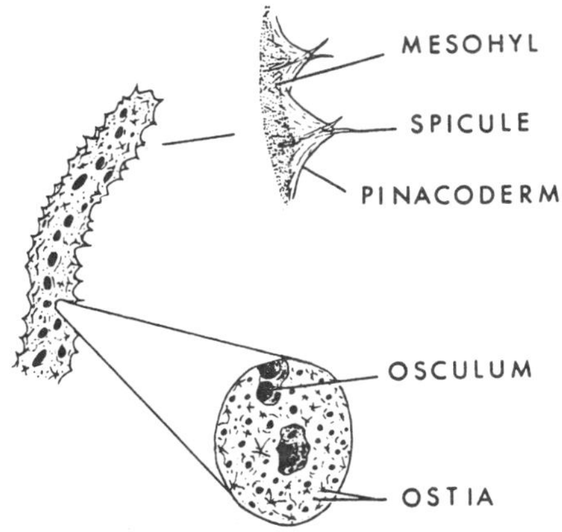

Fig. 2.4. Topography of a freshwater sponge.

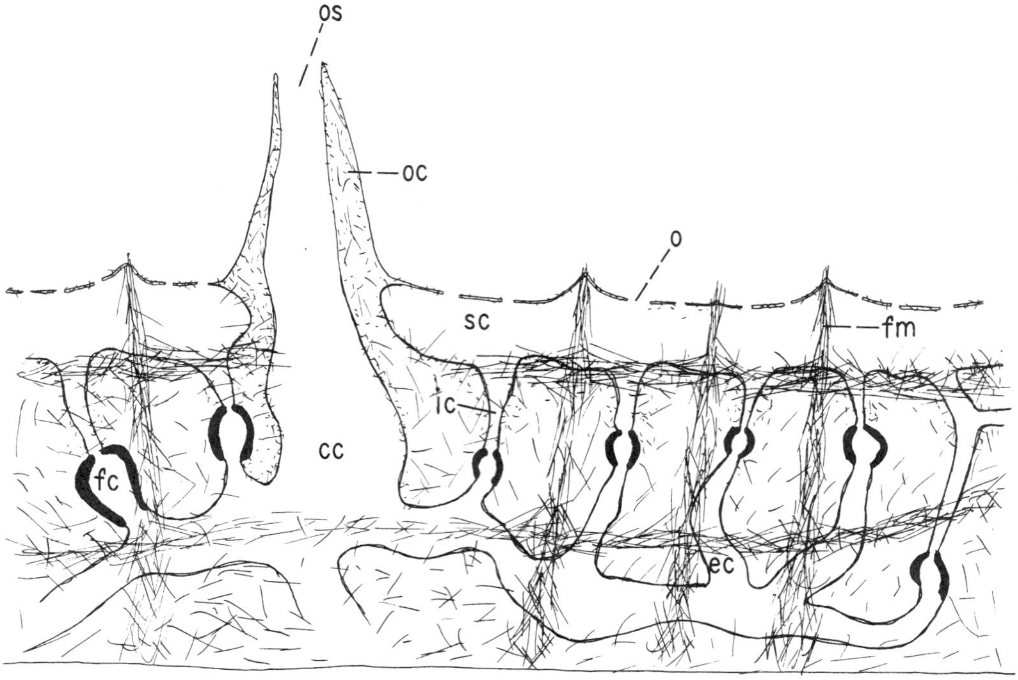

Fig. 2.5. Diagrammatic section of peripheral portion of a typical freshwater sponge. Heavy black lines represent layers of choanocytes lining flagellated chambers. Delicate matrix of spongin not shown. cc, central cavity; ec, excurrent canal; fc, flagellated chamber; fm, fascicle of megascleres; ic, incurrent canal; o, ostium; oc, oscular chimney; os, osculum; sc, subdermal cavity.

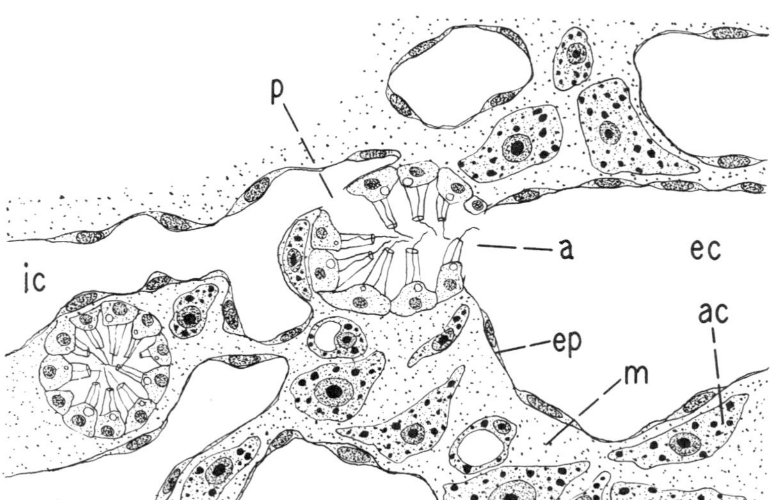

Fig. 2.6. Detailed structure of Spongillidae, semidiagrammatic, showing two flagellated chambers. Spicules and delicate spongin matrix omitted. a, apopyle; ac, amoebocyte; ec, exhalent canal; ep, epithelial cell; ic, inhalent canal; m, mesohyl matrix; p, prosopyle. (Modified from Brien 1936).

tained algal cells (see below). If these same species occur in deep waters or on the undersides of objects, however, there is little chlorophyll, and the coloration may be some shade of dark brown, tan, gray, flesh color, or reddish (rarely). On the other hand, there are a few species that are seldom or never green, even when growing in bright light. Coloration may be modified by the ingestion of food and inorganic particles of a certain shade or by the presence of pigmented metabolic materials in the tissues.

Skeleton. Freshwater sponges contain great numbers of needlelike microscopic spicules that aid in supporting the sponge body. Such spicules are of two general types. The larger spicules, or megascleres (skeletal spicules) (Fig. 2.7), form a reticulate network through the sponge and around and through the many cavities and channels. In addition to many megascleres arranged haphazardly, there are loose overlapping fascicles of megascleres radiating outward from the center or base of the sponge and frequently extending through the dermal layer to the outside. There are also transverse fascicles that run between the radiating fascicles. More or less minute amounts of fibrous and veil-like spongin bind the megascleres together.

The smaller type of spicules, or microscleres (flesh or dermal spicules), are most abundant in the peripheral and dermal portion of the sponge body, although there are also some lining the deeper channels. Microscleres are arranged in a scattered fashion, rarely in fascicles. Microscleres are lacking in many genera of spongillids.

Megascleres and microscleres may be sharp-pointed (amphioxea) or blunt (amphistrongyla) and with few to many spines of variable size, depending on the species. Those genera with microscleres have acerate, stellate, or birotulate (Fig. 2.7) microscleres. Additional types of spicules are found associated with the special reproductive bodies, or gemmules, and these are discussed below.

The siliceous material of sponge spicules is often said to have the composition of "hydrated silica" or "opal." Analyses show about

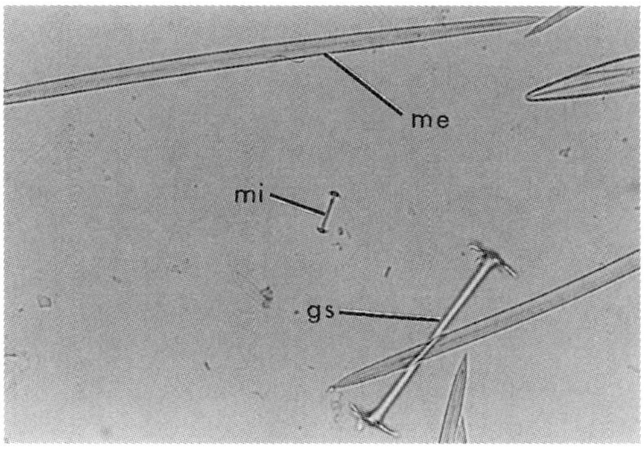

Fig. 2.7. Representative spicules of *Corvomeyenia everetti.* me, megasclere; mi, microsclere; gs, gemmosclere.

92% silicon dioxide, 7% water, and small amounts of magnesium, potassium, and sodium oxides.

Small and simple spicules are formed within single specialized archeocytes, called silicoblasts, but large and complex spicules are formed by two or more silicoblasts. The first sign of a spicule is an extremely thin organic axial thread within the silicoblast, the siliceous material being deposited around it in ever-increasing thickness.

Nutrition. Fundamentally, freshwater sponges efficiently strain minute particles from the water in which they live. Such particulates usually range from colloidal to 50 μm in diameter. The flagella of the choanocytes are directed more or less toward the apopyles of the flagellated chambers. Their spiral or undulating movements collectively produce weak water currents that enter the sponge through the ostia and leave through the oscula. As the water traverses the flagellated chambers, bacteria, phytoplankton, and bits of detritus are caught on filamentous strands of mucous produced on the edge of the collars of the collar cells as well as on the exposed surface of the cells proper. The collar has been likened to a thin pseudopodial membrane. All digestion is intracellular and particles are ingested by the choanocytes and endopinacocytes. Food particles are passed on to the adjacent mesohyl and into the archeocytes, amoebocytes, and lophocytes where digestion is completed, and into vacuolar where there may be considerable storage. Nondigestible material is subsequently egested.

Prosopyles are so small that they frequently become plugged, and in such cases the particles are carried off to one side by slow "flowing" movements of the dermal protoplasm. They may later sink below the surface and be taken up by archeocytes. Various laboratory investigations have shown filtration rates of 0.01 to 0.28 mL of water per second per cubic centimeters of sponge tissue. Clearance rates are in the magnitude of 0.05 cm^3 of food particles per second per gram of sponge material.

It is now thought that the inter- and intracellular "zoochlorellae" of Spongillidae are not essential to sponge metabolism, but there is evidence to show that their presence results in better sponge growth and that they can control hatching rate in gemmules. Such algae are certainly ingested plankton species, including *Pleurococcus*, *Chlorella*, and other minute genera, that are passed from cell to cell and persist, multiply, and are digested very slowly. In dim light or darkness they are digested much more rapidly than in bright light.

Circulation. There is no special circulatory mechanism in sponges. Materials are transferred by diffusion and amoebocytic transport.

Respiration. The exchange of oxygen and carbon dioxide occurs everywhere in exposed cells of sponges, both externally and internally. Diffusion of these dissolved gases is probably rapid and efficient, since all sponge cells are either directly in contact with water or only a few cells distant from it.

Excretion. Dissolved metabolic waste materials simply diffuse into the water surrounding and passing through the sponge body. Granular metabolic materials are usually given off from archeocytes into the excurrent canals but may also be retained by the cells for varying periods of time. Choanocytes and many archeocytes sometimes contain one or two contractile vacuoles, and it is probable that they function mainly in keeping the proper osmotic pressure in the tissues.

Contractile elements, impulse conduction. Although sponges are completely sessile and there is no definite muscle system, a few specialized archeocytes, or myocytes, have the ability to contract or relax. A few bipolar and multipolar "nervous type" cells have been described by some investigators, but nothing is known about their role in the transmission of impulses.

The oscular chimneys are normally kept open by the excurrents, but if environmental conditions become unfavorable, the oscula will slowly close in several minutes. Such contraction is produced by the action of fusiform myocytes around the openings. In *Ephydatia fluviatilis* it has been shown that even under normal circumstances the oscular chimneys (1 to 3 mm high) continuously and slowly change in form, from high to low and wide to narrow. The osculum will close and the chimney will collapse in several seconds if the latter is struck a sharp blow with a needle. Complete recovery takes 20 to 30 seconds. Sticking a needle into the sponge body 2 mm from the chimney base, however, has no effect on the chimney. Some workers report that under unfavorable conditions ostia will slowly close as a result of contractility of the surrounding protoplasm, but there are no special sensory or nerve cells, and there is no coordination and no transmission of impulses (or perhaps transmission for 2 to 3 mm at most).

Reproduction, growth. Many details of syngamic reproduction are still obscure and confused. Nevertheless, sexual reproduction is well developed in freshwater sponges. Some Spongillidae are hermaphroditic, and a single individual may produce eggs and sperm simultaneously, but in others the sexes are separate, or an individual may produce sperm at one time and eggs at another. Perhaps both conditions prevail, depending on the species. One investigator has found that *Spongilla lacustris* may exhibit alternative hermaphroditism, in which a sponge may be exclusively male one year and exclusively female the next year. Other work suggests that females may be extremely rare or transitory, present only long enough to produce ova.

In temperate regions sperm and egg production are thought to occur mostly during July and August, and to a lesser extent in June and September, but in species in southern parts of the United States reproductive activity may be year round. The oogonia are archeocytes that grow by absorbing and ingesting many adjacent cells. By the time the egg is mature it is large and lies loosely in one of the internal cavities of the sponge.

Although earlier investigators thought that sperm developed from archeocytes, more recent work has shown that in at least two species they develop from choanocytes that lose their collars and flagella, become more or less amoeboid, migrate into the mesohyl, and function as spermatogonia. Mature flagellated sperm are formed in abundance in baglike "follicles." These sperm are released into the chambers of the sponge and out through the oscula into the surrounding water. Some of them are eventually drawn into the chambers of the same or other sponges where they fertilize the mature ova.

Several different and peculiar types of early embryology occur in the Spongillidae, but in all of them a flagellated embryo is formed that is released from the parent and swims about for a few hours or days before settling to the substrate and metamorphosing to form a minute young sponge. Growth is mainly a process of proliferation and accretion with accompanying development of additional orifices, channels, and chambers. Usually most of the sponge body dies or becomes dormant in the autumn and partly disintegrates during the winter, particularly in northern latitudes, and although it sometimes does not revive in the following spring, it is more common for the surviving remnant to revive and continue growth during the following season. In this way a specimen may remain more or less intact from year to year, growing in thickness, but with only the outermost layer being alive and active. This persistence of sponges occurs especially on firm, broad substrates and where there is little ice action. Many times, however, a sponge may remain active and green throughout the winter, even under ice. A few species, e.g., *Anheteromeyenia ryderi*, commonly remain in the "vegetative" state throughout the northern winter.

Reduction bodies. Under highly unfavorable conditions a large sponge forms numerous "reduction bodies," and a small sponge may form a single reduction body (see Fig. 1.5B). The cells and mesohyl shrink into the skeleton and the canal system disappears, leaving the outer dermal membrane surrounding a mixture of totipotent cells, mesohyl, choanocytes, and spicules. Under favorable conditions such a reduction body may grow and regenerate a new sponge. Reduction bodies have been observed for only a few species, and it is not known whether they are produced regularly throughout the Spongillidae. Reduction bodies may be produced in the laboratory by placing the sponge in tap water, where presumably there are toxic ions.

Regeneration. Like certain marine species, the Spongillidae have remarkable powers of regeneration. If bits of *Ephydatia* or *Spongilla* are macerated and dissociated by squeezing through a fine mesh net, small clumps of cells will coalesce within a few hours, slowly reconstitute, and within several days grow into a new minute sponge.

Gemmules. The formation of gemmules is one of the most characteristic features of freshwater sponges. These are highly resistant resting stages similar in function to the statoblasts of freshwater ectoprocts. Briefly, a gemmule is a spherical structure (Fig. 2.8) with a secreted thick and vacuolated outer layer, invested spicules (Figs. 2.9, 2.10), and an internal mass of undifferentiated totipotent archeocytes. Gemmule diameter ranges from about 150 to 1000 (m. Coloration may be whitish (Plate 1B) or various shades of yellow or brown. Gemmules are known to be produced by only a small number of marine species, but they are quite simple and have only a single thick enveloping membrane instead of the complex structure found in freshwater species.

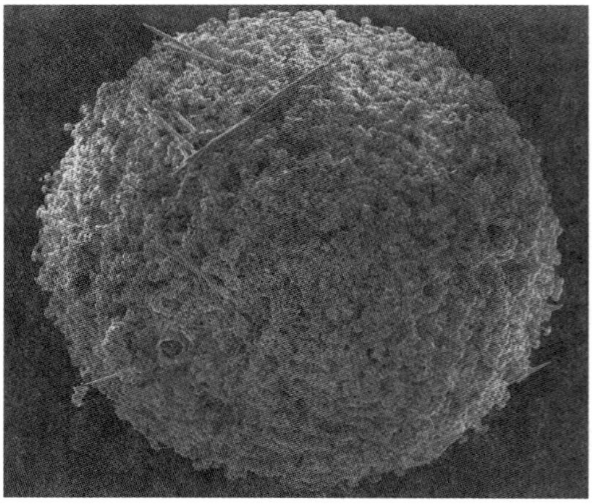

Fig. 2.8. Scanning electron micrograph of a gemmule of *Corvomeyenia everetti* showing embedded gemmoscleres and a few attached megascleres.

Formerly it was thought that gemmules appeared only at the close of the growing season, but it is now known that they may be formed at any time during the growing season. *Spongilla lacustris* is the only common species that regularly forms gemmules in September and October. In this species, and perhaps in others also, the control of gemmulation is influenced by endogenous factors during the summer and then triggered by environmental factors in the autumn. Because of autumn disintegration of superficial material, it is much easier to observe gemmules at that time.

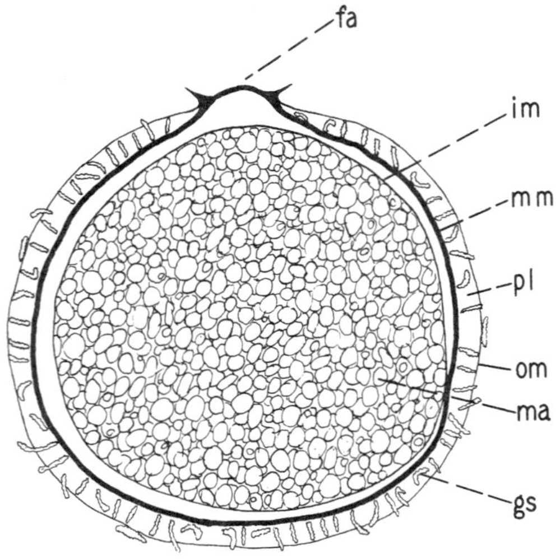

Fig. 2.9. Diagrammatic section of a sponge gemmule, ×100. fa, foraminal area; gs, gemmule spicule; im, inner sclerotized membrane; ma, mesenchymal archeocytes; mm, middle sclerotized membrane; om, outer sclerotized membrane (absent in some species); pl, pneumatic layer (containing spicules).

Anheteromeyenia ryderi, *Eunapius fragilis*, and *Heteromeyenia tubisperma* gemmules undergo diapause which is broken by low temperatures and to some extent drying for one to several months. Gemmules may be deposited in a compact basal layer in a sponge, or they may be scattered throughout the sponge body. The first sign of gemmule development is the grouping of many food-filled archeocytes into discrete masses. A thin inner sclerotized membrane is formed around each mass of archeocytes, followed by a heavier, middle (or outer) sclerotized membrane that has a single small foraminal area. The latter sometimes has a narrow, projecting foraminal tubule (Fig. 2.11). In most species of *Heteromeyenia* the foraminal tube is elongate and provided with variously arranged tendrils (Fig. 2.12). Except for the location of the foraminal aperture or foraminal tubule, the gemmule is covered with a complete, crustlike, thick, honeycombed coating, or pneumatic layer (Fig. 2.10). The thickness of the pneumatic layer is affected by chemical conditions, mainly pH, and conditions associated with season. Sometimes a third thin, outer sclerotized membrane covers the pneumatic layer. *Spongilla lacustris* produces two distinctly different kinds of gemmules: one small, green, and thin walled, and the other large, brown, and thick walled. Each type appears at a different point of the season.

In most species many specialized spicules are imbedded in the pneumatic layer of the gemmule and cover its outer surface (Fig. 2.8). Such gemmule spicules are arranged tangentially or in a heterogeneous manner; they differ markedly from the megascleres and microscleres. Gemmoscleres assume many forms, including thin or stout amphioxea, spiny amphistrongyla, or birotulates with wheel-like ends called rotules that

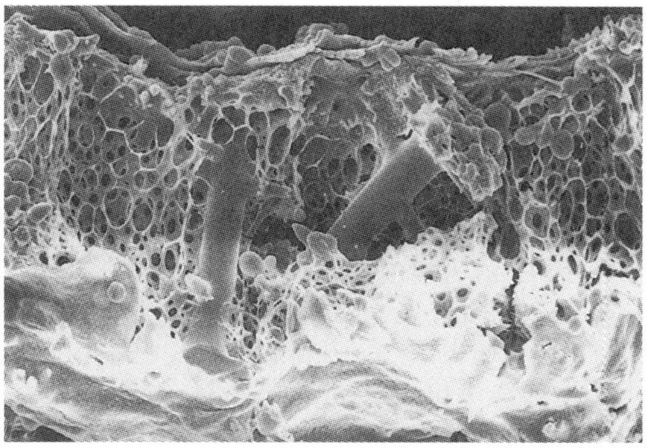

Fig. 2.10. Scanning electron micrograph of a cut through gemmule of *Ephydatia fluviatilis.* Note gemmosclere surrounded by honeycomb matrix.

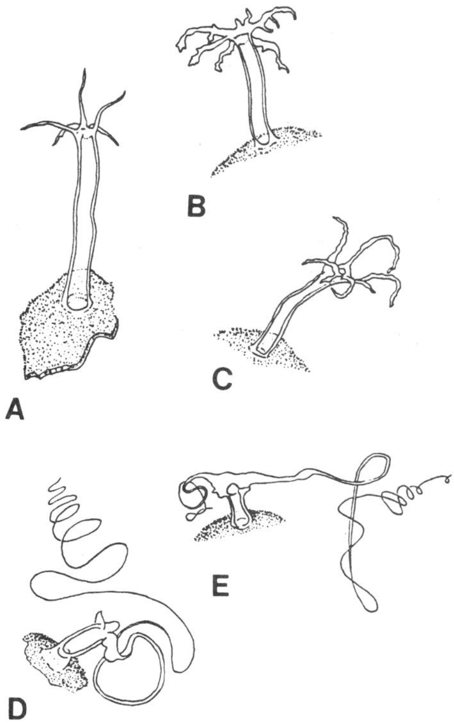

Fig. 2.12. Gemmule foramina of *Heteromeyenia tubisperma* (A–C) and *H. latitenta* (D, E).

may be smooth disks or deeply excised to form spines (Fig. 2.13).

Gemmules may remain attached to the substrate or become freed and rise to the surface or sink to the bottom. They winter over easily, withstand repeated freezings and thawings, and have been found to be viable after being kept dried for one to three years or more.

Germination may occur any time at water temperatures of about 13 to 23°C, but the process requires 2 weeks or more in the autumn and only a few days in the spring. Some investigators maintain that gemmules require a quiescent dormant period of one or more cold months before they will germinate. When a gemmule germinates, the membrane (micropyle) covering the foramen disappears and the inner creamy mass of archeocytes slowly flows out of the foraminal aperture and more or less covers the empty gemmule shell. This process is usually completed in 2 to 6 days. The archeocytes undergo histological differentiation and form the first cells of the several functional cell types. Some spicules, channels, and subdermal spaces may be formed as

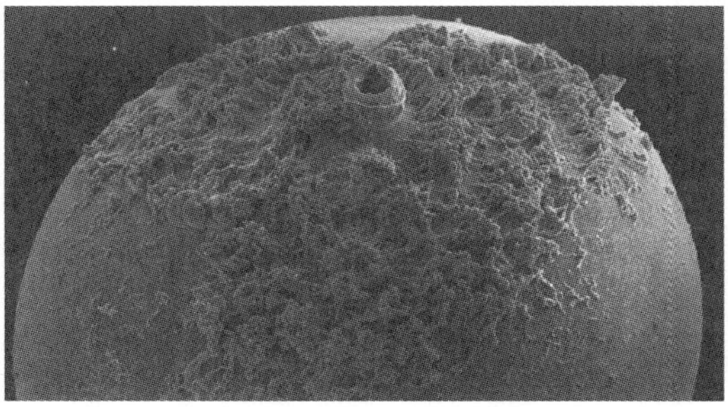

Fig. 2.11. Scanning electron micrograph of *Corvomeyenia everetti* gemmule before gemmosclere layer has been deposited. Foramen near top.

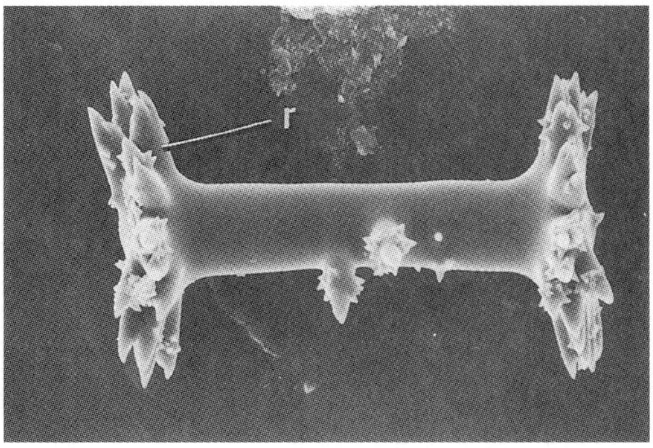

Fig. 2.13. Gemmosclere of *Ephydatia fluviatilis.* r, rotule. Modified from Smith, 1994.

early as the third day, and choanocytes may be functioning as early as the fifth day. Further development and growth result in typical sponge structure.

Gemmules commonly contain zoochlorellae, and it has been found that such zoochlorellae, in both light and dark, initiate gemmule germination at lower temperatures and increase the speed of germination. It is usually difficult to tell whether renewed growth of an old sponge in the spring is initiated by gemmule germination or by resumption of growth by dormant cells, but probably both mechanisms are responsible.

Ecology. Sponges are most common in waters less than 2 m deep. They are seldom found as deep as 4 m but under favorable conditions can exist at considerable depths (Fig. 2.1). Collections from as deep as 50 m have been made. In very favorable habitats sponge growths may almost obscure all suitable substrates. Spongillidae have occasionally been reported to clog water pipes and conduits, but otherwise they are of no economic importance.

They are rarely found on a mud bottom, in silty water, or in small puddles. They are seldom found in extremely rapid waters. In general, individuals in running waters are smaller and more encrusting, and show little or no branching. *Spongilla lacustris* may often be found in both the branching and encrusting forms in the same body of water, but it is thought that the encrusting form is merely the earlier growth form that becomes branching as the sponge matures. Compared with most groups of freshwater invertebrates, sponges are quite sensitive to variations in environmental conditions. *Ephydatia fluviatilis* will tolerate a minor amount of pollution, and a few other species a lesser degree.

Most species are more or less photopositive. *Duosclera mackayi*, *Radiospongilla crateriformis*, *Trochospongilla pennsylvanica*, and *Ephydatia fluviatilis*, however, prefer dim light and are usually on the undersides of objects in clear water.

The most important limiting factors controlling presence or absence of sponge species seem to be calcium, silicon, bound carbon dioxide, and pH (the last as a rough measurement of the general suitability of water). The following species are acid (soft) water forms seldom occurring where the pH is above 7.0 or 7.3: *Spongilla aspinosa*, *Anheteromeyenia argyrosperma*, *Duosclera mackayi*, and *Corvomeyenia everetti*. Alkaline species include *Heteromeyenia tubisperma*, *Ephydatia muelleri*, and *E. fluviatilis*. Several species, such as *Anheteromeyenia ryderi*, *Spongilla lacustris*, *Trochospongilla pennsylvanica*, and *Eunapius fragilis*, tolerate a wide range in bound carbon dioxide and hydrogen-ion concentration, although they fare better under alkaline conditions.

All common species regularly occur in waters having less than 12.0 ppm of calcium. *Spongilla lacustris*, however, has been found where the calcium concentration was as high as 33.0 ppm, and *Heteromeyenia baileyi* where there was 53.4 ppm of calcium. The distribution of *Ephydatia muelleri* and *Eunapius fragilis* appears to be restricted by low calcium concentration. The former has not been reported from waters containing less than 5.6 ppm and the latter 2.1 ppm of calcium. *Duosclera mackayi* and *Corvomeyenia everetti*, on the other hand, appear to grow only where small quantities of calcium are present; the former is restricted to waters having less than 3.2 ppm of calcium, and the latter 2.5 ppm. *Spongilla lacustris* is independent of ordinary calcium concentrations and grows luxuriantly in waters containing from 0.16 to 32.4 ppm.

In view of the fact that sponges require large amounts of silicon for spicule formation, it is surprising to find that most species attain good growth in waters that may contain only a trace of silicon. The only important exception is *Ephydatia muelleri*, which requires a minimum of 1.6 ppm of silicon dioxide. Nevertheless, some species show certain variations in spicule abundance and morphology, depending on the amount of silicon present. *Eunapius fragilis* has a poorly developed skeleton in water containing less than 1.0 ppm of silicon dioxide. In *Spongilla lacustris* both microscleres and megascleres become progressively attenuated in specimens collected in waters ranging from 13.0 ppm of silicon dioxide down to only a trace. In fact, a concentration of 0.56 ppm is the lowest value at which there is full microsclere development; there is no greater development above 0.56, but the microspines are absent below this concentration. In soft-water lakes this species has an abundance of spicules intermediate in size between megascleres and microscleres. *Trochospongilla pennsylvanica* has progressively heavier spicules and closer approximation in the size of the two rotules in lakes of increasing mineralization.

It is common to find three different species in the same lake, less common to find four, and rare to find five or six. Often there are definite associations of certain species, suggesting similar but obscure habitat requirements. The following pairs of species are commonly associated in the same lake or stream: *Ephydatia muelleri* and *Eunapius fragilis*, *Ephydatia muelleri* and *Spongilla lacustris*, *Eunapius fragilis* and *Spongilla lacustris*, and *S. lacustris* and *Trochospongilla pennsylvanica*. There are species clusters that are ecologically restricted to specific habitats displaying certain chemical characteristics. Thus, an acid-water community might include *Spongilla aspinosa*, *Duosclera makayi*, and *Corvomeyenia everetti*. Harrison (1988) has provided a convenient means to determine the general preferences of many species of spongillids for a number of commonly assessed chemical and physical features of freshwater.

Associations and predators. The irregular growth form of many sponges affords a favorable substrate for a wide variety of metazoans, including six orders of insects, crustaceans, water mites, annelids, nematodes, ectoprocts, nemerteans, turbellarians, and even mollusks. Larvae of one insect order, spongilla flies (Sisyridae), pierce the cells and suck out the contents and are entirely dependent on sponges as a food source. A few additional insect larvae also feed on sponge cells, and it is now known that crayfish (especially *Orconectes*) feed on sponges. A protoctoctistan parasite, possibly a microsporidian, lives within the various cell types responsible for producing the gemmule and interferes with gemmule production. The typical result of infestation is malformed gemmules, especially the outer covering.

Dispersal, geographic distribution. Sponges easily spread throughout single drainage systems, and by virtue of their resistant gemmules they are probably transported overland by wind, insects, birds, and mammals including humans. Sponges can attach and grow on other organisms and thus be transported around (Fig. 2.14), especially to deeper zones in waters with fluctuating levels.

Spongilla lacustris, *Eunapius fragilis*, and *Ephydatia fluviatilis* are essentially cosmopolitan and have been found the world over wherever intensive collecting has been done in suitable areas. *Trochospongilla pennsylvanica*, *Anheteromeyenia ryderi*, *Radiospongilla crateriformis*, and probably several other species are widely distributed in the Northern Hemisphere. The four most common American species are *Spongilla lacustris*, *Eunapius fragilis*, *Ephydatia muelleri*, and *Heteromeyenia tubisperma*.

Certain other species, however, are rare and poorly known in the United States. *Spongilla wagneri*, *Spongilla aspinosa*, *Corvomeyenia carolinensis*, *Stratospongilla penneyi*, and *Corvospongilla becki* are each known from only a single or several localities. Within the past 20 to 40 years a few U.S. species have become nearly extinct, chiefly because of pollution.

Collecting. Hand picking in the shallows and using a long-handled rake in deeper waters will produce an abundance of speci-

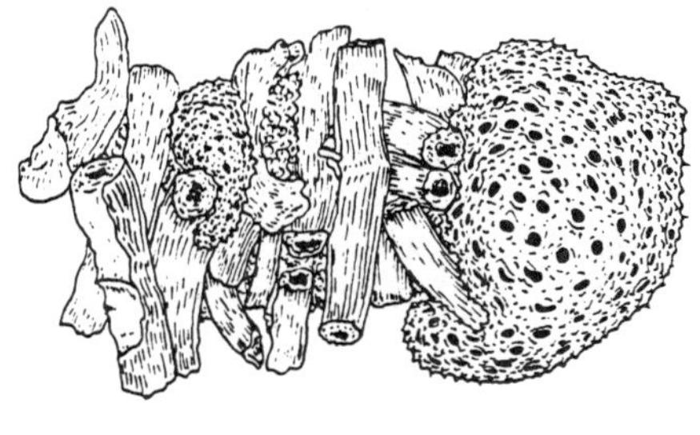

Fig. 2.14. *Eunapius fragilis* attached to caddis fly larvae cases.

mens. Sponges should be handled gingerly because of their fragility. It is important to collect basal, middle, and outer portions of the sponge to capture all the various spicular regions. Individual sponges must be kept separate to avoid mixing of bits of different species.

Culturing. Small *Ephydatia* and *Spongilla* have been kept alive in aquaria up to 5 or 6 months, but these are exceptionally long periods. Usually sponges remain in good condition for only 1 to 3 weeks, then begin to lose color and disintegrate. Death occurs in such short periods even if there is a continuous flow of water through the aquarium.

Some investigators have had success using a dead bacterial suspension for food.

If a small sponge or a bit of a larger one (about a cubic centimeter) is placed cut surface downward on a glass slide submerged under 5 to 10 cm of water, the sponge will often adhere and grow for a time, especially if the water is changed frequently. Never use tap water.

Germination of gemmules and early growth may be easily studied in the laboratory. Gemmules should be brought in and placed on slides or coverslips in shallow pails of water from the natural habitat. Germination will usually begin in 7 to 25 days, but sometimes the process may be hastened by first "vernalizing" the gemmules in water in a refrigerator at 5 to 10°C for a day or two. After germination is completed and the young sponge is securely attached, the slide should be removed to an aquarium in which the water is frequently renewed. Under favorable conditions the sponge may grow and cover as much as half of the slide and live for a month or two.

Preserving, preparing. The identification of freshwater sponges to species has been historically based, for the most part, on the gross morphology of the spicules. Recently, the use of scanning electron microscopy (SEM) has revealed finer details of skeletal structures, which are of further use in taxonomy. Several investigators advocate the use of SEM in future studies of taxonomic relationships of species. Although the application of SEM to systematic investigations of the Spongillidae will eventually become universal, and many of the figures in this chapter are from SEM studies, most individuals interested in merely determining sponge species usually will not have access to such equipment. Light microscopes are more readily available and the characters used to distinguish species in this manual are all detectable using standard light microscopical techniques. Objectives necessary for examination of prepared material are 10× (low power), 40× (high power), and occasionally 100× (oil immersion). Besides spicules, the morphology and arrangement of the gemmules in certain genera provide additional aids to species determination.

Samples of sponges can be dried, which is the traditional method of preservation, or placed directly in 10% buffered formalin or 50% isopropyl or 70% ethyl alcohol. Material preserved in formalin should be diluted in water for a few days and transfered to 50% isopropyl or 70% ethyl alcohol. Once preserved, either way, a small portion of an individual sample is removed and placed in a Kimax or Pyrex test tube. The portion should contain a few gemmules, if present in the sponge, and some of the outer membrane or pinacoderm.

The removed sample is then immersed in about 2 or 3 mL of concentrated nitric acid (HNO_3), which should be carefully pipetted into the test tube. Care must be taken to avoid getting any acid on the skin, clothing, or other surfaces susceptible to the acid's corrosive affects. The acid containing the sample is then heated, while being constantly agitated, preferably over an open flame until moderately boiling, not violently, to dissolve the spongin and gemmule capsule, which contain the various spicules. If possible, the boiling procedure should be carried out under a hood or similar ventilated enclosure for further protection from the boiling acid.

The heated fluid is then allowed to cool before an equal volume of water, ideally distilled, is pipetted into the test tube. The mixture is agitated and set aside to permit the spicules to settle to the bottom. Allow several minutes to elapse to guarantee complete settling of the spicules. At this time a small hot plate should be turned up to about 93°C and a clean slide placed on it. Returning to the preparation and using a new pipette, decant the supernatant from the test tube until only a residue containing the spicules remains. Add another 3 or 4 mL of water and reagitate. The procedure of settling, decanting, and adding water is repeated twice more to insure removal of the acid.

Once the acid has been removed, pipette a sample of the residue from the bottom of the test tube and carefully drop enough spicule-containing fluid onto the slide that will fit within a 22-mm² coverslip. It is important that a specific pipette be used per sample. Do not interchange or reuse pipettes with different samples or the preparation may become contaminated with spicules of other species. When the sample on the slide is dry remove the slide from the hot plate, allow it to cool, and examine it under a dissecting microscope. Remove any grit or foreign matter.

Add mounting medium (any permanent media will do) and cover the preparation with a glass coverslip. Label the slide and place it under a compound microscope. Scan the preparation first under low power, then high power. These are the only magnifications required for most species. For those species in which more detailed scrutiny of spicules is required, place a drop of immersion oil on the coverslip and examine with the oil immersion objective.

When examining spicules under magnification two things must be taken into consideration. One is that despite care used in the preparation of the slide, two or more species may be represented in the sample (Fig. 2.15). This often results when two adjoining species were combined in the same preservative. Also, substrates might contain remnants of the former occurrence of species other than the one present. If multiple species are present in the prepared sample then some initial confusion can occur when trying to identify species. The other consideration is that diatom frustules, like sponge spicules, are siliceous and are common in spicule preparations. There is little similarity between most diatom frustules and spicules of most sponge species; however, to be

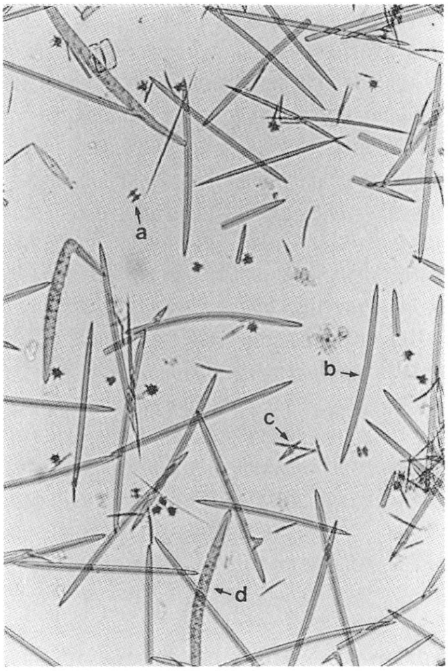

Fig. 2.15. Spicules resulting from two species mixed together: a, *Ephydatia* gemmosclere; b, *Spongilla* megasclere; c, *Spongilla* microscleres; d, *Ephydatia* megasclere.

sure of not confusing the two, any recognized manual including can be consulted for illustrations and descriptions of diatoms.

Spicules that are poorly or incompletely developed can, because of unfavorable environmental conditions, allow proper identification, especially if all potentially occurring forms (megascleres, microscleres, gemmoscleres) are present in the sample. In the event that exact identification is not possible due to spicule malformation, the only solution is to set the material aside until more specimens can be collected at some other date when environmental conditions are improved.

Although the morphology of the spicules is the most important feature of the sponge for identification purposes, the characteristics and arrangement of the gemmules are helpful for determining species as well, particularly in certain genera. The location of gemmules in the sponge is helpful for distinguishing certain genera and, in a few cases, species. It is useful, therefore, to note the arrangement and position of the gemmules. In the genus *Heteromeyenia*, the morphology of the foramen of the gemmule is actually more significant than the spicules in determining the species (Fig. 2.12). To inspect the gemmules of these species, it is necessary to dissect the gemmules from the sponge mass and delicately remove the residual spongin attached to the tendrils, if present, of the foraminal tube. This process can be undertaken only with the aid of a binocular dissecting microscope. For more detailed observations of the foraminal tube of species of *Heteromeyenia*, a small portion of the gemmule containing the foramen can be mounted on a slide in a water-soluble medium, covered with a coverslip, and examined with a compound microscope.

Taxonomy. American freshwater sponges have been studied by only a few investigators, some of the more important early contributors being Potts, Smith, Old, Penney, and Jewell. In recent years Harrison, Poirrier, Reiswig, and Volkmer-Ribeiro have been especially productive in investigating species limits and variations within North American genera. The exact number of species occurring in the United States is unknown, since recent studies have placed in synonymy several formerly recognized taxa. Several recognized genera are monotypic in the United States or contain only a few species. Much remains to be done, however, especially from taxonomic, distributional, physiological, and ecological viewpoints.

Since general growth form is usually so variable within a species, it is a poor character for species identification. The detailed anatomy of gemmules, gemmule spicules, megascleres, and microscleres is essential for accurate identification. There is still considerable disagreement among specialists concerning dividing lines between genera. Indeed, there is evidence of intergeneric or "parazoan" species mixtures in some forms. Such intergeneric hybrids have included a report of a sponge mass consisting of both *Anheteromeyenia* and *Eunapius* spicules, as well as the distinctive gemmules of both, with no intermediate structures. Many collections will produce species of different genera overgrowing one another with some aberrations apparent in the zone of contact. In this manual the generic system of Penney and Racek (1968) has been followed.

Rough average or typical lengths of spicules are given in the key that follows; bear in mind, however, that in certain environments these figures are subject to a variation of plus or minus 30%. The various spicule types for a particular species are roughly drawn to the same scale in Figs. 2.15 to 2.19. Furthermore, the spicule figures represent only a moderate amount of variation, and it must be emphasized that there is a greater degree of variation evident in geographical and environmental populations. The key has been modified in part after Smith (1995).

KEY TO THE GENERA OF FRESHWATER PORIFERA OF THE UNITED STATES

1. Microscleres present 2
Microscleres absent 7

2. Microscleres pointed, with spines or smooth 3
Microscleres minute birotulates or stellate 5

3. Microscleres straight and with knobbed spines that are longer near center of spicule; gemmoscleres, when present, birotulates of two forms depending upon length, grading into one another; sponge growth flat and encrusting to massive; 4 species, widely distributed (Fig. 2.16A–K) . . . . ***Heteromeyenia***
Microscleres gently curved with microspines of even size distributed over the spicule or microspines absent, gemmoscleres, when present, not birotulates 4

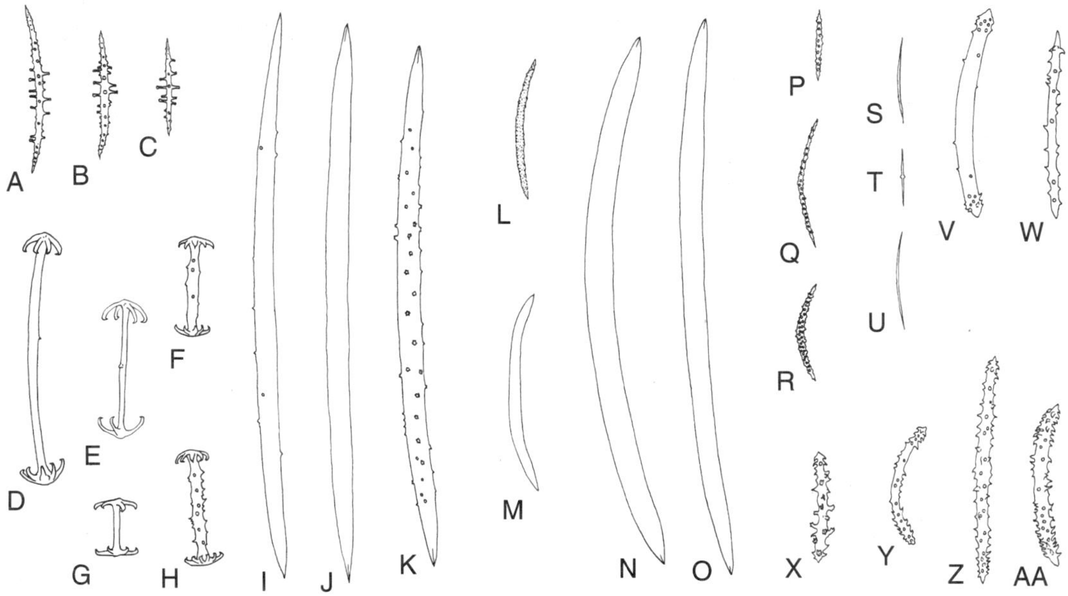

Fig. 2.16. Spicules of freshwater sponges. *Heteromeyenia:* A–C, microscleres; D–H, gemmoscleres; I–K, megascleres. *Stratospongilla:* L, microsclere; M, gemmosclere. *Spongilla:* N, O, megascleres; P–U, microscleres; V–AA, gemmoscleres. (L, M redrawn from Harrison 1979.)

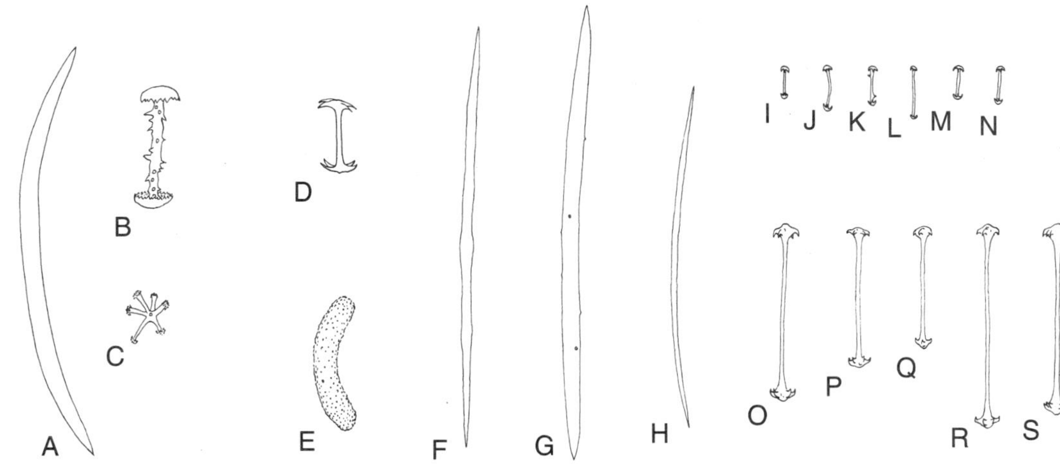

Fig. 2.17. Spicules of freshwater sponges. *Dosilia*: A, megasclere; B, gemmosclere; C, microsclere; *Corvospongilla*: D, megasclere; E, gemmosclere; *Corvomeyenia*: F–H, megascleres; I–N, microscleres; O–S, gemmoscleres. (A, B redrawn from Penney and Racek 1968; R, S redrawn from Poirrier 1978.)

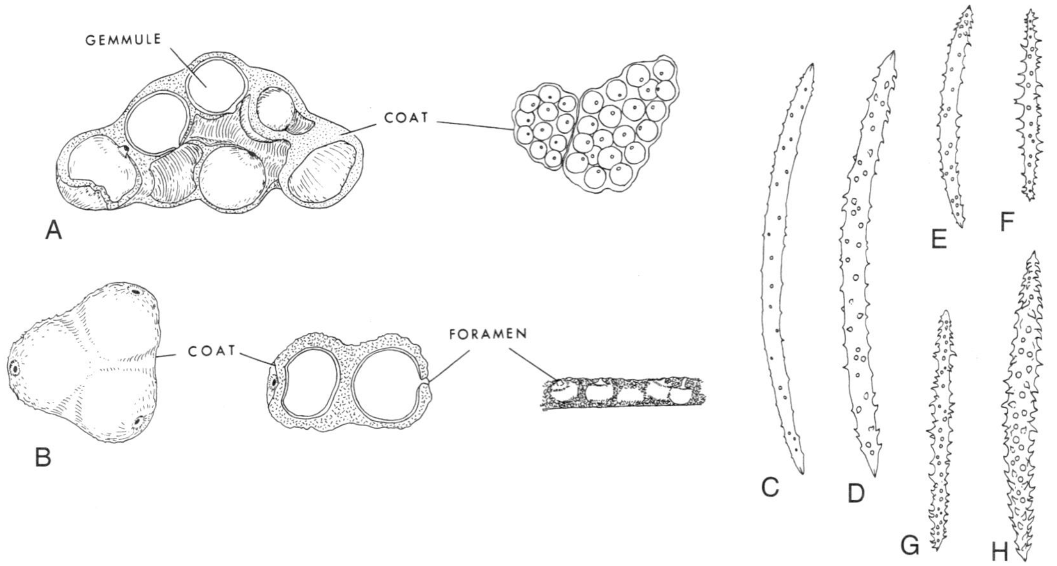

Fig. 2.18. Gemmules and spicules from freshwater sponges. A, gemmules of *Duosclera;* B, gemmules of *Eunapius. Duosclera:* C, D, megascleres; E–H, gemmoscleres.

4. Gemmoscleres amphioxea, smooth or with very small microspines, gently curved or with slightly bent extremities, sponge encrusting; 1 species; FL (Fig. 2.16L, M) ..... ***Stratospongilla***
Gemmoscleres amphistrongyla, coarsely spined; sponge may be incrusting or massive, often forming free, fingerlike extensions; 3 species; widely distributed (Fig. 2.16N–AA) ***Spongilla***

5. Microscleres stellate; 1 species, central and southern U.S. (Fig. 2.17A–C) .... ***Dosilia***
Microscleres birotulate **6**

6. Gemmoscleres, when present, larger birotulates of uniform morphology and between 65 and 110 μm in length, rotules small with short, curved hooks or variably but coarsely spined amphioxea or amphistrongyla about 20 to 60 m in length; sponge growth delicate and encrusting with slender extensions; 3 species, rare in eastern states and Canada (Fig. 2.17F–Q) ***Corvomeyenia***
Gemmoscleres, when present, stout amphistrongyla about 30 to 50 μm in length, with minute spines; 1 species, LA (Fig. 2.17D, E) .... ***Corvospongilla***

7. Gemmules and gemmoscleres absent. [The absence of gemmules and gemmoscleres in species lacking microscleres makes accurate identification even to genus nearly impossible. Nonetheless, the occasion will arise when it is worthwhile to attempt to categorize specimens as much as possible to guide future study. The following selections provide the closest match of various taxa and characteristics of the megasclere.]
   A. Megascleres always smooth .... *Eunapius* and some *Ephydatia* populations
   B. Megascleres smooth to spiny with all degrees of variation .... *Ephydatia*
   C. Megascleres always spiny.
      1. Spines blunt .... *Trochospongilla*
      2. Terminal spines directed toward spicule tips ..... *Anheteromeyenia* and *Radiospongilla*
      3. Some spicules with terminal spines directed towards center .... *Duosclera*

Gemmules and gemmoscleres present **8**

8. Gemmules enclosed in a thick yellowish or reddish-brown coat (Fig. 2.18A, B) , in groups of two or more or forming a contiguous layer at base of sponge; gemmoscleres not birotulates or true gemmoscleres absent **9**
Gemmules not enclosed in coat, scattered throughout sponge **10**

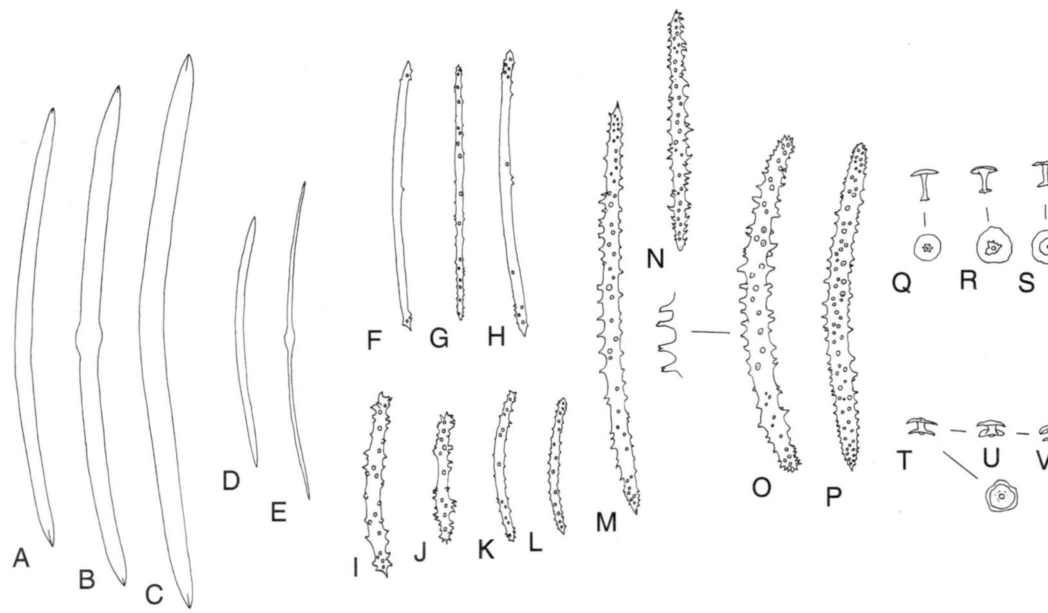

Fig. 2.19. Spicules of freshwater sponges. *Eunapius:* A–E, megascleres; F–N, gemmoscleres. *Trochospongilla:* O–P, megascleres; Q–V, gemmoscleres.

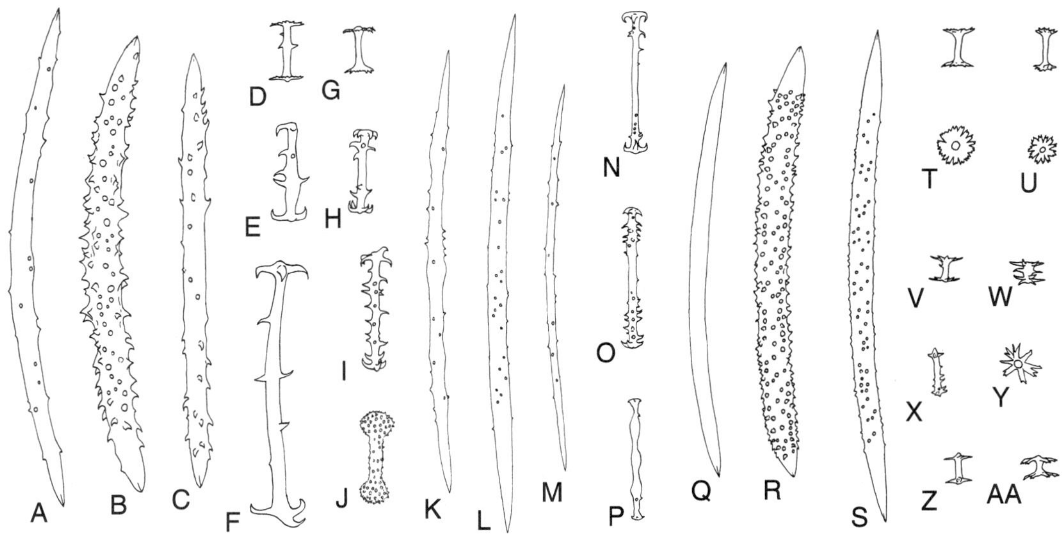

Fig. 2.20. Spicules of freshwater sponges. *Anheteromeyenia:* A–C, megascleres; D–J, gemmoscleres. *Radiospongilla:* K–M, megascleres; N–P, gemmoscleres. *Ephydatia:* Q–S, megascleres; T–AA, gemmoscleres. (J redrawn from Penney and Racek 1968.)

9. Gemmules in groups of a few to several, occasionally forming a basement layer, foramen always directed inward in groups or downward in basement layer; skeletal and gemmule megascleres spined and somewhat transitional; 1 species in the eastern U. S.(Fig. 2.18A, C–H) ***Duosclera***

Gemmules forming a layer at base of sponge or in small groups, foramen of gemmule always directed outside in groups or upward in basement layer; megascleres smooth, gemmoscleres coarsely spined, especially towards tips, occasionally transitional with megascleres; 1 species widespread throughout the U.S. (Figs. 2.18B; 2.19A–N) . . . . ***Eunapius***

10. Spines of megascleres, viewed under high magnification, with blunt spiny tips or crowns; rotules of gemmoscleres with entire margins, not excised to form serrations or spines; widely distributed, 3 species (Fig. 2.19O–V) . . . . ***Trochospongilla***

Spines of megascleres pointed; rotules of gemmoscleres with serrated or spiny (rayed) margins **12**

12. Gemmoscleres with only one form represented; megascleres, if spined, with terminal spines sometimes directed toward tips (genus *Radiospongilla*) or not **13**

Gemmoscleres demonstrating two distinct or a graded series of forms; megascleres always spined and with terminal spines directed toward spicule tips; widely distributed, 3 species (Fig. 2.20A–J) . . . . ***Anheteromeyenia***

13. Gemmosclere length ranging from 75 to 95 μm, rotules with curved hooks; megascleres always with spines; foramen of gemmule situated in a recessed area and encircled by radially arranged gemmoscleres; 2 species, 1 widely distributed species, 1 exotic TX (Fig. 2.20K–P) ***Radiospongilla***

Gemmosclere length much less than 75 μm, rotules without curved hooks; megascleres smooth, weakly spined , or densely and coarsely spined, with all three conditions potentially present; gemmule not as above; 2 species widely distributed (Fig. 2.20Q–AA) . . .***Ephydatia***

REFERENCES

Arndt, W. 1928. Porifera, Schwämme, Spongien. *Tierwelt Dtschl.* 4:1–94.

Berg, K. 1948. Studies on the bottom animals of Esrom Lake. *Mem. Acad. R. Soc. Lett. Danemark, Ser. 9,* 8:1–255.

Bowerbank, J. S. 1863. A monograph of the Spongillidae. *Proc. Zool. Soc. London* (1863): 440–472.

Brien, P. 1932. Contribution A l'étude de la regeneration naturelle chez les Spongillidae. *Arch. Zool. Exp. Gén.* 74:461–506.

——. 1936. La reorganisation de l'éponge après dissociation par filtration et phénomènes d'involution chez *Ephydatia fluviatilis. Arch. Biol.* 48:185–268.

Bronsted, A., and H. V. Bronsted. 1953. The effects of symbiotic zoochlorellae on the germination rate of gemmules of *Spongilla lacustris* (L.). *Vidensk. Medd. Dan Naturhist. Foren. Kobenhavn* 115:133–144.

Courrèges, V. C., and P. E. Fell. 1989. Sexual and asexual reproduction by the freshwater sponge *Anheteromeyenia ryderi*, with emphasis on spermatogenic activity. *Trans. Am. Microsc. Soc.* 108:127–138.

De Vos, L., et al. 1991. *Atlas of Sponge Morphology.* 117 pp. Smithsonian Inst. Press, Washington, DC.

Eshleman, S. K. 1949. A key to Florida's freshwater sponges, with descriptive notes. *Q. J. Fla. Acad. Sci.* 12:35–44.

Fell, P. E. 1990. Environmental factors affecting dormancy in the freshwater sponge *Eunapius fragilis* (Leidy). *Invert. Reprod. Dev.* 18:213–219.

Frost, T. M., and C. E. Williamson, 1980. In situ determination of the effect of symbiotic algae on the growth of the freshwater sponge *Spongilla lacustris. Ecology* 61:1361–1370.

Gilbert, J. J. 1975. Field experiments on gemmulation in the fresh-water sponge *Spongilla lacustris. Trans. Am. Microsc. Soc.* 94:347–356.

Gilbert, J. J., and T. L. Simpson. 1976. Sex reversal in a freshwater sponge. *J. Exp. Zool.* 195: 145–151.

Harrison, F. W. 1971. A taxonomical investigation of the genus *Corvomeyenia* Weltner (Spongillidae) with an introduction of *Corvomeyenia carolinensis* sp. nov. *Hydrobiologia* 38:123–140.

——. 1979. The taxonomic and ecological status of the environmentally restricted spongillid species of North America, V: *Ephydatia subtilis* (Weltner) and *Stratospongilla penneyi* sp. nov. *Hydrobiologia* 65:99–105.

——. 1982. Developmental biology of freshwater sponges. In: *Developmental Biology of Freshwater Invertebrates*, pp. 1–67, edited by F. W. Harrison and R. R. Cowden. Liss, New York.

——. 1988. Utilization of freshwater sponges in paleolimnological studies. *Palaeogeog. Palaeoclimatol. Palaeoecol.* 62:387–397.

Harrison, F. W., and R. R. Cowden (eds.). 1976. *Aspects of Sponge Biology.* 354 pp. Academic, New York.

Harrison, F. W., and M. B. Harrison. 1977. The taxonomic and ecological status of the environmentally restricted spongillid species of North America, II: *Anheteromeyenia biceps* (Lindenschmidt). *Hydrobiologia* 55:167–169.

Harrison, F. W., et al. 1990. The dermal membrane of *Eunapius fragilis.* In: *New Perspectives in Sponge Biology*, pp. 223–227, edited by K. Rützler. Smithsonian Inst. Press, Washington, DC.

Imsiecke, G. 1993. Ingestion, digestion, and egestion in *Spongilla lacustris* (Porifera, Spongillidae) after pulse feeding with *Chlamydomonas reinhardtii* (Volvocales). *Zoomorphology* 113:233–244.

Jewell, M. E. 1935. An ecological study of the fresh-water sponges of northern Wisconsin. *Ecol. Monogr.* 5:461–504.

——. 1939. An ecological study of the fresh-water sponges of Wisconsin, II: the influence of calcium. *Ecology* 20:11–28.

——. 1952. The genera of North American freshwater sponges: *Parameyenia*, new genus. *Trans. Kans. Acad. Sci.* 55:445–457.

Jorgensen, C. B. 1944. On the spicule-formation of *Spongilla lacustris* (L.), I: the dependence of the spicule formation on the content of dissolved and solid silicic acid on the milieu. *Biol. Medd. Kjobenhavn* 19:1–45.

Leveaux, M. 1939. La formation des gemmules chez les Spongillidae. *Ann. Soc. R. Zool. Belg.* 70:53–96.

——. 1941, 1942. Contribution A l'etude histologique de l'ovogenese et de Ia spermatogenese des Spongillidae. *Ann. Soc. R. Zool. Belg.* 72:251–269; 73:33–50.

Lindenschmidt, M. J. 1950. A new species of fresh-water sponge. *Trans. Am. Microsc. Soc.* 69:214–216.

McNair, G. T. 1923. Motor reactions of the fresh-water sponge, *Ephydatia fluviatilis. Biol. Bull.* 44:153–166.

Neidhofer, J. R. 1940. The fresh-water sponges of Wisconsin. *Trans. Wis. Acad. Sci. Arts Lett.* 32:177–197.

Old, M. C. 1931a. A new species of fresh-water sponge. *Trans. Am. Microsc. Soc.* 50:298–299.

——. 1931b. Taxonomy and distribution of the freshwater sponges (Spongillidae) of Michigan. *Pap. Mich. Acad. Sci. Arts Lett.* 15:439–476.

——. 1932a. Environmental selection of the freshwater sponges (Spongillidae) of Michigan. *Trans. Am. Microsc. Soc.* 51:129–137.

——. 1932b. Contribution to the biology of freshwater sponges (Spongillidae). *Pap. Mich. Acad. Sci. Arts Lett.* 17:663–679.

Paduana G. M., and P. E. Fell. 1997. The occurrence of *Corvomeyenia carolinensis* (Harrison, 1971) (Porifera: Spongillidae) in Connecticut. *Hydrobiologia* 350:123–125.

Penney, J. T. 1933a. Reduction and regeneration in fresh water sponges (*Spongilla discoides*). *J. Exp. Zool.* 65:475–495.

——. 1933b. A new fresh-water sponge from South Carolina. *Proc. U. S. Nat. Mus.* 82:1–5.

——. 1954. Ecological observations on the fresh-water sponges of the Savannah River Project area. *Univ. S. C. Publ. Ser. 3*, 1:156–192.

——. 1960. Distribution and bibliography (1892–1957) of the freshwater sponges. *Univ. S.C. Publ. Ser. 3*, 3:1–97.

Penney, J. T., and A. A. Racek. 1968. Comprehensive revision of a world-wide collection of fresh-water sponges (Porifera: Spongillidae). *Bull. U. S. Nat. Mus.* 272:1–184.

Poirrier, M. A. 1974. Ecomorphic variation in gemmoscleres of *Ephydatia fluviatilis* Linnaeus (Porifera: Spongillidae) with comments upon its systematics and ecology. *Hydrobiologia* 44: 337–347.

——. 1977. Systematic and ecological studies of *Anheteromeyenia ryderi* (Porifera: Spongillidae) in Louisiana. *Trans. Am. Micros. Soc.* 96:62–67.

——. 1978. *Corvospongilla becki* n. sp., a new fresh-water sponge from Louisiana. *Trans. Am. Micros. Soc.* 97: 240–243.

Poirrier, M. A., and M. Winter.1990. Gemmule pneumatic layer development in Louisiana popluations of *Spongilla lacustris.* In: *New Perspectives in Sponge Biology*, pp.491–496, edited by K. Rützler, Smithsonian Inst. Press, Washington, DC.

Poirrer, M. A., et al. 1987. Comparative morphology of microsclere structure in *Spongilla alba*, *S. cenota*, and *S. lacustris* (Porifera, Spongillidae). *Trans. Am. Micros. Soc.* 106:302–310.

Potts, E. 1887. Fresh water sponges: a monograph. *Proc. Acad. Nat. Sci. Phila.* (1887): 157–279.

Reiswig, H. M., and A. Ricciardi. 1993a. Resolution of the taxonomic status of the freshwater sponges *Eunapius mackayi*, *E. igloviformis*, and *Spongilla johanseni* (Porifera: Spongillidae). *Trans. Am. Micros. Soc.* 112:262–279.

——. 1993b. Reexamination of *Corvospongilla novaeterrae* (Porifera, Spongillidae), an environmentally restricted freshwater sponge from eastern Canada. *Can. J. Zool.* 71: 1954–1962.

Saller, U. 1990. A redescription of the freshwater sponge *Trochospongilla horrida* (Porifera, Spongillidae). *Abh. Naturwiss. Ver. Hamburg* 31/32:163–174.

——. 1991. Symbiosis of *Spongilla lacustris* (Spongillidae) and green algae: algae uptake, distribution and final whereabouts. In: *Fossil and Recent Sponges*, pp.299–305, edited by J. Reitner and H. Keupp. Springer, Berlin.

Sasaki, N. 1934. Report on the fresh-water sponges obtained from Hokkaido. *Sci. Rep. Tohoku Imp. Univ. (Ser. 4)* 9:219–248.

Schroder, K. 1936. Beitrage zur Kenntnis der Spiculabildung der Larvenspiculation und der Variationsbreite der Gerbstnadeln von Süsswassershwämmen. *Z. Morphol. Oekol. Tiere* 31:245–267.

Simon, L. 1953. Uber die Spezifitdt der Nadeln und die Variabilitdt der Arten bei den Spongilliden. *Zool. Jahrb. AU Allg. Zool. Physiol. Tiere* 64:207–234.

Simpson, T. L., and P. E. Fell. 1974. Dormancy among the Porifera: gemmule formation and germination in fresh water and marine sponges. *Trans. Am. Microsc. Soc.* 93:544–577.

Smith, D. G. 1976. An intergeneric fresh-water sponge mixture (Demospongiae: Spongillidae). *Trans. Am. Microsc. Soc.* 95:235–236.

——. 1994. First report of a freshwater sponge (Porifera: Spongillidae) from the West Indies. *J. Nat. Hist.* 28:981–986.

——. 1995. *Keys to the Freshwater Macroinvertebrates of Massachusetts.* 243 pp. Sunderland, Massachusetts.

Smith, F. 1921a. Data on the distribution of Michigan fresh-water sponges. *Pap. Mich. Acad. Sci. Arts Lett.* 1:418–421.

——. 1921b. Distribution of the fresh-water sponges of North America. *Bull. Ill. Nat. Hist. Surv.* 14:922.

Ungemach, L. F., et al. 1997. Possession and loss of cold tolerance by sponge gemmules: a comparative study. *Invert. Biol.* 116:1–5.

Van Trigt, H. 1919. A contribution to the physiology of the fresh-water sponges. (Spongillidae). *Tijd. Ned. Dierkd. Vereen. (II)* 17:1–220.

Weissenfels, N. 1992a. The influence of an eucaryotic intranuclear call parasite on the production of gemmules in *Ephydatia fluviatilis* (Porifera, Spongillidae). *Zoomorphology* 111: 207–216.

——. 1992b. The filtration apparatus for food collection in freshwater sponges (Porifera, Spongillidae). *Zoomorphology* 112:51–55.

Wierzejski, A. 1935. Süsswasserspongien. Monographische Bearbeitung. *Mem. Acad. Polon. Sci. Lett., Ser. B*, 9:1–242.

3

CNIDARIA

(Hydroids, jellyfish)

Although the Phylum Cnidaria contains in excess of 9000 species, the vast majority are marine. Two classes, the Scyphozoa (true jellyfish) and the Anthozoa (corals, sea anemones, etc.), are exclusively marine, and in the third class, the Hydrozoa, only about 22 species are known from the freshwaters of the United States. These include about 12 hydras, one uncommon "freshwater jellyfish," a parasitic form that is free-living as an adult, a minute polyp with single-celled tentacles, and two species that live in brackish coastal waters but that enter freshwater, a common colonial polyp and a rare solitary protohydroid.

Although not conspicuous, freshwater cnidarians are typical representatives of the fauna of ponds, spring brooks, unpolluted streams and rivers, and lakes. They are all sessile and the polyps may be found attached to stones, twigs, vegetation, or debris, occasionally in enormous numbers. They usually live at depths not exceeding 1.5 m but have been found as deep as 300 m. Populations are most dense during the summer months. Regarding hydras, when there is a food shortage, a polyp often secretes a gas bubble beneath the pedal disk and floats to the surface film, where the bubble bursts, leaving the hydra hanging downward from the surface film.

General characteristics. Freshwater cnidarians, especially hydras (Fig. 3.1), should be familiar to all who have had an elementary biology or zoology course. Like other cnidarians, the polyps are radially symmetrical, the main body, or column, being an elongated cylinder from less than 1 to over 25 mm long. Attachment to the substrate is effected by a pedal disk that consists largely of special secretory cells. The distal end of the column has a circlet of tentacles whose length varies from less than one-half to five times the length of the column. Usually there are 5 to 8 tentacles, sometimes over 20 in certain species. The end of the column in the center of the circlet of tentacles is raised and domelike to form a hypostome that bears a single opening, the combined mouth and anus. The single, continuous, internal body cavity is a gastrovascular cavity; it usually continues out into the hollow tentacles.

A phenomenon characterizing cnidarians is the shift between two functional forms during the life history of many living species. The phenomenon is called alternation of generations and although it is not unique to cnidarians the transformation between a sessile biology and a motile biology is not demonstrated in other phyla as dramatically as it is in the Cnidaria. The process is exhibited best in the Class Hydrozoa, to which all freshwater cnidarians belong, and involves both a complete reversal in orientation of the animal and a shift between asexual and sexual methods of reproduction. Generally, a sessile polyp with the oral (mouth) pole fac-

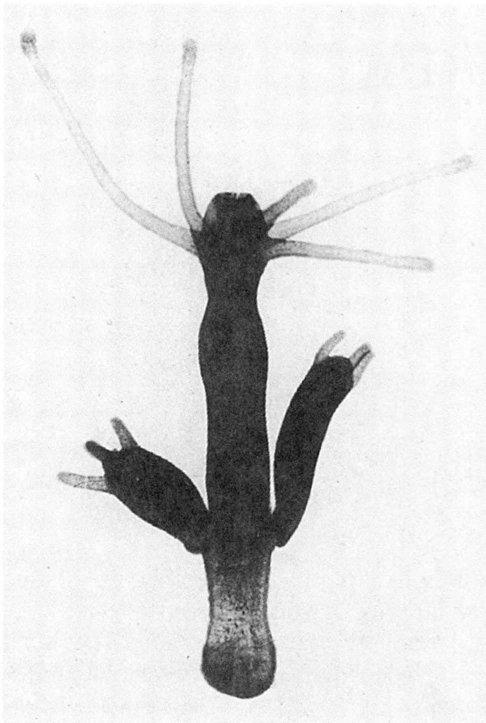

Fig. 3.1. Hydra with two buds (partially contracted, ×12). (Courtesy General Biological Supply House, Inc., Chicago, IL.)

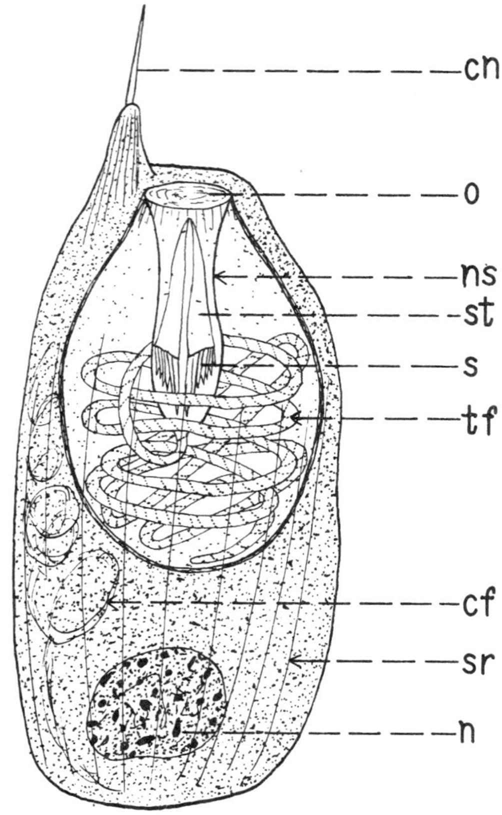

Fig. 3.2. Undischarged stenotele nematocyst within its cnidoblast. cf, cnidoblast filament; ns, nematocyst shaft (butt); o, operculum; s, spines; sr, supporting rods of cnidoblast; st, stylet; tf, tubular filament (actually much longer than shown).

ing up develops asexual buds that give rise to a motile, swimming egg-forming medusa in which the oral surface is pointed downward.

Many groups of cnidarians have reduced or lost one of the two phases. Such is the case with freshwater cnidarians and only one species living in the United States, *Craspedacusta sowerbii*, retains the medusa stage. All other species have lost the medusa stage of their life history, and all reproductive processes, whether sexual or asexual, have been shifted to the polyp.

Because of the disparate biology and structure of the freshwater cnidarians occurring in the United States, each genus is introduced and discussed.

Nematocysts. All cnidarians have minute stinging capsules, or nematocysts, imbedded in certain epidermal cells (Fig. 3.2). Each nematocyst contains a coiled capillary thread, which, on proper stimulation, can be "exploded" and extruded. Nematocysts function in food getting and protection by means of their entangling, adhesive, and paralyzing properties. Nematocysts are especially abundant on the tentacles and distal part of the column.

A nematocyst is a spherical to elongated, capsular, sclerotized organoid ranging from 5 to 25 μm in length in freshwater species. It is contained within a special epidermal cell, the cnidoblast. Inside of the nematocyst is a minute, coiled, hollow thread, or tubule, whose base is attached to the inner distal wall of the nematocyst. A lid, or operculum, covers the distal opening of the capsule, and a tiny, triggerlike bristle, or cnidocil, projects into the surrounding water from the surface of the cnidoblast near the edge of the operculum. In the Phylum Cnidaria as a whole there are many different types of nematocysts, but in the freshwater hydras there are only four types as follows: Desmonemes (volvents) are small, pyriform, or spherical nematocysts. The tubule forms only one loop, is closed at the distal end, and is usually unarmed but many have extremely fine spinules (Fig. 3.3). Desmonemes appear to function mainly in entangling and wrapping around bristles of the prey. Stenoteles (penetrants) are large, spherical nematocysts. The coiled tubule is open at the tip and has a definite butt, or enlarged basal portion, which bears three spines and three spiral rows of minute spinules (Fig. 3.3). Exploding stenoteles pierce the prey and inject a paralyzing material. There are two types of isorhiza (glutinant) nematocysts, but both are characterized by a tubule that is open at the tip and by the absence of a butt. Holotrichous isorhizas are larger and have the tubule covered with minute spinules (Fig. 3.3), which function mainly in defense. Atrichous isorhizas are smaller and have a smooth tubule (Fig. 3.3), which appear to function mainly in locomotion by discharging and sticking when the tentacles are placed in contact with a substrate.

Nematocysts have an unusual mode of origin and migration. They are formed within special interstitial cnidoblast cells of the epidermis. Most cnidoblasts first appear in the distal third of the column. Before their contained nematocysts are fully formed, many cnidoblasts pass through the mesoglea and gastrodermis into the gastrovascular cavity. Here they circulate briefly before being taken up again by gastrodermal cells, especially in the tentacles. Then they are transported back through the mesoglea to their final location in the epidermis. In hydras, the cnidoblasts migrate as amoebae through the epidermis into the tentacles where they become positioned for action. Every day, hundreds or thousands of cnidoblasts make this journey to replace the used nematocysts.

According to one theory, when the projecting cnidocil is touched, the operculum springs open and the nematocyst tubule instantaneously turns inside out as it whips out of the capsule. Expulsion is facilitated by high pressure maintained within the cell. The base of the tubule emerges first and the rest follows from base to tip. Following expulsion, the internal coiled tubule disintegrates. Discharged nematocysts are lost and are of no further use to the hydra.

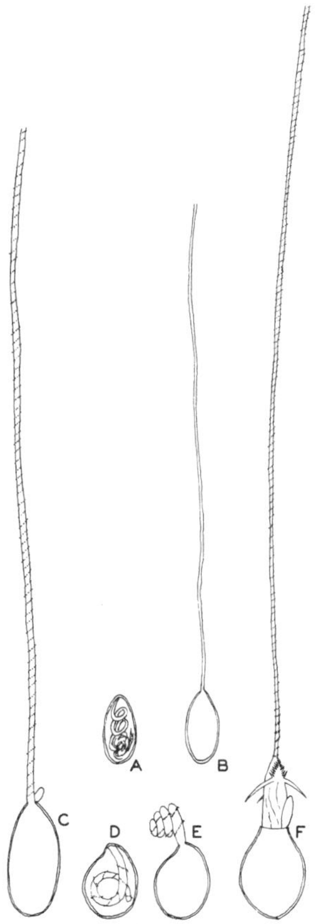

Fig. 3.3. Typical hydra nematocysts. A, undischarged atrichous isorhiza; B, discharged atrichous isorhiza; C, discharged holotrichous isorhiza; D, undischarged desmoneme; E, discharged desmoneme; F, discharged stenotele. A–E, ×2000; F, ×1000. (Cnidoblasts not shown.)

It is thought that stenotele and desmoneme discharge is usually dependent on a combined chemical and mechanical stimulus. The juice and body fluids of the normal prey will not in themselves discharge nematocysts nor will a fine glass rod touched to the cnidocils. In addition, the nematocysts of satiated animals do not react when food organisms touch the tentacles. This indicates that the nervous system is also involved with expulsion of nematocysts.

The physiological mechanisms of nematocyst action are poorly understood, but it is thought that the discharge is produced by an increased pressure within the capsule caused by increased permeability of the capsule wall. Perhaps a chemical stimulus acts first by reducing surface tension and lowering the resistance of the cnidoblast and cnidocil to mechanical stimuli.

One of the most unusual phenomena centering around nematocyst biology is the use of nematocysts by certain species of *Microstomum* for their own protection and food getting. When one of these turbellarians feeds on a hydra only the desmonemes and atrichous isorhizas are digested, and the other two types of nematocysts are passed through the mesenchyme of the *Microstomum* and into the epidermis. Here they are correctly oriented by associated phagocytes and discharged when touched by other animals.

Hydra. Hydras are solitary polyps and are the most commonly encountered freshwater cnidarians. Hydras have a well-developed column that ranges from 1 to 25 mm long. The tentacles are well developed and number between 5 and 12, depending on species and age. Although some hydras have a characteristic green or brown color, others have a variable coloration, depending on the kind and amount of ingested food. Thus, they may be various shades of translucent whitish, grayish, tan, or red.

The body wall is simply constructed and typical of all cnidarians. Externally it consists of a layer of epidermis, and internally there is a layer of gastrodermis lining the gastrovascular cavity. Several different histological types of epidermal and gastrodermal cells can be distinguished. Between these two epithelia is a very thin, secreted, noncellular, cementlike mesoglea.

Hydras are typically sessile creatures (Fig. 3.4); when they are undisturbed the column is extended, the tentacles are spread out, and there is no movement visible to the naked eye. Occasionally, however, the tentacles wave about or the whole animal contracts and expands unaccountably. Hydras are also capable of moving slowly on the substrate, but the exact method of movement has not been observed. Occasionally they exhibit somersaulting or inchworm-like looping movements. Mechanically, they attach the distal end of the body to the substrate and pull themselves along by contractions and elongations. The central part of the basal disk may secrete a small, gas bubble that enables the hydra to rise to the surface and remain suspended at the surface film in quiet waters. Budding is the common asexual method of reproduction. Such buds usually originate at the junction of the gastric and stalk regions. A bud first appears as a hemispherical outpouching that later elongates, becomes cylindrical, and develops tentacles. Subsequent buds appear spirally. The gastrovascular cavity of the budding individual and parent are continuous, but soon the base of the bud pinches off completely and the new individual becomes separated and independent. When food is plentiful a new bud may be produced every 2 or 3 days.

Both transverse and longitudinal fission are occasionally observed in hydras, but these methods of reproduction are not normal and usually follow injuries or depression periods. Powers of regeneration and susceptibility to grafting are well developed, and a large amount of experimental work has been done on these phases of hydra biology.

Syngamic reproduction is usually limited to the autumn months when water temperatures are dropping, but in a few species it may appear in spring or summer when temperatures are highest or when the habitat is drying up. In hermaphroditic species the testes are distal and the ovaries proximal; sometimes ovaries and testes mature simultaneously, sometimes at different periods. In dioecious species the ovaries or testes may be distributed throughout most of the distal half or three-quarters of the column. In general, males far outnumber females, in striking contrast to the situation in most freshwater invertebrate groups.

A gonad consists of an accumulation of epidermal interstitial cells within the epidermis, but in a developing ovary these cells fuse or engulf each other, leaving a single,

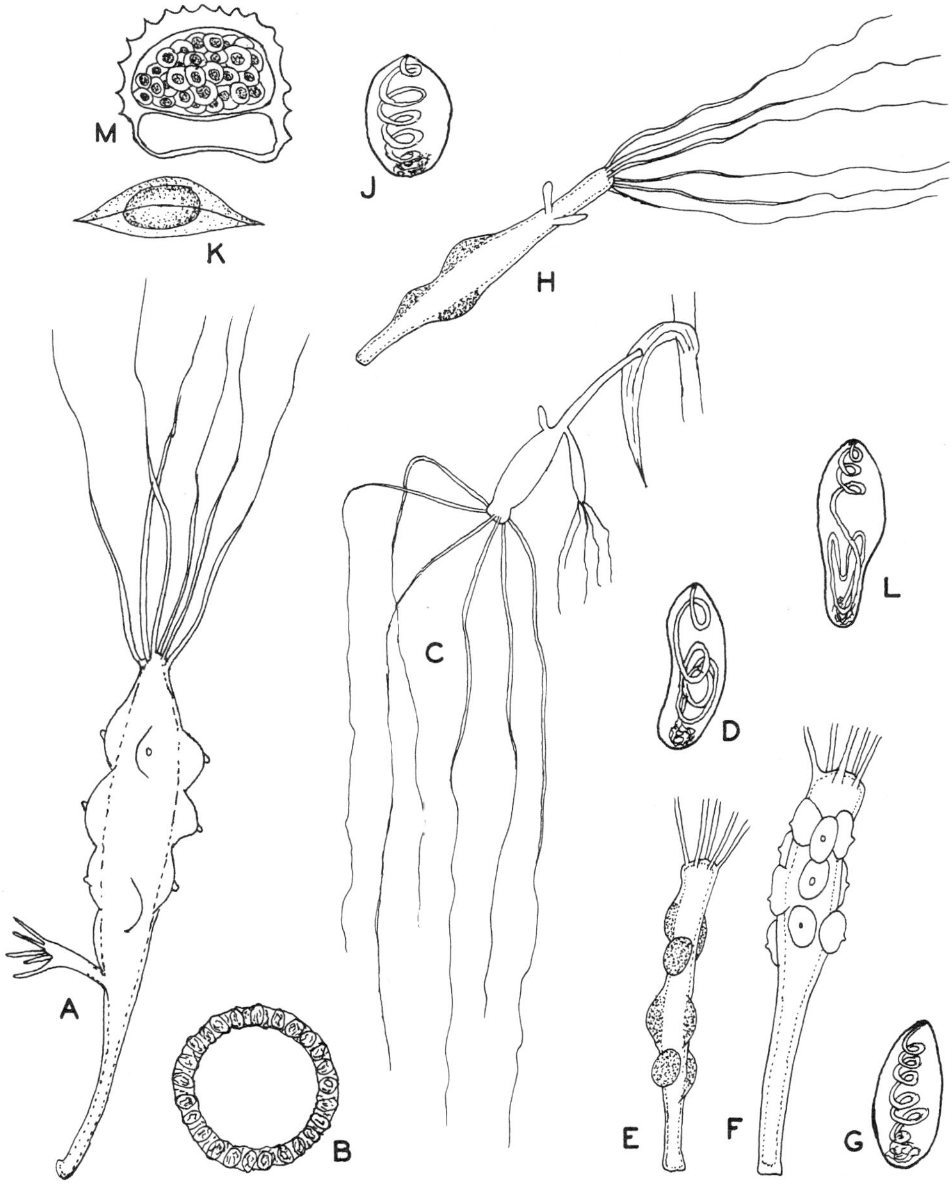

Fig. 3.4. Structure of hydras. A, male *Hydra cauliculata*, ×10; B, optical section of thecated embryo of *H. cauliculata;* C, *H. oligactis*, ×5; D, holotrichous isorhiza of *H. oligactis;* E, male *H. oligactis* with testes, ×4; F, male *H. pseudoligactis* with testes, ×4; G, holotrichous isorhiza of *H. pseudoligactis;* H, sexually mature *H. utahensis*, showing basal ovaries and distal testes, ×7; J, holotrichous isorhiza of *H. utahensis;* K, theca and contained zygote of *H. utahensis;* L, holotrichous isorhiza of *Hydra viridissima;* M, cross section of theca of *Hydra hadleyi.* (A and B modified from Hyman 1938; C–E modified from Hyman 1930; F–K modified from Hyman 1931b.)

large, food-filled ovum. A mature testis is usually hemispherical or helmet-shaped. It usually has a "nipple" (Fig. 3.4A), and many of the sperm escape to the outside through the tip of the nipple. Ovaries are hemispherical or bulbous swellings. When an egg is mature, the covering epidermis ruptures and withdraws, forming a small cup or cushion under the egg. If not fertilized promptly, the egg dies. Sperm may come from the same or a different individual.

About a day after fertilization the embryo secretes a sclerotized yellowish shell, the embryonic theca. In some species when the theca is being formed it is somewhat sticky, and at this stage the whole embryo drops off the parent or is fastened to the substrate by bending movements of the parent. It sticks to the substrate while the theca hardens. In most species the embryo remains on the parent longer and is spherical. Such thecated embryos (Fig. 3.4K) may be spherical or planoconvex, spinous or smooth. The diameter ranges from 0.40 to 0.60 mm. It is assumed that they are highly resistant to adverse environmental conditions and endure drying and freezing temperatures. After a dormant period of 3 to 10 weeks or more, the theca softens and splits, permitting the exit of an embryo that already has a gastrovascular cavity and tentacles.

Little is known about the length of life under normal conditions. Under favorable conditions in the laboratory, however, hydras live for 3 to 12 months or more. It is generally considered that they are immortal because new growth compensates for tissue lost during budding.

Craspedacusta. Although hydras are by far the most familiar freshwater hydrozoans, there is one American species of "jellyfish," *Craspedacusta sowerbii.* Unlike large marine jellyfish (scyphozoa), however, it has a velum and its life history includes a colonial (but simple) polyp.

Craspedacusta sowerbii was first found in England in 1880; more recently it has been noted many times in Europe and on other continents but only rarely in Africa. It has been reported from about 60 localities in the United States since 1908, when it was discovered in this country. It has been found mostly in small lakes, ponds, fishponds, and old water-filled quarries, especially between July and October. In addition, there are a few reports for rivers, sluggish streams, and large reservoirs (e.g., Lake Mead).

A mature medusa stage of *C. sowerbii* (Fig. 3.5) has the typical umbrella or bell shape, manubrium, and four radial canals and one circular canal forming the gastrovascular cavity. The diameter of the bell is about 5 to 22 mm, and the circumference bears 50 to 500 solid tentacles of varying length and arranged in three to seven series, depending on the medusa size. Only short tentacles are involved in feeding; long tentacles are presumably stabilizers for swimming. The bell margin also bears many statocysts, often 60 to 100 or more. The velum is rather thick. Nematocysts are especially abundant around the mouth, at the edge of the bell, and on the tentacles. *Craspedacusta* actively swims up, down, or sideways in a dancing sort of movement produced by varying contractions of the bell. Food consists of zooplankters, chiefly in the 0.2- to 2.0-mm range. It is not thought to be an important predator on very small fish. Crayfish are the only important group that feed on the medusae, especially when the latter are at rest on the substrate.

Most *Craspedacusta* populations appear to consist of all males or all females, the occurrence of mixed populations having been observed only a few times. The four gonads consist of convoluted or flaplike masses on the subumbrellar surface of the bell in the vicinity of the radial canals. Both eggs and sperm are released into the surrounding water, where fertilization occurs.

The zygote develops into a simple, branching, colonial hydroid 2 to 8 mm long, usually consisting of two to four individuals (seldom up to 10) (Fig. 3.6). Single individuals are usually less than 2 mm long, and tentacles are absent. The hydroid creeps about slowly on the substrate, feeding on small metazoans. The region around the mouth bears clumps of nematocysts. The body secretes a sticky, mucous material to which debris adheres and more or less covers the body. During the winter months the hydroids contract and secrete a sclerotized protective coat.

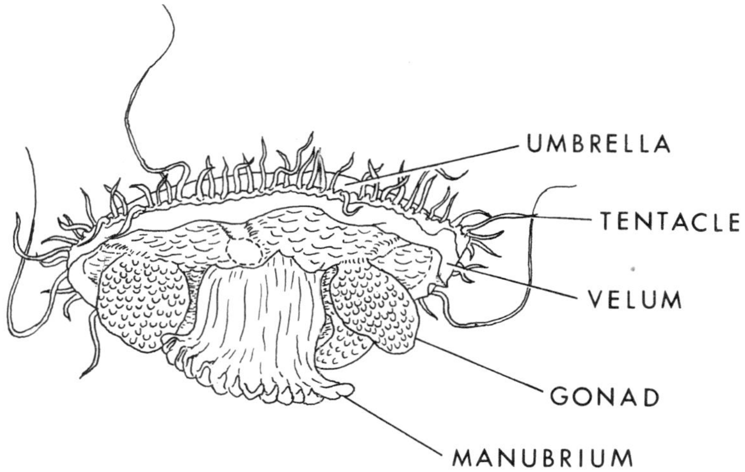

Fig. 3.5. Medusa of *Craspedacusta sowerbii*, ×5.

The hydroid stage has three methods of asexual reproduction. Ordinary budding is similar to that of the hydras, and new individuals may remain attached or constrict off from the parent (Fig. 3.7). Nonciliated planulalike buds (frustules) may also be produced. These constrict off, elongate, and creep about slowly on the substrate for a time, and then develop into new polyps. The third method involves the production of medusa buds. When these break free from the parent polyp the bell diameter is about 0.4 mm, and there are only eight tentacles,

Fig. 3.6. Polyp cluster of *Craspedacusta sowerbii*, ×100.

but rapid growth and development follow, resulting in the large, sexually mature medusae. Both polyps and medusae feed on a wide variety of small invertebrates.

The occurrence of the medusae is highly sporadic and unpredictable. In some years they may be found in abundance in certain ponds, but in subsequent years they may be absent, only to reappear several years later or in several successive years. In other ponds they have been found only once and have never reappeared, in spite of careful searching. Production of the medusa appears to reach its peak in late summer, usually following a sudden warm spell. There is some evidence to show that the medusae are positively phototropic.

The hydroid generation is only occasionally noticed, probably chiefly because of its small size and inconspicuous nature. It is possible that the hydroids live undetected for long periods, only occasionally giving rise to the medusoid generation. For many years the hydroid was called *Microhydra ryderi*, and it was not until 1924 that it was definitely associated with *Craspedacusta sowerbii*.

The freshwater jellyfish was introduced from China into the United States around the turn of the century. It has been reported from nearly all states except the northern most and is found in rivers, natural lakes, and ponds, and in quarry ponds where pop-

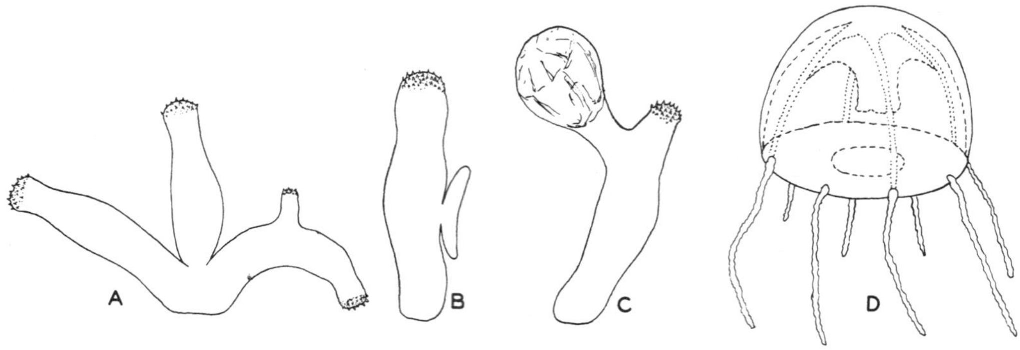

Fig. 3.7. Stages in the life history of *Craspedacusta sowerbii*. A, typical colony of hydroids, ×12; B, hydroid with a maturing planulalike bud, ×15; C, hydroid with a maturing medusa bud, ×15; D, newly released medusa with eight tentacles, ×100, (A and D greatly modified from Payne 1924.)

ulations of the medusa stage can be enormous. The species has been found living in water-treatment facilities as well. On a worldwide basis, *C. sowerbii* appears to be chiefly confined to a broad band between 45° north and 45° south latitude.

Several related species of freshwater limnomedusae have been reported from other continents, including one species from China, two or three in Africa, seven in India, and one in Trinidad. All of these species have their closest relatives in marine situations, and most invasions of freshwaters appear to transpire along tropical shores.

Calpasoma. The tiny polyps of *Calpasoma dactyloptera* have at times been considered variants of *C. sowerbii.* Indeed, the two species are very similar and an intermediate form has been reported at least once. Experimental evidence has shown, however, that the two forms are not transformational and represent separate and distinct species. *Calpasoma* is among the smallest of cnidarian genera, the maximum size reported is 200 μm in length. The only described species, *C. dactyloptera* (Fig. 3.8), forms a solitary polyp and has tentacles that are about 150 μm in length when fully extended. The tentacles are arranged in two poorly defined rows and are unique among cnidarians in that they consist of a single cell containing a vacuole. Nematocysts of the same type found in *C. sowerbii* are distributed throughout the tentacle or tentaculocyte (Fig. 3.8A).

Reproduction in *C. dactyloptera* is by frustule formation. Tiny frustules formed on the epidermis migrate away from the parent and give rise to a new polyp.

Calposoma dactyloptera has been found rarely in the wild and most research on the species has been performed on laboratory cultures. The species seems to be often found in the company with *C. sowerbii* and apparently prefers aquatic plants for a substrate.

Cordylophora. A colonial brackish-freshwater cnidarian is *Cordylophora caspia.* It is a profusely branching, mosslike hydrozoan (Fig. 3.9) superficially similar to the marine *Obelia* and its many relatives. Some taxonomists believe this species should be called *Cordylophora lacustris.* During the summer months brackish-water colonies attain a height of 20 to 100 mm from the substrate, but in freshwaters growths are usually smaller. Some freshwater colonies have been reported to reach 70 mm but usually the colonies seldom exceed 30 mm in height. Two kinds of polyps make up such colonies; hydranths (Fig. 3.10) are spindle-shaped feeding individuals with a prominent manubrium and 10 to 20 thin, scattered tentacles, and sporosacs are ovoid reproductive polyps.

It is notable that only a few sporosacs are produced on colonies growing in strictly freshwaters. A single female sporosac usually produces 6 to 20 ova in brackish waters, but only 3 to 6 in freshwaters. Mature ova are retained on the sporosacs, but the sperm

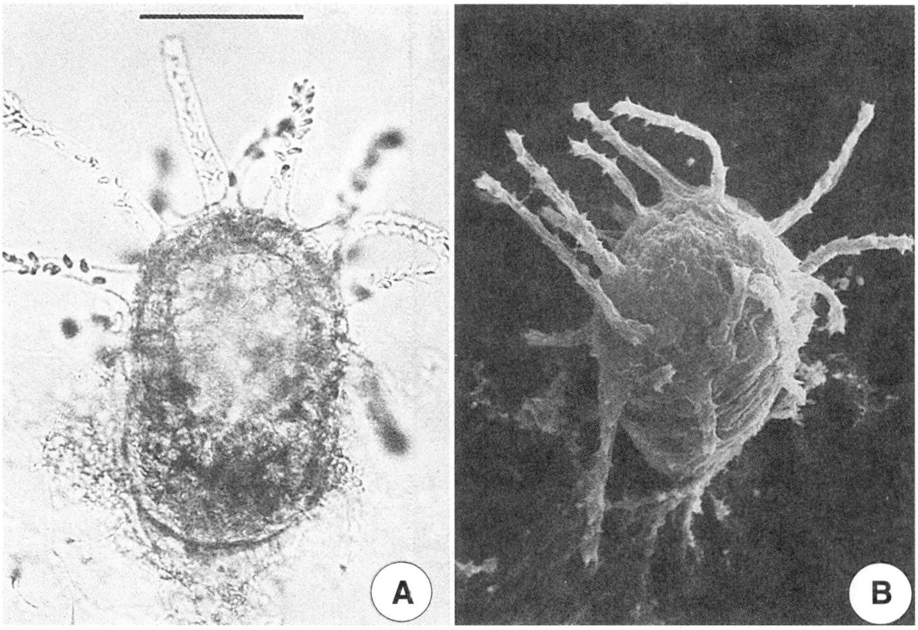

Fig. 3.8. Polyps of *Calpasoma dactyloptera*, × 200. A, photomicrograph of live animal showing nematocysts in tentacles; B, scanning electron micrograph of specimen. Scale, 100 μm. (From Campbell and Rahat 1975, with permission.)

from male sporosacs are released into the water, and after swimming about for a short time they penetrate through the covering of the female sporosacs and fertilize the eggs. The embryo develops on the sporosac and forms a free-swimming ciliated planula that is released from the colony. The planula attaches to the substrate and by growth and differentiation gives rise to a new colony. Thus, there is no true medusoid generation in *Cordylophora*. The sporosacs themselves represent the abortive medusa generation. In the autumn the hydranths disintegrate and only a few horizontal and more basal upright portions of the colony winter over.

Although *Cordylophora caspia* is a brackish water species it has become well established and persistent in freshwater containing high conductivity. The species has been found well inland in large rivers and lakes and has even been recorded in cave pools. The species occurs in brackish inlets and estuaries from New Jersey to the Gulf of the St. Lawrence, but has also been found in inland rivers in several states. This species is probably cosmopolitan but has also been regarded as an introduced species in North America. It is widely distributed in northern Europe and has been found in numerous other parts of the world.

Protohydra. An unusual type of hydroid is *Protohydra leuckarti* (Fig. 3.11). Although not a true freshwater species, it occurs in the mud and sand of brackish estuaries, swamps, and creeks, some of which are almost fresh. It has the general appearance of a hydra but has no tentacles. The body is usually less than 2 mm long but may attain a maximum of 5 mm. The species exhibits dioecious syngamic reproduction and also reproduces by transverse and longitudinal fission. *Protohydra* has been reported several times along the New England coast and from numerous places in Great Britain and northwestern Europe.

The writer (RWP) has seen several related minute forms in interstitial psammolittoral and stream bottom waters. These were very short and had only three or four blunt tentacles.

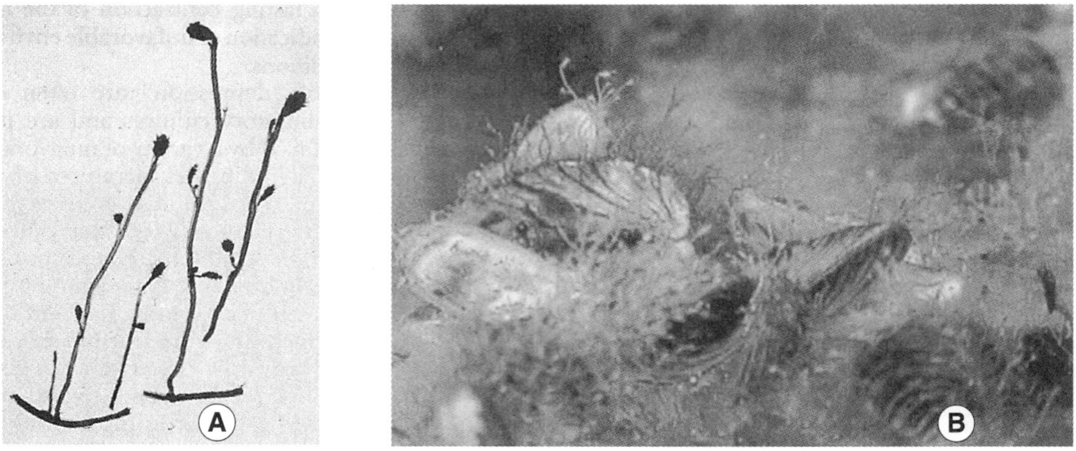

Fig. 3.9. Colonies of *Cordylophora caspia*. A, individual polyps ×5; B, colonies growing on zebra mussels, *Dreissena polymorpha*, ×1.5. (B © Nadine Folino, with permission.)

Polypodium. *Polypodium hydriforme* is probably the most bizaare of the freshwater cnidarians and one of the most unusual species of all Cnidaria. The species spends almost all of its pre-adult life as a parasite of the eggs of acipenseriform fishes. The adult is a free-living, solitary polyp (Fig. 3.12). It has a shortened column, a well-developed hypostome, and either 12 or 24 tentacles. The polyp is very small, the column reaching about 2 mm in length, but the tentacles are considerably longer. The species contains nematocysts of one type, atrichous isorhiza, which are represented in three size classes. The tentacles are divided into two functional forms, a walking form with the largest nematocysts and a feeding form with medium-sized nematocysts. As implied by the nomenclature of the tentacles, *Polypodium* is fairly mobile and actively hunts for its food. Adults of *P. hydriforme* feed on oligochaetes and turbellarians.

The adult reproduces by asexual binary fission during which an equal number of tentacles is dedicated to each new polyp, and by sexual methods. Adults may be dioeceous or hermaphroditic. Formation of gonads is entodermal with either two complex female gonads or four male gonads or both appearing in a specimen. Gametes produced by gonads develop into a planulalike larva, which presumably finds a way into the gonads of female fishes. Somehow, it is not yet known, the eggs of the fish become infected with *Polypodium* larvae. The larva develops into a stolon (Fig. 3.13) that eventually occupies the entire egg and kills it. Curiously, this intracellular form is turned inside out. Perhaps this allows the gastrodermis, now on the outside, to more easily absorb the nutrients of the sturgeon egg. When the fish lays its eggs, the infected dead eggs rupture as polyps along the stolon begin to break free.

Polypodium hydrifome was introduced into North America from Russia sometime before 1974. It is beginning to show up in several Midwestern states.

Feeding, digestion. Cnidarians are strictly carnivorous and eat many kinds of small metazoans, including cladocerans (but not ostracods), copepods, insects, annelids, and turbellarians. Occasionally freshwater cnidarians become a hatchery nuisance by killing fish fry, but *Hydra oligactis* is probably the only species large enough to be of any real importance. When a food organism is swimming or creeping about and by chance comes in contact with the tentacles or column, it is paralyzed and killed in a few seconds by the nematocysts that are discharged from those parts of the column or tentacles contacted. The organism is then

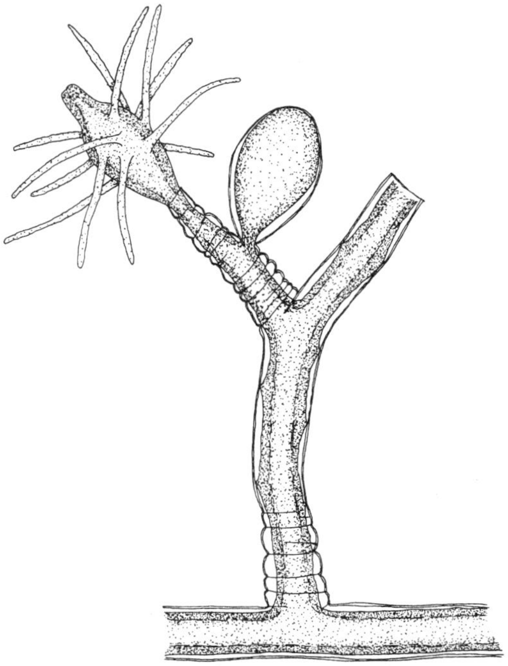

Fig. 3.10. Small portion of colony of *Cordylophora lacustris*, ×15, showing one hydranth and one immature gonophore.

Fig. 3.11. *Protohydra leuckarti*, ×35.

manipulated so that it is taken into the mouth. The latter is variously described as star-shaped or circular; both it and the column are capable of great distension, and a cladoceran four times the diameter of the column may be ingested. When the usual food supply is insufficient, hydras may feed on the organic material of the substrate, but this is certainly not a normal means of feeding. The feeding reaction in hydras (writhing and twisting of the tentacles toward the mouth) is induced by reduced glutathione (and perhaps by other substances). Presumably this is the key mechanism in the feeding process of hydras.

Digestion in cnidarians is both extra- and intracellular. Preliminary extracellular digestion occurs in the gastrovascular cavity, resulting in the maceration of the food organisms into a mush within several hours. These small particles are then taken up by pseudopodial action of many of the gastrodermal cells. Further digestion is intracellular. Only the gastrodermal cells in the upper part of the column appear to function in digestion; those in the more basal part of the column (stalk) are highly vacuolated and have no digestive power. Excess food is stored in the gastrodermal cells as fat and glycogen. Indigestible material is voided from the mouth.

Hydra viridissima is called the green hydra because of the abundance of algal cells (probably mostly *Chlorella*) in the gastrodermis. Such living algae probably have a true symbiotic relationship with the hydra, but dead and disintegrating cells are digested and used as food. The algae are passed from one generation to the next in the egg cytoplasm.

Physiology. Circulation of materials within the gastrovascular cavity is effected by movements of the whole animal and by beating of the flagella of some of the gastrodermal cells. Such cells usually have one or two flagella.

Respiration occurs through the general body surface. None of the hydrozoans are active in the presence of reduced quantities of dissolved oxygen.

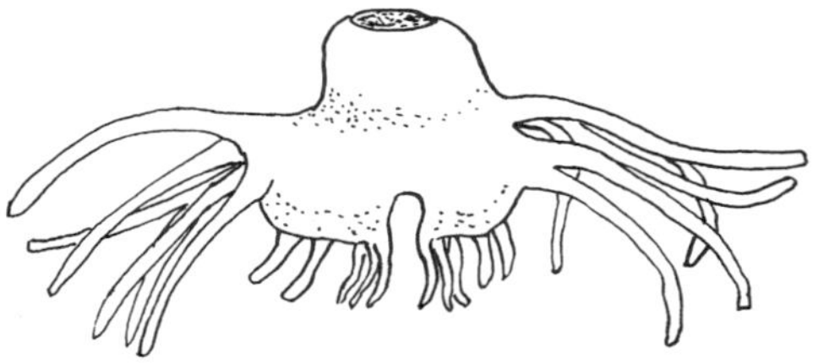

Fig. 3.12. An adult *Polypodium hydriforme*, ×25. (Redrawn from Raikova 1973.)

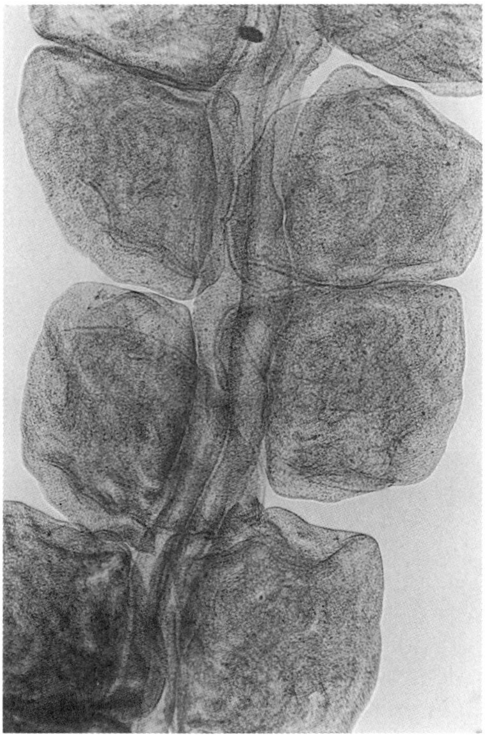

Fig. 3.13. Stolon of *Polypodium hydriforme* with several inverted polyps, ×50.

There seems to be no special excretory mechanism, and it is presumed that metabolic wastes simply diffuse out into the surrounding water. A few workers, however, have reported a small "aboral" or "excretory" pore in the center of the basal disk. Osmoregulation seems to be a function of the individual cells.

When strongly stimulated, and under adverse environmental conditions, hydras contract markedly and assume a barrellike shape with small, budlike contracted tentacles. Though there are no special bands or sheets of muscle tissue to produce such contractions and elongations, many of the epidermal cells and some gastrodermal cells have one or a few threadlike myonemes associated with their basal portions. The longitudinal fibers are restricted to the epidermis and the circular fibers to the gastrodermis; the former are more abundant.

Elongated sensory cells are numerous in the epidermis of the tentacles and oral region. They are generalized receptors for touch, temperature, and substances in solution, and pass basally into one or more processes of the nervous system, which is a nerve net ramifying everywhere at the base of the epidermis. The neurones are bi- and multipolar ganglion cells and their neurites, and the entire system has the appearance of a continuous network. There are no coordinating or relay centers, and stimuli simply diffuse throughout the nerve net in all directions.

Ecology of *Hydra*. Although hydrids are typical members of the littoral and shallow stream associations, they have occasionally been reported from waters as deep as 40 to 350 m. It is not unusual to find them in swift streams or in the wave-swept rocky littoral of large lakes. Unless there are large pieces of debris present, however, they do not occur on fine, muddy bottoms. The greatest populations are usually found in late spring and early summer, followed by a marked decrease in July or August. Sometimes there is also an autumn maximum in September or October. Populations may be vigorous in midwinter, but are usually scanty by the time of the ice melt in late winter.

Gonad production is often induced in 2 or 3 weeks by lowered laboratory temperatures, usually 10 to 15°C, but often a culture becomes sexual without any apparent stimulus. Medium-hard waters with pH readings of 7.6 to 8.4 seem to be especially favorable for most species. Although little is known about minimal oxygen concentrations, about 3.0 mL/L seems to be a typical value. Hydras do not occur in oxygen-deficient depths of lakes. In general, any lasting contraction of the tentacles is an indication of unfavorable environmental conditions. Periods of "depression" are often observed in laboratory cultures and are presumably induced by a variety of unfavorable conditions, including temperatures above 20°C, lack of sufficient dissolved oxygen, toxic salts, transfer to clean water, and accumulation of metabolic materials.

Commensals, enemies. Commensal or parasitic ciliates, such as *Kerona* and *Trichodina*, are common on the external surface of hydras, where they feed on small bits of food. They do not discharge the nematocysts, although many free-living ciliates, both smaller and larger than *Kerona* and *Dichodina*, do easily discharge them.

Carnivorous enemies are not abundant or important but include a few turbellarians, nemertines, crustaceans, snails, and aquatic insects. *Anchistropus minor* is a cladoceran that feeds exclusively on hydras. Epidemics of an external amoeboid parasite, *Hydramoeba hydroxena*, sometimes decimate vigorous hydra growths.

Geographic distribution, dispersal. *Hydra viridissima* is cosmopolitan. *Hydra oligactis* is widely distributed in the Northern Hemisphere. The other species of hydras occurring in the United States, however, have not been reported from continents other than North America, although there are similar species in Eurasia. About 25 species are recorded in the literature as occurring outside of the continent of North America, but about half of these are so sketchily described as to be unrecognizable. Europe has eight species. Six American hydras are known from single localities, and several others have been reported from a few scattered areas. Undoubtedly, when more intensive collecting is done, especially in the western half of the country, the geographical ranges of most species will be greatly extended.

Hydras attached to mollusk shells may be easily distributed through single drainage systems, and thecated embryos are probably easily blown or transported overland. It is also possible that hydra are transported by birds when pondweeds get tangled about their legs.

Collecting. Unless hydras are exceedingly abundant, they may be collected most easily by bringing small objects and aquatic plants into the laboratory where they may be picked off. If the containers are placed in semidarkness, the hydras will migrate to the surface film as the dissolved oxygen in the water decreases. The grayish or brownish thecated embryos should be sought for on aquatic vegetation or debris with the low powers of a binocular dissecting microscope.

Other species are normally collected with other benthic organisms. Seeking individual species may be a challenge because of the extreme small size of several of the species of freshwater cnidarians.

Culturing. Specimens keep well in aquaria in moderate light if the temperature is below 20°C and if they are well fed with cladocerans, copepods, or brine shrimp. Three or four daphnids per day per hydra will produce prolific budding. A fraction of the aquarium water should be changed frequently, and only pond water should be used, not tapwater. Other species such as *Calpasoma* have been maintained in laboratory cultures for years by providing aquatic vegetation and small annelids. A popular medium for maintaining freshwater cnidarians is the "M" solution of Lenhoff and Brown (1970).

Periods of depression may be overcome by changing the culture water. If necessary, hand feed depressed individuals by holding (with fine forceps) a freshly killed daphnid in contact with the hydra.

Preserving, preparing. The best method for identifying hydras and polyps of other species is to examine both live and dead material. To kill hydras and other species in an expanded condition they should first be gently scrubbed or squirted off their normal substrate into a dish. After they are fully expanded in about 2 mL of water, dash a copious quantity of hot Bouin's fixative into the dish. After a few minutes place the specimens in cold Bouin's fluid for 10 hours. Hance (1957) gives an alternative excellent method for killing hydras in an expanded condition. Preserving specimens nevertheless usually results in loss of the natural body and tentacle proportions. Freshwater cnidarians can also be relaxed by placing some menthol crystals on the water surface followed by monitoring. When the organisms are relaxed they can be fixed directly in 5% formalin. If done judiciously excellent preparations can be achieved. Jellyfish should be fixed directly in buffered 5% formalin.

Nematocysts provide key characters for identification. They must be observed in the living state by viewing a tentacle or polyp slightly compressed under a cover glass.

Dissociation of the tissues to permit an examination of the cells may be accomplished by placing hydras in a few drops of Hertwig-Schneider's fluid on a slide for 15 to 20 minutes. This solution consists of one part 0.02% osmic acid and four parts 5.0% glacial acetic acid. Dilute Bouin's fluid and 1% nitric acid are also fairly useful for dissociation.

Slide preparation is the best means of examining fixed polyps. Temporary whole mounts can be prepared with glycerol while semipermanent mounts can be set up with Hoyers medium or CMCP. For permanent preparations wash the specimens in several changes of alkaline 30% alcohol, stain, dehydrate, and mount permanently.

Taxonomy. All freshwater cnidarians are currently placed in the Class Hydrozoa. Most of the basic information on our American hydras has been gained through the extensive and careful investigations of Libbie H. Hyman. The consolidation of *Chlorohydra* and *Pelmatohydra* under the single genus *Hydra* for all American species follows her work and that of later investigators, notably Campbell (1983). The species level taxonomy of *Hydra* remains unsettled at best.

There are two approaches for *Hydra* identification. One is the use of more traditional keys (Hyman, in Edmondson 1959; Forrest 1963; earlier editions of this book), which are species by species. An alternative method proposed by Campbell (1983, 1987) utilizes a conservative approach by recognizing four groups defined by specific characters, each group containing a few to several variable but presumably related species.

Although the systematics of the freshwater jellyfishes are not completely worked out (Rahat and Campbell 1974), most recent works (Petersen 1990) use the family Olindiadidae. The same holds true for the family Clavidae, which includes *Cordylophora caspia* (=*lacustris*). The jellyfish species *Craspedacusta sowerbii* was, in the past, also known as *Microhydra ryderi* (named for the polyp stage). Older references list the latter, but it has since been synonymized with *C. sowerbii* by most authors. Currently, neither *Polypodium* nor *Calpasoma* can be assigned to any current family with certainty.

KEY TO GENERA OF FRESHWATER CNIDARIA OF THE UNITED STATES

1. Animal a solitary or colonial polyp 2
   Animal a medusa (jellyfish) . . . . **OLINDIADIDAE, *Craspedacusta***
2. Animal a solitary polyp 3
   Animal joined with others to form a cluster or colony, either by means of a direct connection or through a common structure 6
3. Polyp typically with well-developed column and well-developed tentacles (Fig. 3.1), usually several millimeters in length; pigments consisting of gold, brown, and green; common throughout the U.S., about 16 species . . . . **HYDRIDAE, *Hydra***
   Polyp minute, column short, 3 mm or less in length; with or without tentacles; white or translucent 4

4. Polyp with compressed column, 1 or 2 mm in length; tentacles in clusters of 6 or 12 (Fig. 3.12); 1 species, CA, MI, MO, ND . . . . ***Polypodium***
Polyp with extended column, tentacles minute or absent **5**
5. Polyp less than 1 mm in length, mouth encircled by delicate tentacles (Fig. 3.8) which are usually not evident in death; rare, 1 species . . . . ***Calpasoma***
Polyp to 3 mm in length, without tentacles (Fig. 3.11); in brackish estuaries, swamps, and creeks, some of which are almost fresh, rare along the New England coast; 1 species **PROTOHYDRIDAE, *Protohydra***
6. Colony brushy in appearance, with many stolons giving rise to polyps and sporosacs; tentacles scattered over surface of manubrium (Fig. 3.10); 1 species in tidal freshwater of eastern states, inland in AR, CA, IL, KY, LA, OH, OK, TN, TX, WA . . . . . . **CLAVIDAE, *Cordylophora***
Polyps without tentacles, nematocysts evident under high magnification, polyp clusters containing from a few to 14 individuals (Fig. 3.6); species presence also indicated by the occurrence of planktonic jellyfish (medusae) (Fig. 3.5); 1 species distributed throughout the U.S. in natural and artificial habitats. . . . . ***Craspedacusta***

REFERENCES

Acker, T. S. 1976. *Craspedacusta sowerbii*: an analysis of an introduced species. In: *Coelenterate Ecology and Behavior*, pp. 219–226, edited by G. O. Mackie, Plenum, New York.

Acker, T. S., and A. M. Muscat. 1976. The ecology of *Craspedacusta sowerbii* Lankester, a freshwater hydrozoan. *Am. Midl. Nat.* 95:323–336.

Ahmad, M. F., et al. 1987. A new species of *Limnocnida* (Limnomedusae, Cnidaria) from a freshwater aquarium in India. *Hydrobiologia* 144:33–36.

Augustin H, et al. 1987. A sewage plant as a remarkable new habitat of the fresh-water polyp *Craspedacusta sowerbii* (Hydrozoa: Coelenterata). *Limnologica* 18:225–226.

Berg, K. 1948. Studies on the bottom animals of Esrom Lake. *Mem. Acad. R. Soc. Lett. Danemark*, Ser. 9, 8:1–255.

Bryden, R. R. 1952. Ecology of *Pelmatohydra oligactis* in Kirkpatricks Lake, Tennessee. *Ecol. Monogr.* 22:45–68.

Campbell, R. D. 1983. Identifying hydra species. In: *Hydra: Research Methods*, pp. 19–28, edited by H. M. Lenhoff. Plenum, New York.

——. 1987. A new species of hydra (Cnidaria: Hydrozoa) from North America with comments on species clusters within the genus. *Zool. J. Linn. Soc.* 91:253–263.

Campbell, R. D. and M. Rahat. 1975. Untrastructure of neamtocytes and one-celled tentacles of the freshwater coelenterate, *Calpasoma dactyloptera*. *Cell Tiss. Res.* 159:445–457.

Dejdar, E. 1934. Die Süsswassermeduse *Craspedacusta sowerbii* Lankester in monographischer Darstellung. *Z. Morphol. Oekol. Tiere* 28:595–691.

DeVries, D. R. 1992. The freshwater jellyfish *Craspedacusta sowerbyi:* A summary of its life history, ecology, and distribution. *J. Freshwat. Ecol.* 7:7–16.

Dodson, S. I., and S. D. Cooper. 1983. Trophic relationships of the freshwater jellyfish *Craspedacusta sowerbii* Lankester 1880. *Limnol. Oceanogr.* 28:345–351.

Edmondson, W. T. (Ed.) 1959. *Ward and Whipple's Fresh-water Biology*. 2nd ed. 1248 pp. Wiley, New York.

Ewer, R. F. 1947. On the function and mode of action of the nematocysts of *Hydra*. *Proc. Zool. Soc. London* 117:365–376.

——. 1948. A review of the Hydridae and two new species of *Hydra* from Natal. *Proc. Zool. Soc. London* 118:226–244.

Forrest, H. 1963. Taxonomic studies on the hydras of North America, VIII: description of two new species, with records and a key to the North American hydras. *Trans. Am. Microsc. Soc.* 82:6–17.

Grayson, R. F. 1971. The freshwater hydras of Europe, 1: a review of the European species. *Arch. Hydrobiol.* 68:436–449.

Hadley, C. E., and H. Forrest. 1949. Taxonomic studies on the hydras of North America, 6: description of *Hydra hymanae*, new species. *Am. Mus. Novit.* 1423:1–14.

Hance, R. T. 1957. Mass fixation of hydra. *Trans. Am. Micros. Soc.* 76:387–389.

Hoffman, G. L. 1999. *Parasites of North American Freshwater Fishes*, 2d ed. 539 pp. Comstock, Ithaca, NY.

Holstein, T. 1995. *Cnidaria: Hydrozoa.* Süßwasserfauna von Mitteleuropa. 1/2. 110 pp. Gustav Fischer, Stuttgart.

Holstein, T., and P. Tardent. 1984. An ultrahigh-speed analysis of exocytosis: nematocyst discharge. *Science* 223:830–833.

Hyman, L. H. 1929. Taxonomic studies on the hydras of North America, 1: general remarks and description of *Hydra americana*, new species. *Trans. Am. Microsc. Soc.* 48:242–255.

——. 1930. Studies on hydras, II: the characters of *Pelmatohydra oligactis* (Pallas). *Trans. Am. Microsc. Soc.* 49:322–333.

——. 1931a. Studies on hydras, III: rediscovery of *Hydra carnea*. L. Agassiz (1850) with a description of its characters. *Trans. Am. Microsc. Soc.* 50:20–29.

——. 1931b. Studies on hydras, IV: description of three new species with a key to the known species. *Trans. Am. Microsc. Soc.* 50:302–315.

——. 1938. Taxonomic studies on the hydras of North America, V: description of *Hydra caulicculata*, n. sp., with notes on other species, especially *Hydra littoralis*. *Am. Mus. Novit.* 1003: 1–9.

Jones, C. S. 1941. The place of origin and the transportation of cnidoblasts in *Pelmatohydra oligactis* (Pallas). *J. Exp. Zool.* 87:457–476.

——. 1947. The control and discharge of nematocysts in hydra. *J. Exp. Zool.* 105:25–60.

Kepner, W. A., et al. 1935. The manipulation of hydra's nematocysts by Microstomum. *Science* 82:621.

Kepner, W. A., et al. 1951. The discharge of nematocysts of the hydra, with special reference to the penetrant. *J. Morphol.* 88:23–48.

Lenhoff, H. M. 1968. Behavior, hormones, and *Hydra*. *Science* 161:434–442.

Lenhoff, H. M., and R. D. Brown. 1970. Mass culture of *Hydra:* an improved method and its application to other aquatic invertebrates. *Lab. Anim.* 4:139–154.

Lytle, C. F. 1982. Development of the freshwater medusa *Craspedacusta sowerbii*. In: *Developmental Biology of Freshwater Invertebrates*, pp. 129–150, edited by F. W. Harrison and R. R. Cowden., Liss, New York.

Matthews, D. C. 1966. A comparative study of *Craspedacusta sowerbii* and *Calpasoma dactyloptera* life cycles. *Pacific Sci.* 20:246–259.

McClung, G. D., et al. 1978. First records of *Cordylophora lacustris* Allman, 1871 (Hydroida: Clavidae) in Texas with morphological and ecological notes. *Southwestern Nat.* 23: 363–370.

Miller, D. E. 1936. A limnological study of *Pelmatohydra*, with special reference to their quantitative seasonal distribution. *Trans. Am. Microsc. Soc.* 55:123–193.

Moore, T. E. 1954. Notes on fresh-water jellyfish, and new records for Illinois. *Nat. Hist. Misc. (Chicago Acad. Sci.)* 132:14 pp.

Mueller, J. F. 1950. Some observations on the structure of hydra, with particular reference to the muscular system. *Trans. Am. Microsc. Soc.* 69:133–147.

Pauly, R. 1902. Untersuchungen uber den Bau und die Lebensweise der *Cordylophora lacustris* Allman. *Jena. Z. Naturwiss.* 36:737–780.

Payne, F. 1924. A study of the freshwater medusa, *Craspedacusta ryderi*. *J. Morphol.* 38:387–430.

Pennak, R. W. 1956. The freshwater jellyfish *Craspedacusta* in Colorado with some remarks on its ecology and morphological degeneration. *Trans. Am. Microsc. Soc.* 75:324–331.

Petersen, K. W. 1990. Evolution and taxonomy in capitate hydroids and medusae (Cnidaria: Hydrozoa). *Zool. J. Linn. Soc.* 100: 101–231.

Rahat, M., and R. D. Campbell. 1974. Three forms of the tentacled and non-tentacled fresh-water coelenterate polyp genera *Craspedacusta* and *Calpasoma*. *Trans. Am. Microsc. Soc* 93: 235–241.

Raikova, E. V. 1973. Life cycle and systematic position of *Polypodium hydriforme* Ussov (Coelenterata), a cnidarian parasite of the eggs of Acipenseridae. *Publ. Seto Mar. Biol. Lab.* 20: 165–173.

Rowan, W. 1930. On a new *Hydra* from Alberta. *Trans. R. Soc. Can., Sect. 5, Biol. Sci.* 24:165–170.

Welch, P. S., and H. A. Loomis. 1924. A limnological study of *Hydra oligactis* in Douglas Lake, Michigan. *Trans. Am. Microsc. Soc.* 43: 203–235.

4

PLATYHELMINTHES

(Free-living flatworms)

Considering the Phylum Platyhelminthes as a whole, only a small fraction of the described species are free-living in freshwater habitats. There are several parasitic groups and a large cluster of provisional orders of free-living forms. Many of the species of flatworms are marine and a few (about 10 in the United States) are terrestrial in warm, damp habitats or parasitic on marine invertebrates; the remainder are freshwater forms. Many species of land planarians have been reported from the United States, all belonging to either *Bipalium* or *Rhynchodemus*, and all are introduced from their original extensive ranges in southeastern Asia and Japan. They are large planarians, usually 60 to 350 mm long, customarily found in the warm damp soil of greenhouses, but one has manged to become established outdoors.

The free-living flatworms are divided into about 13 orders. Several are exclusively marine or have only a few freshwater species. The orders Tricladida, Catenulida, Macrostomida, Typhloplanida, Kalyptorhynchia, and Dalyellioida have many freshwater and some terrestrial representatives.

There is one record of a microscopic acoel flatworm being found in U.S. freshwater (Mirror Lake, New Hampshire), and two other rare species are known from Germany and Poland. These worms are only about 0.5 mm long and lack a digestive cavity. Probably all are recent immigrants from ancestors in the marine environment. Careful collecting should reveal additional freshwater Acoela, especially in rivers and lakes not far from the seacoast.

Freshwater flatworms are found everywhere, usually on or closely associated with a substrate. The larger species are sometimes confused with small leeches, which they resemble somewhat in shape and color; the microscopic species resemble large ciliates in size, shape, and general habits.

General characteristics. All freshwater flatworms are more or less elongated. They may be highly flattened and leaflike, cylindrical, or spindle-shaped. Almost invariably, however, there is some evidence of flattening on the ventral surface. In many species the anterior end is differentiated and sufficiently specialized to resemble a "head."

Tricladida are usually 5 to 30 mm long, but the other orders are seldom more than 4 mm long, the majority being microscopic and 300 to 1200 μm long. Because of their habit of reproducing by fission, some small species customarily occur in chains of several individuals, or zooids. The general surface of the body is usually more or less covered with cilia, and the term "flatworms" was coined by Ehrenberg in 1831, when he detected the minute vortical currents of water generated at the anterior end by ciliary action. In the familiar species of planarians, to which every zoology student is exposed in college, cilia are absent from the dorsal surface and edges of the body.

The body cavity of flatworms is a typical gastrovascular cavity having but a single external opening, which functions as both mouth and anus. In the Tricladida the gastrovascular cavity has three main irregular branches, one median and anterior and the others posterior and lateral (Fig. 4.1). The other orders (Fig. 4.2) have a single median gastrovascular chamber, but in a few representatives its outline is more irregular, owing to small lateral caeca.

In a few flatworms the mouth is terminal or subterminal. More characteristically it is ventral and somewhere between one-fifth and three-fifths of the way back from the anterior end (Fig. 4.2). A protrusible or nonprotrusible muscular pharynx surrounds that portion of the gastrovascular cavity just inside the mouth. If protrusible, the pharynx often lies in a separate pharyngeal chamber.

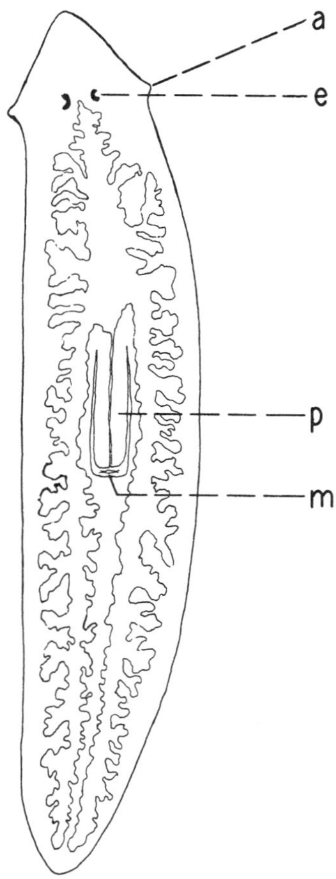

Fig. 4.1. Gross anatomy of a typical planarian, showing the three divisions of the gastrovascular cavity, ×15. a, auricle; e, eyespot; in, mouth (on ventral surface of body); p, pharynx (lying in pharyngeal chamber).

Almost all species are hermaphroditic and reproduce sexually. One or two genital pores are found on the ventral surface, usually posterior to the mouth.

A wide range of coloration occurs in freshwater flatworms, although the brilliant patterns of marine species are never approached. The more familiar planarians (Tricladida) are various shades of gray, brown, or black; often there are dorsal stripes, spots, or mottlings. The ventral surface, however, is usually lighter in color or gray and without a color pattern. Some species have a highly variable coloration. *Girardia tigrina*, for example, may be dark yellow, olive, brown, or brownish black, and the pattern may range from lightly mottled to dense coloration (Fig. 4.3). The lighter shades usually develop in laboratory cultures. Cave species and a few noncave species are whitish. The other four orders are characteristically grayish or colorless, but a few are brilliant yellow, gold, orange, red, or rose. Other species are a bright green, owing to internal symbiotic zoochlorellae. In all naturally light-colored flatworms the body may take on a variety of darker shades, depending on the particular contents of the gastrovascular cavity. Two darkly pigmented eyespots are usually present near the anterior end, but in some species eyespots are absent, and in a few others they are numerous.

Epidermal and subepidermal glands secrete an abundance of mucus that usually covers the body. Minute rodlike rhabdites are produced in the epidermis and subepidermal tissues of many genera, and these, on being extruded to the surface of the body, disintegrate and produce an additional adhesive mucus.

Thin layers of circular and oblique, and a thicker layer of longitudinal muscle fibers lie below the basement membrane of the epidermis. The cells that line the gastrovas-

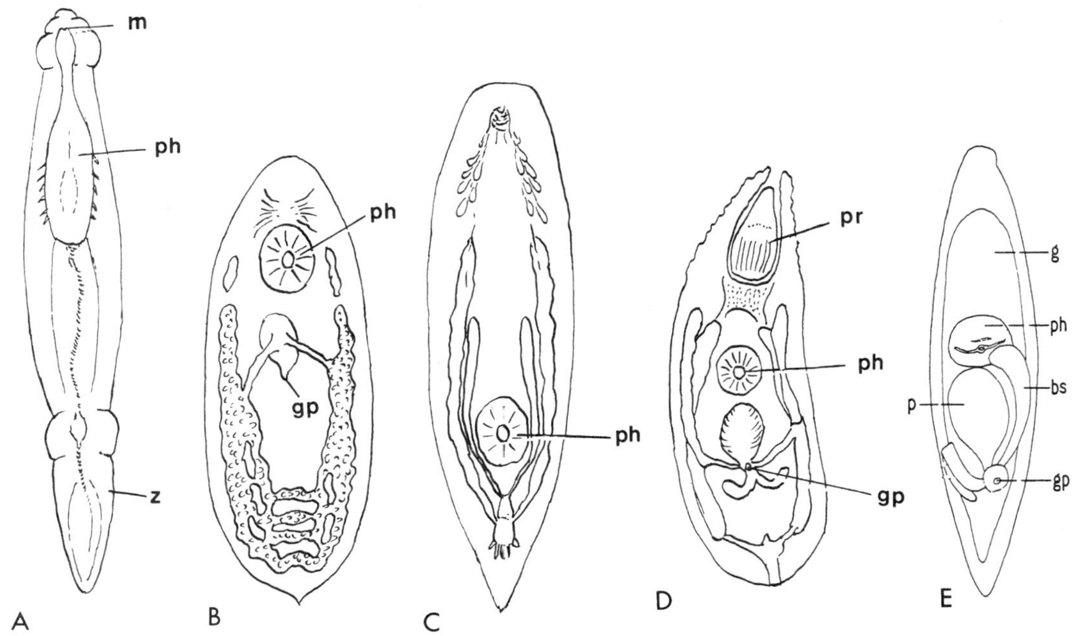

Fig. 4.2. Structure of some representative planarian worms, various magnifications. A, Catenulida; B, *Phaenocora* (Typhloplanoida); C, *Olisthanella* (Typhloplanoida); D, *Polycystis* (Kalyptorhynchia); E, *Castrada* (Typhloplanoida). bs, bursa copulatrix; g, gut; gp, gonopore; m, mouth; p, penis bulb; ph, pharynx; pr, proboscis; z, zooid. (A and C redrawn from Cannon and Faubel 1988; B and D redrawn from Hyman 1951a).

cular cavity are large and phagocytic. Between the inner gastrodermis and the outer muscle fibers and epidermis the body consists largely of an undifferentiated mass of parenchyma cells. A few muscle fibers are found in the parenchyma, but otherwise it appears to function as a loose packing material between the excretory and reproductive organs. In the smaller flatworms the parenchyma is often highly vacuolated, glandular, and watery.

Locomotion. Triclad flatworms cannot swim but are capable of locomotion on a substrate or on the underside of the surface film of water. Smooth gliding movements are produced by the action of cilia on the thin coat of mucus secreted onto the substrate by cyanophilous glands in the ventral parenchyma. Faint waves of muscular contractions pass longitudinally down the body and supplement ciliary action. On strong stimulation, planarians may crawl actively by means of pronounced waves of muscular contraction that pass from anterior to posterior. A few species have anterior adhesive organs, which may be used as a leech uses its anterior sucker for locomotion. Catenulids, macrostomids, typhloplanids, lecithoepitheliates, and prolecithorans move on a substrate as do the triclads, but many are also capable of swimming about freely in the water, owing to their abundant cilia. When conditions are unfavorable, when the water is too warm, or when organsisms are fully fed, planarians are not active but remain more or less contracted for days at a time.

Feeding, digestion. The most familiar type of feeding occurs in the planarians. Here the pharynx is long, highly muscular, and protrusible through the mouth. When the presence of food (usually living, dead, or crushed animal matter) is detected in the water, the animals quickly move toward it and place the ventral surface of the body in

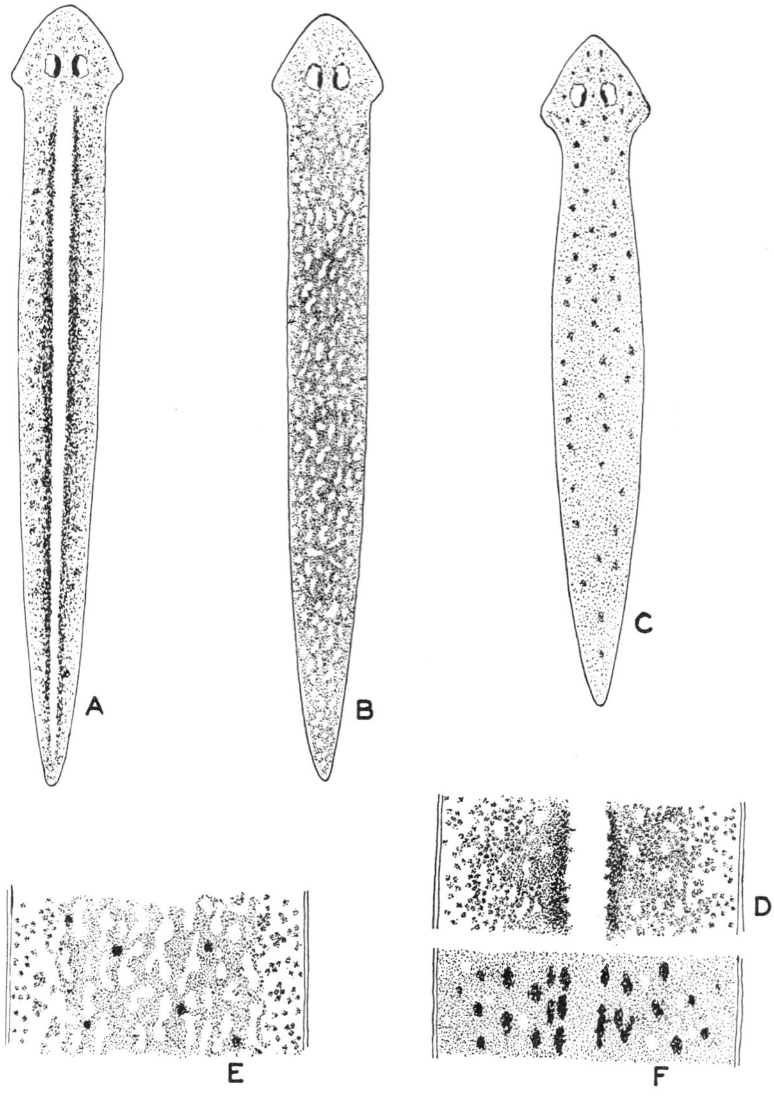

Fig. 4.3. Variations in color pattern of *Girardia tigrina*. A, striped form; B, spotted form; C, dark spots on uniform brown background; D–F, enlarged views of similar patterns. (From Hyman 1939a.)

contact with it. The pharynx is then extruded through the mouth and its tip is placed against the food (Fig. 4.4). Sometimes the pharynx may be extended to as much as half of the body length. There appears to be no appreciable secretion of digestive juices from the tip of the pharynx, and apparently only soft or disintegrating tissues are capable of being sucked up into the main gastrovascular cavity by the muscular action of the pharynx. The gastrovascular cavity becomes filled with fluid and small bits of tissues after 30 to 80 minutes of feeding. There is no good evidence for extracellular digestion in planarians. Instead, individual particles are ingested by pseudopodial action of the large gastrodermal cells, and digestion is intracellular. Evidence based on the small amount of work that has been done on 'microturbellarians,' however, indicates that

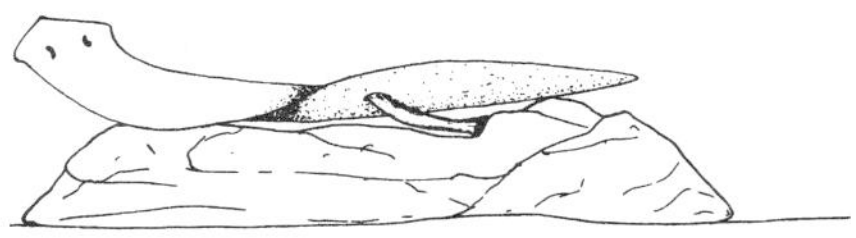

Fig. 4.4. Planarian with extruded pharynx, just beginning to feed on a bit of food.

they probably have considerable extracellular digestion.

Some species of *Phagocata* (Tricladida) are interesting in that they are polypharyngeal (Fig. 4.5). In 'microturbellarians' the pharynx is short and generally not greatly protrusible, but the normally small mouth may be expanded to a marked degree during feeding.

Although some flatworms are sometimes said to be vegetarians, the great majority of flatworms are zoophagous and feed on small living invertebrates. Experimental work has shown that oligochaetes and chironomids are often the preferred items. Nevertheless, dead animal matter may often be ingested, and the microscopic species subsist on fresh corpses of larger metazoans as well as on whole living or dead Protoctista, rotifers, nematodes, and gastrotrichs. Small individuals, such as *Catenula* and *Macrostomum*, are especially abundant in decaying organic matter and perhaps are essentially omnivorous. Some species of *Phaenocora* and *Macrostomum*, however, prefer animal tissues and eject plant material from the pharynx. *Mesostoma* are voracious predators on dipteran larvae, especially mosquitoes, and have been considered as possible control agents. Many substrate species prefer a diatom diet.

Some species are so voracious that they will exterminate other species in a mixed culture. Cannibalism has occasionally been observed, especially in crowded cultures where there are injured specimens.

A few 'microturbellarians' have been found to form an adhesive layer or mass on the substrate by the discharge of rhabdites at the anterior end. Such mucus with its entangled detritus, bacteria, algae, and protozoans is then gathered into a ball and ingested. Some investigators suggest that this method of feeding may be more common than is suspected. Planarians also exude rhabdites, which form a mucus on coming in contact with water. Small crustaceans get caught in this material, as well as in the usual slime track, and are thus easy prey.

Most Dendrocoelidae have an anterior, adhesive organ that is used in feeding as well as anchorage (Fig. 4.6). Such species pounce on small crustaceans and hydras and grasp them firmly by bringing the lateral edges of the sucker together. The prey is then held against the substrate in the pharyngeal region while feeding proceeds. *Procotyla fluviatilis* feeds chiefly on living crustaceans; it cannot be cultured on raw meat. In general, the larger species of 'microturbellarians' feed on living plankton and substrate crustaceans (up to 10/day) and annelids, but

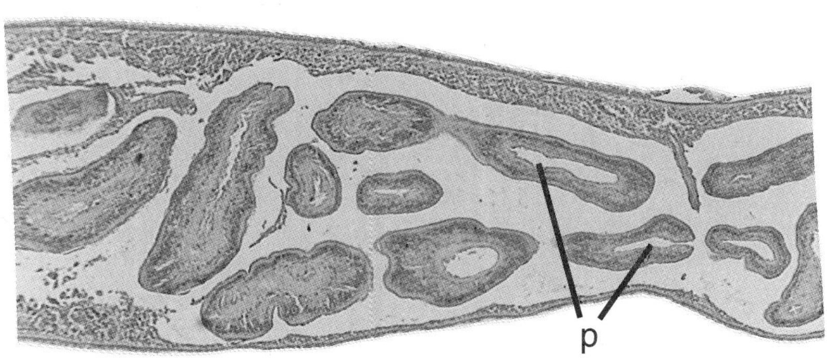

Fig. 4.5. Section through *Phagocata woodworthi* showing polypharenges, ×60. p, pharenges.

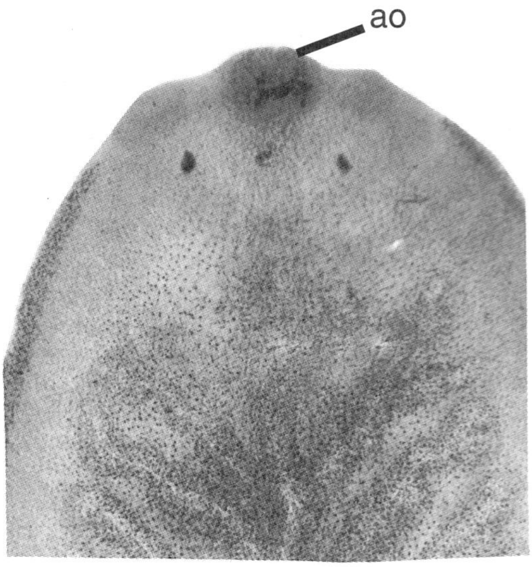

Fig. 4.6. Dorsal view of head of *Procotyla fluviatilis*, with adhesive organ (ao) retracted.

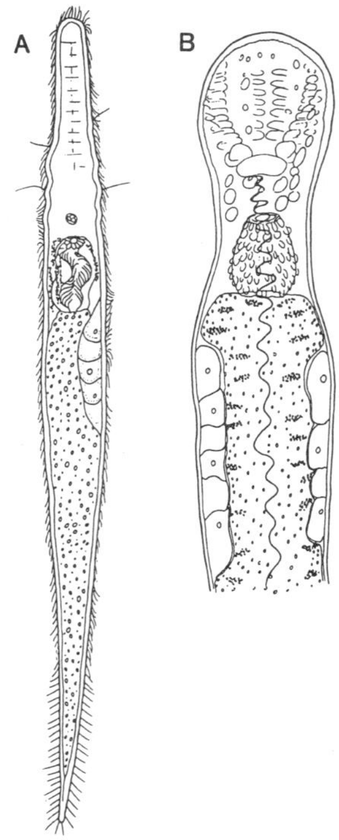

Fig. 4.7. Two undescribed species of *Rhynchoscolex*, showing the range of shape in the proboscis. (From Kolasa et al. 1987, with permission.)

dead asellids and gastropods are common foods also.

Many species of 'microturbellarians' distributed among genera of different families have a stationary proboscis (Fig. 4.7) or one that can be everted. These structures are presumably involved with several functions but are probably important in locating food.

Microstomum and a few other genera have become curiously adapted for retaining the undischarged nematocysts of ingested hydras. Within 10 to 20 hours after ingestion, the nematocysts are found distributed over the body surface between the epidermal cells. Such nematocysts are retained as defense devices for the remainder of the life of the flatworm.

There is no special circulatory system in flatworms. Materials are distributed throughout the gastrovascular cavity by means of general muscular and body movements. The tissues are nowhere thick or bulky, and diffusion of materials from cell to cell is easily facilitated.

Respiration. The respiratory exchange of oxygen and carbon dioxide occurs through the general epidermis and perhaps to a slight extent through the gastrodermis also. Many 'microturbellarians' are characteristic of habitats where there is much decay and low concentrations of dissolved oxygen; some are able to withstand anaerobic conditions for variable lengths of time, although there is little definite information on this phase of flatworm biology in the literature. Planarians (triclads), however, all require high concentrations of oxygen and occur only in well-oxygenated habitats.

Symbiotic zoochlorellae occur in the parenchyma and gastrodermis of many 'microturbellarians,' although they are not necessarily consistently present in a particular species. Photosynthesis releases oxygen to the flatworm tissues, while the worm contributes carbon dioxide and nitrogenous compounds for algal metabolism. Algae may be passed on to the succeeding generation in the flatworm egg.

Excretion. The flame bulb, or protonephridial, system of flatworms has a variable pattern. There may be few to many flame bulbs, but their ducts usually empty into two longitudinal collecting ducts (Fig. 4.8). *Stenostomum* and *Catenula* are exceptional in having a single median collecting duct and a double ciliated cell (Type 1). Collecting ducts open to the ventral surface by means of one or two pores, which are variable in position. Macrostomids contain a Type 2 system involving a single terminal cell with several cilia, while the remaining groups have a Type 3 system, which contains several flame cells on perikaryon.

As in certain other phyla, the excretory process is a function of the ducts rather than the flame bulbs themselves. Nevertheless, the physiological significance of the whole flame bulb system appears to vary from one species to another. Sometimes the system has an important excretory function; sometimes it is apparently unimportant. In many species it is of great importance in osmoregulation; in other species it appears to be relatively unimportant. Recent study has revealed that in at least *Geocentrophora*, a link exists between the ciliary pits on the surface, normally considered to be sensory structures only, and excretory canals. Thus, the flame bulb system may not act alone as an excretory system.

Gyratrix hermaphroditus occurs in freshwater, brackish, and marine habitats. In the former it has a highly developed flame bulb system. In brackish habitats the system is less highly developed, with the bladder and ampullae lacking. In marine habitats, however, it is poorly developed or absent.

It is probable that a large fraction of the

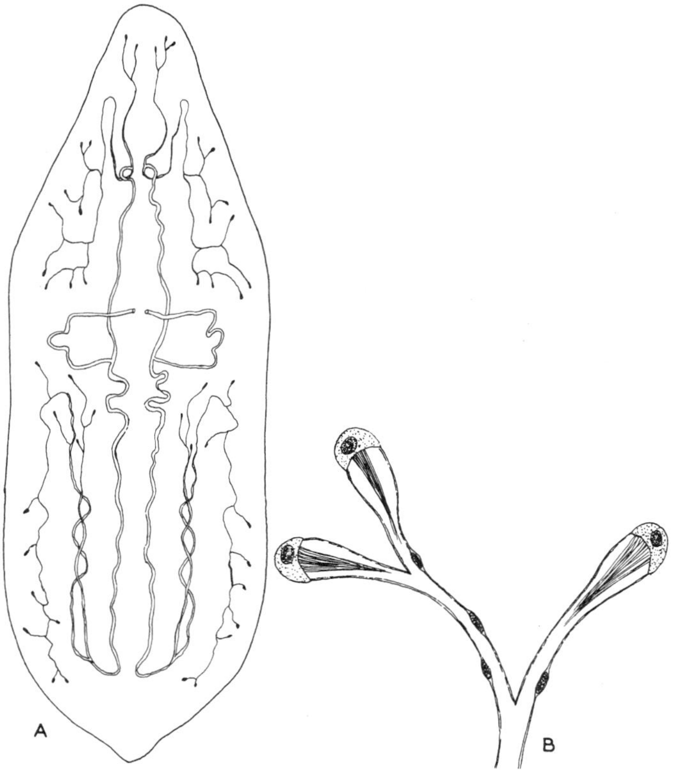

Fig. 4.8. Diagrams of protonephridial structures. A, protonephridial system of *Mesostoma* (the two proximal pores are located just inside the mouth); B, three flame bulbs at terminal endings of tubules.

excretory materials is disposed of through the general epidermal surface and also perhaps to a slight extent through the gastrodermis into the gastrovascular cavity. Some excretory granules are commonly retained and stored in the tissues until the animal dies.

Nervous system. The brain consists of two more or less well-defined lobes, or ganglia, near the anterior end. Usually there are two prominent longitudinal nerve trunks, one originating in each ganglion, but often there are one to several additional, smaller trunks. Many cross commissures and small lateral nerves originate from the longitudinal trunks and innervate all of the organs and tissues. An abundance of sensory fibers are given off from the brain and innervate the anterior end (Fig. 4.10).

Sensory receptors. Eyespots are the most highly developed sense organs, especially in planarians. Several photoreceptors have been identified in free-living flatworms. A typical eyespot is shaped like a beam or a pigmented cup containing translucent light-sensitive receptor cells, which are attached to nerve fibers at the outer portion of the cup. Other kinds of ocelli occur in macrostomids and proseriates among others. Some spring and most cave species of Tricladida have no eyespots. Most 'microturbellarians' have one pair, but some are blind, and a few have multiple eyespots (Figs. 4.10, 4.23A, E).

Such eyespots are relatively simple in structure and are sometimes unpigmented and called light-refracting bodies rather than eyes. In *Stenostomum*, for example, the so-called eyespot is colorless and consists of a few to many basal refractive mitochondria

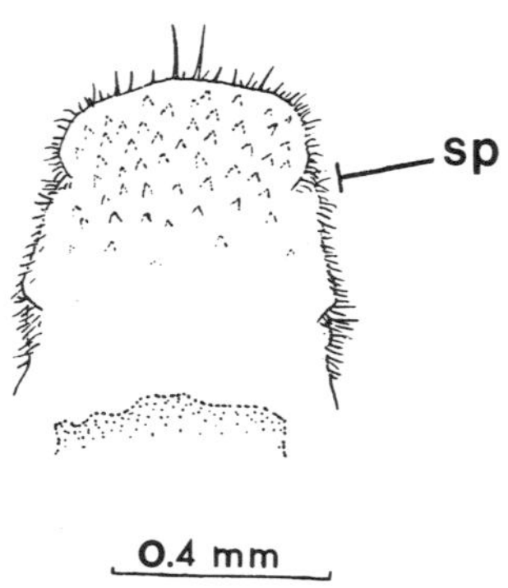

Fig. 4.9. Sensory pits (sp) on the head of *Bothrioplana semperi*, ×50.

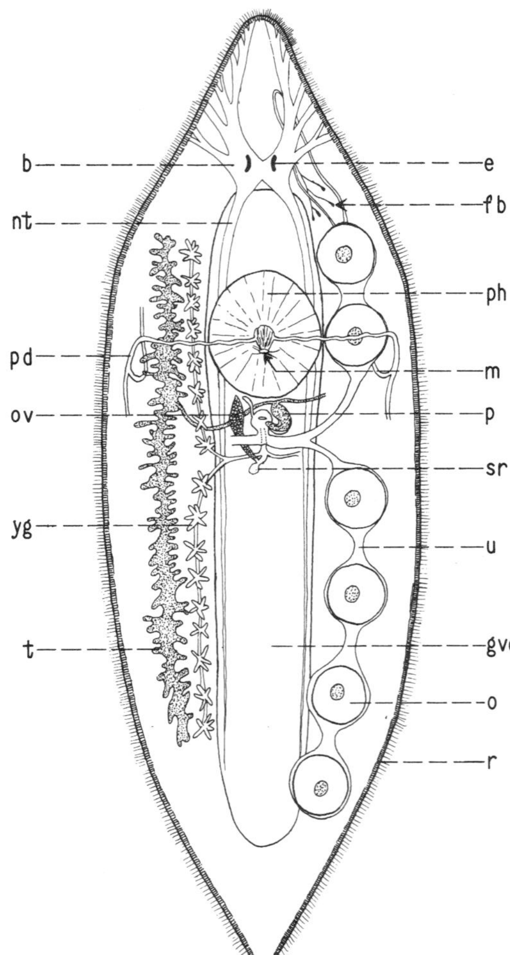

Fig. 4.10. Structure of *Mesostoma ehrenbergi* Focke, ×20, ventral view. b, brain; e, eyespot; fb, flame bulb (distributed over entire body); gvc, gastrovascular cavity; m, mouth; nt, longitudinal nerve trunk; o, ovum; ov, ovary; p, penis (projected into genital atrium); pd, protonephridial duct; ph, pharynx; r, rhabdites in epidermis; sr, seminal receptacle; t, testes (shown only on left side of body); yg, yolk gland. (Modified from Ruebush 1940.)

within cells. Actually, a visual function has not yet been definitely established for this structure. Although eyespots maybe well developed, most flatworms are photonegative, especially in moderate to strong light, and as a consequence are usually found in shaded areas or under objects.

Some species have anterior sensory pits, often ciliated (Fig. 4.9), which are presumably chemoreceptors, but assist in osmoregulation and respiration. They may be rounded, oblong, slitlike, and shallow or deep (Figs. 4.9, 4.20). Larval *Rhynchoscolex* and adult *Catenula*, acoels, and *Otomesostoma* have a median anterior statocyst (Figs. 4.18C, D, 4.19A), but none of the other American freshwater genera have such a structure. It appears likely that a statocyst has arisen independently in each taxon. In addition to the general coat of short cilia, many have a few long sensory cilia, especially at the anterior and posterior ends (Figs. 4.12, 4.13). These move independently of each other and of the short cilia. In some species they are relatively stiff and are called "spines."

Other receptors, mostly in the epidermis, function in the detection of currents, in olfaction, and in response to contact. The pharynx and anterior end are especially sensitive.

Reproduction, development. With reference to their reproductive habits, there are three general groups of flatworms: (1) species that reproduce only asexually, by budding or fission; (2) species that reproduce solely by sexual methods; and (3) species that are capable of reproducing sexually or asexually, depending on the genetic strain, physiological strain, and prevailing ecological conditions.

Fission does not occur in the majority of triclads, though certain common species are in the third category. Among these latter species the relative frequency of sexual and asexual reproduction varies greatly from one species to another. In *Girardia tigrina*, for example, there appear to be at least two physiological varieties. One variety reproduces only asexually. Under appropriate conditions of temperature and food supply, individuals pinch in at about midlength on each edge. The pinching in proceeds rapidly until the two halves are completely separated. Then the anterior end of the posterior individual reconstitutes a head, and the posterior end of the anterior individual forms a tail. External factors may change the rate of fission or inhibit it, but they do not change the reproduction to the sexual type. Fission occurs only above 10°C, with frequency increasing to a maximum at 25 to 28°C. Such fission strains have been kept as long as 5 years in the laboratory without development of sex organs. In other physiological races of *G. tigrina* there may be an alternation of reproductive methods according to water temperature. Sex organs may develop during the winter and early spring, and egg capsules are deposited in May and June. As the water temperature increases, however, the reproductive organs degenerate, and in July, August, and September reproduction may be entirely by fission. By late autumn the sex organs are again beginning to develop. If the water temperature remains low throughout the year, however, sexual reproduction may be continuous.

It has also been suggested that there are still other strains of *G. tigrina* that reproduce sexually regardless of normal seasonal temperature variations. *Girardia dorotocephala* and a few other triclads also appear to exist in several physiological varieties comparable to the situation in *G. tigrina*. Nevertheless *G. dorotocephala* is rarely observed to reproduce sexually under natural conditions.

Procotyla fluviatilis is incapable of asexual reproduction; it is sexual and deposits egg capsules during the winter months. Asexual reproduction is also unknown in *Cura foremanii*, but it reproduces sexually at all seasons. In *Phagocata vernalis* and *P. velata*, however, sexual reproduction is rare and apparently occurs only during the winter months in permanent bodies of water.

In fission the posterior part of the body tears itself away; each of the two resulting individuals then grows and regenerates the lost parts. In fragmentation the animal breaks into several parts, each of which en-cysts in a slime mass. After an inactive period, a small individual emerges from the cyst, becomes active, and grows to normal size. Triclads have seldom been observed dividing, and there is some evidence to show that the process occurs at night and is completed within 30 minutes to several hours.

Another form of asexual reproduction in flatworms occurs in *Schmidtea polychroa* in which both parthenogenetic and sexual individuals occur. In parthenogenetic animals, sperm only stimulate egg development. It has been shown, however, that nonsexual reproduction in this species reduces fertility.

The seasonal reproductive habits of 'microturbellarians' are poorly known, but most of the genera reproduce solely by the sexual method. *Catenula*, *Rhynchoscolex*, *Suomina*, *Microstomum*, and *Stenostomum* are the only genera capable of reproducing asexually. In these forms fission proceeds rapidly and often along several transverse planes so that there may be a series of 2 to 8 zooids in a chain (Fig. 4.11). Occasionally, a few species in this group of five genera have been observed to develop sex organs and reproduce sexually. At least one species of *Phaenocora* spends about 7 months in a diapause egg stage.

In 'microturbellarian' fission a full complement of organs is usually developed before separation. Under favorable conditions divisions may occur as frequently as every 5 to 10 days in triclads, but in 'microturbellarians' the time interval is much shorter.

All American freshwater flatworms are hermaphroditic, and the male and female organ systems are unusually complex in many genera. They are not present during periods of fission but develop later, presumably from undifferentiated parenchyma cells. The male organs sometimes mature before the female organs. Various types of reproductive systems are diagrammed in Figs. 4.10, 4.12, 4.13 and only a typical planarian arrangement will be described here (Fig. 4.15).

In the female reproductive system the two ovaries are on either side of the body near the anterior end. A longitudinal oviduct (or ovovitelline duct) proceeds posteriorly from each ovary and collects yolk cells from numerous lateral yolk glands. The two ovovitelline ducts unite near the midline, posterior to the mouth, and enter a common cavity, the female atrium. A saclike bursa copulatrix connects with the female atrium by means of a bursal canal.

The testes are small spherical bodies arranged roughly in two longitudinal rows. Each one connects with one of the two vasa deferentia by means of a minute duct. As the vasa deferentia proceed posteriorly they are widened and convoluted to form spermiducal

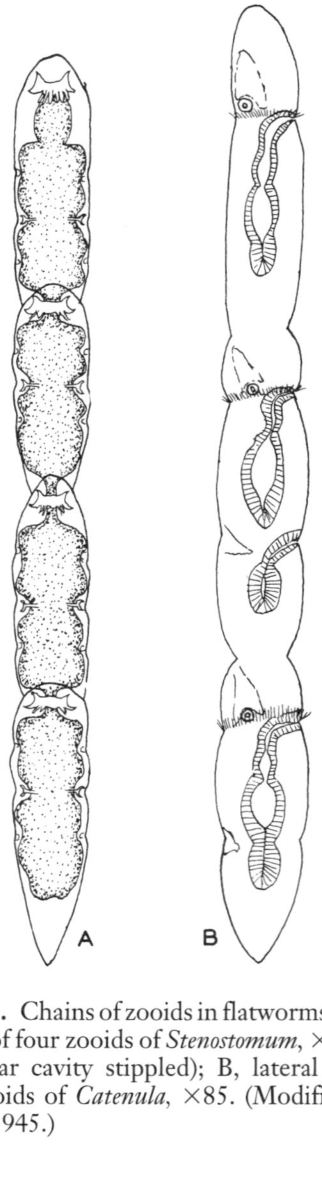

Fig. 4.11. Chains of zooids in flatworms. A, dorsal view of four zooids of *Stenostomum*, ×24 (gastrovascular cavity stippled); B, lateral view of three zooids of *Catenula*, ×85. (Modified from Marcus 1945.)

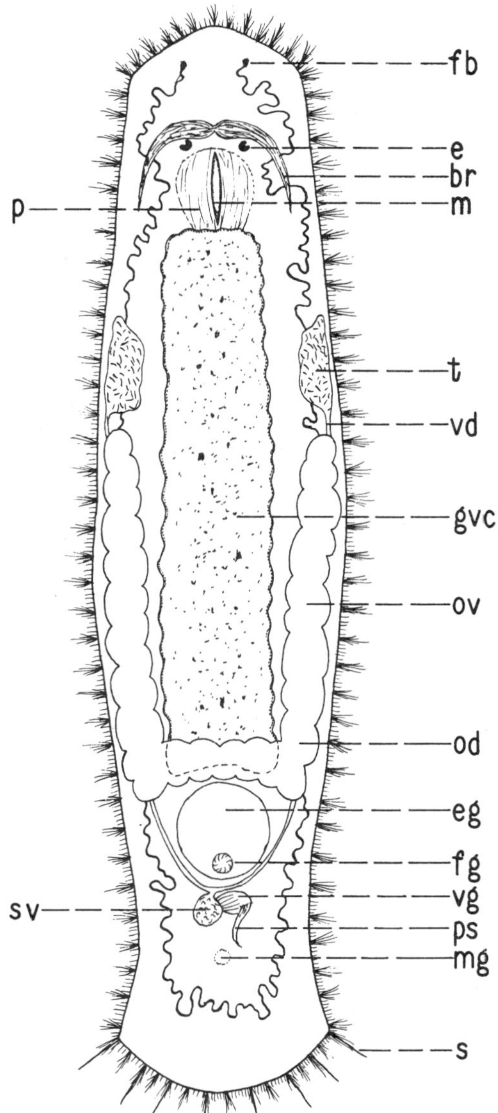

Fig. 4.12. Structure of *Macrostomum appendiculatum* (Fabr.), ×140, a species common everywhere in fresh, brackish, and salt waters. br, brain; e, eyespot; eg, egg; fb, flame bulb; fg, female genital pore; gvc, gastrovascular cavity; m, mouth; mg, male genital pore; od, oviduct; ov, ovary; p, pharynx; ps, penis stylet; s, spine; sv, spermiducal vesicle; t, testes; vd, vas deferens; vg, prostatic vesicle (Modified from Ferguson 1939.)

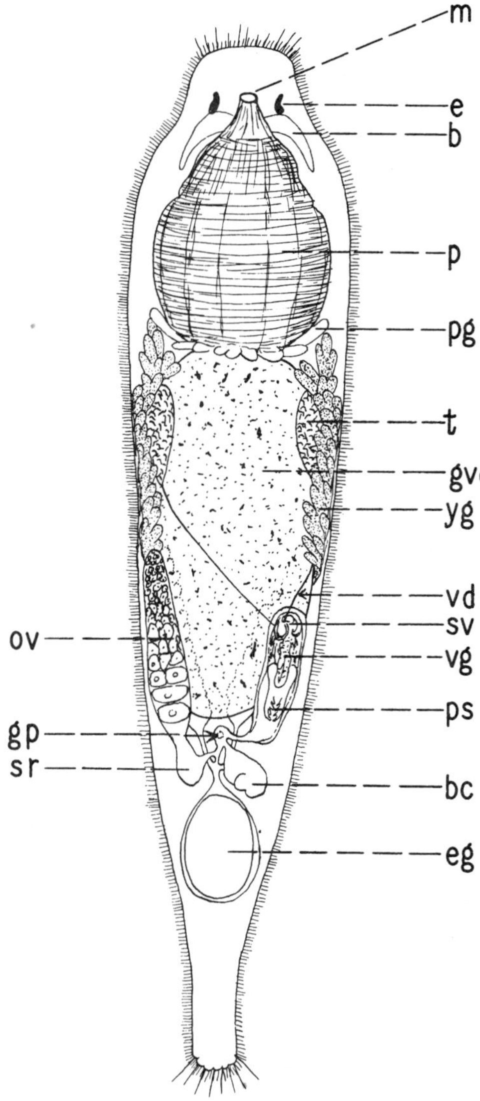

Fig. 4.13. Structure of *Dalyellia*, ×70, diagrammatic. b, brain; bc, bursa copulatrix; e, eyespot; eg, egg; gp, genital pore; gvc, gastrovascular cavity; m, mouth; ov, ovary; p, pharynx; pg, pharyngeal gland; ps, penis stylet; sr, seminal receptacle; sv, spermiducal vesicle; t, testes; yg, yolk glands; vd, vas deferens; vg, prostatic vesicle. (Modified from Ruebush and Hayes 1939.)

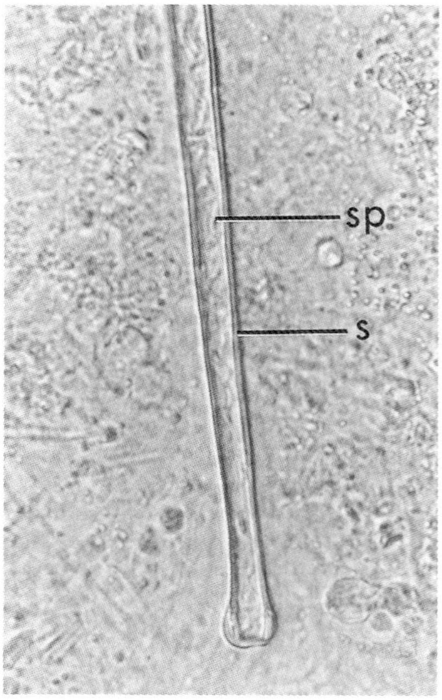

Fig. 4.14. Style of male reproductive system of *Macrostomum tubum*, ×100. s, style; sp, sperm within style.

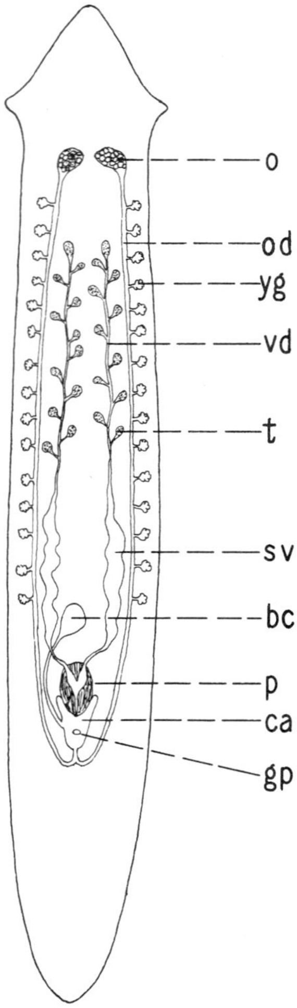

Fig. 4.15. Diagrammatic ventral view of a typical triclad reproductive system, ×16. (The relationships of the male and female atria are distinct only in a lateral view.) bc, bursa copulatrix; ca, common atrium; gp, genital pore; o, ovary; od, ovovitelline duct; p, penis; sv, spermiducal vesicle; t, testis, vd, vas deferens; yg, yolk gland.

vesicles where ripe sperm are stored. The spermiducal vesicles, either with or without uniting into a common duct, enter the basal portion of the highly muscular penis, and the tip of the latter projects into a cavity, the male atrium. Male and female atria unite ventrally to form a common atrium, which opens externally as a common genital pore.

Additional cement glands, copulatory glands, and other accessory reproductive structures (e.g., Fig. 4.17E, F) occur in many flatworms. Other orders often have the distal end of the penis modified to form a stylet (Fig. 4.14), and sometimes the basal part is a prostatic vesicle, which secretes a granular material vital to the sperm. Many 'microturbellarians' have two genital pores.

As is the case in most hermaphroditic animals, two flatworms usually copulate and exchange sperm. The penis of each animal is elongated and protruded from its genital pore and through the genital pore of the other animal, the sperm being discharged into the bursal canal in planarians. Both animals exude quantities of mucus from the genital pore region during copulation. On completion of copulation, sperm are stored in the bursa copulatrix for a short time and then leave and migrate up the oviducts to the region of the ovaries. After a variable interval, the eggs are fertilized as they leave the ovaries.

As the zygote moves down the oviduct it accumulates yolk cells, and in the male atrium aggregates of zygotes and yolk cells become surrounded by a proteinaceous capsule. Such an egg capsule usually contains 2 to 20 zygotes and thousands of yolk cells. It leaves the body through the genital pore, and in *Girardia*, *Cura*, and a few other species it is attached to the substrate by means of a stalk (Fig. 4.16A). Each worm may deposit a series of egg capsules during the breeding season.

Hypodermic impregnation appears to be the rule in *Stenostomum*. A single egg is ripened in each ovary during the reproductive season. It is fertilized inside the body and is liberated to the outside by a rupturing of the body wall.

More commonly, however, 'microturbellarians' copulate and exchange sperm in the usual fashion. Fertilized eggs are usually laid singly, each enclosed in a capsule or shell that is sometimes stalked (Fig. 4.16B). Many of the Typhloplanidae produce two kinds of eggs. Thin-shelled "summer" eggs hatch promptly after deposition, but the larger, thick-shelled "winter" eggs have delayed hatching and may be dormant during the winter; they are capable of withstanding unfavorable environmental conditions. Winter eggs may be extruded from the parent body in the customary way or they may be liberated on the death and disintegration of the parent. Some species are regularly viviparous; the summer eggs hatch within the body and the young break out of the posterior region of the parent to the outside. In 'microturbellarians' other than the Typhloplanidae only eggs of the winter type are produced.

Self-fertilization is rare in the Tricladida and has been definitely demonstrated only for *Cura foremanii*. Certain species sometimes produce sterile egg capsules. Some investigators think that summer eggs of typhlopanids are self-fertilized.

Egg capsules of triclad flatworms are 2 to 4 mm in diameter and light-colored when recently deposited, but soon become dark brown or blackish. Although they are resistant to low temperatures, they cannot withstand isolated drying. Depending on the season, an egg capsule may winter over or hatch in about 2 weeks. The several worms that emerge are only 1 to 3 mm long. Development is direct and there are no special larvae stages in the American freshwater species. With the exception of reproductive structures, the newly hatched worm often has the full complement of organ systems.

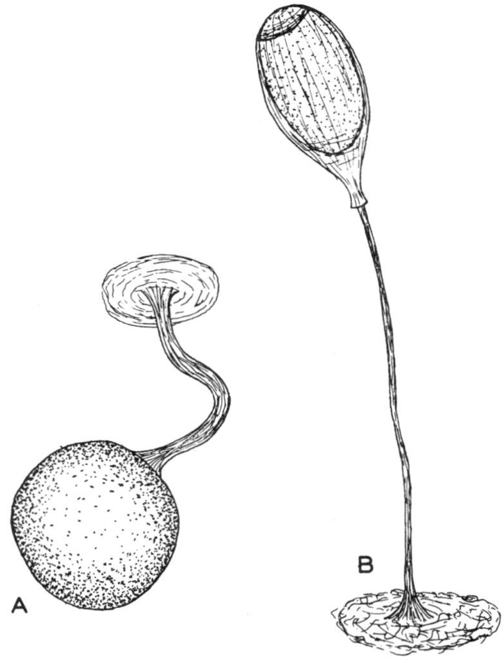

Fig. 4.16. Flatworm egg capsule and flatworm encapsulated egg. A, stalked planarian egg capsule, ×7; B, stalked encapsulated egg of *Gyratrix*, ×100 (note operculum at apex).

Phagocata velata and *P. vernalis* have an unusual habit of fragmenting, especially when their habitat becomes warm in late spring and summer. When a worm reaches a length of 12 to 15 mm, the internal organs degenerate and the whole worm fragments into numerous small bits. Each piece secretes mucus, which hardens to form a small cyst, and a whole miniature individual is reconstituted within the cyst. Encystment of a whole mature animal has also been reported. Such cysts may hatch in a few weeks or may winter over.

Length of life is known chiefly from laboratory cultures. Sexually reproducing species live for only a few weeks or months, usually dying in summer or autumn after producing winter eggs. A few species, however, may survive a year or more by encysting during unfavorable conditions. Species capable of fission can presumably continue indefinitely under appropriate conditions. Such laboratory cultures have been kept going for up to 11 years.

The larger species of planarians, especially *Girardia*, have long been classical objects for regeneration and grafting experiments. A great deal of such work was done in the United States, especially between 1910 and 1930. If extended specimens are carefully cut transversely into two or several pieces with a razor blade, a large percentage of the pieces will regenerate into whole new worms, the original longitudinal axis being maintained. Worms cut in two along the median line regenerate respective right and left halves to form two new worms. Partial longitudinal cuts at the anterior end often produce worms with two or more complete heads. Regeneration potentialities are most highly developed in the anterior half or three quarters of the body. In spite of this regeneration capacity, planarians are rather delicate, and when handled roughly in the laboratory they easily develop breaks and irregularities in the body.

Ecology. Springs, spring brooks, ditches, marshes, pools, ponds, lakes, and caves all have their characteristic flatworm populations. Most species, and especially triclads, are photonegative and are found under objects or in debris during the daytime. Flatworms thrive on any kind of substrate where there is an appropriate food supply. Although they are most characteristic of the shallows, a few species have been collected from lake bottoms as deep as 100 m.

Triclads are typical of brooks and streams; the other four orders are never abundant in such habitats. Rheophile forms are also usually cold stenotherms; common species are *Cura foremanii*, *Girardia dorotocephala*, *Polycelis coronata*, *Phagocata velata*, and *P. morgani*. The last-named species is restricted to water between 4 and 14°C. *Procotyla fluviatilis* occurs in all kinds of habitats, including running waters, standing waters, and even brackish areas. In general, specimens from running waters attain greater size than those from standing waters. *Girardia tigrina* and *Phagocata vernalis* are both standing-water triclads, but whereas the former species is a eurytherm, the latter is active only during the cold months in temperate zone ponds and ditches.

The nontriclads are especially abundant during the warm months in pools and ponds containing quantities of algae, other vegetation, and organic debris. In a single lake, where many microhabitats or niches exist, as many as 30 species have been recorded, especially in hard-water lakes and ponds. Almost any stagnant pool will contain one or more species of *Dalyellia*, and usually one or more additional species in each of *Stenostomum*, *Macrostomum*, *Catenula*, *Microstomum*, and *Mesostoma*. Such species are eurythermal since they occur in habitats having a wide daily and seasonal temperature range. *Gyratrix*, *Phaenocora*, and *Otomesostoma* are also chiefly eurythermal. *Rhynchomesostoma rostratum* and certain species of *Macrostomum*, *Dalyellia*, and *Castrada*, however, are some of the few genera known to be cold stenotherms. Species composition in a small pond is usually a dynamic process with succession often occurring during a season. The dominant species groups of the 'microturbellarian' fauna may change up to four times a year. Some species spend as much as 8 months per year in diapause.

Temporary streams and ponds can also support a diverse flatworm fauna. Many species of *Mesostoma* and *Hymanella retenouva* and *P. velata* are common in temporary ditches, streams, and ponds in the northeast.

Kolasa et al. (1987) have recently demonstrated the presence of a rich 'microturbellarian' fauna on the bottom and in the interstitial waters of streams and springs in New York state. Although the worms were never abundant, these investigators found 33 species of which 5 were undescribed and 8 were new to North America. Twelve species of *Stenostomum* accounted for the great majority of individuals collected.

The great majority of flatworm genera occurring in freshwaters are restricted to that environment. *Macrostomum* is the most distinct exception, with many species in brackish and marine environments. As implied in a previous section, *Gyratrix hermaphroditus* is one of the very few euryhaline species; it is found in freshwater, brackish, and marine environments. There is some evidence to show that this taxon actually consists of a large cluster of incipient species.

In those 'microturbellarian' genera occurring in both standing and running waters there is often a correlation of size and activity with habitat. Still waters are characterized by larger, more sedentary species, whereas the small, faster-moving species are more abundant in waters where there is a current, however slow.

Some species of *Pseudophaenocora* thrive in the presence of only a trace of oxygen, and many other small species occur in water where the concentrations are as low as 5 to 40% saturation. Most triclads are restricted to water whose oxygen concentration does not fall below 70% saturation.

Hymanella retenuova is the only planarian known to produce a thick-walled egg capsule capable of withstanding drought. In addition, however, *Phagocata velata* and *P. fawsetti* undergo multiple fission during drought; each fragment secretes a slime layer that hardens into a resistant cyst.

Enemies, parasites. Flatworms can be an important element in the diet of other animals. Study has shown that in certain environments stonefly larvae prey heavily on flatworms and tend to select the species they attack. Odonata nymphs consume a few planarians, and members of the other orders appear to be taken occasionally by nematodes, annelids, and a few crustaceans and aquatic insects.

Internal parasites of flatworms include ciliates, gregarines, and (rarely) larval mermithid nematodes. Some flatworms may be found in the brood pouch of isopods and amphipods, but they are probably commensals rather than true parasites.

Geographical distribution. So far as flatworms are concerned, that portion of the United States west of the Great Plains is almost a terra incognita. A few triclad records have been published for this region, but 'microturbellarians' have been even more greatly neglected. East of the Mississippi the triclads have been extensively studied, especially by Hyman, and although the nontriclads have been studied to a considerable degree by Kepner, Ferguson, Ruebush, and Kolasa, an enormous amount of work yet remains to be done. The number of microscopic species appears to be very large, and each new investigation turns up new species. As shown in the key, some triclads are common and widely distributed, whereas others are known from only one or two localities. Cave species are highly endemic, each form being restricted to a single cave or cave system, or to a group of caves in the same general area and always in unglaciated regions.

Although many 'microturbellarians' have thus far been reported from only one or several collecting areas, a few appear to be common and essentially cosmopolitan in the Holarctic region. Typical examples are *Macrostomum appendiculatum*, *Stenostomum leucops*, *Castrella truncata*, *Phaenocora unipunctata*, and *Bothromesostoma personatum*.

Little definite information is available concerning the dissemination of flatworms. *Girardia dorotocephala* is generally distributed in spring brooks and marshy springs, yet it probably is unable to migrate actively throughout a single drainage system because of its ecological preference for headwaters. The same is true of *Phagocata morgani*, which is widely distributed east of the Mississippi. *Girardia tigrina*, however, is never found in springs or spring brooks, but it does occur in a wide variety of other lotic and lentic habitats and presumably can migrate actively through drainage systems over wide areas. *Procotyla fluviatilis* and *Cura foremanii* also probably spread by active migration. Species such as *Phagocata vernalis*, however, which occur in temporary ponds and ditches, must rely entirely on passive means of distribution.

Even less is known about the dissemination of microscopic species, but the mechanisms of distribution must be fairly efficient, since the same species may be found everywhere in suitable habitats, even in temporary ponds and puddles that are dry during the late summer and early autumn. Undoubtedly the cyst stages, winter eggs, and perhaps even egg capsules are transported from place to place by natural agencies and larger animals that are capable of moving or flying overland from one pond, lake, or stream to another. *Oligochoerus*, a genus of freshwater acoel flatworm, has been found in several European freshwater canals, rivers, and lakes. A single record of a freshwater acoel has been made in the United States, but the identity remains undetermined.

Collecting. Specimens should be removed from the substrate with a certain amount of care, because the animals are easily damaged during handling. Removal of specimens from vegetation or vegetative debris can be best accomplished by gently shaking the material in clean jar containing water from the collection locality. A sharp-edged tool can be used to lift the worm from firm substrates. The specimen is then placed in a clean jar of water. The water containing the collected animals must be kept cool or cold. If the water warms too much or stays warm for too long the worms will die and decompose rapidly. Cooling by refrigeration will keep the worms alive for an extended period of time, even in the absence of feeding, thus allowing some flexibility in scheduling investigations.

Spring brooks maybe "baited" for planarians with cubes of raw, lean beef. After 15 minutes the meat should be examined and the worms shaken off into a container of water. Otherwise they may be washed or picked off the undersides of objects with a large pipette. If planarians are abundant in watercress, the plants should be placed in containers in the laboratory. As oxygen decreases, the worms will migrate to the top, where they may be picked off with a glass rod or pipette. Planarians may be taken from pond and lake substrates by placing the bait in a glass jar with a well-perforated lid. Retrieve the jar after 24 hours.

Because of their small size, the other orders are more difficult to obtain and concentrate in numbers. The uppermost 1 or 2 cm of bottom debris as well as algae and rooted vegetation from any body of water will yield these forms if such material is placed in wide-mouthed jars or aquaria in the laboratory. In 6 to 12 hours many of the specimens will migrate to the surface film and may be picked up with a pipette. Such containers should also be allowed to stand undisturbed for 2 to 20 days. As decomposition increases, more individuals and additional species will be found.

Sometimes a Birge cone net drawn through dense vegetation will take large numbers. Small species are common inhabitants of the undersurface of lily pads. The Bou-Rouch or comparable sampler is useful for collecting interstitial species (see Appendix).

Culturing. Planarians may be kept in standing water, provided the water does not become too warm and is changed every other day. Tapwater is often toxic. Enameled pans and glass or crockery containers are all suitable. Since the animals are photonegative, it is best to put objects into the container, such as pebbles or bits of broken flowerpots, to provide shaded places. Also, the whole container may be covered. Cubes of fresh meat (especially beef liver), earthworm fragments, or chopped mealworms are suitable foods and should be supplied once or twice a week. *Procotyla fluviatilis*, however, requires small living crustaceans. After 2 or 3 hours any remaining food fragments should be removed, and if the cultures are in standing water, the water should be changed. Planarians will live for 3 to 12 weeks without any food but they get progressively smaller.

Most planarians are seldom found reproducing sexually, but if sexual specimens of *Girardia tigrina* are collected in the field, they will usually continue producing egg capsules in the laboratory. *Procotyla fluviatilis* is especially useful for a study of sexual reproduction because of its translucency. Its sexual period extends from autumn through winter and sometimes into early spring. Nevertheless, *Cura foremanii* is about the only species that can really be depended on to produce egg capsules regularly under laboratory conditions.

Temporary cultures of 'microturbellarians' may develop and be kept for several weeks in the original containers of decaying aquatic vegetation and debris in the laboratory. More or less permanent cultures may be maintained in petri dishes or fingerbowls in a variety of ways. *Stenostomum* and certain other genera may be cultured in boiled wheat, rice, or rye grain infusions similar to those used for other micrometazoans and protoctistans. Subcultures should be started every 5 or 6 weeks. Pablum or cultures of similar foods are often successful. Some cold stenothermal species of *Macrostomum* can be cultured only in a refrigerator. Food may be crushed crustaceans, chopped mealworms, or chopped aquatic oligochaetes.

Preparing, preserving. Identification of flatworms to family, genus, and even species should be made on living specimens whenever possible. Shape, and especially the shape of the anterior end, is highly variable in preserved specimens, even though carefully prepared. Therefore, although it is customary to preserve specimens in the field during most types of surveys, especially when conducting quantitative studies, it is far more useful to maintain collected material of flatworms alive for examination. If placed directly into preservative of any kind, extensive contraction and distortion will occur, obliterating any evidence of natural form and shape important in most flatworm groups for accurate identification.

Live worms must be examined using a dissecting microscope with a range of magnification of 10× to 20×, or 30×. For viewing, specimens should be placed in a small, clean container with just enough water from the collection site to cover the animals. Although the worms move actively about, the external features required for identification should be nevertheless clearly evident. Manipulation of the specimens is sometimes necessary to examine specific structures. This is best undertaken with the assistance of thin blunt probes. Smaller species of nontriclad flatworm worms can be further examined by placing them alive on a slide with water and gently covering them with a coverslip. If available, methyl cellulose, a viscous fluid that is marketed as Planaslo, mixed with the water on the slide will slow the worms, thus facilitating examination. Prepared slides should be studied using a compound microscope with a strong light source. Some workers have suggested preparing squash mounts to expose certain features of the reproductive system. This procedure is not necessary for any genera in this key.

Acceptable methods of preservation vary according to taxonomic representation. The most effective means of preserving triclad flatworms is to replace the water in the container holding the worms with a 2 to 5% nitric acid solution. The acid solution will kill the animals in a somewhat flattened extended state. Immediately after the worms have been killed, the acid solution should be decanted off and specimens placed in FAA (formalin–acetic acid–alcohol) prepared following the formula provided in the last chapter. Nontriclad worms should be narcotized prior to preservation. A simple method for doing this is to immerse the worms in a cold 10% ethyl alcohol solution. Narcotization takes less than a minute and worms should be promptly flooded with hot FAA to prevent deterioration. Worms fixed in this manner will undergo some contraction and shrinkage but other methods available, though possibly producing slightly better preparations, are more complicated and require fairly toxic ingredients. A recently discussed method suggests quick freezing animals in CO_2 ice and fixing in 70% ethyl alcohol at about −6 to −9°C.

Following fixation in FAA for a few days, specimens can be transferred to either 70% ethyl or 50% isopropyl alcohol for permanent storage. If histological examination is planned, then the procedure should be undertaken soon after fixation, since alcohol tends to clear tissues over time. Another method of permanently maintaining specimens is to whole mount animals on slides. Clearing is accomplished by dehydrating the worms through a series of ethyl alcohols of increasing strength until reaching 95% ethyl alcohol. This is followed by treating the specimens with absolute alcohol and subsequently with xylene. When thorougly cleared the specimen can be mounted on a slide with a petroleum-base mounting medium and covered with a coverslip. Preserved specimens and slides should be labeled for future study.

There are several other suitable methods of killing and fixing planarians, but all methods should begin with the animals extended in a very small amount of water. Hot saturated mercuric chloride in 0.9% sodium chloride is good. Two percent nitric acid for 1 minute followed by 70% alcohol is also fairly good. Two percent nitric acid may be dropped on the planarians, and then they should be flooded with hot saturated mercuric chloride in 0. 9% sodium chloride. A biological supply company advocates narcotizing in strychnine water for a few minutes and then flooding with hot Gilson's fluid. The latter solution consists of the following: 5 g of mercuric chloride, 5 mL 80% nitric acid, 1 mL of glacial acetic acid, 25 mL 70% alcohol, and 220 mL of water; filter after 3 days.

Species identification of almost all non-triclads and species identification of some triclads are entirely dependent on serial sections, and especially on the detailed anatomy of the reproductive system of the mature individual. Chromosome structure and number is also used to separate, taxonomically, certain species of flatworms. However, both reproductive morphology and karyotype features require specialized methodologies to resolve and both involve equipment and training not available to most users of this key. Yet, investigators who have access to standard histological apparatus and stains may want to supplement information presented in this manual to corroborate identifications. It is unfortunate that such work can be done only by the experienced specialist, and the considerable effort involved in making species identifications will probably always maintain a dearth of investigators.

Because of their greater translucency, 'microturbellarians' are advantageously studied alive microscopically in a hollow-ground slide or by placing them between two square coverslips sealed at the edges with petroleum jelly. Such a mount may be turned over on a slide and both sides of the specimens examined. Slight pressure compresses the worms so that they move about only slowly. If the preparation is carefully made, even an oil immersion lens (×1000) may be used.

Flatworm fixation usually takes 30 to 60 minutes. If the fixative contains mercuric chloride, the specimens should be washed several times in 50% alcohol containing a little tincture of iodine. Pigmented specimens may be bleached in undiluted hydrogen peroxide. Seventy percent alcohol and dioxan are suitable preservatives. Stained whole mounts of rhabdocoels are sometimes useful, orange G and Delafield's hematoxylin being suitable stains.

Taxonomy. Table 4.1 Shows the arrangement of orders, families, and genera used in this manual. The higher taxonomy generally follows Rieger et al. (1991). Some categorical names have been adjusted to provide a more traditional and practical approach.

The terms Turbellaria, Neorhabdocoela, and Rhabdocoela are now believed to represent unnatural groups of free-living flatworms. Although all have been used, and will be probably continue be used to define classes and subclasses of flatworms, their application to taxonomic categories should be discouraged. I (DGS) have therefore avoided their use as taxonomic categories throughout this chapter.

The flatworm fauna of the United States is poorly known, especially in the western half of the country, and therefore a species name should never be given to a specimen simply because it is the only one previously reported from a particular region. Several genera are sorely in need of revision. Study on North American species certainly lags far behind that in Europe (Kolasa et al. 1985).

Complicating the issue is the fact that many of our species have a long history of taxonomic confusion and synonomy, chiefly as the result of variable characters, poor fixation, and identification by inexperienced workers. The section of the key that includes the 'microturbellarian' genera has been revised considerably. Although still utilizing some of Ruebush's (1941) characters, it has been updated according to information developed over the last several years. The triclad section has also been greatly revised in-

Table 4.1. ORDERS, FAMILIES, AND GENERA OF FREE-LIVING PLATYHELMINTHES IN THE UNITED STATES

Order Aceola
Order Catenulida
- Catenulidae
  - *Catenula, Dasyhormus, Suomina*
- Coohordariidae
  - *Chordarium*
- Stenostomidae
  - *Stenostomum, Myostenostomum, Rhynchoscolex*

Order Macrostomida
- Macrostomidae
  - *Macrostomum*
- Microstomidae
  - *Microstomum*

Order Dalyellioida
- Provorticidae
  - *Provortex, Pilgramilla*
- Dalyelliidae
  - *Castrella, Dalyellia, Gieysztoria, Microdalyellia*

Order Typhloplanoida
- Typhloplanidae
  - *Amphibolella, Ascophora, Bothromesostoma, Castrada, Dochmiotrema, Krumbachia, Limnoraunis, Mesostoma, Olisthanella, Opistomum, Phaenocora, Prorhynchella, Protoascus, Pseudophaenocora, Rhynchomesostoma, Strongylostoma, Typhloplana, Typhlopanella*

Order Kalyptorhynchia
- Gyratricidae
  - *Gyratrix*
- Polycystidae
  - *Polycystis, Opisthocystis*
- Koinocystidae
  - *Microkalyptorhynchus*

Order Prorhynchida
- Prorhynchidae
  - *Geocentrophora, Prorhynchus*

Order Prolecithophora
- Plagiostomidae
  - *Hydrolimax*

Order Proseriata
- Otomesostomidae
  - *Otomesostoma*
- Bothrioplanidae
  - *Bothrioplana*

Order Tricladida
- Dugesiidae
  - *Cura, Girardia, Schmidtea*
- Planariidae
  - *Hymanella, Phagocata, Paraplanaria, Polycelis, Siedlia*
- Dendrocoelidae
  - *Dendrocoelopsis, Kenkia, Procotyla, Rectocephala, Sphalloplana*

corporating information in Kenk (1972, 1989), Kawakatsu and Mitchell (1981), and Vries and Sluys (1991). Both the triclad and 'microturbellarian' sections are constructed with practicability in mind, without strict attention to fundamental generic criteria. Table 4.1 is an outline of the taxonomic relationships recognized in this manual.

There are many specific issues regarding certain genera appearing in this volume. One regards *Mesostoma* and *Bothromesostoma*. *Bothromesostoma* is defined by a combination of features detectable only by histology. Also, at least one species contains only one these features and otherwise appears closely related an otherwise typical *Mesostoma* species. Because of these problems, *Bothromesostoma* is not included in the key. Only one unambiguous species, *B. personatum*, is known from the United States. The status of families and genera of triclad worms continues to be debated. *Dugesia* has been split into three genera (formerly subgenera), of which two, *Girardia* and *Schmidtea*, occur in the United States. Kawakatsu and Mitchell (1998) have raised *Seidlia* to genus level, which includes two species in the United States. Unfortunately, *Seidlia* and *Polycelis* can be clearly separated by histological examination only. Other large genera will be almost certainly be split eventually.

Another issue concerns the genera of the Dalyelliidae and the Provorticidae. There seems to be uncertainty regarding the validity or presence in the United States of certain genera and the occurrence of unassigned populations makes the problem more complicated. I (DGS) have included probably more genera in the key than are recognized or present in the United States.

KEY TO THE FAMILIES AND GENERA OF FREE-LIVING FLATWORMS OCCURRING IN THE UNITED STATES

1. Pharynx directed anteriorly, mouth terminal or slightly subterminal (Figs. 4.2A; 4.18B) **2**
   Pharynx directed ventrally or posteriorly, if pharynx not clearly evident, mouth ventral (Figs. 4.2B–D; 4.3) **10**
2. Eyes distinct, each constituting a concave mass or connected eyespots; posterior produced to a point in most genera; egg capsules often present; typically swimming forms (Fig. 4.13) **DALYELLIDAE, PROVORTICIDAE, 3**
   Eyes absent or if present, not forming a concave cup, worms pigmentless or mottled on dorsal surface, egg capsules not produced; creeping forms **8**
3. Each eye consisting of two pigmented spots connected by a pigmented band (Fig. 4.17A); body usually darkly pigmented with blue or black; about 2 species .... ***Castrella***
   Each eye a concave mass; with green, brown, or no pigment **4**
4. Worm pigmentless, about 0.5 mm in length; posterior rounded (Fig. 4.17B); at least 1 species ***Provortex***
   Worm pigmented, posterior pointed, length greater than 0.5 mm **5**
5. Copulatory organ a tube without a spinose structure; 1 species .... ***Pilgramilla***
   Spinose copulatory aparatus present (Fig. 4.17E, F) **6**
6. Cuticular aparatus of male copulatory organ consisting of a collar with stout spines at the distal end (Fig. 4.17E), one pair of dark eyes; several species .... ***Gieysztoria***
   Cuticular aparatus of male copulatory organ consisting of a long tongue and two spinous processes (Fig. 4.17F) **7**
7. Mature animals 2 to 3 mm long; 1 to many egg capsules present and scattered throughout the body; with greenish symbionts and brown pigment (Fig. 4.13); several species in the U.S. ***Dalyellia***
   Mature animals only 1 to 2 mm long; only one egg capsule at a time; rarely more; green symbionts restricted to intestine (Fig. 4.17C); about 10 or 15 species in the U.S. ***Microdalyellia***

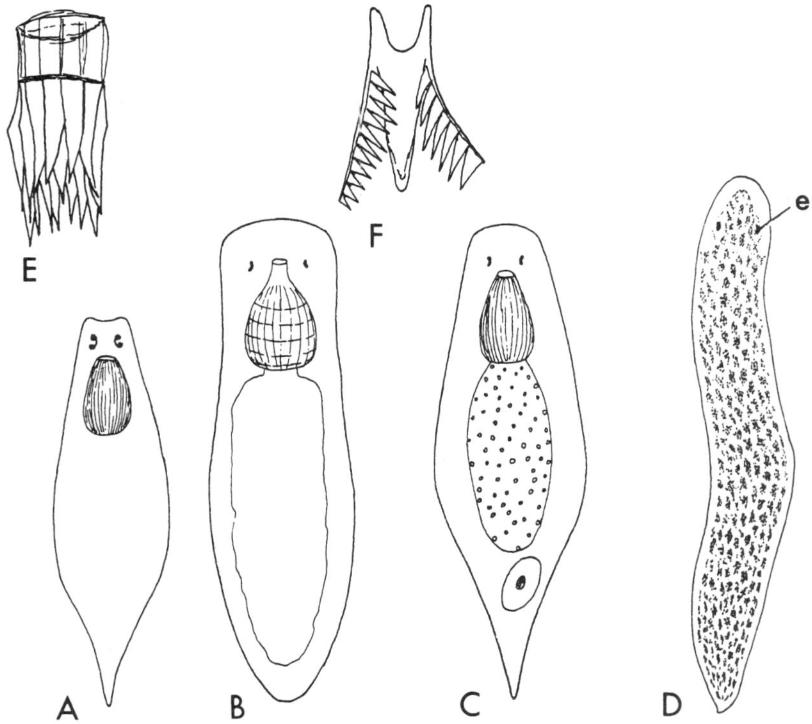

Fig. 4.17. Structure of Dalyellidae, Provorticidae, and Plagiostomidae. A, *Castrella*, ×60, showing pharynx and eyespots; B, *Provortex*, ×125, showing eyespots, pharynx, and gastrovascular cavity; C, *Microdalyellia*, ×50, showing eyespots, pharynx, gastrovascular cavity, zoochlorellae; D, *Hydrolimax grisea*, ×20, showing eyespots (e); E, cuticular apparatus of male copulatory organ of *Gieysztoria;* F, cuticular apparatus of male copulatory organ of *Dalyellia* and *Microdalyellia*. (A and C modified from Ruebush 1941; B modified from Ruebush 1935b.)

8. Worms without pigment, flat, rarely exceeding 3 mm; anterior margin wider than rest of body. **PRORHYNCHIDAE, 9**

Worms with dark, mottled pigment dorsally, plump, reaching 15 mm in length; anterior end not widened (Fig. 4.17D); 1 species, CT, MA, NC, NJ, PA **PLAGIOSTOMIDAE, *Hydrolimax***

9. Eyes absent, body narrow throughout length; penis stylet straight (Fig. 4.18A); 1 widespread species .... ***Prorhynchus***

A pair of weak close-set eyes; body rather wide (Fig. 4.18B); penis stylet bent at right angle; 1 common and several rare species .... ***Geocentrophora***

10. Largest worms less than 1 mm to just over 3 mm in length; plump, cylindrical or flat **11**

Largest worms over 5 mm in length; typically flat **42**

11. Worm not forming chains of zooids; a statocyst present **12**

If statocyst present then worm consisting a chain of zooids, otherwise statocyst absent **13**

12. Worms microscopic, less than 1 mm in length; without distinct gut (Fig. 4.18C); at least one undetermined genus, NH .... **ACOELA**

Worms up to 3 mm in length; a saccate gut extending anteriorly and posteriorly (Fig. 4.18D); 1 widespread species .... **OTOMESOSTOMIDAE, *Otomesostoma***

13. Animals forming chains of zooids (Figs. 4.11; 4.19B) **14**

Animal comprising a single individual **20**

14. With or without a statocyst at the anterior end of each zooid; principal zooid containing an anterior, ciliated constriction (Fig. 4.19A) **15**

Without a cephalic constriction or statocyst **17**

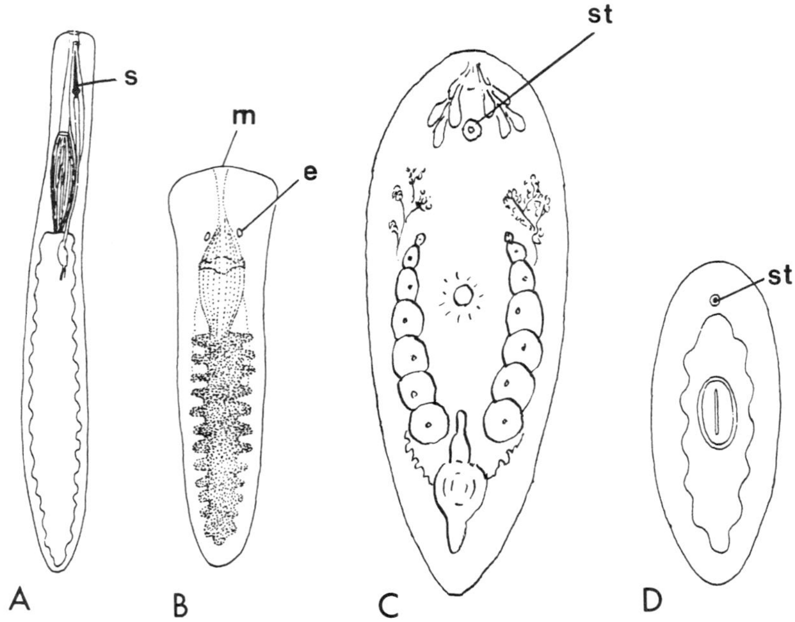

Fig. 4.18. Structure of Prorhynchidae, Otomesostomidae, Acoela. A, *Prorhynchus stagnalis*, ×50, showing reproductive style (s); B, *Geocentrophora applanata*, showing location of mouth (m) and eyespot (e); C, Acoela, ×100, showing statocyst; D, *Otomesostoma*, ×35, showing statocyst (st). (C redrawn from Cannon and Faubel 1988.)

15. Brain pronouncedly anterior to mouth, anterior lobe bulblike ; 1 species **CHORDARIIDAE, *Chordarium***
Anterior not so distinguished . . . . **CATENULIDAE (in part), 16**

16. Intestinal lumen continuous through zooids; 1 species . . . . ***Dasyhormus***
Intestinal lumen not continuous, statocyst present (Fig. 4.19A); 5 species . . . . ***Catenula***

17. With a narrow or swollen proboscislike structure at the anterior end (Figs. 4.7, 4.19B); 2 described species . . . . **STENOSTOMIDAE, *Rhynchoscolex***
Without such a structure **18**

18. With a preoral pouch at the anterior end of the intestine; without obvious ciliated bands on anterior surface (Fig. 4.19C); many common species **MICROSTOMIDAE, *Microstomum***
Without such a pouch; with a ciliated pit on either side of the anterior end **STENOSTOMIDAE, 19**

19. Intestine with anterior muscular constriction; 1 species . . . . ***Myostenostomum***
Without such a constriction; very common and many species . . . . ***Stenostomum***

20. Anterior with a protrusible proboscis in addition to ventral pharynx (Figs. 4.2D; 4.20A), or a specialized ciliated constriction, pit, or retractable structure **21**
Without a proboscis or ciliated features **27**

21. With a protrusible proboscis **22**
With ciliated pits, ring, or sensory structure **24**

22. With a distinct posterior penis stylet (Fig. 4.20A), anterior end narrow with black eyes; white and 2 mm long; a difficult complex of similar species . . . . . . **GYRATRICIDAE, *Gyratrix***
Without a posterior stylet . . . . **POLYCYSTIDAE, 23**

23. Up to 2 mm long; colorless anteriorly, reddish posteriorly; proboscis opening terminal (Fig. 4.2D); 1 rare species . . . . ***Polycystis***

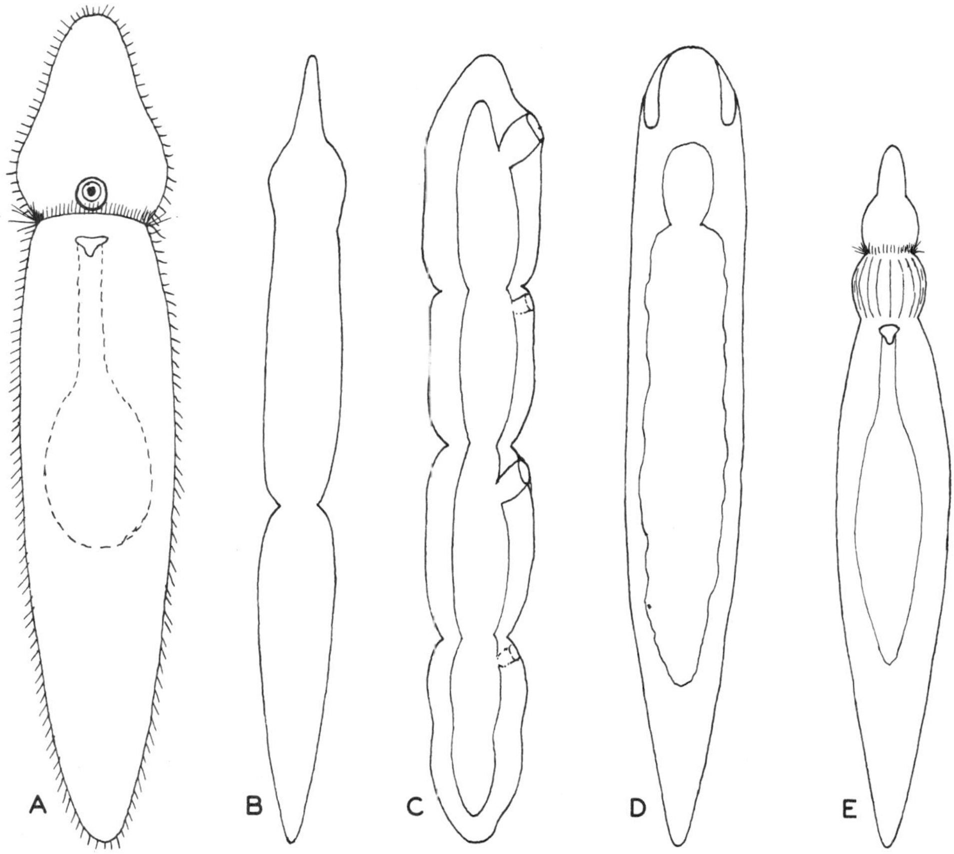

Fig. 4.19. Structure of Catenulida and Macrostomida. A, single zooid of *Catenula*, ×230, showing statocyst, mouth, and gastrovascular cavity; B, two zooids of *Rhynchoscolex*, ×55; C, lateral view of four zooids of *Microstomum*, ×35; D, dorsal view of single zooid of *Stenostomum*, ×220; E, ventral view of *Suomina*, ×200, showing mouth, gastrovascular cavity, and ciliated anterior transverse groove. (A modified from Heinlein and Wachowski 1944.)

Less than 2 mm long; white to gray coloration; proboscis opening subterminal; 1 rare species **_Opisthocystis_**

24. With a ciliated ring constricting the anterior end (Fig. 4.19E) **CATENULIDAE (in part), _Suomina_**

Without a ciliated ring . . . . **.25**

25. With a retractile and sensory anterior end (Fig. 4.20D); 1 species **TYPHLOPLANIDAE (in part), _Rhynchomesostoma_**

Without such a structure but with a median anterior sensory pit . . . . **.26**

26. With a pair of anterolateral ciliated pits (Fig. 4.20B); 1 species **TYPHLOPLANIDAE (in part), _Prorhynchella_**

Without anterolateral ciliated pits (Fig. 4.20C); 1 species **KOINOCYSTIDAE, _Microkalyptorhynchus_**

27. Worms with a broad spatulalike truncate posterior . . . . **.28**

Worms without broad truncate posterior . . . . **.30**

28. Posterior margin slightly concave or straight; reproductive organs evident on either side of intestine (Fig. 4.12); several species . . . . **.MACROSTOMATIDAE, _Macrostomum_**

Posterior margin convex with a small median tail, organs not so apparent **TYPHLOPLANIDAE (in part), 29**

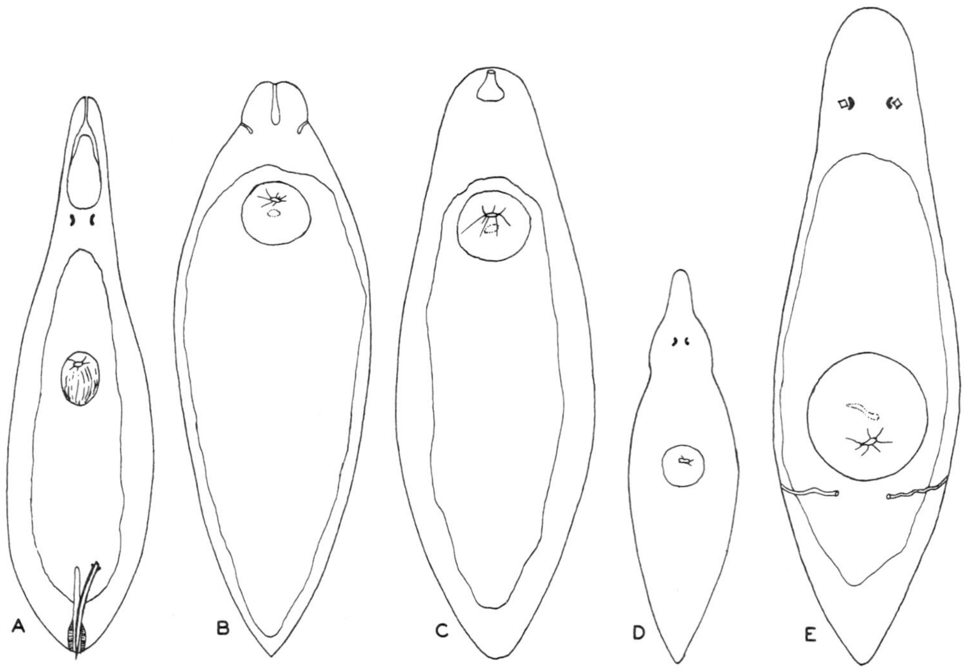

Fig. 4.20. Structure of Kalyptorhynchia and Typhloplanoida. A, ventral view of *Gyratrix hermaphroditus*, ×40, showing proboscis, eyespots, mouth, pharynx, gastrovascular cavity, and penis stylet; B, ventral view of *Prorhynchella*, ×100, showing anterior sensory pit, anterolateral ciliated pits, mouth, pharynx, and gastrovascular cavity; C, ventral view of *Microkalyptorhynchus*, ×80, showing anterior sensory pit, mouth, pharynx, and gastrovascular cavity; D, ventral view of *Rhynchomesostoma*, ×70, showing eyespots, mouth, and pharynx; E, ventral view of *Krumbachia minuta*, ×180, showing eyespots, mouth, pharynx, gastrovascular cavity, protonephridial tubules, and protonephridial pores. (B modified from Ruebush 1939; C modified from Ruebush 1935a; D modified from Ruebush 1941; E modified from Ruebush 1938.)

29. Mature animals up to 4.8 mm long; without green symbionts; 1 species in sulfur springs ***Pseudophaenocora***
Mature animals 1 to 3 mm long (Fig. 4.2B); often with green symbionts in body; about 5 species . . . . ***Phaenocora***

30. Pharynx directed posteriorly or positioned near anterior of body **31**
Pharynx near middle of body . . . . **TYPHLOPLANIDAE (in part), 33**

31. Pharynx directed posteriorly **32**
Pharynx near anterior end of body (Fig. 4.21C); 1 rare species . . . . ***Protoascus***

32. Without anterior pairs of sensory pits; distinct penis style posteriorly (Fig. 4.22D); 1 species ***Opistomum***
With anterior sensory pits (Figs. 4.9; 4.22E), no posterior penis style; 1 rare species **BOTHRIOPLANIDAE, *Bothrioplana***

33. Separate excretory openings present on ventral surface or at gonopore; pharynx in posterior half of body; worms 3 mm or less in length (three difficult genera) **34**
Separate excretory openings not present, ducts empty to mouth; worms 1 to 5 mm in length **36**

34. Pharynx in posterior third of body (Figs. 4.20E; 4.21A); 3 uncommon species . . . ***Krumbachia***
Pharynx just posterior of midpoint **35**

35. Eyes typically present; 1 species . . . . ***Olisthanella***
Eyes typically absent (Fig. 4.21B); body translucent; rare, 1 species . . . . ***Amphibolella***

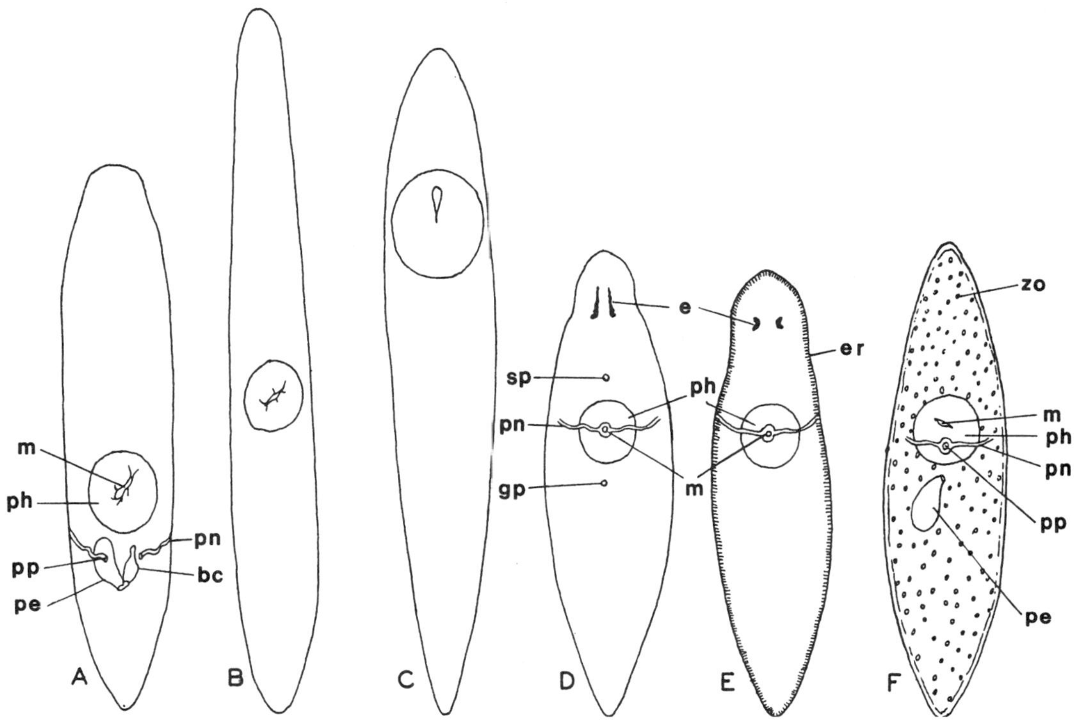

Fig. 4.21. Structure of Typhloplanoida. A, ventral view of *Krumbachia virginiana*, ×30, showing mouth (m), pharynx (ph), penis (pe), bursa copulatrix (bc), protonephridial tubules (pn), and protonephridial pores (pp); B, ventral view of *Amphibolella*, ×125; C, *Proascus wisconsinensis*, × 95; D, ventral view of *Mesostoma macroprostatum* (and *Bothromesostoma*), ×11, showing eyespots (e), pore of sensory pouch (sp), protonephridial tubules (pn), mouth (m), pharynx (ph), and common genital pore (gp); E, ventral view of *Strongylostoma*, ×50, showing eyespots (e), protonephridial tubules, mouth (m), pharynx (ph), and epidermal rhabdites (er); F, ventral view of *Typhloplana*, ×65, showing zoochlorellae (zo), mouth (m), pharynx (ph), protonephridial tubules (pn), protonephridial pore (pp), and penis (pe). (A, D, E, and F modified from Ruebush 1941; B modified from Kepner and Ruebush 1937.)

36. Eyes present, even if deeply placed; if eyes absent often with green symbionts or dark pigment; gonopore common with mouth or separate **37**
Eyes absent; single gonopore well separated from pharynx **40**

37. Green pigment (symbionts) present; about 1 to 1.5 mm long **38**
Without green pigment; worms reaching 5 mm in length **39**

38. Green pigment distinct (Fig. 4.21F); a large saccate bursa copulatrix absent; 1 species ***Typhloplana***
Green pigment dispersed among yellow and white pigment; a large saccate bursa copulatrix present; several species . . . . ***Castrada***

39. Eyes red; worms 1 to 1.5 mm long (Fig. 4.21E); 2 species . . . . ***Strongylostoma***
Eyes, when evident, black, worms up to 5 mm long; typically darkly pigmented (Fig. 4.22B, C) but some species without pigment (Fig. 4.7); 9 or 10 species, many widespread, often found in temporary ponds . . . . ***Mesostoma***

40. Gonopore well posterior **41**
Gonopore near pharynx; 1 species . . . . ***Ascophora***

41. A large and long yolk gland situated over the intestine; 1 species . . . . ***Limnoraunis***
No such large organ over intestine; 1 species . . . . ***Dochmiotrema***

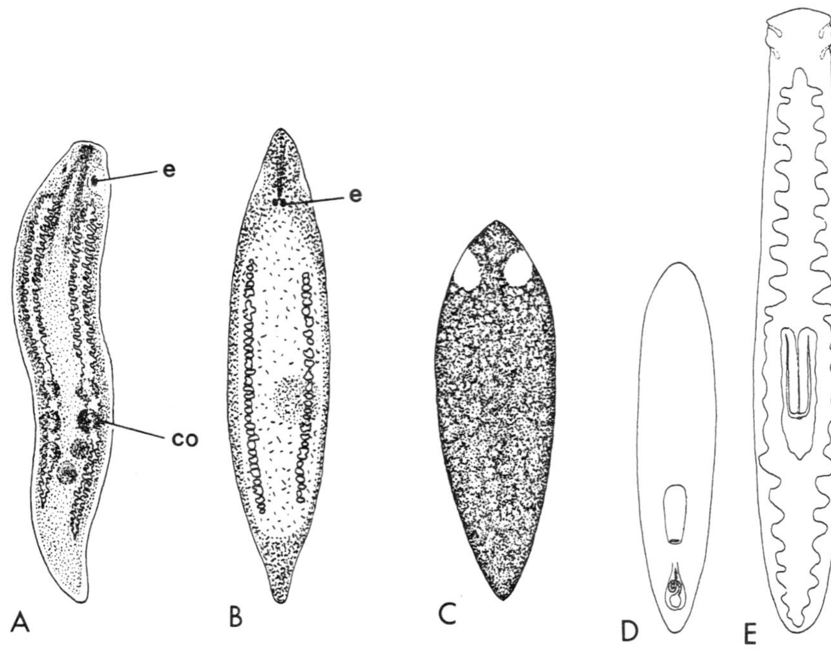

Fig. 4.22. Structure of Typhloplanidae and Bothrioplanidae, dorsal views. A, *Mesostoma macroprostatum*, ×25, showing eyespots (e) and cocoon (co); B, *M. curvipenis*, ×50, showing eyespots (e); C, *M. columbianum*, X 50; D, *Opistomum*, ×16; E, *Bothrioplana semperi*, ×16.

42. Pharynx bulbous, directed ventrally; an anterior ventral pit sometimes present (Figs. 4.21D; 4.22A); a few species reaching 10 mm in length (see 39, includes *Bothromesostoma*) **TYPHLOPLANIDAE, *Mesostoma***

Pharynx plicate, directed posteriorly (Figs. 4.1; 4.4) **43**

43. Eyes many and scattered along the anterior margin (Fig. 4.23A–C); 3 species, western states and MA . . . . **PLANARIIDAE (in part), 44**

Eyes paired or scattered, but never along anterior margin, or eyes absent **45**

44. Male atrium small and without a thick surrounding musculature (requires histological preparation); 1 species, CO, ID, MT, NM, SD, UT, WY . . . . ***Polycelis***

Male atrium large and with a thick, muscular coat; with or without cylindrical gland beneath male atrium (requires histological preparation); 2 species, CA, MA, NV . . . . ***Seidlia***

45. Median anterior margin with an adhesive organ, ranging from either a slight glandular swelling to an elongate tube; eyes present or absent . . . . **DENDROCOELIDAE (in part), 46**

Without such a structure, eyes present; rarely absent **51**

46. Eyes absent; caves, cave springs, and deep lakes **47**

Eyes present **49**

47. One species known only from Lake Tahoe, CA, NV . . . . ***Dendrocoelopsis* (in part)**

Occurring in springs and caves **48**

48. With an anterior protruding tubular adhesive organ (Fig. 4.24A); 3 species, MO, OR ***Kenkia***

With an anterior cuplike adhesive organ (Fig. 4.22B–H); 18 species distributed generally in nonglaciated regions . . . . ***Sphalloplana***

49. Pigmented **50**

Not pigmented (Figs. 4.6; 4.23F); 2 species, 1 common in eastern half of U.S. . . . . . . ***Procotyla***

50. Color black; adhesive organ an invagination of the anterior margin (Fig. 4.23G); 1 species, DC ***Rectocephala***

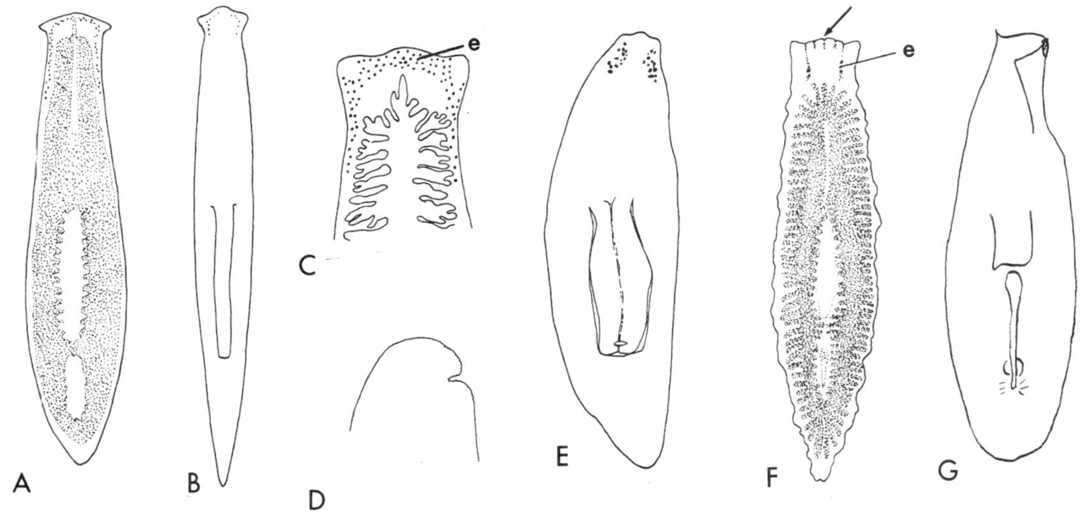

Fig. 4.23. Structure of Planariidae and Dendrocoelidae, dorsal views. A, *Siedlia remota*, ×10; B, *Polycelis coronata*, ×10; C, head of *P. coronata* showing eyespots (e); D, lateral view of head of *Dendrocoelopsis americana*, showing adhesive organ; E, *D. americana*, ×18; F, *Procotyla fluviatilis*, ×15, showing adhesive organ (arrow) and eyespots (e); G, *Rectocephala exotica*, ×20. (A from Smtih 1988, with permission; B and C modified after Hyman 1931a; G redrawn from Hyman 1953).

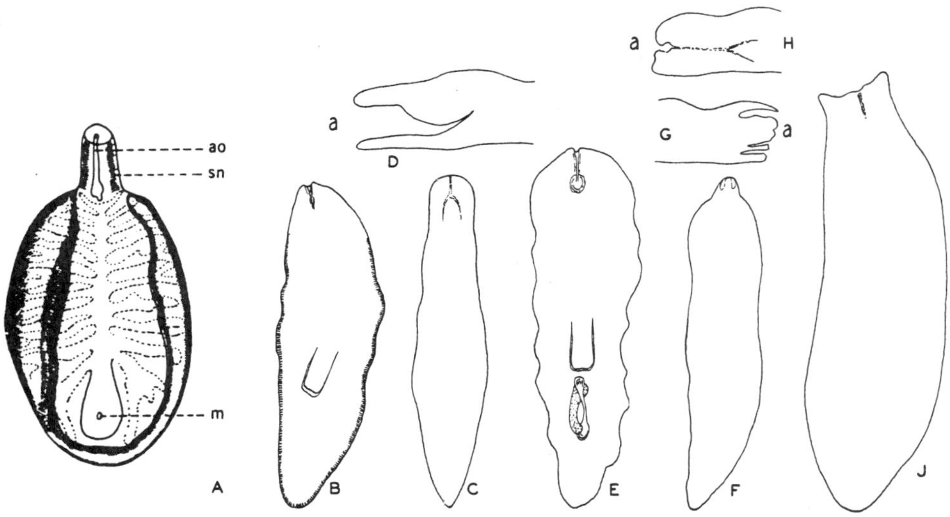

Fig. 4.24. Structure of Kenkiidae, dorsal views unless otherwise indicated. A, *Kenkia rhynchida*, ×11; B, *Sphalloplana virginiana*, ×4; C, *S. buchanani*, ×4; D, longitudinal section of anterior end of *S. buchanani*, showing the invaginated adhesive gland; E, *S. hubrichti*, ×3 (penis stippled); F, *S. pricei*, ×5, G, longitudinal section of anterior end of *S. pricei*, showing partly protruded adhesive gland; H, same, showing retracted adhesive gland; J, *S. percoeca*, ×6, a, anterior end. ao, adhesive organ; m, mouth; sn, snout. (A, C, D, and F–H modified from Hyman 1937; B and E modified from Hyman 1945; J modified from Buchanan 1936.)

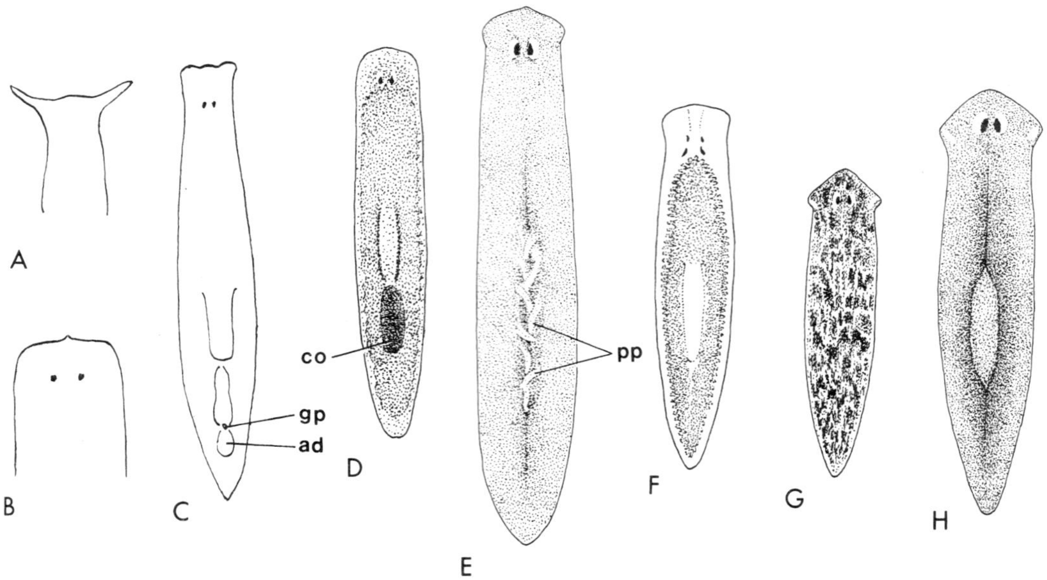

Fig. 4.25. Structure of Planaridae and Dugesiidae, dorsal views. A, head of *Phagocata bursaperforata*, ×20; B, head of *P. velata*, ×20, showing eyespots; C, *Paraplanaria dactyligera*, ×15, showing location of gonopore (gp) and adenodactyl (ad); D, *Hymanella retenouva*, ×15, showing cocoon (co); E, *Phagocata woodworthi*, × 15, showing pharenges (pp); F, *P. morgani*, ×15; G, *Girardia tigrina*, ×15; H, *Cura foremanii*, ×15. (A and B modified from Kenk 1972.)

Color light gray or brown, adhesive organ a weak swelling (Fig. 4.23D, E); 1 species, MT, OR, WA . . . . *Dendrocoelopsis* **(in part)**

51. Without eyes **52**

With eyes **53**

52. Without obvious auricles (Fig. 4.6), 2 species, one rare eyeless form in VA **DENDROCOELIDAE (in part),** ***Procotyla***

With long filiform auricles (Fig. 4.25A); one rare eyeless species, AL, GA **PLANARIIDAE (in part),** ***Phagocata*** **(in part)**

53. Auricles weak and anterior to eyes (Fig. 4.25B), or absent; few species polypharyngeal; white or pigmented . . . . **PLANARIIDAE (in part), 54**

Auricles present at level with or slightly behind eyes (Fig. 4.1), usually with light patches; always monopharyngeal; always darkly pigmented. . . . . **DUGESIIDAE, 56**

54. With a muscular pouch posterior to gonopore (Fig. 4.25C); 2 species in eastern states ***Paraplanaria***

Without such a muscular pouch **55**

55. With a broad anterior margin, no trace of auricles; adults typically with a large reddish egg capsule behind pharynx (Fig. 4.25D); 1 species in vernal ponds in eastern and gulf states ***Hymanella***

Anterior margin often with weak auricles, a slight constriction behind; no such egg capsule (Fig. 4.25E, F); about 20 species widespread throughout the U.S. . . . . ***Phagocata*** **(in part)**

56. Auricles low, rounded lobes (Fig. 1.11A); eyes well anterior of auricles; 1 species, NY ***Schimdtea***

Auricles triangular lobes; eyes at level of or just anterior of auricles **57**

57. Worms either with light dashes behind auricles and a mottled pharynx or without light dashes and unpigmented or faintly pigmented pharynx (Figs. 4.3; 4.25G); 4 species, 2 very common and widespread. . . . . ***Girardia***

Worms with light dashes behind auricles (Fig. 4.25H), pharynx unpigmented; 1 common species in eastern states . . . . ***Cura***

REFERENCES

Armitage, M. J., and J. O. Young. 1991a. Predators and planariid competitors of the triclad *Phagocata vitta* (Dugés). *Hydrobiologia* 211: 43–50.

——. 1991b. The feeding biology of the triclad *Phagocata vitta* (Dugés): a laboratory study. *Hydrobiologia* 215:63–72.

Ax, P., and J. Dörjes. 1966. *Oligochoerus limnophilus*, nov. spec., ein kaspisches Faunenelement als erster Süsswasservertreter der Flatworms Acoela in Flussen Mitteleuropas. *Int. Rev. gesamten Hydrobiol.* 51:15–44.

Ball, I. R., et al. 1981. The planarians (Turbellaria) of temporary waters in eastern North America. *Life Sci. Contr. R. Ont. Mus.* 127:1–27.

Bauchhenss, J. 1971. Die Klein turbellarien Frankens. Ein Beitrag zur Systematik und Ökologie der Turbellaria excl. Tricladida in Silddeutschland. *Life Sci. Contr. R. Ont. Mus.* 56:609–666.

Blaustein, L. 1990. Evidence for predatory flatworms as organizers of zooplankton and mosquito community structure in rice fields. *Hydrobiologia* 199:179–191.

Bronsted, H. V. 1955. Planarian regeneration. *Biol. Rev.* 30:65–126.

Buchanan, J. W. 1936. Notes on an American cave flatworm, *Sphalloplana percoeca* (Packard). *Ecology* 17:194–211.

Cannon, L. R. G. 1986. *Turbellaria of the World: A Guide to Families and Genera.* 136 pp. Queensland Museum, Brisbane.

Cannon, L. R. G., and A. Faubel. 1988. Turbellaria. In: *Introduction to the Study of Meiofauna*, pp. 273–282, edited by R. P. Higgins and H. Thiel. Smithsonian Institution Press, Washington, DC.

Carpenter, J. H. 1970. *Geocentrophora cavernicola* n. sp. (Turbellaria, Alloeococla): first cave allococoel. *Trans. Am. Microsc.* Soc. 89:124–133.

Castle, W. A., and L. H. Hyman. 1934. Observations on *Fonticola velata* (Stringer), including a description of the anatomy of the reproductive system. *Trans. Am. Microsc. Soc.* 53:154–171.

Chandler, C. M. 1966. Environmental factors affecting the local distribution and abundance of four species of stream-dwelling triclads. *Invest. Ind. Lakes Streams* 7:1–56.

Chandler, C. M., and J. T. Darlington. 1991. A survey of the epigean planarians (Turbellaria: Tricladida: Paludicola) of Alabama. *J. Freshwater Ecol.* 6:155–160.

Chodorowski, A. 1959. Ecological differentiation of Turbellaria in Harsz-Lake. *Pol. Arch. Hydrobiol.* 6:33–73.

——. 1960. Vertical stratification of Turbellaria species in some littoral habitats of Harsz Lake. *Pol. Arch. Hydrobiol.* 8:153–163.

Curtis, W. C. 1902. The life history, the normal fission, and the reproductive organs of *Planaria maculata. Proc. Boston Soc. Nat. Hist.* 31: 515–559.

Ehlers, U. 1991. Comparative morphology of statocysts in the Plathelminthes and the Xenoturbellida. *Hydrobiologia* 227:263–271.

Faubel, A., and J. Kolasa. 1978. On the anatomy and morphology of the freshwater species of Acoela (Turbellaria): *Limnoposthia polonica* (Kolasa et Faubel, 1974). *Bull. Acad. Pol. Sci. Ser. Sci. Biol. Classe II* 26:393–397.

Ferguson, F. F. 1939–1940. A monograph of the genus *Macrostomum* O. Schmidt 1848, Parts I to VIII. *Zool. Anz.* 126:7–20; 127:131–144; 128:49–68; 188–205, 274–291; 129:21–48, 120–146, 244–266.

——. 1954. Monograph of the macrostomine worms of Turbellaria. *Trans. Am. Microsc. Soc.* 73:137–164.

Ferguson, F. F., and W. J. Hayes, Jr. 1941. A synopsis of the genus *Mesostoma* Ehrenberg 1835. *J. Elisha Mitchell Sci. Soc.* 57:1–52.

Folsom, T. C., and H. F. Clifford. 1978. The population biology of *Dugesia tigrina* (Platyhelminthes: Turbellaria) in a thermally enriched Alberta, Canada lake. *Ecology* 59: 966–975.

Hayes, W. J., Jr. 1941. Rhabdocoela of Wisconsin, I: morphology and taxonomy of *Protoascus wisconsinensis* n. g., n. sp. *Am. Midl. Nat.* 25: 388–401.

Heinlein, E., and H. E. Wachowski. 1944. Studies on the flatworm *Catenula virginia. Amer; Midl. Nat.* 31:150–158.

Heitkamp, U. 1978. Speziationsprozesses bei *Gyratrix hermaphroditus* Ehrenberg, 1831 (Turbellarias, Kalyptorhynchia). *Zoomorphologie* 90:227–251.

——. 1988. Life-cycles of microturbellarians of pools and their strategies of adaptation to their habitats. *Prog. Zool.* 36:449–456.

Hyman, L. H. 1925. The reproductive system and other characters of *Planaria dorotocephala*. *Trans. Am. Microsc. Soc.* 44:51–89.

——. 1928. Studies on the morphology, taxonomy, and distribution of North American triclad Turbellaria, 1: *Procotyla fluviatilis*, commonly but erroneously known as *Dendrocoelum lacteum*. *Trans. Am. Microsc. Soc.* 47:222–255.

——. 1931a. Studies on triclad Turbellaria, Ill: on *Polycelis coronata* (Girard). *Trans. Am. Microsc. Soc.* 50:124–135.

——. 1931b. Studies on triclad Turbellaria, IV: recent European revisions of the triclads, and their application to the American forms, with a key to the latter and new notes on distribution. *Trans. Am. Microsc. Soc.* 56:316–335.

——. 1931c. Studies on triclad Turbellaria, V: descriptions of two new species. *Trans. Am. Micros. Soc.* 50:336–343.

——. 1937. Studies on triclad Turbellaria, VIII: some cave planarians of the United States. *Trans. Am. Microsc. Soc.* 36:457–477.

——. 1938. North American Rhabdocoela and Alloeocoela, 11: rediscovery of *Hydrolimax grisea* Haldeman. *Am. Mus. Novit.* 1004:1–19.

——. 1939a. North American triclad Turbellaria, IX: the priority of *Dugesia* Girard 1850 over *Euplanaria* Hesse 1897 with notes on American species of *Dugesia*. *Trans. Am. Microsc. Soc.* 58:264–275.

——. 1939b. North American triclad Turbellaria, X: additional species of cave planarians. *Trans. Am. Microsc. Soc.* 58:276–284.

——. 1945. North American triclad Turbellaria, XI: new, chiefly cavernicolous, planarians. *Am. Midl. Nat.* 34:475–484.

——. 1951a. North American triclad Turbellaria, XII: synopsis of the known species of freshwater planarians of North America. *Trans. Am. Microsc. Soc.* 70:154–167.

——. 1951b. *The Invertebrates: Platyhelminthes and Rhynchocoela: The Acoelomate Bilateria*, Vol. II, 550 pp. McGraw-Hill, New York.

——. 1953. North American triclad Turbellaria, 14: a new, probably exotic, dendrocelid. *Am. Mus. Novit.* 1629:1–6.

Janssen, C. N., and A. Faubel. 1992. Revision of the subfamily Mesostominae, Rhabdocoela, Platyhelminthes. *Mitt. hamb. Zool. Mus. Inst.* 89:7–47.

Jenkins, M. M., and H. P. Brown. 1963. Cocoon production in *Dugesia dorotocephala* (Woodworth) 1897. *Trans. Am. Microsc. Soc.* 82: 167–177.

Karling, T. G. 1963. Die Turbellarien Ostfennoskandiens, V: Neorhabdocoela. 3. Kalyptorhynchia. *Fauna Fenn.* 17:1–59.

Kawakatsu, M., and R. W. Mitchell. 1981. Redescription of *Kenkia rhynchida*, a troglobitic planarian from Oregon, and a reconsideration of the family Kenkiidae and its genera (Turbellaria, Tricladida, Paludicola). *Ann. Zool. Japonenses* 54:125–141.

——. 1998. Redescription of a North American freshwater planarian, *Seidlia remota* (Smith, 1988), with taxonomic notes on *Seidlia* and *Polycelis* species from the Far East and Central Asia (Turbellaria, Seriata, Tricladida, Paludicola). *Bull. Fuji Women's College* 36:95–110.

Kawakatsu, M., et al. 1989. Freshwater planarians of Colorado, with special reference to their vertical distribution in the Rocky Mountain National Park region. *Nat. Hist. Invent. Colorado* 10:1–26.

Kenk, R. 1935. Studies on Virginian triclads. *J. Elisha Mitchell Sci. Soc.* 51:79–126.

——. 1941. Induction of sexuality in the asexual form of *Dugesia tigrina* (Girard). *J. Exp. Zool.* 87:55–70.

——. 1970. Freshwater triclads (Turbellaria) of North America, 11: new or little known species of *Phagocata*. *J. Exp. Zool.* 83:13–34.

——. 1972. Freshwater planarians (Turbellaria) of North America. *Biota Freshwater Ecosystems, E.P.A. Identification Manual No.* 1: 1–8 1.

——. 1973a. Freshwater triclads (Turbellaria) of North America, V: the genus *Polycelis*. *Smithson. Contrib. Zool.* 135:1–15.

——. 1973b. Freshwater triclads (Turbellaria) of North America, VI: the genus *Dendrocoelopsis*. *Smithson. Contrib. Zool.* 138:1–16.

——. 1974. Index of the genera and species of the freshwater triclads (Turbellaria) of the world. *Smithson. Contrib. Zool.* 183:1–90.

——. 1975a. Freshwater triclads (Turbellaria) of North America, VII: the genus *Macrocotyla*. *Trans. Am. Microsc. Soc.* 94:324–339.

——. 1975b. Freshwater triclads (Turbellaria) of North America, VIII: *Dugesia arizonensis*, new species. *Proc. Biol. Soc. Wash.* 88:113–120.

——. 1977a. Freshwater triclads (Turbellaria) of North America, X: three species of *Phagocata* from the Eastern United States. *Proc. Biol. Soc. Wash.* 89:645–652.

——. 1977b. Freshwater triclads (Turbellaria) of North America, IX: the genus *Sphalloplana*. *Smithson. Contrib. Zool.* 246:1–38.

——. 1982. Freshwater triclads (Turbellaria) of

North America, XIII: *Phagocata hamptonae*, new species, from Nevada. *Proc. Biol. Soc. Wash.* 95:161–166.

——. 1984. Freshwater triclads (Turbellaria) of North America, XV: two new subterranean species front the Appalachian region. *Proc. Biol. Soc. Wash.* 97:209–216.

——. 1987. Freshwater triclads (Turbellaria) of North America, XVI: more on subterranean species of *Phagocata* of the eastern United States. *Proc. Biol. Soc. Wash.* 100:664–673.

——. 1989. Revised list of the North American freshwater planarians (Platyhelminthes: Tricladida: Paludicola). *Smithson. Contrib. Zool.* 476:1–10.

Kepner, W. A., and T. K. Ruebush. 1935. *Microrhynchus virginianus* n. gen. n. sp. *Zool. Anz.* 111:257–261.

——. 1937. *Amphibolella virginiana* n. sp. *Zool. Anz.* 118:103–107.

Kepner, W. A., et al. 1934. Observations upon *Rhynchomesostomum rostratum*. *Zool. Anz.* 107: 188–192.

Kepner, W. A., et al. 1937. *Castrada virginiana* n. sp. *Zool. Anz.* 119:307–314.

Kolasa, J., and A. Faubel. 1974. A preliminary description of a freshwater Acoela (Turbellaria): *Oligochoerus polonicus* nov. spec. *Boll. Zool.* 41:81–85.

Kolasa, J., et al. 1985. New records for two mosquito predators (Turbellaria: *Mesostoma*) in the northeastern United States. *Entomophaga* 30:83–85.

Kolasa, J., et al. 1987. Microturbellarians from interstitial waters, streams, and springs in southeastern New York. *North Am. Benthol. Soc.* 6:125–132.

Kromhout, G. A. 1943. A comparison of the protonephridia in freshwater, brackish-water, and marine specimens of *Gyratrix hermaphroditus*. *J. Morphol.* 72:167–169.

Luther, A. 1955. Die Dalyelliiden (Turbellaria Neorhabdocoela). Eine Monographie. *Acta Zool. Fenn.* 87:1–337.

——. 1960. Die Turbellarien Ostfermoskandiens, 1: Acoela, Catenulida, Macrostomida, Lecithoepitheliata, Prolecithophoraund, Proseriata. *Fauna Fenn.* 7:1–154.

——. 1962. Die Turbellarien Ostfennoskandiens, IV: Neorhabdocoela 1. *Fauna Fenn.* 12: 1–7 1.

——. 1963. Die Turbellarien Ostfennoskandieiis, IV: Neorhabdocoela 2. *Fauna Fenn.* 16: 1–163.

Maguire, B. 1963. Population regulation by ectocrines in *Cura foremanii* (Turbellaria). *Proc. XVI Int. Cong. Zool.* 1:64.

Marcus, E. 1945. Sobre *Catenulida brasileiros*. *Univ. Sao Paulo, Bol. Fac. Filos. Cienc. Lett. Zool. No.* 10:8–100.

——. 1946. Sobre *Turbellarias limnicos* brasileiros. *Univ. Sao Paulo, Bol. Fac. Filos. Cienc. Lett. Zool. No.* 11:5–253.

Mitchell, R. W. 1968. New species of *Sphalloplana* (Turbellaria; Paludicola) from the caves of Texas and a reexamination of the genus *Speophila* and the Family Keiikiidae. *Ann. Speleol.* 23:597–620.

Nuttycombe, J. W. 1956. The *Catenula* of the eastern United States. *Am. Midl. Nat.* 55: 419–433.

Nuttycombe, J. W., and A. J. Waters. 1938. The American species of the genus *Stenostomum*. *Proc. Am. Philos. Soc.* 79:213–284.

Pörner, H. 1966. Die rhabdocoeliden Turbellarien der Gewdsservonjenaund Umgebung. *Limnologica* 4:27–44.

Raikova, O. I., et al. 1995. Ultrastructure of the ciliary pits in the *Geocentrophora* group (Platyhelminthes, Lecithoepitheliata). *Hydrobiologia* 305:169–175.

Ramalingham, S., and P. Randall. 1984. An improved fixation technique for the rhabdocoel *Mesostoma* Ehrenberg 1835. *Can. J. Zool.* 62: 1893–1894.

Redfield, E. S. P. 1915. The grasping organ of *Dendrocoelum lacteum*. *J. Anim. Behav.* 5:375–380.

Reynoldson, T. B. 1958. The quantitative ecology of lake-dwelling triclads in northern Britain. *Oikos* 9:94–138.

——. 1964. Evidence for intra-specific competition in field populations of triclads. *J. Anim. Ecol.* 33:187–201.

——. 1967. A key to the British species of freshwater triclads. *Sci. Publ. Freshwater Biol. Assoc.* 23:1–28.

Reynoldson, T. B., and L. S. Bellamy. 1971. Intraspecific competition in lake-dwelling triclads. *Oikos* 22:315–328.

——. 1973. Interspecific competition in lake-dwelling triclads. *Oikos* 24:301–313.

Reynoldson, T. B., and A. D. Sefton. 1976. The food of *Planaria torva* (Müller) (Turbellaria-Tricladida), a laboratory and field study. *Freshwater Biol.* 6:3 83–393.

Rhode, K. 1991. The evolution of protonephridia of the Platyhelminthes. *Hydrobiologia* 227: 315–321.

Rieger, R. M., et al. 1991. Platyhelminthes: Turbellaria. In: *Microscopic Anatomy of the Invertebrates*, Vol. 3, pp. 7–140, edited by F. W. Harrison and B. J. Bogitsh. Wiley–Liss, New York.

Riser, N. W., and M. P. Morse, eds. 1974. *Biology of the Turbellaria.* 530 pp. McGraw–Hill, New York.

Rixen, J.-U. 1961. Kleinturbellarien aus dem Literal der Binnengewässer Schleswig-Holsteins. *Arch. Hydrobiol.* 57:464–538.

Ruebush, T. K. 1935a. The genus *Olisthanella* in the United States. *Zool. Anz.* 112:129–136.

——. 1935b. The occurrence of *Provortex affinis* Jensen in the United States. *Zool. Anz.* 111: 305–308.

——. 1937. The genus *Dalyellia* in America. *Zool. Anz.* 119:237–256.

——. 1938. *Krumbachia minuta* n. sp. (Turbellaria Rhabdocoela). *Zool. Anz.* 122:260–265.

——. 1939. A new North American rhabdocoel Turbellarian, *Prorhynchella minuta* n. gen., n. sp. *Zool. Anz.* 127:204–209.

——. 1940. *Mesostoma ehrenbergii wardi* for the study of the Turbellarian type. *Science* 91: 531–532.

——. 1941. A key to the American freshwater Turbellarian genera, exclusive of the Tricladida. *Trans. Am. Microsc. Soc.* 60:29–40.

Ruebush, T. K., and W. J. Hayes, Jr. 1939. The genus *Dalyellia* in America, II: a new form from Tennessee and a discussion of the relationships within the genus. *Zool. Anz.* 128:136–152.

Sanders, O., and R. M. Sanders. 1970. Webs secreted by planarians. *J. Biol. Psych.* 12:65–70.

Schwank, P. 1980. Neue limnische Turbellarien. *Arch. Hydrobiol.* 88:463–490.

Schwartz, S. S., and P. D. N. Hebert. 1982. A laboratory study of the feeding behavior of the rhabdocoel *Mesostoma ehrenbergii* on pond Cladocera. *Can. J. Zool.* 60:1305–1307.

Shinn, G. L. 1993. Formation of egg capsules by flatworms (Phylum Platyhelminthes). *Trans. Am. Microsc. Soc.* 112: 18–34.

Sluys, R., et al. 1995. Taxonomy, ecology, and karyology of a new species of *Phagocata* from Spain, with a discussion on the phylogenetic systematics of the genus *Phagocata* s.l. (Platyhelminthes, Tricladida, Paludicola). *Can. J. Zool.* 73:557–568.

Smith, D. G. 1988. A new, disjunct species of triclad flatworm (Turbellaria: Tricladida) from a spring in southern New England. *Biol. Bull.* 175:246–252.

——. 1991. On some larger turbellarian worms (Platyhelminthes) living in temporary fresh waters of southern New England. *Hydrobiologia* 220:255–265.

——. 1998. *Mesostoma* (Platyhelminthes: Rhabdocoela, Typhloplanidae) of Mona Island, Puerto Rico, with a description of a new species. *Invert. Biol.* 117:101–108.

Smith, D. G., and C. C. Menzie. 1994. *Hydrolimax grisea* (Platyhelminthes: Palgiostomidae) in southern New England, with notes on the anatomy of the pharynx. *Trans. Am. Microsc. Soc.* 113:258–262.

Sopott-Ehlers, B. 1991. Comparative morphology of photoreceptors in free-living plathelminths: a survey. *Hydrobiologia* 227:231–239.

Strayer, D. 1985. The benthic micrometazoans of Mirror Lake, New Hampshire. *Arch. Hydrobiol./Suppl.* 72:287–426.

Timoshkin, O. A. 1991. Turbellaria Lecithoepitheliata: morphology, systematics, phylogeny. *Hydrobiologia* 227:323–332.

Von Graf, L. 1911. Acoela, Rhabdocoela, und Alloeocoela des Ostens der Vereinigten Staaten von Amerika. *Z. Wiss. Zool.* 99:1–108.

——. 1913. Turbellarias, II. Rhabdocoelida. *Das Tierreich* 35:1–484.

Vries, E. J. de, and R. Sluys. 1991. Phylogenetic relationships of the genus *Dugesia* (Platyhelminthes, Tricladidia, Paludicola). *J. Zool. Lond.* 223: 103–116.

Weinzierl, R. P., et al. 1999. High fecundity and low fertility in parthenogenetic planarians. *Invert. Biol.* 118:87–94.

Young, J. O. 1970. British and Irish freshwater microturbellarians: historical records, new records and a key for their identification. *Arch. Hydrobiol.* 67:210–241.

——. 1975. The population dynamics of *Phaenocora typhlops* (Vejdovsky) (Turbellaria: Neorhabdocoela) living in a pond. *J. Anim. Ecol.* 44:251–262.

Young, J. O., and T. B. Reynoldson. 1965. A laboratory study of predation on lake-dwelling triclads. *Hydrobiologia* 26:307–313.

5

NEMERTEA

(Proboscis worms)

Most freshwater biologists have never had the experience of observing living freshwater nemerteans. One reason is that these organisms are seldom sought intentionally and are usually overlooked or disregarded in collections made with other taxa uppermost in mind. Another reason is their extremely "spotty" and local abundance. Weedy ponds, masses of filamentous algae, the undersides of lily pads, and the general substrate of littoral areas are preferred habitats, but often nemerteans may be taken from only one restricted area in a pond or stream and are apparently absent elsewhere. A third reason they are often overlooked is their superficial resemblance to oligochaetes: they have a pseudosegmental appearance owing to the many gonads and intestinal diverticula, especially when these diverticula are full of food.

The phylum is well represented by marine forms, there being only a handful of freshwater and terrestrial forms known and these are largely confined to southern latitudes. Freshwater nemerteans have been badly neglected in the United States, but presumably there is only one native genus, *Prostoma* (Fig. 5.1). The first form discovered in the United States was described in 1850 as *Emea rubrum*, which has had a long history of taxonomic and nomenclatorial confusion. It is now called *Prostoma graecense* and is identified by the presence of a small pit at the anterior margin of the head (Fig. 5.2) This species is widely distributed in the United States, as well as worldwide. A second species is *Prostoma eilhardi.* It is likewise distributed worldwide, although it is rarely found in the United States. Unfortunately, serial sections are necessary for the separation of these two species. *P. graecense* has a well-developed ciliated esophagus; *P. eilhardi* has an indistinct, unciliated esophagus. These features are affected by method of killing, preservation, and preparation and can be difficult to assess. Five other species of *Prostoma* are known from other countries, chiefly from single localities. In addition, there are seven other monospecific freshwater genera, each known from a single locality, including Java, Germany, Brazil, Chile, China, and New Zealand.

One of the monotypic forms is *Apatronemertes albimaculosa*, known only from aquaria in Germany and the United States. The native range of the species remains unknown but is presumably somewhere in the tropics. In the United States, *A. albimaculosa* has been found in aquaria associated with tropical plants raised in Florida and it is likely that the species has become established in that state.

General characteristics. *Prostoma graecense* is elliptical in cross section, somewhat flattened ventrally, up to 30 mm long, and 0.6 to 2.0 mm. in diameter. Coloration is highly variable. Young and well-fed specimens are whitish or pale yellow, but mature

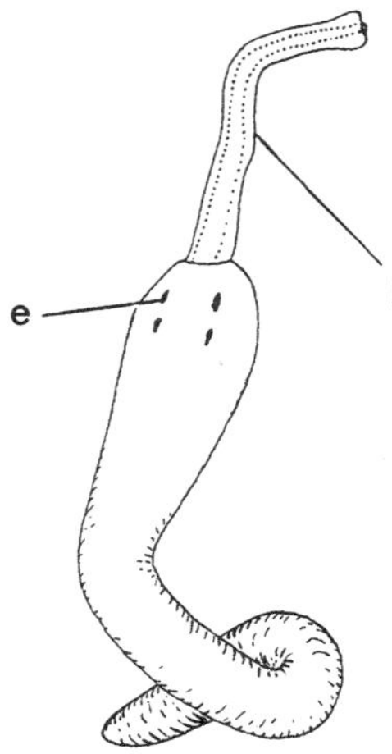

Fig. 5.1. A specimen of *Prostoma* sp. ×10, with proboscis extended. e, eyespot; pr, proboscis.

specimens are usually yellowish red, orange, or deep red. There is also a rare green variety. The anterior end is rounded and without a definite head; the posterior end is tapered to a point. Three pairs of ocelli are usually present, but occasionally there are two or four pairs. They are arranged in two longitudinal rows on the dorsal surface of the anterior end and are sometimes placed asymmetrically. The body is smooth, unsegmented, and covered with a columnar ciliated epithelium.

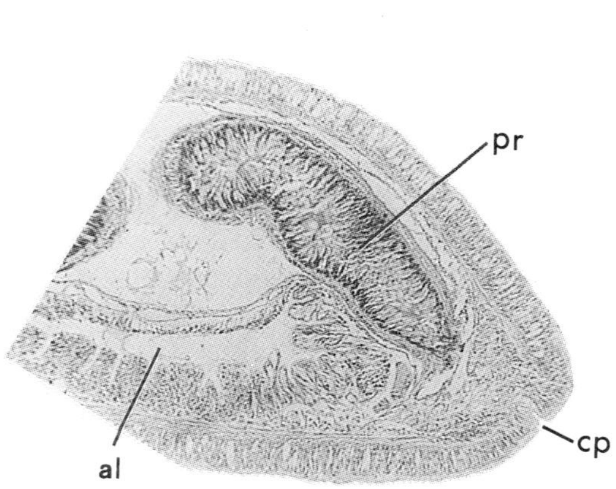

Fig. 5.2. Longitudinal section of the head of *Prostoma graecense*, showing small anterior cephalic pit. al, alimentary canal; cp, cephalic pit; pr, proboscis.

Apatronemertes is much larger, reaching 120 mm in length. It is a bright red worm and the head contains distinct lateral sensory slits (Fig. 5.3). Eyes have not been detected. Unlike *Prostoma*, the mouth is separate from the proboscis pore and located in a ventral position (Fig. 5.4).

Owing to a well-developed longitudinal and circular muscle system, these worms are highly contractile. A smooth, gliding type of locomotion is produced by cilia acting on a copious slime track secreted by the epidermis.

Digestive system, feeding. The mouth opens into a buccal cavity, which is followed by an esophagus, stomach, intestine, and terminal anus. The entire length of the long intestine has pairs of lateral diverticula, and from the anterior end of the intestine a pair of pyloric caeca extend forward. The digestive tract has no musculature of its own.

The proboscis apparatus (Fig. 5.5) is a unique feature of nemerteans. A fluid-filled longitudinal cavity, the rhynchocoel, runs throughout most of the body length dorsal to the digestive tract. The muscular wall of the rhynchocoel is called the proboscis sheath. Lying within the rhynchocoel is a tubular muscular proboscis, which can be protruded anteriorly and withdrawn back into the rhynchocoel by a long retractor muscle. The proboscis is very long (Fig. 5.1), sometimes two or three times the body length, and when not in use it is strongly contracted and coiled within the rhynchocoel. The rhynchodaeum (Fig. 5.5) is the cavity of the proboscis and it has an opening to the outside, either in common with the mouth or separate. In *Prostoma* there are actually three parts to the proboscis. Anteriorly it is a thick-walled tube. Farther back is a middle bulbous portion bearing a functional sclerotized stylet set in a heavy base along with two lateral pockets containing accessory stylets in various stages of development (Fig. 5.6). The functional stylet is frequently lost and is then replaced by another from the lateral pockets. Posterior to the bulb is a blind tube. When the proboscis is protruded, the bulb and posterior blind tube are everted through the anterior thick-walled tube and

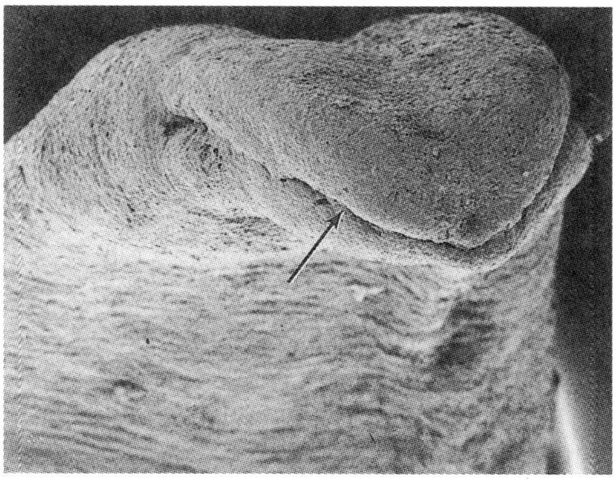

Fig. 5.3. Head of *Apatronemertes albimaculosa*, Arrow, sensory slit.

out of the mouth in such a fashion that the stylet is at the tip of the extended proboscis. Proboscis eversion is effected by muscle contraction exerting pressure on the fluid in the rhynchocoel.

The proboscis functions in defense, food getting, and food handling. While *Prostoma* creeps about on the substrate its extended proboscis probes and darts constantly. If the proboscis comes in contact with a small metazoan, the stylet may be used to stab the animal repeatedly. A toxic material is thought to be secreted from the posterior chamber of the proboscis and discharged

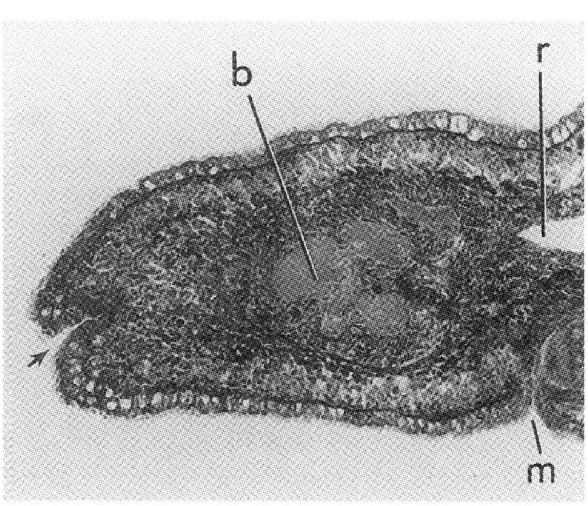

Fig. 5.4. Longitudinal section of head of *Apatronemertes albimaculosa* b, brain; m, mouth; r, rhynchocoel, arrow denotes sensory slit.

through the stylet apparatus. The proboscis is also coiled around the prey and quantities of sticky mucus are secreted, thus further immobilizing the prey. Once grasped, the food can be additionally manipulated by the proboscis. The mouth and buccal cavity are capable of great distension, and most food organisms are ingested intact. If the prey is too large, however, only the body juices are sucked out. *Prostoma* is chiefly carnivorous and feeds on all kinds of small metazoans, and especially on small oligochaetes. It may also feed on recently dead animals, and a few times it has been observed ingesting organic detritus.

Preliminary digestion occurs in the stomach and further digestion in the intestine and its diverticula (Fig. 5.5). Specimens may be kept for 6 months or more without food, but the metabolic requirements during this period may produce a shrinkage to one-tenth or less of normal size.

General anatomy, physiology. Below the epidermis *Prostoma* has a thick connective tissue dermis and well-defined layers of longitudinal and circular muscle. Visceral organs are separated by mesenchymal connective tissue.

A closed circulatory system is present, the main vessels being a dorsal longitudinal contractile vessel and two lateral longitudinal contractile vessels. The blood is clear, sometimes faintly yellowish reddish. Possibly it contains hemoglobin.

Excretion is effected by a protonephridial system consisting of many branched tubules with clusters of terminal flame bulbs and several nephridiopores along each side of the body. Presumably this apparatus also functions in osmoregulation. The protonephridial system is much more extensive than in marine nemertines.

The brain (Figs. 5.4, 5.5) is dark-colored and four-lobed and surrounds the esophagus. Two prominent lateroventral nerve cords pass from it and extend to the posterior end. Smaller paired nerves supply the head, proboscis, and proboscis sheath. The ocelli lie beneath the epidermis and are irregular cup-shaped masses of black, brown, or reddish pigment granules. Anterior sen-

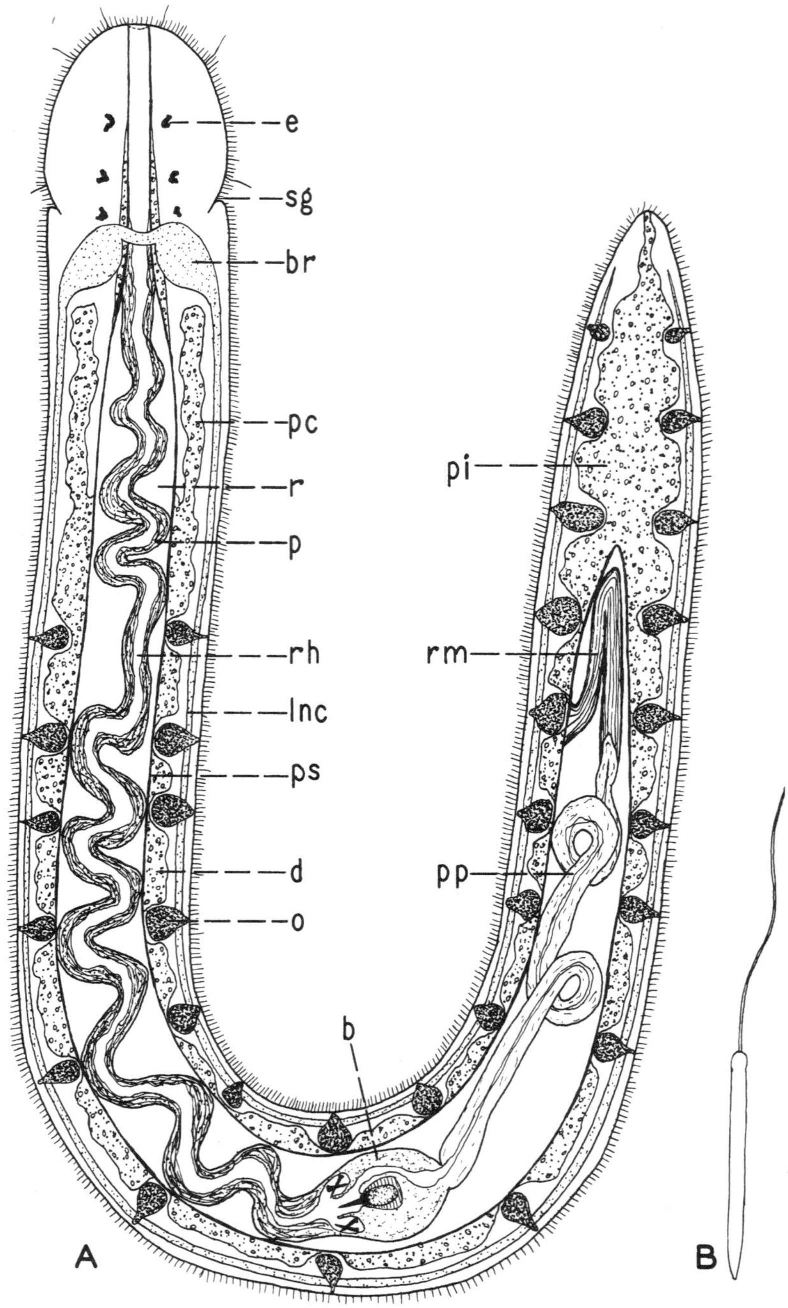

Fig. 5.5. Structure of *Prostoma graecense*, semidiagrammatic. A, general anatomy, dorsal view, ×16; B, with protruded proboscis, ×1.5. b, bulbous portion of proboscis; br, brain; d, diverticulum of intestine; e, ocellus; lnc, lateral nerve cord; o, ovotestis; p, anterior portion of proboscis; pc, pyloric caecum; pi, posterior portion of intestine; pp, posterior portion of proboscis; ps, proboscis sheath; r, rhynchocoel; rh, rhynchodaeum; rm, retractor muscle of proboscis; sg, sensory groove.

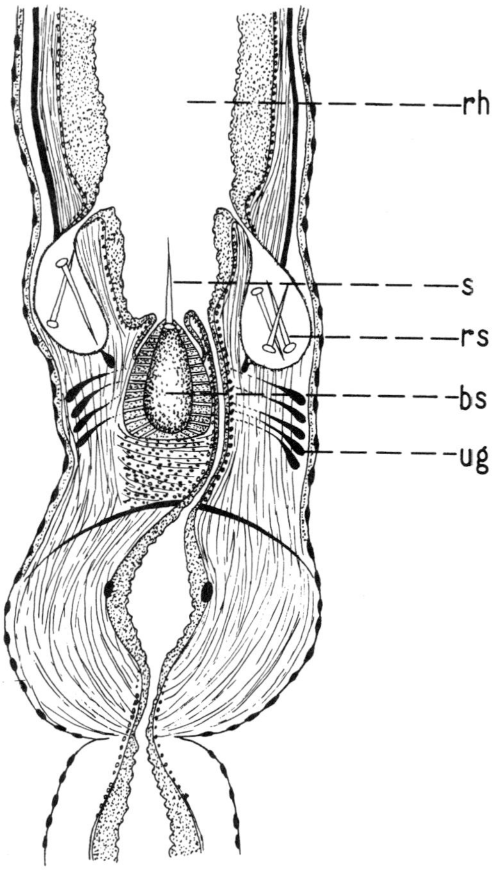

Fig. 5.6. Bulbous portion of proboscis and adjacent structures of *Prostoma*. bs, base of stylet; rh, rhynchodaeum; rs, accessory stylet; s, functional stylet; ug, unicellular glands. (Greatly modified from Reisinger 1926.)

sory grooves are presumably chemotactile, and sensory cells with long cilia are especially abundant at the extremities.

There are no special respiratory structures or adaptations. It is possible that some respiratory exchange occurs through the wall of the digestive tract in addition to the general body surface.

Reproduction, life history. *Prostoma* and *Apatronemertes* are hermaphroditic and protandric (not confirmed in *Apatronemertes*); saclike ovotestes alternate with the intestinal diverticula. Each ovotestis opens to the outside by a lateral genital pore and produces many sperm but only a single ripe egg. In *Prostoma* the additional oocytes degenerate. Breeding may occur at any time, but most often between May and December, provided the water temperature is above 10°C. A sexually mature individual secretes a mucous sheath around itself, releases sperm and a double row of eggs into the sheath, and then crawls out of the sheath. Not infrequently two individuals, one of which is at the time functionally male and the other female, occupy the same mucous sheath, and cross-fertilization occurs. Contact with water causes dissolution of the germinal vesicles around the eggs and they become fertilized. Development is direct, and under laboratory conditions (25°C) the young emerge in about 5 days. There is no special larval stage. Young worms leave the mucous mass and become independent. Newly hatched worms frequently swim about freely above the substrate.

Almost nothing is known concerning the details of the life history of *Apatronemertes*. I (DGS) have observed animals containing large encapsulated eggs and developed oviducts collected in aquaria in July.

Prostoma has considerable powers of regeneration. The whole posterior portion of the body may be regenerated provided the remaining anterior portion contains at least half of the esophageal region. A new proboscis may be regenerated if the old one is lost.

Ecology. *Prostoma graecense* can be most easily found in autumn, when it is relatively abundant. It occurs only in well-oxygenated shallow and littoral standing waters and sluggish streams, especially in masses of filamentous algae, on rooted aquatics, or in the debris of the general substrate. No explanation is known for its localized distribution in such places. It is especially active at night, when it does most of its feeding and egg deposition. It is highly thigmotactic and reacts positively to water currents. With the approach of autumn and freezing temperatures *Prostoma* migrates to deeper waters and returns to the shallows in spring.

Under adverse environmental conditions, such as insufficient oxygen, high temperature, abnormal substrate, and lack of food, *Prostoma* forms resting cysts by rounding up within a secreted outer layer of mucus, which soon hardens and becomes covered with detritus. The worm may remain alive within such a cyst for a few days to a few weeks, emerging when ecological conditions are again appropriate. Cysts cannot withstand complete drying and are of limited value in dissemination.

Prostoma is preyed upon by larger worms and carnivorous insects and crustaceans.

Apatronemertes has been collected in close association with various aquatic plants where it lives among the roots and stems. In the United States it seems to be most frequently found among beds of tropical *Vallisnera* and *Echinodoris*, plants distributed widely throughout the country by the aquarium trade.

Culturing, preserving. *Prostoma* may be cultured in an aquarium that has a thin layer of soil on the bottom and plenty of rooted aquatics. Excessive bacterial decomposition inhibits nemerteans and should be avoided. Feed the worms bits of liver, oligochaetes, including the genera *Dero*, *Limnodrilus*, and *Stylaria*, or small metazoans. Apparently, insect larvae and ciliates are rejected.

Because it is so highly contractile, *Prostoma* should always be anesthetized before it is killed. Chloretone and other solutions may be used with varying success. Kill in 80% alcohol or hot FAA. Preserve in 70% alcohol.

Taxonomy. About 650 valid species of nemerteans are known. The great majority are marine, but a few occur on land in freshwaters. The genus *Prostoma* is placed in the order Hoplonemertea and is essentially cosmopolitan. *Apatronemertes* is in the order Heteronemertea.

The heavy reliance on body color, number of reserve stylets, length of rhynchocoel, and number of ocelli and size in earlier studies rendered the taxonomy of *Prostoma* confused and difficult. Currently, species of *Prostoma* are distinguished by anatomical features of the head and anterior alimentary canal. The recognition of these characters is easily obscured by poor preservation, extreme contraction during preservation, and clearing caused by long storage in alcohol. *Prostoma graecense* has also gone under the generic designations of *Emea*, *Tetrastemma*, and *Stichostemma* and under the species designations of *asensoriatum*, *aquarium*, and *rubrum*. Only carefully prepared histological sections can resolve species identity.

Apatronemertes and *Prostoma* probably live together only in Florida. They can be easily distinguished by their size differences, pigment, and the presence or absence of eyes (see above). Besides *Apatronemertes*, it is probable that other tropical genera as well as European species of *Prostoma* have been introduced into this country on imported aquatic plants. Furthermore, it is quite possible that additional native species await discovery.

REFERENCES

Böhmig, L. 1898. Beiträdge zur Anatomie und Histologie der Nemertmen. *Z. Wiss. Zool.* 64:479–564.

Child, C. M. 1901, The habits and natural history of *Stichostemma*. *Am. Nat.* 35:975–1006.

Coe, W. R. 1943. Biology of the nemerteans of the Atlantic coast of North America. *Trans. Conn. Acad. Arts Sci.* 35:129–328.

Gibson, R., and J. Moore. 1978. Freshwater nemerteans: new records of *Prostoma* and a description of *Prostoma canadiensis* sp. nov. *Zool. Anz.* 201:77–85.

——. 1976. Freshwater nemerteans. *Zool. J. Linn. Soc.* 58:177–218.

Gibson, R., and Q. Sang. 1991. A new freshwater heteronemertean from the Zhujiang (Pearl River), People's Republic of China. *Hydrobiologia* 220:167–178.

Gibson, R., and J. O. Young. 1974. First records of *Prostoma* (Hoplonemertea) from East Africa. *Zool. Anz.* 193:103–109.

Loden, M. S. 1974. Observations on the freshwater nemertean, *Prostoma rubrum* (Leidy), in Alabama. *Am. Midl. Nat.* 91:434–435.

Montgomery, T. 1896. *Stichostemma aesensoriatum* n. sp., a freshwater nemertean from Pennsylvania. *Zool. Anz.* 19:436–438.

Moore, J., and R. Gibson. 1985. The evolution and comparative physiology of terrestrial and freshwater nemerteans *Biol. Rev.* 60: 257–312.

Reisinger, E. 1926. Nemertini. *Biol. Tiere Deutschl.* 7:1–24.

Stiasny-Wijnhoff, G. 1938. Das *Prostoma* Dugés, eine Gattung von Süsswasser-Nemertinen. *Arch. Neerl. Zool. Suppl.* 3:219–230.

Wilfert, M., and R. Gibson. 1974. A new genus of hermaphroditic freshwater heteronemertean (Nemertinea). *Z. Morph. Tiere* 79:87–112.

6

GASTROTRICHA*

In company with protoctists, rotifers, nematodes, and small oligochaetes, the Gastrotricha are a part of the characteristic assemblage of microorganisms on aquatic vegetation and on the debris (periphyton) forming the substrates of standing waters. They are also commonly a part of the upper interstitial, or meiobenthos. Formerly, it was thought that gastrotrichs were almost restricted to freshwaters, but they are now known to be abundant in marine habitats. The phylum consists of only two orders; the Order Macrodasyida is predominantly marine, and the Order Chaetonotida is predominantly freshwater.

General characteristics. Most freshwater gastrotrichs range from 100 to 300 μm in length, although a few are as small as 70 μm or as large as 760 μm. The body (Fig. 6.1) is short to long and wormlike, with a flat ventral surface. Head and arched trunk regions are obvious, and often there is a more or less distinct neck. Two toelike projections, forming the furca, are commonly present at the posterior end.

The margin of the head is variable, even within the same genus. Sometimes the entire margin is smooth, sometimes it has a median anterior lobe and one or two pairs of variously developed lateral lobes. Tufts of sensory cilia are customarily situated in the depressions between lobes. The mouth is terminal, subterminal, or ventral, and is often surrounded by short delicate bristles. In freshwater species the anus lies in a dorsal position at the base of the furca; rarely is it terminal or subterminal.

Locomotor cilia occur chiefly on the ventral surface of the head and trunk. Most commonly there are several patches on the head and two longitudinal bands running throughout most of the body length (Fig. 6.3H, J). Gastrotrichs appear to have only monociliated epithelial cells.

Coloration is variable, usually light gray to reddish brown, and mostly determined by the contents of the digestive tract. A starved gastrotrich is grayish and translucent.

Internal organization is relatively uncomplicated. The digestive tract is tubular and without diverticula. The space between the digestive tract and thin body wall is a pseudocoel and it contains the simple muscle, protonephridial, nervous, and reproductive systems (Fig. 6.1). There are no special circulatory or respiratory structures.

Many workers have discussed the phylogenetic relationships of gastrotrichs with the Nematoda, Rotatoria, Kinorhyncha, Turbellaria, and other groups, but the Gastrotricha are here considered sufficiently distinctive to constitute a separate phylum.

*Prepared with the assistance of Jacek Kisielewski.

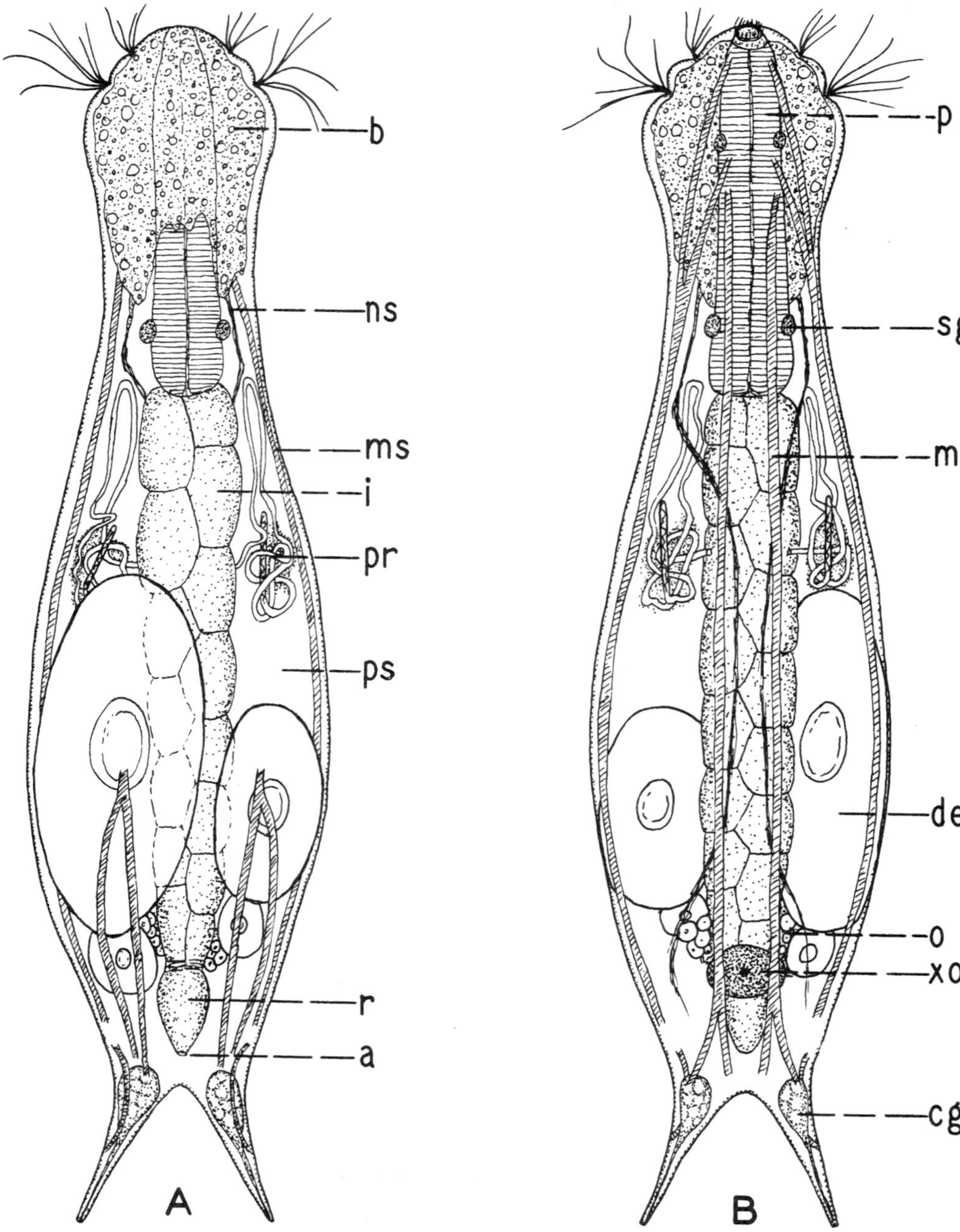

Fig. 6.1. Diagrammatic internal structure of a typical gastrotrich. A, dorsal; B, ventral. a, anus; b, brain; cg, cement gland; de, developing egg; i, intestine; ms, muscle strand; ns, longitudinal nerve strand; o, ovary; p, pharynx; pr, protonephridium; ps, pseudocoel; r, rectum; sg, salivary gland; xo, "X organ." (Greatly modified from Zelinka 1889.)

Cuticular structures. The general body surface in gastrotrichs is covered with a cuticle secreted by the underlying syncytial hypodermis. In the great majority of species, however, the cuticle is not simple and thin but patterned and with a variety of outgrowths, especially dorsally and laterally (Figs. 6.2, 6.7, 6.8).

In a few species the cuticle is locally thickened to form variously arranged plates, but more commonly there are series of cuticular scales that usually overlap more or less. Both plates and scales may have small keels or spines; the latter range from very short to exceedingly long and robust. In the absence of plates and scales, spines are often attached directly to the thin cuticle. Sometimes spines are barbed and have a triradiate or platelike base. Many species have midventral scales or plates. Size and shape of plates, scales, and spines commonly vary within the same individual, depending on the location on the body. A few simple setae are often present (Figs. 6.5, 6.6), especially near the anterior and posterior ends of the body. The cephalic shield is a thickened cuticular plate covering the anterior end of the head above the mouth (Fig. 6.3G). Sometimes there are less distinct special plates posterior to the mouth and laterally on the head.

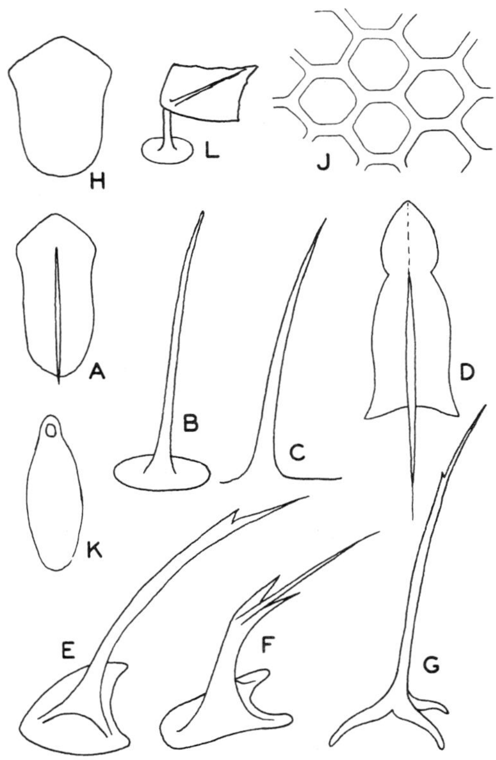

Fig. 6.2. Cuticular structures of Gastrotricha. A–G, *Chaetonotus;* H and J, *Lepidodermella;* K, *Polymerurus;* L, *Aspidiophorus.* H and J, simple cuticular plates; A, keeled plate; B, D–F, spined plates; C, cuticular spine; K and L, pedunculated scales. (J from Brunson 1950.)

Locomotion. Beating movements of the ventral cilia against the substrate produce a characteristic smooth, graceful, gliding type of location that is sometimes quite rapid; presumably the lateral tufts of head cilia are also of some importance. Such gliding locomotion is especially notable in the Chaetonotidae. The long spines of the Dasyclytidae are used in springing and leaping. Although the spines are not supplied with special muscles, they are moved vigorously toward the plane of the body when the arched body is straightened. Most gastrotrichs may temporarily leave the substrate and swim about in the water with the ventral and head cilia, but they seldom go far above the substrate. *Polymerurus* demonstrates a peculiar sinuous activity that is a defensive reflex that superficially resembles nematode movement.

A cement gland is usually found in the base of each process of the furca, with its duct opening at or near the tip. The adhesive secreted material is used for temporary attachment to objects. A similar arrangement is found in rotifers. In the marine order Macrodasyida there is often an abundance of small lateral and posterior adhesive tubules. In certain Chaetonotida that have adopted a "semipelagic" lifestyle the adhesive tubes are lost.

Feeding, digestive system. Gastrotrichs browse about on the substrate, ingesting bacteria, algae, small protoctists, and organic detritus. Currents induced by head cilia aid in concentrating and bringing food particles to the mouth.

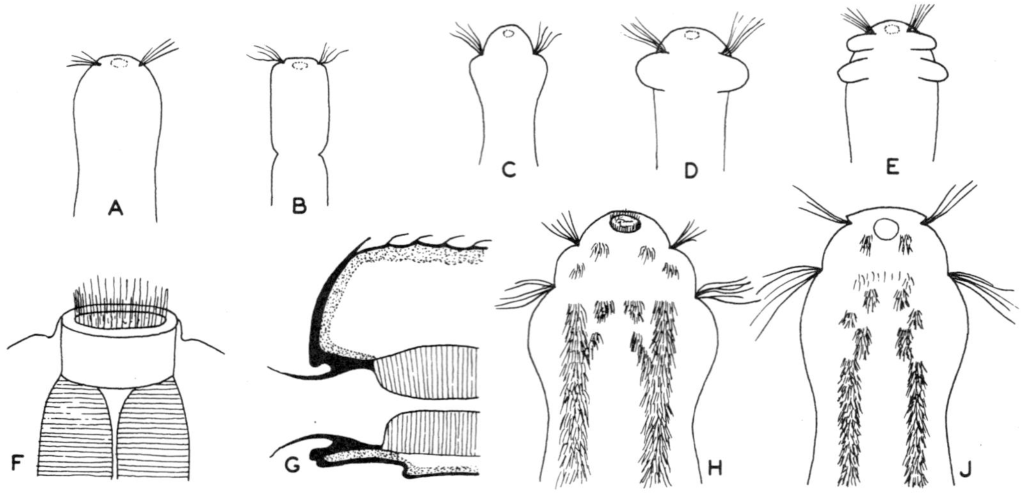

Fig. 6.3. Structure of head region of Gastrotricha. A–E, dorsal view of typical heads, showing general shape and head lobes; F, semidiagrammatic ventral view of mouth region, showing circular collar, oral bristles, and anterior portion of pharynx; G, diagrammatic longitudinal section of anterior end, showing oral bristles, oral cavity, anterior portion of pharynx, and the cephalic shield; H, ventral view of anterior end of a typical *Chaetonotus*, showing arrangement of cilia; J, ventral view of anterior end of a typical *Ichthydium*, showing arrangement of cilia. (A–E modified from Brunson 1950; F–J modified from Remane 1935–1936.)

Often the mouth opening is surrounded by a projecting cuticular collarlike lip bearing the oral bristles. The mouth cavity opens directly into the pharynx, which is remarkably similar to the nematode pharynx. It is a long, cylindrical, muscular organ, usually slightly larger toward the inner end and sometimes with one or two indistinct bulbs. Its cavity is more or less triquetrous and lined with thin cuticle. Four small unicellular salivary glands are closely associated with the pharynx, one pair anteriorly and one posteriorly.

The intestinal wall consists of only four longitudinal rows of cells. There are no caeca, although a short rectum is distinguishable just before the anus. One investigator has reported that reserve food materials are stored in special subepidermal cells, which break loose and float about in the pseudocoel.

Excretion. The Chaetonotidae, including the great majority of freshwater species, all have a single pair of flame bulbs at about midlength. The two protonephridial tubules are considerably coiled in the pseudocoel and open ventrally near the midline. Undoubtedly, the protonephridial system functions in both excretion and osmoregulation. Although there is some disagreement, it is probable that the other families of the Chaetonotida also have a protonephridial system. Some investigators claim that the marine species of Chaetonotidae have a protonephridial system; others maintain that it is absent. In the many marine order Macrodasyida, the system is lacking, but at least one freshwater species contains a protonephridial system. It is possible that the intestine of all gastrotrichs has an accessory excretory function.

Muscle system. The longitudinal muscle strands are quite prominent, especially ventrally. Usually there are about six pairs, with origins and insertions on the body wall. Their contraction produces shortening, curving, and partial rolling up of the body. A few delicate circular fibers in the body wall have recently been demonstrated for a few species, but they are thought to be of little importance in movement. In addition to the radial pharyngeal musculature, there are a few circular fibers on the outside of the pharynx and in the region of the rectum.

Nervous system. A large, dorsal, lobed, saddle-shaped brain covers the anterior portion of the pharynx, and two lateral nerve strands extend almost to the posterior end of the body. Fine branches innervate the body wall and viscera.

Sensory receptors. One or two pairs of pigmented" eyespots" have been seen on the lateral surfaces of the brain in a few species, but their function is questionable. The tufts of cilia on the head are thought to function in touch, current detection, and the detection of dissolved materials. They are inserted in special thick glandular cells of the hypodermis. Pairs of anterior and posterior tactile bristles are common.

Reproduction, life history. Freshwater gastrotrichs are all parthenogenetic females, males being unknown. Recently, however, atypical sperm have been seen in the genital area of several species. The two parts of the ovary are lateral to the intestine. Each portion consists of a few oocytes and eggs in various stages of development. The oviduct is so skimpy that some investigators doubt its existence as a definite duct. The genital pore is ventral. A single, ventral, saclike "X organ" lies just inside the genital pore; possibly it represents a vestigial copulatory bursa.

A gastrotrich usually produces one to five eggs during its lifetime. A mature egg is quite large, and before leaving the parent it distends the body greatly. Nevertheless, it is plastic, and as it emerges from the genital pore it is momentarily constricted into a dumbbell shape. When the egg comes in contact with the water an outer shell hardens. Sometimes the shell bears spines or other protuberances. Eggs are usually attached to some small object on the substrate.

Brunson (1949) distinguishes two definite types of eggs, differing slightly in size. Tachyblastic eggs begin cleavage as soon as deposited and hatch 12 to 70 hours later. Opsiblastic eggs (Fig. 6.4) are slightly larger and are produced in old cultures and when environmental conditions become unfavorable. They have a heavier shell and survive desiccation, freezing, and unusually high temperatures. The duration of dormancy is variable, depending on ecological conditions, but it is known to last as long as 2 years. Opsiblastic eggs are usually the last ones laid by a particular gastrotrich, but there are some exceptions.

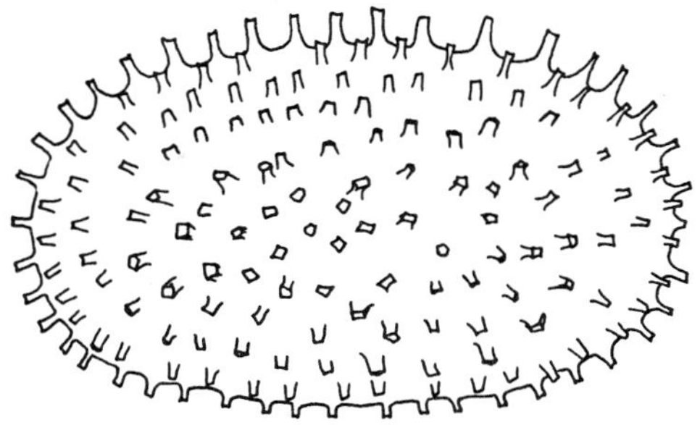

Fig. 6.4. Opsiblastic egg of *Chaetonotus*. (Redrawn from Zelinka 1889.)

Gastrulation begins at the 32-cell stage. Development is direct. On emergence from the egg a gastrotrich is quite large, and during the ensuing several hours it grows by 20 to 30% to attain mature size. As compared with the Macrodasyida, the Chaetonotida are composed of relatively few cells; perhaps the number is constant within each species. Little is known about the normal length of life; individuals in cultures live for 3 to 22 days, depending on the species and ecological conditions. Like the situation in rotifers, the cuticle is never shed during the life history.

Ecology. Sixteen genera of gastrotrichs are restricted to freshwaters: *Anacanthoderma*, *Arenotus*, *Dasydytes*, *Haltidytes*, *Kijanebalola*, *Lepidochaetus*, (Fig. 6.5), *Marinellina*, *Neogossea*, *Ornamentula*, *Polymerurus*, *Proichthydioides*, *Proichthydium*, *Redudasys*, *Setopus*, *Stylochaeta*, and *Undala*. Six genera occur in both marine and freshwaters, or brackish (*Dichaetura*) and freshwaters. The marine and freshwater genera are *Aspidiophorus*, *Chaetonotus*, (Fig. 6.5), *Ichthydium*, *Heterolepidoderma*, *Lepidodermella*. Only a very few species are known to occur in both types of environments, and some investigators believe that these are all actually pairs of varieties. About 35 genera are restricted to marine habitats.

Gastrotrichs are typical of puddles, marshes, wet bogs, and the shallow litoral,

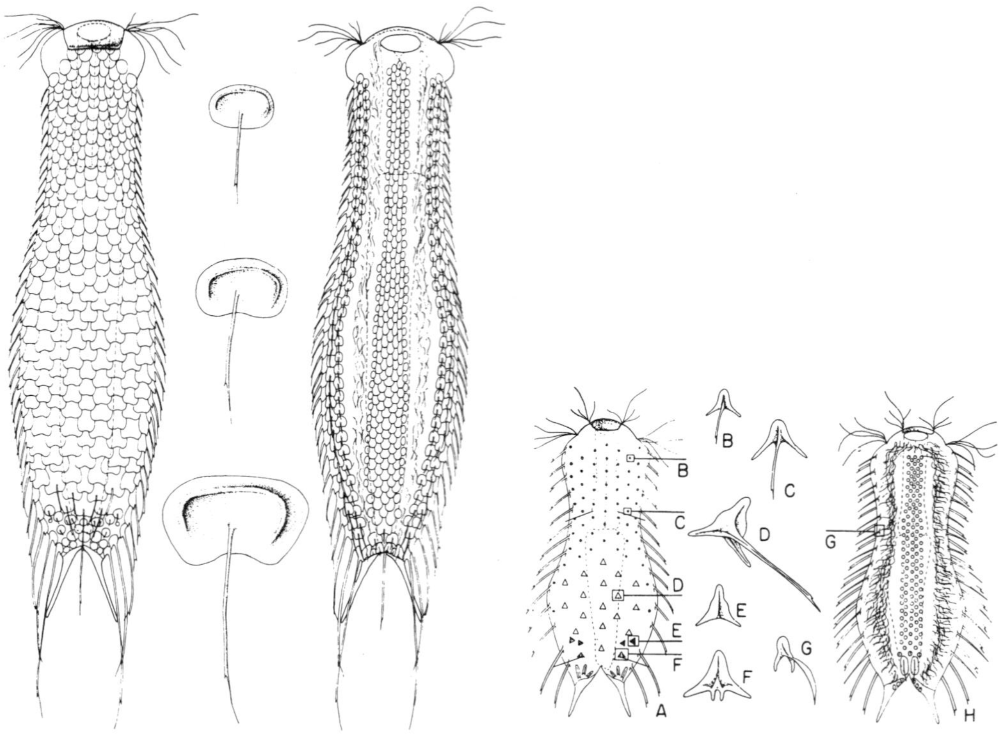

Fig. 6.5. Detailed structure of *Lepidochaetus* (left) and *Chaetonotus* (right). Note ventral ciliation patterns. (From Balsamo 1983, revised by Kisielewski 1991.)

especially where there is detritus and decaying material. Old laboratory protoctist cultures or aquaria with disintegrating aquatic plants commonly have gastrotrich populations. They are uncommon in running waters generally and perhaps absent from swift streams.

Most species remain closely associated with the substrate, although the Neogosseidae and Dasydytidae may swim about above it. Their occurrence in the plankton of the shallows, however, is atypical and fortuitous. Little is known about the depth distribution and concentrations of freshwater species, although *Chaetonotus* has been found in bottom deposits as deep as 36 in. Recently it has been found that the capillary water of sandy beaches a meter or two from the water's edge supports a considerable population of gastrotrichs.

Although maximum numbers are usually found in summer and autumn, gastrotrichs may be collected at any time of the year. Other than opsiblastic eggs, no special desiccation or overwintering mechanisms are known.

Many species may be collected from habitats where there is much decay, and where the dissolved oxygen content consequently may be less than 1 ppm. Minimal oxygen concentrations are not definitely known, however, although it is probable that gastrotrichs may withstand at least temporary anaerobiosis.

Geographical distribution, dispersal. Except for some areas in Europe, South America, the Middle East, and three or four areas in the United States, there has been almost no intensive, systematic collection of gastrotrichs. Most records in North America are the result of incidental or superficial studies, often associated with collections of other groups of freshwater invertebrates. As a result, we know little about the geographical distribution of individual species. Nevertheless, the greatest number of species are known from North America (66) (Kisielewski 1999). Most freshwater genera, however, appear to be cosmopolitan and occur the world over in suitable habitats; these are *Aspidiophorus*, *Chaetonotus*, *Dasydytes*, *Ichthydium*, *Lepidodermella*, *Neogossea*, and *Polymerurus*. Overland dispersal is undoubtedly effected by the resistant opsiblastic eggs.

Enemies. In addition to the toll taken by browsing insects and crustaceans, gastrotrichs are also devoured by amoeboid protozoans, hydras, oligochaetes, turbellarians, and perhaps by tardigrades and nematodes.

Collecting. Sandy beach washings, and rinsings of aquatic vegetation often yield good collections. Specimens may be concentrated by straining the wash water through fine bolting silk. For bottom species the thin layer of surface mud and debris should be carefully scooped up and placed in a suitable container with water. After standing for a time in the laboratory, material at the debris and water interface should be sucked up with a long pipette and examined under the high power of the binoculars.

Culturing. Old protoctistan cultures are sometimes suitable if gastrotrichs are seeded into them, and one investigator recommends using two drops of uncooked egg yolk in 100 mL of spring or pond water. In either case it is best to add a little bottom debris to the culture.

One-tenth percent malted milk in well water is especially useful for pure cultures. After the water is brought to a boil, malted milk powder is added, and the solution is then further boiled for about a minute. Let stand a week before inoculating with gastrotrichs.

Regardless of the culture method used, about half the volume of fluid should be renewed every 2 to 3 weeks.

Preserving, preparing. Two percent osmic acid is an excellent fixative. Place the living specimens in a drop of water on a slide and invert for 5 to 10 seconds over the mouth of the bottle of osmic acid. Saturated mercuric chloride, 10% formalin, and Bouin's fluid are alternative fixatives that often give good results. Specimens may be stored in 70% alcohol or 5% formalin. If glycerin is to be used for mounting, transfer the gastrotrichs to a concave watch glass containing 5% glycerin in 50% alcohol. Cover loosely to keep out dust and permit evaporation. After about a week the water and alcohol will have evaporated and the specimens will be in concentrated glycerin. Any of the several mounting methods for rotifers may be used (page 155). Glycerin is excellent for permanent storage.

Some species require narcotization before fixing, and a wide variety of substances may be tried with varying success, including a concentrated aqueous menthol solution, chloretone, 2% benzamine lactate, 2% butyn, and 2% hydroxylamine hydrochloride.

Scales, plates, and spines may be isolated by the judicious use of dilute acetic acid. Stain with fuchsin. Oil immersion is essential for critical work and species identification. An Irwin loop is ideal for sorting and handling individual gastrotrichs.

Taxonomy. Only two investigators have made important taxonomic contributions to the biology of North American gastrotrichs. Stokes published the results of his New Jersey observations in 1887 and 1888, and Brunson began his series of papers in 1947. Obviously an enormous amount of work remains to be done in this country. Some of the most important European investigations are those of Zelinka (1889), Grunspan (1910), and Remane (1935–1936). Although concerned with faunae outside of North America, the many studies by J. Kisielewski provide information on certain genera and species occurring in the United States. The most reliable and significant characteristics for distinguishing genera are the nature of the furca and the presence or absence and detailed structure and arrangement of cuticular plates, scales, and spines.

The majority of American species are unreported or undescribed. As a consequence, almost every study, however superficial, turns up new species. It would be fruitless to construct a key to the approximately 45 American species known to date.

About 250 species have been described from the freshwaters of the world, and all are currently placed in 22 genera. The key that follows includes only nine genera; one of these has not yet been reported from the United States (*Dichaetura*), but they are widely distributed and will undoubtedly be found here eventually. Three other rare freshwater genera are not included in the key: *Proichthyidium* has been collected only once in the Argentine, and one species each of *Redudasys* and *Arenotus* have been found in Brazilian psammon.

Remane (1935–1936) gives evidence for grouping *Dasydytes*, *Setopus*, *Anacanthodenna*, and so on, into a single genus, *Dasydytes*, but recent study supports the division of *Dasydytes* into several genera. *Lepidoderma* is preoccupied and *Lepidodermella* is used instead.

KEY TO GENERA OF GASTROTRICHA OF THE UNITED STATES*

1. Caudal furca present (Figs. 6.6, 6.7) **2**
True caudal furca absent, but sometimes with rudimentary protuberances, long spines, or setose styloid processes (Fig. 6.8D–J) **8**

2. Caudal furca not forked; cuticle smooth, or with spines, plates, simple scales, or spined scales or plates; head usually with two to four tufts of cilia; cephalic shield present or absent; usually with one pair of posterior bristles (Figs. 6.6B–F; 6.8A, B); two longitudinal bands of cilia; common and widely distributed; includes about 90% of the known freshwater species **CHAETONOTIDAE, 3**
Caudal furca forked; cuticle covered with spined plates; head without tufts of cilia but with many single cilia; cephalic shield absent; four to many posterior bristles (Fig. 6.6A); several rare and poorly known species reported from brackish water in Europe; not known in the U.S. . . . . . . **ICHAETURIDAE, *Dichaetura***

3. Caudal furca very long and jointed; body usually with spines or spines and scales (Fig. 6.6E); several species . . . . . . ***Polymerurus***
Caudal furca not especially long, not jointed **4**

4. Cuticle smooth, without plates, scales, or spines, but sometimes with setae (Fig. 6.6C, D); about 25 species some of which are marine . . . . . . ***Ichthydium***
Cuticle not smooth **5**

5. Body covered with complex stalked plates, each consisting of a circular basal plate, a short stalk, and a rhombic end plate (Fig. 6.6F); about 15 species; rare in the U.S. . . . . . ***Aspidiophorus***
Body with simple plates, scales, or spines **6**

6. Body with spines or spined scales (Fig. 6.7); very common; over 170 species worldwide ***Chaetonotus***
Body without spines or scales but covered with plates **7**

7. With flat, broad plates or scales (Fig. 6.8A, B); about 15 species . . . . . . ***Lepidodermella***
With small, keeled plates (Fig. 6.8C); several uncommon species; rare in the U.S. ***Heterolepidoderma***

8. Head with two club-shaped tentacles; body with long spines; two long spines or two styloid processes at posterior end of body; several uncommon species (Fig. 6.8D); rare in the U.S. **NEOGOSSEIDAE, 9**
Head without tentacles; body with long spines, arranged singly or grouped; ventral ciliation in two longitudinal bands or two longitudinal rows of tufts; head ciliation usually with three ventral transverse bands of cilia (Fig. 6.8E–J); about 25 species in several genera, all but one either absent or very rare in the United States, two genera keyed out below **DASYDYTIDAE, 10**

9. Posterior truncate with postero-lateral projections with long spines (Fig. 6.8D); 2 species, LA ***Neogossea***
Posterior rounded with median spines and setae, longest at the midline (Fig. 6.7G); rare ***Kijanebalola***

10. Posterior with papilla-like bristled protrusions (Fig. 6.8F, G); rare, LA . . . . . . ***Stylochaeta***
Posterior without such structures; common and widespread . . . . . . ***Dasydytes***

*Revised by Jacek Kisielewski.

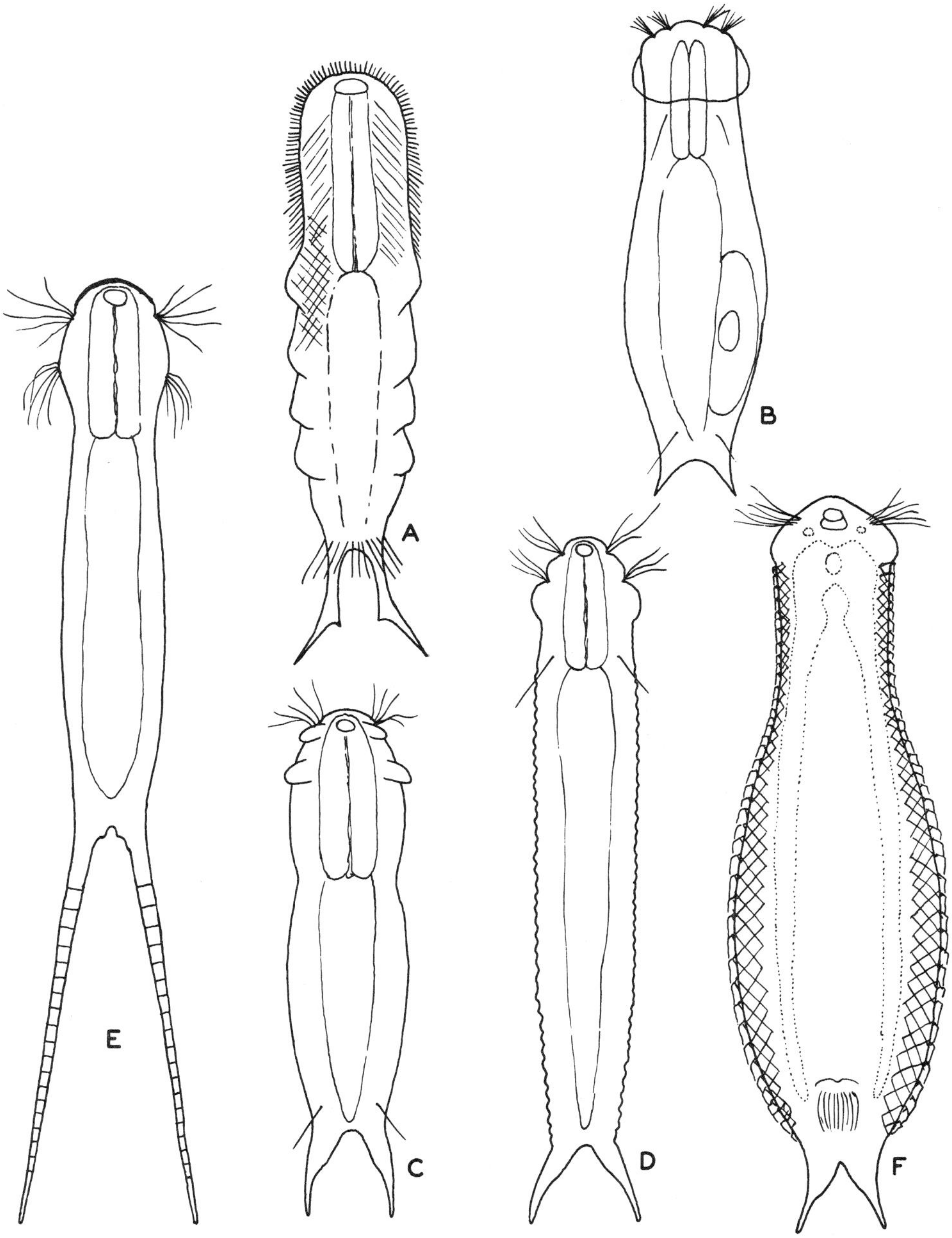

Fig. 6.6. Structure of Gastrotricha. A, *Dichaetura;* B, *Ichthydium*, showing cephalic shield; C and D, *Ichthydium;* E, *Polymerurus;* F, ventral view of *Aspidiophorus*, showing arrangement of cilia. (A modified from Metschnikoff 1865; B modified from Konsuloff 1913; C modified from Brunson 1949; D and E modified from Brunson 1950; F modified from Saito 1937.)

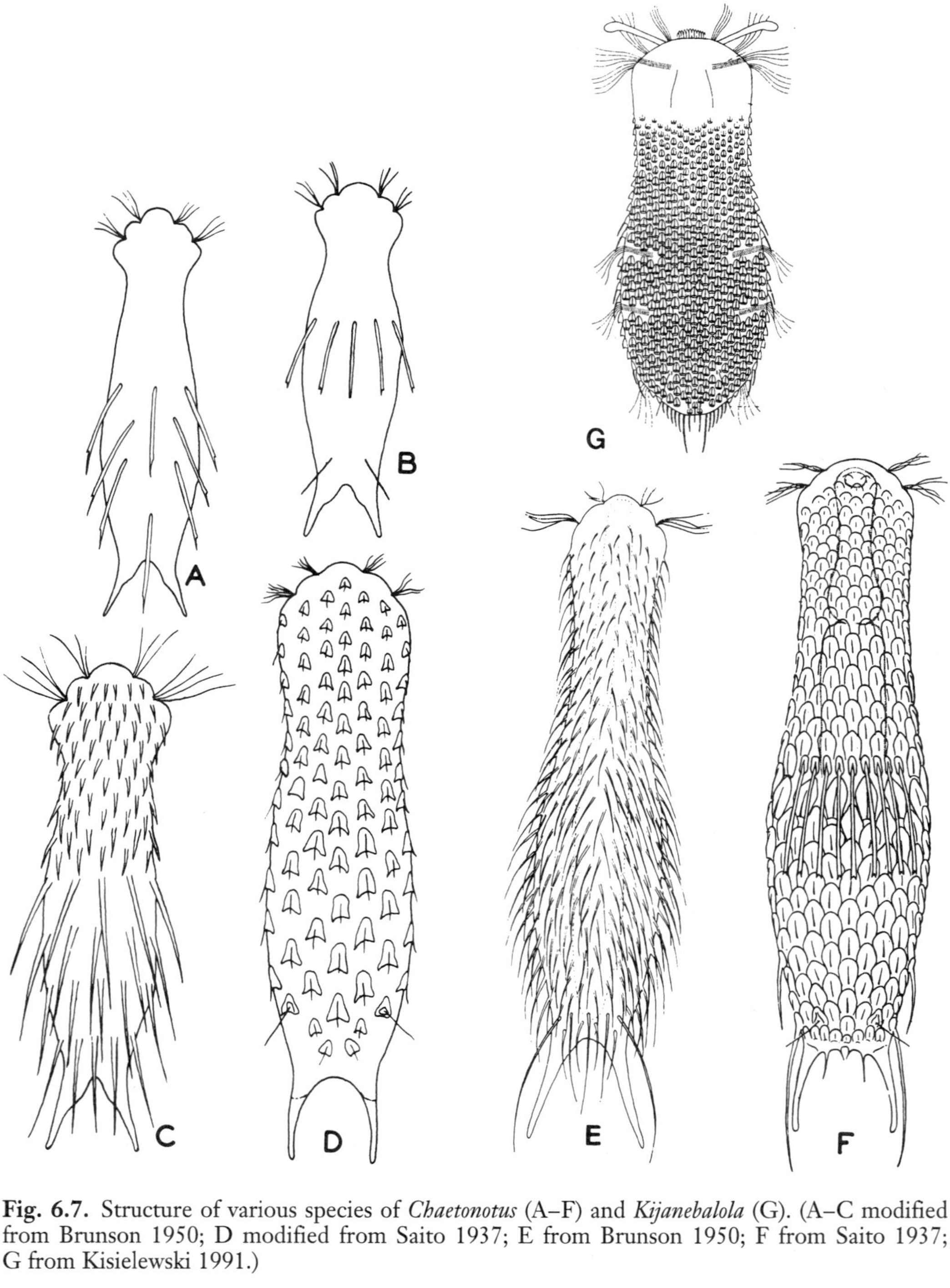

Fig. 6.7. Structure of various species of *Chaetonotus* (A–F) and *Kijanebalola* (G). (A–C modified from Brunson 1950; D modified from Saito 1937; E from Brunson 1950; F from Saito 1937; G from Kisielewski 1991.)

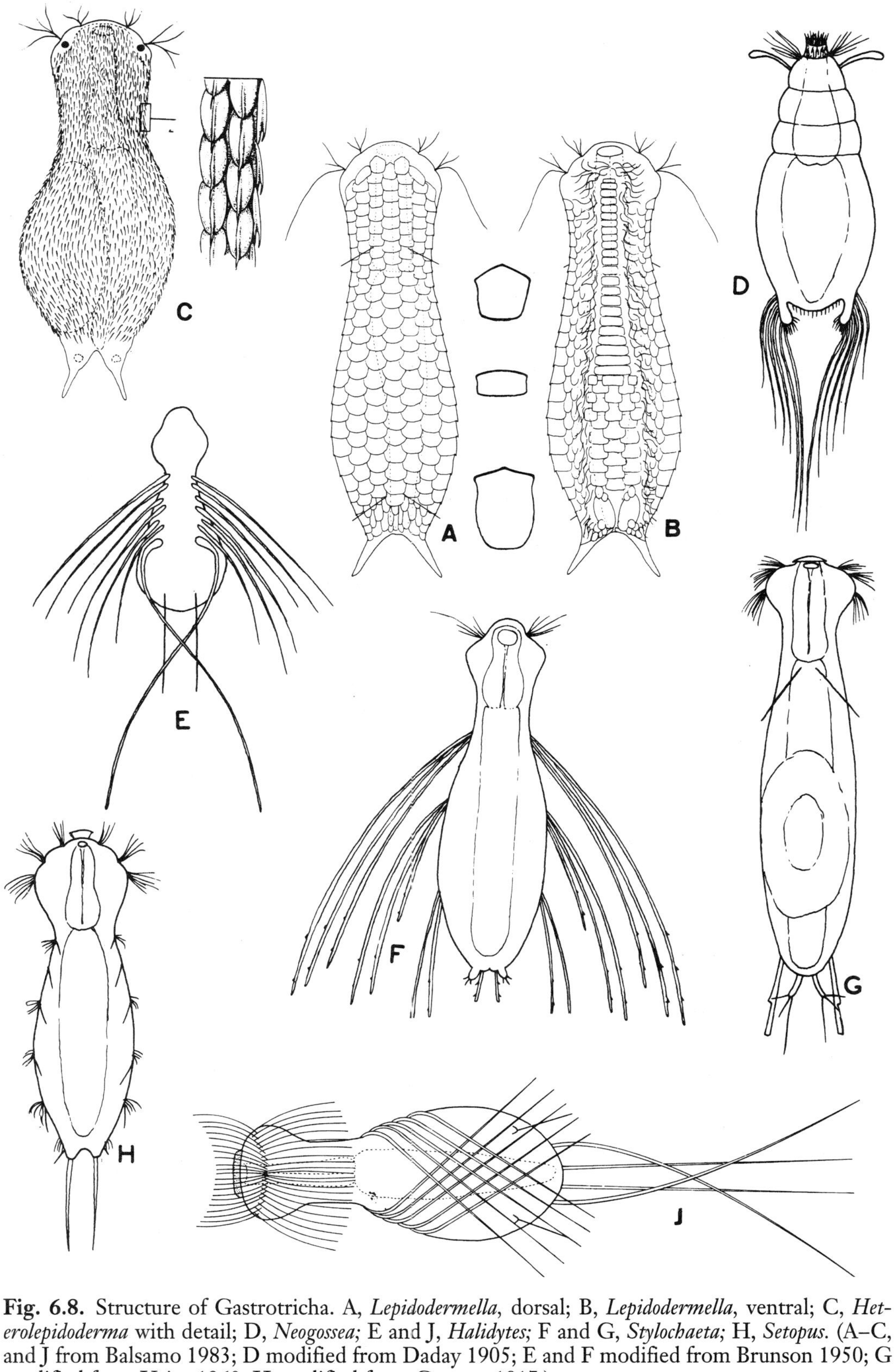

Fig. 6.8. Structure of Gastrotricha. A, *Lepidodermella*, dorsal; B, *Lepidodermella*, ventral; C, *Heterolepidoderma* with detail; D, *Neogossea;* E and J, *Halidytes;* F and G, *Stylochaeta;* H, *Setopus.* (A–C, and J from Balsamo 1983; D modified from Daday 1905; E and F modified from Brunson 1950; G modified from Voigt 1960; H modified from Greuter 1917.)

REFERENCES

Balsamo, M. 1977. Prime recerche sui Gastrotrichi dulciacquicoli Italiani. *Att. Soc. Toscana Sci. Nat.* 84:87–150.

——. 1983. Gastrotrichi (Gastrotricha). *Guide Riconosc. Spec. Anim. Acque Int. Ital.* 20:1–92.

Bennett, L. W. 1979. Experimental analysis of the trophic ecology of *Lepidodermella squammata* (Gastrotricha: Chaetonotoidea). *Trans. Am. Microsc. Soc.* 98:254–260.

Brunson, R. B. 1947. Gastrotricha of North America, II: four new species of *Ichthyidium* from Michigan. *Pap. Mich. Acad. Sci. Arts Lett.* 33:59–62.

——. 1948. *Chaetonotus tachyneusticus:* a new species of gastrotrich from Michigan. *Trans. Am. Microsc. Soc.* 67:350–351.

——. 1949. The life history and ecology of two North American gastrotrichs. *Trans. Am. Microsc. Soc.* 68:1–20.

——. 1950. An introduction to the taxonomy of the Gastrotricha with a study of eighteen species from Michigan. *Trans. Am. Microsc. Soc.* 69:325–352.

Daday, E. 1905. Untersuchungen uber die Susswasser-Mikrofauna Paraguays. *Zoologica* 18:72–86.

Davidson, D. B. 1938. A new species of gastrotrichan: *Chaetonotus ribustus,* new species. *Am. Mus. Novit.* 972:1–6.

De Beauchamp, P. M. 1934. Sur la morphologie et l'ethologie des *Neogossea* (gastrotriches). *Bull. Soc. Zool. Fr.* 58:331–342.

Faucon, A. S., and W. D. Hummon. 1976. Effects of mine acid on the longevity and reproductive rate of the Gastrotricha *Lepidodermella squammata* (Dujardin). *Hydrobiologia* 50:265–269.

Greuter, R. 1917. Beitradge zur Systematik der Gastrotricha in der Schweiz. *Rev. Suisse Zool.* 25:35–70.

Grunspan, T. 1910. Die Süsswassergastrotrichen Europas: Einezusarnmenfassende Darstellungihrer Anatomie, Biologie und Systematik. *Ann. Biol. Lacustre* 4:211–365.

Kisielewski, J. 1981. Hermaphroditism of freshwater gastrotrichs in natural conditions. *Bull. Acad. Pol. Sci., Ser. Biol.* 29:167–172.

——. 1982. Gastrotrichia of two complexes of peat hags near Siedlce. *Fragm. Faun., Warszawa* 27(4):39–57.

——. 1986. Freshwater Gastrotricha of Poland. *Frag. Faun.* 30 (9–16):139–295.

——. 1987. Two new interesting genera of Gastrotricha (Macrodasyida and Chaetonotida) from the Brazilian freshwater psammon. *Hydrobiologia* 153:23–30.

——. 1991. Inland-water Gastrotrichia from Brazil. *Ann. Zool., Warszawa* 43, Suppl. 2. 168 pp.

——. 1999. A preliminary study of the inland-water Gastrotrichia of Israel. *Israel J. Zool.* 45:135–157.

Konsuloff, S. 1913. Notizen uber die Gastrotrichen Bulgariens. *Zool. Anz.* 43:255–260.

Krivanek, R. C., and J. O. Krivanek. 1959a. A new and a redescribed species of *Neogossea* (Gastrotrichia) from Louisiana. *Trans. Am. Microsc. Soc.* 77:423–428.

——. 1959b. A new species of *Stylochaeta* (Gastrotrichia) from Louisiana. *Trans. Am. Microsc. Soc.* 78:77–81.

Metschnikoff, E. 1865. Uber einige wenig bekannte niedere Tierformen. *Zeitsch. Wiss. Zool.* 4:450–465.

Mola, P. 1932. Gastrotricha delle acque dolci italiane. *Int. Rev.* 26:397–423.

Packard, C. E. 1936. Observations on the Gastrotricha indigenous to New Hampshire. *Trans. Am. Microsc Soc.* 55:422–427.

Remane, A. 1927. Beitradge zur Systematik der Susswassergastrotrichen. *Zool. Jahrb. Abt. Syst.* 53:269–320.

——. 1935–1936. Gastrotricha. *Klassen u. Ordnungen des Tierreich* 4 (Abt. 2, Buch 1, Teil 2, Lfg. 1–2):1–242.

Ruttner-Kolisko A. 1955. *Rheomorpha neiswestnovae* und *Marinellina flagellata,* zwei phylogenetisch interssante Wurmtypen aus dem Süswasserpsammon. *Öster. Zool. Z.* 6:55–69.

Saito, I. 1937. Neue und bekannte Gastrotrichen der Umgebung von Hiroshima (Japan). *J. Sci. Hiroshima Univ., Ser. B, Div. 1,* 5:245–265.

Stokes, A. C. 1887. Observations on *Chaetonotus. The Microscope* 7:1–9, 33–43.

——. 1888. Observations on a new *Dasydytes* and a new *Chaetonotus. The Microscope* 8:261–265.

Voigt, M. 1960. Gastrotricha. *Die Tierwelt Mitteleur.* 1:4a:1–74.

Weiss, M. J., and D. P. Levy. 1979. Sperm in "parthenogenetic" freshwater gastrotrichs. *Science* 205:302–303.

Zelinka, C. 1889. Die Gastrotrichen: Fine monographische Darstellung ihrer Anatomie, Biologie und Systematik. *Z. Wiss. Zool.* 49: 209–384.

7

ROTIFERA

(Wheel animals)

If we were to designate a single major taxonomic category that is most characteristic of freshwaters, it could only be the Phylum Rotifera. The rotifers are one of the few groups that have unquestionably originated in freshwaters, and it is here that they have attained their greatest abundance and diversity. For this reason we have chosen to give more detailed consideration to rotifers than to any other freshwater taxon. Probably 2500 species have been described, but less than 5% of these are found in marine and brackish waters. Only two plankton species occur in the mid-Atlantic.

Except for occasional new species descriptions, the basic taxonomy of the Rotifera is well established. Recent emphasis has centered on experimental work, especially biochemical phenomena, clone genetics, species competition, selective feeding, and sensory physiology.

Rotifers were first studied and described by Leeuwenhoek in 1703, and since his time they have become classical objects for study by the amateur microscopist and professional hydrobiologist alike. They are essentially all microscopic, the length range for the phylum being about 40 μm to 2.5 mm, but the great majority are between 100 and 500 μm long. These fascinating creatures have long been called Rotifera, or rotifers, because in some species the disklike ciliated anterior end, or corona, has a fancied and illusory resemblance to a pair of revolving wheels owing to the synchronized beating of the coronal cilia. Rotifers occur in an endless variety of aquatic and semiaquatic habitats, including the limnetic and deepest regions of the largest lakes and the smallest puddles. They are found in damp soil and vegetable debris, in mosses that may be wetted or dampened only occasionally, in the interstices between the sand grains of lake beaches, from the Arctic and Antarctic to the tropics, in small rock depressions, and even in cemetery flower urns and eave troughs. About 20 species of bdelloids have been found in Antarctica, but the total bdelloid fauna there is probably much larger.

The vast majority of rotifers encountered under natural conditions are females. Males are definitely known for relatively few species; they are much smaller than the females, degenerate, and seldom live for more than 2 or 3 days. The biology of males is discussed, but most of the remarks in this chapter concern only female rotifers.

General characteristics. The Phylum Rotifera (Figs. 7.1–7.3) is now divided into three classes, the Seisonidea, the Bdelloidea, and the Monogononta. The Seisonidea includes only a single genus, *Seison*, occurring as commensals on marine Crustacea; the ovaries have no vitellaria, males are well developed, and the corona is rudimentary. Almost all Bdelloidea (Fig. 7.3) occur in freshwaters; the ovaries have vitellaria, males are

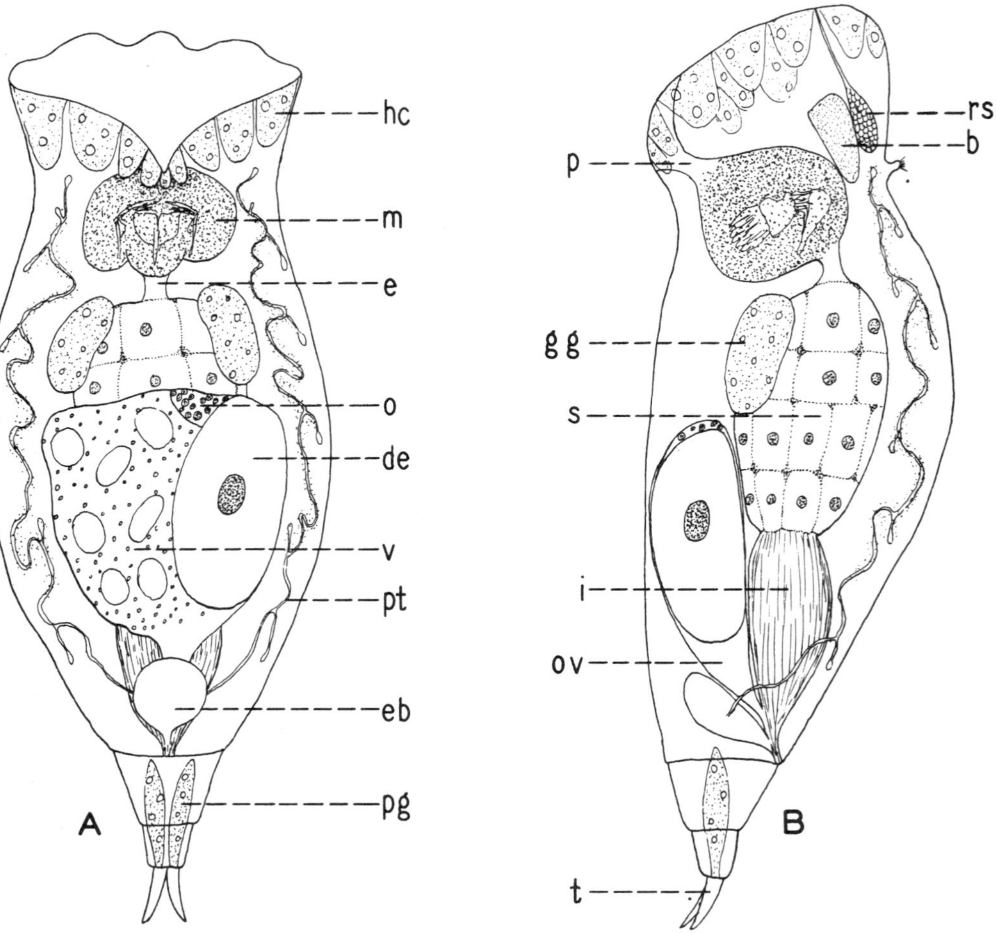

Fig. 7.1. Diagrammatic structure of *Epiphanes senta* (Müller), ×400. A, ventral view; B, lateral view. (Musculature and coronal ciliation not shown; see Fig. 7.2.) b, brain; de, developing egg; e, esophagus; eb, excretory bladder; gg, gastric gland; hc, hypodermal cell of head; i, intestine; in, mastax; o, ovary; ov, oviduct; p, pharynx; pg, pedal gland; pt, protonephridial tubule; rs, retrocerebral sac; s, stomach; t, toe; v, vitellarium.

unknown, reproduction is exclusively by parthenogenesis, and the corona is well developed. The Bdelloidea are appropriately named since the body is highly contractile and they creep about on the substrate in a leechlike manner.

The members of the Class Monogononta constitute about 90% of the known species of rotifers and are characterized by a single ovary, by a mastax that is not ramate, and by the presence or absence of a lorica or a secreted tube. Males are known for some species and not for others; they are small and degenerate. This class is divided into three orders: Ploima, Flosculariacea, and Collothecacea.

The Order Ploima (Figs. 7.1, 7.2) includes most of the free-swimming, limnetic plankton, and littoral species. These rotifers typically have a posterior "foot" and two "toes." The corona is not especially large. The Order Flosculariacea includes some species that are free swimming and many more that are sessile as adults. The foot, when present, has no toes. Often there is a secreted gelatinous envelope or case. The

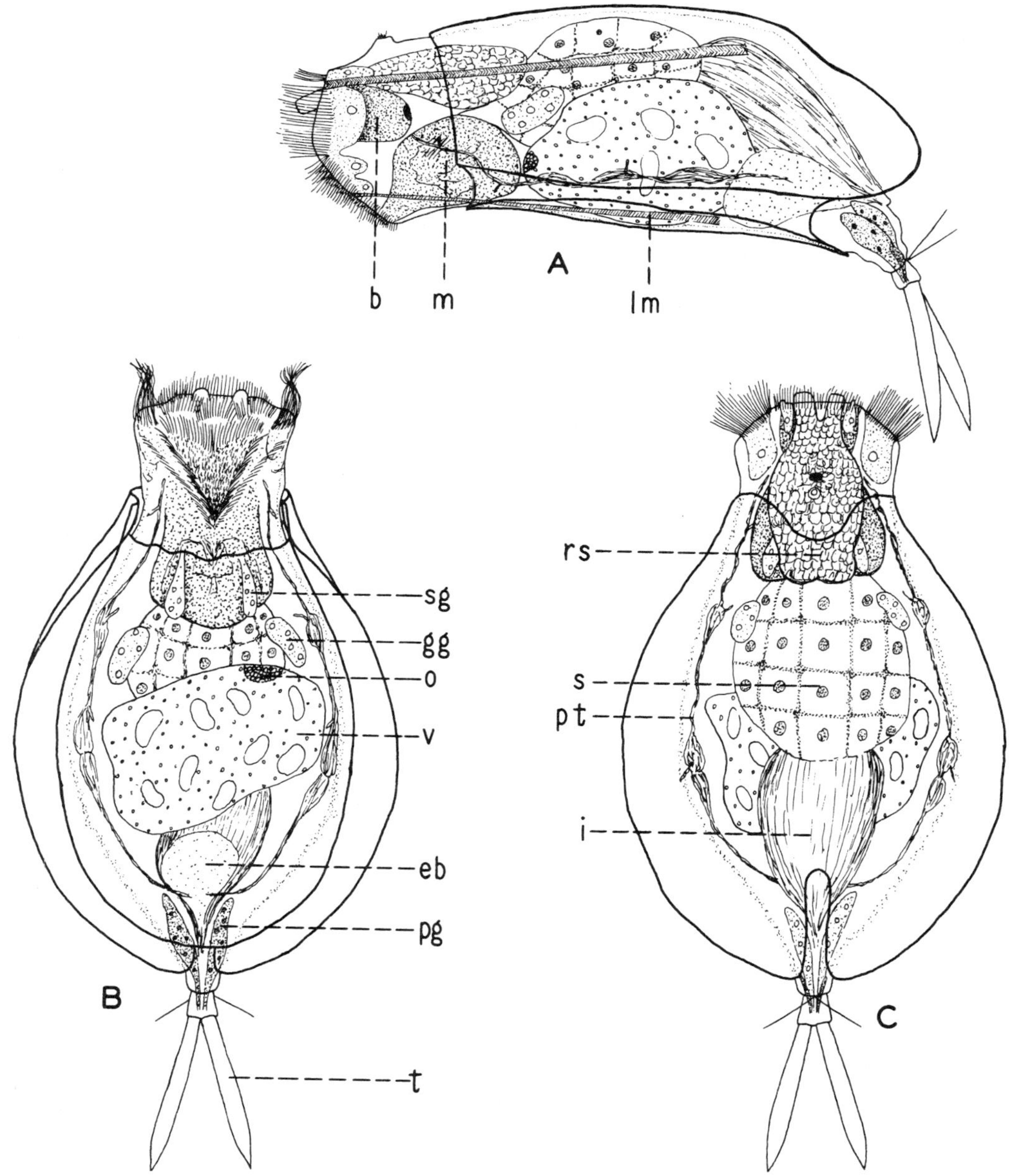

Fig. 7.2. Structure of *Euchlanis dilatata* Ehr., ×320, a typical loricate rotifer, semidiagrammatic. A, lateral; B, ventral; C, dorsal. b, brain; eb, excretory bladder; gg, gastric gland; i, intestine; lm, longitudinal muscle; m, mastax; o, ovary; pg, pedal gland; pt, protonephridial tubule; rs, retrocerebral sac; s, stomach; sg, salivary gland; t, toe; v, vitellarium. (Modified from Myers 1930.)

corona is not especially large. The Order Collothecacea includes those rotifers with a very large, lobed corona and a mouth that is centrally located at the lower end of a funnellike infundibulum; they are usually solitary and sessile.

As a group, the rotifers display an amazing range of morphological variations and adaptations. Yet the great majority have several fundamental features in common. The body is usually elongated and cylindrical, though the complete range of shape extends

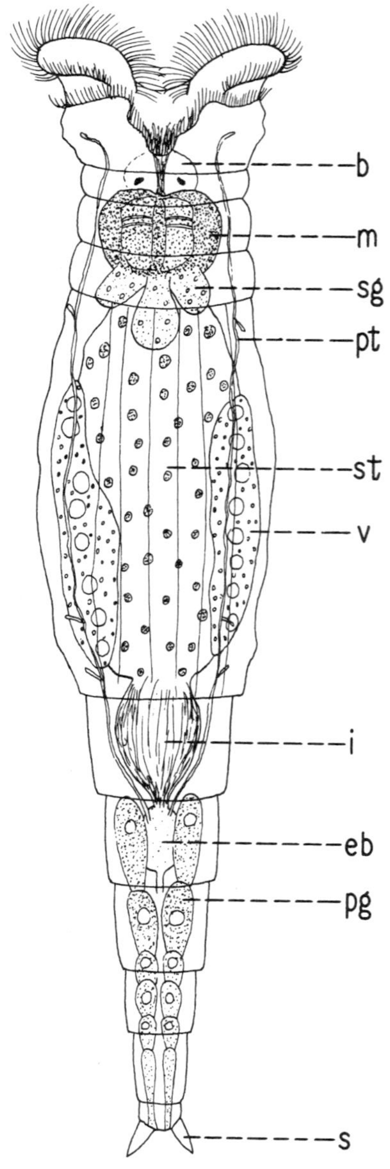

Fig. 7.3. Ventral view of *Philodina*, a typical bdelloid rotifer, ×375, semidiagrammatic. b, brain; eb, excretory bladder; i, intestine; in, mastax; pg, pedal gland; pt, protonephridial tubule; s, spur (toes are completely retracted); sg, salivary gland; st, stomach; v, vitellarium. (Modified from Hickernell 1917.)

from linear attenuation to spherical. Usually three main body regions may be distinguished: head, trunk, and foot. The head is more or less distinctly set off from the large trunk region, although there is rarely an elongated neck region. In the bdelloid rotifers the body appears to be composed of segments, but these are only superficial annuli indicating the zones of folding and telescoping of the cuticle when the animal contracts; typically there are 15 to 18 such annuli. There is no true segmentation in rotifers. The posterior tapering foot usually consists of two or more superficial segmentlike parts, or it may be highly retractile and annulated or creased; sometimes it is absent. When present, the foot usually has two (sometimes none, one, three, or four) terminal "toes."

The general body surface of a rotifer is covered with a cuticle that is secreted by a syncytial hypodermis. The hypodermis forms a thin layer under most of the cuticle, but at the anterior end it is thick, cushionlike, and inwardly lobed. In some rotifers the cuticle is very thin and flexible, but in many genera a portion of the cuticular surface is thickened, more or less rigid, and is termed a lorica. The lorica may be poorly developed and consists of several thin, elastic, platelike sections of the trunk cuticle, or it may be thick, rigid, boxlike, sculptured, and inelastic and may involve the entire surface of the trunk, much of the foot, and even a part of the head. Between these two extremes, every gradation in lorica development is to be found, sometimes even within a single genus (e.g., *Cephalodella*).

The periphery and more or less of the corona surface are ciliated, but the density and pattern of ciliation are highly variable. In general, these cilia serve the double function of locomotion and feeding.

The mouth may be anterior and in the center of the corona, peripheral, subterminal, or ventral. The anus, or cloacal aperture, is typically located dorsally at the base of the foot.

The mastax is a structure peculiar to the digestive system of rotifers, and no comparable device is known elsewhere in the animal kingdom, with the possible exception of the marine Gnathostomulida. Superficially, it has the appearance of a bulbous swelling lying between the pharynx proper and the esophagus. It consists of a complicated arrangement of muscles that activate a set of translucent sclerotized jaws (collectively called the trophi), which are used to seize, tear, grind, or macerate the food (Figs. 7.8–7.10).

True coloration of the tissues is usually grayish, yellowish, or sometimes a pink or bluish cast. The apparent coloration, however, is more often tan, green, or brownish, owing to the varied contents of the digestive tract and stored excretory granules.

Corona and its ciliation. The most primitive type of corona is probably that of the creeping and browsing notommatid Ploima. It consists merely of an anterior, ventral, or oblique surface more or less densely covered with cilia (Figs. 7.4, 7.6); the buccal area, which surrounds the mouth, is always densely ciliated, but this buccal field maybe extended so that it covers some or all of the corona (Fig. 7.4A). From this simple type, all the more complicated and intergrading arrangements are presumed to be derived. In addition to the dense, short cilia of the coronal surface, for example, there may be a peripheral wreath of large, strong cilia (Fig. 7.4B–E). On the other hand, with the exception of the buccal and peripheral areas, some genera, such as *Eosphora* (Fig. 7.39K), have little or no ciliation on the coronal surface. *Epiphanes* and many other genera have a small buccal field, a strong peripheral wreath, and several transverse rows of long cilia and clumps of cilia located on protuberances on the corona. *Synchaeta* has several long, sensory coronal bristles in addition to ordinary cilia (Fig. 7.4F). In many Notommatidae, *Synchaeta*, and a few other genera the corona bears a pair of lateral ciliated earlike processes, or auricles (Fig. 7.39D). In the Bdelloidea and Flosculariacea the surface of the corona is usually bare, but it has two peripheral concentric wreaths of long, strong cilia with a densely ciliated furrow between (Figs. 7.42–7.46). In *Stephanoceros* and *Collotheca* the corona is a greatly concave infundibulum and its upper edge is folded and developed into prominent lobes or arms. These protuberances bear long, stiff setae. The

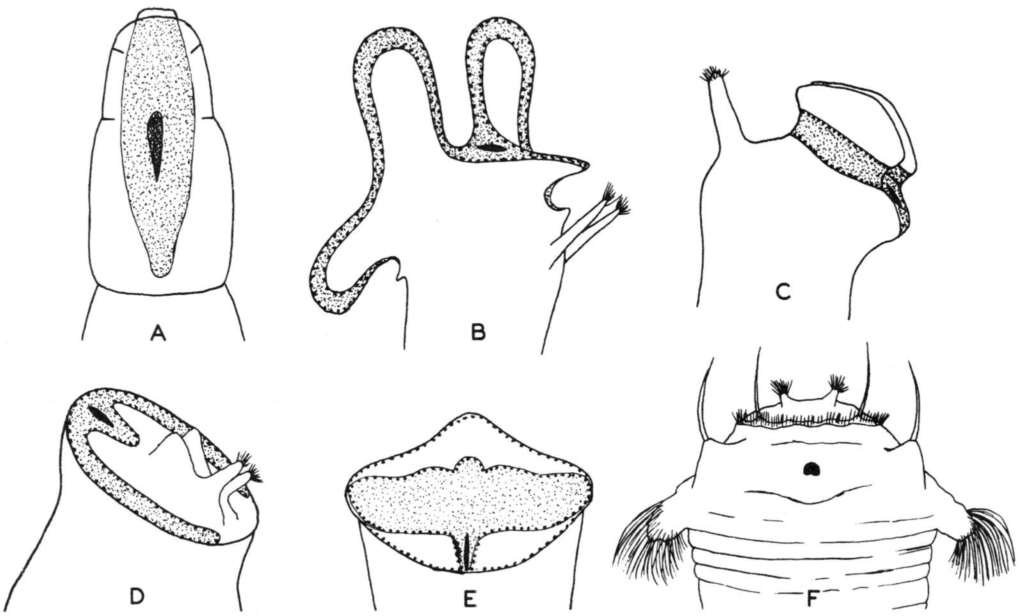

Fig. 7.4. Examples of coronal ciliation. Membranelles and large cilia shown by large dots, smaller cilia by small dots. A, *Dicranophorus*, ventral; B, *Floscularia*, lateral, showing the ventral pellet-making concavity; C, *Philodina*, lateral; D, *Conochilus*, lateral; E, *Cyrtonia*, ventral; F, *Synchaeta*, dorsal.

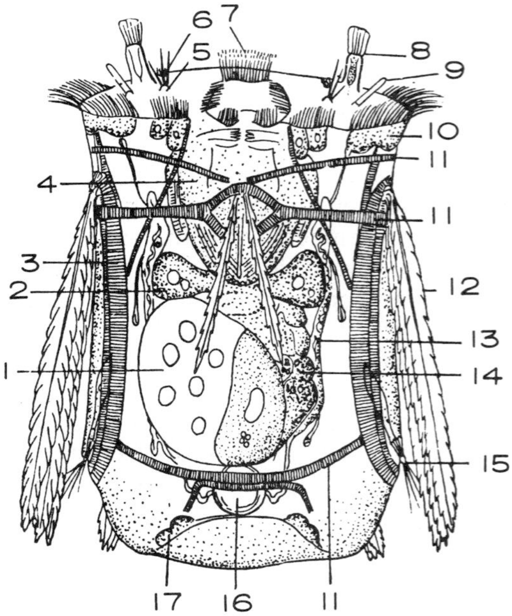

Fig. 7.5. *Polyarthra vulgaris*, showing lateral locomotor appendages. (Modified from various sources.)

mouth, at the bottom of the infundibulum, is surrounded by bands of cilia. In *Acyclus*, *Cupelopagis*, and *Atrochus* there are no coronal cilia. In addition to these several types of corona, there are other arrangements and intergradations involving sparse to dense ciliation.

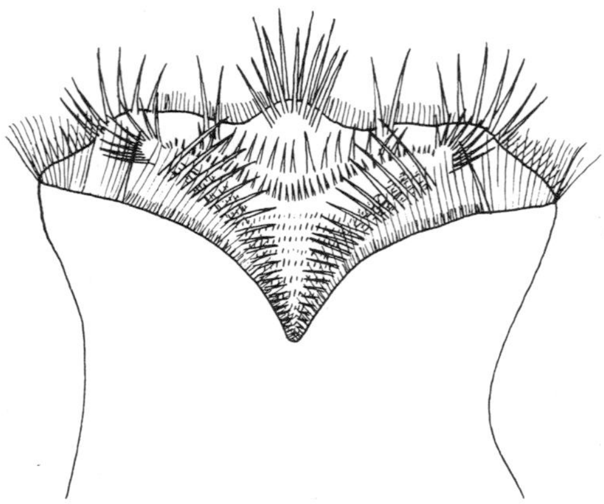

Fig. 7.6. Coronal ciliation of *Epiphanes senta*. (Modified from de Beauchamp 1909.)

Locomotion. Movements through the water are mostly dependent on the beating of the peripheral cilia, and many plankton and limnetic species remain in permanent suspension throughout life without ever coming in contact with a substrate. Such locomotion is often a combination of twisting on the longitudinal axis and spiral movements of the whole animal. A few plankton genera, such as *Filinia*, *Hexarthra*, and *Polyarthra*, also move by sudden jerks and leaps, owing to sudden beating movements of their long appendages (Fig. 7.5). Most of the free swimming, nonplankton species having a foot and toes are able to creep about and browse as a result of the beating of the coronal cilia against the substrate combined with a pushing action of the toes. Toes are useful in steering during both creeping and swimming.

In the Bdelloidea (Fig. 7.7), however, the corona is retracted during creeping and in such a manner as to produce an anterior terminal suckerlike proboscis that functions in the leechlike movements. The toes of bdelloids are usually everted only during active creeping (Figs. 7.44C, F; 7.45K; 7.46C).

Most rotifers with a foot and toes have pedal glands. The Ploima usually have two pedal glands in the base of the foot or in the posterior part of the trunk, whereas the Bdelloidea have from 4 to 20 such glands. Pedal glands open via ducts at the tips of the toes, at the base of the toes, or, in some Bdelloidea, also into the spurs. The secretion is used for anchoring the animal to the substrate temporarily by the tips of the toes or spurs.

Most of the Collothecacea and Flosculariacea are sessile as adults, but their newly hatched young swim about actively for a short time before attaching to the substrate.

Tube construction. A wide variety of protective tubes are constructed or secreted by many genera of sessile rotifers, and even by a few free-swimming species. In Ptygura the tube is usually gelatinous, sometimes flocculose, colorless to brown, and it appears to be secreted by the general body surface; in many species it is more or less covered with bits of adherent debris, and in one species the tube is supplemented with fecal pel-

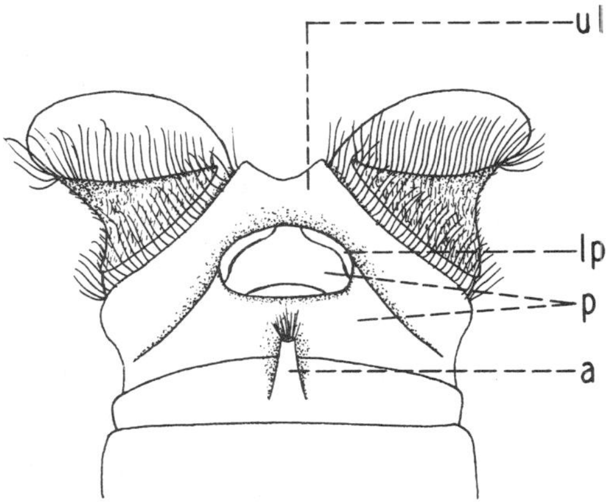

Fig. 7.7. Dorsal view of anterior end of *Macrotrachela*, a typical bdelloid rotifer. a, antenna; lp, lamella of proboscis; p, proboscis; ul, upper lip.

lets. *Stephanoceros* secretes a thin, hyaline tube, and most species of *Limnias* form a thin, firm, and sometimes annulated tube.

Several species of *Floscularia* construct their tubes of small pellets of debris, and it is a fascinating process to watch. The newly settled young *Floscularia* secretes a rough gelatinous tube that is to form the base of the main tube. On the ventral side of the body, just posterior to the mouth, there is a conical, ciliated depression into which a large gland secretes gelatinous material. As an accessory part of the coronal cilia action, small bits of detritus are collected and concentrated in the ciliated cup. This detritus is thoroughly mixed with the gelatinous secretion, and by means of ciliary action in the cup the mass is rotated on its axis and is formed into a hard symmetrical pellet having the shape of a pistol bullet. By appropriate movements of the head and trunk, the pellet is pushed into place at the top of the tube with the tip pointed outward. The upper edge of the tube is always level with the pellet-making cup, and as the animal elongates, it continuously manufactures pellets so that growth closely corresponds with the length of the tube (Edmondson 1945).

According to the specific constituents of the pellets, the tubes are usually yellow, green, or brown, although in one species they are almost colorless and composed largely of the gelatinous secretion. Another species of *Floscularia* constructs its tube out of fecal pellets.

Mastax and trophi. The anatomy and action of the mastax and its contained trophi are so variously modified for special feeding habits that it is best to consider these structures before discussing feeding and the digestive system. Food is brought into the mouth by ciliary action, down the short pharynx, and into the mastax. Only the anterior, dorsal, or anterodorsal part of the mastax contains a cavity through which the food passes. The greater, basal part of this organ consists of minute muscles and the sclerotized trophi, whose movements the muscles bring about. In traversing the mastax, however, the food must pass through the moving, distal, toothlike portions of the trophi.

The trophi consist of one median piece and three paired lateral pieces, all of which are subject to morphological variation and specialization. The fulcrum is basal and median and serves as an attachment for the two rami; these three pieces collectively form the incus. The two unci are usually toothed to a varying degree, and each of these pieces is attached laterally to a manubrium. An uncus and its associated manubrium collectively form a malleus. Thus, the trophi consist of one incus (composed of one fulcrum and two rami) plus two mallei (each composed of one manubrium and one uncus) (Fig. 7.8A, B). Eight basic, but variable, types of mastax are recognized according to the relative development and specialization of their parts. These are characterized briefly below, and their occurrence in the various families of rotifers is given in the taxonomic outline on pages 158–159.

Malleate (Fig. 7.8B). Specialized for horizontal grinding of plankton and particulate detritus. Feebly prehensile.

Virgate (Fig. 7.10). The fulcrum is long and expanded at the base for the attachment of the hypopharynx, a powerful domelike mastax muscle; the upper (dorsal) end of the hypopharynx is free and covered with cuticle, and when it contracts it acts as a piston and sucks food into the mouth and mastax

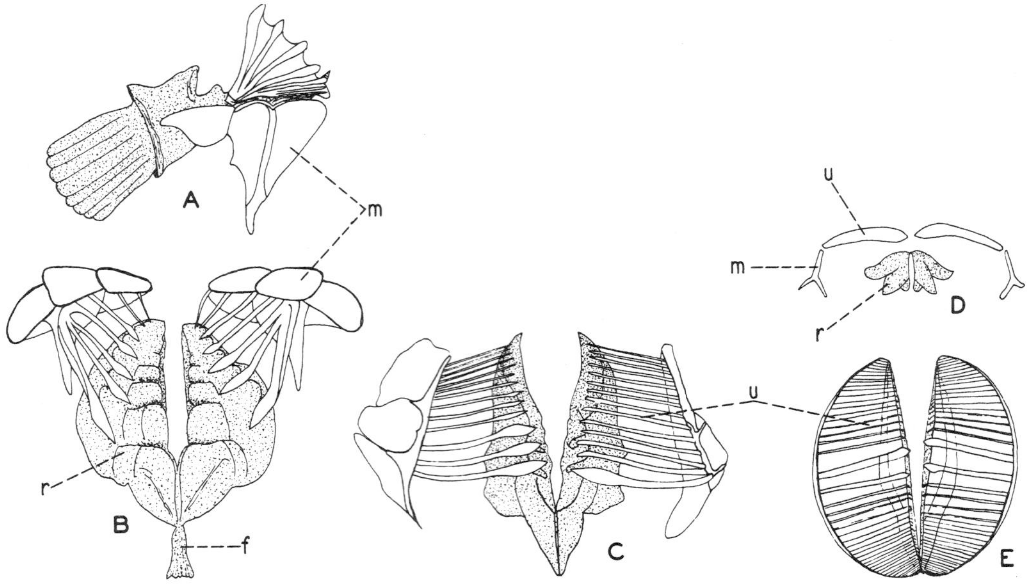

Fig. 7.8. Typical trophi. A, left side view of malleate trophi of *Epiphanes senta;* B, anterior view of trophi of *E. senta;* C, anterior view of malleoramate trophi of *Floscularia;* D, optical section of ramate trophi of a bdelloid rotifer; E, anterior view of ramate trophi of a bdelloid rotifer. Inci are stippled. f, fulcrum; m, manubrium; r, ramus; u, uncus. (Modified from de Beauchamp 1909.)

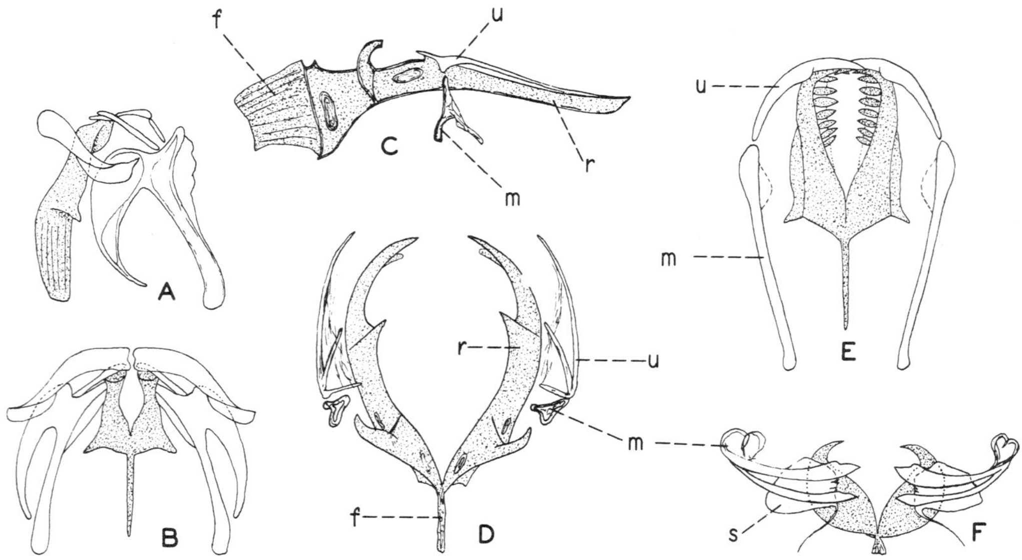

Fig. 7.9. Typical trophi. A, lateral view of cardate trophi of *Lindia;* B, ventral view of trophi of *Lindia;* C, lateral view of incudate trophi of *Asplanchna;* D, anterior view of trophi of *Asplanchna;* E, anterior view of forcipate trophi of *Dicranophorus;* F, anterior view of uncinate trophi of *Stephanoceros.* Inci are stippled. f, fulcrum; m, manubrium; r, ramus; s, subuncus; u, uncus. (A and B modified from Harring and Myers 1922; C–F modified from de Beauchamp 1909; E modified from Harring and Myers 1928.)

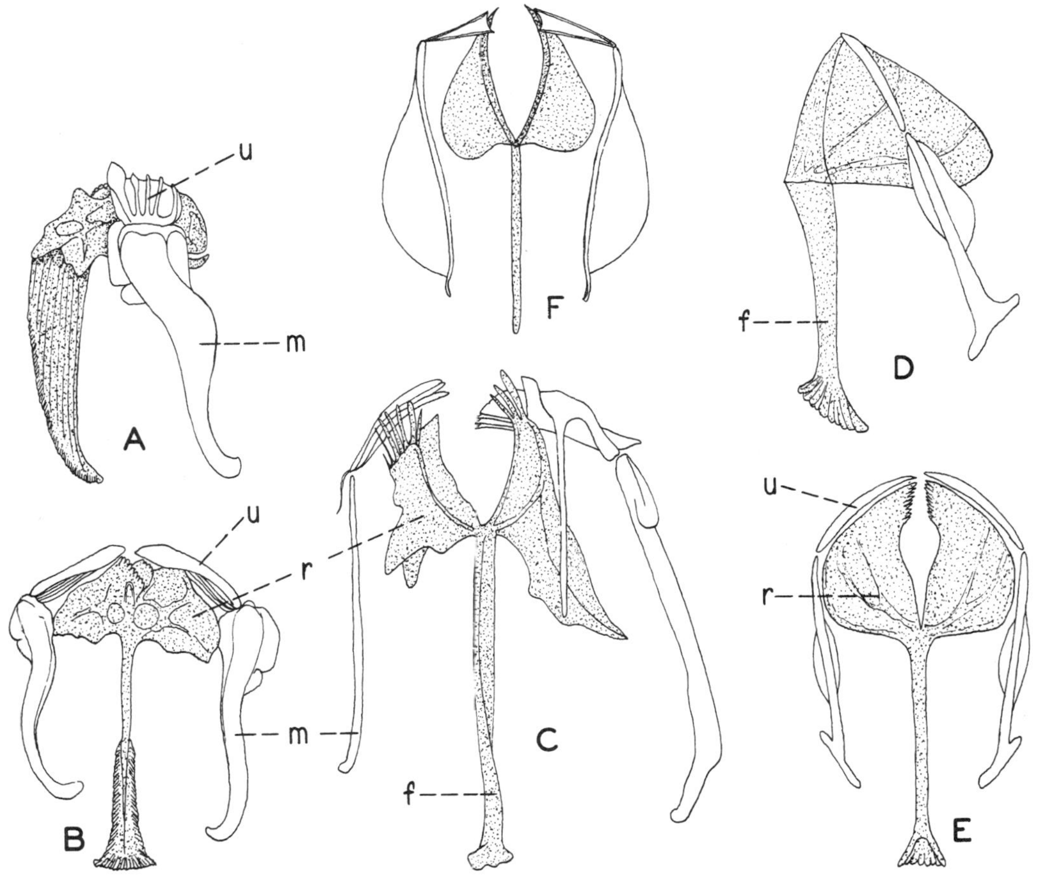

Fig. 7.10. Typical virgate trophi. A, lateral view of trophi of *Notommata;* B, ventral view of trophi of *Notommata;* C, ventral view of trophi of *Trichocerca;* D, lateral view of trophi of *Cephalodella;* E, ventral view of trophi of *Cephalodella;* F, ventral view of trophi of *Synchaeta.* Inci are stippled. f, fulcrum; m, manubrium; u, uncus; r, ramus. (A and B modified from Harring and Myers 1922; C modified from de Beauchamp 1909; D and E modified from Harring and Myers 1924.)

cavity. Rotifers with a virgate mastax commonly extract the fluids of plant cells and the body fluids of microscopic animals by this suction mechanism. In addition to suction, the trophi of the virgate mastax may be used for biting or nibbling preliminary to ingestion and the grinding action. Plankton, periphyton, and particulate detritus are also used as the chief food by many rotifers with a virgate mastax.

Cardate (Fig. 7.9A, B). This type of mastax occurs only in the Subfamily Lindiinae. Like the virgate type, the cardate mastax functions by suction, but the whole mastax is able to oscillate on a transverse axis, and a complicated epipharynx supports the mouth. The Lindiinae are only incidentally predatory, the chief food being periphyton and particulate detritus.

Forcipate (Fig. 7.9E). The trophi of this type of mastax are elongated, strongly compressed dorsoventrally, and adapted for protrusion through the mouth and the capture and tearing of the prey (Protoctista and micrometazoa). The trophi can sometimes be protruded for half their length.

Incudate (Fig. 7.9C, D). The trophi are prehensile and all parts are reduced except the pincerlike protrusile rami. The mastax musculature is likewise reduced. Food consists mostly of zooplankton.

Ramate (Fig. 7.8D, E). This mastax is specialized for the grinding of plankton, periphyton, and detritus. All parts of the trophi are reduced or vestigial except the semicircular unci, which have opposing linear teeth. The fulcrum is usually absent.

Malleoramate (Fig. 7.8C). This type is much like the ramate mastax, but the first few ventral teeth are larger, more highly developed, and detached, while the remaining teeth are progressively smaller striations. The fulcrum is present but small.

Uncinate (Fig. 7.9F). Specialized for laceration of ingested plankton, periphyton, and detritus. An intermediate piece, or subuncus, facilitates greater movement of the unci on the rami.

The trophi of all types of mastax are minutely sculptured, ridged, and barred. These complex markings can be discerned to advantage only with the oil immersion lens and they are mostly omitted in the trophi figures in this chapter.

Food, feeding. In both sessile and free-swimming nonpredatory rotifers the coronal cilia are of primary importance in creating localized currents that guide and concentrate the loose periphyton, small plankton organisms, and detritus at the region of the mouth. Indeed, the great majority of species are omnivorous and ingest all organic particles of the appropriate size. Common examples are *Cephalodella*, *Filinia*, *Keratella*, *Lecane*, *Proales*, *Euchlanis*, *Epiphanes*, *Brachionus*, the Bdelloidea (e.g., *Philodina*, Figs. 7.3, 7.11), and most sessile species. *Abrochtha intermedia* has a well developed trochus but does not seem to feed by filtration but rather sweeps the substrate with its cilia, picking up food particles.

Predatory species probably detect their prey by touch or by biochemical stimuli. The following common genera feed on other rotifers and all kinds of small Metazoa, either in the plankton or on a substrate: *Asplanchna*, (Fig. 7.12) *Dicranophorus*, *Ploesoma*, *Synchaeta*, and *Trichocerca*. A careful investigation has shown that *Asplanchna* will usually ingest all plankton organisms larger than 15 μm.

Many notommatids, as well as a few genera in other families, feed mostly on the fluid contents of filamentous algal cells. *Cupelopagis*, *Acyclus*, and *Atrochus* are all sessile and without coronal cilia. They have a large funnellike corona, or infundibulum, however, and when the prey wander into this cavity, they are quickly enclosed and ingested. A few free-living rotifers have highly specialized food habits. *Acyclus inquietus*, for example, lives among *Sinantherina* colonies and eats the motile young. *Dicranophorus isothes* is thought to subsist entirely on small Cladocera, and *D. thysanus* feeds on dead copepods, Cladocera, and oligochaetes.

Digestive system. Most Flosculariacea, Bdelloidea, and Ploima have similar digestive systems (Fig. 7.13A, B, D–F). The mouth opens into a narrow, ciliated pharynx which leads to the cavity of the mastax. A variable number of small salivary glands are closely associated with the mastax. After the food has been macerated in the mastax, it passes into a short to long esophagus, which leaves the dorsal or posterodorsal surface of the mastax. The esophagus extends to the large, thick-walled, ciliated stomach, where most of the digestion and absorption are thought to occur. Usually there is a pair of oval or bean-shaped gastric glands on the anterior margin of the stomach. The intestine may be distinctly or indistinctly set off from the stomach. It is much smaller, sometimes very narrow, thin-walled, and often ciliated. The cloaca is short and seldom ciliated; it opens to the outside dorsally at the base of the foot.

The anterior end of the Collothecacea digestive tract is much different (Fig. 7.13C). The buccal area is situated at the base of the large funnellike infundibulum. Food taken into the mouth at the bottom of the infundibulum passes through a narrow esophageal tube, which hangs freely in a very large cavity, the proventriculus. The small mastax is situated at the bottom of the proventriculus.

Digestion is regularly extracellular, but in a few genera, such as *Chromogaster*, *Asco-*

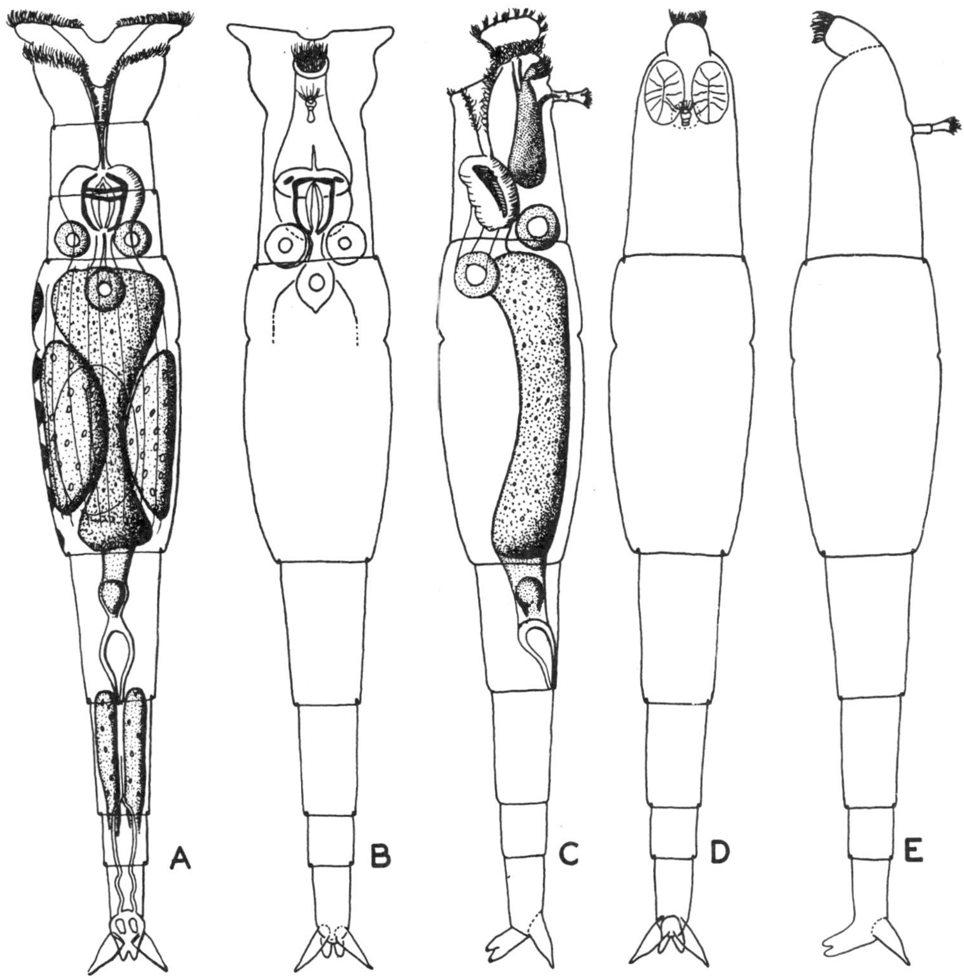

Fig. 7.11. Structure of *Philodina*, diagrammatic. A, ventral view; B, dorsal view with most internal organs not shown; C, lateral view with reproductive system not shown; D, dorsal view with corona retracted; E, lateral view with corona retracted. (From Calaway 1968, by permission.)

morpha, and a few others, it is clearly intracellular. No gastric glands are present in these rotifers, but the stomach has large gastric ceca that fill much of the pseudocoel.

Although many rotifers have true symbiotic zoochlorellae in the stomach wall, it is now believed that some instances previously reported as symbiosis are in reality ingested algal cells that persist for several days during their very slow intracellular digestion in the cells of the stomach wall.

In members of the bdelloid family Habrotrochidae there is no stomach cavity and the stomach is a continuous syncytial mass. As food leaves the esophagus it enters the stomach protoplasm in the form of food pellets, or food vacuoles, which circulate about and become digested.

Respiration. Most littoral and plankton rotifers have rather high oxygen requirements, but certain genera are capable of withstanding anaerobic conditions for short periods and very low concentrations of dissolved oxygen (0.1–1.0 ppm, for example) for extended periods.

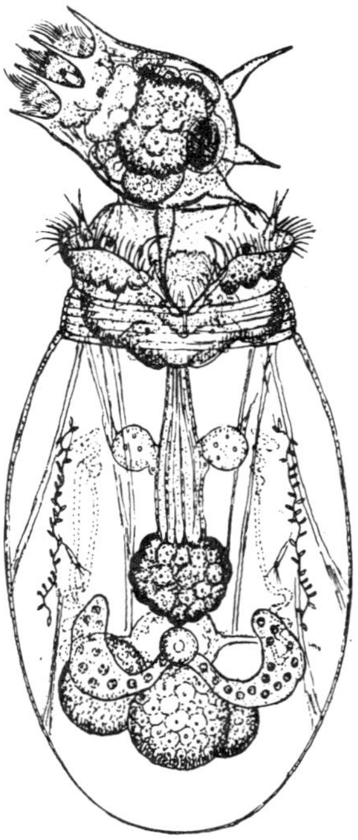

Fig. 7.12. *Asplanchna* about to ingest a *Brachionus.* (From Halbach 1971.)

Limnetic plankton genera, such as *Asplanchna*, *Filinia*, *Polyarthra*, and *Keratella*, commonly occur in the oxygen-poor hypolimnion of lakes during midsummer and midwinter. Rotifers living at a depth of a few centimeters in the interstices of sandy beaches likewise are in a region of low oxygen. Trickling filters of sewage plants and oxygen-deficient bottom mud and hay infusions characteristically contain species of *Lecane*, *Lepadella*, and many bdelloids. Undoubtedly the creation of water currents by ciliary action ensures an adequate supply of oxygen to rotifers under most circumstances.

Osmotic control, excretion. Like most small freshwater organisms, rotifers are readily permeable to water, but the internal osmotic pressure is kept constant by the action of the protonephridial system (Fig. 7.14). There are from four to 50 symmetrically arranged flame bulbs throughout much of the body. These are connected to a long, convoluted collecting tubule on each side, a part of the tubule being thin walled, another part thick walled and glandular. The two collecting tubules empty into an excretory bladder that has a very short duct leading to the ventral side of the cloaca. Excess water and probably a certain amount of metabolic waste are absorbed from the fluid of the pseudocoel by the flame bulbs and wall of the collecting ducts and are excreted into the cavity of the ducts. The weak pressure produced by the beating of the tufts of cilia in the flame bulbs creates a slight current, which aids in carrying the excretory fluid through the tubules to the bladder. Here it is stored only temporarily, since the bladder normally contracts and empties its contents to the outside through the cloacal aperture (anus) one to six times per minute. In a few Ploima and in typical Bdelloidea there is no separate bladder; the collecting tubules merely empty into the thin-walled posterior part of the cloaca, which is modified into a bladder.

Old rotifers are usually distinguished by their darker pigmentation, most of which consists of accumulated excretory granules retained in the tissues.

Muscles. Both smooth and striated muscle fibers occur in rotifers (Fig. 7.15). The latter is often associated with the rapid movement of long spines and special appendages, as in *Polyarthra* and *Hexarthra.*

Muscles are always arranged in small, bandlike groups of fibers, never in flat sheets. The circular muscle system consists of 4 to 15 bands attached just beneath the hypodermis. These bands are typically complete in the Monogononta, but in the Bdelloidea the bands usually have a ventral gap. One set of posterior longitudinal muscle bands are inserted in the foot or posterior of the trunk. An anterior set of longitudinal bands have their insertions in the coronal region and their origins near the middle of the trunk. *Testudinella* has short dorsoventral muscles, no circular ones. *Cupelopagis* has a remarkably complex system of muscles. A few delicate visceral and cutaneovisceral muscles help move and hold the internal organs in place.

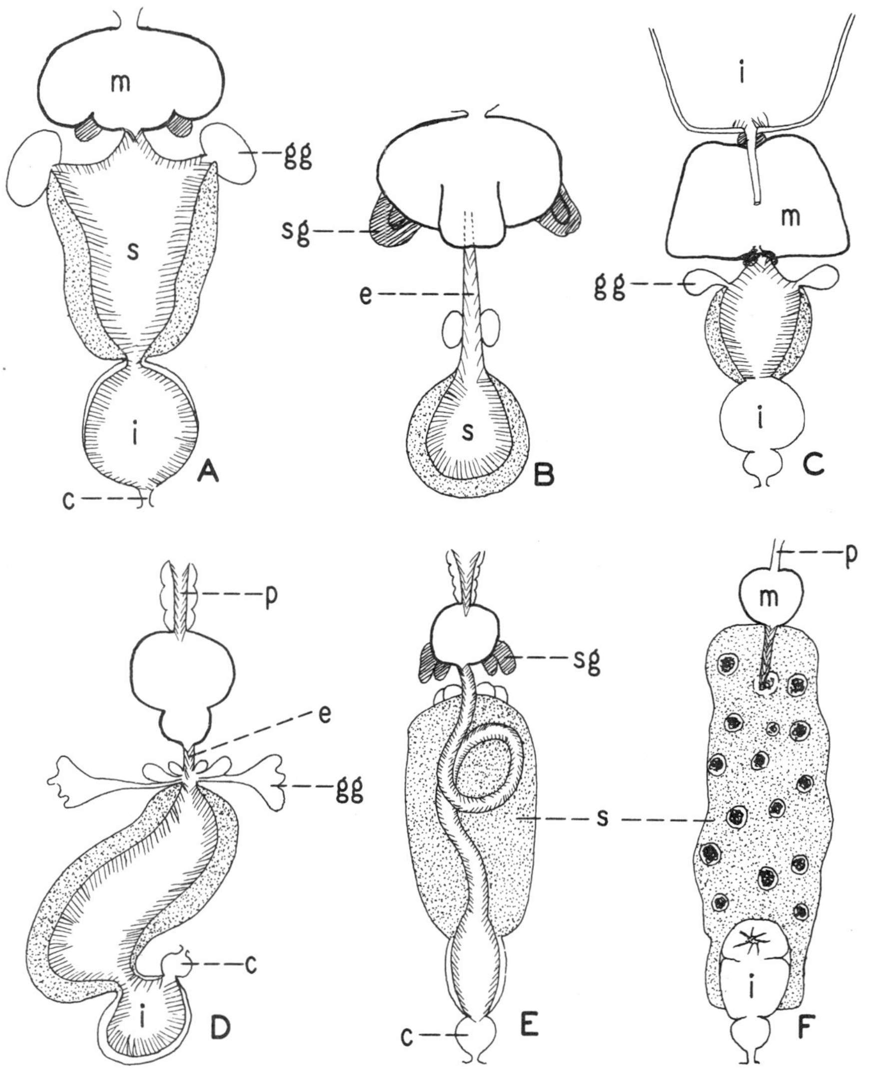

Fig. 7.13. Diagrams of typical rotifer digestive systems. A, *Epiphanes;* B, *Asplanchna;* C, *Stephanoceros;* D, *Testudinella;* E, *Mniobia;* F, *Habrotrocha.* c, cloaca; e, esophagus; gg, gastric gland; k, intestine; i, intestine; m, mastax; p, pharynx; s, stomach; sg, salivary gland. (A–E greatly modified from de Beauchamp 1909.)

Nervous system. The arrangement of the rotifer nervous system is nonlinear. The largest mass of nervous tissue is the sac-shaped cerebral ganglion, or "brain," lying on the dorsal surface of the mastax and esophagus and sometimes more or less obscured by the hypodermal syncytium of the corona. By means of fine paired nerve fibers it is connected with the mastax ganglion (sometimes absent) on the ventral surface of the mastax and with the caudal ganglion in the foot region. A few additional fine connecting fibers and associated nerve cell bodies innervate the sensory areas, muscles, and viscera (Fig. 7.16).

Sense organs. Perhaps the most conspicuous sensory area is the cervical eyespot. Characteristically it lies at the lower end of the brain or is embedded in the brain and

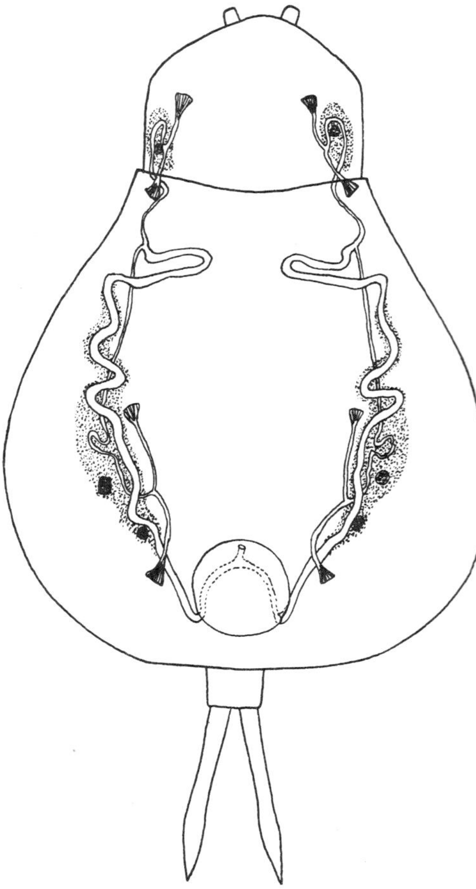

Fig. 7.14. Protonephridial system of *Euchlanis*. (Modified from Stossberg 1932.)

consists of a cup- or bowl-shaped mass of red pigment granules containing a refractile globule. Occasionally it is double. Many rotifers have two additional frontal eyespots located on the corona and widely separated. Eyespots are lacking in the adults of most sessile species, but they are present in the active immature stages.

A small, median, dorsal, setose antenna is usually present; rarely it is paired. In addition, the Monogononta have a pair of lateral antennae in the posterior third of the body and often a minute caudal antenna on the dorsal surface of the foot. Special tufts of cilia, membranelles, setae, and sensory protuberances or papillae are of common occurrence on the corona of many Monogononta.

For many years the retrocerebral organ of rotifers went almost unnoticed. It consists of two parts: a pigmented retrocerebral sac situated dorsal or posterior to the brain and opening on two coronal papillae by a paired duct, and a small paired subcerebral gland on either side of the sac with ducts alongside those of the sac (Fig. 7.17). Either of these two components may be lacking, but usually it is the latter. "Bacteroids" often occur in the retrocerebral sac and occasionally in the subcerebral glands. A variety of functions have been postulated for the retrocerebral organ, none of them conclusive. The most plausible suggestion is a special sense organ.

Reproduction. The Monogononta have a simple, ventral, elongated, saclike reproductive system (Figs. 7.1, 7.2). just within the distal end of the sac is a small ovary consisting of a clump of developing oocytes. The vitellarium is a syncytium with a few large nuclei; it occupies half or more of the capacity of the sac, and as eggs in the ovary mature one at a time and move downward slightly, the vitellarium contributes a large mass of yolk to the egg so that it becomes relatively large. On maturing, the egg passes down the short oviduct, into the cloaca, and out the cloacal aperture. The egg is quite elastic and is momentarily constricted and elongated as it passes through these small ducts.

The Bdelloidea reproductive system is similar to that of the Monogononta except that it is V or Y shaped and has two ovaries and two vitellaria (Fig. 7.18). The two oviducts are very delicate or rudimentary.

Males are unknown in the bdelloids, and reproduction is solely by parthenogenesis. In the Monogononta, however, males have been reported or described for those species whose biology has been carefully worked out, and it is thought that males occur regularly in this group, although there are many genera in which males have never been seen, and perhaps in these groups reproduction is also exclusively parthenogenetic. It is commonly stated that males are rare, but this belief has arisen simply because the occurrence of males in most species is restricted to only

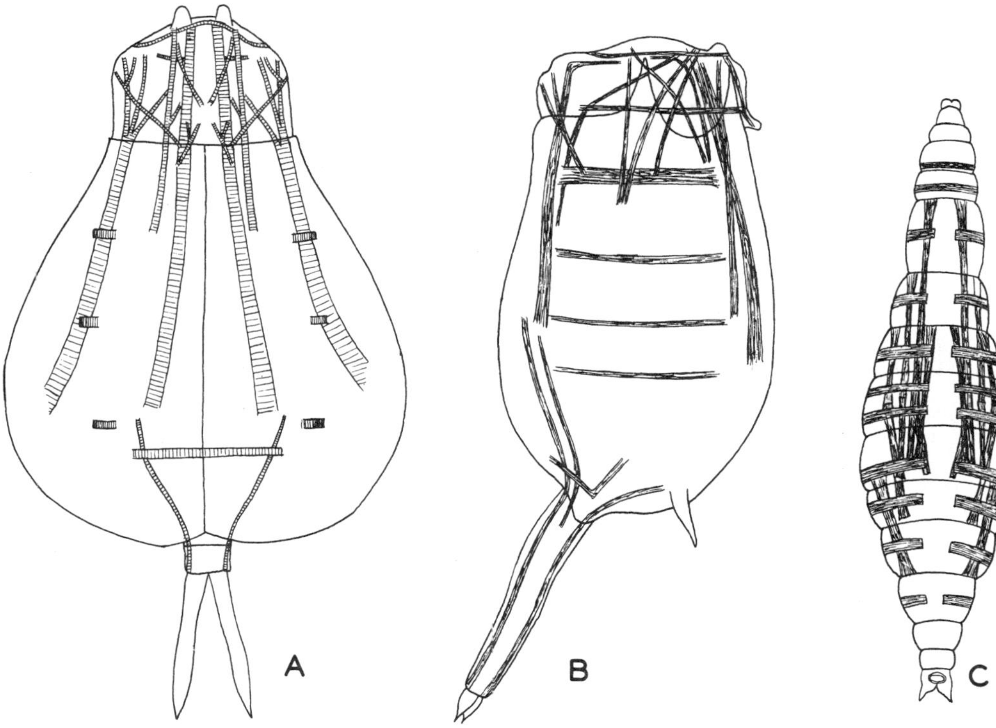

Fig. 7.15. Rotifer musculature, semidiagrammatic. A, *Euchlanis*, dorsal; B, *Brachionus*, lateral; C, typical bdelloid, ventral. (A and B modified from Stossberg 1932.)

one to three weeks during the year. If collections are made during the brief time of male occurrence, males can be found in abundance.

Ploimate rotifers have two distinct types of females, which, with the exception of two or three species, are morphologically indistinguishable. During the greater part of the year the females reproduce exclusively by parthenogenesis. These females are said to be amictic, and both their body cells and their amictic eggs all have the diploid number of chromosomes (Fig. 7.19). Their eggs undergo only one, nonreductional division during maturation in the ovary. The second type of female is called mictic. Such females appear only at critical times of the year, especially when there are marked changes in the factors of the environment (see page 144). The eggs of a mictic female undergo the usual double meiotic division and are therefore haploid. In contrast to amictic females, the mictic females are capable of being fecundated by the males. If fecundation does occur, the fertilized eggs (also called winter eggs, or resting eggs) that they deposit are thick-walled and highly resistant to adverse environmental conditions. If, however, a mictic female is not impregnated, she deposits eggs that promptly hatch into males. Fundamentally, a resting egg is a potential male-producing egg that has been fertilized and that, on hatching, produces an amictic female.

One mictic female may produce both fertilized eggs and male eggs, but males and females, or fertilized eggs and females, are never produced by the same parent. In fact, mictic females are commonly seen carrying both male eggs and fertilized eggs at the posterior end of the body. Fecundation of an amictic female by a male has no effect. The two kinds of female are always physiologically distinct, not interchangeable. A typical sequence of generations is shown in Fig. 7.20. Under natural conditions, there are

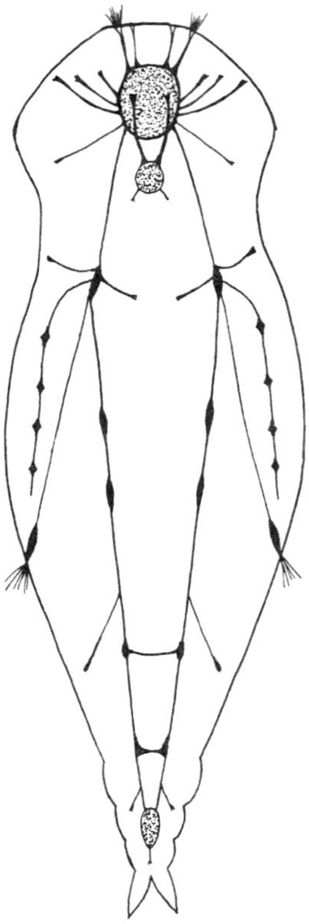

Fig. 7.16. Diagram of rotifer nervous system. (Greatly modified from Remane 1932.)

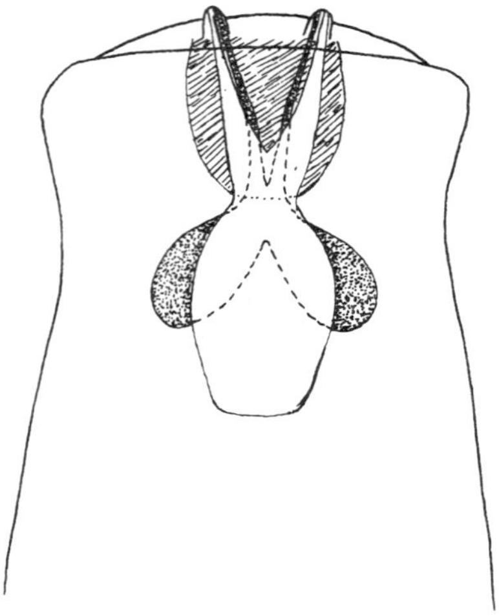

Fig. 7.17. Diagram of a typical retrocerebral organ (dorsal view). Retrocerebral sac clear, subcerebral glands stippled, and brain with parallel diagonal lines.

usually only one or two mictic generations per year, whereas there may be up to 20 or 40 amictic generations. The female that hatches from a resting egg is always amictic, but the following generation may be amictic or mictic.

The number of amictic eggs that a female may produce during her lifetime is highly variable and depends on both the species and environmental conditions. Specific laboratory cultures have shown that *Brachionus calyciflorus* averages only 3.6 eggs and *Testudinella elliptica* only 5.0, whereas *Epiphanes senta* averages 45.4 and *Proales sordida* 24.3. Little is known about the number of eggs that a mictic female produces, but presumably it does not differ greatly from the number produced by an amictic female.

Some rotifers are ovoviviparous, the eggs being retained in the oviduct or pseudocoel until hatched. These include two species of Notommatidae, many bdelloids, *Asplanchna*, *Conochilus*, *Rhinoglena*, and a few others. Their young are usually released through the cloaca, but many bdelloids are known to release the young through ruptures in the body wall. In *Asplanchna* the male may fecundate its own mictic mother before it is released from the body of the mother.

Amictic, or summer, eggs are usually released from the body of the female and hatch within a day or two. The summer eggs of some plankton species contain large oil droplets, which facilitate suspension and flotation. In the following genera, however, the summer eggs are usually carried loosely attached to the posterior end of the body until hatching: *Polyarthra*, *Pompholyx*, *Keratella*, *Hexarthra*, *Brachionus*, and *Filinia*.

Resting (or winter, or fertilized) eggs are heavy, thick-shelled, and often sculptured. On release from the female they usually sink to the bottom, but in the autumn they may sometimes be found floating on the surface or blown ashore. They are remarkably resistant to desiccation, high temperatures, low temperatures, and adverse chemical conditions for long periods. Many species winter over in the resting egg stage. Unlike summer eggs, the resting eggs require a la-

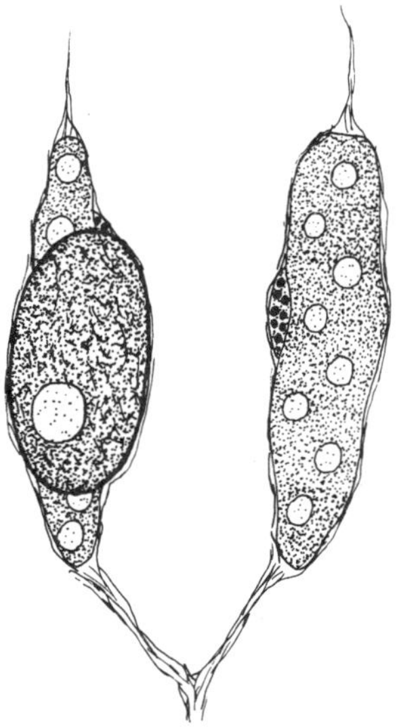

Fig. 7.18. Diagram of typical bdelloid reproductive system, showing two vitellaria, two ovaries, two vestigial oviducts, and one developing egg.

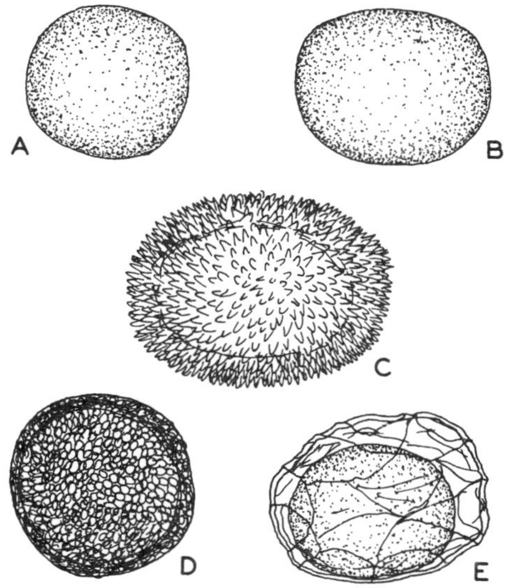

Fig. 7.19. Rotifer eggs. A and B, amictic eggs; C– E, fertilized (resting) eggs.

tent period of several weeks to several months before they can hatch. Hatching is stimulated by combinations of changes in temperature, osmotic pressure, water chemistry, aeration, and even algae. In spite of some textbook statements to the contrary, resting eggs are known for less than 1% of all rotifer species.

A curious fourth type of rotifer egg has recently been reported from cultures of *Keratella quadrata* and *K. hiemalis.* It is an unfertilized "pseudosexual resting egg" produced by parthenogenesis in the absence of males. The morphology and development are similar to those of ordinary resting eggs.

Figure 7.20 summarizes the complicated and interrelated reproductive events as they have been worked out for a few species of ploimate rotifers.

Biology of males. With few exceptions, male rotifers are minute, degenerate, and short-lived. Commonly, they are about a third as long as the female. The digestive tract is absent or vestigial, although a few species have a mastax and stomach. Mouth and anus are never present. The males have no well developed lorica or spines, and the corona is always anterior and well supplied with cilia. They are very rapid swimmers, never sessile.

The simple reproductive system occupies most of the pseudocoel. The testis is pear-shaped or globular and is held in place by a small strand supposed to be a remnant of the digestive tract. The vas deferens is short to long and sometimes has two or four associated prostate glands. Some species have no true penis, the vas deferens being everted through a small ciliated orifice during copulation. Other species have a well-developed penis that is everted continuously; sometimes it is larger than the foot. In a few males the whole posterior end of the body is tube-like and specialized for copulation.

Males of plankton species are the most strongly reduced. Some are not much more than a sperm sac surrounded by protoplasm and a cuticle, and having an anterior tuft of cilia.

Males are ready for copulation within an hour after hatching. They are extremely active, especially in the presence of females, and swim about erratically until they come in contact with one. Copulation and sperm transfer occur either through the cloacal

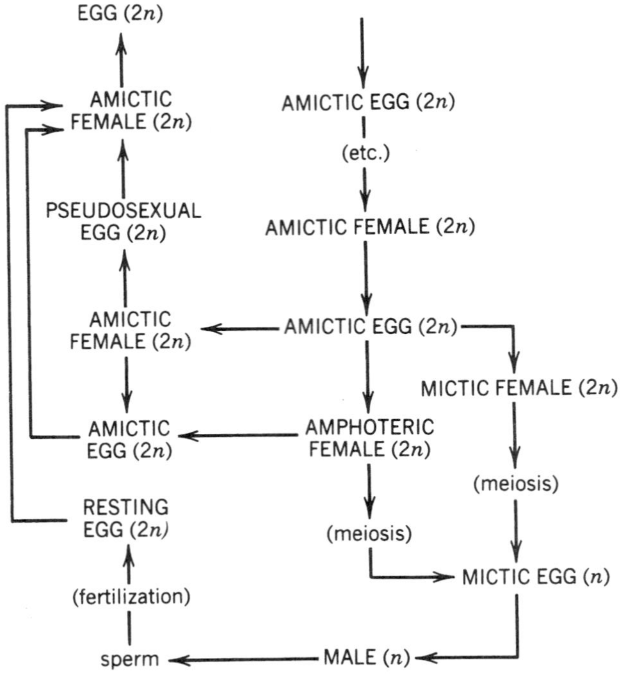

Fig. 7.20. Diagrammatic sequence of generations and chromosome complements in a typical ploimate rotifer.

aperture of the female or through the body wall (*Asplanchna*) (Fig. 7.21), and the male dies immediately thereafter. If a female is not located, the male lives for only 2 or 3 days; *Asplanchna* males, however, are known to live for as long as 4 to 7 days.

Males have been reported for only about 10% of the species of Monogononta. Few have been carefully studied and figured. Males have been reported for only three or four species in the large genus *Lecane*, and there are only three or four records for the family Trichocercidae.

In the so-called perennial species of rotifers, males usually appear in spring and autumn; in "summer" species they appear in autumn; and in "winter" species they ordinarily occur in spring.

Development and longevity. The rotifer zygote undergoes unequal and regular cleavage, and the gastrula consists of an outer layer of cells (ectoderm) enclosing an inner cell mass (mesoderm and endoderm). Cell division and differentiation proceed rapidly, with each organ developing from a definite cell complex. When all organs have received their full complement of cells, cell division ceases.

Most plankton and littoral species grow rapidly during the several hours after hatching, but thereafter very slowly if at all. Many sessile species, however, grow throughout life, chiefly as a result of the elongation of the posterior end. Although rotifers have no ecdyses, the mature adult may be three to five times as large as the newly hatched young.

Although there may be considerable size variations among the individuals of a single species, it has been found, for the several species carefully studied, that there is an approximately constant number of cells in all adults of the same species. In *Epiphanes senta*, for example, there are about 959 nuclei. Each organ in the great majority of specimens has the same number of nuclei, though a few tissues, especially the vitellarium, show slight variations in the number of nuclei.

Length of life from hatching until death, is highly variable. *Epiphanes senta* averages

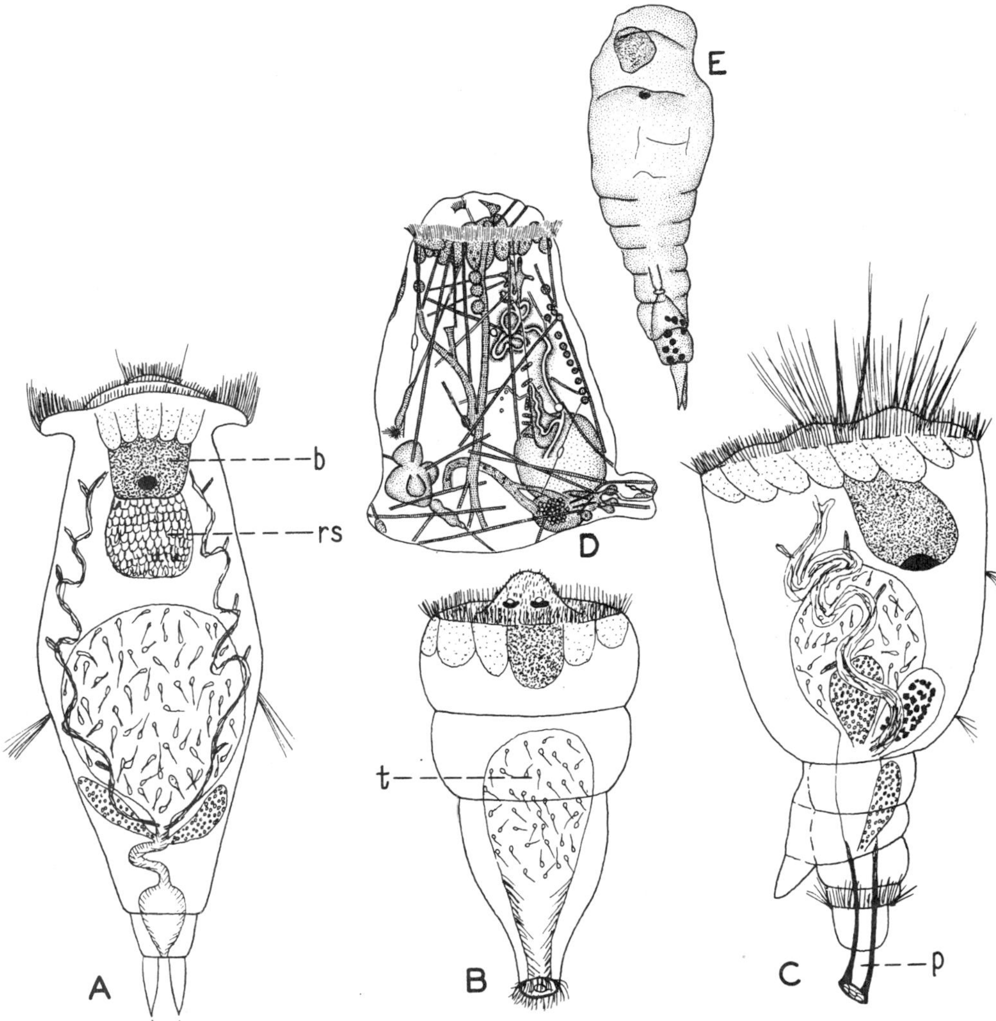

Fig. 7.21. Typical male rotifers, semidiagrammatic and with musculature omitted (except in D). A, dorsal view of *Notommata;* B, dorsal view of *Filinia;* C, lateral view of *Brachionus;* D, *Asplanchna;* E, Lecane. b, brain; p, penis; rs, retrocerebral sac; t, testes (sperm sac). (A–C greatly modified from Wesenberg-Lund 1923; D from Sudzuki 1956; E from Sudzuki 1959.)

about 8 days, *Lecane inermis* 7.4 days, *Brachionus calyciflorus* 6 days, and *Proales decipiens* only 5.5 days. At the other extreme, *Euchlanis incisa* averages 21 days, *Keratella quadrata* 22 days, *Cupelopagis vorax* 6 weeks, and many bdelloids are active for 3 to 6 weeks.

As might be expected, the ability to regenerate lost or damaged organs is poorly developed. Amputation of the coronal lobes of *Stephanoceros* results in regeneration only in immature specimens; regeneration is lacking or very slow in full-grown individuals.

Syngamic reproduction. Since about 1904 there has been a vigorous and many-sided controversy over the relative significance of physiological, ecological, and genetic factors in the production of mictic females and males, and bisexual reproduction. Scores of papers dealing with a wide variety of field observations and controlled laboratory experiments have appeared. Reduced to simplest terms, the chief problem may be stated as follows: Most ploimate rotifers reproduce by a series of parthenogenetic generations throughout most of the year; yet, suddenly, mictic females and males may appear in the population, and resting eggs are produced, usually in the autumn, but sometimes in the spring, or both. How may this phenomenon be explained? What factors are responsible for male production?

Most of the important observations and experimental work centering around this involved problem have stemmed from the laboratories of Shull, Whitney, Tauson, Wesenberg-Lund, and (recently) Gilbert. Their many papers should be consulted for a complete account. For our purposes, however, it is sufficient to summarize the present status of the matter and to mention some of the more significant items relating to sexual periodicity that have appeared as side issues in these studies.

To begin, it is obvious that species differ considerably from one another with respect to the nature and degree of ecological factors that instigate their bisexual periods. Furthermore, some species behave differently from one body of water to another and from one year to another. In one lake, for example, a species may regularly produce males in the autumn; in another lake it may have both spring and autumn bisexual periods; in a third lake the occurrence of bisexual periods may be sporadic. *Polyarthra*, for example, sometimes produces males at almost any time of the year, in both ponds and lakes.

In general, small populations of rotifers usually remain parthenogenetic and bisexuality appears during population increases. External factors, rather than genetic makeup, control sexual periodicity, but these factors are nonspecific. Careful experimental work indicates that the chief factors responsible for the appearance of males in a population are (1) a change in the type of food (green to nongreen, and vice versa), (2) an increase in food supply, and (3) a decrease in food supply. Other factors that are important in instigating male production are crowding, addition of fresh culture medium, addition of distilled water, making the culture medium more alkaline, changing the temperature (rarely), and starving (rarely). Sometimes, however, even if measurable conditions remain unaltered, males unaccountably appear in a population.

The solution of the problem must, of course, be sought in a study of the amictic female, for whether she produces eggs that will hatch into more amictic females or eggs that will hatch into mictic females is a matter determined by her physiology. The decision as to which type of egg she will deposit is made in the 3-hour period before she releases the egg from her oviduct. Eggs hatching into mictic females are slightly smaller than eggs hatching into amictic females, and several investigators contend that in this difference lies the crux of the whole problem. The production of a large or small egg might very well depend on the available amount of yolk in the ovary, which, in turn, might indirectly depend on environmental conditions. For example, the appearance of mictic females is in part brought on by a change in the photoperiod, by greater population densities, and by minute quantities of vitamin C in the water. This matter is further complicated by the fact that some species produce polymorphic resting eggs that may show a considerable range of size, shell sculpture, and color.

Epiphanes senta invariably has a sexual period under natural conditions, but experimentally this species has been carried through more than 500 successive parthenogenetic generations; such old cultures are characterized by a considerably lowered vigor and reproductive rate. *Proales decipiens*, however, has been carried through more than 250 successive parthenogenetic generations with no changes in vigor or fecundity, and it is interesting to note that the male is unknown in this species. Such facts appear to constitute some evidence that periodic bisexuality may have a rejuvenating effect on rotifer populations through the associated meiotic and fertilization processes.

Periodicity. Like so many other plankton and littoral organisms, rotifers exhibit periodic and sometimes quite striking cycles of abundance during the year, a fact familiar to anyone who has done seasonal plankton work. Various investigators have designated certain species as monocyclic, dicyclic, polycyclic, or acyclic and perennial, according to whether their annual population curves have one, two, several, or no pronounced peaks. *Kellicottia longispina* and *Conochilus unicornis* are usually considered monocyclic, although they are both perennial in certain lakes. *Euchlanis dilatata* commonly has an autumn maximum. *Brachionus angularis* and *Keratella cochlearis* are often considered dicyclic, with spring and autumn maxima, but are sometimes perennial. Other species are more variable. *Polyarthra* is dicyclic (summer and autumn), polycyclic, or perennial. *Keratella quadrata* may be most abundant in spring, summer, or autumn, or it may be perennial. *Asplanchna priodonta* is variously considered monocyclic, dicyclic, or perennial.

There has been an unfortunate tendency on the part of some investigators to generalize on the periodicity of a species on the basis of their observations of a few large lakes during only 1 or 2 years. As emphasized by Pennak (1949), however, the cycles of abundance for plankton species are highly variable within each species, variable from year to year within a single lake, and especially variable from one small lake to another.

Variation, cyclomorphosis. Like the cladocerans, many species of plankton rotifers exhibit a remarkable range of variation in size and lorica development from one habitat to another. The most striking examples of these local "races" occur in *Keratella* and *Brachionus*. In *Brachionus calyciflorus* (Fig. 7.22), for instance, all degrees of development in the four posterior spines have been reported, ranging from long and well developed to very short or absent (Fig. 7.32B–D). At least in part, the production of spines in this species is induced by starvation, low temperatures, and substance(s) elaborated into the water by the predatory *Asplanchna*. Comparable variations in posterior spine development occur in *B. quadidentata* (Fig. 7.32H), *B. caudatus* (Fig. 7.32F), and *Platyias patulus;* to some extent the anterior spines are also variable. Until about 1915 these variations were often described as distinct species, and taxonomists have had to contend with the resultant confusing synonymy. Now, however, they are recognized as forms or races comprising relatively few species. An extreme example is shown by *Keratella cochlearis*, which supposedly exhibits at least 13 "forms" (see Fig. 7.26).

Variations in the size of races from different localities are also striking. Ahlstrom (1940) gives the average width of the lorica of *Brachionus calyciflorus* for 21 localities all over the world, and the variation is 124 to 300 μm. This is a remarkable range of av-

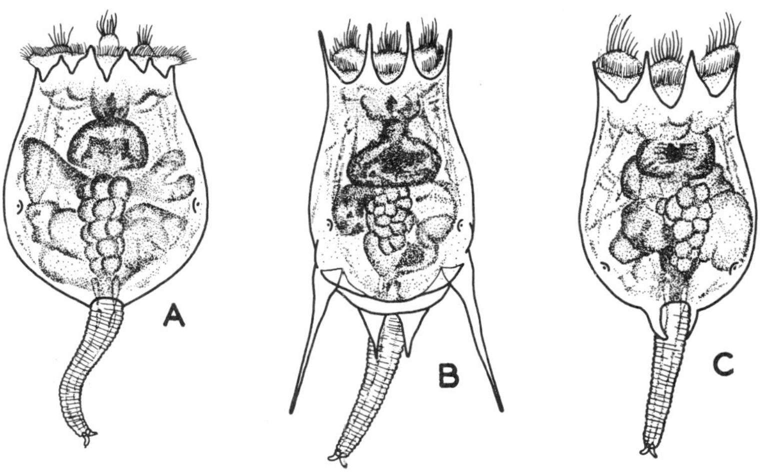

Fig. 7.22. A, *Brachionus rubens;* B, *B. calyciflorus* with large posterolateral spines; C, *B. calyciflorus* without posterolateral spines. (From Halbach 1971.)

erage size for a species having determinate cell division.

Superimposed on the confusion of varieties in some loricate plankton species is an annual cyclomorphic series, that is, a sequence of generations in the same strain marked by progressive modifications in morphology through a long parthenogenetic series. In *Keratella cochlearis* (Figs. 7.25A–D, 7.26), for example, the lorica has a long posterior spine during the cold months followed by a gradual reduction in length until July and August when there is a very short spine or no spine. At this point syngamic reproduction often occurs, followed by a gradual reversal in spine length during the remainder of the year. Such well-defined cycles usually occur in small lakes that become quite warm in summer. In Lake Michigan there is no cycle, and in other bodies of water only a partial cycle. Sometimes the two extreme lorica forms have no transitional forms. The literature contains scores of papers dealing with *K. cochlearis* cyclomorphosis and each study seems to show results that are more or less different from others.

In *Keratella quadrata* (Figs. 7.25H–L) a form with long posterior spines initiates the cycle. The spines are progressively reduced in successive generations until a form with very short spines or no spines is dominant. Syngamic reproduction may then occur, and the cycle starts all over again with a long-spined form. But often this species is long-spined the year round even in small ponds.

From these examples, and judging from the occurrence of two or three widely differing varieties in the same population, it is obvious that cyclomorphosis is a variable phenomenon within a single species. Certainly it cannot be ascribed to an inherent mechanism or simply to changes in environmental factors, but rather to a complicated interaction between organism and environment. In fact, the problems of cyclomorphosis, population fluctuations, local races, and bisexual reproduction all seem to be inextricably bound up with each other and with a multitude of ecological factors, Although this complex is a most difficult one, it is probably capable of solution by means of careful field and laboratory observation and experimentation. Gilbert (1973), for example, has shown that the lateral humps of male and female *Asplanchna sieboldi* protect them from cannibalism.

The species problem. Without detailing the evidence here, information is accumulating at an increasing rate that the genetics of single species and related subspecies is a complex matter. For example, some species show evidence of two or more genetic "clones" living in a single body of water. While these clones differ imperceptibly in their detailed morphology, they do show clear distinctions in such items as their feeding rates, reproductive habits, geographic distribution, seasonal periodicity, longevity, and biochemistry. Thus, for all practical purposes, they should be classified as good and separate species. But most rotiferologists have long been staunch "lumpers," and it will take a long time to generalize and clarify most species and subspecies problems. See below under Taxonomy for more comment.

General ecology. As emphasized earlier, rotifers occur in an extreme variety of habitats. They are poorly represented only in hot and cold springs, rushing streams, brackish waters, and extremely saline waters. Some common species are usually present in the same pond year after year, but sometimes unaccountably disappear for one or more years, then suddenly reappear.

Approximately 75% of the known species occur in the littoral areas of lakes and ponds, but only about 100 species are limnetic or planktonic enough in their habits to be completely emancipated from a substrate. The Bdelloidea are most characteristic of mosses and sphagnum. These habitats may be submerged, emergent, or terrestrial and only occasionally wet. Only a few species of the following genera of bdelloids regularly occur in the littoral zone: *Rotaria*, *Embata*, *Philodina* (Fig. 7.23), and *Dissotrocha;* two or three other genera are uncommon in littoral waters. *Abrochtha* appears to be found only in temporary pools that can remain dry for

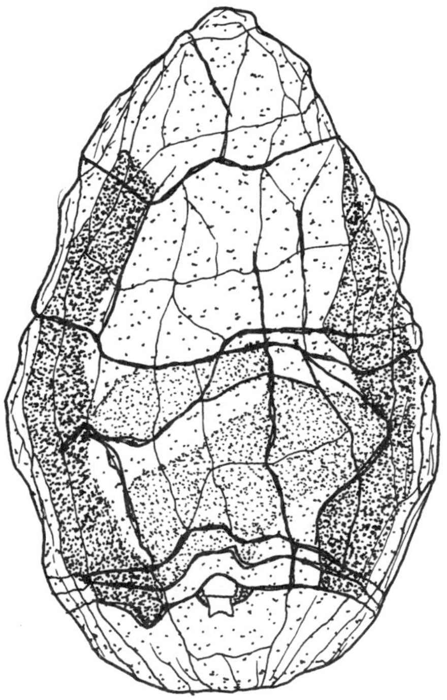

Fig. 7.23. Desiccated *Philodina.* (Modified from Hickernell 1917.)

long periods. On the other hand, only a few ploimate species occur commonly in moss and only if it is kept wet.

Increasingly, the psammon and hyporheos of lakes and rivers are being considered important habitats of rotifers. In European lakes, over 11 families have representative species occurring in the lake psammon. Generally, the interstices of lotic environments contain a greater diversity of rotifers than that of lentic environments.

A few free-living species are highly specific in their habitat preferences. *Acyclus inquietus* is invariably found among colonies of *Sinantherina. Collotheca algicola* appears to be restricted to clusters of *Gloeotrichia* filaments, and one or two other sessile species are also common on this alga. *Brachionus plicatilis* and *B. pterodinoides* are confined to extremely alkaline ponds and lakes, especially in the western states. *Proales rheinardti* is unusual in that it occurs in mountain springs and among *Fucus* in marine bays and inlets! Only two genera, *Seison* and *Zelinkiella,* are known to be restricted to marine habitats.

Brackish waters contain many euryokous freshwater species, as well as some species that are restricted to this habitat. The genus *Synchaeta* is remarkable for its many brackish and estuarine species.

Rotifers serve as food to a host of predators, including a few protoctistans, *Craspedacusta,* the larger cladocerans, many cyclopoid and calanoid copepods, *Chaoborus,* and *Mysis.* Even very small fish may capture rotifers.

Although the Rotifera are primarily free-living organisms, a few species have become specialized for a parasitic existence. *Notommata trypeta* is parasitic in colonies of the alga *Gomphosphaeria. Proales parasitica* is a common parasite inside *Volvox* colonies. A few species of *Albertia* are ecto- and endoparasites of aquatic oligochaetes. *Drilophaga judayi* has been found only in the free-living state, but a second species in this genus is parasitic on *Lumbriculus,* and a third on *Erpobdella.* Aside from such true parasites, many free-living rotifers live as commensals or epizoics on a wide variety of freshwater invertebrates, especially crustaceans and insects. Common examples include certain species of Bdelloidea, *Pleurotrocha, Lepadella,* and *Testudinella.* On the other hand, some of the larger plankton species often have the pseudocoel packed with ramifying filaments or sausage-shaped bodies, especially during the warm months. These are parasitic fungi that occasionally become epidemic.

Judging from the large plankton literature, it is obvious that most plankton communities average between 40 and 500 rotifers per liter, with populations in excess of 1000 per liter being unusual. The waters of mountain lakes and large oligotrophic lakes, however, sometimes contain less than 20 rotifers per liter, especially during the cold months. The most dense plankton rotifer population ever recorded from unpolluted waters is 5800 specimens per liter, cited by Pennak (1949).

Within ordinary limits, the density of natural rotifer populations appears to be governed by the amount of available food. The maximum density of rotifer populations appears to be closely correlated with the relative amount of available substrate and exposed surfaces. Using 5800 rotifers per liter as a maximum for plankton, the corresponding maximum for sessile species inhabiting finely divided littoral plants is about 25,000 rotifers per liter (Edmondson 1946). Furthermore, rotifers inhabiting the wet interstices of sandy beaches are reported to attain a maximum density of 1.155 million individuals per liter of damp sand (Pennak 1940). The greater the surface, therefore, the greater the population. Presumably this phenomenon is associated with more food, more attachment space, more nooks for protection, and reduced predation loss. Below the water's edge the sand contains a negligible rotifer population, but the top 3 or 4 cm of sand extending from the water's edge as far as 3 or 4 m landward is a favorable habitat. Although many of the sandy beach rotifers occur regularly in the littoral, about 45 psammobiotic species are known to be restricted to the sand habitat, especially species of *Dichocerca*, *Encentrum*, *Lecane*, and *Wierzejskiella*.

In contrast to littoral forms, the limnetic, or open-water, plankton genera occur over a wide depth range, even as deep as 100 or 200 m. *Asplanchna*, *Filinia*, *Polyarthra*, and *Keratella* are all good examples. If open-water limnetic plankton samples from many lakes are analyzed statistically, the following rotifers will be most abundant and generally distributed: *Keratella cochlearis*, *K. quadrata*, *Kellicottia longispina*, *Filinia longiseta*, *Asplanchna* spp., and *Conochilus unicomis*. *Brachionus* spp., *Synchaeta* spp., and *Conochiloides* spp. will be next most common. As shown by Halbach (1973), positive association coefficients are involved between most of these species cooccurrences.

Those pond and lake plankters having eyespots exhibit, in varying degrees, a 24-hour cycle of vertical movements in accordance with the variations in subsurface illumination. Beginning late in the afternoon, a slow drift upward begins, and maximum abundance occurs near the surface in the early morning. Around dawn a reverse drift downward begins, until maximum concentrations in the deep waters are attained about noon. Such movements are also of general occurrence among copepods and cladocerans, but in rotifers they are, of course, much less pronounced: Average amplitudes of only 1 to 3 m are common and amplitudes as high as 8 or 10 m are very unusual.

Unfortunately, many papers contain generalizations on rotifer ecology that are based on "controlled" conditions in the laboratory, and do not apply to natural field conditions. Experimental work on rotifers has burgeoned during the past 20 years, and it is difficult to generalize. Results are often contradictory and confusing because of reliance on manipulating a single ecological factor at a time. Food clearance rates, competition with other rotifer species and with cladocerans, biochemical control of polymorphism, cannibalism phenomena, selective feeding, food selection, predation effects, induction of parthenogenetic and bisexual reproduction, and sensory physiology are all complicated phenomena that await additional critical research.

Ecology of sessile rotifers. The work of Edmondson (1944, 1945) has contributed much information to the ecology of sessile rotifers (Floscularidae and Collothecidae). For the most part, sessile species are never present in quantities below a temperature of 15°C, the largest populations always being present at temperatures about 20°C. They are usually most plentiful in late spring, especially in small bays and littoral nooks in ponds and lakes. Although any substrate may be used, sessile species are by far the most abundant on submerged aquatic vegetation. *Utricularia vulgaris* var. *americana* has the most extensive, dense, and varied rotifer fauna of all species. Other favored plants are *Myriophyllum*, sphagnum, potamogetons, *Anacharis*, and *Vallisneria*.

Some species occur on a wide variety of plants, but others exhibit striking preferences. Thus *Collotheca gracilipes* is almost invariably on *Utricularia vulgaris americana*, *Rygura melicerta* inhabits *Gloeohichia* and *Coleochaete*, whereas *Collotheca campanulata* and *Ptygura barbata* are usually found on epiphytic algae. *Cupelopagis vorax*, however, is restricted to broad, flat leaf surfaces. Edmondson concluded that species distribution is determined mostly by water chemistry and available plant substrates.

Hydrogen-ion effect. Harring and Myers (1928) first pointed out some striking correlations between the pH of a body of water and the composition of its ploimate rotifer fauna. In general, alkaline waters (above pH 7.0) contain relatively few species but large numbers of individuals, whereas acid waters (below pH 7.0) contain large numbers of species and few individuals. In the hard-water lakes near Madison, Wisconsin, Myers identified 100 species in a week of collecting, but he found the same number of species in only 1 hour of collecting in a small acid lake in northern Wisconsin (personal communication). In the alkaline Yahara basin of southern Wisconsin he collected 138 species, but in the acid waters of Mount Desert Island in Maine he collected 497 species (Myers, 1931, 1933, 1934a).

As the result of his extensive field collecting, Myers distinguished three ecological groups of species with reference to their occurrence in waters of various hydrogen-ion concentrations. (1) Alkaline water species are those that are confined to waters having pH readings above 7.0. *Asplanchna*, *Asplanchnopus*, *Mytilina*, *Brachionus*, *Filinia*, *Lacinularia*, *Sinantherina*, *Eosphora*, and *Notholca* are all alkaline genera, but there are also species in other genera that are restricted to alkaline waters. (2) Transcursion species are those that occur in both alkaline and acid waters. By far the large majority of rotifers belong in this category. (3) Acid water species are those that are confined to lakes, ponds, and bogs having an acid reaction. There are no genera restricted to acid waters, only certain species in genera that also have transcursion and alkaline representatives. The following genera contain numerous acid water species: *Cephalodella*, *Lepadella*, *Lecane*, *Monostyla*, *Trichocera*, and *Dicranophonts*.

Most genera contain representative species in each of these three groups, but a specialist can tell at a glance whether a particular collection came from an acid or alkaline body of water. Even though an acid lake and an alkaline lake may be separated by a ridge of land only 50 inches wide, the rotifer faunas are completely different.

In a study of many areas in Wisconsin and New England, Edmondson (1944) lists 8 alkaline, 36 transcursion, and 12 acid species of sessile rotifers. In an investigation of many ponds and lakes (mostly acid) of the Pocono Plateau in Pennsylvania, Myers (1942) found 4 alkaline species, 313 transcursion, and 75 acid water species. In a study of the acid waters of the Adirondacks, Myers (1937a) found 42 acid water species and 171 transcursion species. On the other hand, he found (1936) a total of only 106 alkaline and transcursion species in the alkaline waters of southern California, even though he collected almost daily for 3 months.

To be sure, pH 7.0 is by no means a sharp dividing line between acid or alkaline water species; it is only an approximation. Some acid species, for example, are found in waters giving readings as high as pH 7.1 or 7.2, and alkaline species likewise occur as low as pH 6.9 or 6.8. The hydrogen-ion concentration itself cannot be a governing factor in ecological distribution, since pH is a reflection of a whole complex of physical and chemical variables, and actually the limiting factor may be one or more of these ecological variables. Total dissolved salts, calcium, and bicarbonate, for example, have been suggested as being fundamentally important. Acid water rotifers can easily be transferred to alkaline waters where they will live, and vice versa, but their parthenogenetic eggs will not hatch in the unfavorable environment.

The preceding remarks relating to pH refer only to ploimate rotifers; they do not apply to the Bdelloidea, which are apparently all transcursional and capable of tolerating a relatively wide range of ecological conditions.

Desiccation. Even certain modern textbooks include the statement or inference that all rotifers can be dried to form a resistant cyst, which when returned to water will excyst and become active again. In the first place, true desiccation is restricted to the majority of Bdelloidea and does not occur at all in the much larger ploimate group. Second, in the great majority of bdelloids desiccation involves drying and shrinking of the body contents and cuticle without the formation of a true cyst. Only a few bdelloids are known to form true secreted cysts; when subjected to drying, these species secrete a surrounding gelatinous material that quickly hardens and becomes resistant.

Bdelloids typically inhabit moss on old walls, bark, or ground; they are much more abundant on emergent sphagnum than on submerged sphagnum. Such habitats are wet only intermittently from rains, wave action, or high water, and the process of desiccation allows the rotifer to tide over the frequent unfavorable periods of dryness. Very few truly aquatic bdelloids are able to undergo desiccation.

Bdelloids that are isolated in a little water on a smooth, bare surface are unable to undergo desiccation and they die. They must be associated with irregular surfaces and interstices where drying occurs more gradually and where there is a protective effect, such as in debris, sand, or among moss or sphagnum. When drying begins, the bdelloid draws both ends of the body into the central portion of the trunk and then puckers the two ends as though they were drawn up by pursestrings (Fig. 7.23). This globular body mass then becomes wrinkled and loses much water. The pseudocoel is almost nonexistent, and the volume of the animal is only one-third or one-fourth of its normal size.

A remarkable rearrangement of chromatin occurs during desiccation. Instead of remaining in its usual large karyosome mass in the center of the cell, it fragments, becomes highly dispersed, and adheres to the inner surface of the cell membrane. Presumably this adjustment facilitates suspended animation and the utilization of oxygen at a greatly retarded rate.

Desiccated rotifers may be kept dry as dust for months or even years. There is one record of specimens being revived from moss kept dry for 27 years. When the dried rotifers are placed in water, they quickly absorb water and become extended and normally active. Sometimes complete recovery occurs in only 10 minutes, but under other circumstances it may take a whole day.

Desiccation always produces some mortality, even under favorable conditions, and only a fraction of a group of desiccated rotifers are viable and can be revived. A main cause of mortality among bdelloid cysts is believed to be oxidation caused by high temperatures and humidity. Rapid drying is injurious, and, in general, the longer the interval of desiccation, the lower will be the percentage of recovery. Nevertheless, even under natural conditions, bdelloids are able to survive several alternating periods of activity and desiccation.

Dispersal, geographic distribution. Overland transport of resting eggs or desiccated bdelloids is easily effected by winds and animals, and many rotifers are potentially cosmopolitan. By "cosmopolitan" we mean that similar habitats all over the world have similar rotifer faunas, and whether a given species is present in a particular body of water depends partially on geographic location, and partially on the precise complex of ecological conditions prevailing there. As Jennings (1903) has said, "The problem of the distribution of the Rotifera, is, then, a problem of the means of distribution." The same species of rotifers are found in limnetic areas, in moss, in sandy beaches, in sphagnum bogs, in comparable farm ponds, and so on, the world over. Several species of *Lecane* are found on several continents and the genus occurs throughout the world. Even the "rare" species are being reported from localities everywhere that appropriate habitats prevail and where the collector is diligent. Additionally, long-term study has shown and will show the existence of overlooked species previously unreported on a specific continent. Nevertheless, there are many species with restricted distributions, for example, the species of *Notholca* endemic to Lake Baikal in Russia. Similarly, there are 46 described species of *Brachionus;* only 22 are cosmopolitan, and the others are confined to single continents, especially in the Southern Hemisphere. None are endemic to Europe.

Pseudoploesoma formosum and *Trichocerca platessa* are common in the acid waters of the northeastern states but have never been collected elsewhere in the world. Many such similar instances are difficult to explain on the basis of our present knowledge of rotifer ecology. Presumably, however, such species do not have the proper distribution mechanisms to overcome their restricted ranges.

In spite of rigorous climatic isolation, rotifers have been remarkably successful in colonizing Antarctic and Subantarctic areas. DePaggi and Koste (1984), for example, list 27 Monogononta and 16 Bdelloidea as occurring south of 60° south latitude.

Collecting. Plankton species are easily taken with a townet, thrownet, or a fine dipnet with a vial or centrifuge tube fastened to the lower open end. Vegetated littoral areas are much richer collecting grounds, especially for browsing species, but nets are of little use there. Instead, masses of the plant material should be carefully placed in large screw-top jars along with water from the same habitat. The jars should be no more than half filled with plant material. When the jars are brought to the laboratory, they should be uncapped and placed in a north window. As the oxygen slowly decreases the rotifers will gather at the top of the water and usually toward the lighted side, where they may be pipetted out. Sessile species may be found with the dissecting binoculars on bits of plants. Such temporary culture jars will keep well for a week or more if the lowermost third of the water is changed daily.

Bdelloids may be brought to the laboratory in handfuls of wet or dry moss, which may be stored dry. When the material is to be studied it should be soaked for a few hours in water, then washed vigorously several times. The wash water and sediment should be kept in covered watch glasses or petri dishes for weeks or months and examined periodically under the microscope. Some bdelloids appear promptly, others not for 2 or 3 weeks.

Interstitial species may be washed out of streamside and lakeside sand and gravel.

Culturing. Many rotifer culture methods are suggested in the literature. Some methods are better for certain species than others, and an investigator who wishes to culture a previously uncultured species should try a variety of methods and modifications to get the best results. Species of *Lecane* have been raised in 0.1% malted milk in pond water, the solution being changed daily. The same genus has been cultured in rye extract prepared by grinding 20 grains of rye in a mortar and boiling in 100 mL of water for 20 minutes; the rotifers should be kept in drops of this solution and changed daily. Beef cube extract prepared by boiling one cube in 400 mL of tapwater is sometimes good. Dried breakfast oats may be used by boiling 8 to 30 flakes for 3 minutes in 100 mL of water, filtering, and using after 24 hours. *Epiphanes* and certain other pond genera may be cultured in hay infusions that have been inoculated with a suitable protoctist or algal food organism. "Stable tea" is prepared by boiling 800 mL of horse manure in 1000 mL of water for 1 hour; cool, strain, dilute with two parts boiled rain water, inoculate with a food organism, and age a week or 10 days before inoculating with the rotifer. Bdelloids may be cultured easily in 0.1% nonfat dried milk in tap or pond water. Sessile species are extremely difficult to culture.

Aloia and Moretti (1973) describe a pure culture sequence of *Aerobacter–Paramecium–Asplanchna*. Many modern laboratory cultures of rotifers are maintained with pure cultures of algae used for food.

The most important precaution to be observed in culturing rotifers is to change the culture solution frequently and prevent excessive bacterial activity. Several species, especially *Lecane*, have been successfully cultured in sterile media containing various mixtures of liver extract, vitamins, salts, and peptones.

Preparing, preserving. Since the identification of loricate rotifers depends largely on the details of the lorica, they are conveniently killed and fixed by placing directly in 10% formalin. This causes the soft parts of the animal to contract strongly, leaving the lorica well exposed for easy study.

The illoricate, free-swimming Ploima, however, require a much different technique. The identification of these forms depends largely on anatomical details that can be studied only in an uncontracted, natural state. As a consequence, these rotifers must be narcotized before killing and fixing. The following method is adapted from Myers (1942) and should be performed under dissecting binoculars. With a fine pipette place the rotifers in about 3 mL of water in a concave watch glass. If the specimens are from acid waters, add three or four drops of 1% Neo-Synephrin hydrochloride; repeat at 5-minute intervals until the rotifers are all lying on the bottom and ciliary action has about stopped. Kill and fix by adding three drops of 0.5 or 1.0% osmic acid from a dropping bottle and quickly stirring into the water. Pour the contents of the watch glass into a small vial, and after 5 minutes decant or pipette off most of the fluid; then add water. Repeat the decantation and addition of wash water three times at 15-minute intervals. Another useful method for illoricate rotifers involves narcotizing and killing with 0.04% procaine hydrochloride, although it takes 10 to 15 hours of contact!

If the rotifers are from alkaline waters, a 5% solution of Novocain or cocaine hydrochlorate in 50% methyl alcohol gives better results than the Neo-Synephrin. Usually, however, these narcotics are difficult to obtain, even for such scientific purposes, and other more readily available substitutes must be used. Chloretone and 2% benzamine lactate are useful, and the writer has had considerable success with 2% butyn and 2% hydroxylamine hydrochloride.

The osmic acid solution will last much longer without deteriorating if it is dissolved in 1% chloroplatinic acid (platinic chloride) and kept in a dropping bottle that is painted black. If rotifers are exposed to osmic acid too long, they will blacken but they can be bleached in weak potassium hydroxide.

Many illoricate species, especially Notommatidae and bdelloids, are so delicate and sensitive that narcotization is usually not successful. Ricci and Melone (2000) suggest studying the specimens alive in a drop of water on a slide without a coverslip, permitting the specimens to move about freely in a fully expanded form. Otherwise, it usual to flood the specimens with an equivalent amount of boiling water in a watch glass; this treatment kills a large fraction of the animals in a well-expanded and lifelike condition. Formalin or other fixative should then be added.

Hanley (1949) strongly recommends the following procedure. Narcotize by adding several drops of 2% benzamine hydrochloride to a small amount of water containing the rotifers, the time being variable from one species to another. Kill in 10% formalin. Wash at least six times in 5% formalin to remove all the benzamine hydrochloride, which might later crystallize out. After killing and fixing, all rotifers can be stored in 2 to 5% formalin containing 2% glycerin. Many workers add a little eosin, as this colors the rotifers and makes them easier to study and handle.

Preserved specimens can be studied and kept for several days in a shallow, hollow ground slide, provided the coverslip is sealed with petroleum jelly. Permanent mounts are best made with glycerin or glycerin jelly. In either case the rotifers should be slowly run up to pure glycerin. This is done by placing the preserved specimens in 5% glycerin in a depression embryology slide or concave watch glass and leaving it loosely covered for about a week so that all the water evaporates. In this way the specimens are penetrated slowly and partially cleared, and they retain their turgidity and natural shape.

Individual specimens in glycerin should never be picked up with a pipette since they usually stick to the inner surface of the glass and are easily lost. Instead, they may be readily picked out of glycerin under the dissecting binoculars with an Irwin loop. Specimens handled in this way may be conveniently set aside while other preparations are being made. Glycerin jelly slides are made by placing a melted drop on a slide, touching the drop with the tip of the bristle or loop holding the rotifer, and quickly orienting the specimen under the binoculars before putting on the round coverslip. The glycerin jelly should be allowed to set for several hours. Then the excess may be trimmed from the edge of the coverslip and several coats of Murrayite applied to its edge with a camel's hair brush and turntable at 24-hour intervals.

Fluid glycerin mounts are much more difficult to prepare but are superior for some purposes. Detailed directions for one method are given in Harring and Myers (1928). An alternative method is as follows: Three or four bits of passe partout or gummed paper are fixed symmetrically to a 1 × 3 slide and spaced about 10 or 12 mm apart to act as supports for the coverslip and thus prevent crushing of the specimen. The rotifer is then placed in a droplet of glycerin in the center of the slide and midway between the supports. The coverslip is next lowered into place very carefully, and when a camel's hair brush of Murrayite is touched to the edge of the coverslip, capillarity will draw the Murrayite between the coverslip and slide and it will surround the glycerin droplet and completely fill the capillary space. Such slides should be set aside to dry for several hours and then given several rings of Murrayite at 24-hour intervals for greater permanence. Incidentally, as of this writing, Murrayite is almost impossible to obtain. Alternatives, such as various types of fingernail polish, are usually suitable.

Trophi mounting. Since genus and species identifications are often dependent on the anatomical details of the trophi, it is sometimes essential that permanent or semipermanent mounts of these jaws be made. The following method, requiring some skill and practice and much patience, is modified from Myers (1937a). Place a drop of 1:10 Clorox, or similar caustic alkali, just inside the concavity of a shallow concavity slide, and place a similar drop just outside the concavity. Then place a 22-mm-square coverslip on the outside drop and push it over the concavity until it is almost in contact with the inside drop. Next place the rotifer in the inner drop by means of a bristle or Irwin loop. If the coverslip is pushed slowly over the inner drop, the rotifer will be drawn under. And by working the coverslip over the concavity by short pushes, and adding small quantities of solution after each advance, the rotifer will be forced into the acute angle formed by the edge of the concavity and the under surface of the coverslip. The edges of the coverslip should then be carefully dried and painted with petroleum jelly or Murrayite. The position of the rotifer should be noted, and in about a half-hour it will all have dissolved except the trophi. Such trophi slides will last for several to many months, and if a petroleum jelly seal is used, the trophi may be oriented by tapping or moving the coverslip slightly.

De Smet (1998) has reviewed several methods and proposed a rapid method for use in scanning electron microscopy with excellent results (Fig. 7.47). Animals are dissolved in 0.5 μL of hypochlorite on a coverslip that is fixed to a slide with glycerin. After a few minutes the trophi are exposed. Gradually remove the hypochlorite with a pipette and replace with distilled water; about 10 sessions are necessary. Allow the preparation to dry while adjusting the trophi with a dissecting needle. Either mount in glycerin jelly or sputter coat, depending on viewing preference.

Taxonomy. Rotifer taxonomy owes much to Harring's painstaking "Synopsis of the Rotatoria," which appeared in 1913, bringing order out of chaotic nomenclature. As has been pointed out several times, the rotifer fauna of the world is enormous, with certain genera reputed to contain nearly 500 species. In the genus *Lecane*, about 90 species are reported from North America.

The establishment of new species is becoming ever more demanding of precise and detailed descriptions. Specifically, the description of the carona and trophi requires both SEM and light microscope methods to accurately identify species-specific features and reduce the chance for overlooking variation. New methods of preparing specific feeding and locomotor structures are being developed and will need to be employed before a new species achieves general acceptance. In spite of the efforts of specialists, some outdated terminology still persists in the literature.

The arrangement of genera and families used here is greatly modified from the plan given by Remane in Das Tierreich (1929–1933). It is based primarily on the fundamental structure and modifications of the mastax, in addition to other obvious morphological features. Recently, several classifications of the Rotifera have been offered that rely heavily on character polarities and computer-generated estimates of phylogenetic distance. A few include the Rotifera with the parasitic Anthocephala as a single phylum. The arrangement followed here is after Melone et al. (1998). Table 7.1 includes the genera given in the key that follows.

The key includes all freshwater genera likely to be encountered in the United States. It excludes genera that are exclusively or nearly always amphibious. The key to the Bdelloidea is modified from Ricci and Melone (2000) and uses simple and relatively easy to observe features of adult specimens. A few modifications have been added following data in Vidrine et al. (1985). This key does not include about 30 rare freshwater genera that have been found only on other continents and that may eventually be found in the United States. Numbers of species are only approximations for the numbers reported or likely to be reported for the United States.

Table 7.1. The Classes, Orders, Families, and Genera of Rotifera Occurring Fresh Waters of the United States

Class Seisonidea

Class Bdelloidea (ramate mastax)

- Philodinidae
  - *Dissotrocha, Embata, Macrotrachela, Mniobia, Philodina, Pleuretra, Rotaria*
- Habrotrochidae
  - *Ceratotrocha, Habrotrocha, Scepanotrocha*
- Philodinavidae
  - *Abrochtha, Henoceros, Philodinavus*
- Adinetidae
  - *Adineta, Bradysclea*

Class Monogononta

- Order Flosculariacea (malleoramate mastax)
  - Flosculariidae
    - *Beauchampia, Floscularia, Lacinularia, Limmas, Octotrocha, Pseudoecistes, Ptygura, Sinantherina*
  - Conochilidae
    - *Conochilus, Conochiloides*
  - Hexarthridae
    - *Hexarthra*
  - Testudinellidae
    - *Pompholyx, Testudinella*
  - Filniidae
    - *Filindia, Tetramastix*
  - Trochosphaeridae
    - *Trochosphaera*
- Order Collothecacea (uncinate mastax)
  - Collothecidae
    - *Collotheca, Stephanoceros*
  - Atrochidae
    - *Acyclus, Atrochus, Cupelopagis*

Order Ploima

Notommatidae

Notommatinae (virgate or virgate forcipate mastax)

Cephalodella, Drilophaga, Enteroplea, Eosphora, Eothinia, Itura, Monommata, Notommata, Pleurotrocha, Resticula, Rousseletia, Scaridium, Sphyrias, Taphrocampa, Tylogrocha

Tetrasiphoninae (modified virgate mastax)

Tetrasiphon

Proalidae (modified malleate mastax)

Bryceella, Proales, Proalinopsis

Lindiidae (cardate mastax)

Lindia

Birgeidae (aberrant mastax, with a pair of pseudunci)

Birgea

Dicranophoridae (forcipate mastax)

Albertia, Aspelta, Dicranophorus, Dorria, Encentrim, Erignatha, Myersinella, Pedipartia, Streptognatha, Wierzejskiella

Synchaetidae (virgate or virgate-forcipate mastax)

Polyarthra, Ploesoma, Pseudoploesoma, Synchaeta

Microcodonidae (forcipate mastax)

Microcodon

Gastropodidae (virgate mastax)

Ascomorpha, Chromogaster, Gastropus

Trichocercidae (virgate mastax)

Ascomorphella, Elosa, Trichocerca

Asplanchidae

Asplanchna, Asplanchnopus, Harringia

Epiphanidae (malleate mastax)

Cyrtonia, Epiphanes, Mikrocodides, Rhinoglena

Brachionidae (malleate mastax)

Anuraeopsis, Argonotholca, Brachionus, Kellicottia, Keratella, Notholca, Platyias

Euchlanidae (malleate mastax)

Beauchampiella, Dipleuchlanis, Euchlanis, Tripleuchlanis

Mytilinidae (malleate mastax)

Lophoocharis, Mytilina

Trichotriidae (malleate mastax)

Macrochaetus, Trichotria

Colurellidae (malleate mastax)

Colurella, Lepadella, Paracolurella, Squatinella

Lecanidae (malleate mastax)

Lecane (Monostyla)

KEY TO GENERA OF FRESHWATER ROTIFERA OF THE UNITED STATES

1. With a single ovary (Figs. 7.1; 7.2); mastax not ramate; lorica or secreted tube present or absent **Class MONOGONONTA, 3**

With paired ovaries (Figs. 7.3; 7.11); mastax ramate (Fig. 7.8E); without a lorica or secreted tube **2**

2. Ovaries without a vitellarium; male well developed; corona disk rudimentary; one genus and several species commensal on marine Crustacea . . . . **Class SEISONIDEA**

Each ovary with a vitellarium (Fig. 7.18); males absent; reproduction entirely by parthenogenesis; corona usually well developed; body cylindrical, highly contractile, and telescopic; free-swimming or with leechlike creeping movements; with a dorsal proboscis; toes, when present, usually retracted within foot except when creeping; two nonretractile spurs usually present (Figs. 7.44G; 7.45J, K); about 380 species worldwide. . . **Class BDELLOIDEA, 94**

3. Female almost invariably free-swimming or creeping; lorica present or absent; without a secreted tube; corona not especially large; foot, when present, typically with two toes; solitary **5**

Female usually sessile; without a lorica but often in a secreted tube; corona large; foot, when present, long and annulated and without toes but with a terminal attachment disk; solitary or colonial **4**

4. Corona surrounded by two concentric wreaths of cilia with a ciliated furrow between (Fig. 7.4B, D); cilia of outer wreath always shorter than inner; mouth not central; usually with one or two well-developed antennae; mastax malleoramate (Fig. 7.8C); solitary or colonial **Order FLOSCULARIACEA, 81**

Corona very large, without such a double wreath of cilia; without well-developed antennae; mouth central; mastax uncinate or poorly developed; buccal field horseshoe shaped and located at the bottom of a large, concave infundibulum (Fig. 7.43); solitary, often sessile **Order COLLOTHECACEA, 90**

5. Foot and toes always absent **6**

Foot always present; toes usually present **20**

6. With 12 transparent, movable, sword- or blade-shaped lateral appendages (Fig. 7.24B); body short and more or less cylindrical; about 10 variable species, of which several are common plankton forms . . . . **Order PLOIMA, *Polyarthra***

Without 12 such appendages **17**

7. With six stout, muscular, setose appendages; body conical; with a double ciliary wreath and a ciliated groove between (Fig. 7.24A); the "jumping rotifer"; usually in small alkaline lakes and ponds during the summer months; several species **Order FLOSCULARIACEA, *Hexarthra***

Without such appendages **8**

8. Without a lorica but with three or four long, movable spines (Fig. 7.24E, F) **Order FLOSCULARIACEA, 9**

Lorica present or absent; spines, if present, rigid and confined to anterior and posterior margins of lorica **10**

9. With two lateral and one posterior spine (Fig. 7.24E); several common limnetic species, one very common species . . . . ***Filinia***

With two unequal lateral and two unequal posterior spines (Fig. 7.24F); one uncommon plankton species; sometimes included in *Filinia* . . . . ***Tetramastix***

10. With a spinous lorica composed of two plates immovably fused laterally **Order PLOIMA, 11**

Lorica absent, or present and not spinous **14**

11. With one very long posterior spine and four or six anterior spines, of which three are very long (Fig. 7.24H, J); two limnetic species, one of which is very common . . . . ***Kellicottia***

With one, two, or no posterior spines, and with six short to medium anterior spines; highly variable plankton and limnetic species **12**

12. Dorsal surface of lorica with a pattern of polygonal facets (Figs. 7.25A–L; 7.26; 7.27); cyclomorphic; about 20 species, of which several are very common . . . . ***Keratella***

Dorsal surface of lorica plain, pustulose, or longitudinally striated; usually in ponds **13**

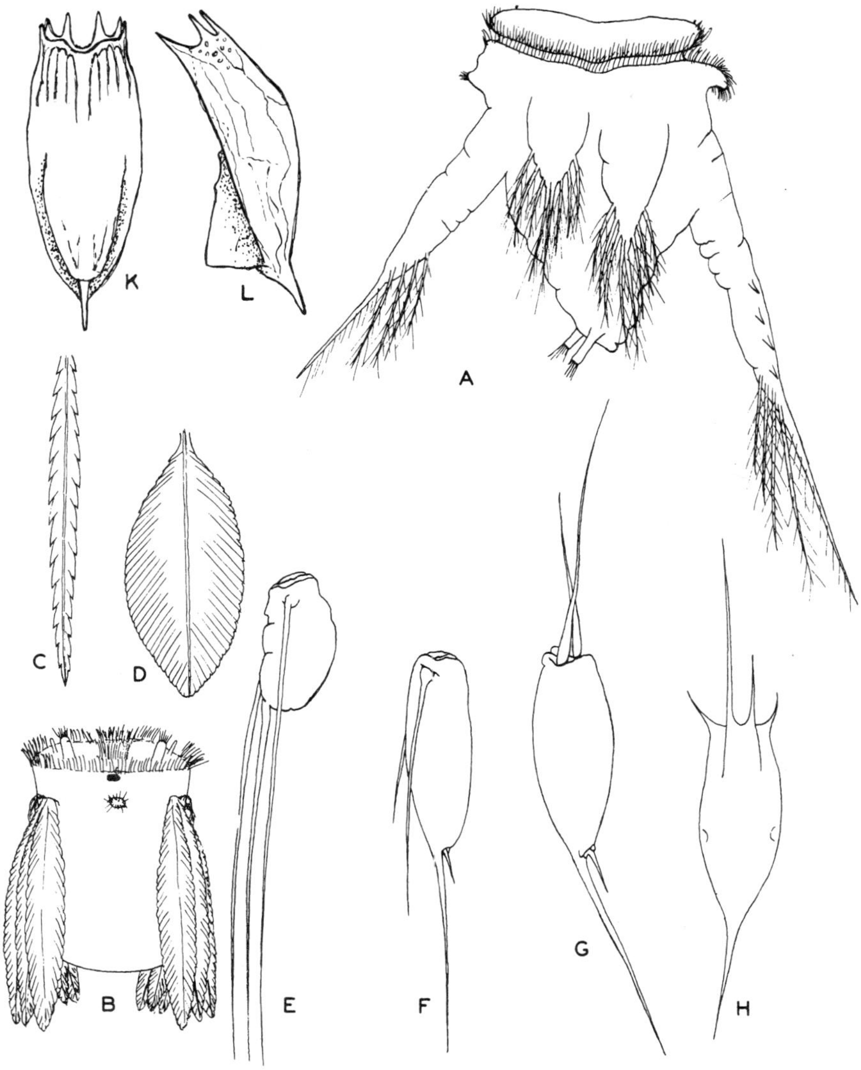

Fig. 7.24. Structure of Ploima and Flosculariacea. A, right side of *Hexarthra;* B, dorsal view of *Polyarthra;* C and D, single appendages of *Polyarthra;* E, lateral view of *Filinia;* F, *Tetramastix opoliensis* with anterior spines in normal position; G, *T. opoliensis* with anterior spines retracted forward; H, dorsal view of lorica of *Kellicottia bostoniensis;* J, dorsal view of *K. longispina;* K, ventral view of *Argonotholca;* L, side view of *Argonotholca.* (K and L modified from Donner 1966.)

13. With a broad ventral bulbous keel (Fig. 7.24K, L); rare in coastal waters, common in brackish areas . . . . ***Argonotholca***
Without such a keel (Fig. 7.25M, N); about 10 species . . . . ***Notholca***

14. Large transparent species, usually 400 to 2000 μm long **15**
Small, more opaque species, less than 200 μm long **16**

15. Sac-shaped (Fig. 7.29), with a well-developed corona; intestine and anus lacking; mastax incudate; vitellarium horseshoe-shaped or globose (Fig. 7.28A, B); often viviparous; predatory; morphology variable; body sometimes with humps or winglike processes; about 6 very common plankton species . . . . **Order PLOIMA, *Asplanchna***

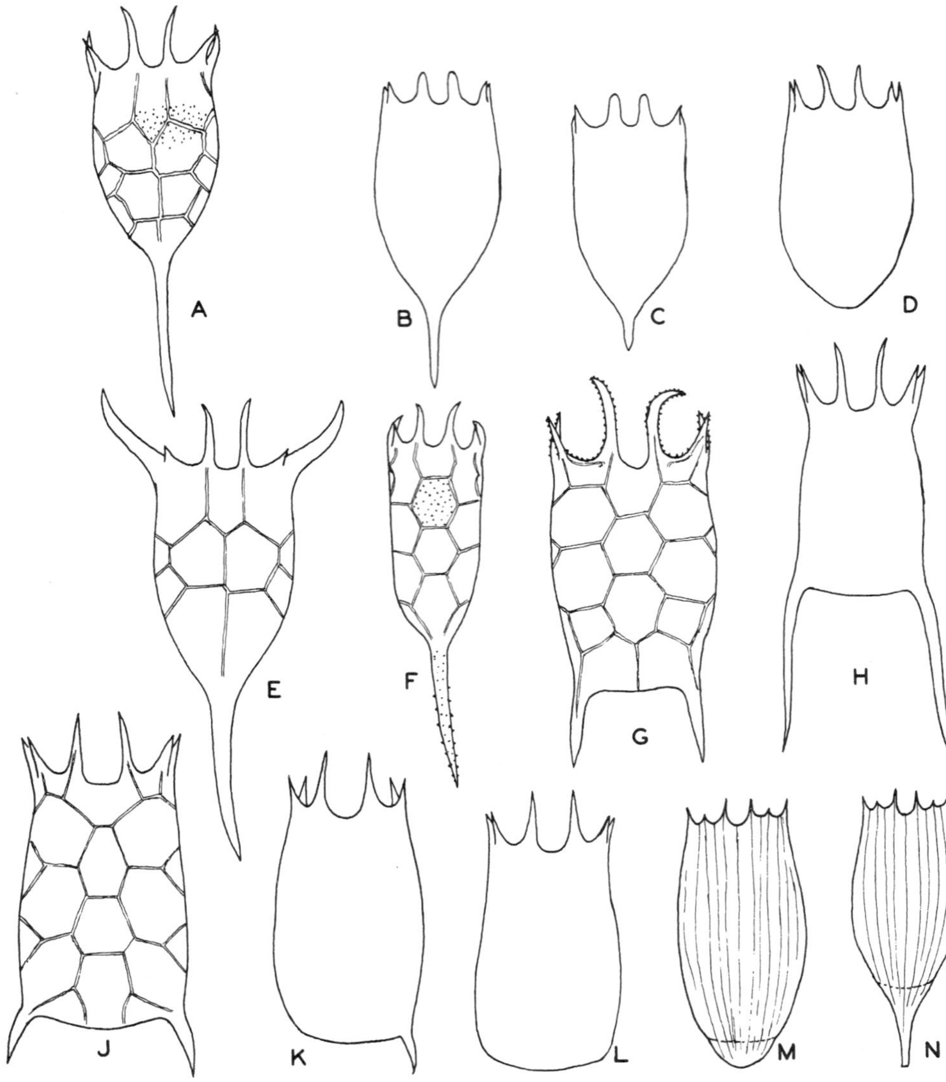

Fig. 7.25. Lorica of typical species of *Keratella* and *Notholca* (dorsal views). A–D, *Keratella cochlearis;* E, *K. taurocephala* (an acid water species); F, *K. gracilenta;* G, *K. serrulata* (common in acid waters); H–L, *K. quadrata;* M and N, *Notholca.* Dorsal plates shown only for A, E–G, and J; small amount of pustulation shown only in A and F. (A–L modified from Ahlstrom 1943.)

Spherical and without a true corona, the body being divided into unequal hemispheres by a ciliary band; intestine and anus present; mastax malleoramate (Fig. 7.28C); up to 750 μm in diameter; very rare in small polluted ponds; more likely to be found in southern states; 2 species . . . .**Order FLOSCULARIACEA, *Trochosphaera***

16. With a thin lorica, or lorica absent; sac-shaped; with a very large, lobed stomach filling much of the body cavity; anus absent; usually densely colored or opaque (Fig. 7.28D); mastax virgate; several plankton species . . . .**Order PLOIMA, *Ascomorpha***

With a thick lorica; not sac-shaped **17**

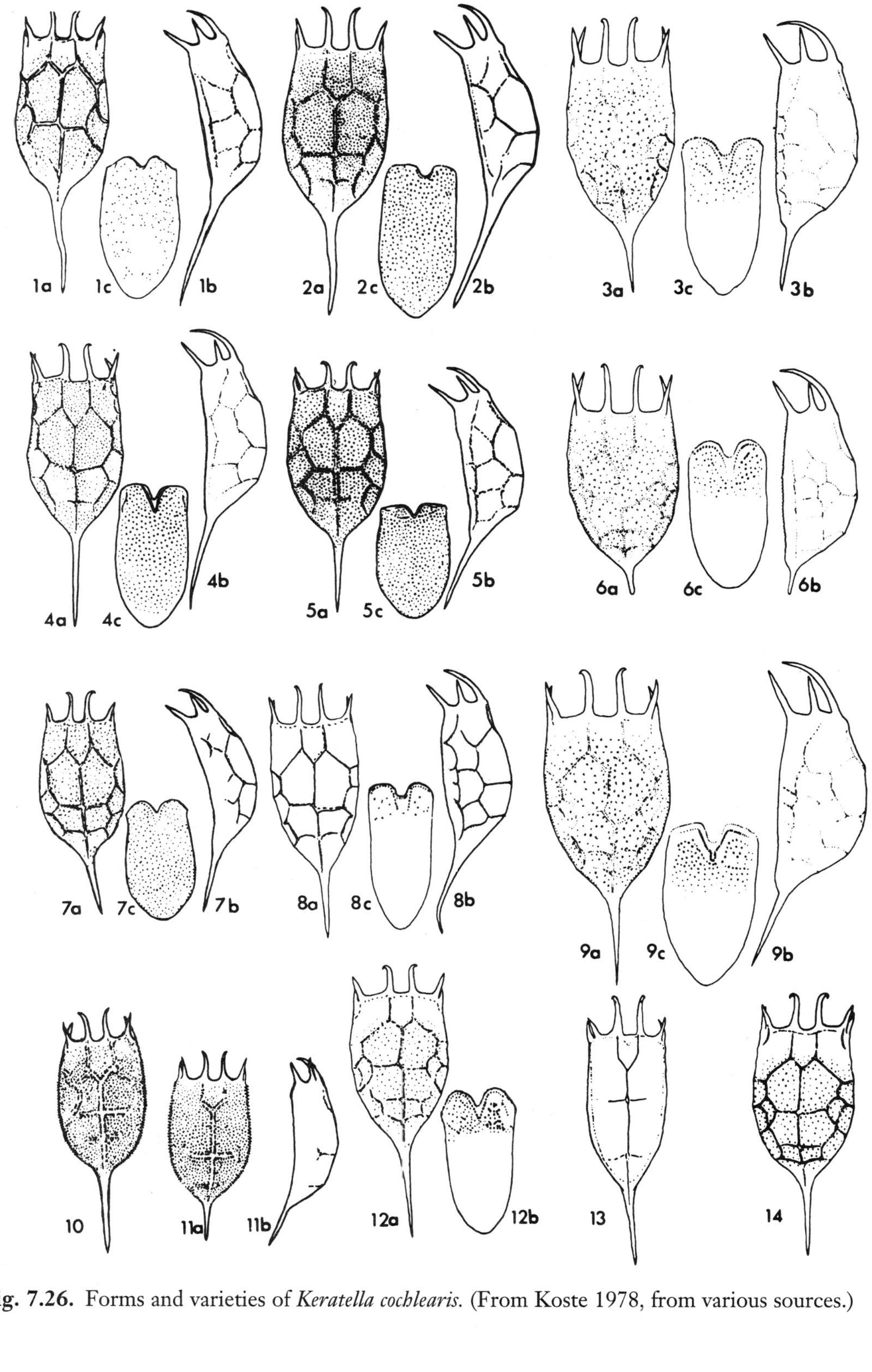

Fig. 7.26. Forms and varieties of *Keratella cochlearis.* (From Koste 1978, from various sources.)

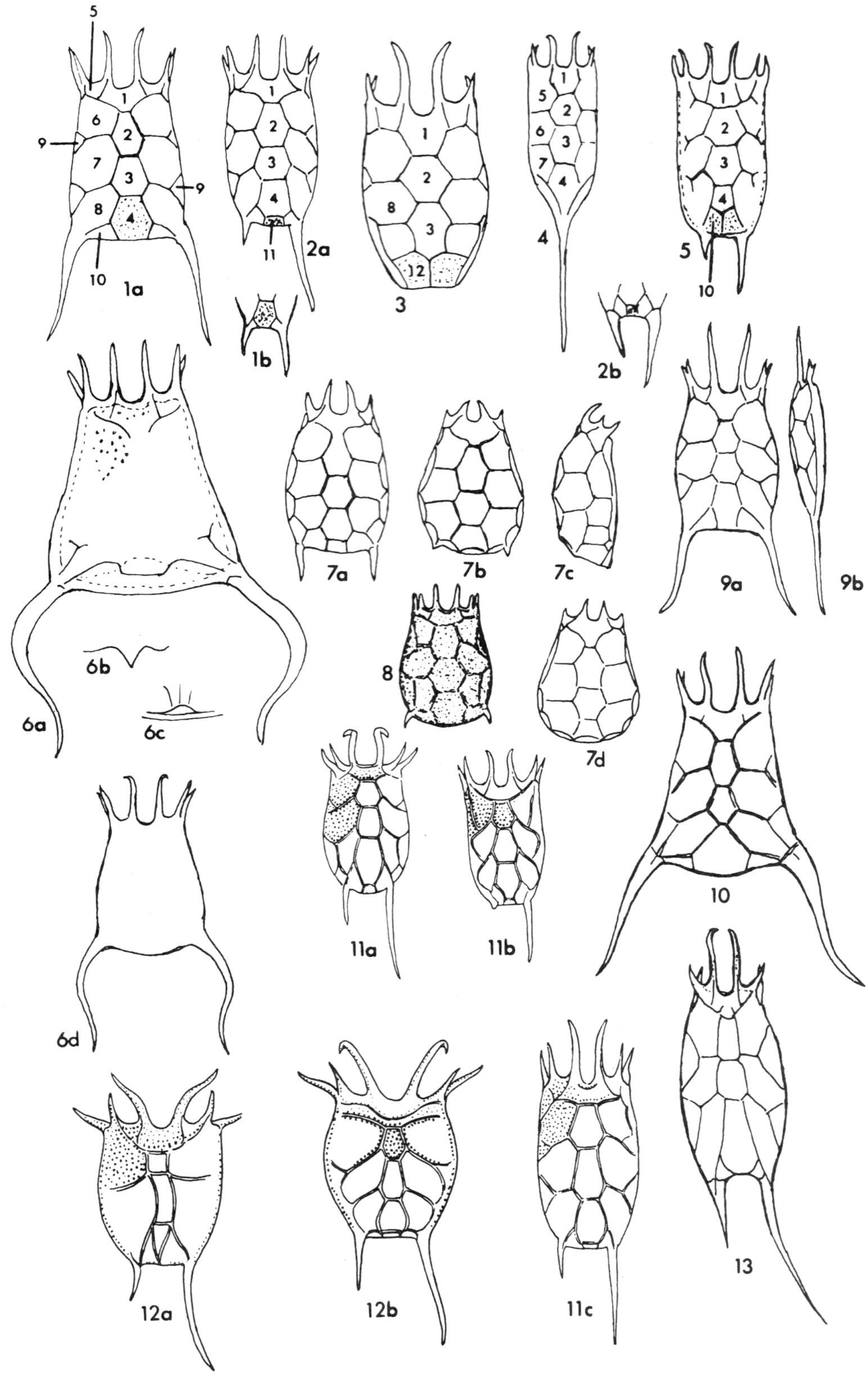

Fig. 7.27. Various species and varieties of *Keratella.* (From Koste 1978, from various sources.)

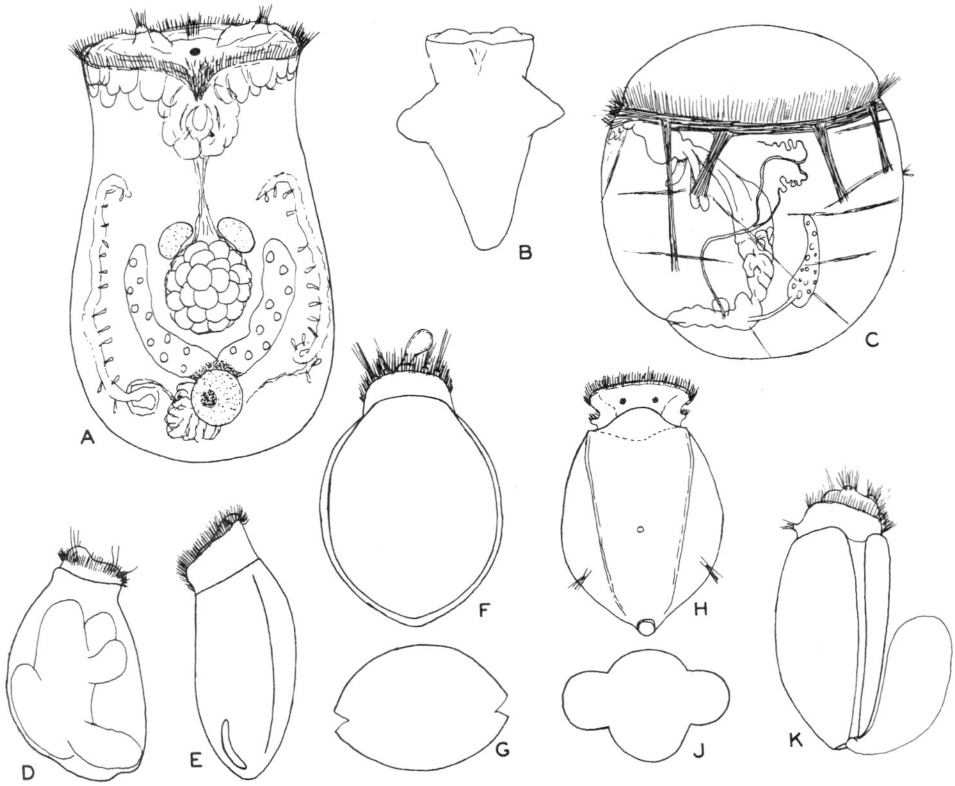

Fig. 7.28. Structure of Ploima and Flosculariacea. A, ventral view of *Asplanchna* (muscle fibers not shown); B, *Asplanchna* with winglike processes; C, lateral view of *Trochosphaera;* D, lateral view of *Ascomorpha* showing lobed stomach; E, lateral view of *Elosa worralli;* F, dorsal view of *Chromogaster;* G, cross section of *Chromogaster;* H, dorsal view of *Pompholyx;* J, cross section of *Pompholyx;* K, lateral view of *Anuraeopsis fissa* carrying egg.

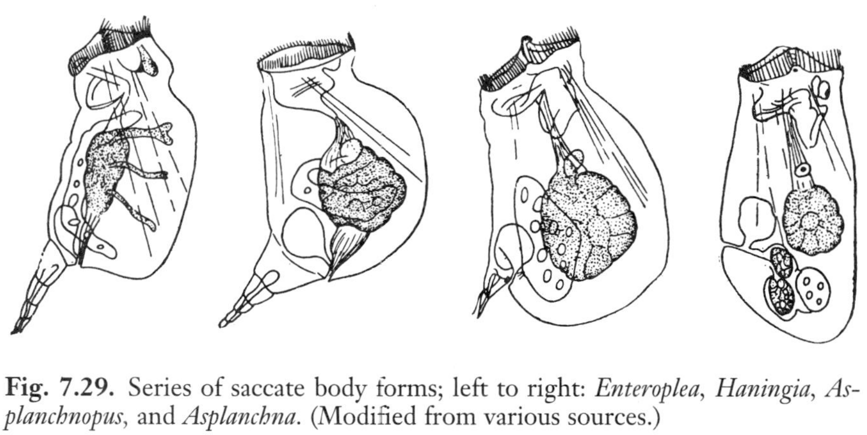

Fig. 7.29. Series of saccate body forms; left to right: *Enteroplea, Haningia, Asplanchnopus,* and *Asplanchna.* (Modified from various sources.)

17. Lorica composed of a single cylindrical piece, trilobate in section and with a crescentic opening posteriorly on the left side (Fig. 7.28E); trophi asymmetrical; one uncommon species **Order PLOIMA, *Elosa***

Lorica composed of a dorsal and a ventral plate (Fig. 7.28F–K); trophi symmetrical **18**

18. Corona with one to three fingerlike processes in addition to ciliated papillae; lorica composed of two convex plates fused laterally (Fig. 7.28F, G); mastax virgate; probably two species **Order PLOIMA, *Chromogaster***

Corona without fingerlike processes **19**

19. Lateral edges of dorsal and ventral plates of lorica closely confluent; flat or four-lobed in cross section; two frontal eyespots (Fig. 7.28H, J); mastax malleoramate; several species, including only two in the U.S. .... **Order FLOSCULARIACEA, *Pompholyx***

Lateral edges of dorsal and ventral plates of lorica connected by infolded cuticle; ventral plate flat, dorsal plate arched (Fig. 7.28K); mastax malleate; 1 common and variable species; several very rare species .... **Order PLOIMA, *Anuraeopsis***

20. With a well-developed rigid lorica **21**

Lorica absent, or present and poorly developed and flexible .... **Order PLOIMA, 43**

21. Foot attached to the ventral surface; lorica composed of one rigid piece, never dorsoventrally flattened; foot usually annulated (Fig. 7.30A, B); mastax virgate ..... **Order PLOIMA, 22**

Foot terminal or subterminal **24**

22. Surface of lorica plain; body laterally compressed; lorica with a small ventral opening for the foot; foot with one or two toes (Fig. 7.30A); 3 species .... ***Gastropus***

Surface of lorica marked with ridges or vesicles; body not laterally compressed; lorica usually open along the midventral line; foot with two toes (Fig. 7.30B, C) **23**

23. Corona with frontal palps; foot annulated; lorica open along the midventral line (Fig. 7.30B); mastax virgate and adapted for prehension; about 4 species .... ***Ploesoma***

Corona without frontal palps but with two unique juxta-buccal protuberances; foot composed of two telescoping segments and emerging from an oval opening in the lorica (Fig. 7.30C); mastax virgate but not adapted for prehension; 1 rare species in acid waters ***Pseudoploesoma***

24. Body cylindrical, more or less curved, and asymmetrical; lorica a single cylindrical piece, often with teeth and longitudinal grooves or ridges; toes spinelike, of unequal length, and with several small spinules at their base (Fig. 7.30D–J); creeping or free-swimming; mastax virgate; trophi asymmetrical (Fig. 7.10C); common; many species ... **Order PLOIMA, *Trichocerca***

With another combination of characters; rarely asymmetrical; toes not spinelike **25**

25. Foot long, retractile, annulated, and terminating in a tuft of cilia; dorsal and ventral plates of lorica completely fused laterally; greatly flattened dorsoventrally and sometimes nearly circular (Fig. 7.31B); mastax malleoramate; 120 to 300 μm long; about 15 littoral species **Order FLOSCULARIACEA, *Testudinella***

Foot ending in one or two .... **Order PLOIMA, 26**

26. Entire margin spiny; lorica heavy, with 4 to 10 long dorsal spines and six posterior spines; foot short (Fig. 7.31A); mastax malleate; about 6 littoral species .... ***Macrochaetus***

Entire margin of body not spiny **27**

27. Head bearing a wide, circular shield; lorica cylindrical or pyriform, often with one or two median dorsal spines, or several spines on posterior margin of lorica; sometimes with a spine at the base of the toes (Fig. 7.31C, D); mastax malleate; about 10 uncommon species ***Squatinella***

Head without a wide, circular shield **28**

28. Usually with two spines at base of foot, or rarely with two posterior dorsolateral spines on the lorica; lorica composed of one boxlike piece, usually thick and with large facets (Fig. 7.31E, K); mastax malleate; about 8 species .... ***Trichotria***

Without two such spines **29**

29. Body moderately flattened dorsoventrally; dorsal and ventral plates of lorica completely fused laterally; anterior dorsal margin of lorica usually with four or six spines; posterior margin with or without spines; two toes; mastax malleate **30**

With another combination of characters; mastax usually malleate or submalleate; never dorsoventrally flattened and never with four or six spines on anterior dorsal margin of lorica **31**

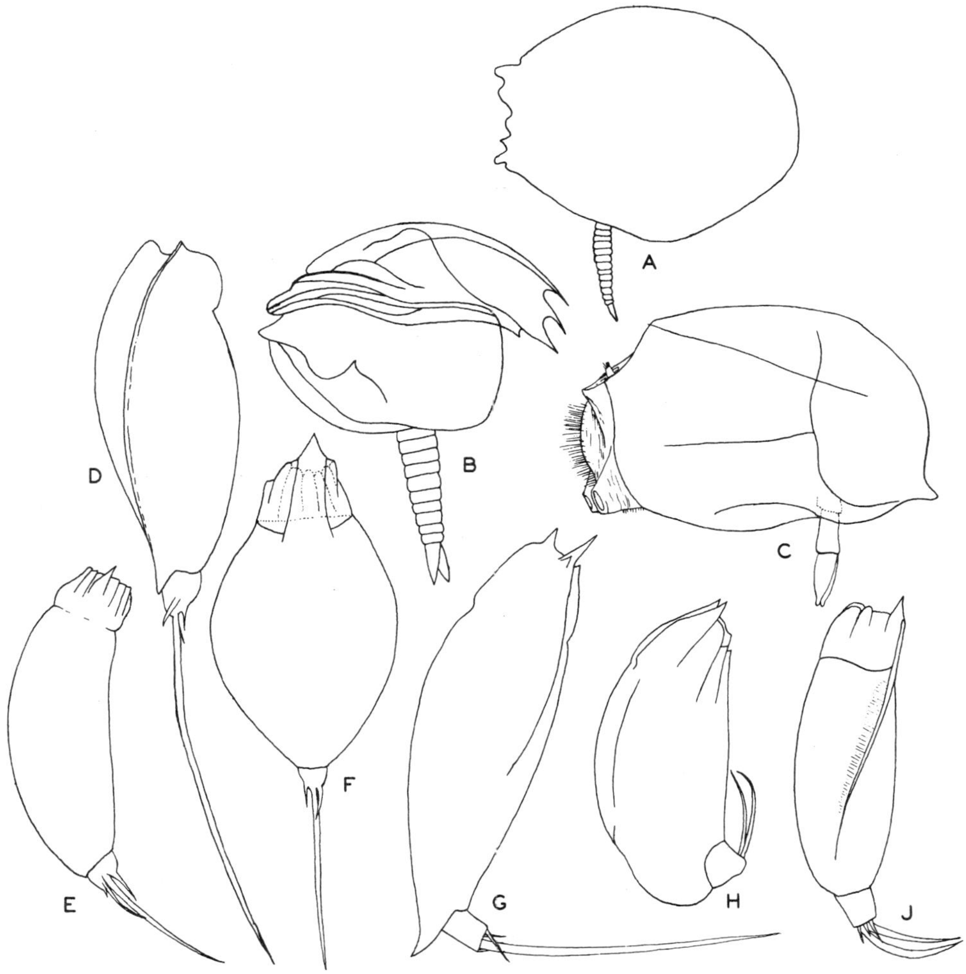

Fig. 7.30. Structure of ploimate rotifers. A, lateral view of *Gastropus;* B, lateral view of *Ploesoma;* C, lateral view of *Pseudoploesoma formosum;* D to J, common species of *Dichocerca* (contracted). (B modified from Wesenberg-Lund 1930; C modified from Myers 1938.)

30. Foot segmented and retractile (Fig. 7.32N); three species .... ***Platyias***
Foot long, annulated, retractile, not segmented (Fig. 7.32K); about 20 plankton species, some of which are common and highly variable, especially in hard waters; sometimes in dense populations .... ***Brachionus***

31. Strongly compressed laterally; lorica composed of two lateral plates; frontal head hood present; toes long and tapering; lorica open along anterior, ventral, and posterior margins (Fig. 7.31 F, G); small species **32**
Not strongly compressed laterally; lorica of different construction; frontal hood absent **33**

32. Terminal foot segment long (Fig. 7.31F); three rare species .... ***Paracolurella***
Terminal foot segment short; about 20 species, some common (Fig. 7.31G, H) ..... ***Colurella***

33. Foot long; toes long; lorica without spines or plates **34**
Foot short; toes short to long **35**

34. Lorica vase shaped and thin; toes shorter than rest of body (Fig. 7.33B); mastax virgate; 1 species .... ***Scaridium***
Lorica pear shaped; toes about as long as rest of body (Fig. 7.33A); mastax malleate; 1 rare species .... ***Beauchampiella***

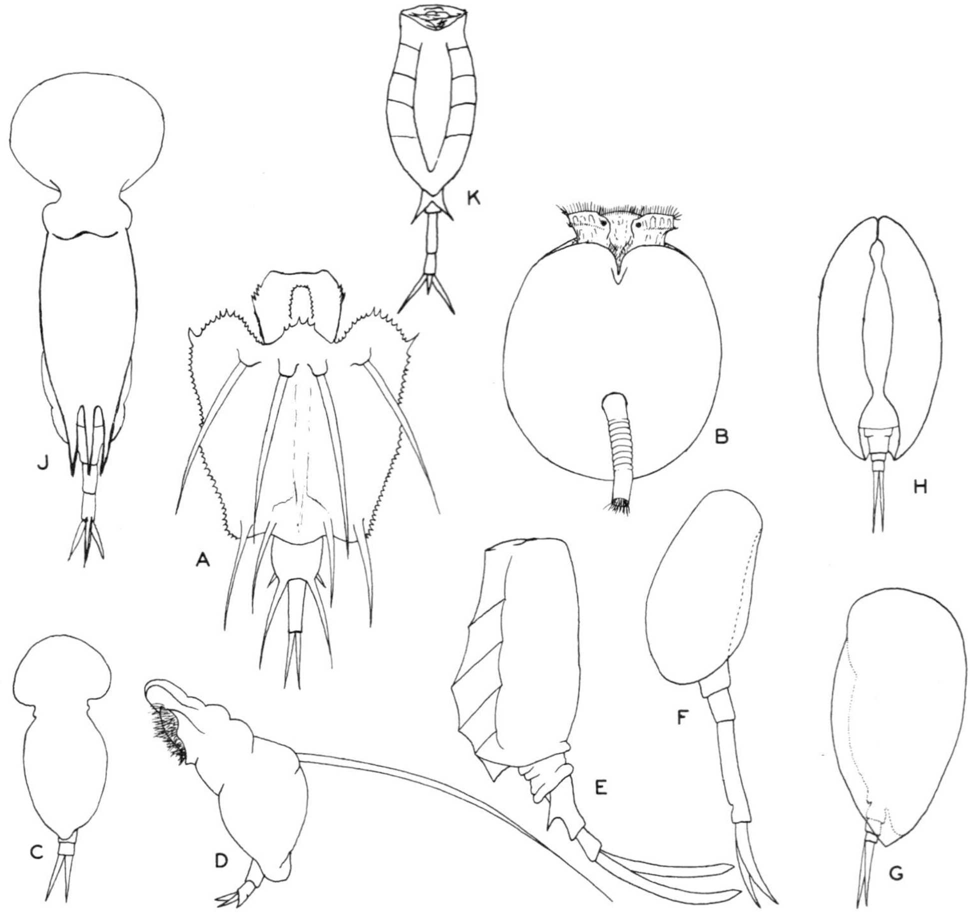

Fig. 7.31. Structure of Colurellidae, Trichotriidae, Testudinellidae, and Brachionidae. A, dorsal view of contracted *Macrochaetus;* B, ventral view of *Testudinella;* C, dorsal view of *Squatinella;* D, lateral view of a spined species of *Squatinella;* E, lateral view of contracted *Trichotria;* F, lateral view of contracted *Paracolurella;* G, lateral view of contracted *Colurella;* H, ventral view of contracted *Colurella;* J, dorsal view of *Squatinella;* K, dorsal view of *Trichotria.* (D modified from Hauer 1935; F modified from Myers 1934a.)

35. Lorica composed of a ventral plate and two lateral plates, the edges of the latter forming two dorsal ridges; often with anterior and posterior spines (Fig. 7.33C–E); about 10 littoral species, usually in hard waters . . . . ***Mytilina***
Lorica constructed differently **36**

36. Lorica composed of one piece, capacious, rigid, boxlike, and with a prominent dorsal keel or low ridge extending the entire length of the lorica (Fig. 7.33F, G); several littoral species ***Lophocharis***
Lorica usually composed of a dorsal and a ventral plate **37**

37. Dorsal and ventral plates rigidly united at the edges; with an anterior opening for the protrusion of the head and a posterior opening through which the foot projects; foot well developed (Fig. 7.33H–K); about 35 very small species, 60 to 180 μm long; littoral . . . . ***Lepadella***
Dorsal and ventral plates not united at the edges; foot not well developed; associated with the substrate but occasionally found in the plankton **38**

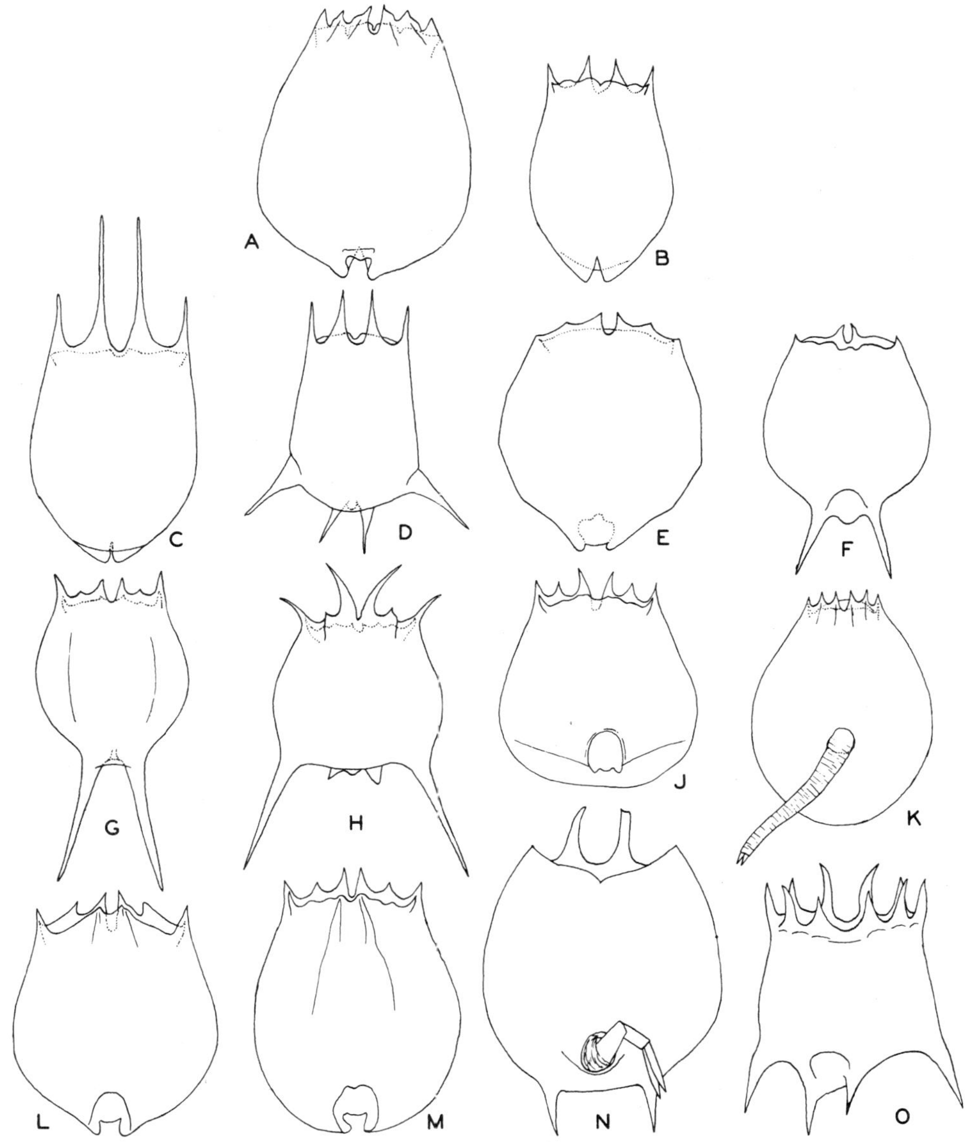

Fig. 7.32. Structure of lorica of some common species of *Brachionus* and *Platyias*. A dorsal view of *Brachionus plicatilis*, a species characteristic of highly alkaline waters; B, ventral view of *B. calyciflorus;* C and D, dorsal views of *B. calyciflorus;* E, dorsal view of *B. angularis;* F, ventral view of *B. caudatus;* G, dorsal view of *B. havanaensis;* H, dorsal view of *B. quadridentata;* J, ventral view of *B. quadridentata;* K, ventral view of *B. pterodinoides*, a species characteristic of highly alkaline waters; L, dorsal view of *B. rubens;* M, ventral view of *B. urceolaris;* N, ventral view of *Platyias quadricornis;* O, ventral view of *P. patulus.* (Redrawn and modified from Ahlstrom 1940.)

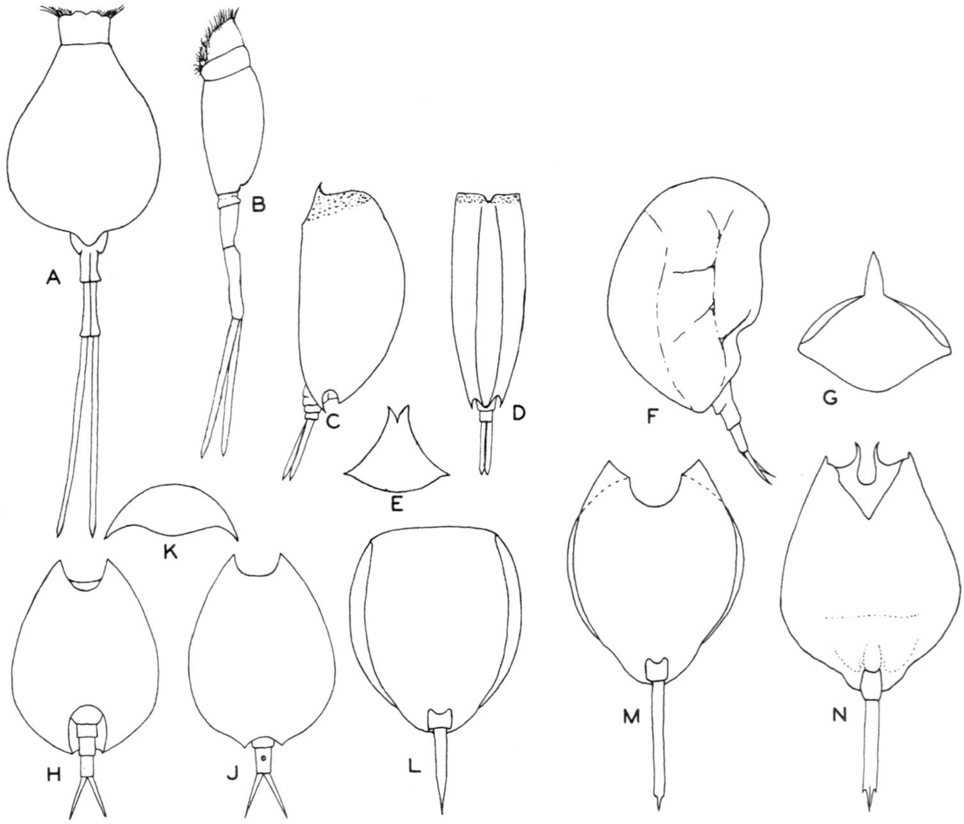

Fig. 7.33. Structure of ploimate rotifers. A, contracted *Beauchampiella eudactylota;* B, lateral view of *Scaridium longicaudum;* C, lateral view of contracted *Mytilina;* D, dorsal view of contracted *Mytilina;* E, cross section of *Mytilina;* F, lateral view of contracted *Lophocharis;* G, cross section of *Lophocharis;* H, ventral view of contracted *Lepadella;* J, dorsal view of contracted *Lepadella;* K, cross section; L–N, ventral views of three contracted species of *Monostyla* (=*Lecane*).

38. With a single long toe; strongly compressed dorsoventrally and oval to ovate in outline (Fig. 7.33L–N); about 25 small littoral species; usually united with *Lecane* .... ***Monostyla***
With two toes **39**

39. Prismatic or spindle-shaped; toes tapering, usually curved; lorica thin and poorly developed (Fig. 7.34A–C); mastax virgate; more than 100 littoral species .... ***Cephalodella***
More or less flattened dorsoventrally; toes long and slightly fusiform or parallel sided and straight; lorica rigid and well developed (Fig. 7.34D–K); mastax malleate or submalleate **40**

40. Dorsal plate much smaller than ventral plate (Fig. 7.34D); cuticle connecting the dorsal and ventral plates in the form of a deep groove dividing the body cavity into two unequal portions (Fig. 7.34E); 1 species .... ***Dipleuchlanis***
Dorsal plate the same size or larger than the ventral plate **41**

41. Foot with two rudimentary segments, of which only the posterior is movable; toes parallel sided; strongly compressed dorsoventrally (Fig. 7.34H–K); many small littoral species; (Figs. 7.35, 7.36) 150 to 250 μm long; includes *Monostyla* .... ***Lecane***
Foot with two or three obscure segments; toes usually slightly fusiform; ventral plate flat; dorsal plate arched and convex, sometimes keeled (Fig. 7.2) **42**

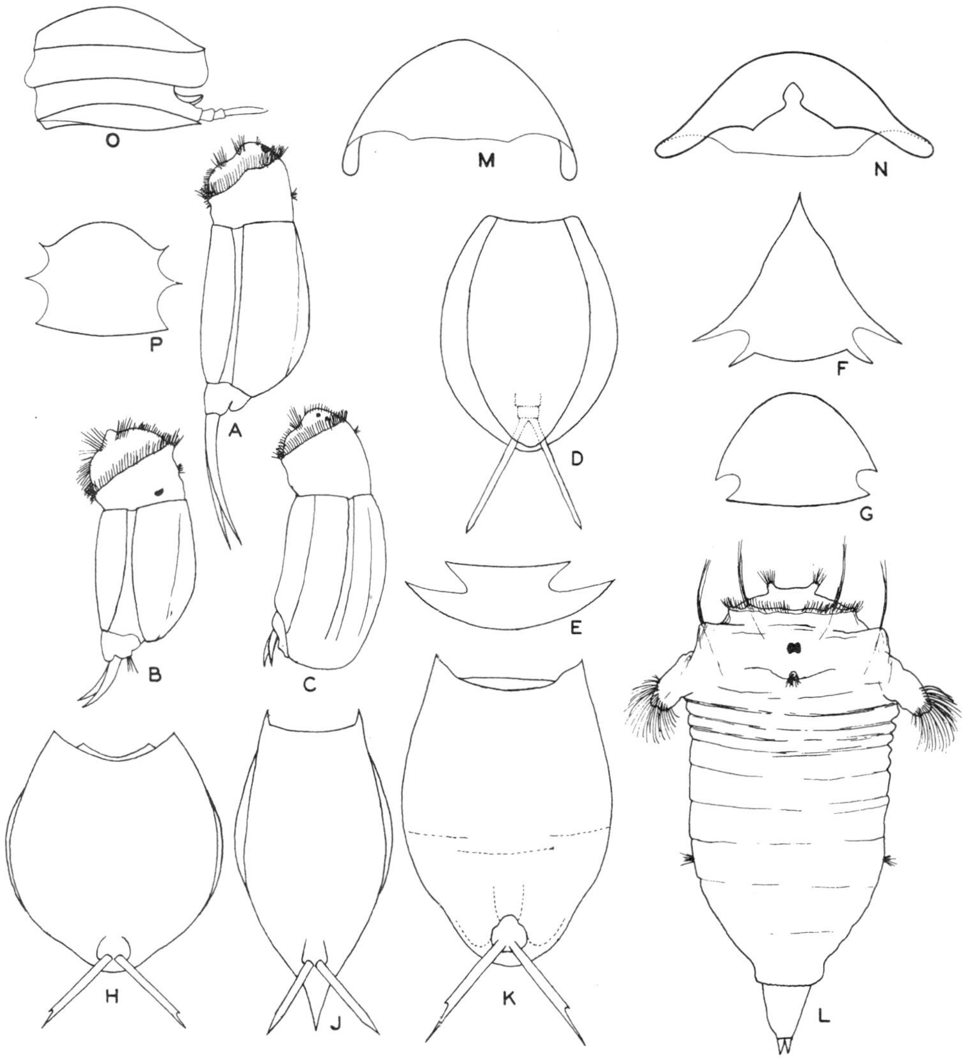

Fig. 7.34. Structure of ploimate rotifers. A–C, three typical species of *Cephalodella;* D, dorsal view of retracted *Dipleuchlanis;* E, cross section of *Dipeuchlanis;* F and G, cross sections of *Euchlanis;* H–K, ventral views of typical species of *Lecane* (contracted); L, dorsal view of *Synchaeta;* M, cross section of *Euchlanis;* N, anterior edge of *Euchlanis;* O, lateral view of lorica of *Tripleuchlanis;* P, cross section of *Tripleuchlanis.* (D–G modified from Myers 1930; H–K modified from Harring and Myers 1926; M-P from Myers 1930.)

42. Lateral sulci very wide, with a stiff flange (Fig. 7.34O, P); in brackish waters. . . ***Tripleuchlanis***
Lateral sulci without a flange (Figs. 7.2; 7.34F, G); about 15 large species, some very common ***Euchlanis***

43. Foot rudimentary; body saccate (Fig. 7.37G); one species parasitic in *Volvox* colonies ***Ascomorphella***
Foot not rudimentary; body not saccate; free-living. . . . . **.44**

44. With four long, prominent sensory bristles on the corona; body conical; toes small (Fig. 7.34L); mastax virgate; about 15 plankton species, several of which are common; some species occur in salt and brackish water . . . . ***.Synchaeta***
Without four such bristles . . . . **.45**

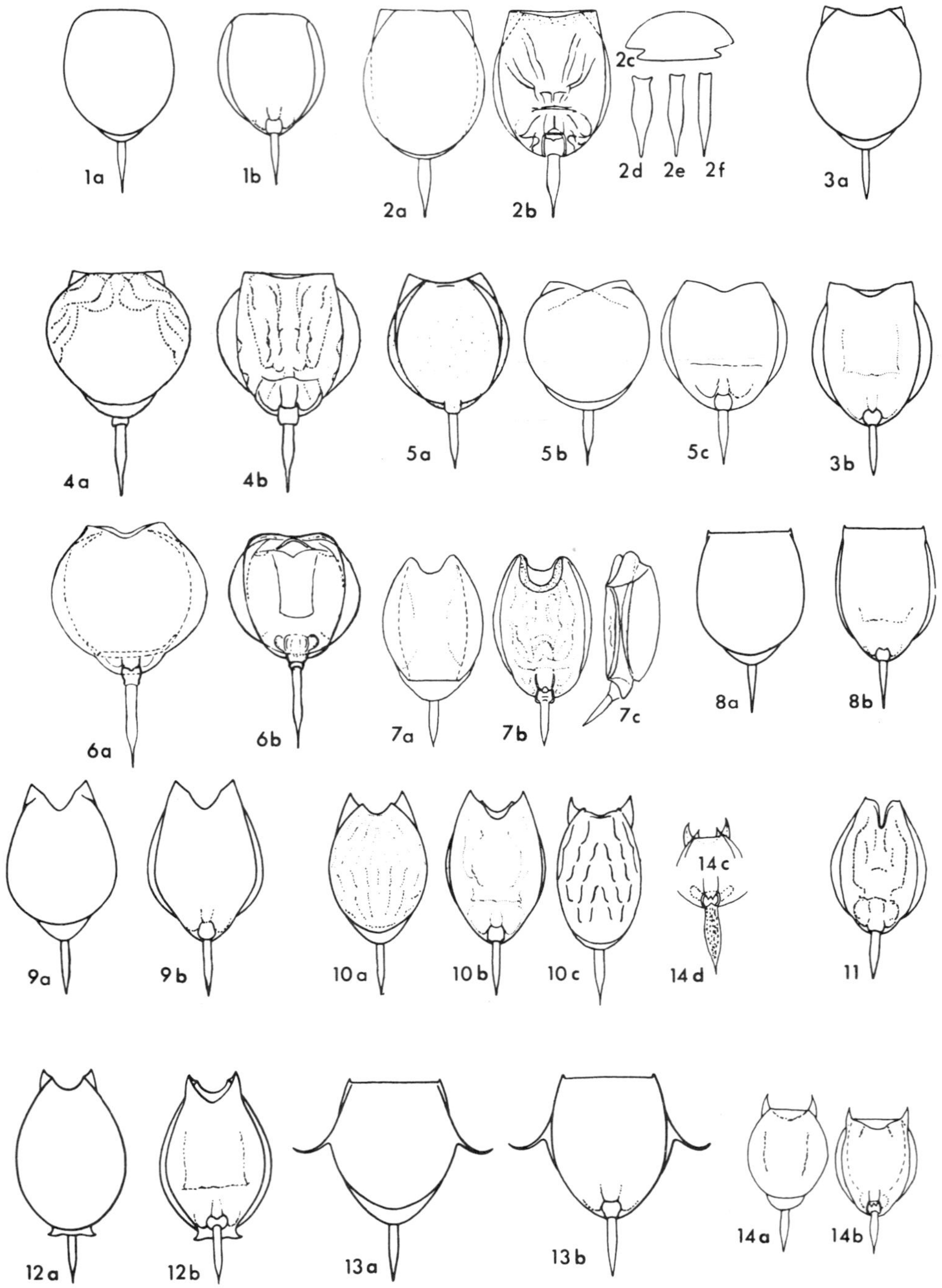

Fig. 7.35. Various species of *Lecane* (=*Monostyla*). (From Koste 1978 after various other authors.)

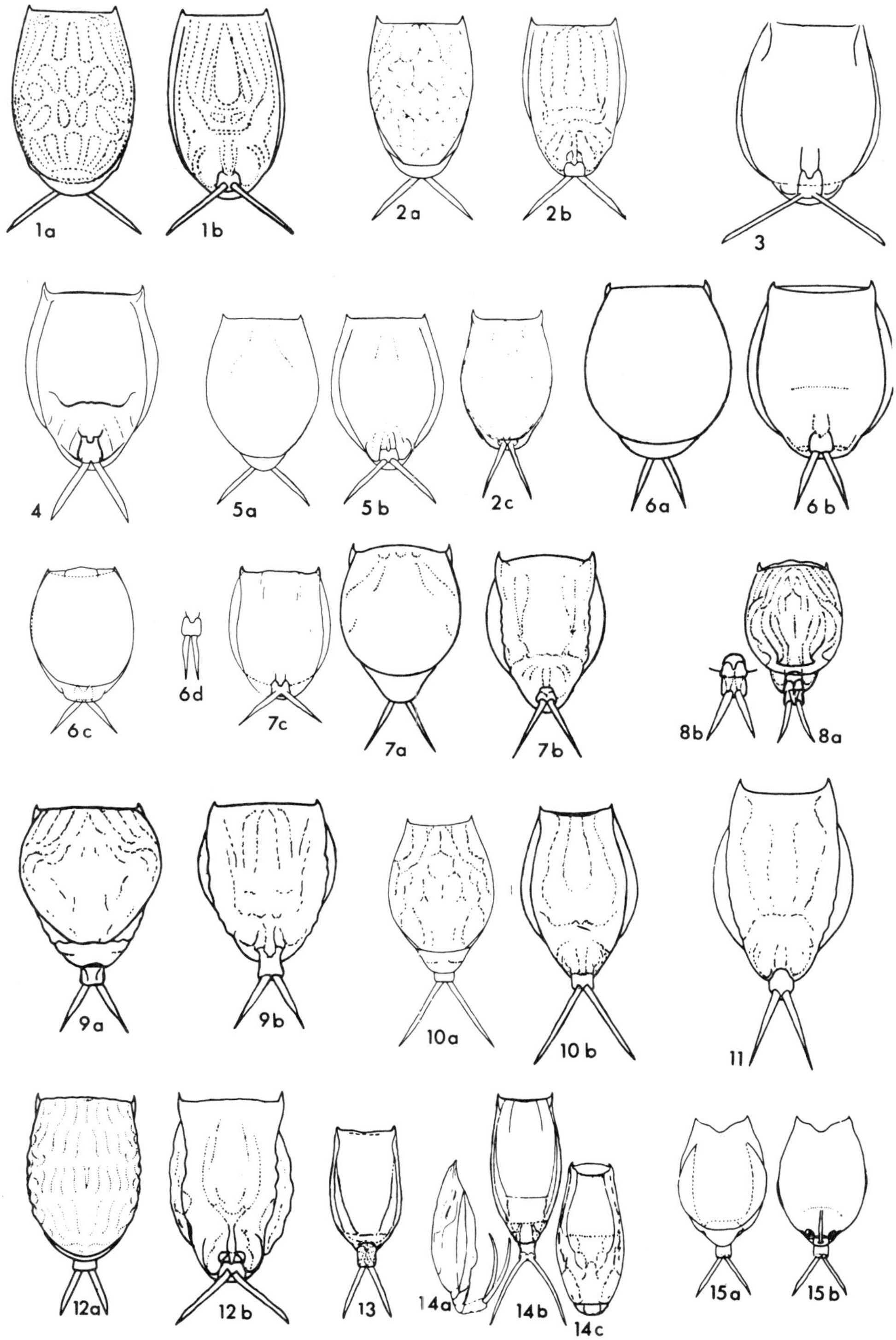

Fig. 7.36. Various species of *Lecane* (From Koste 1978 after various other authors.)

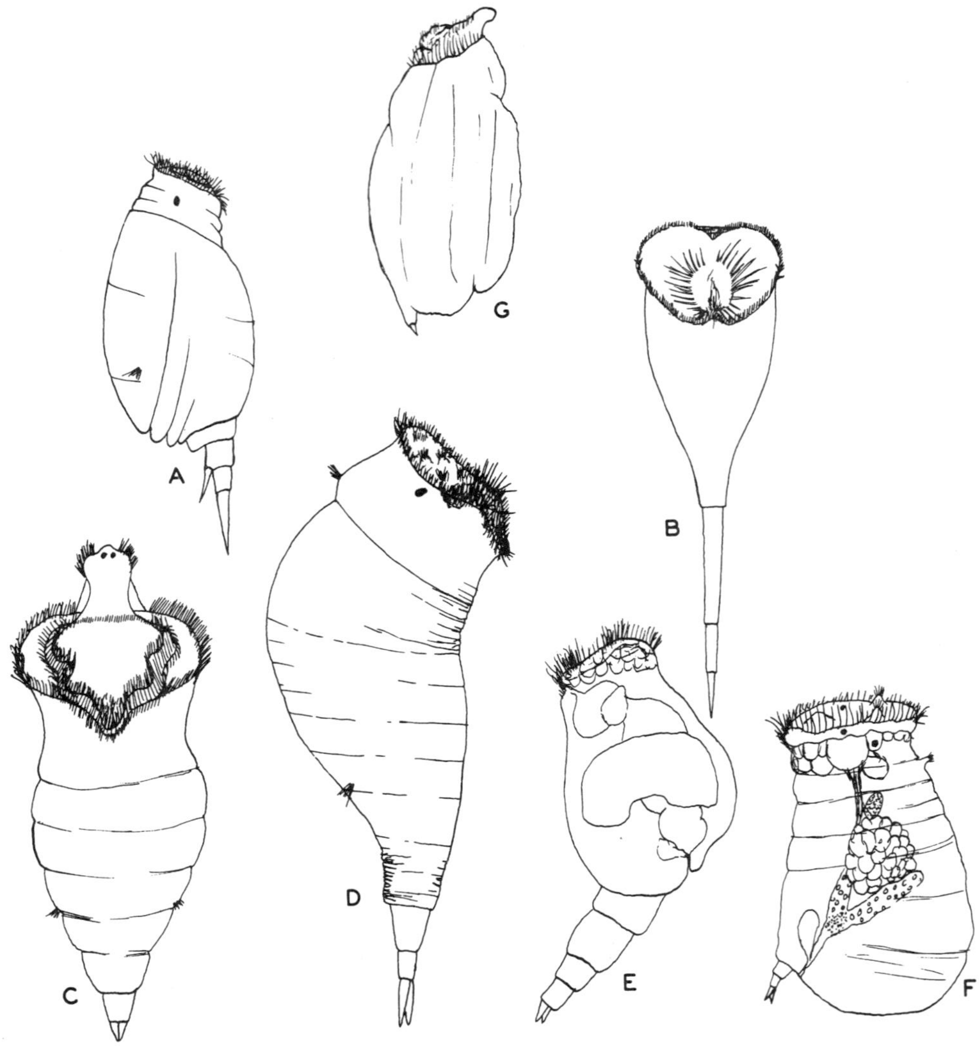

Fig. 7.37. Structure of ploimate rotifers. A, *Mikrocodides chlaena;* B, *Microcodon clavus;* C, ventral view of *Rhinoglena frontalis;* D, *Cyrionia cuba;* E, *Haningia;* F, *Asplanchnopus;* G, lateral view of *Ascomorphella volvocicola.*

45. Foot very long, about half of total length; a single toe; corona flat and circular (Fig. 7.37B); mastax forcipate; a single, uncommon species, 200 μm long .... ***Microcodon***
Foot much shorter almost invariably with two toes **46**

46. Corona complex, with an outer band of cilia and an inner band of cilia, sometimes also with accessory rows of cilia and ciliated protuberances (Figs. 7.1; 7.37A, C, D); mastax malleate; large species **47**
Corona simple, with a single peripheral band of cilia; general surface of corona without cilia, with few scattered cilia, or with cilia more or less covering the corona; corona occasionally with two slight sensory protuberances **50**

47. With a single toe; sometimes a small spur at the base of the toe; foot thick (Fig. 7.37A); 1 U.S. species .... ***Mikrocodides***
With two toes **48**

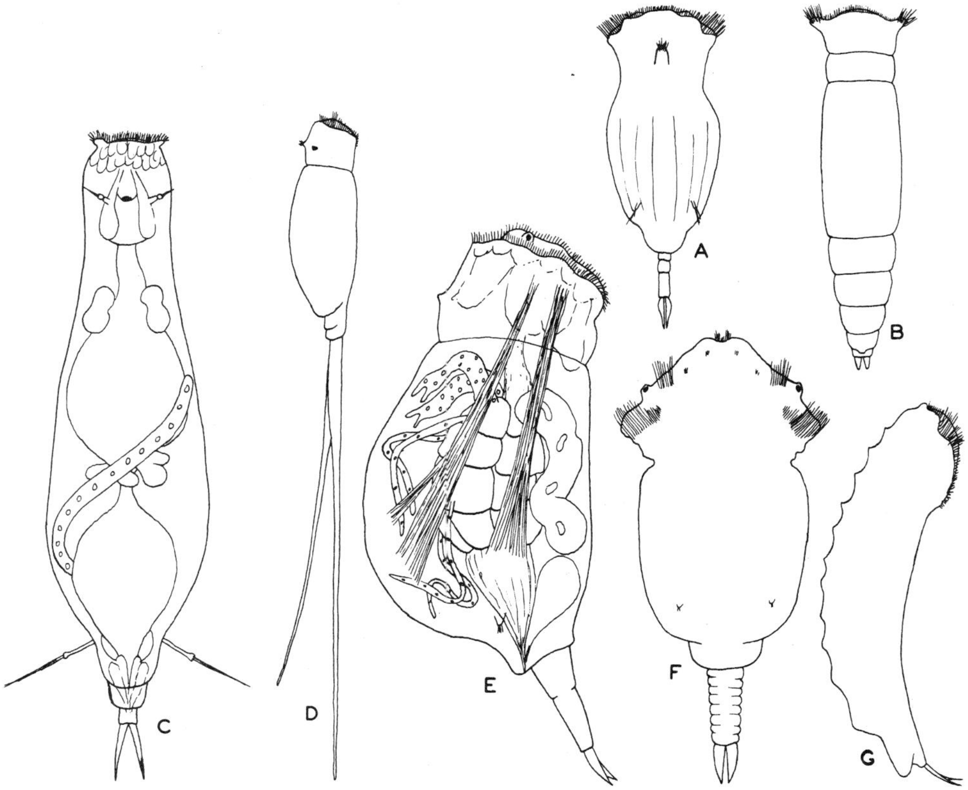

Fig. 7.38. Structure of Notommatidae. A, *Birgea enantia;* B, *Lindia;* C, *Tetrasiphon hydrocora;* D, *Monommata;* E, *Enteroplea lacustris;* F, *Sphyrias lofuana;* G, *Taphrocampa.* (A–C modified from Harring and Myers 1922; D–G modified from Harring and Myers 1924.)

48. With a large dorsal proboscis at the anterior end bearing two eyespots (Fig. 7.37C); one littoral species, 270 to 400 μm long . . . . ***Rhinoglena***
Without such a proboscis **49**

49. Eyespot absent; body not strongly tapered posteriorly (Fig. 7.1); common during spring in stock ponds and puddles containing much organic matter; probably 5 species; 300 to 600 μm long . . . .***Epiphanes***
Eyespot present; body strongly tapered posteriorly (Fig. 7.37D); one littoral species; 250 to 300 μm . . . . ***Cyrtonia***

50. Large, globular species; mastax incudate (Fig. 7.37E, F); carnivorous **51**
Smaller species, more or less opaque; mastax not incudate; mostly illoricate or weakly loricate (see couplet 61) **52**

51. Intestine and anus absent (Fig. 7.37F); three rare littoral species; up to 1 mm long ***Asplanchnopus***
Intestine and anus present (Fig. 7.37E); 2 rare littoral species . . . .***Harringia***

52. Mastax forcipate; trophi strongly compressed dorsoventrally and adapted for protrusion and tearing prey (Fig. 7.9E); corona oval and ventral, or subcircular and strongly oblique; marginal cilia short, with the exception of two lateral, auriclelike tufts of cilia; rostrum large; buccal field evenly ciliated; retrocerebral sac present **73**
Mastax not forcipate; corona frontal or oblique **53**

53. Mastax virgate or virgate–forcipate, adapted for sucking out the contents of plant cells and animals (Fig. 7.10); a pair of lateral auricles often present; retrocerebral sac present **54**
Mastax not virgate or virgate–forcipate; retrocerebral sac rudimentary or absent **55**

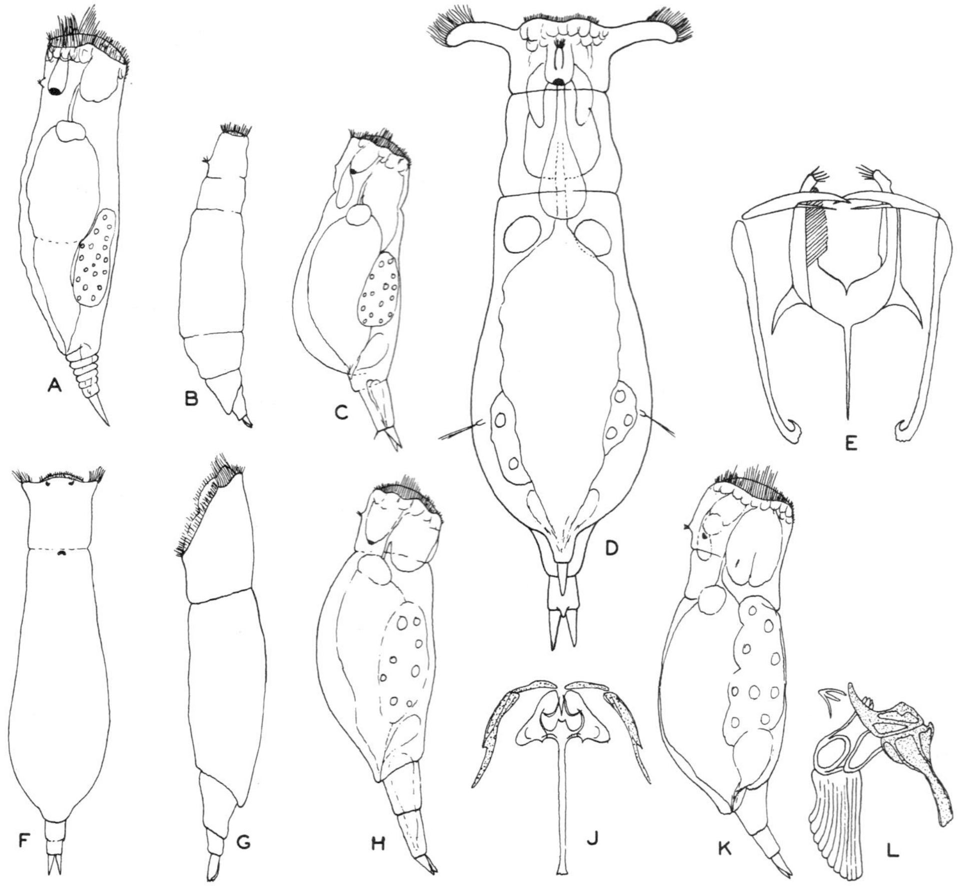

Fig. 7.39. Structure of Notommatidae and Dicranophoridae. A, *Tylotrocha monopus;* B, *Drilophaga;* C, *Rousseletia corniculata;* D, dorsal view of *Notommata;* E, ventral view of *Itura* trophi; F, *Itura;* G, *Dorria dalecarlica;* H, *Pleurotrocha;* J, ventral view of *Pleurotrocha trophi* (mallei stippled); K, *Eosphora;* L, lateral view of *Eosphora* trophi (malleus stippled). (A, C, H, and J modified from Harring and Myers 1924; B, D, K, and L modified from Harring and Myers 1922; E and F modified from Harring and Myers 1928; G modified from Myers 1933.)

54. Lateral antennae located near base of foot, long, knobbed, and with excessively long setae; ovary long, slender, and ribbonlike; auricles absent; dorsal antenna double and very long (Fig. 7.38C); corona oblique; total length 750 to 1000 μm; 1 rare species in acid waters ***Tetrasiphon***

Lateral antennae not located near base of foot; ovary not long, slender, and ribbonlike; auricles present or absent; dorsal antenna not especially long; corona frontal or oblique **57**

55. Mastax modified malleate; unci adapted for crushing and grinding; corona oblique, with two lateral tufts of cilia but no auricles (Fig. 7.40D–F); total length 90 to 400 μm **71**

Mastax not modified malleate or malleate–virgate; corona oblique or frontal; auricles present or absent **56**

56. Body cylindrical and tapering posteriorly, sometimes indistinctly annulated; foot not set off sharply from body (Fig. 7.38B); mastax cardate (Fig. 7.9A); total length 250 to 1200 μm; about 10 species ....***Lindia***

Foot slender and sharply set off from the wide body (Fig. 7.38A); trophi aberrant, with a pair of pseudunci; 240 to 275 μm long; 1 rare littoral species ....***Birgea***

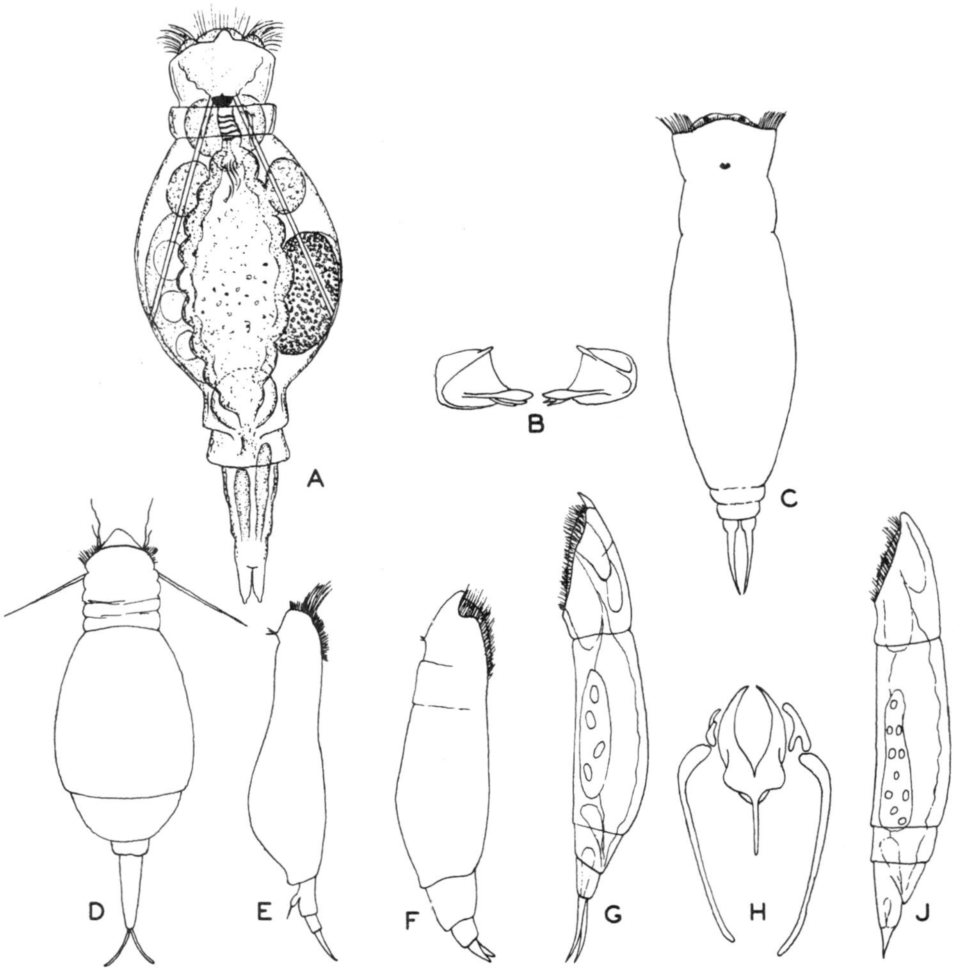

Fig. 7.40. Structure of Proalidae and Notommatidae. A, *Resticula;* B, frontal view of unci of *Eothinia;* C, *Eothinia;* D, *Bryceella;* E, *Proalinopsis;* F, *Proales;* G, *Aspelta;* H, trophi of *Aspelta;* J, *Pedipartia gracilis.* (E and F modified from Harring and Myers 1922; B and C modified from Harring and Myers 1924; G and H modified from Harring and Myers 1928; J modified from Myers 1936; A from Donner 1970.)

57. Toes longer than rest of body, usually unequal in length (Fig. 7.38D); about 20 species, usually in acid waters . . . . ***Monommata***
Toes short **58**

58. Gastric glands long, ribbonlike, and bifurcated; with four slender accessory gastric appendages; body short and saclike (Fig. 7.38E); one uncommon littoral species; 600 μm long ***Enteroplea***
Without such glands **59**

59. Head very broad and truncate anteriorly; two eyespots widely separated on protuberances; foot annulated (Fig, 7.38F); one carnivorous littoral species . . . . ***Sphyrias***
With other characters **60**

60. Body marked with transverse folds or annuli; elongated and spindle-shaped; foot rudimentary (Fig. 7.38G); several littoral species; 200 μm long . . . . ***Taphrocampa***
Body not so marked **61**

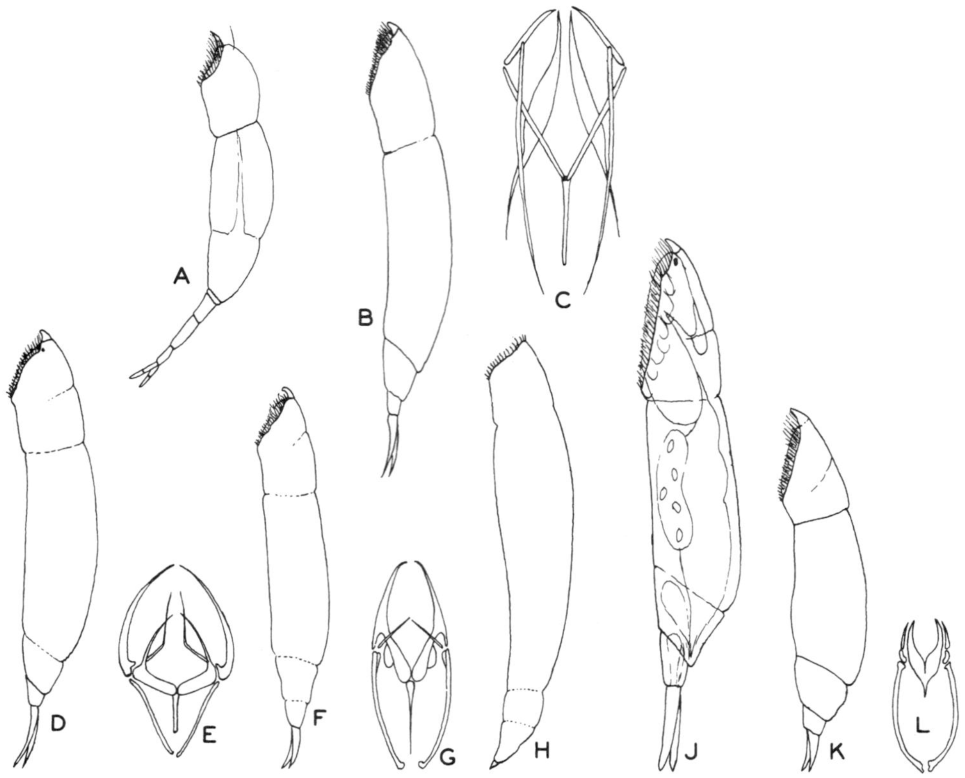

Fig. 7.41. Structure of Dicranophoridae. A, *Wierzejskiella;* B, *Streptognatha lepta;* C, trophi of *S. lepta;* D, *Erignatha;* E, trophi of *Erignatha;* F, *Myersinella;* G, trophi of *Myersinella;* H, *Albertia;* J, *Dicranophorus;* K *Dicentrian;* L, trophi of *Encentrum.* (B–L modified from Harring and Myers 1928.)

61. Toes relatively long, pointed, and usually curved; foot rudimentary and unjointed; body prismatic or spindle shaped (Fig. 7.34A–C); retrocerebral sac absent; lorica present, more or less sclerotized and composed of a dorsal and a ventral plate; both rami and unci of very simple construction (Fig. 7.10D, E); more than 50 U.S. species, some of which are very common ***Cephalodella***

Toes comparatively short; foot present and distinct; retrocerebral sac present; lorica absent; rami and unci more complicated **62**

62. Foot unsegmented, though sometimes annulated (Fig. 7.39A–C) **63**

Foot segmented (Fig. 7.39D–K) **65**

63. With a single toe; retrocerebral sac absent (Fig. 7.39A); trophi highly specialized virgate; 1 rare species in ponds and reservoirs; 250 μm long . . . . ***Tylotrocha***

With two toes; retrocerebral sac present or absent; trophi not highly specialized virgate . . . **64**

64. Retrocerebral sac absent; eyespot absent; foot short (Fig. 7.39B); one species ectoparasitic on oligochaetes, a second ectoparasitic on *Erpobdella*, and one other (U.S.) possibly free-living; often covered by a gelatinous secretion . . . . ***Drilophaga***

Retrocerebral sac present; eyespot present; foot long (Fig. 7.39C); one littoral species; 130 μm long . . . . ***Rousseletia***

65. With a projecting auricle on each side of the head, each auricle bearing long cilia; trophi usually asymmetrical (Figs. 7.10A, B; 7.39E) **66**

Without definite auricles but often with a tuft of long cilia in the usual location of an auricle; trophi symmetrical **67**

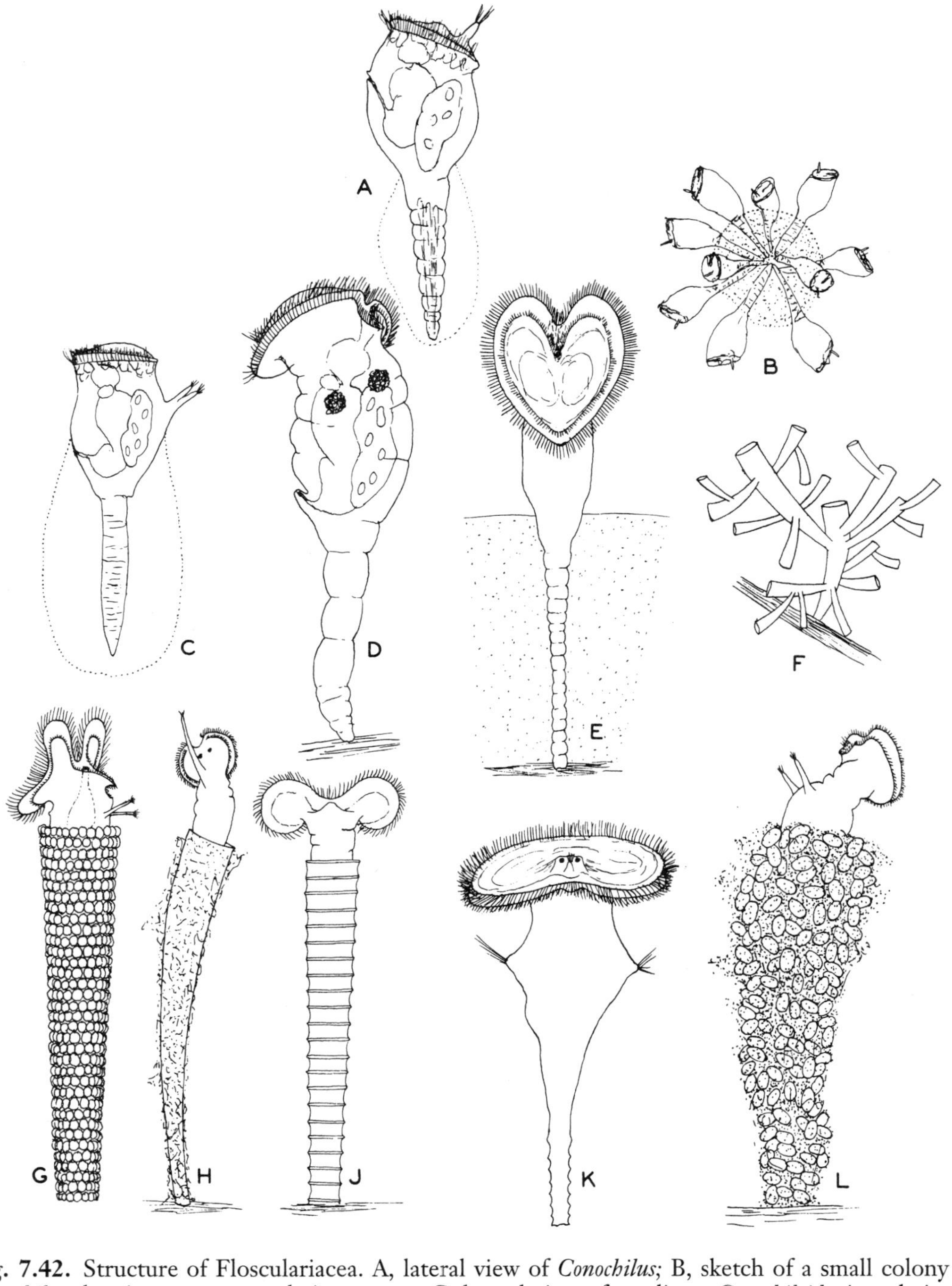

Fig. 7.42. Structure of Floscularíacea. A, lateral view of *Conochilus;* B, sketch of a small colony of *Conochilus* showing common gelatinous case; C, lateral view of a solitary *Conochiloides* in gelatinous case; D, lateral view of *Sinantherina;* E, *Lacinularia* in gelatinous matrix of colony; F, cluster of empty *Floscularia* tubes; G, lateral view of *Floscularia* (note pellet-making concavity on ventral surface of head); H and J, dorsal views of *Limnias;* K, *Pseudoecistes rotifer;* L, *Ptygura.* (All greatly modified from various sources.)

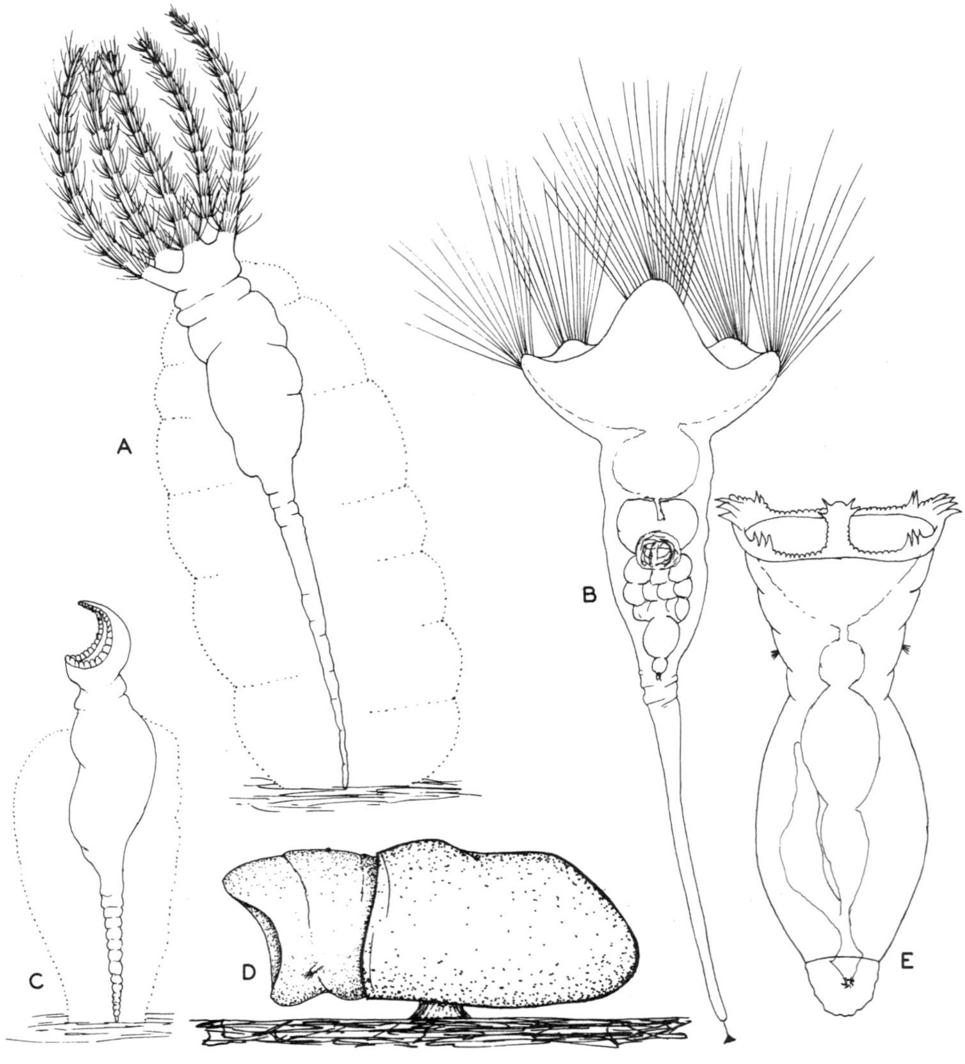

Fig. 7.43. Structure of Collothecacea. A, *Stephanoceros;* B, dorsal view of *Collotheca* (gelatinous tube omitted); C, lateral view of *Acyclus inquietus;* D, lateral view of *Cupelopagis vorax;* E, dorsal view of *Atrochus tentaculatus.*

66. Rami roughly hemispherical (Fig. 7.10A, B); with a cervical eyespot only (Fig. 7.39D); total length 100 to 1000 μm, usually more than 300 μm; a few species covered by a gelatinous secretion; many littoral species, some very common and spectacular . . . . ***Notommata***

Rami lyrate, broad, and strongly divergent at the base, continuing as parallel rods toward the tips (Fig. 7.39E); with a cervical eyespot and two frontal eyespots (Fig. 7.39F); about 4 uncommon littoral species . . . . ***Itura***

67. Eyespot absent (Fig. 7.39G); in mountain brooks among submerged moss; one species; 200 μm long . . . . ***Dorria***

One or three eyespots present; not in mountain brooks in moss **68**

68. Retrocerebral sac absent (Fig. 7.39H); rami without teeth (Fig. 7.39J); toes separate or fused; total length 110 to 250 μm; 8 littoral species . . . . ***Pleurotrocha***

Retrocerebral sac present (Fig. 7.39K); rami with one or more teeth; toes separate. **69**

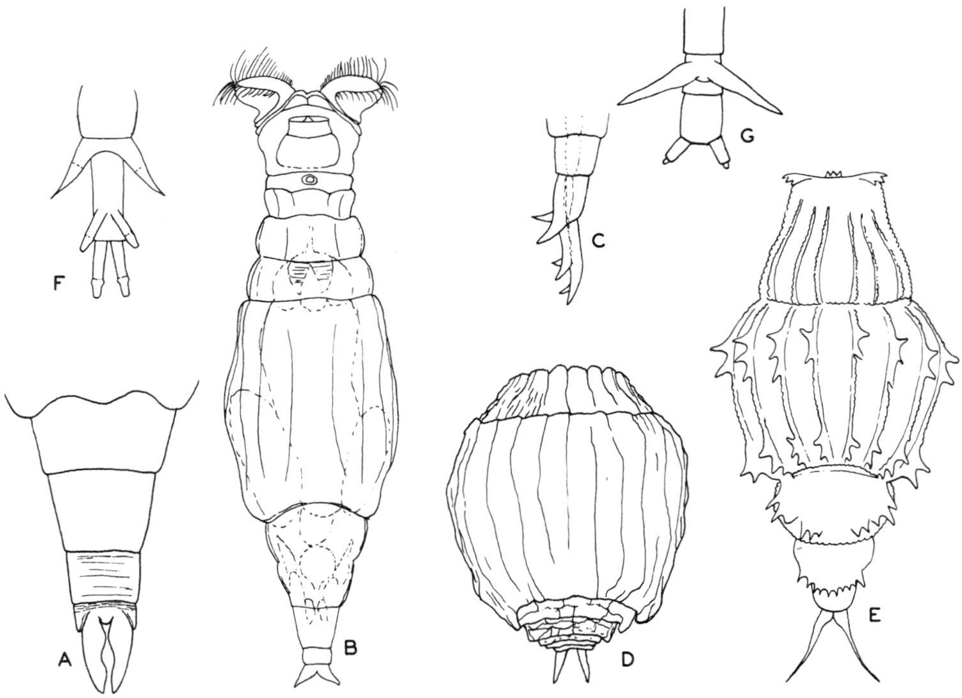

Fig. 7.44. Structure of Bdelloidea. A, posterior end of *Embata* while swimming (toes retracted); B, dorsal view of *Philodina* (toes retracted); C, foot of *Philodina* (toes protruded); D, contracted *Philodina;* E, dorsal view of partly contracted *Dissotrocha;* F, dorsal view of foot of *Philodina* (toes protruded); G, toes and spurs of *Embata.* (A modified from Murray 1906; B and E modified from Murray 1911; C and D modified from Hickernell 1917.)

69. Each uncus with a single simple tooth; rami bent at right angles and with one or two teeth at midlength and many small ones more distally (Fig. 7.39L); total length 300 to 500 μm; about 5 littoral species . . . . *Eosphora*

One or both unci with a single main tooth plus accessory teeth (Fig. 7.40B); rami more or less bent at right angles but with a different arrangement of teeth; total length 170 to 400 μm **70**

70. With one eyespot, or eyespot absent (Fig. 7.40A); each uncus with one main tooth and with accessory teeth on either the left or right uncus; probably 6 littoral species . . . . ***Resticula***

With three eyespots (Fig. 7.40C); both unci with one main tooth and accessory teeth (Fig. 7.40B); about 6 littoral species . . . . ***Eothinia***

71. With a very long stylet on each side of the head (Fig. 7.40D); two littoral and sandy beach species; 130 μm long . . . . ***Bryceella***

Without such stylets **72**

72. Tail region or basal foot segment with a knoblike papilla bearing a tuft of setae or a spine (Fig. 7.40E); 5 littoral species . . . . ***Proalinopsis***

Without such a tuft of setae or spine (Fig. 7.40F); one species with a single toe; about 30 littoral species, some quite common . . . . ***Proales***

73. Trophi asymmetrical (Fig. 7.40H) **74**

Trophi symmetrical (Fig. 7.41C, E, G, L) **75**

74. Corona ventral; retrocerebral sac present; toes long (Fig. 7.40G); about 12 species, usually in acid waters and sphagnum bogs . . . . ***Aspelta***

Corona oblique; retrocerebral sac absent; toes short (Fig. 7.40J); 1 rare species known from sandy beaches . . . . ***Pedipartia***

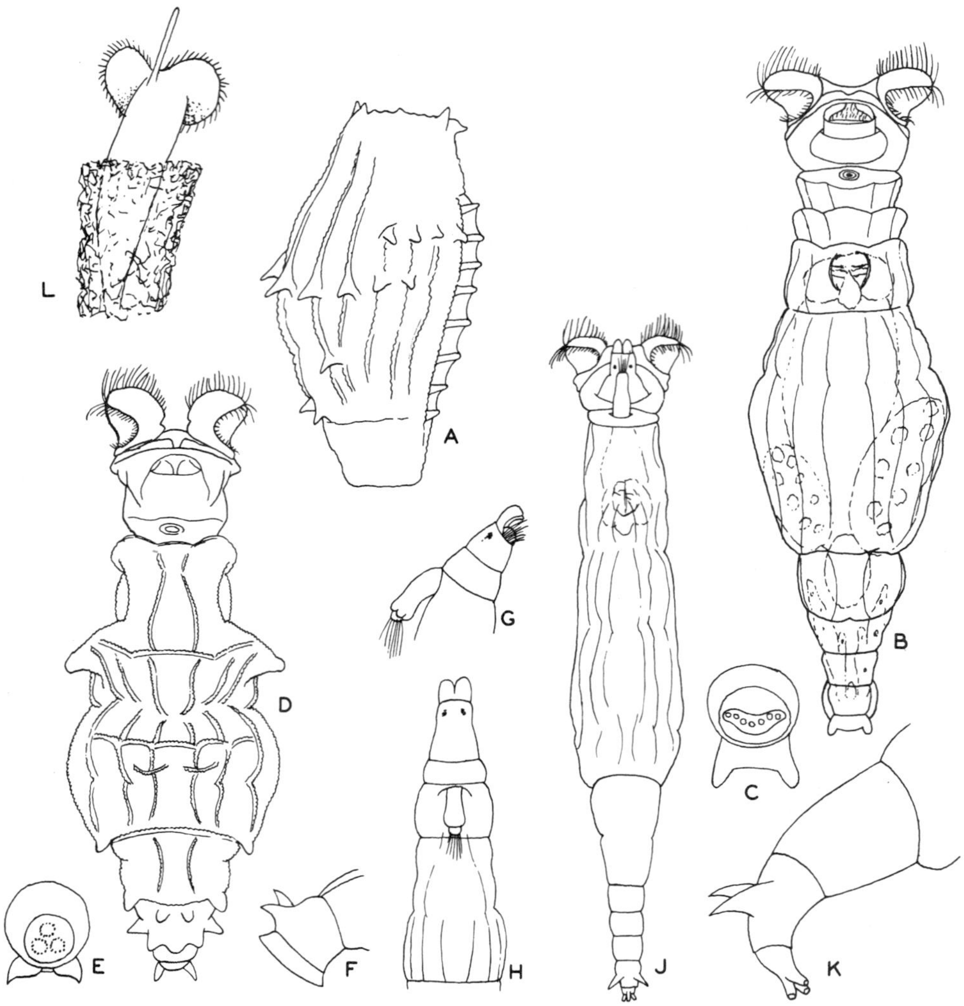

Fig. 7.45. Structure of Bdelloidea. A, lateral view of contracted *Pleuretra;* B, dorsal view of *Mniobia* (toes retracted); C, ventral view of spurs and disc of *Mniobia;* D, dorsal view of *Macrotrachela* (toes retracted); E, end view of foot of *Macrotrachela* (toes retracted); F, lateral view of foot of *Macrotrachela* (toes retracted); G, lateral view of head of creeping *Rotaria;* H, dorsal view of anterior end of creeping *Rotaria;* J, dorsal view of swimming *Rotaria* (toes shown extruded); K, lateral view of foot of *Habrotrocha* (toes extruded); L, distal end of *Beauchampia crucigera.* (A–J modified from Murray 1911.)

75. Foot one-fourth to one-third the total length, composed of one to four segments (Fig. 7.41A); 3 uncommon species; most common in sandy beaches . . . . ***Wierzejskiella***
Foot shorter **76**

76. Incus Y-shaped and composed of rods; with a long naviculoid sclerite attached to the tip of each uncus (Fig. 7.41C); a single acid water species . . . . ***Streptognatha***
With incus and unci of different construction **77**

77. Rami slender and bent at nearly a right angle at midlength and terminating in long, slender, single teeth (Fig. 7.41E); about 5 littoral species . . . . ***Erignatha***
Rami not slender and bent at nearly a right angle at midlength; retrocerebral sac present or absent; corona oblique or ventral **78**

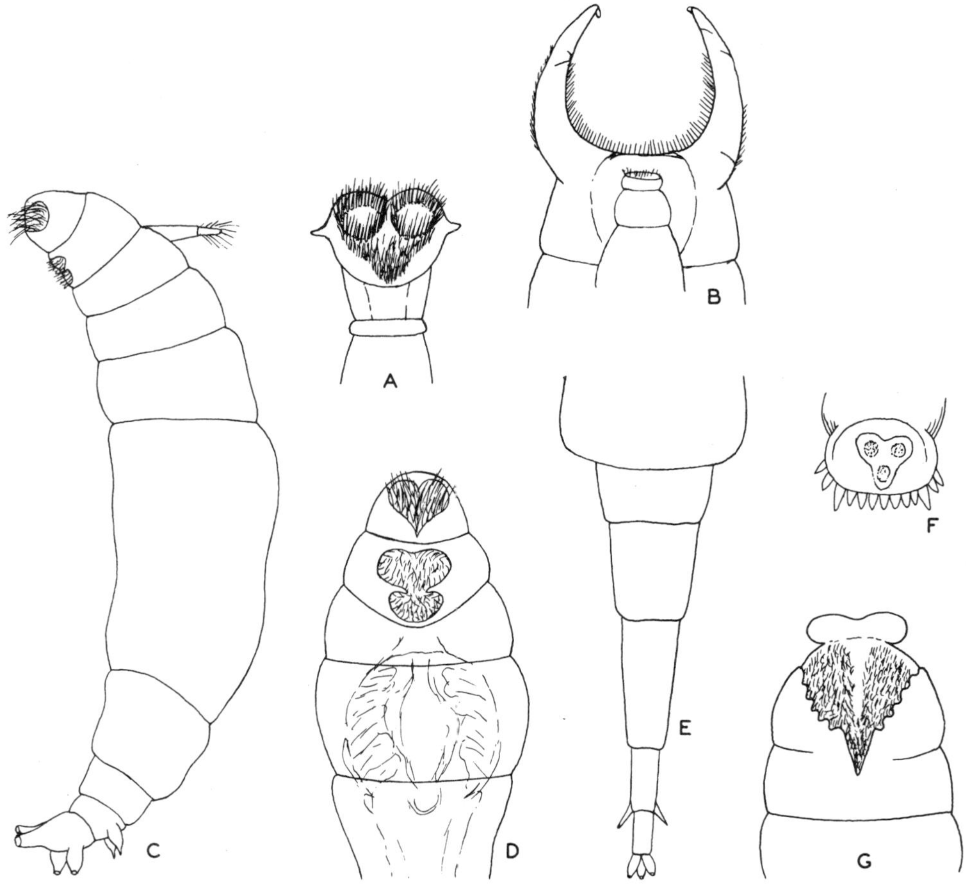

Fig. 7.46. Structure of Bdelloidea. A, anterior end of *Scepanotrocha;* B, dorsal view of anterior end of *Ceratotrocha;* C, *Philodinavus paradoxus* with protruded toes; D, ventral view of anterior end of *P. paradoxus;* E, posterior end of *Adineta* with protruded toes; F, ventral view of posterior end of *Bradyscela clauda* showing papilliform lobes and region of three retracted toes; G, ventral view of anterior end of *B. clauda.* (A modified from Bryce 1910; C and D modified from Murray 1906.)

78. With a posterior projection at the external angles of the rami (Fig. 7.41G); retrocerebral sac absent; corona oblique or ventral (Fig. 7.41F); 2 sandy beach species; 130 μm long **_Myersinella_**

Without such projections on the rami **79**

79. Toes minute and fused, or lacking; corona oblique; eyespots lacking; rostrum minute; retrocerebral sac absent (Fig. 7.41H); several species, most or all of which are ecto- or endoparasites of oligochaetes .... **_Albertia_**

Toes prominent, not fused; corona oblique or ventral; eyespots present or absent; rostrum conspicuous; retrocerebral sac present or absent; not parasitic **80**

80. Corona essentially ventral; two frontal eyespots; retrocerebral sac present and large (Fig. 7.41J); rami often with shearing teeth; unci robust (Fig. 7.9); about 50 littoral species **_Dicranophorus_**

Corona oblique; one eyespot, or eyespots absent; retrocerebral sac present or absent (Fig. 7.41K); rami never with shearing teeth; unci weak and needlelike (Fig. 7.41L); about 15 species; chiefly littoral .... **_Encentrum_**

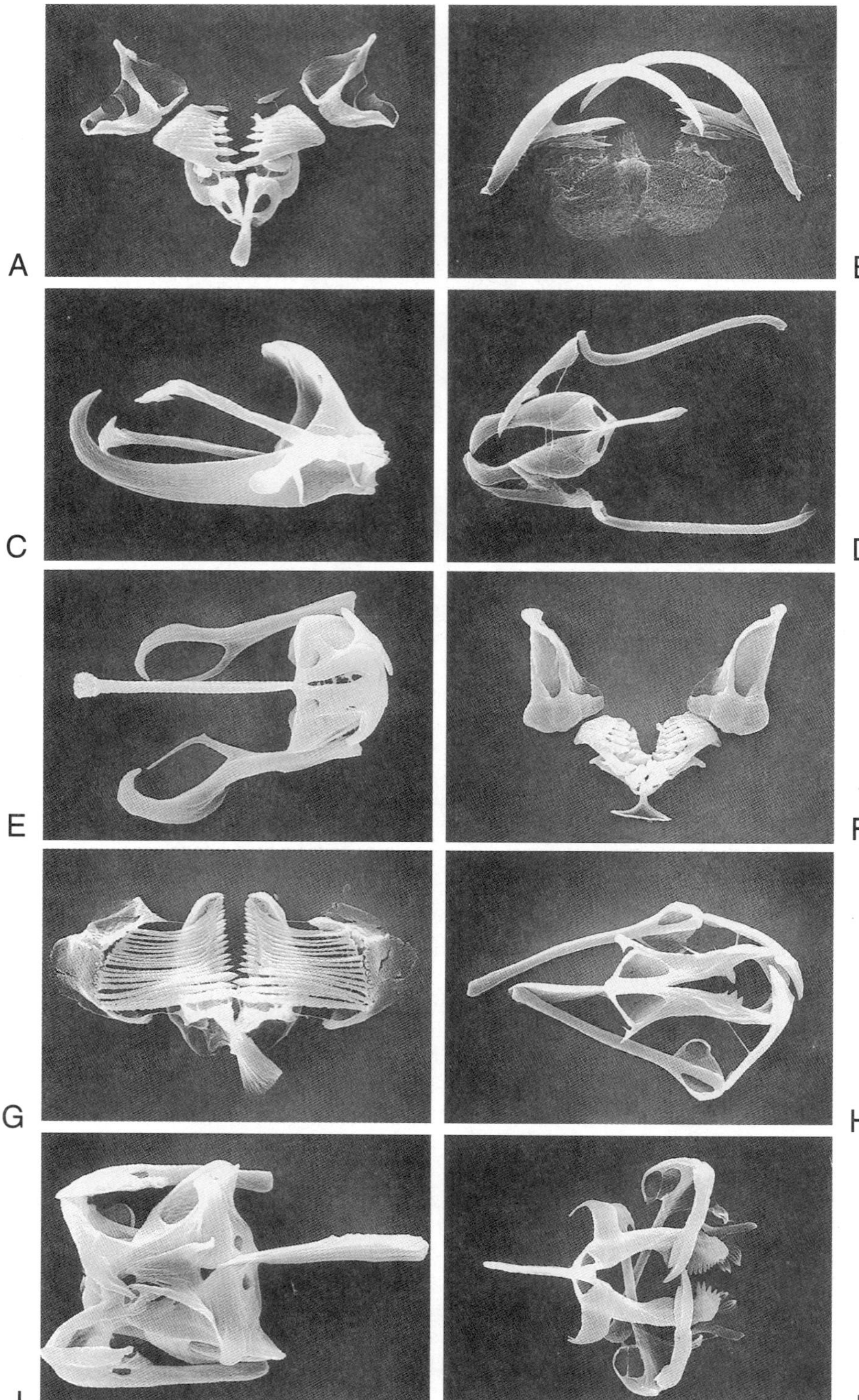

Fig. 7.47. Trophi of various rotifer genera. A, *Epiphanes;* B, *Cupelopagis;* C, *Ascomorpha;* D, *Aspelta;* E, *Cephalodella;* F, *Lepadella;* G, *Testudinella;* H, *Dicranophorus;* I, *Eosphora;* J, *Itura.* (From De Smet 1998, with kind permission from Kluwer Academic Publishers.)

81. With a ventral gap in the ciliary wreath; mouth on the corona, near the dorsal edge; gelatinous case (Fig. 7.42A–C); limnetic **82**
Ciliary wreath with a dorsal gap or without a gap; mouth ventral to the corona; eyespots usually absent (Fig. 7.42D–L); adults usually sessile **83**
82. With one or two antennae on the corona; colonies composed of radiating individuals inhabiting coherent gelatinous tubes; 2 common species (Fig. 7.42A, B) . . . . ***Conochilus***
With one or two ventral antennae below the corona; solitary or colonial; probably four limnetic species, of which two are common (Fig. 7.42C) . . . . ***Conochiloides***
83. Adults in spherical, sessile colonies attached to aquatic plants **84**
Adults solitary or colonial, but colonies never spherical **85**
84. Colonies without tubes; trunk usually with two to four opaque protuberances; corona roughly kidney-shaped (Fig. 7.42D); eggs on ovifera; about 5 species . . . . ***Sinantherina***
Colonies with adhering gelatinous tubes containing eggs; trunk without protuberances; corona heart-shaped and with the sinus ventral (Fig. 7.42E); about 10 species . . . . ***Lacinularia***
85. Corona with eight prominent lobes and one small median dorsal lobe; in a gelatinous tube; 1 poorly known species . . . . ***Octotrocha***
Corona with less than eight lobes, or not lobed. **86**
86. Corona four-lobed; two large ventral antennae; with a tube composed of gelatinous material, pellets of debris, or fecal pellets (Fig. 7.42F, G); about 6 species . . . . ***Floscularia***
Corona two-lobed and elliptical, circular, or kidney-shaped **87**
87. Tube chitinoid, often opaque and more or less covered with debris; corona of two distinct lobes, or nearly circular; dorsal antenna short to long (Fig. 7.42H, J); about 6 species . . . ***Limnias***
Tube not chitinoid, or tube absent **88**
88. Tube absent (Fig. 7.42K); able to leave substrate and swim about; 1 rare species. ***Pseudoecistes***
Tube present **89**
89. Dorsal antenna small or apparently absent; ventral (lateral) antennae short to long; corona a wide oval or nearly circular, indistinctly bilobed; tube made of debris, fecal pellets, or gelatinous and more or less covered with debris (Fig. 7.42L); one species free-swimming; 20 species; 200 to 1000 μm long . . . . ***Ptygura***
Dorsal antenna very long; corona of two distinct lobes; tube irregular, translucent, and gelatinous; one rare species up to 500 μm long . . . . ***Beauchampia***
90. With setae but no cilia on the corona; sessile; in a gelatinous tube; foot terminated by a long, nonretractile peduncle, ending in an adhesive disk **91**
Without setae or cilia on the corona; gelatinous case present or absent **92**
91. Corona drawn out into long, pointed arms, which bear whorls of setae (Fig. 7.43A); 2 species ***Stephanoceros***
Corona circular, lobed, or pointed (Fig. 7.43B); setae not arranged in whorls; some species free-swimming; many species . . . . ***Collotheca***
92. Body long, with a long, annulated, tapering stalk and with or without a gelatinous case; corona produced into a large dorsal lobe, and bordered by a thin, festooned membrane (Fig. 7.43C); lives in *Sinantherina* colonies; 1 rare species . . . . ***Acyclus***
Body shorter, without a gelatinous case; foot absent. **93**
93. Body saclike, with a flat attachment disk on the ventral surface; corona a large concave chamber (Fig. 7.43D); usually on broad, flat leaves of aquatic plants; up to 800 μm long; 1 rare species ***Cupelopagis***
Body spindle-shaped; corona with numerous small peripheral tentacles (Fig. 7.43E); 1 species ***Atrochus***
94. Mastax deep in the body **95**
Mastax superficial, near surface **105**
95. Trochi present **96**
Trochi absent; with 2 spurs and 3 toes (Fig. 7.46E); about 10 species in shallow littoral ***Adineta***
96. Stomach with fecal pellets, usually with several teeth **97**
Stomach without fecal pellets, usually only a few teeth **98**
97. Trochi partially covered by a hood or lobe (Fig. 7.46A); several uncommon species, usually in mosses . . . . ***Scepanotrocha***
Trochi not covered by hood; many species . . . . ***Habrotrocha***

98. Toes absent 99
Toes present (Fig. 7.45B, C), about 50 species .... ***Mniobia***
99. Adults viviparous (containing live young) **100**
Adults oviparous **102**
100. Rostrum visible (often with eyes present) (Fig. 7.45G–J), about 15 species.......... ***Rotaria***
Without visible rostrum **101**
101. Integument sculptured (Fig. 7.44E), 2 or 3 teeth; about 5 species .... ***Dissotrocha***
Integument smooth, 2 teeth; about 3 species ectocommensal on many aquatic invertebrates ***Embata***
102. Foot with 3 toes **103**
Foot with 4 toes **104**
103. Corona with lateral wings (Fig. 7.46B), several rare species .... ***Ceratotrocha***
Corona without lateral wings (Fig. 7.45D–F), with about 80 species .... ***Macrotrachela***
104. Integument sculptured (Fig. 7.45A), about 4 rare species in mosses .... ***Pleuretra***
Integument smooth (Fig. 7.44B, C); about 20 species .... ***Philodina***
105. Trochi present; 1 species, temporary pools .... ***Abrochtha***
Trochi absent **106**
106. With 2 short spurs (Fig. 7.46C, D); 1 species not yet reported from the U.S. ... ***Philodinavus***
With 1 spur and 4 big toes; 2 species, not yet reported from the U.S. but probably present ***Henoceros***

REFERENCES

Ahlstrom, E. H. 1940. A revision of the rotatorian genera *Brachionus* and *Platyias* with descriptions of one new species and two new varieties. *Bull. Am. Mus. Nat. Hist.* 77:143–184.

——. 1943. A revision of the rotatorian genus *Keratella* with descriptions of three new species and five new varieties. *Bull. Am. Mus. Nat. Hist.* 80:411–457.

Aloia, R. C., and R. L. Moretti. 1973. Sterile culture techniques for species of the rotifer *Asplanchna*. *Trans. Am. Microsc. Soc.* 92:364–371.

Bartos, E. 1948. On the Bohemian species of the genus *Pedalia* Barrois. *Hydrobiologia* 1:63–77.

——. 1951. The Czechoslovak Rotatoria of the Order Bdelloidea. *Vestn. Cesk. Spol. Zool.* 15: 241–500.

Beach, N. W. 1960. A study of the planktonic rotifers of the Ocqueoc River system, Presque Isle County, Michigan. *Ecol. Monogr.* 30:339–357.

Bérzins, B. 1951. On the collothecacean Rotatoria: with special reference to the species found in the Aneboda district, Sweden. *Ark. Zool.* 1:565–592.

Brakenhoff, H. 1937. Zur Morphologie der Bdelloidea. *Zool. Jahrb. Abt. Anat. Ont. Tiere* 63:125–182.

Bryce, D. 1910. On a new classification of the bdelloid Rotifera. *J. Quek. Micros. Club.* 11: 61–92.

Buchner, H. 1941. Experimentelle Untersuchungen uber den Generationswechsel: der Rddertiere, II. *Zool. Jahrb. Abt. Allg. Zool. Physiol. Tiere* 60:279–344.

Budde, E. 1925. Die parasitischen Räddertiere mit besonderer Berücksichtigung der in der Umgegend von Minden 1: W. beobachteten Arten. *Z. Morphol. Oekol. Tiere* 3:706–784.

Calaway, W. T. 1968. The Metazoa of waste treatment processes: rotifers. *J. Water Pollut. Control Fed.* 40:R412–R422.

Campbell, R. S. 1941. Vertical distribution of the plankton Rotifera, in Douglas Lake, Michigan, with special reference to depression individuality. *Ecol. Monogr.* 11:1–19.

Carlin, B. 1939. Über die Rotatorien einiger Seen bei Aneboda. *Medd. Lunds Univ. Limnol. Inst.* 2:1–68.

——. 1943. Die Planktonrotatorien des Motalastram. *Medd. Lunds Univ. Limnol. Inst.* 5:1–255.

Cori, C. 1925. Zur Morphologie und Biologie von *Apsilus vorax* Leidy. *Z. Wiss. Zool.* 125: 557–584.

De Beauchamp, P. M. 1909. Recherches sur les rotiféres: les formations tégumentaires et Pappareil digestif. *Arch. Zool. Exp. Gen.* (4) 10: 1–410.

——. 1912. Instructions for collecting and fixing rotifers in bulk. *Proc. U.S. Nat. Mus.* 42:181–185.

——. 1928. Coup d'oeil sur les recherches récentes relatives aux rotiferes et sur les méthodes qui leur sont applicables. *Bull. Biol. Fr. Belg.* 62:51–125.

de Paggi, S. J., and W. Koste. 1984. Checklist of the rotifers recorded from Antarctic and subantarctic areas (Rotatoria). *Senckenbergiana Biol.* 65:169–178.

De Smet, W. H. 1998. Preparation of rotifer trophi for light and scanning electron microscopy. *Hydrobiologia* 387/388:117–121.

Dobers, E. 1915. Über die Biologic der Bdelloidea. *Int. Rev.* 7 (Suppl. 1):1–128.

Donner, J. 1951. Erste Übersicht über die Rotatorienfauna einiger Humusboden. *Oesterreich. Zool.* Z. 3:175–240.

——. 1966. Ordnung Bdelloidea. Rotatoria, Radertiere. *Bestimmungsbucher Bodentauna Eur.* 6:1–297.

——. 1970. Die Räddertierbestande submerser Moose der Salzach und anderer Wasser-Biotope des Flussgebietes. *Arch. Hydrobiol. Suppl.* 36:109–254.

——. 1972. Die Räddertierebestande submerser Moose und weiterer Merotope im Bereich der Staurdume der Donau an der deutsch-Österreichischen Landesgranze. *Arch. Hydrobiol. Suppl.* 44:49–114.

Dumont, H. J., and J. Green (eds.). 1980. Rotatoria: Proceedings of the 2nd International Rotifer Symposium. *Hydrobiologia* 73:1–263.

Edmondson, W. T. 1936. Fixation of sessile Rotatoria. *Science* 84:444.

——. 1939. New species of Rotatoria, with notes on heterogonic growth. *Trans. Am. Microsc. Soc.* 58:459–472.

——. 1940. The sessile Rotatoria of Wisconsin. *Trans. Am. Microsc. Soc.* 59:433–459.

——. 1944. Ecological studies of sessile Rotatoria, Part 1: factors affecting distribution. *Ecol. Monogr.* 14:3–166.

——. 1945. Ecological studies of sessile Rotatoria, Part II: dynamics of populations and social structures. *Ecol. Monogr.* 15:141–172.

——. 1946. Factors in the dynamics of rotifer populations. *Ecol. Monogr.* 16:357–372.

——. 1948. Ecological applications of Lansing's physiological work on longevity in Rotatoria. *Science* 108:123–126.

——. 1949. A formula key to the rotatorian genus *Ptygura*. *Trans. Am. Microsc. Soc.* 68:127–135.

——. 1960. Reproductive rates of rotifers in natural populations. *Mem. Ist. Ital. Idrobiol.* 12: 21–77.

——. 1965. Reproductive rate of planktonic rotifers as related to food and temperature in nature. *Ecol. Monogr.* 35:61–111.

Finesinger, J. E. 1926. Effects of certain chemical and physical agents on fecundity and length of life, and on their inheritance in a rotifer *Lecane* (*Distyla*) *inermis* (Bryce). *J. Exp. Zool.* 44:63–94.

Gilbert, J. J. 1970. Monoxenic cultivation of the rotifer *Brachionus calyciflorus* in a defined medium. *Oecologia* 4:89–101.

——. 1973. The adaptive significance of polymorphism in the rotifer *Asplanchna humps* in males and females. *Oecologia* 13:135–146.

——. 1974. Dormancy in rotifers. *Trans. Am. Microsc. Soc.* 93:490–513.

——. 1975. Polymorphism in the rotifer *Asplanchna sieboldi:* variability in the body-wall—outgrowth response to dietary tocopherol. *Physiol. Zool.* 48:404–419.

——. 1976a. Sex-specific cannibalism in the rotifer *Asplanchna sieboldi*. *Science* 194:730–731.

——. 1976b. Polymorphism in the rotifer *Asplanchna sieboldi:* biomass, growth, and reproductive rate of the saccate and campanulate morphotypes. *Ecology* 57:542–551.

——. 1985. Competition between rotifers and *Daphnia*. *Ecology* 66:1943–1950.

Gilbert, J. J., et al. 1979. Taxonomic relationships of *Asplanchna brightwelli*, *A. intermedia*, and *A. sieboldi*. *Arch. Hydrobiol.* 87:224–242.

Halbach, U. 1970a. Die Ursachen der Temporal variation von *Brachionus calyciflorus* Pallas (Rotatoria). *Oecologia* 4:262–318.

——. 1970b. Einfluss der Temperatur auf die Populations dynamik des planktischen Rddertieres *Brachionus calyciflorus* Pallas. *Oecologia.* 4:176–207.

——. 1971. Zum Adaptivwert der zyklomorphen Dornenbildung von *Brachionus calyciflorus* Pallas (Rotatoria), 1: Rduber-Beute-Beziehung in Kurzzeit-Versuchen. *Oecologia* 6:267–288.

——. 1973. Quantitative Untersuchungen zur Assoziation von planktischen Rotatorien in Teichen. *Arch. Hydrobiol.* 71:233–254.

Hanley, J. 1949. The narcotization and mounting of Rotifera. *Microscope* 7:154–159.

Harring, H. K. 1913. Synopsis of the Rotatoria. *Bull. U.S. Nat. Mus.* 81:1–226.

——. 1916. A revision of the rotatorian genera *Lepadella* and *Lophocharis* with descriptions of five new species. *Bull. U.S. Nat. Mus.* 51: 527–568.

Harring, H. K., and F.J. Myers. 1922. The rotifer fauna of Wisconsin. *Trans. Wis. Acad. Sci. Arts Lett.* 20:553–662.

——. 1924. The rotifer fauna of Wisconsin, II: a revision of the notommatid rotifers, exclusive of the Dicranophorinae. *Trans. Wis. Acad. Sci. Arts Lett.* 21:415–549.

——. 1926. The rotifer fauna of Wisconsin, III: a revision of the genera *Lecane* and *Monostyla. Trans. Wis. Acad. Sci. Arts Lett.* 22:315–423.

——. 1928. The rotifer fauna of Wisconsin, IV: the Dicranophorinae. *Trans. Wis. Acad. Sci. Arts Lett.* 23:667–808.

Hauer, J. 1935. Rotatorien aus dem Schluchsee-moor und seiner Umgebung. *Verh. Naturwiss. Ver.* 31:47–130.

Hickernell, L. M. 1917. A study of desiccation in the rotifer *Philodina roseola*, with special reference to cytological changes accompanying desiccation. *Biol. Bull.* 32:343–407.

Hlava, S. 1904. (Monograph of the Melicertidae.) *Arch. Prir. Prozk. Cech.* 13:1–79.

Hudson, C. T., and P. H. Cosse. 1889. *The Rotifera; or Wheel-Animalcules, Both British and Foreign.* 2 vols. 272 pp. London.

Jacobs, M. H. 1909. The effects of desiccation on the rotifer *Philodina roseola. J. Exp. Zool.* 6:207–265.

Jennings, H. S. 1900. Rotatoria of the United States, with special reference to those of the Great Lakes. *Bull. US. Fish Comm.* 19:67–105.

——. 1903. Rotatoria of the United States, II: a monograph of the Rattulidae. *Bull. U.S. Fish Comm.* (1902):273–352.

Jennings, H. S., and R. S. Lynch. 1928a. Age, mortality, fertility, and individual diversities in the rotifer *Proales sordida* Gosse, I. *J. Exp. Zool.* 50:345–407.

——. 1928b. Age, mortality, fertility, and individual diversities in the rotifer *Proales sordida* Gosse, II: *J. Exp. Zool.* 51:339–381.

King, C. E. 1967. Food, age, and the dynamics of a laboratory population of rotifers. *Ecology* 48:111–128.

Koste, W. 1978. *Rotatoria. Die Radertiere Mitteleuropas. Monogononta.* 2 vols. 673 pp. text and 234 plates. Gebrud. Borntraeger, Berlin.

Lansing, A. I. 1942. Some effects of hydrogen ion concentration, total salt concentration, calcium and citrate on longevity and fecundity of the rotifers. *J. Exp. Zool.* 91:195–212.

Maly, E. J. 1975. Interactions among the predatory rotifer *Asplanchna* and two prey, *Paramecium* and *Euglena. Ecology* 56:346–358.

Melone, G., et al. 1998. Phylogenetic relationships of phylum Rotifera with emphasis on the families of Bdelloidea. *Hydrobiologia* 387/388: 101–107.

Melone, G., and C. Ricci. 1995a. Rotatory apparatus in bdelloids. *Hydrobiologia* 313/314: 91–98.

——. 1995b. An uncommon bdelloid rotifer: *Abrochtha intermedia* (de Beauchamp, 1909). *Boll. Zool.* 62:29–34.

Miller, H. M. 1931. Alternation of generations in the rotifer *Lecane inermis* Bryce: life histories of the sexual and non-sexual generations. *Biol. Bull.* 60:345–381.

Montgomery, T. H. 1903. On the morphology of the rotatorian family Floscularidae. *Proc. Acad. Nat. Sci. Phila.* 55:363–395.

Murray, J. 1906. Some Rotifera of the Scottish Lochs. *Trans. R. Soc. Edin.* 45:151–191.

——. 1911. Rotifera Bdelloidea. *Proc. R. Irish Acad.* 31:1–20.

Myers, F. J. 1930. The rotifer fauna of Wisconsin, V: the genera *Euchlanis* and *Monornmata. Trans. Wis. Acad. Sci. Arts Lett.* 25:353–411.

——. 1931. The distribution of Rotifera on Mount Desert island. *Am. Mus. Novit.* 494: 1–12.

——. 1933. The distribution of Rotifera on Mount Desert Island, III: New Notommatidae of the genera *Pleurotrocha, Lindia, Eothinia, Proalinopsis,* and *Encentrum. Am. Mus. Novit.* 660:1–18.

——. 1934a. The distribution of Rotifera on Mount Desert Island, Part V: a new species of the Synchaetidae and new species of Asplanchnidae, Trichocercidae, and Brachionidae. *Am. Mus. Novit.* 700:1–16.

——. 1934b. The distribution of Rotifera on Mount Desert Island, Part VII: New Testudinellidae of the genus *Testudinella* and a new species of Brachionidae of the genus *Trichotria. Am. Mus. Novit.* 761:1–8.

——. 1936. Psammolittoral rotifers of Lenape and Union lakes, New Jersey. *Am Mus. Novit.* 830:1–22.

——. 1937a. A method of mounting rotifer jaws for study. *Trans. Am. Microsc. Soc.* 56:256–257.

——. 1937b. Rotifera from the Adirondack region of New York. *Am. Mus. Novit.* 903:1–17.

——. 1938. New species of Rotifera from the collection of the American Museum of Natural History. *Am. Mus. Nov.* 1011:1–17.

——. 1941. *Lecane curvicornis* var. miamiensis, new variety of Rotatoria, with observations on the feeding habits of rotifers. *Not. Nat.* 75:1–8.

——. 1942. The rotatorian fauna of the Pocono plateau and environs. *Proc. Acad. Nat. Sci. Phila.* 94:251–285.

Nachtwey, R. 1925. Untersuchungen (ber die Keimbahn, Organogenese und Anatomic von *Asplanchna priodonta Gosse. Z. Wiss. Zool.* 126: 239–492.

Nauwerck, A. 1978. Notes on the planktonic rotifers of Lake Ontario. *Arch. Hydrobiol.* 84: 269–301.

Örstan, A. 1998. Factors affecting long-term survival of dry bdelloid rotifers: a preliminary study. *Hydrobiologia* 387/388:327–331.

Pawlowski, L. K. 1938. Materialen zur Kenntnis der mossbewohnenden Rotatorien Polens. 1. *Ann. Mus. Zool. Pol.* 13:115–159.

——. 1970. Les Rotifères de le Rivière Grabia. *Soc. Sci. Lodz, Pr. Wydz. III–Nauk. Mat. Ptyzr.* 110: 1–127.

Pax, F., and K. Wulfert. 1941. Die Rotatorien deutscher Schwefelquellen und Thermen. *Arch. Hydrobiol.* 38:165–213.

Pejler, B. 1956. Introgression in plankton Rotatoria with some points of view on its causes and conceivable results. *Evolution* 10:246–261.

——. 1957. On variation and evolution in planktonic Rotatoria. *Zool. Bidr. Uppsala* 32:1–66.

——. 1962. Morphological studies on the genera *Notholca*, *Kellicottia* and *Keratella* (Rotatoria). *Zool. Bidr. Uppsala* 33:295–309.

Pejler, B., et al. (eds.). 1983. Biology of rotifers. *Hydrobiologia* 104:1–396.

Pennak, R. W. 1940. Ecology of the microscopic Metazoa inhabiting the sandy beaches of some Wisconsin lakes. *Ecol. Monogr.* 10:537–615.

——. 1949. Annual limnological cycles in some Colorado reservoir lakes. *Ecol. Monogr.* 19: 233–267.

Pourriot, R. 1965. Recherches sur l'écologiedes rotifères. *Vie Milieu, Suppl.* 21:1–224.

Preissler, K. 1977. Do rotifers show "avoidance of the shore"? *Oekologia* 27:253–260.

Rauh, F. 1963. Untersuchungen uber die Variabilität der Radertiere, III: Die experimentelle Beeinflussung der Variation von *Brachionus calyciflorus* und *Brachionus capsuliflorus. Z. Morphol. Oekol. Tiere* 53:61–106.

Remane, A. 1929a. Intrazellulare Verdauung bei Rädertieren. *Z. Vgl. Physiol.* 11: 146–154.

——. 1929b. Rotatoria. *Tierwelt Nord Ostsee.* 7:1–156.

——. 1929–1933. Rotatorien. *Klassen Ordnungen Tier Reichs 4 (Abt. 2, Buch 1) Lief.* 1–4:1–576.

Ricci, C. N. 1987. Ecology of bdelloids: how to be successful. *Hydrobiologia* 147:117–127.

Ricci, C., and G. Melone. 1998. The Philodinavidae (Rotifera Bdelloidea): a special family. *Hydrobiologia* 385:77–85.

——. 2000. Key to the identification of the genera of bdelloid rotifers. *Hydrobiologia* 418: 73–80.

Rousselet, C. F. 1902. The genus *Synchaeta:* a monographic study, with descriptions of five new species. *J. R. Microsc. Soc.* (1902):269–290, 393–411.

Rudescu, L. 1960. Rotatoria. *Fauna Rep. Pop. Rom.* 2(2):1–1192.

Ruttner-Kolisko, A. 1946. Über das Auftreten unbefruchteter "Dauereier" bei *Anuraea aculeata* (*Keratella quadrata*). *Oesterreich. Zool. Z.* 1:179–181.

——. 1969. Kreuzungsexperimente zwischen *Brachionus urceolaris* und *Brachionus quadridentatus*, ein Beitrag zur Fortpflanzungsbiologie der heterogonen Rotatoria. *Arch. Hydrobiol.* 65:397–412.

——. 1972. Das Zooplankton der Binnengewässer, I Teil, III. Rotatoria. *Die Binnengewässer* 26:99–234.

——. 1983. The significance of mating processes for the genetics and for the formation of resting eggs in monogonont rotifers. *Hydrobiologia* 104:181–190.

Salt, G. W., et al. 1978. Trophi morphology relative to food habits in six species of rotifers (Asplaitchniclae). *Trans. Am. Microsc. Soc.* 97: 469–485.

Schmid-Araya, J. M. 1998. Rotifers in interstitial sediments. *Hydrobiologia* 387/388:231–240.

Seehaus, W. 1930. Zur Morphologie der Rädertiergattung *Testudinella* Bory de St. Vincent. *Z. Wiss. Zool.* 137:175–273.

Segers, H. 1996. The biogeography of littoral *Lecane* (Rotifera). *Hydrobiologia* 323:169–197.

——. 1998. An analysis of taxonomic studies on Rotifera: a case study. *Hydrobiologia* 387/398: 9–14.

Sladecek, V. 1983. Rotifers as indicators of water quality. *Hydrobiologia* 100:169–201.

Stemberger, R. S., and J. J. Gilbert. 1985. Body size, food concentration, and population growth in planktonic rotifers. *Ecology* 66: 1151–1160.

Stossberg, K. 1932. Zur Morphologie der Räddertiergattungen *Euchlanis, Brachionus* und *Rhinoglena. Ecology.* 142:313–424.

Sudzuki, M. 1955. Life history of some Japanese rotifers, 1: *Polyarthra trigla* Ehrenberg. *Sci. Rep. Tokyo Kyoiku Daigaku, Sect.* B, 8:41–63.

——. 1956. (On the general structure and the seasonal occurrence of the males in some Japanese rotifers, IV.) (*Zool. Mag.*) 65:1–6.

——. 1959. (On the general structure and the seasonal occurrence of the males in some Japanese rotifers, VIII.) (*Zool. Mag.*) 68:1–7.

Turner, P. N., and H. L. Taylor. 1998. Rotifers new to Florida. *Hydrobiologia* 387/388:55–62.

Valkanov, A. 1936. Beitrag zur Anatomie und Morphologie der Rotatoriengattung Trochosphaera Semper. *Trav. Soc. Bulg. Sci. Nat.* 17: 177–195.

Varga, L. 1966. Kerekesférgek I. Rotatoria I. *Fauna hung.* 80:1–144.

Viaud, G. 1940. Recherches experimentales sur le phototropisme des rotifères. *Bull. Biol. Fr. Belg.* 74:249–308; 77:68–93, 224–242.

Vidrine, M. F., et al. 1985. Free-swimming colonial rotifers (Monogononta: Flosculariacea: Flosculariidae) in southwestern Louisiana rice fields. *Freshwater Invert. Biol.* 4:187–193.

Wallace, R. L. 1977. Distribution of sessile rotifers in an acid bog pond. *Arch. Hydrobiol.* 79:478–505.

Weber, E. F. 1898. Faune rotatorienne du bassin de Leman. *Rev. Suisse Zool.* 5:263–785.

Wesenberg-Lund, C. 1923. Contributions to the biology of the Rotifera, I: the males of the Rotifera. *Det. Kgl. Vidensk. Selsk. Skr. Naturv. Math.* (8) 4:189–345.

——. 1930. Contributions to the biology of the Rotifera, II: the periodicity and sexual periods. *Det. Kgl. Vidensk. Selsk. Skr. Naturv. Math.*(9)2: 1–230.

Williams, L. G. 1966. Dominant planktonic rotifers of major waterways of the United States. *Limnol. Oceanogr.* 11:83–91.

Wiszniewski, J. 1934. Les rotiferes psammiques. *Ann. Mus. Zool. Pol.* 10:339–399.

Wulfert, K. 1938. Die Rädertiergattung *Cephalodella* Bory de St. Vincent. *Arch. Naturgesch.* 7:137–152.

——. 1964. Unsere gegenwärtige Kenntnis der Rotatoriengattung *Macrochaetus* Perty 1850. *Limnologica* 2:281–309.

——. 1965. Revision der Rotatorien-Gattung *Platyias* Harring 1913. *Limnologica* 3:41–64.

Zacharias, O. 1902. Zum Kapital der "wurstförmigen Parasiten" bei Rädderthieren. *Zool. Anz.* 25:647–649.

8

NEMATODA

(Roundworms)*

Almost any collection of sand, mud, debris, or vegetation from the bottom or wet margins of a pond, lake, brook, or river will contain small roundworms, or nematodes, sometimes in great abundance. Yet, as a group, the freshwater species are poorly known, and few American investigators have been concerned with them since the extensive taxonomic and ecological studies of N. A. Cobb, of the U. S. Department of Agriculture, whose many papers appeared during the first third of the 20th century. Nematodes have been generally avoided by aquatic biologists because of their small size and difficulties of identification. Although some freshwater nematodes can be readily assigned to the proper genus, there are a great many undescribed species in most genera.

Most of the available information about the Phylum Nematoda is concerned with the thousands of parasitic and predatory species on plants and animals, some of which are of extreme importance to the health and economic well being of man. Less than 0.1% of all nematode literature citations are concerned with free-living aquatic species. Probably no more than 150 free-living species have been reported from the freshwaters of the world, but this figure must represent only a small fraction of the actual number of species living in this environment. They occur in such widely differing habitats as pools of polar ice, hot springs, puddles, the bottoms of streams, and the deepest and largest lakes. As of 1967, about 500 species of freshwater nematodes were reported from Europe.

The account of the Nematoda as given in this chapter is written specifically from the standpoint of the free-living freshwater species of the United States, but most of the generalizations also apply to the phylum as a whole.

General characteristics. Nematodes occurring in the substrates of freshwaters are quite small animals with a body length of usually 1 to 2 mm, rarely less than 1 mm. A few families have species that reach several centimeters in length.

The majority of aquatic species are translucent and their internal morphology is easily seen under the microscope. Most of the coloration of nematodes is determined by the contents of the digestive tract and intestinal cell inclusions. In cross section the nematode body is more or less circular. The body tapers anteriorly and posteriorly, more so posteriorly, with the tail often elongated and filiform. The majority of species have three unicellular glands in the tail. Their secretion is carried to the posterior end by one to three minute ducts, and the flow to the outside is regulated by a special pore, or

*Prepared with the assistance of Reyes Peña Santiago.

spinneret, at the tip of the body (Figs. 8.5C, F; 8.8A, B, D). This secretion is sticky and is used for temporary attachment to objects. Some nematodes have a pair of lateral caudal pores, or phasmids, of uncertain function (Fig. 8.8C), although they are probably chemoreceptors.

The mouth is terminal and is usually surrounded by six lips, sometimes fused to form only three, that are more or less distinct. The lip region may be continuous with the body or set off by a constriction. Ventrally, many species have an excretory pore. The transverse slitlike anus is subterminal.

The outermost layer of the body is a nonliving, secreted, multilayered, proteinaceous cuticle, which is variously marked, sculptured, or scaly. Sometimes the transverse striations are composed of lines or rows of dots so fine that they may be detected only with critical illumination and an oil immersion lens. At the other extreme, the cuticle may have prominent transverse grooves, rows of dots, or even scalelike folds (Fig. 8.9C). Some species (e.g., *Plectus*) have longitudinal striations or long, low, keel-like wings, or alae, which appear to aid in stiffening the animal. In the males of a few freshwater species there is a caudal bursa composed of two more of less prominent alae (Fig. 8.1B). Setae are sometimes present, especially at the anterior end; these are simply elongated projections of the general cuticular exoskeleton. The cuticle extends inward and lines the mouth cavity, esophagus, vulva, and posterior portion of the digestive tract.

The hypodermis, or subcuticle, is usually a syncytial tissue; it secretes the cuticle and lies just below it. This layer is thin except dorsally, ventrally, and laterally, where it is greatly thickened and protruded inward to form the four chords (Fig. 8.2).

The remainder of the body wall consists of an innermost layer of muscle cells comprising four groups, or fields, of cells separated by the four chords; there is no circular muscle. In meromyarian nematodes there are usually only two to four cells per field, but in polymyarian nematodes each field has numerous cells. A muscle cell is long and spindle-shaped, and contains numerous ribbon-shaped muscle fibrils. The fibrils may all lie at the base of the cell and adjacent to the hypodermis (platymyarian arrangement), or the fibrils may extend varying distances up the sides of the cells and partially enclose the sarcoplasm (coelomyarian arrangement) (Fig. 8.2). In nematodes, the neuromuscular connections are made by the extensions of the muscle cells rather than by neurons.

The body cavity of a nematode is properly called a pseudocoel and contains only a few connective tissue cells. There is no true mesodermal epithelium covering the digestive tract. All support of internal structures is derived from fluid that is maintained under pressure. Locomotion requires the combination of the hydrostatic nature of the cavity and the longitudinal muscles. Movement is accomplished by the nearly constant and rapid whiplike movements in a dorsoventral plane. No changes occur in body diameter and proportions. These thrashing, serpentine movements usually produce little or no locomotion in water alone, but debris and vegetation afford sufficient friction to result in forward locomotion. A few species move along the substrate like an inchworm, and the rare free-swimming forms are able to move through the water just above the substrate.

A nematode grows by shedding the cuticle four times, including the cuticular lining of the stomodeum and proctodeum.

Digestive system. The mouth is followed by the stoma forming the stomodeum. The stoma maybe a more or less narrow, cylindrical structure, but often, depending on feeding habit, becomes wider and is thus referred to as a buccal cavity. The buccal cavity can be conical, funnel-like, tubular, or barrel-shaped with stiffening rods or plates; rasps and special teeth, or onchia, are common. In some herbivorous (including plant parasitic) and carnivorous species the pharynx is supplied with a solid or hollow stilettolike protrusile spear (Figs. 8.3E, 8.9).

The structure of the pharynx is an important taxonomic character. It is circular in cross section, is lined with cuticle, is com-

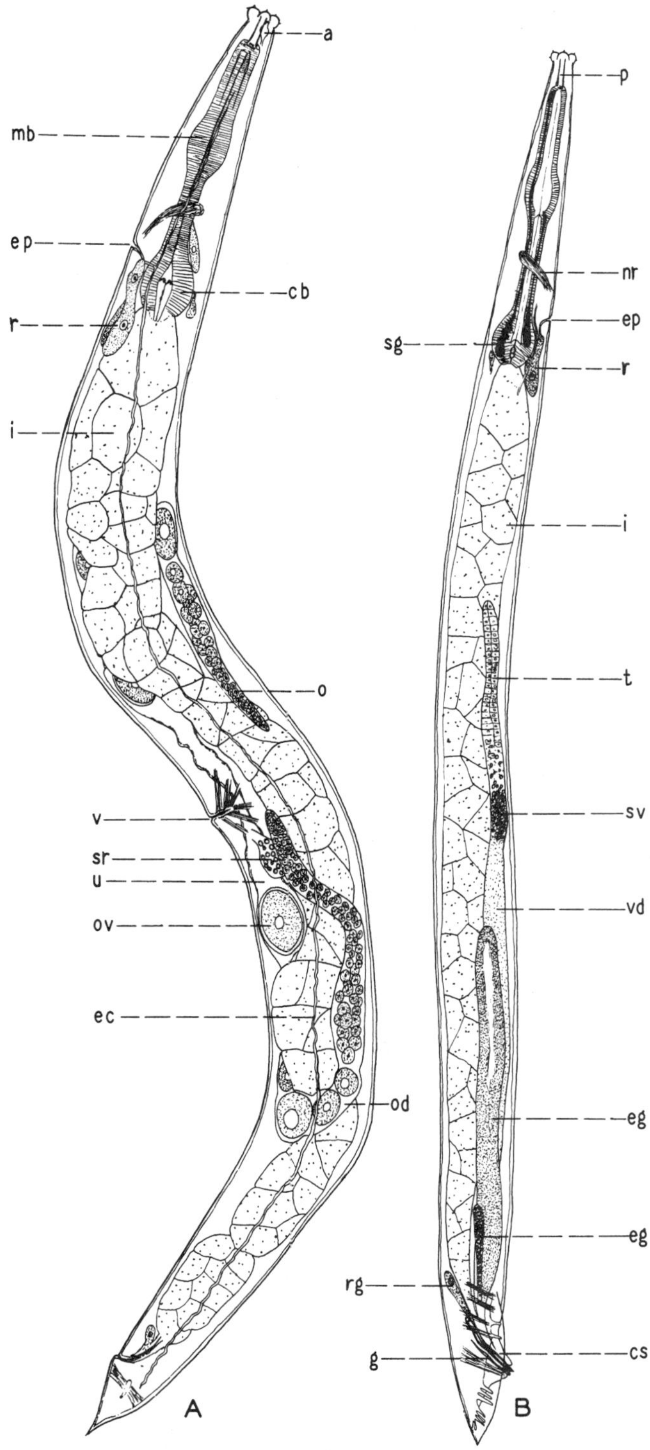

Fig. 8.1. Structure of *Rhabditis* (Rhabditidae), a typical freshwater nematode, ×95. A, female; B, male. A genital bursa is shown in the male, but this structure is rare in free-living species. a, amphid; cb, cardiac bulb of esophagus; cs, copulatory spicule; ec, excretory canal; eg, ejaculatory gland; ep, excretory pore; g, gubernaculum; i, intestine; mb, middle bulb of esophagus; nr, nerve ring; o, ovary; od, oviduct; ov, mature fertilized ovum; p, pharynx; r, renette; rg, rectal gland; sg, salivary gland (glands and ducts solid black; present in both sexes but shown only in figure of male); sr, seminal receptacle; sv, seminal vesicle; t, testis; u, uterus; v, vulva and genital pore; vd, vas deferens. (Modified from Chitwood 1937, by permission.)

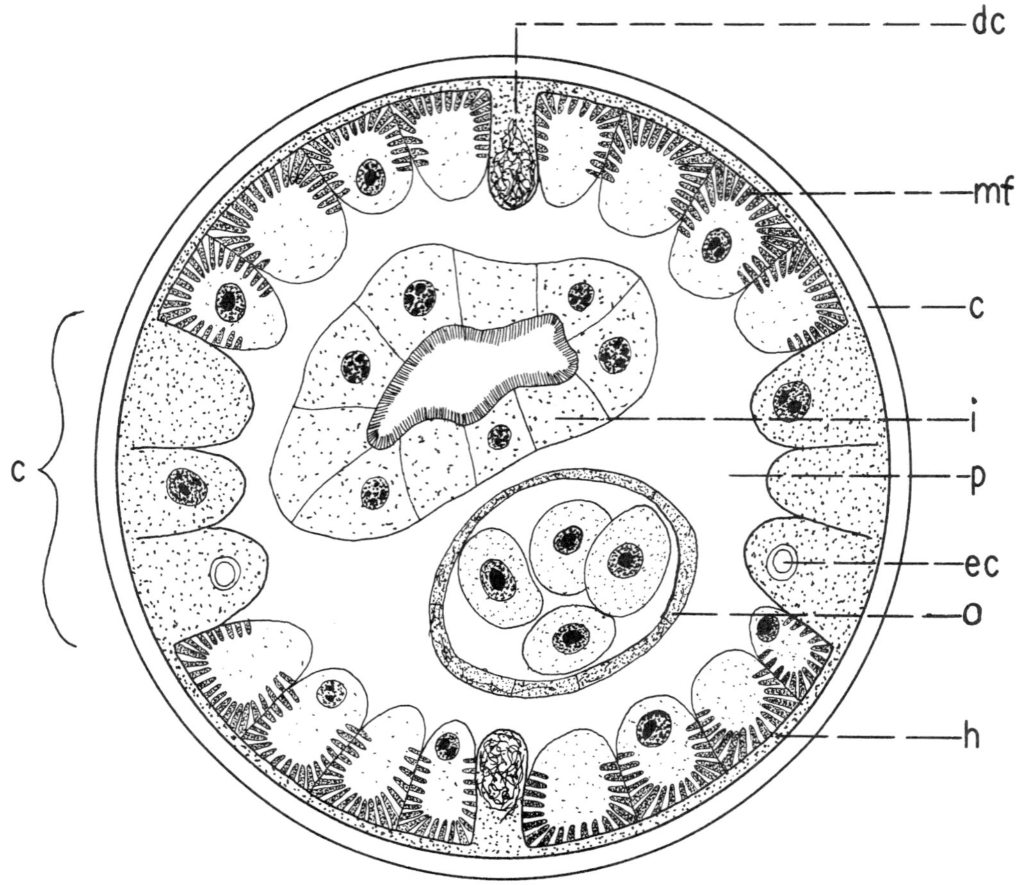

Fig. 8.2. Diagrammatic cross section of a typical female nematode. c, cuticle; dc, dorsal chord of hypodermis containing dorsal nerve cord; ec, excretory canal; h, hypodermis; i, intestine, showing rodlike layer on inner surface; lc, lateral chord of hypodermis; mf, muscle fibers of muscle cell; o, oviduct with four developing eggs; p, pseudocoel.

posed mostly of radial muscle fibers, epithelium, and has a small triquetrous cavity. Three unicellular salivary glands are usually embedded in the pharyngeal wall; their ducts open into the lumen of the stoma or esophagus. In some species the pharynx is elongated, plain, and more or less cylindrical, but more commonly there are one or two ellipsoid or spherical bulbar swellings (Fig. 8.3B–D). If a single bulb is present (Fig. 8.3B), it is usually located at the posterior end of the esophagus and is termed the cardiac bulb and often has a valve, whereas the second (middle) bulb with a valve (Fig. 8.3D) is located somewhere near midlength. In many species a simple swelling occurs (Fig. 8.3C) but does not represent a true bulb. The pharynx is a suctorial and peristaltic organ, with the bulbs serving to increase the pumping action.

The pharynx is separated from the intestine by a valvelike structure called the cardia. The long intestine extends from esophagus to rectum and is devoid of ceca or loops in freshwater species. In some nematode groups (e.g., Dorylaimida, Mononchida) the intestine becomes a complex organ with tubercles and large accessory cells. It consists of a single layer of epithelial cells, though rarely there is a flimsy covering of connective tissue or muscle fibers. The inner surface of the intestine is lined with a layer of

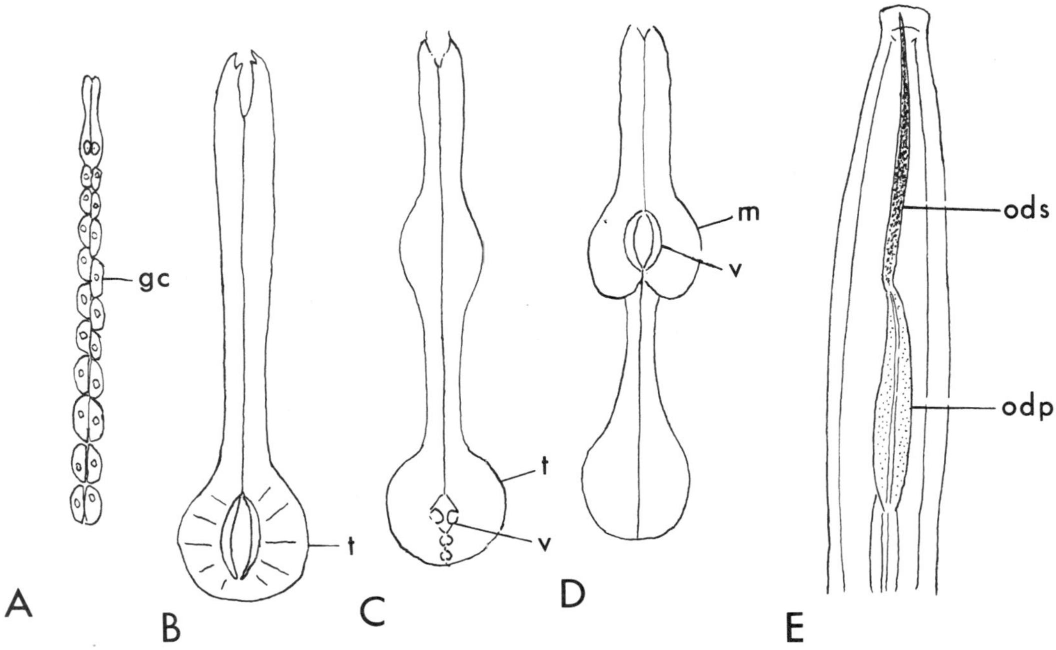

Fig. 8.3. Esophagus morphology and specialization for plant parasitism. A, glandular, Mermithidae; B, with a terminal bulb, *Microlaimus* (Desmodoridae); C, simple middle swelling and terminal valvular bulb, *Rhabditis* (Rhabditidae); D, middle bulb with valve and terminal bulb, *Diplogaster* (Diplogasteridae); E, anterior end of *Longidoris* (Longidoridae) showing two part spear for attacking plants. gc, gland cell; odp, odontophore; ods, odontostyle; t, terminal bulb; v, valve. (Redrawn or modified after Maggenti 1981.)

curious minute rodlike outgrowths, or microvilli, constituting the brush border (Fig. 8.2). These are sometimes mistaken for cilia, but true cilia (with basal bodies) are unknown in this phylum. Most intestinal digestion is thought to occur in the anterior half of this organ.

The rectum is separated from the intestine by a constricting muscle ring. The rectum is almost always a short cuticularized tube opening through the anus. Several unicellular rectal and anal glands are characteristic of this region of the body. The rectum and the male reproductive pore form a cloaca.

Feeding. All varieties of food habits are represented in free-living freshwater nematodes. Some feed only on dead plant material, others on living hydrophytes, and still others only on dead animal material. A fourth group are detritus feeders, consuming both dead plant and dead animal material. Herbivorous species are variously specialized for biting and chewing living plant tissues, and others are equipped with a rigid hollow stylet that pierces plant cells and through which the protoplasm is sucked by the action of the esophagus. Similarly, the predaceous and carnivorous forms, such as *Mononchus*, *Dorylaimus*, *Nygolaimus*, and *Actinolaimus*, show a variety of anatomical modifications for feeding. The lips may be specialized for seizing or tearing the prey, and the pharynx is often partially eversible and supplied with onchia or rasps for crushing, rasping, or macerating. Many carnivorous species have the hollow stylet type of feeding. The prey consists of Protoctists and all types of small Metazoa, including oligochaetes, rotifers, gastrotrichs, tardigrades, and other nematodes.

Material digested in the alimentary canal is absorbed into the fluid of the surrounding hemocoel and carried to all the internal organs and body wall by circulation resulting from the incessant body movements. There is no special circulatory system.

Excretion. The excretory system presents a varied picture in the phylum as a whole. In its simplest condition there is a single ventral excretory cell, or renette, which opens through an excretory pore on the midventral line in the region of the esophagus by way of a short to long duct (Fig. 8.4). In other nematodes two lateral excretory canals are embedded in the lateral chords of the hypodermis throughout most of the body length. They are connected anteriorly and ventrally by a transverse canal, thus forming an H or U shape. A very short duct connects the transverse duct with the excretory pore. There may or may not be two special cells (the renette) associated with the transverse duct (Fig. 8.1).

In a few genera, including *Dorylaimus*, no excretory system has been found. Considerable excretion through the digestive tract may occur in all nematodes.

Nervous system. A ring of nerve fibers and associated indistinct ganglia encircle the esophagus. From this mass, small nerves are given off that innervate sensory structures in the cephalic region. Four or more longitudinal nerves extend posteriorly in the hypodermis, of which the ventral nerve is usually the most prominent and may be ganglionated along its length. Especially at the extremities of the body, there are scattered small ganglia and numerous branch nerves that connect the longitudinal nerves with each other.

Sense organs. Nematodes have several kinds of receptors: mechanoreceptors, chemoreceptors, and photoreceptors. The cephalic lips display specific patterns among nematodes and are believed to be mechanoreceptors. The somatic setae, often present between anterior and posterior ends, are probably also tactile. Paired eyespots occur in some species of *Punctodora* and *Monhystera*, as well as in a few other genera. Such eyespots are small clumps of reddish, violet, or blackish granules embedded in the pharynx or between the pharynx and body wall. A minute lenslike device may be present in front of each clump of pigment, or it may be absent. There is some difference of opinion as to whether there are nervous connections between eyespots and nervous system.

A peculiar feature of nematodes is the presence of two cephalic sense organs called amphids (Fig. 8.6) which are generally considered to be chemoreceptors. They open on the lateral surface of the body and are situated in the lip region or slightly behind the lips. Externally, an amphid may have the shape of a narrow slit, an oval, a circle, or spiral grooves and ridges on the cuticle. Internally, it consists of a small pouch (fovea), ducts, and receptors. Some nematodes have a pair of caudolateral pores called phasmids of uncertain function but are possibly chemoreceptors.

Reproduction, development. Most species of nematodes are dioecious and reproduction is syngamic, but in many species males are rare or unknown, and reproduction is usually parthenogenetic, as in *Rhabdolaimus*, *Monhystera*, and *Alaimus*. Hermaphroditic species occur in *Mononchulus*, *Aphanolaimus*, *Ironus*, *Tobrilus*, *Mononchus*, and a few other genera but the condition is considered rare. The female is usually the larger, but the sexes can be distinguished by the sex characters.

In general, the reproductive systems are simpler and less extensive than those of parasitic species and produce relatively few gametes. In the mature female the genital pore, or vulva, can be seen as a transverse slit on the ventral midline near the middle of the body. There is a short muscular vagina, which may connect with a single thin-walled, tapering reproductive tubule or, more typically, may branch to form two tubules, one anterior and one posterior. Beginning at the vagina, each tubule consists of a wide uterus, an oviduct, and a fine terminal ovary composed of very small oocytes. Sometimes the uterus has a small associated seminal receptacle (Fig. 8.1). A reproductive tubule may

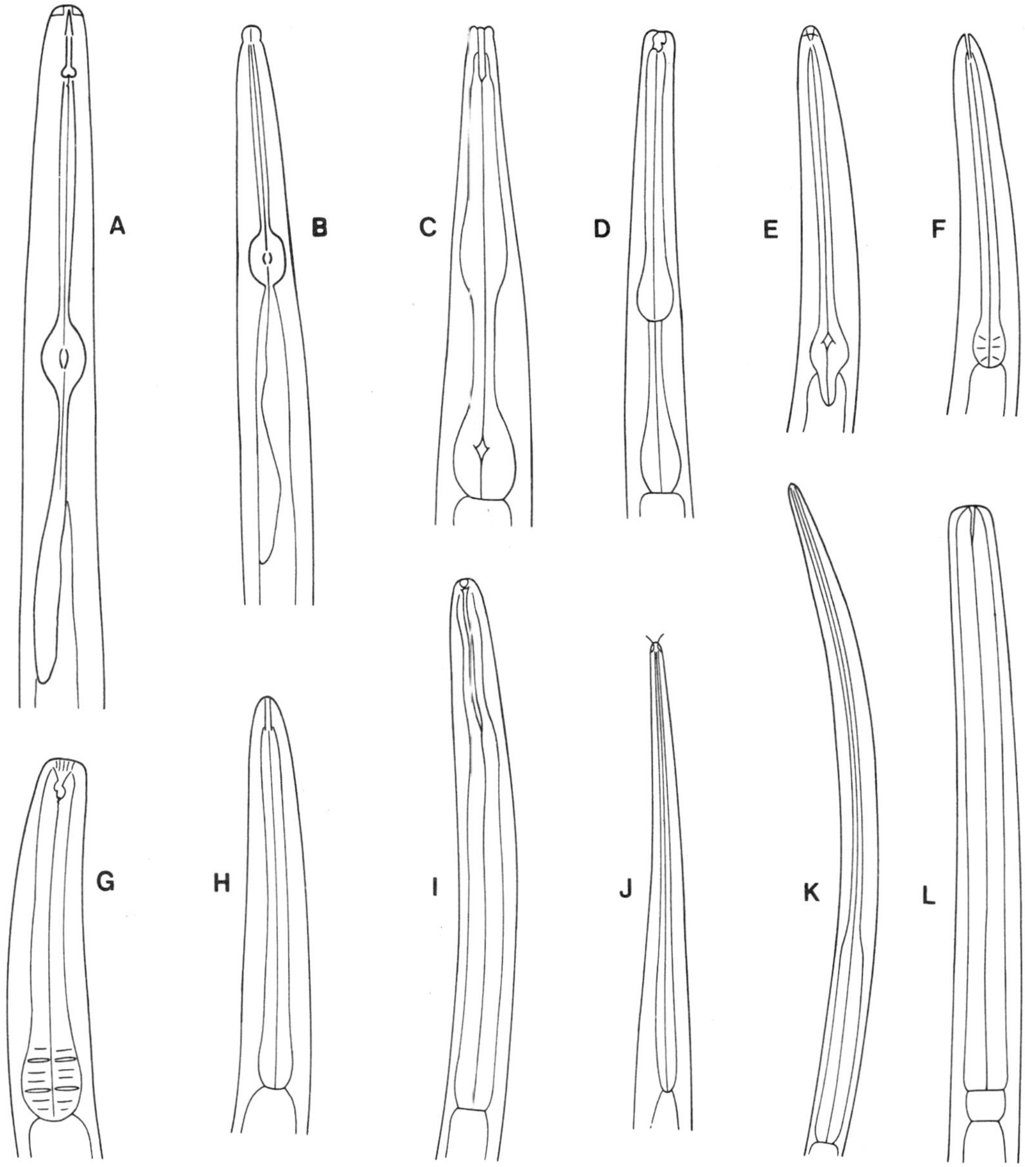

Fig. 8.4. Esophagus shapes. A, *Hirschmanniella* (Tylenchidae); B, *Aphelenchoides* (Aphelenchoididae); C, *Rhabditis* (Rhabditidae); D, *Butlerius* (Diplogasteridae); E, *Plectus* (Plectidae); F, *Leptolaimus* (Leptolaimidae); G, *Achromadora* (Achromadoridae); H, *Cylindrolaimus* (Diplopeltidae); I, *Ironus* (Ironoidae); J, *Aphanolaimus* (Leptolaimidae); K, *Alaimus* (Alaimidae); L, *Tripyla* (Tripylidae). (Modified from Ferris et al. 1973.)

be straight or curved. Thus, the reproductive system of a female nematode may have one of a combination of four possible configurations: double- or single-tubed and straight or curved (Fig. 8.9D).

The male reproductive system is ordinarily composed of two opposed testes (Fig. 8.16), a posterior and an anterior branch. Commonly, one branch is reduced. Beginning basally, each tubule consists of a long vas deferens, short seminal vesicle, and thin testis (Fig. 8.1B). Ejaculatory glands some-

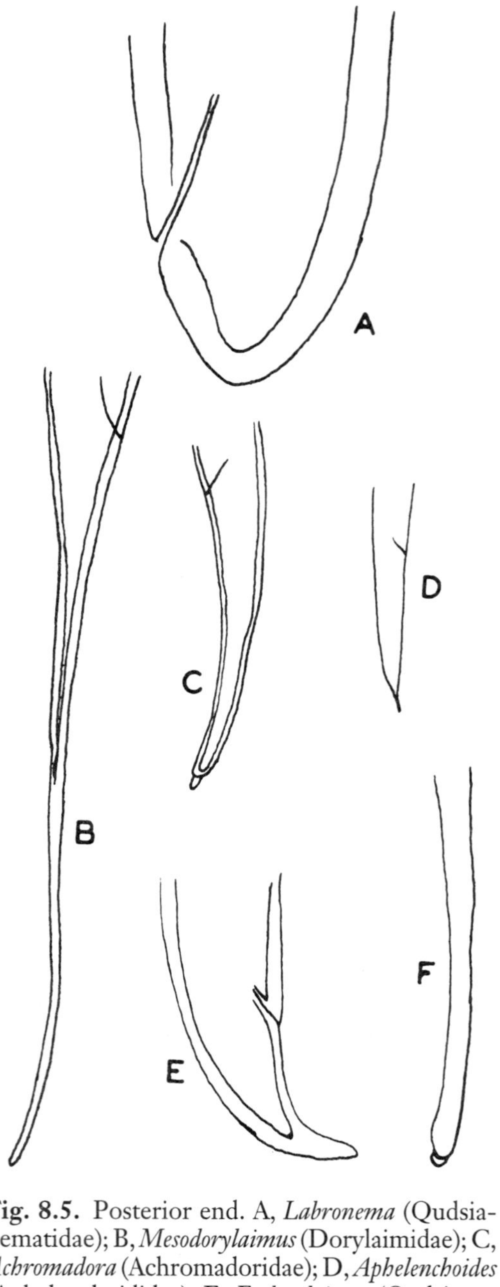

Fig. 8.5. Posterior end. A, *Labronema* (Qudsianematidae); B, *Mesodorylaimus* (Dorylaimidae); C, *Achromadora* (Achromadoridae); D, *Aphelenchoides* (Aphelenchoididae); E, *Eudorylaimus* (Qudsianematidae); F, *Labronema* (Qudsianematidae). (Modified from Ferris et al. 1973.)

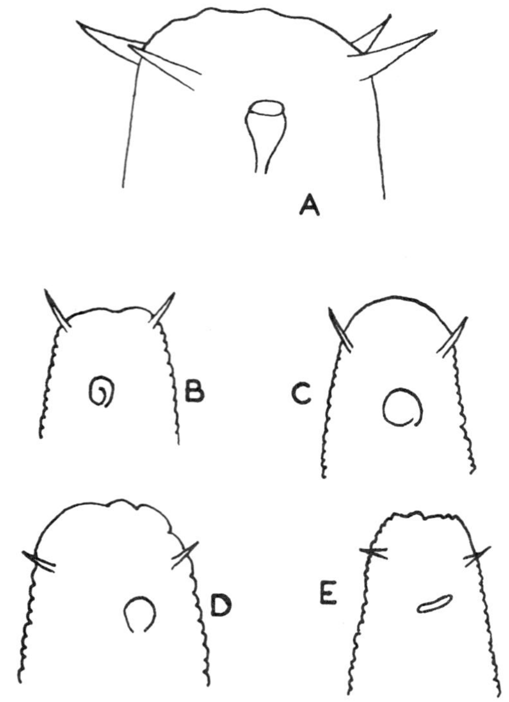

Fig. 8.6. Amphid shapes. A, *Tobrilus* (Tobrilidae); B, *Achromadora* (Achromadoridae); C, *Prodesmodora* (Desmodoridae); D, *Plectus* (Plectidae); E, *Anaplectus* (Plectidae). (Modified from Ferris et al. 1973.)

times open into the basal part of the vas deferens. Since the rectum of the male is common to both the reproductive and digestive ducts, it is more appropriately called a cloaca. A pair of sclerotized copulatory spicules are contained within a delicate pouch on the dorsal side of the cloaca. During copulation these spicules are extruded through the cloacal aperture, or anus, of the male and into the vulva (Fig. 8.7). They are moved back and forth by muscles so that the vulva and vagina are kept open and so that sperm are more easily propelled into the vagina. The gubernaculum is a sclerotized thickening in the dorsal wall of the male cloaca that acts as a spicular guide and prevents the spicules from rupturing the spicular pouch and cloaca when exerted.

A clasping device, or copulatory bursa (Fig. 8.1B), is common in parasitic species but is rare in the males of freshwater species. It consists of two thin, transparent, longitudinal flaps of cuticle near the posterior end. It is widest opposite the anus and may or may not be long enough to encompass the posterior tip of the animal.

Fertilization occurs internally in the upper end of the uterus and the egg shell is

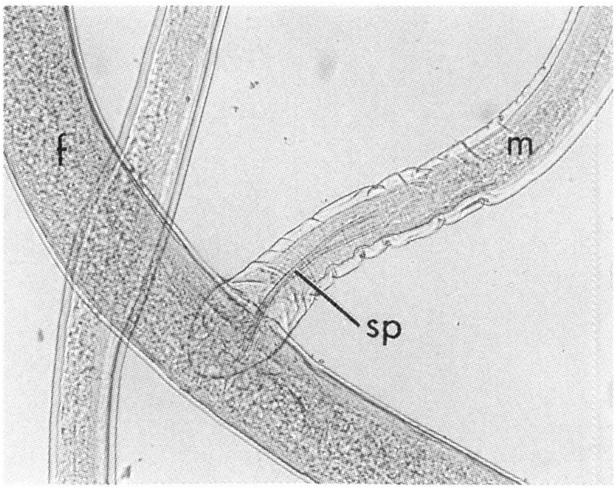

Fig. 8.7. Copulating mermithids. f, female; m, male; sp, spicule.

laid down in the lower uterus. The fertilized eggs usually leave the body before the onset of segmentation, but some species are viviparous or ovoviviparous. The embryo is formed and hatches in a few hours to a few weeks, depending on the particular species and ecological conditions. The newly hatched young are fully developed except for size, reproductive system, and certain specializations of the lips, pharynx, and esophagus. It is thought that there is almost always a series of four molts (five instars) in the complete life history. In addition to the general body covering, a molt also includes shedding of the cuticular lining of pharynx, esophagus, vulva, and rectum. The secondary sex characters are not apparent until after either the penultimate molt or the last molt. Little is known about the longevity of free-living nematodes, but males are thought to be much shorter lived than females.

Ecology. Among all metazoan phyla, the Nematoda are perhaps one of the most abundant groups on Earth and the most highly adaptable from ecological and physiological standpoints. Nematodes are known from virtually every habitat in the sea, freshwater, and on land; some are generalists, but many are very specialized forms. Furthermore, some species show an amazing ability to thrive in a wide variety of habitats and a wide range of ecological conditions. The same species, for example, may be found from the tropics to the subarctic, from warm springs to inland saline lakes to cold apine lakes, and on many types of substrates.

Many different nematode taxa are found in freshwater. Some of them are characteristic or almost exclusive from this habitat, such as the species of the genera *Cryptonchus*, *Chronogaster*, *Anonchus*, *Mononchus*, and *Dorylaimus*. Others occur in both freshwater and terrestrial soils, for example, in the genera *Prismatolaimus*, *Tobrilus*, and *Mesodorylaimus*. The species of the genera *Oncholaimus* and *Chromadorita* are both marine and freshwater nematodes; and species of *Monhystera* even occur in terrestrial soils, freshwater, and marine habitats. Moreover, terrestrial soils and freshwaters share many nematode species as the result of mixing. The origin of freshwater nematodes must be looked for in marine forms that conquered first the sediments of inland waters and then probably the terrestrial soils.

Freshwater nematode distribution has been barely explored even though all evidence suggests that the fauna of freshwaters is rich and ecologically diversified. In lakes in Germany species are abundant at nearly all depths, with the highest concentration in the littoral zone. Study has revealed that up to 59% of the littoral and 75% of the profundal fauna is made up of nematodes and over 70 species have been recorded in one lake. In a spring in Spain 45 species have been found, and distribution in these environments is controlled by mineral content, with chloride concentration being the most significant limiting factor. As pointed out long ago by Cobb, nematode populations often reach tremendous numbers both in terrestrial soils and in the substrates of bodies of waters, where they are of great mechanical importance in continuously working over and mixing soils is much the same way as oligochaetes. Even in the wet sand of lake beaches near the water's edge, which is not an especially favorable habitat for nematodes, up to eight adult nematodes occur per milimeter of sand. Some of the most dense populations, especially species of *Mononchus*, *Ironus*, and *Tripyla*, occur in the sand and gravel filter beds of sewage disposal plants.

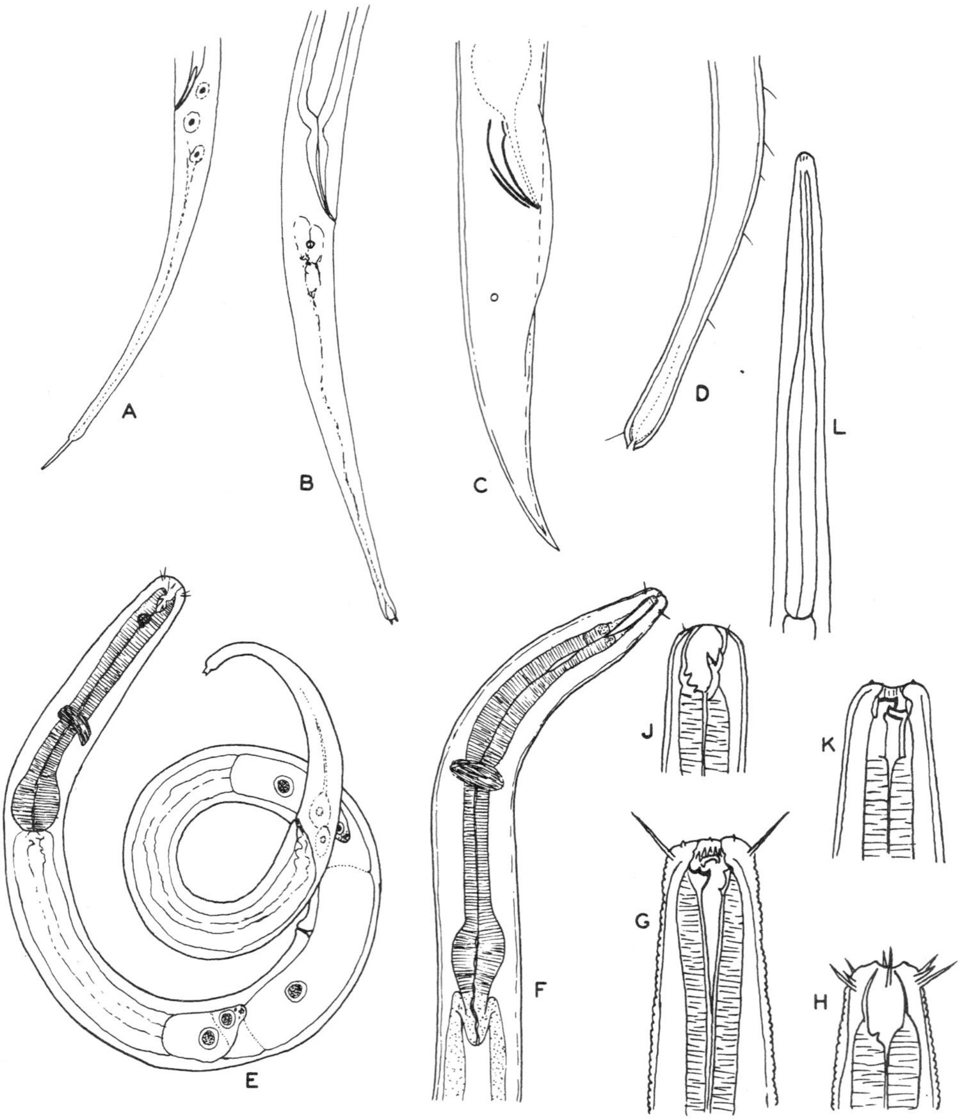

Fig. 8.8. Structure of nematodes. A, posterior end of *Rhabdolaimus* (Rhabdolaimidae), showing caudal glands and spinneret; B, posterior end of *Aphanolaimus* (Leptolaimidae), showing caudal glands, spinneret, and phasmid; C, posterior end of male *Tylenchus* (Tylenchidae), showing phasmid; D, posterior end of *Tobrilus* (Tobrilidae), showing spinneret; E, female *Achromadora* (Achromadoridae), ×350; F, anterior end of *Plectus* (Plectidae); G, anterior end of *Punctodora* (Chromadoridae); H, anterior end of *Prismatolaimus* (Prismatolaimidae); J, anterior end of *Butlerius* (Diplogasteridae); K, anterior end of *Mononchoides* (Diplogasteridae); L, anterior end of *Nygolaimus* (Nygolaimidae).

A wide variety of species occur around the roots of aquatic plants. In natural freshwater habitats most specimens are confined to the uppermost 5 cm of the substrate, with the highest concentration in the top 2 cm. Unlike most freshwater invertebrate taxa, the (micro) nematodes are ecologically tied to the substrate and do not swim freely in the water. They are chiefly members of the meiobenthos and, in addition to sewage disposal habitats, they attain their greatest densities in soft muddy substrates of lakes,

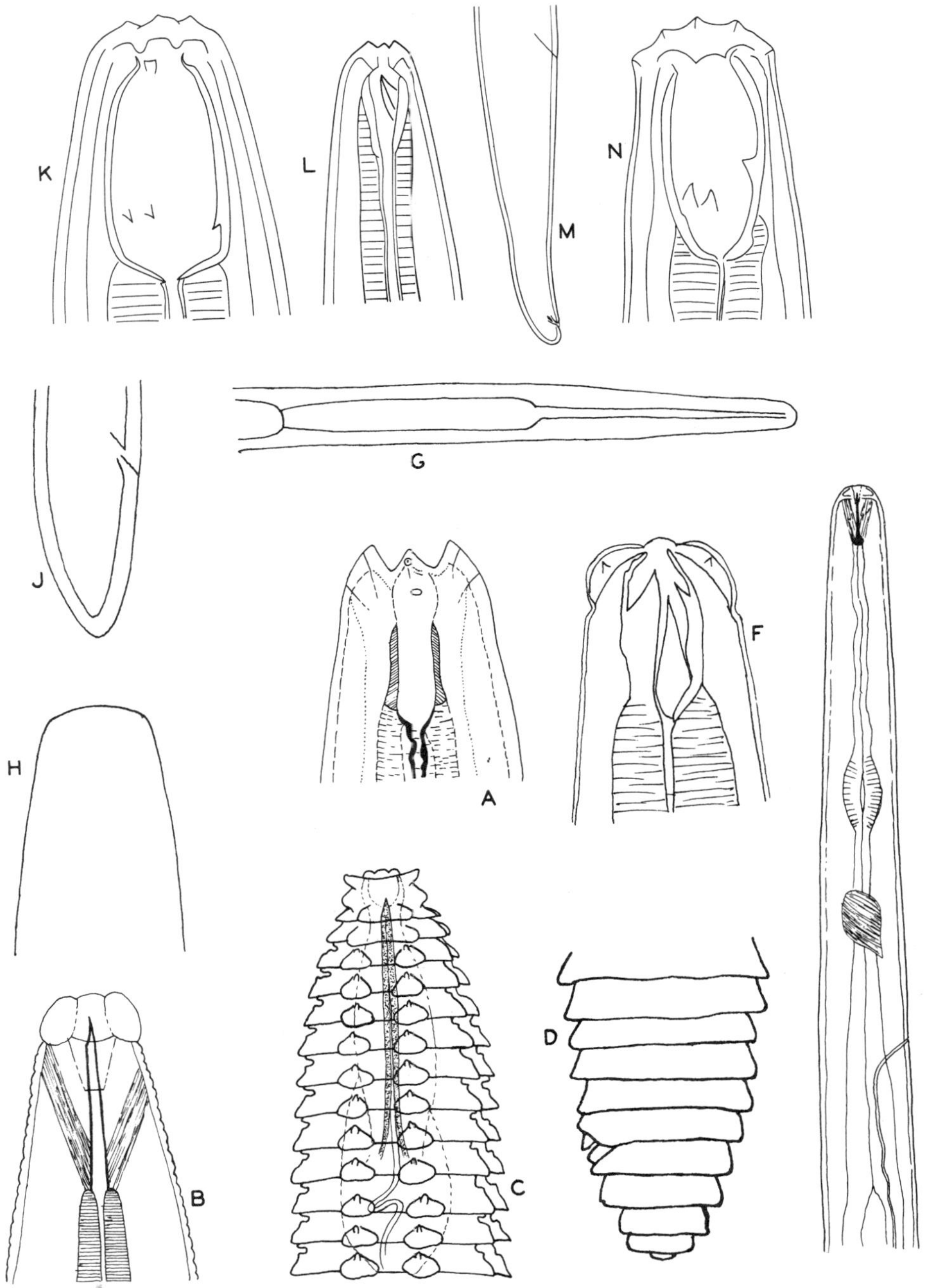

Fig. 8.9. Structure of nematodes. A, anterior end of *Panagrolaimus* (Panagrolaimidae) B, anterior end of *Aphelenchus* (Aphelenchidae); C, anterior end of *Criconema* (Criconematidae); D, posterior end of *Criconemoides* (Criconematidae); E, anterior end of *Tylenchus* (Tylenchidae); F, anterior end of *Oncholaimus* (Oncholaimidae); G, anterior end of *Thornia* (Qudsianematidae); H, anterior end of *Thornia* (Qudsianematidae); J, posterior end of female *Thornia* (Qudsianematidae); K, anterior end of *Anatonchus* (Mononchidae); L, anterior end of *Mononchulus* (Mononchulidae); M, posterior end of female *Mononchulus* (Mononchulidae); N, anterior end of *Miconchus* (Mononchidae). (A modified from Thorne 1937; B modified from Chitwood 1939; C modified from Cobb 1915; G–J modified from Ferris et al. 1973; K–N from Ferris et al. 1973.)

ponds, and streams, often in excess of 100,000 specimens/m^2 and to a depth of 2 cm. Such nematode assemblages commonly consist of 5 to 10 species, altough some thorough studies report up to 150 species in one stream substrate.

Although populations of aquatic nematodes can be maintained indefinitely in low concentrations of dissolved oxygen, 2–10% saturation, for example, it is thought that active nematodes are capable of withstanding anaerobic conditions for only one to several weeks at a time. The eggs, however, are highly resistant and may remain viable after many months in the absence of oxygen and after repeated freezing and thawing. Because of the ability of many nematodes to withstand near anaerobic conditions, they are often an important component of the recovery zone fauna below sewage outfalls and can be ordinarily found in wastewater treatment facilities. As much as 75% of the nematode fauna occurring in a river with wastewater discharge can be traced to the source of the wastewater. There seems to be very little information about encystment and desiccation among aquatic nematodes, and it is not known how important these mechanisms are for dissemination; however, the eggs of soil and parasitic nematodes are highly resistant to desiccation (even for periods up to 10 or 20 years) and to high and low temperatures.

Collecting, preparing, preserving. Perhaps the chief reason for the general neglect of the nematodes by investigators in aquatic zoology is the fact that they almost invariably occur in quantities of bottom mud and fine organic debris from which it is difficult to separate them. Also, preparation and preservation requires some skill and patience. Southey (1985) provided an interesting review of methods and techniques customarily used in general nematology that can serve to handle freshwater specimens.

Nematodes can be collected from both sediments and organic debris and concentrated in clean water by different methods. The Baermann (1917) funnel technique is the simplest, but produces good results for qualitative studies and needs only basic materials. The equipment consists of a funnel (made of plastic or glass) with a piece of rubber tubing attached to the stem and closed by a spring or screw clip. The funnel is placed in a support and almost filled with tapwater. About 100 mL of sediment, with or without organic material, is placed in a square of butter muslin, which is folded to enclose the material, then gently submerged in the water in the funnel. Active nematodes pass through the cloth and sink to the bottom of the funnel stem. After some hours, or overnight, a small quantity of water containing the nematodes is decanted and examined.

Extraction methods, in general, yield a quantity of living nematodes whose presistent movements make their study difficult under the microscope. One of the best methods to kill nematodes and avoid coiling is to heat the water containing them to 60–65°C for a few minutes.

Several fixatives have been recommended for studying nematode morphology, but the most widely used is FA 4:1 [formalin (=40% formaldehyde) 10 mL; glacial acetic acid 1 mL; distilled water up to 100 mL], but 4% formaldehyde also gives good results. Heating the fixative to 60–65°C is advised.

Some morphological features of the nematodes become obscure after fixation, and can be cleared by processing in different media, mainly glycerol; however, glycerin penetrates the nematode cuticle very slowly, so the process needs some time to be completed. A good and rather fast method involves transferring fixed specimens to hot lactophenol (lactic acid 20 mL, phenol 20 mL, glycerine 40 mL, distilled water 20 mL) and leaving it there until the liquid becomes thick. The specimens are then transferred to a warm muxture of 75% glycerol and 25% lactophenol and kept warm for 5 minutes. The nematodes are finally transferred to pure glycerol and mounted in the same medium. Permanent slides with coverslips should be ringed with Murrayite, "Zut," or similar cement. Many aquatic nematodes may be cultured in the laboratory in 3% agar in tapwater with a little added soil and debris.

Nematode morphology can be made with low- and high-power objectives of the mi-

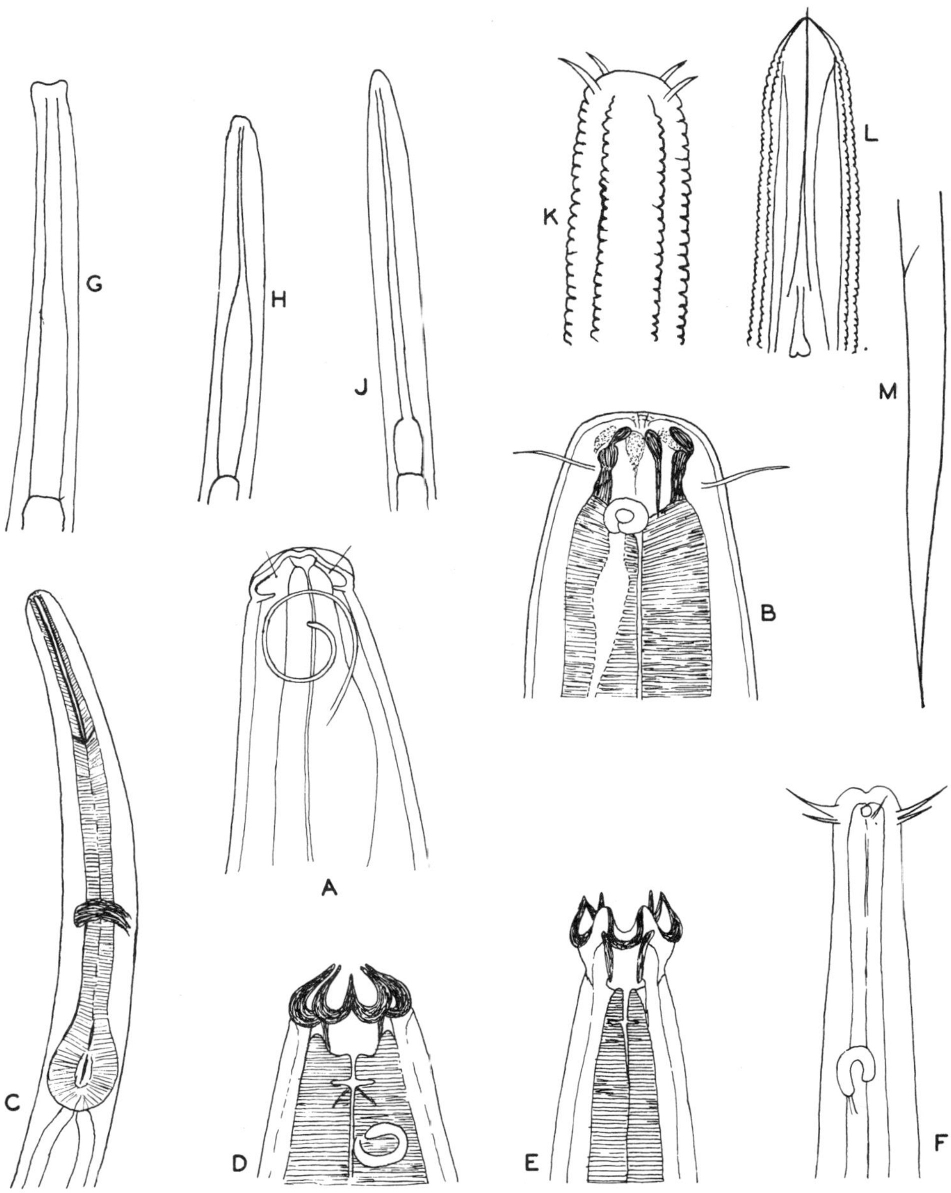

Fig. 8.10. Anterior end of nematodes. A, *Aphanolaimus* (Leptolaimidae); B, *Anonchus* (Leptolaimidae); C, *Rhabdolaimus* (Rhabdolaimidae); D, *Euteratocephalus* (Teratocephalidae); E, *Teratocephalus* (Teratocephalidae) (annulations not shown); F, *Bastiana* (Bastianiidae) (annulations not shown); G, *Labronema* (Qudsianematidae); H, *Eudorylaimus* (Qudsianematidae); J, *Aulolaimoides* (Aulolaimidae); K, *Atylenchus* (Tylenchidae); L, *Hemicycliophora* (Criconematidae); M, posterior end of *Tylenchus* (Tylenchidae). (A, C–E modified from Cobb 1914; B modified from Cobb 1913; F modified from Chitwood 1937; G–M modified from Ferris et al. 1973.)

croscope, but critical illumination and the oil immersion objective (×1000) are usually essential to study more precise details. Identification of nematode taxa requires considerable experience, especially at he genus and species level.

Taxonomy. This volume treats the Nematoda as a phylum and the higher classification adopted is that presented in Roberts and Janovy (2000). The number of genera truly "aquatic" is probably impossible to assess because so many genera occur in both water and soil and in between. As many as 1094 genera of soil and freshwater are recognized. Many other genera are normally parasitic on plants and not actually "free-living" and one large family of very conspicuous nematodes, the Mermithidae, is free-living only in the adult stage, the juveniles being parasitic of insects. To complicate matters, several estuarine forms can venture into and live in freshwater.

Freshwater families, genera, and species are found among all major groups of nematodes. The arrangement of families included in the key is provided in Table 8.1.

Several lists of genera of freshwater nematodes of North America are available, the most comprehensive and recent being that of Esser and Buckingham (1987), who list 133 genera. Other lists collectively add 16 or 17 genera and if the Mermithidae are included, then almost 170 genera have been reported from the United States. Because of the enormous number of genera and the difficulty in identifying these taxa the key treats nematodes to the family level only. The key includes 55 families. Many of these families are represented in U.S. freshwaters by only a single genus. In these cases the family characters used in the key are specific to that genus. It is very possible that previously unrecorded described genera or newly discovered genera will broaden the definition of certain families, making the key less useful. Identifying nematodes, as has been stated already, is a difficult task and with the increasing number of taxa being discovered regularly it may eventually become necessary to devise computerized keys to handle the multitude of characters and variations used to diagnose genera and species of freshwater nematodes.

The key is mainly based on the general works of Andrássy (1976) and Lorenzen (1994). Those readers interested in the identification of nematodes at generic level (and even specific level in some cases) may consult the following works: Jairajpuri and Ahmad (1992) for members of the orders Dorylaimida and Triplonchida; Jairajpuri and Khan (1982) for Mononchida; Andrássy (1984) for Monhysterida, Chromadorida and Rhabditida in partim; Andrássy (1983) for Rhabditida; and Siddiqi (1986) and Hunt (1993) for Tylenchida. Even at the family level, the nematodes are difficult to identify. Using the key effectively requires an understanding of nematode terminology.

KEY TO THE ORDERS AND FAMILIES OF FRESHWATER NEMATODES OF THE UNITED STATES*

1. Pharynx usually consisting of slender but muscular anterior part and expanded cylindrical posterior portion, sometimes completely cylindrical (Fig. 8.4H–K), rarely slender and ending in a basal bulb (Figs. 8.3B; 8.4A–G) or simply glandular (Fig.8.3A); amphid fovea a wide pouch beneath the cuticle (Fig. 8.6A), opening in a transverse slit at the lip region base; male genital papillae in a ventromedian series **2**

Without such combination of characters, i.e., pharynx generally with one or two bulbs (Figs. 8.3D; 8.4D), amphids very small and undistinguishable or big and circular or spiral (Fig. 8.6B–D), male without a ventromedian series of genital papillae. **7**

*Prepared by Reyes Peña Santiago.

Table 8.1 Families of Nematoda Occurring in Fresh Waters of the United States

- Class Enoplea
  - Order Enoplida
    - Alaimidae
    - Ironoidae
    - Oncholaimidae
    - Oxystominidae
    - Tobrilidae
    - Tripylidae
  - Order Triplonchida
    - Diphtherophoridae
  - Order Dorylaimida
    - Actinolaimidae
    - Aporcelaimidae
    - Aulolaimoididae
    - Belondiridae
    - Campydoridae
    - Dorylaimidae
    - Leptonchidae
    - Longidoridae
    - Desmodoridae
    - Ethmolaimidae
    - Haliplectidae
    - Leptlaimidae
    - Microlaimidae
    - Odontolaimidae
    - Plectidae
    - Prismatolaimidae
    - Rhabdolaimidae
    - Teratocephalidae
  - Order Monohysterida
    - Diplopeltidae
    - Monhysteridae
    - Mydonomidae
    - Nordiidae
    - Nygellidae
    - Nygolaimidae
    - Qudsianematidae
    - Tylencholaimidae
  - Order Mononchida
    - Bathyodontidae
    - Mononchidae
    - Mononchulidae
  - Order Mermithida
    - Mermithidae
- Class Rhabditea
  - Order Chromadorida
    - Achromadoridae
    - Aulolaimidae
    - Bastianiidae
    - Chromadoridae
    - Chronogasteridae
    - Cyatholaimidae
    - Xyalidae
  - Order Rhabditida
    - Bunonematidae
    - Cephalobidae
    - Diplogasteridae
    - Cylindrocorporidae
    - Panagrolaimidae
    - Rhabditidae
  - Order Tylenchida
    - Aphelenchidae
    - Aphelenchoididae
    - Criconematidae
    - Hoplolaimidae
    - Tylenchidaee

2. Cephalic setae absent, caudal glands almost always absent **3**
Cephalic setae present, caudal glands present . . . . **ORDER ENOPLIDA (in part), 10**

3. Stoma armed with a stylet or spear (sometimes a mural tooth) of variable morphology (Figs. 8.9B; 8.11F) **4**
Stoma unarmed, without stylet **5**

4. Spear complex, asymmetrical, and generally connected to accessory elements (Fig. 8.15M); excretory pore present; prerectum absent; male without adanal pair of genital papillae. . . . . . . **ORDER TRIPLONCHIDA, DIPHTHEROPHORIDAE**
Spear not so complex, almost always symmetrical; excretory pore absent, prerectum present; male with adanal pair of genital papillae. . . . . **DORYLAIMIDA, 14**

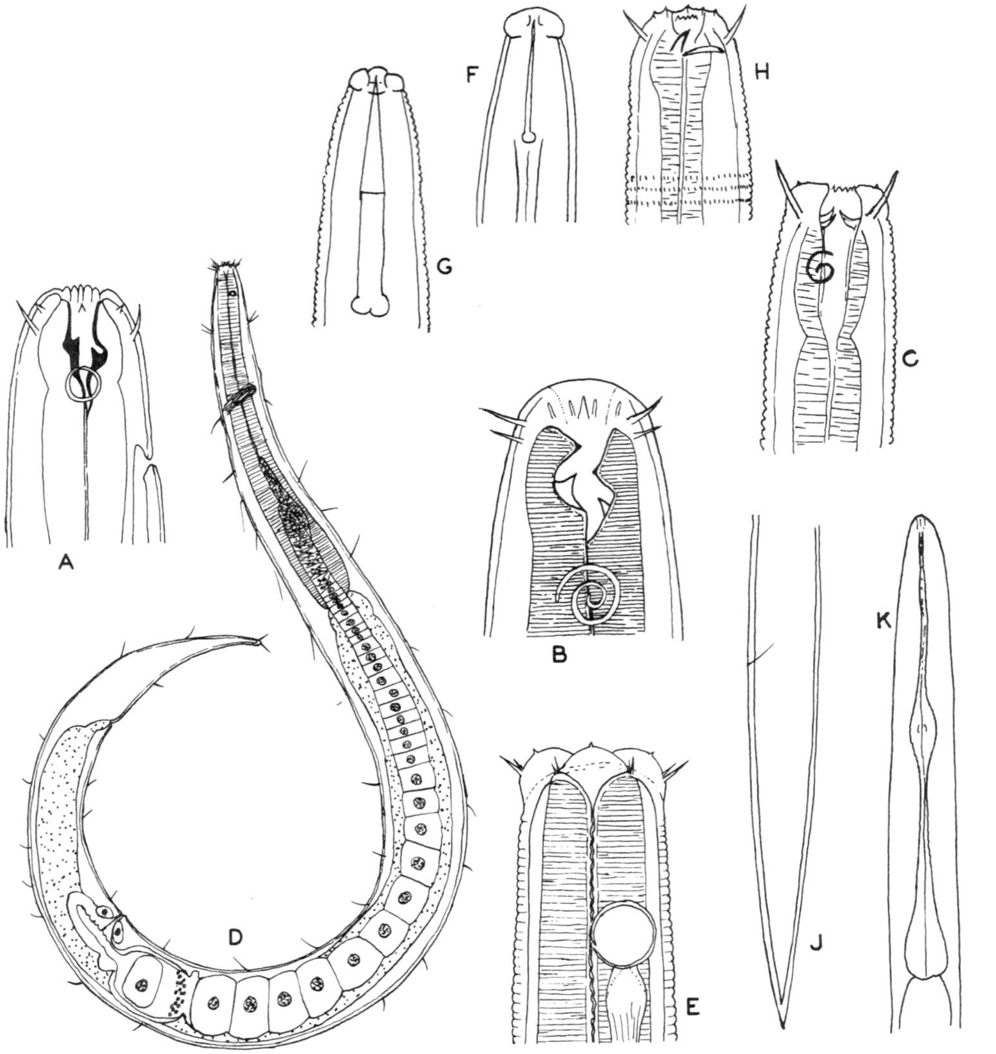

Fig. 8.11. Structure of nematodes. A, anterior end of *Microlaimus* (Desmodoridae); B, anterior end of *Achromadora* (Achromadoridae); C, anterior end of *Ethmolaimus* (Ethmolaimidae); D, female *Monhystera* (Monhysteridae), ×110; E, anterior end of *Monhystera* (Monhysteridae); F, anterior end of *Aphelenchoides* (Aphelenchoididae); G, anterior end of *Hirschmanniella* (Tylenchidae); H, anterior end of *Chromadorita* (Chromadoridae); J, posterior end of female *Hirschmanniella* (Tylenchidae); K, anterior end of *Tylenchus* (Tylenchidae). (A modified from Chitwood 1937; B and D modified from Cobb 1914; E modified from Micoletzky 1925; C, and F–K modified from Ferris et al. 1973.)

5. Stoma very wide, usually a buccal cavity with sclerotized walls and with one or more teeth (Figs. 8.9K, L; 8.14E–G) . . . . **ORDER MONONCHIDA, 27**
   Stoma quite narrow and small, poorly differentiated **6**
6. Stoma tubular and straight or lacking; esophagus glandular (Fig. 8.3A); intestine absent or reduced; larger nematodes, almost always 10 mm or more in length.
   ORDER MERMITHIDA, MERMITHIDAE
   Stoma very short; esophagus tubular and muscular (Fig. 8.4K); intestine present; smaller nematodes, almost always under 2 mm long (Figs. 8.12F, 8.13L)
   ORDER ENOPLIDA, ALAIMIDAE

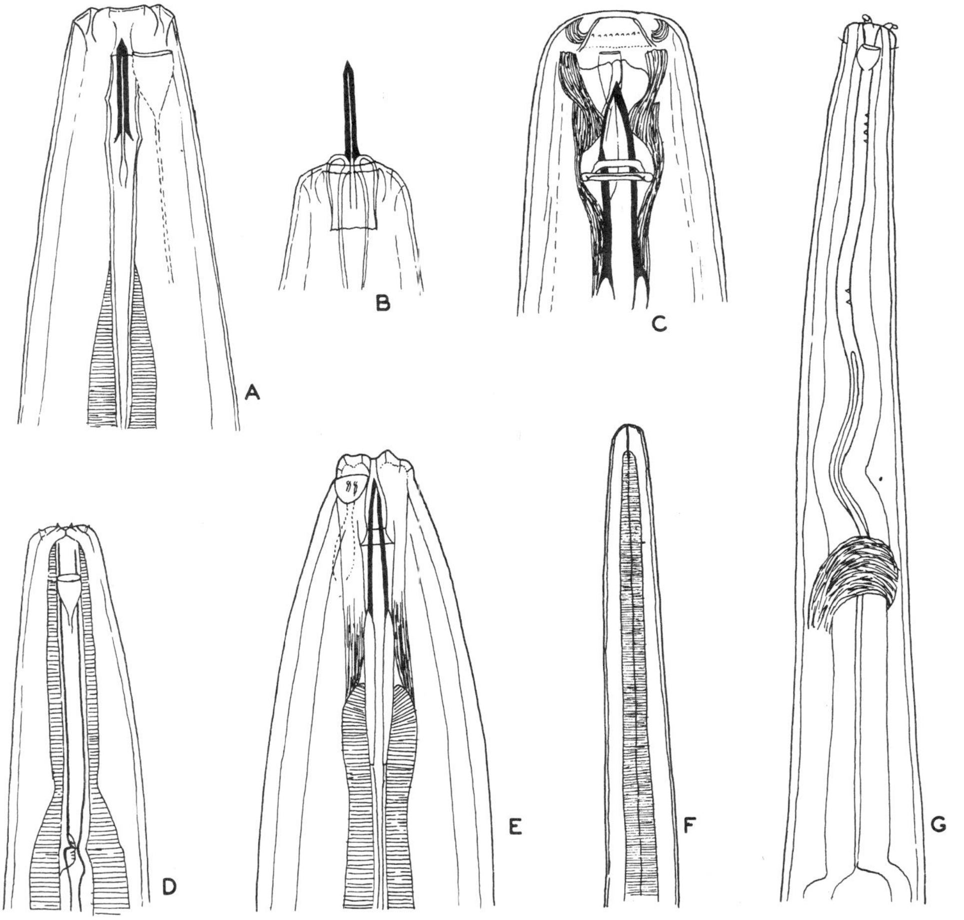

Fig. 8.12. Anterior end of nematodes. A, *Nygolaimus* (Nygolaimidae); B, *Nygolaimus* (Nygolaimidae) with protruded spear; C, *Actinolaimus* (Actinolaimidae); D, *Cryptonchus* (Cryptonchidae); E, *Dorylaimus* (Dorylaimidae); F, *Alaimus* (Alaimidae); G, *Ironus* (Ironoidae). (A and B modified from Thorne 1930; C and D modified from Cobb 1913; E modified from Micoletzky 1925; F and G modified from Cobb 1914.)

7. Cephalic and somatic setae present; amphids postlabial, well developed, circular, spiral, or slit-like 8
   Cepahlic and somatic setae absent; amphids labial, very small and porelike 9
8. Ovaries usually reflexes; amphids usually spiral or derived (Figs. 8.8E; 8.13M, R, S) **ORDER CHROMADORIDA, 29**
   Ovaries always outstretched; amphids usually circular . . . .**ORDER MONHYSTERIDA, 44**
9. Buccal (stomal) cavity unarmed, pharyngeal median bulb ususally lacking **ORDER RHABDITIDA, 46**
   Buccal cavity armed with a stylet, pharyngeal median bulb always present (Figs. 8.9B, C; 8.11 F, G, K) . . . .**ORDER TYLENCHIDA, 51**
10. Stoma wide (a buccal cavity) with sclerotized walls; one large dorsal tooth and two or more sub-ventral teeth (Fig. 8.9F); pharynx beginning at proximal end of stoma **ONCHOLAIMIDAE**
   Stoma narrower, teeth in other arrangement; pharynx beginning at the distal end of the stoma **11**

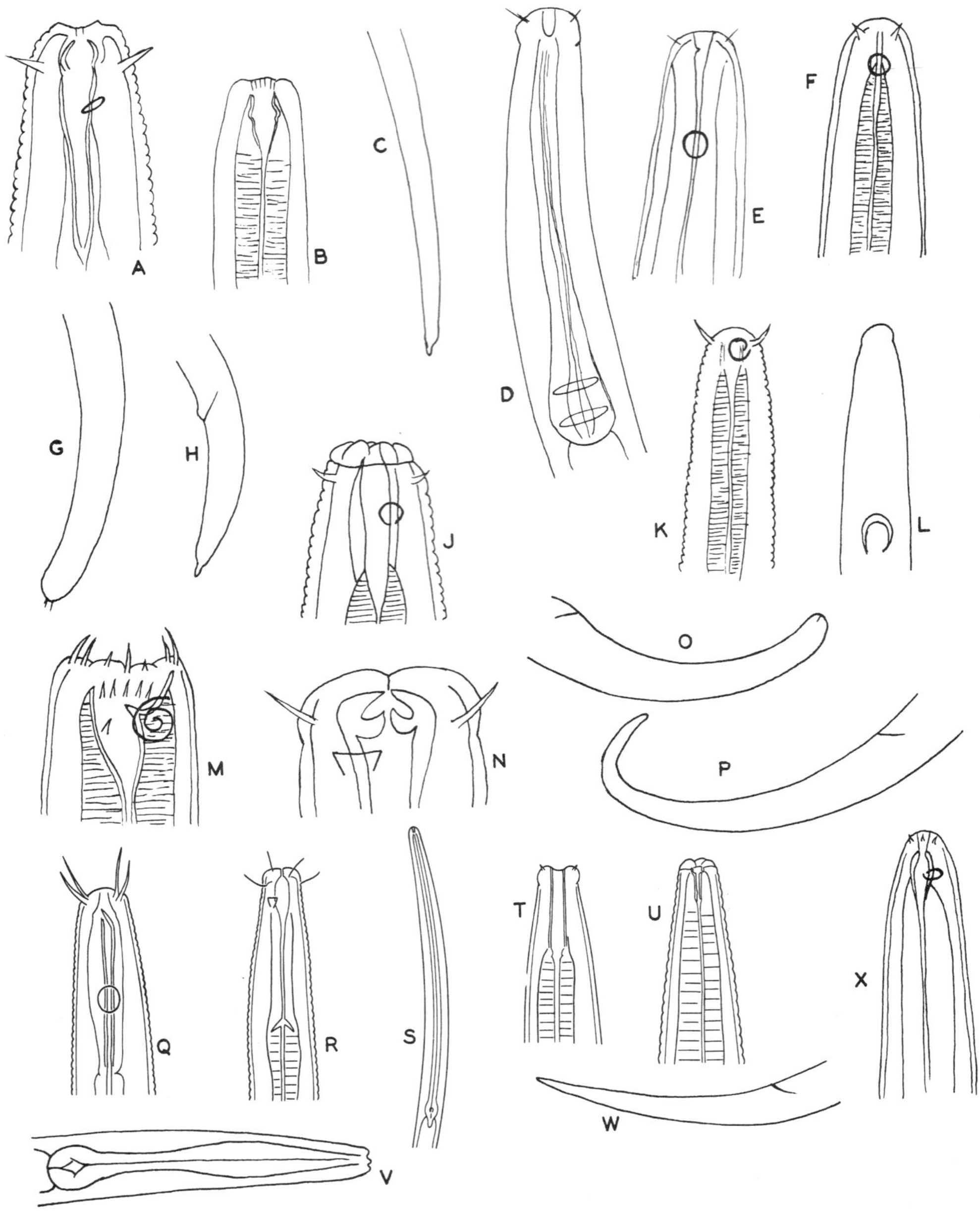

Fig. 8.13. Anterior and posterior ends of nematodes. A, *Anaplectus* (Plectidae); B, *Monochromadora* (Rhabdolaimidae); C, *Monochromadora* (Rhabdolaimidae), D, *Prodesmodora* (Desmodoridae); E, *Monhystrella* (Monhysteridae); F, *Cylindrolaimus* (Diplopeltidae); G, *Plectus* (Plectidae); H, *Plectus* (Plectidae) female; J, *Plectus* (Plectidae); K, *Paraphanolaimus* (Leptolaimidae); L, *Amphidelus* (Alaimidae); M, *Paracyatholaimus* (Cyatholaimidae); N, *Ironus* (Ironoidae); O, *Cylindrolaimus* (Diplopeltidae); P, *Cryptonchus* (Cryptonchidae); Q, *Odontolaimus* (Odontolaimidae); R, *Chronogaster* (Chronogasteridae); S, *Chronogaster* (Chronogasteridae); T, *Rhabditis* (Rhabditidae); U, *Acrobeloides* (Cephalobidae); V, *Acrobeloides* (Cephalobidae); W, *Eucephalobus* (Cephalobidae); X, *Goffartia* (Diplogasteridae). (All modified from Ferris et al. 1973.)

11. Stoma practically absent; amphids often enlarged longitudinally; body very slender **OXYSTOMINIDAE**
Stoma present; amphids small; body stouter **12**
12. Stoma tubular with parallel walls, long (Figs. 8.4I; 8.12G; 8.13N) . . . . **IRONOIDAE**
Stoma funnel or barrel-shaped **13**
13. Stoma quite narrow; male ventromedian supplements small, papilliform (Figs. 8.4L; 8.14B, C) **TRIPYLIDAE**
Stoma well developed; male ventromedian supplements large, vesiculate (Figs. 8.6A; 8.8D; 8.14A) . . . . **TOBRILIDAE**
14. Feeding apparatus provided with mural tooth (Fig. 8.15A) **15**
Feeding apparatus provided with odontostyle (Fig. 8.3E) **17**
15. Basal part of pharynx small, with a strongly developed triquetrous chamber (Fig. 8.15A) **CAMPYDORIDAE**
Basal expanded part of pharynx fairly long and without triquetrous chamber **16**
16. Basal expanded part of the pharynx enclosed in thick sheath of spiral muscles (Fig. 8.15B, C) **NYGELLIDAE**
Basal expanded part of the pharynx not enclosed in muscular sheath (Figs. 8.8L; 8.12A, B) **NYGOLAIMIDAE**

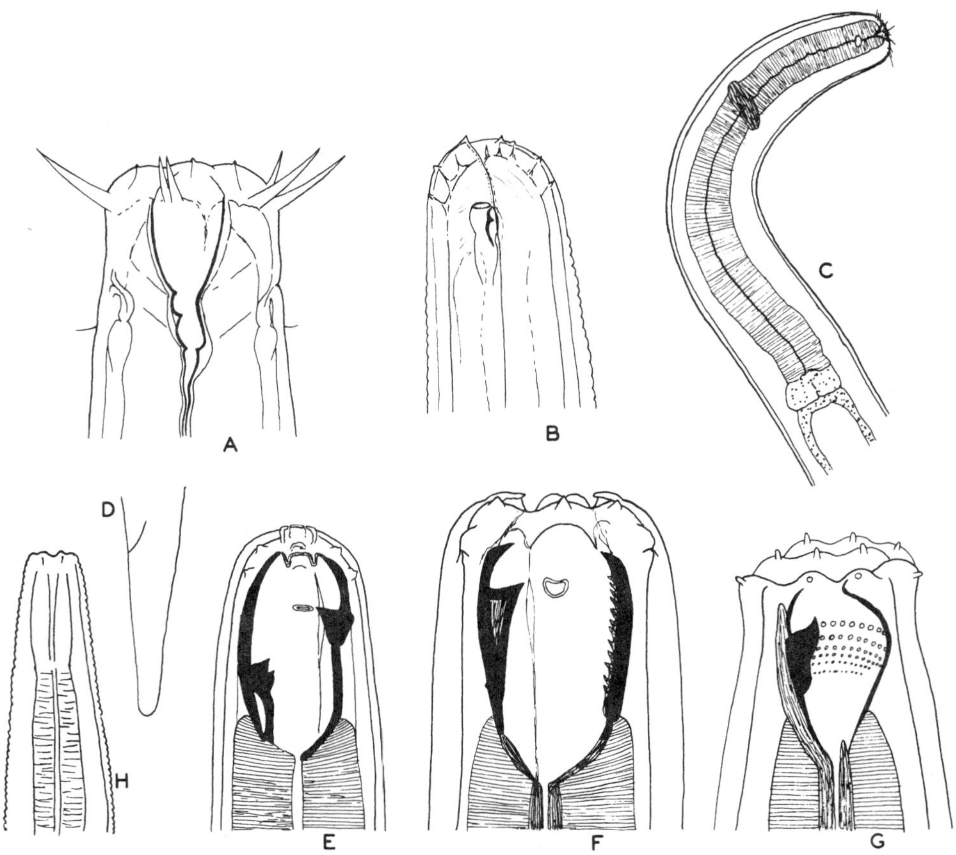

Fig. 8.14. Anterior and posterior ends of nematodes. A, *Tobrilus* (Tobrilidae); B and C, *Tripyla* (Tripylidae); D, *Cephalobus* (Cephalobidae); E, *Mononchus* (Mononchidae); F, *Prionchulus* (Mononchidae); G, *Mylonchulus* (Mononchidae); H, *Cephalobus* (Cephalobidae). (A and B modified from Micoletzky 1925; C modified from Cobb 1914; E–G modified from Cobb 1917; D and H modified from Ferris et al. 1973.)

17. Stoma strongly sclerotized, provided with plate or basketlike structures, frequently accompanied by large onchia with or without denticles (Fig. 8.12C) .... **ACTINOLAIMIDAE**
Stoma usually thin-walled, without onchia or denticles **18**
18. Odontostyle very long and attenuated (Fig 8.3E); only three pharyngeal glands **LONGIDORIDAE**
Odontostyle comparatively much smaller; five pharyngeal glands **19**
19. Expanded part of the pharynx enclosed in spiral muscular sheath (Fig. 8.15D) **BELONDIRIDAE**
Expanded part of the pharynx not enclosed in spiral muscular sheath **20**
20. Inner cuticle layer coarsely striated, provided with abundant radial striae; expanded part of the pharynx usually up to one-third of the total neck length **21**
Inner cuticle layer not striated, radial striae absent; expanded part of the pharynx almost always more than one-third of the total neck length **23**
21. Pharynx consisting of three sections, two separated by a constriction (Fig. 8.15I, J) **AULOLAIMOIDIDAE**
Pharynx consisting of two sections **22**
22. Stylet usually well developed; expanded part of the pharynx about one third of the total neck length (Fig. 8.15E) .... **TYLENCHOLAIMIDAE**
Stylet rather slender, attenuated, sometimes solid, needlelike; pharyngeal bulb usually pyriform (Fig. 8.15F, G) .... **LEPTONCHIDAE**
23. Stylet somewhat irregular, asymmetrical; enclosed in a guide ring (Fig. 8.15H) **MYDONOMIDAE**
Stylet simple, symmetrical (with or without guide rings) **24**
24. Stylet with wide aperture occupying usually more than one-half of the total length of stylet; guiding ring not sclerotized, plicated (Fig. 8.15K) .... **APORCELAIMIDAE**
Stylet with smaller aperture, usually one-third or less of the total length of stylet; guiding ring sclerotized and sometimes double (Fig. 8.15L) **25**
25. Odontostyle attenuated, usually longer than the lip region width (Fig. 8.15L) **NORDIIDAE**
Odontostyle not attenuated and usually about as long as the lip region width **26**
26. Large and stout nematodes, usually with long filiform tail exhibiting sexual dimorphism (Figs. 8.5B; 8.12E) .... **DORYLAIMIDAE**
Medium-sized nematodes, with short tail similar in sexes (Figs. 8.5A, E, F; 8.9G, H, J; 8.10G, H) **QUDSIANEMATIDAE**
27. Stoma tubular, surrounded by the distal part of the pharynx **28**
Stoma spacious, barrel-shaped, not surrounded by the pharynx (Figs. 8.9K, N; 8.14E–G) **MONONCHIDAE**
28. Mouth cavity with minute denticles or unarmed .... **BATHYODONTIDAE**
Mouth cavity armed with a large subventral tooth (Fig. 8.9L, M) .... **MONONCHULIDAE**
29. Buccal cavity funnellike, armed with teeth, one dorsal, two subventral **30**
Buccal cavity tubular, very rarely armed with teeth **35**
30. Cuticle punctuated (with dots or adornments) **31**
Cuticle not punctuated **34**
31. Always only a single anterior testis; amphids usually much wider than long (Figs. 8.8G; 8.11H) **CHROMADORIDAE**
Always two testes; amphids hardly wider than longer **32**
32. Cephalic sensilla situated in two circles at the same level and not separated (Fig. 8.13M) **CYATHOLAIMIDAE**
Cephalic sensilla in two circles at different level **33**
33. Ovaries situated on the same side of the intestine; males very rare (Figs. 8.4G; 8.5C; 8.6B; 8.8E; 8.11B) .... **ACHROMADORIDAE**
Gonads situated on different sides of the intestine; males very frequent (Fig. 8.11C) **ETHMOLAIMIDAE**
34. Only the anterior testis is present; ovaries reflexed (Figs. 8.6C; 8.11A; 8.13D) **DESMODORIDAE**
Two testes present (Fig. 8.16F); ovaries linear (Fig. 8.16B) .... **MICROLAIMIDAE**

35. Labial region with six deep cuticularized incisions and six extensions (Fig. 8.10D,E) **TERATOCEPHALIDAE**
No incisions or extensions in the labial region **36**
36. Anterior portion of the pharynx consisting of a cuticularized tube not surrounded by muscles; posterior portion shorter and strongly muscular (Fig. 8.10J) . . . .**AULOLAIMIDAE**
Pharynx with different morphology **37**
37. Amphids always dorsally spiral, with the canalis always ventral to the lateral line of the body (Fig. 8.8H) . . . .**PRISMATOLAIMIDAE**
Amphids with different morphology **38**
38. Dorsal tooth eversible, spearlike (Fig. 8.13Q) . . . .**ODONTOLAIMIDAE**
Dorsal tooth, if present, not eversible **39**
39. Pharynx ending in a basal bulb with valvular aparatus **40**
Pharyngeal bulb without valvular aparatus **41**
40. Females with only one anterior ovary (Fig. 8.13R, S) . . . .**CHRONOGASTERIDAE**
Females with two ovaries (Figs. 8.4E; 8.6D, E; 8.8F; 8.13A, G, H, J) . . . .**PLECTIDAE**
41. Anterior part of the pharynx weakly muscular, a small median bulb present; body annulated (Fig. 8.16C) . . . . **HALIPLECTIDAE**
Pharynx with other morphology **42**
42. Amphids nonspiral, pocketlike (Figs. 8.8A; 8.10C; 8.13B, C) . . . .**RHABDOLAIMIDAE**
Amphids spiral **43**
43. A circle of six long anterior cephalic setae (Fig. 8.10F) . . . .**BASTIANIIDAE**
Anterior cephalic setae usually invisible or hardly detectable (Figs. 8.4F, J; 8.8B; 8.10A, B; 8.13K) . . . . **LEPTOLAIMIDAE**
44. Buccal cavity narrow and tubular, or very small, without teeth; amphids spiral (Figs. 8.4H; 8.13F, O) . . . .**DIPLOPELTIDAE**
Buccal cavity funnel-like, occasionally with toothlike formations; amphids circular **45**
45. Anterior ovary lying to the right of the intestine; cuticle always smooth (Figs. 8.11D, E; 8.13E) **MONHYSTERIDAE**
Anterior ovary lying to the left of the intestine; cuticle always striated . . . .**XYALIDAE**
46. Stoma composed of small rhabdions, generally small and straight; pharynx without median bulb; female gonad mostly unpaired; male without bursa **47**
Stoma long, tubular or spacious, containing denticles or teeth; pharynx often with median swelling (bulb); ovaries generally two; bursa present, sometimes reduced **48**
47. Ovary with double flexure in the postvulval part; stoma quite straight, buccal rings (rhabdions) equally cuticularized (Figs. 8.13U–W; 8.14D, H) . . . . **CEPHALOBIDAE**
Ovary without double flexure in the postvulvar part; buccal rings unevenly cuticularized (Fig. 8.9A) . . . . **PANAGROLAIMIDAE**
48. Pharynx with two bulbs: a median, muscular, and a terminal glandular bulb (Fig. 8.3D; 8.4D); stoma usually carrying large teeth; bursa mostly reduced **49**
Pharynx with only one terminal basal bulb (Fig. 8.3B, C); stoma tubular without prominent teeth; bursa generally strongly developed **50**
49. Buccal tube extremely long and slender, one-fourth of the total neck length or more; anterior part of the pharynx unusually strong, bulblike over its entire length (Fig. 8.16D) **CYLINDROCORPORIDAE**
Buccal cavity not so long and slender; anterior part of the pharynx divided, consisting of a more slender part and a swelling (Figs. 8.4D; 8.8J, K; 8.13X) . . . .**DIPLOGASTERIDAE**
50. Body strikingly asymmetrical, one side ornamented with bumps or protuberances (Fig. 8.16E) **BUNONEMATIDAE**
Body bilaterally symmetrical, without protuberances (Figs. 8.1; 8.4C; 8.13T) **RHABDITIDAE**
51. Dorsal pharyngeal gland opening into lumen of pharynx within the median bulb and in front of valves; bursa rarely present, bearing papillae, if bursa absent, there are small genital papillae; pharyngeal median bulb usually quadrangular (Fig. 8.4B) **52**
Dorsal pharyngeal gland opening into lumen of pharynx behind base of spear; bursa present (only exceptionally reduced), no genital papilla; pharyngeal median bulb more rounded (Fig. 8.4A) **53**

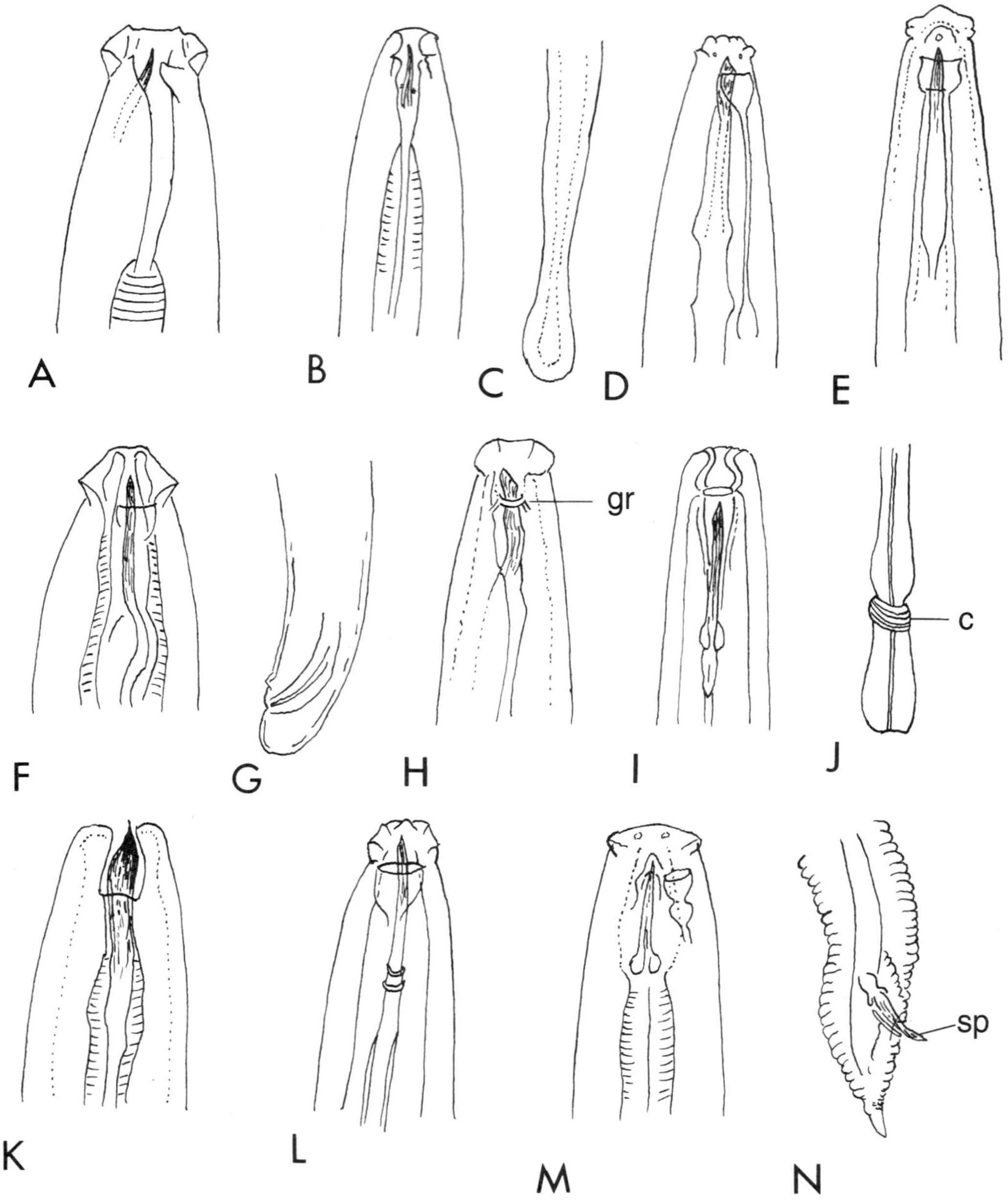

Fig. 8.15. Anterior and posterior ends of nematodes. A, Campydoridae; B, Nygellidae; C, Nygellidae; D, Belondiridae; E, Tylencholaimidae; F, Leptonchidae; G, Leptonchidae; H, Mydonomidae; I, Aulolaimoididae; J, esophagus of Aulolaimoididae; K, Aporcelaimidae; L, Nordiidae; M, Diphtherophoridae; N, Hoplolaimidae. c, constriction; gr, guide ring; sp, spicule. (A–N redrawn from Jairajpuri and Ahmad 1992, after various sources.)

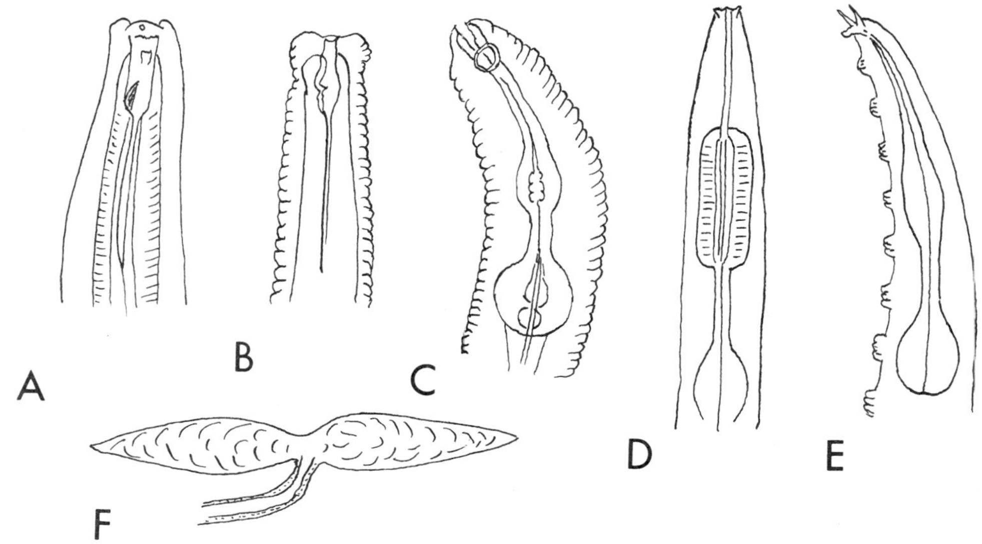

Fig. 8.16. Anterior and posterior ends and testes of nematodes. A, Bathyodontidae; B, Microlaimidae; C, Haliplectidae; D, Cylindrocorporidae; E, Bunonematidae; F, testes of most nematodes. (A–E redrawn from Goodey and Goodey 1963; F redrawn from Malakhov 1994.)

52. Bursa present, well developed, provided with large papillae; female tail short and broadly rounded (Fig. 8.9B) . . . . **APHELENCHIDAE**
Bursa absent or if present in the form of a very small rudiment on tail tip; female tail generally pointed (Figs. 8.4B; 8.5D; 8.11F) . . . . **APHELENCHOIDIDAE**

53. Median bulb strikingly large; spear unusually long and strong; annulation of cuticle mostly very coarse; males rare or rudimentary (Figs. 8.9C; D; 8.10L) . . . . **CRICONEMATIDAE**
Median bulb normally developed, spear of moderate length; annulation of cuticle fine; males normally developed **54**

54. Bursa encircling tail or reaching nearly tail tip, phasmids on bursal flaps; ovaries usually two; tail blunt (Fig. 8.15N) . . . . **HOPLOLAIMIDAE**
Bursa shorter, generally small and adanal, without phasmids; ovaries generally unpaired; tail mostly filiform, pointed (Figs. 8.4A; 8.8C; 8.9E; 8.10K, M; 8.11G, J, K) **TYLENCHIDAE**

REFERENCES

Adamson, M. L. 1987. Phylogenetic analysis of the higher classification of the Nematoda. *Can. J. Zool.* 65:1478–1482.

Andrássy, I. 1976. *The Evolution as a Basis to the Sistematization of Nematodes.* 288 pp. Pitman, London.

——. 1983. *A Taxonomic Review of the Suborder Rhabditina (Nematoda: Secernentia).* 241 pp. Éditions de l'ORSTOM, Paris.

——. 1984. *Klasse Nematoda: (Orduneng Monhysterida, Desmoscolecida, Araeolaimida, Chromadorida, Rhabditida).* 509 pp. Gustav Fischer, Stuttgart.

——. 1999. A census of genera and subgenera of free-living nematodes. *J. Nematode Morph. Syst.* 2:45–68.

Aumann, J. 1993. Chemosensory physiology of nematodes. *Fundam. Appl. Nematol.* 16:193–198.

Baermann, G. 1917. Eine einfache Methode zur Auffindung von Ankylostomum (Nematoden) Larven in Erdproben. *Geneesk. Tijdschr. Ned.-Indië* 57:131–137.

Baylis, H. A., and R. Daubney. 1926. *A Synopsis of the Families and Genera of Nematoda.* 277 pp. British Museum of Natural History, London.

Bird, A. F. 1971. *The Structure of Nematodes.* 318 pp. Academic, New York.

Chatterji, R. C, 1935. Permanent mounts of nematodes. *Zool. Anz.* 109:270.

Chitwood, B. G. 1931. A comparative histological study of certain nematodes. *Z. Morph. Oekol. Tiere* 23:237–284.

Chitwood, B. G., and M. B. Chirwood. 1930. A technique for the embedding of nematodes. *Trans. Am. Microsc. Soc.* 49:186–187.

——. 1974. *Introduction to Nematology.* 315 pp. University Park Press, Baltimore, MD.

Cobb, M. V. 1915. Some fresh-water nematodes of the Douglas Lake region of Michigan, USA. *Trans. Am. Microsc. Soc.* 34:21–47.

Cobb, N. A. 1913. New nematode genera found inhabiting fresh water and non- brackish soils. *J. Wash. Acad. Sci.* 3:432–444.

——. 1914. The North American free-living freshwater nematodes. *Trans. Am. Microsc. Soc.* 33:69–134.

——. 1914–1935. *Contributions to the Science of Nematology.* 490 pp. Baltimore, MD. (Issued in 26 parts, bound in one volume; some parts are reprints of previously published articles, others are original.)

——. 1917. The mononchs: a genus of freeliving predatory nematodes. *Soil Sci.* 3:431–486.

——. 1918a. Estimating the nema population of soil. *USDA Bur. Plant Ind. Agric. Tech. Circ.* 1: 1–48.

——. 1918b. Filter-bed nemas: nematodes of the slow sand filter beds of American cities (including new genera and species) with notes on hermaphroditism and parthenogenesis. *Contrib. Sci. Nematol.* 7:189–212.

——. 1935. A key to the genera of free-living nemas. *Proc. Helminth. Soc. Wash.* 2:1–40.

Coomans, A. 1979. The anterior sensilla of nematodes. *Rev. Nématologie* 2:259–283.

Esser, R. P., and G. R. Buckingham. 1987. Genera and species of free-living nematodes occupying fresh-water habitats in North America. In: *Vistas on Nematology: A Commemoration of the Twenty-fifth Anniversary of the Society of Nematologists*, pp. 477–487, edited by J. A. Veech and D. W. Dickson, Society of Nematologists, Hyattsville, MD.

Ferris, V. R., et al. 1973. Genera of freshwater nematodes (Nematoda) of eastern North America. *Biota Freshwater Ecosystems Identification Manual* 10: 1–38.

Filipjev, I. N. 1934. The classification of the free-living nematodes and their relation to the parasitic nematodes. *Smithson. Misc. Coll.* 89:1–63

Gerber, K., et al. 1987. A comprehensive catalogue of plant parasitic nematodes associated with aquatic and wetland plants. *Univ. Florida Agric. Exp. Sta. Bull.* 871:1–170.

Goodey, T., and J. B. Goodey. 1963. *Soil and Freshwater Nematodes.* 544 pp. Wiley, New York.

de Grisse, A. 1969. Redescription ou modification de quelques techniques utilisées dans l'étude des nématodes phytoparasitaires. *Med. Rijks. Landbouwwetenschappen Gent* 34:351–369.

Grundmann, A. W. 1955. Improved methods for preparing and mounting nematodes for study. *Turtox News* 33:152–153.

Hoeppli, R. J. C. 1926. Studies of free-living nematodes from the thermal waters of Yellowstone Park. *Trans. Am. Microsc. Soc.* 45:23–255.

Hunt, D. J. 1993. *Aphelenchida, Longidoridae and Trichodoridae: Their Systematics and Bionomics.* 352 pp. CAB International., Wallingford, UK.

Jairajpuri, M. S., and W. Ahmad. 1992. *Dorylaimida. Free-living, Predacious and Plant-Parasitic Nematodes.* 458 pp. E. J. Brill, Leiden.

Jairajpuri, M. S., and W. U. Khan. 1982. *Predatory Nematodes (Mononchida) with Special Reference to India.* 131 pp. Associated, New Delhi.

Linford, N. B. 1937. The feeding of some hollow-stylet nematodes. *Proc. Helminth. Soc. Wash.* 4: 41–46.

Lorenzen, S. 1994. *The Phylogenetic Systematics of Freeliving Nematodes.* 383 pp. The Ray Society, London.

Maggenti, A. 1981. *General Nematology.* 372 pp. Springer , New York.

Malakhov, V. V. 1994. *Nematodes. Structure, Development, Classification, and Phylogeny.* 286 pp. Smithsonian Institution Press, Washington, DC.

Meyl, A. 1960. Freilebenden Nematoden. *Tierwelt Mitteleuro.* 1:(5a):1–164.

Micoletzky, H. 1925. Die freilebenden Süsswasserund Moornematoden Dänemarks. *Det. Kgl. Danske Vidensk. Selsk. Skr. Naturv. Math. Afd.* 10:57–310.

Nicholas, W. L. 1984. *The Biology of Free-Living Nematodes*, 2d ed. 251 pp. Clarendon Press, Oxford, UK.

Ocana, A., and R. Morales. 1992. The influence of ionic composition on the distribution of nematode species in springs of the province of Granada (Spain). *Hydrobiologia* 237: 81–92.

Prejs, K. 1970. Some problems of the ecology of benthic nematodes (Nematoda) of Mikolajskie Lake. *Ekol. Polska* 18:225–242.

Raski, D. J., and A. R. Maggenti. 1983. Tylenchidae: morphological diversity in a natural, evolutionary group. In: *Concepts in Nematode Systematics*, pp. 131–142, edited by A. R. Stone, H. M. Platt, and L. F. Khalil. Academic, London.

Roberts, L. S., and J. Janovy, Jr. 2000. *Foundations of Parasitology*, 6th ed. 670 pp. McGraw-Hill Higher Education, Boston.

Schneider, W. 1937. Freilebende Nematoden der Deutschen Limnologischen Sundaexpeclition nach Sumatrajava, und Bali. *Arch. Hydrobiol.* 15 (Suppl):30–108.

Seinhorst, J. W. 1962a. On the killing, fixation and transferring to glycerin of nematodes. *Nematologica* 8:29–32.

——. 1962b. Modifications of the ellutriation method for extracting nematodes from soil. *Nematologica* 8:117–128.

Siddiqi, M. R. 1986. *Tylenchida. Parasites of Plant and Insects.* 645 pp. CAB, Slough, UK.

Southey, J. F. 1985. Laboratory methods for work with plant and soil nematodes. Great Britain Agriculture, Fisheries and Food Reference Book 402:1–202.

Stefanski, W. 1938. Les nematodes libres des lacs Tatra Polanaises, leur distribution et systematique. *Arch. Hydrobiol.* 33:585–687.

Strayer, D. L., and E. B. O'Donnell. 1992. The hyporheic nematode community of some streams in southeastern New York state, U. S. A. *Stygologia* 7:143–148.

Tarjan, A. C. et al. 1977. An illustrated key to nematodes found in fresh water. *J. Water Poll. Control Fed.* 49:2318–2337. (Key to genera reproduced on World Wide Web; url: *http://ianrwww.unl.edu/ianr/plntpath/nematode/key/nemakey.htm.*)

Thorne, G. 1930. Predaceous nemas of the genus *Nygolaimus* and a new genus *Sectonema*. *J. Agric. Res.* 41:445–466.

——. 1937. A revision of the nematode family Cephalobidae Chirwood and Chitwood. *Proc. Helminthol. Soc. Wash.* 4:1–16.

——. 1939. A monograph of the nematodes of the superfamily Dorylaimoidea. *Capita Zool.* 8:1–261.

Traunspurger, W. 1996. Distribution of benthic nematodes in the littoriprofundal and profundal of an oligotrophic lake (Königssee, National Park Berchtesgaden, FRG). *Arch. Hydrobiol.* 135:557–575.

Yeates, G. W., et al. 1993. Feeding habits in soil nematode families and genera: an outline for soil ecologists. *Nematologica* 25:315–331.

Zullini, A. 1982. Nematodi (Nematoda). Guide per il riconoscimento delle specie animali delle acque interne italiane. *Consig. Naz. Ricerche, Italy* 17:1–117.

9

NEMATOMORPHA

(Horsehair worms, gordian worms)

Because of their fancied resemblance to animated horsehairs, members of the Phylum Nematomorpha have long been known as "horsehair worms," and because of their habit of becoming inextricably entangled in masses of two to many individuals (Fig. 9.1) they are also commonly called "gordian worms," after the Gordian knot episode of antiquity. Sometimes a single worm may be found tightly coiled around a small twig.

The adults move clumsily by slow undulations or writhe about in puddles and in the shallows of marshes, ponds, lakes, and streams. Sometimes they occur on the shore near the water's edge, especially if the substrate is wet. The immature stages are internal parasites of insects, usually crickets, terrestrial beetles, and grasshoppers, but occasionally aquatic insects.

General characteristics. Adult gordian worms range from about 10 to 70 cm in length with a diameter of 0.3 to 2.5 mm. The body diameter is usually constant except at the extremities. However, in many species the diameter decreases in either direction from the midpoint of the body, giving the body a tapered appearance. Coloration includes opaque yellowish, gray, tan, dark brown, and blackish and can include spots or blotches. There is considerable variation in size and coloration within single species. Males are consistently smaller than females and often exhibit slight coiling of the posterior end. The two sexes may be found in about equal numbers or decidedly unequal numbers.

The small mouth is terminal or subterminal, and sometimes absent. The cloacal aperture is terminal, subterminal, or ventral. Sexes of several genera can be distinguished by the structure of the posterior end. Females of *Paragordius* have three lobes. Males of *Gordius*, *Parachordodes*, *Paragordius*, and *Gordionus* have two lobes. In females of *Gordius*, *Gordionus*, and *Parachordodes* the posterior end is simple and rounded, and this is also the situation in both sexes for *Pantachordodes*, *Chordodes*, and *Pseudochordodes*. In some species the posterior end of the female is slightly bulbous.

These worms were not separated from the nematodes until 1886, and they are often confused with terrestrial mermithid nematodes. Externally, however, Mermithidae are easily differentiated by their pointed ends, tapered body, smooth cuticle, and the location of the anus.

Cuticle. The gordian cuticle is unusually complex, lamellated, and fibrous (Figs. 9.2, 9.6). Areoles are irregular, rounded, or polygonal thickened surface areas of the cuticle that are present in most species. Sometimes they are poorly differentiated or flat and project only slightly from the general surface of the cuticle; in other species they may be papillate to a varying degree. Often the areoles bear small pore canals, granules, setae, or filamentous processes. The furrows

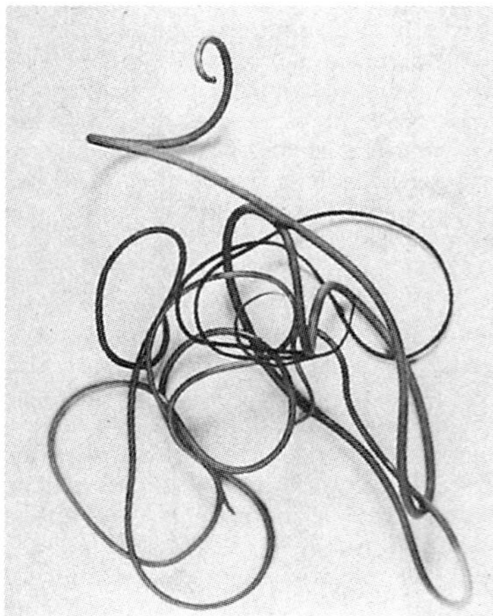

Fig. 9.1. Three female gordians, ×1.

or flat areas between areoles likewise may contain setae, granules, papillae, or filamentous processes. In the cloacal region of the male there are often special tracts of setae, thorns, or adhesive wartlike structures that presumably aid in copulation.

Internal anatomy, physiology. In addition to cuticle, the body wall consists of a flat to columnar hypodermis and an innermost layer of longitudinal muscle fibers (Fig. 9.3). There are no circular fibers, and consequently the diameter of the body cannot be varied. Locomotion consists of localized bending and writhing movements produced by contractions of the longitudinal muscle fibers in one or two quadrants at a time. Males are able to "swim" in an undulatory manner, but such locomotion is not to be compared with the rapid wriggling of small nematodes.

Particulate food is never ingested at any time during the life history, the digestive tract being degenerate and completely functionless. A pharynx is usually differentiated but has no cavity, and the midgut is a minute narrow tubule. The posterior end of the digestive tract is a cloaca into which the two genital ducts open.

The gonads consist of a pair of long, cylindrical bodies that fill most of the pseudocoel in mature animals. Ripe eggs and sperm pass posteriorly through short paired oviducts

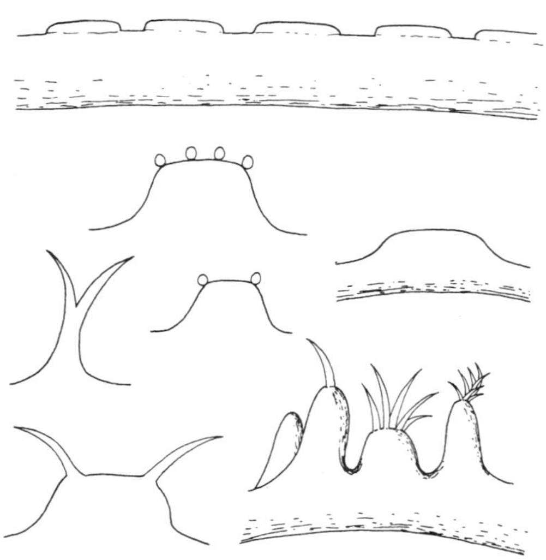

Fig. 9.2. Various types of areoles, diagrammatic. (Partly modified from Müller 1927.)

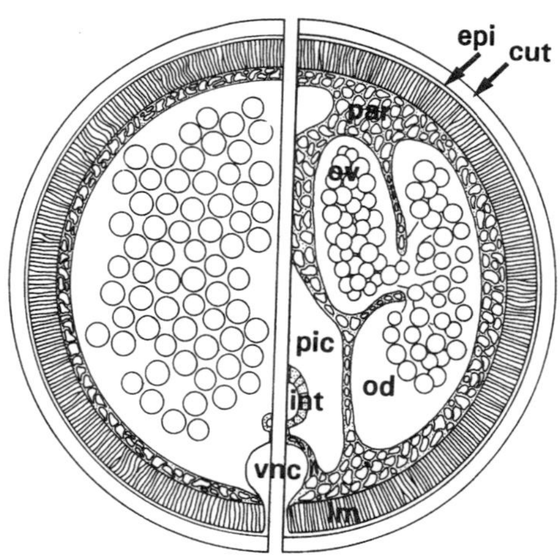

Fig. 9.3. Cross section of a typical female gordian, ×60, right side, early reproductive state; left side, late reproductive state with many oocytes filling pseudocoel. cut, cuticle; epi, epicuticle; int, intestine; lm, longitudinal muscle; od, oviduct; par, parenchyma; pic, periintestinal cavity (pseudocoel); vnc, ventral nerve cord. (After Schmidt-Rhaesa 1997, © Spektrum Akademischer Verlag, Heidelberg, Berlin with permission.)

and sperm ducts, respectively, into the cloaca, and thence to the outside. In the female there is a median ventral seminal receptacle extending forward from the anterior end of the cloaca.

A cerebral mass, or brain, encircles the digestive tract at the anterior end, and a longitudinal ventral nerve cord extends throughout the body. One or two ocelli have been described at the anterior end of some species, but otherwise sensory receptors are limited to tactile areas near the brain and perhaps also to some of the cuticular outgrowths on the general body surface.

There are no special circulatory, respiratory, or excretory structures. The residual pseudocoel space between body wall and visceral organs is filled with loose mesenchyme.

Reproduction, life history. Probably depending on the species and ecological conditions, copulation usually occurs in spring, early summer, or autumn. Although it is not uncommon to find up to 10 or 20 worms aggregated in a single writhing mass in the springtime, only one female and one male are involved in a copulation. A male coils the posterior tip of his body around the posterior end of the female and deposits sperm at her cloacal opening. The sperm then migrate into the female cloaca, and may be stored for a short time in the seminal receptacle. Fertilization is internal, and presumably the eggs are fertilized just before being extruded from the cloacal opening of the female. Eggs are deposited in long gelatinous strings that swell and may exceed the size of the parent. Such masses contain up to several million eggs. Males die after copulation and females after egg deposition.

Depending on water temperature, the incubation period is usually 15 to 80 days. The resulting larva (Fig. 9.4) is unlike the adult and has a very brief free-swimming existence. It is cylindrical, often annulated, only about 250 μm long, and divided into a trunk and large muscular presoma. The anterior end of the presoma has an eversible proboscis bearing three long stylets and three circlets of hooks (Fig. 9.5). The anterior portion of the trunk contains a large glandular mass that opens at the tip of the presoma via a long duct. Posteriorly the trunk contains a blind nonfunctional intestine that opens to the outside by means of a pore.

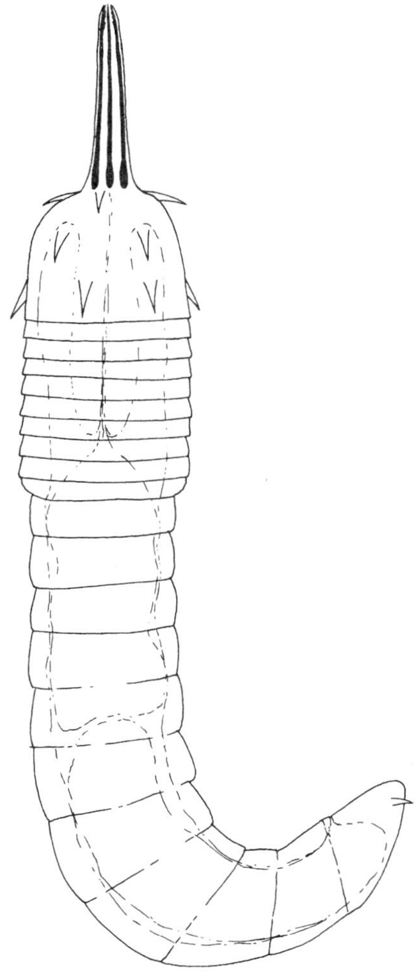

Fig. 9.4. Diagram of a typical gordian larva, ×500, with protruded proboscis.

The immediate and ultimate fates of this larval stage have been the subject of a long and confusing controversy, and even the most recent investigations do not seem to have cleared up the problem. Within 24 hours after hatching, the larva is usually thought to encyst on vegetation or any other convenient substrate at or near the water's

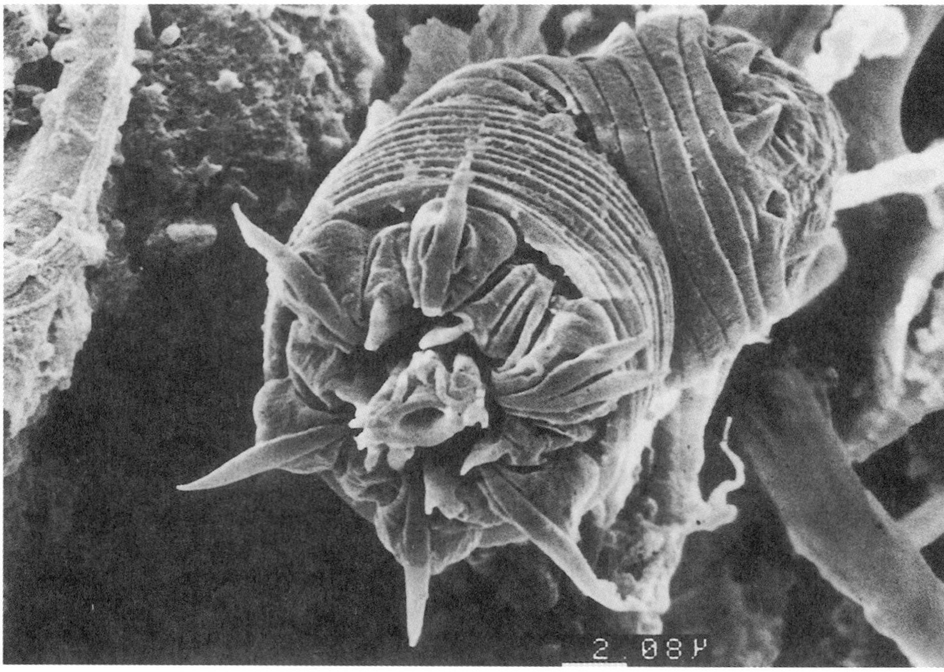

Fig. 9.5. SEM micrograph of anterior end of *Chordodes* larva. Note proboscis in center surrounded by hooks and spines. (After Bohall et al. 1997.)

edge. The colder the water, the more readily encystment occurs. Such cysts may remain viable for 2 months in water and up to 1 month in a moist habitat such as grass and other vegetation along the waterside. The latter condition may result when the water level of a stream or pond is falling.

Under favorable circumstances, a wide variety of crickets, grasshoppers, and terrestrial and aquatic beetles and other aquatic insects accidentally ingest these cysts along with the vegetation to which they are attached. Within the digestive tract the cyst wall disintegrates, and the released larva bores its way through the wall of the intestine and into the hemocoel of the host. Large numbers of cysts are ingested by abnormal hosts (snails, oligochaetes, aquatic insects, fish, etc.), but in these hosts the larva dies or reencysts in the tissues instead of developing further. If such an abnormal host is eaten by an omnivorous insect that is a normal host, the contained larva then burrows through the wall of the digestive tract to the hemocoel of the latter and undergoes further development. There is also the possibility (but one that is losing favor) that motile larvae actively penetrate through the body wall of almost any small aquatic or waterside animal but develop further only within such insects as crickets, grasshoppers, and beetles.

Within the host the worm digests and absorbs the surrounding tissues as a source of nourishment. Metamorphosis and development to the adult stage are completed in several weeks to several months, by which time the worm is a tightly coiled mass in the hemocoel. There may be one to several worms per host. Sometimes gordians overwinter in the host body.

If the host falls into water or becomes wetted at a time when the contained worm is mature or nearly mature, the latter breaks through the body wall and is henceforth free-living. There is some evidence that hosts tend to seek water when their worms are ready to emerge. Newly released worms that do not have access to water soon die. If emergence from the host occurs in spring or summer, copulation usually soon follows. Sometimes emergence occurs in the autumn and then the worms remain solitary, customarily hibernating in waterside grass and roots, and become active and copulate in the water in the following spring. Length of the complete life cycle is variable, ranging from about 2 to 15 months.

Unlike most parasites, gordians generally have little host specificity, and the same species may develop in a variety of grasshoppers or beetles. A few gordians have even been reported emerging from caddis flies and dragonflies. Adults have occasionally been recorded as human parasites, but such cases are undoubtedly fortuitous and produced by accidental ingestion of worms in food or water.

Ecology. Little is known about gordian ecology. The worms occur in suitable habitats wherever appropriate hosts are found, from the tropics to cold latitudes, and to altitudes above timberline in mountainous areas. Some species are widely distributed throughout North and South America, and a few occur in both Europe and the United States or are Holarctic. The genus *Gordius* is essentially cosmopolitan. The most common and widely distributed species in the United States are *Gordius robustus* and *Paragordius varius*.

There seem to be few habitat preferences, and the same species may be found in both running and standing waters. Occasionally, they are on the shore just above the water's edge but more usually in water 2 to 20 cm deep. Besides small puddles, habitats can also include artificial water bodies, such as swimming pools, stock tanks, watering bowls for pets, and drainage ditches. Specimens have even been observed in washing machines and toilets, sometimes causing great consternation for the discoverer.

A few observations indicate that the worms are more active at night, but there is no evidence for a true negative phototaxis. Males are much more active than females and move about in a serpentine manner for hours in the water and on the bottom. Females are sluggish and move about but little.

Preparing. Seventy to 90% alcohol is a suitable fixative and preservative, but for more critical work hot saturated mercuric chloride containing 5 to 10% acetic acid is much better, especially if histological sections are contemplated.

For a study of the extremities, cut anterior to the cloacal aperture and posterior to the dark ring near the anterior end; mount in lactophenol, clarite, or balsam after dehydrating. Small bits of cuticle should be sliced from the middle third of the body with a sharp razor blade by stretching the worm firmly over the left forefinger. Such pieces should be placed in a drop of glycerin for about an hour. Then gently scrape off the soft tissues from the underside with needles. Rinse the cleaned cuticle in a change of glycerin and mount, outer surface up, in glycerin jelly on a slide. Lactophenol and clarite mounts are often preferable for certain cuticular structures.

Taxonomy. The Phylum Nematomorpha is divided into the Nectonematoidea and Gordioidea. The former category contains a single marine species whose larvae are parasitic in crustaceans. All the other species are freshwater, totaling about 110 for the world, and belong in the Gordioidea. This category consists of 5 families and about 16 genera. About 12 species in the following taxa occur in the United States: Chordodidae (*Chordodes*, *Pantachordodes*, *Pseudochordodes*, *Euchordodes*), Parachordodidae (*Parachordodes*, *Gordionus*, *Paragordius*), and Gordiidae (*Gordius*).

Some generic characters and most species characters are based largely on the surface pattern and sculpturing of the cuticle. For studying some of these details it is often necessary to use oil immersion.

Only a small amount of work on American gordians has been published since the papers of Montgomery and May, and the taxonomy and distribution of our species are still very poorly known. Many of the descriptions published prior to 1910 were quite brief and did not include sufficient fine details, and for this reason some are not trustworthy. Miralles, in a series of papers, has redescribed some of the species, but detailed studies of the kind by Chandler and Wells (1989), Smith (1994), and Schmidt-Rhaesa et al. (1998) using large series of both sexes and SEM will be necessary to clarify the morphological limits of most species and even several genera. Especially, the nature of the cuticle and a redefinition of its various manifestations will be required in order to provide more meaningful discussion of morphology and structure of freshwater nematomorphs.

Currently, seven genera comprising about 17 species are known from the United States (including Alaska). Recent research leaves unresolved the relationships of certain genera (*Gordionus* and *Parachordodes*, *Neochordodes* and *Pseudochordodes*), but all are included in the key. Families are not used, following the recommendations of Schmidt-Rhaesa (1997). The several families in use likely do not characterize natural groups. For example, the anatomy of *Paragordius varius* is very unlike that of many other genera, but is included in either the Chordodidae or Parachordodidae.

As a preamble to using the key, it is useful to note certain features of the posterior end of the worm or worms. Noting the morphology of the posterior will reveal the sex of the animal and assist in determining the genus. Males of several genera (*Gordionus*, *Gordius*, *Parachordodes*, *Paragordius*) have a bilobed posterior, while females of these genera have a simple rounded border with a terminal cloaca (*Gordionus*, *Gordius*, *Parachordodes*) or a trilobed posterior (*Paragordius*). Posterior lobes of males of *Parachordodes* and *Gordionus* have distinct, divergent rows of bristles on either side of the cloaca, while *Gordius* males have a distinct crescent anterior to the cloaca. Specimens with a swollen posterior and either a terminal cloaca or a subterminal cloaca and a slight swelling to either side are females and males, respectively, of *Chordodes*, *Pseudochordodes*, or *Neochordodes*.

KEY TO GENERA OF NEMATOMORPHA OF THE UNITED STATES

1. Cuticular areoles absent in one or both sexes, if evident, simple and weakly defined elevations only, polygonal or ovoid; bilobed posterior of male (Fig 9.6D) with a transverse crescent just anterior of cloaca; four species, one common and widespread, the others rare in eastern states **_Gordius_**

Cuticular areoles, simple or complex, present and well defined; without a crescent on posterior ventral surface of male **2**

2. Posterior with elongate lobes, if bilobed with rows of tubercles on either side of cloaca (Fig. 9.8A), or posterior (female) with three lobes (Fig. 9.6B); one wide-ranging species **_Paragordius_**

Posterior with two lobes, but not as above, or without lobes **3**

3. Cuticle with one type of areole only, simple elevations (Fig. 9.7C, D) **4**

Cuticle with complex areoles of different size and structure (Fig. 9.7D–H) **5**

4. Areoles crowded on the surface; interareolar spaces narrow, sometimes with spines or bristles (Fig. 9.7C); lobes of male with spines and bristles (Fig. 9.7G); about three rare species scattered around the U.S. . . . . **_Gordionus_**

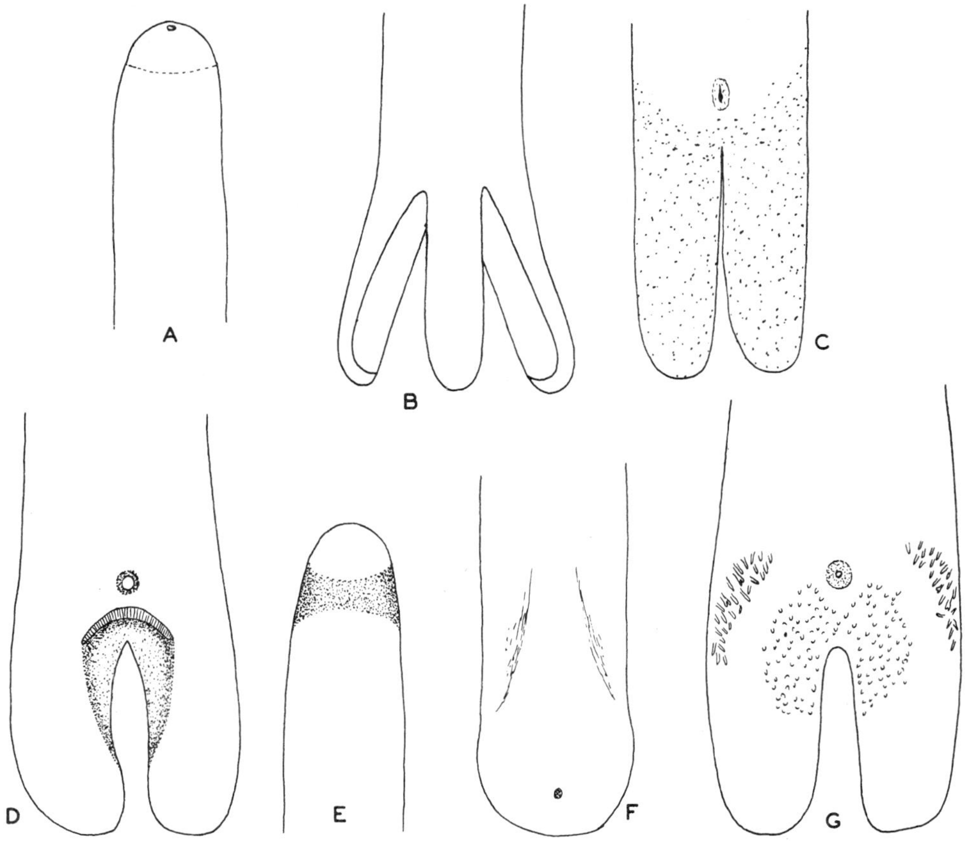

Fig. 9.6. Extremities of Nematomorpha. A, anterior end of *Paragordius;* B, posterior end of female *Paragordius;* C, posterior end of female *Neochordodes;* D, posterior end of male *Gordius;* E, anterior end of *Gordius;* F, posterior end of female *Gordius;* G, posterior end of male *Gordionus.* (After Poinar and Doelman 1974.)

Areoles single or conjoined clusters of two or three, large, flat, and ovoid, occasionally with conical elevations (Fig. 9.7D); interareolar spaces expansive; female posterior with two lobes (Fig. 9.6C) at least one species, may include others presently in *Pseudochordodes*, rare, CA ***Neochordodes***

5. Areoles very complex and variable (Fig. 9.7F, G), typically with spiny crowns and well developed or weak processes, giving cuticle a granular appearance; male posterior very weakly bilobed (Fig. 9.8B); one widely distributed species . . . . ***Chordodes***

Areoles of two types, without dense spines and processes on surfaces, not granular appearing **6**

6. Smaller type of areole most common, somewhat ovoid; larger areole bean-shaped with a central pore canal (Fig. 9.7E); posterior similar to *Gordionus;* about three rare species generally in more western parts of the country . . . . ***Parachordodes***

Smaller type of areole most common, somewhat polygonal, occasionally with a pair of small tubercles on surface; larger areoles commonly paired, separated by a narrow furrow with pore canal (Fig. 9.7H); posterior rounded in male (Fig. 9.8C); four rare species, AR, CA, MT, NE, NV, TX, WO . . . . ***Pseudochordodes***

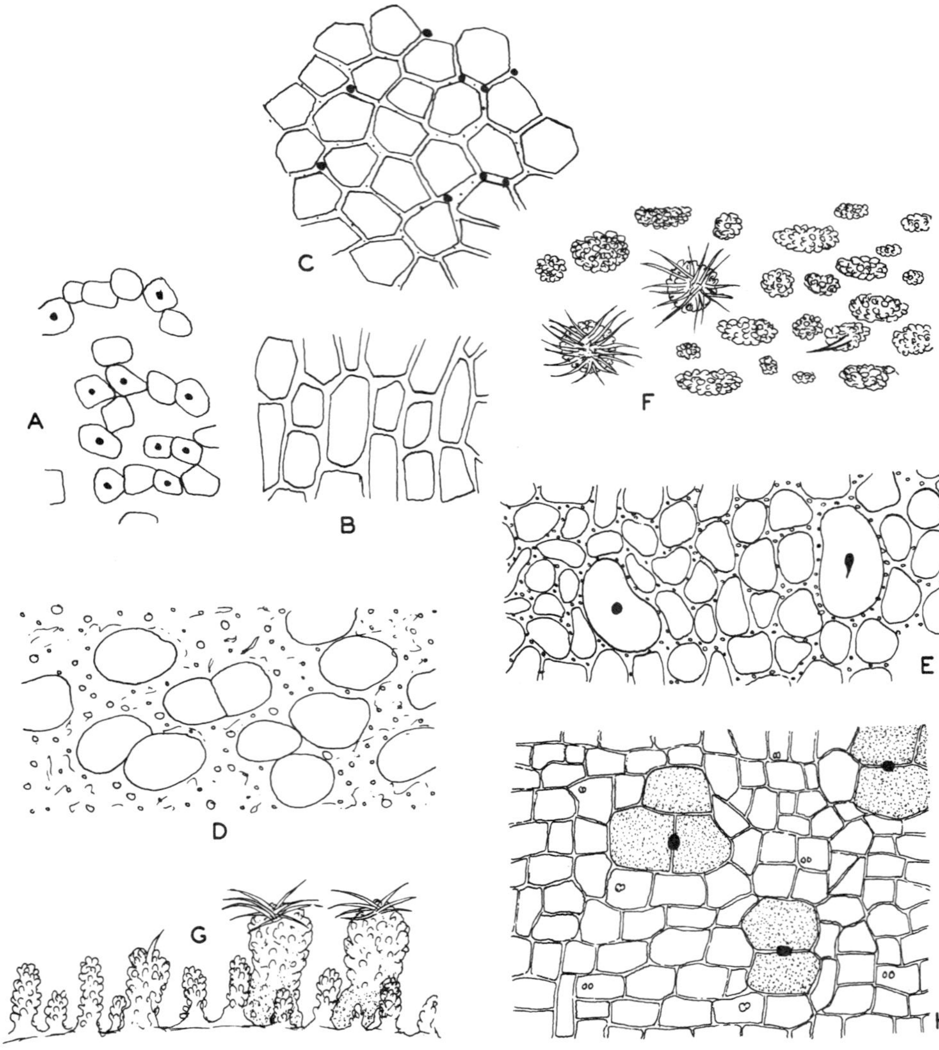

Fig. 9.7. Areoles and other cuticular structures of Nematomorpha, semidiagrammatic. (Pore canals shown in solid black.) A, *Paragordius;* B, *Gordius;* C, *Gordionus;* D, *Neochordodes;* E, *Parachordodes;* F, surface view of *Chordodes* areoles; G, side view of *Chordodes* areoles; H, *Pseudochordodes.* (E modified from Heinze 1935a; F and G modified from Camerano 1897; H modified from Carvalho 1942.)

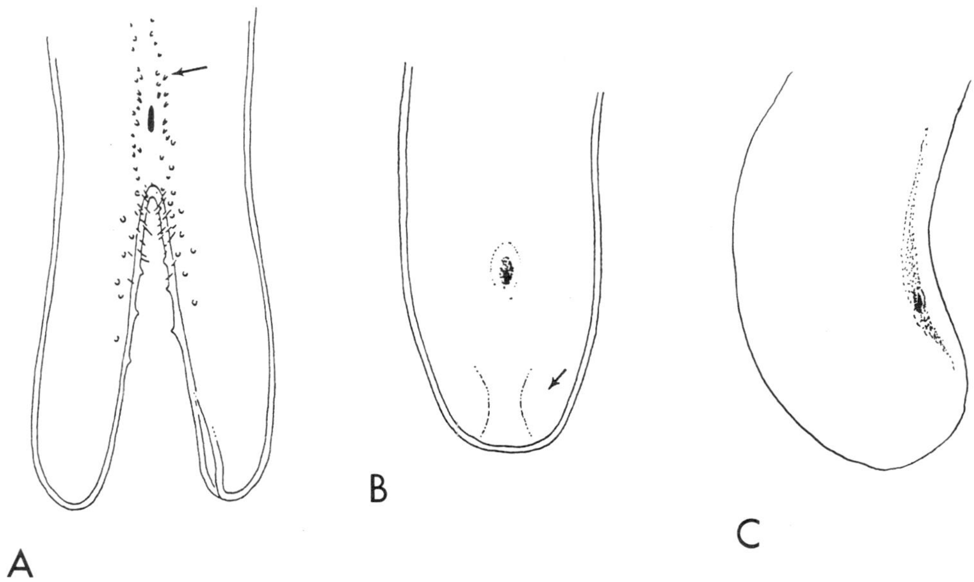

Fig. 9.8. Extremities of Nematomorpha. A, male posterior of *Paragordius*, arrow denotes tubercles; B, male posterior of *Chordodes*, arrow denotes lobes; C, male posterior of *Pseudochordodes*. (C redrawn from Miralles and Villalobos 1995.)

REFERENCES

Bohall, P. J., et al. 1997. External morphology of larvae of *Chordodes morgani* (Nematomorpha). *Invert. Biol.* 116:26–29.

Camerano, L. 1897. Monografia dei Gordii, *Mem. R. Acad. Sci. Torino* (2) 47:339–419.

——. 1915. Revisione dei Gordii. *Mem. R. Acad. Sci. Torino* 66:1–66.

Carvalho, J. C. M. 1942. Studies on some Gordiacea of North and South America. *J. Parasitol.* 28:213–222.

Chandler, C. M. 1985. Horsehair worms (Nematomorpha, Gordioidea) from Tennessee, with a review of the taxonomy and distribution in the United States. *J. Tenn. Acad. Sci.* 60:59–62.

Chandler, C. M., and M. R. Wells. 1989. Cuticular features of *Chordodes morgani* (Nematomorpha) using scanning electron microscopy. *Trans. Am. Microsc. Soc.* 108:152–158.

Conn, D. B. 1994. Fine structure of the oocytes and female reproductive tissues of *Gordius* sp. (Nematomorpha). *Trans. Am. Microsc. Soc.* 113: 420–421 (abstract).

Dorier, A. 1930. Recherches biologiques et systématiques sur les Gordiacés. *Trav. Lab. Hydrobiol. Piscicult. Grenoble* 22:1–180.

Heinze, K. 1935a. Über das genus *Parachordodes* Camerano 1897 nebst allgemeinen Angaben über die Familie Chordodidae. *Z. Parasitenk.* 7:657–678.

——. 1935b. Über Gordiiden. *Zool. Anz.* 111: 23–32.

——. 1937. Die Saitenwürmer (Gordioidea) Deutschlands: Eine systematisch-faunistische Studie über Insektenparasiten aus der Gruppe der Nematomorpha. *Z. Parasitenk.* 9:263–344.

——. 1941. Saitenwurmer oder Gordioidea. *Tierwelt Dtschl.* 39:1–78.

——. 1952. Über Gordioidea, eine systematische Studie über Insektenparasiten aus der Gruppe der Nematomorpha. *Z. Parasitenk.* 15:183–202.

Miralles, D. A. B. de, and L. C. de Villalobos. 1993. Redescription de *Gordius agasizzi* Montgomery, 1898, de *Pseudochordoes occidentalis* comb. n. y nuevo registro para *P. bedriagae* (Camerano, 1896) (Gordioidea, Nematomorpha). *Iheringia* 75:107–111.

——. 1995. A review of the Gordiacea (Nematomorpha) in the collection of the California Academy of Sciences with the description of a new species. *Graellsia* 51:17–22.

Miralles, D. A. B. de, et al. 1997. Una nueva especie y una neuva combinación de gordiaceos (Nematomorpha). *Neotropica* 43:53–56.

Montgomery, T. H., Jr. 1898a. The Gordiacea of certain American collections with particular reference to the North American fauna. *Bull. Mus. Comp. Zool. Harvard Univ.* 32: 23–59.

——. 1898b. The Gordiacea of certain American collections, with particular reference to the North American fauna, II. *Proc. Calif. Acad. Sci.* (3)1:333–344.

Müller, G. W. 1927. Über Gordiacean. *Z. Morphol. Oeokol. Tiere* 7:134–219.

Poinar, G. O., and J. J. Doelman. 1974. A reexamination of *Neochordodes occidentalis* (Mont.) comb. n. (Chordodidae: Gordioidea): larval penetration and defense reaction in Culex pipiens L. *J. Parasit.* 60:327–335.

Redlich, A. 1980. Description of *Gordius attoni* sp. n. (Nematomorpha, Gordiidae) from northern Canada. *Can. J. Zool.* 58:382–385.

Schmidt-Rhaesa, A. 1997. *Nematomorpha.* Süßwasserfauna von Mitteleuropa. 4/4. 124 pp. Gustav Fischer, Stuttgart.

Schmidt-Rhaesa, A., et al. 1998. Scanning electron microscopy and intraspecific variation in *Euchordoes nigromaculatus* from New Zealand. *J. Helminth.* 72:65–70.

Schuurmans Stekhoven, J. H., Jr. 1943. Contribution a l'etude des gordiides de la fauna Belge. *Bull. Mus. R. Hist. Nat. Belg.* 19:1–28.

Smith, D. G. 1991. Observations on the mophology and taxonomy of two *Parachordodes* species (Nematomorpha, Godioida, Chordodidae) in southern New England (USA). *J. Zool. Lond.* 225:469–480.

——. 1994. A reevaluation of *Gordius aquaticus difficilis* Montgomery, 1898 (Nematomorpha, Gordioidea, Gordiidae). *Proc. Acad. Nat. Sci. Phila.* 145:29–34.

10

TARDIGRADA

(Water Bears)*

Although tardigrades are strictly aquatic animals, they are seldom abundant in the usual types of habitats harboring freshwater invertebrates. They are not plankton organisms; they are collected in aquatic mosses and algae, on rooted aquatics, or in the mud and debris of puddles, ponds, and lakes. A large number of species are on terrestrial wet mosses, liverworts, and certain angiosperms with a rosette growth form. The animals are active in the droplets and film water. They also occur in the capillary water between the sand grains of sandy beaches up to several cm from the water's edge.

Mature individuals range from 85 to 1200 μm in length but are usually less than 500 μm long. The cylindrical body, the four pairs of stumpy lateroventral legs, and a deliberate " pawing" sort of locomotion produce a fancied resemblance to a miniature bearlike creature, therefore the Tardigrada have come to be called "water bears."

The great majority of species have been reported from lichens, turfs, leaf-litter, liverworts, and terrestrial and freshwater mosses. All other species have been reported from marine habitats, mostly interstitial.

Nearly all of what we know about tardigrade biology has been contributed through the efforts of British, German, Italian, and French investigators. Water bears have been badly neglected in the United States, even though we undoubtedly have a rich fauna. Within the past 20 years, however, there has been increased interest in tardigrades by American investigators. About 140 species have been reported from this country, and this figure includes marine, moss, and freshwater forms. About 850 species have been described the world over.

General characteristics. Tardigrades are fascinating objects for electron and low-power microscopy investigations. The body consists of a distinct head and four indefinite "body segments" (Figs. 10.1–10.3). Each leg is generally armed with four claws or two pairs of double claws. Crawling and creeping are more or less sluggish, the claws being used for clinging to vegetation and debris. The mouth is anterior, subterminal, or ventral, and the cloacal aperture, or anus, lies between the bases of the last legs. The entire surface of the body is covered with a complex, multilayered cuticle secreted by the thin epidermis, which consists of polygonal cells. Depending on the species, the cuticle may be variously thickened and marked. In the Heterotardigrada there are also various spines, papillae, and setae. Only liquid materials are ingested, and the anterior part of the digestive tract has a sucking pharynx and a pair of long, piercing stylets.

Some hermaphroditic species are known, but generally the sexes are separate. In some genera the great majority of individuals are

*Prepared with the assistance of Giovanni Pilato

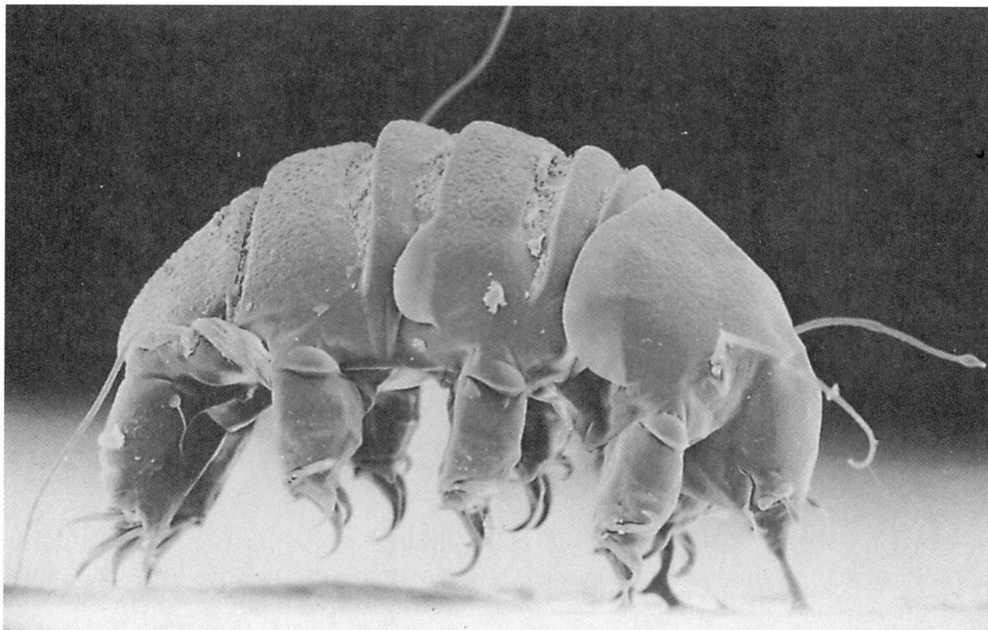

Fig. 10.1 Scanning electron micrograph of *Novechiniscus*, ×600, a semiaquatic species. (Courtesy R. O. Schuster, University of California, Davis.)

known. Usually there are 6 to 12 ecdyses in the life history.

A remarkable feature of most aquatic and semiaquatic freshwater tardigrades is their ability to tide over dry periods in a shriveled, rounded, cryptobiotic state. When normal moisture conditions return, the animal resumes its normal appearance and activity.

Coloration is variously produced by the pigmentation of the epidermis, dissolved materials in the body fluid, reserve food bodies, and the contents of the digestive tract. Gray, yellow brown, reddish, green, and brown are common colors. Some species are relatively dark and opaque, especially among older individuals; others are translucent and pale.

The tardigrades are coelomate animals. During early embryology only anlagen of the mesodermic coelomic spaces are formed; the main body cavity of adults is an hemocoel. There are no special circulatory or respiratory systems. Materials diffuse and circulate about easily in the body fluid, and respiration occurs through the general body surface.

Feeding, digestive system. Tardigrades are mostly plant feeders. The cellulose wall of moss leaflet cells and algal filaments are pierced with the stylets and the fluid contents are sucked out by pharyngeal pumping action. Occasionally, however, the body fluids of small metazoans, such as nematodes and rotifers, are used as food. *Milnesium tardigradum* is said to be chiefly carnivorous.

The portion of the digestive system extending from the esophagus is called the bucco-pharyngeal apparatus (Fig. 10.4). It consists of the mouth cavity (peribuccal structures may be present: peribuccal lamellae, papulae); the buccal tube (which in some genera is rigid and in others has a caudal flexible portion); and the muscular, or sucking, pharynx (also called the pharyngeal bulb). The pharynx has thick walls composed of radial muscle fibers. Its small triquetrous cavity is lined with cuticle, which can be sclerotized to form continuous bars or series of sclerotized individual pieces called "placoids" (Figs. 10.5, 10.6).

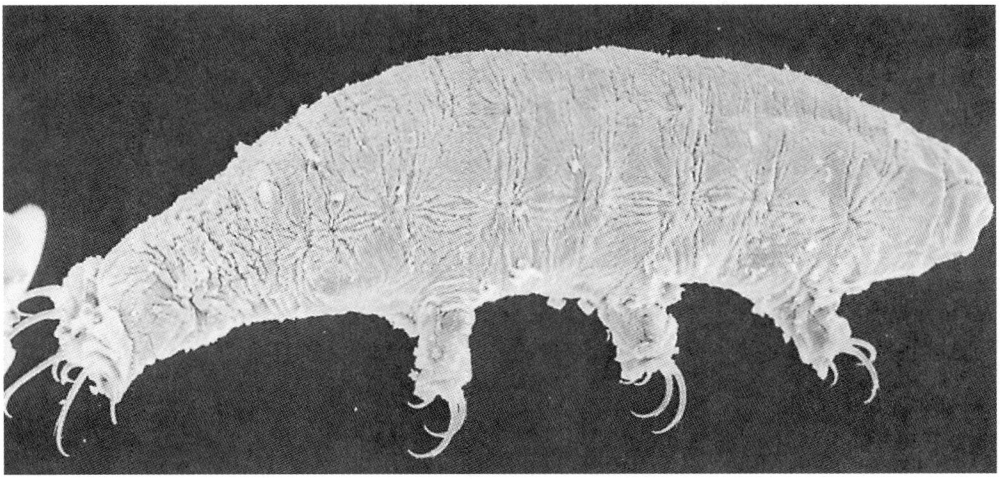

Fig. 10.2 Scanning electron micrograph of *Dactylobiotus*, ×150. (Courtesy of R. O. Schuster, University of California, Davis.)

The two stylets lie at an angle on either side of the buccal tube and the sharp tips project into the anterior end of the mouth cavity through special lateral slots, or stylet sheaths. Generally, a transverse support extends from the buccal tube to the basal portion of each stylet. Several pairs of small muscles extending from the base of the stylets to the buccal tube and the sucking pharynx are capable of withdrawing and protruding the stylets a considerable distance out of the mouth.

A pair of large salivary glands open into the mouth cavity. They also function in the secretion of new stylets and stylet supports at molting and for this reason are sometimes called stylet glands. Some investigators believe that they have an excretory function, especially in the Heterotardigrada. A few small unicellular glands occur near the mouth. The function of their secretion is unknown. They are ingrowths of the epidermis.

The esophagus ranges from long to very short. It empties into the capacious stomach, or midgut, which is usually without diverticula but may have slight lateral caeca. The hindgut, or rectum, is short and has a small cavity. In the Class Eutardigrada it is a true cloaca (Fig. 10.4). The anus or cloacal aperture is a longitudinal or transverse slit.

Digestion in the anterior portion of the stomach is acid; posteriorly it is alkaline. Reserve food may be stored in epidermal cells or in special reserve food bodies (storage cells) that float about in the body cavity. Feces are often released into the old cuticle during molting, especially in *Echiniscus*.

Excretion. In the Eutardigrada there are three small glands situated at the junction of midgut and hindgut, each of which consists of only three cells. One gland is dorsal and two are lateral. All supposedly have an excretory function, and perhaps they are also osmoregulatory. The paired glands are commonly called Malpighian tubules, even if the assumed homology with the arthropodan Malpighian tubules is questionable. Malpighian tubules are absent in the Heterotardigrada; in the Echiniscidae ventral organs at the level of the second and third pairs of legs, composed of a few cells and convoluted canals, are present, which probably perform osmoregulatory and excretory functions. Some excretory material is stored in the epidermal cells, some excretory granules are left within the old exoskeleton at ecdysis, and some are given off at the anterior end when the stylet mechanism is shed just preceding an ecdysis.

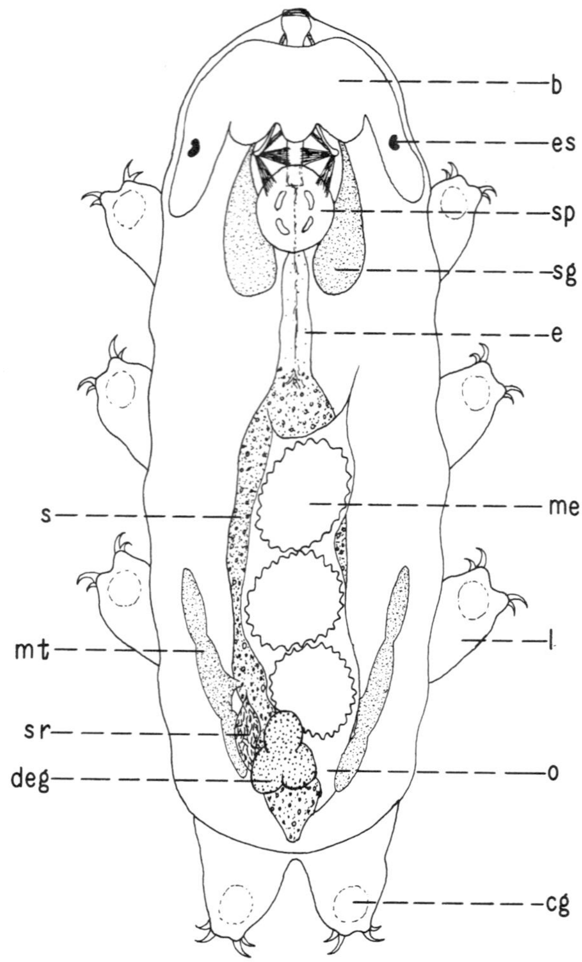

Fig. 10.3 Semidiagrammatic dorsal view of *Macrobiotus*, ×220; trunk and leg muscle not shown. b, brain; cg, claw gland; deg, dorsal excretory gland; e, esophagus; es, eyespot; 1, leg; me, mature egg in saclike ovary; mt, Malpighian tubule; o, oviduct; s, stomach; sr, seminal receptacle; sg, salivary gland; sp, sucking pharynx. (Greatly modified from Marcus 1928c.)

Muscle system. In addition to the pharynx and the stylet muscles, a tardigrade usually has from about 40 to 140 long, thin body muscles, the number depending on the species. Each such muscle is merely a single fibrillar cell with one nucleus, or several such cells. Most origins and insertions are on the body wall (Fig. 10.7). Contractions of the dorsal and ventral longitudinal muscles bring about a slight shortening or curvature of the body. Each leg is moved by a set of muscles originating on the dorsal and ventral body wall and inserting near the tip of the leg. The body wall is devoid of circular muscle fibers, but the turgor of the body fluid presumably acts as an antagonist to the longitudinal and leg muscles. Muscles extending from the body wall to the cloaca facilitate egg deposition and defecation. Longitudinal fibers associated with the stomach have been observed in a few species.

Nervous system. The large dorsal brain commonly has two long lateral lobes and two or three median lobes; it covers much of the buccal tube (Fig. 10.8). Two broad circum-

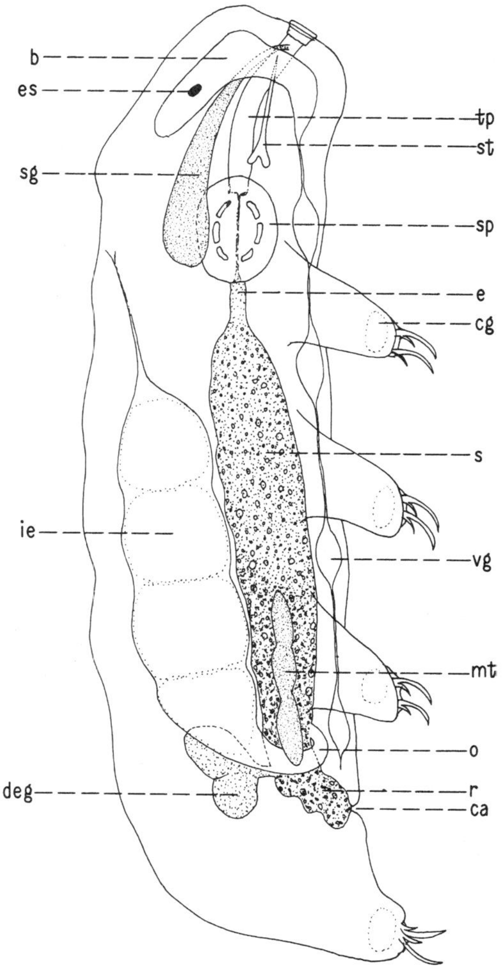

Fig. 10.4 Semidiagrammatic lateral view of a typical eutardigrade, ×220; all musculature omitted. b, brain; ca, cloacal aperture; cg, claw gland; deg, dorsal excretory gland; e, esophagus; es, eyespot; ie, immature egg in saclike ovary; mt, Malpighian tubule; o, oviduct; r, rectum; s, stomach; sg, salivary gland; sp, sucking pharynx; st, stylet; tp, buccal tube; vg, ventral ganglion.

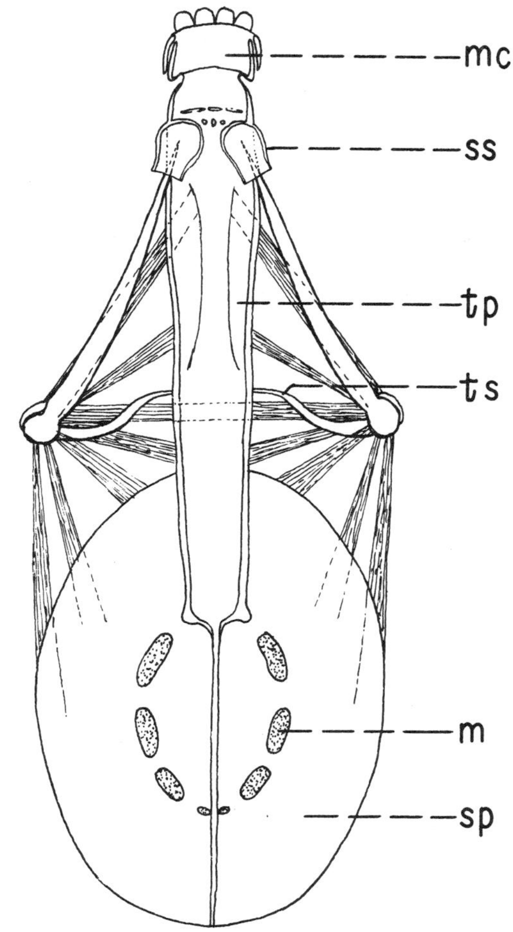

Fig. 10.5 Semidiagrammatic dorsal view of buccal apparatus and associated musculature of *Macrobiotus;* stylets are shown retracted. m, macroplacoids; mc, mouth cavity; sp, sucking pharynx; ss, stylet sheath; tp, buccal tube; ts, transverse stylet support.

pharyngeal connectives pass around the tube and unite with a subpharyngeal ganglion. In the trunk there is a chain of four conspicuous ventral ganglia united by two longitudinal nerve strands. Frequently these ganglia are slightly bilobed. Paired nerve strands originating in the brain and ganglia innervate all parts of the body. Some of the longer strands, especially those extending into the legs, have small terminal ganglionic masses.

Sense organs. The great majority of tardigrades have a pair of eyespots in the lateral lobes of the brain. Each consists of a cup-shaped mass of black or red pigment granules.

The Heterotardigrada have some structures to which a sensory function has been attributed (head cirri, cephalic papillae, clava, filaments). The Eutardigrada generally lack sensory appendages but have internal cephalic structures described as "sensory fields" covered by cuticle that exhibit variable permeability and thus chemosensory function.

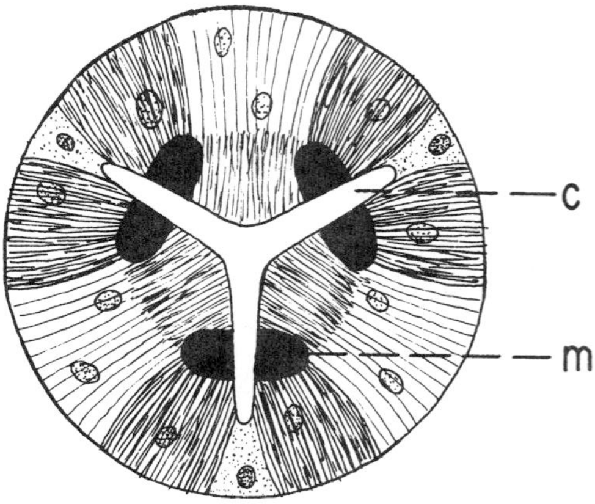

Fig. 10.6 Diagrammatic cross section of a typical tardigrade sucking pharynx showing radial muscle fibers. c, triquetrous cavity; m, macroplacoid.

Cuticular structures. Freshwater species usually have a thicker cuticle than marine species, and semiaquatic species have a thicker cuticle than aquatic species. The surface may be smooth or variously sculptured, granular, or papillate. Some species of Eutardigrada have transverse grooves or folds, but have dorsal armorlike, thickened plates (Fig. 10.12B–G). Many marine species are nothing short of spectacular in the development of very long cirri, lateral trunk expansions, dorsoventral flattening, padded claw tips, great caudal processes, and other embellishments.

Reproduction, growth. Few species are hermaphroditic. Males are unknown in most species of *Echiniscus* and *Pseudechiniscus*, which reproduce exclusively by parthenogenesis. Some species show sexual dimorphism in the shape and cuticular structure of the last legs, in the size and curvature of the claws on the front legs, and in the smaller size of the males.

The gonad is an unpaired sac dorsal to the digestive tract and with one or two attachment fibers at the anterior end. In the male there are two vasa deferentia that curve around each side of the intestine and have a small swollen portion that acts as a seminal vesicle. The female and the hermaphroditic forms have a single gonoduct on the right side of the intestine. On the other side a blind seminal receptacle may sometimes be found in the females of some Eutardigrada and in most marine Heterotardigrada. In the Eutardigrada the genital ducts open into the rectum, which is therefore more appropriately a cloaca. In the Heterotardigrada there is a separate preanal genital pore different in shape in males and females.

In one of the true aquatic species of Eutardigrada "external fertilization" has been observed. One or more males clamber about on the body of a female before she is ready to release her eggs but after the old cuticle has become loosened in preparation for a molt (Fig. 10.9). The males release sperm through the cloacal aperture of the old cuticle, and the eggs are fertilized when they are released into the cavity between the new and old cuticle. In typical semiaquatic species fertilization is "internal" and does not necessarily occur just before an ecdysis. Sperm enter the genital pore or cloacal aperture and fertilize the eggs before their release from the body of the female.

Most female Echiniscoidea produce only 2 to 6 eggs, but in other groups a female produces anywhere from 1 to 30 eggs. Sometimes the whole complement may be released in 15 to 30 minutes. In *Macrobiotus* as well as other genera the eggs are deposited freely and singly, or in groups. In other freshwater forms they are contained in the newly shed cuticle. Free eggs have processes for attachment to the substrate, and appear faceted, spinous, tuberculate, or variously sculptured. Eggs released into the old cuticle usually have a smooth surface. Inside the true shell a thin membrane surrounds the developing embryo.

The young emerge from the egg by rupturing the shell with their stylets. Newly hatched individuals are one-third to one-fifth the size of mature specimens, and all growth occurs by enlargement of cells already present. Aside from their smaller size, immature tardigrades may be recognized by their underdeveloped reproductive system, slight differences in the details of the bucco-pharyngeal apparatus, and, in some Heterotardigrada, by the lower number of claws per leg.

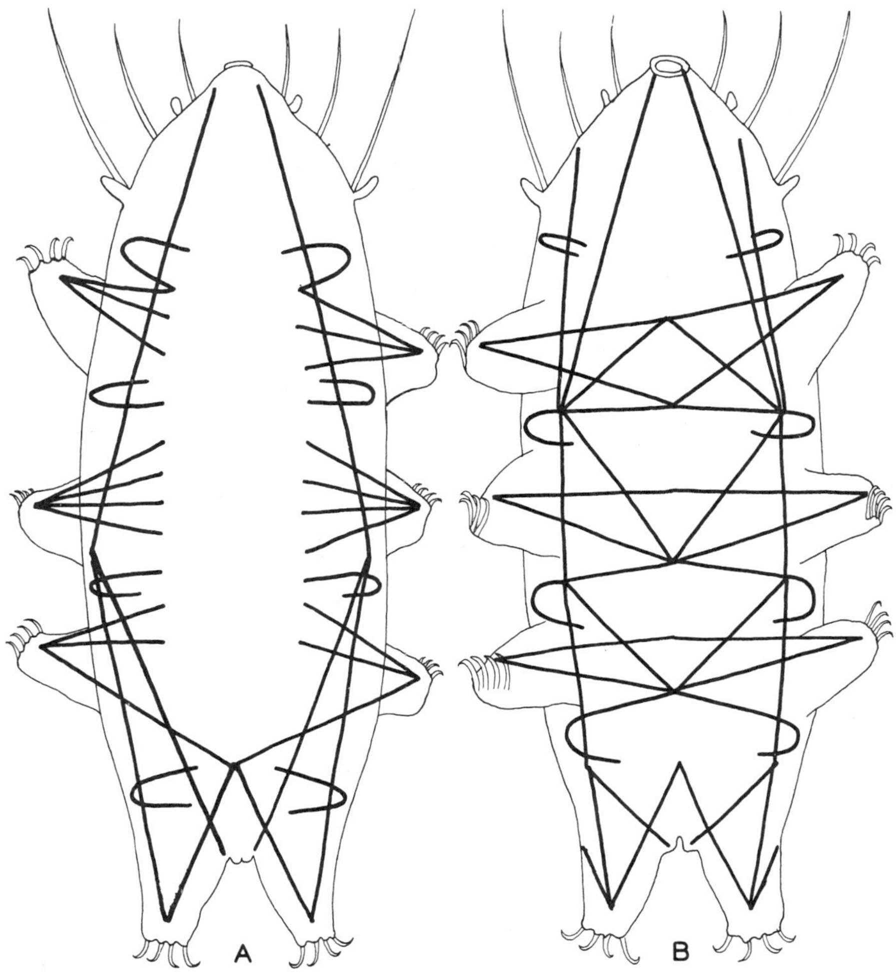

Fig. 10.7 Trunk and leg musculature of *Echiniscus*, ×350, semidiagrammatic. A, dorsal view; B, ventral view. (Greatly modified from Marcus 1928c.)

Most investigators indicate 6 to 12 molts during the life history, and sexual maturity is usually attained sometime after the second or third molt. The old cuticle is usually shed after a period of growth, but it may also be shed as a response to hunger or unfavorable environmental conditions. Several days before a molt the sclerotized portions of the buccal–pharyngeal apparatus are ejected ("simplex stage") and new ones (attached to the old muscles) promptly begin to be formed by the salivary glands and the pharynx. A day or two before a molt, the animal shrinks temporarily and slightly. Then the old cuticle ruptures at the anterior end and the animal crawls out. The lining of the rectum and the claws are lost at molting, and new claws are initially secreted by special claw glands in the tip of each leg.

There appears to be considerable variation in longevity, assuming that the animal is continuously active. Two months and 24 months are perhaps the extremes.

General ecology. Some species (*Hypsibius dujardini* and *H. convergens*) are truly aquatic and some are both aquatic and semiaquatic, but the majority are semiaquatic only and occur chiefly on terrestrial mosses and lichens. Active tardigrades can usually

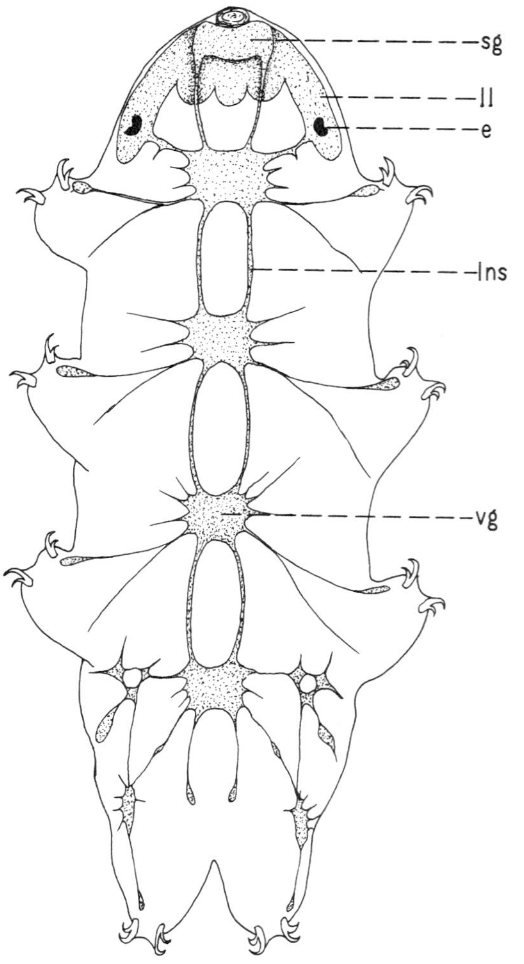

Fig. 10.8 Ventral view of nervous system of *Macrobiotus*, ×220. e, eyespot; ll, lateral lobe of brain; lns, longitudinal nerve strand; sg, subpharyngeal ganglion; vg, ventral ganglion. (Modified from various sources.)

be found if one searches diligently in masses of filamentous algae, washings of higher aquatic plants, or in the debris on the bottom of ponds and pools. The semiaquatic species, however, are active only when the plant has been wetted by rain or when it is splashed at streamside. Many mosses, liverworts, lichens, turfs, leaf-litter, and rosette angiosperms contain large tardigrade populations.

Mosses on rocks or on the bark of trees contain tardigrades, even though they may be wet enough to allow them to be active for less than 5% of the time. Concentrations up to 22,000 dry and inactive cryptobiotic tardigrades per gram of dry moss have been reported. Mosses with especially thick cellulose cell walls, such as *Polytrichum*, usually do not harbor tardigrades. Presumably their stylets cannot penetrate such walls. Favorable portions of sandy beaches sometimes contain 300 to 400 tardigrades/10 cm^3 of wet sand. Sometimes they occur as deep as 15 cm in the sand.

Most species appear to be eurythermal and are normally active anywhere between near-freezing temperature and 25 to 30°C. Some forms have been found in warm springs where they are continuously exposed to temperatures of 40°C. Although aquatic species are generally characteristic of the littoral, there are records of collections as deep as 100 m on lake bottoms. Some semiaquatic species occur from the tropics to the Arctic and Antarctic and from sea level to mountain tops up to 6000 m high. The annual period of activity varies correspondingly, from the year round to only a week or two.

Little is known about minimal oxygen requirements, but it is usually inferred that tardigrades cannot endure low dissolved oxygen concentrations as do many other small aquatic metazoans.

Aquatic species are usually most abundant between January and May, and sometimes there is a second maximum in the autumn months. Moss dwellers do not have pronounced seasonal maxima. Individuals are active as long as the moss is damp.

Anhydrobiosis. When a bit of moss dries, many of the associated Tardigrada have the unusual ability to assume an inactive cryptobiotic state that may persist for as long as 4 to 7 years. When normal ecological conditions again prevail, the tardigrades come out of their quiescent state and resume normal activity. Most marine species and true aquatic freshwater species apparently do not have the ability to assume an cryptobiotic condition.

Under drying conditions the head, posterior end, and legs are retracted, and the whole body becomes more or less rounded into a "tun." Considerable water is lost and

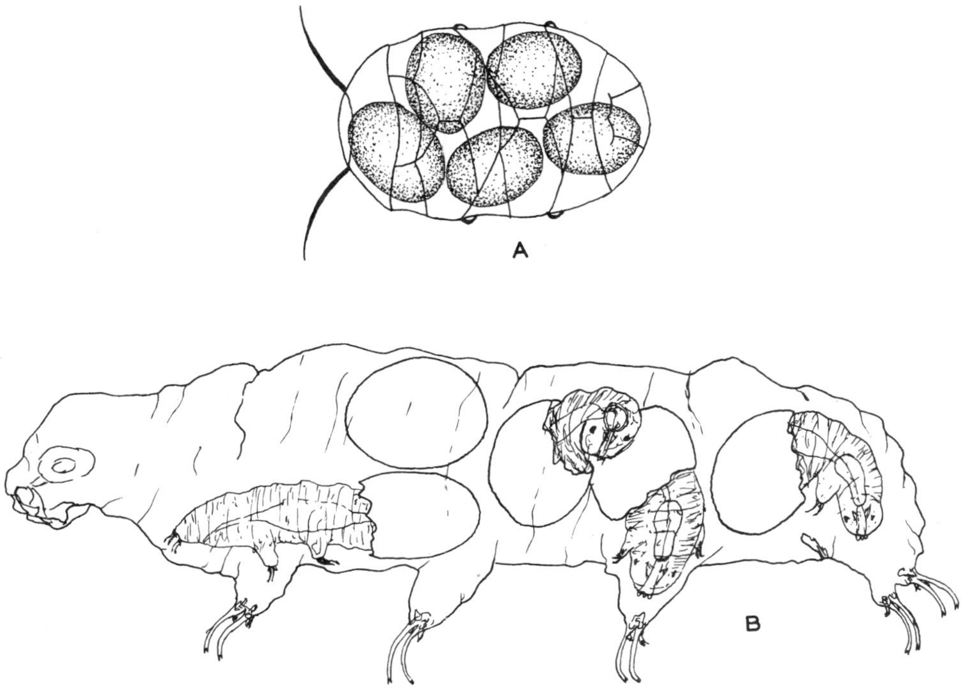

Fig. 10.9 Cast tardigrade exoskeletons containing eggs. A, dorsal view of *Pseudechiniscus;* B, lateral view of *Milnesium tardigradum* (four of five eggs in process of hatching). (A modified from Ramazotti and Maucci 1983.).

the anhydrobiotic animal is therefore shriveled and wrinkled (Fig. 10.10). Internal visceral structures are not greatly modified. Although drying is almost always responsible for bringing on cryptobiosis, a lack of sufficient dissolved oxygen and perhaps other unfavorable conditions may also be responsible. Tuns can be formed only when the drying process occurs slowly and when the relative humidity is high.

Anhydobiotic tardigrades are highly resistant to abnormal environmental conditions. Some remain viable even after being kept experimentally in strong brine solutions, at 100°C for 6 hours, and at −190°C for as long as 20 months.

Revival from anhydrobiosis takes but a short time when individuals are properly wetted. Usually the interval is 4 minutes to several hours, and occasionally a day or two. The animal absorbs water, swells, and promptly becomes active. Revival is more rapid at higher temperatures, with higher dissolved oxygen, and in animals having more stored food material. Moss species usually go through numerous alternating periods of activity and cryptobiosis during their life cycle, depending on the nature of their habitat. One author states that the life cycle may be as long as 60 years, including cryptobiotic periods. Laboratory animals can easily be dried and revived 10 times or more.

Cysts. Damage, hunger, and other abnormal ecological conditions sometimes bring about the formation of resistant cysts, especially in the true aquatic species of *Dactylobiotus.* The animal contracts inside the old wrinkled cuticle and forms a dark-colored, thick-walled inner cyst (Fig. 10.11). The internal organs undergo a variable degree of degeneration. When ecological conditions are favorable the animal becomes reconstituted, the cyst wall ruptures, and the tardigrade emerges.

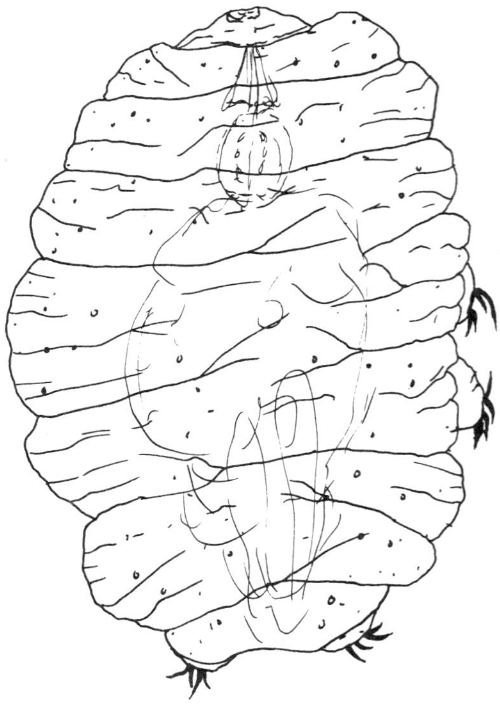

Fig. 10.10 Dorsal view of a tardigrade "tun" in the anhydrobiotic state, ×300.

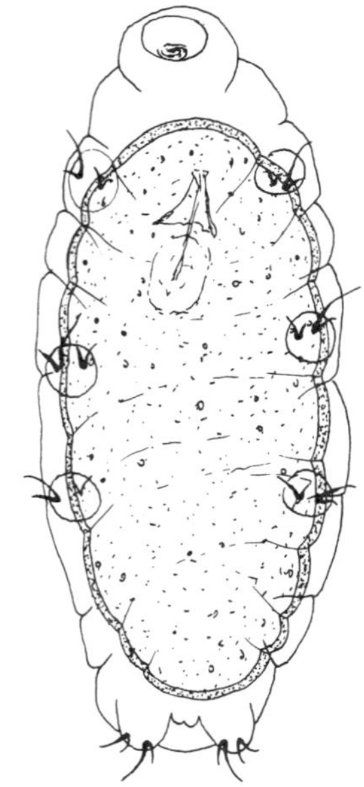

Fig. 10.11 Cyst of *Hypsibius*, semidiagrammatic, ×75. (Greatly modified from Marcus 1928a.)

Anoxybiosis. A third type of inactive condition in tardigrades is called anoxybiosis. It usually comes about when there is insufficient oxygen. The body swells and becomes rigid and turgid. There are no movements, and it is difficult to decide whether the animal is alive. After a maximum of about 5 days in this condition the animal dies, but if food and oxygen are supplied promptly the animal again resumes normal activity. Sometimes individuals emerging from anabiosis go into the state of anoxybiosis.

Crowe (1975) recognizes two other kinds of latency among tardigrades, both of which have been only superficially investigated. One is cryobiosis, which is closely associated with tun formation and is a state of suspended animation assumed on freezing and other exceptionally low temperatures. The second is osmobiosis, a cryptobiotic or tun state induced by exceptionally high osmotic pressures when the animals are placed in abnormal salt concentrations.

Everitt (1981) worked on tardigrade populations of an Antarctic pool that was frozen solid for 8 months each year. The animals (*Hypsibius*) overwintered in a "cryptobiotic state" and quickly became active in an algal mat when the pond thawed. Densities as high as 470 animals per gram of algae were found.

Geographical distribution, dispersal. Many species of tardigrades are widely distributed. Some are Holarctic, generally distributed in the Northern Hemisphere, or even cosmopolitan (*Macrobiotus hufelandi*, *Hypsibius oberhauseri*, *Milnesium tardigradum*). Other species, of course, have been reported only from single mainland or island localities. It seems that the tropics contain notably few species and few individuals.

Passive distribution is effected by currents, animals (usually insects), waves, and wind, especially in the anhydrobiotic and cyst stages. There is considerable positive evidence for wind as an important distributive agency.

Collecting. Almost any bit of suitable moss will yield tardigrades. If the moss is wet, simply rinse it out in water and examine the washings. Masses of filamentous algae and rooted aquatics may be treated similarly. Dry mosses and lichens should be soaked for 30 minutes to several days. At intervals rinse such plant material, and the tardigrades will be found as they emerge from cysts and the anabiotic state.

Culturing. Tardigrades are not easy to culture, and little actual culture work seems to have been done. They may be kept in masses of filamentous algae or wet moss for a few days to weeks. Change a large fraction of the water every three or four days.

Preserving, preparing. Unlike many other small freshwater metazoans, tardigrades are easy to fix in extended form. A large percentage will be suitable when killed and fixed in 5% formalin or 85% alcohol. Especially fine specimens may be secured by inducing anoxybiosis under a coverslip sealed with petroleum jelly before killing and fixing. This procedure usually requires 10 minutes to 3 hours. Some investigators recommend the use of one of a variety of narcotics before fixation. Fixatives that are particularly useful for staining and histological work are osmic acid, saturated mercuric chloride, and 2% acetic acid. Piercing the body wall with a fine needle permits better staining and dehydration. Simple unstained semipermanent mounts may be made in glycerin jelly. The sclerotized parts of the buccal apparatus maybe isolated on a slide by the judicious use of 10% potassium hydroxide.

Taxonomy. The phylogenetic affinities of tardigrades are highly debatable. Some workers considered the tardigrades to be closely related to nematodes, but they are coelomate animals. Often they are placed near the mites in the Phylum Arthropoda, especially because of their indistinct segmentation, four pairs of legs, and piercing stylets. Many workers place them near the Onychophora.

About 60 genera are recognized in the whole phylum, of which 40 contain a total of about 750 aquatic and semiaquatic freshwater species, the others being marine. Thirty of these genera have been reported from the United States.

The chief taxonomic structures used in generic and specific determinations include details of the buccal apparatus, armor, claws, cirri, filaments, and eggs. *Echiniscus*, *Pseudechiniscus*, *Macrobiotus*, *Isohypsibius*, *Diphascon*, and *Hypsibius* are large genera that are often subdivided into species groups or subgenera for taxonomic convenience. Most of the earlier descriptions are insufficient, owing to the false supposition that the structures of the claws alone were sufficient and trustworthy characters. The present systematic arrangement of the tardigrades is based on the taxonomic criteria of Pilato (1969, 1975). The works of Bertolani (1982) and Ramazotti and Maucci (1983) represent the most complete references to date.

The following key was kindly prepared by Giovanni Pilato and includes all valid aquatic and semiaquatic genera known from the United States. *Thulinia*, recently reported form the United States (Strayer et al. 1994), has been added to the key. The key to certain Hypsibiidae makes reference to claws of specific types, for example, *oberhaeuseri* or *Ramazzottius* type. There are three types: In the *oberhaeuseri* type the inner claws are short and stout while in the outer claws the basal portion is long and articulated with the short, secondary branch (Fig. 10.14K). In the *Isohypsibius* type the the claws are slender, the basal portion not so elongate and articulated, and the secondary branch forms a strong angle with the main branch. In the *Hypsibius* type the basal portion is also short and not articulated and the secondary branch more continuously curves but does not form a distinct angle.

Although there has been a recent increase in interest in tardigrades in North America, little is known of the distribution of most genera in the United States and specific range information is not given. For similar reasons, with respect to species in each genus, numbered references contain usually only general information on the genus identified.

KEY TO FAMILIES AND GENERA OF TARDIGRADA OF THE UNITED STATES

1. Head with anterior cirri and lateral filaments; four separate but similar claws on each leg (Fig. 10.12B–G) .... **Class HETEROTARDIGRADA, ECHINISCIDAE, 2**

Head without cirri and lateral filaments (Fig. 10.14N–P); each leg with two double claws or unlike pairs of claws, or with two simple claws (Fig. 10.14C–E, K–M) **EUTARDIGRADA, 8**

2. Segmental plates I–IV unpaired; median plates 1 and 2 bar-shaped (Fig. 10.1) . ***Novechiniscus***

Body with thick dorsal paired plates, variously sculptured; median plates not bar-shaped (Fig. 10.12D–G) **3**

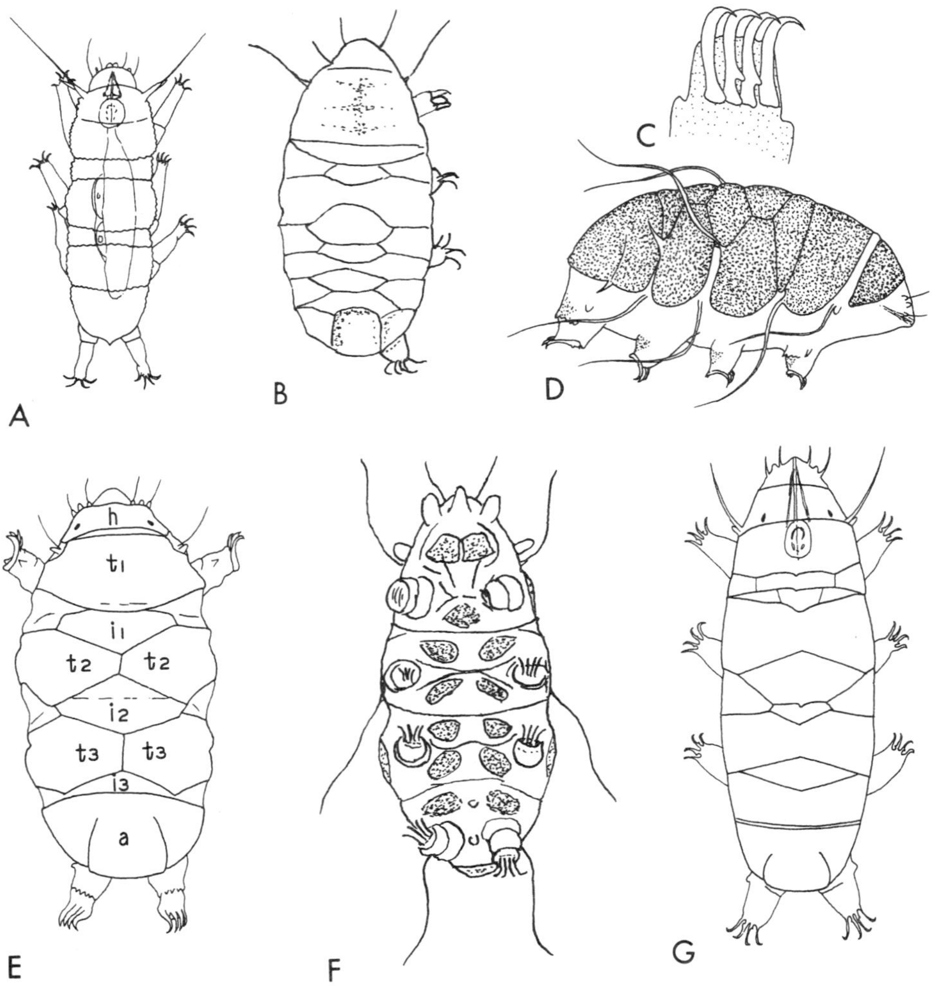

Fig. 10.12 Structure of Heterotardigrada. A, dorsal view of *Oreella*, ×280, a rare genus, not yet found in the United States; B, *Bryodelphax*, ×410; C, claws of *Echiniscus;* D and E, typical species of *Echiniscus*, ×240; F, ventral view of *Testechiniscus*, ×440; G, dorsal view of *Pseudechiniscus*, ×340. a, anal plate; h, head plate; i1, i2, and i3, median plates (or intercalaries); t1, first trunk plate; t2, second trunk plate; t3, third trunk plate (divided). (B redrawn from Pilato 1974; F redrawn from Kristensen 1987.)

3. Three trunk plates (Fig. 10.12B, D–F) **4**
Four trunk plates (Fig. 10.12G) **7**
4. Median plates 1and 2 divided, median plate 3 undivided (Fig. 10.12B); ventral plates present or absent . . . . ***Bryodelphax***
Median plates 1 and 2 undivided (but can show a transverse fold) **5**
5. Ventral plates absent **6**
Ventral plates present . . . . ***Testechiniscus***
6. Dorsal plates not very distinct; anal plate divided into three parts by sutures; mid-dorsal appendage, if present, located between median plate 2 and the segmental plate 3 ***Hypechiniscus***
Dorsal plates very distinct; anal plate notched but not divided; no medial dorsal appendage evident (Fig. 10.12D, E) . . . . ***Echiniscus***
7. Cirrus A are normal filaments (Fig. 10.12G) . . . . ***Pseudechiniscus***
Cirrus A horn-shaped . . . . ***Cornechiniscus***
8. Each leg with two long, slender claws and two short, heavy claws (Fig. 10.14C); pharyngeal bulb elongated and without placoids (Fig. 10.14A, B); mouth surrounded by six prominent papillae . . . . **MILNESIIDAE,** ***Milnesium***
Each leg usually with two double claws (Fig. 10.14D, E, K, L) or rarely single claws (Fig. 10.14M); pharyngeal bulb with or without placoids **9**
9. Double claws of each leg symmetrical with respect to the median plane of the leg (Figs. 10.13B, D; 10.14D, E); buccal tube with a longitudinal anteroventral lamina (Fig. 10.13A, C, E) **MACROBIOTIDAE, 10**
Double claws of each leg asymmetrical with respect to the median plane of the leg (Figs. 10.13F, G; 10.14K, L), or single claws (Fig. 10.14M) **15**
10. Caudal portion of the buccal tube flexible and with a spiral thickening (Fig. 10.13E) ***Pseudodiphascon***
Buccal tube rigid without a spiral thickening **11**
11. A longitudinal ridge present on anterodorsal portion of the buccal tube . . . . ***Adorybiotus***
No dorsal ridge on anterior portion of buccal tube **12**
12. Secondary branch of each claw reduced and forming an angle almost 90° with main branch (Fig. 10.13D) . . . . ***Dactylobiotus***
Both branches of each claw with tips close together **13**
13. Main and secondary branches of each claw diverging almost from base of claw (Fig. 10.14D); a marginal hook present on ventral lamina of buccal tube (Fig. 10.13C) . . . . ***Murrayon***
Main and secondary branch of each claw joined for almost half of length (Fig. 10.14E); no hook present on the margin of ventral lamina **14**
14. Peribuccal lamellae present (Fig. 10.13A); buccal tube width variable . . . . ***Macrobiotus***
Peribuccal papulae instead of lamellae present; buccal tube narrow . . . . ***Minibiotus***
15. The two branches of each double claw solidly joined to each other up to the base of the claw (Fig. 10.14L), or single claws (Fig. 10.14M), with or without a minute basal spur **CALOHYPSIBIIDAE, 16**
The main branch of each claw joined to the basal portion far from the base of the claw (Figs. 10.13F, G; 10.14K) . . . . **HYPSIBIIDAE, 19**
16. Claws double **17**
Claws single or with a minute basal spur **18**
17. Hind legs with one or two claws, secondary branch reduced; buccal tube with ventral lamina ***Parhexapodibius***
Hind legs with two double claws, secondary branch not especially reduced; buccal tube without ventral lamina . . . . ***Calohypsibius***
18. Claws single or with an additional minute basal spur; hind legs with claws ***Haplomacrobiotus***
Claws single; hind legs reduced and without claws . . . . ***Haplohexapodibius***
19. Bucco-pharyngeal tube with a flexible caudal portion and spiral thickening (Fig. 10.13H) . . **20**
Buccal tube rigid **25**
20. Pharyngeal bulb without placoids or with one placoid in each series **21**
Pharyngeal bulb with more than one placoid in each series **22**

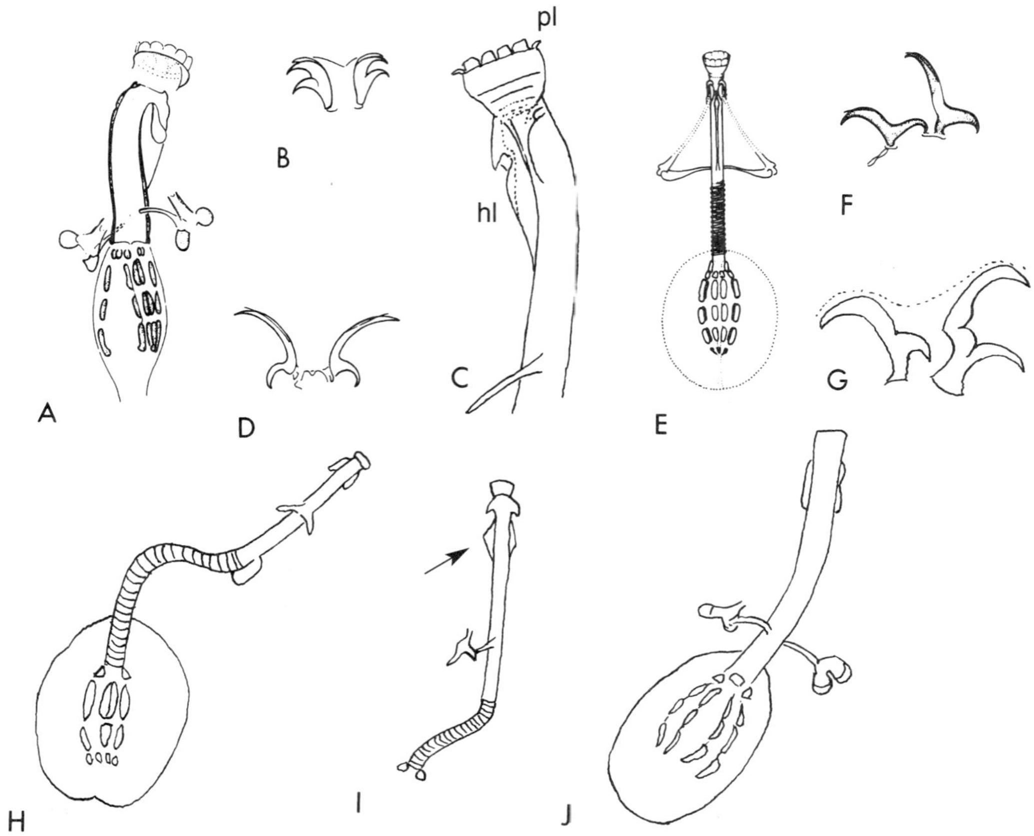

Fig. 10.13 Bucco-pharyngeal structure and claws of tardigrades. A, bucco-pharyngeal apparatus of *Macrobiotus;* B, claws of *Macrobiotus;* C, bucco-pharyngeal apparatus of *Murrayon;* D, claws of *Dactylobiotus;* E, bucco-pharyngeal apparatus of *Pseudodiphascon;* F, claws of *Isohypsibius* type; G, claws of *Hypsibius* type; H, bucco-pharyngeal apparatus of *Diphascon;* I, bucco-pharyngeal tube of *Hebesuncus;* J, bucco-pharyngeal apparatus of *Isohypsibius.* hl, hooked lamella; pl, peribuccal lamella. (A, B, D, F from Schuster et al. 1980; H, I redrawn from Pilato 1987; others redrawn from various sources.)

21. Stylet supports absent; rigid portion of bucco-pharyngeal tube very short, about 15% of total length of the tube . . . . ***Astatumen***

Stylet supports present, rigid portion of the bucco-pharyngeal tube clearly more than 15% of the length of the tube . . . . ***Itaquascon***

22. Anterior portion of the buccal tube with a dorsal and ventral apophysis (for insertion of stylet muscles) and shaped like a triangle (Fig. 10.13H, I, arrow) **23**

Anterior portion of buccal tube with a dorsal and ventral apophysis and appearing as a wide, flat ridge **24**

23. Dorsal and ventral apophyses for the stylet muscles symmetrical in the frontal plane (Fig. 10.13H) . . . . ***Diphascon***

Dorsal and ventral apophyses asymmetrical in the frontal plane (Fig. 10.13I) . . . . . ***Hebesuncus***

24. Furcae of the stylets with postero-lateral processes thickened and rounded at their apices (Fig. 10.14F) . . . . ***Mesocrista***

Furcae of the stylets with postero-lateral processes spoonlike and tapering at their apices (Fig. 10.14G) . . . . ***Platicrista***

25. Buccal tube with an antero-ventral lamina . . . . ***Doryphoribius***

Buccal tube without ventral lamina **26**

26. Peribuccal lamellae present **27**

Peribuccal lamellae absent **28**

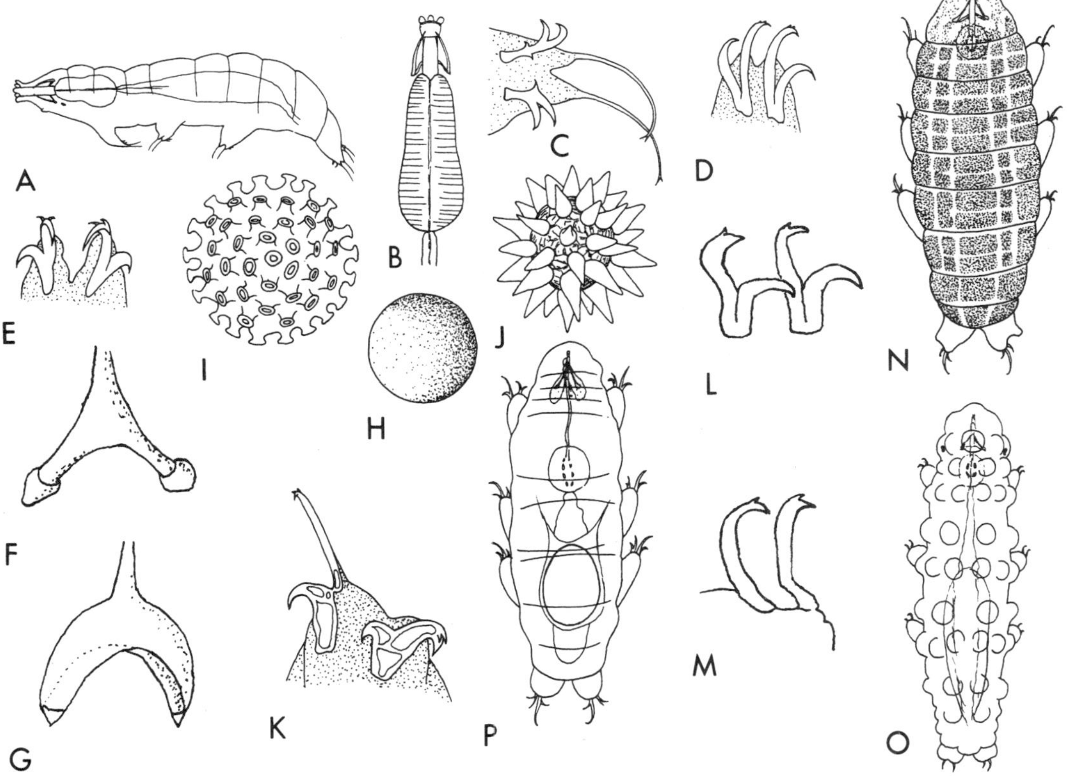

Fig. 10.14 Structure of tardigrades. A, *Milnesium tardigradum*, ×75; B, bucco-pharyngeal apparatus of *Milnesium;* C, claws of *Milnesium;* D, claws of *Murrayon;* E, claws of *Macrobiotus;* F, stylet furcae of *Mesocrista;* G, stylet furcae of *Platicrista;* H, egg of *Hypsibius,* I and J, eggs of *Macrobiotus;* K, claws of *Ramazzottius* type; L, claws of *Calohypsibius* type; M, single claws of Calohypsibiidae; N, dorsal view of *Ramazzottius;* O, dorsal view of *Isohypsibius* (see also Fig. 1.8C); P, dorsal view of *Diphascon;* N–P ×220. (F, G redrawn from Pilato 1987; others modified from various sources.)

27. Peribuccal lamellae numerous (almost 30) . . . . ***Pseudobiotus***
Peribuccal lamellae fewer, about 12 in number . . . . ***Thulinia***

28. Claws of the *Ramazzottius* type (Fig. 10.14K) . . . . ***Ramazzottius***
Claws of different type **29**

29. Claws of *Isohypsibius* type (Fig. 10.13F); anterior portion of buccal tube with dorsal and ventral apophysis produced as a longitudinal ridge (Fig. 10.13J) . . . . ***Isohypsibius***
Claws of *Hypsibius* type; anterior portion of buccal tube with dorsal and ventral apophysis produced as a hook . . . . ***Hypsibius***

REFERENCES

Baumann, H. 1921. Beiträge zur Kenntnis der Anatomie der Tardigraden (*Macrobiotus hufelandi*). *Z. Wiss. Zool.* 68:637–652.

——. 1922. Die Anabiose der Tardigraden. *Zool. Jahrb. Abt. Syst. Geogr. Biol. Tiere* 45:501–556.

Beasley, C. W. 1990. Tardigrada from Gunnison Co., Colorado, with a description of a new species of *Diphascon. Southwestern Nat.* 35:302–304.

Bertolani, R. 1982. Tardigradi. *Guide per il riconoscimento delle specie animali delle acque interne italiane.* C.N.R. AQ/1/68 15:1–104.

Christenberry, D., and R. P. Higgins. 1979. A new

species of *Pseudodiphascon* (Tardigrada) from Alabama. *Trans. Am. Microsc. Soc.* 98:508–514.

Crowe, J. H. 1975. The physiology of cryptobiosis in tardigrades. *Mem. 1st. Ital. Idrobiol. Suppl.* 32:37–59.

Crowe, J. H., and R. P. Higgins. 1967. The revival of *Macrobiotus areolatus* Murray (Tardigrada) from the cryptobiotic state. *Trans. Am. Microsc. Soc.* 86:286–294.

Dastych, H. 1980. *Niesporczaki (Tardigrada) Tatrzanskiego Parku Narodowego.* 232 pp. Monografie Fauny Polski 9. Polska Akad. Nauk, Warszawa.

Dewel, R. A., et al. 1992. Unusual cuticle-associated organs in the heterotardigrade, *Echiniscus viridissimus. J. Morph.* 212:123–140.

Doyère, L. 1840, 1842. Mémoire sur less Tardigrades. *Ann. Sci. Nat. Zool.* (2) 14:269–361; 17:193–205; 18:5–35.

Everitt, D. 1981. An ecological study of an Antarctic freshwater pool with particular reference to Tardigrada and Rotifera. *Hydrobiologia* 83:225–239.

Guidetti, R. 1998. Two new species of Macrobiotidae (Tardigrada: Eutardigrada) from the United States of America, and some taxonomic considerations of the genus *Murrayon. Proc. Biol. Soc. Wash.* 111:663–673.

Higgins, R. P. 1959. Life history of *Macrobiotus islandicus* Richters with notes on other tardigrades from Colorado. *Trans. Am. Microsc. Soc.* 78:137–154.

——. (ed.). 1975. International symposium on tardigrades. *Mem. 1st. Ital. Idrobiol. Suppl.* 32: 1–469.

Kathman, R. D. 1990. Some tardigrades from Colorado, with a description of a new species of *Macrobiotus* (Macrobiotidae: Eutardigrada). *Proc. Biol. Soc. Wash.* 103:300–303.

Kathman, R. D., and C. W. Beasley. 1997. Synonymy of *Platicrista cheleusis* (Tardigrada: Eutardigrada: Hypsibiidae). *Proc. Biol. Soc. Wash.* 110:558–559.

Kristensen, R. M. 1987. Generic revision of the Echiniscidae (Heterotardigrada), with a discussion of the origin of the family. In: *Biology of Tardigrades,* pp. 261–335, edited by R. Bertolani. Selected Symposia and Monographs, U.Z.I., Mucchi, Modena.

Marcus, E. 1928a. Spinnentiere oder Arachnoidea, IV: Bartierchen (Tardigrada). *Tierwelt Dtschl.* 12:1230.

——. 1928b. Zur Ökologie und Physiologie der Tardigraden. *Zool. Jahrb. Abt. Allg. Zool.* 44: 323–370.

——. 1928c. Zur vergleichenden Anatomie und Histologie der Tardigraden. *Zool. Jahrb. Abt. Allg. Zool.* 45:99–158.

——. 1929. Zur Embryologic der Tardigraden. *Zool. Jahrb. Abt. Anat.* 50:333–384.

——. 1936. Tardigrada. *Das Tierreich* 66:1–340.

May, R.-M. 1948. La vie des Tardigrades. *Hist. Nat.* 8: 1–133.

Morgan, C. I., and P. E. King. 1976. *British Tardigrades.* 132 pp. Academic, London.

Müller, J. 1935. Zur vergleichenden Myologie der Tardigraden. *Z. Wiss. Zool.* 147:171–204.

Murray, J. 1907a. Scottish Tardigrada, collected by the Lake Survey. *Trans. R. Soc. Edin.* 45:641–668.

——. 1907b. Encystment of Tardigrada. *Trans. R. Soc. Edin.* 837–854.

——. 1910. Tardigrada. *Sci. Rep. Br. Antarctic Exped.* 1907–1909 1:81–185.

Peterson, B. 1951. The tardigrade fauna of Greenland: a faunistic study with some few ecological remarks. *Medd. Grønland* 150:1–94.

Pigon, A., and B. Weglarska. 1955. Rate of metabolism in tardigrades during active life and anabiosis. *Nature (London)* 176:121–122.

Pilato, G. 1969. Evoluzione e nouve sistemazione degli Eutardigradi. *Boll. Zool.* 36:327–345.

——. 1974. Tre nouve specie di tardigradi muscicoli di Cina. *Animalia* 1:59–68.

——. 1975. On the taxonomic criteria of the Eutardigrada. *Mem. 1st. Ital. Idrobiol. Suppl.* 32: 277–304.

——. 1982. The systematics of Eutardigrada: a comment. *Z. Zool. Syst.* 20:271–284.

——. 1987. Revision of the genus *Diphascon* Plate, 1889, with remarks on the subfamily Itaquasconinae (Eutardigrada, Hypsibiidae). In: *Biology of Tardigrades,* pp. 337–357, edited by R. Bertolani, Selected Symposia and Monographs, U.Z.I., Mucchi, Modena.

Ramazzoti, G., and W. Maucci. 1983. Il Phylum Tardigrada. *Mem. 1st. Ital. Idrobiol. Dott. Marco Marchi* 41:1–1012.

Riggin, G. T. 1962. Tardigrada of southwest Virginia: with the addition of a description of a new marine species from Florida. *Tech. Bull. Va. Agric. Exp. Sta.*152:1–145.

Rudescu, L. 1964. Tardigrada. *Fauna Rep. Pop. Rom.* 4:1–401.

Schuster, R. O., et al. 1980. Systematic criteria of the Eutardigrada. *Trans. Am. Microsc. Soc.* 99:284–303.

Schuster, T. R., and A. A. Grigarick. 1965. Tardigrada from Western North America: with emphasis on the fauna of California. *Univ. Calif. Publ. Zool.* 76:1–67.

Strayer, D. L., et al. 1994. Tardigrades from shallow groundwaters in southeastern New York, with the first record of *Thulinia* from North America. *Trans. Am. Microsc. Soc.* 113:325–332.

Walz, B. 1975. Modified ciliary structures in receptor cells of *Macrobiotus hufelandi* (Tardigrada). *Cytobiology* 11:181–185.

——. 1978. Electron microscopic investigations of cephalic sense organs of the tardigrade *Macrobiotus hufelandi*. C.A.S. Schultze. *Zoomorphologie* 89:1–19.

11
ENTOPROCTA
(Banded moss animals)

The Entoprocta is the most poorly represented invertebrate group in freshwater. In fact only one species of this small otherwise marine phylum is known from freshwater environments. The Entoprocta are small colonial or solitary sessile invertebrates with anatomical characteristics linking them to a number of otherwise very differently constructed invertebrate groups.

The phylum to which the Entoprocta has been most often compared is the Ectoprocta. The ectoprocts have often been combined with entoprocts as the Bryozoa, but each group represents very different kinds of organisms. The Entoprocta lack a coelom and the tentacles form a ring enclosing both the mouth and the anus. A specialized vestibule functions as both a marsupium and a food-processing structure.

Entoprocts are either colonial or solitary and individual zooids are rarely more than 5 mm in size. Colonies comprise individuals that are joined by a stolon or are simply branched from a single zooid. Noncolonial species are either free-living or commensal with other organisms. Only a single species of freshwater entoproct is currently recognized, *Urnatella gracilis*. This species was described by the master microscopist and zoologist of the 19th century Joseph Leidy, who discovered and described many aquatic taxa from eastern Pennsylvania during his career.

General characteristics. *Urnatella gracilis* bears close resemblance to marine representatives of the phylum but contains certain unique morphological features. Colonies of *U. gracilis* consist of few to many individual zooids, all arising from a single stalk rather than a sessile stolon (Fig. 11.1). An entire colony may reach 3 or 4 mm in height but may branch many millimeters laterally. The individual zooid consists of a terminal calyx, housing most of the organs, a stalk, and a pedal disk or footplate. The calyx is a saclike structure with several upwardly directed tentacles. The epithelium of the medial surfaces of the tentacles is ciliated, elsewhere covered with cuticle, which extends around the rest of the calyx. The U-shaped gut is differentiated into stomach, intestine, and rectum, the latter occupying a raised anal cone.

The flexible stalk is the main support for the zooid and consists of several sclerotized bands separated by narrow, liable constrictions. Muscle within the stalk can bend and straighten the individual to adjust its position relative to the surrounding aquatic environment. The pedal disk is an expanded portion of the stalk and secures the animal or colony to the substrate. Adherence is presumably the result of secretion of adhesive material from the pedal disk. Both the stalk and the pedal disk are covered by a continuation of the cuticle that covers the calyx.

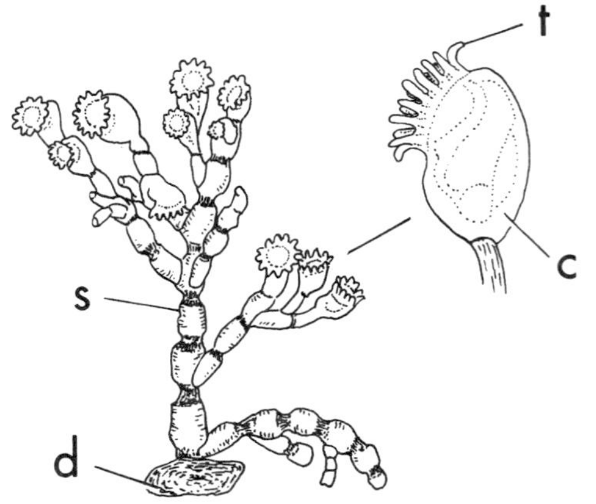

Fig. 11.1. Colony of *Urnatella gracilis*, ×30, enlargement of calyx to right. c, calyx; d, disk; s, stalk; t, tentacle.

Food and Feeding. *Urnatella gracilis* is a filter feeder. Cilia on the tentacles and vestibule of the calyx create feeding currents. Water containing suspended particulate material and minute plankton is drawn into the cup of the calyx. Particles and consumable plankton are removed and passed into the mouth and stomach, assisted by mucus, which traps the food. During inactivity or when the animal is overcome by excessive suspended material the tentacles fold over the calyx.

Reproduction. Unlike marine species, the freshwater entoproct is capable of several modes of reproduction. The typical ciliated larva produced by its marine relatives has been abandoned and sexual reproduction in *U. gracilis* has been reduced to a rare occurrence. When undertaken, fertilized eggs from hermaphroditic adults produce a small, short-lived larva that settles and grows into an individual zooid.

Reproduction in this species is predominately asexual. The colony increases size by budding new individuals from germination zones along the stalk. Following the active growing season, usually above 20°C, the colonies degenerate and only the stalks survive. Dormant stalks serve as hibernacula, generating new calyces when favorable conditions resume. If the stalk becomes unrooted, it cannot reattach even if it can generate new calyces.

An important mechanism for dispersal and the founding of new colonies involves the asexual budding of stolonlike stalks called creeping stolons. These small propagules contain small muscular suction disks to attach to the substrate while germinating secondary stalks and calyces. The developing young colony cements itself to the substrate at the site of the disk and becomes a mature, permanent colony.

Ecology. *Urnatella gracilis* seems to prefer warmer environments and does not achieve luxuriant growth at temperatures less than 15°C. Optimal temperatures appear to be above 20°C.

Throughout its range, *U. gracilis* has been most frequently reported from moving water, either large rivers or canals, but records from standing water, especially large lakes, are known. *Urnatella gracilis* is rather tolerant of a moderate range of alkalinity and turbidity. Little preference exists for substrate beyond a firm surface. Specimens have been collected on rocks, organic debris, vegetation, concrete, and even trash. A common substrate for this species is the shells of live mollusks (Fig. 11.2). Freshwater mussels and pleurocerid snails can support dense colonies under favorable conditions.

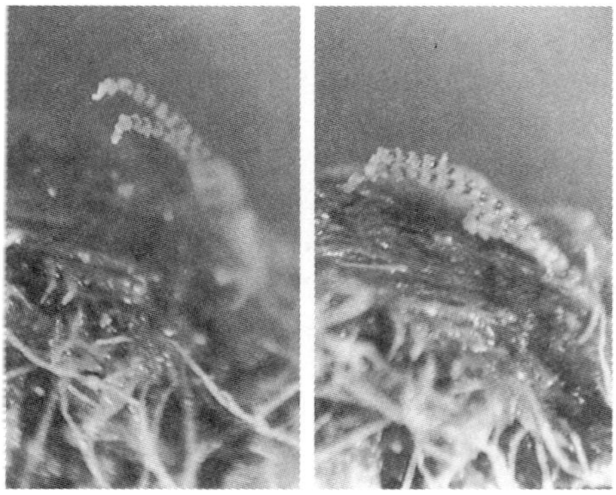

Fig. 11.2. Colonies of *Urnatella gracilis* attached to the shell of the freshwater mussel, *Popenaias popei*. Calyx is missing.

Distribution. *Urnatella gracilis* probably occurs throughout the United States. Populations have been found in several southwestern states—California, Nevada, Texas, and Oklahoma. East of the Mississippi, the species has been collected in many states from Ohio east to New England and south to the Gulf of Mexico.

Culturing and preserving. Emschermann (1987) developed an effective method of culturing *U. gracilis* that involves maintaining colonies in cages in a constant circulation system. Water temperature above 20°C promotes the most prolific growth, temperatures below this value usually result in stagnant growth or degeneration. Quiescent stalks can maintain viability up to 3 years when stored in dishes at 0 to 4°C.

Specimens can be preserved by placing them directly in 10% formalin, but extreme contraction will occur. Narcotization with menthol crystals followed by fixation in 10% formalin will result in nicely relaxed samples. Fixed colonies can be stored in alcohol whole or stained in methylene blue, hematoxylin, or other stains, cleared, and mounted on slides.

Taxonomy. Emschermann (1987) has determined that only one species is known throughout the world, although populations from Asia require further study. *Urnatella gracilis* has been placed in marine families, but is currently believed to represent a monotypic family, the Urnatellidae. Beyond morphological and reproductive traits, *U. gracilis* appears to have a distinctly different karyotype than at least a few marine species.

REFERENCES

Backus, B. T. 1986. Karyotype of the fresh-water entoproct *Urnatella gracilis*. *Trans. Am. Microsc. Soc.* 105:73–75.

Curry, M. G., B. Everitt, and M. F. Vidrine. 1981. Haptobenthos on shells of living freshwater clams in Louisiana. *Wasmann J. Biol.* 39: 56–62.

Emschermann, P. 1987. Creeping propagation stolons: an effective propagation system of the freshwater entoproct *Urnatella gracilis* Leidy (Barentsiidae). *Arch. Hydrobiol.* 108:439–448.

Eng, L. L. 1977. The freshwater entoproct, *Urnatella gracilis* Leidy, in the Delta–Mendota canal, California. *Wasmann J. Biol.* 35:196–202.

Hull, H. C., et al. 1980. Occurrence of *Urnatella gracilis* Leidy in the Tampa bypass canal, Florida. *Florida Scientist* 43: 12–14.

King, D. K., et al. 1988. Morphology and ecology of *Urnatella gracilis* Leidy, (Entoprocta), a freshwater macroinvertebrate from artificial riffles of the Tombigbee River, Mississippi. *J. Freshwater Ecol.* 4:351–360.

Smith, D. G. 1980. *Goniobasis virginica* (Gastropoda: Pleuroceridae) in the Connecticut River. *Nautilus* 94:50–54.

Weise, J. G. 1961. The ecology of *Urnatella gracilis* Leidy: Phylum Entoprocta. *Limnol. and Oceanogr.* 6:228–230.

12

ECTOPROCTA

(Moss animals)

Unpolluted and unsilted waters, especially ponds and the shallows of lakes and slow streams, characteristically contain colonies of Ectoprocta (Plate 2.3). To the casual observer such colonies are often mistaken for a mat of moss, and certainly some of the common species do have a superficial plant-like appearance. A large, encrusting, sessile colony may consist of thousands of individuals covering an area in excess of a square meter, and a massive gelatinous species may attain the size of a grapefruit or a small watermelon. Usually ectoprocts occur on the undersides of logs and stones, or on twigs and other objects where the light is dim.

Ectoprocta are more generally distributed and obvious in saltwater habitats than in freshwaters, and about 4000 marine species have been described the world over, as contrasted with about 50 freshwater species. Only about 25 species have been reported from North American freshwaters.

General characteristics. The unit of organization is a microscopic, more or less cylindrical zooid (Figs. 12.1, 12.2). Superficially it resembles certain cnidarian polyps, and it wasn't until 1830 that the Ectoprocta were distinguished from the Cnidaria as a separate phylum. The body wall, or cystid, of the zooid consists of a thin living layer and a secreted dead outer layer. The Y-shaped digestive tract is suspended in the extensive coelomic cavity. The mouth is situated on a large, distal, circular, oval, or horseshoe-shaped lophophore bearing numerous ciliated tentacles, but the anus is located posterior to or below the lophophore. Sometimes the lophophore and its tentacles are collectively called the tentacular crown. The lophophore, its tentacles, the base of the lophophore, and all of the viscera constitute a polypide (collectively, the protruded living parts of an individual zooid). Typically the zooids or individuals of a colony are all connected in a branched twig-like manner. The coelom is more or less continuous from one polypide to another, depending on the relative development of septa.

Only the tentacular crown and distal portion of the body are exposed to the water, the rest of the body and the whole colony being invested by a nonliving outer protective layer, which is secreted by the body wall. This protective layer may be thin to massive, corneous, cuticular, gelatinous, delicate, or tough, depending on the particular species and variety. According to the mode of growth and type of outer layer, there are three general but intergrading growth forms: (1) branching and threadlike, (2) matlike or crustlike, and (3) gelatinous and massive. In addition, *Plumatella casmiana* and, to a greater extent, *P. nitens* and *P. fungosa* occasionally form "honeycomb" colonies 2 to 20 mm or more thick (Fig. 12.3). In the older portions of colonies the outer layer is commonly

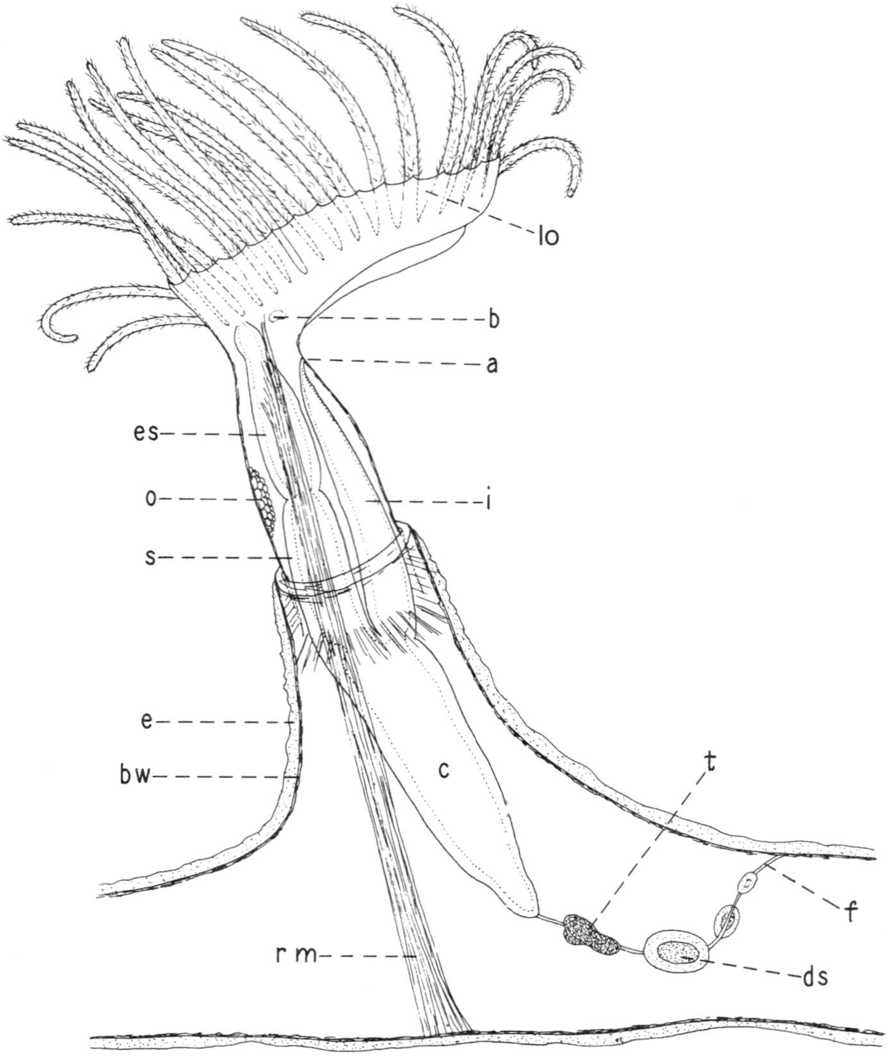

Fig. 12.1. Structure of a *Plumatella* zooid, diagrammatic. Tentacles and retractor muscle shown for only one side. a, anus; b, brain; bw, body wall; c, caecum; lo, lophophore; ds, developing statoblasts (3); e, cuticula; es, esophagus; f, funiculus; i, intestine; o, ovary; rm, retractor muscle; s, stomach; t, testis.

opaque, brownish, and more or less covered with detritus, algae, and microzoa. But the living portions of the zooid are delicate, translucent, and only slightly pigmented. Internal structures can usually be observed only in young individuals.

When undisturbed, the ectoproct tentacular crown is extended through the distal aperture of the outer layer. It is large and beautifully expanded, and a colony has the appearance of a patch of tiny and delicate flowers, but at the slightest disturbance the tentacular crowns are retracted within the cystids in a flash. After a few minutes they are again slowly extruded and expanded.

Growth and proliferation of a colony is a matter of simple asexual budding. Sexual reproduction, when evident, is restricted to a few weeks during the year. Each individual is hermaphroditic, and early development

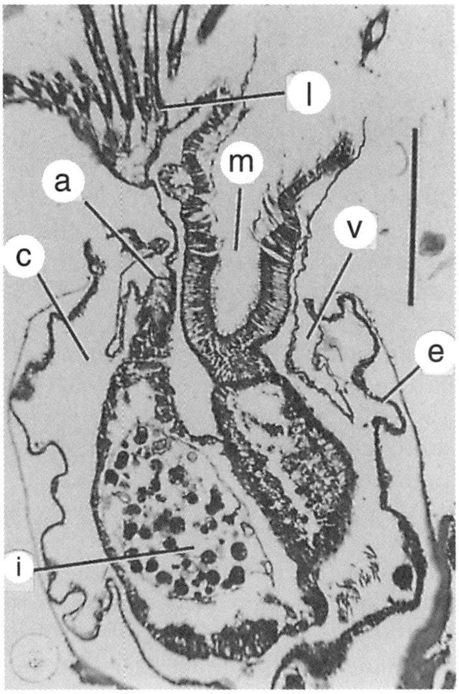

Fig. 12.2. Longitudinal section through *Plumatella recluse*. a, anus; c, coelom; e, epidermis; i, intestine; l, lophophore; m, mouth; v, vestible. Scale, 0.5 mm. (From Smith 1992.)

takes place within the coelom. On being released to the outside, the ciliated embryo soon becomes fixed to the substrate and gives rise to a new colony.

Statoblasts are asexual internal buds whose functional cells are enclosed by two tightly fitting convex sclerotized and sculptured valves (Figs. 12.6–12.9). On being released from the colony, statoblasts are distributed at random by currents and winds; they may float or remain on the bottom. A statoblast is highly resistant to drying, cold, and other adverse environmental conditions, but under appropriate conditions it germinates to produce a new colony.

The Phylum Ectoprocta is divided into three classes with living examples: the Phylactolaemata, the Stenolaemata, and the Gymnolaemata. All species in the first of these classes are restricted to freshwaters, but the much larger Class Gymnolaemata is exclusively marine except for about 10 fresh-

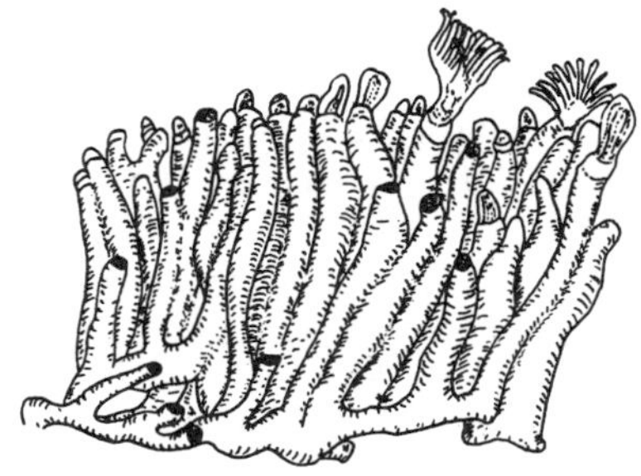

Fig. 12.3. Bit of a massive colony of *Plumatella nitens*, ×5.

water species worldwide. Four of the five orders in this class are marine, and in the fifth order, the Ctenostomata, only three American species occur in strictly freshwaters. The Stenolaemata are all marine. Unlike many marine species, freshwater ectoprocts exhibit no polymorphism.

Locomotion. Although ectoproct colonies are generally sessile, a few freshwater species are capable of sluggish creeping, pseudopodlike movements (Fig. 12.4). *Cristatella mucedo* is the most active form, and a colony may progress as much as 1 to 10 cm/day. The small colonies of *Lophopus crystallinus* and *Lophopodella carteri*, as well as young colonies of *Pectinatella magnifica*, also move, but usually less than 2 cm/day. The mechanism of colony locomotion is obscure, but ciliary action on the tentacles, the musculature of the body wall, and perhaps polypide retractor muscle action are probably of some importance.

Feeding, digestive system. When the lophophore is expanded under normal conditions, the tentacles are almost motionless,

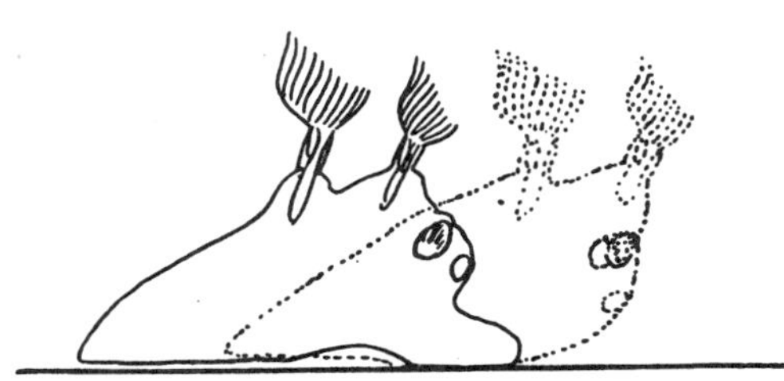

Fig. 12.4. Pseudopodal movement of *Lophopus crystallinus*. (Modified from Marcus 1934.)

but their abundant cilia beat constantly and create miniature currents along the lophophore that bring algae, protoctist, micrometazoa, and detritus to the region of the mouth. Bacteria can also be an important source of food and likely supplements other resources. If large and inedible particles reach the mouth, the individual tentacles may bend over in such a way as to brush the particles away and alter the ciliary currents. In all Phylactolaemata the base of the lophophore is surrounded by a membranous cup, or calyx, which is festooned or scalloped; presumably this structure increases the efficiency of feeding. The distal end of the animal may extend upward from the substrate or downward from the underside of an object, but feeding goes on effectively regardless of position. Phylactolaemata have a flaplike epistome over the mouth. It may cover the mouth or aid in rejecting or warding off undesirable particles. In most freshwater ectoprocts, the lophophore is produced into a horseshoe shape, a manifestation that increases the size of the structure and presumably increases the effective feeding area.

The anterior portion of the esophagus is slightly modified to form a somewhat muscular pharynx. Food passes through the mouth (Fig. 12.2) and pharynx and into the esophagus, and when a sufficient quantity has accumulated in the esophagus, the esophageal valve relaxes and the food passes into the stomach, where most of the digestion occurs. Below the stomach and corresponding to the basal leg of the Y is a long, baglike absorptive organ, the cecum. Connecting the lower end of the cecum and the body wall is a long strand of tissue, the funiculus. The tubular intestine passes from the upper end of the cecum to the anus; its terminal portion is slightly modified to form a rectum. The anus is located just below the lophophore. The feces are released as pellets, sometimes surrounded by a peritrophic membrane. The Phylactolaemata alimentary canal is ciliated only around the mouth and in the pharynx, but in the Gymnolaemata cilia extend as far as the upper portion of the stomach.

Much of the polypide coloration is determined by the specific contents of the alimentary canal. By changing the diet experimentally it is possible to get yellow, green, and brown coloration all in the same day.

Circulation. There is no circulatory mechanism in ectoprocts, but materials are distributed quickly and efficiently via the coelomic fluid owing to the movements of the alimentary canal and retractions and extrusions of the tentacular crown.

Respiration. No special respiratory devices are present. The greater portion of the carbon dioxide and oxygen exchange undoubtedly occurs generally through the lophophore tissues.

Excretion. No excretory system occurs in the ectoprocts, although some investigators claim otherwise. Coelomic amoebocytes containing engulfed excretory granules supposedly make their way through the body wall to the outside.

In *Cristatella*, *Lophopus*, *Plumatella*, and possibly some other genera the epidermal cells of the lophophore apparently accumulate excretory substances and liberate them to the outside. In addition, quantities of metabolic wastes are undoubtedly retained in many of the body tissues until the animal dies and disintegrates. A true intertentacular organ is unknown in freshwater ectoprocts.

A few Phylactolaemata, including *Hyalinella*, *Lophopus*, *Cristatella*, and *Lophopodella*, sometimes undergo partial disintegration similar to true "brown-body" formation in marine forms, especially under unfavorable conditions. It is thought that this phenomenon is at least partially excretory in nature. The polypides (especially older individuals) first retract permanently and then begin to undergo degeneration. The lophophore, tentacles, digestive tract, and nervous system degenerate (Fig. 12.14). Such a body remains unchanged, or it may further disintegrate into small fragments that circulate about in the coelom. Sometimes, in *Plumatella casmiana*, a new polypide is then regenerated from an internal bud on the body wall; it comes to occupy the space vacated by the old polypide. Sometimes the same zooid may undergo "brown-body" formation. In spite of certain textbook statements to the contrary, a "brown-body" has never been seen to regenerate a new cystid.

Muscle system. The most prominent muscles are the two retractors, which are somewhat stringy and not at all compact. They originate at the base of the cystid and are inserted on the base of the lophophore and uppermost portion of the digestive tract, one on each side (Fig. 12.1). Their contraction pulls the whole distal end of the polypide back into the coelom and cystid. Some short indistinct fibers connect the upper part of the polypide with the nearby body wall, and a few associated fibers aid in rotating the tentacular crown and depressing the lobes.

The thin body wall contains a few loosely arranged circular and longitudinal fibers, which are effective in everting the polypide. Other indistinct muscles elevate the epistome, move the alimentary canal slightly, and act as dilators and sphincters.

Nervous system. A single large ganglion, or "brain," is situated at the distal end between the mouth and anus. From it a series of flimsy fibers with a few associated nerve cell bodies radiate out to all tissues and organs. There are no proven nervous connections between adjacent zooids. Special sense organs are absent, but ectoprocts are remarkably sensitive to light and other stimuli, certain epidermal cells being specialized as receptors.

Reproduction, development. Growth of a colony is an asexual form of proliferation, often dichotomous, whereby a part of the body wall grows outward and eventually forms a new zooid. In this way, beginning with a single zooid in the spring, the resulting colony may consist of thousands of living and dead individuals by autumn. Developing *Pectinatella*, *Cristatella*, and *Lophopodella* colonies often undergo fission, thus giving rise to two or more new colonies (Fig. 12.14).

The period of sexual reproduction in a colony probably lasts about 1 to 4 weeks, during the warmer months in northern latitudes, and occurs simultaneously with statoblast production. Individual zooids are hermaphroditic in all species. The single ovary develops from the peritoneum of the body wall near the distal end of the cystid, and the testis develops from the peritoneum of the funiculus. Unlike the situation in the great majority of hermaphroditic animals, there seems to be no current evidence for cross fertilization in freshwater Ectoprocta. Mature eggs and sperm are produced either simultaneously or in rapid succession. Ordinarily only a single ovum matures at a time and is fertilized by sperm in the coelomic fluid just as it leaves the ovary. The zygote then becomes implanted in a special pouchlike ooecium, which develops from the body wall in the immediate vicinity of the ovary. Here the embryo develops into a pear-shaped, oval, or spherical ciliated larva 1 to 2 mm long containing one to five immature polypides (Fig. 12.5). A *Cristatella* larva, however, may have up to 25 polypides or polypide buds. This larva breaks out of the parent through a special large pore at the base of the lophophore, or it escapes when

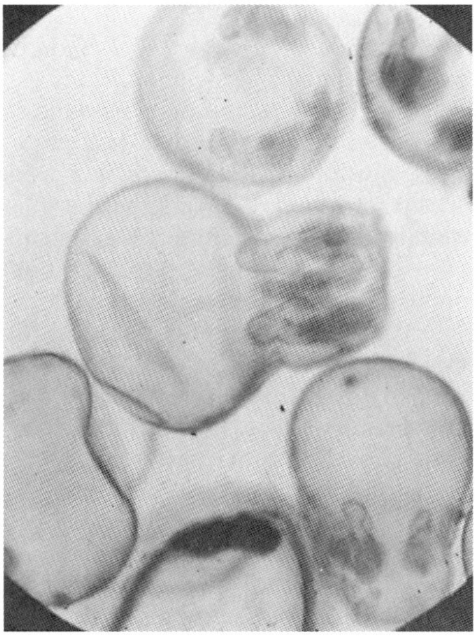

Fig. 12.5. Ciliated free-swimming larvae of *Plumatella*, ×27. Center embryo contains three developing zooids.

the parent degenerates (Fig. 12.7). It is free-swimming for only a few minutes or for as long as 20 to 24 hours and then settles on the substrate, evaginates, and eventually gives rise to a mature colony. Sexual reproduction occurs within only a small portion of the colonies within a population but is considered, in *Cristatella*, at least, to be an important aspect of reproduction.

Little specific information is available concerning the natural longevity of individual zooids, but presumably the great majority live for 3 to 8 weeks in temperate climates. Large, old colonies contain many dead zooids in the parts of the colony that were first formed. Although colonies customarily die out in the late summer or autumn, there are records of *Paludicella*, *Fredericella*, and *Lophopus* colonies being active throughout the winter, especially if it is mild and the water temperature does not drop below 2°C.

Statoblasts. The production of highly resistant statoblasts is a unique feature of freshwater Phylactolaemata. They function mainly in tiding the species over unfavorable environmental conditions and in geographic dissemination.

Statoblasts develop by asexual budding from the funiculus, and in some zooids several stages of development may be found simultaneously. A mature statoblast is roughly biconvex, but the central portion is considerably thickened into a capsule, owing to its contained mass of undifferentiated germinative cells. The peripheral portion (annulus, or float) usually consists of air cells and is relatively thin; sometimes it bears spiny, barbed, or hooked process (Fig. 12.6). Except for the germinative cells, a statoblast therefore consists of two sclerotized, dead valves that are tightly attached to each other in the region of the annulus. Early stages in statoblast development are almost smooth; older ones are strongly ridged or mammillate and tan, brown, or black. Usually each zooid produces 2 to 8 statoblasts, although more than 20 per zooid have been reported.

The literature distinguishes many types of statoblasts. Floatoblasts are common to all phylactolaemates except *Fredericella* and

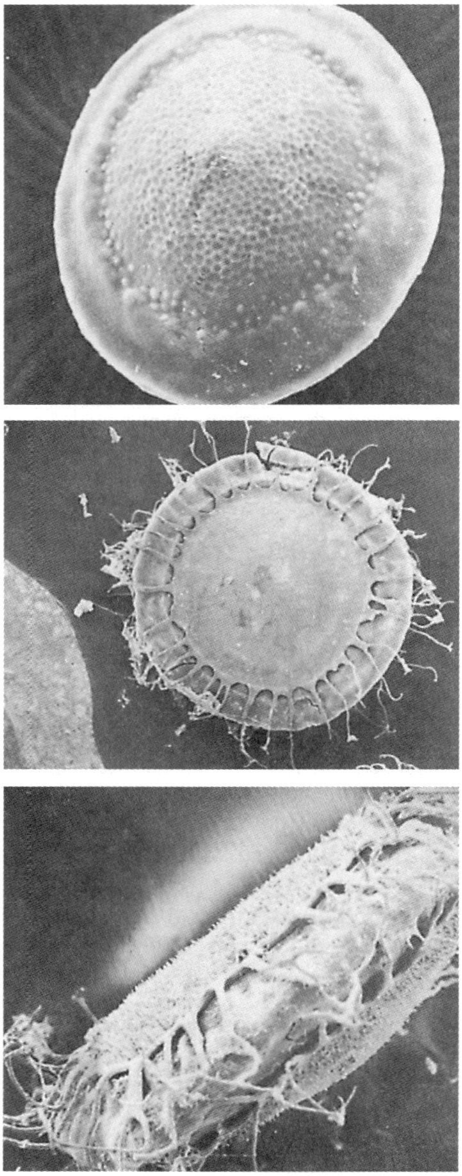

Fig. 12.6. Ectoproct statoblasts. Top, *Plumatella orbisperma*, ×160; middle, *Cristatella mucedo* spinoblast, ×40; bottom, edge view of *C. mucedo* spinoblast, ×75, showing suture zone. (SEM photos courtesy of K. S. Rao.)

exhibit broad biological and morphological diversity (Figs. 12.16B, C; 12.17B, C, E). They are freed after the cystid disintegrates or are released from the living zooid by means of a hyaline collarlike protrusion (Fig. 12.10).

Floatoblasts of most species have no peripheral processes; they are annulated when mature and are 0.2 to 1.0 mm long. These floatoblasts generally float, but in certain species they sink.

There are several specialized kinds of floatoblasts besides the common form described above. Spinoblasts are free but are provided with spined, barbed, or hooked processes. Length ranges from 0.7 to 1.3 mm. *Cristatella*, *Lophopodella*, *Lophopus*, and *Pectinatella* produce spinoblasts (Figs. 12.6, 12.8, 12.11). Spinoblasts do not float unless first dried, suggesting an adaptation to drought conditions. A leptoblast is a fragile floatoblast that germinates without a diapause. A pyknoblast is thick-walled (Fig. 12.7, top). *Plumatella casmiana* produces leptoblasts, pyknoblasts, and an "immediate form that germinates only after a diapause." All other species have a single morphotype.

Sessoblasts are sessile, fixed, or cemented to the outer layer wall; an annulus may be present or absent, well developed or reduced. The length is 0.3 to 0.6 mm. *Plumatella* and *Stephanella* are currently the only genera believed to produce sessoblasts (Fig. 12.9). The question of sessoblasts being a feature of *Hyalinella* remains unresolved. The three species of *Fredericella* produce piptoblasts (Fig. 12.7, middle), which are neither cemented to the substrate nor capable of floating. They lack an annulus and usually sink to the bottom when the outer layer disintegrates.

Formerly, it was thought that statoblast production is limited to late summer and autumn just before the colonies die off, but it is now well established that statoblasts may also be regularly produced during the earlier portion of the growing season. In the tropics statoblasts usually appear with the onset of hot weather. Sexual reproduction and statoblast production may occur simultaneously in the same zooid; there is little evidence for alternation of the two methods.

On disintegration of the zooids, the floatoblasts are released to the vagaries of currents, waves, and wind. Some of them sink or remain on the bottom. Others float to the surface, and under appropriate conditions may be found in such abundance as to form windrows on the shore, especially in spring and autumn. Sessoblasts, however, remain attached to the substrate until the last bit of old outer layer has disintegrated. A few species actively release statoblasts (Fig. 12.10).

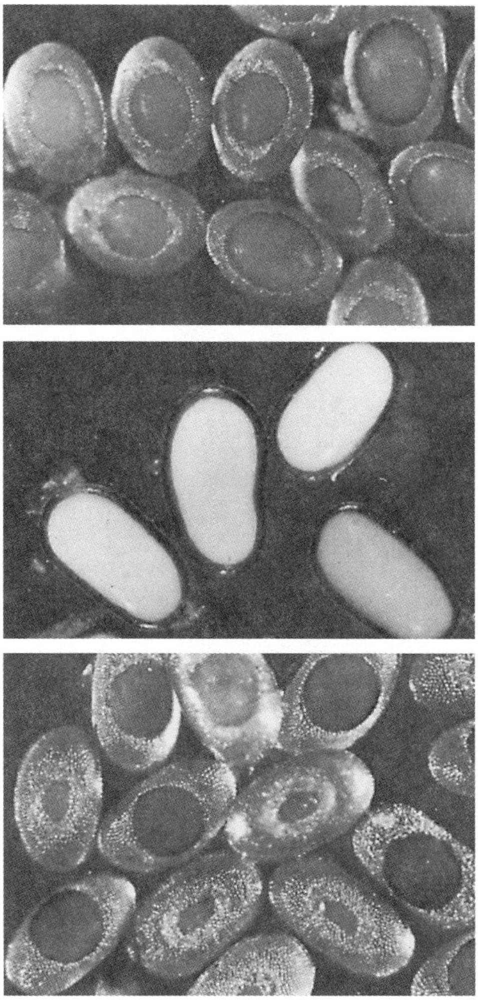

Fig. 12.7. Ectoproct statoblasts, all ×60. Top, pyknoblasts of *Plumatella casmiana;* top middle, piptoblasts of *Fredericella indica;* bottom, floatoblasts of *Plumatella emarginata.* (All but bottom courtesy of T. S. Wood.)

The length of the statoblast dormant period is highly variable, depending on the species, the individual, temperature, and other environmental conditions. In high latitudes there is usually only one generation

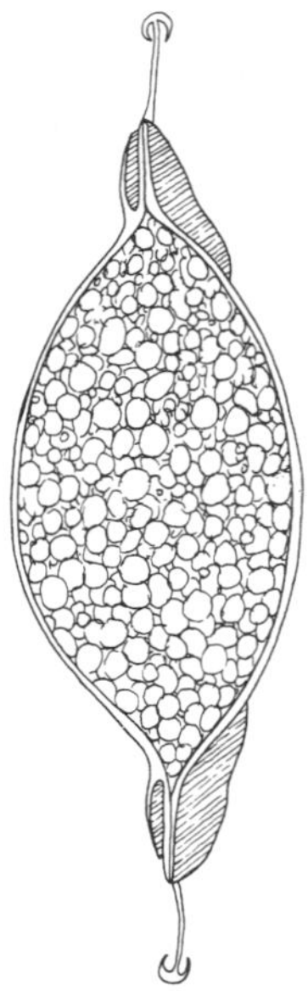

Fig. 12.8. Diagrammatic section of statoblast of *Pectinatella magnifica;* showing the two valves, hooked spines, central capsule filled with germinative cells, and float.

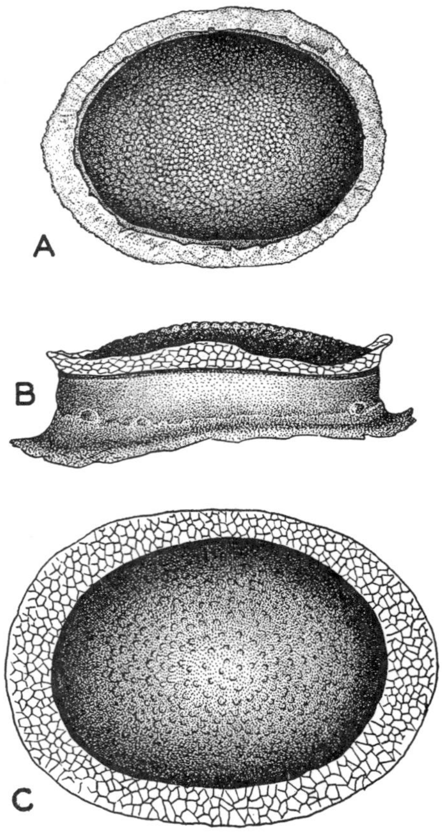

Fig. 12.9. Sessoblasts of *Plumatella*, ×120. A, unattached surface view of sessoblast of *P. emarginata;* B, side view of sessoblast of *P. "repens"* (note sclerotized lamella that resembles the float of floatoblasts); C, unattached surface view of same. (From Rogick 1940a.)

per year, and the majority of statoblasts winter over and germinate the following spring, but south of the 50th parallel it appears that the growing season is often sufficiently long for two or three main crops of statoblasts. Colonies mature in the spring, and their statoblasts hatch out and produce a second group of mature colonies by late summer or early autumn. These individuals, in turn, give rise to statoblasts that winter over. In spite of their ability to survive drying and freezing temperatures, statoblasts have a very high natural mortality rate.

Under natural conditions the period of statoblast dormancy or desiccation extends from the end of one growing season to the beginning of the next, which is usually 4 to 9 months, and most workers are agreed that a majority of statoblasts of all species are viable for the duration of such periods.

The natural period of dormancy and germination is usually 30 to 150 days, the chief governing factors being low temperatures and desiccation. Germination begins in the spring when the water reaches 10 to 19°C or more, depending on the species. The first evidence of germination is the gradual separation of the two statoblast valves along the equator, a process that may take 1 to 8 days. A mass of undifferentiated yolk cells slowly grows and protrudes from between the valves. These cells are whitish or grayish, very contractile, and they adhere to the substrate. The primary zooid meanwhile develops between the valves, hidden from view. It utilizes the abundant stored food of the undifferentiated yolk cells, and as it enlarges it protrudes, pushes the valves farther apart, and absorbs and incorporates more and more of the yolk mass, so that the latter soon be-

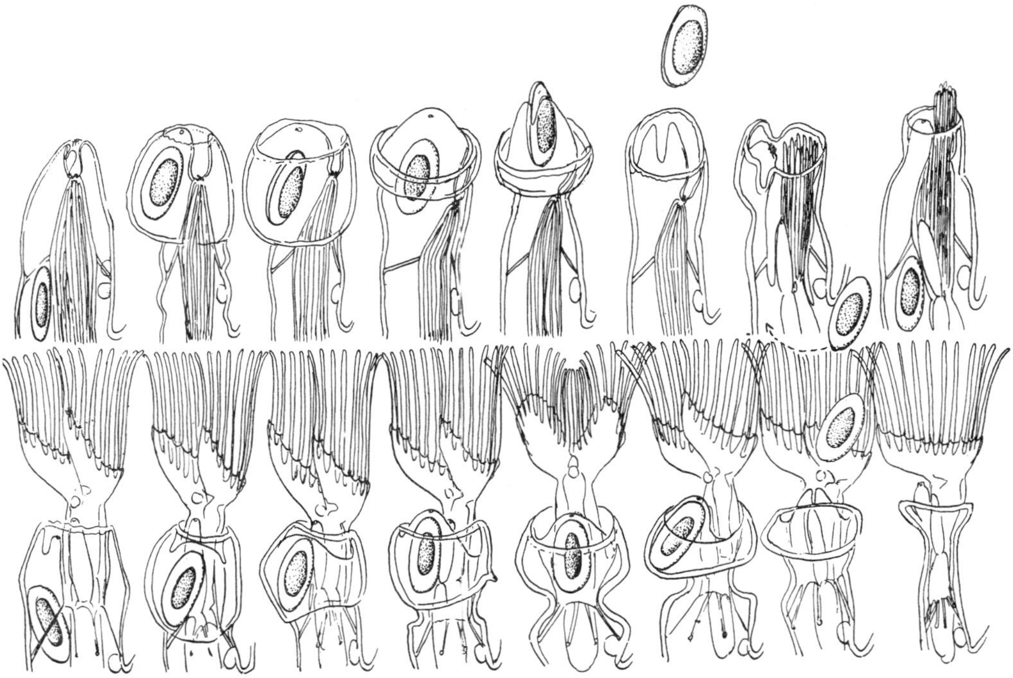

Fig. 12.10. Sequence of release of two statoblasts in *Plumatella.* (Revised from Marcus 1942.)

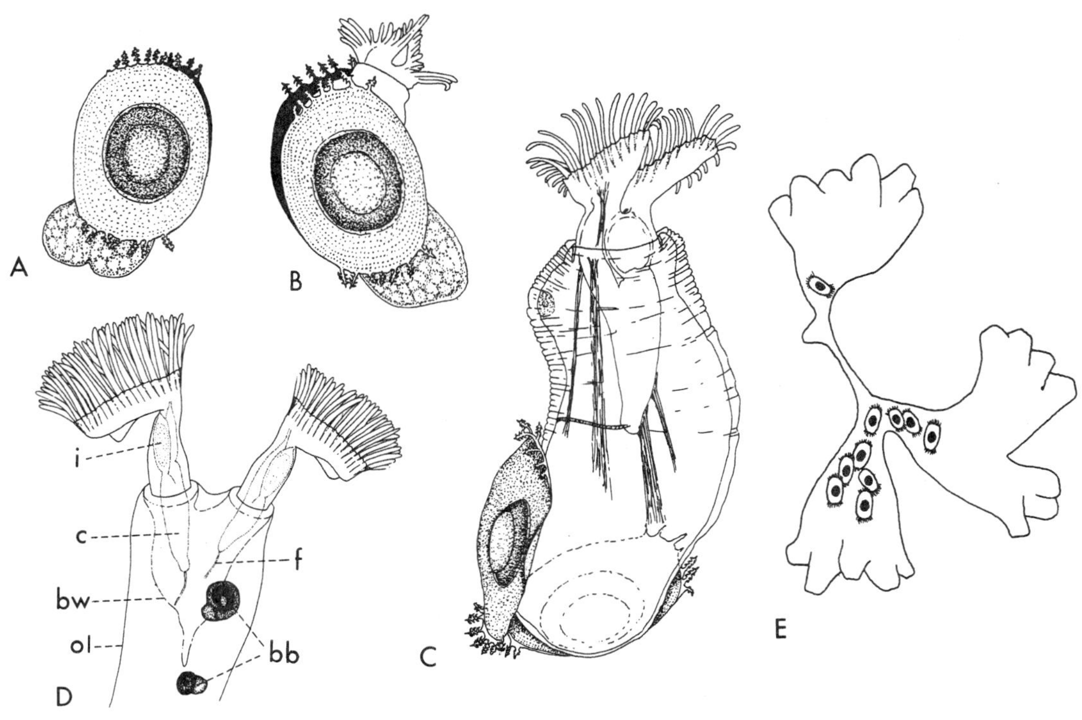

Fig. 12.11. Biology of *Lophopodella carteri,* ×33. A, statoblast valves separating and mass of undifferentiated yolky cells extruded; B, primary polypide emerging; C, primary zooid almost mature (note small bud on left side near distal end); D, semidiagrammatic view of a small portion of a *Lophopodella carteri* colony, ×25, showing two "brown bodies." bb, "brown bodies"; bw, body wall; c, cecum; ol, outer layer; f, funiculus; i, intestine; E, diagram of dividing colony, ×5. Polypides are shown contracted; 10 statoblasts present. (Modified or redrawn from Rogick 1935, 1938.)

comes an indistinguishable part of the zooid. As growth proceeds, additional zooids appear by budding, and eventually a colony results.

A large amount of experimental laboratory work has been conducted on the viability of statoblasts, but the results vary so much from one species to another and from one investigator to another that it is difficult to generalize. Most workers, however, seem to agree that natural freezing and normal winter temperatures have little effect on the percentage of later germination, but little is known about the lethal effects of unusually low temperatures.

Viability of statoblasts dried and kept at room temperature differs widely from one species to another. *Hyalinella*, *Pectinatella*, and *Plumatella* statoblasts apparently cannot survive desiccation beyond 10 months. *Lophopodella* statoblasts have been observed to germinate after being dry for 4 years. In general, wet storage in the refrigerator produces a lower statoblast mortality than drying at room temperature for the same length of time. Statoblasts die within a few days when kept in a laboratory desiccator.

When statoblasts are placed in water and germinated in the laboratory the period of incubation is highly variable. Some individuals and species germinate in 4 to 10 days; others require 1 to 4 months for germination. Much information exists suggesting that germination is temperature dependent. The lower the temperature, the lower the rate of germination.

Hibernacula. The gymnolaemate Paludicellidae and Victorellidae do not form statoblasts, but rather hibernacula, which are similar to statoblasts in function. A hibernaculum is a specially modified external winter bud. Like statoblasts, they consist of masses of cells protected by a thick sclerotized coat (Fig. 12.12).

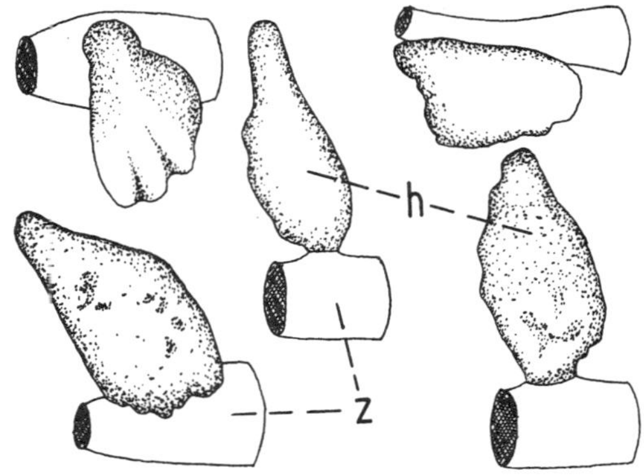

Fig. 12.12. Hibernacula of *Paludicella articulata*, ×50. h, hibernaculum; z, bits of old outer layer. (Modified from Rogick 1935.)

Ecology. Quiet ponds, backwaters, bays, and slow streams, especially where there are sunken logs, twigs, rocks, and aquatic vegetation, are ideal places for Ectoprocta. Sandy and pebbly bottoms are poor habitats, but certain species of *Plumatella*, *Fredericella*, and *Paludicella* are often found in turbulent waters. Although they are common in stagnant waters, ectoprocts are never found under markedly polluted conditions and only sparingly where the quantity of dissolved oxygen falls below 30% saturation. Silting also discourages establishment and growth, and strongly acid waters and bogs almost never contain ectoprocts. *Lophopodella carteri* is probably the species that is most tolerant of decay and stagnation.

Light is not necessary for development and growth, as shown by the occurrence of ectoprocts in water pipes, closed conduits, and even artesian water. On the other hand, they are seldom present in bright light, although *Cristatella mucedo* occurs regularly on the upper sides of objects in the shallows. All other species found in shallow water are attached to the undersides or shaded sides of rocks, logs, vegetation, and boards.

Most Ectoprocta are collected in waters less than 1 m deep, but they commonly occur at considerably greater depths, and if appropriate collecting equipment is used, almost all of our common species may be taken in lakes at depths up to 5 or 10 m. *Paludicella* has been taken as deep as 36 m and *Fredericella* up to 214 m. *F. indica* is sometimes found in marl lakes; it also occasionally grows up from sandy substrates in an arborescent form.

Very few freshwater ectoprocts are limited to any specific kind of habitat. Certain species of *Plumatella* seem to have restricted habitat preferences. *Plumatella reticulata* and *emarginata* appear to be found only in large and small streams, respectively. *Plumatella repens*, *P. nitens*, and *P. recluse* are almost always found in standing water. *Stephanella* and *Sineportella*, two recently discovered genera in North America, are known only from moving water. Up to eight species have been found in a single lake. *Victorella pavida* is a rare species of Gymnolaemata that is largely limited to brackish waters but will enter tidal freshwater.

Most American species attain their greatest abundance in the summer when the water temperature reaches 19 to 23°C and begin to die off in the autumn when the water cools to 10°C. *Pectinatella*, however, tolerates a much smaller range of temperature and usually dies when the water temperature goes below 16°C. *Fredericella* and *Lophopus*, on the other hand, are unusually tolerant and do well anywhere between 2 and 23°C. *Plumatella* is sometimes collected at water temperatures as high as 37°C.

Economic importance. Occasional summer plagues of large floating gelatinous colonies of *Pectinatella magnifica* clog the screens of water intakes and the grates of hydroelectric plants so effectively that it requires a full-time worker to keep them clear.

In communities using unfiltered and unchlorinated water, ectoprocts sometimes become established in the pipes, and dead and living masses clog the valves and meters. Occasionally, fishing trap nets accumulate large growths. The usual offenders are *Paludicella*, *Fredericella*, and *Plumatella*.

Ectoprocta are an element in the diet of many freshwater invertebrates, including planarians, snails, oligochaetes, Hydrachnidia, crustaceans, and especially Trichoptera and chironomids. Statoblasts and colonies are often eaten by small fish. The outer layer forms an excellent substrate for a variety of commensals and browsers, especially ciliates and rotifers.

A possible beneficial effect is the observed capacity of at least one species, *Lophopodella carteri*, to inhibit the settling of larvae of the invasive zebra and quagga mussels (Dreissenidae).

Geographical distribution, dispersal. *Paludicella articulata*, *Plumatella "repens"*, *Fredericella indica*, *Plumatella emarginata*, and *Hyalinella punctata* are the only known truly cosmopolitan species; they occur in suitable habitats all over the world. At the other extreme, some North American species have greatly restricted and often puzzling distributions. *Plumatella orbisperma*, *P. nitens*, *P. recluse*, and *Sineportella forbesi* are Nearctic; *Pottsiella erecta* is Nearctic and Neotropical; *Stephanella* is in eastern North America and Japan; and *Cristatella mucedo* and *Plumatella fungosa* are circumpolar. *Lophopus crystallinus* is so rare that its native range can only be guessed. *Pectinatella magnifica* is probably native only to North America but has been recently introduced into Japan and Germany. *Plumatella vaihiriae* is an exotic native to Australian regions while *Lophopodella carteri* is believed to be imported from Asia.

Unquestionably the paucity of records for some of the American species is a reflection of the fact that little systematic and intensive collecting has been done. Future collecting, especially in the South, and west of the Mississippi River, will probably show that some of our "rare" species are actually widely distributed.

Chiefly by virtue of their resistant statoblasts and hibernacula, ectoprocts are one of the few freshwater groups that are clearly capable of being transported overland from one body of water to another by animals. There are definite records of statoblast occurrence in mud on the feet and feathers of waterfowl, and in the fur of mammals. Some statoblasts are capable of germinating after passing through the alimentary canal of waterfowl, turtles, frogs, salamanders, and fish (see Bushnell 1973).

Collecting. Bits of colonies, sessoblasts, and hibernacula should be carefully removed from their substrate with forceps, spatula, and scalpel, but if the colonies are on small pieces of wood, twigs, or stones, they may be brought to the laboratory intact. A dredge or double rake is necessary for collecting below a depth of 1 m. Floating statoblasts may often be obtained in abundance at the right season by skimming the surface of the water with a net or by scooping them up from windrows.

Culturing. Ectoprocta are among the most difficult of all freshwater invertebrates to culture in the laboratory. If intact colonies are used, they may sometimes be kept alive in finger bowls for 5 to 15 days, provided the water is replaced frequently with fresh pond water, preferably from their natural habitat. Sometimes a small quantity of fresh yeast is beneficial to the culture.

Some investigators report partial success using water from aquatic plant aquariums along with a bit of the organic debris from the bottom of such aquariums. Such debris usually contains an abundance of suitable microorganisms for food. Both water and debris should be changed every 2 or 3 days.

Wood (1971) presents a simple and highly successful method of culturing in the laboratory. Statoblasts are first allowed to germinate in petri dishes. Once the ancestrulae (first individuals) are formed, they stick to the glass, and the petri dish is then placed in an aquarium where it is held upside down by Plexiglas strips. Fecal pellets and detritus fall away and leave the colony clean. The most reliable food source is the suspended organic matter normally present in a large, thriving aquarium. For a good review of culturing ectoprocts see Wood (1989).

Preparing, preserving. Ectoprocts are highly sensitive to foreign substances, and if ordinary methods of fixation are used the tentacular crown contracts into the cystid so strongly as to make the whole animal useless for study. Consequently, some method of narcotization must first be used so that the animals can be killed and fixed with the lophophore expanded.

The colonies and a part of their substrate should be placed in a finger bowl one-third to one-half full of water from their habitat, and the subsequent narcotization must be done very gradually and patiently. If chloretone, menthol, or chloral hydrate is used, drops of saturated solution should be added slowly over a period of 1 to 3 hours. When the animals no longer respond appreciably to jarring or touching, most of the fluid should be withdrawn with a large pipette. Then Bouin's fixative or 5% formalin should be poured over the animals, and if the narcotization has been successful, they will be killed in a beautifully expanded condition.

An especially effective method involves menthol. Place the colonies and a bit of their substrate in a finger bowl of water. Sprinkle a $\frac{1}{4}$ teaspoonful of menthol crystals on top of the water, cover the finger bowl with a glass plate, and let stand for several hours. If allowed to remain for too long, the animals will die and quickly decompose. When fully narcotized, withdraw most of the fluid from the finger bowl, add 5% formalin, and let stand for 1 day.

Ectoprocta may be mounted either unstained or stained with hematoxylin or borax carmine. Before mounting or staining, however, the outer layer should be gently brushed as free of debris as possible. If glycerin jelly mounts are to be made, the animals must first be run up into glycerin. Place them in a watch glass containing a solution of 5% glycerin in 50% alcohol and cover with a loose cover to keep out dust and permit evaporation of the alcohol and water. When nearly pure glycerin remains, they are ready for mounting. For permanent mounts the stained specimen must be dehydrated through an alcohol series, cleared in xylene, and mounted in balsam or some synthetic medium.

Taxonomy. After a hiatus of about 30 years, following the significant work of Rogick and Bushnell, a recent surge of systematic work on North American freshwater Ectoprocta has emerged. Investigators, notably Wood, Mukai, Backus, and Smith, have recognized several taxa new to either science or North America and refined the taxonomic system of the Phylactolaemata. Several new species of *Plumatella* have been introduced to the literature that demonstrate the diversity and broad ecological range of this genus. These authors have also noted the decline of some species while at the same time documented the spread of alien species into North America. Nevertheless, many areas in the United States remain unsurveyed and the potentially rich ectoproct faunae of many regions, such as the Pacific northwest and the southeastern states, have hardly been examined.

Most of our species are distinctive, especially if both the colonies and their statoblasts are available for identification. Indeed, the genera are easily distinguished with the naked eye or hand lens. It is most useful, if possible, to bring the specimens into the laboratory alive. Examination of the fully extended lophophore can resolve questions of tentacle number, lophophore size, and shape, and provide the examiner the rare treat of witnessing one of the most exquisite creatures inhabiting freshwater. The most common and widely distributed genus, *Plumatella,* is quite variable, and identifying species can be a challenge. On the other hand, many genera are monotypic as understood but future study may split some populations taxonomically. Previous editions and earlier works listed *Stollela* (with two species), but the genus is now generally considered a peculiar growth form of *Plumatella "repens"*.

KEY TO THE GENERA OF FRESHWATER ECTOPROCTA OF THE UNITED STATES

1. Lophophore circular and without an epistome (Fig. 12.13); colony covered with a corneous or membranous cuticle; stoloniferous and threadlike; statoblasts absent; hibernacula produced (Fig. 12.12) . . . . **Class GYMNOLAEMATA 2**

Lophophore horseshoe-shaped, oval, or circular, and with an epistome (mouth flap); colonies not threadlike; statoblasts produced. . . . . **Class PHYLACTOLAEMATA, 5**

2. Lophophore with over 15 tentacles; colony conspicuously branched, cystids corneous and darkly pigmented . . . . **PALUDICELLIDAE, 3**

Lophophore with 8 tentacles; colony branched or not; cystids pliable and membranous, somewhat translucent . . . . **VICTORELLIDAE, 4**

3. Colony consisting of a series of club-shaped cystids arranged end to end and separated from each other by partitions; 16 to 18 tentacles; cystid openings on upper surface and four-sided (Fig. 12.13G); zooids about 2 mm long; colony recumbent or partially erect (Figs. 12.13B, C); standing and slowly flowing waters; reported from scattered localities chiefly east of the Mississippi River but probably cosmopolitan in the United States; collected as far north as Alaska . . . . ***Paludicella***

Colony consisting of a stolon from which single individuals arise; cystid openings terminal, and pentagonal (Fig. 12.13H); 19 to 21 tentacles; zooids about 1.5 mm long (Fig. 12.13D); on upper surface of stones in rapid water; reported from streams in CT, LA, NC, OH, PA, VA, TN, and TX . . . . ***Pottsiella***

4. Colony branching, adherent and repent (Fig. 12.13E); zooids consisting of narrow basal portion expanding to a broad sac containing the polypide (Fig. 12.13F); inland fresh water; 1 extremely rare species, IL . . . . ***Sineportella***

Colony rarely branching in freshwater, typically with single upright zooids arising from stolon-like strand (Fig. 12.13A), if branching then with zoids arising directly from parent zooid; 1 brackish-water species that enters freshwater in southeastern states . . . . ***Victorella***

5. Statoblasts all free, annulated, and with processes (Figs. 12.11; 12.14); colony soft, gelatinous, transparent, without septa, and not dendritic (Figs. 12.14B, E; 12.15H, J) **6**

Often with two types of statoblasts; one attached and without float cells (sessoblasts); the other unattached, funicular, and annulated (floatoblasts) (except *Fredericella*); statoblasts without processes; colony most often dendritic, with either tubular opaque or swollen gelatinous cystid **9**

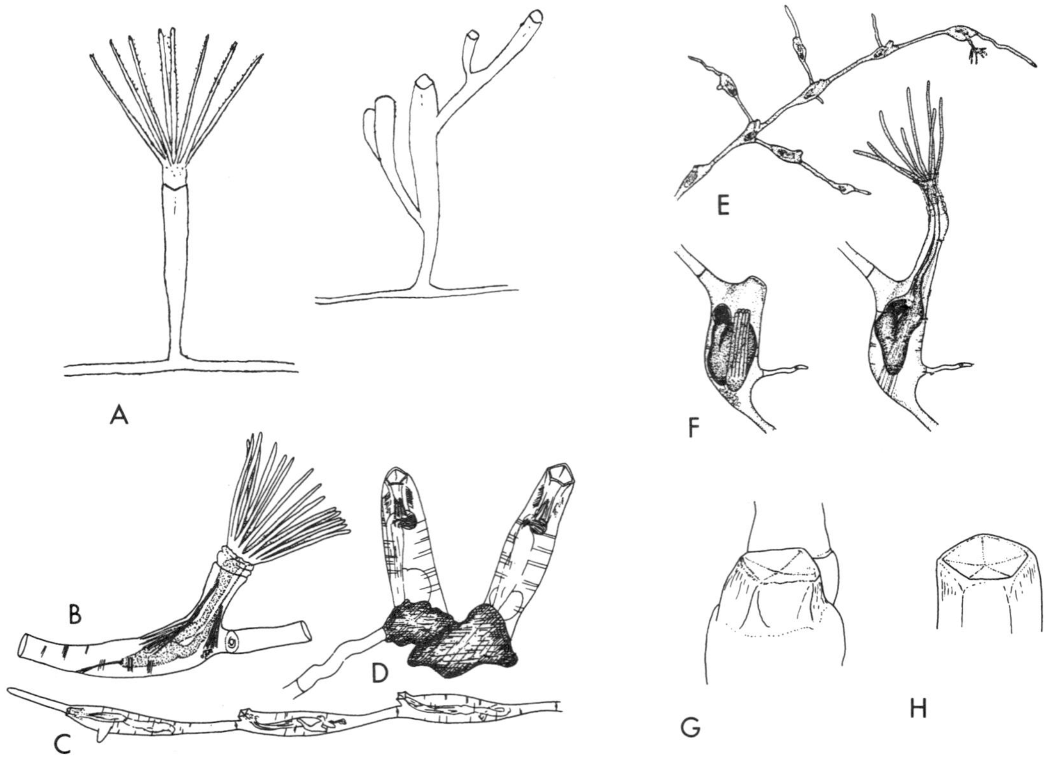

Fig. 12.13. Structure of freshwater Gymnolaemata. A, small portion of colonies of *Victorella pavida*, ×32; B, zooid of *Paludicella articulata*, ×33; C, portion of a young colony of *P. articulata* showing three retracted individuals and two buds; D, young colony of *Pottsiella erecta*, ×33, showing two hibernacula; E, colony of *Sineportella forbesi*, ×40; F, zooids (right extended) of *S. forbesi*, ×50; G, apex of withdrawn zooid of *Paludicella articulata*, ×50; H, apex of withdrawn zooid of *Pottsiella erecta*, ×50. (A modified after Porrier and Mulino 1977; B and C modified from Rogick 1940a; D modified from Kraepelin 1887; E, F from Wood and Marsh 1996b.)

6. With a transient secretion elaborated by flat base of colony; colony elongated and slowly motile (Fig. 12.14B); statoblasts with two circlets of hooked spines; dorsal circlet with 10 to 34 hooks and ventral circlet with 20 to 50 hooks (Figs. 12.14C, D); zooids on upper surface and arranged marginally in symmetrical fashion; all polypides retract into a common cavity; usually 80 to 90 tentacles; long dimension of colony sometimes as much as 10 to 30 cm, but more typically 2 to 6 cm; in standing waters or slow currents, on upper side of submerged objects and vegetation; widely distributed but not common . . . .**CRISTATELLIDAE, *Cristatella***

Lobular gelatinous cystid present throughout life of colony and sometimes very well developed; colony not usually motile; statoblasts never with two circlets of hooked spines 7

7. Colonies in the form of rosettes on gelatinous base or surrounding a gelatinous core; very young colonies motile; colonies commonly investing twigs and other submerged objects, especially in shaded places in quiet waters (Fig. 12.14E); 60 to 84 tentacles; epistome red in fresh animals; free statoblasts with single row of marginal anchorlike spines (Fig. 12.8); widely distributed in U.S., but uncommon west of the Mississippi Valley.
PECTINATELLIDAE, *Pectinatella*

Colonies in the form of small gelatinous masses with radiate branching; a few to 70 zooids per colony; epistome not red; free statoblasts pointed or with hooked spines at each end (Figs. 12.11; 12.15D, E) . . . .**LOPHOPODIDAE, 8**

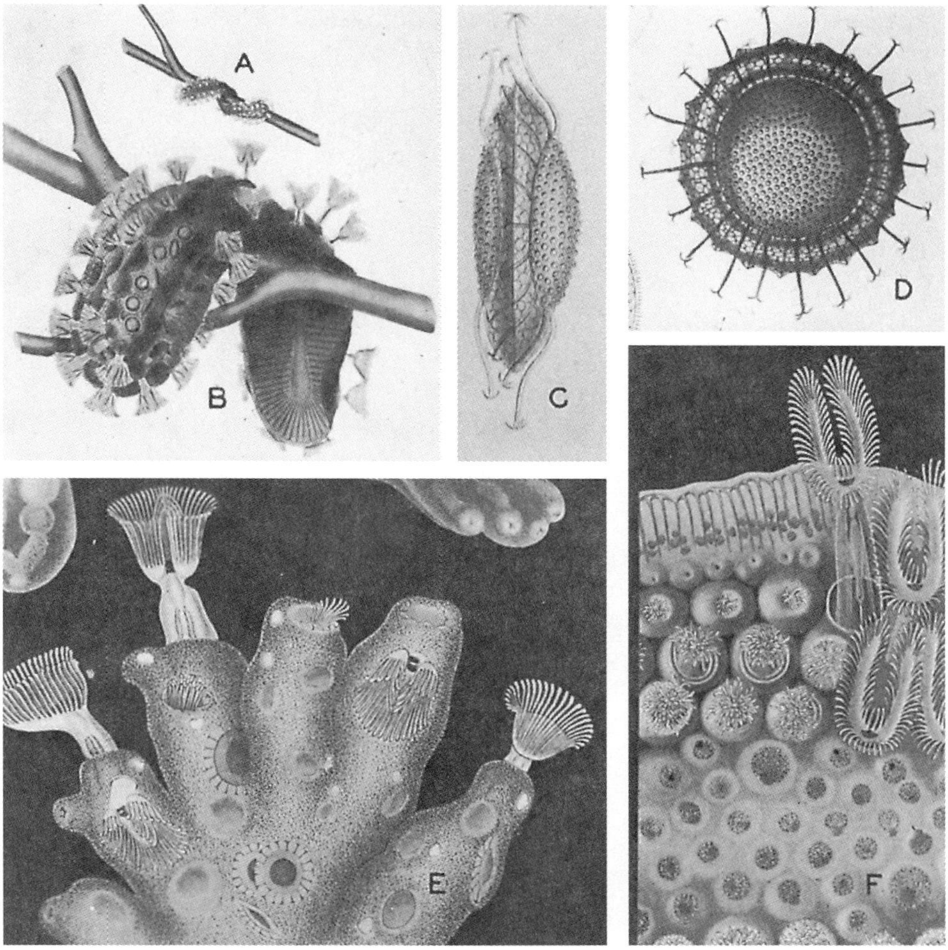

Fig. 12.14. Structure of *Cristatella* and *Pectinatella*. A, colony of *Cristatella mucedo*, ×5; B, colony of *C. mucedo*, ×40; C, edge view of *C. mucedo* statoblast, ×50; D, surface view of *C. mucedo* statoblast; E, rosette of *Pectinatella magnifica;* F, edge of colony of *Cristatella mucedo* showing three rows of marginal polypides. (A–D from Allman 1856; E and F from Kraepelin 1887.)

8. Free statoblasts broadly oval, with one or several hooked polar spines (Fig. 12.11); cystid translucent; colonies small, circular to oval, and mound-shaped when only a few polypides are present, but radiating lobate when many polypides are present; average colony 5 mm wide, 8 mm long, and with 20 to 45 polypides; 52 to 82 tentacles; infrequently reported from northeastern quarter of the United States ....***Lophopodella***

   Free statoblasts elongated, attenuated to a point at each pole (Figs. 12.15D, E); colony sac-shaped, erect, often lobed, and usually less than 1 cm across; cystid gelatinous (Figs. 12.15H, J); with about 60 tentacles; usually attached to vegetation in standing water; very rare in eastern states .... ***Lophopus***

9. Lophophore circular or elliptical; colony tubular, branched, and antlerlike (Fig. 12.15B); outer layer usually opaque and brown; all statoblasts (piptoblasts) free, without annulus (Fig. 12.15A, C); no floatoblasts; 3 species, 1 very common
FREDERICELLIDAE, *Fredericella*

   Lophophore horseshoe-shaped even if very small; floatoblasts present; sessoblasts also present in most species **10**

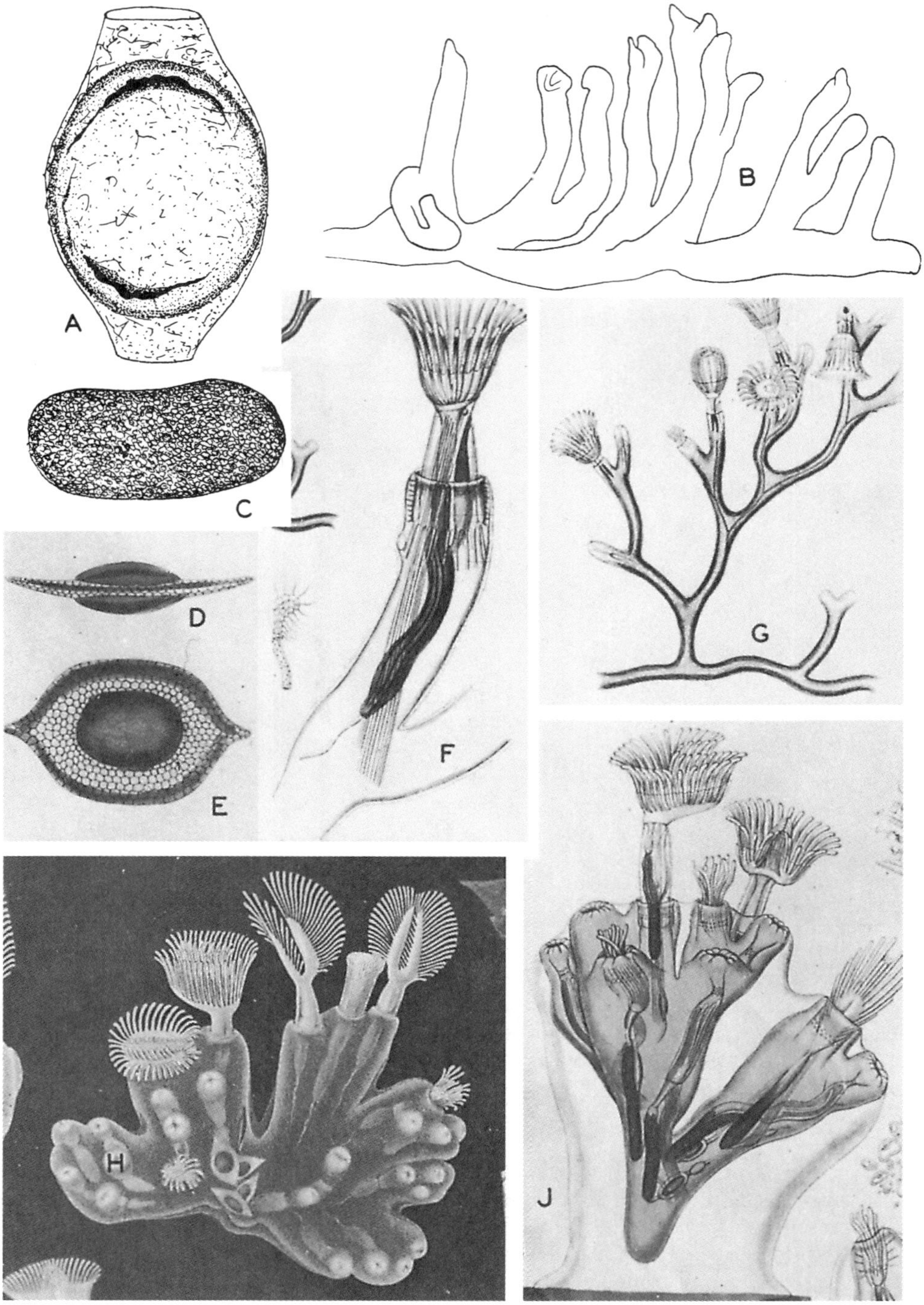

Fig. 12.15. Structure of Plumatellidae. A, cuticula enclosing piptoblast of *Fredericella australiensis*, ×100; B, branching in colony of *F. australiensis*, ×12 (polypides shown retracted); C, piptoblast of *F. indica*, ×85; D, edge view of statoblast of *Lophopus crystallinus*, ×40; E, surface view of same statoblast; F, zooid of *Fredericella indica*, ×15; G, small colony of *F. indica*, ×3; H and J, small colonies of *Lophopus crystallinus*, ×7. (A and B modified from Rogick 1945; C modified from Rogick 1937; D–G, and J from Allman 1856; H from Kraepelin 1887.)

10. Cystid very soft, highly transparent, and branching, with several lateral branches and very erect zooids (Fig. 12.16A); tentacle number less than 50; floatoblasts circular (Fig. 12.16B), sessoblast with one surface containing raised tubercular processes; 1 very rare species, MA . . . **STEPHANELLIDAE, *Stephanella***

Cyst ranging from soft and transparent to firm and opaque; branches repent or compact; floatoblast circular (rarely) or elliptical, sessoblast without tubercular processes on either surface **PLUMATELLIDAE, 11**

11. With soft, hyaline, gelatinous cystid; thickly branched, often completely covering the substrate (Fig. 12.16D); 1 common and widespread species . . . . ***Hyalinella***

With thick and somewhat gelatinous branches (Fig. 1.11b) or thin, firm, and sometimes horny branches (Fig. 12.17); outer layer usually long and tubular, pale and transparent to dark brown; in masses up to 5 cm high or closely adherent to substrate (Fig. 12.17); many species in a variety of habitats, widely distributed with several common species . . . . ***Plumatella***

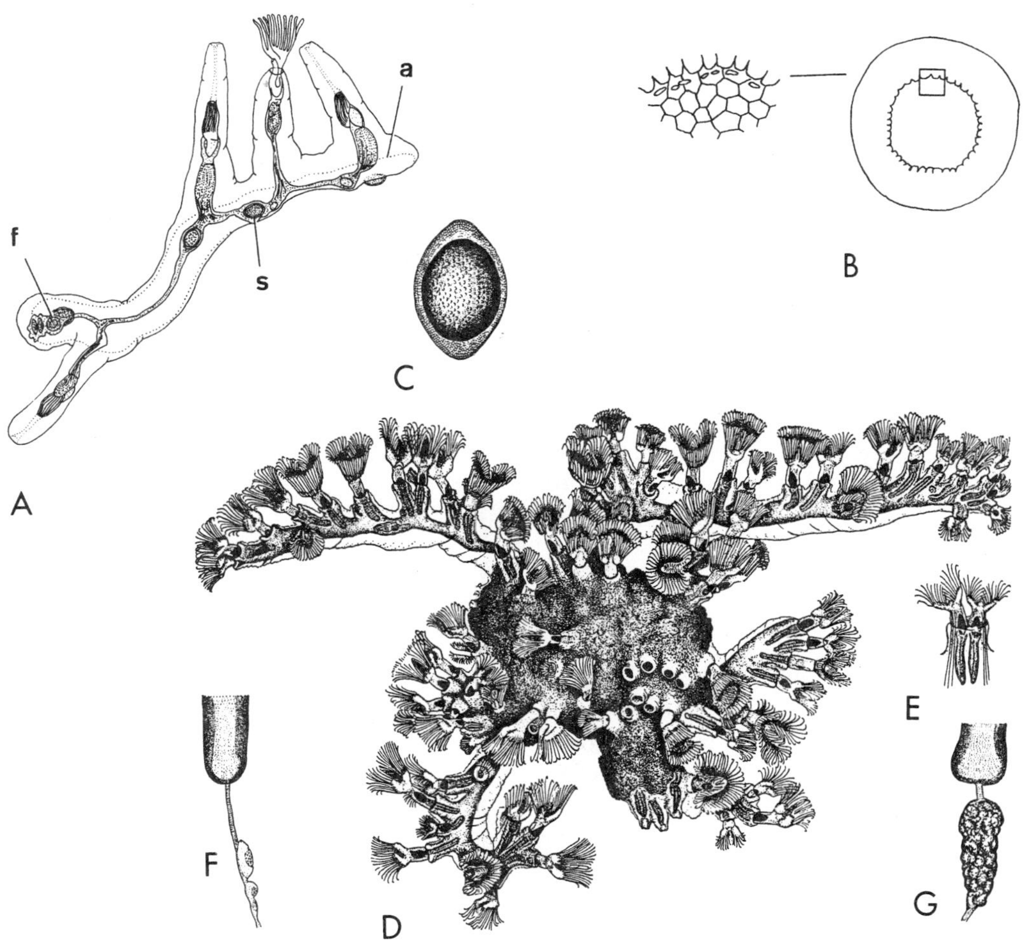

Fig. 12.16. Structure of Plumatellidae. A, Colony of *Stephanella hina;* B, floatoblast of *S. hina*, enlargement of box to left; C, floatoblast of *Plumatella vaihiriae*, ×60; D, entire colony of *Hyalinella punctata*, ×6; E, twin polypides of *H. punctata;* F, G, immature sperm mass on funiculus of *H. punctata*. a, attachment site; f, floatoblast; s, sessoblast. (A from Smith 1988; D–G from Rogick 1935.)

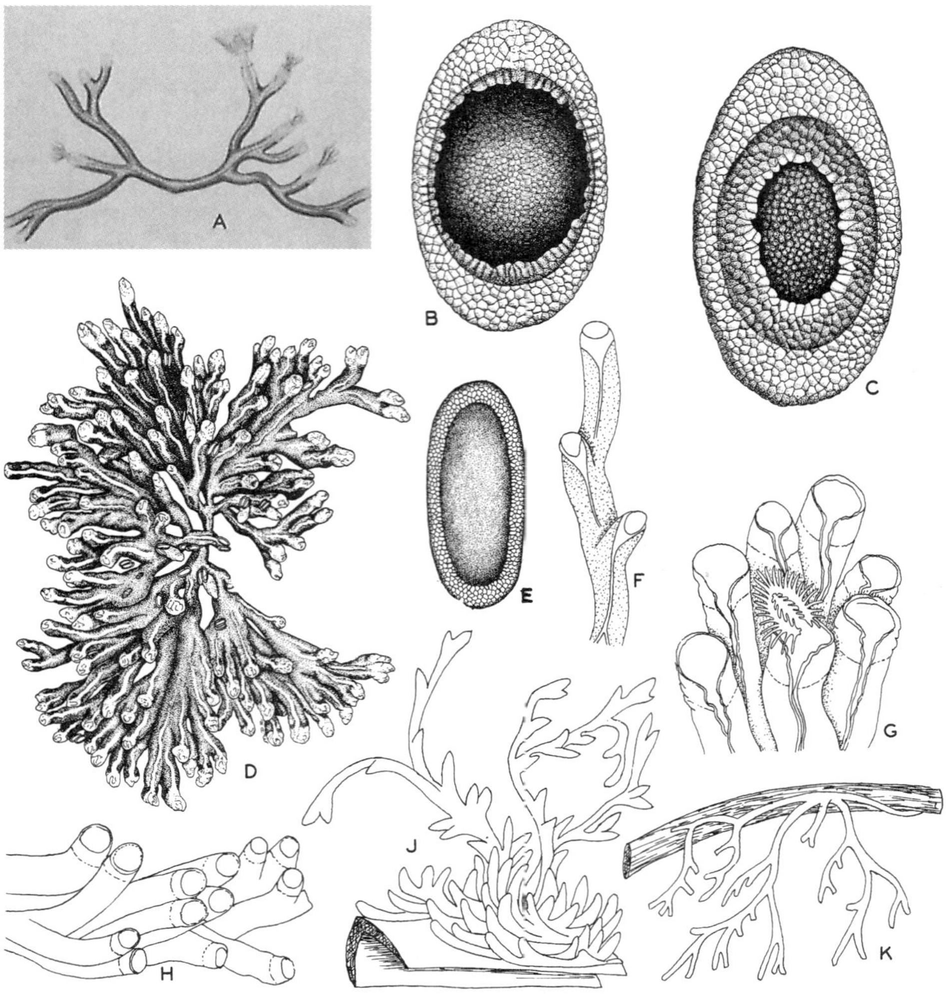

Fig. 12.17. Structure of *Plumatella*, A, small colony of *Plumatella "repens"*, ×4; B and C, opposite surfaces of floatoblasts of *P. "repens"*, ×100; D, colony of *P. casmiana*, ×5; E, floatoblast of *P. casmiana*, ×65; F, diagram of small portion of colony of *P. "repens"* (polypides all retracted); G, small portion of colony of *P. emarginata* (only one polypide partially expanded); H, small portion of colony of *P. "repens"* (all polypides retracted); J, diagram of small portion of colony of *P. "repens"* (all polypides retracted); K, diagram of small portion of colony of *P. fruticosa* (all polypides retracted). (A from Allman 1856; B–D from Rogick 1941; E from Rogick 1943; F modified from Newcomer 1950; G and H modified from Rogick 1935; J modified from Rogick 1940a.)

REFERENCES

Agrawal, V., and K. S. Rao. 1981. Studies on the body wall of some phylactolaematous Ectoprocta, 1: *Plumatella casmiana* Oka. *Bioresearch* 5:41–46.

Allman, J. 1856. *A Monograph of the Fresh-water Polyzoa*. 119 pp. Ray Society, London.

Brown, C. J. D. 1933. A limnological study of certain fresh-water Polyzoa with special reference to their statoblasts. *Trans. Am. Microsc. Soc.* 52:271–316.

Bushnell, J.H. 1966. Environmental relations of Michigan Ectoprocta, and dynamics of natural populations of *Plumatella repens. Ecol. Monogr.* 36:95–123.

——. 1968. Aspects of architecture, ecology, and zoogeography of freshwater Ectoprocta. *Atti Soc. Ital. Sci. Nat. Mus. Civ. St Nat. Milano* 108:129–151.

——. 1973. The freshwater Ectoprocta: a zoogeographical discussion. In: *Living and Fossil Bryozoa: Recent Advances in Research*, pp. 503–521, edited by E. P. Larwood. Academic, London.

Bushnell, J. H., and K. S. Rao. 1974. Dormant or quiescent stages and structures among the Ectoprocta: physical and chemical factors affecting viability and germination of statoblasts. *Trans Am. Microsc.* Soc. 93:524–543.

——. 1979. Freshwater Bryozoa: micro-architecture of statoblasts and some Aufwuchs animal associations. *Syst. Assoc Spec, Vol. "Advances in Bryozoology"* 13:75–91.

Dendy, J. W. 1963. Observations on bryozoan ecology in farm ponds. *Limnol. Oceanogr.* 8: 478–482.

Dehdsahti B., and D. W. Blinn. 1986. A bryozoan from an unexplored cave at Montezuma Well, Arizona. *Southwest. Nat.* 31:557–558.

Everitt, B. 1975. Fresh-water Ectoprocta: distribution and ecology of five species in southeastern Louisiana. *Trans. Am. Microsc. Soc.* 94: 130–134.

Ingold, J. L., et al. 1984. Ecology and population genetics of the freshwater bryozoan, *Pectinatella magnifica* Leidy. *J. Freshwater Ecol.* 2:499–508.

Kraepelin, K. 1887. Die deutschen Süssasserbryozoen. Eine Monographie, I: Anatomisch-systematischer Teil. *Abh. Natw. Vereins Hamburg* 10:1–168.

Lacourt, A. W. 1968. A monograph of the freshwater Bryozoa- Phylactolaemata. *Uitg. Rijksmus. Nat. Hist. Leiden* 93:1–159.

Lauer, T. E., et al. 1999. Evidence of recruitment inhibition of zebra mussels (*Dreissena polymorpha*) by a freshwater bryozoan (*Lophopodella carteri*). *J. N. Am. Benthol. Soc.* 18:406–413.

Marcus, E. 1934. Über *Lophopus crystallinus* (Pall.). *Zool. Jahrb. Abt. Anat. Ontog. Tiere* 58: 501–606.

——. 1942. Skôbre Bryozoa do Brasil II. Univ. Sao Paulo. *Bol. Zool.* 6:57–96.

Mukai, H. 1974. Germination of the statoblasts of a freshwater Bryozoan, *Pectinatella gelatinosa. J. Exp. Zool.* 187:27–40.

——. 1999. Comparative morphological studies on the statoblasts of lower phylactolaemate bryozoans, with discussion on the systematics of Phylactolaemata. *Sci. Rep. Fac. Educ. Gunma Univ.* 46:51–91.

Mundy, S. F. 1980. A key to the British and European freshwater bryozoans. *Sci. Publ. Freshwater Biol. Assoc. U. K.* 41:1–31.

Newcomer, W. S. 1950. Additions to the description of *Plumatella repens furcifer Jullien. Am. Midl. Nat.* 44:205–210.

Oda, S. 1959. Germination of the statoblasts in freshwater Bryozoa. *Sci. Rep. Tokyo Kyoiku Daigaku, B,* 135:90–135.

Okamura, B., and L. A. Doolan. 1993. Patterns of suspension feeding in the freshwater bryozoan *Plumatella repens. Biol. Bull.* 184:52–56.

Poirrier, M. A., and M. M. Mulino. 1977. Effects of environmental factors on the distribution and morphology of *Victorella pavida* (Ectoprocta) in Lake Pontchartrain, Louisiana, and vicinity. *Chesapeake Sci.* 18:347–352.

Raddum, G. G., and T. M. Johnsen. 1983. Growth and feeding of *Fredericella sultana* (Bryozoa) in the outlet of a humic acid lake. *J. Freshwater Ecol.* 2:499–508.

Ricciardi A., and T. S. Wood. 1992. Statoblast morphology and systematics of the freshwater bryozoan *Hyalinella orbisperma* (Kellicott, 1882). *Can. J. Zool.* 70:1536–1540.

Richelle E., et al. 1994. Bacterial feeding by the freshwater bryozoan *Plumatella fungosa* (Pallas, 1768). *Hydrobiologia* 291:193–199.

Rogick, M. D. 1935. Studies on fresh-water Bryozoa, II: the Bryozoa of Lake Erie. *Trans. Am. Microsc. Soc.* 54:245–263.

——. 1937. Studies on fresh-water Bryozoa, VI: the finer anatomy of *Lophopodella carteri* var. *typica. Trans. Am. Microsc. Soc.* 56:367–396.

——. 1938. Studies on fresh-water Bryozoa, VII: on the viability of dried statoblasts of *Lophopodella carteri* var. *typica. Trans. Am. Microsc. Soc.* 57:178–199.

——. 1940a. Studies on fresh-water Bryozoa, IX: additions to New York Bryozoa. *Trans. Am. Microsc. Soc.* 59:187–204.

——. 1940b. Studies on fresh-water Bryozoa, XI: the viability of dried statoblasts of several species. *Growth* 4:315–322.

——. 1941. Studies on fresh-water Bryozoa, X: the occurrence of *Plumatella casmiana* in North America. *Trans. Am. Microsc. Soc.* 60:211–220.

——. 1943. Studies on Fresh-water Bryozoa, XIV: the occurrence of *Stollela indica* in North America. *Ann. N.Y. Acad. Sci.* 45:163–178.

——. 1945. Studies on fresh-water Bryozoa, XV: *Hyalinella punctata* growth data. *Ohio J. Sci.* 45:55–79.

Rogick, M. D., and C. J. D. Brown. 1942. Studies on fresh-water Bryozoa, XII: a collection from various sources. *Ann. N.Y. Acad. Sci.* 43: 123–144.

Rogick, M. D., and H. van der Schalie. 1950. Studies on fresh-water Bryozoa, XVII: Michigan Bryozoa. *Ohio J. Sci.* 50:136–146.

Smith, D. G. 1985. *Lophopodella carteri* (Hyatt), *Pottsiella erecta* (Potts), and other freshwater Ectoprocta in the Connecticut River (New England, U. S. A.). *Ohio J. Sci.* 85:67–70.

——. 1988. *Stephanella hina* (Ectoprocta: Phylactolaemata) in North America, with notes on its morphology and systematics. *J. N. Am. Benthol. Soc.* 7:253–259.

——. 1992. A new freshwater moss animal in the genus *Plumatella* (Ectoprocta: Phylactolaemata: Plumatellidae) from New England (U. S. A.). *Can. J. Zool.* 70:2192–2201.

Toriumi, M. 1956. Taxonomical study on freshwater Bryozoa, XVII: general consideration: interspecific relation of described species and phylogenetic consideration. *Sci. Rep. Tohoku Univ. Ser.* 4, *Biol.* 22:57–88.

——. 1971. Additional observations on *Plumatella repens* (L.) (a fresh-water bryozoan), IV: reexamination on the field materials of *P. repens* and *P. fungosa*. *Bull. Mar. Biol. Sta. Asamuschi* 14:117–126.

Uotila L., and J. Jokela. 1995. Variation in reproductive characteristics of colonies of the freshwater bryozoan *Cristatella mucedo*. *Freshwater Biol.* 34:513–522.

Wood, T. S. 1971. Laboratory culture of freshwater Ectoprocta. *Trans. Am. Microsc. Soc.* 90: 92–94.

——. 1988. *Plumatella reticulata* sp. nov. in Ohio (Bryozoa: Phylactolaemata). *Ohio. J. Sci.* 88: 101–104.

——. 1989. Ectoproct bryozoans of Ohio. *Bull. Ohio Biol. Surv.* 8:1–70.

——. 1996. *Plumatella nitens*, a new species of freshwater bryozoan from North America (Ectoprocta: Phylactolaemata), previously misidentified. *Hydrobiologia* 328:147–153.

Wood, T. S., and B. T. Backus. 1992. Differentiation of North American and European forms of *Fredericella sultana* (Blumenbach) (Ectoprocta: Phylactolaemata). *Hydrobiologia* 237:185–193.

Wood, T. S., and T. G. Marsh. 1996a. The sinking floatoblasts of *Lophopodella carteri* (Bryozoa: Phylactolaemata). In: *Bryozoans in Space and Time*, pp. 383–389, edited by D. P. Gordon et al. NIWA Research, Wellington, N.Z.

——. 1996b. *Sineportella forbesi*, a new victorellid bryozoan from Illinois (Ectoprocta: Ctenostomata). *J. N. Am. Benthol. Soc.* 15:610–614.

——. 1999. Biofouling of wastewater treatment plants by the freshwater bryozoan, *Plumatella vaihiriae* (Hastings, 1929). *Water Res.* 33:609–614.

13

ANNELIDA

(Polychaetes, Aeolosomatidae, aquatic oligochaetes, Branchiobdellida, leeches)

Typical segmented worms, constituting the Phylum Annelida, are constructed on a tube-within-tube plan. The body wall is soft, muscular, and covered with a very thin cuticle. The specialized digestive tract has a terminal mouth and anus and is supported in the coelom by thin transverse septa that mark the internal segmental divisions. Unlike the condition in arthropods, segmentation is homonomous. There is an extensive closed circulatory system, becoming somewhat open in leeches, and a few to most segments have a pair of long, tubular nephridia, or excretory organs. The two ventral nerve cords and their segmental ganglia are fused along the median line; the brain typically consists of a pair of cerebral ganglia, and these connect with the nerve cord by two circumpharyngeal connectives. Anterior eyespots are present in some genera.

Five taxonomic groups of annelids are represented in freshwaters. Oligochaeta, or earthworms, have many freshwater representatives; the Hirudinea, or leeches, are dominantly a freshwater group; the Polychaeta, however, which are so varied and richly represented in saltwater, occur in freshwater only as a few rare and sporadic species. In the past, another class, the Archiannelida, was recognized, comprising several unrelated minute annelids demonstrating reduced or modified annelid features. Recent study has revealed that these disparate taxa are most likely highly specialized polychaetes adapted for living in interstitial environments. The Branchiobdellida are highly modified commensals or parasites of crayfishes and the Aeolosomatidae, traditionally placed with the Oligochaeta, is increasingly considered to comprise a distinct group within the Annelida. These five groups are easily distinguished from each other, and because they are so different they are considered in separate sections of this chapter.

There is a trend to divide annelids into two major groups, one containing the architypical Polychaeta, recognized as a separate class, and another containing all remaining annelids that possess a clitellum. The clitellum is a glandular swelling of the epidermis of several genital segments and functions as a source of specific secretions necessary for the completion of sexual reproduction. Thus, all annelids with a clitellum are placed in the Class Clitellata. The traditional classes Oligochaeta, Branchiobdellida, and Hirudinea are now treated as orders. The Aeolosomatidae, formerly in the Oligochaeta, is now generally considered to be a nonclitellate group. The polychaetes are dioecious while the clitellates are hermaphroditic. Most aquatic clitellates primarily reproduce asexually by budding; sexual reproduction, usually rare in many species, is syngamic. Cross fertilization is the rule.

Polychaeta have muscular, paired, lateral projections from the body wall, called parapodia; these are absent in the other groups occurring in freshwater. Groups of setae accompany the parapodia of polychaetes, and setae are also present in the Oligochaeta, though not in the Hirudinea and branchiobdellids. Polychaetes usually have accessory tentacles or other appendages at the anterior end but these are absent in the other two groups. Leeches are dorsoventrally flattened and have an oral and a caudal sucker; the body segments are subdivided into superficial annuli. None of these features occur in the oligochaetes, polychaetes, and branchiobdellids, although the branchiobdellids have a caudal attachment disk structure. A key to the principal groups of Annelida is provided below.

KEY TO MAJOR GROUPS OF ANNELIDS

1. Worms without reproductive organs or reproductive organs associated with other organs or structural systems; prostomium with cilia, tentacles, scales, or papillae; with setae; aquatic **2**
   Worms with distinct reproductive organs and a body wall swelling in the vicinity of the gonads (clitellum); prostomium without elaboration, simple or with suckerlike organs or a single proboscis; with or without setae; aquatic to semiterrestrial **3**
2. Worms minute or large, with lateral, setiferous swellings or lobes; prostomium with fleshy elaboration . . . .**POLYCHAETA 270**
   Worms minute; without lateral extensions; prostomium with cilia only **AEOLOSOMATIDAE, *Aeolosoma* 275**
3. Prostomium not fused with subsequent segments or absent; setae present **OLIGOCHAETA 276**
   Prostomium fused with anterior segments to form an attachment structure; setae absent **4**
4. Worms small to large, more than 20 body segments; body faintly to darkly pigmented **HIRUDINEA 301**
   Worms small, less than 20 body segments; not pigmented; exclusively commensal with crayfish **BRANCHIOBDELLIDA 296**

POLYCHAETA

Polychaetes are almost exclusively marine annelids, and an amazing variety of species and enormous numbers of individuals are found in many types of marine habitats. Some species (especially Nereidae) are euryhaline and occur also in estuaries and brackish waters, but less than 50 species of polychaetes are known to be restricted to the freshwaters of the world.

Polychaete worms are very unique among annelids in general. They are believed to demonstrate the most generalized and primitive traits of the Annelida and exhibit anatomical features that easily distinguish them from other annelids. The prostomium of polychaetes bears specialized structures involved with sensory functions, food capture, respiration, and defense. These structures may be in the form of fleshy or fine tentacular processes, scaly lobes, or palps. Many species of the Nereidae (Fig. 13.1A) contain a powerful eversible pharynx with biting jaws. Supplemental teeth called paragnaths or papillae are sometimes present on the pharynx. The body of the polychaete worm consists of many segments, each with a pair of lateral fleshy extensions or parapodia (Figs. 13.1B, 13.2). The degree of development of the parapodia varies among particular families and genera, but in all cases setae are present on the apices of the parapodia. Like the prostomium, the parapodia provide many functions for the organism, including locomotion, respiration, sensory reception, and defense.

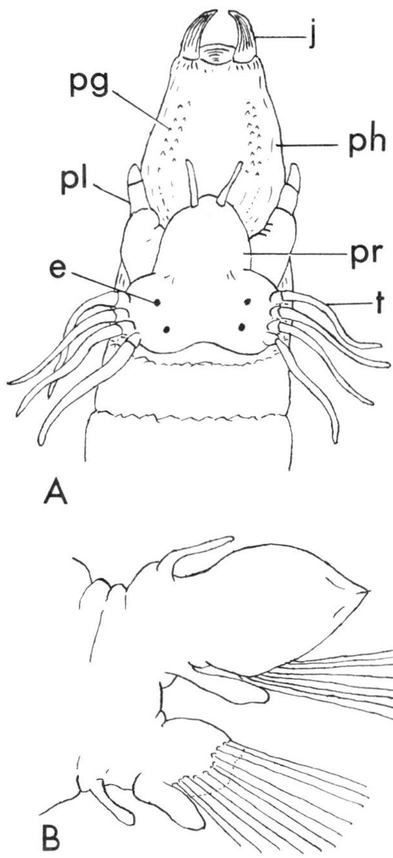

Fig. 13.1. Features of *Nereis*, A, head and associated structures; B, parapodium. e, eye; j, jaw; pg, paragnath; ph, pharynx; pl, palp; pr, prostomium; t, tentacle. (A, B composed from various sources.)

Fig. 13.2. *Marenzellaria viridis* from the Connecticut River, ×1.

The colonization of inland waters by a small number of polychaetes is by no means restricted to one family or a few genera. Eight families are represented, and all species have close marine relatives. In addition to being found in the United States, freshwater polychaetes have been reported sporadically from Yugoslavia, Lake Baikal in Asia, Viet Nam, Gabon, Java, Sumatra, China, Great Britain, Trinidad, Curaçao, and other islands in the Caribbean Sea, and Brazil. With few exceptions, they have been collected from freshwater rivers, lakes, and springs within 20 miles of the seacoast, and this seems to be an indication that polychaetes, in spite of their phylogenetic antiquity, are relative newcomers to freshwaters. Presumably, the physiological and morphological adjustments necessary for the marine–brackish–freshwater transition constitute almost insurmountable barriers. Perhaps the most striking incidence of freshwater polychaete distribution is the occurrence of two species in Lake Baikal, a body of water that is geologically ancient and more than 1600 km from the nearest saltwater.

Most freshwater polychaetes look like the familiar *Nereis* of invertebrate zoology courses. About 13 species have been reported from the United States. *Nereis limnicola* (Fig. 13.3A), one of the Nereidae, was first collected in 1895 from Lake Merced near San Francisco, a body of freshwater less than 0.4 km from the ocean and formed in very recent geological times. It may be taken as far north as Oregon. *N. succinea* is widely distributed in brackish to fresh waters on both east and west coasts. Another nereid, *Lycastoides alticola* (Fig. 13.3B), was described from a single specimen collected in 1892 in a mountain stream at an altitude of 2100 m in "Sierra Laguna, Lower California." It has apparently not been collected since. *Laeonereis culveri* and *Marenzellaria viridis* (Figs. 13.2, 13.3C) occur in brackish to fresh waters along the east and gulf coasts of the United States.

Among tube-building polychaetes, a single serpulid, *Ficopomatus enigmaticus* (Fig. 13.3D), is known from marine to brackish and fresh waters near Oakland, California, and *F. miamiensis* from a few streams entering the

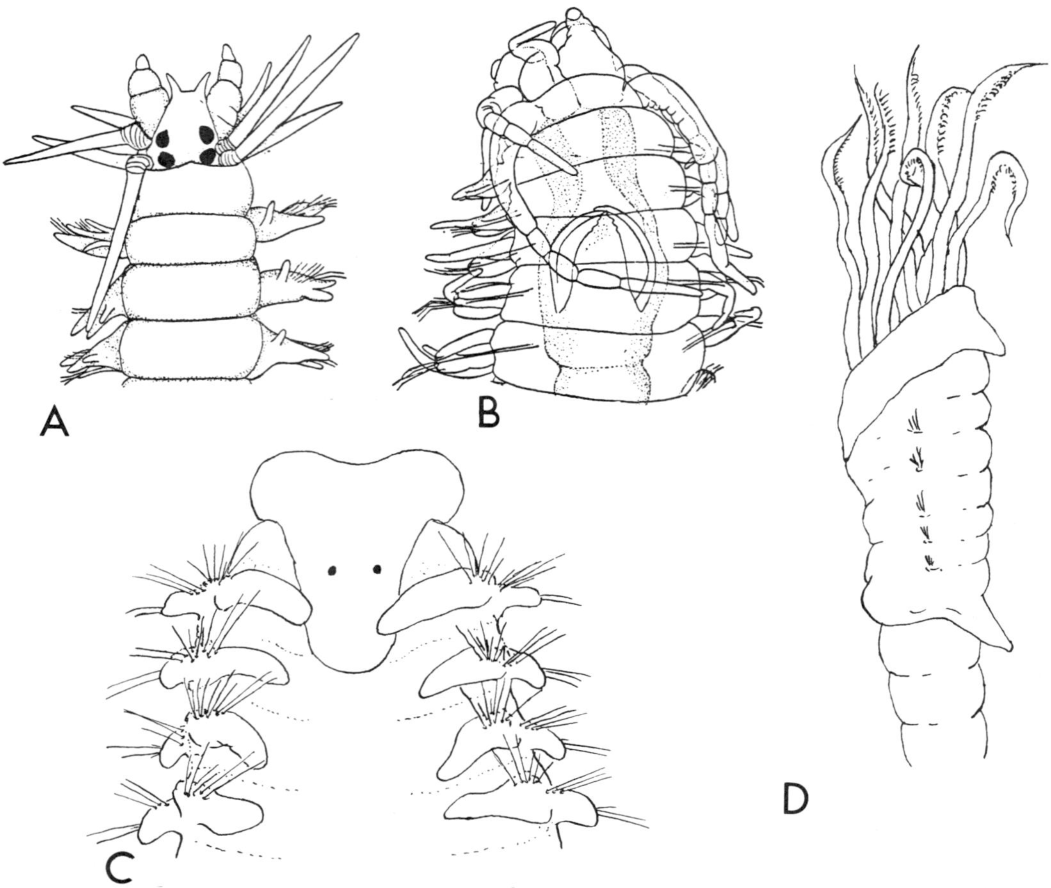

Fig. 13.3. Representative freshwater polychaetes. A, *Nereis limnicola;* B, *Lycastoides alticola;* C, *Marenzellaria viridis;* D, *Ficopomatus enigmaticus,* (A, B from Johnson 1903; D redrawn from Foster 1972.)

Gulf of Mexico. Both build a calcareous tube. In 1859 and in several later papers Joseph Leidy described an interesting sabellid polychaete, *Manayunkia speciosa* (Fig. 13.4), from the Schuylkill River near Philadelphia; it was later found in other areas in southeastern Pennsylvania and New Jersey. In 1939 Krecker described another species in the same genus, *M. eriensis* Krecker, from mud samples taken at a depth of 17 m in Lake Erie, near Put-in-Bay, Ohio. More recently this species has been considered a synonym of *M. speciosa.* It has since been found to be widely distributed in the Great Lakes, Georgia, New York, Oregon, California, Alaska, New England, North Carolina, South Carolina, and the Gulf Coast. This is undoubtedly the polychaete most likely to be found by the freshwater biologist, and undoubtedly it is generally distributed in the United States.

Manayunkia speciosa is 3 to 5 mm long and inhabits a tube built of mud or sand and mucus. The head bears two large lateral lophophorelike structures, each of which has about 18 long, ciliated tentacles used in feeding. There is a pair of eyes near the median line of the head, and also a linear group of pigmented spots on each half of the lophophore, which are presumed to be light sensitive. Each of the first 11 of the 12 trunk segments has a pair of reduced, knoblike parapodia bearing a bundle of 4 to 10 setae. The circulatory system is well developed, and

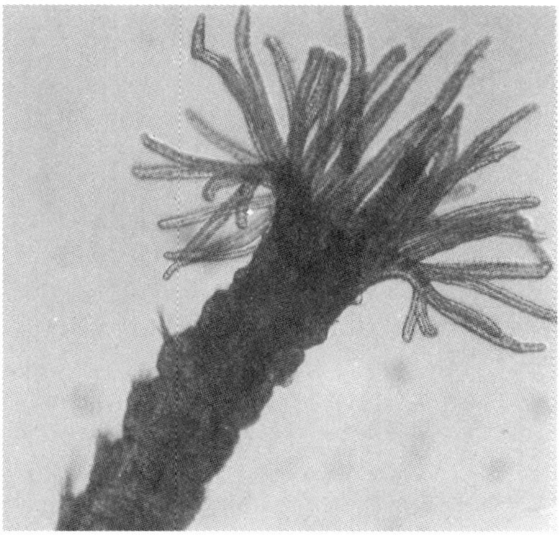

Fig. 13.4. *Manayunkia speciosa*, ×30, a freshwater polychaete, photographed from life.

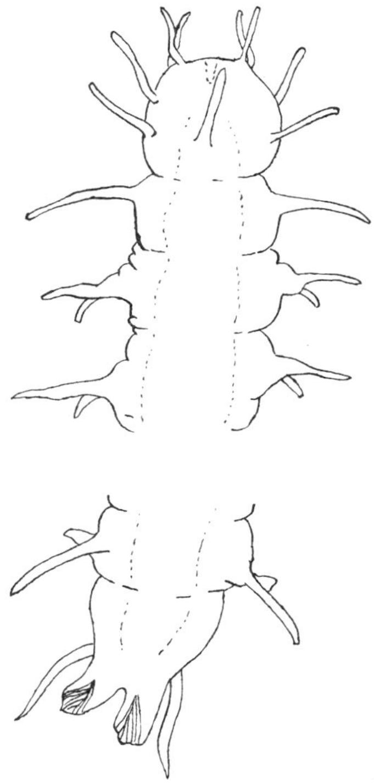

Fig. 13.5. *Hesionides riegerorum*, anterior and posterior portions. (Modified from Westheide 1979.)

the bright green chlorocruorin of the blood colors the whole body. Leidy showed that development is direct and that there is no free-swimming trochophore larva. The large size and lateral swellings of the sixth trunk segment may be an indication that these polychaetes also reproduce by budding. See Plate 5 for whole organism.

Most recently, a small (1.3-mm) polychaete, *Hesionides riegerorum* (Fig. 13.5) has been described from the sandy beaches of a North Carolina river; it has very close marine relatives. In addition to the species just mentioned, 5 or 10 additional polychaetes of marine families have occasionally been found in brackish waters in estuaries. It is uncertain whether any of these can reproduce in freshwaters.

One of the most remarkable phylogenetic discoveries among freshwater invertebrates is the existence of a species of groundwater polychaete. *Troglochaetus beranecki* (Fig. 13.6) was first described from underground waters of caverns in Switzerland (Delachaux 1921), and more recently this worm has been taken in Hungary, Romania, Austria, and Germany. In 1969 RWP found what is thought to be the same species in the interstitial water 20 cm below the surface of gravels of mountain streams in Colorado at altitudes of 2640 and 3050 m. More recently, hundreds of specimens have been found in similar Colorado habitats. Recently, the genus has been found elsewhere in the United States. *Troglochaetus* averages only 630 μm in length. There are two blunt prostomial tentacles and seven pairs of parapodia. Each parapodium has a dorsal and a ventral fascicle of long, thin setae that are easily broken off during field collecting. There are no segmental septa. The species is closely related to several marine genera, especially the littoral genus *Thalassochaetus*, from the North Sea, but family placement is uncertain.

Many of the species of polychaetes collected in freshwater are actually estuarine forms that undoubtedly reproduce in saltwater. Sexes are typically separate and adults spawn gametes, which combine to form a planktonic larva. Very little is known about

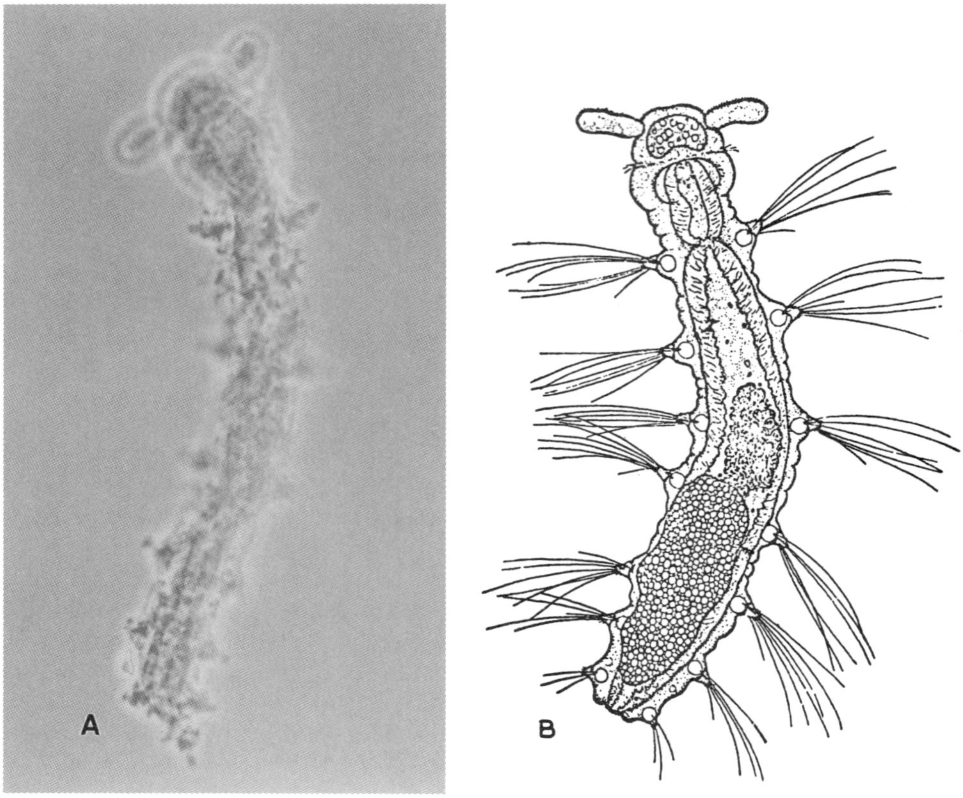

Fig. 13.6. *Troglochaetus beranecki*, ×100. A, phase contrast photograph; B, showing general anatomy. (A, specimen courtesy of D. L. Strayer; B, from Andrassy 1956.)

the reproductive biology of those species (about 4 or 5) that are found exclusively in freshwater.

The key treats only those genera that have actually been collected in freshwater. It incorporates data from several sources, including Smith (1964), Foster (1972), and Klemm (1985). Nearly all genera have only a single species occurring in freshwater.

KEY TO THE FRESHWATER POLYCHAETA OF THE UNITED STATES

1. Worm microscopic, less than 2 mm in length; of interstitial facies; with reduced parapodia and simple round head with fleshy appendages **2**

Worm macroscopic, 3 to 5 mm or larger; parapodia reduced (tubicolous species) or well developed; head of complex structure, with many tentacles, or combination of palps and tentacles **3**

2. Head with several cirri (Fig. 13.5); with about 13 segments; 1.3 mm in length; 1 species near coast; NC . . . . **HESIONIDAE, *Hesionides***

Head with two lateral fleshy antennae (Fig. 13.6) with about 7 segments, less than 1 mm in length; occurring well inland in interstitial and groundwater habitats; one rare species, widely distributed in the northern U.S. . . . . ***Troglochaetus***

3. Sedentary, tube constructing worm; parapodia reduced, simple setae bearing lobes; anterior with well developed hood or collarlike structure containing many tentacles (Fig. 13.3D) or without tentacles **4**
Burrowing or crawling worms; anterior with a pair of tentacles **6**
4. Tube calcareous and with a lid (operculum); in brackish freshwater along western and southeastern coasts; 2 species . . . .**SERPULIDAE, *Ficopomatus***
Tube delicate, flexible, and composed of particles, without operculum; with 2 distinct eyes **5**
5. Worm 10 to 30 mm; prostomium shield-shaped (Fig 13.3C); anterior parapodia divided at base, with large dorsal notopodial lobes (Fig. 13.2); 1 species along the eastern coast **SPIONIDAE, *Marenzellaria***
Worm less than 5 mm in length; prostomium with two clusters of long tentacles (Fig. 13.4); 1 species with a wide but spotty distribution, typically in large lakes and rivers; exclusively freshwater . . . .**SABELLIDAE, *Manayunkia***
6. Parapodia broad with two major rami, each with a cluster of setae; eyes present **NEREIDAE (in part), 7**
Parapodia with only one distinct lobe and one set of setae; eyes absent (Fig. 13.3B); 1 very rare species; mountainous CA . . . .**NEREIDAE (in part), *Lycastoides***
7. Eversible pharynx lacking paragnaths or papillae; 1 uncommon species; FL, NC ***Stenoninereis***
Eversible pharynx with either paragnaths or papillae **8**
8. Eversible pharynx with paragnaths (Fig. 13.1A); about 2 species entering freshwater on both coasts with a landlocked population in CA . . . .***Nereis***
Eversible pharynx with papillae; 1 species entering freshwater along the east coast ***Laeonereis***

AEOLOSOMATIDAE

The Aeolosomatidae have been traditionally placed with the Oligochaeta, "Archiannelida," and other groups within the Annelida. The archiannelids are no longer considered a natural group and the anatomy and biology of the aeolosomatids is clearly nonoligochaete in nature and purportedly very primitive among annelids. The aeolosomatids are peculiar aquatic annelids found in freshwater environments around the world.

The aeolosomatids are small worms ranging from 1 to 10 mm in length (Fig. 13.7). The body is typically translucent except for the presence of colored globules occurring in the epidermis (Figs. 13.8, 13.9D). The anatomy is very simple in construction. There are no septa separating segments, the body wall musculature is weak, and the nervous system is entirely intraepidermal. The head is large and rounded and contains a large ventral mouth invested by a broad patch of cilia. A large hollow proboscis extending over the mouth contains the brain (Figs. 13.8, 13.9D). The anterior margin of the head is ringed by small bristles presumably tactile in function. Segmentation is evidenced by periodic groups of long hairlike setae (two ventral and two dorsal pairs) along the length of the animal (Fig. 13.9A–C). The gut is not partitioned into functional regions and ends with a terminal anus.

The aeolosomatids reproduce primarily by asexual means involving fission or the splitting off of a new worm. A bud is formed posteriorly (Figs. 13.7, 13.10) and the organ and tissue systems at the anterior extremity of the bud differentiate or become more complex. Sexual reproduction is apparently rare. Some species of *Aeolosoma* are known to encyst when the water temperature is 6°C or lower. A translucent, hardened coat of mucus is formed around the tightly coiled worm. Emergence from the cyst occurs at about 20°C.

Locomotion is accomplished by two means. The first is a form of swimming that involves the cilia on the head; the gliding motion of the worm is reminiscent of free-

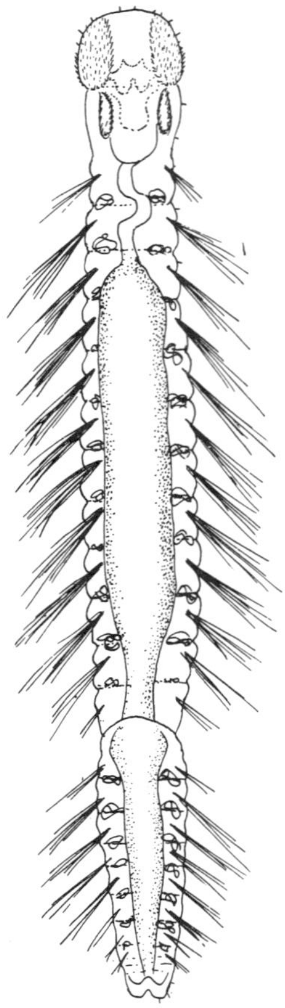

Fig. 13.7. Dorsal view of *Aeolosoma*, showing separating bud, ×50. (Modified from Marcus 1944.)

living flatworms. Otherwise, the aeolosomatids crawl in a similar fashion to other annelids. The setae coordinate with the body wall muscles to produce a form of peristalsis, resulting in the incremental forward lurches customary of earthworms.

Feeding has been studied in some detail. Food is acquired by an interesting process including the head muscles, cilia, and marginal bristles. Particulate debris are swept toward the mouth by the cilia. Larger particles are surrounded by the flattened head, which then seals the particles by securing the head margins to the substrate using the bristles and internal musculature to create the seal. The interior of the head is then lifted creating a sudden vacuum that dislodges the particles, which are then swept into the mouth by the cilia.

The Aeolosomatidae occur in all types of freshwater environments. They have been collected in the profundal of lakes and the interstices of streambeds and groundwater. All species are placed in the single genus *Aeolosoma.* There are about 5 species native to North America and at least one of these, *A. variegatum*, is cosmopolitan. A sixth species presumably native to the Indian subcontinent, *A. bengalense*, has been widely cultured and sold as an experimental animal and is probably established in the United States. Features and length of the setae are the usual means of distinguishing species.

OLIGOCHAETA

Aquatic oligochaetes have the same fundamental structure as the common terrestrial earthworms, and the anatomy, physiology, and behavior of the latter are so familiar to everyone who has been exposed to an elementary course in college zoology that it would be superfluous to repeat these details here. Instead, the various specializations and peculiarities of aquatic oligochaetes will be emphasized.

General characteristics. Representatives of 9 families of oligochaetes occur in the freshwaters of the United States. The Naididae, Tubificidae, North American Opistocystidae, and Parvidrilidae are strictly aquatic. The Opistocystidae are extremely rare. The Lumbriculidae are semiaquatic. Some species of Haplotaxidae and Enchytraeidae are aquatic, semiaquatic, or amphibious and occur especially in debris at the edges of streams, ponds, or in marshes and soggy ground; other species of these two families, however, are essentially terrestrial and never occur in water. The Sparganophilidae has one aquatic genus occurring in the United States. Several other families of oligochaetes are essentially or fully terrestrial, but a few have species normally occurring in freshwater. Among the North American terrimegadril Opisthopora (terrestrial families) one exotic species of the otherwise terres-

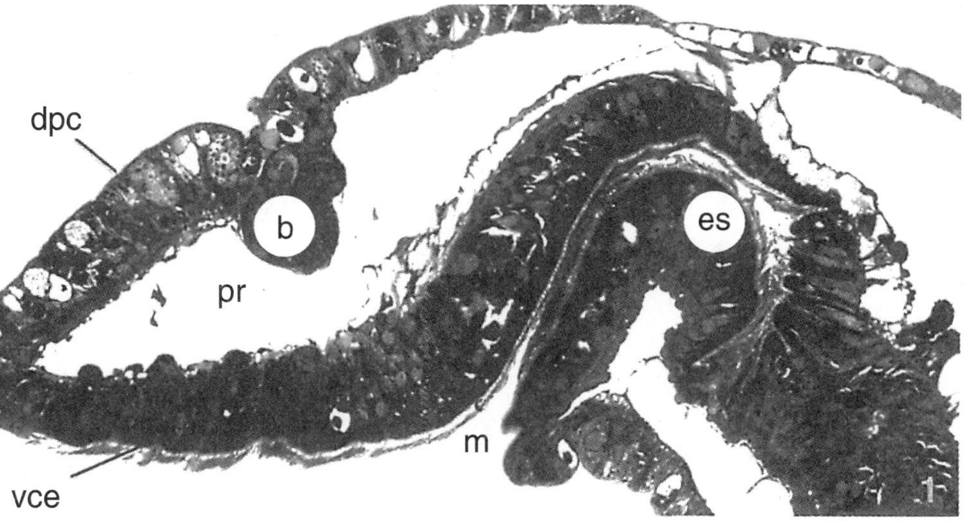

Fig. 13.8. Longitudinal section of *Aeolosoma*. b, brain; dpc, dorsal pigment cells; es, esophagus; m, mouth; pr, prostomium; vce, ventral ciliated epidermis. (Modified from Potswald 1971.)

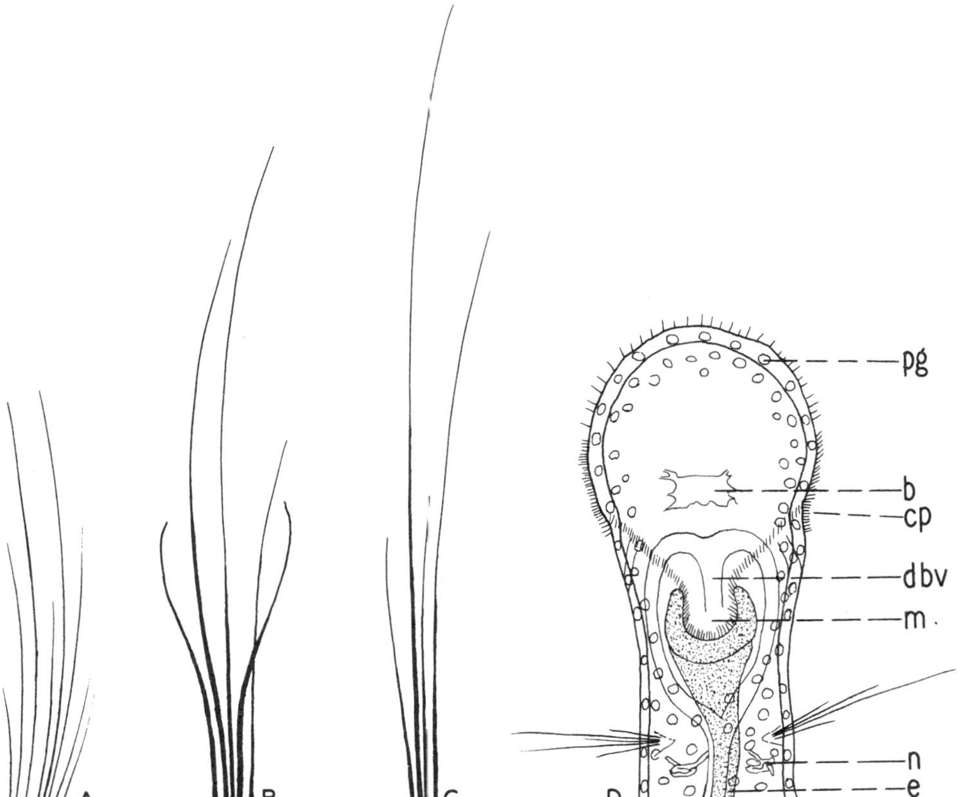

Fig. 13.9. Structure of *Aeolosoma*, A, setal bundle of *Aeolosoma hemprichi;* B, setal bundle of *A. leidyi;* C, setal bundle of *A. headleyi;* D, ventral view of anterior end of *A. leidyi*, b, brain; cp, ciliated pit; dbv, dorsal blood vessel; e, esophagus; m, mouth; n, nephridium; pg, pigment globules.

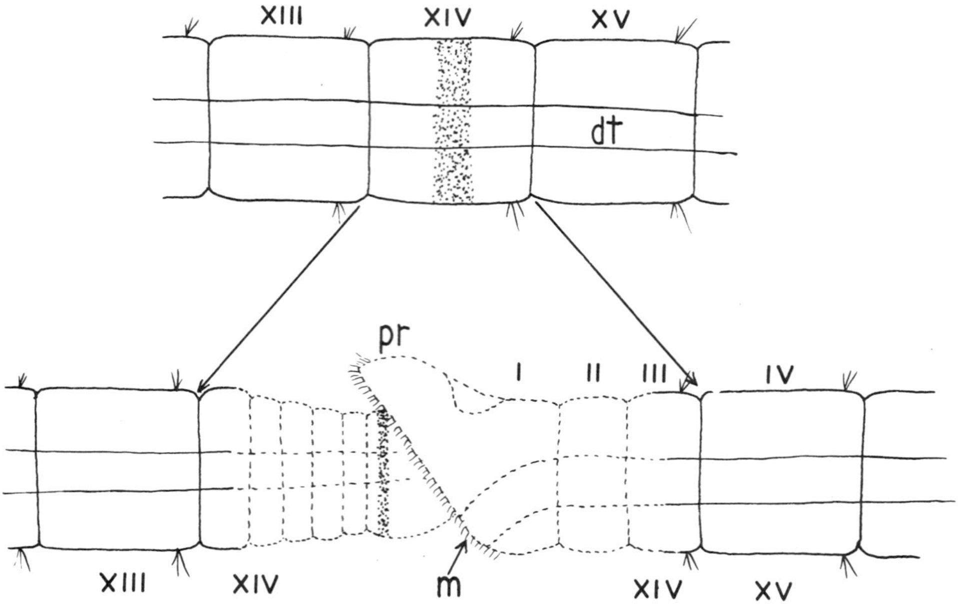

Fig. 13.10. Budding in *Aeolosoma*, showing segmental derivatives in daughter zooids. dt, digestive tract; m, mouth; pr, prostomium. (Modified from Marcus 1944.)

trial family Lumbricidae (a species of the genus containing *Eiseniella)* is regularly found in freshwaters.

The great majority of true aquatic species are common in the mud and debris substrate of stagnant pools and ponds and in streams and lakes everywhere. They are sometimes especially abundant in masses of filamentous algae. Tubificidae occur even in the deepest parts of large lakes, but the shallows down to a depth of a meter are the usual habitat for the great majority of species.

Compared with amphibious and terrestrial forms, the true aquatic oligochaetes are much more delicately constructed and small, the usual length range being 1 to 30 mm. The body wall is thin, and the internal organs can be easily seen in living specimens.

The prostomium is a presegmental lobe at the extreme anterior end of the body; it projects in a rooflike fashion above the mouth and is sometimes elongated into a proboscis. In locating various external and internal structures of oligochaetes it is customary to designate the body segments serially by Roman numerals, beginning at the anterior end. Thus, the first complete body segment immediately behind the prostomium is designated as I.

The number of segments is quite variable, even in mature individuals, and commonly there is a ±25% variation from the average for various specimens of a particular species. The threadlike Haplotaxidae have the largest number of segments, sometimes up to 500. At the opposite extreme the Naididae usually have between 7 and 40 segments. The other families have an intermediate number of segments, tubificids, for example, usually consist of 40 to 200 segments.

The chitinoid setae are arranged in four bundles, two dorsolateral and two ventrolateral. Setae are never found on segment I, but usually begin on II, although in some genera they are absent from several successive segments, either dorsally, ventrally, or both. Detailed structure of the setae and their arrangement are important as taxonomic characters. A bundle may consist of from 1 to as many as 20 setae, and the number in a particular bundle may be constant from one specimen to another within a species, or it may vary. Setae may be long or short (Fig. 13.11), straight or curved, or even sigmoid;

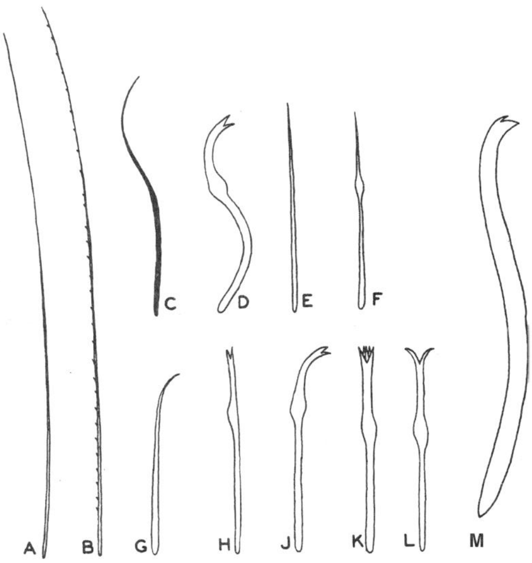

Fig. 13.11. Typical setae of aquatic Oligochaeta. A and B, hair; C–L, aciculate; B, serrate; C and D, sigmoid; D, H, and J, bifurcate; G, simple; J, hooked; K, pectinate; L, biuncinate; M, unequal bifurcate.

their structure may be long, pointed, and hairlike or short, thick, and truncated. The tips of setae can be blunt, single-pointed, bifurcate, or pectinate. A plethora of nomenclature is used to describe the multitude of forms and expressions of setae (also called chaetae), with authorities following different systems over the years. Long, thin, hairlike setae are referred to as capilliform, aciculate, or simply hair setae, the latter term used in this manual. Shorter aciculate setae are often called crotchets when they demonstrate some tip structure. In the Naididae the dorsal crotchet setae are called needles if they differ from the ventral setae. Herein they are all simply termed setae.

The tip morphology of setae is very diverse, and correct determining of the tip structure is critical for accurate species identification. Single-toothed or blunt tips are uncinate or simple; bifurcate or bifid tips have two teeth, which may vary in relative size and direction. Pectinate tips have a comblike series of teeth, typically with the two outer teeth better developed, whereas in palmate tips all the teeth are similar (Fig. 13.11).

Locomotion. Usually locomotion is a crawling movement on or in the superficial layers of the substrate; it is similar to that of the earthworm and involves contractions of the muscular body wall and obtaining a purchase with the setae. A few naidids, such as *Pristina* and *Stylaria*, are effective swimmers that move along just above the substrate in a serpentine manner. *Lumbriculus* undergoes a combination of rapid directional reversals and rotating or helical swimming movements presumably to avoid predation.

Feeding. The great majority of aquatic oligochaetes obtain nutriment by ingesting quantities of the substrate after the manner of the earthworm, the organic component being digested as it passes through the alimentary canal. Sometimes the food is ingested at a depth of 2 or 3 cm below the water–substrate interface. Under some circumstances the food may consist largely of filamentous algae, diatoms, or simply miscellaneous plant and animal detritus. Carnivorous forms are poorly known. *Chaetogaster* feeds on microcrustacians, insect larvae, protoctists, and other oligochaetes. The author (DGS) has observed *Stylaria* feeding on freshwater sponges.

Respiration. Because the body wall is thin and often well supplied with capillaries, most of the carbon dioxide and oxygen exchange occurs through the general body surface. In most naidids and in some tubificids water is taken into the anus and passed forward for a variable distance by antiperistalsis and ciliary action, and these processes are thought to constitute an accessory respiratory mechanism.

Dero has handsome ciliated gills surrounding the anal region (Fig. 13.16D). The posterior extremity has the form of a shallow cup, and the gills originate within the cup or from its margin. *Branchiura* has a fingerlike dorsal and ventral gill on each of the posterior segments. All oligochaete gills are well supplied with blood vessels. Gilled species remain quietly in position with the anterior portion of the body hidden in the substrate and in a tube constructed of fine debris and projecting up from the substrate for a variable distance. The posterior end projects vertically out of the tube and into the water. When disturbed, the worm withdraws into its tube and the substrate in a fraction of a second.

Most tubificids also build tubes, the projecting posterior end of the animal being waved about vigorously in the water, presumably to circulate the water and make more oxygen available to the body surface. Tubificids are commonly red, owing to dissolved hemoglobin in the blood.

Epidermal sensory structures. Many species of Naididae have a pair of eyespots (Fig. 13.16A). They are entirely epidermal, the overlying cuticle being unmodified. Each eye consists of a transverse row of five or six small visual cells and a large number of pigment granules that cover the sensory cells medially and posteriorly. Other epidermal sense organs, consisting of either single cells or aggregates and often arranged symmetrically, occur over the general body surface of all oligochaetes but are especially concentrated at the anterior end. Sensory hairs are common in the Naididae, and in *Slavina appendiculata* they are borne on prominent sensory papillae (Fig. 13.16B). Several aquatic oligochaetes have a long proboscis projecting from the prostomium (Fig. 13.16C) that is likely sensory in function. Epidermal sensory cells are generally assumed to be sensitive to tactile, thermal, and chemical stimuli.

Reproduction. Asexual reproduction is widespread among freshwater oligochaetes and all families subjected to experimental study demonstrate some form of asexual reproduction. Asexual methods vary and include fission by architomy (fragmentation followed by regeneration) or, more commonly, paratomy (formation of buds followed by fragmentation). In typical paratomous species the budding segment grows and undergoes repeated transverse divisions that result in four or more anterior segments of a new posterior worm and several new posterior segments of the anterior parent worm. These regions rapidly become differentiated. The two animals remain attached for a variable length of time, and the more posterior individual may in turn produce a third individual posterior to its budding zone. This process may be repeated so that in some species there may be chains of as many as four to eight zooids in various stages of development. After an interval, however, two adjacent and mature zooids separate and become completely independent individuals. Under favorable conditions most naidids reproduce a new individual every 2 or 3 days. In *Dero* it is unusual to find a chain of more than two zooids. In general, zooid chains are longest in spring and autumn.

The budding zone is localized in one of the segments toward the posterior end. The location of the budding segment is usually constant within a single individual, but under experimental conditions it is possible to alter its position slightly; favorable temperature, food, and oxygen conditions will move the zone anteriorly a segment or two; unfavorable conditions may move it posteriorly a segment or two. On the other hand, the specific location of the fission zone varies greatly from one individual to another within a species. Thus, in *Nais elinguis*, the budding segment varies from 12 to 20, and in *Stylaria fossularis*, it varies from 8 to 23 but is usually 18.

Parthenogenesis also occurs and both meiotic and ameiotic patterns of cell division result in formation of diploid, triploid, or aneuploid clones. Sexual reproduction certainly occurs and sometimes may be very common, especially in *Chaetogaster*. The presence of sexual organs is essential for positive identification of many genera and species of aquatic oligochaetes, especially the Enchytraeidae.

All oligochaetes are hermaphroditic (Fig. 13.12A). Syngamic reproduction is essentially similar to the familiar story in the earthworm. The epidermis in the reproductive zone of the adult worm is modified to form a clitellum in which are found several kinds of secretory cells involved with different aspects of copulation. The clitellum is well developed in terrestrial families, less so in aquatic families. It is reduced to simple lateral swellings in the tiny aquatic *Parvidrilus*.

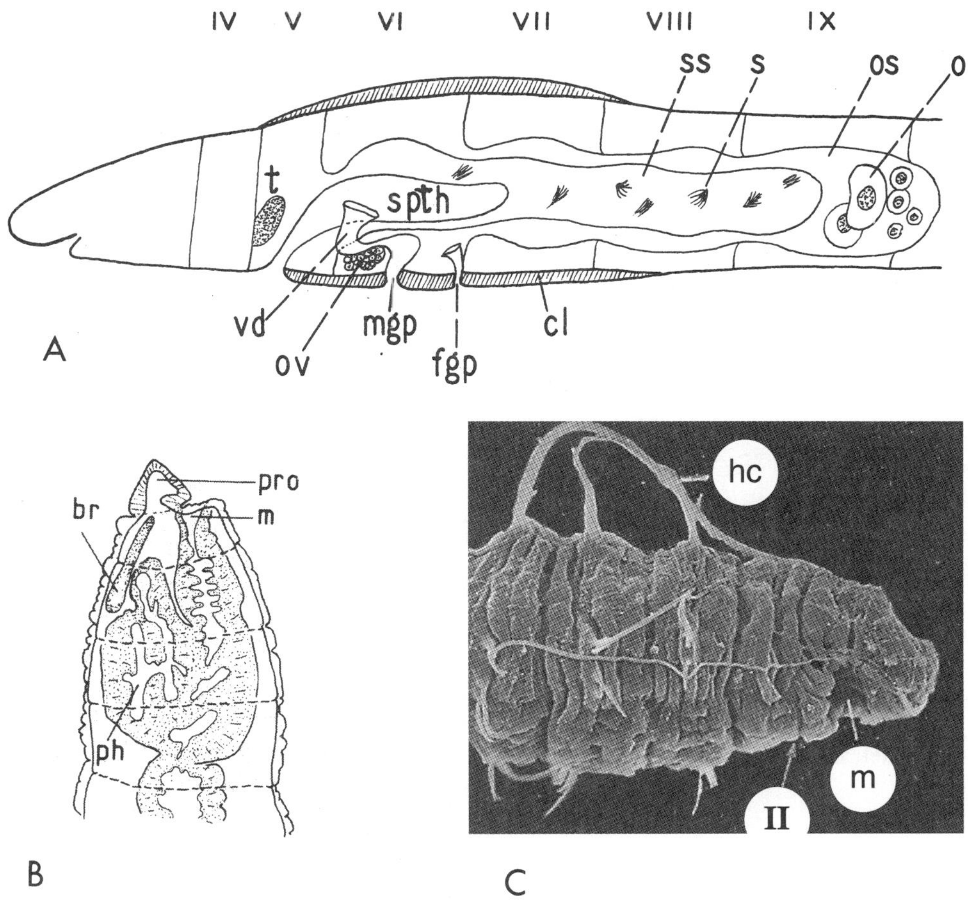

Fig. 13.12. Structure of aquatic Oligochaeta. A, diagram of typical Naididae sex organs; B, anterior of *Teneridrilus;* C, anterior of *Parvidrilus,* a minute interstitial genus. br, brain; cl, clitellum; fgp, female genital pore; hc, hair chaetae; m, mouth; mgp, male genital pore; o, ovum; os, ovisac; ov, ovary; ph, pharynx; pro, prostomium; s, sperm; spth, spermatheca; ss, sperm sac; t, testis; vd, vas deferens. Roman numerals indicate segments. (B from Erséus et al. 1990; C from Erséus 1999.)

Embryos containing cocoons (Fig. 13.13), produced by secretions of the clitellum, are deposited on rocks, vegetation, and debris, especially in the late summer and early autumn when a large percentage is consumed by small fish. Although many fertilized eggs may be deposited into a cocoon, only a few develop to juveniles and exit the cocoon. Syngamic reproduction has often been reported as early as July but probably follows an asexual cycle, as is the case with many other freshwater invertebrates. Under rare circumstances reproduction in isolated tubificids indicates self-fertilization. Life cycles of macroscopic species usually take 1 or 2 years.

The capacity to regenerate lost or damaged anterior or posterior segments can be quite high in aquatic oligochaetes. The cellular phenomenon associated with regeneration in oligochaetes is in certain ways similar to that of platyhelminth worms. Regeneration appears to be best developed in the Naididae and Tubificidae. In these two families the ability to regenerate new body regions is quite high. Terrestrial families have much less regenerative ability.

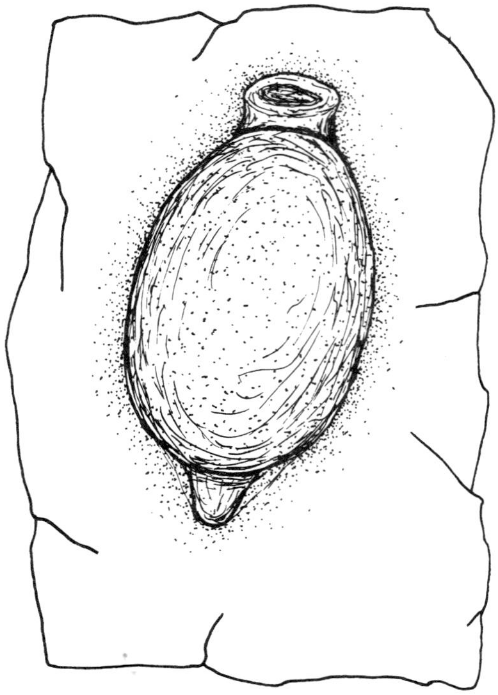

Fig. 13.13. Cocoon of *Stylaria* deposited on a bit of debris, ×60.

Ecology. Aquatic oligochaetes occupy a niche equivalent to that occupied by terrestrial species. They feed on bottom mud and mix it much as earthworms effectively mix the surface layers of garden and meadow soils and in this manner are important in aquatic ecosystems, cycling nutrients in the top layers of the substrate.

Temperature is not usually a limiting factor, but it often determines the relative abundance of oligochaetes. Some amphibious species, especially enchytraeids and lumbriculids, are notable because they are found on the shores of cold mountain streams and brooks originating from snowbanks and glaciers. Many microannelids occur in hyporheic, phreatic, and psammolittoral situations as a part of the meiobenthos. The recently discovered Parvidrilidae is known only from the interstices of groundwaters. Freshwater oligochaetes are also well represented in cave environments. The Naididae, Tubificidae, and especially Lumbriculidae have genera with cave-adapted species and several are known from only a single cave or well.

Tubificidae can be dominant at depths exceeding 1 m in poorly oxygenated water, and in the deep waters of lakes there are sometimes more than 8000 individuals/m^2. They are also well represented in CO_2-rich spring-fed waters, and can develop large populations in fish hatcheries. Because tubificids have been identified as intermediate hosts for "whirling disease" causing myxozoans, their presence in hatcheries is a cause for concern. By far the most concentrated populations are found in streams and rivers that are polluted with sewage. Here tubificids occur in dense waving masses (Fig. 13.14). In fact, many species of Tubificidae are usually considered an "indicator" of organic pollution, especially where the water is between 10 and 60% saturated with oxygen. Certain *Tubifex*, *Limnodrilus*, *Nais*, and possibly enchytraeids can be inhabitants of organically polluted waters as well as sewage settling tanks and trickling filters. *Tubifex* often exhibits negative aerotaxis; that is, it actively migrates from regions of high dissolved oxygen to those of low oxygen.

Most of the true aquatic species are able to thrive in low concentrations of dissolved oxygen, and many are also able to withstand the complete absence of oxygen for extended periods. Tubificids, for example, can endure the summer and winter periods of stagnation in lakes. Nevertheless, populations cannot be maintained indefinitely in the absence of oxygen. Dausend (1931) found that only one-third of the specimens of *Tubifex* that he used were able to survive anaerobic conditions for 48 days at 0 to 2°C and at higher temperatures the fraction was progressively smaller. Other workers report the survival of very small percentages of *Tubifex* populations after 120 days under anaerobic conditions. In general, the lower the oxygen concentration the more the animal projects from its tube and the more vigorous are the waving aeration movements of the posterior part of the body, but when the oxygen concentration nears complete exhaustion most tubificids become relatively quiescent.

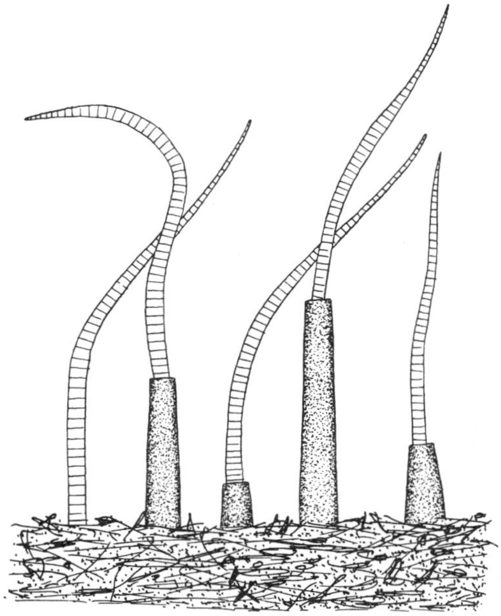

Fig. 13.14. *Tubifex*, a typical tube builder, ×3.

More diverse oligochaete faunas can be found in deep water where oxygen is in good supply, such as in recently created artificial reservoirs. The writer (DGS) has found that the deep, well-oxygenated water of the Quabbin Reservoir in Massachusetts supports a rich oligochaete assemblage, including genera of the Tubificidae, Naididae, and Enchytraeidae.

Forested wetlands, semipermanent ponds, and wet meadows can support large populations of lumbriculids, especially *Lumbriculus variegatus*, enchytraeids, a few tubificids, and the opisthoporan species *Eiseniella tetraeda* and *Sparganophilus eiseni*. Of the several species of luminescent freshwater invertebrates, one, an enchytracid annelid, has been reported from the shores of a river in Russia in concentrations of 100 to 250/m^2.

Freshwater oligochaetes are quite tolerant of low salt conditions characteristic of upper estuaries, and many genera among the Tubificidae, Naididae, and Enchytraeidae have been collected in association with estuarine organisms. A few "freshwater" genera also have saltwater representatives, especially *Tubifex*, *Spirosperma*, and *Clitellio*.

Dispersal, geographic distribution. Oligochaetes undoubtedly migrate easily throughout whole drainage systems, but passive and accidental overland transport presumably is most effective during the cocoon stage in those species that have syngamic reproduction. The importance of aquatic vascular plants for arboreta and the aquarium trade has resulted in the importation and establishment of some species in the United States.

Juveniles of *Tubifex* are known to produce temporary protective tests composed of mucus, sand particles, and organic detritus. Several Naididae are known to produce cocoons in times of environmental stress. It is during these resistant stages that individuals may be easily transported from place to place.

The aquatic oligochaete fauna of the United States has recently been the object of many regional and national surveys. However, the freshwater oligochaetes of certain parts of the country, such as the deep south and southwest, remain poorly known. Collections in the southeast and Midwest have revealed many new distribution records. As currently understood, the freshwater tubificate faunas of the United States east of the Rocky Mountains seem to be geographically rather uniform. Florida is currently known to contain over 100 species in 32 genera of naidids and tubificids, and the Great Lakes region contains just under 100 species in 36 genera of naidids and tubificids. Regarding faunas of specialized habitats, the oligochaetes of groundwaters have come under increased scrutiny, and as many as 21 species have been collected from the ground waters of glaciated New York. In 10 Illinois springs, 33 species have been found, many considered rare.

Many species once believed rare and considered to represent introductions have recently been shown to have wider distributions and are no longer assumed to be exotic. Nevertheless, introductions have occurred as they have for many aquatic invertebrates; especially, the tropical aquatic plant trade has most likely introduced many exotic species to the United States. Many species simply are found around the Northern Hemisphere. Several species of *Chaetogaster*, *Nais*, *Dero*, *Tubifex*, and *Slavina appendiculata* and *Eiseniella tetraedra* are examples of common and cosmopolitan species. The Lumbriculidae are widespread but most surface water species are found in temperate and cold–temperate zones; the family seems to be especially well represented in western states.

Collecting. If bottom material is finely divided, a sieve or strainer may be used for field collecting. If the straining process is too vigorous, the more delicate species will be damaged and fragmented. Otherwise, the bottom debris should be brought into the laboratory, where it may be examined immediately or allowed to stand until the worms can be seen at the surface of the debris or crawling up the sides of the container. Masses of filamentous algae can be rinsed or washed out with a stream of water. The small species can best be found with the aid of a dissecting microscope. Amphibious and large species can usually be picked up with tweezers by sorting stream and lakeside debris.

Culturing. The success of a laboratory culture often depends on the amount and kind of water supply and the available oxygen, but many species can be grown in their natural substrate when it is kept in shallow parts into which water drips or slowly flows. Water should never be changed rapidly. Naidids often become abundant in old protoctistan cultures.

Preparation. For gross structure and even for many fine details of small oligochaetes there is no substitute for the examination of living specimens. Nevertheless, preserved specimens are unavoidable for lengthy examinations and comparisons. Perhaps the greatest difficulty encountered in working with Naididae, Tubicifidae, and Lumbriculidae is the usual strongly contracted and distorted condition, which makes careful anatomical work impossible. Chloretone, chloroform vapors, and magnesium sulfate have all been suggested as narcotic agents, but almost invariably the posterior end of the worm being narcotized disintegrates long before the anterior end has become quiet. Placing live worms in carbonated soda or beer can produce well relaxed specimens in a short period. Some success with 2% hydroxylamine hydrochloride has been realized. Any methods of relaxation require close monitoring of the specimens to avoid deterioration. Regardless of the method used, it is almost impossible to kill *Dero* with expanded gills. Most investigators advocate killing and fixing small oligochaetes without the use of a narcotic. Some tubificids will die in an extended condition if diluted fixatives are used. If many specimens are being fixed, they may be placed in a shallow dish, and then the water is poured off. The worms thereupon elongate and crawl about, and if they are flooded with hot Schaudinn's fixative, a good percentage of the worms will remain elongated and in good condition.

Small annelids are best fixed and preserved in the field with 5% formalin. If necessary, Amman's lactophenol is a good clearing agent. It consists of 40 g of carbolic acid, 40 mL of lactic acid, 80 mL of glycerol, and 40 mL of water. Rinse preserved worms in water, place on a slide, add a few drops of Amman's fluid, and then a circular coverslip. In 30 minutes to a day or two the worms will be sufficiently cleared to distinguish setae and other details. For permanent mounts, replace Amman's fluid with polyvinyl lactophenol or warm glycerine jelly and ring the coverslip with Murrayite. Permanent glycerin jelly mounts are generally satisfactory, either with or without staining. Over time, most temporary or semipermanent mounts involving glycerin or its derivatives will clear the specimen completely, even if it is stained. Once cleared, specimens become fragile, flacid, and brittle and will no longer store or stain well. Stained mounts should be made permanent by dehydration followed by clearing in xylene. Mounting media can include natural or synthetic resins. Using this technique will prevent loss of stain. Good stains include hematoxylin and methylene blue (see Appendix). Microdriles should be overstained then destained in acid alcohol (1% glacial acetic acid in 70% ethyl alcohol).

It is best to set some specimens aside for different preparations; some for permanent stained mounts, some for semipermanent cleared mounts, and some for histology or dissection, if required. Large amphibious or terrestrial species are best killed and preserved in an extended condition in accordance with standard methods. If the specimens are to be sectioned, they should be first kept in water or on wet filter paper until the digestive tract is free of grit.

Another method involves utilizing live specimens. Place one in water on a slide with a circular coverslip. Then remove most of the water at the edge of the coverslip with blotting paper and allow the mount to dry. The worm will be compressed, and then it will burst and flatten, leaving the body wall and setae nicely in place. If a drop of glycerin is placed at the edge of the coverslip, capillarity will draw it under, and such a slide can be made semipermanent by ringing with Murrayite.

Similar preparations of larger dead or preserved specimens of Tubificidae and some Lumbriculidae may be made if they are left in water for a day or two until they begin to decompose and are soft enough for setae preparations. The number, arrangement, and structural details of the setae are essential for the identification of Naididae, Tubificidae, and some Lumbriculidae, but whole mounts do not often allow an examination of the finer details. These drastic methods obviously obliterate the specimen but will fully exhibit the setae. Ultimately, if the detailed examination is desired, then examination using SEM techniques produce the best results.

Taxonomy. In addition to setae, other characters of taxonomic significance in small species are the location of the budding zone, gill structure, shape and size of the prostomium, and the arrangement of certain blood vessels. Body length is of little value, since preserved specimens exhibit varying degrees of contraction. In addition, the published literature is often confusing because it is not made clear whether lengths refer to a single anterior zooid or a whole chain of zooids. Many diagnostic features of aquatic oligochaetes, particularly Naididae and Tubificidae, are present only in adult specimens. The location and number of genital pores, the morphology and position of the modified or genital setae if produced, and the structure of ancillary reproduction structures such as penial sheaths are observable in sexually mature worms.

Most major taxonomic studies in which new taxa are described rely on histological preparations. Both longitudinal and transverse stained sections are usually considered necessary to clearly demonstrate internal characteristics.

Research on North American freshwater oligochaetes has been invigorated by the monographic studies of Brinkhurst, Jamieson, Hiltunen, and their colleagues. Several subsequent investigations have addressed the relationships of specific genera and added new taxa. The increase in study of groundwater and phreatic habitats has revealed new forms and will certainly lead to the discovery of additional species and genera in the future.

Unfortunately, the aquatic Enchytraeidae remains an obscure and difficult group taxonomically and although the family can be well represented in aquatic oligochaete samples, there remains no practical way to distinguish genera and species. Identification of species may require the development of rapid protein staining methods.

A systematic arrangement of the freshwater Oligochaeta of the United States is given below. The system of higher taxa and genera follows Barnes and Harrison (1992) and Kathman and Brinkhurst (1999), respectively. The family Parvidrilidae was recently described by Erséus (1999). Currently, only one species of the Haplotaxidae is believed to occur in the United States. *Metataxis brinkhursti*, once listed as a haplotaxid, is now placed among the otherwise terrestrial Lumbricina (Kathman and Brinkhurst 1999). It thus remains unclear whether this species is actually aquatic. Certain taxa assigned to this genus in the past are presently considered synonyms or have affinities with other groups. Among the Naididae, *Homochaeta* has not been confirmed in North America and is not included. *Arcto-*

drilus and *Alexandrovia* (Tubificidae) have not been reported from the contiguous United States and are omitted. Among the Lumbriculidae not included are *Rhynchelmis* and is its allies and *Styloscolex* from localities northwest of the contiguous United States. For a listing of Enchytraeidae known from North America consult Tynen (1975). The remaining 54 genera listed in Table 13.1 are included in the key.

The families of freshwater oligochaetes are distinguished primarily on the basis of internal anatomy. Certain external characteristics can help separate many specimens into at least family level groups. The three most difficult families are, of course, the largest and these are the Lumbriculidae, Naididae, and Tubificidae. The lumbriculids always have just two setae per cluster and they are bifurcate or simple. Many surface water lumbriculids can reach 300 mm in length and some species have a proboscis. Eyes are never present. The naidids are the smallest in size, about 1 to about 20 mm but usually much less. They can have eyes and multiple setae of different varieties, and several species possess a proboscis. Most genera have the dorsal setae beginning after seg-

Table 13.1. Orders, Families, and Genera of Oligochaetes Occurring in Fresh Waters of the United States

- Order Tubificida
  - Enchytraeidae
  - Naididae
    - *Allonais, Amphichaeta, Arcteonais, Bratislavia, Chaetogaster, Dero, Haemonais, Nais, Ophidonais, Paranais, Piguetiella, Pristina, Pristinella, Ripistes, Slavina, Speciaria, Stephensoniana, Stylaria, Uncinais, Vejdovskyella*
  - Tubificidae
    - *Aulodrilus, Bothrioneurum, Branchiura, Haber, Ilyodrilus, Isochaetaetidea, Limnodrilus, Phallodrilus, Potamothrix, Psammoryctides, Quistadrilus, Rhizodrilus, Rhyacodrilus, Spirosperma, Tasserkidrilus, Telmatodrilus, Teneridrilus, Tubifex, Varichaetadrilus*
  - Opistocystidae
    - *Crustipellis, Opistocysta, Trieminentia*
  - Parvidrilidae
    - *Parvidrilis*
- Order Lumbriculida
  - Lumbriculidae
    - *Eclipidrilus, Kincaidiana, Lumbriculus, Phagodrilus, Spelaedrilus, Stylodrilus, Tenagodrilus, Trichodrilus*
- Order Haplotaxida
  - Haplotaxidae
    - *Haplotaxis*
- Order Opisthopora
  - Cohort Aquamegadrili
    - Spargonaphilidae
      - *Sparganophilus*
  - Cohort Terrimegadrili
    - Lumbricidae
      - *Eiseniella*

ment II. The tubificids never have eyes or a proboscis, are slightly larger than naidids, and contain a variety of multiple setae, and the dorsal setae always begin at segment II. Most genera of the aquatic families Lumbriculidae, Naididae, and especially Tubificidae are diagnosed by reproductive anatomy, and determining identity of a specific genus using external features alone is often difficult and in some genera seemingly impossible. External characters such as setae, position of reproductive pores, and prostomial development overlap among many genera and the reader is cautioned that in several areas of the key, more than one genus can be determined for a specific specimen.

KEY TO FAMILIES AND GENERA OF FRESHWATER OLIGOCHAETA OF THE UNITED STATES

1. Worms large and robust, diameter 3 or more mm; posterior quadrangular in transverse section; with a thick clitellum **2**

Worms can be long but are always thin, diameter 2 mm or less, and delicate; posterior segments round; with a poorly defined clitellum **3**

2. Anus terminal; prostomium clearly differentiated from first segment (Fig. 13.15A); 1 aquatic species, widespread . . . . **LUMBRICIDAE, *Eiseniella***

Anus dorsal; prostomium partially fused with first segment (Fig. 13.15B); at least 1 aquatic species in the U.S. . . . . **SPARGANOPHILIDAE, *Sparganophilus***

3. Worm minute, about 1 mm in length; all setae beginning on segment III, ventral and dorsal hair setae present (Fig. 13.12C); 1 one rare species, AL . . . . . . **PARVIDRILIDAE, *Parvidrilus***

Worm minute to elongate, reaching 400 mm in a few species; at least ventral setae beginning on segment II; ventral hair setae absent **4**

4. Worm very long and thin, reaching 400 mm in length; only one bifurcate seta per bundle; 1 very rare species in scattered locations . . . . **HAPLOTAXIDAE, *Haplotaxis***

Worm up to 300 mm in length, but usually much less; multiple setae per bundle in at least anterior and ventral portion of animal **5**

5. Dorsal and ventral setae in uniform groups of 2 or 4 and usually similar throughout worm; setae short and stout, bifurcate or simple **6**

Setae 2 or more per bundle (Figs. 13.18A; 13.19B, G), at least ventrally and anteriorly, but number, type, and size varies, rarely similar dorsally (if present) and ventrally (Fig. 13.18C, D), setae simple (Fig. 13.18N), bifurcate (Fig. 13.18C, D, H, L), pectinate or palmate (Figs. 13.18 O–Q; 13.21G–J); hair setae present (Fig. 13.19B) or absent **20**

6. Four setae per bundle, all simple and similar; body usually less than 10 mm but stout; whitish or pale . . . . **ENCHYTRAEIDAE (in part)**

Two setae per bundle throughout entire worm, dorsal and ventral setae similar and hair setae absent (except 1 species of *Stylodrilus*, hairlike seta but with thickened base), setae either bifurcate or simple **7**

7. Worm up to 100 mm in length and thin, usually darkly pigmented (in surface water species) and soft-bodied; with either two simple or two bifurcate setae per bundle **LUMBRICULIDAE, 8**

Worm usually 10 mm or less in length but relatively stout, typically pale or whitish and with a thick cuticle rendering the worm stiff, turgid; two simple setae per bundle **ENCHYTRAEIDAE (in part)**

8. With a proboscis **9**

Without a proboscis **11**

9. Worm under 30 mm in length, with a short proboscis; genital pores on X through XII; from ground water habitats; 1 species, TN . . . . ***Trichodrilus* (in part)**

Worms up to 50 mm; with a long proboscis **10**

10. Anterior setae bifurcate, posterior simple; 1 species in northwestern states ***Kincaidiana* (in part)**

All setae simple, about 3 species scattered throughout U.S. . . . . ***Eclipidrilus* (in part)**

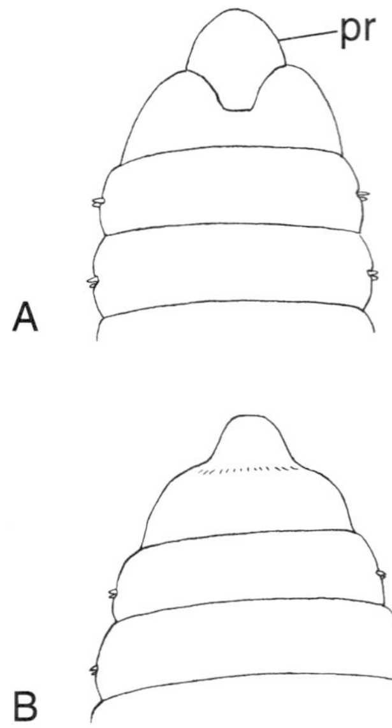

Fig. 13.15. Anterior segments of Opisthopora. A, *Eiseniella;* B, *Sparganophilus,* pr, prostomium.

11. Worms limited to groundwater or cave environments; usually small, less than 50 mm in length (most species below 10 mm); pigmentation absent or reduced; setae simple, somewhat hair-like, and delicate **12**
Worms from surface waters; up to 300 mm in length but usually smaller; pigmented; setae robust **15**
12. With 4 pairs of reproductive pores on X; 1 species in VA cave . . . . ***Spelaedrilus***
Less than 4 pairs of pores on X **13**
13. A deep ventral cavity present containing penial organ; known only from groundwaters in AL; 1 species . . . . ***Tenagodrilus***
Without such a cavity **14**
14. Reproductive pores commencing on IX, 2 species, AL, OR, TN, WV . . . ***Stylodrilus* (in part)**
Reproductive pores commencing on X; 2 species from DC, IL, WV . . . ***Trichodrilus* (in part)**
15. Worm up to 300 mm in length; anterior segments enclosing large, muscular pharynx; setae simple and sigmoid, especially posterior; 1 species in the northwestern U.S. . . . . . ***Phagodrilus***
Worm less than 300 mm, rarely over 100 mm; without enlarged pahrynx; setae bifurcate or simple throughout **16**
16. With external fixed pair of penes on ventral surface at X; 1 widespread species ***Stylodrilus* (in part)**
Without fixed external penes **17**
17. Setae simple **18**
Seate bifurcate (Fig. 13.11M) **19**
18. Worm less than 20 mm in length; reproductive pores on VIII through IX; 1 species from Lake Tahoe, CA . . . . ***Kincaidiana* (in part)**
Worm 20 mm or more in length; reproductive pores on VIII (rarely) to XII; 4 species scattered around the U.S. . . . . ***Eclipidrilus* (in part)**
19. Worm reaching 100 mm in length; darkly pigmented and with large blind blood vessels posteriorly; 2 very common and widespread species, possibly synonymous . . . . ***Lumbriculus***
Worm less than 50 mm in length; pale; without distinct posterior blood vessels; 1 rare species, CA . . . . ***Stylodrilus* (in part)**
20. Worms small, rarely exceeding 12 mm; with a proboscis in combination with posterior processes, 2 lateral and 1 medial (Fig. 13.17A, B); very rare . . . . **OPISTOCYSTIDAE, 21**
Worms small to 50 mm in length; without combination of proboscis and posterior appendages **23**

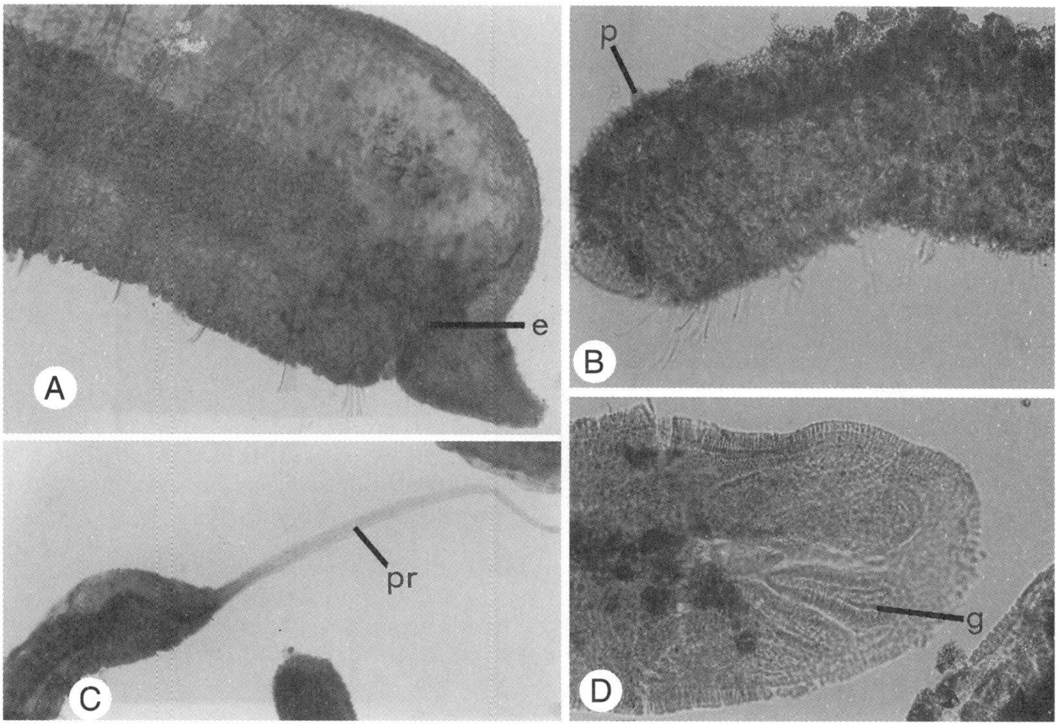

Fig. 13.16. Structure of Naididae. A, anterior of *Uncinais;* B, anterior of *Slavina;* C, anterior of *Stylaria;* D, posterior of *Dero,* e, eye; g, gill; p, papilla; pr, proboscis.

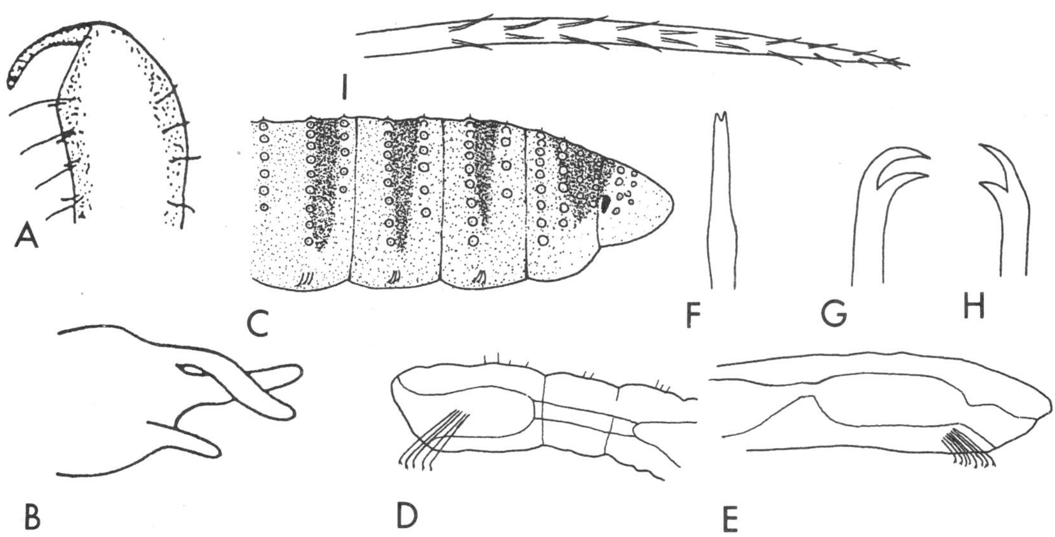

Fig. 13.17. Structure of Opistocystidae and Naididae. A, anterior of *Crustipellis;* B, posterior of *Crustipellis;* C, anterior end of *Ophidonais serpentina;* D, anterior end of *Chaetogaster diastrophus;* E, anterior end of *C. limnaei;* F, dorsal seta of *O. serpentina;* G, tip of seta of *Chaetogaster limnaei;* H, terminal bifurcation of seta of *C. diaphanus;* I, dorsal seta of *Vejdovskyella*. (A, B from Milligan 1997.)

21. Up to 12 mm in length, with 50 segments; 1 species, FL . . . . *Trieminentia*
Less than 10 mm; with 30 or less segments 22

22. Minute worm, less than 3 mm long; dorsal setae including hispid (serrate) hairs; 1 species, FL, LA, MI, NJ, NC . . . . *Crustipellis*
Worm exceeding 5 mm; dorsal setae without hispid hairs; 1 species in NJ, PA . . . . *Opistocysta*

23. Body papillate, roughened, or obscured with foreign matter (Fig. 13.16B) 24
Body smooth or without adherent material 27

24. With eyes; 1 common and widespread species . . . . **NAIDIDAE (in part),** ***Slavina***
Without eyes 25

25. Body wall glandular and covered with foreign matter; ventral setae bifurcate, upper tooth much larger and longer than ventral; 2 species in the eastern U.S.
NAIDIDAE (in part), ***Stephensoniana***
Body wall distinctly papillate, with some attached material; ventral setae simple or bifurcate, posteriorly the ventral tooth largest and very broad (Figs. 13.21N; 13.22J)
TUBIFICIDAE (in part), 26

26. Prostomium and segment I retractile (retracted in preserved specimens); most species with only a single seta per bundle on posterior segments; 4 species, 2 very widespread
Spirosperma
Prostomium and segment I not clearly retractile; multiple ventral setae posteriorly; 1 common and widespread species . . . . *Quistadrilus*

27. Specimens with anterior proboscis or several terminal appendages, gills or dorsal setae beginning after segment XV or dorsal setae absent . . . . **NAIDIDAE (in part), 28**
Specimens without any of these features 34

28. With a proboscis (Figs. 13.16C; 13.18G, K; 13.19E, F) 29
Without a proboscis 32

29. Dorsal setae beginning on segment II (Fig. 13.18G, K); several species distributed throughout the U.S. . . . . *Pristina*
Dorsal setae beginning on segment IV or VI 30

30. Dorsal hair setae in bundles of more than 3 (up to 18) 31
Dorsal hair setae in bundles of 1–3 (Fig. 13.19G); 1 widespread species . . . . *Stylaria*

31. Ventral setae missing in segments IV and V; 1 species in northeastern states . . . . *Ripistes*
Ventral setae complete; 1 widespread species . . . . *Arcteonais*

32. With posterior gills enclosed in a hood (fossa) (Figs. 13.16D; 13.18J); many species throughout the U.S. . . . . *Dero*
Without posterior gills 33

33. Without dorsal setae (Fig. 13.17D, E); several species, some commensal with other invertebrates or vertebrates, widely distributed . . . . *Chaetogaster*
With dorsal setae absent before XVIII; 1 species in the eastern half of the U.S. . . . . *Haemonais*

34. Dorsal setae beginning in segment II 44
Dorsal setae beginning in segment III or later . . . . **NAIDIDAE (in part), 35**

35. Dorsal setae beginning in III 36
Dorsal setae beginning in segment V or VI 37

36. Dorsal setae on III elongate (Fig. 13.18B); 1 species, FL, NJ . . . . *Bratislavia* **(in part)**
Dorsal setae on III not elongate; 2 widespread species . . . . *Amphichaeta*

37. Dorsal setae beginning on segment V; no hair setae; 4 species in coastal freshwater
Paranais
Dorsal setae beginning on VI; with or without hair setae 38

38. Hair setae absent 39
Hair setae present (Fig. 13.19B, D) 40

39. Dorsal setae straight and simple or with slight terminal indentation (Fig. 13.17F); 1 widespread species . . . . *Ophidonais*
Dorsal setae slightly curved and distinctly bifurcate (Fig. 13.19J); 1 widespread species (Fig. 13.16A) . . . . *Uncinais*

40. Dorsal hair setae hispid (serrate) (Figs. 13.11B; 13.17I); 2 widely distributed species
Vejdovskyella
Dorsal hair setae not hispid 41

41. Up to 6 dorsal hair setae per bundle; 1 widespread species . . . . *Specaria*
Never more than 3 hair setae per bundle dorsally . . . . **(3 similar genera), 42**

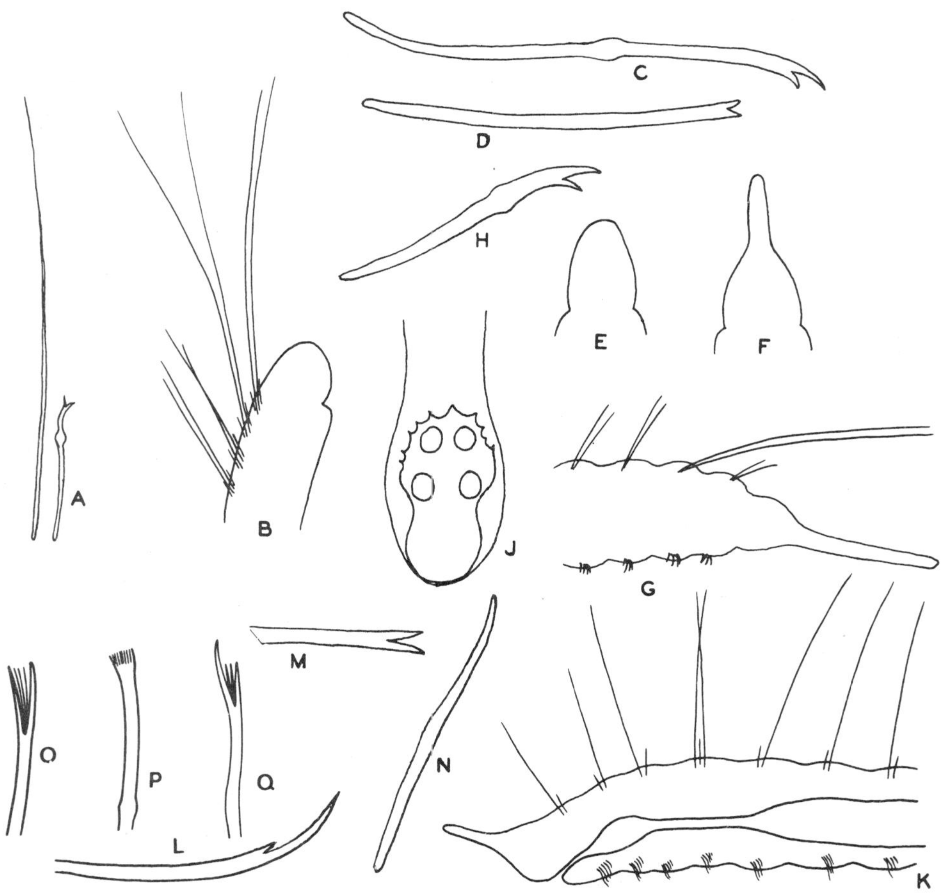

Fig. 13.18. Structure of Naididae. A, dorsal bundle of *Pristina osborni;* B, anterior end of *Bratislavia bilongata;* C, ventral seta of *Pristina aequiseta;* D, dorsal seta of *P. aequiseta;* E and F, extreme variations in anterior end of *P. breviseta;* G, anterior end of *P. longiseta;* H, thick seta of VI of *Nais pardalis;* J, branchial apparatus of *Dero nivea;* K, anterior end of *Pristina plumaseta;* L, tip of ventral seta of *Chaetogaster setosus;* M, dorsal seta of *Nais elinguis;* N, dorsal seta of *N. simplex;* P, and Q, examples of pectinate setae (tips). (B modified from Chen 1944; G modified from Marcus 1943; K modified from Leidy 1883; C, D, H, J modified from Brinkhurst and Jamieson 1971; O–Q modified from Hiltunen and Klemm 1980.)

42. Dorsal setae usually pectinate, less so bifurcate, with 1 very prominent tooth; 2 species generally distributed . . . . ***Allonais***
Dorsal setae usually bifurcate, less so pectinate; without a prominent tooth **43**

43. Ventral setae of II–V different from those posterior to VI; large genus with several common and widely distributed species . . . . ***Nais***
Ventral setae similar throughout body (Fig. 13.19K); 2 species generally distributed in the northeastern U.S. . . . . ***Piguetiella***

44. Dorsal setae bundle usually composed of 1–4 long hair setae and distinct, elongate, needlelike setae, simple or with nearly parallel teeth, with apex or teeth pointed upward; body length about 10 mm or less; two very similar genera with indeterminate number of species **45**
Dorsal setae typically curved apically with teeth oblique to shaft (Figs. 13.21E; 13.22K–M), hair setae present or absent; body usually well over 10 mm in length
TUBIFICIDAE (in part), 46

Fig. 13.19. Structure of Naididae. A, dorsal view of anterior end of *Nais communis;* B, dorsal bundle of setae of *N. communis;* C, ventral bundle of setae of *N. communis;* D, dorsal bundle of setae of *N. elinguis;* E and F, anterior end of *Stylaria lacustris;* G, dorsal bundle of setae of *S. lacustris;* H, ventral setae of *Dero* II, VI, and X; J, seta of *Uncinais uncinata* II; K, seta of *Piguetiella michiganensis* II; L, ventral setae of *Bratislavia unidentata* II and VI. (H–L from Hiltunen and Klemm 1980.)

45. Dorsal setae commonly with 1 simple needle per bundle, occasionally a long, very thin hair seta; at least 1 species in the eastern U.S. . . . . ***Bratislavia*** **(in part)**
Dorsal setae bundles with 1–3 needle like setae, simple, bifurcate, or weakly pectinate, and 1–4 hair setae; several obscure species, some possibly not valid, widespread in the eastern half of the U.S. . . . . ***Pristinella***

46. Hair setae absent **47**
Hair setae present (Fig. 13.20B) **56**

47. Dorsum of prostomium with a median pit and canal; modified ventral (genital) setae, if present, with severly bent, beaklike tip; 1 widespread species . . . . ***Bothrioneurum***
Without a prostomial pit; modified ventral setae, if present, otherwise **48**

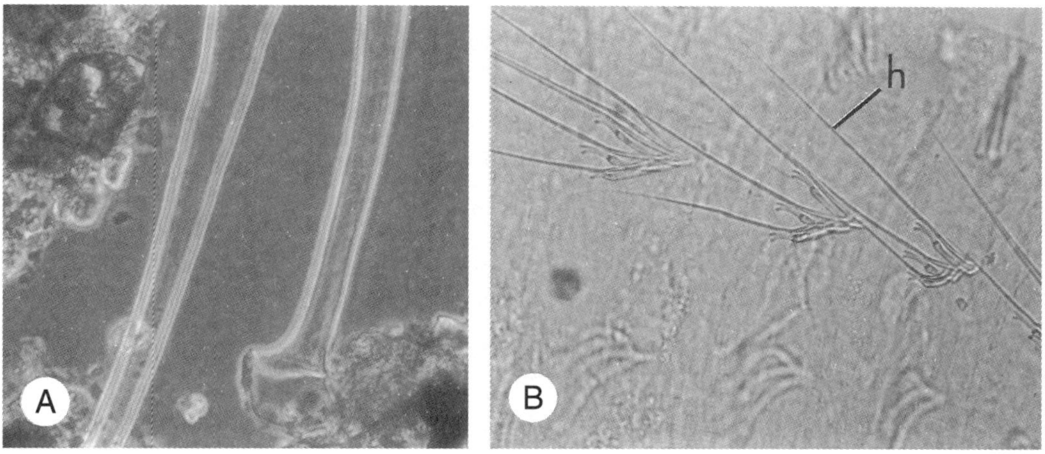

Fig. 13.20. Structure of Tubificidae. A, penial sheaths of *Limnodrilus;* B, dorsal setae clusters of *Tasserkidrilus,* h, hair seta.

48. With simple setae anteriorly or posteriorly 49
No simple setae 50

49. Anterior dorsal setae simple, posteriorly becoming dome tipped with peripheral minute teeth; 1 species .... ***Telmatodrilus***
Posterior ventral and dorsal setae simple; 1 species in the Great Lakes .... ***Phallodrilus***

50. Mature worm with modified ventral setae from IX– XI, broad terminally, knoblike or spatulate (Fig. 13.22O); 1 rare and discontinuous species .... ***Rhizodrilus***
Mature worm without penial setae or penial setae otherwise 51

51. Mature specimens with a pair of tubular short or long penial sheaths present in region of X and visible in cleared and weakly stained specimens (Fig. 13.20A) 52
Without tubular penial sheaths 53

52. Penial sheaths long and folded at base, or short and only slightly longer than broad, with upper margin flared; 3 species scattered about the U.S. .... ***Varichaetadrilus* (in part)**
Penial sheaths long, smooth, and narrow, with upper margin flared, scalloped, or otherwise modified (Fig. 13.20A); many species in the U.S., a few very common .... ***Limnodrilus***

53. Mature worms with long, modified ventral setae on X, the distal portion fluted and hollow (Figs. 13.21P; 13.22A, B, N) 54
Mature worm without modified ventral (genital) setae or modified setae otherwise 55

54. With strongly curved ventral setae in posterior portion of body or not; penes not typically everted in preserved specimens; 2 species, 1 common and widespread .... ***Isochaetides***
Without strongly curved ventral setae; penes typically everted in preserved specimens; 1 species in the Great Lakes region .... ***Potamothrix* (in part)**

55. Setae bifurcate throughout; 1 very scattered species .... ***Rhyacodrilus* (in part)**
Setae diverse, especially posteriorly, either pectinate (Fig. 13.22F, H) or widened distally (Figs. 13.21K, L; 13.22G, H), or reduced to one with curved teeth (Fig. 13.21M); many species generally distributed .... ***Aulodrilus* (in part)**

56. Worm with large, eversible, and lacunar pharynx filling anterior coelom (Fig. 13.12B); 4 species largely restricted to the west coast .... ***Teneridrilus***
Worm without enlarged pharynx; linear in coelom 57

57. Anterior and dorsal setae pectinate 58
Anterior and dorsal setae bifurcate 65

58. Penial sheaths, if present, very short, somewhat conical, and inconspicuous, not tubular (Fig. 13.22C) (identification of following genera requires adult worm) 59
With distinct tubular penial sheaths; upper (proximal) end often reflected over shaft or flared (2 very similar genera) 64

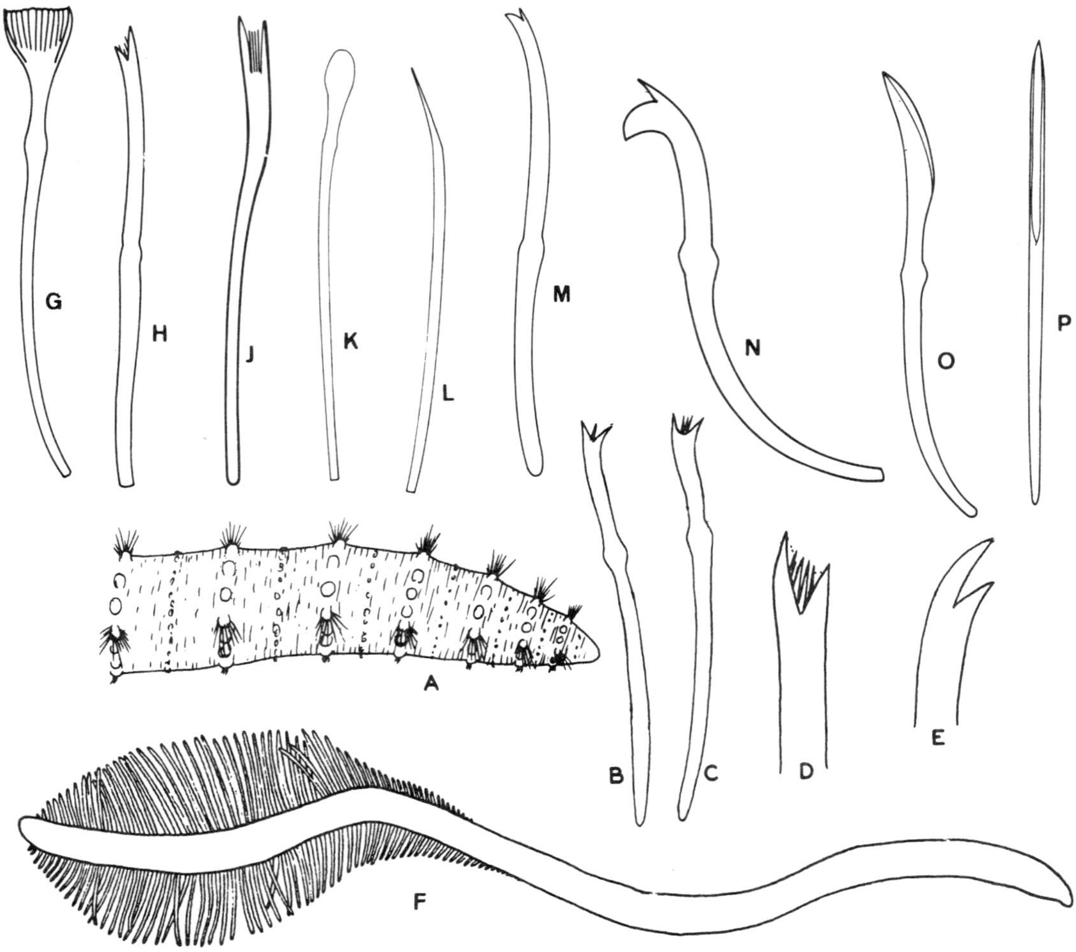

Fig. 13.21. Structure of Tubificidae. A, anterior end of typical tubificid; B and C, dorsal pectinate setae of *Tubifex tubifex;* D, distal end of dorsal pectinate seta of *Ilyodrilus;* E, tip of seta of *Limnodrilus;* F, lateral view (diagrammatic) of *Branchiura sowerbyi*, ×2, showing dorsal and ventral gills; G, anterior dorsal pectinate seta of *Psammoryctides;* H and J, anterior dorsal pectinate seta of *Rhyacodrilus;* K, facial view of posterior dorsal crotchet seta of *Aulodrilus pigueti;* L, lateral view of same; M, ventral crotchet seta of *Aulodrilus;* N, posterior dorsal crotchet seta of *Tubifex harmani;* O, seta of *Potamothrix;* P, *Psammoryctides* spermathecal setae of segment X. (G–P from Stimpson et al. 1975.)

59. Worm without modified ventral (genital) setae on X or XI; 3 species, 2 on the west coast, 1 widespread . . . . ***Ilyodrilus***
With modified ventral (genital) setae on X or XI or both **60**

60. Modified ventral setae on XI, elongate and bifurcate; 2 species restricted to few localites in western states . . . . ***Varichaetadrilus* (in part)**
Modified ventral setae otherwise **61**

61. Several modified ventral setae on XI, tips blunt and convergent; 4 species generally distributed ***Rhyacodrilus* (in part)**
Modified setae on VIII, X, or XI, fluted and hollow (3 superficially similar genera) **62**

62. Modified setae with broad, opened tips (Fig. 13.22A, B); 3 species widespread in the north central U.S. . . . . ***Potamothrix* (in part)**
Modified setae narrow to tips **63**

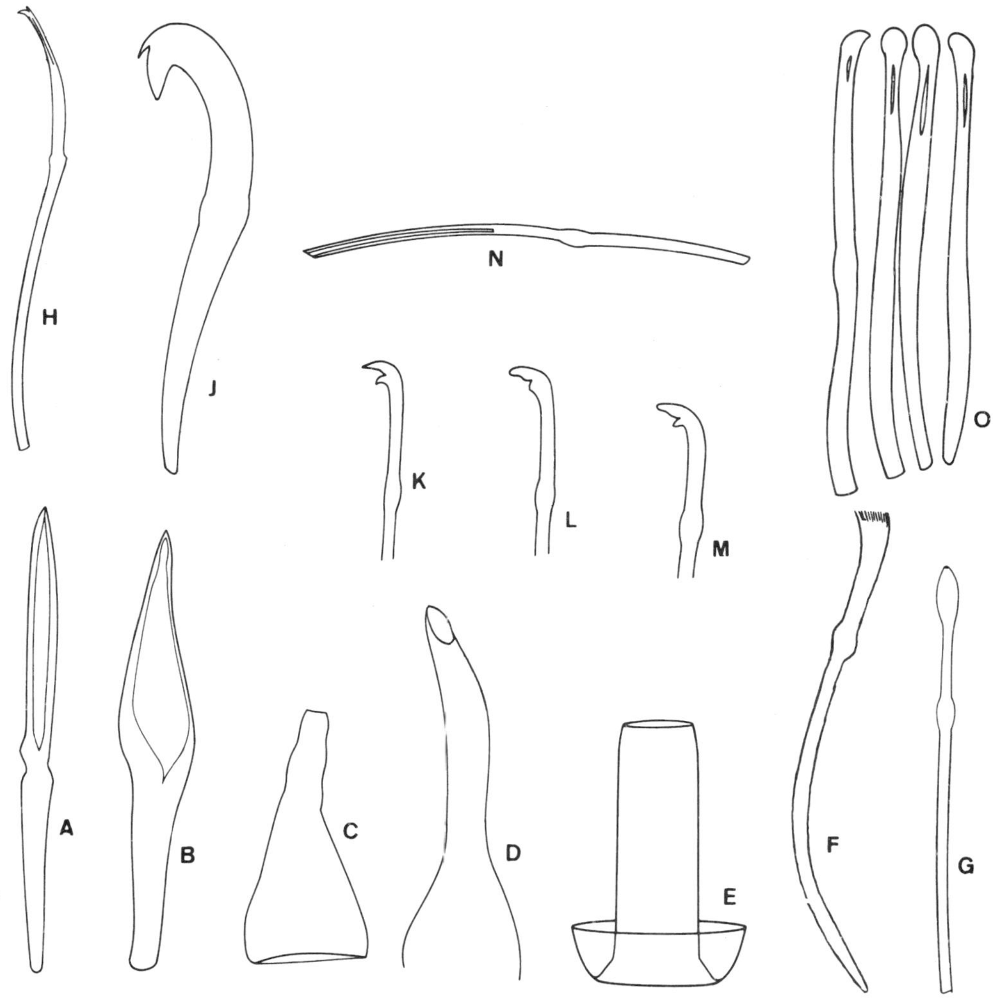

Fig. 13.22. Structure of Tubificidae. A and B, spermathecal setae of *Potamothrix;* C, penis sheath of *Ilyodrilus templetoni;* D and E, same for *Tubifex;* F, posterior dorsal crotchet seta of *Aulodrilus americanus;* G and H, aspects of posterior dorsal crotchet seta of *Aulodrilus limnobius;* J, posterior dorsal seta of *Isochaedites curvisetosus;* K to M, variations in tip of ventral setae of *Limnodrilus udekemianus;* N, spermathecal seta of *Isochaetides freyi;* O, knobbed setae in XI of *Rhizodrilus lacteus,* (Chiefly from Stimpson et al. 1975.)

63. Modified setae on X (Fig. 13.21P); 3 species discontinuous in the U.S. ..... ***Psammoryctides***
Modified setae on X and XI; 1 species in coastal freshwater of eastern states .... ***Haber***

64. Penial sheaths short, only slightly longer than wide (Fig. 13.22D, E); many species, several widespread and common .... ***Tubifex***
Penial sheaths distinctly longer than width; 2 species with a mostly northern distribution ***Tasserkidrilus***

65. Ventral setae 4 or less; modified (genital) ventral setae unknown, but probably blunt tipped and convergent; 1 species known only from Lake Tahoe, CA .... ***Rhyacodrilus* (in part)**
Ventral setae 4 or more, up to 16; modified ventral setae otherwise **66**

66. Dorsal setae becoming oar-shaped (Fig. 13.21K) or with accessory tooth (trifurcate) posteriorly; 2 widespread species .... ***Aulodrilus* (in part)**
Dorsal setae bifurcate throughout; 1 species in Great Lakes region and eastward ***Potamothrix* (in part)**

BRANCHIOBDELLIDA

In these commensals of crayfishes there is always a distinct head region. They are so similar to leeches and oligochaetes that they have been at different times considered a leech family or a higher taxon within either the Hirudinea or Oligochaeta. From a morphological standpoint, however, the Branchiobdellida form a connecting link between typical oligochaetes and true leeches.

The body (Fig. 13.23) ranges from 1 to 12 mm in length and consists of head, trunk, and a posterior glandular attachment disk. The cylindrical head consists of four fused segments, which are sometimes superficially annulated. A prostomium is absent. The mouth is surrounded by the peristomium and epidermal gland cells used for attachment, the region is referred to as the anterior attachment site. The peristomium may also contain a series of small projections or lips. Just inside the mouth is a pair of sclerotized jaws. Anterior as well as posterior attachment is accomplished by gland cells of two types, an adhesive secreting form and a releasing form.

The trunk consists of 11 segments, although the last three are often indistinct and somewhat fused into the base of the attachment disk. Trunk segments are usually subdivided externally by transverse sulci. Setae are absent, and there are only two pairs of nephridia. In several genera the trunk segments are dorsally thrown into folds, ridges, or digitate processes. The anus is on the dorsal side of segment X. In most species a spermatheca (Fig. 13.24) opens on the ventral midline of V, and there is a male genital pore on the ventral midline of VI. There is a pair of testes in each of V and VI. At maturity, the testes disassociate into clumps of spermatids, which, with developing and mature spermatozoa, fill the coelomic cavities of V and VI in American members of the group. A pair of ovaries is present in VII, and there are two lateroventral female genital pores. Large eggs may be seen in sexually mature specimens. The glandular clitellum forms around V, VI, and VII. Cocoons are formed, but almost nothing is known about the embryological development.

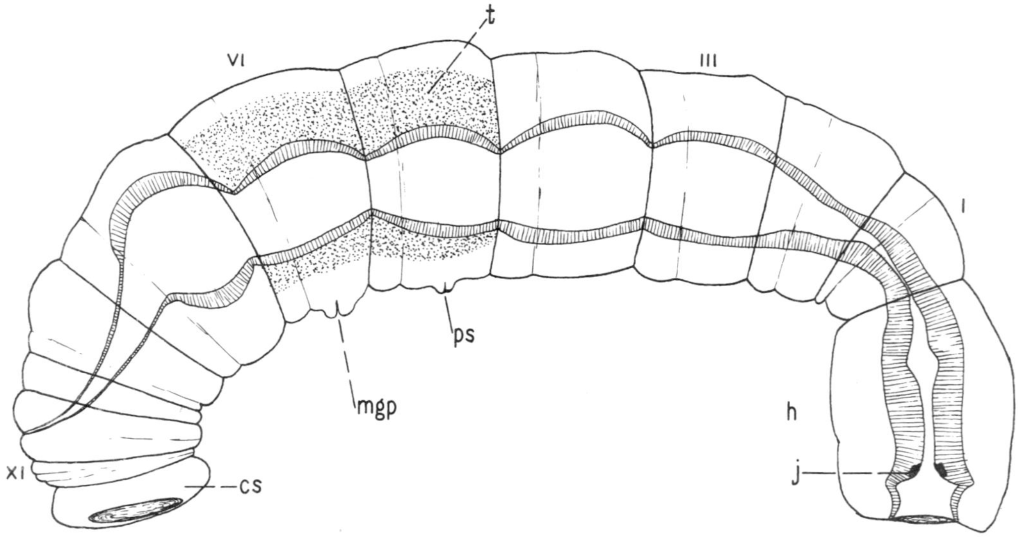

Fig. 13.23. Diagram of *Cambarincola*, lateral view, showing digestive tract, ×70. cs, caudal sucker; h, head; j, jaw; mgp, male genital pore; ps, pore of spermatheca; t, testes (shaded areas).

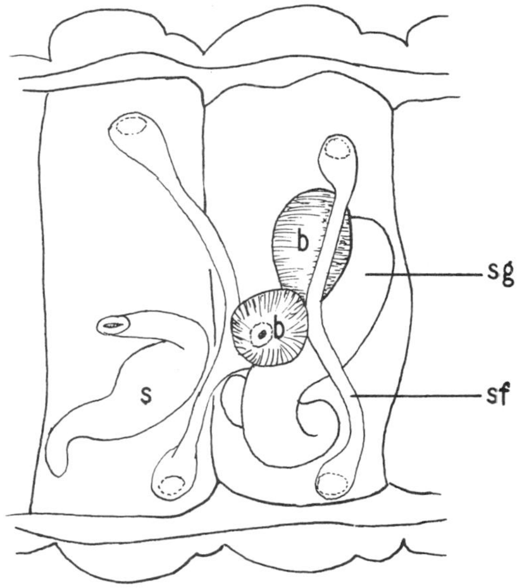

Fig. 13.24. Ventral view of reproductive organs of V and VI of *Ankyrodrilus* (testes not shown). b, bursa; sf, sperm funnel; sg, seminiducal gland; s, spermatheca. (Modified from Holt 1965.)

Branchiobdellids in the United States are always associated with cambarid and astacid crayfish, except for some species of *Cambarincola* that are found on cavernicolous isopods (*Caecidotea*) and on the portunid crab *Callinectes sapidus* in brackish and fresh water. They occur on the gills, on the inner surface of the gill chambers, and on the general anterior external surface of the body and appendages. Sometimes a single crayfish may harbor as many as 200 or 300 worms. It is thought that the worms are essentially commensals since their digestive tracts usually contain only diatoms, other algae, and organic debris. Nevertheless, some species almost surely feed on blood of the host taken from gills or intersegmental membranes, and the species found on *Caecidotea* apparently feed on the eggs of the host. Branchiobdellids have survived up to 240 days in the laboratory, separated from their hosts, but have never been observed to produce cocoons under such conditions.

Except for *Xironogiton*, species of branchiobdellids reported from western crayfish (*Pacifastacus*) in Pacific drainages have thus far not been found on eastern crayfish (Cambaridae), and vice versa; otherwise, there seems to be no host specificity. One species of *Cambarincola*, for example, has been reported from at least 11 species of Cambaridae. Nevertheless, most species are known only from small geographic areas, often only one state or one cave. Sometimes, when a crayfish carries two or more species of branchiobdellids, the worms are distributed in a specific and localized manner. In a study by Gelder and Smith (1987), for example, the crayfish *Cambarus bartoni* had *Bdellodrilus* only in the branchial chambers, *Cambarincola* only on the oral and ventral surfaces of the body, and *Xironogiton* only on the chelipeds.

A convenient place to find branchiobdellids is in the debris at the bottom of jars of preserved crayfish. Some genera, however, which inhabit the gills beneath the carapace do not become dislodged and must be removed manually. If living specimens are collected, they are best killed and fixed in a solution consisting of 4 parts formalin and 96 parts of 70% alcohol. Some workers relax living worms with magnesium chloride. Either stained or unstained specimens may be dehydrated and mounted permanently on slides in a suitable medium.

The Branchiobdellida has been the object of much study recently. Several key papers have elucidated aspects of functional anatomy, revised old and described new taxa, and provided additional data on species distributions. Most of the approximately 155 described species (Gelder 1996b) occur only in the United States. The species in the United States are distributed among 14 genera and 4 families. The key does not distinguish among families; one must consult Gelder (1996b) for this information. The systematic arrangement follows Holt (1986). The key was originally prepared by P. C. Holt and has been modified slightly.

Unfortunately, the determination of genera in this key is based mostly on differences in the anatomy of the male reproductive system (Figs. 13.25, 13.26). In many cases this requires careful observation of the internal reproductive organs, which can be done with whole mounts or by dissection of large

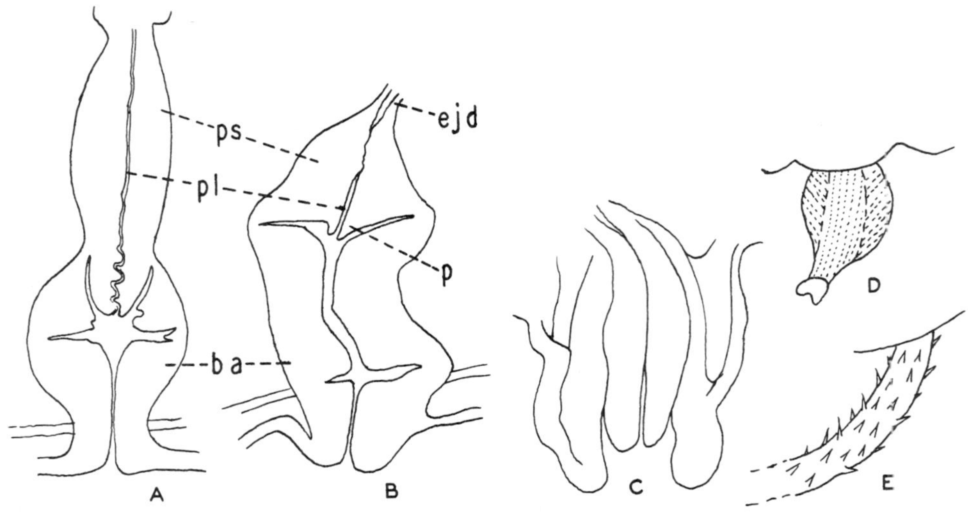

Fig. 13.25. Male copulatory structures in Branchiobdellida. A, male copulatory apparatus in *Triannulata;* B, male copulatory apparatus in *Magmatodrilus;* C, protrusible penis in shape of a muscular cone; D, eversible penis with internal strands; E, cuticular tube type of penis. ba, bursal atrium; ejd, ejaculatory duct; p, penis; pl, lumen of penis; ps, penial sheath. (A and B modified from Holt 1974; C–E courtesy of Perry C. Holt.)

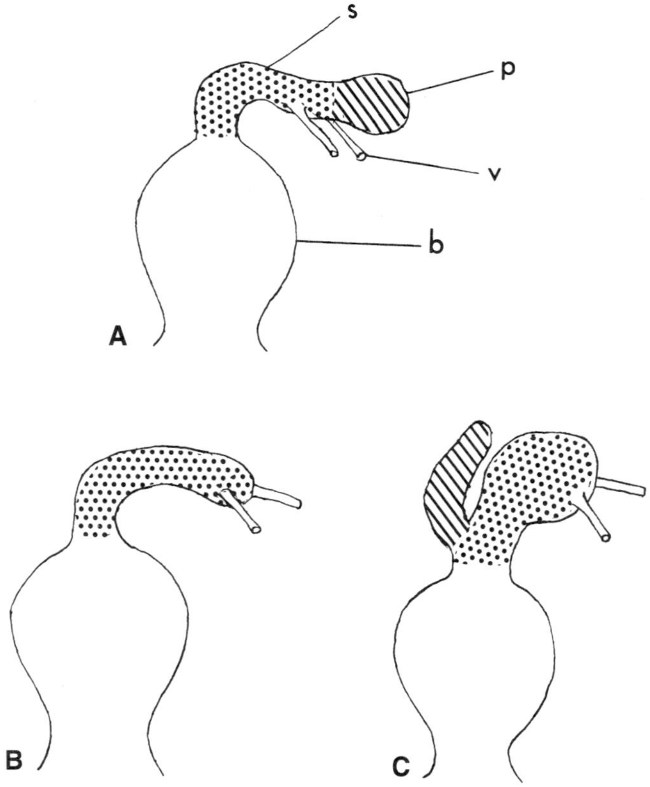

Fig. 13.26. Schematic of male reproductive systems in the Branchiobdellida. b, bulb (penial); p, prostate; s, seminiducal gland; v, vas deferens.

worms. Even so, considerable experience and effort is required. Students of the Branchiobdellida utilize the terms ectal and ental to denote whether a structure is positioned toward an extremity, typically outer or inner. The terms proximal and distal refer to the "starting point" and "end point" of a structure, respectively, and are more widely used in biology. To avoid confusion, proximal and distal are used parenthetically with ental and ectal when referring to the location or point of attachment of a specific structure. Otherwise, see Holt (1960b, 1986) and Holt and Opell (1993) for methods and definitions of anatomical terms. Four families are represented in the United States: the Branchiobdellidae, Bdellodrilidae, and Xironodrilidae without a prostate (Fig. 13.26B), and the Cambarincolidae with a variously developed prostate in addition to the seminiducal gland (Fig. 13.26A, C). Lengths given are of preserved specimens and are approximate; extended living worms may be two or three times as long.

KEY TO GENERA OF BRANCHIOBDELLIDA OF THE UNITED STATES

1. Body dorsoventrally flattened (depressed); paired nephridiopores on dorsal surface of segment III **2**

Body cylindrical (terete); single nephridiopore opening on dorsal surface of segment III **4**

2. Seminiducal gland not bifurcated entally (proximally) **3**

Seminiducal gland with ental (proximal) bifurcation (prostate) (Fig. 13.26C); vasa deferentia enter seminiducal gland near ectal (distal) end; spermatheca prominent (Fig. 13.24); body 2 to 3 mm long; 2 species in VA and TN . . . . . ***Ankyrodrilus***

3. Vasa deferentia enter tubular seminiducal gland near ectal (distal) end (Fig. 13.26B); spermatheca disproportionately small; 2 to 8 mm long (Fig. 13.28B); 5 species on crayfishes of both eastern and western mountains . . . . . ***Xironogiton***

Vasa deferentia enter tubular or subspherical seminiducal gland basally; spermatheca prominent; up to 6 mm long; 5 species on eastern crayfish . . . . . ***Xironodrilus***

4. Penis eversible, often with a distinct cuticularized tube (Figs. 13.25D, E) **5**

Penis a protrusible muscular cone (Fig. 13.25C); with a prostate (Fig. 13.26A) **12**

5. Eversible penis a cuticular tube, often very long, with or without spines (Fig. 13.25E) **6**

Eversible penis with internal strands (Fig. 13.25D), or an epithelial or muscular wall or both **7**

6. Vasa deferentia enter seminiducal gland ectad (distal) to ental (proximal) end of latter; prostate absent (Fig. 13.26B); penis without spines; about 2 mm long; 1 species in GA ***Cronodrilus***

Vasa deferentia enter ental (proximal) end of seminiducal gland; prostate present, often rudimentary (Fig. 13.26A); penis with spines; 1 to 3 mm long; 3 species in TN and KY ***Oedipodrilus***

7. With cylindrical projections borne dorsally on transverse folds of segments II to VIII (Fig. 13.27B); upper lip with four tentacles; 3 mm long; 2 species in ID, OR, UT, WA, WY ***Ceratodrilus***

Without such dorsal projections **8**

8. With pairs of large, clear glands in lateral body wall of segments I to IX (Fig. 13.27E); upper jaw a longitudinal ridge fitting concavity of lower jaw; bifid spermatheca; gill-inhabiting delicate worms; 2 to 3 mm long; 1 species in eastern states . . . . . ***Bdellodrilus***

Without lateral glands; spermatheca not bifid **9**

9. Penis and ejaculatory duct a continuous heavily muscular eversible tube (Fig. 13.25A); 4 to 5 mm long (Fig. 13.28C); 1 species in CA, OR, WA . . . . . ***Triannulata***

Penis and ejaculatory duct distinct regions; ejaculatory duct not eversible **10**

10. Bursal atrium long and heavily muscular (Fig. 13.25B); penial sheath and eversible penis short; seminiducal gland very long with prominent anterior deferent lobe; prostate absent (Fig. 13.28D); about 3 mm long; 1 species in CA . . . . . ***Magmatodrilus***

Bursal atrium shorter than penial sheath; noneversible penial sheath **11**

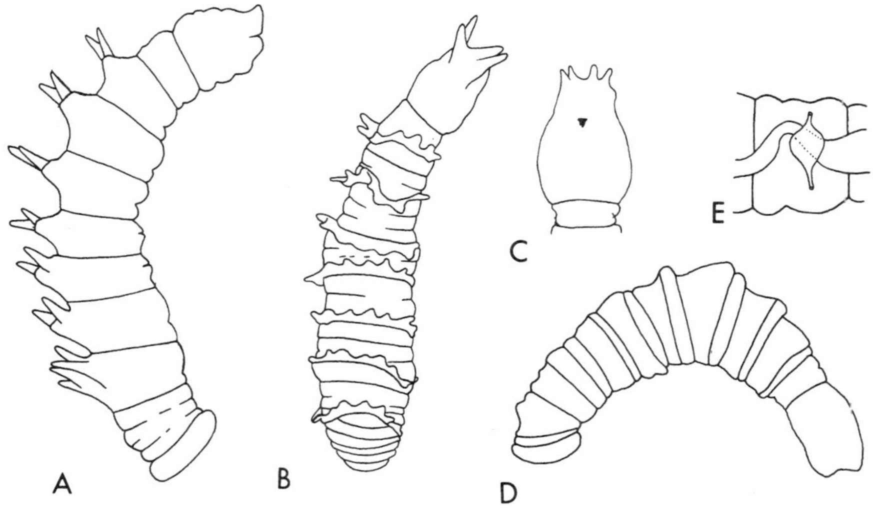

Fig. 13.27. Structure of Branchiobdellida. A, lateral view of *Pterodrilus*, ×85; B, dorsal view of *Ceratodrilus*, ×35; C, dorsal view of head of *Cambarincola;* D, lateral view, *Cambarincola*, ×24; E, lateral view of fourth trunk segment of *Bdellodrilus* showing lateral gland. (B modified from Hall 1915; D modified from Goodnight 1941.)

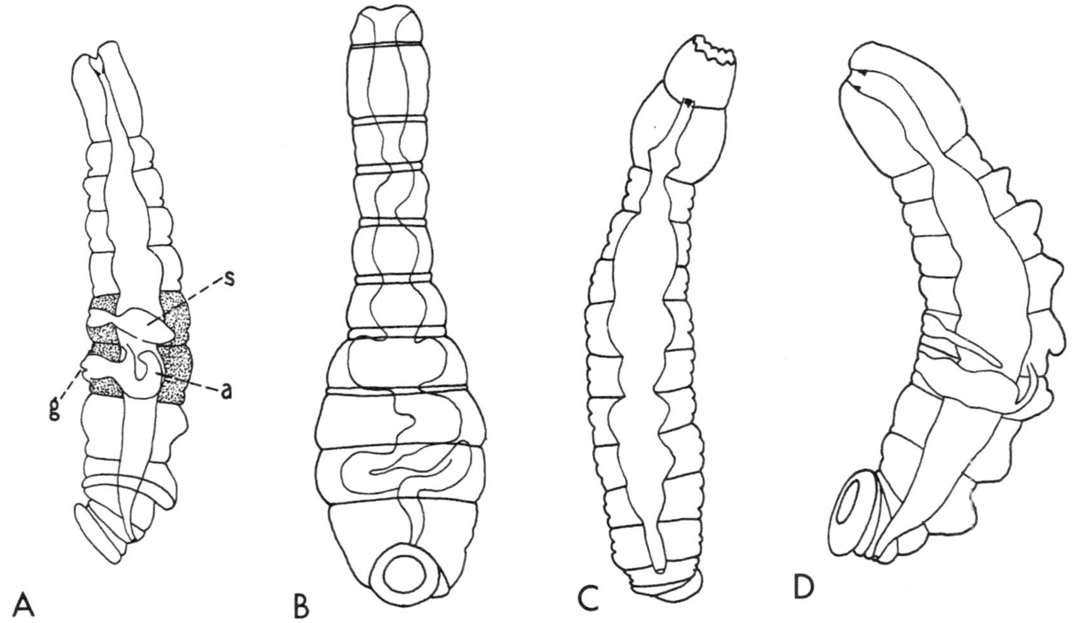

Fig. 13.28. Structure of Branchiobdellida. A, *Cambarincola*, lateral view, ×30; B, ventral view of *Xironogiton*, ×18; C, dorsal view of *Triannulata*, ×21; D, lateral view of *Magmatodrilus*, ×30. a, accessory sperm duct; g, male genital pore; s, spermatheca; testes shown as shaded area in A. (A, C, D modified from Goodnight 1940; B redrawn from Moore 1898.)

11. Seminiducal gland small and slender, subequal to ejaculatory duct in length and diameter; prostate short; penis cuticular, enclosed when retracted by muscular extension of ejaculatory duct, which projects into bursal atrium; 2 to 3 mm long; 1 species in TN ...... ***Tettodrilus***
Seminiducal gland various but always larger than ejaculatory duct; prostate present or absent; penis eversible, cuticular with strands connecting its outer everted surface to wall of penial sheath; 1 to 5 mm, long; 6 species with scattered distributions .... ***Sathodrilus***

12. Spermatheca absent; bursa asymmetrically subglobular; about 1.5 min long; 3 species; IN, KY, MI, TN .... ***Ellisodrilus***
Spermatheca present (Fig. 13.28A) **13**

13. Large worms; without dorsal appendages, though peristomial lobes or tentacles may be present; dorsal ridges, if present, not confined to segment VIII; 2.5 to 8 mm long; about 50 species in the U.S.; generally distributed .... ***Cambarincola***
Small worms; usually with finger- or fanlike projections on some dorsal ridges (Fig. 13.27A), but sometimes a single prominent dorsal ridge on segment VIII; 1 to 2 mm long; 9 species; uplands of eastern United States and Ozark Mountains .... ***Pterodrilus***

HIRUDINEA (Leeches)

The leeches are predominantly a freshwater group, although there are many marine as well as numerous terrestrial species, the latter mostly in the tropics. About 60 freshwater species are known from the United States, and they are common inhabitants of ponds, marshes, lakes, and slow streams, especially in the northern half of the country. To the layman, leeches are collectively known as "bloodsuckers," but many forms are predators and scavengers, and only a small minority of U.S. species will take blood from warm-blooded animals. Unlike the great majority of freshwater invertebrates, leeches are often brightly colored and patterned, and in some species the coloration is highly variable. The length of mature leeches ranges from minute species only 5 mm long to some of the giant *Haemopis* that have been reported to reach a length of 45 cm when extended and swimming.

Research in recent years has revealed that the two main branches of leeches, the Rhynchobdellida and Arhynchobdellida, have evolved several means to extract blood from mammalian hosts. A wide variety of biochemical tools are produced, including vasodilators, spreading agents, proteolytic enzymes, and platelet-formation inhibitors. In addition, many kinds of symbiotic bacteria have been discovered in organs of the digestive system that conserve or help break-down ingested blood. Because of the obvious pharmacological value of leeches, more than 90% of recent leech literature citations have been of a physiological or biochemical nature.

General characteristics. Unlike oligochaetes, the Hirudinea are dorsoventrally flattened (Fig. 13.29). The mouth is surrounded by an oral sucker that may be large, or small, fused with the anterior end, and liplike. The caudal sucker usually faces ventrally and is much larger, discoid, powerful, and expanded over a short central attachment pedestal. The anus is dorsal and in front of the sucker. The surface of the body may be smooth, wrinkled, tuberculate, or papillated. Although most genera have characteristic shapes, it should be borne in mind that leeches are highly muscular and contractile, and the body outline may vary greatly with locomotion, reaction to stimuli, and method of killing and fixation.

From the standpoint of basic body segmentation, external appearances are deceiving. All leeches are composed of only 34 true segments, but each segment is subdivided into a definite and constant number of superficial annuli. In the middle portion of the body a segment typically has 2 (rarely), 3 (Fig. 13.35), 5, or 6 annuli, but in some of the Piscicolidae (fish leeches) there are 7, 12,

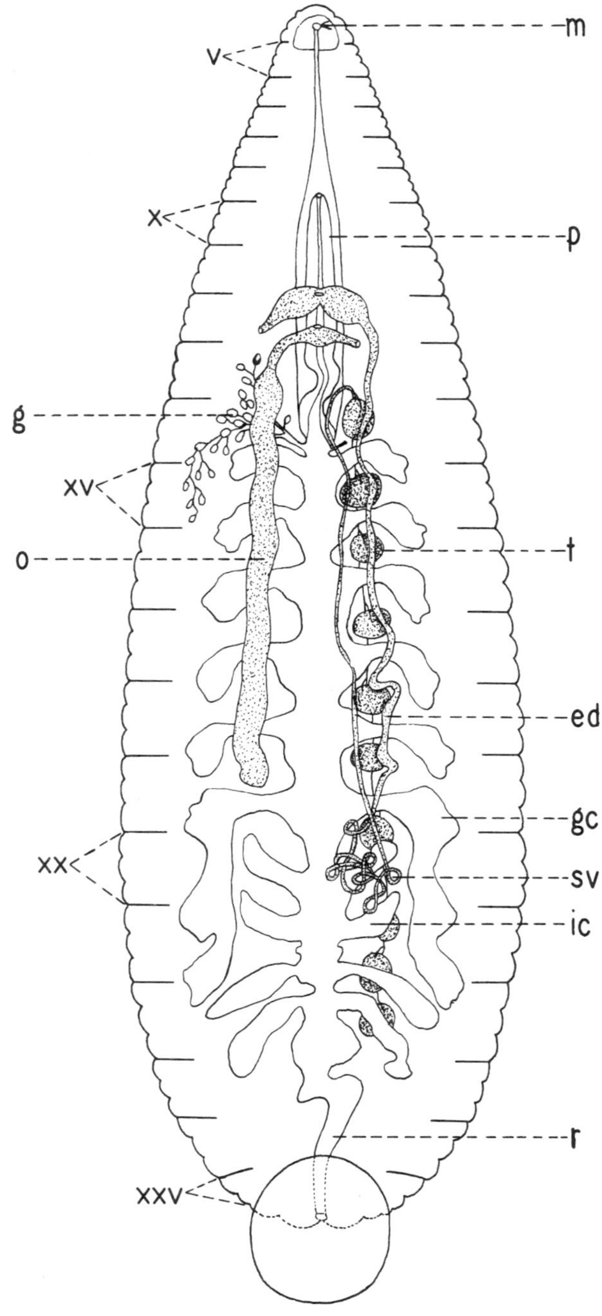

Fig. 13.29. Ventral diagram of *Glossiphonia complanata*, ×10, showing annulations, reproductive system, and digestive system. Reproductive system stippled; female organs shown only on the left; male organs shown only on the right. ed, ejaculatory duct; ic, intestinal caecum (second); g, salivary glands; gc, gastric cecum (sixth); m, mouth; o, ovary; p, proboscis; r, rectum; t, testis; sv, seminal vesicle. (Modified from Harding and Moore 1927.)

or 14 annuli. Toward the extremities of the body annulation is incomplete so that there are fewer annuli per segment than in the midbody segments. In *Haemopis marmorata*, for example, midbody segments all have 5 annuli, but segments I through VII are represented by 1, 1, 1, 2, 2, 3, and 4 or 5 annuli, respectively; and XXIII through XXVII are represented by 5, 4, 3, 2, and 1 annuli. Furthermore, in all leeches there is considerable fusion of the true segments of the body at the extremities, and especially at the posterior end. In *H. marmorata* segments XXVIII through XXXIV are all fused to form the caudal sucker.

The true body segmentation has been precisely established by a study of the nervous system. Each of the 34 segments is represented by a ganglion and paired nerves, but these structures are crowded together and partially fused at the anterior and posterior ends.

Some investigators distinguish the following major body regions in leeches: head (I–VI), preclitellar (VII–IX), clitellar (X–XIII), postclitellar (XIV–XXIV), anal (XXV–XXVII), and caudal sucker (XXVIII–XXXIV). The degree of differentiation of these regions is highly variable from one species to another.

The Class Hirudinea is divided into two orders. Each member of the Order Rhynchobdellida has a small, porelike mouth in the oral sucker through which a muscular proboscis can be protruded; jaws are absent; the blood is colorless; and there are three or six or more annuli per segment in the middle of the body. Members of the Order Arhynchobdellida have a medium to large mouth, no proboscis, jaws present or absent, red blood, and usually five annuli per segment in the middle of the body.

The Order Rhynchobdellida is subdivided into two American families. The Glossiphoniidae generally have a flattened body (Fig. 13.30), the anterior sucker more or less fused with the body, one to four pairs of eyes, and usually three annuli in each of the midbody segments. The American Piscicolidae, on the other hand, are somewhat cylindrical, with a distinct anterior sucker, one or two pairs of eyes, and usually six or more annuli per midbody segment; these leeches are almost invariably found as parasites on the body of fish or crustaceans.

The Order Arhynchobdellida contains three families in North America (Fig. 13.30). The Family Erpobdellidae have three or four pairs of eyes (never arranged in an arch), no jaws, and no gastric caeca. The Hirudinidae and Haemopidae, however, have five pairs of eyes (forming a regular arch), toothed jaws, and one to several pairs of gastric ceca.

Color. When collecting specimens in the field, notes on coloration should always be taken, since leeches quickly fade in preservatives, with only red, brown, and black pigments persisting. Uniform coloration is a rare condition, most leeches being brightly mottled, spotted, or segmentally striped; sometimes the color pattern is arranged transversely, sometimes longitudinally (Plate 6). The ventral surface is almost always paler than the dorsal. The Glossiphoniidae range from translucent and pinkish to an opaque and brightly colored condition. *Glossiphonia complanata*, for example, is dull green to brownish, with six longitudinal rows of yellowish spots and two dark longitudinal paramedian lines. *Macrobdella decora*, the American medicinal leech, is very showy; the general coloration is green with obscure longitudinal stripes or short median lines; in the midline are bright orange or reddish spots, while the marginal spots are black. The ventral surface is rich orange, either plain or spotted with black. *Placobdella parasitica* is richly and variably colored, and it is difficult to find two specimens with exactly the same coloration and intensity. The general ground color is some shade of greenish brown, but the body is variously spotted, striped, or blotched with yellow and orange. A histological study of this species has revealed three types of pigment granules: (1) a pale yellowish granular reflecting substance, (2) chromatophores that range from black (contracted) to green (expanded), and (3) reddishbrown chromatophores that form a fine network in the expanded state. Leeches show no background color adaptation, but usually become paler in darkness

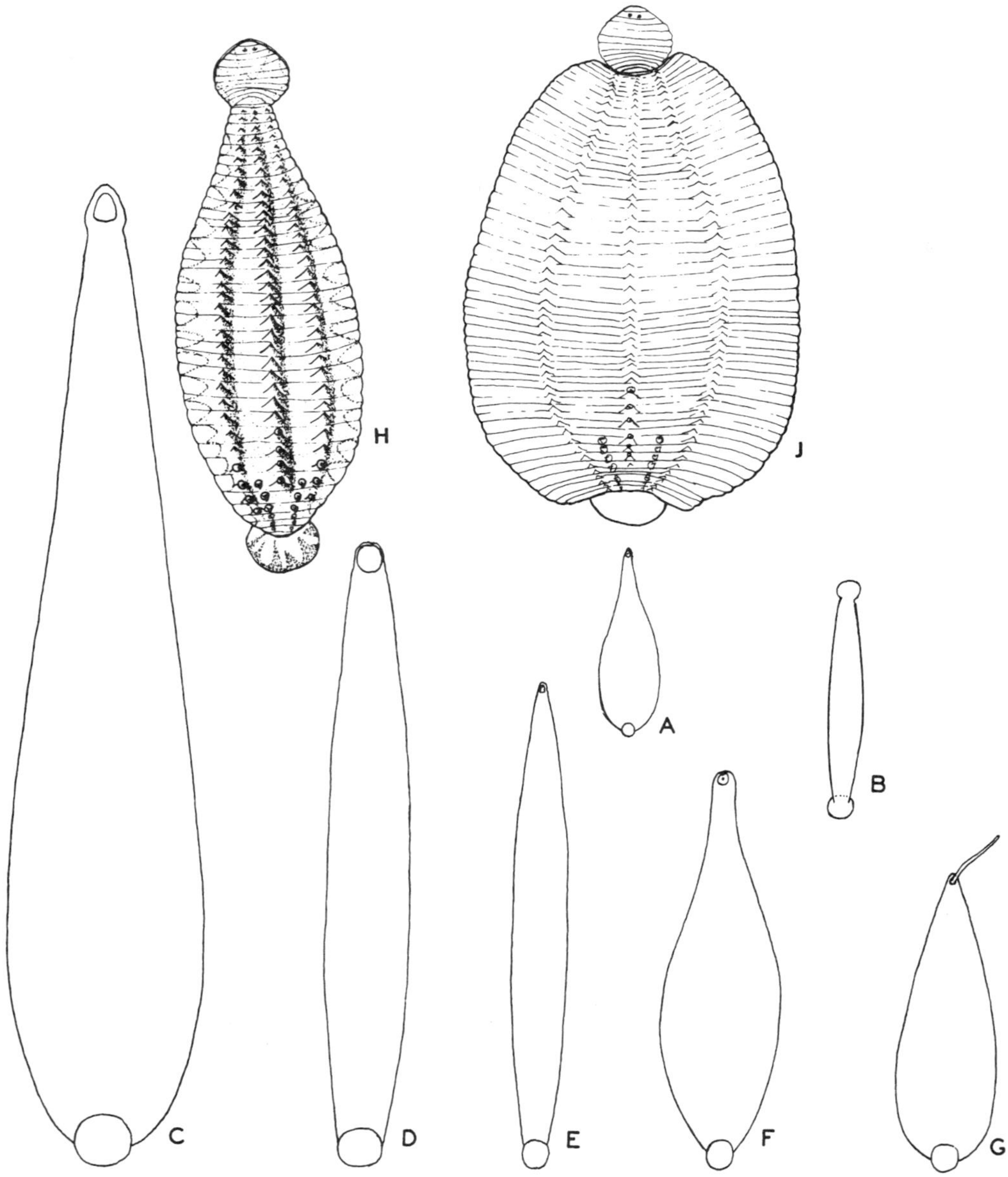

Fig. 13.30. Relative shape and size of typical American leeches, ×1. A, *Helobdella;* B, *Piscicola;* C, *Haemopis;* D, *Macrobdella;* E, *Erpobdella;* F, *Placobdella;* G, *Glossiphonia*, with extruded proboscis; H, *Placobdella montifera*, normal shape; J, same, contracted. (H and J from Klemm 1972.)

and darker under illumination. Color changes are thought to be governed through the central nervous system.

Locomotion. All leeches can move about on the substrate with characteristic creeping, looping, or inchworm movements, by getting a firm grip alternately with the oral and caudal suckers. The body shape changes greatly during such locomotion. The Hirudinidae and Erpobdellidae are excellent swimmers, moving along rapidly with graceful up-and-down or side-to-side undulations. The Glossiphoniidae are poor swimmers, however, and when disturbed they usually roll up like a pillbug and fall to the substrate.

Feeding, digestive system. The structure of the digestive system is partially correlated with food habits. In the Rynchobdellida there are no jaws, and the mouth is normally a small pore in the oral sucker. The anterior portion of the digestive tract contains an elongated, muscular, protrusible proboscis (Fig. 13.31) that lies in a proboscis sheath and is thrust forward and out of the mouth during feeding. There is a very short esophagus at the base of the proboscis, but the unicellular salivary glands in this region have long ducts that carry their secretion into the proboscis. In some rhynchobdellids there are, in addition to salivary glands, flat saclike structures called esophageal organs or mycetomes (Fig. 13.32). These structures contain symbiotic bacteria and although their function has not been established, it is believed they are involved with predigestion of ingested blood. The large stomach extends approximately through the middle third of the body and has from 1 to 10 pairs of lateral gastric ceca (often branched), in which quantities of ingested food may be stored for long periods prior to digestion. The gastric ceca are capable of great distension, and after a large meal a leech may weigh two to four times as much as it did before the meal. Digestion occurs in the intestine, and this part of the digestive tract is also supplied with ceca; usually, however, there are only four small pairs. The rectum is an unspecialized part of the digestive tract just before the anus.

The Glossiphoniidae have a wide variety of food habits. Many species are scavengers, feeding on dead animal matter, but more commonly they are carnivorous and feed on snails, insects, oligochaetes, and other small invertebrates. Some species have a notable preference for aquatic insects. *Glossiphonia complanata* and *Helobdella stagnalis* are almost always associated with snails on which they feed and are consequently commonly called "snail leeches." *Placobdella* and *Helobdella* are often temporary parasites on fish, frogs, and turtles, and they have occasionally been reported as taking a blood meal from humans, especially from an open abrasion on the skin. Except during the breeding season, the Pis-

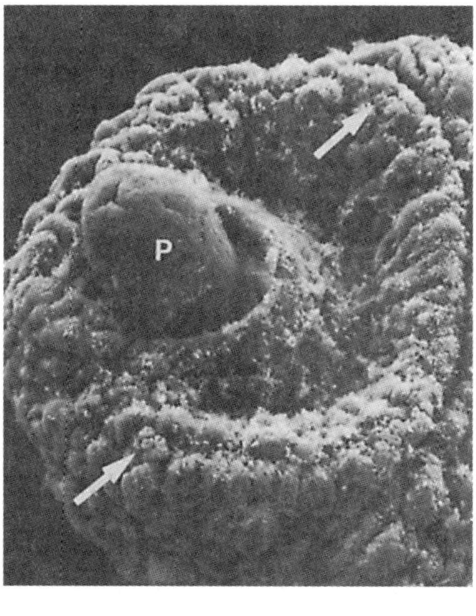

Fig. 13.31. Proboscis (p) of *Desserobdella phalera.* (From Jones and Woo 1990.)

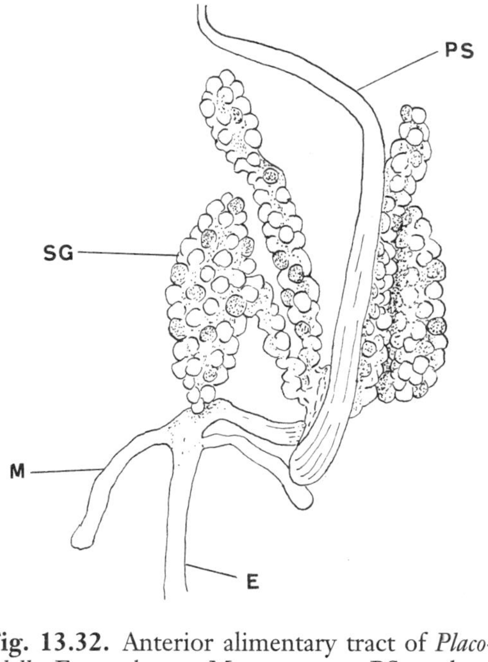

Fig. 13.32. Anterior alimentary tract of *Placobdella,* E, esophagus; M, mycetome; PS, proboscis sheath; SG, salivary gland.

cicolidae are almost invariably found as parasites on fish or crustaceans; they feed on the surface mucus as well as on blood and tissue fluids.

The fact that certain Rynchobdellida are capable of taking blood from vertebrates does not seem to have been satisfactorily explained, especially since these leeches have no teeth with which to make an incision in the host epidermis. In certain highly specialized South American glossiphoniids, a powerful protein is produced called hementin. The substance interferes with fibrin production, which in turn prevents clotting. Other glandular secretions have been identified that disrupt other blood chemicals and break down cutaneous tissues of the host. All of these products are secreted at the tip of the proboscis. North American species of bloodsucking glossiphoniids probably have similar mechanisms at their disposal to facilitate feeding.

The general plan of the Arhynchobdellida digestive system is similar to that of the Rynchobdellida, but there are several striking differences. For one thing, there is no proboscis, and in its place is a strong, muscular, sucking pharynx. The Hirudinidae and some Haemopidae, or true bloodsuckers, have three sharp-toothed jaws behind the large mouth (Figs. 13.33, 13.34). These are used in making incisions in the skin of the host. Some Haemopidae have no teeth or only vestigial teeth and normally feed on small invertebrates or dead animal matter, though *Haemopis marmorata* will occasionally attack humans. The latter species is called the "horse leech" because it commonly lives in ponds and troughs, where it attacks drinking cattle and horses. The Erpobdellidae never have true jaws. They are scavengers or feed mostly on small invertebrates, which they sometimes swallow whole; occasionally they suck blood from fish, frogs, and humans, especially at bleeding epidermal abrasions. None of the American Arhynchobdellida have intestinal ceca. The Hirudinidae and Haemopidae have one to several pairs of gastric ceca, but the Erpobdellidae have none.

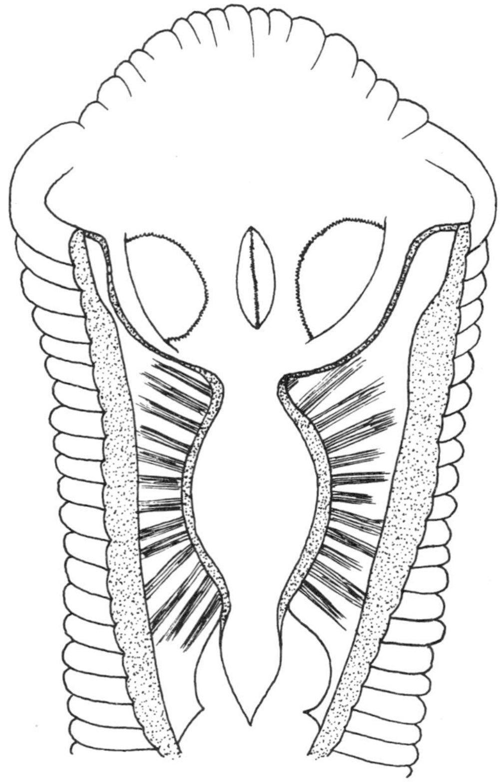

Fig. 13.33. Diagrammatic ventral dissection of the anterior end of *Hirudo*, Stippling denotes cut edges of oral chamber, pharynx, and body wall.

Macrobdella and *Philobdella* are the only common American leeches that regularly take human blood. Like all other bloodsuckers, they attach to the host with the caudal sucker and explore with the anterior end until a suitable spot is located, especially where the skin is thin. The oral sucker is then attached tightly and three fine painless incisions are made by back-and-forth rotary motions of the jaws. Sufficient blood is taken to distend the stomach and its ceca so greatly that the leech may be five times as heavy as it was when it began feeding. Much of the fluid of the ingested blood is excreted by the nephridia during feeding, the stored organic components, both cellular and noncelluar, being considerably concentrated. When the leech has filled its digestive tract, it leaves the host voluntarily, but the incisions keep on bleeding for a variable time because of the persistence of small quantities of the salivary anticoagulant, hirudin, in the adjacent tissues.

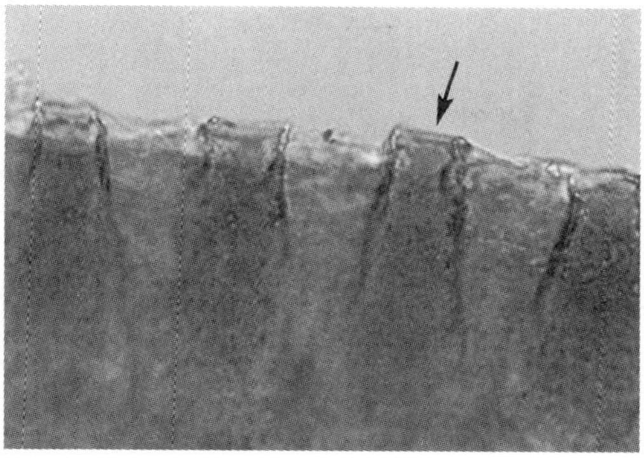

Fig. 13.34. Jaw of *Haemopis marmorata*. Arrow, tooth. (From Smith and Hanlon 1997.)

True bloodsucking leeches require only an occasional full meal, the stored blood being digested and utilized very slowly. Specimens have been kept for more than 2 years without feeding.

Free-living species may feed actively at temperatures as low as 2°C, but food intake rate usually stabilizes between 7 and 12°C. A "typical" small carnivorous leech requires only 1 to 4 chironomid larvae per day, for example. Predaceous species are usually nocturnal feeders.

Circulatory system. Unlike the condition in oligochaetes, where there is a spacious coelom regularly divided by transverse septa, the coelom of leeches is largely obliterated and reduced by greatly developed muscular, botryoidal, and parenchymatous tissues. The remaining sinuslike cavities consist of several longitudinal canals interconnected by smaller branch canals. The nerve cord, larger blood vessels, and reproductive organs are located in these sinuses.

The main portion of the circulatory system of typical Rynchobdellida consists of a large dorsal longitudinal blood vessel and a large ventral longitudinal blood vessel, which are connected near the two extremities by a series of convoluted branch vessels. The muscular wall of the anterior portion of the dorsal vessel acts as a heart. Near the posterior end the dorsal vessel is greatly expanded into an intestinal blood sinus that envelops the intestine and its caeca, and muscular contractions of the intestine are thought to aid in forcing the blood anteriorly out of this sinus and through the dorsal vessel proper. All the body tissues are well supplied with capillaries. In addition to the blood circulatory system, the coelomic sinuses contain a lymphlike fluid that circulates slowly. The two systems are unconnected, blood flowing in one and lymph in the other, both fluids are colorless. In some of the Piscicolidae the lymphatic system has a series of paired lateral pulsating vesicles that force the lymph around the body.

In the other two orders the true blood vascular system has disappeared, and the system of coelomic sinuses alone remains in a modified form. It contains blood that is red because of dissolved hemoglobin.

Respiration. Some marine leeches have special lateral gill-like structures, but none of the American freshwater species have any special respiratory structures. The oxygen and carbon dioxide exchange readily occurs through the general body surface since there is a rich network of capillaries just below the epidermis. Many species attach to the substrate with one or both suckers and undulate the body, probably to facilitate respiration. Little is known about the ability of active leeches to tolerate low concentrations of dissolved oxygen, but many species will live through several days without free oxygen. They do not appear to be able to tolerate highly acid conditions or decomposition gases at low oxygen concentrations.

Excretion. There are never more than 17 pairs of nephridia and these are all segmentally arranged in the middle body segments. The internal openings, located in the sinus system, are simple ciliated funnels in some species and simple to complex, variable "ciliated organs" in others. The paired external openings, or nephridiopores, are in the ventral furrows between annuli.

Nervous system. The brain is in the region of segments V and VI and consists of the fused ganglia and basal portions of the nerves of the first six body segments. There is a small suprapharyngeal mass, a small connective passing around each side of the pharynx, and a very large subpharyngeal mass. The posterior ganglionic mass, in segments XXV and XXVI, represents the more or less fused ganglia of XXV through XXXIV. The ventral nerve cord connects brain and posterior ganglionic mass, and each of VII through XXIV has a ganglion with three pairs of nerves.

Sense organs. Typical leeches have a variable number of small papillae arranged in transverse rows or extending completely around the body (Fig. 13.35). These are of two general types. Ordinary cutaneous papillae are tactile and possibly chemical receptors. Sensillae, however, are more specialized, light-colored or whitish, and sensitive to light and mechanical impulses. Although there may be numerous transverse rows or rings of cutaneous papillae in a single segment, there is never more than one row or ring of sensillae per segment. Usually a ring of sensillae consists of a dorsal transverse row of six or eight sensillae and a ventral row of six sensillae. Such a ring of sensillae is often called the neural ring, or sensory ring, and is always situated on the neural or sensory annulus of every somite. Anteriorly there may be sensillae surrounding the mouth (Fig. 13.31, arrows) and in some leeches these are able to detect vibrations of prey. The term is usually applied to the annulus and not to the sense organs themselves.

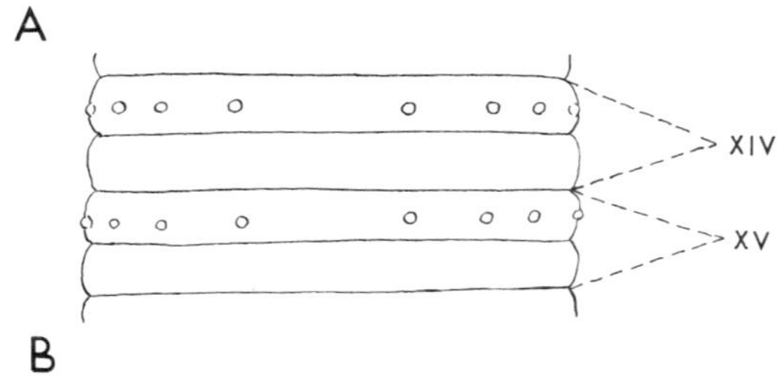

Fig. 13.35. Arrangement of sensillae and papillae in typical triannulate segments (e.g., *Placobdella*, A) and the biannulate *Oligobdella*, B. Sensillae are shown with central dots; all other areas are cutaneous papillae. Note that sensillae occur only on the middle annulus of each segment and that they are symmetrically arranged.

True paired eyes are found at the anterior end of the body. Each eye consists of a clump of dark-colored pigment granules, often cup-shaped, and an adjacent sensory area. Glossiphoniidae have one to four pairs of eyes: Piscicolidae one or two or more, Erpobdellidae three or four, and Hirudinidae five. A few leeches have eyespots on the caudal sucker or body somites.

Reproduction. In a very few leeches there is a common genital pore for both sets of reproductive organs, but usually there are two genital pores located on the median ventral line in XI and XII and separated by one to five or more annuli, the male genital pore always being more anterior than the female (see Fig. 13.41). The female reproductive system is much simpler than that of the male. It consists of two saclike ovaries of variable size and two very short, narrow oviducts that unite to form a vagina. The vagina, in turn, opens to the outside through the female genital pore. In the Hirudinidae and Haemopidae the female reproductive system also has an albumen gland, vaginal cecum, and muscular vagina.

Usually there are 5 to 12 pairs of segmentally arranged spherical testes, but in some of the Erpobdellidae they may be minutely subdivided like linear bunches of grapes. On each side of the body the testes are connected to a long vas deferens by means of short vasa efferentia. Each vas deferens has a modified seminal vesicle region that, in turn, becomes an ejaculatory duct. The two ejaculatory ducts unite to form a median atrium that is often complex and glandular.

The relative development and size of the

seminal vesicles and ejaculatory ducts vary greatly from one species to another. They may be short to long, straight or much convoluted, and small to massive. In the seminal vesicles sperm become cemented together and stored for future use.

An eversible muscular bursa is associated with the male atrium. In the Glossiphoniidae it is rudimentary, but in the other families it is generally larger and functional. In the Hirudinidae and Haemopidae, however, the bursa region contains a coiled filamentous tubular penis, often having a complicated structure.

As the sperm leave the seminal vesicle and pass down the ejaculatory ducts into the region of the atrium, they are formed into spermatophores (except in the Hirudinidae and Haemopidae) for conveyance to another individual during copulation. These cylindrical or double club-shaped structures, which have a chitinoid covering, consist of two more or less adherent tubes containing sperm bundles (Fig. 13.36). Spermatophores are formed by glandular areas of the atrium and adjacent portions of the ejaculatory ducts.

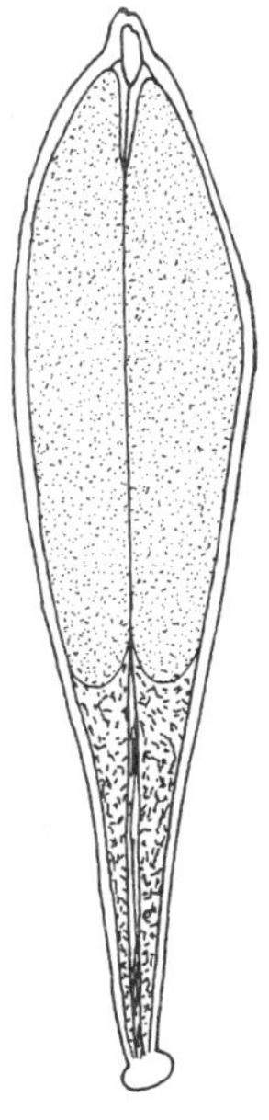

Fig. 13.36. Spermatophore of *Glossiphonia*, showing two oval bundles of sperm. (Modified from various sources.)

Like most hermaphroditic animals, an individual leech does not fertilize itself, but copulates with another individual and exchanges sperm. In the Rhynchobdellida and Erpobdellidae copulating pairs implant spermatophores on the ventral or dorsal surface of each other, usually in the clitellar region, but sometimes only one member of a copulating pair releases spermatophores. The sperm issue out of the spermatophores and penetrate the superficial tissues to the sinuses of the recipient leech with the aid of local histolysis. Soon the sperm reach the ovaries in the ventral sinuses and fertilize the eggs. In the Hirudinidae and Haemopidae the penis of each member of a copulating pair is extruded so that sperm are deposited at the genital pore or in the vagina of the recipient.

The clitellum of a leech is relatively inconspicuous, but in the breeding season it secretes ringlike structures that slip off the anterior end of the body to form cocoons, just as in the earthworm. As the cocoon passes the female genital pore it receives and encloses from one to several fertilized eggs. The Piscicolidae and most Arhynchobdellida fasten their cocoons to the substrate, but some of the arhynchobdellids superficially bury their cocoons in soft substrates, often in or on exposed mud along the shores of bodies of water. Cocoons are usually 2 to 15 mm long, more or less oval, and flattened on the attachment surface. The outer surface may be gelatinous, spongy, tough and membranous, horny, or chitinoid (Fig. 13.37). After a variable period of growth, the young leeches emerge from the cocoons and become independent.

The Glossiphoniidae do not form true cocoons but either carry their fertilized eggs in membranous capsules on the ventral surface of the body or attach them to a firm substrate. After hatching, the young attach to the body of the parent on the ventral surface. They are attached by means of mucous threads and probably feed on mucus until they attain quite a large size, at which time they leave the parent.

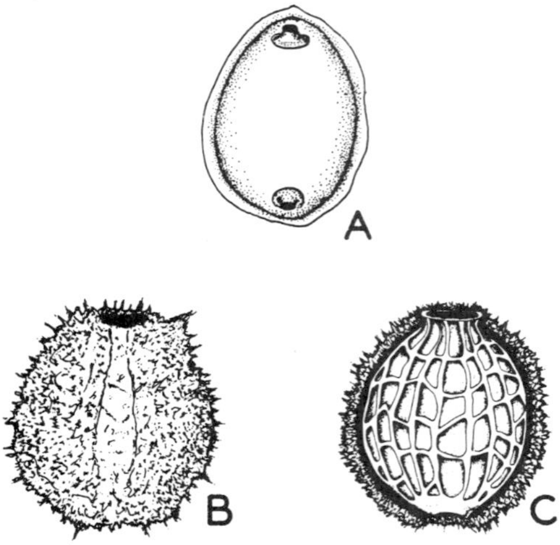

Fig. 13.37. Leech cocoons. A, *Erpobdella;* B, complete cocoon of *Cystobranchus;* C, cocoon of *Cystobranchus* with spongy layer removed from upper surface. (B and C modified from Hoffmann 1955.)

Copulation usually begins in the spring when temperatures are changing and after a good meal has been obtained; most cocoons are deposited between May and August. Some species produce batches of eggs for periods of 5 to 6 months, and empty cocoons may be found as late as October or November.

Little is known about the length of life of leeches. *Macrobdella decora* takes 2 to 3 years to mature, certain other Hirudinidae and Haemopidae may require 4 to 5 years, and it is claimed that some specimens may live for 10 to 15 years.

Ecology. Leeches abound in warm, protected shallows where there is little wave action and where plants, stones, and debris afford concealment. In such habitats population densities of 700 or more leeches per square meter of bottom have been reported. Exceptional densities exceed 50,000/m^2. Superficial daylight examination of a likely body of water may be disappointing, but bear in mind that leeches are chiefly nocturnal, and they are hidden under stones, vegetation, and debris during the hours of daylight. The great majority of specimens are collected between the water's edge and depths of 2 m, but a few species have been collected as deep as 50 m. Leeches require substrates to which they can adhere and consequently are uncommon on pure mud, clay, or sand bottoms.

Few species and few individuals occur in strongly acid lakes, and they are generally absent from sphagnum bogs. There is some evidence indicating that low calcium content of waters may be a limiting factor in leech distribution. They are notably absent from soft-water high mountain lakes, but this may be due to food shortages in such waters.

Unlike most groups of freshwater invertebrates, there is very little habitat preference among leeches. The same species (*Helobdella stagnalis,* for example) may occur in lakes, ponds, springs, slow streams, marshes, and on all kinds of suitable substrates. A few species of Erpobdellidae and Glossiphoniidae may be collected in swift streams, and some of them even tolerate some degree of pollution.

The leech fauna of intermittent ponds persists because some species are able to tide over dry periods by burrowing into the mud bottom, losing considerable weight, and constructing a small mucus-lined cell. Leeches have been known to estivate in this manner for more than 4 weeks at a time. Winters are spent in a lethargic dormant state buried in the upper parts of the substrate but below the frost line. During summer stagnation, leeches often migrate to shallow waters near shore where there is more dissolved oxygen.

Leeches serve as food for dytiscid larvae, caddis larvae, and odonate nymphs, and fish, but the predation rate is generally low.

Although American leeches are all considered aquatic organisms, there are some that are actually amphibious. *Haemopis,* for example, commonly crawls out of the water and wanders about on the moist shore and ground, sometimes as much as a quarter of a mile from the water, where it feeds on living and dead invertebrates, especially earthworms.

Semipermanent and permanent parasites,

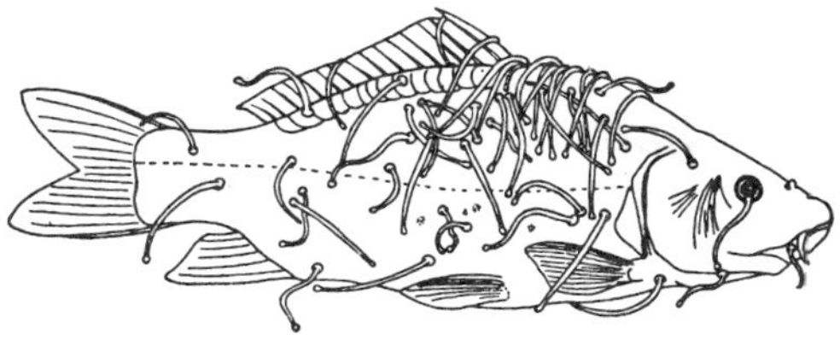

Fig. 13.38. A small carp heavily infected with *Piscicola,* × 0.5.

such as the Piscicolidae and *Placobdella parasitica,* remain on the body of the host except during the breeding season when they become free-living long enough to deposit their eggs on the substrate. But some of the fish leeches are completely dependent and deposit their cocoons on the host fish.

The true fish leeches (Piscicolidae) are almost exclusively found on fish (Fig 13.38). Only during reproductive periods or other rare periods do they leave their hosts. Fish parasitic glossiphoniids are not so obligated and can be often found free-living. Piscicolids and fish parasitic glossiphoniids are apparently not prey specific, since one species can be found on several fish species. In fact, both fish parasitic glossiphoniids and piscicolids can be found on the same fish species. The anadromous shortnose sturgeon, *Acipenser brevirostrum,* can host as many as six species of leeches—two glossiphoniids and four piscicolids.

Economic importance. The common medical practice of bloodletting as treatment for a wide variety of human ills supported a considerable leech industry in 18th and 19th centuries Europe. *Hirudo medicinalis* was raised commercially in so-called leech ponds, especially in France, Hungary, and Russia, and it is estimated that 25 million leeches were used in France in 1846 and 7 million in London hospitals in 1863.

Leeching never attained such great importance in the United States, and between 1840 and 1890 it is doubtful if more than 1.5 million leeches were used in any one year, including both the imported and native species (usually *Macrobdella decora*). The common European bloodsucker, *Hirudo medicinalis,* has been introduced at various times in the past but apparently without success. The maximum wholesale price was about 100 dollars per thousand. By 1890 the use of leeches had dwindled greatly, and by the turn of the century it was difficult to find a drugstore that still carried leeches in stock. Bloodletting by the use of leeches is chiefly confined to a few older European elements of our population. The practice is mostly resorted to as a home remedy for boils, contusions, and abscessed teeth. Very recently, however, several large American centers have gone to the use of *Hirudo* for removing blood from areas where tissue has been transplanted or reattached in microsurgery. In England, there is a large leech-producing firm that keeps a stock of 30,000 leeches, which are worth about five dollars each. In addition to the anticoagulant hirudin, other organic substances are derived from salivary secretions of the animals and are used for a variety of biochemical research.

Leeches sometimes become so abundant in northern lake resort areas as to constitute a nuisance for bathers. Disturbances in the water, such as splashing and currents, are known to attract bloodsuckers quickly. Unfortunately, there are usually no effective remedial measures that can be taken to reduce leech populations, but if the water level of a lake can be lowered at least a meter in the late fall, dormant leeches in the exposed muddy shallows will be killed by temperatures of −6°C or less. Leeches may be temporarily controlled in localized bathing beach areas by applying 100 lb of powdered lime per acre per day in the shallows.

Hook and line fishermen commonly use a wide variety of leeches as bait, the most common species probably being *Nephelopsis obscura,*

With reference to fish populations, the Glossiphoniidae, Hirudinidae, and Erpobdellidae can be dismissed as being of no real importance as parasites. Heavy infestations of Piscicolidae, however, have sometimes caused considerable mortality in both hatcheries and the natural habitat. Meyer (1969) summarizes the generally unsuccessful methods for treating fish infestations.

Dispersal, geographic distribution. Since many leeches temporarily attach to or definitely parasitize ducks and shorebirds, they are easily transported overland from one body of water or drainage system to another. In fact, the leeches are one of the new freshwater groups in which such transport is definitely known to occur. Some investigators report the recovery of viable cocoons from duck feces. Within a single drainage system, passive dispersal by attachment to fish is undoubtedly of great importance.

In addition, leeches actively and readily spread from place to place and colonize suitable habitats within individual drainage systems. Peculiar seasonal upstream mass migrations of great numbers of leeches have been reported several times. As shown in the key, some American leeches are known from only one or two states, whereas others are generally distributed in suitable habitats. One factor that may control the distribution of leeches and explains why some species appear to have restricted habitats or geographical ranges is that many species are somewhat host or prey specific. Thus, *Desserobdella cryptobranchii* is found only in areas containing the hellbender, *Cryptobranchus alleghenienisis*, its host. Other examples include *Oligobdella biannulata*, which parasitizes salamanders of the central Appalachian uplands, and *D. picta*, a parasite of amphibians of northern wetlands. *Erpobdella montezuma* feeds exclusively on the amphipod *Hyalella montezuma* in Montezuma Well, Arizona. Two species of *Glossiphonia* occur throughout Europe and much of Asia; *Helobdella stagnalis* and *Theromyzon tessulatum* are widely distributed in the north and south temperate zones.

Collecting, preparing. Usually specimens are collected individually with forceps by turning over rocks, logs, and debris in the shallows. A small dipnet should be carried to capture the rapid swimmers and to retrieve sluggish glossiphoniids as they roll up and fall to the bottom. Some collectors have had success by placing the body of a dead and bleeding small animal as bait in the water where there is a slight current. If leeches are kept in aquariums, they must be provided with food, especially the Hirudinidae, which otherwise have a habit of crawling out and wandering about on the floor.

For close examination in the living state leeches should be narcotized. Carbonated water, chloretone, chloroform vapors, chloral hydrate, nicotine sulfate, and magnesium sulfate are all variously successful for different species. The author (DGS) has had exceptional success with beer. Specimens to be killed and preserved should receive similar treatment until they no longer respond to pinching. If carefully done, exceptionally good specimens may require 1 to 5 hours or more for narcotization. They are then placed in a flat dish and flooded with the killing agent, which may be any one of a variety of standard fixatives: Schaudinn's fluid, 0.5% chromic acid, and 2% formalin in 50% alcohol are all suitable. If an acid-containing fixative is used, care must be exercised to remove all acid after fixation; otherwise, the connective tissues swell greatly. Permanent preservation should be in 80% alcohol or 5% formalin. Preserved leeches should be straight, moderately extended, undistorted, and not too hard or too soft.

Bennike (1943) advises narcotizing leeches in 15% alcohol for 15 minutes. Then they should be removed, stretched to normal size, and placed in 10% formalin or 70% alcohol. If they bend or contract, they have not been kept in the weak alcohol long enough.

Laboratory cultures. With the exception of the Piscicolidae, laboratory cultures can be easily maintained, especially since leeches need to be fed only infrequently. It is most important that the culture water be kept cool and clean. Although leeches are tolerant of a wide range of temperatures, they do not usually survive in the laboratory in direct sunlight or water that is too warm. A wide variety of common substances are toxic, even in minute quantities; copper and chlorine are especially harmful. Dead leeches, feces, and unused food material should always be removed promptly. Water should be changed as soon as it shows signs of becoming foul.

Small Glossiphoniidae may be kept in finger bowls or similar dishes of pond water containing a few shoots of *Anacharis* or other aquatic plants. Pieces of shell or pebbles afford additional concealment and protection. Small balanced aquariums also make excellent containers for leeches in the laboratory. Living snails should be added occasionally for food. Large bloodsucking glossiphoniids should be kept in larger containers and require a feeding of turtle, frog, or toad blood every month or two. Glossiphoniids thrive and readily reproduce under suitable conditions, but care should be taken to prevent overcrowding.

Erpobdellidae are very active, especially at night, and require large containers with tightly fitting covers or screens so that they will not crawl out. Stones and pieces of shell are a suitable substrate for the deposition of cocoons. Earthworms, immature insects, and ground fresh meat are appropriate foods for maintaining satisfactory cultures.

Hirudinidae may be kept in large, low aquariums or earthenware jars, but similar precautions must be taken against wandering. They do not require vegetation, but since they are somewhat amphibious there should be room for them to crawl up on the side of the container above the water level, and it is well to have a sloping bank of sandy soil at one end with a partial cover of moss, stones, or sticks. A vertebrate blood meal should be supplied at intervals of about 6 months, but the diet may be varied with frog eggs, immature insects, and earthworms. Cocoons are mostly deposited in wet soil just above the water's edge.

Taxonomy. The northeastern states and the Great Lakes region have been the center of systematic research on North American hirudineans since the beginning of the 20th century. Nearly all described species were originally collected east of the Mississippi River and north of the Gulf Coast states. The leech fauna of the western states, the Gulf Coast, and the biotically rich Ozarkian region remains poorly known. It seems likely that with serious investigation these parts of the country will yield new taxa.

External key characters are used wherever possible, but in a few couplets it has been necessary to use internal anatomical features that require dissection. Dissection usually involves a longitudinal incision along the dorsal midline in the anterior portion of the animal. Lengths are for average specimens when at rest. The 20 genera listed in this key are believed to be a complete list for the United States. The classification follows Sawyer (1986) and Barta and Sawyer (1990).

KEY TO FAMILIES AND GENERA OF HIRUDINEA OF THE UNITED STATES

1. Mouth a small pore in the oral sucker through which a muscular proboscis can be protruded; jaws absent; blood colorless . . . . **Order RHYNCHOBDELLIDA, 2**

Mouth medium to large, without a proboscis; jaws present (Figs. 13.33, 13.34, 13.39G) or absent; blood red . . . . **Order ARHYNCHOBDELLIDA, 3**

2. Body flat and much wider than head, never cylindrical; oral sucker ventral and more or less fused to body (a few species of *Placobdella* with distinct head); young clinging to ventral side of adult; poor swimmers in strictly freshwater . . . . **GLOSSIPHONIIDAE, 4**

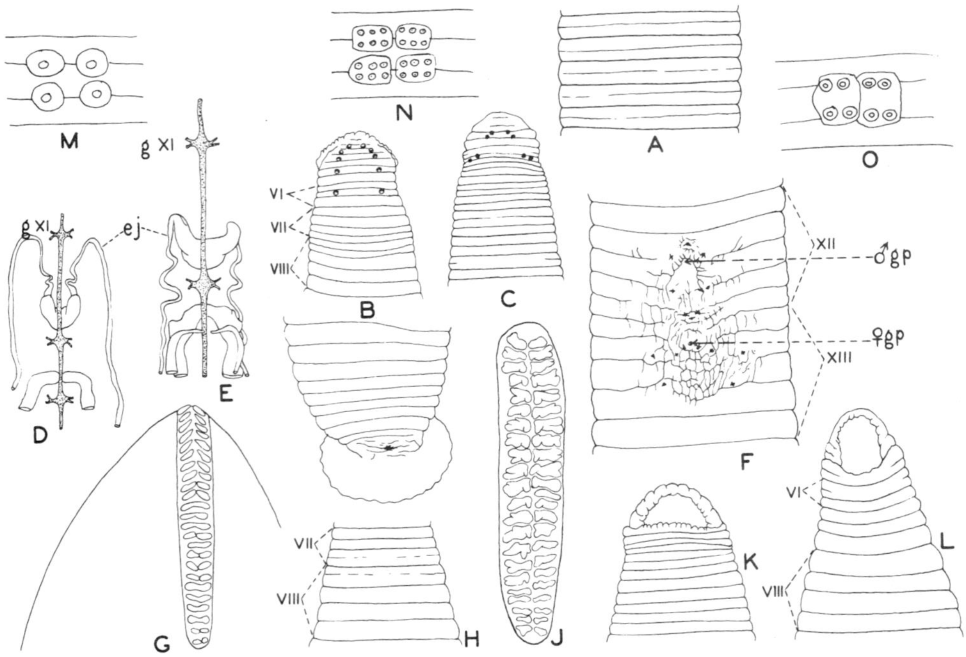

Fig. 13.39. Structure of Arhynchobdellida. A, typical midbody segment of *Nephelopsis obscura*, showing subdivided annuli; B, anterior and posterior ends of *Macrobdella decora;* C, anterior end of *Erpobdella parva;* D, dorsal view of part of reproductive system of *E. parva;* E, dorsal view of part of reproductive system of *Mooreobdella fervida;* F, ventral view of XII and XIII of *Philobdella gracilis*, showing copulatory gland region and genital pores; G, edge view of median jaw of *P. gracilis;* H, segments VII and VIII of *Haemopis marmorata;* J, edge view of teeth of median jaw of *H. terrestris;* K, ventral view of anterior end of *H. plumbea;* L, ventral view of anterior end of *H. grandis;* M, N, and O, arrangement of copulatory gland pores in *Macrobdella decora*, *M. sestertia*, and *M. ditetra*, respectively. ej, ejaculatory duct; gXI, eleventh ganglion; gp, gonopore; nerve tissue stippled. (B–D, K, and L modified from Nachtreib et al. 1912; E–G, and J modified from Moore 1901; M–O modified from Klemm 1972.)

Body cylindrical, usually divided into a narrow anterior and a wide posterior region; oral sucker distinctly separated from body (Fig. 13.42); 6 to 30 mm long; almost invariably found as parasites on fishes in freshwater and brackish water; fish leeches . . . . **PISCICOLIDAE, 13**

3. Five pairs of eyes forming a regular arch (Fig. 13.39B); jaws typically present and toothed, but may be toothless or absent; one to several pairs of large gastric caeca; most species highly pigmented; medium to very large size, to 200 mm in length **16**

Three or four pairs of eyes (rarely absent) never arranged in a regular arch (Fig. 13.39C); jaws absent; gastric caeca absent; body linear; pigmented or not; occasionally over 100 mm in length but usually much less, 20 to 50 mm. **ERPOBDELLIDAE, 18**

4. Posterior sucker distinctly set off from body by a pedicle and forming a deep cup (Fig. 13.43B), with or without marginal denticles; semipermanent parasites of fishes or amphibians; 3 or 4 species; generally rare . . . . ***Actinobdella***

Posterior sucker without a distinct pedicle, margin papillate but without denticles; predatory or parasitic **5**

5. Eyes coalesced or almost touching, only one large obvious pair but eyes may be compound, having elements of additional eyes; with enlarged head or not **6**

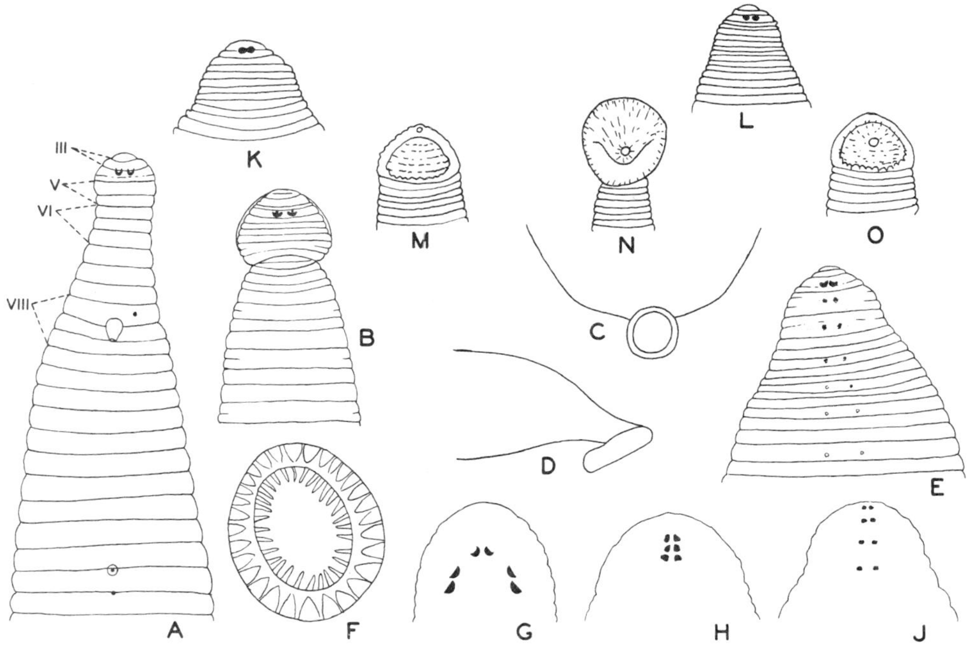

Fig. 13.40. Structure of Rhynchobdellida. A, dorsal view of anterior end of *Helobdella stagnalis* also showing position of genital pores on ventral surface; B, dorsal view of anterior end of *Placobdella montifera;* C, ventral view of posterior end of *Actinobdella pediculata* (annulations not shown); D, lateral view of same; E, dorsal view of anterior end of *P. hollensis* (note transitions between well-developed pair of eyes and sensillae); F, caudal sucker of *Actinobdella inequiannulata;* G, eyes of *Alboglossiphonia heteroclita;* H, eyes of *Glossiphonia complanata;* J, eyes of *Theromyzon;* K and L, anterior end of *Placobdella,* showing eyes fused and adjacent; M to O, ventral views of anterior suckers of Rhynchobdellida showing various locations of the porelike mouth. (A, C–E modified from Nachtreib et al. 1912; B and F modified from Moore 1901; K–O modified from Klemm 1972.)

Eyes separated by more than the diameter of the eye, 1, 3, or 4 distinct pairs, not compound; head not set off from body **8**

6. Eyes closely set, sigmoid and compound (Figs. 13.30H, J; 13.44E, F), appearing to have elements of secondary eyes; mouth pore on anterior rim of sucker; head set off from body (2 species) or not; large species reaching 70 mm, parasitic of reptiles; at least 9 species; generally widespread and common, a few species rare .... ***Placobdella***

Eyes coalesced; head not set off from body; mouth pore in sucker disk; adults generally less than 25 mm **7**

7. Large, distinct posterior sucker; midbody segments biannulate (Fig. 13.35B); less than 10 mm; parasitic on salamanders; 1 species known only from mountains of NC, SC ... ***Oligobdella***

Small posterior sucker; typically with white patches on head, occasionally on mid body, and anal region (Fig. 13.43C); adults 5 to 25 mm; parasitic of primarily amphibians and fish, a few species specialists for specific hosts; 4 uncommon species .... ***Desserobdella***

8. Single pair of eyes; parasites or predators of invertebrates **9**

Three or four pairs of eyes; bird and molluscivorous leeches **11**

9. Gonopores united on single annulis; 1 uncommon species (Fig.44C), MI .... ***Marvinmeyeria***

Gonopores separated by one annulis **10**

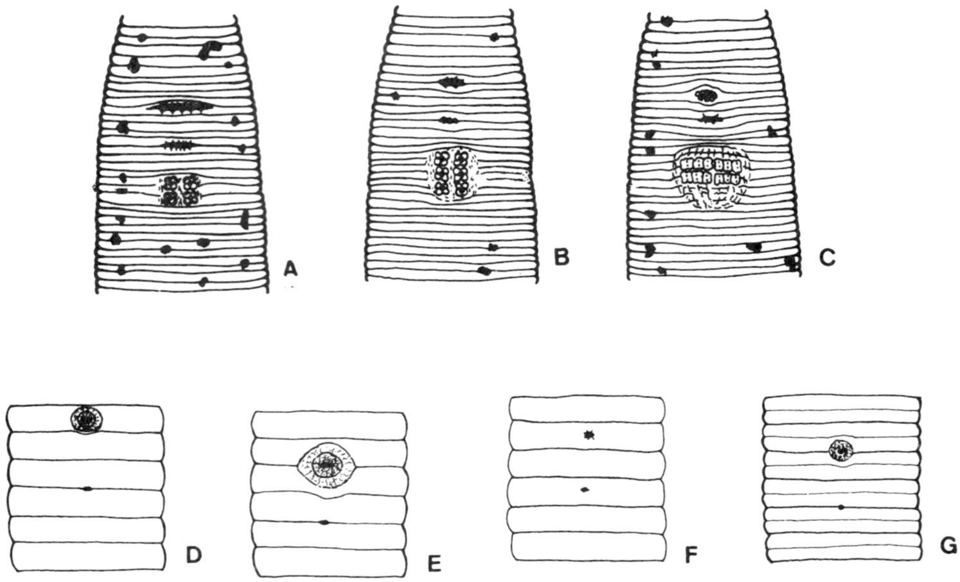

Fig. 13.41. External reproductive structures. A–C show the male genital pore, female genital pore, and copulatory glands along the midline for *Macrobdella decora*, *M. diplotertia*, and *M. sestertia*, left to right; D–G show relative positions of male genital pore (anterior) and female genital pore (posterior) for *Erpobdella punctata*, *E. punctata*, *Mooreobdella bucera*, and *Nephelopsis obscura*, left to right. (Modified from Klemm 1972.)

10. Body terete, pale (Fig. 13.43D), translucent; 1 widely distributed but uncommon species ***Gloiobdella***

Body wide and flattened, pigmented; about 5 species, most common and widespread genus of glossiphoniids . . . . ***Helobdella***

11. With three pairs of eyes **12**

With four pairs of eyes (Fig. 13.44A); body soft and translucent, to 30 mm in length; bird leeches; 2 or 3 rare species in northern states . . . . ***Theromyzon***

12. Eyes in parallel rows; pigment with 2 distinct paramedial dark stripes (Fig. 13.44D); adults about 20 mm; parasitic and predacious, chiefly on mollusks; 2 species, one very common and widespread . . . . ***Glossiphonia***

Eyes in weak triangular pattern, middle pair farthest apart (Fig. 13.44B); without pigment; adults less than 10 mm; 1 uncommon species in eastern states . . . . ***Alboglossiphonia***

13. Caudal sucker flattened, as wide or wider than the widest part of body, often with ocelli; vesicles on margin of trunk; suckers distinctly set off from body (Figs. 13.42A–C, G, H); 2 pairs of eyes or none **14**

Caudal sucker concave, weakly developed, and narrower than widest part of body, without ocelli; vesicles absent; 1 pair of eyes **15**

14. Vesicles small, difficult to see on preserved leeches; body weakly or not divided into anterior and posterior parts (Fig. 13.42C) . . . . ***Piscicola***

Vesicles large, obvious after preservation; body strongly divided into anterior and posterior parts (Fig. 13.42A, B); uncommon . . . . ***Cystobranchus***

15. Body slim; colorless or with scattered pigment cells; 24 to 26 mm long; widely distributed ***Myzobdella***

Body thicker; with six longitudinal rows of brown-black pigments (Fig. 13.43A); 6 to 8 mm long; scattered distribution in eastern half of United States . . . . ***Piscicolaria***

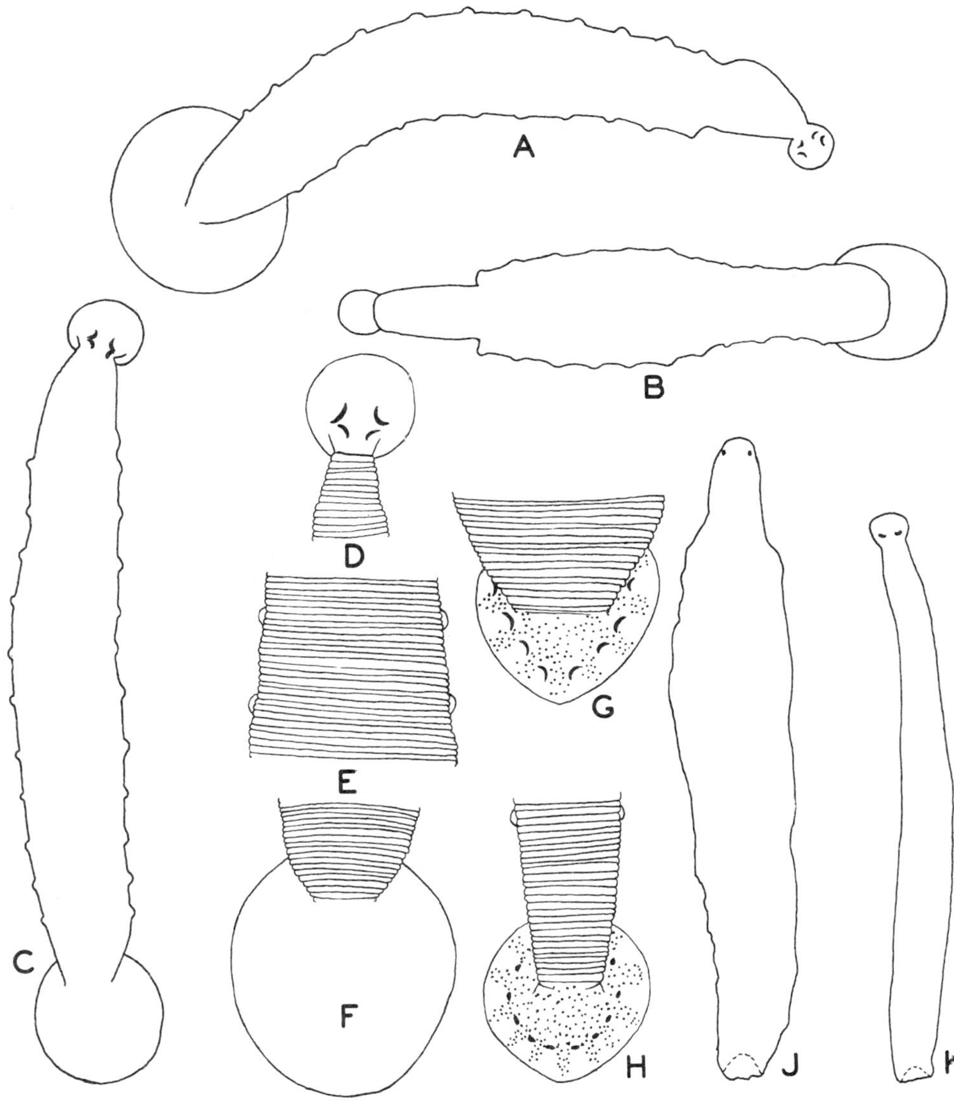

Fig. 13.42. Structure of Piscicolidae. A, *Cystobranchus verrilli,* ×10; B, *Cystobranchus vivida,* ×4; C, *Piscicola punctata,* ×6.5; D–F, anterior, middle, and posterior portions of *P. punctata;* G, posterior end of *P. salmositica;* H, posterior end of *P. milneri;* J, *Myzobdella lugubris,* ×6.5; K, *Illinobdella moorei,* ×8. (A, C, J, and K modified from Meyer 1940; D–F, and H modified from Meyer 1946a; B modified from Moore 1898; G modified from Meyer 1946b.)

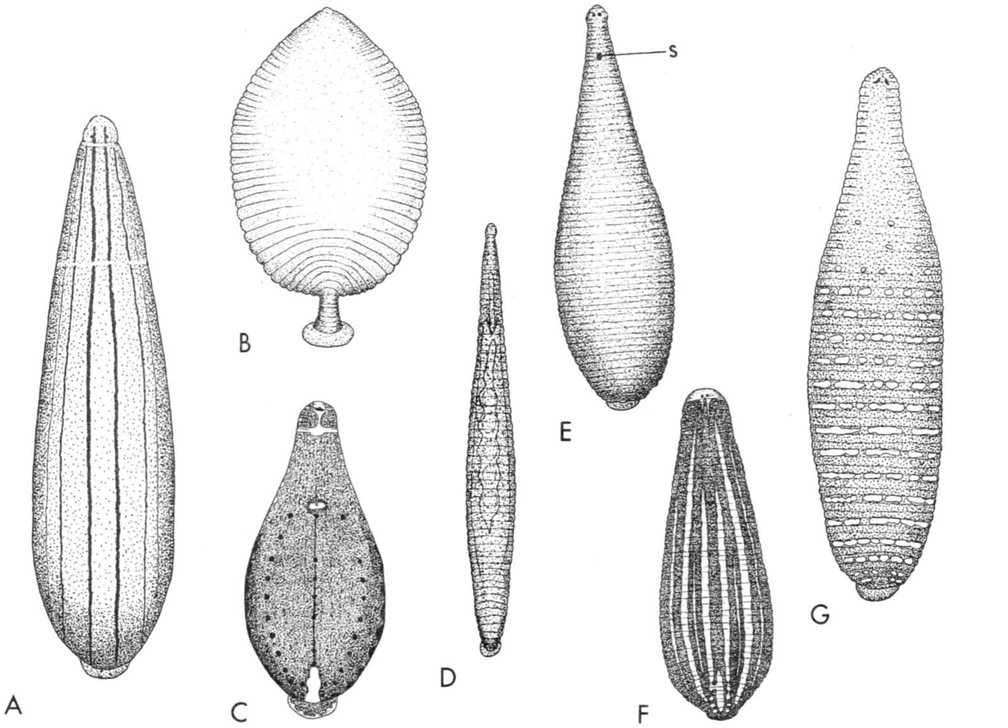

Fig. 13.43. Rhynchobdellid leeches. A, *Piscicolaria reducta*, ×5; B, *Actinobdella pediculata*, ×2; C, *Desserobdella phalera*, ×6; D, *Gloiobdella elongata*, ×4; E, *Helobdella stagnalis*, ×4; F, *H. fusca*, ×4; G, *H. transversa*, s, scute. (A–G from Sawyer 1972.)

16. Ventral surface with gonopores only; large species, often mottled, very dark or with a median stripe (Fig. 13.45F, G) . . . . **HAEMOPIDAE, *Haemopis***

Ventral surface with either porous, glandular region around gonopores or patch of glandular pores posterior to gonopores; usually with bright-colored markings (Fig. 13.45D, E) **HIRUDINIDAE, 17**

17. Copulatory glands behind gonopores on ventral surface (Figs. 13.39M–O; 13.41A–C); 4 species with scattered distributions (Fig. 13.45D) . . . . ***Macrobdella***

Copulatory glands in a depression in vicinity or surrounding gonopores (Fig. 13.39F); 2 species generally in southeastern states or Mississippi River valley (Fig. 13.45E) . . . . . . ***Philobdella***

18. With four eyes arranged in an anterior horizontal fashion, and four posterior eyes similarly arranged in a horizontal line; feeds on aquatic invertebrates, also a scavenger; up to 100 mm long; dorsally with blotches (Fig. 13.45C); 1 species in northern states . . . . ***Nephelopsis***

Eyes fewer in number or in a different arrangement (bell curve) **19**

19. Sperm (ejaculatory) ducts forming an anterior loop (requires dissection) (Fig. 13.39D); pigment often marked by longitudinal dark bands of some sort (Fig. 13.45A); 3 or 4 pairs of eyes, or eyes absent; 10 to 100 mm long; 1 very common and 3 uncommon species generally in northern half of the U.S. . . . . ***Erpobdella***

Sperm (ejaculatory) ducts lateral in direction, not forming loops (Fig. 13.39E); pigment, if present, typically uniform (Fig. 13.45B), blotchy, or rarely (1 species) with longitudinal bands; eyes 3 pairs; adults not exceeding 55 mm; 5 species ranging from the east coast to the interior, most records in the eastern U.S. . . . . ***Mooreobdella***

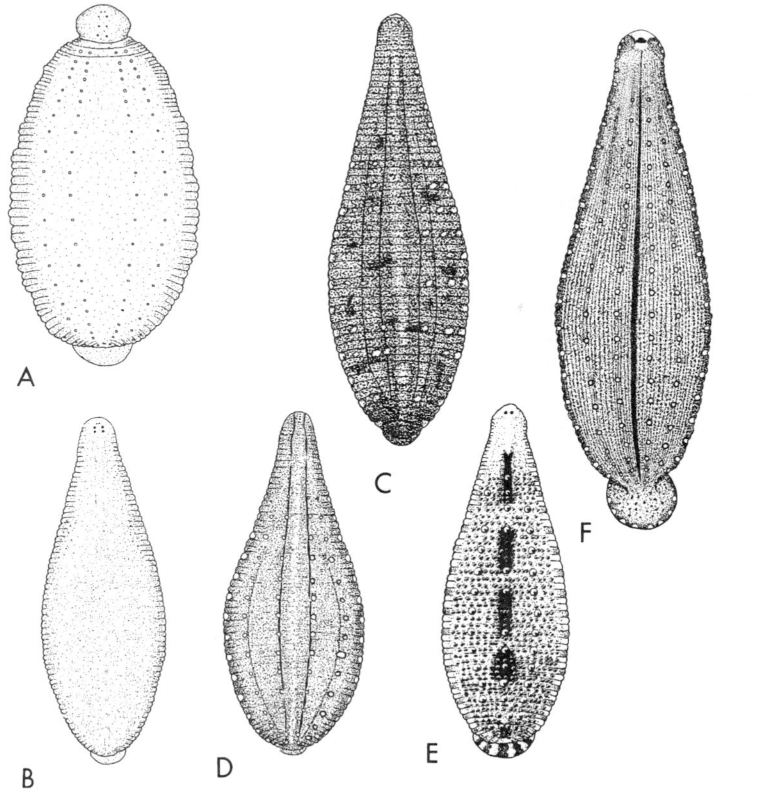

Fig. 13.44. Rhynchobdellid leeches. A, *Theromyzon*, ×3; B, *Alboglossiphonia heteroclita*, ×7; C, *Marvinmeyeria lucida*, ×4; D, *Glossiphonia complanata*, ×4; E, *Placobdella ornata*, ×1; F, *Desserobdella picta*, ×3. (A–E from Sawyer 1972; F from Barta and Sawyer 1990.)

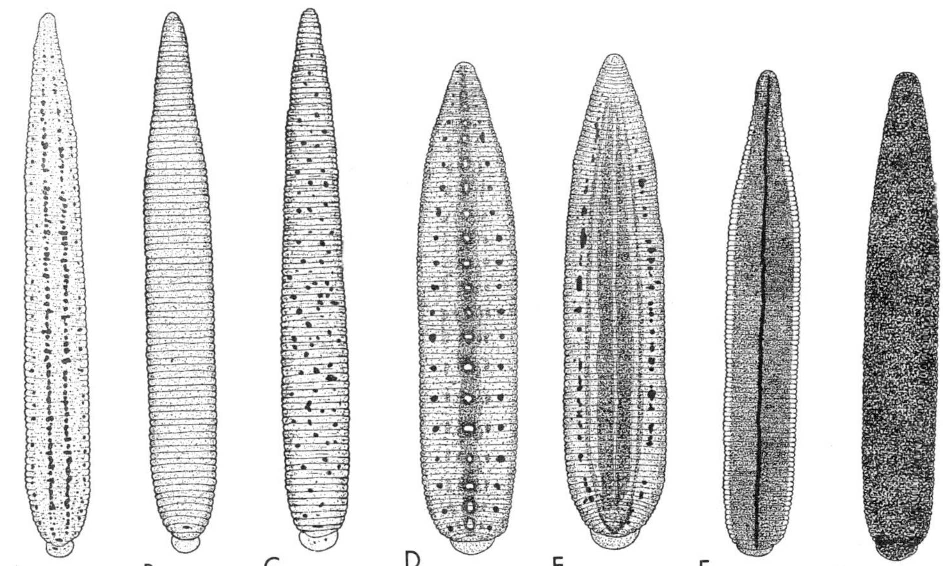

Fig. 13.45. Arhynchobdellid leeches. A, *Erpobdella punctata*, ×2; B, *Mooreobdella fervida*, ×5; C, *Nephelopsis obscura*, ×1; D, *Macrobdella decora*, ×1; E, *Philobdella gracilis*, ×1; F, *Haemopis terrestris*, ×1; G, *H. marmorata*, ×1. (A–G from Sawyer 1972.)

REFERENCES

Andrassy, I. 1956. *Troglochaetus beranecki* Delachaux, ein Repräsentant der für die Fauna Ungarns neuen Tierklasse Archiannelida. *Ann. Hist. Nat. Mus. Nat. Hung.* 7:371–375.

Appleby, A. G., and R. O. Brinkhurst. 1970. Defecation rates of three tubificid oligochaetes found in the sediment of Toronto Harbour, Ontario. *J. Fish. Res. Bd. Can.* 27:1971–1982.

Aston, R. J. 1968. The effect of temperature on the life cycle, growth and fecundity of *Branchiura sowerbyi* (Oligochaeta: Tubificidae). *J. Zool. London* 154:29–40.

——. 1973. Tubificids and water quality: a review. *Environ. Pollut.* 5:1–10.

Barnes, R. D., and F. W. Harrison. 1992. Introduction. In: *Microscopic Anatomy of the Invertebrates, Vol. 7: Annelida*, pp. 1–9, edited by F. W. Harrison and S. L. Gardiner. Wiley-Liss, New York.

Barta, J. R., and R. T. Sawyer. 1990. Definition of a new genus of glossiphoniid leech and a redescription of the type species, *Clepsine picta* Verrill, 1872. *Can. J. Zool.* 68:1942–1950.

Becker, C. D., and M. Katz. 1965. Distribution, ecology, and biology of the salmonid leech, *Piscicola salmositica* (Rhynchobdellae: Piscicolidae). *J. Fish. Res. Bd. Can.* 22:1175–1195.

Bennike, S. A. B. 1943. Contributions to the ecology and biology of the Danish freshwater leeches (Hirudinea). *Fol. Limn. Scandinavica* 2: 1–109.

Bilello, A. A., and H. E. Potswald. 1974. A cytological and quantitative study of neoblasts in the naid *Opidonais serpentina* (Oligochaeta). *Wilhelm Roux' Arch.* 174:234–249.

Blinn, D. W. 1988. Intraspecific discrimination of amphipod prey by a freshwater leech through mechanoreception. *Can. J. Zool.* 66: 427–430.

Blinn, D. W., and R. W. Davies. 1989. New distribution records for leeches (Hirudinoiodea) in Arizona. *Southwestern Nat.* 34:431–432.

Blinn, D. W., et al. 1986. Surface sensilla on the predaceous fresh-water leech *Erpobdella montezuma:* possible importance in feeding. *Trans. Am. Microsc. Soc.* 105:21–30.

——. 1987. Specialized pelagic feeding by *Erpobdella montezuma* (Hirudinea). *Hoarctic Ecol.* 10:235–240.

Bogatov, V. V., et al. 1980. The luminescence of freshwater Oligochaeta. *Hydrobiol. J.* 16:50–51. (Translated from Russian.)

Bousfield, E. C. 1887. The natural history of the genus *Dero. J. Linn. Soc. (Zool.)* 20:91–107.

Brandt, E. 1978. Anpassungen von *Tubifex tubifex* Willer (Annelida, Oligochaeta) an die Temperature, den Sauerstoffgehalt und den Ernährungszustand. *Arch. Hydrobiol.* 84:302–338.

Brinkhurst, R. O. 1964. Studies on the North American aquatic Oligochaeta, 1: Naididae and Opistocystidae. *Proc. Acad. Nat. Sci. Philadelphia* 116:195–230.

——. 1966. The Tubificidae (Oligochaeta) of polluted waters. *Proc. Int. Assoc. Theor. Appl. Limnol.* 16:854–859.

——. 1970. Distribution and abundance of tubificid (Oligochaeta) species in Toronto harbour, Lake Ontario. *J. Fish. Res. Bd. Can.* 27: 1961–1969.

——. 1985. The generic and subfamilial classification of the Naididae (Annelida: Oligochaeta). *Proc. Biol. Soc. Wash.* 98:470–475.

——. 1986. Guide to the freshwater aquatic microdrile oligochaetes of North America. *Can. Spec. Publ. Fish. Aquatic Sci.* 84:1–259.

——. 1988. A taxonomic analysis of the Haplotaxidae (Annelida, Oligochaeta). *Can. J. Zool.* 66:2243–2252.

Brinkhurst, R. O., and K. A. Coates. 1985. The genus *Paranais* (Oligochaeta: Naididae) in North America. *Proc. Biol. Soc. Wash.* 98:303–313.

Brinkhurst, R. O., and D. G. Cook (eds.). 1980. *Aquatic Oligochaete Biology*, 530 pp. Plenum, New York.

Brinkhurst, R. O., and B. G. M. Jamieson. 1971. *Aquatic Oligochaeta of the World*, 860 pp. University of Toronto Press, Toronto.

Brinkhurst, R. O., and M. J. Wetzel. 1984. Aquatic Oligochaeta of the world: supplement. A catalogue of new freshwater species, descriptions, and revisions, *Can. Tech. Rep. Hydrogr. Ocean Sci.* 44:1–101.

Bunke, D. 1967. Zur Morphologie und Systematik der Aeolosomatidae Beddard 1895 und Potarnodrilidae nov. fam. (Oligochaeta). *Zool. Jahrb. Abt. Syst. Oekol. Geogr. Tiere* 94:187–368.

Chen, Y. 1940. Taxonomy and faunal relations of the limnetic Oligochaeta of China. *Contrib. Bio. Lab. Sci. Soc. China, Zool. Sec.* 14:1–132.

Christensen, B. 1994. 1. Annelida-Clitellata. In: *Reproductive Biology of Invertebrates, VI (B)*, pp.1–23, edited by K. G. Adiyodi and R. G. Adiyodi. Oxford & IBH, New Delhi.

Cook, D. G. 1975. Cave-dwelling aquatic Oligochaeta (Annelida) from the Eastern United States. *Trans. Am. Microsc. Soc.* 94:24–37.

Cook, D. G., and R. O. Brinkhurst. 1973. Marine flora and fauna of the northeastern United States. Annelida: Oligochaeta. *NOAA Tech. Rpt. NMFS CIRC-374.* 23 pp.

Dausend, K. 1931. Über die Atmung der Tubifiziden. *Z. vergl. Physiol.* 14:557–608.

Davies, R. W., and R. P. Everett. 1975. The feeding of four species of freshwater Hirudinoidea in southern Alberta. *Proc. Int. Assoc. Theor. Appl. Limnol.* 19:2816–2827.

Davies, R. W., et al. 1982. Passive dispersal of four species of freshwater leeches (Hirudinoidea) by ducks. *Freshwater Invert. Biol.* 1:40–44.

——. 1985. *Erpobdella montezuma* (Hirudinoidea: Erpobdellidae), a new species of freshwater leech from North America. *Can. J. Zool.* 63: 965–969.

Delachaux, T. 1921. Un Polychete d'eau douce cavernicole: *Troglochaetus beranecki* nov. gen. nov. spec. (Note Prél). *Bull. Soc. Sci. Nat. Neuchatel* 45:3–11.

Erséus, C. 1999. *Parvidrilus strayeri*, a new genus and species, an enigmatic interstitial clitellate from underground waters in Alabama. *Proc. Bio. Soc. Wash.* 112:327–337.

Erséus, C. et al. 1990. Redefinition of *Teneridrilus* Holmquist (Oligochaeta: Tubificidae), with description of two new species from North America. *Proc. Bio. Soc. Wash.* 103:839–846.

Famme, P., and J. Knudsen. 1985. Aerotaxis by the freshwater oligochaete *Tubifex* sp. *Oecologia* 65:599–601.

Foster, N. 1972. Freshwater polychaetes (Annelida) of North America, *Biota Freshwater Ecosystems Identification Manual* 4:1–15.

Gates, G. E. 1954. Exotic earthworms of the United States. *Bull. Mus. Comp. Zool. Harv. Univ.* 111:219–258.

Gelder, S. R. 1980. A review of the symbiotic Oligochaeta (Annelida). *Zool. Anz.* 204:69–81.

——. 1996a. Description of a new branchiobdellidan species, with observations on three other species, and a key to the genus *Pterodrilus* (Annelida: Clitellata). *Proc. Biol. Soc. Wash.* 109:256–263.

——. 1996b. A review of the taxonomic nomenclature and a checklist of the species of the Branchiobdellidae (Annelida: Clitellata). *Proc. Biol. Soc. Wash.* 109:653–663.

Gelder, S. R., and R. C. Smith. 1987. Distribution of branchiobdellids (Annelida: Clitellata) in northern Maine, USA. *Trans. Am. Microsc. Soc.* 106:85–88.

Goodnight, C. J. 1940. The Branchiobdellidae (Oligochaeta) of North American crayfishes. *Ill. Biol. Monogr.* 17:1–75.

Goodnight, C. J. 1941. The Branchiobdellidae (Oligochaeta) of Florida. *Trans. Am. Microsc. Soc.* 60:69–74.

Greene, K. L. 1974. Experiments and observations on the feeding behavior of the freshwater leech *Erpobdella octoculata* (L.) (Hirudinea: Erpobdellidae). *Arch. Hydrobiol.* 74:87–99.

Gruffydd, L. D. 1965. Notes on a population of the leech, *Glossiphonia heteroclita*, infesting *Lymnaea periger. Ann. Mag. Nat. Hist.* 8:144–150.

Hall, M. C. 1914. Description of a new genus and species of the discodrilid worms. *Proc. U.S. Nat. Mus.* 48:187–193.

Harding, W. A. and J. P. Moore. 1927. Hirudinea. *Fauna British India.* 302 pp.

Hartman, O. 1938. Brackish and freshwater Nereidae from the northeast Pacific with the description of a new species from central California. *Univ. Calif. Publ. Zool.* 43:79–82.

——. 1959. Capitellidae and Nereidae (marine annelids) from the Gulfside of Florida, with a review of freshwater Nereidae. *Bull. Mar. Sci. Gulf Carib.* 9:153–168.

Herlant-Meewis, H. 1950.Cyst-formation in *Aeolosoma hemprichi* (Ehr.). *Biol. Bull.* 99:172–180.

Herrmann, S. J. 1970. Systematics, distribution, and ecology of Colorado Hirudinea. *Am. Midl. Nat.* 83:1–37.

Hiltunen, J. K., and D. J. Klemm. 1980. *A Guide to the Naididae (Annelida: Clitellata: Oligochaeta) of North America*, EPA Publ. 600/4–80–031, 47 pp.

Hoffman, R. L. 1963. A revision of the North American annelid worms of the genus *Cambarincola* (Oligochaeta: Branchiobdellidae). *Proc. U.S. Nat. Mus.* 114 (3470):271–371.

Hoffmann, J. 1955. Quelques caractères éthologiques de la Piscicolidae: *Cystobranchus respi-*

rans Troschel. *Arch. Inst. Grand-Ducal Luxemb. Sect. Sci. Nat. Phys. Math.* (*New ser.*) 22:223–225.

Holmquist, C. 1973. Fresh-water polychaete worms of Alaska with notes on the anatomy of *Manayunkia speciosa* Leidy. *Zool. Jahrb. Syst.* 100:497–516.

Holt, P. C. 1960a. On a new genus of the Family Branchiobdellidae (Oligochaeta). *Am. Midl. Nat.* 64:169–176.

——. 1960b. The genus *Ceratodrilus* Hall (Branchiobdellidae, Oligochaeta) with the description of a new species. *Va. J. Sci.* 11:53–77.

——. 1965. On *Ankyrodrilus*, a new genus of branchiobdellid worms (Annelida). *Va. J. Sci.* 16:9–21.

——. 1967a. *Oedipodrilus oedipus*, n, g., n. sp. (Annelida, Clitellata: Branchiobdellida). *Trans. Am. Microsc. Soc.* 86:58–60.

——. 1967b. Status of genera *Branchiobdella* and *Stephanodrilus* in North America with description of a new genus (Clitellata: Branchiobdella). *Proc. U.S. Nat. Mus.* 124:1–10.

——. 1968a. The Branchiobdellida: epizootic annelids. *The Biologist* 50:79–94.

——. 1968b. New genera and species of branchiobdellid worms (Annelida: Clitellata). *Proc. Biol. Soc. Wash.* 81:291–318,

——. 1968c. The genus *Pterodrilus* (Annelida: Branchiobdellida). *Proc. U.S. Nat. Mus.* 125:1–44.

——. 1973a. Branchiobdellids (Annelida: Clitellata) from some eastern North American caves, with descriptions of new species of the genus *Cambarincola*. *Int. J. Speleol.* 5:219–256.

——. 1973b. Epigean branchiobdellids (Annelida: Clitellata) from Florida. *Proc. Biol. Soc. Wash.* 86:79–104.

——. 1974. The genus *Xironogiton* Ellis, 1919 (Clitellata: Branchiobdellida). *Va. J. Sci.* 25: 5–19.

——. 1981. A resume of the members of the genus *Cambarincola* (Annelida: Branchiobdeflida) from the Pacific drainage of the United States. *Proc. Biol. Soc. Wash.* 94:675–695.

——. 1986. Newly established families of the Order Branchiobdellida (Annelida: Clitellata) with a synopsis of the genera. *Proc. Biol. Soc. Wash.* 99:676–702.

Holt, P. C., and B. D. Opell. 1993. A checklist of and illustrated key to the genera and species of the Central and North American Cambarincolidae (Clitellata: Branchiobdellida). *Proc. Biol. Soc. Wash.* 106:251–295.

Hovingh, P., and D. J. Klemm. 2000. *Erpobdella lahontana* (Annelida: Hirudinea: Arhynchobdellida: Erpobdellidae), a new species of freshwater leech from North America. *Proc. Biol. Soc. Wash.* 113:155–161.

Howmiller, R., and M. S. Loden. 1976. Identification of Wisconsin Tubificidae and Naididae. *Trans. Wis. Acad. Sci. Arts Lett.* 64:185–197.

Jennings, J. B., and V. M. Van Der Lande. 1967. Histochemical and bacteriological studies on digestion in nine species of leeches (Annelida: Hirudinea). *Biol. Bull.* 133:166–183.

Johnson, H. P. 1903. Freshwater nereids from the pacific coast and Hawaii with remarks on freshwater Polychaeta in general. In: *Mark Anniversary Volume*, pp. 205–223. Henry Holt, New York.

Jones, S. R. M., and P. T. K. Woo. 1990. Redescription of the leech *Desserobdella phalera* (Graf, 1899) n. comb. (Rhynchobdellida: Glossiphoniidae), with notes on its biology and occurrence on fishes. *Can. J. Zool.* 68: 1951–1955.

Kaster, J. L. 1979. Morphological development and adaptive significance of autotomy and regeneration in *Tubifex tubifex* Müller. *Trans. Am. Microsc. Soc.* 98:473–477.

——. 1980. The reproductive biology of *Tubifex tubifex* Müller (Annelida: Tubificidae). *Am. Midl. Nat.* 104:355–364.

Kaster, J. L., and J. H. Bushnell. 1981a. Cyst formation by *Tubifex tubifex* (Tubificidae). *Trans. Am. Microsc. Soc.* 100:34–41.

——. 1981b. Occurrence of tests and their possible significance in the worm, *Tubifex tubifex* (Oligochaeta). *Southwest Nat.* 26:307–310.

Kathman, R. D., and R. O. Brinkhurst. 1999. *Guide to the Freshwater Oligochaetes of North America*, 264 pp. Aquatic Resources Center, College Grove, TN.

Klemm, D. J. 1972. Freshwater leeches (Annelida: Hirudinea) of North America. *Biota Freshwater Ecosystems Identification Manual* 5:1–53.

——. 1975. Studies on the feeding relationships of leeches (Annelida Hirudinea) as natural associates of mollusks. *Sterkiana* 58:1–51.

——. 1977. A review of the leeches (Annelida: Hirudinea) in the Great Lakes Region. *Mich. Academician* 9:397–418.

——. 1982. Leeches (Annelida: Hirudinea) of North America. *EPA Publ.* 600/3-82-025, 177 pp.

——. (ed.). 1985. *A Guide to the Freshwater An-*

nelida (Polychaeta, Naidid and Tubificid Oligochaeta, and Hirudinea of North America), 198 pp. Kendall/Hunt, Dubuque, IA.

——. 1995. *Identification Guide to the Freshwater Leeches (Annelida: Hirudinea) of Florida and Other Southern States.* 82 pp. Florida Department Environ. Protection, Tallahassee, FL.

Klemm, D. J., and J. K. Hiltunen. 1992. The freshwater Annelida (Polychaeta, naidid and tubificid Oligochaeta, and Hirudinea) of the Great Lakes region: an overview. In: *Proceedings 1991 Midwest Pollution Control Biologists Meeting—Environmental Indicators: Measurement and Assessment Endpoints*, pp. 35–48, edited by T. P. Simon and W. S. Davis. USEPA, Chicago, IL.

Kutschera, U. 1987. Notes on the taxonomy and biology of leeches of the genus *Helobdella* Blanchard 1896 (Hirudinea: Glossiphoniidae). *Zool. Anz.* 219:321–323.

——. 1988. A new leech species from North America, *Helobdella californica* nov. sp. (Hirudinea: Glossiphoniidae). *Zool. Anz.* 220:173–178.

Learner, M. A., et al. 1978. A review of the biology of British Naididae (Oligochaeta) with emphasis on the lotic environment. *Freshwater Biol.* 8:357–375.

Leidy, J. 1883. *Manayunkia speciosa. Proc. Philadelphia Acad. Nat, Sci.* 1883:204–212.

Loden, M. S. 1981. Reproductive ecology of Naididae (Oligochaeta). *Hydrobiologia* 93:115–123.

Maloney, S. D., and C. M. Chandler. 1976. Leeches (Hirudinea) in the Upper Stones River drainage of Middle Tennessee. *Am. Midl. Nat.* 95:42–48.

Mann, K. H. 1962, *Leeches (Hirudinea); Their Structure, Physiology, Ecology and Embryology*, 201 pp. Pergamon, New York.

Manoleli, D. G., et al. 1998. *Haemopis caeca* (Annelida: Hirudinea: Arhynchobdellida: Haemopidae), a new species of troglobitic leech from a chemoautotrophically based groundwater ecosystem in Romania. *Proc. Biol. Soc. Wash.* 111:222–229.

Marcus, E. 1944. Sôbre Oligochaeta limnicos do Brasil, *Univ. Sao Paulo, Bol. Fac. Filos, Cien Lett. Zool.* 8:5–135.

Marcus, E. 1943. Sobre Naididae do Brasil. *Univ. Sao Paulo, Bol. Fac. Filos, Cien. Lett. Zool.* 7:1–247.

Meyer, F. P. 1969. A potential control of leeches. *Prog. Fish Cult.* 31:160–163.

Meyer, M. C. 1940. A revision of the leeches (Piscicolidae) living on fresh-water fishes of North America. *Trans. Am. Microsc. Soc.* 59:354–376.

——. 1946a. Further notes on the leeches (Piscicolidae) living on fresh-water fishes of North America. *Trans. Am. Microsc. Soc.* 65:237–249.

——. 1946b. A new leech, *Piscicola salmositica* n. sp. (Piscicolidae), from steelhead trout (*Salmo gairdneri gairdneri* Richardson, 1838). *J. Parasitol.* 32:467–473.

Miller, J. A. 1929. The leeches of Ohio. *Contrib. Franz Theodore Stone Lab* 2:1–38.

Milligan, M. R. 1997. *Identification Manual for the Aquatic Oligochaeta of Florida*, Vol. 1: *Freshwater Oligochaetes*, 187 pp. Florida Department Environ. Protection, Tallahassee, FL.

Moore, J. P. 1895. The anatomy of *Bdellodrilus illuminatus*, an American discodrilid. *J. Morphol.* 10:497–540.

——. 1898. The leeches of the U.S. National Museum. *Proc. U.S. Nat. Mus.* 21:543–563.

——. 1901. The Hirudinea of Illinois. *Bull. Ill. State Lab. Nat. Hist.* 5:479–547.

——. 1912. Leeches of Minnesota. *Geol. Nat. Hist. Surv. Minn. Zool.* 5, 111:64–150.

——. 1923. The control of blood-sucking leeches, with an account of the leeches of Palisades Interstate Park. *Roosevelt Wild Life Bull.* 2:1–53.

Nachtreib, H. F., et al. 1912. The leeches of Minnesota. *Geol. Nat. Hist. Surv. Minn. Zool. Ser.* 5:1–150.

Oosthuizen, J. H., and R. W. Davies. 1992. Redescription of the duck leech, *Theromyzon rude* (Baird, 1869) (Rhynchobdellida: Glossiphoniidae). *Can. J. Zool.* 70:2028–2033.

——. 1993. A new species of *Theromyzon* (Rhynchobdellida: Glossiphoniidae), with a review of the genus in North America. *Can. J. Zool.* 71:1311–1318.

Pennak, R. W. 1971. A fresh-water archiannelid from the Colorado Rocky Mountains. *Trans. Am. Microsc. Soc.* 90:372–375.

Peterson, D. L. 1983. Life cycle and reproduction of *Nephelopsis obscura* Verrill (Hirudinea: Erpobdellidae) in permanent ponds of northwestern Minnesota. *Freshwater Invert. Biol.* 2:165–172.

Poe, T. P., and D. C. Stefan. 1974. Several environmental factors influencing the distribution of the fresh-water polychaeta, *Manayunkia speciosa* Leidy. *Chesapeake Sci.* 12:235–237.

Potswald, H. E. 1971. A fine structural analysis of the epidermis and cuticle of the oligochaete

Aeolosoma bengalense Stephenson. *J. Morph.* 135:185–211.

Reynolds, J. W. 1977. *The Earthworms (Lumbricidae and Sparganophilidae) of Ontario*, 141 pp. Life Sci. Misc. Pub., R. Ont. Mus.

Reynoldson, T. B. 1947. An ecological study of the enchytraeid worm population of sewage bacteria beds. *J. Anim. Ecol.* 16:26–37.

Richardson, L. R. 1969. A contribution to the systematics of the hirudinid leeches, with description of new families, genera and species. *Acta Zool. Acad. Sci. Hung.* 15:97–149.

Rupp, R. W., and M. C. Meyer. 1954. Mortality among brook trout, *Salvelinus fontinalis*, resulting from attacks of freshwater leeches. *Copeia* 1954:294–295.

Sapkarev, J. A. 1967–1968. The taxonomy and ecology of leeches (Hirudinea) of Lake Mendota, Wisconsin. *Trans. Wis. Acad.* 56:225–253.

Sawyer, R. T. 1970. Observations on the natural history and behavior of *Herpobdella punctata* (Leidy) (Annelida: Hirudinea). *Am. Midl. Nat.* 83:65–80.

——. 1972. North American freshwater leeches, exclusive of the Piscicolidae, with a key to all species. *Ill. Biol. Monogr.* 46:1–154.

——. 1986. *Leech Biology and Behaviour*, 3 vols. 1100 pp. Oxford Univ. Press, Oxford, UK.

Sawyer, R. T., and R. M. Shelley. 1976. New records and species of leeches (Annelida: Hirudinea) from North and South Carolina. *J. Nat. Hist.* 10:65–97.

Schmelz, R. M. 1996. Species separation and identification in the Enchytraeidae (Oligochaeta, Annelida): combining morphology with general protein data. *Hydrobiologia* 334:31–36.

Singer, R. 1978. Suction-feeding in *Aeolosoma* (Annelida). *Trans. Am. Microsc. Soc.* 97:105–110.

Smith, D. G. 1980. New records of leeches (Annelida: Hirudinea) from the shortnose sturgeon (*Acipenser brevirostrum*) in the Connecticut River. *Proc. Helminth. Soc. Wash.* 47:147–148.

——. 1985. Notes on leeches (Hirudinea) of the Connecticut Lakes region of northern New Hampshire. *Proc. Helminth. Soc. Wash.* 53:294–295.

Smith, D. G., and S. Hanlon. 1997. *Macrobdella sestertia* (Hirudinea: Hirudinidae) in Maine and a key to the hirudiniform leeches of Maine. *Northeast. Nat.* 4:231–236.

Smith, R. I. 1942. Nervous control of chromatophores in the leech *Placobdella parasitica. Physiol. Zool.* 15:410–417.

Smith, R. I. (ed.). 1964. *Keys to Marine Invertebrates of the Woods Hole Region*, Systematics-Ecology Program, Marine Biological Laboratory, Woods Hole, Massachusetts. Contrib. 11. 208 pp.

Soos, A. 1965. Identification key to the leech (Hirudinoidea) genera of the world, with a catalogue of the species, 1: Family: Piscicolidae. *Acta Zool. Acad. Sci. Hung.* 11:417–463.

——. 1966a. On the genus *Glossiphonia* Johnson, 1816, with a key and catalogue to the species. *Ann. Hist. Nat. Mus. Nat. Hung. Zool.* 58:271–279.

——. 1966b. Identification key to the leech (Hirudinoidea) genera of the world, with a catalogue of the species, III: Family: Erpobdellidae. *Acta Zool. Acad. Sci. Hung.* 12:371–407.

——. 1969. Identification key to the leech (Hirudinoidea) genera of the world, with a catalog of the species, V: Family: Hirudinidae. *Acta Zool. Acad. Sci. Hung.* 15:151–201.

Spencer, D. R. 1976. Occurrence of *Manayunkia speciosa* (Polychaeta: Sabellidae) in Cayuga Lake, New York, with additional notes on its North American distribution. *Trans. Am. Microsc. Soc.* 95:127–128.

——. 1980. The aquatic Oligochaeta of the St. Lawrence Great Lakes region. *Aquatic Oligochaete Biol.* 1:115–164.

Stimpson, K. S., et al. 1975. Distribution and abundance of inshore oligochaetes in Lake Michigan. *Trans. Am. Microsc. Soc.* 94:384–394.

——. 1982. A guide to the freshwater Tubificidae of North America. *EPA Publ*, 600/3-82-033.

Strayer, D., and E. Bannon-O'Donnell. 1988. Aquatic microannelids (Oligochaeta and Aphanoneura) of underground waters of southeastern New York. *Am. Midl. Nat.* 119:327–335.

Taylor, W. D. 1980. Observations on the feeding and growth of the predaceous oligochaete *Chaetogaster langi* on ciliated Protozoa. *Trans. Am. Microsc. Soc.* 99:360–367.

Tynen, M. J. 1975. A checklist and bibliography of the North American Enchytraeidae (Annelida: Oligochaeta). *Syllogeus* (9):1–14.

Walker, J. G. 1970. Oxygen poisoning in the annelid *Tubifex tubifex*, I: response to oxygen exposure. *Biol. Bull.* 138:235–244.

Watermolen, D. J. 1996. Notes on the leech *Desserobdella picta* (Hirudinea: Glossiphoniidae). *J. Freshwater Ecol.* 11:211–217.

Webb, D. W., et al. 1998. The macroinvertebrate biodiversity, water quality, and hydrogeology of ten karst springs in the Salem Plateau section of Illinois, USA. In: *Studies in Crenobiology*, pp.39–48, edited by L. Botosaneanu. Backhuys, Leiden.

Weigl, A. M. 1994. Ultrastructure of the adhesive organs in branchiobdellids (Annelida: Clitellata). *Trans. Am. Microsc. Soc.* 113: 276–301.

Westheide, W. 1979. *Hesionides riegerorum* n. sp., a new freshwater polychaete from the United States. *Int. Rev. Gesamten Hydrobiol.* 64:273–280.

Young, J. O. 1983. A stimulus for egg production in *Glossiphonia complanata*. (L.) (Hirudinoidea: Glossiphoniidae). *Freshwater Invert. Biol.* 8:112–115.

Young, J. O., and S. M. Spelling. 1986. The incidence of predation on lake-dwelling leeches. *Freshwater Biol.* 16:465–477.

Young, W. 1966. Ecological studies of the Branchiobdellida (Oligochaeta). *Ecology* 47:571–577.

Plate 1. Top, *Anheteromeyenia ryderi*, a freshwater sponge, from life; bottom, large gemmule field and reduction bodies of the sponge *Radiospongilla crateriformis*.

Plate 2. Top left, polyps of *Cordylophora caspia*, a colonial cnidarian, from life; top right, the jellyfish *Craspedacusta sowerbii*, from life; bottom left, the ectoproct *Pectinatella magnifica*, filmed underwater; bottom right, the ectoproct *Cristatella mucedo* covering rooted vegetation. (Bottom photos © E. J. Klekowski, with permission.)

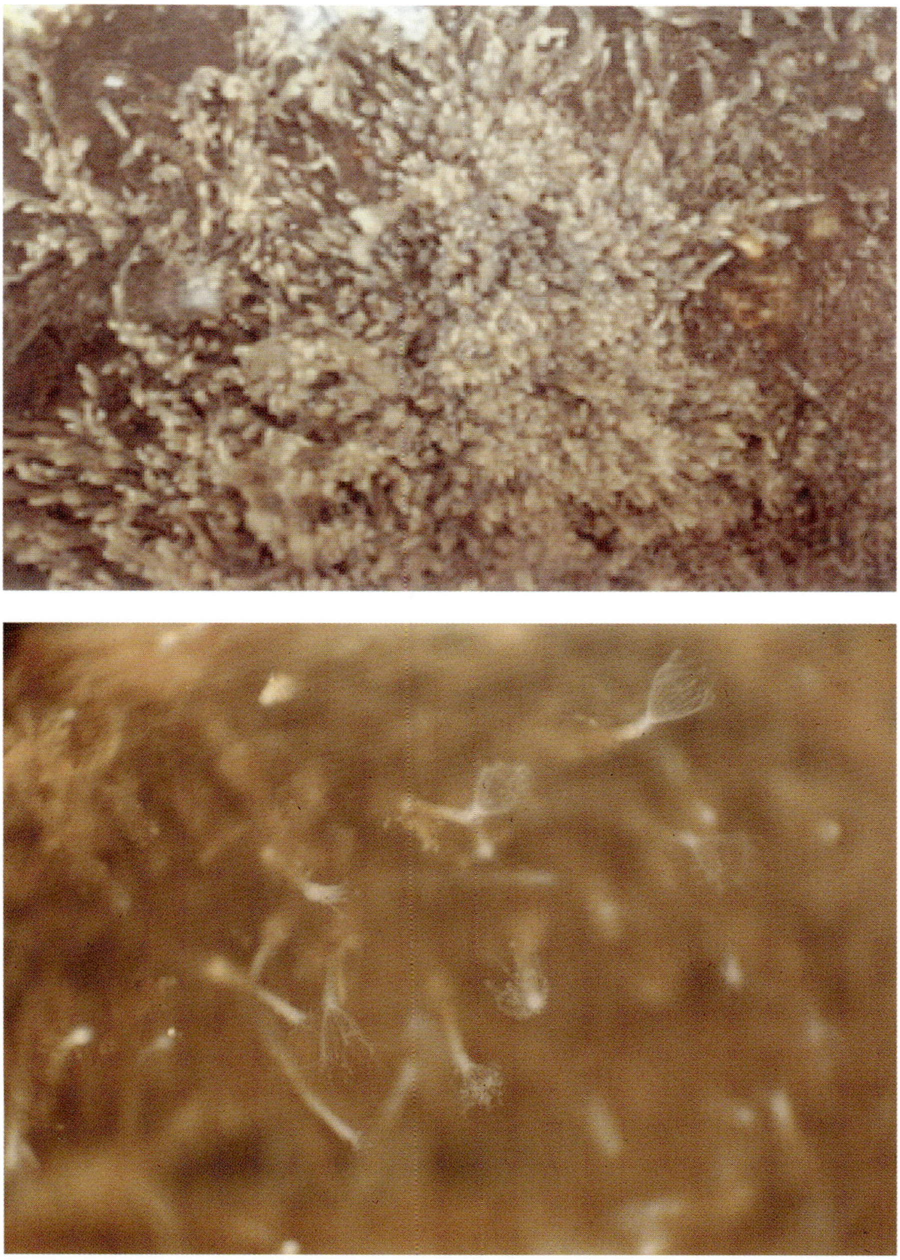

Plate 3. Ectoproct colonies. Top, large colony of *Plumatella emarginata*; bottom, several lophophores of *Pottsiella erecta* zooids. (Top photo © Terry Marsh, with permission.)

Plate 4. Scanning electron micrographs of rotifers. Top, *Rotaria* sp., a bdelloid; bottom, *Testudinella patina*, a monogonont. (Courtesy G. Melone.)

Plate 5. The freshwater polychaete *Manayunkia speciosa*, composed from videotape, from life. (A. MacConnell, A. Wier, S. Werle.)

Plate 6. Representative rhynchobdellid (Glossiphoniidae) leeches. Top left, *Helobdella fusca*; top right, *Placobdella papillifera*; bottom left, *Placobdella parasitica*; bottom right, *Placobdella picta*. (Top photos © William Moser, with permission; bottom left © Donald Klemm, with permission.)

Plate 7. Common mussels of the Mississippi River drainage system. A, *Pyganodon grandis*; B, *Quadrula metanerva*; C, *Fusconaia ebena*; D, *Cyprogenia aberti*; E, *Obovaria olivaria*. (From Harris and Gordon 1990, *Arkansas mussels*, Arkansas Game and Fish Commision, Little Rock, with permission.)

Plate 8. Common mussels of the Mississippi River drainage system. A, *Amblema plicata*; B, *Leptodea fragilis*; C, *Obliquaria reflexa*; D, *Ellipsaria lineolata*; E, *Potamilus purpuratus*. (From Harris and Gordon 1990, *Arkansas mussels*, Arkansas Game and Fish Commision, Little Rock, with permission.)

Plate 9. Some bivalve and gastropod mollusks of the eastern United States. A, *Ligumia nasuta*; B, *Alasmidonta varicosa*; C, *Goniobasis virginica*; D, *Viviparus georgianus*. (A, B by Ethan Nedeau.)

Plate 10. Mantle displays of freshwater mussels (Unionidae). Top, *Villosa vibex*; bottom, *Lampsilis fasciola*. (Top photo © Linda Richardson, with permission; bottom photo courtesy Shane Hanlon.)

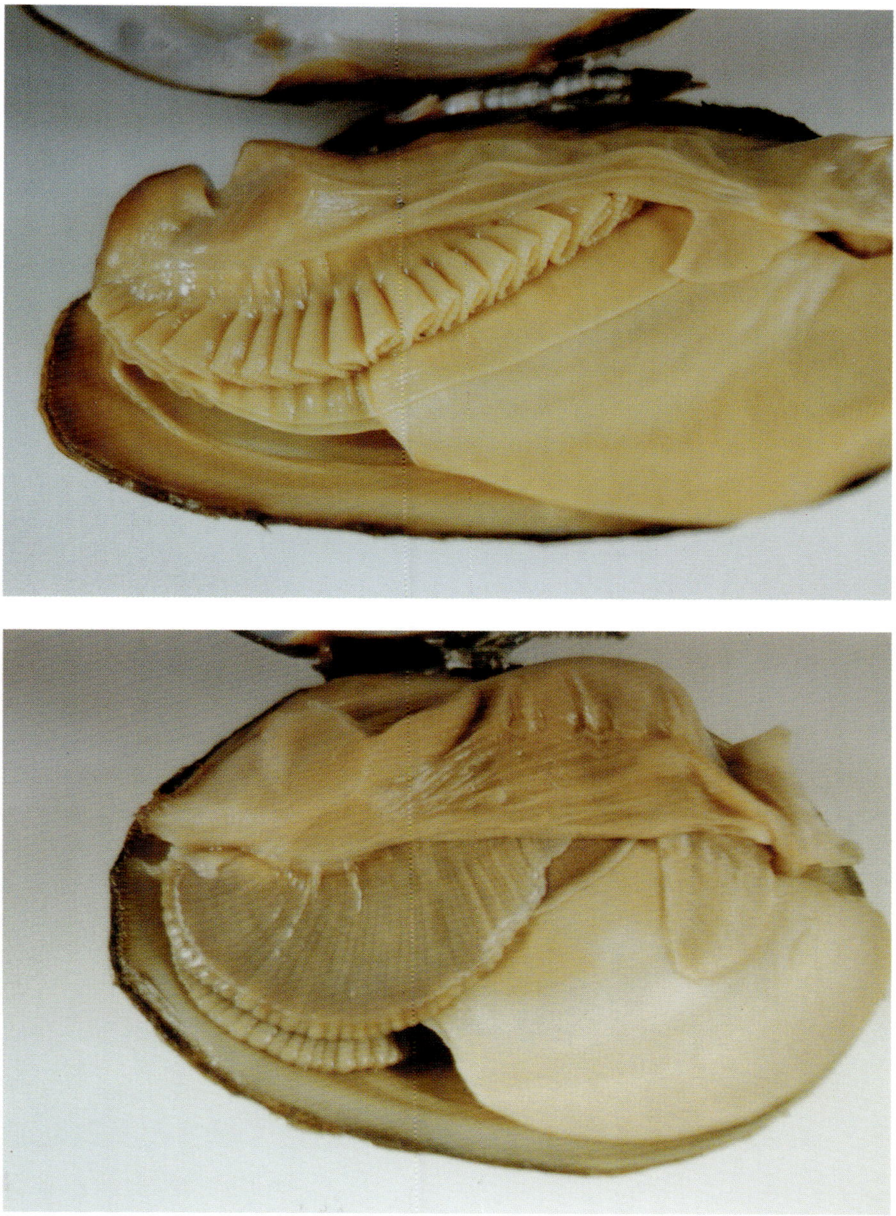

Plate 11. Specialized marsupial gills (demibranchs) of mussels. Top, *Ptychobranchus fasciolaris*; bottom, *Glebula subrotunda*.

Plate 12. Branchiopoda. Top, the spinicaudate clam shrimp *Eulimnadia texana*; bottom, the notostracan *Triops longicaudatus*. (Both photos courtesy Naida Zucker.)

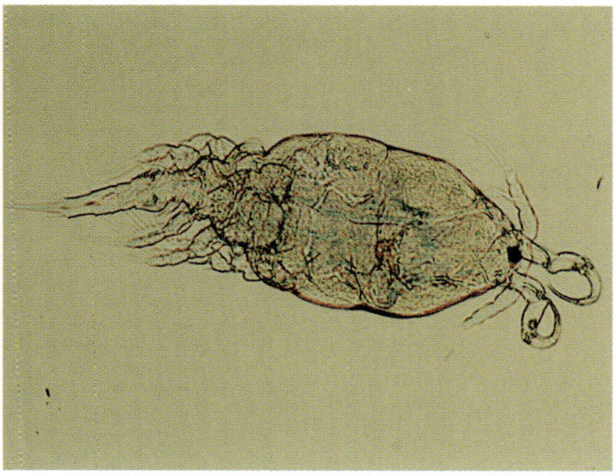

Plate 13. Some specialized crustaceans. Top, *Antrolana* lira, a cave adapted isopod, from life; bottom left, the groundwater amphipod *Stygobromus tenuis*, from life; bottom right, a parasitic copepod *Ergasilus nerka*, stained specimen. (Top photo © 1994 Susan Middleton and David Liittschwager, with permission.)

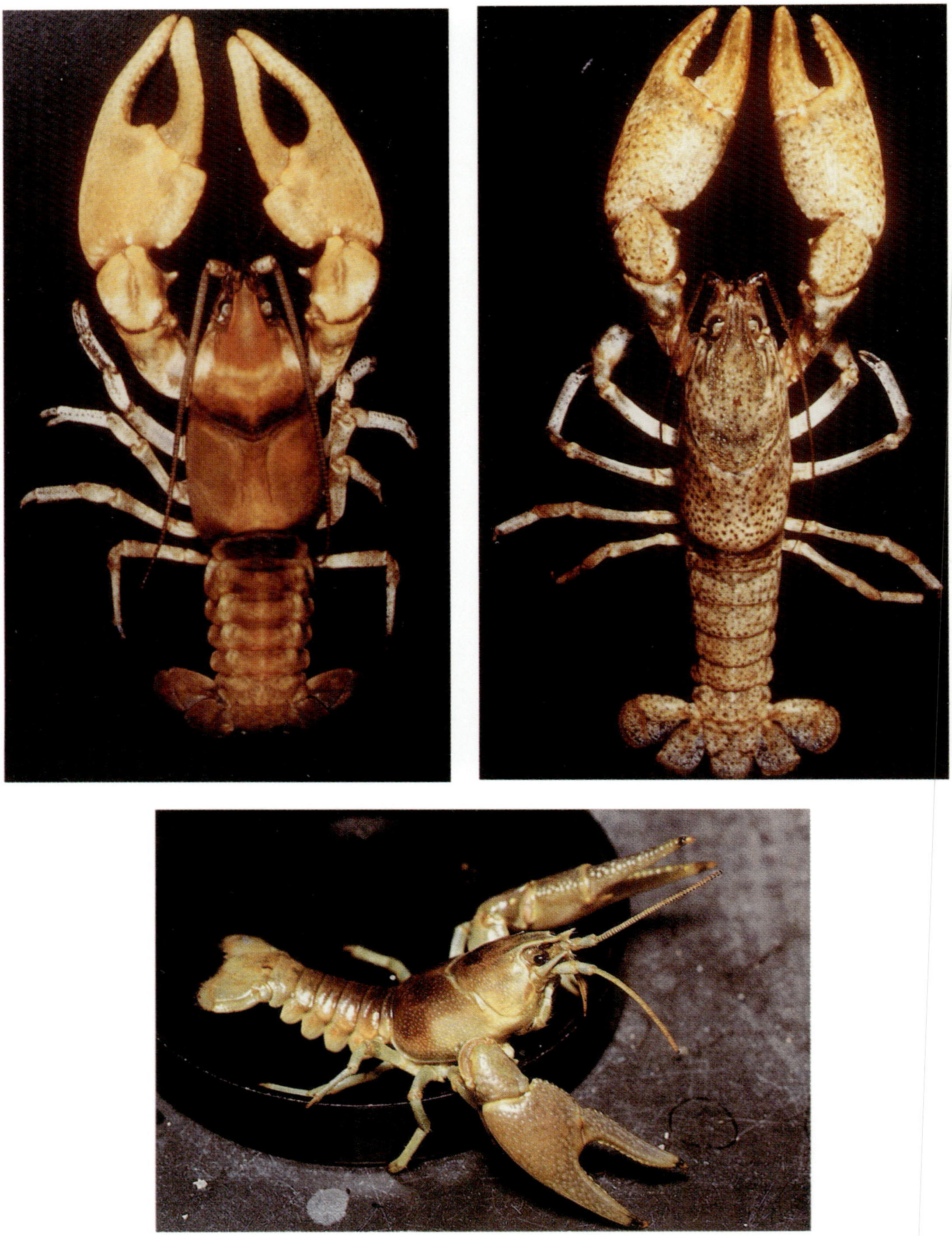

Plate 14. Representative crayfish. Top left, *Cambarus longirostris*; top right, *Orconectes compressus*; bottom *Orconectes rusticus*. (Top photos © Geunter Schuster, used with permission; bottom from D. G. Smith 1994, Exotic Aquatic Invertebrates, Massachusetts Wildlife 44: 17–20, photo by Bill Byrne.)

Plate 15. Representative crayfish. Top, *Cambarus dubuis*; bottom, *Troglocambarus maclanei*. (Top photo by Dan Duerson, courtesy G. Schuster; bottom from C. E. Mohr and T. L. Poulson. 1966. *The Life of the Cave*. McGraw-Hill Book Co., with permission.)

Plate 16. Some exotic invertebrates. Top, the "Polynesian moss animal" *Plumatella vaihiriae* (Ectoprocta); middle, zebra mussels *Dreissena polymorpha* (left two specimens) and an asian clam *Corbicula fluminea* (right specimen); bottom, *Platychirograpsus spectabilis* (Crustacea), the saber crab. (Top photo © Terry Marsh, with permission; middle photo from D. G. Smith 1994, Exotic Aquatic Invertebrates, Massachusetts Wildlife 44: 17–20, photo by Bill Byrne; bottom photo © Wayne Price, with permission.)

14

MOLLUSCA

(Gastropods, pelecypods)

The Mollusca are probably the most recognizable invertebrate organisms to scientists and nonscientists alike. They have provided a food source for nearly every culture since the beginning of civilization, and their colorful and ornate shells have been sought after and treasured. Some early civilizations even utilized mollusk shells as currency.

Beyond their obvious importance in human history, mollusks have played a far more significant role in metazoan evolution and have become one of the most successful animal groups on earth. Mollusks include the largest and most intelligent examples of invertebrates and they have gained entry into nearly every habitat in the aquatic biosphere. A significant reason for this success is the plasticity apparently built into some of their distinctive traits, namely a muscular foot, a multifunctional mantle, and a hard shell.

The basic mollusk is a composite of several fundamental and discrete features. The most readily apparent is the shell, a hard structure composed of layers of crystallized and mineralized calcium carbonate cemented together by bands of complex proteins (conchiolin). The shell is secreted and lined by a sheet of glandular epithelium called the mantle. The mantle also houses such organs as gills and structural glands and can be regionally modified to serve a number of auxiliary functions. A distinct head is present in most mollusks containing sensory structures and a file-like rasping tool called a radula. The radula is situated behind the mouth and encased in a thick bulb of muscle, the buccal capsule. Forming the base of the mollusk is a powerful foot, primarily involved with locomotion, but variously modified for different functions in the most highly specialized mollusks.

The invasion of freshwater has been accomplished by two classes, the Gastropoda and the Pelecypoda. The gastropods are the ubiquitous snails and are by far the most successful of the two classes in freshwater. The two major groups of gastropods, the operculate snails and the air-breathing snails, contain many families of freshwater species and the air-breathing forms are far better represented in freshwater and on land than in saltwater. The pelecypods, although reasonably well accounted for in freshwater, are the most part recent colonizers of freshwater habitats and most groups with freshwater representatives have many more species occurring in marine or brackish-water environments.

GASTROPODA (Snails, limpets)*

Only 15 out of a total of the approximately 300 families of gastropods are partially or wholly found in freshwaters of the United States. Nevertheless, almost every type of freshwater habitat, from the smallest ponds and streams to the largest lakes and rivers, has its characteristic population of snails. Only the coldest high altitude lakes, grossly polluted waters, and acid lakes are predictably without snails. Some localities may yield only one or two species, whereas others may contain dozens. Sometimes the collector needs to hunt assiduously to find a few specimens; under other circumstances they may be present in enormous numbers.

Snails are animals of the substrate and are found creeping about on all types of submerged surfaces, chiefly in water 10 cm to 2 m deep. Substrate types selected by gastropod species include clean cobble, silt and detritus, living hydrophytes, hydrophyte detritus, and allochthonous organic matter.

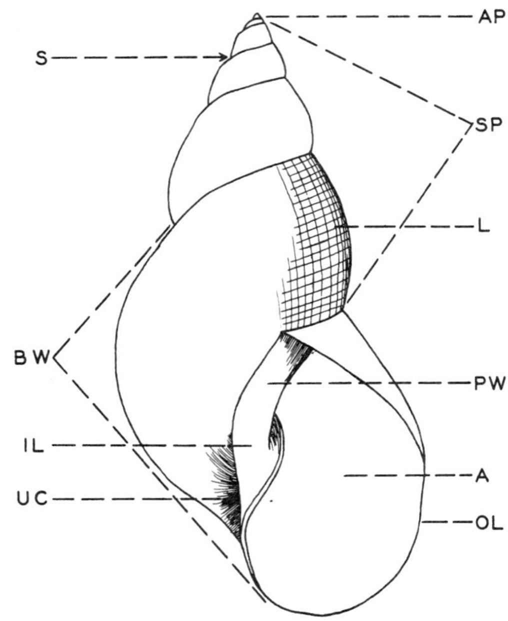

Fig. 14.1. Anatomical features of a typical *Lymnaea* shell, ×3. A, aperture; AP, apex; BW, body whorl; IL, inner lip reflected over columella; L, spiral and growth lines; OL, outer lip; PW, parietal wall; S, suture; SP, spire; UC, umbilical chink. (Modified from Baker 1928a.)

General characteristics. All freshwater gastropods of the United States have a spiral or discoidal coiled shell (Fig. 14.1). However, in several American families and genera (the limpets), the shell is in the form of a very low cone. The range in the maximum dimensions of the shell (length or width) of the mature animal is about 2 to 70 mm.

The foot is the muscular portion of the animal that projects from the shell (Figs. 14.2, 14.3). For the most part it is inconspicuously colored, usually being grayish, brownish, or blackish, and often flecked or mottled with yellowish or whitish. The ventral surface of the foot is flat and there is a more or less prominent head at the anterior end. The foot consists mainly of muscle tissue, as well as much of the nervous system. The rest of the internal organs form the visceral hump that lies within the shell. Forming a lining for the shell and enclosing the viscera is the mantle, a thin layer of specialized tissue. The shell grows by the addition of the shell material secreted by the edge of the mantle (collar) at the orifice of the shell.

The head bears a pair of tentacles, which range in shape from short, blunt, or conical to long and filiform. Eyes are on or near the base of the tentacles. The mouth is on the ventral surface of the head at the extreme anterior end. In most species the mouth is in contact with the substrate, but in the Viviparidae, Pleuroceridae, Hydrobiidae, and Valvatidae it is at the end of a muscular proboscis, or rostrum. In the Ampullariidae and Neritidae this rostrum is divided into two long, tentaclelike lobes. In the Lymnaeidae the head itself is widened into two flat lateral lobes, constituting the velum, a structure that persists from the larval stage.

Freshwater gastropods have either internal gills that are specialized folds of the mantle for aquatic respiration (Prosobranchia)

*This section was prepared with the assistance of Eileen Jokinen, with additional input from Robert Hershler.

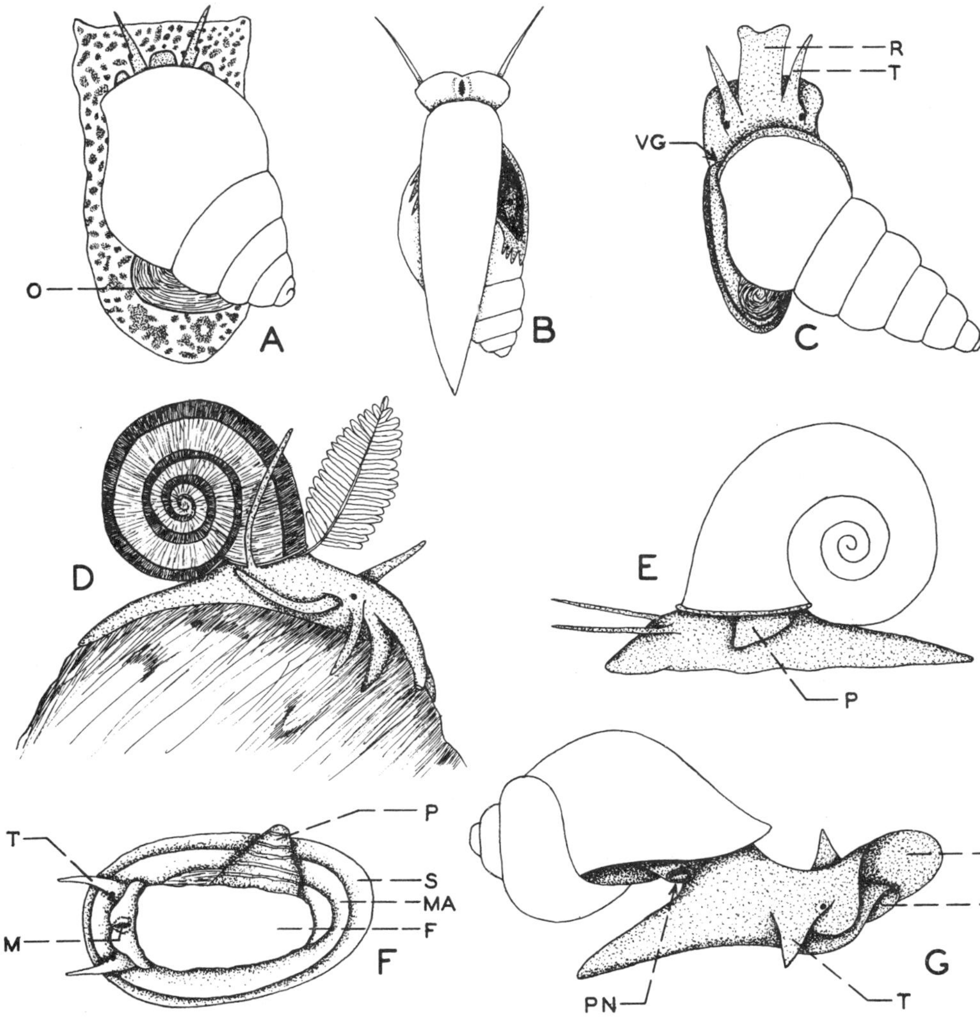

Fig. 14.2. External features of typical Gastropoda. A, female *Campeloma*, ×1.2; B, ventral view of *Physa*, ×5; C, *Pomatiopsis*, ×5; D, *Valvata*, ×5, showing plumose left external gill, rudimentary right gill, verge, tentacles, rostrum, and divided foot; E, *Helisoma*, ×5; F, ventral view of *Ferrissia*, ×5; G, *Lymnaea*, ×1.5. F, foot; M, mouth; MA, mantle; O, operculum; P, pseudobranch; PN, pneumostome; R, rostrum; S, shell; T, tentacle; V, velum; VG, vertical groove of foot. (A–F modified from Baker 1928a; G modified from Baker 1911.)

or an internal air-filled "lung" consisting of a special portion of the mantle cavity surrounded by highly vascularized mantle (Pulmonata). The respiratory opening (pneumostome) to either the internal gills or lung is usually situated on the right side at the edge of the shell. In those aquatic snails with a lung, the edge of the mantle is often elongated into a respiratory siphon through which air is obtained at the surface film.

The body of a snail is securely fastened to the shell internally by a large columellar muscle. Contraction of this muscle folds the foot and withdraws it and the head from view. Prosobranch snails with gills have a discoidal sclerotized (rarely calcareous) operculum attached dorsally on the posterior part of the foot, and when the animal is withdrawn into the shell the operculum fits into the opening and effectively closes it.

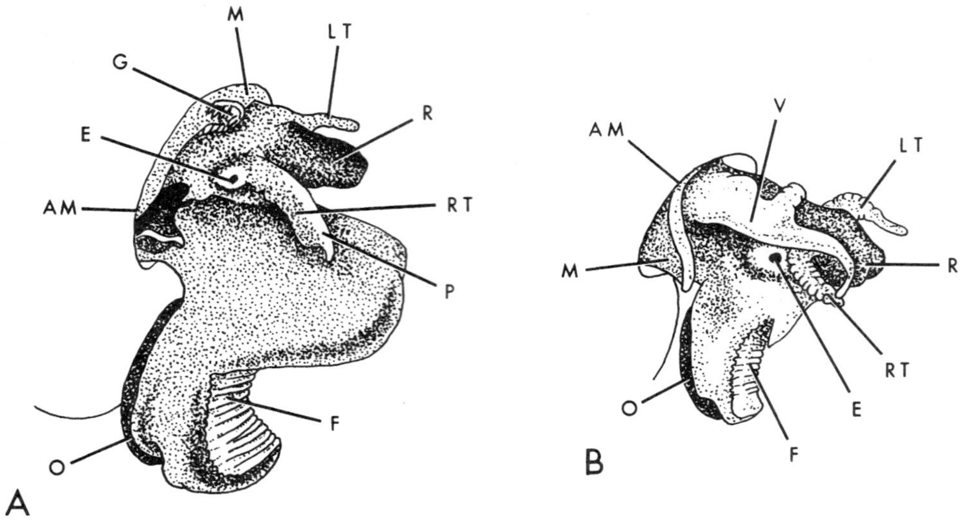

Fig. 14.3. Heads of gastropods. A, *Viviparus georgianus;* B, *Amnicola limosa.* AM, aperture margin; E, eye; F, foot; G, gill; LT, left tentacle; M, mantle; O, operculum; P, penis; R, rostrum; RT, right tentacle; V, verge.

Figure 14.4 shows the most common types of opercula.

In some prosobranch families portions of the male reproductive structures are external. In the Hydrobiidae (Fig. 14.3B) and Valvatidae a projecting male copulatory organ of variable structure and size, the verge, occurs more or less laterally in the region of the neck. In the Viviparidae the penis is attached to the right tentacle (Fig. 14.3A), forming a verge, which is much larger than the left tentacle.

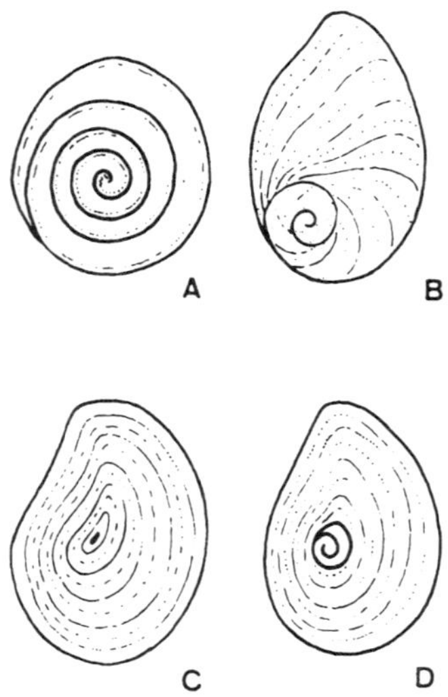

Fig. 14.4. Types of opercula. A, multispiral; B, paucispiral; C, concentric; D, concentric with spiral nucleus. (From Burch 1982.)

Shell. Compared with marine and tropical land snails, freshwater snail shells are a somewhat drab-appearing lot. Nevertheless, they exhibit a wide degree of variation and expression (Fig. 14.5). A few species are rich green or yellow with red tinges or bright color bands, but the great majority are nondescript gray, tan, brown, blackish, or "horn colored," sometimes with indistinct bands or a mottled appearance. The shell proper is covered with a thin organic periostracum in which many of the pigments are deposited. The periostracum has the important function of protecting the underlying chalky white calcium carbonate from erosion.

In most of the common and widespread species the surface of the shell is superficially smooth, but if it is examined closely, fine longitudinal growth lines and spiral sculpture may be distinguished. Sometimes these lines can be detected only with the aid of a lens. At the other extreme are a few spe-

Fig. 14.5. Representative prosobranch shells. A, *Bellamya japonica*, ×1.5 (Viviparidae); B, *Tulotoma magnifica*, ×1 (Viviparidae); C, *Valvata tricarinata*, ×5 (Valvatidae); D, *Neritina reclivata*, ×5 (Neritidae); E, *Pomacea canaliculata*, ×0.5 (Ampullaridae); F, *Leptoxis praerosa*, ×2 (Pleuroceridae); G, *Io fluvialis*, ×0.5 (Pleuroceridae); H, *Goniobasis virginica*, ×1 (Pleuroceridae); I, *Pomatiopsis lapidaria*, ×6 (Pomatiopsidae). (A after Smith 2000a; B after Hershler et al. 1990b; C after Clarke 1973; F after Dillon and Ahlstedt 1997.)

cies whose shells are pronouncedly costate, ridged, carinate, or tuberculate.

When a shell is held with the opening toward the observer and with the tip upward, the shell is said to be dextral if the opening is on the observer's right and sinistral if on the left. The great majority of species are dextral. Some families are sinistral and some genera contain both dextral and sinistral species, but only in a very few species are both dextral and sinistral specimens known.

The important features of a typical gastropod shell are shown in Fig. 14.1. The opening of the shell is called the aperture, and the main portion of the shell above the aperture is the spire. The first large coil, or whorl, of the shell is the body whorl. The nucleus, or protoconch, is the small round knob of from $1\frac{1}{4}$ to $1\frac{1}{2}$ whorls with or without distinct sculpturing at the apex of the shell. The whorls are coiled around a central axis, or columella. An inner lip at the columellar margin of the aperture is often present and reflected over the columellar region. The inner surface of the shell immediately adjacent to the inner lip is termed the parietal wall. The outer lip, or peristome, lies on the opposite side of the aperture; it is often reflected back on itself in mature animals. The umbilicus, if present, is a small chink or narrow slit between the reflected inner lip and the body whorl. It may be a definite hole one whorl deep; or it may be so well developed that all of the larger whorls revolve around a thin-walled hollow tube occupying the position of the typical solid columella. In many instances the umbilicus is sealed and the shell is said to be imperforate.

Considerable variations in size, shape, shell thickness, and surface markings occur within many genera, and sometimes even within a single species. These variations are particularly striking in the Pleuroceridae. In some cases variations within a species are clearly correlated with certain ecological features of the environment. Thickness of the shell, for example, may be directly related to the amount of calcium in the water or the rate of growth. In highly alkaline waters, especially in the western states, many species show increased tendencies toward being ridged, plicate, or rugose. Specimens in such waters are also generally smaller than those in other habitats. In species on wave-beaten shores and in rapid streams there is usually an increase in the relative size of the aperture and a decrease in the length of the spire.

Locomotion. In spite of many close observations, the detailed mechanism of the usual gliding movements is not well understood. As a snail moves along on the underside of the surface film or on the substrate, it leaves a familiar "slime track," a thin, flat ribbon of mucus secreted by the ventral surface of the foot, especially near the anterior end. The secretion of mucus is apparently necessary for locomotion, perhaps only as a lubricant, but it is not definitely known whether the actual movement of the snail is entirely due to indistinct waves of muscular contraction on the ventral surface of the foot, to ciliary action in the same area, or to both.

Another method of locomotion has been called "hunching." It involves obvious muscular contractions of the foot and a jerky pulling forward of the shell. Such movements are not normal, however, and usually occur when the snail is entangled in vegetation or is out of water.

A third method of locomotion, called "spinning," is utilized chiefly by some species of *Physa*, *Lymnaea*, and *Helisoma;* it occurs when the snail is moving through the water rather than over the substrate. For moving upward in this manner the snail osmotically decreases its specific gravity. First, a thread of mucus is fastened to the substrate, and the snail moves upward with the lateral margins of the foot brought together as a vertical mucous thread is extended behind. Similarly, a snail at the surface film or on some object above the bottom may fasten a thread and move downward, trailing a mucous track. The same thread may be used a number of times by the same or other individuals, but it soon becomes brittle and breaks. For the most part, those snails that spin have slender, tapered, or a pointed foot.

Feeding, digestive system. The great majority of freshwater gastropods are primarily vegetarians. The living algae, bacteria, and fungi that cover most submerged surfaces form the chief food, but dead plant material is frequently ingested, and occasionally dead animal material is eaten by some species. Many pulmonates are good scavengers and are generally omnivorous. There are even a few records of lymnaeids feeding on living animals. When first deposited, mucous trails are sticky and accumulate debris and microscopic organisms, and as snails move about they incidentally utilize it and its contents as food. The entire foot and head areas are chemoreceptive for taste but are decreasingly sensitive laterally and posteriorly. Food in the form of animal tissues elicits responses much more quickly than plant tissues. It has been clearly shown that the higher the temperature, the greater the activity and food intake.

Just inside the dorsal part of the mouth is a set of one, two, or three small sclerotized jaws (Fig. 14.6A–D), used in cutting off bits of food. One jaw is dorsal and (or) two are lateral. Immediately behind the jaws the digestive tract is swollen to form a large buccal mass. The ventral portion of this organ is greatly thickened and contains stout, movable buccal cartilages to which muscles are attached. Dorsally, the buccal cartilages and their muscles are covered by the radula, one of the most characteristic features of the

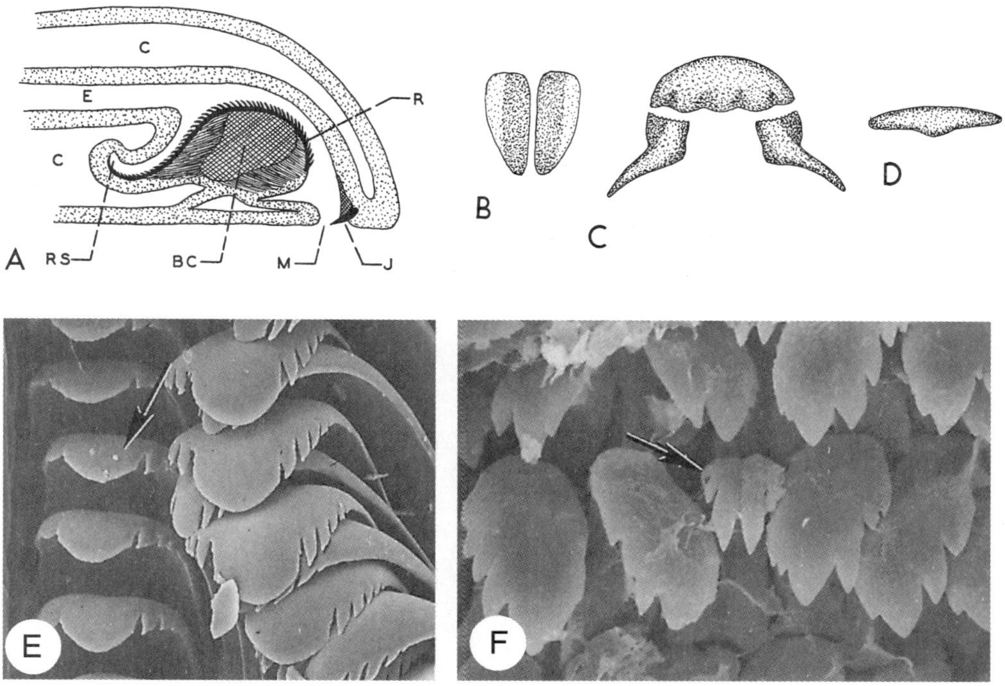

Fig. 14.6. Features of the feeding system. A, diagrammatic longitudinal section through anterior end of a snail; B, sclerotized jaws of *Pleurocera;* C, same for *Lymnaea;* D, single dorsal jaw of *Aplexa;* E, portion of tooth row in *Bellamya;* ×50; F, portion of tooth row of *Ferrissia;* ×100. BC, buccal cartilages; C, body cavity; E, esophagus; J, sclerotized jaw; M, mouth; R, radula; RS, radula sac. Arrows in E, F denote central tooth. (E from Smith, 2000a; F from Smith and Brousseau, 1996).

Gastropoda. It is essentially a longitudinal, toothed, straplike, chitinoid structure, which occupies a position analogous with that of the human tongue. The radula is moved back and out of the mouth very rapidly by means of the underlying cartilages and muscles, and in this way pieces of food are thoroughly scrapped from the substrate and subsequently ground between it and the roof of the buccal cavity. The radula wears away as a result, especially at the anterior end, but it is continuously replaced since it is formed in a radular sac at the posterior end of the buccal mass and grows outward much like the human fingernail.

The teeth of the radula are fastened to the flat radular membrane in transverse rows. The number of teeth in a transverse row ranges from 7 to 175 (Figs. 14.6E, F; 14.7). The number of transverse rows is usually large, some species having more than 150. Thus, the total number of teeth in a radula may range into the thousands. The morphology and the size of the teeth vary greatly from one family and genus to another (Fig. 14.7), but all teeth are fundamentally similar; each has a base of attachment and a reflected portion bearing the cutting points, or cusps.

In each transverse row three general types of teeth are usually recognized: (1) a single, distinctive, median tooth called the central tooth, (2) a series of lateral teeth on each side of it, and (3) a series of marginal teeth located outside of the laterals. Thus, in those families having seven teeth in a transverse series, they are named in order: second marginal, first marginal, lateral, central, lateral, first marginal, and second marginal. In the other families there are correspondingly more lateral and marginal teeth, and the morphological gradations between adjacent

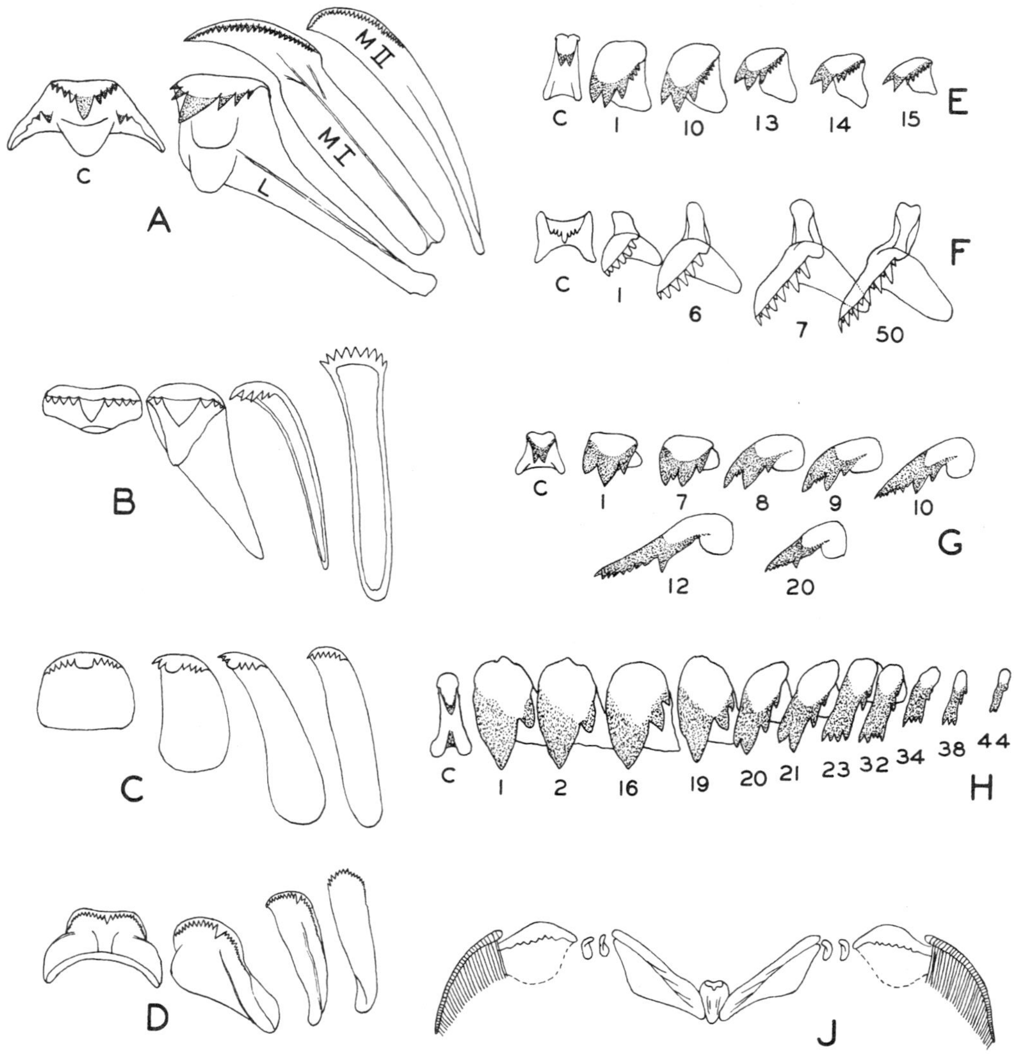

Fig. 14.7. Radular teeth of typical gastropods. A, *Amnicola* (Hydrobiidae); B, *Goniobasis* (Pleuroceridae); C, *Viviparus* (Viviparidae); D, *Valvata* (Valvatidae); E, *Ferrissia* (Ancylidae); F, *Physa* (Physidae); G, *Helisoma* (Planorbidae); H, *Lymnaea* (Lymnaeidae); J, *Neritina* (Neritidae). C, central tooth; L, lateral tooth; MI and MII, first and second marginal teeth; teeth designated by number are a few laterals and marginals selected from the complete series. (A modified from Baker 1928a; B–G redrawn from Baker 1928a; H modified from Baker 1911.)

teeth are very gradual. In the Lymnaeidae it is customary to designate a series of transitional teeth that lie between typical laterals and marginals. The differences in radular structure occurring among various groups of gastropods is apparently functional. In the case of *Physa* (a physid) and *Pseudosuccinea* (a lymnaeid) living in the same habitat, differences in food preference have been observed and these can be correlated with structural differences in the radula and other parts of the alimentary system.

The esophagus, which leaves the buccal mass, is quite long (Fig. 14.8); it passes from the foot into the visceral mass contained within the shell, where it may or may not be dilated to form a crop. A pair of salivary glands may be found along the esophagus or

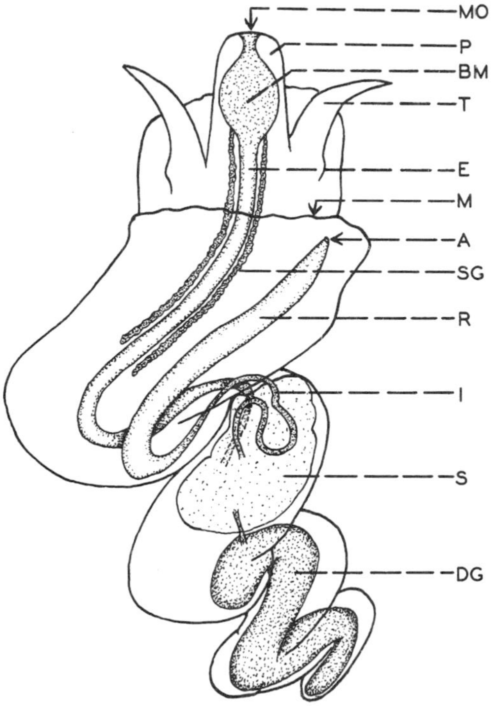

Fig. 14.8. Diagrammatic dorsal view of head and visceral mass of *Pleurocera*, showing location of digestive organs. A, anus; BM, buccal mass; DG, digestive gland; E, esophagus; I, intestine; M, edge of mantle; MO, mouth; P, proboscis; R, rectum; S, stomach; SG, salivary gland; T, tentacle. (Modified from Magruder 1935.)

on the crop. Shortly behind the crop is the dilated stomach, which may have a gizzard associated with it. In some genera the gizzard contains sand, presumably to macerate the food further. The stomach is followed by the long intestine, whose posterior end is dilated to form the rectum. The anus opens headward into the mantle cavity near the edge of the mantle and shell. A very large digestive gland, the so-called liver, empties into the stomach.

Circulatory system. The heart is situated in the lower part of the visceral hump, usually on the left side. In mature individuals with thin shells and in immature individuals the beating may be seen through the shell. The heart consists of one auricle (rarely two) and a ventricle. The ventricle pumps the blood through a short aortic trunk to all parts of the body via a series of arteries and capillaries. From the capillaries the blood passes into spaces, or sinuses, in the tissues collectively called the hemocoel. From the hemocoel it passes into veins and thence back to the auricle.

The blood of most gastropods contains a dissolved, complex, copper-containing compound, hemocyanin, which, like hemoglobin, is capable of transporting oxygen. Oxidized hemocyanin imparts a definite bluish color to the blood; in the reduced state it is colorless. In the Planorbidae, however, hemoglobin is the respiratory pigment and the tissues appear red.

Respiration. In the Ampullariidae, Bithyniidae, Viviparidae, Valvatidae, Pleuroceridae, Thiaridae, Hydrobiidae, and Neritidae respiration is strictly aquatic and occurs through an internal primary gill, or ctenidium, to which the surrounding water has easy access. This structure consists of a series of narrow, flat leaflets well supplied with blood and arranged like the teeth of a comb. Except for the Valvatidae, the ctenidium is located on the surface of the mantle within the mantle cavity. In some species there are over 100 leaflets in the ctenidium. In the Valvatidae the ctenidium is external and plumose (Fig. 14.2D). The gilled snails may obtain a small amount of oxygen through the general body surface.

Most pulmonate snails, members of the Physidae, Lymnaeidae, Planorbidae, Ancylidae, Lancidae, and Acroloxidae do not have primary gills, but obtain oxygen through a "lung," or pulmonary cavity. This is an air-filled, saclike, highly vascularized portion of the mantle cavity that occupies as much as one-half of the body whorl. The pneumostome, a small opening to the pulmonary cavity, is situated near the edge of the mantle, above the foot. In the Lymnaeidae and Physidae the edge of the mantle is frequently drawn out into a long, muscular respiratory siphon around the pneumostome. It is probable that the mantle border is especially significant in all snail respiration. At variable intervals most pulmonates come to the surface of the water where the pneumostome is brought into contact with the atmosphere and a fresh supply of air is taken into the pulmonary cavity. In a very quiet laboratory the opening of the pneumostome at the surface may be heard as a faint clicking sound. The migrations to the surface by spinning or on emergent vegetation appear to be largely governed by temperature and the amount of dissolved oxygen in the water. Under both natural and experimental conditions migrations may occur every few minutes to several hours.

However, many pulmonates rarely or never come to the surface for air. This is particularly true of certain species of *Lymnaea*, *Physa*, and *Helisoma*, which may pass their entire life cycle without access to the surface. Some pulmonates have been collected in water more than 15 m deep, and Forel collected *Lymnaea* at a depth of 250 m in Lake Geneva, Switzerland. Certainly under such circumstances there are no migrations to the surface for air. Although there seems to be little specific information available (Cheatum 1934), it appears that pulmonates that occur at great depths and some that remain submerged for protracted periods in the shallows fill the pulmonary cavity with water and use it as a gill. In many species, however, it is well established that all oxygen absorption may occur through the general body surface in both immature and mature individuals. Indeed, in the limpets the respiratory cavity is vestigial. Individuals in this family, as well as in the Planorbidae, have a secondary gill, the pseudobranch. This is a short, blunt to long, conical projection from the dorsal or lateral portion of the foot near the edge of the shell. It is highly vascularized and undoubtedly functions as an accessory gill (Fig. 14.2F).

Under experimental conditions it has been found that few snails can tolerate anaerobic conditions more than 48 hours.

Excretion. The renal organ is spongy and of varying size and shape. It is situated near the heart and respiratory cavity or gill, and its small duct opens near the anus. Its primary function is the removal of nitrogenous wastes, including ammonia and urea.

Muscles. Aside from the foot, which is composed mostly of muscle tissue, there are several important free muscles. The columellar muscle is attached to the shell internally and serves to withdraw the animal. Sets of muscles protract, retract, or depress the buccal mass. Those parts of the reproductive system directly concerned with copulation are usually supplied with protractors and retractors.

Nervous system. The greater portion of the gastropod nervous system (the "brain") usually consists of nine large ganglia, eight of which are paired and in close proximity. They are connected with each other by commissures and are arranged around the esophagus just behind the buccal mass. Large branching nerves originating in these ganglia innervate all parts of the body. Several small accessory ganglia are associated with some of the sense organs.

Sensory areas and organs. The eyes, situated at the base of the tentacles, are well developed, although little is known concerning the powers of vision. The ability to detect certain substances in solution (chemoreception) is probably centered in the osphradium, a small specialized area of the mantle cavity. The sense of hearing, or perhaps more appropriately the sense of equilibrium and the ability to detect vibrations, is centered in two statocysts. Each of these is a minute sac closely associated with the central nervous system. Each statocyst consists of a fluid in which are suspended a variable number of calcareous bodies called statoliths. Although the general body surface reacts to touch, the tentacles are especially sensitive. It seems logical to assume that taste is centered in the mouth region where this sense has been attributed to Semper's organ.

Reproduction. The anatomy of the reproductive system varies widely in the freshwater gastropods and is of increasing taxonomic significance, especially for the separation and identification of difficult species having nearly identical shells. The Physidae, Lymnaeidae, Planorbidae, Ancyli-

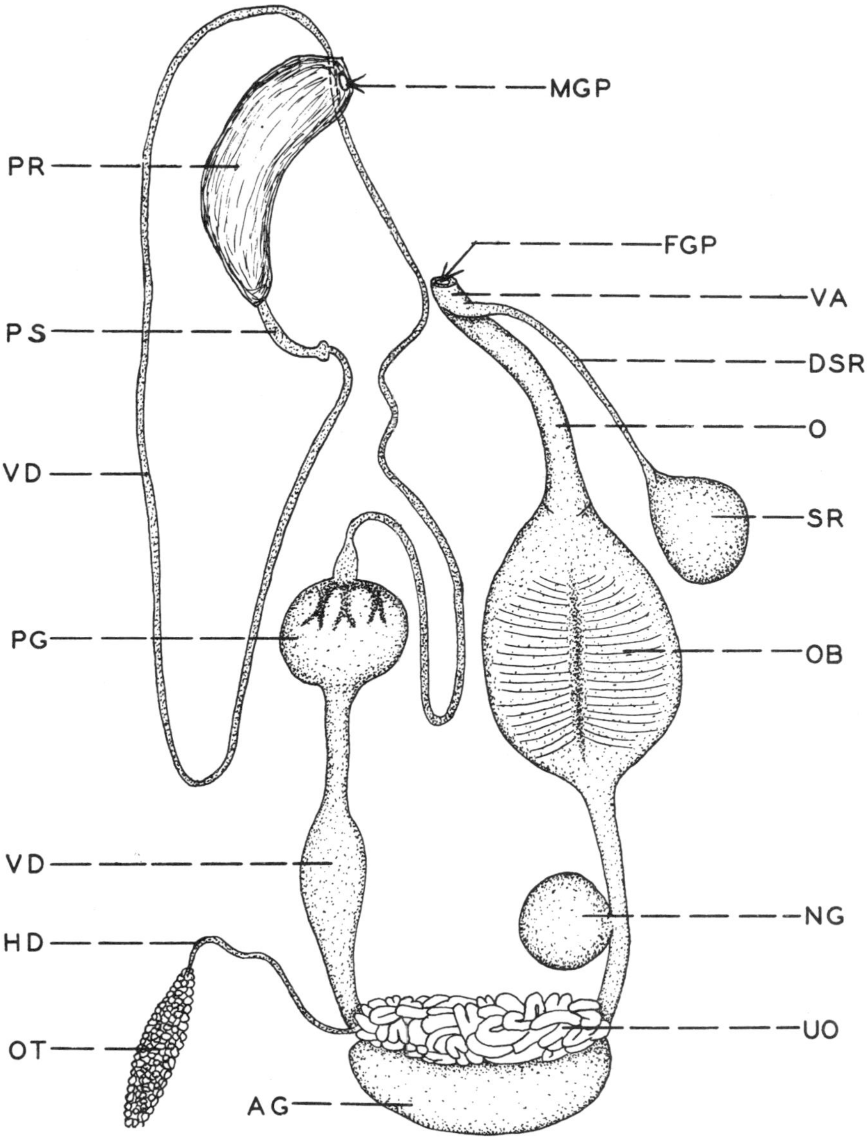

Fig. 14.9. Reproductive system of *Lymnaea stagnalis;* protractors and retractors of preputium and penis sheath not shown. AG, albumen gland; DSR, duct of seminal receptacle; FGP, female genital pore; HD, hermaphroditic duct; MGP, male genital pore; NG, nidamental gland; O, oviduct; OB, oviducal bulb; OT, ovotestis; PG, prostate gland; PR, praeputium; PS, penis sheath; SR, seminal receptacle; UO, uterine portion of oviduct; VA, vagina; VD, vas deferens. (Modified from Baker 1911.)

dae, Valvatidae, Acroloxidae, and Lancidae are hermaphroditic, both the male and female reproductive organs being in the same individual (Fig. 14.9). In all other freshwater families the sexes are separate. The basal portions of the ducts lie in the foot and body whorl; the distal organs, however, occur in the smaller whorls, with the ovotestis, ovary, or testis being more or less embedded in the digestive gland at the tip of the spire.

Except in sinistral species, the male genital pore is usually located near the base of the right tentacle. It is commonly situated at the end of a muscular, protrusible, intromittent copulatory organ, or penis, which is withdrawn into the body except during copulation. In the Viviparidae the right tentacle is modified as a penis sheath (Figs. 14.3A; 14.10). In the Hydrobiidae and Valvatidae the copulatory organ, usually called a verge (Fig. 14.3B), remains protruded and cannot be retracted. Sometimes the verge is long and thin; in other genera it is blunt, lobed, or divided. There is no male copulatory organ in the Pleuroceridae and the male genital pore lies at the edge of the mantle cavity. The female genital pore is unspecialized and usually lies at the base of the neck near the pulmonary aperture or at the edge of the mantle cavity.

Available information indicates that a variety of conditions exist regarding fertilization in the hermaphroditic freshwater Gastropoda. In some species isolated individuals produce young; under such circumstances both mature sperm and mature eggs are produced simultaneously, and self-fertilization without self-copulation occurs in the reproductive tract. In other cases two individuals may copulate and exchange sperm, each animal acting as male and female during the process; this is thought to be the common condition. In still other instances, however, the ovotestis may produce eggs at one time and sperm at another, and therefore an individual may act as either male or female during copulation, but not both.

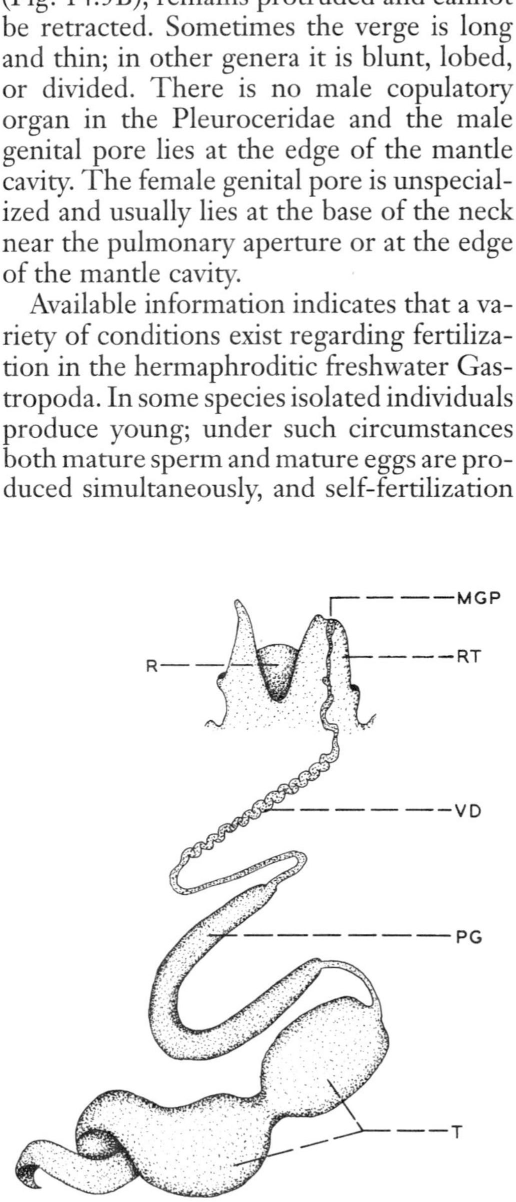

Fig. 14.10. Reproductive system of male *Campeloma integrum*. R, rostrum; RT, right tentacle; see legends of foregoing figures for other structures. (Modified from Baker 1928a.)

Parthenogenesis is uncommon, but it is known to occur in native *Campeloma*, especially in the northern states, a species of *Tryonia*, and the recently introduced *Potamopyrgus*. The introduced Thiaridae, native to Asia, are normally parthenogenic.

Oviposition usually occurs in the spring, although it may continue into the summer and early fall. Some species produce few eggs; others hundreds at a time. Eggs may be encapsulated singly, in pairs, or in high numbers. The egg mass capsule may be chitinous or gelatinous. The amphibious Ampullaridae lay eggs on emergent plants in a complex ropelike and chambered capsule.

The early developmental stages occur within the egg mass, and by the time the young snail leaves, it has taken on the basic morphological features of the adult, and the shell has one to two whorls. The Viviparidae and Thiaridae are ovoviviparous, the individuals being well developed at birth.

It is thought that in the great majority of species the usual length of life is 9 to 15 months. Some species, however, have two or three generations per year, especially in southern states, and notably among limpets and *Physa*. Length of life is often determined by periods of enforced estivation. Some of the Lymnaeidae, Planorbidae, *Viviparus*, and *Goniobasis* are known to live as long as 2 to 4 years, but this longevity is conditional on habitat and latitude.

Predators. Perhaps the greatest natural enemies of snails are fishes. It has been estimated that about 20% of our freshwater fish species feed to a greater or lesser extent on mollusks. Observations on snail dispersion in a small pond, for example, have shown that "refuges" such as sunken boats can harbor numbers of larger adults, whereas more open and exposed areas rarely contain large individuals. For only a few predators, however, do snails form a significant portion of the diet; some of these are the suckers, perch, sheepshead, pumpkinseed, and whitefish. Some ducks, shorebirds, and occasional amphibians eat snails. Among the invertebrates their most important predators are leeches, beetles and "snail-eating" fly larvae, Hemiptera and Odonata nymphs, and crayfish. In the laboratory, ostracods will attack small snails and destroy their eggs. The introduced Asian mystery snails (*Bellamya*) are sometimes eaten by humans.

Many species of snails serve as the intermediate hosts of digenetic trematodes. The larval stages typically encountered are the miracidia, sporocysts, redia, and cercaria. These various stages in the trematode's life history can be abundant in the vascular spaces, tissues, and pericardium. Many families of trematodes can be involved, including all species infecting humans while in the adult stage. The exotic tropical snails *Tarebia* and *Biomphalaria* can serve as intermediate hosts for *Paragonimus*, the human lung fluke, and *Schistosoma*, the blood fluke, respectively. Swimmers' itch, a common irritant to humans, is the cercaria of a common species of *Schistosoma* hosted by native snails. The reproductive capacity may be greatly decreased due to damage to the organs, and the digestive system may also be severely damaged.

Ecology. Most species of gastropods are not strong indicators of successional stages of lake history. Regarding many prosobranch families, geography is as important as habitat, if not more so, in determining snail distribution. Nevertheless, certain ecological factors of the environment have a pronounced influence on habitat selection and biological activities. Physical parameters, including substrate type and volume, can have a noticable affect on diversity. Shale and clay are preferred substrates, as are those of organic origin. Beds of aquatic plants in lotic situations are occupied by many species. Inland saline lakes are generally without snail species.

One of the most important factors in snail distribution is the amount of dissolved salts in the water, especially calcium carbonate, which is the essential material for shell construction. This aspect of growth, the adding of new shell material, especially by the juvenile, is likely the most sensitive part of the life history, as has been demonstrated for *Helisoma*. It is generally true that soft waters with high amounts dissolved organic matter contain few species and individuals, whereas hard waters contain many species and individuals. The majority of Lymnaeidae, for example, occur in water high in carbonates (more than 15 ppm of bound carbon dioxide). A few species, however, are striking in their ability to thrive in soft waters low in carbonates. The Valvatidae can tolerate waters containing less than 8 ppm but prefer higher concentrations. *Campeloma* has been found living in water containing only 1 ppm. It is difficult to understand how calcium carbonate can be absorbed and utilized at such low concentrations. It is very likely that these conditions are ephemeral, since recruitment would likely fail at this low level. Many species are, of course, adapted to a wide range of carbonate content; the common *Amnicola limosa* and *Physa heterostropha* have been collected from lakes having a calcium content ranging from 1 to 30 ppm; and other species undoubtedly have still wider limits. On substrates composed of mixed high- and low-carbonate rubble, the majority of individual snails will select the high-carbonate pieces.

There is some evidence to show that calcium is stored as concretions in connective tissues of the mantle edge, pulmonary region, and the foot. In general, the ionic uptake mechanisms have a high affinity for calcium, and, at least in several species investigated, the calcium content of the blood remains remarkably constant, in spite of varying ecological factors. Furthermore, in some species there is no relation between the calcium:tissue ratio and the level of environmental calcium. As a limiting factor, however, it appears that calcium levels may be instrumental in preventing development of the gastropod egg.

Hydrogen-ion concentration is closely associated with, and partly determined by, the carbon dioxide content, and lakes low in carbonates are usually toward the acid side of the scale (below pH 7.0), whereas those high in carbonates are almost always alkaline. It follows, then, that the great majority of species and the largest numbers of individuals occur under alkaline conditions. All the Valvatidae and many Lymnaeidae, for example, are confined to waters having pH readings of 7.0 or above. Nevertheless, there are some striking exceptions in other families. *Ferrissia parallelus* has been found in waters ranging from pH 6.0 to 8.4, and *Amnicola limosa* from pH 5.7 to 8.3. The genus *Lyogyrus* along the Atlantic coast appears to have adapted to nearly the lowest extreme tolerable among gastropods, pH 5.4. In general, however, snails are uncommon in lakes and streams whose surface waters are more acid than pH 6.2. A major restricting aspect of low pH is its interference with successful reproduction. *Amnicola limosa* will survive in low pH habitats. However, reproduction ceases below 5.0. It also appears that the tolerance limit of the offspring, more so than the adult, is what determines successful habitation. Hatching success seems to decline precipitously below pH 6.0.

Dissolved oxygen is another important limiting factor, some pulmonates and all gilled species requiring rather high concentrations. For this reason severely polluted rivers and the deeper parts of lakes that become oxygen deficient during the summer and winter are usually devoid of gastropods. Some limpets seem to be found only where the water remains almost saturated. For common species of *Physa*, 2 ppm is about the limiting level for dissolved oxygen.

The great majority of species and individuals occur in the shallows, especially in water less than 3 m deep. At greater depths river and lake faunas tend to become more and more depauperate. The fact that the shallows are a favorite habitat is probably correlated with the abundance of food in this zone. Although a few genera, such as *Lymnaea*, occur in a wide variety of habitats, many forms are restricted to particular types of substrates. *Pleurocera* is usually found on rocky or sandy shoals, the Viviparidae are most common on sandy bottoms, and the Ampullariidae are mud lovers. *Physa* occurs in greatest abundance where there is a moderate amount of aquatic vegetation and organic debris, and it is rare among dense mats of vegetation. The Hydrobiidae are usually found among aquatic plants, sometimes in enormous numbers. Throughout the United States and especially in the arid southwest, springs and wells support both high numbers and diversity of hydrobiid snails. The "watercress snails," genus *Fontigens*, are often found in dense mats of watercress and other vegetation in cool springs and their outlet streams. Swift streams with sand or gravel bottoms and wave-swept beaches are generally poor places for finding snails. Nevertheless, in certain parts of the United States swift streams with boulder or talus substrates can support pleurocerid, hydrobiid, and ancylid species.

Large bodies of water usually have many species and small bodies have few, the reason presumably being that the former have a greater variety of specific subhabitats suitable for the individual species. In large water bodies with adequate habitat diversity, many species can seemingly coexist. A lake in Connecticut supports 17 species in five families living in the near shore zone. Cohabitation is possible to some extent by differences in recruitment periods and food preference. Subterranean and even interstitial habitats support snail populations, primarily hydrobiids. Other families with representatives living in cave waters include the Physidae and the Pleuroceridae. Typically, the cave-dwelling species are albinistic and eyeless.

Although some gastropods may be active down to the freezing point of water, such conditions are not favorable, and perennially cold lakes and streams and bodies of water in high, mountainous areas contain few species and individuals. At the other extreme, only a few U. S. species can tolerate a continuous temperature of 40°C. *Physa* and certain western hydrobiids are examples of snails that can tolerate temperatures in excess of 30°C. Some lake pulmonates exhibit seasonal migrations correlated with temperature. In the fall they move into deeper waters and in spring into the shallows.

Hibernation, aestivation. Populations in shallow ponds that freeze solid are able to overwinter by settling into the mud and debris on the bottom and hibernating in the frozen substrate. In intermittent streams and ponds that are dry for a short period in the summer many pulmonates (e.g., *Planorbula*, *Aplexa*) settle into and among vegetation and organic debris up to a depth of several inches and estivate during the unfavorable period. The most effective seal and protection is afforded by a mud bottom with a high percentage of clay. In *Stagnicola* the formation of an epiphragm has been studied closely. This structure is usually produced when the snail is embedded in the mud. It is a sheet of mucus that is formed just within the aperture by the foot. On drying, it hardens and forms an effective seal that aids in protecting the animal during the drought period. Several successive epiphragms may form as the snail retreats further into the shell. This ability may account for the presence of *Stagnicola* in some waters to the exclusion of other genera. Viable estivating *Planorbis* and *Lymnaea* have been kept in the laboratory for more than 3 years. Some juvenile *Lymnaea* usually move out onto the adjacent shore and estivate in detritus for 3 to 4 months of cold or dry conditions. Young *Stagnicola* have been observed to even migrate up into trees to estivate. Mature *Lymnaea*, however, usually move into deeper water and rarely leave the water.

Overwintering nonhibernating gastropods usually do not ingest food, and consequently they usually suffer loss of weight. Nevertheless, some physids have been observed feeding under ice.

Geographic distribution, dispersal. The freshwater gastropod fauna of the United States is rich in species (about 500), rich in individuals, and rich in diversity of form. Examples of statewide studies carried out during the 20th century in the northern United States include Wisconsin, which contains 95 species; Indiana with 99 species; New York, after several exhaustive surveys, producing 61 species; 142 from Michigan; but only 23 in North Dakota. The glaciated and calcium-poor coastal New England states have the fewest species of the region, about 35. The Midwestern states of Missouri and Kansas have 56 and 21 (probably low), respectively. In the west, California has records of 67 species, Texas has over 100 species, and New Mexico 21. In the southeast, Florida contains 83 species. It is estimated that more than 90% of the species and half of the genera found in North American waters are endemic.

Some genera, such as *Stagnicola*, *Helisoma*, *Gyraulus*, and *Ferrissia*, occur from coast to coast. Certain species in these genera have very high dispersal rates and are good colonizers, i.e., supertramps. A few of these species seem to appear before any others when artificial habitats are formed (e.g., reservoirs and quarry ponds). As a whole, pulmonates, because of their physiology, are better dispersers than prosobranchs. The Pleuroceridae are largely restricted to the Mississippi River system and eastward. Pleurocerid genera would be considered among the poorest dispersers among the freshwater gastropods and demonstrate very high endemism.

Concerning other prosobranchs, native species of Viviparidae are not found west of the Mississippi drainage, but two imported species are found in California. The Neritidae, represented by a single species, *Neritina reclivata*, occurs only in Alabama, Florida, and along the Gulf coast. Freshwater species of the Ampullaridae are found only in Alabama, Florida, Georgia, and Texas, and all but one are introduced.

Some genera are similarly restricted. *Lanx* occurs only in the west coast states, *Aplexa* in the northern states, *Lyogyrus* in the Atlantic coast states, and *Tulotoma* only in the Alabama River and its tributaries. Perhaps the most restricted genera of all, *Gyrotoma*, *Amphigyra*, and *Neoplanorbis*, are (were) confined to the Coosa River, Alabama. *Bithynia tentaculata*, a small euryokous species in northeastern United States, has a mysterious history in the country. Many believe it was introduced from Europe, but there are some indications that it is native to at least some watersheds in western New York. Many species in the Mississippi and other eastern drainages undoubtedly would find suitable ecological conditions in many areas west of the Continental Divide, but their rate of geographic spread across the central plains and over the Divide since the last glaciation has been slow, except where humans have made intentional transplants.

Over the past 150 years, America's freshwaters have been degraded by industrial pollution (organic and inorganic), domestic sewage, agrichemicals, water diversion, impoundment of small and large rivers, destruction of wetlands, landfill leachates, and exotic species. This problem is especially acute in the southeastern region of the United States where many endemic forms are known to have existed. A classic example is the Coosa River, Alabama, where *Amphigyra*, *Neoplanorbis*, and *Gyrotoma* once thrived but have undoubtedly become extinct. The information on geographic ranges of unusual species in this chapter should not, therefore, be accepted at face value.

The present geographic distribution of many American freshwater snails is extremely complicated, poorly understood, and beyond the scope of this volume. It involves, among other things, a careful consideration of past geological, geographic, and climatic changes. Presumably it is relatively easy for a species to migrate slowly and extensively throughout an entire, connected drainage system, provided favorable environmental conditions exist. Movements between normally isolated bodies of water, or between inaccessible parts of the same drainage system, however, are necessarily passive and dependent on outside agencies. Occasional floodwaters may leave their normal channels and carry eggs or adults to distant places. There are numerous instances recorded in the literature where small snails have been found on the feathers or in mud on the feet and legs of ducks and shorebirds, and most authorities are agreed that this is a most important means of increasing the geographical range.

Several species native to Europe, Asia, and South America have been introduced into the United States by human immigrants, the aquarium trade, and ship ballast. Some have "escaped" from confined habitats to favorable natural habitats. At least some *Bithynia tentaculata* populations were supposedly imported to the Great Lakes region from northern Europe in the late 1870s and have spread widely throughout the St. Lawrence River system. *Tarebia* and *Melanoides*, which are native to the Far East and western Pacific islands, have been introduced into Florida, where they are now locally abundant. Two ampullariid genera, *Marisa* and *Pomacea*, originally from South America, and the tropical planorbid genus *Biomphalaria*, an intermediate host for the blood fluke *Schistosoma* sp., have become established in several southern states. Exotic Viviparidae have been in the United States since the beginning of the 20th century. The genus *Bellamya* (*Cipangopaludina* in the previous literature), originally from Asia, is now known in many parts of the country and at least one European viviparid, *Viviparus viviparus*, is believed to be present in the eastern United States. Very recently, a tiny hydrobiid from New Zealand, *Potamopyrgus* sp., has been released into interior North America. Other introduced freshwater species have been collected in various parts of the country, and the list will undoubtedly grow with time.

Collection, preparation. Collecting apparatus depends on the type of environment being visited. In the shallows of lakes and streams simple hand picking may be used. In deeper waters a long-handled net or tin strainer is effective. In water more than a meter or two deep it is usually necessary to work from a boat with a dredge or grab. Bottom debris and vegetation should be sifted and examined on shore. Limpets should be carefully removed from the substrate with a small knife blade. Uncontaminated wide-mouth bottles may be used to transport specimens to the laboratory. Leave an air pocket at the top and keep the specimens cool.

For anatomical work the animal should be narcotized prior to preservation. Especially in the Hydrobiidae, where species identification is usually based on the morphology of the undistorted verge, it is essential to narcotize the animal and every effort should be taken to ensure that relaxed specimens are available for examination. Remove enough water from the collection locality so that the specimens are just covered. The most effective method is to sprinkle menthol or magnesium sulfate crystals on the surface of the collection water and wait until the snails become very sluggish (which may be a whole day later). Some pulmonates are very difficult to narcotize and *Helisoma* simply retracts and dies. Typically, osmotic processes will begin to break down and the animal will swell. It is then time to kill and fix the specimens in 10% buffered formalin. Following fixation, about 2 days to a week, depending on size of the animal, the specimens should be washed in water to remove the formalin and then transferred to 75% ethyl alcohol or 60% isopropyl alcohol. The specimens should never be left in formalin, because it makes the specimens brittle and eventually damages the shell.

Either freshly killed or alcohol-preserved specimens may be used for dissection. The foot and the attached visceral hump should be kept intact by carefully pulling the latter from the shell with a pin or hairpin. If the specimen is small, the shell may be broken away from the underlying parts bit by bit, or the shell may be dissolved in dilute hydrochloric or sulfuric acid. The only instruments necessary for the dissection itself are a fine scalpel, forceps, scissors, and needles. First, the foot should be pinned down in a small dish (Stender dish, see Appendix) containing a thick layer of black paraffin on the bottom. Beginning at the head, a median, dorsal, longitudinal cut should be made through the body wall; it should be carried well into the mantle area, using great care not to injure the underlying parts. The dorsal flaps of the body wall and mantle may then be pinned back, thus exposing most of the viscera.

The radula may be isolated and cleaned by simply placing the whole buccal mass in cold 10% potassium hydroxide for several hours to a day or by heating the solution for a few minutes. A quicker method of preparing the radula for mounting involves dropping a few drops of household bleach on the isolated buccal mass. Within a few minutes the radula will be free of tissues and is ready for further processing. After a thorough rinsing in dilute acetic acid and twice in distilled water, the radula should be placed between two slides with a strip of paper on each side of the radula to prevent crushing. It is wise to mount the radula with the teeth up and to make several transverse cuts so the surface of the teeth can be more closely examined at the ends of the cuts. The two slides should be bound together with string. While in this flattened position, the radula should be dehydrated in an alcohol series and cleared. One slide is then removed and the radula is covered with mounting fluid and a coverslip. It is sometimes advantageous to stain it in strong chromic acid or carmine. The teeth at the anterior end of the radula are too worn to be diagnostic. In small species, the radula, once removed from the surrounding tissue, can be air dried on a slide, covered with any of several semipermanent or permanent media, and covered with a coverslip.

If the dry shell is to be preserved, the animal should first be pulled out and then the shell cleaned. It is sometimes effective to simply allow the animal to die and decompose. The viscera can be easily washed out of the shell, and the shell dosed in alcohol for a few days to eliminate odor. Larger specimens should be rubbed internally with a bit of sponge on a wire and then rinsed with a syringe. The external surface is often encrusted with algae, debris, and calcium carbonate or iron oxide deposits, and the appearance may be much improved by scrubbing with a stiff toothbrush and water, or by placing the shell in a weak solution of oxalic acid for a half-hour or longer. Finally the shell should be well rinsed and dried before storage. It is not necessary to remove the animal from the shell in the minute species, but the whole specimen should be kept in 70% alcohol for a day or two and then simply allowed to dry. Some workers place the live snail in a stoppered shell vial in the freezing compartment of the refrigerator. After it is frozen solid, thaw it; the whole visceral mass then can be easily pulled out of the shell. Opercula should be cleaned in oxalic acid, dried, and then placed in the shell whose aperture is plugged with cotton.

Jokinen (1992) has provided a timely and thorough discussion on protocols and proper storage of gastropod specimens.

Taxonomy. Two major groups of gastropods occur in freshwaters of the United States. The first comprises the operculate snails, generally placed in the Prosobranchia and characterized by the presence of an operculum, a streptoneurous (torted) nerve arrangement and a true gill attached to the mantle cavity wall, and a heart possessing either a single or two auricles (Neritidae). All members contain a primary (1°) gill and are dioecious, rarely parthenogenetic (Thiaridae, Viviparidae, Hydrobiidae). The pulmonates or nonoperculate snails, consisting of the Physidae, Lymnaeidae, Lancidae, Planorbidae, Acroloxidae, and Ancylidae, have a highly vascularized mantle cavity capable of absorbing oxygen, a heart with a single auricle, and a radula consisting of many teeth in a transverse row. The visceral nerves do not cross (detorted), but form a simple loop. All are hermaphroditic.

Enormous strides in the taxonomy of the Hydrobiidae have been made in recent years. The hydrobiids are clearly the most diverse of freshwater gastropods in North America. Fortunately, several features of the reproductive and intestinal anatomy are useful in identifying genera and even species, and several well-illustrated studies exist enabling researchers to determine the species of snail at hand. The Pleuroceridae, however, are still very problematical. These moderate- to large-sized snails display a myriad of shell ornamentation and share somewhat similar karyotypes, and so far, little anatomical information has proved useful in separating taxa. Radular differences have been used for distinguishing *Leptoxis* from other genera, but apparently only a few species have been examined and the character should be used with caution. Therefore, shell features continue to be largely used to differentiate genera and species. Burch (1982) lists some 200 species and subspecies described in the United States; 111 of these are in *Goniobasis* alone. Recent efforts to resolve the taxonomic problems of the Pleuroceridae are concentrating on DNA sequences, but many populations of these snails are extinct or nearly extinct and many questions regarding systematic affinity may go forever unresolved.

A list of the families and genera recognized in the key is provided below (Table 14.1). It has been constructed using information in a number of sources, more importantly Burch (1982), Jokinen (1992), and Kabat and Hershler (1993).

The key to the Lymnaeidae is modified from Jokinen (1992), which follows a more liberal approach to lymnaeid systematics in recognizing several genera. Within the Physidae and Planorbidae, the genera *Physa* and *Helisoma* are often distinguished from *Physella* and *Planorbella*, respectively, but many workers do not consider the differences sufficient to recognize two genera in each case. A broad concept of *Physa* and *Helisoma* is adopted here. The genus *Bellamya* includes the long recognized genus *Cipangopaludina*. The key that follows includes essentially all the known genera occurring in the United States. Only freshwater forms are indicated, those strictly brackish in occurrence have been omitted.

KEY TO FAMILIES AND GENERA OF FRESHWATER GASTROPODA OF THE UNITED STATES

1. Without an operculum, and mantle cavity a lung and indicated by a small lateral opening in the mantle, or with a small pseudobranch outside the mantle cavity; pulmonates **62**
   With an operculum (Fig. 14.4); gill present in mantle cavity; mantle open anteriorly; prosobranchs **2**
2. Shell globose, with a very low spire, up to 20 mm in length (Fig. 14.5D); operculum calcareous and with processes on lower margin; 1 species in FL, GA, MS . . . . **NERITIDAE,** ***Neritina***
   Shell variously shaped, but if shaped like *Neritina* then less than 5 mm in length **3**

Table 14.1. Families and Genera of Freshwater Gastropods of the United States

Subclass Prosobranchia	
Neritidae	*Neritina*
Valvatidae	*Valvata*
Viviparidae	*Bellamya, Campeloma, Lioplax, Tulotoma, Viviparus*
Ampullaridae	*Marisa, Pomacea*
Thiaridae	*Melanoides, Tarebia*
Pleuroceridae	*Goniobasis, Gyrotoma, Juga, Leptoxis, Lithasia, Pleurocera*
Bithyniidae	*Bithynia*
Pomatiopsidae	*Pomatiopsis*
Hydrobiidae	*Amnicola, Antrobia, Antrorbis, Antroselates, Aphaostracon, Balcanorbis, Birgella, Cincinnatia, Clappia, Cochliopina, Dasyscias, Fluminicola, Fontigens, Gillia, Heleobops, Holsingeria, Hoyia, Lepyrium, Littoridinops, Lyrogyrus, Notogillia, Phreatoceras, Phreatodrobia, Potamopyrgus, Pristinicola, Probythinella, Pyrgophorus, Pyrgulopsis, Rhapinema, Somatogyrus, Spilochlamys, Stiobia, Stygopyrgus, Taylorconcha, Texapyrgus, Tryonia*
Subclass Pulmonata	
Lymnaeidae	*Acella, Bulimnea, Fossaria, Lymnaea, Pseudosuccinea, Radix, Stagnicola*
Lancidae	*Fisherola, Lanx*
Acroloxidae	*Acroloxus*
Physidae	*Alpexa, Physa*
Planorbidae	*Amphigyra, Biomphalaria, Drepanotrema, Gyraulus, Helisoma, Menetus, Neoplanorbis, Planorbula, Promenetus, Vorticifex*
Ancylidae	*Ferrissia, Hebetancylus, Laevapex, Rhodacmea*

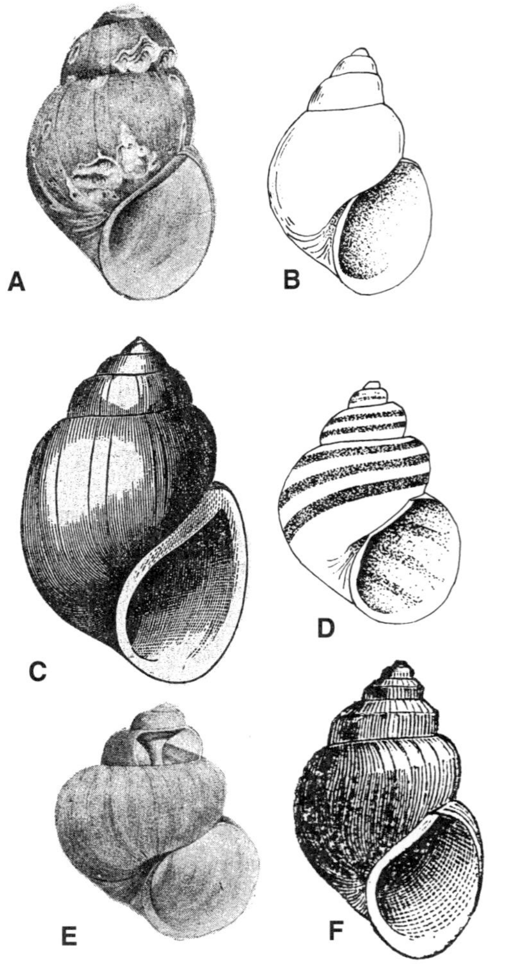

Fig. 14.11. Shells of Viviparidae. A, *Campeloma decisum*, worn specimen, ×1; B, *C. decisum*, ×1; *C. crassula*, ×1; D, *Viviparus georgianus*, ×1; E, *V. intertextus*, ×1; F, *Lioplax subcarinata*, ×1. (A, E from Clarke 1973; C, F from Tryon 1873.)

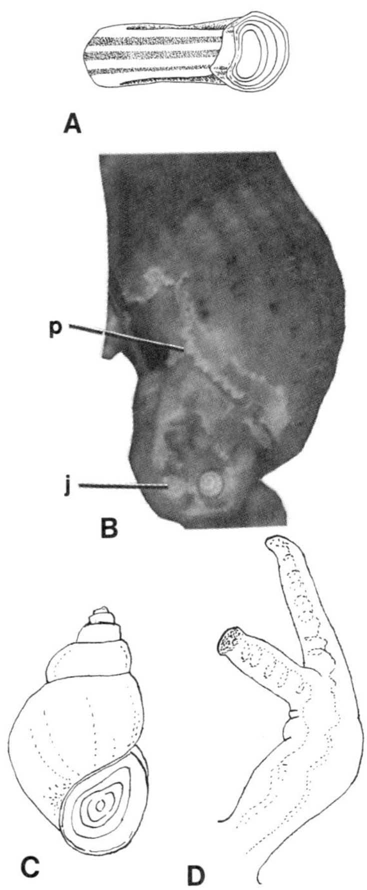

Fig. 14.12. Exotic tropical prosobranchs. A, *Marissa cornuaurietus* with operculum in place; B, dissected preserved specimen of *Tarebia granifera*, ×4; C, shell of *Bithynia tentaculata* with operculum in place, ×3.5; D, verge of *B. tentaculata*. p, mantle papillae; j, juvenile. (C redrawn from Berry 1943; D redrawn from Macan 1960.)

3. Operculum concentric (Fig. 14.4C) with or without spiral nucleus, calcareous (rare) or corneous; most species large, to 60 mm in length 4
   Operculum paucispiral or multispiral (Fig. 14.4A, B, D), corneus, adult shells usually small (less than 10 mm), unusually to 60 mm **11**
4. Operculum calcareous, with a spiral nucleus (Fig. 14.4D); male with a verge; (Fig. 14.12D); shell under 20 mm in length; 1 introduced species (Fig. 14.12C), widespread in St. Lawrence River system and adjacent drainages connected by canals . . . . . **BITHYNIIDAE, *Bithynia***
   Shell over 20 mm in length, globose, operculum corneous; male without a verge **5**
5. Body whorl considerably larger than spire (Figs. 14.5E; 14.12A), which is low to planospiral; mantle with a siphonal groove in addition to gill; amphibious and oviparous **AMPULLARIDAE, 6**
   Body whorl not considerably larger than spire, spire elevated, conical (Figs. 14.2A; 14.11); males with enlarged tentacle (containing penis) (Figs. 14.3A; 14.10); aquatic and ovoviviparous . . . .**VIVIPARIDAE, 7**

6. Shell planospiral (Fig. 14.12A); 1 introduced species, FL .... ***Marisa***
Shell low conical (Fig. 14.5E); 2 introduced, 1 native species, AL, FL, GA, TX ..... ***Pomacea***
7. Operculum with spiral nucleus (Fig. 14.4D), shell carinate, with raised spiral lines (Fig. 14.11F), or unsculptured, aperture margin sinuous; 6 species, scattered in the eastern half of U.S. ***Lioplax***
Operculum concentric throughout, shell tuberculate or unsculptured, aperture margin straight **8**
8. Shell with midwhorl carina interrupted to form tubercles (Fig. 14.5B), aperture somewhat compressed; 1 rare species, AL .... ***Tulotoma***
Shell without discrete sculpture, aperture circular to broadly lacrimate **9**
9. Shell brown, surface smooth, or with very subtle low carina (Fig. 14.5A), or faintly malleated, to 60 mm in length; 2 introduced species, scattered about mainly in the northern half of U.S. ***Bellamya***
Shell with shades of yellow, green, red, and blended with brown, without carina or malleations, with or without color bands, shell length usually less than 45 mm **10**
10. Shell yellow, yellowish brown, reddish-brown, without spiral color bands, aperture generally narrower than long (Fig. 14.11A–C); males typically rare in several species; 8 species, most found in southeastern sates, a few more widespread .... ***Campeloma***
Shell greenish or yellowish brown to dark brown, reddish color bands present (Fig. 14.5D) or absent; aperture as wide or wider than long; males always present; 3 species, plus possibly one exotic, eastern half of U.S. .... ***Viviparus***
11. Shell spire depressed (Fig. 14.5C); operculum circular and multispiral (Fig. 14.4A); gill external (Fig. 14.2D); hermaphroditic, right side of animal with a long slender penial organ diminishing with age, pallial tentacle (cleaning filament) present (Fig. 14.2D); about 10 species generally distributed in a variety of habitats but usually large lakes **VALVATIDAE,** ***Valvata***
Shell spire high conical to planospiral; operculum paucispiral or with multispiral nucleus; sexes separate, right side of male with a verge (penial organ) of variable form or verge absent, pallial tentacle absent **12**
12. Shell small, rarely over 10 mm in length, rarely decollated or with strong sculpture; males with a verge (Fig. 14.3B) **21**
Shell medium to large, intact specimens usually well over 20 mm, often decollated and usually with well marked sculpture; verge absent **13**
13. Mantle edge digitate (Fig. 14.12B); shell with high acute spire, aperture lacrimate, lower margin evenly curved; parthenogenic and ovoviviparous (Fig. 14.12B) ...... **THIARIDAE, 14**
Mantle edge not digitate; shell globose to high conical, spire often decollated, aperture elliptical to lacrimate, lower margin slightly to strongly angulate, or caniculate; dioecious and oviparous .... **PLEUROCERIDAE, 15**
14. Shell surface distinctly nodular or granular in appearance (Fig. 14.12B); 1 introduced species, FL, TX .... ***Tarebia***
Shell surface with spiral lines that intersect longitudinal ribs; 2 introduced species, AR, FL, TX ***Melanoides***
15. Shell large, to 50 mm in length, whorls with variously developed spines (Fig. 14.5G), rarely absent; aperture strongly caniculate; 1 species, TN, VA .... ***Io***
Shell 30 mm or less in length, sculptured, sometimes strongly, but without projecting spines, aperture angulate to weakly caniculate **16**
16. Aperture with a sinus above, shells somewhat globose with variously developed raised spiral sculpture (Fig. 14.13A, B); many described species, all presumably extinct, AL ***Gyrotoma***
Aperture without a sinus, shells globose to conical, sculpture ranging from absent to elaborate and complex **17**
17. Specimens from west of the Continental Divide; 8 species, CA, NV, OR, WA .... ***Juga***
Specimens from east of the Continental Divide **18**
18. Median cusp of lateral radular tooth with a broadly rounded edge; shell low conical or ovate (Fig. 14.13C), over 20 species, eastern U.S. .... ***Leptoxis***
Median cusp of lateral radular tooth with pointed or straight edge; shell ranging from ovate to high conical; 3 highly convergent genera, in some cases impossible to separate **19**

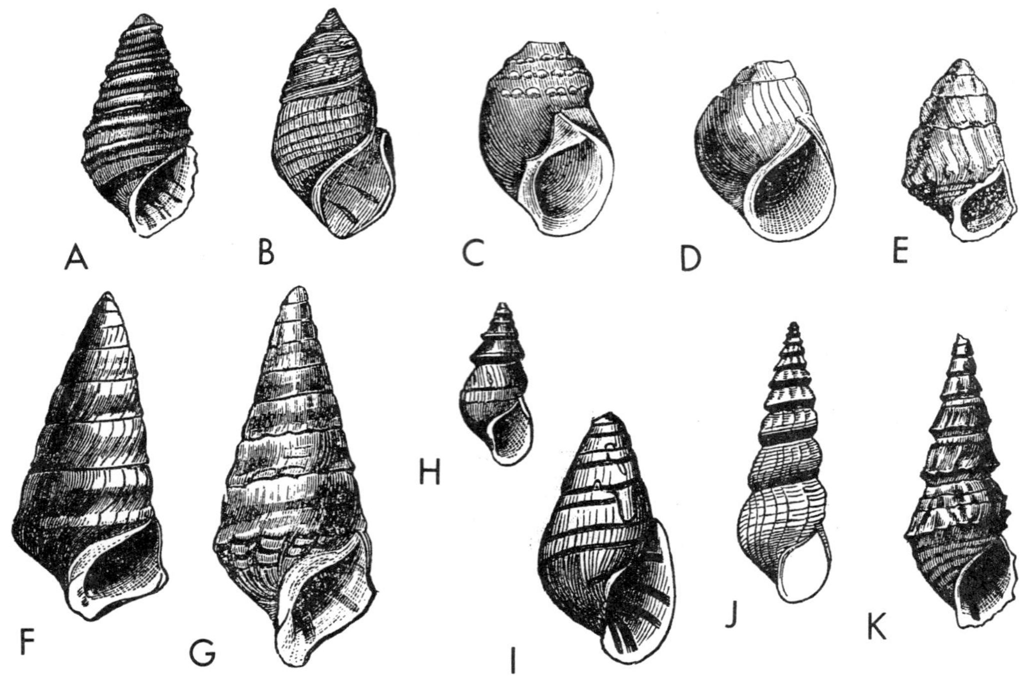

Fig. 14.13. Representative Pleuroceridae, all ×1. A, *Gyrotoma cariniferum;* B, *G. pyramidatum;* C, *Leptoxis plicata;* D, *Lithasia subglobosa;* E, *Pleurocera alveare;* F, *P. canaliculatum;* G, *P. moniliferum;* H, *Goniobasis acutocarinata;* I, *G. ampla;* J, *G. arachnoidea;* K, *G. boykiniana.* (A–K after Tryon 1873.)

19. Shell broadly to rarely high conical, usually with strongly shouldered or buttressed body whorl (Fig. 14.13D), sculpture absent or consisting of thick, heavy interrupted carinae; 10 species of the interior U.S. . . . . ***Lithasia***
Shell high conical, rarely low conical with or without acute spire (often decollated), sculpture variable or absent **20**

20. Aperture typically caniculate (Fig. 14.13E–G), spire usually high and acute, whorls flattened; 21 species, interior U. S. and St. Lawrence River system . . . . ***Pleurocera***
Aperture angulate, spire elevated but usually not acute (Fig. 14.13H–K), whorls generally rounded; over 60 described species, many probably invalid, distributed throughout the eastern U.S. (*Elimia* used in some literature) . . . . ***Goniobasis***

21. Shell high conical (Figs. 14.2C), up to 10 mm in length; foot with vertical groove; verge an elongate lobe with lateral thickening (Fig. 14.18A); usually amphibious but sometimes found on submerged substrates; 3 species in coastal CA and 3 species in eastern and southern states **POMATIOPSIDAE, *Pomatiopsis***
Shell high conical to planospiral, rarely reaching 10 mm, usually much smaller; foot without a vertical groove; verge variable; aquatic . . . . **HYDROBIIDAE, 22**

22. Shell with elongate, detached body whorl; verge a simple lobe (Fig. 14.16A); 1 species in TX wells . . . . ***Phreatoceras***
Shell distinctly coiled **23**

23. Shell planospiral **24**
Shell with raised spire **27**

24. Shell with enlarged body whorl (Fig. 14.19D); verge (Fig. 14.15C) a truncated lobe with subapical patch of glands; 1 rare species, AL . . . . ***Lepyrium***
Shell without enlarged body whorl; in phreatic waters **25**

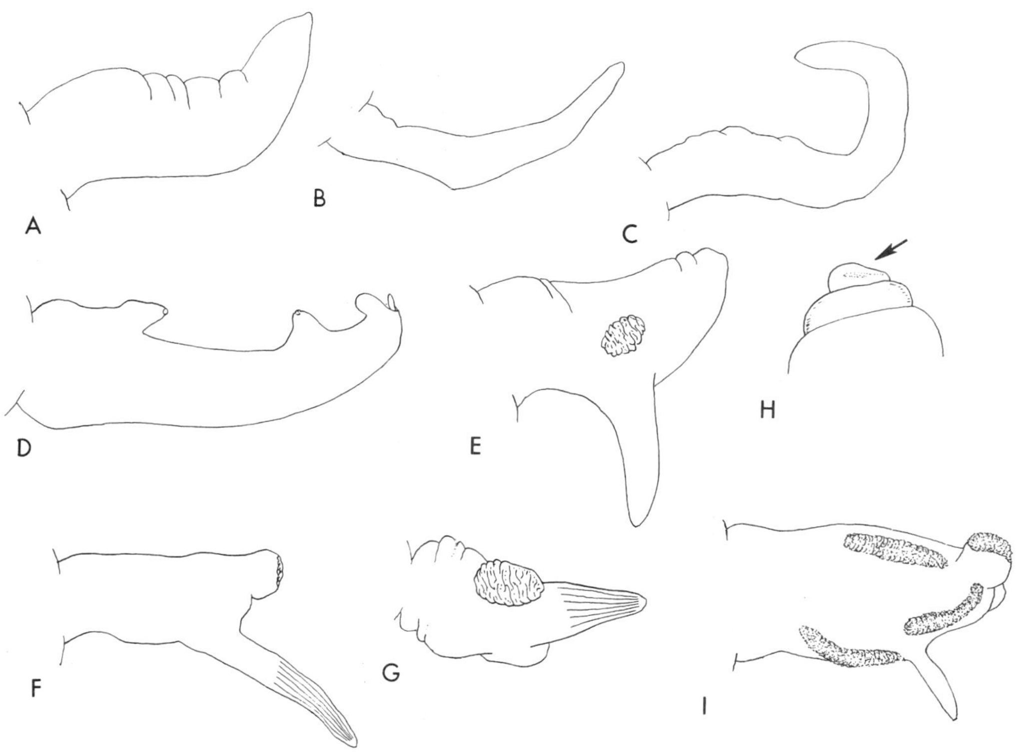

Fig. 14.14. Verge and shell characters of Hydrobiidae. A, *Fluminicola;* B, *Taylorconcha;* C, *Pristinicola;* D, *Tryonia* (Arizona); E, *Pyrgulopsis* (New Mexico); F, *Pyrgulopsis* (California); G, *Pyrgulopsis* (New Mexico); H, *Probythinella;* arrow denotes apical whorls; I, *Pyrgulopsis* (Arizona). (A redrawn from Hershler and Frest 1996; B, C redrawn from Hershler et al. 1994; D, I redrawn from Hershler and Landye 1988; E, G redrawn from Taylor 1987; F redrawn from Hershler 1995.)

25. Known only from a cave in AL; verge an elongate, simple lobe (Fig. 14.17E); 1 species ***Antrorbis***
Known only from wells in southern Texas **26**
26. Verge with pediculate gland (Fig. 14.16B); 1 species .... ***Balcanorbis***
Verge a simple lobe (similar to Fig. 14.16A); 8 species .... ***Phreatodrobia***
27. Shell with several color bands and occasionally raised spiral lines on whorls, verge a simple elongate lobe (Fig. 14.16E); 1 species, TX .... ***Cochliopina***
Shell without color bands **28**
28. Specimens from west of the Continental Divide (northern CA, ID, NV, OR, UT, WA, or WY) **29**
Specimens from elsewhere **33**
29. Shell broadly conical to high conical, occasionally with spiral carina, aperture entire (Fig. 14.20J, L); verge a broad, blunt lobe with distinct penial branch and glandular lobe (Figs. 14.14F; 14.15B); at least 15 species .... ***Pyrgulopsis* (in part)**
Shell low conical with enlarged body whorl or high conical with flat whorls; verge a simple, elongate wrinkled lobe **30**
30. Shell high conical with flattened whorls, body whorl not enlarged; verge narrow and U-shaped (Figs. 14.14A–C; 14.15G) or males absent **31**
Shell globose or low conical (Fig. 14.20I), with inflated body whorl; verge bent, but not U-shaped, often broad, especially at base (Fig. 14.14A) **32**

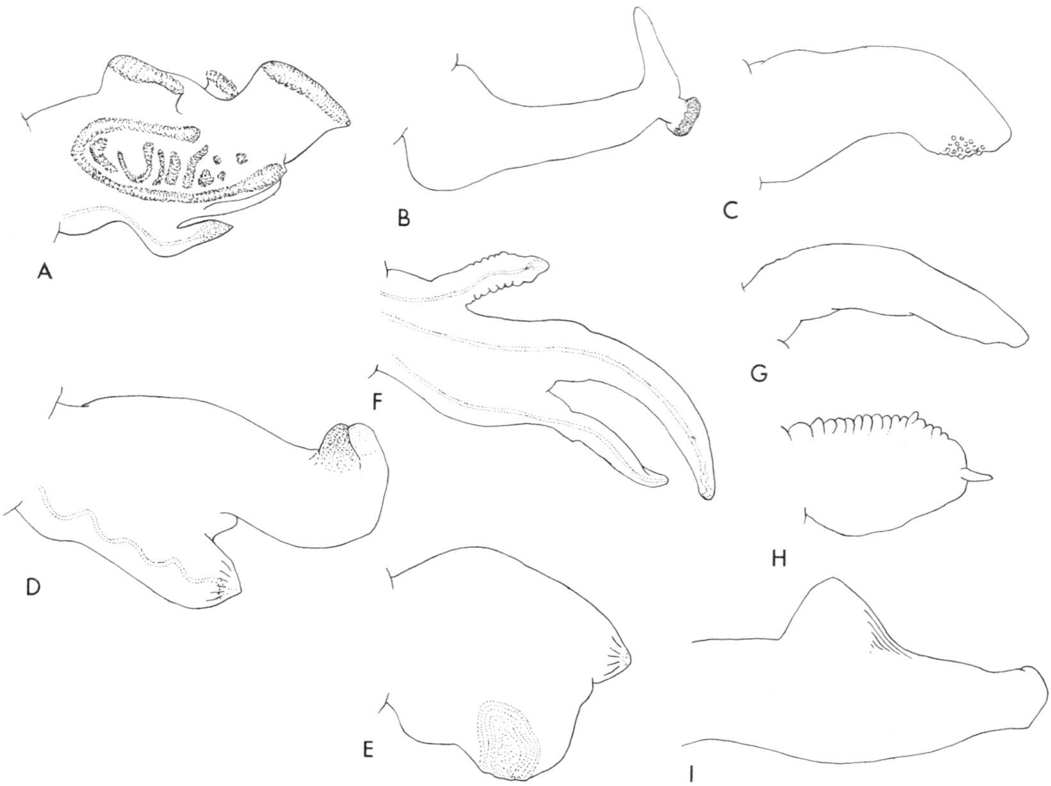

Fig. 14.15. Verge characteristics of Hydrobiidae. A, *Paludina;* B, *Pyrgulopsis;* C, *Lepyrium;* D, *Birgella;* E, *Birgella;* F, *Fontigens;* G, *Somatogyrus;* H, *Gillia;* I, *Probythinella.* (A composed from Berry 1943, and Hershler and Thompson 1996; B, D, F, I redrawn from Berry 1943; C, E, H redrawn from Thompson 1984b.)

31. Shell tapered, aperture; males in normal ratio; 1 species, ID, OR, WA .... ***Pristinicola***
Shell acuminate, occasionally carinate (Fig. 14.19E); males rare or absent, if present verge strongly U-shaped (Fig. 14.17D); 1 introduced species, ID, recently in upper Missouri River system, MT .... ***Potamopyrgus***

32. Shell reaching 4 mm in length; known only from southern ID; 1 species .... ***Taylorconcha***
Shell up to 12 mm in length; widespread; 10 species .... ***Fluminicola***

33. Specimens from southwest; AR, southern CA, NM, or TX **34**
Specimens from the Mississippi River drainage and east **37**

34. Animal of subterranean facies, without eyes or pigment; shell with raised spiral lines on whorls, less than 2 mm in length **36**
Epigean, with pigment and eyes; shell without spiral lines, typically greater than 2 mm in length **35**

35. Shell broadly conical, body whorl somewhat enlarged (Fig. 14.20J); occasionally carinate; verge a broad, blunt organ with distinct penial tube and glandular lobe (Fig. 14.14E, G, I); at least 15 species .... ***Pyrgulopsis* (in part)**
Shell high conical, turreted (Fig. 14.20K), occasionally ribbed; verge a slender lobe with terminal papilla and lateral, cone-like glands (Fig. 14.14D); about 7 species, AR, southern CA, NM, west TX, southern NV .... ***Tryonia* (in part)**

36. Shell with strong spiral lines; verge an elongate lobe with single lateral tuberous gland (Fig. 14.16G); 1 species, TX .... ***Texapyrgus***
Shell with weak spiral lines (Fig. 14.19C); verge an elongate lobe with two tubelike lateral glands (Fig. 14.16D); 1 species, TX .... ***Stygopyrgus***

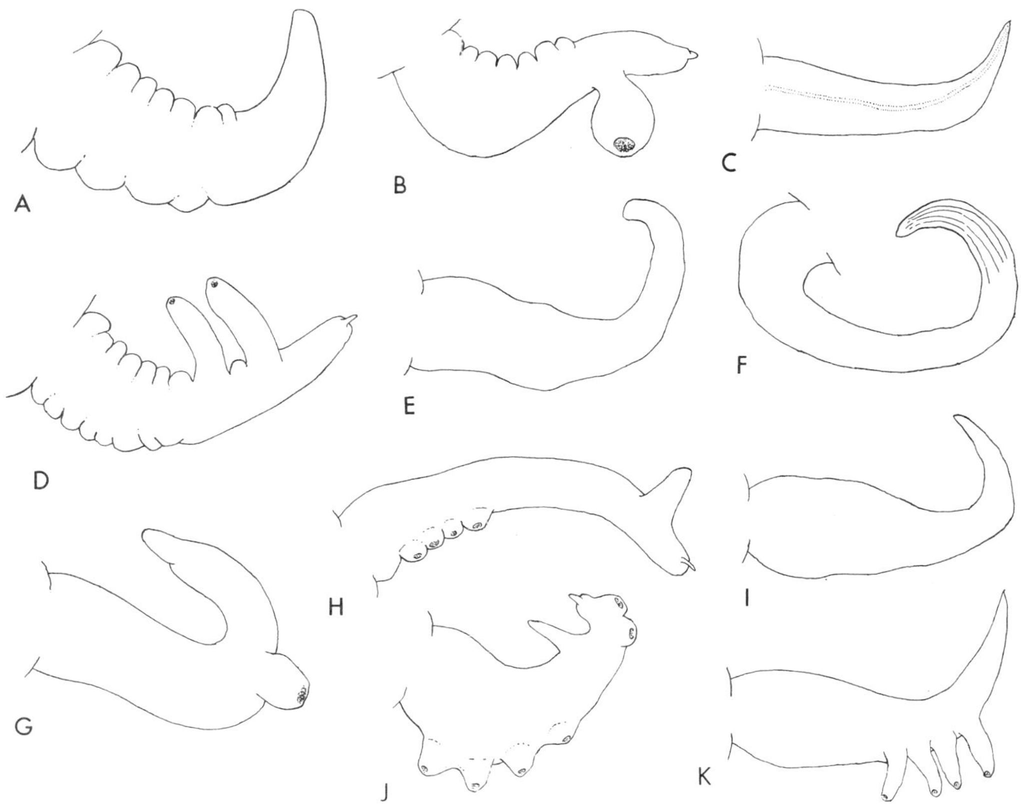

Fig. 14.16. Verge characteristics of Hydrobiidae. A, *Phreatoceras;* B, *Balcanorbis;* C, *Antrobia;* D, *Stygopyrgus;* E, *Cochliopina;* F, *Antroselates;* G, *Texapyrgus;* H, *Heleobops;* I, *Aphaostracon;* J, *Heleobops;* K, *Aphaostracon.* (A–K redrawn from Hershler and Thompson 1992.)

37. Verge unknown; shell high conical, about 3 mm in length; 1 poorly defined species, possibly misidentified; known only from deep waters of Lake Michigan, WI . . . . ***Hoyia***
Shell likely similar, but not from same habitat **38**
38. Verge containing 2 or 3 discrete accessory lobes and 3 ducts (Fig. 14.15F); shell length 2 to 6 mm, broadly conical to turreted (Fig. 14.20D); species typically associated with springs and spring fed streams or cave waters (unpigmented); 9 species, a few common and widely distributed . . . . ***Fontigens***
Verge with 1 or no accessory lobe, 2 or less ducts; epigean or phreatic **39**
39. Operculum paucispiral (Fig. 14.4B) or with multispiral nucleus (Fig. 14.4D); verge bifurcate with 2 ducts, proximal bifurcation truncate, glands absent (Fig. 14.18C, E) **40**
Operculum paucispiral; verge with lobes or glands, but unlike above, with a single duct **43**
40. Eyeless and without pigment; in cave environments **41**
With eyes and pigment; epigean **42**
41. Shell with distinct spiral sculpture, 1 species known only from a cave in FL . . . . ***Dasyscias***
Shell without sculpture; 2 or 3 poorly known species in caves in MO and KY
***Amnicola* (in part)**
42. Shell small, usually less than 3 mm in length (Fig. 14.20G); operculum usually with multispiral nucleus (Fig. 14.4D); central radular tooth with only 1 basocone on each side; 4, possibly more species generally in eastern states and Great Lakes region . . . . ***Lyogyrus***
Shell 4 to 5 mm in length; operculum paucispiral (Figs. 14.4B; 14.20H); central tooth with 2 or more basocones; 1 common and widespread species, several others poorly known or restricted in range . . . . ***Amnicola* (in part)**

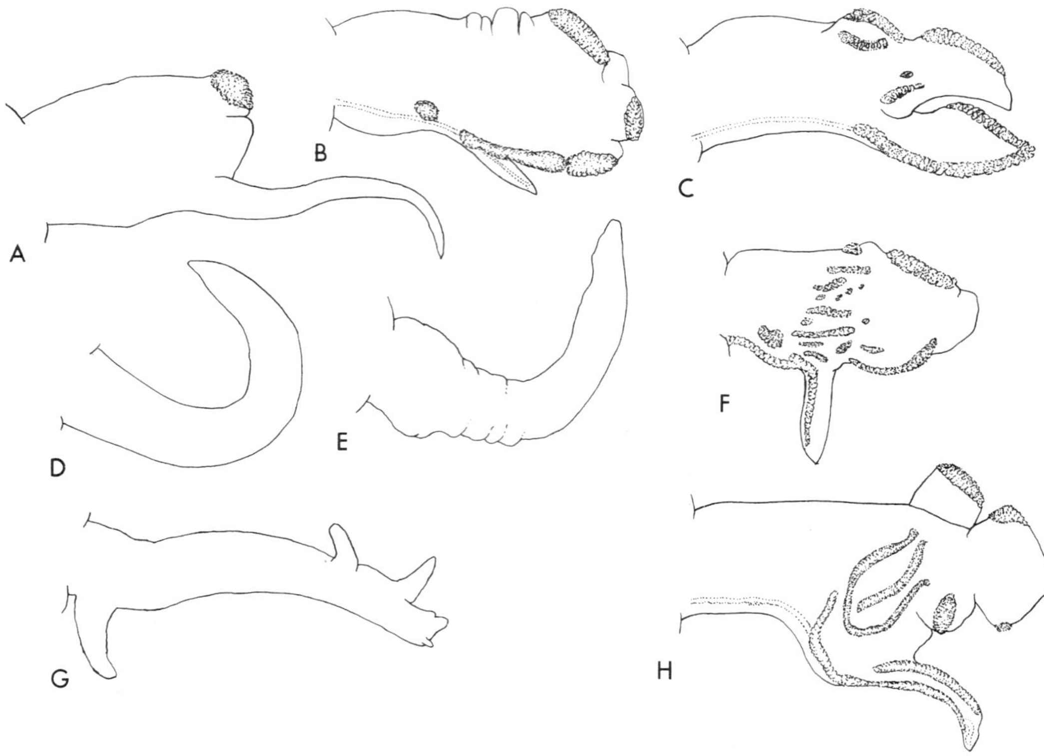

Fig. 14.17. Verge characteristics of Hydrobiidae. A, *Rhapinema;* B, *Stiobia;* C, *Spilochlamys;* D, *Potamopyrgus;* E, *Antrorbis;* F, *Notogillia;* G, *Tryonia* (Florida); H, *Cincinnatia.* (A, C, F–G, H redrawn from Thompson 1968; B redrawn from Thompson and McCaleb 1978; E redrawn from Hershler and Thompson 1990.)

43. Recorded only from caves; generally pigmentless and eyeless; verge a simple elongate lobe (Fig. 14.16C, F) **44**
Recorded only from epigean waters, pigmented and with eyes; verge otherwise **46**

44. Shell turreted (Fig. 14.19A); known only from a cave in Virginia; 1 species . . . . . ***Holsingeria***
Shell globose or low conical; known elsewhere **45**

45. Known only from a cave in MO; 1 species (Fig. 14.19B) . . . . ***Antrobia***
Known only from caves in IN and KY; 1 species . . . . ***Antroselates***

46. Apical whorl of shell sunken below subsequent whorls giving shell a truncated appearance (Figs. 14.14H; 14.20C), verge bilobed, one lobe a cone at the midpoint of the verge (Fig. 14.15I); 1 common and widespread species . . . . ***Probythinella***
Apical whorl at or above level of subsequent whorl; verge otherwise **47**

47. Verge consisting of a simple lobe, without fleshy glands, with a terminal penial opening or papilla, or verge (males) absent **48**
Verge a broad appendage with accessory glands as patches or lateral cones, penis terminal or branched from main lobe **51**

48. Shells small, about 2 or 3 mm in length; 2 species, very rare in AL, possibly extinct . . . ***Clappia***
Shells larger, 3.5 to 8 mm in length **49**

49. Shell acuminate (Fig. 14.19E), occasionally carinate; males rare or absent, if present verge strongly U-shaped; 1 introduced species, Lake Ontario . . . . ***Potamopyrgus* (in part)**
Shell globose, males always present **50**

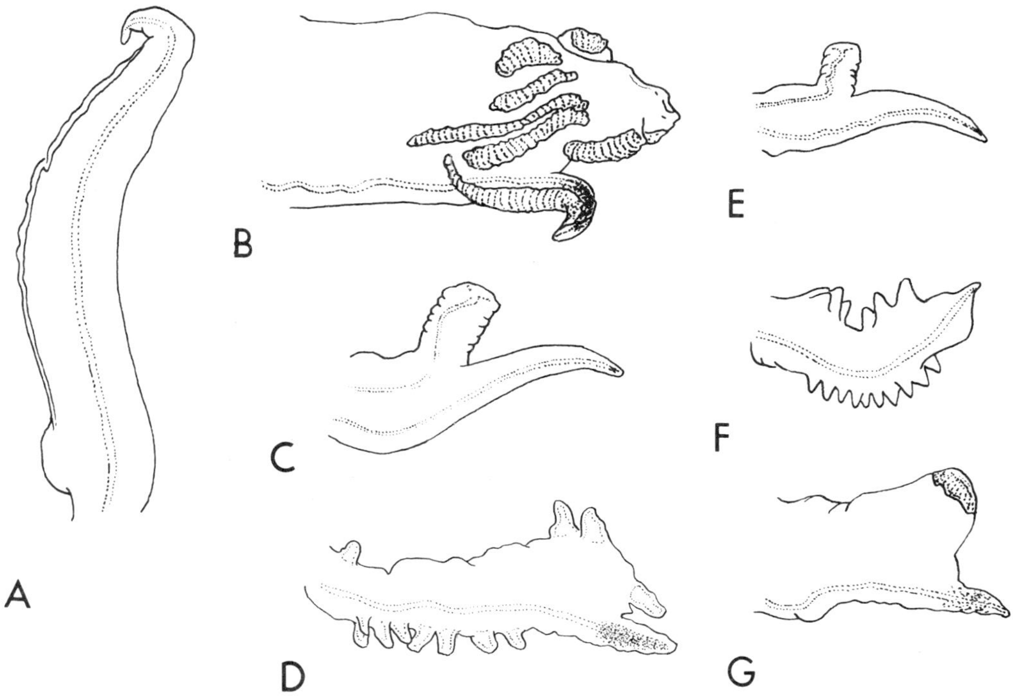

Fig. 14.18. Verge characteristics of Pomatiopsidae and Hydrobiidae. A, *Pomatiopsis;* B, *Cincinnatia;* C, *Amnicola;* D, *Pyrgophorus;* E, *Lyogyrus* (and *Dasyscias*); F, *Littoridinops;* G, *Pyrgulopsis.* (A redrawn from Smith 1986a; D from Smith and Brousseau 1996; F redrawn from Smith, 1987.)

50. Shells 3.5 to 6 mm in length; verge a simple flat lobe (Fig. 14.15G); about 35 species, most poorly known, generally in eastern states . . . . ***Somatogyrus***

Shells 6 to 8 mm in length; verge a short broad lobe with lateral folding and a distinct terminal papilla (Fig. 14.15H); 1 species in eastern states . . . . ***Gillia***

51. Verge a complex organ comprising a separate penial appendix and a broad lobe containing a single or several glands **52**

Verge an elongate lobe with lateral glandular cones or papillae, penis terminal **59**

52. Lobe of verge with several marginal papillate glands (Fig. 14.18D), shell conical and often with raised spiral line and conical spines, 2 species in coastal TX and southern FL ***Pyrgophorus***

Lobe of verge with glandular pads or patches, not papillate; shell without spiral line or spines **53**

53. Verge with several glandular patches on margins and face and often extending onto the penis itself **54**

Verge with a single or few terminal or lateral glands; penis without obvious glands **57**

54. Shell with weak spiral keels (Fig. 14.19F); glands on verge restricted to periphery (Fig. 14.17B), known only from a single spring in AL; 1 species . . . . ***Stiobia***

Shell without keels; glands on face of verge as well as periphery **55**

55. Penial lobe blade like, as large as glandular lobe (Fig. 14.17C); shell with open umbillicus; 3 species in GA, FL . . . . ***Spilochlamys***

Penial lobe smaller (Figs. 14.15A; 14.17F, H; 14.18B), more lateral in position **56**

56. Shell globose, with closed umbilicus; 2 species, AL, GA, FL . . . . ***Notogillia***

Shell conical, umbilicus open (Fig. 14.20E, F); many species, the status of which unclear, includes *Paludina;* distributed throughout the eastern U.S. ***Cincinnatia***

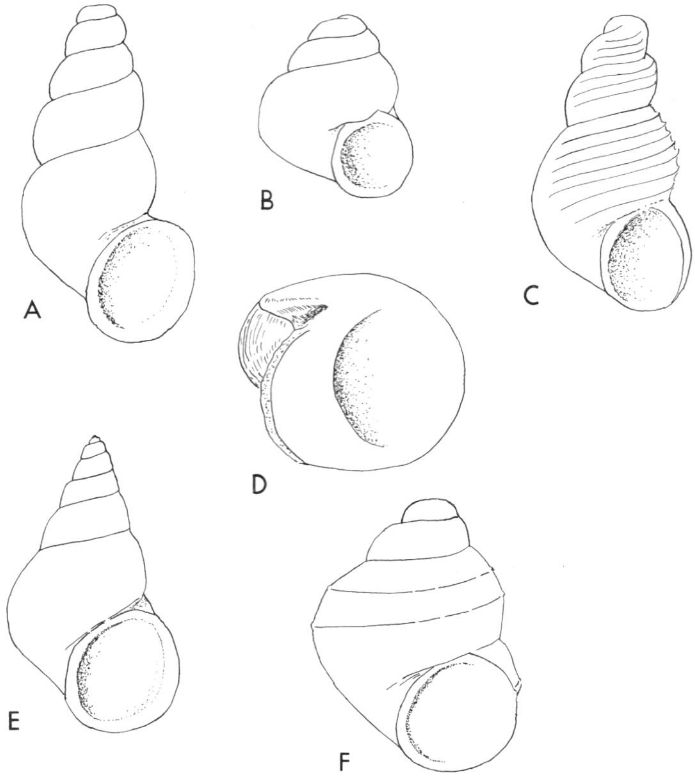

Fig. 14.19. Shells of Hydrobiidae. A, *Holsingeria;* B, *Antrobia;* C, *Stygopyrgus;* D, *Lepyrium;* E, *Potamopyrgus;* F, *Stiobia.* (A–C redrawn from Hershler and Holsinger 1990; D redrawn from Thompson 1984a; E adapted from several sources; F redrawn from Thompson and McCaleb 1978.)

57. Shell globose, reaching 9 mm in length, with large body whorl and aperture; verge a broad lobe with a single gland bearing extension or appendage and conelike penis (Fig. 14.15D, E); 1 species widespread in Mississippi River and Great Lakes region .... ***Birgella***

Shell globose to conical, aperture not exceptionally large; penial portion of verge extending beyond glandular portion **58**

58. Penial portion of verge greatly elongate (Fig. 14.17A); shell globose; 1 species in AL, GA, FL ***Rhapinema***

Penial portion of verge not so elongate, terminal gland typically pediculate (Figs. 14.15B; 14.18G); shell conical, rarely carinate; about 10 species in the Mississippi River drainage and eastward .... ***Pyrgulopsis* (in part)**

59. Shell conical with rather flat whorls (Fig. 14.20B) **60**

Shell with well-rounded whorls (Fig. 14.20K) **61**

60. Verge with terminal papilla and various marginal mound or conelike glands (Fig. 14.16H, J); 2 species, coastal FL, MD .... ***Heleobops***

Verge with many marginal papillae, typically overlapping, on one side, a few on the opposed side (Fig. 14.18F); 2 species, at or near coast along eastern seaboard and Gulf coast ***Littoridinops***

61. Shell occasionally with low ribs; verge elongate with usually 1 large and a few smaller glandular cones on opposite margins (Fig. 14.17G); 2 species, FL .... ***Tryonia* (in part)**

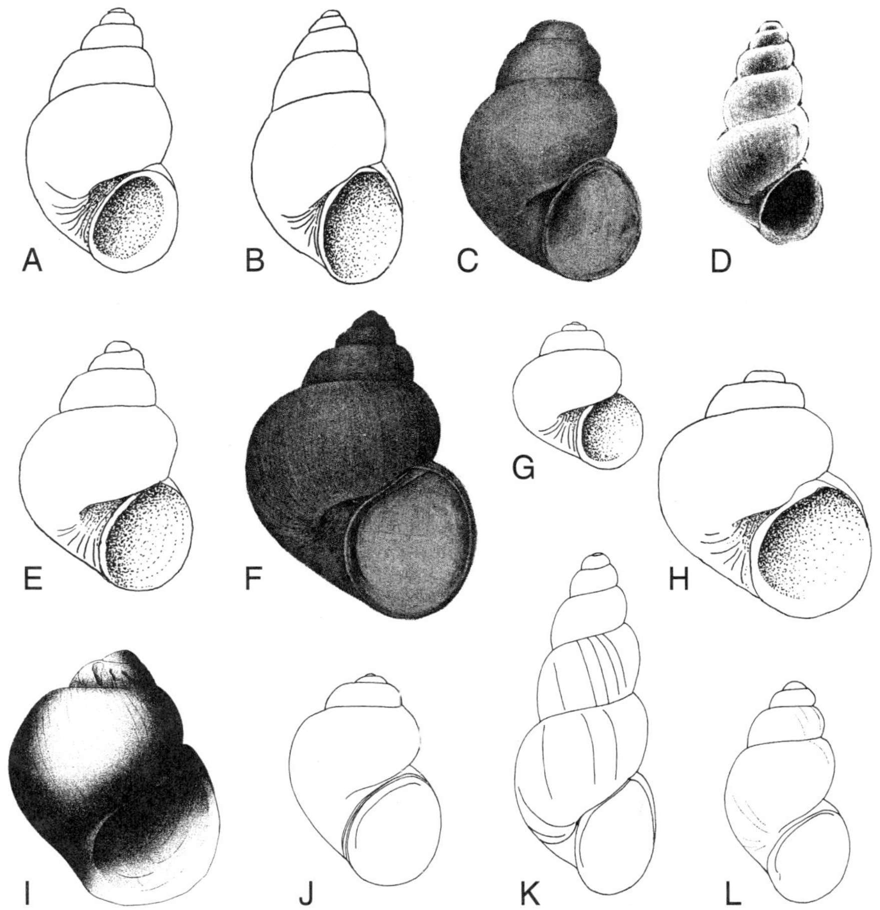

Fig. 14.20. Shells of Hydrobiidae. A, *Pyrgulopsis lustrica*, ×5; B, *Littoridinops tenuipes*, ×5; C, *Probythinella lacustris*, ×4; D, *Fontigens nickliniana;* ×6; E, *Cincinnatia winkleyi*, ×4; F, *Paludina cincinnatiensis*, ×4; G, *Lyogyrus grana*, ×5; H, *Amnicola limosa*, ×4; I, *Fluminicola seminalis*, ×4; J, *Pyrgulopsis metcalfi*, ×6; K, *Tryonia gilae*, ×6; L, *Pyrgulopsis bernardina*, ×6. (C, F from Clarke 1973; D from Hershler et al. 1990a; I from Hershler and Frest 1996; J–L from Taylor 1987.)

Whorls without ornamentation; verge an elongate, acuminate lobe with none to several marginal papillae along one side (Fig. 14.16I, K); 9 species, FL . . . . ***Aphaostracon***

62. Shell without any vestige of a spiral, in the form of a limpet or cap-shaped **63**

Shell with a spire, even if reduced or depressed (planospiral) **70**

63. Specimens from west of the Continental Divide **64**

Specimens from east of the Continental Divide . . . .**ANCYLIDAE (in part), 67**

64. Apex of shell sharply produced to a curved posteriorly directed point (Fig. 14.21A, B); 1 species in high altitude lakes, CO, MT . . . .**ACROLOXIDAE, *Acroloxus***

Apex of shell not produced to a curved point **65**

65. Apex of shell to right of midline, on posterior half of shell (Fig. 14.21C, E); about 4 species generally distributed . . . .**ANCYLIDAE (in part), *Ferrissia***

Apex of shell at center, slightly or clearly anterior (Fig. 14.21F) . . . .**LANCIDAE, 66**

66. Apex clearly anterior of center of shell; 1 species, ID, OR . . . .***Fisherola***

Apex at or slightly anterior to center of shell (Fig. 14.21F); 4 species, CA, OR . . . .***Lanx***

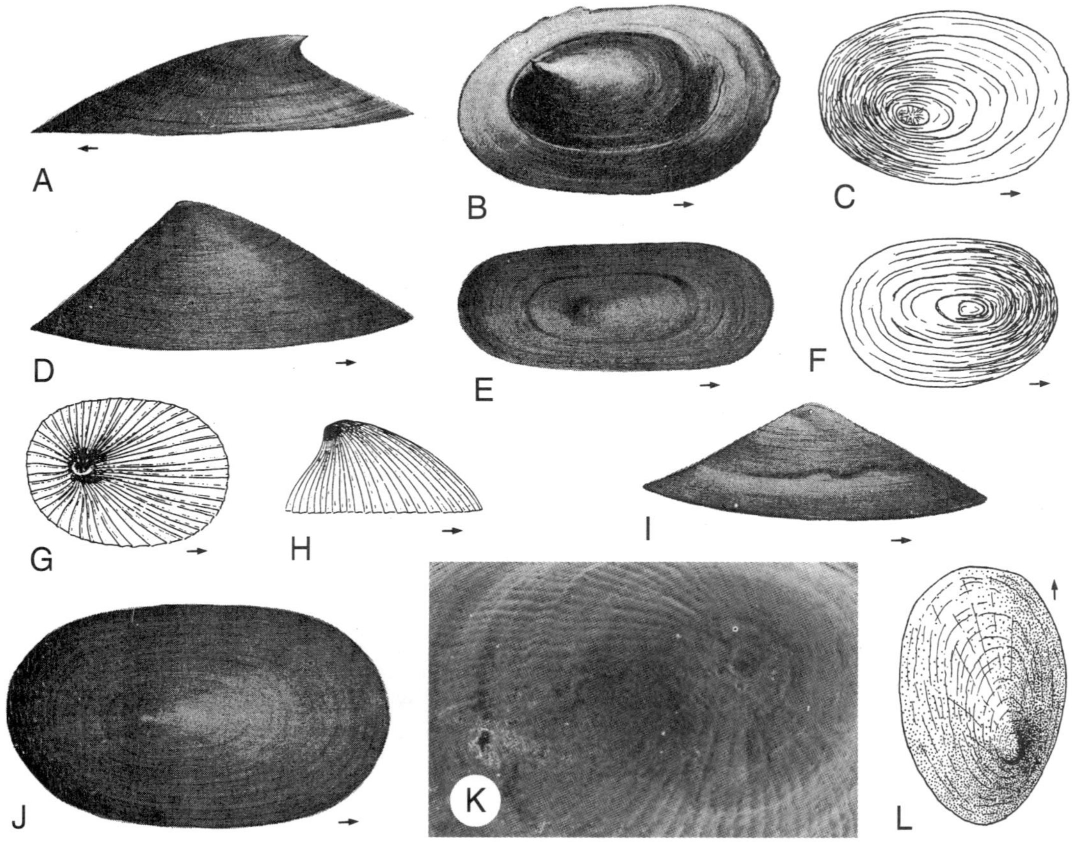

Fig. 14.21. Representatives of the Acroloxidae, Ancylidae, and Lancidae. A, *Acroloxus coloradensis*, lateral view; B, *A. coloradensis*, dorsal view; C, *Ferrissia rivularis*, dorsal view; D, *F. rivularis*, lateral view; E, *F. parallela*, dorsal view; F, *Lanx newberryi*, dorsal view; G, *Rhodacmea filosa*, dorsal view; H, *R. filosa*, lateral view; I, *Laevapex fuscus*, lateral view; J, *L. fuscus*, dorsal view; K, apex of *Ferrissia* spp.; L, *Hebetancylus excentricus*. Dorsal view. Arrows indicate direction of anterior. (A, B, E, I, J after Clarke 1973; G, H from Burch 1979; K from Smith and Brousseau 1996; L from Basch 1963.)

67. Shell elevated, apex near midline, with a shallow depression (Fig. 14.21G, H); 3 species, in drainages of the eastern Mississippi River system . . . . ***Rhodacmea***
Shell low to moderately elevated, apex slightly to eccentrically to right of midline, apex without a depression **68**

68. Shell low and somewhat broad (Fig. 14.21I, J), nearly ovate, apex near or just right of midline, not pronounced; tentacles with a black core; found in quiet, lentic waters; 2 species, widely distributed . . . . ***Laevapex***
Shell low to moderately high, apex clearly right of midline, generally pronounced; tentacles white **69**

69. Shell apex usually with fine radiating lines (Fig. 14.21K), apex variously to right of mid line; 6 species, several very common and widespread . . . . ***Ferrissia***
Shell apex without fine radiating lines, apex distinctly right of mid line (Fig. 14.21L); 1 species, GA, FL, TX . . . . ***Hebetancylus***

70. Shell planospiral . . . . **PLANORBIDAE, 81**
Shell with raised spire **71**

71. Aperture on left side of shell (Fig. 14.22A–D) . . . . **PHYSIDAE, 72**
Aperture on right of shell . . . . **LYMNAEIDAE (in part), 74**

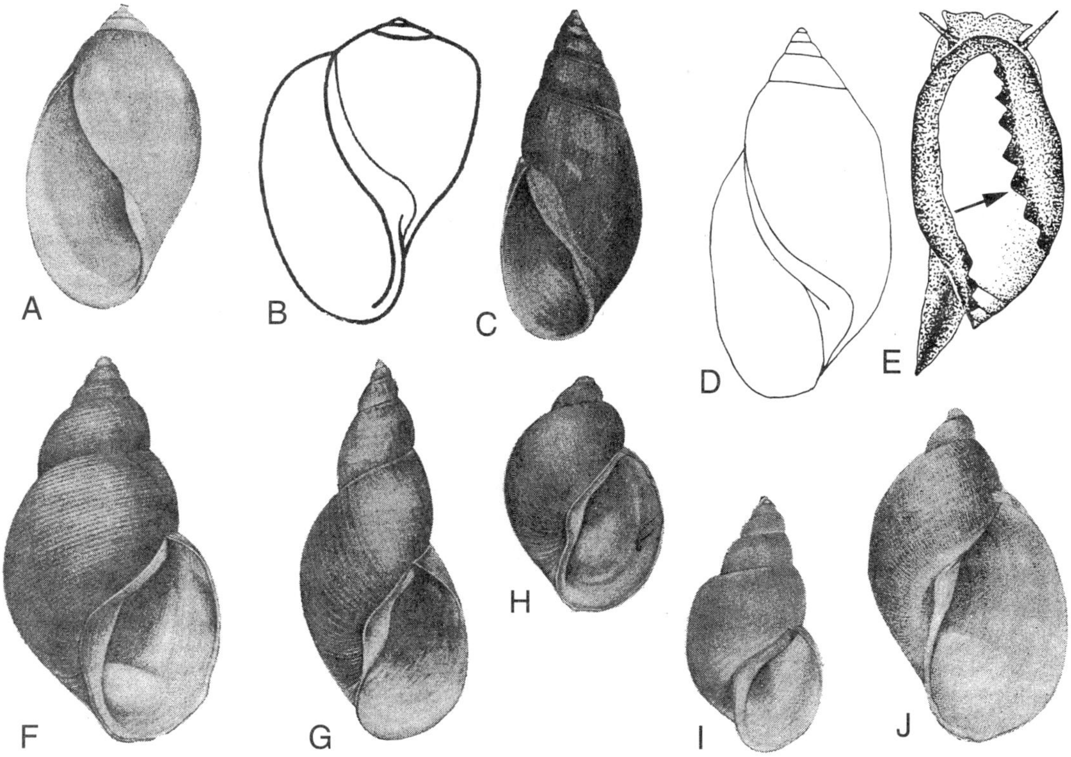

Fig. 14.22. Representative Physidae and Lymnaeidae. A, *Physa gyrina*, ×3; B, *P. ancillaria*, ×3; C, *Aplexa elongata*, ×3; D, *Stenophysa marmorata*, ×3; E, *S. marmorata*, arrow denotes mantle margin serration, ×3; F, *Stagnicola caperata*, ×3; G, *S. elodes*, ×3; H, *S. catascopium*, ×3; I, *Fossaria obrussa*, ×3; J, *Pseudosuccinea columella*, ×3. (A, C, F–J from Clarke 1973; B from Jokinen 1983; D, E from Richards 1964.)

72. Shell fusiform, with a high spire and narrow (Fig. 14.22C), compressed aperture, surface glossy; usually associated with temporary waters; 1 widespread species . . . . ***Aplexa***
Shell globose to conical, aperture wide, spire low to elevated (Fig. 14.22A, B, D), surface glossy or not; in most aquatic habitats **73**

73. Several species extremely common and widespread, mantle edge with fingerlike extensions on at least one side of the mantle; nearly 30 described species, many ranging throughout most of the United States . . . . ***Physa***
Exotic, tropical species in TX; mantle edge serrated (Fig. 14.22E), extending beyond aperture; 2 species . . . . ***Stenophysa***

74. Aperture much longer than spire **75**
Aperture length equal to or less than aperture **77**

75. Aperture lacrimate, shell with a tapered spire (Figs. 14.22J; 14.23A), fine spiral lines present on body whorl; 1 species, widely distributed . . . . ***Pseudosuccinea***
Shell not as above **76**

76. Aperture large, broad, and ear-shaped (Fig. 14.23B); 1 introduced species, scattered throughout the U.S. . . . . ***Radix***
Aperture large and rounded (Fig. 14.22H), but not ear-shaped, or narrow; 2 or 3 species in northern states . . . . ***Stagnicola* (in part)**

77. Shell and aperture exceptionally narrow, appearing needle-like (Fig. 14.23C); 1 species, northeastern quarter of the U.S. . . . . ***Acella***
Shell not as above **78**

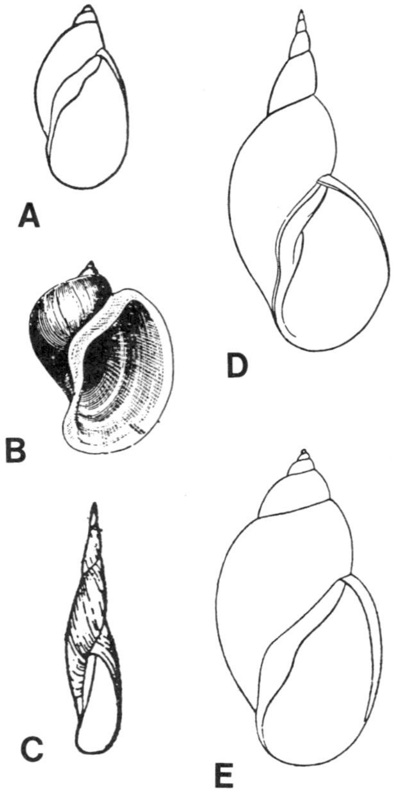

Fig. 14.23. Shell form in Lymnaeidae. A, *Pseudosuccinea columella*, ×3; B, *Radix auricularia*, ×1.3; C, *Acella haldemani*, ×2; D, *Lymnaea stagnalis*, ×1.5; E, *Bulimnea megasoma*, ×1.5. (A–E from Walker 1918.)

78. Shell thin and light colored, fragile, and with a very acute spire, length to 30 mm (Fig. 14.23D); 1 species, discontinuous in northern states . . . . ***Lymnaea***
Shell otherwise **79**

79. Shell large, in excess of 30 mm in length, robust and broad, spire not acute (Fig. 14.23E); 1 species in interior states and St. Lawrence River drainage . . . . ***Bulimnea***
Shell not as above **80**

80. Shell with surface microsculpture consisting of lines or crenulation (Fig. 14.22F, G), whorls rounded, length usually over 15 mm; over 20 species, predominately in northern states ***Stagnicola* (in part)**
Shell without microsculpture, whorls often shouldered, length generally less than 15 mm (Fig. 14.22I); about 23 species, several very common and widespread . . . . ***Fossaria***

81. Whorls increasing in size gradually, aperture width about one-half or less of shell width . . . **84**
Body whorl much larger than preceding whorls, aperture width greater than half the width of the shell **82**

82. Specimens from west of the Continental Divide, body whorl not keeled, shell greater than 5 mm in diameter (Fig. 14.25D); 2 species, CA, NV, OR . . . . ***Vorticifex***
Specimens east of the Continental Divide, aperture width greater than two-thirds of shell width or body whorl keeled (Fig. 14.25E, F); shell minute **83**

83. Aperture greater than two-thirds width of shell; 1 species, AL . . . . ***Amphigyra***
Aperture less than two-thirds width of shell, body whorl keeled; 4 species, AL . . ***Neoplanorbis***

84. Shell with lamellae (teeth) a short distance within aperture (Figs. 14.24A, B; 14.25A); 2 species, 1 eastern, 1 western U.S. . . . . ***Planorbula***
Shell without lamellae or teeth **85**

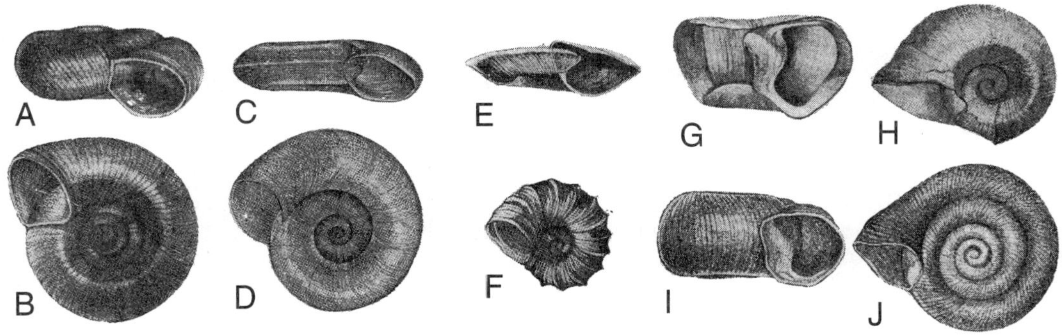

Fig. 14.24. Representative Planorbidae. A, *Planorbula armigera*, ×3; B, *P. armigera*, top view, ×4; C, *Gyraulus deflectus*, ×4; D, *G. deflectus*, top view, ×4; E, *Promenetus exacuous*, ×5; F, *Menetus cristata*, ×7; G, *Helisoma anceps*, ×2; H, *H. anceps*, top view, ×2; I, *Helisoma campanulatum*, ×3; J, *H. campanulatum*, top view, ×3. (A–J from Clarke 1973.)

85. Maximum frontal height (including aperture) more than one-half shell diameter, aperture height usually (except a few *Menetus*) greater than whorl height; specimens typically greater than 6 mm in diameter (Fig. 14.24G–J); about 20 species, widely distributed .... ***Helisoma***
Maximum frontal height (including aperture) less than one-half shell diameter, aperture height not more than whorl height; shell usually 6 mm or less in diameter .... **.86**
86. Distribution in southernmost states, subtropical .... **.87**
Distribution in central and northern half of the U.S. .... **.90**
87. Aperture markedly deflected (Fig. 14.24C, D), compressed; shell not keeled and flat, without spiral lines; 1 species, widely distributed .... ***Gyraulus* (in part)**
Aperture not deflected, shell extremely flat and keeled or not, with spiral lines or not ..... **.88**
88. Aperture extending below body whorl (Fig. 14.25B, C); 1 species, widely distributed ***Menetus* (in part)**

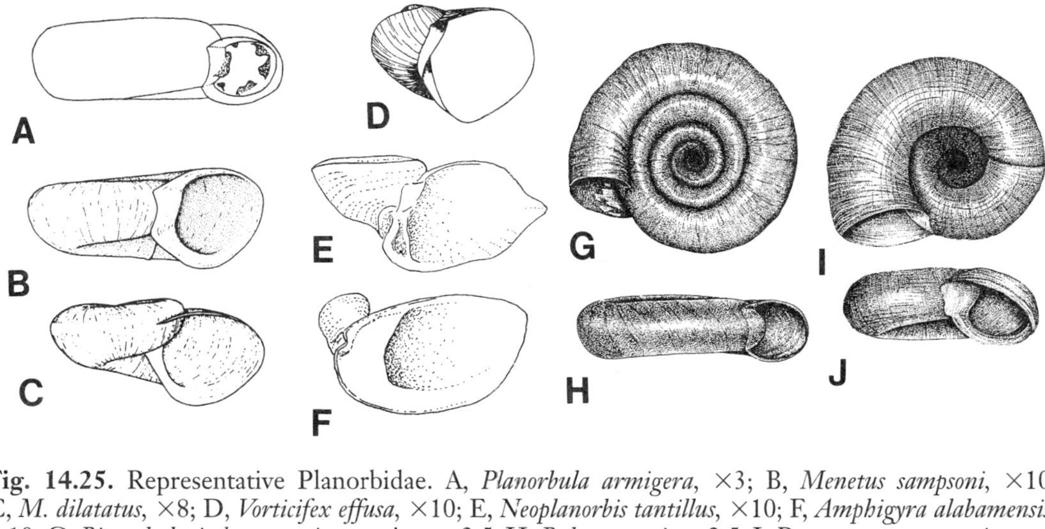

Fig. 14.25. Representative Planorbidae. A, *Planorbula armigera*, ×3; B, *Menetus sampsoni*, ×10; C, *M. dilatatus*, ×8; D, *Vorticifex effusa*, ×10; E, *Neoplanorbis tantillus*, ×10; F, *Amphigyra alabamensis*, ×10; G, *Biomphalaria havanensis*, top view, ×2.5; H, *B. havanensis*, ×2.5; I, *Drepanotrema aeruginosum*, top view, ×5; J, *D. aeruginosum*, ×5. (B, C from Burch 1979; E, F redrawn from Burch 1979; G–J from Malek 1985, courtesy of the Pan American Health Organization <www.paho.org>.)

Aperture not extending below body whorl **89**

89. Shell either extremely flat and keeled (Fig. 14.25I, J) or with raised spiral lines; 3 species, AR, FL, TX . . . . ***Drepanotrema***

Shell otherwise (Fig. 14.25G, H); 2 species, introduced into AR, FL, LA, TX ***Biomphalaria***

90. Shell with any or combination of deflected aperture, raised costae (Fig. 14.24F), or greater than 4 whorls; 5 species, generally confined to northern states . . . . ***Gyraulus* (in part)**

Shell lacking any of the above features; carinate or not **91**

91. Aperture extending considerably below body whorl (Fig. 14.25C); 4 species, generally eastern states but 1 western species . . . . ***Menetus* (in part)**

Aperture only slightly deeper than body whorl (Fig. 14.24E); shell flat; 2 species in central and northern states . . . . ***Promenetus***

PELECYPODA (Clams, Mussels)

Unlike the Gastropoda, the bivalve mollusks are all aquatic, and although they occur in nearly all types of freshwater habitats, they are most abundant and varied in our larger rivers.

General characteristics. The pearly mussels (Unionoida, Fig. 14.26) of our inland waters range from about 30 to 250 mm in length. The fingernail, pea, or marsh clams and false mussels (Veneroida) are much smaller, rarely exceeding 30 mm in length. The two valves of the shell are securely attached to each other dorsally by an elastic hinge ligament, and under natural conditions the valves gape slightly to permit the protrusion of the muscular hatchet- or ax-shaped foot (Fig. 14.27) at the anteroventral margin and the inhalant and exhalant openings at the posterior margin. There are no tentacles, head, or eyes. During normal conditions, the animal lies obliquely with the ventral half (or more) hidden in the substrate; in some species the shell may be entirely hidden from view with only the mantle openings appearing at the surface. Byssate pelecypods are attached to a firm surface and the entire animal is exposed.

Locomotion. Unattached clams and mussels move over and through the substrate by means of a series of contractions of the intrinsic muscle fibers of the foot and protractor and retractor muscles attached to the shell. The foot is first elongated and thrust forward in the substrate; then the distal portion of the foot becomes more or less swollen transversely so that it obtains a purchase, and at the same time the entire foot is shortened; as a result the main portion of the body and its enclosing shell are pulled forward slightly. These repeated muscular contractions therefore result in locomotion that is essentially a series of short "hunching" movements. Some species are known to move several feet per hour and the resulting long, troughlike tracks of larger mussels may often be seen on sandy lake bottoms where there is little wave action. The stimuli responsible for extensive movements of clams are not generally known, but it is thought that stagnation and fall in water level are of primary importance. Thin-shelled species are generally more active than heavy-shelled species.

Shell, muscles. Mussel shells exhibit a variety of shapes. Some are decidedly elongated or oval; others are subcircular, rhomboidal, quadrate, trapezoidal, or subtriangular (Fig. 14.26). At the dorsal margin of each valve (Fig. 14.28), just anterior to the hinge ligament, is a raised area, the umbo or beak, which denotes the point at which growth began in the juvenile mussel. In the larger river mussels this area is often eroded. Surrounding each umbo and extending out to the edges of the valves is a series of concentric lines, which are formed at the edge of the shell as the animal grows. At intervals

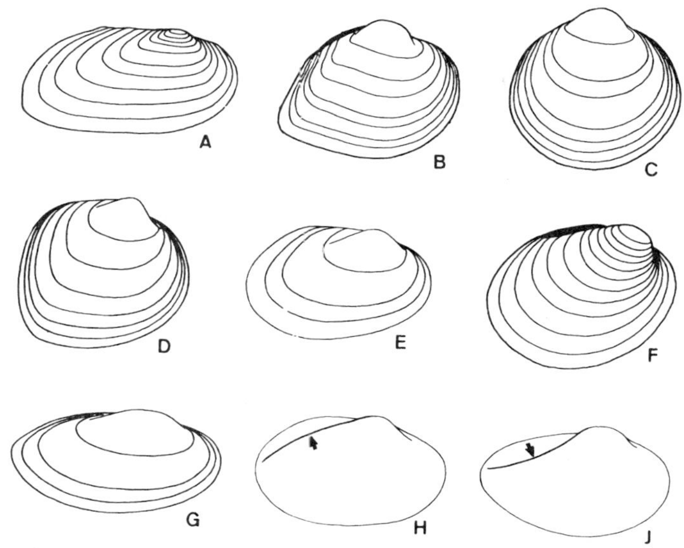

Fig. 14.26. Typical unionid shell shapes. A, rhomboidal or elongated; B, triangular; C, circular; D, quadrate; E and F, oval; G, elliptical; H, posterior ridge convex; J, posterior ridge concave. (Modified from Burch 1973.)

these rings are more closely grouped, forming a more or less distinct ridge. In many species it can be shown that these markings clearly denote limits of annual growth and winter rest periods, especially if an adequate sample from a particular locality is studied. The annual growth ring picture is often confused, however, by the presence of accessory rings formed in response to a variety of temporary unfavorable environmental conditions, such as lack of food, low oxygen supply, and fall in water level. In addition to these concentric markings, many genera bear tubercles, raised lines, knobs, or wrinkles (costae).

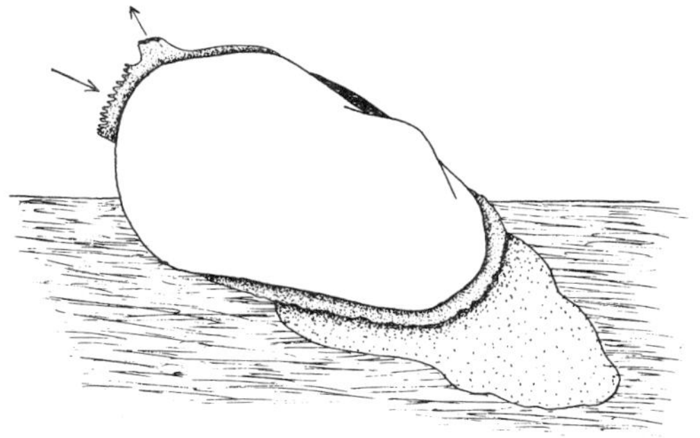

Fig. 14.27. Female *Lampsilis siliquoidea* in natural position in river bottom, ×0.5. Arrows indicate currents of water. (Modified from Baker 1928a.)

External coloration is whitish to brown in the Sphaeriidae and the shells are rather drab. Shells of the Unionoida can be brightly colored (Plates 7–9) and can exhibit yellows, greens, reddish brown to brown or black. There may be color rays as well. The false mussels (Dreissenidae) often have zigzag markings on the shell. The internal surface of the shell is thickly coated with nacre, or mother of pearl, which ranges in color from silvery white through pink to dark purple; it is composed of extremely thin alternating laminae of calcium carbonate and an organic substance. Between the nacre and the periostracum is the prismatic layer; it consists of minute, closely packed, prismlike blocks of calcium carbonate (Fig. 14.29). The important features of the inner surface of a typical unionid shell are shown in Fig. 14.30. Near the dorsal margin there is usually a series of projecting and interlocking hinge teeth, which aid in keeping the two

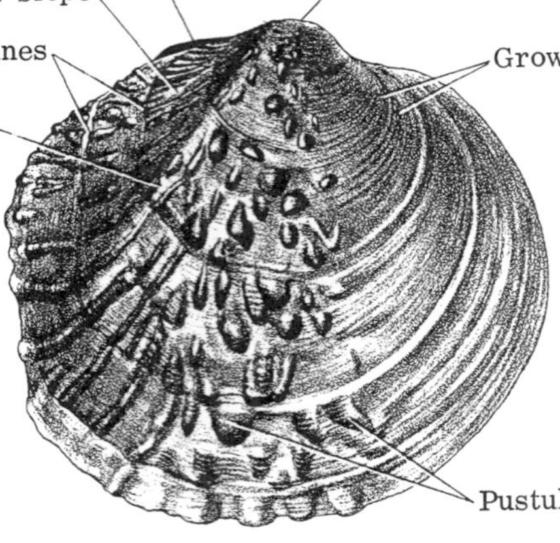

Fig. 14.28. Shell structure of *Cyclonaias.* (From Burch 1973.)

valves in juxtaposition. In the Unionidae the pseudocardinal teeth are in the anterior part of the shell below the umbones, and the long, narrow, ridgelike lateral teeth are more posterior and below the region of the ligament. Either one set or both are absent in certain genera. In the Veneroida there are obvious lateral teeth posterior and usually anterior to the true cardinals.

The hinge teeth serve as a fulcrum, and the springlike hinge ligament normally keeps the shell slightly open. When the foot and mantle are withdrawn, however, the shell may be tightly closed by the contraction of the two large but short transverse muscles fastened to the inner surface of the valves, the anterior and posterior adductors. In addition to the scars produced by the attachment of the adductor muscles, each valve typically contains the scars of the three foot muscles. Above, and more or less closely associated with the posterior adductor scar, is a small scar produced by the attachment of the posterior retractor of the foot. The anterior adductor has a comparable associated scar for the anterior retractor of the foot, as well as an additional scar for the protractor muscle of the foot. There are also some small and indistinct scars in the cavity of the umbo from muscles that aid in holding the visceral mass in place.

Except where muscles are attached, the inner surface of the shell is completely lined with a sheet of glandular tissue, the mantle. This structure is securely fastened near the edges of the valves along the muscular pallial line. Such an attachment prevents foreign particles from getting between mantle and shell. An indentation in the pallial line may be present near the posterior end of the shell; it is called the pallial sinus and is produced by intrinsic muscles that retract siphons. The entire lateral surface of the mantle secretes the nacre, but the periostracum and prismatic layer are formed only by the border of the mantle at the periphery of the shell.

Gross visceral anatomy. When one valve of a mussel is removed, the underlying mantle can be seen covering the viscera. At the posterior end of a typical animal the edges of the two lobes of the mantle are modified to form slitlike to round tubular openings. In most Unionoida the uppermost opening is subdivided by a fusion of the mantle lobes. The two openings produced are the supra-anal and anal (exhalant) openings (Fig. 14.32). In more primitive examples (Margaritiferidae, some Unionidae), the two openings are not distinguished. In neither case are true siphons formed. However, at least one genus (*Potamilus*) produces, during the juvenile stage, an extensible ex-

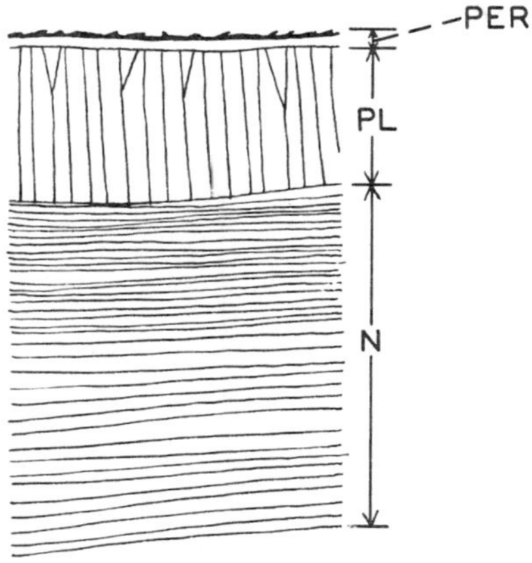

Fig. 14.29. Small section of mussel shell. PER, periostracum; PL, prismatic laver; N, nacre.

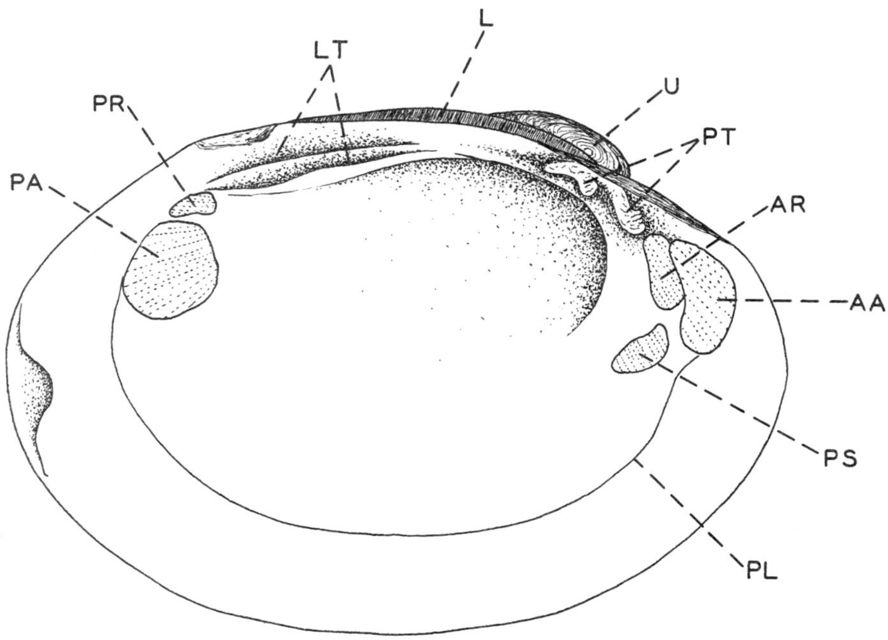

Fig. 14.30. Diagrammatic view of inner surface of left valve of *Lampsilis*, ×1. AA, anterior adductor muscle scar; AR, anterior retractor muscle scar; L, ligament; LT, lateral teeth; PA, posterior adductor muscle scar; PL, pallial line; PR, posterior retractor muscle scar; PS, scar of protractor muscle of foot; PT, pseudocardinal teeth; U, umbo.

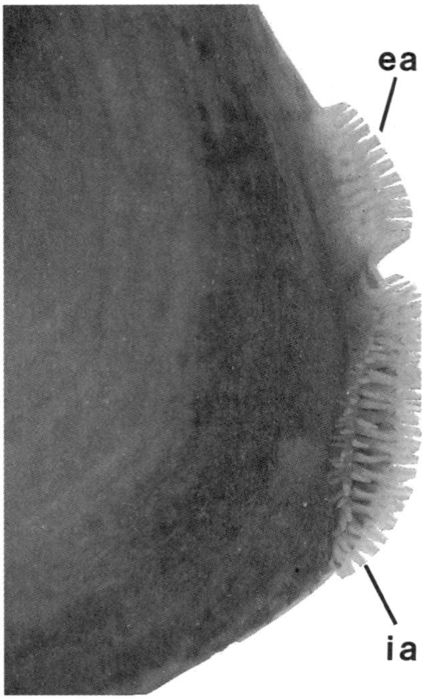

Fig. 14.31. Apertural region of *Elliptio complanata.* ea, exhalant aperture; ia, inhalant aperture.

halant "pseudosiphon" of unknown function. Below the exhalant aperture there is no fusion of the mantle edges to form a tubular siphon. Instead, the mantle edges only project slightly and form an inhalant aperture or opening (Fig. 14.32). The margins around the apertures are always papillate (Fig. 14.31) or crenulate. The Veneroida produce tubular siphons by means of fusion of the mantle edges. These siphons are reduced in the genus *Pisidium.* In a few genera (*Musculium*), the siphons produce long, cylindrical tubes (Fig. 14.33).

When the mantle is removed from one side, the two long, flat demibranchs of that side are exposed to view. Each demibranch is composed of two lamellae that are united at the lower but not the upper margin, the two demibranchs together being roughly W-shaped in cross section and compose, in entirety, the gill. The outer demibranch is connected at the outside and top to the mantle, and the inner demibranch is variously attached at the inside to the outer surface of the main median visceral mass; the base of the inner lamella of the outer demibranch and the base of the outer lamella of the inner demibranch are attached to each other. In most freshwater pelecypods, the cavity between the two lamellae of each demibranch is subdivided by vertical partitions into many narrow chambers or water tubes, closed below but opening above into a longitudinal space along the dorsal portion of the demibranch called the suprabranchial chamber. Only in the Margaritiferidae are vertical water tubes absent. Communication with the suprabranchial chamber is as in other mussels. At the posterior end of the animal the suprabranchial chambers are fused to form a single cavity, the cloacal chamber. Near the anterior end of the demibranchs and surrounding the small slitlike mouth are two flat labial palps on each side. The upper lip is continuous with the left and right outer palps, and the lower lip is continuous with the left and right inner palps. The greater portion of the foot lies between

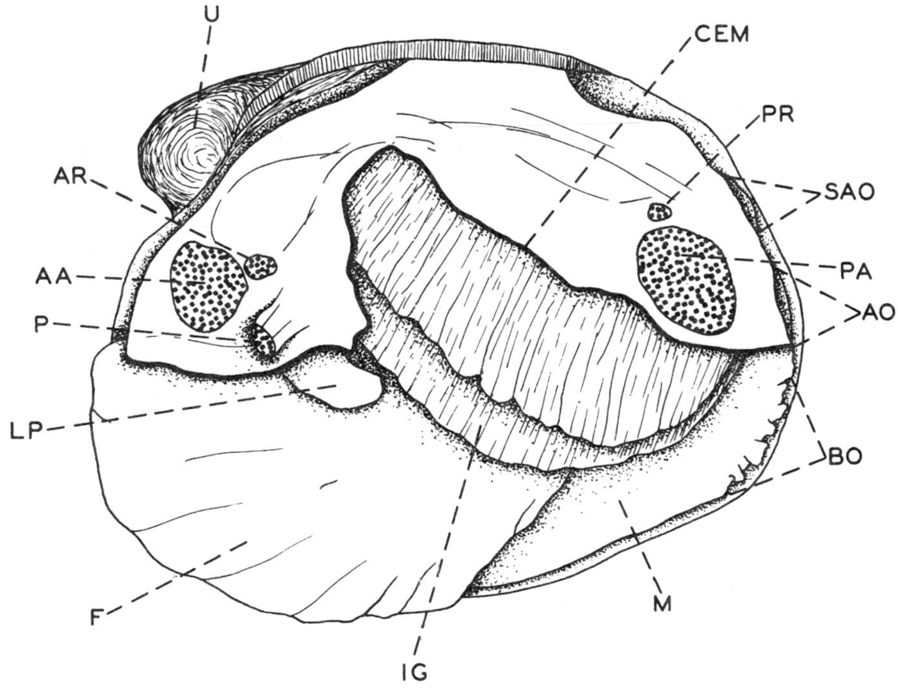

Fig. 14.32. Diagrammatic partial dissection of *Fusconaia*, ×1.4. AA, anterior adductor muscle; AO, anal opening; AR, anterior retractor muscle; BO, branchial opening; CEM, cut edge of mantle; F, foot; IG, inner demibranch; LP, labial palps; M, mantle; P, protractor muscle of foot; PA, posterior adductor muscle; PR, posterior retractor muscle; SAO, supra-anal opening; U, umbo. (Modified from Lefevre and Curtis 1912.)

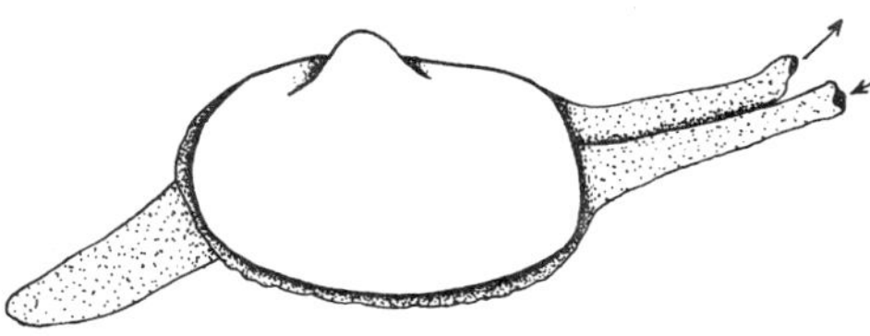

Fig. 14.33. *Musculium*, showing branchial and anal siphons, ×2. Arrows indicate currents of water. (Modified from Baker 1928a.)

the demibranchs in the anterior part of the mantle cavity. It tapers posteriorly, and the median visceral mass of the posterior part of the animal is confined to the dorsal part of the mantle cavity.

In byssate species, within the Dreissenidae, the foot is reduced and plays a very minor role in locomotion. Specialized glands within the foot of these species produce a fibrous byssus (Fig. 14.34) that attaches to a substrate and essentially renders the animal stationary.

Feeding, digestive system. The food of a clam or mussel consists chiefly of fine organic detritus dislodged from the substrate and momentarily in suspension. Bacterial picoplankton and microphytoplankton are also eaten, but large planktonic organisms are not utilized. The inner surface of the mantle, the demibranchs, and the visceral mass are covered with cilia that beat in such a coordinated way as to draw a stream of water in through the inhalant aperture or siphon. This water passes into the mantle cavity and enters the many minute openings, or ostia, of the demibranch lamellae into the demibranch chambers, then upward to the suprabranchial chambers, and finally posteriorly through the cloacal chamber and out

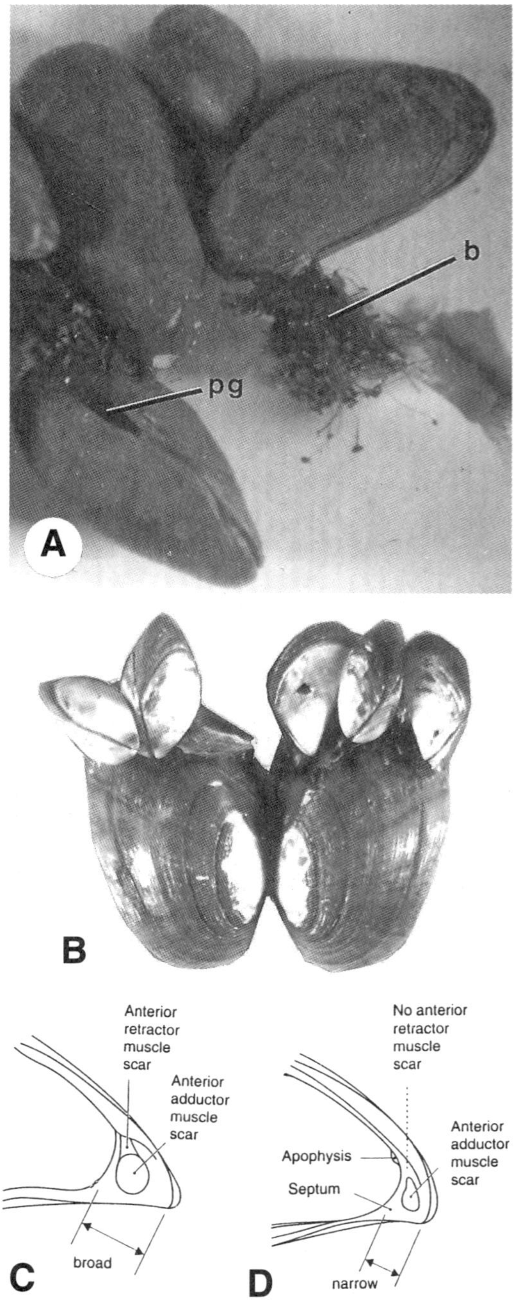

Fig. 14.34. Structure and biology of Dreissenidae. A, cluster of *Mytilopsis leucophaeata*, ×2; B, *Dreissena polymorpha* attached to a unionid mussel shell, ×1; C, anterior of shell of *Dreissena;* D, anterior of shell of *Mytilopsis.* b, byssus; pg, pedal gape. (C, D modified from MacNeill 1992.)

the exhalant opening. As this continuous stream of water passes over the outer surface of the demibranchs, ingested food particles become entangled in their mucous covering. The mucus with its contained food is driven by ciliary action along the edge of the demibranchs toward the mouth, where it is picked up by cilia on the labial palps and directed into the mouth in a constant ribbon-like stream.

Sphaeriidae burrowing well below the mud surface must rely completely on organic detritus for their food. Indeed, it has recently been established that subsurface mud inhabiting sphaeriids have an interstitial suspension-feeding mechanism. As they burrow through the mud, organic particles and bacteria are placed in suspension momentarily, to be filtered out promptly and taken into the digestive tract.

Although some inorganic silt may be mixed with the organic food taken into the mouth, a large portion of the inedible material is separated out beforehand. The mechanisms involved in this selection and separation have been the subject of a long controversy. It appears that the labial palps are most important, and that by some means much of the indigestible material is separated out on their surface and carried to the ventral edges where it drops off and is carried backward by ciliary action and expelled between the valves just below the inhalant siphon. It is likely that some additional selection and separation of food and inorganic debris is effected by the edge of the inhalant siphon and the mouth. The entire digestive tract is imbedded in the main visceral mass. The mouth opens into a short esophagus that leads to the more or less bulbous stomach situated in the anterior dorsal portion of the foot. The stomach is surrounded by a large digestive gland (Fig. 14.35). The long, narrow intestine has several coils behind and below the stomach before it proceeds to the dorsal region and runs posteriorly above the posterior adductor muscle and opens into the mantle cavity through the anus just above the exhalent opening of the mantle cavity.

The anterior part of the intestine of the Unionidae has a lateral diverticulum, or

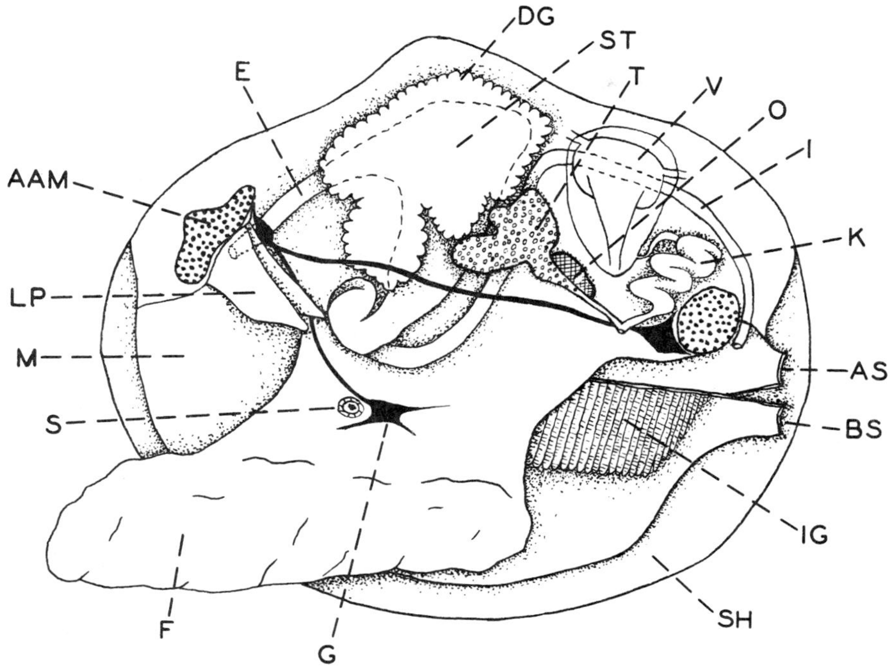

Fig. 14.35. Diagrammatic longitudinal section of *Musculium*, ×12. AAM, anterior adductor muscle; AS, anal siphon; BS, branchial siphon; DG, digestive gland; E, esophagus; F, foot; G, pedal ganglion; I, intestine; IG, inner demibranch; K, kidney; LP, labial palp; M, mantle; O, ovary; S, statocyst; SH, shell; ST, stomach; T, testis; V, ventricle of heart.

groove, containing the curious crystalline style. This is a cylindrical structure of a hyaline, milky, or brownish color and of a dense, gelatinous consistency. During normal feeding activities it is revolved on its long axis by the ciliary epithelium of the style sac. The anterior end of the style usually projects into the stomach, where it rubs against a shield and is constantly being eroded away. This erosion liberates a polysaccharide-digesting enzyme and is thought to function also in separating food from foreign particles. The style disappears during periods of starvation, but is regenerated as a response to ingestion. It is formed only slowly at low temperatures.

Circulation. The circulatory system consists of a single ventricle, two auricles, blood vessels, and spaces or sinuses in the tissues. Dorsally and posterior to the stomach is the thin-walled pericardial cavity, containing the heart and through which the posterior part of the intestine passes longitudinally. The ventricle is an oval organ surrounding the intestine, and the two flat auricles lie at the sides of the ventricle and open into it just below the middle of each side.

Arising from the ventricle are an anterior and a posterior artery whose branches supply all of the soft parts of the animal; the mantle and demibranchs are supplied with a particularly large quantity of blood. From the smaller arteries the blood passes into numerous sinuses in the tissues, and thence into veins and back to the auricles. The blood is colorless or slightly bluish owing to dissolved hemocyanin, a respiratory pigment. The heart rate is usually less than 20 beats per minute, but a rate as high as 100 beats per minute has been recorded.

Respiration. Although the entire body surface is in contact with water and probably functions in respiration, the greater portion of the oxygen–carbon dioxide exchange undoubtedly occurs in the mantle and demibranchs. Hibernating bivalves rely heavily on the edges of the mantle for respiration.

Excretion. The two kidneys, or organs of Bojanus, lie in the visceral mass immediately below or behind the pericardial cavity (Fig. 14.35). Each kidney is a dark-colored more or less convoluted tubule with the opening at one end in the pericardial cavity and the opening at the other (external) in a suprabranchial chamber.

Nervous system. Unlike snails, mussels have a simple nervous system, with only three pairs of ganglia (Fig. 14.35). The cerebropleural ganglia lie immediately posterior to the anterior adductor muscle, one on each side of the esophagus, and are connected by a short commissure passing over the esophagus; the pedal ganglia are deeply imbedded in the foot and are more or less fused; the visceral ganglia lie just anterior to the posterior adductor and are also partially fused. Connecting the cerebropleural ganglia with the pedal and visceral ganglia are two pairs of long commissures. Each ganglion is the source of nerve fibers to the adjacent organs.

Sense organs. Maintenance of equilibrium is effected by a pair of minute statocysts situated near the pedal ganglia. Each statocyst (Fig. 14.35) is an ovoid or spherical fluid-filled cavity lined with sensory cells and containing a spherical concretion, the statolith.

The osphradia are two small areas of specialized epithelium on the roof of the cloacal chamber. Their function has never been successfully established, but most writers assume that they are useful in detecting dissolved foreign materials in the water.

In addition to being sensitive to touch, the projecting edges of the mantle are generally capable of detecting pronounced changes in light intensity, and the valves can be caused to shut by casting a shadow or bright light on the mantle. Some species of the Unionidae have a distinct pigment spot near the inhalant aperture that may be photoactive.

Reproduction. Among the Veneroida, the Sphaeriidae and *Corbicula* (Corbiculidae) are uniformly hermaphroditic and ovoviviparous, while the Dreissenidae are dioecious. In the Margaritiferidae and Unionidae the sexes are usually separate, but several species are sporadically or consistently hermaphroditic. The ovaries and testes are imbedded in the upper part or visceral portion of the foot and are connected with the suprabranchial chambers by short ducts.

Polymesoda (Corbiculidae) and *Rangia* (Mactridae) almost certainly reproduce in saltwater and are not discussed here.

The details of reproduction differ markedly in all three families. In the Sphaeriidae it has been suggested that self-fertilization can occur in the reproductive ducts. The zygotes pass into the suprabranchial chambers and then downward into the water tubes of the inner pair of demibranchs. Here they are brooded while they develop and grow, and the demibranchs, becoming greatly distended, are then called marsupia. An adult may contain from 1 to an astounding 60 young in various stages of development. When released from the marsupia, the immature individuals are fully formed, having the shell and all morphological features of adults; sometimes they are surprisingly large, often one-quarter to one-third as long as the parent. For a while, the juveniles of *Musculium* and a few *Sphaerium* contain a byssus. A byssus is retained in adult *Eupera.* Reproduction is thought to continue throughout the year, although very few young are released during the winter months. Many species exhibit a wide range of phenotypic plasticity in the features of their life history.

The genus *Corbicula* undergoes a reproductive pattern very similar to that of the Sphaeriidae. The clams deposit eggs into demibranch marsupia where they are presumably fertilized and develop quickly. A larval stage is bypassed and within a short time, "spat" or juveniles a few millimeters in length are released. The juveniles produce a mucilaginous thread that aids in anchoring the young clam. This thread is retained after maturity and is considered by some to be a byssus. Growth is rapid and maturity can be reached in the same season. *Corbicula* can produce enormous numbers of offspring, the estimates ranging as high as 70,000 per season.

The Dreissenidae are largely marine pelecypods, but several species have become physiologically adapted to freshwater. The freshwater species, including the two presently established in the United States, have retained their marine ancestors' reproductive traits. Eggs and sperm are spawned into the water and following fertilization the embryo develops into a veliger larva. The larva lives among the plankton for a short while and then settles onto a firm substrate and begins to acquire adult features, including the byssus. Adult mussels produce enormous numbers of gametes. An adult female can produce tens of thousands of eggs during her lifetime.

In the Margaritiferidae and Unionidae, the unfertilized eggs pass into the suprabranchial chamber and then downward into the water tubes, or spaces, of the demibranchs. Depending on the species, all four demibranchs, the two outer demibranchs, or only special parts of the outer demibranchs (Fig. 14.36) are utilized. Sperm of the male are apparently swept out of the animal through the exhalant aperture into the surrounding water. Some of the sperm are then drawn fortuitously into the inhalant aperture of a mature female in the vicinity; they pass through the ostia of the demibranch lamellae and fertilize the eggs in the water tubes. The embryos are retained only for the early

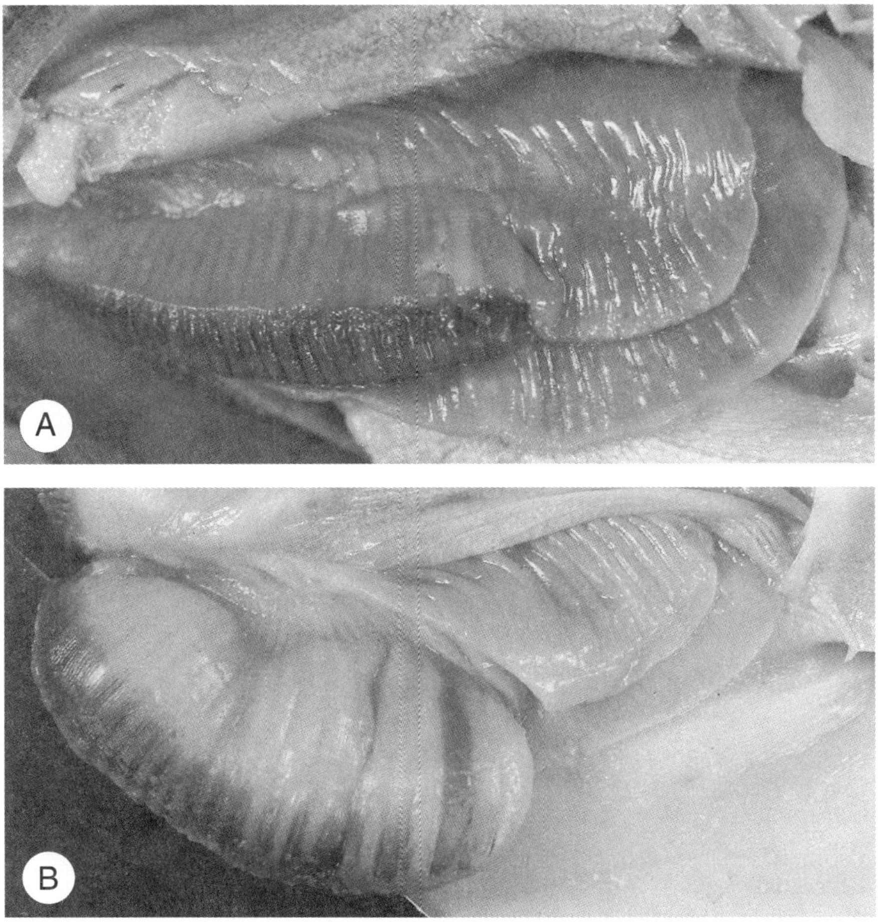

Fig. 14.36. Outer demibranch of female mussels. A, *Lampsilis perovalis;* B, *Lampsilis siliquoidea.* (From Haag et al. 1995.)

stages of development, but growth is pronounced, and the demibranchs (marsupia) soon become variously distended. The number of embryos present at one time ranges from several thousand in the smaller lampsiline species to more than 3 million in some margaritiferids.

Depending on the species, breeding may begin in the first to eighth year of life. Gametogenesis may continue throughout the year or be restricted to certain months of the year. The period of gravidity varies among mussel genera and even species and is affected by latitudinal distribution. Northern populations tend to show a specific seasonality, with the gravid period extending only a few months (*Elliptio*) to most of the winter (e.g., *Lampsilis*, *Ligumia*, *Potamilus*). Larvae are typically released in the spring but some species release in the fall. In southern species, there seems to be little seasonality with respect to gametogenesis and gravidity. Larvae are released over periods of months or within a short time span. As with other aspects of reproduction in southern parts of the country, larval release in not seasonally dependent.

All American freshwater mussels (Unionoida) produce a physiologically and anatomically unique type of larva, the glochidium (Fig. 14.38), which becomes a temporary and obligatory parasite on a fish. The mature glochidium, which is released from the marsupium of the female mussel, ranges from 0.05 to 0.50 mm in diameter. Superficially, it appears to be a small edition of an adult in its general morphology. There are two chitinoid valves joined at their bases along a hinge, lined by a layer of cells comprising the larval mantle. The valves are held together by a single adductor muscle. The mantle bears several tufts of sensory hairs. Usually a larval thread extends from the two valves that is thought to be useful in becoming attached to the host fish (Fig. 14.39A). The general shape of the valves may be simple and oval or rounded, roughly triangular, or ax-head shaped (in *Potamilus* only, Fig. 14.38L). In several genera the apex of each valve bears a hooklike projection that is spiny on the surface (Figs. 14.38A–D, 14.39A).

In the Margaritiferidae and many Unionidae the glochidia leave the female via the suprabranchial chambers and exhalant aperture; they may be free or more or less embedded in mucus. In other Unionidae (e.g., *Lampsilis*, *Villosa*, *Truncilla*, Fig. 14.40), irregular masses called conglutinates, break through temporary openings in the water tubes, usually at their base. In some forms (*Lampsilis* spp., *Strophitus*), specialized encapsulated structures are released into the environment, occasionally of exceptional appearance (Fig. 14.37), which apparently lure potential host fish into biting and releasing hundreds of glochidia. In most cases, the glochidia become scattered and eventually sink to the bottom where they remain

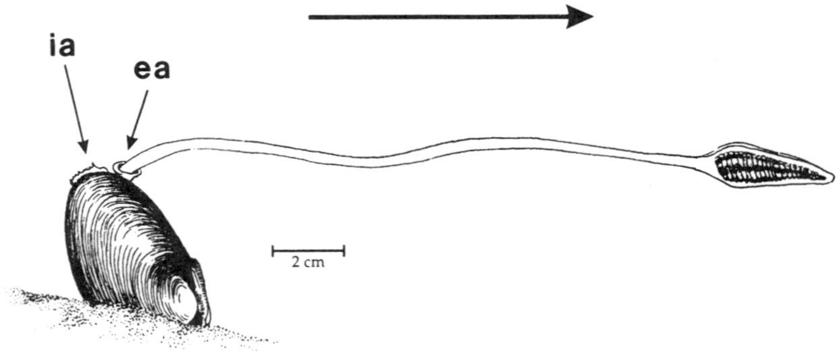

Fig. 14.37. "Super" conglutinate produced by *Lampsilis perovalis*. ea, exhalant aperture; ia, inhalant aperture, arrow denotes direction of current. (From Haag et al. 1995.)

with the valves gaping upward. If they do not come in contact with the body of a fish they usually die within several days. Fish become infected by brushing against the bottom, by stirring up the bottom with fin movements, and by taking in water that contains the glochidia in temporary suspension. In the lampsiline group of Unionidae, special modifications of the marsupium and mantle edge have evolved to increase likelihood of attachment to hosts. The larvae are exceedingly sensitive and clamp tightly to the superficial tissues immediately on contacting the fish. The hookless species usually become attached to the host's gills whereas hooked species attach to the fins or general body surface. Some forms are host specific and infect only one species of fish; others are capable of infecting several to many species of fish. Often, the glochidia become attached to an unsuitable host and are eventually discarded.

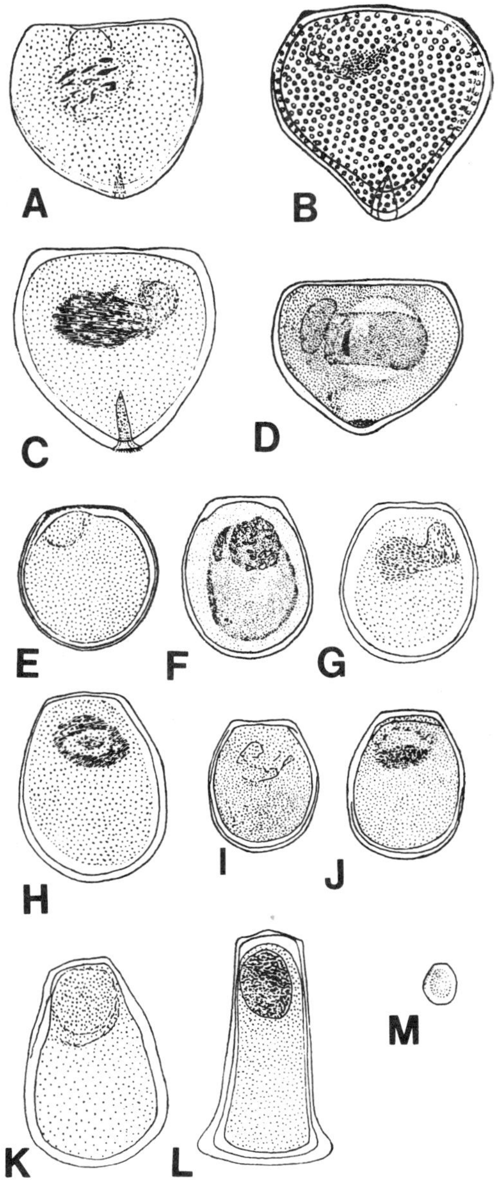

Fig. 14.38. Typical mussel glochidia, face views. A–D, triangular hooked glochidia; E–K, circular to ovoid–elongate glochidia; L, *Potamilus* type; M, Margaritiferidae type. All × 100. (From Baker 1928a.)

Within a short time the tissues of the fish begin to grow over the glochidium, and in a day or two the latter is completely encased and forms a small projecting nodule. This parasitic stage usually lasts from 10 to 30 days, although some records indicate as long as 190 days; undoubtedly water temperatures influence the duration of the parasitic period.

During encystment the glochidial structures change markedly, and most of the adult organ systems become more or less definitely formed. The single adductor muscle is replaced by two; the mouth and intestine are present; the demibranchs, kidneys, heart, and foot are represented by functional rudiments; and the mantle has taken on a more definitive structure. Usually, however, there is little increase in size. There is evidence that the speed at which this process occurs and overall success of metamorphosis is tied closely with temperature and certain ambient chemical parameters coupled with the degree of development of immunological responses within the host specimen. Fish that have previously served as hosts for the same mussel species can acquire a resistance to future infestation.

At the close of the parasitic period the young mussel breaks out of its cyst, falls to the bottom, and enters a period appropriately called the juvenile stage (Fig. 14.39B). The organ systems tend to develop rapidly, and the beginning of the adult shell appears beneath the glochidial shell. The foot is long and ciliated and provides a means of loco-

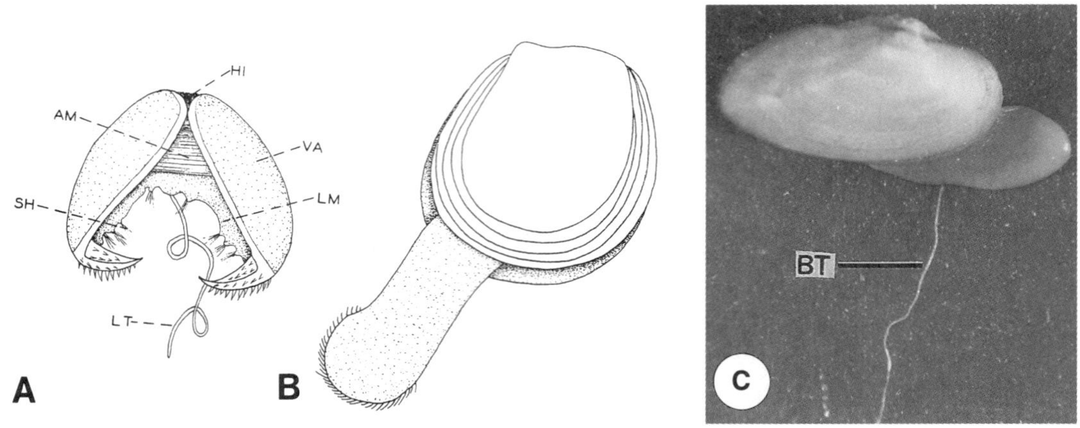

Fig. 14.39. Immature Unionidae. A, end view of glochidium of *Anodonta*, ×130; B, lateral view of early juvenile stage of *Actinonaias*, ×150; C, late juvenile of *Elliptio*. AM, adductor muscle; BT, byssal thread; HI, hinge; LM, larval mantle; LT, larval thread; SH, tuft of sensory hairs; VA, valve. (A modified from Tucker 1928; B modified from Lefevre and Curtis 1912.)

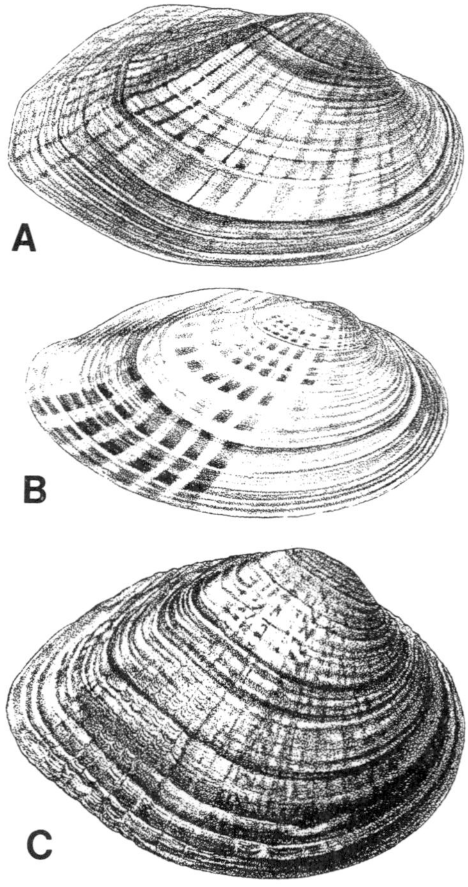

Fig. 14.40. Shells of Unionidae. A, *Lampsilis*, ×1; B, *Villosa*, ×0.6; C, *Truncilla*, ×0.6. (A–C from Burch 1973.)

motion. A long, sticky, hyaline thread, or byssus (Fig. 14.39C), is secreted by the posterior part of the foot in many species; it apparently facilitates attachment to the substrate and is usually shed in a few weeks. The smaller juvenile mussels burrow into the bottom and for this reason are seldom collected. As growth proceeds, however, the adult habits are assumed and the organs take on their adult characteristics. Strictly speaking, the juvenile stage lasts from 1 to as much as 8 years, that is, until the animal is sexually mature.

A few species are known to be able either to complete development to the juvenile stage in the brood chamber or to leave the parent as a glochidium and parasitize a fish in the usual manner. Amphibians have been implicated as hosts for unionoid mussels. One species, *Simpsoniconcha ambigua*, regularly parasitizes *Necturus*, the mud puppy. Artificial infection of mussel glochidia has been successfully undertaken on larval salamanders (*Ambystoma tigrinum*, *Eurycea bislineata*)

Even though more than 3000 glochidia have been found parasitizing a single adult fish, there is little evidence that they do any direct harm in natural habitats. Natural infections are believed to be very light. Nevertheless, heavy infections (100 or more) have been shown to kill 30-mm fingerling trout by causing secondary bacterial infections.

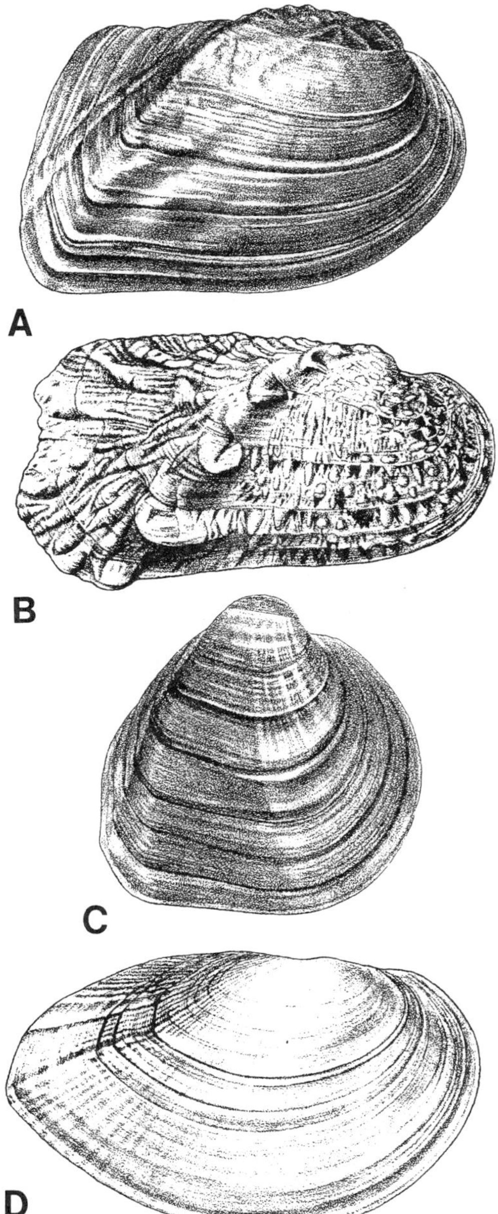

Fig. 14.41. Shells of Unionidae. A, *Plectomerus;* B, *Quadrula;* C, *Pleurobema;* D, *Utterbackia.* (A–D from Burch 1973.)

Growth, longevity. The period of growth extends from about April through September, and the actual rate of growth is dependent on numerous factors, including temperature, food supply, current, and the specific chemical nature of the water. Growth is certainly more seasonal in northern climates.

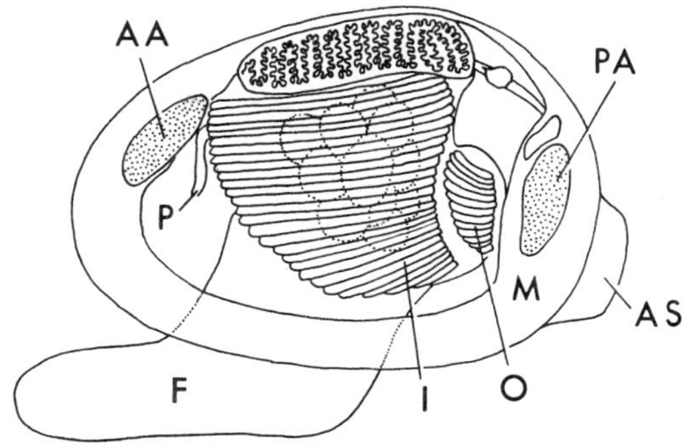

Fig. 14.42. Anatomy of *Pisidium.* AA, anterior adductor muscle; AS, anal siphon; F, foot; I, inner gill (containing young); M, mantle; O, outer gill; PA, posterior adductor muscle.

Mussels usually grow to a length of 30 to 80 mm in two growing seasons. Some species of mussels are short-lived, not exceeding 8 or 9 years of age, others, like *Anodonta,* can live for 10 to 15 years. The longest-lived species, and arguably the longest lived of all freshwater invertebrates, are those in the Margaritiferidae, which have been estimated to live from 25 to 65 years in the United States. Most Sphaeriidae are thought to live no longer than 12 to 18 months. At least one species of *Sphaerium* has a complete life cycle in 1 month or less. At the other extreme, some sphaeriids live as long as 3 years.

Predators and parasites. The Sphaeriidae (Fig. 14.43) and *Corbicula* (Fig. 14.44A) are eaten regularly by numerous fishes, including the gizzard shad, whitefish, buffalo, sucker, redhorse, sturgeon, and perch, and soft-shelled species (e.g., *Anodonta, Pyganodon, Utterbackia,* Fig. 14.41) of mussels are eaten by catfish (Fig. 14.45) and sheepshead. Shortnose sturgeon (*Acipenser brevirostrum*) intestines from the Connecticut River sometimes contain large numbers of immature *Elliptio.* Muskrats are undoubtedly the most important mammalian enemy of mussels. They usually select thin-shelled forms and either let them die on shore and eat them when the shell gapes, or force the shell open with their teeth. In the United States, mussels are also occasionally eaten by mink, otter, raccoon, turtles, hellbenders (*Cryptobranchus*), and most recently, humans.

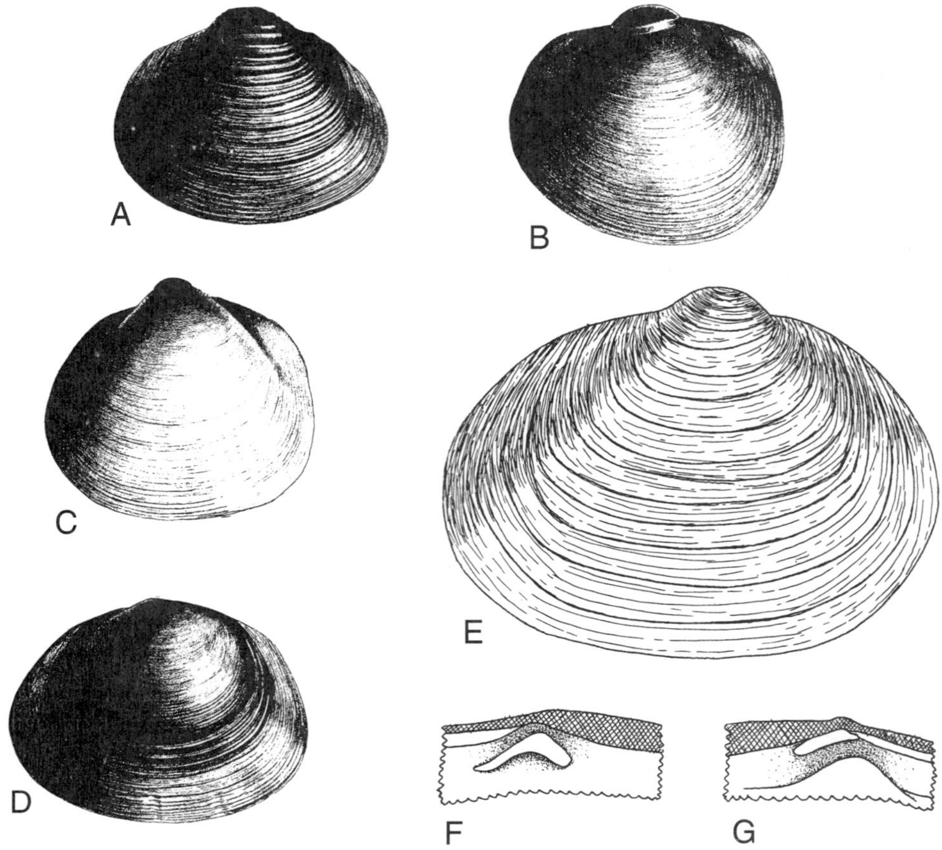

Fig. 14.43. Structure of Sphaeriidae. A, left valve of *Sphaerium*, ×4; B, left valve of *Sphaerium*, ×4; C, left valve of *Musculium*, ×3; D, left valve of *Sphaerium*, ×3; E, right valve of *Sphaerium simile*, ×5; F, cardinal tooth of right valve of *S. simile*; G, cardinal teeth of left valve of *S. simile*. (A–D from Burch 1973; F and G modified from Baker 1928a.)

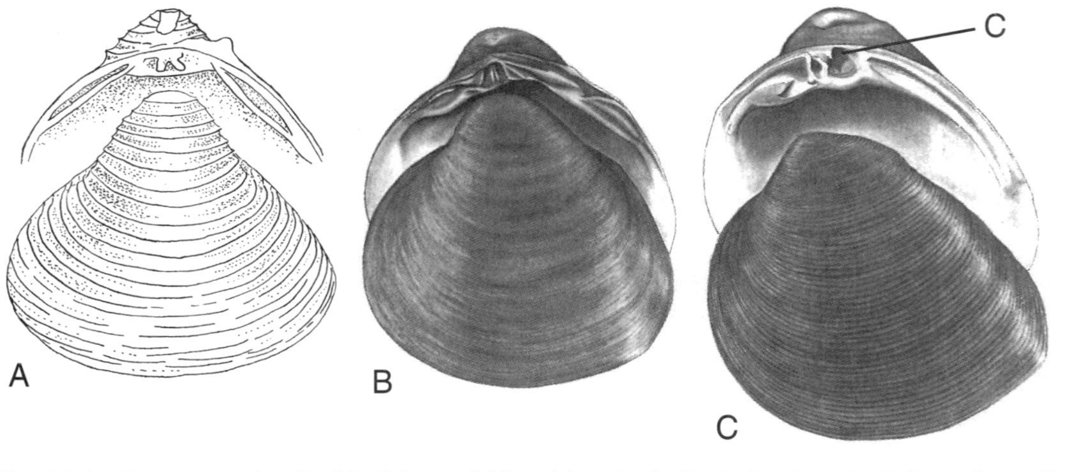

Fig. 14.44. Representative Corbiculidae and Mactridae. A, *Corbicula fluminea;* B, *Polymesoda caroliniana;* C, *Rangia cuneata.* C, chondrophore (B, C from Heard 1982.)

Fig. 14.45. View of the bottom of the Connecticut River showing large bed of *Elliptio complanata.* In the background is a large channel catfish, *Ictalurus punctatus,* an important predator of freshwater bivalves. (© E. Klekowski, with permission.)

The branchial and mantle cavities are often infested with parasitic or commensal Hydrachnidia (*Unionicola*). The chironomid *Baeoctenus bicolor* apparently lives and feeds on the demibranchs of some mussel species. The larval and adult stages can damage demibranch and mantle tissues. The adults lay eggs within the tissues. Larval digenetic trematodes (e.g., Allocreadiidae: cercaria, metacercaria) infest sphaeriids and possibly unionoids. Freshwater mussel shells harbor, by means of nacrezation, cercaria of other trematodes (e.g., Bucephalidae). These parasites may be responsible for the production of small irregular pearls and blisters on the inner surface of the host shell, and sometimes the nacre may be discolored. Aspidobothrid trematodes (e.g., *Cotylaspis, Aspidogaster*) are sometimes found in the branchial, intestinal, and pericardial cavities.

Ecology. Freshwater bivalves occur in all types of unpolluted habitats but they are most abundant and varied in the larger rivers (Fig. 14.45). Small creeks and spring brooks are devoid of mussels, although they contain Sphaeriidae. Similarly, small lakes, especially seepage lakes, contain few species, and large drainage lakes contain many.

Like snails, bivalves are most abundant in the shallows, especially in water less than 2 m deep. In the largest rivers and some lakes, however, unionoid mussels occur as deep as 7 m. The record depth collection is *Anodonta,* which was taken at 30 m in Lake Michigan. The specimens were 7 to 14 years old but less than 53 mm long. Sphaeriidae are regularly found in great abundance on some lake bottoms more than 30 m below the surface. One of the deepest records is of *Pisidium conventum* dredged from 219 m. These small bivalves, especially the immature ones, are often active burrowers and have been taken up to 25 cm deep in soft substrates. Under such conditions the sphaeriids are said to be in a "resting stage."

The occurrence of the largest species and largest numbers of mussels in rivers where there is a good current has not yet been satisfactorily explained from an ecological standpoint. It is possible that abundant food supply, the specific type of substrate, an abundance of oxygen, water chemistry, or a combination of factors may be significant. River genera have thicker shells than lake genera, and individuals of a single species inhabiting both types of environments usually have heavier shells in running waters. Also, in a single species, flatter and less inflated varieties may occur in upper tributaries, and more swollen varieties in the lower parts of a river system. Mussels inhabiting exposed lake shoals may be somewhat stunted and more heavy-shelled than specimens from protected waters. Greater numbers of species are usually found in the lower waters of river systems rather than in the small tributaries. Although not definitive, investigation has revealed that lotic populations of species do segregate on the basis of pH, hydrology, and physical features of a river or stream basin and tend to aggregate in areas protected or shielded from the dislocation activities of very high flow. In large rivers that drain to the sea, certain species prefer riverine sections close to tidal influence. The greater abundance of species in large, shallow lakes is probably due to the fact that such environments have a greater variety of subhabitats or niches than small lakes. Although sphaeriids are common, there are no mussels in high mountain lakes, probably chiefly because these lakes do not contain the proper host fishes for the glochidia and are nutrient poor. Additionally, such lakes are usually poor in calcium and could not support mussels anyway.

Bare rock bottoms and shifting sands and mud, with their accompanying high turbidities, are unsuitable for bivalves, whereas stable gravel, sand, and substrates composed of sand or gravel mixed with other materials support the largest populations, some such beds having more than 50 mussels per square meter. Though soft mud bottoms are generally uninhabited, there are some common "mud-loving" species such as *Anodonta* and *Pyganodon.* Customarily, mussels inhabit substrates free of rooted vegetation, but there are numerous exceptions. The Sphaeriidae are less specific in their occurrence and are found on all types of bottoms except clay and rock; in favorable situations populations of more than 10,000 individuals per square meter may be counted.

To a large extent the geographical distribution of the freshwater bivalves appears to be determined by the chemical features of waters. Sustained populations of anodontines are rarely found in acid waters (pH below 7.0) or in waters having a bound carbon dioxide content of less than 15.0 mg/L. Adults of some species are able to tolerate acid events of pH 3.0, but over time lose the periostracum and undergo physiological changes, especially in blood chemistry. The glochidia, however, very likely cannot survive pH values this low. Decidedly alkaline waters and an abundance of calcium carbonate for shell construction favor large numbers of individuals and species. The Sphaeriidae, on the other hand, have become adapted to a wider range of conditions, and some species are common in lakes having a pH as low as 6.0 and a bound carbon dioxide content of only 2.0 mg/L. It is difficult to understand how a calcium carbonate shell can be secreted and maintained under such unfavorable conditions. Salts appear to be actively absorbed directly from the water. Biomass calculations sometimes show that the calcium in unionid shells may be 50% as much as the total calcium dissolved in the water of a lake in which the clams occur. A few Sphaeriidae species are indicators of lake chemistry trophic conditions, but, at least in the Sphaeriidae, there is little correlation between $CaCO_3$ in the shell and water hardness of the habitat. Potassium seems to be relatively toxic, and waters having more than 7 mg/L of potassium are generally devoid of mussels.

Because of the apparent disappearance of mussels from the shallows during the winter months, it was formerly inferred that they migrated to deeper waters in the fall. It is now believed, however, that migrations are unimportant and that most individuals simply burrow more deeply into the bottom of their summer habitat so that only the inconspicuous edges of the apertures project. The apertures are opened infrequently, and the animals remain in a dormant state throughout the winter. Ten degrees C is thought to be an average critical hibernation temperature.

Many Sphaeriidae are known to burrow into the drying substrate in times of drought. In vernal pools in northern states, a few species in the genera *Sphaerium* and *Musculium* are able to withstand seasonal drying by lowering metabolism and either respiring by periodic minute shell gapes or withstanding hypoxia. The ubiquitous *Pisidium casertanum* has been often observed by myself (DGS) living in damp soils above the waterline of streams and appears to be somewhat amphibious. Among the Unionidae, species of *Uniomerus*, *Carunculina*, *Ligumia*, and *Amblema* have all been found buried in substrates that were moist or dry for several months.

The Family Unionidae is believed to have evolved in the North American continent and, more specifically, in the general area of the Mississippi drainage basin, for it is here that the greatest diversity of species occurs. There appear to be fewer taxa in Europe and parts of Asia, but the unionid fauna of Asia remains very poorly known. The Margaritiferidae are widely distributed throughout the Northern Hemisphere, and the Sphaeriidae are worldwide. The other native families mentioned in this chapter are predominantly estuarine or marine, only a handful of species having invaded brackish waters and the coastal areas of rivers.

Two families contain exotic species that have become well established in North America, the Dreissenidae and Corbiculidae. Both families are predominately estuarine but contain species that have made the successful transition to freshwater. *Corbicula fluminea* is native to large rivers in eastern China and entered the United States around 1930. *Dreissena*, which hails from northern Caspian Sea region of central and eastern Europe, appeared in about 1985. Both genera have spread rapidly throughout the country, *Corbicula* largely through the southern and central states, *Dreissena* through the northern and central states. In their own way, each genus has produced a multitude of problems for both native aquatic faunas and industries that require free-flowing water. An enormous amount of research has followed in their wake, primarily of the sort that seeks to control or prevent the species in their new habitats. Neither species seems to have been checked in its colonization of new waters, and barriers previously believed significant in arresting their spread (e.g., temperature, calcium concentration) are apparently being overcome. In the Connecticut River, Connecticut, *C. fluminea* appeared in 1991 (Fig. 14.46). Water temperature in the Connecticut River approaches 0.0°C for several months during winter; however, *Corbicula* is surviving in spite of the long-held belief that sustained near zero temperature conditions eliminate the species. A similar situation has been reported in Ohio.

Dreissena, with its ability to disperse during the veliger stage and its unique byssal form of attachment, poses all manner of

Fig. 14.46. A handfull of *Corbicula fluminea* collected in a few minutes.

threats to humans and animals alike. It has the unsavory habit of attaching to native mussel species (Fig. 14.34B), eventually weighing down the host and denying it access to its food. Like *Corbicula, Dreissena* will invade and live within water distribution pipes of municipal and industrial water users, causing fouling and clogging problems in the systems.

By migrating naturally and extensively into suitable habitats, a species may, over a long period of time, come to occupy much of a large drainage area. More rapid migration is undoubtedly effected by fishes carrying the glochidial stage of mussels, and by birds and large insects transporting Sphaeriidae and immature Unionidae overland in mud or debris on their legs and feet and in the feathers. Whereas most migrations have occurred over long periods of geological history and drainage evolution, other forms of migration have certainly been relatively rapid. The removal of a barrier such as ice sheets characteristic of glacial episodes has allowed colonization of entire drainage systems within a few thousand years, and the construction of bypasses of dams and locks can allow upstream colonization within a hundred years.

Pollution, silting of rivers, shell harvesting, and the construction of dams are decidedly inimical to our American mussel (unionoid) fauna, as are the byssus-producing dreissenids. Predation by mammals that feed on the rarest species exacerbates the problem. Even with the advent of water treatment, the byproducts of treatment are toxic for a certain downstream distance, especially to glochidia. In fact, the glochidia and juveniles of mussels are among the least tolerant of any organism living in freshwater. Juveniles are very susceptible to depressed oxygen as well as pH and other toxins. Many of the less common species are fast disappearing, and reduced populations appear to be unsuccessful in increasing their present restricted geographical distribution without human intervention. Gradual declines have certainly occurred since at least the Industrial Revolution, and single events involving the die-off of parts of or entire populations have been documented since the 1970s. Moreover, freshwater mussels once supported a bustling pearl button industry and in the early 1950s, the cultured pearl industry of Japan found that small pellets from American thick-shelled mussels formed excellent nuclei for cultured pearls. Although the industry remains small, the overharvesting of shells in certain areas has also been a cause for decline. Williams et al. (1992) estimate that about 7% of the native mussel fauna is extinct and several authorities have maintained that between 40 and 50% of the fauna is in need on conservation.

Recent surveys, compared with early surveys, give some idea of the rate at which our fauna is becoming depauperate. The Tennessee River system has been hard hit in terms of stream alteration. Currently, the system supports less than 50 species contrasted with 100 species before the 1936 Tennessee Valley Authority impoundment period began. A study by Parmalee and Hughes (1993) of the recently created Tellico dam has shown that 90% of the mussel fauna was eliminated and replaced by a much lower diversity of silt- and mud-tolerant species.

Matteson and Dexter (1966) report a "drastic decline" in the mussel fauna of the Big Vermilion River between 1918 and 1962. Isom and Yokley (1968) report only 48 species in the Duck River, Tennessee, as contrasted with 63 reported originally by Ortmann. In the Illinois River, Starrett (1971) found only 24 species in 1966–1969 (5 species represented by single specimens) as contrasted with 45 in 1906–1912. Suloway (1981) found only 15 unionids in the Kankakee River, Illinois, in contrast to 29 species reported in 1906. Only three mussel beds in the Kentucky River are now said to be commercially valuable. Channalization and dredging of several of the major rivers east of the Mississippi River flowing into the Gulf of Mexico have seen major declines in population sizes. Statewide, 14 species are now extinct in Ohio (Watters 1995), and 18 species are presumably gone from Illinois (Cummings and Mayer 1997).

In light of the serious nature of the decline of freshwater mussels, research on life-history characteristics and conservation has undergone a dramatic upswing. Important recent reviews (Hoggarth 1992; Watters 1994) on the life history of freshwater mussels have formed the basis for continuing and future research. Since these reviews, well over 20 studies have been published (e.g., Bruenderman and Neves 1993; Gordon and Layzer 1993; Michaelson and Neves 1995; Haag and Warren 1997; Howells 1997) on the life history and host fish biology, especially for endangered species. In recent years, a broad field of restorative biology has developed specifically for freshwater mussels. Attempts to transplant cultured animals into former habitats has been met with some success, and the experimental introductions have produced useful biological information. A recent survey identified 24 U. S. scientists involved in some capacity of restoration biology of freshwater mussels.

Collection, preparation. Little special apparatus is necessary for field collecting. The larger species may easily be handpicked in the shallows; small forms may be washed from vegetation or sifted from bottom debris. A long-handled rake may be used for collecting mussels in water of moderate depths, and dredges are necessary for sampling in deep lakes. A viewing box that blocks out the light and allows clear views of the bottom of shallow water is very useful in locating mussels. Modern survey techniques are increasingly relying on the use of SCUBA and snorkeling to locate animals.

Because of the decline of freshwater mussels and the prohibition many states place on collecting, the collection and killing of live specimens is discouraged. Typically, if live animals are about, then shells of the deceased should be easily found. By placing a live animal in a clear, clean container and allowing it to remain undisturbed, the mantle structures will be exposed by the animal and observed or photographed. Animals should always be returned to the substrate in the manner they were found. If preservation is necessary and permitted, then every effort should be taken to carefully preserve the animal and in a manner in which its use can be maximized. Thus, any mussel to be preserved should be narcotized. Several reagents have been tried with various success. The simplest and probably best chemical is powdered menthol, which is sprinkled over the surface of the container housing the animal. When the animal is anesthetized, and prior to death, 10% buffered formalin should be injected hypodermically into the pericardial cavity. This will cause expansion of the mantle tissues to a lifelike position and fill the vascular system with fixative, insuring proper fixation. Simply placing a live, active animal in alcohol or formalin renders the specimen nearly useless because of extreme contraction and insufficient penetration of preservative. Following fixation, the shell and animal can be kept together in 70% ethyl alcohol or if separated, the shell dried and stored, the animal maintained in 70% ethyl alcohol. Fresh dry shells can be stored without treatment or treated to prevent flaking and excessive cracking by brushing a Permount–paraffin mix thinned in xylene directly on the periostracum. Shells should be kept in closed containers to keep out dust and pests.

The veneroids can be preserved the same way as freshwater mussels. As in mussels, narcotization will enable fixation with the organ systems in a more lifelike manner, especially the siphonal region which can help in specimen identification and reveal characters of systematic value. Fixation can be accomplished by simply immersing the narcotized specimen in 10% formalin. Shells of most veneroids, especially sphaeriids, are very delicate and in time, will dissolve in fluid. Thus, shell material is probably best preserved dry. Storage of dry material should be in enclosed vials or boxes to prevent pest and dust damage.

Preservation of the animal should be in 70 or 80% alcohol, not formalin.

Taxonomy. In keeping with the rise of biological research on freshwater clams and mussels, taxonomic investigations have increased as well. The large, native, and richly diverse freshwater mussels of the Order Unionoida have received much attention, especially in the area of clarification of species limits among the numerous genera. Investigations have concentrated on both traditional methods involving analysis of shell characters or on a rapidly emerging discipline, DNA sequence analysis. Unfortunately, anatomical studies lag behind both of these methods. Nevertheless, much disagreement remains on the number of valid species existing or formerly existing in the United States. The number of recognized species appears to be growing, starting with just over 220 listed by Burch (1975a) to nearly 300 in more recent works. In many cases, however, the true number may never be known due to the relatively rapid rate of extinction occurring among these animals.

The Sphaeriidae have not experienced the intensity of taxonomic study as have the unionoids. Most species are still known or recognized by shell characters only. Natural groupings appear to be established among genera and within the large genus *Pisidium*. However, many species-level taxa need to be reexamined anatomically to confirm or invalidate their placement within a specific genus or subgenus. It is believed by some workers that several recognized species of *Pisidium* are conspecific with other species, but there has been little effort to fully investigate their relationships. It is probable that the actual number of species is somewhat lower than found in the most recent accounts. Herrington (1962), Burch (1975a), and Mackie (1986) list between 37 and 39 species of sphaeriids within the United States.

Because of their well-documented impact on biological systems and water-dependent industries, the exotic species, although few in number, have been allocated more than their share of taxonomic scrutiny. In fact, U.S. populations of the Dreissenidae and the Corbiculidae (*Corbicula*) have probably undergone taxonomic study equal to that in their home territories. The result of these studies has revealed that at least two species of *Dreissena* and one of *Mytilopsis* presently occur in the United States. The number of species of *Corbicula* living in the country is not certain. Generally, one species, *C. fluminea*, is considered to be present, but some data suggest that an additional species is present. Table 14.2 gives the genera presented in the key and their higher classifi-

Table 14.2. Classification of Freshwater Pelecypods of the United States

Order Unionoida
- Margaritiferidae
  - *Margaritifera, Margaritinopsis, Pseudunio*
- Unionidae
  - *Actinonaias, Alasmidonta, Amblema, Anodonta, Anodontoides, Arcidens, Cyclonaias, Cyprogenia, Cyrtonaias, Dromus, Ellipsaria, Elliptio, Elliptoideus, Epioblasma, Fusconaia, Glebula, Gonidea, Hemistena, Lampsilis, Lasmigona, Lemiox, Leptodea, Lexingtonia, Ligumia, Medionidus, Megalonaias, Obliquaria, Obovaria, Plectomerus, Plethobasis, Pleurobema, Popenaias, Potamilus, Ptychobranchus, Pyganodon, Quadrula, Quincuncina, Simpsonaias, Strophitus, Toxolasma, Tritogonia, Truncilla, Uniomerus, Utterbackia, Villosa*

Order Veneroida
- Dreissenidae
  - *Dreissena, Mytilopsis*
- Corbiculidae
  - *Corbicula, Polymesoda*
- Mactridae
  - *Rangia*
- Spaheriidae
  - *Eupera, Musculium, Pisidium, Sphaerium*

cation, which follows Burch (1975a,b) and Smith (2000b) for native species.

The key is designed to accommodate live mussels as well as preserved specimens, particularly in the portions dealing with the Unionoida. The use of characters evident in living individuals of unionoids eliminates the necessity of killing the animals, a practice now prohibited in most states. The combination of live animals, which can be returned to the wild, and the collection of the shells in the same habitat should permit identification to genus in nearly all instances.

Examining live sphaeriids will also greatly facilitate identifying genera when using the key. The shells of remaining veneroids is all that is necessary to identify the genus.

KEY TO FAMILIES AND GENERA OF FRESHWATER PELECYPODA OF THE UNITED STATES

1. Hinge teeth consisting of central cardinals beneath beaks and lateral teeth on either side (Figs. 14.43, 14.44A), or with lateral teeth insconspicuous and beaks anterior, pointed, and with a distinct shelflike myophore within; adults rarely over 25 mm in length, with (Fig. 14.34) or without a byssus; true siphons present . . . . **Order VENEROIDA, 2**

Hinge teeth consisting of psedocardinal teeth just anterior of beaks and laterals just posterior, or hinge teeth reduced to absent, beaks central or subcentral, never with a myophore; adults between 25 and 250 mm in length, without a byssus in adult; true siphons absent **Order UNIONOIDA, 10**

2. Shell equilateral or inequilateral, but without eccentrically placed beaks; adult with a well-developed foot and with or without a simple byssus **4**

Shell beaks eccentrically placed at anterior end (Fig. 14.34); adults with a reduced foot and a complex byssus (Fig. 14.34) . . . . **DREISSENIDAE, 3**

3. Shell without an apophysis (Fig. 14.34C), zigzag markings persistent on adult periostracum; inland freshwater; 2 species introduced to Great Lakes region, spreading rapidly ***Dreissena***

Shell with an apophysis beneath the myophore (Fig. 14.34D); zigzag markings absent on periostracum of adults (Fig. 14.34); upper estuaries and tidal freshwater; 2 species in eastern states, Gulf Coast, and possibly Mississippi River . . . . ***Mytilopsis***

4. Adult shell subtriangular with elevated beaks, with two cardinal teeth in each valve or a deep pocket (chondrophore); adults 10 to 60 mm in length **5**

Adult shell round, oval, or quadrate with low beaks (Fig. 14.43), with one or two cardinal teeth in one valve, one in the other and without chondrophore; adults 2 to 15 mm in length **SPHAERIIDAE, 7**

5. Area of cardinal teeth with a deep pocket (chondrophore), posterior lateral tooth much longer than anterior (Fig. 14.44C); 1 species in tidal freshwater and estuaries, eastern states and Gulf of Mexico . . . . **MACTRIDAE, *Rangia***

Area of cardinal teeth without chondrophore, lateral teeth equal or subequal (Fig. 14.44A, B) . **CORBICULIDAE, 6**

6. Restricted to tidal marshes, rarely in freshwater; shell without strong ridges (Fig. 14.44B); 1 species in southeastern and Gulf Coast states . . . . ***Polymesoda***

Widespread (introduced) in the interior and coastal U.S., periostracum yellowish in the young, becoming black in older individuals, shell with coarse ridges (Fig. 14.44A); 1 or 2 species ***Corbicula***

7. Beaks posterior to center (Fig. 14.49); animal with a single siphon, outer gill as high or smaller than inner demibranch or outer demibranch absent (Fig. 14.42); juveniles loosely arranged in marsupial demibranch; about 24 species, widely distributed . . . . ***Pisidium***

Beaks at or near center of shell; animal with two siphons, outer demibranch equal to inner demibranch **8**

8. With only one cardinal tooth in each valve, adult with a byssus; periostracum typically blotchy; 1 species in southeastern states . . . . ***Eupera***

With two cardinal teeth in one valve, adult without a byssus; periostracum color uniform . . . **9**

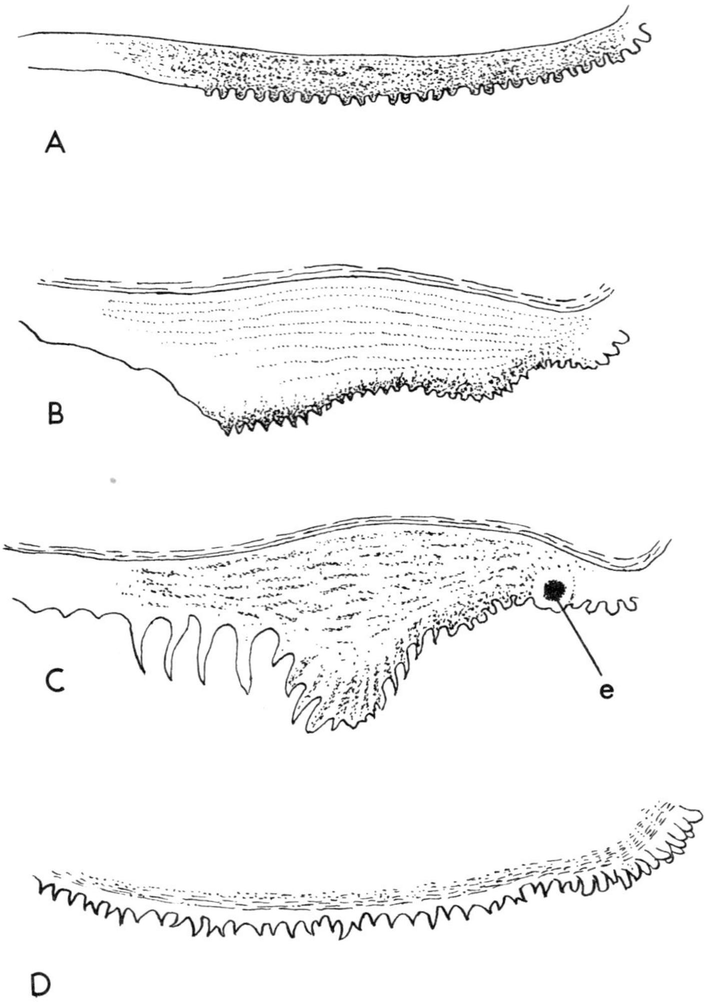

Fig. 14.47. Mantle structure of female Unionidae. A, *Ligumia;* B, *Lampsilis teres;* C, *Lampsilis straminea;* D, *Glebula subrotunda.* e, eyespot. (Drawn from anesthetized and preserved animals.)

9. Shell delicate and usually with embryonic caps or prominences in beaks (Fig. 14.43B, C); anterior lateral teeth without tubercles; siphons free for most of length (Fig. 14.33), many juveniles in demibranchs; 4 species distributed throughout the U.S. .... ***Musculium***

Shell strong and without caps (Fig. 14.43A, D, E); anterior lateral teeth with tubercles; few to several juveniles in demibranchs; siphons fused for most of length; 8 species distributed throughout the U.S. .... ***Sphaerium***

10. Adult shell somewhat kidney-shaped or arcuate (Fig. 14.50A, B), without external color markings (e.g., rays) anytime during life, midinterior of fresh shells with distinct scars (Fig. 14.50B); inhalant aperture with branched papillae .... **MARGARITIFERIDAE, 11**

Adult shell variable in shape, mid-interior of shell without scars; color markings often present, especially in younger shells; inhalant aperture with unbranched papillae (Fig. 14.31) ...... **UNIONIDAE, 13**

11. Shell with corrugation or plication posteriorly (Fig. 14.50C), 2 species, AL, AR, LA ***Pseudunio***

Shell without surface sculpture **12**

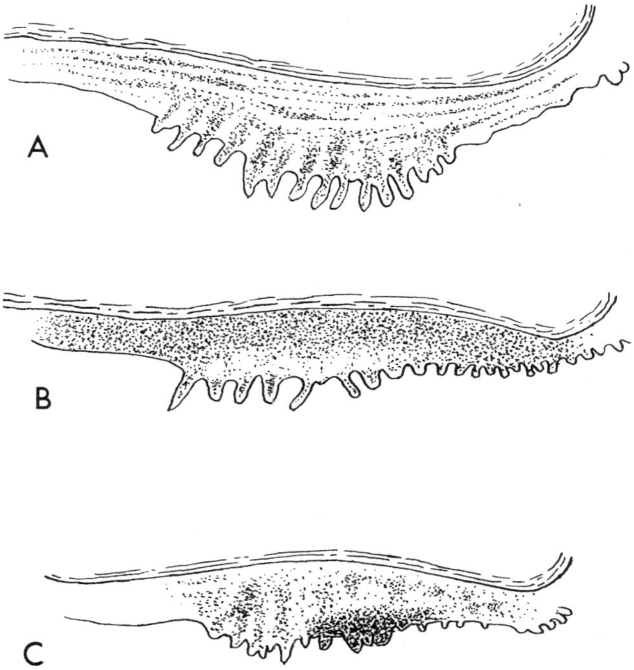

Fig. 14.48. Mantle structure of female Unionidae. A, *Villosa vibex;* B, *V. villosa;* C, *V. lienosa.* (Drawn from anesthetized and preserved animals.)

12. Anterior cusp of left valve produced (Fig. 14.50D), interlamellar junctions of demibranchs generally discontinuous; 1 species (Fig. 14.52A), northeast U.S. . . . . ***Margaritifera***
Anterior cusp of left valve reduced (Fig. 14.50E), interlamellar junctions of demibranchs generally continuous; 2 species (Fig. 14.50A, B); Mississippi and northwest coastal drainages ***Margaritinopsis***

13. Shell without hinge teeth, dorsal and ventral shell margins steeply convergent anteriorly (Fig. 14.54H); 1 species known only from CA, OR, WA . . . . ***Gonidea***
Shell with or without hinge teeth, dorsal and ventral margins slightly convergent or otherwise, never steeply convergent; occurring east of the Continental Divide (one exception, *Anodonta*) **14**

14. Shell without hinge teeth, or with pseudocardinals only, or both pseudocardinal and lateral teeth (least frequent condition); glochidia (where known) with spiny ventral hooks (Fig. 14.39A); marsupium entire outer demibranch and swollen when gravid but undifferentiated externally; mantle margin of female plain; exhalant aperture smooth or rarely papillate (2 genera) **15**
Shell with pseudocardinal and lateral teeth, rarely one or other reduced; glochidia without hooks; marsupium either outer demibranch only or both inner and outer, marsupial demibranch not greatly swollen when gravid or markedly differentiated externally from inner demibranch; mantle margin of female plain or elaborated chromatically and anatomically; exhalant aperture papillate, occasionally appearing crenulate **25**

15. Hinge teeth absent, without even a trace or with only an uneven ridge anterior to beaks, shell usually fragile **16**
Hinge teeth present in some form, at least vestigial to well-developed pseudocardinal evident; shell fragile to strong **20**

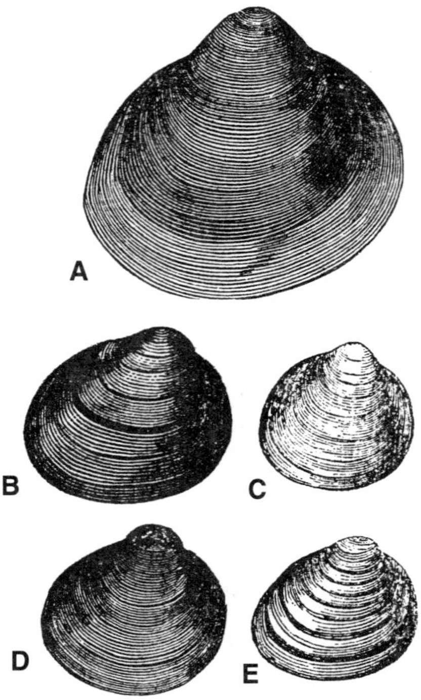

Fig. 14.49. Representative *Pisidium*. A, *Pisidium varaibile*, ×4; B, *P. dubium*, ×2; C, *P. equilaterale*, ×4; D, *P. ferrugineum*, ×10; E, *P. adamsi*, ×2. (From Prime 1865.)

16. Shell large, over 50 mm in length; exhalant aperture crenulate or smooth; periostracum not rayed on midportion of shell **17**
Shell small, usually less than 40 mm in length; periostracum often rayed throughout (Fig. 14.51H); exhalant aperture papillate; 2 species widespread in the interior U.S. ***Anodontoides***

17. Beak sculpture with double-looped bars **18**
Beak sculpture with concentric bars; possibly a few species in northeast states ***Pyganodon* (in part)**

18. Beaks not elevated above hinge line (Fig. 14.41D); about 4 species in southeastern and Gulf coast drainages, ranging into TX and the Mississippi River system . . . . ***Utterbackia***
Beaks elevated above hinge line **19**

19. West of the Continental Divide; about 6 species (Fig. 14.54G), northwestern states and southern CA, AR . . . . ***Anodonta***
East of the Continental Divide; about 3 species (Fig. 14.52D), both sides of the Appalachian Mountains (includes 1 *Anodonta*) . . . . ***Pyganodon* (in part)**

20. Hinge teeth represented only by a thickening of the hinge anterior of the beak; beak sculpture with coarse concentric ridges; glochidia in transverse gelatinous cylinders within demibranch; 2 species, 1 very widespread throughout the U.S. east of the Continental Divide (Fig. 14.52F) . . . . ***Strophitus***
Hinge teeth represented by either reduced or strong pseudocardinals; beak sculpture with low concentric ridges or double-looped; glochidia loosely packed in gravid demibranch **21**

21. Shell usually small, less than 50 mm in length, somewhat delicate, lateral teeth absent except one species (*Alasmidonta*), beak sculpture concentric or double-looped **22**
Shell larger, from 50 mm to over 200 mm, fragile to thick and strong; lateral teeth present or absent, beak sculpture double-looped **24**

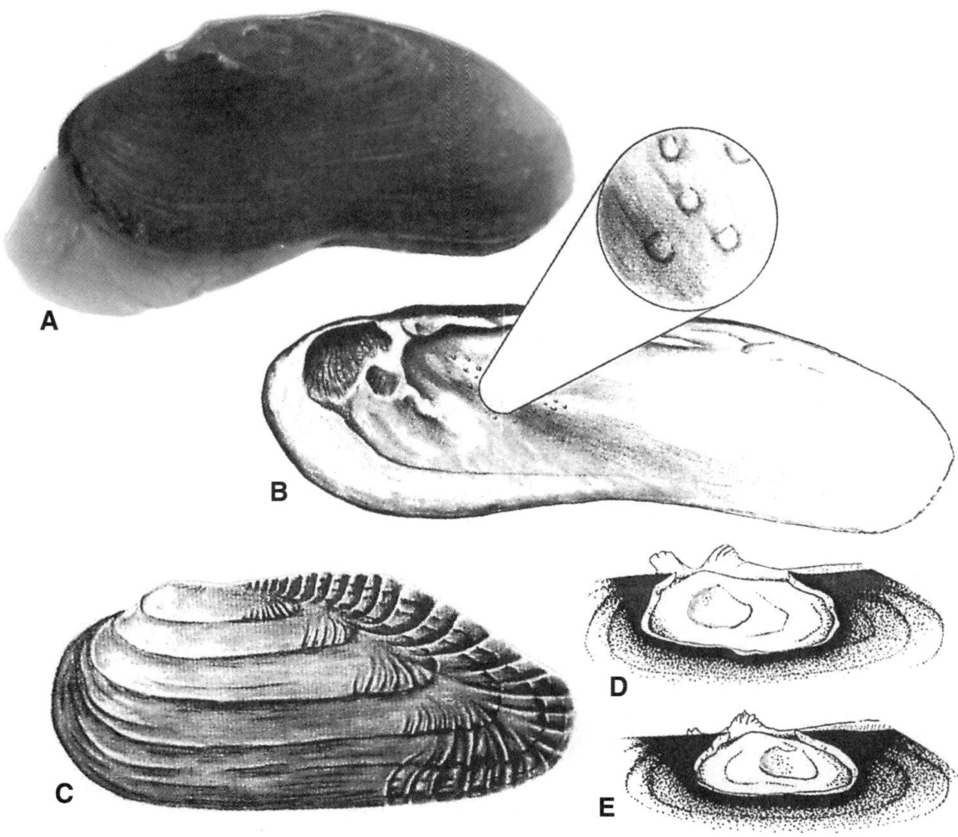

Fig. 14.50. Representative Margaritiferidae. A, *Margaritinopsis falcata*, preserved; B, *M. monodonta*, showing mantle attachment scars; C, *Pseudunio hembeli;* D, hinge line of left valve of *Margaritifera;* E, hinge line of left valve of *Margaritinopsis.* (B from Smith 1983; D, E modified from Smith 1980.)

22. Pseudocardinal teeth weak, shell elongate, posterior slope of shell without ridge **23**
Pseudocardinal teeth weak to strong, shell not elongate, posterior often truncate and with a ridge, sometimes strong (Fig. 14.53A–E); about 10 species, east and west of the Appalachian Mountains (includes *Pegias* of many authors) .... ***Alasmidonta***

23. Beak sculpture double-looped, shell without rays; found under large flat rocks in large streams; 1 species in Mississippi River drainage (Fig. 14.53I) .... ***Simpsoniconcha***
Beaks with low concentric ridges, shell occasionally with broad rays; found in open water; 1 very rare species, upper Mississippi River system .... ***Hemistena***

24. Shell ornamented with strong ridges on shell surface, beaks high and pronounced (Fig. 14.53J, K); exhalant aperture weakly papillate; 2 species scattered in interior drainages ***Arcidens***
Shell with posterior costae on posterior slope or without ornamentation, beaks low, not prominent (Fig. 14.53F–H); exhalant aperture crenulate to smooth; about 5 species generally west of the Appalachian Mountains .... ***Lasmigona***

25. Shell ovate, elliptical, to rarely quadrate, sometimes alate, strong or delicate; with few exceptions without surface sculpturing; periostracum brightly colored, varying from yellow to rarely black, often with greenish rays or other markings; female with only outer demibranch gravid, marsupium greatly swollen and differentiated externally from rest of demibranch and inner demibranch (Fig. 14.36), mantle edge before inhalant aperture broader than elsewhere and usually modified (Figs. 14.47; 14.48); shells commonly sexually dimorphic (Fig. 14.54E, F) **26**

Fig. 14.51. Representative Unionidae, Mississippi River drainage. A, *Quadrula quadrula;* B, *Cyclonaias tuberculata;* C, *Fusconaia flava;* D, *Elliptio dilatata;* E, *Pleurobema sintoxia;* F, *Ligumia subrostrata;* G, *Leptodea fragilis;* H, *Anodontoides ferussacianus;* I, *Lampsilis higginsi;* J, *Obliquaria reflexa;* K, *Plethobasis cyphus;* L, *Actinonaias ellipsiformis.* (A–L from *Freshwater Mussels of the Upper Mississippi River,* 63 pp., WI Dept. Nat. Res.)

Shell ovate, quadrate to elongate, strong to massive; surface sculpturing present or not; periostracum can be bright and rayed but is usually drab or dark colored; female with both inner and outer or just outer demibranchs gravid and forming thin pads without external modification, mantle edge plain; shells rarely sexually dimorphic **45**

26. Shell large, to 200 mm in length, dishlike, and frequently alate, usually compressed, inflated in a few species, hinge teeth lamellar, especially pseudocardinals; glochidia very small (less than 90 μm) or ax-shaped; mantle margin plain to slightly crenulate **27**

Shell small to large, ovate to quadrate but never alate, hinge teeth thick and coarse to lamellar; glochidia over 100 μm and never ax-shaped; mantle margin expanded and slightly to highly modified **28**

27. Shell typically strong, one species fragile, compressed to inflated; periostracum yellowish-green to black; glochidia ax-shaped (Fig. 14.38L); 4 species, widely distributed in the interior U.S. ***Potamilus***

Shell usually thin, fragile, and compressed (Fig. 14.51G); periostracum bright yellow to yellowish-green; glochidia not ax-shaped; 3 species widely scattered in the interior U.S. ***Leptodea***

28. Shell with surface sculpture, either thick tubercles, deep grooves or ridges, or costae on posterior portion of shell **29**

Shell surface smooth **34**

29. Shell elliptical, with thin gently curved hinge, sculpture confined to wrinkles or costae on posterior of shell **30**

Shell high, ovate, with thick, sharply curved hinge, sculpture consisting of tubercles, grooves, or ridges **31**

30. Marsupium occupying entire outer demibranch, produced into folds along ventral margin (Plate 11); mantle margin anterior to inhalant aperture with a short row of delicate papillae; 1 species east-central Mississippi River system . . . . ***Ptychobranchus* (in part)**

Marsupium restricted to middle of demibranch, greatly extended when gravid; mantle in front of inhalant aperture broad with, reportedly, filiform projections along margin; at least 5 species in Gulf coast and Mississippi River drainage systems (Fig. 14.54B) . . . . . . ***Medionidus***

31. Shell surface with a few large tuberculate prominences on middle of disk (Fig. 14.51J); marsupium in middle 2/3 of demibranch; 1 species widely distributed in Mississippi River system ***Obliquaria***

Surface sculpture and marsupium otherwise **32**

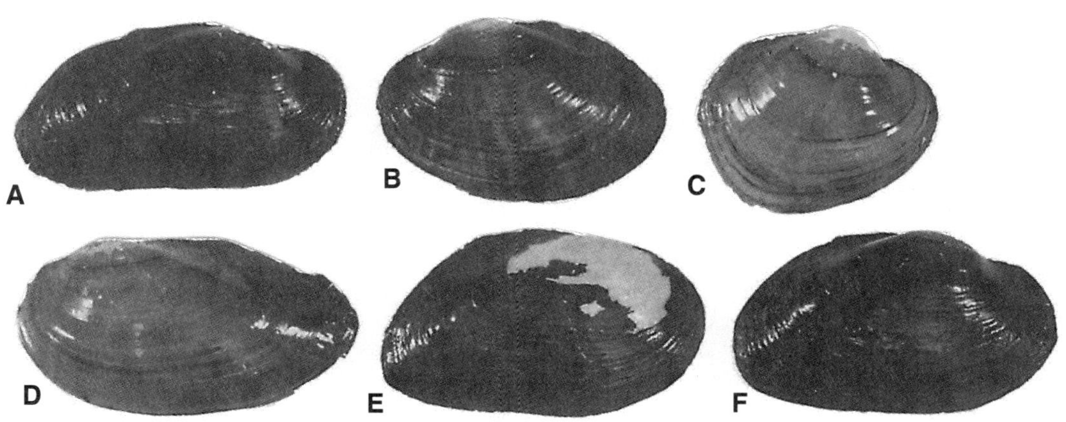

Fig. 14.52. Representative Margaritiferidae and Unionidae, Atlantic coast dranaiges. A, *Margaritifera margaritifera;* B, *Lampsilis radiata;* C, *Lampsilis cariosa;* D, *Pyganodon cataracta;* E, *Elliptio complanata;* F, *Strophitus undulatus.* (From *Freshwater Mussels of New England,* U.S. Fish Wildlife Ser., Concord, NH.)

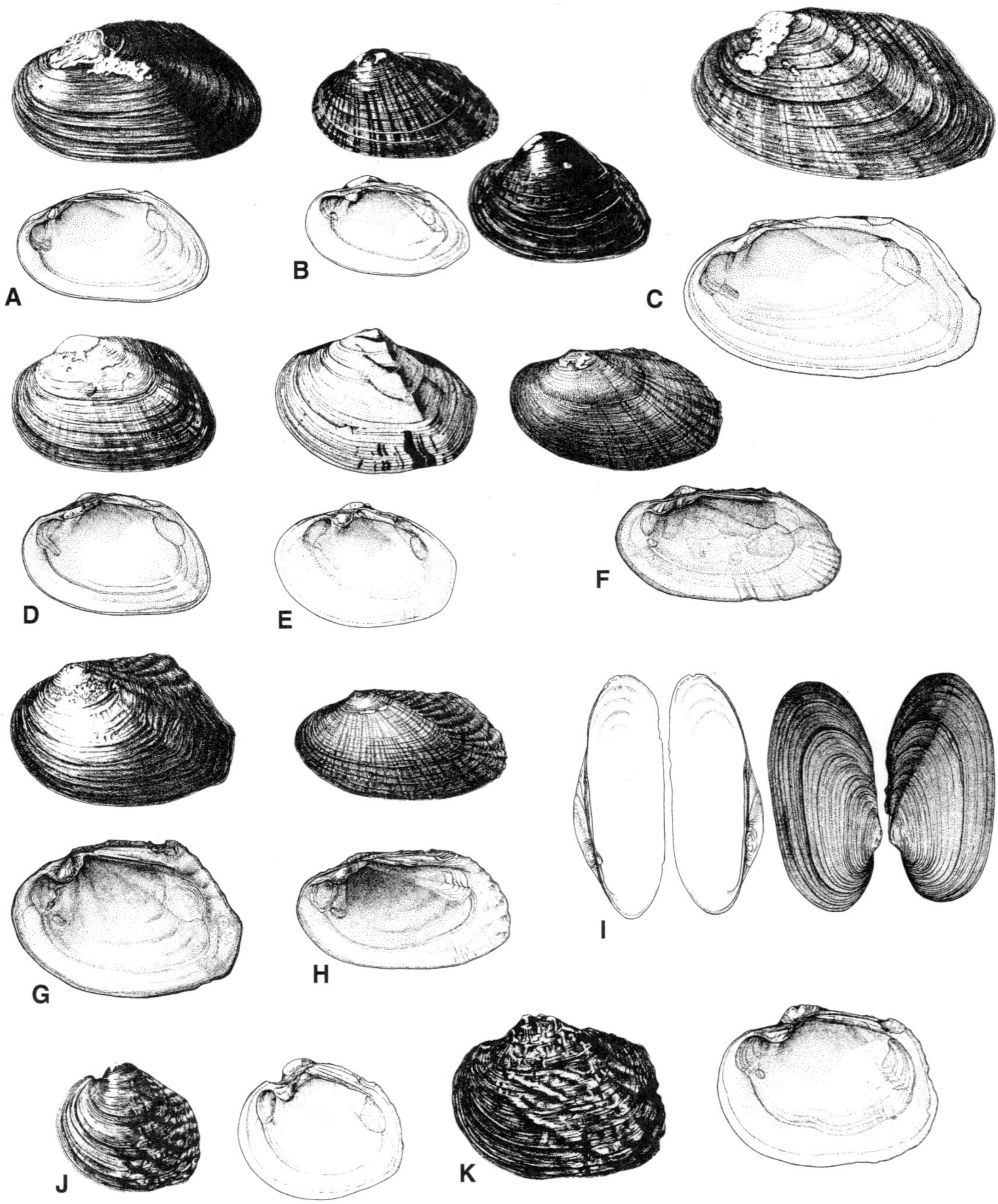

Fig. 14.53. Left and right valves of some Unionidae with "hooked glochidia." A, *Alasmidonta heterodon;* B, *A. undulata;* C, *A. varicosa;* D, *A. viridis;* E, *A.* (*Pegias*) *fabula;* F, *Lasmigona compressa;* G, *L. complanata;* H, *L. costata;* I, *Simpsoniconcha ambigua;* J, *Arcidens wheeleri;* K, *A. confragosa.* (A–E, J, K from Clarke 1981; F–I from Clarke 1985.)

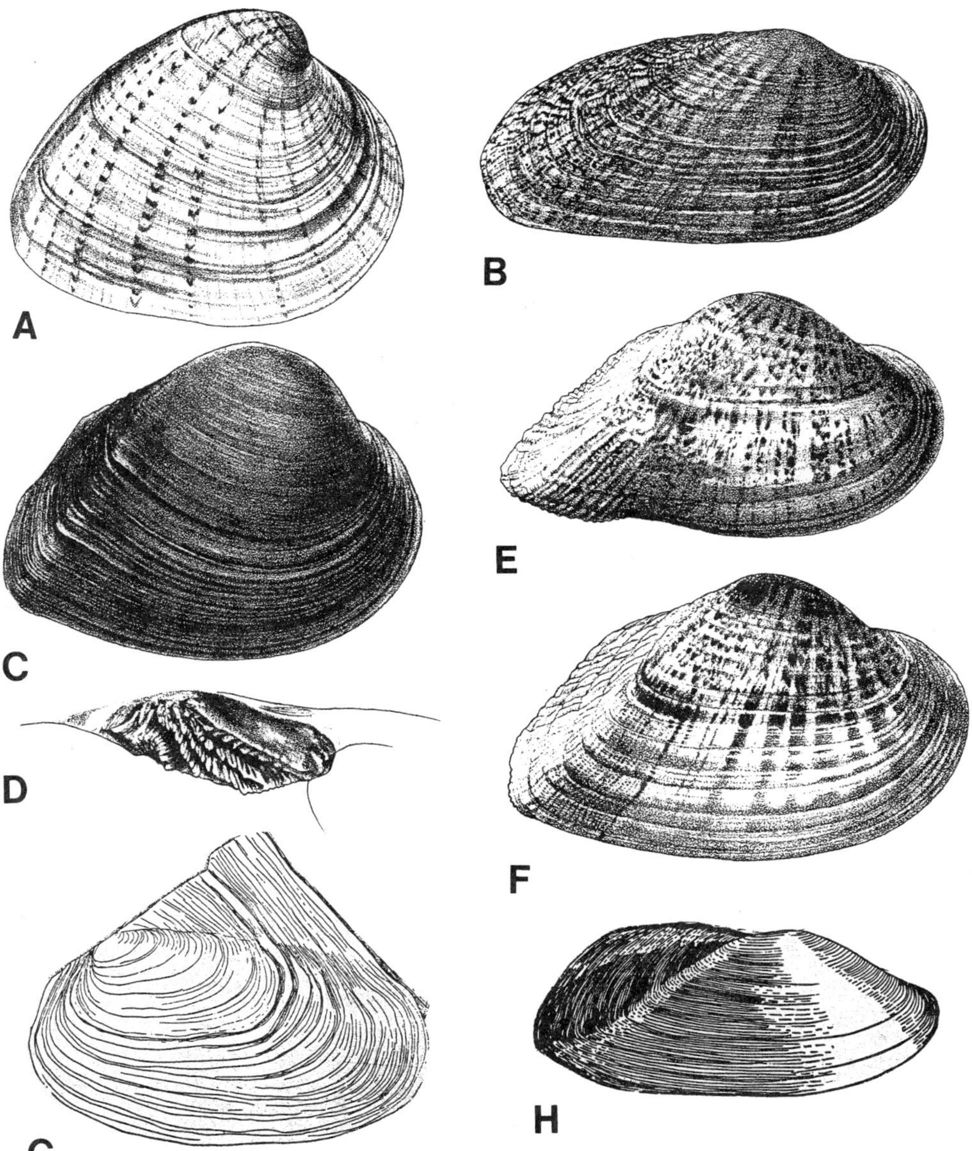

Fig. 14.54. Representative unionid genera. A, *Ellipsaria;* B, *Medionidus;* C, *Glebula;* D, pseudocardinal tooth of *Glebula;* E, female *Plagiola;* F, male *Plagiola;* G, *Anodonta;* H, *Gonidea.* (A–F from Burch 1973; G, H modified from Keep 1935.)

32. Shell with distinct sulcus on surface, bordered by tubercles (Plate 7D); gravid marsupium in anterior part of demibranch, extending well beyond ventral margin and curled posteriorly; 2 species, west and east central Mississippi River system .... ***Cyprogenia***

Shell without sulcus; marsupium on posterior half of demibranch and extending ventrally somewhat but not curled **33**

33. Shell with deep grooves radiating out onto posterior surface, ventral margin straight; marsupium restricted to posterior half of demibranch; 1 species in Tennessee River system ***Lemiox***

Shell with strong ridges running parallel with growth lines, ventral margin rounded; marsupium occupying most of demibranch and becoming folded in older females; 1 species in east-central Mississippi River system .... ***Dromus***

34. Shell quadrate, periostracum with broad, broken color rays; marsupium occupying entire outer demibranch, produced into folds along ventral margin (Plate 11); mantle margin anterior to inhalant aperture lamellate with a short row of delicate pointed papillae; shell without sexual dimorphism; 4 species in the interior United States .... ***Ptychobranchus* (in part)**
Shell ovate to elliptical or rhomboid, sexual dimorphism pronounced or not; marsupium and mantle edge otherwise **35**

35. Shell nut-shaped, tumid, ovate to elliptical (Plate 7E), with swollen or elevated beaks, hinge moderately to very thick; shell exhibiting weak sexual dimorphism; gravid marsupium fan-shaped (Plate 11) **36**
Shell subovate, elliptical, rhomboid, to elongate, with or without strong posterior ridge, hinge thin to thick, sexual dimorphism usually well marked but in some genera weak; gravid marsupium occupying all of outer demibranch or variously restricted to posterior part but not fan-shaped **38**

36. Shell round–ovate, rarely elliptical (Plate 7E); mantle edge in front of inhalant aperture reported to be thick, crenulate, and pigmented with a dark streak; about 6 species, distributed throughout the interior U.S., Gulf Coast, and St. Lawrence River system .... ***Obovaria***
Shell triangular-ovate; mantle edge in front of inhalant aperture lamellate with fine serrate denticles (Fig. 14.47D) **37**

37. Shell with a dark, patternless periostracum (Fig. 14.54C); 1 species in Gulf Coast region ***Glebula***
Shell periostracum with dark, broken rays on a lighter background (Fig. 14.54A); 1 species in the interior U.S. .... ***Ellipsaria***

38. Shell rarely subovate (1 species), usually elongate (Fig. 14.51F); mantle edge in front of inhalant aperture in female uniformly broad and either light or dark colored, with many fleshy, annulated papillae and with or without a dark pigment spot near aperture (Fig. 14.47A); 5 species scattered in the central and eastern U.S. .... ***Ligumia***
Shell not elongate, mantle edge otherwise **39**

39. Shell small, rarely over 40 mm in length, elliptical, with a dark rayless periostracum; mantle edge in front of inhalant aperture of female with a few papillae and, distal from the aperture, a large fleshy lobe or caruncle; 8 species, widely distributed in the central and eastern U.S. ***Toxolasma***
Shell larger, elliptical, rhomboid, or subovate, periostracum typically with color rays on at least part of shell; mantle without fleshy caruncle **40**

40. Shell usually less than 60 mm in length, surface of posterior half sulcate or with one or two radiating ridges, sometimes with fine wrinkles or costae; sexual dimorphism strongly pronounced; mantle in front of inhalant aperture in female forming a broad, thick lobe, inner edge papillate; many described species, most extinct; interior U.S. .... ***Epioblasma***
Shell and mantle otherwise **41**

41. Sexual dimorphism poorly expressed; female mantle margin in front of inhalant aperture not strongly differentiated; marsupium occupying one half to all of outer demibranch **42**
Sexual dimorphism distinct; female mantle strongly differentiated; marsupium occupying posterior half of demibranch only **44**

42. Shell rhomboidal, somewhat inflated, entire outer demibranch marsupial; found only in southeastern NM, TX; 1 species .... ***Cyrtonaias***
Shell elliptical or subovate, compressed, rarely inflated, marsupium occupying less than all of demibranch; widely distributed genera **43**

43. Shell strong but not heavy, posterior and ventral margin forming a sharp angle (Fig. 14.40C); marsupium in posterior half of demibranch; 4 species in the interior and southern U.S. ***Truncilla***
Shells fragile to heavy; posterior margin rounded to ventral margin; marsupium occupying most of outer demibranch; about 4 species, widely distributed in the interior U.S. ***Actinonaias***

44. Shell rarely exceeding 70 mm in length, usually thin; mantle margin on front of inhalant aperture of female subtly pigmented, with a medial expansion forming a lobe with several large, digitiform papillae (Fig. 14.48A–C), elsewhere along margin few to many small papillae present; female displays while lying on the substrate (Plate 10); many described species from the Atlantic and Gulf coasts as well as the interior U.S (Fig. 14.40B) .... ***Villosa***

Shell up to 150 mm in length, occasionally thin but usually thick to massive; mantle margin in front of inhalant aperture highly pigmented, increasing in width anteriorly and forming a free lobe, the edge with minute (rarely) to long ribbonlike or falciform extensions, a dark pigment spot often present near inhalant aperture (Fig. 14.47B, C); female displays from a partially buried position (Plate 10); about 20 or more species, distributed throughout the U.S. east of the Continental Divide (Figs. 14.40A; 14.51I; 14.52B, C) . . . . ***Lampsilis***

45. Shell heavy to massive and usually large, exceptionally reaching 250 mm in length, surface with discrete sculpturing, but without spiny projections; most genera with inner and outer demibranchs marsupial; exhalant aperture with very short and inconspicuous mantle fusion forming separate anal and supra-anal openings, or separation absent **46**

Shell thin to heavy, but not massive, rarely over 150 mm in length, most genera much smaller; shell smooth, rarely with spiny projections or costae on posterior slope; most genera with outer demibranchs marsupial only; separation of exhalant aperture into anal and supra-anal openings made by conspicuous fusion, rarely inconspicuous or absent **54**

46. Shell heavy and massive, surface with long, wavelike and uninterrupted subparallel ridges extending to posterior margin; both inner and outer demibranchs marsupial **47**

Shell solid, but not massive, ridges, if present, interrupted and accompanied by chevron-shaped bars, or with small to large nodules, grooves, or fine costae; either inner and outer or only outer demibranchs marsupial **48**

47. Sculpturing covering entire shell; 2 species in Gulf Coast and Mississippi River drainages ***Megalonaias***

Sculpturing absent from anterior of shell; 2 species in Gulf Coast and Mississippi River drainages (Plate 8A) . . . . ***Amblema***

48. Shell surface dominated by nodules or costae or both on posterior ridge **49**

Shell surface dominated by ridges or bars, which are interupted, sometimes forming a chevron-shaped pattern **52**

49. Shell with uneven perimeter, ventral margin often emarginate or interrupted by a bulge, not sexually dimorphic **50**

Shell elongate-quadrate, ventral margin sharply curved downward at junction with posterior ridge, sexually dimorphic; both inner and outer demibranchs marsupial; 1 species widespread in the interior U.S. . . . . ***Tritogonia***

50. Shell ovate to triangulate, sculpture present on middle and posterior surface; only outer demibranchs marsupial **51**

Shell broadly ovate, nodular sculpture present on entire shell, or represented by a few rows of heavy prominent nodules (Figs. 14.41B; 14.51A), or nodules very reduced; a sulcus sometimes present bordered by nodules; inner and outer demibranchs marsupial; between 9 and 18 recognized species scattered about the interior and Gulf Coast of the U.S. . . . ***Quadrula***

51. Shell ovate, beak subcentral, not elevated (Fig. 14.51B); anal aperture not separated from supra-anal by mantle fusion; 1 species in the interior U.S. . . . . ***Cyclonaias***

Shell triangular, beak anterior and pronounced (Fig. 14.51K); anal aperture separated from supra-anal by a short fusion, which is occasionally absent; 3 species in Gulf Coast and interior drainages . . . . ***Plethobasis***

52. Shell about 50 mm in length, with a series of chevron-shaped bars, often interrupted, with apex of chevron on posterior ridge; dorsal margin curved, nacre white; 3 species, Gulf Coast of AL, FL, to GA, TX . . . . ***Quincuncina***

Shell over 100 mm in length, with intersecting bars and ridges; dorsal margin straight, nacre entirely or partly purple **53**

53. Posterior ridge low, purple nacre limited to outer perimeter; exhalant aperture not subdivided; 1 species, Gulf Coast of AL, FL, into GA . . . . ***Elliptoideus***

Posterior ridge distinct (Fig. 14.41A), nacre entirely purple; exhalant aperture subdivided; 1 species, Gulf Coast to southern Mississippi River drainage . . . . ***Plectomerus***

54. Shell subovate to triangular, with a distinct posterior ridge and often a sulcus in front of the ridge (Fig. 14.51C), both inner and outer demibranchs marsupial and typically bright red in live or fresh animal; exhalant aperture subdivided by a very short mantle connection, which is sometimes absent; 13 species, most in the interior U.S., a few in Gulf Coast region and on southern Atlantic slope . . . . ***Fusconaia***

Shell high, elliptical, triangular, to quadrate and elongate, posterior ridge may be present but without sulcus, beaks either eccentrically placed or low; only outer demibranchs marsupial, white to red colored; exhalant aperture subdivided by a short but conspicuous connection, rarely absent **55**

55. Shell high, triangular or elliptical, with swollen eccentric beaks (Figs. 14.41C; 14.51E), rarely quadratotruncate with low beaks, but never quadrate or elongate; periostracum yellowish to dark brown and often with color rays of some sort; gravid demibranchs white to pinkish orange to bright orange; two very similar genera **56**

Shell quadrate, elliptical, or elongate, but not high; beaks may be elevated but never eccentric, periostracum typically dark in adults; gravid demibranch white **57**

56. Gravid demibranch orange or red; 2 species, central Appalachian region . . . . ***Lexingtonia***

Gravid demibranch white to rarely pink; over 20 species, many very similar, distributed mostly in Gulf Coast region, a small number of species ranging to southern Atlantic slope and into Mississippi River system . . . . ***Pleurobema***

57. Shell elongate–quadrate, periostracum light to dark brown; anal opening papillate; known only from Rio Grande River system, NM, TX, 1 species . . . . ***Popenaias***

Shell variably quadrate to elliptical to elongate, periostracum dark brown to black in older individuals; anal opening crenulate or papillate; widely distributed **58**

58. Shell quadrate, with a satiny, not shiny, periostracum, pseudocardinal teeth not strongly developed; exhalant aperture subdivided by a long connection, as long as anal opening, anal margin crenulate; 5 species ranging from southern Atlantic coast through Gulf Coast drainages to CO, TX, north into the Mississippi River system . . . . ***Uniomerus***

Shell quadrate, elliptical, or elongate (Figs. 14.51D; 14.52E), periostracum brown to black in older specimens, typically shiny, pseudocardinal teeth well developed, serrate; exhalant aperture subdivided by a short connection, anal opening papillate (Fig. 14.31); over 30 recognized species distributed mainly along the Atlantic coast, several ranging to the Gulf Coast and into the Mississippi and St. Lawrence River systems . . . . ***Elliptio***

REFERENCES

Aho, J. 1966. Ecological basis of the distribution of the littoral freshwater molluscs in the vicinity of Tampere, South Finland. *Ann. Zool. Fenn.* 3:287–322.

Aldridge, D. W. 1983. Physiological ecology of freshwater prosobranchs. In W. D. Russell-Hunter (ed.), *The Mollusca*, Vol. 6, *Ecology*, pp. 329–358, edited by W. D. Russel-Hunter. Academic, Orlando, FL.

Allen, W. R. 1921. Studies of the biology of freshwater mussels: experimental studies of the food relations of certain Unionidae. *Biol. Bull.* 40:210–242.

Arey, L. B. 1932. The formation and structure of the glochidial cyst. *Biol. Bull.*. 62:212–221.

Baker, F. C. 1911. The Lymnaeidae of North and Middle America. *Spec. Publ. Chica. Acad. Sci.* 3:1–539.

——. 1916. The relation of mollusks to fish in Oneida Lake. *N.Y. State Coll. For. Tech. Publ.* 4:1–366.

——. 1918a. The productivity of invertebrate fish food on the bottom of Oneida Lake, with special reference to mollusks. *N.Y. State Coll. For. Tech. Publ.*. 9:1–233.

——. 1918b. The relation of shellfish to fish in Oneida Lake, New York. *N. Y. State Coll. For. Circ.* 21:1–34.

——. 1928a. The fresh water Mollusca of Wisconsin, Part I: Gastropoda. *Bull. Wis. Acad. Sci., Arts and Lett.* 70:1–507.

——. 1928b. The fresh water Mollusca of Wisconsin, Part II: Pelecypoda. *Bull. Wis. Acad. Sci., Arts and Lett.* 70:1–495.

——. 1945. *The Molluscan Family Planorbidae.* 530 pp. Univ. Illinois Press, Urbana, IL.

Basch, P. F. 1959. The anatomy of *Laevapex fuscus*, a freshwater limpet (Gastropoda: Pulmonata). *Misc. Publ. Mus. Zool. Univ. Mich.* 108:1–56.

——. 1963. A review of the recent freshwater limpet snails of North America (Mollusca: Pulmonata). *Bull. Mus. Comp. Zool.* 129:401–461.

Bedford, J. W., et al. 1968. The freshwater mussel as a biological monitor of pesticide concentrations in a lotic environment. *Limnol. Oceanogr.* 13:118–126.

Berg, K., and K. W. Ockelmann. 1959. The respiration of freshwater snails. *J. Exp. Biol.* 36:690–708.

Berry, E. G. 1943. The Amnicolidae of Michigan: distribution, ecology, and taxonomy. *Mus. Zool. Univ. Mich. Misc. Publ.* 57:1–68.

Boss, K.J. 1974. Oblomovism in the Mollusca. *Trans. Am. Microsc. Soc.* 93:460–481.

Bovbjerg, R. V. 1968. Responses to food in lymnaeid snails. *Physiol. Zool.* 4:412–423.

——. 1975. Dispersal and dispersion of pond snails in an experimental environment varying to three factors, singly and in combination. *Physiol. Zool.* 48:203–215.

Boycott, A.E. 1936. The habitats of fresh-water Mollusca in Britain. *J. Anim. Ecol.* 5:116–186.

Brown, C. J. D., et al. 1938. The size of certain naiades from western Lake Erie in relation to shoal exposure. *Am. Midl. Nat.* 19:682–701.

Bruenderman, S A., and R. J. Neves. 1993. Life history of the endangered fine-rayed pigtoe *Fusconaia cuneolus* (Bivalvia: Unionidae) in the Clinch River, Virginia. *Am. Malacol. Bull.* 10:83–91.

Bryce, G. W. 1970. Rediscovery of the limpet, *Acroloxus coloradensis* (Basommatophora: Acroloxidae), in Colorado. *Nautilus* 83:105–108.

Buchanan, A. C. 1980. Mussels (Naiades) of the Meramec River basin, Missouri. *Missouri Depart. Conserv. Aquatic Ser.* 17:1–67.

Burch, J. B. 1972. Freshwater sphaeriacean clams (Mollusca: Pelecypoda) of North America. *Biota of Freshwater Ecosystems, Identification Manual* 3:1–31.

——. 1973. Freshwater Unionacean clams (Mollusca: Pelecypoda) of North America. *Biota of Freshwater Ecosystems, Indentification Manual* 11:1–176.

——. 1975a. *Freshwater Sphaeriacean Clams (Mollusca: Peletypoda) of North America.* 99 pp. Malacological Publications, Hamburg, MI.

——. 1975b. *Freshwater Unionacean Clams (Mollusca: Pelecypoda) of North America.* 204 pp. Malacological Publications, Hamburg, MI.

——. 1979. Genera and subgenera of recent freshwater gastropods of North America (north of Mexico). *Malacol. Rev.* 12:96–99.

——. 1982. *Freshwater snails (Mollusca: Gastropoda) of North America.* EPA Report 600/3-82-026, 294 pp.

Burky, A. J. 1971. Biomass turnover, respiration, and interpopulation variation in the stream limpet *Ferrissia rivularis* (Say). *Ecol. Monogr.* 41:235–251.

——. 1983. Physiological ecology of freshwater bivalves. In: *The Mollusca,* Vol. 6: *Ecology,* pp. 281–327, edited by W. D. Russell-Hunter. Academic, Orlando, FL.

Burky, A. J., et al. 1979. The ratio of calcareous and organic shell components of freshwater sphaeriid clams in relation to water hardness and trophic conditions. *J. Moll. Stud.* 45:312–321.

Carriker, M. R. 1946. Morphology of the alimentary system of the snail *Lymnaea stagnalis appressa* Say. *Trans. Wis. Acad. Sci. Arts Lett.* 38:1–88.

Chamberlin, R. V., and D. T. Jones. 1929. A descriptive catalog of the Mollusca of Utah. *Bull. Univ. Utah* 19:1–203.

Cheatum, E. P. 1934. Limnological investigations on respiration, annual migratory cycle, and other related phenomena in fresh-water pulmonate snails. *Trans. Am. Microsc. Soc.* 53: 348–407.

Churchill, E. P., Jr., and S. I. Lewis. 1924. Food and feeding in fresh-water mussels. *Bull. U. S. Bur. Fish.* 39:437–471.

Clampitt, P. T. 1970. Comparative ecology of the snails *Physa gyrina* and *Physa integra* (Basommatophora: Physidae). *Malacologia* 10: 113–151.

Clarke, A. H. (ed.). 1970. Papers on the rare and endangered mollusks of North America. *Malacologia* 10:1–56.

——. 1973. The freshwater mollusks of the Canadian interior basin. *Malacologia* 13:1–509.

——. 1979. Sphaeriidae as indicators of trophic lake stages. *Nautilus* 94:178–184.

——. 1981. The Tribe Alasmidontini (Unionidae: Anodontinae), Part 1: *Pegias, Alasmidonta,* and *Arcidens. Smithson. Contrib. Zool.* 326:1–101.

——. 1985. The Tribe Alasmidontini (Unionidae: Anodontinae), Part II: *Lasmigona* and *Simpsonaias. Smithson. Contrib. Zool.* 399:1–75.

Clarke, A. H., and C. O. Berg. 1959. The freshwater mussels of central New York. *Mem. Cornell Univ. Agr. Exp. Sta.* 367:1–79.

Coker, R. E. 1919. Freshwater mussels and mussel industries of the United States. *Bull. U. S. Bur. Fish.* 36:13–89.

Coker, R. E., et al. 1921. Natural history and propagation of fresh-water mussels. *Bull. U.S. Bur. Fish.* 37:76–181.

Counts, C. L., III. 1986. The zoogeography and history of the invasion of the United States by *Corbicula fluminea* (Bivlvia: Corbiculidae). *Am. Malacol. Bull. Spec. Ed.* 2:7–39.

Cummings, K. S., and C. A. Mayer. 1992. Field guide to freshwater mussels of the midwest. *Ill. Nat. Hist. Surv. Man.* 5:1–194.

——. 1997. Distributional checklist and status of Illinois freshwater mussels (Mollusca: Unionacea). In: *Conservation and Management of Freshwater Mussels II: Initiatives for the Future*, pp. 129–145, edited by K. S. Cummings et al. Proc. UMRCC Symposium, 16–18 October 1995, St. Louis, Missouri.

Cummings, K. S., et al. (editors). 1997. *Conservation and Management of Freshwater Mussels II: Initiatives for the Future*, 293 pp. Proc. UMRCC Symposium, 16–18 October 1995, St. Louis, Missouri.

Cvancara, A. M. 1983. *Aquatic Mollusks of North Dakota*. North Dakota Geol. Surv. Reports of Investigations 78. 141 pp.

Davis, G. M., and M. Mazurkiewicz. 1985. Systematics of *Cincinnatia winkleyi* (Gastropoda: Hydrobiidae). *Proc. Acad. Natl. Sci. USA* 137:28–47.

Davis, G. M., and M. McKee. 1989. A new species of *Heleobops* (Prosobranchia: Hydrobiidae: Littorininae) from Maryland. *Proc. Natl. Acad. Sci.* 141:213–249.

Dawson, J. 1911. The biology of *Physa*. *Behavior Monogr.* 1:1–120.

Dazo, B. C. 1965. The morphology and natural history of *Pleurocera acuta* and *Goniobasis livescens* (Gastropoda: Ceritheacea: Pleuroceridae). *Malacologia* 3:1–80.

DeWitt, R. M. 1954. Reproduction, embryonic development, and growth in the pond snail, *Physa gyrina* Say. *Trans. Am. Microsc. Soc.* 73:124–137.

Dillon, R. T., Jr. 1991. Karyotypic evolution in pleeurocerid snails, II: *Pleurocera*, *Goniobasis*, and *Juga*. *Malacologia* 33:339–344.

Dillon, R. T., Jr., and S. A. Ahlstedt. 1997. Verification of the specific status of the endangered Anthony's river snail, *Athearnia anthonyi*, using allozyme electrophoresis. *Nautilus* 110:97–101.

Dundee, D. S. 1977. Catalog of introduced molluscs of eastern North America (north of Mexico). *Sterkiana* 55:1–37.

Dussart, G. B. J. 1979. *Sphaerium corneurn* (L.) and *Pisidium* spp. Pfeiffer: the ecology of freshwater bivalve molluscs in relation to water chemistry. *J. Moll. Stud.* 45:19–34.

Fichtel, C., and D. G. Smith. 1995. The freshwater mussels of Vermont. *Vermont Fish and Wildlife Dep. Tech. Rpt.* 18:1–54.

Foster, T. D. 1932. Observations on the life historv of a fingernail shell of the genus *Sphaerium*. *J. Morphol.* 53:473–497.

Fuller, S. L. H., and I. Brynildson. 1985. *Freshwater Mussels of the Upper Mississippi River.* 63 pp. Wisconsin Dept. Nat. Res., Madison.

Gale, W. F., and R. L. Lowe. 1971. Phytoplankton ingestion by the fingernail clam, *Sphaerium transversum* (Say) in Pool 19, Mississippi River. *Ecology* 52:507–513.

Gillespie, D. M. 1968. Population studies of four species of molluscs in the Madison River, Yellowstone National Park. *Limnol. Oceanogr.* 14:101–114.

Gilmore, R. J. 1917. Notes on reproduction and growth in certain viviparous in mussels of the family Sphaeriidae. *Nautilus* 31:16–30.

Goodrich, C. 1922. The Anculosae of the Alabama River drainage. *Mus. Zool. Univ. Mich. Misc. Publ.* 7:1–57.

——. 1924. The genus *Gyrotoma*. *Mus. Zool. Univ. Mich. Misc. Publ.* 12:1–32.

——. 1941. Distribution of the gastropods of the Cahaba River, Alabama. *Mus. Zool. Univ. Mich. Occas. Pap.* 428:1–30.

Goodrich, C., and H. van der Schalie. 1939. Aquatic mollusks of the Upper Peninsula of Michigan. *Misc. Publ. Mus. Zool. Univ. Mich.* 43:1–45.

——. 1944. A revision of the Mollusca of Indiana. *Am. Midl. Nat.* 32:257–326.

Gordon, M. E., and J. B. Layzer. 1989. Mussels (Bivalvia: Unionoidea) of the Cumberland River. *U. S. Fish and Wildlife Service Biol. Rep.* 89(15):1–99.

——. 1993. Glochidial host of *Alasmidonta atropurpurea* (Bivalvia: Unionoidea, Unionidae). *Trans. Am. Microsc. Soc.* 112:145–150.

Gordon, M. E., et al. 1978. *Baeoctenus bicolor* (Diptera: Chironomidae) parasitic in unionid bivalve molluscs, and notes on other chironomid–bivalve associations. *J. Fish. Res. Bd. Can.* 35:154–157.

Goudreau, S. E., et al. 1993. Effects of wastewater treatment plant effluents on freshwater mollusks in the upper Clinch River, Virginia, USA. *Hydrobiologia* 252:211–230.

Green, R. H. 1980. Role of a unionid clam population in the calcium budget of a small Arctic lake. *Can. J. Fish. Aquat. Sci.* 37:219–224.

Greenaway, P. 1971. Calcium regulation in the freshwater mollusc, *Limnaea stagnalis* (L.) (Gastropoda: Pulmonata), 1: the effect of internal and external calcium concentration. *J. Exp. Biol.* 54:199–214.

Gunter, G. 1936. Radular movements in gastropods. *J. Wash. Acad. Sci.* 26:361–365.

Haag, W. R., and M. L. Warren, Jr. 1997. Host fishes and reproductive biology of 6 freshwater mussel species from the Mobile Basin, USA. *J. North Am. Benthol. Soc.* 16:576–585.

——. 1999. Mantle displays of freshwater mussels elicit attacks from fish. *Freshwater Biol.* 42:35–40.

Haag, W. R., et al. 1995. An extraordinary reproductive strategy in freshwater bivalves: prey mimicry to facilitate larval dispersal. *Am. Midl. Nat.* 34: 471–476.

——. 1999. Host fishes and host-attracting behavior of *Lampsilis altilis* and *Villosa vibex* (Bivalvia: Unionidae). *Am. Midl. Nat.* 141:149–157.

Hanna, G. D. 1966. Introduced mollusks of western North America. *Occas. Pap. Calif. Acad. Sci.* 48:1–108.

Harman, W. N. 1972. Benthic substrates: their effect on fresh-water Mollusca. *Ecology* 53:271–277.

Harman, W. N., and C. O. Berg. 1970. Freshwater Mollusca of the Finger Lakes region of New York. *Ohio J. Sci.* 70:146–170.

——. 1971. The freshwater snails of central New York with illustrated keys to the genera and species. *Search, Agriculture* 1:1–68.

Harman, W. N., and J. L, Forney. 1970. Fifty years of change in the molluscan fauna of Oneida Lake, New York. *Limnol. Oceanogr.* 15:454–460.

Harris, J. L., and M. E. Gordon. 1990. *Arkansas Mussels.* 32 pp. Arkansas Game and Fish Commission, Little Rock.

Hartfield, P. D., and R. G. Rummel. 1985. Freshwater mussels (Unionidae) of the Big Black River, Mississippi. *Nautilus* 99:116–119.

Heard, R. W. 1982. *Guide to Common Tidal Marsh Invertebrates of the Northeastern Gulf of Mexico.* 82 pp. Mississippi Alabama Sea Grant Consortium, MASGP-79-004.

Heard, W. H. 1962. Distribution of Sphaeriidae (Pelecypoda) in Michigan, USA. *Malacologia* 1:139–160.

——. 1965. Comparative life histories of North American pill clams (Sphaeriidae: *Pisidium*). *Malacologia* 2:381–411.

——. 1970. Reproduction of fingernail clams (Sphaeriidae: *Sphaerium* and *Musculium*). *Malacologia* 10:421–455.

——. 1975. Sexuality and other aspects of reproduction in *Anodonta* (Pelecypoda: Unionidae). *Malacologia* 15:81–103.

——. 1979. Identification manual of the freshwater clams of Florida. *Florida Dept. Environ. Regul., Tech. Ser.* 4(2):1–83.

Henderson, J. 1924. Mollusca of Colorado, Utah, Montana, Idaho, and Wyoming. *Univ. Colo. Studies* 13:65–223.

——. 1929. Non-marine Mollusca of Oregon and Washington. *Univ. Colo. Stud.* 17:47–190.

——. 1936a. Mollusca of Colorado, Utah, Montana, Idaho, and Wyoming: supplement. *Univ. Colo. Stud.* 23:81–145.

——. 1936b. The non-marine Mollusca of Oregon and Washington. Supplement, *Univ. Colo. Stud.*. 251–280.

Herrington, H. B. 1962. A revision of the Sphaeriidae of North America (Mollusca: Pelecypoda). *Univ. Mich. Mus. Zool. Misc. Publ.* 118:1–74.

Hershler, R. 1989. Springsnails (Gastropoda: Hydrobiidae) of Owens and Amargosa River (exclusive of Ash Meadows) drainages, Death Valley system, California–Nevada. *Proc. Biol. Soc. Wash.* 102:176–248.

——. 1994. A review of the North American freshwater snail genus *Pyrgulopsis* (Hydrobiidae). *Smithson. Contrib. Zool.* 554:1–115.

——. 1995. New freshwater snails of the genus *Pyrgulopsis* (Rissoidea: Hydrobiidae) from California. *Veliger* 38:343–373.

Hershler, R., and T. J. Frest. 1996. A review of the North American freshwater snail genus *Fluminicola* (Hydrobiidae). *Smithson. Contrib. Zool.* 583:1–41.

Hershler, R., and J. R. Holsinger. 1990. Zoogeography of North American hydrobiid cavesnails. *Stygologia* 5:5–16.

Hershler, R., and J. J. Landye. 1988. Arizona Hydrobiidae (Prosobranchia: Rissoacea). *Smithson. Contrib. Zool.* 459:1–63.

Herschler, R., and G. Longley. 1986. Phreatic hydrobiids (Gastropoda: Prosobranchia) from the Edwards (Balcones Fault Zone) aquifer, south-central Texas. *Malacologia* 27:127–172.

Herschler, R., and F. G. Thompson. 1987. North American Hydrobiidae (Gastropoda: Rissoacea): redescription and systematic relationships of *Tryonia* Stimpson, 1965 and *Pyrgulopsis* Call and Pilsbry, 1886. *Nautilus* 101:25–32.

——. 1990. *Antrorbis breweri*, a new genus and species of hydrobiid cavesnail (Gastropoda) from Coosa River basin, northeastern Alabama. *Proc. Biol. Soc. Wash.* 103:197–204.

——. 1992. A review of the aquatic gastropod subfamily Cochliopinae (Prosobranchia: Hydrobiidae). *Malacol. Rev., Suppl.* 5:1–140.

——. 1996. Redescription of *Paludina integra* Say, 1821, type species of genus *Cincinnatia* (Gastropoda: Hydrobiidae). *J. Moll. Stud.* 62: 33–55.

Hershler, R., et al. 1990a. A revision of the North American freshwater snail genus *Fontigens* (Prosobranchia: Hydrobiidae). *Smithson. Contrib. Zool.* 509:1–49.

——. 1990b. Rediscovery of *Tulotoma magnifica* (Conrad) (Gastropoda: Viviparidae). *Proc. Biol. Soc. Wash.* 103:815–824.

——. 1994. Two new genera of hydrobiid snails (Prosobranchai: Rissoidea) from the northwestern United States. *Veliger* 37:221–243.

Hoff, C. C. 1940. Anatomy of the ancylid snail, *Ferrissia tarda* (Say). *Trans. Am. Microsc. Soc.* 59:224–242.

Hoggarth, M. A. 1992. An examination of the glochidia–host relationships reported in the literature for North American species of Unionacea (Mollusca: Bivalvia). *Malacol. Data Net* 3: 1–30.

Hornbach, D. J., and D. L. Childers. 1987. Life-history variation in a stream population of *Musculium partumeium* (Bivalvia: Pisidiidae). *J. North Am. Benthol. Soc.* 5:263–271.

Howells, R. G. 1997. New fish hosts for nine freshwater mussels (Bivalvia: Unionidae) in Texas. *Texas J. Sci.* 49:255–258.

Howells, R. G., et al. 1996. *Freshwater Mussels of Texas.* 218 pp. Texas Parks and Wildlife, Austin.

Hubendick, B, 1951. Recent Lymnaeidae: their variation, morphology, taxonomy, nomenclature, and distribution. *K. Svenska Vetensho. Handl.* 3:1–225.

Hudson, R. G., and B. G. Isom. 1984. Rearing juveniles of the freshwater mussels (Unionidae) in a laboratory setting. *Nautilus* 98:129–137.

Hunter, R. D. 1990. Effects of low pH and low calcium concentration on the pulmonate snail *Planorbella trivolvis:* a laboratory study. *Can. J. Zool.* 68:1578–1583.

Imlay M. J. 1973. Effects of potassium on survival and distribution of freshwater mussels. *Malacologia* 12:97–113.

Isom, B. G. 1969. The mussel resources of the Tennessee River. *Malacologia.* 7:397–425.

——. 1986. Historical review of Asiatic clam (*Corbicula*) invasion and biofouling of waters and industries in the Americas. *Am. Malacol. Bull. Spec. Ed.* 2:1–6.

Isom, B. G., and R. G. Hudson. 1982. In vitro cultures of parasitic freshwater mussel glochidia. *Nautilus* 96:147–151.

Isom, B. G., and P. Yokley. 1968. The mussel fauna of Duck River in Tennessee, 1965. *Am. Midl. Nat.* 80:34–42.

Johnson, R. I. 1978. Systematics and zoogeography of *Plagiola* (+*Dysnomia* + *Epioblasma*), an almost extinct genus of freshwater mussels (Bivalvia: Unionidae) from Middle North America. *Bull. Mus. Comp. Zool.* 148:239–320.

——. 1980. Zoogeography of North American Unionacea (Mollusca: Bivalvia) north of the maximum Pleistocene glaciation. *Bull. Mus. Comp. Zool..* 149:77–189.

Jokinen, E. H. 1978, The aestivation pattern of a population of *Lymnaea elodes* (Say) (Gastropoda: Lymnaeidae). *Am. Midl. Nat.* 100:43–53.

——. 1982. *Cipangopaludina chinensis* (Gastropoda: Viviparidae) in North America. *Nautilus* 96:89–95.

——. 1983. The freshwater snails of Connecticut. *Bull. State Geol. Nat. Hist. Surv. (Connecticut)* 109:1–83.

——. 1985. Comparative life history patterns within a littoral zone snail community. *Verh. Int. Verein. Limnol.* 22:3292–3299.

——. 1987. Structure of freshwater snail communities: species–area relationships and incidence categories. *Am. Malacol. Bull.* 5:9–19.

——. 1992. The freshwater snails (Mollusca: Gastropoda) of New York State. *Bull. N. Y. State Mus.* 482:1–112.

Jones, R. O. 1949. Propagation of fresh-water mussels. *Prog. Fish-Cult.* 12:13–25.

Jones, W. C., and B. A. Branson. 1964. The radula, genital system, and external morphology in *Mudalia potosiensis* (Lea) 1841 (Gastropoda: Prosobranchiata: Pleuroceridae) with life history notes. *Trans. Am. Microsc. Soc.* 83:41–62.

Kabat, A. R., and R. Hershler. 1993. The prosobranch snail family Hydrobiidae (Gastropoda:

Rissoidea): review of classification and supraspecific taxa. *Smithson. Contrib. Zool.* 547:1–94.

Keep, J. 1935. *West Coast Shells.* 350 pp. Stanford Univ. Press, Palo Alto, CA.

Kesler et al. 1986. Trophic preferences and feeding morphology of two pulmonate snail species from a small New England pond, U.S.A. *Can. J. Zool.* 64:2570–2575.

Kunz, G. F. 1898. The freshwater pearls and pearl fisheries of the United States. *Bull. U. S. Fish Comm.* 17:373–426.

Lauritsen, D. D. 1986. Filter-feeding in *Corbicula fluminea* and its effect on seston removal. *J. North Am. Benthol. Soc.* 5:165–172.

Lefevre, G., and W. C. Curtis. 1912. Studies of the reproduction and artificial propagation of freshwater mussels. *Bull. U. S. Bur. Fish.* 30: 105–201.

Leonard, A. B. 1959. Handbook of gastropods in Kansas. *Univ. Kansas State Biol. Surv. Misc. Publ.* 20:1–224.

Lopez, G. R., and I. J. Holopainen. 1987. Interstitial suspension-feeding by *Pisidium* spp. (Pisidiidae: Bivalvia): a new guild in the lentic benthos. *Bull. Am. Malacol. Union* 5:21–29.

Macan, T. T. 1950. Ecology of fresh-water Mollusca in the English Lake District. *Anim. Ecol.* 19:124–146.

——. 1960. A key to the British fresh- and brackish-water gastropods. *Freshwater Biol. Assoc. Sci. Publ.* 13:1–44.

Machin, J. 1975. Water relationships. In: *Pulmonates,* Vol. 1: *Functional Anatomy and Physiology,* pp. 105–164, edited by V. Fretter and J. Peake. Academic, London.

Mackie, G. L. 1979. Dispersal mechanisms in Sphaeriidae (Mollusca: Bivalvia). *Bull. Am. Malacol. Union* (1979):17–21.

——. 1981. Nearctic freshwater Sphaeriacea (Bivalvia). *Bull. Am. Malacol. Union* 50:49–52.

——. 1986. Adaptations of Pisidiidae (Heterodonta: Corbulacea) to freshwater habitats. *Am. Malacol. Bull. Spec. Ed.* 2:223–229.

Mackie, G. L., et al. 1980. *A Guide to Freshwater Mollusks of the Lauretian Great lakes with Special Emphasis on the Genus Pisidium.* EPA Report 600/3-80-068, 144 pp.

MacNeill, D. B. 1992. *Identification of Juvenile Dreissena polymorpha and Mytilopsis leucophaeata.* NY Sea Grant and Cornell Coop. Exten. pamphlet.

Magruder, S. R. 1935. The anatomy of the freshwater prosobranchiate gastropod, *Pleurocera canaliculatum undulatum* (Say). *Am. Midl. Nat.* 16:883–912.

Mäkäla, T. P., and A. O. J. Oikari. 1992. The effects of low pH on the ionic balance in freshwater mussel *Anodonta anatina* L. *Ann. Zool. Fennici* 29:169–175.

Malek, E. A. 1985. Snail hosts of schistosomiasis and other snail-transmitted diseases in tropical America: a manual. *Pan Am. Health Organ., Sci. Publ.* 478:1–325.

Malone, C. R. 1965. Dispersal of aquatic gastropods via the intestinal tract of water birds. *Nautilus* 78:135–139.

Mathiak, H. A. 1979. *A River Survey of the Unionid Mussels of Wisconsin.* 76 pp. Sand Shell Press, Horizon, WI.

Matteson, M. R., and R. W. Dexter. 1966. Changes in pelecypod populations in Salt Fork of Big Vermilion River, Illinois, 1918–1962. *Nautilus* 79:96–101.

McKee, P. M., and G. L. Mackie. 1983. Respiratory adaptations of the fingernail clams *Sphaerium occidentale* and *Musculium securis* to ephemeral habitats. *Can. J. Fish. Aquatic Sci.* 40:783–791.

McMahon, R. F. 1975. Effects of artificially elevated water temperatures on the growth, reproduction and life cycles of a natural population of *Physa virgata* Gould. *Ecology* 56: 1167–1175.

——. 1982. The occurrence and spread of the introduced Asiatic freshwater clam, *Corbicula fluminea* (Muller), in North America: 1924–1982. *Nautilus* 96:134–141.

——. 1983. Physiological ecology of freshwater pulmonates. In: *The Mollusca,* Vol. 6, *Ecology,* edited by W. D. Russell-Hunter, pp. 359–430. Academic, Orlando.

McNeil, C. W. 1963. Winter survival of *Stagnicola palustris nuttalliana* and *Physa propinqua. Ecology* 44:187–191.

Meier-Brook, C. 1969. Substrate relations in some *Pisidium* species (Eulamellibranchiata: Sphaeriidae). *Malacologia* 9:121–125.

Michaelson, D. L., and R. J. Neves. 1995. Life history and habitat of the endangered dwarf wedgemussel *Alasmidonta heterodon* (Bivalvia: Unionidae). *J. North Am. Benthol. Soc.* 14:324–340.

Mikkelsen, P. S. 1985. A rapid method for slide mounting of minute radulae, with a bibliography of radula mounting techniques. *Nautilus* 99:62–65.

Monk, C. R. 1925. The anatomy and life-history of a fresh-water mollusk of the genus *Sphaerium*. *J. Morphol. Physiol.* 45:473–501.

Murphy, G. 1942. Relationship of the fresh-water mussel to trout in the Truckee River. *Calif. Fish Game* 28:89–102,

Murray, H. D., and E. C. Roy, 1968. Checklist of fresh-water and land mollusks of Texas. *Sterkiana* 30:25–42.

Nalepa, T. F., and D. W. Schloesser (eds.). 1993. *Zebra Mussels—Biology, Impacts, and Control.* 810 pp. CRC Press, Boca Raton, FL.

Nekrassow, A. D. 1928. Vergleichende Morphologie der Laiche von Süsswasser-Gastropoden. *Z. Morphol. Oekol. Tiere* 13:1–35.

Nelson, T. C. 1918. On the origin, nature, and function of the crystalline style of lamellibranchs. *J. Morphol.* 31:53–111.

Neves, R. J. 1987. Recent die-offs of freshwater mussels in the United States: an overview. In: *Proc. Workshop on Die-offs of Freshwater Mussels in the United States*, pp. 7–18, edited by R. J. Neves. U. S. Fish and Wildlife Service, Davenport, IA.

Neves, R. J., and M. C. Odom. 1989. Muskrat predation on endangered freshwater mussels in Virginia. *J. Wildl. Manage.* 53:934–941.

Noland, L. E., and E. Reichel. 1943. Life cycle of *Lymnaea stagnalis* completed at room temperature without access to air. *Nautilus* 57:8–13.

Oesch, R. D. 1984. *Missouri Naiades—A Guide to the Mussels of Missouri.* 270 pp. Missouri Dept. Conserv., Jefferson City.

Ortmann, A. E. 1919. A monograph of the Naiades of Pennsylvania, Part III: systematic account of the genera and species. *Mem. Carnegie Mus.* 8:1–384.

Parmalee, P. W., and M. H. Hughes. 1993. Freshwater mussels (Mollusca: Pelecypoda: Unionidae) of Tellico Lake: twelve years after impoundment of the Little Tennessee River. *Ann. Carnegie Mus.* 62:81–93.

Perera, G., and J. G. Walls. 1996. *Apple Snails in the Aquarium.* 121 pp. T.F.H., Neptune City, NJ.

Pilsbry, H. E. 1934. Review of the Planorbidae of Florida, with notes on other members of the family. *Proc. Acad. Natl. Sci. USA* 86:29–66.

Pip, E. 1987. Species richness of freshwater gastropod communities in central North America. *J. Moll. Stud.* 53:163–170.

Prime, T. 1865. Monograph of American Corbiculidae (Recent and fossil). *Smithson. Misc. Coll.* 145:1–80.

Reigle, N. J. 1967. An occurrence of *Anodonta* (Mollusca, Pelecypoda) in deep water. *Am. Midl. Nat.* 78:530–531.

Richards, C. S. 1964. Studies on Puerto Rican Physidae. *Pub. Health Rpts.* 79:1025–1029.

Roberts, A. D., and M. C. Barnhart. 1999. Effects of temperature, pH, and CO_2 on transformation of the glochidia of *Anodonta suborbiculata* on fish hosts and in vitro. *J. North Am. Benthol. Soc.* 18:477–487.

Roscoe, E. J., and S. Redelings, 1964. The ecology of the fresh-water pearl mussel *Margaritifera margaritifera* (L.). *Sterkiana* 16:19–32.

Ross, L. F., and A. D, Harrison. 1977. Effects of environmental calcium deprivation on the egg masses of *Physa marmorata* Guilding (Gastropoda: Physidae) and *Biomphalaria glabrata* Say (Gastropoda: Planorbidae), *Hydrobiologia* 55: 45–48.

Russell-Hunter, R. D., and W. W. Lull. 1977. Physiologic and environmental factors influencing the calcium-to-tissue ratio in populations of three species of freshwater pulmonate snails. *Oecologia* 29:205–218.

Russell-Hunter, R. D., et al. 1984. Overwinter tissue degrowth in natural populations of freshwater pulmonate snails (*Helisoma trivolvis* and *Lymnaea palustris*). *Ecology* 65:223–229.

Shaw, M., and G. L. Mackie. 1989. Reproductive success of *Amnicola limosa* (Gastropoda) in low alkalinity lakes in south-central Ontario. *Can. J. Fish. Aquat. Sci.* 46:863–869.

Sheehan, R. J., et al. 1989. Fate of freshwater mussels transplanted to formerly polluted reaches of the Clinch and North Fork Holston Rivers, Virginia. *J. Freshwater Ecol.* 5: 139–149.

Shoup, C. S. 1943. Distribution of fresh-water gastropods in relation to total alkalinity of streams. *Nautilus* 56:130–134.

Smith, D. G. 1980. Anatomical studies on *Margaritifera margaritifera* and *Cumberlandia monodonta* (Mollusca: Pelecypoda: Margaritiferidae). *Zool. J. Linn. Soc.* 69:257–270.

——. 1983. On the so-called mantle muscle scars on shells of the Margaritiferidae (Mollusca, Pelecypoda), with observations on mantle-shell attachment in the Unionoida and Trigonioida. *Zool. Scr.* 12:67–71.

——. 1985. Recent range expansion of the freshwater mussel *Anodonta implicata* and its relationship to clupeid fish restoration in the Connecticut River system. *Freshwater Invert. Biol.* 4:105–108.

——. 1986a. The rediscovery of *Pomatiopsis lapidaria* in New England. *Malacol. Rev.* 19:115–116.

——. 1986b. The stomach anatomy of some eastern North American Margaritiferidae (Unionoida: Unionacea). *Am. Malacol. Bull.* 4:13–19.

——. 1987. The genus *Littoridinops* (Mesogastropoda: Hydrobiidae) in New England. *Veliger* 29:442–444.

——. 1996. A method for preparing freshwater mussels (Mollusca: Unionoida) for anatomical study. *Am. Malacol. Bull.* 13:125–128.

——. 1999. Differences in siphonal anatomy between *Dreissena polymorpha* and *D. bugensis* (Mollusca: Dreissenidae) in Lake Ontario. *Am. Midl. Nat.* 141:402–405.

——. 2000a. Notes on the taxonomy of introduced *Bellamya* (Archaeagastropoda: Viviparidae) species in northeastern North America. *Nautilus* 114:31–37.

——. 2000b. The systematics and distribution of the Margaritiferidae. In: *Ecological Studies*, Vol. 145: *Evolution of the Freshwater Mussels—Unionoida*, pp. 33–49, edited by G. Bauer and K. Wächtler. Springer, Berlin.

Smith, D. G., and D. J. Brousseau. 1996. Survey of the freshwater mollusks of St. John, U. S. Virgin Islands, with descriptions and anatomical details of two species. *Carib. J. Sci.* 32:33–42.

Sparks, B. L., and D. L. Strayer. 1998. Effects of low dissolved oxygen on juvenile *Elliptio complanata* (Bivalvia: Unionidae). *J. North Am. Benthol. Soc.* 17:129–134.

Spidle, A. P., et al. 1994. Identification of the Great lakes quagga mussel as *Dreissena bugensis* from the Dnieper River, Ukraine, on the basis of allozyme variation. *Can. J. Fish. Aquat. Sci.* 51:1485–1489.

Starrett, W. C. 1971. A survey of the mussels (Unionacea) of the Illinois River, a polluted stream. *Bull. Ill. Nat. Hist. Surv.* 30:266–403.

Stern, E. M. (Undated). An illustrated key to the freshwater mussels (Bivalvia: Unionacea) of Wisconsin. *Univ. Wisconsin, Stevens Point, Reps. Mus. Nat. Hist* 20:1–75.

Strayer, D. L. 1983. The effects of surface geology and stream size on freshwater mussel (Bivalvia, Unionidae) distribution in southeastern Michigan. *Freshwater Biol.* 13:253–264.

——. 1993. Macrohabitats of freshwater mussels (Bivalvia: Unionacea) in streams of the northern Atlantic Slope. *J. North Am. Benthol. Soc.* 12:236–246.

——. 1999. Use of flow refuges by unionid mussels in rivers. *J. North Am. Benthol. Soc.* 18:468–476.

Strayer, D. L., and K. J. Jirka. 1997. The pearly mussels of New York State. *New York State Mus. Mem.* 26:1–113.

Suloway, L. 1981. The unionid (Mollusca: Bivalvia) fauna of the Kankakee River in Illinois. *Am. Midl. Nat.* 105:233–239.

Surber, T. 1912. Identification of the glochidia of freshwater mussels. *U. S. Bur. Fish. Doc.* 771:1–10.

Taylor, D. W. 1981. Freshwater mollusks of California: a distributional checklist. *Calif. Fish Game* 67:140–163.

——. 1987. Fresh-water molluscs from New Mexico and vicinity. *Bull. New Mexico Bur. Mines Miner. Res.* 116:1–50.

Taylor, R. W. 1980. *A Survey of the Freshwater Mussels of the Ohio River from Greenup Locks and Dam to Pittsburgh, PA.* 69 pp. Rpt. to U. S. Army Corps Engineers, Huntington, West Virgina.

——. 1987. Fresh-water molluscs from New Mexico and vicinity. *Bull. New Mexico Bur. Mines* 116:1–50.

Thompson, F. G. 1968. *The Aquatic Snails of the Family Hydrobiidae of Peninsular Flordia.* 268 pp. Univ. Florida Press, Gainesville, Flordia.

——. 1979. North American freshwater snail genera of the hydrobiid subfamily Lithoglyphinae. *Malacologia* 25:109–141.

——. 1984a. *Freshwater Snails of Florida: A Manual for Identification.* 94 pp. Univ. Florida Press, Gainesville.

——. 1984b. North American freshwater snail genera of the hydrobiid subfamily Lithoglyphinae. *Malacologia* 25:109–141.

Thompson, F. G., and R. Hershler. 1991. Two new hydrobiid snails (Amnicolinae) from Florida and Georgia, with a discussion of the biogeography of freshwater gastropods of South Georgia streams. *Malacol. Rev.* 24:55–72.

Thompson, F. G., and J. E. McCaleb. 1978. A new freshwater snail from a spring in eastern Alabama. *Am. Midl. Nat.* 100:350–358.

Tryon, G. W., Jr. 1873. Land and freshwater shells of North America, Part IV: Strepomatidae (American melanians). *Smithson. Misc. Coll.* 16:1–435.

Tucker, M. E. 1928. Studies on the life cycles of two species of fresh water mussels belonging to the genus *Anodonta. Biol. Bull.* 54:117–127.

Van Cleave, H. J. 1940. Ten years of observation on a fresh-water mussel population. *Ecology* 21:363–370.

Van der Schalie, H. 1938. The naiad fauna of the Huron River, in southeastern Michigan. *Mus. Zool. Univ. Mich. Misc. Publ.* 40:1–83.

——. 1940. Aestivation of fresh-water mussels. *Nautilus* 53:137–138.

——. 1945. The value of mussel distribution in tracing stream confluence. *Pap. Mich. Acad. Sci. Arts Lett.* 30:355–373.

——. 1970. Hermaphroditism among North American freshwater mussels, *Malacologia* 10:93–112.

Van der Schalie, H., and A. Van der Schalie. 1950. The mussels of the Mississippi River. *Am. Midl. Nat.* 44:448–466.

Vanderploeg, H. A., et al. 1995. From picoplankton to microplankton: temperature-driven filtration by the unionid bivalve *Lampsilis radiata siliquoidea* in lake St. Clair. *Can. J. Fish Aquat. Sci.* 52:63–74.

Vidrine, M. F. 1993. *The Historical Distributions of Freshwater Mussels in Louisiana.* 225 pp. Gail Q. Vidrine Collectables, Eunice, LA.

Walker, B. 1918. A synopsis of the classification of the freshwater Mollusca of North America, north of Mexico, and a catalogue of the more recently described species, with notes. *Mus. Zool. Univ. Mich. Misc. Publ.* 6:1–213.

Wallace, C. 1992. Parthenogenesis, sex and chromosomes in *Potamopyrgus. J. Moll. Stud.* 58: 93–107.

Watters, G. T. 1993. *Responses to the Freshwater Mussel Reproduction and Propagation Survey.* No pagination. Ohio Dept. Nat. Res., Columbus, OH.

——. 1994. An anotated bibliography of the reproduction and propagation of the Unionoidea. *Ohio Biol. Surv., Misc. Contrib.* 1:1–158.

——. 1995. *A Guide to the freshwater Mussels of Ohio*, 3d ed., rev. 122 pp. Ohio Division Wildlife, Columbus.

——. 1997. Glochidial metamorphosis of the freshwater mussel *Lampsilis cardium* (Bivalvia: Unionidae) on larval tiger salamanders, *Ambystoma tigrinum* ssp. (Amphibia: Ambystomidae). *Can. J. Zool.* 75:505–508.

Wenz, W. 1938–1944. *Gastropoda*, Band 1: *Prosobranchia und. Allgemeiner Teil.* 1639 pp. Koeltz, Koenigstein-Taunus, Germany.

Wicklow, B. J., and P. M. Beiseim. 1998. *Life History Studies of the Squawfoot Mussel Strophitus undulatus in the Piscataquog River Watershed, New Hampshire.* Freshwater Mussel Symposium, Columbus, Ohio. Program and Abstracts, p. 41.

Williams, D. D., et al. 1992. Conservation status of freshwater mussels of the United States and Canada. *Fisheries* 18:6–22.

Winslow, M. L. 1926. A revised check list of Michigan Mollusca. *Mus. Zool. Univ. Mich. Occas. Pap.* 181:1–28.

Wood, D. H. 1982. The aquatic snails (Gastropoda) of the Savannah River plant, Aiken, South Carolina. *Publ. DOEs Savannah R. Plant, Nat. Environ. Res. Park,* 46 pp.

Wood, E. M. 1974. Development and morphology of the glochidium larva of *Anodonta cygnea* (Mollusca: Bivalvia). *J. Zool. London* 173:1–13.

Wu, S.-K. et al. 1997. Missouri aquatic snails. *Missouri Depart. Conserv., Nat. Hist. Ser.* 5:1–97.

Wurtz, C. B. 1956. Fresh-water mollusks and stream pollution. *Nautilus* 69:96–100.

Zaranko, D. T. et al. 1997. Another exotic mollusc in the Laurentian Great lakes: the New Zealand native *Potamopyrgus antipodarum* (Gray 1843) (Gastropoda, Hydrobiidae). *Can. J. Fish. Aquat. Sci.* 54:809–814.

15

ARACHNIDA

(Water mites, spiders)

The Class Arachnida consists chiefly of terrestrial and parasitic members, including such familiar "arthropods" as spiders, scorpions, ticks, and mites. All are placed in the Phylum Chelicerata, a phylum distinguished by the possession of six pairs of uniramous appendages, tagmata consisting of an anterior prosoma containing all the appendages, and, generally, a posterior opisthosoma, which may be subdivided. All truly aquatic forms belong to the Subclass Acari.

Only the water mites, Hydrachnidia (or Hydrachnellae), among arachnids, have become generally adapted to freshwaters. The Hydrachnidia comprise a subcohort of the Order Prostigmata (=Trombidiformes). They are restricted to freshwater habitats and are further characterized by immature forms that differ markedly from adults in form and habits. Furthermore, considering the animal kingdom as a whole, the Hydrachnidia is one of the few characteristic freshwater groups, being almost restricted to freshwaters; only a few species are found in brackish and salt waters, and none are terrestrial.

Other arachnids that occur in freshwater include the Family Halacaridae, a largely marine group with about 30 freshwater species. Although somewhat similar to other water mites, the halacarids exhibit several characters distinguishing them from the Hydrachnidia. The soil mites, or Orbatei, have a few aquatic species, and certain spiders, Aranae, are associated with freshwater.

HYDRACHNIDIA

Although Hydrachnidia are found in almost all types of freshwater habitats, they are most abundant and characteristic of streams, rivers, ponds, and the littoral region of lakes, especially where there are quantities of rooted aquatic vegetation. Their bright colors, globular to ovoid shape, and clambering and swimming habits identify them unmistakably.

General characteristics. Superficially, water mites appear to be minute spiders, but they differ most significantly from the true spiders in having the cephalothorax and abdomen fused into a single mass, all evident segmentation being lost (Fig. 15.1). A decided minority of genera depart from the usual globular or ovoid shape and are dorsoventrally or laterally compressed. Most Hydrachnidia, and especially the more primitive genera, are soft bodied, but in many forms the cuticle is leathery or thickened to form a series of heavily sclerotized plates. Sometimes the body is almost completely enclosed by two plates, one dorsal and one ventral, whereas in other genera there are numerous smaller plates. The surface of the body is smooth, finely marked, granular, striated, or papillated. A variable number of setae are borne on the body. Total length usually ranges between 0.4 and 3.0 mm.

The six pairs of arachnid appendages are always present, the most conspicuous being

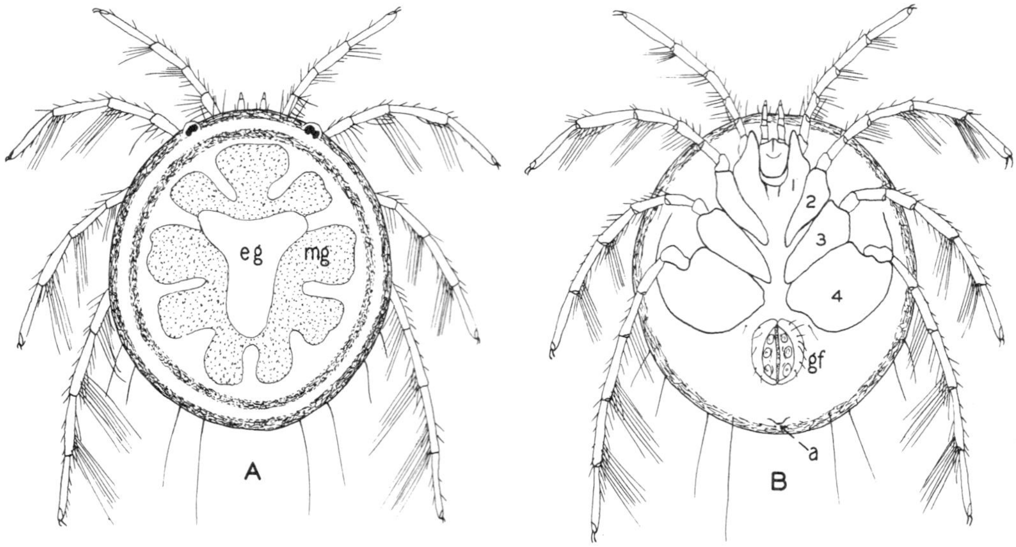

Fig. 15.1. *Mideopsis orbicularis* (Müller) female, ×40. A, dorsal; B, ventral. 1, 2, 3, and 4, first, second, third, and fourth epimera; a, anus; eg, excretory gland; gf, genital field; mg, midgut. (Modified from Soar and Williamson 1925–1929.)

the last four pairs, or legs. Each leg is long and consists of six segments that are variously supplied with spines, setae, and long hairs. The last segment usually bears two terminal movable claws. Each of the eight legs originates from a sclerotized ventral plate called a coxal plate, or epimera. At least some of the epimera are adjacent and more or less fused.

The first two pairs of appendages are borne on an anterior and ventral headlike structure called the capitulum, or maxillary organ. The mouth is located at the anterior end of the capitulum, often at the extremity of a straight or curved rostrum of variable length. The ventral surface of the capitulum is termed the maxillary shield. The chelicerae (usually called mandibles in the Hydrachnidia) are small and, except during feeding, are kept withdrawn into the capitulum. In a few genera they are stilettolike or saberlike (Fig. 15.7B), but usually there is a separate, terminal, curved, clawlike segment (Fig. 15.7A). The five-segmented pedipalps (or "palps") are inserted laterally or at the anterolateral angles of the maxillary shield and are customarily curved ventrally.

The only other conspicuous external feature is the genital field, a group of closely associated structures located on the ventral median line between or behind the epimera and sometimes at the extreme posterior end of the body. The genital pore is a small longitidinal slit in the center of the field; it is flanked by two plates that bear few to many curious knoblike or cuplike acetabula of unknown function. Sometimes there is a pair of longitudinal movable or immovable genital flaps that may or may not cover the acetabula (Figs. 15.1B; 15.7C).

The two pigmented eyes are near the anterior margin of the body and are usually widely separated, but in a few genera they lie close together along the median line. Each eye is characteristically paired. The so-called anus is an inconspicuous pore near the posterior end; sometimes it is borne on a special small plate. Two minute spiracles are located on the dorsal surface of the capitulum near its base and are therefore hidden from view by the body and can be seen only when the capitulum is dissected off.

In contrast to freshwater invertebrates in general, water mites are particularly striking because of their coloration. The majority of species are some shade of red or green, but many others are blue, yellow, tan, or brown. Excellent pictures of mites can be found at several internet sites under "water mites." The basic body coloration is caused by pigmentation of the body wall, but yellow, brown, or blackish markings and patterns, especially in dorsal view, are produced by the coloration of internal organs, especially the digestive tract and excretory gland (Fig. 15.1A). *Axonopsis complanata* is one of the most spectacular species: The appendages and general ventral surface are green; the epimera and much of the dorsal surface are blue; the capitulum, genital field, and central part of the dorsal surface are yellowish; and the large excretory gland produces a brown dorsal Y-shaped area. Some species of *Arrenurus* have equally striking combinations of green, blue, and orange. Stream species of water mites are characteristically brownish. Coloration of some species is somewhat variable, especially in *Limnesia*, and different specimens of the same species taken at the same place and time may exhibit various shades of red or green. These variations can be at least partially ascribed to differences in age and varying contents of the digestive tract and excretory gland.

Locomotion. Swimming is effected by relatively uncoordinated flailing movements of the legs, and the ability to swim is variously developed from one genus to another. The best swimmers have rows of long swimming hairs on the legs, especially the more posterior pairs (Fig. 15.1). At the other extreme are many species that are strictly creepers and crawlers on the bottom or among vegetation; their legs are supplied with only short spines and setae. Swimming ability is lost in many stream forms. Irrespective of the degree of swimming development, Hydrachnidia are clearly associated with the substrate and rarely stray far from it. Specific gravity of the body is usually high, and unless they cling to some object, they quickly fall to the bottom when not in motion. Certain *Unionicola* species are the only American water mites that stray far enough from vegetation and a substrate to be considered an accidental plankton organism. In addition to locomotion, the legs are used periodically for cleaning debris from the body and from each other.

Food, feeding. The great majority of water mites are carnivorous or parasitic, their food consisting chiefly of microcrustaceans (Fig. 15.3), small insects, various kinds of worms, or host tissues. Different species of the same genus might have completely different diets. For example, *Arrenurus hamrumi* feeds exclusively on ostracods while *A. vivida* preys on chironomid insects. Sometimes the more active and voracious species are cannibalistic. Sluggish species usually feed on dead animals, and a few investigators have found that they may utilize vegetable material and detritus, thus making them omnivores.

The prey is seized and held with the palps, with or without the aid of the legs, and the body of the prey is quickly pierced with the mandibles. Only fluids are ingested, and it is probable that secretions of the several pairs of salivary glands are injected into the prey, where they predigest some of the more solid tissues so that they may be sucked up along with the body fluids. The exoskeleton and indigestible remnants of the prey are then discarded.

Internal anatomy, physiology. Ingested fluids pass through the small cavity of the suctorial pharynx and into the midgut, the latter being a capacious, lobed, digestive and absorptive organ filling one-third to one-half of the hemocoel. The midgut usually has about 12 caeca, although there may be as many as 30 in some species. In all the Hydrachnidia whose anatomy has been carefully studied, it has been found that the midgut and its caeca end blindly, and there is no connection with the "anus." Presumably, indigestible food residues are simply regurgitated.

A heart and blood vessels are lacking, but the blood is circulated about freely in the hemocoel by the movements of the muscles and digestive tract.

Some Hydrachnidia have a tracheole system (not visible in preserved material). Each

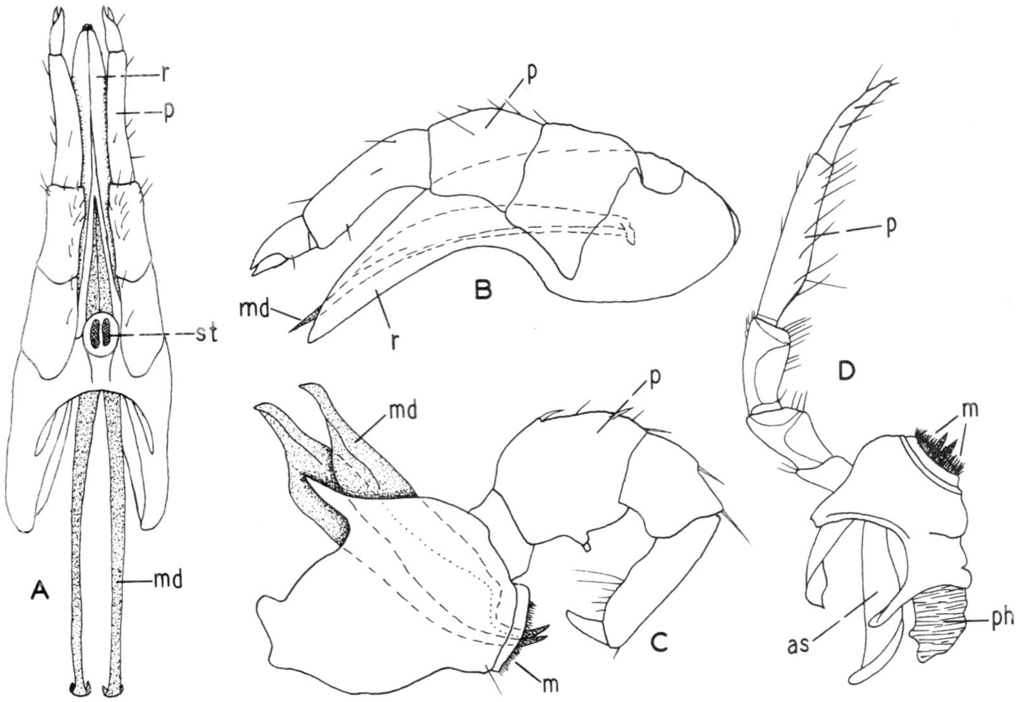

Fig. 15.2. Hydrachnidia capitulum and associated structures. A, dorsal view of capitulum of *Hydrachna;* B, lateral view of capitulum of *Hydrachna* (only left palp shown); C, lateral view of capitulum of *Tyrrellia* (only right palp shown); D, lateral view of capitulum of *Eylais* (only right palp shown). as, air sac; m, mouth (with projecting mandibles); md, mandibles (shaded); p, palp; ph, pharynx; r, rostrum; st, stigma (cross-hatched). (A and D modified from Viets 1936; C modified from Marshall 1940a.)

of the two minute prostigmata opens into an elongated air sac (Fig. 15.2D). At the posterior end of this sac is a single, short tracheole trunk, which branches into many fine, threadlike tracheoles ramifying and anastomosing to all parts of the body.

Although the tracheole system is always filled with a gas (air), no adult water mite has ever been observed taking air at the surface film in the manner of many aquatic insects. The prostigmata arc is thought to be covered with gas-permeable but waterproof membranes. When larvae or developing nymphs parasitize insects that leave the water and become aerial, however, it is probable that they take in air through the spiracles. In some species there seems to be no morphological link between the tracheole system and the prostigmata, which leaves the prostigmata with unknown functions. There is also the possibility that air could be taken into the tracheole system when the mandibles and a good portion of the capitulum are thrust into the air-containing tissues of aquatic plants. Nymphs have been observed in this attitude, but it is doubtful whether adults do it, with the possible exception of females of a few species that may assume the attitude as a preliminary to oviposition. Furthermore, mites are found in many habitats where there are no rooted aquatics.

Aside from the question of the physiological significance of the water mite tracheole system and the source of its contained gas, all investigators seem to agree that most of the oxygen utilized in metabolism is absorbed through the general body surface and

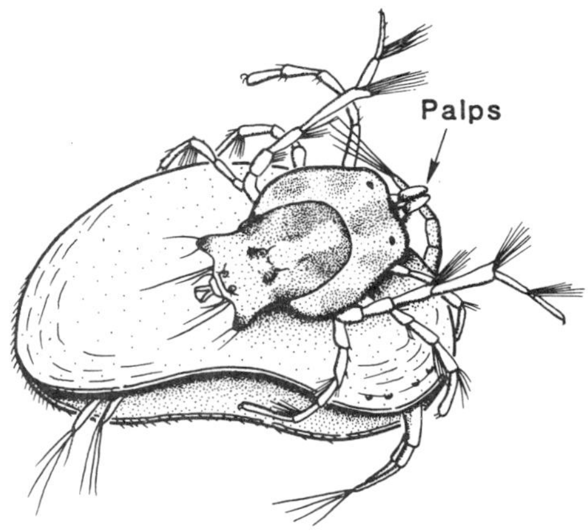

Fig. 15.3. *Arrenurus* feeding on an ostracod. (Redrawn from Gledhill 1985.)

that carbon dioxide diffuses outward through it. Those species with a thick body wall usually have scattered thin porous areas that presumably facilitate the gaseous exchange. Although many species survive temporary conditions where the dissolved oxygen concentration may be below 1 ppm, no species are known to remain active under anaerobic conditions.

The excretory gland is a large dorsal T- or Y-shaped organ (Fig. 15.1A) that connects with the ventral "anus," or "excretory pore," by a duct. Many typical mites have seven pairs of oval or pear-shaped epidermal glands opening over the dorsal and lateral surfaces of the body, and it is possible that these structures have an accessory excretory function.

The nervous system consists of a large central ganglionic mass with little indication of the method of fusion. It is perforated by the pharynx and has radiating nerves. Each of the two lateral eyes consists of two elements that may be clearly separated or more or less fused. An element consists of a pigmented bulb and cuticular lens. Many species have an additional fifth, or median, eye. Other senses are poorly developed; the palps are tactile and probably detect substances in solution, and most of the body setae are probably tactile. Chemosensory properties have been detected in the setae of the palps and segments of the first two walking legs.

In the male there are two testes, two vasa deferentia, and a single penis, which can be extruded through the genital pore. The female has two fused ovaries, two oviducts, and a single vagina, or uterus.

Reproduction. Sometimes the two sexes can be distinguished only by slight differences in the anatomy of the genital field, palps, third legs, and fourth legs, but more commonly sexual dimorphism is pronounced. Males frequently have much larger fourth epimera, and genital fields that differ considerably from those of females. The third and fourth legs of males are sometimes specialized for copulation and the transfer of spermatophores; often the fourth legs are elongated, and in a few genera the fourth segment is flattened and platelike (Fig. 15.4C) or the terminal segment is roughly sickle-shaped (Fig. 15.4B). The third legs may be modified for clasping the female (Fig. 15.4A), and sometimes the distal segments form a club for sperm transfer. In *Arrenurus* sexual dimorphism is extreme, the

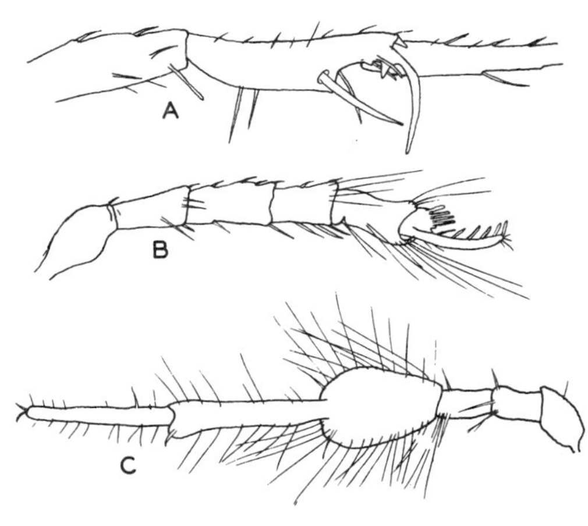

Fig. 15.4. Modified legs of male Hydrachnidia. A, middle segments of third legs of *Hydrochoreutes ungulatus;* B, fourth leg of *Forelia;* C, fourth leg of *Acercus.* (A modified from Marshall 1927; B modified from Marshall 1935; C modified from Marshall 1937.)

males being characterized by a large posterior prolongation, or cauda (Figs. 15.7D, E). Unfortunately, there is no single general anatomical character or several characters that may be used in distinguishing the sexes throughout the order.

Judging from the few species whose reproductive habits have been carefully described, it appears that Hydrachnidia exhibit a wide variety of clasping and spermatophore transfer habits. Occasionally the male is carried on the back of the female for a varying time. The males of a few species drag the female about in a haphazard manner. In some genera the genital pores of the two copulating individuals are almost in direct contact during spermatophore transfer, but more commonly the bodies are placed at various angles to each other (often at right angles), while the male securely holds the first two pairs of female legs with his fourth legs. Regardless of the specific copulatory position, the male picks up the packetlike spermatophores from the tip of his extruded penis with his third legs and transfers them to the region of the female genital pore. Spermatophore transfer takes from less than a minute to as long as an hour.

Statements concerning the breeding season vary widely. Some authorities state that breeding and egg deposition may occur at any time during the year; others maintain that eggs are usually deposited from May through July. It therefore seems logical to assume that the breeding season may vary widely from one species to another and also that it may vary within a single species according to latitude, temperature, and season.

As the female releases the eggs from her genital pore they are fertilized by the sperm contained in the spermatophores previously placed there by the male. Fertilized eggs are rarely deposited singly; usually they are extruded in groups of 20 to 400 onto stones, vegetation, and debris. Each egg may have its own individual gelatinous covering, or the whole group of eggs may be embedded in a common gelatinous mass. Eggs are usually red. *Hydrachna* oviposits in the tissues of aquatic plants.

Development, life history. Depending on temperature and the particular species, the eggs develop into inactive prelarvae within the egg in 1 to 6 weeks. After emergence, the active larvae have functional mouthparts, small epimera, three pairs of legs, and no genital field (Figs. 15.5A, B). After a short free-swimming period, the larvae become attached to aquatic insects with their capitulum, and assume a parasitic existence. Historically, Wesenberg-Lund (1919) favored the theory of some broad preference by larvae for certain insect hosts; thus, he stated that the Limnocharidae usually parasitize hydrometrids, that Eylaidae mostly parasitize aerial insects, that Hydryphantidae usually occur on culicids, and that several other families appear to be almost restricted to insects that do not leave the water. In view of more recent ecological observations, it is possible that his conclusions may be a reflection of restricted mite and insect faunas in the ponds he studied. It would be highly desirable to test this question experimentally to find out definitely whether a particular larval mite has any specific host preference. It is significant that Ephemeroptera and hydrophilid beetles are rarely parasitized and that Trichoptera are uncommon hosts. Dytiscid beetles are considered rare hosts for mites, but a recently discovered family, the Apheviderulicidae, utilizes these insects. The usual hosts are Plecoptera, Odonata, Diptera, and Hemiptera. Chironomid Diptera are the most frequent hosts. The larval mite may come in contact with its host on the substrate or at the surface film. A single insect may carry over 150 mite parasites. There is some uncertainty about the age of the host infected. Forbes and Baker (1990) believe associations that form between *Arrenurus* and larval odonates occur during the last instar of the host. Anderson and Anderson (1998) have observed that *Arrenurus* is not selective of late instars, however.

After a variable feeding period on the host, the larva begins to metamorphose to the inactive protonymphal stage. Feeding ceases and the larva shrinks from its exoskeleton, which then becomes a loose, bag-

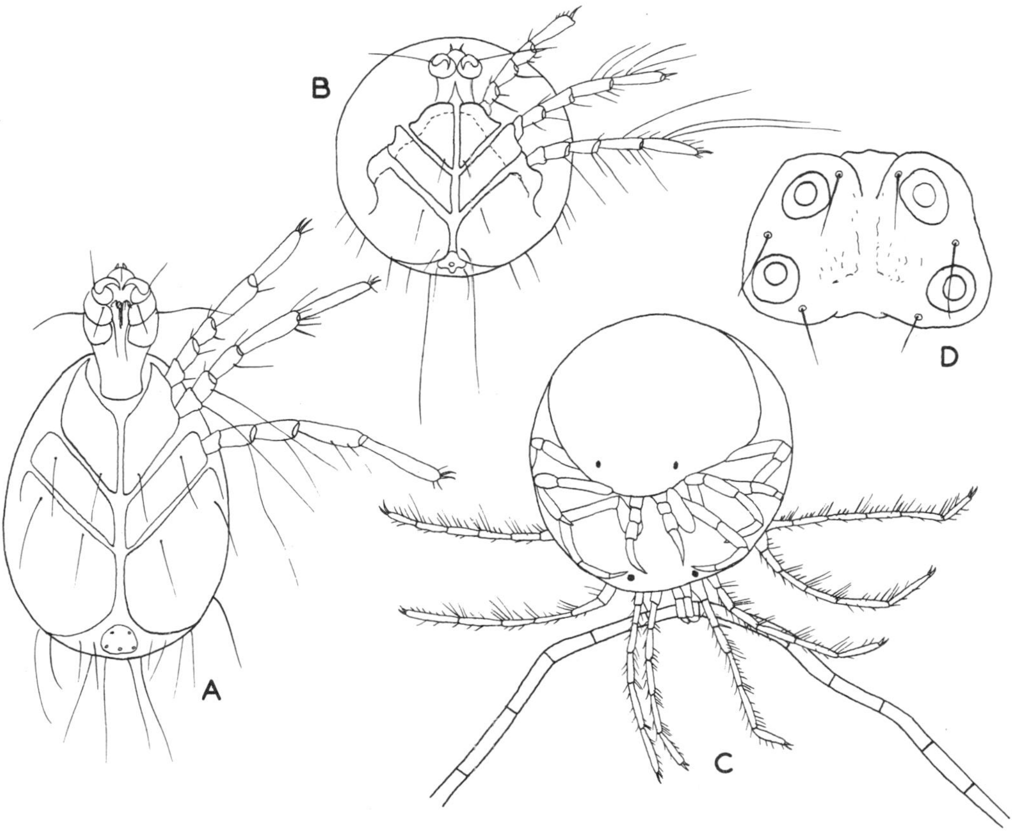

Fig. 15.5. Immature Hydrachnidia. A, larva of *Arrenurus*, ×260; B, larva of *Mideopsis*, ×130; C, adult *Piona* developing within nymphal exoskeleton, attached to filamentous alga, ×40; D, genital field of *Limnesia* nymph. (A and B modified from Soar and Williamson 1925–1929; C modified from Uchida; D redrawn from Viets 1936.)

like puparium (still attached to the host). The contained larva loses its definitive features and becomes more or less reconstituted to form the deutonymph stage. Soon the deutonymph ruptures the puparium and escapes into the water. The period of inactivity between cessation of feeding by the larva and the escape of the nymph is comparable to a pupal stage. The active deutonymph stage lasts for periods ranging from 5 days to 6 months.

The deutonymph is considerably larger than the larva and almost as large as the adult. It has the same general structures and proportions as the adult, but there is no genital opening in the genital field. Unlike the larva, the deutonymph is free-living and has the same carnivorous food habits as the adult.

At the end of a variable period of activity, the deutonymph attaches itself to an algal filament, rooted aquatic, or bottom debris with its mouthparts and enters a second pupal stage, or imagochrysalis. Again, the nymphal exoskeleton becomes a baglike puparium with the developing adult within (Fig. 15.5C). This quiescent period usually lasts only 7 to 10 days, after which the imagochrysalis exoskeleton ruptures and the sexually mature adult emerges.

By way of recapitulation, the stages in the life history of a typical water mite may be outlined as follows: egg, inactive prelarva, active larva, inactive protonymph (or nymphochrysalis), active deutonymph, inactive imagochrysalis, active adult. Some authors use other terminologies for the various stages. The active larval and parasitic larval stages are morphologically indistinguishable.

If the larva attaches to an insect that is aquatic until the termination of the nymphophan, the developed nymph simply breaks out of its puparium into the water. If the insect host is an adult beetle or hemipteran, however, it may leave the water intermittently with its attached larval parasite exposed to the air. Furthermore, the host may metamorphose to an adult that leaves the water and is aerial for the rest of its life (Odonata, Diptera, and Plecoptera, for example). In such instances the fully developed nymph breaks out of the puparium and may fall into the water fortuitously, or it may reenter the water when its host oviposits in the water. Nymphs that fall to the ground from an aerial host are presumed to die very shortly.

Figure 15.6 shows some of the common variations in mite life histories. There are many exceptions to the generalized type of life history outlined in the preceding paragraphs. A few water mites deposit their eggs in the tissues of sponges or the gelatinous matrix of colonial Protoctista. Others are thought to leave the insect host while still in the larval stage and attach to the substrate for the transformation to the nymph. Certain stream species of *Piona* and *Limnesia* do not have a parasitic stage but remain in the gelatinous egg mass until they become nymphs. The species of *Unionicola* and *Najadicola* have unusual life histories. The majority pass their entire lives in the tissues or cavities of freshwater mussels (Unionoida) or snails, but a few species of *Unionicola* as adults either are found in mussels or are free-swimming. One or two species are parasitic of sponges. Sponge-inhabiting forms release eggs in the sponge. Larvae emerge in the spring and infest chironomid insect larvae.

In mussel-inhabiting species, the entire life history can be tied closely to the preferred host mussel species or species group. A community of mussels may have a specifically coevolved community of parasitic mite species associated with it. In dense concentrations of mussels there may be as many as five mite species in a single mussel. Within the mussel, each mite species selects a different site to lay eggs. The larvae emerge and feed on the host tissue. There is generally no permanent damage to the mussel, al-

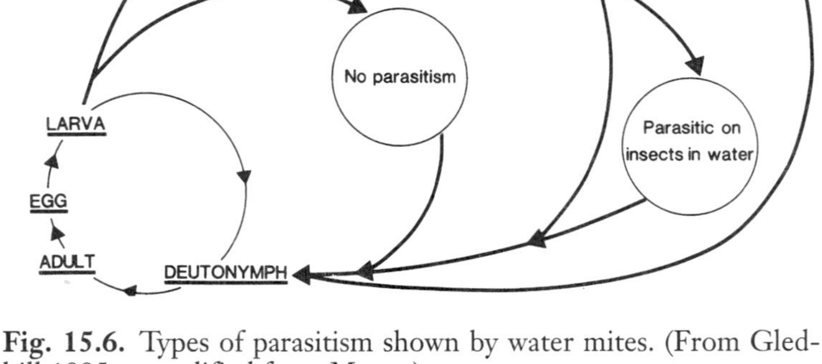

Fig. 15.6. Types of parasitism shown by water mites. (From Gledhill 1985 as modified from Mager.)

though specimens with heavy infestations show signs of stress. Complete life histories are known for relatively few water mites, and it is likely that many additional variations, exceptions, and complications will appear as further studies are made.

Almost nothing is known about the duration of the life cycle or number of generations per year, but it is thought that the great majority of species have a life span of less than 1 year. Overwintering probably occurs most frequently in the adult stage, although parasitic larvae, resting stages, and deutonymphs have also been reported.

Ecology. Unlike most groups of freshwater invertebrates, water mites may be collected in all seasons, even under ice, but the greatest numbers of adults are to be found in late spring and early autumn.

They are invariably most abundant in species and individuals among the heaviest growths of rooted aquatics where the water is only a meter or two deep. Populations in excess of 2000 mites per square meter of bottom have been reported from such habitats. For some unknown reason, however, water mites are rare in beds of *Chara* and *Elodea*. Acid lakes, though superficially an ideal collecting ground, contain very few Hydrachnidia. Wave-beaten sand and gravel shores are also poor places to collect. A study of European water mites revealed that the majority of species collected (~40%) favored organically rich substrates and roots of plants, the fewest (~15%) being confined to sandy substrates.

Many species are capable of clambering and swimming about with surprising agility, whereas others, devoid of swimming hairs, are sluggish bottom creepers. These two extremes are all gradations in crawling, swimming, and clambering ability. Mites are active only during the hours of daylight, and sometimes they remain fixed in the same position all night long.

As the water deepens offshore, the fauna becomes progressively depauperate, and even though Hydrachnidia may be collected from the bottom at depths of 10 to 100 m, such populations are usually scanty and composed of only one to several species. There is some evidence for migration into the weedy shallows in the late spring and back to deeper waters in the autumn, when vegetation disappears, but this subject needs further investigation.

Careful collecting in a favorable lake or weedy pond commonly yields as many as 20 to 40 species. Young (1969) collected 99 species in one county in Colorado, of which 72 were new state records and 3 were new for North America. Angelier et al. (1985) found 86 species in running waters of the Pyrenees, and Modlin and Gannon (1973) collected 32 species in the St. Lawrence Great Lakes.

Some genera, notably *Thyas*, *Hydryphantes*, *Acercus*, *Piona*, and *Limnesia*, are characteristic of temporary ponds. Probably they are capable of withstanding periods of drought by being buried in wet mud and debris of the substrate.

Brook and stream species are often conspicuously modified for withstanding the force of the current. Dorsoventral flattening, small size, strong claws, and the absence of swimming hairs are the usual adaptations. *Mideopsis* is an example of a flattened form that can be found scurrying about river bottoms in large numbers. *Mideopsis*, however, retains long swimming hairs and is a capable swimmer. Stream species produce comparatively small numbers of eggs. Few species may be found in both running and standing waters.

Recent studies of the interstitial and phreatic waters of stream gravels have shown a new habitat for small species of Hydrachnidia and Halacaridae. Streams associated with springs may contain up to 20 species. Sometimes these populations of interstitial mites are remarkably dense and diverse. Especially in Europe and Japan, new genera and species are being described with great frequency.

Coloration is sometimes correlated with the background, though there is little evidence that the resemblance has any protective value.

Geographical distribution, dispersal. Although the geographical distribution and ecological affinities are well known for many species in Europe, collecting in the United States has been so highly regional and generally neglected that we have little reliable information on the total distribution of most species. The majority of species reported from the United States appear to be restricted to this country, and many are known only from a single or few localities and the rare genus, *Najadicola*, appears to be restricted to North America. Many other American species are cosmopolitan and widely distributed throughout the Northern and even the Southern hemispheres. It is striking that in each of about half of the genera occurring in the United States one species (and only one) is widely distributed and common in both North America and Europe. Some of these species are *Piona rotunda, Hydryphantes ruber, Hydrodroma despiciens* (Fig. 15.7), *Lebertia porosa, Hygrobates longipalpis, Hydrochoreutes ungulatus*, and *Mideopsis orbicularis* (Fig. 15.1).

Dispersal is both active and passive, and those species that are parasitic on insects are undoubtedly easily distributed overland to new drainage systems.

Enemies. Hydrachnidia are preyed on by a wide variety of aquatic invertebrates, especially cnidarians and carnivorous insects. They are not ordinarily an important element in the diet of fishes, but occasionally a fish stomach is found in which mites predominate. Several new species have been described from the stomach contents of trout.

Collection, preparation. A successful way to collect quantities of mites is to stir up bottom debris and catch the suspended material with a small dipnet. Hand picking and washing out vegetation in an enameled pan are also useful techniques. Bringing live freshwater sponges into the laboratory is an effective way to collect many species of mites. A light placed along the edge of the container will attract emerging mites. Barr (1979) describes collecting mites at night in traps with chemoluminescent bait. During July and early August the large majority of specimens collected are immature. Living specimens may be kept in the laboratory and fed microcrustaceans or nonpredatory aquatic insects, but crowding is disastrous, since mites are often cannibalistic.

Because they are so active, it is best to examine living specimens in a compression slide or by anesthetizing and relaxing them with chloretone, in water saturated with chloroform vapor, or adding Alka-Seltzer. A binocular microscope, an opaque white background, and a bright spotlight are essential.

A good killing and preserving agent consists of 5 parts by volume of glycerin, 1 part glacial acetic acid, and 4 parts water, other-

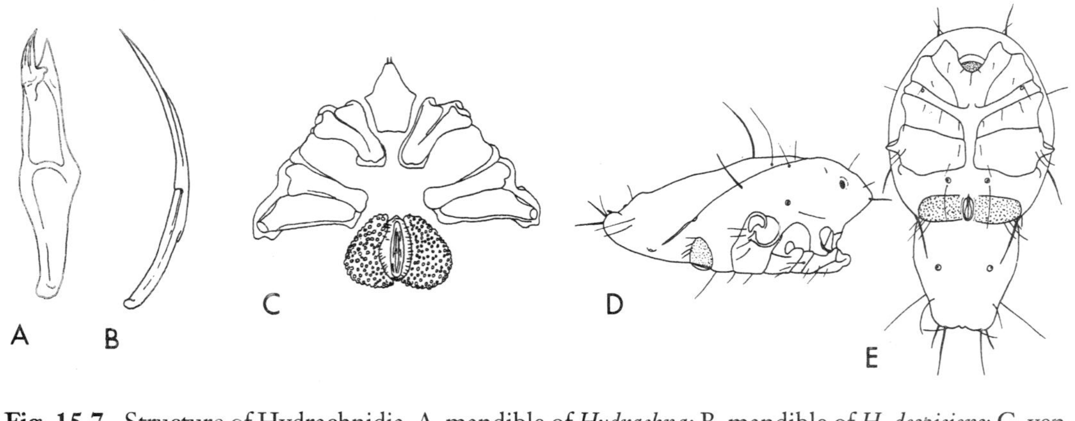

Fig. 15.7. Structure of Hydrachnidia. A, mandible of *Hydrachna;* B, mandible of *H. despiciens;* C, ventral plates of *Hydrodroma despiciens;* D, lateral view of male *Arrenurus*, ×37; E, ventral view of male *Arrenurus*. (C modified from Marshall 1940a; D, E modified from Lavers 1945.)

wise known as Koenike's solution or GAW. Wolcott recommends the following solution: 2 parts glycerin, 3 parts water, 2 parts glacial acetic acid, and 1 part absolute alcohol. Another preservative that maintains flexibility and ease of mounting is a mixture of 50% vinegar and 50% glycerine.

Suitably translucent whole mounts may sometimes be made by cutting a small incision in the dorsal body wall and pressing out the contents, but it is first necessary to clear dense and hard species with acetic corrosive to render them sufficiently translucent for work with the compound microscope. Acetic corrosive consists of equal parts (by weight) of glacial acetic acid, chloral hydrate, and water. For critical study and accurate identification, the mouthparts, capitulum, and legs should be dissected off and mounted separately. Some workers prefer permanent Euparal or similar mounting media (see Lavers 1945); others use glycerin or glycerin jelly. Polyvinyl alcohol and Hoyer's fluid are excellent mounting media that also have clearing properties. See Mitchell and Cook (1952) for a further discussion of making slide mounts.

Taxonomy. Most of all the published information on our American species is based on the work of a small number of investigators. Among the classical literature, Wolcott, whose publications extend from 1899 to 1918; Ruth Marshall, whose numerous papers appeared between 1903 and 1944; Cook, beginning in 1953; and Mitchell, also beginning in 1953, have provided a wealth of information on water mite taxonomy. More recently, I. M. Smith and M. Vidrine have contributed much information on the taxonomy of water mites. The potential number of species in the United States must be large. Worldwide, there are about 300 genera and 5000 species. Regarding the United States, Marshall reported 140 species and varieties from Wisconsin alone. K. H. Viets (1936) lists 444 species from Germany, and Soar and Williamson (1925–1929) list 247 for the British Isles. Mitchell (1954a) lists 291 species from North America, and Crowell (1961) lists 457 taxa from North America. B. P. Smith (1990) suggests that nearly a 1000 occur in Canada alone, so the number of species occurring in the United States is probably more than 1000. In fact, Vidrine (1996) lists 69 American species in *Unionicola* alone! K. O. Viets (1987) lists 4000 species and subspecies worldwide. Very little is known about the Hydrachnidia of phreatic waters, rapid streams, high altitudes, states west of the Great Plains, and southeastern states.

Identification to genus and species is based primarily on the detailed adult anatomy of the epimera, genital field, capitulum, and palps. The taxonomic status and limits of several families and genera are still controversial. Future study will undoubtedly provide better definitions of Hydrachnidia families. Most investigators believe that the fauna is highly diverse and that many new genera and families will be regularly discovered or recognized. About 39 families are currently recognized in the United States, containing over 100 genera. Many characteristics defining water mite families and genera are subtle and often convergent, masking otherwise significant features defining individual taxa. Determining genera requires extensive (expert) knowledge of the morphology and biology of water mites and adult specimens (sometimes of each sex) are generally necessary for identification. Furthermore, acarologists specializing in Hydrachnidia have yet to develop a systematized method for identifying families that is useful to the nonexpert. Thus, unlike in previous editions, a key has not been constructed for these organisms. The 39 families occurring in the United States have been listed in Table 15.1 along with the number (in parentheses) of genera assigned to each. Numbers of genera in the more diverse families are probably not exact and are subject to interpretation by different investigators. Several new genera were added to the U. S. fauna in the 1990s.

The most complete key to adult mite genera is found in Cook (1974a); however, since his publication many new genera and families have been described. B. P. Smith (1990) has provided an artificial key to genera of

Table 15.1. Families of Freshwater Mites in the United States (with Number of Genera in Parentheses)

Acalyptonotidae (2)	Limnocharidae (3)
Anisitsiellidae (6)	Mideidae (1)
Apheviderulicidae (1)	Mideopsidae (1)
Arenohydracaridae (1)	Momoniidae (3)
Arrenuridae (1)	Neoacaridae (2)
Athienemanniidae (3)	Nudomideopsidae (3)
Aturidae (3)	Omartacaridae (1)
Bogatiidae (1)	Oxidae (2)
Chappuisididae (1)	Piersigiidae (1)
Eylaidae (1)	Pionidae (11)
Feltriidae (1)	Rhynchohydracaridae (1)
Hydrachnidae (1)	Rutripalpidae (1)
Hydrodromidae (1)	Sperchontidae (2)
Hydrovolziidae (1)	Stygothrombidiidae (1)
Hydryphantidae (24)	Teutoniidae (1)
Hygrobatidae (5)	Thermacaridae (1)
Krendowskiidae (2)	Torrenticolidae (4)
Laversiidae (1)	Uchidastygacaridae (3)
Lebertiidae (2)	Unionicolidae (2)
Limnesiidae (9)	

northeastern North America. The recently described family Apheviderulicidae, including a single genus known only from California in the United States, is based on larvae and deutonymph stages only. It appears to be most closely related to the Piersigiidae. Mites recovered from bivalve and gastropod mollusks and sponges are *Unionicola* (Unionicolidae) or *Najadicola* (Pionidae) and regarding these genera, keys to the subgenera and species can be found in Vidrine (1996). Smith and Cook (1994) provide a key to the Neomamersinae (Limnesiidae). A few of the families have specialized life styles or morphologies. The Stygothrombidiidae and Wandesiinae have elongate bodies and live in the interstices of the hyporheos. The Thermacaridae is unique to thermal (hot) springs in western states. The Oxidae have their legs grouped close together around the anterior end. The Hydrachnidae contains the single, very common genus *Hydrachna* (Fig. 15.7), which is typically bright red and an active swimmer.

HALACARIDAE

The Halacaridae are predominantly a marine family of predaceous mites in the Acari, and although several of the 30 genera are found in freshwaters, the Halacaridae are not usually considered as belonging in the Hydrachnidia. The family has about 50 freshwater species worldwide.

Morphologically, halacarids are distinguished from Hydrachnidia by the fact that their palps have only three or four segments. In addition, the rostrum and palps project well beyond the anterior end of the body. Body length is less than 0.5 mm. The exoskeleton is leathery and reinforced with four dorsal plates and four ventral plates. There are no separate epimera, and the legs are attached laterally. These mites cannot swim and are usually found creeping about on aquatic plants; on the bottom of ponds, lakes, and streams; in the interstitial waters of sandy and gravel substrates of freshwaters; and in subterranean waters. They appear to be especially common in hyporheic waters where they may outnumber hydrachnids. Sometimes populations reach densities of 5000 individuals per square meter. They may be found from the sea's edge to 5000 m in altitude. Halacarids are unusual in that some forms, especially *Limnohalacarus* and *Soldanellonyx*, frequently are found as members of the lake and pond pleuston association.

In general, halacarids are seldom collected, and their U.S. distribution and ecology are poorly known. They have poor dispersal capabilities and they prefer weakly acidic water. The following genera are either known to occur here or are likely to be taken: *Copidognathus*, *Halacarus*, *Hamohalacarus*, *Limnohalacarus*, *Lobohalacarus*, *Parasoldanellonyx*, *Porohalacarus* (Fig. 15.9), *Porolohmannella*, and *Soldanellonyx* (Fig. 15.8). *Halacarus* and *Copidognathus* are two genera in transition between fresh and salt waters. They are taken in wells and springs near marine shores.

Until much more information and field collecting records are available for the United States, it seems fruitless to draw up a key based on our fragmentary data.

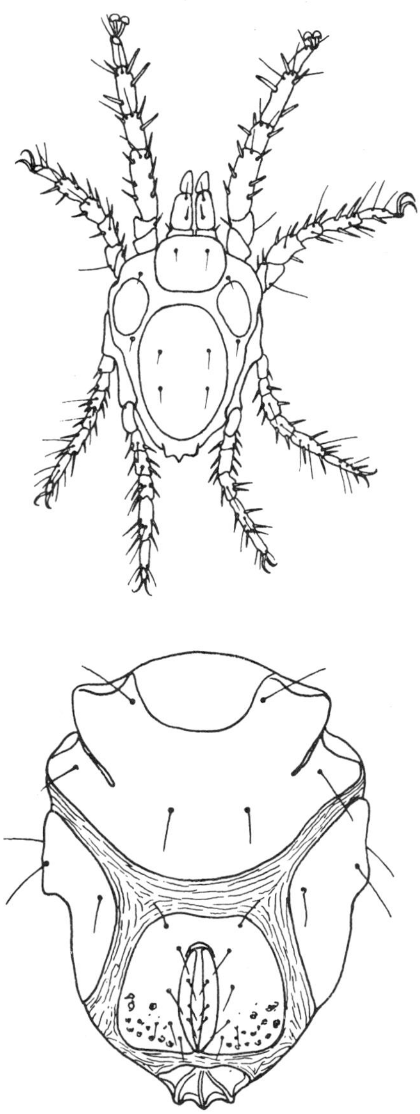

Fig. 15.8. *Soldanellonyx chappuisi.* upper, dorsal view, ×85; lower, ventral view (appendages not shown). (From Imamura 1968.)

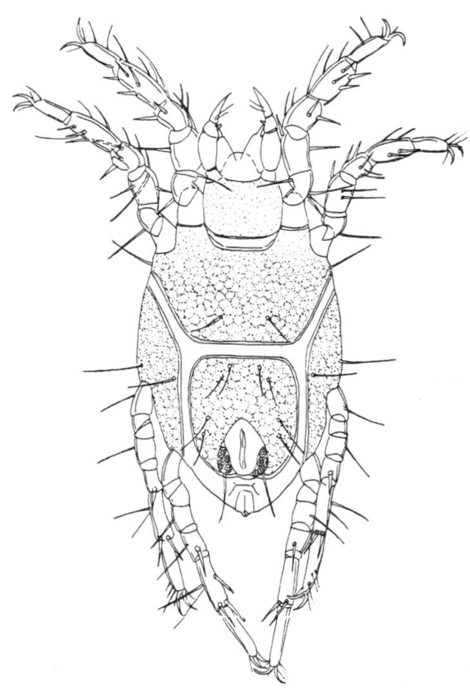

Fig. 15.9. Ventral view of *Porohalacarus*, ×275. (From Ramazzotti and Nocentini 1960.)

ORIBATEI

The Order Oribatida (soil mites) includes mostly terrestrial families, but at least three families have genera with aquatic species in the United States. A few species scattered throughout the remaining families are semi-aquatic or live on moist soil or on *Sphagnum* and other hydrophytes that may be damp or intermittently submerged.

Morphologically, the Oribatei is separated most obviously from the Halacaridae and the true water mites by the modified "tear-drop" shape of the body owing to the broad triangular gnathosoma, or capitulum. Oribatids are also more coriaceous and dark colored.

Undoubtedly, *Hydrozetes* (Figs. 15.10; 15.11) (in the Family Hydrozetidae) is the most common truly aquatic North American genus. It occurs chiefly in small permanent ponds where the mites are restricted to crevices on submerged logs, twigs, and debris (Fig. 15.12). Sometimes it occurs in the pleuston, especially when it carries an air plastron. In addition, *Hydrozetes* can swallow a small bubble into the midgut from the plastron, which, by levitation, brings the mite to the surface, especially at night and when there are mechanical stimuli. In the spring it sometimes forms enormous dense aggregations, as large as 2 mm thick, 10 mm wide, and 2 m long! Populations are scattered during the winter months, and individuals appear to exist in a kind of cold narcosis state. Food consists of wood fibers, fungi, algae, and detritus. *Hydrozetes* may be easily cultured in the laboratory, where the life cycle usually lasts 30 to 80 days. Instars include larva, protonymph, deutonymph, imagochrysalis, and adult (Fig. 15.13). Damselfly nymphs (especially *Enallagma* and *Tetragonewia*) are important predators on *Hydrozetes*.

Other aquatic genera present in the United States are *Mucronothrus* (Trhypochthoniidae) (Fig. 15.14A, B) and *Malaconothrus* and *Trimalaconothrus* (Malaconothridae). *Mucro-*

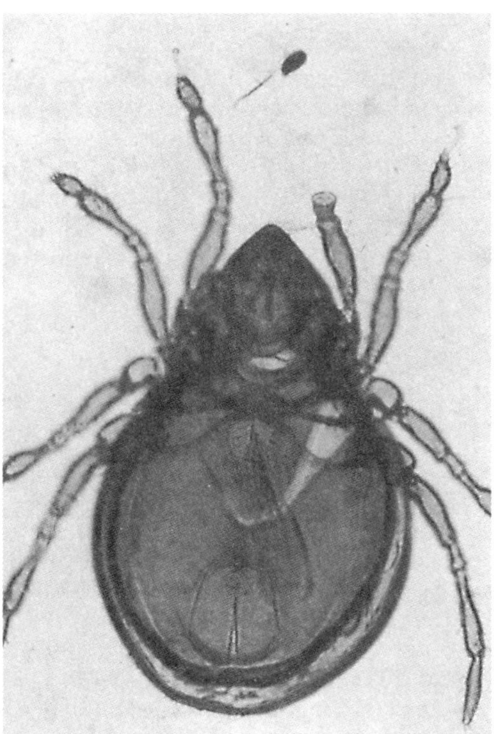

Fig. 15.10. Adult female *Hydrozetes*, 400 μm long (Courtesy of J. Bushnell and D. Burford.)

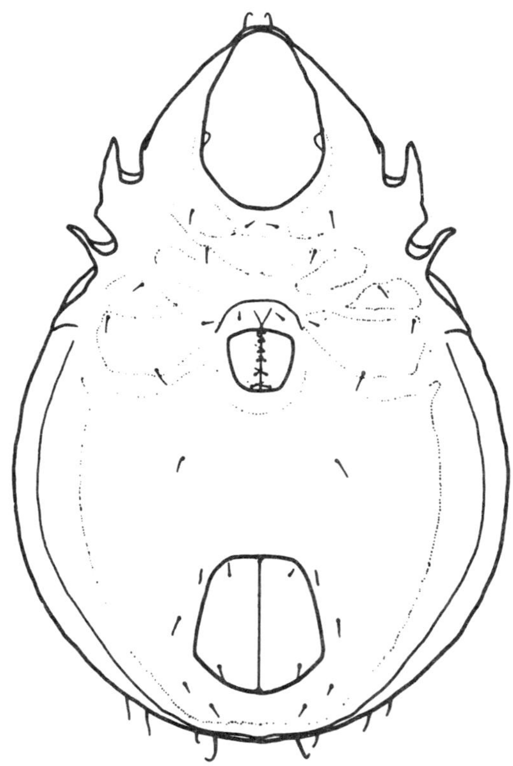

Fig. 15.11. Adult female *Hydrozetes*, ventral, legs omitted, 400 μm long. (Courtesy of J. Bushnell and D. Burford.)

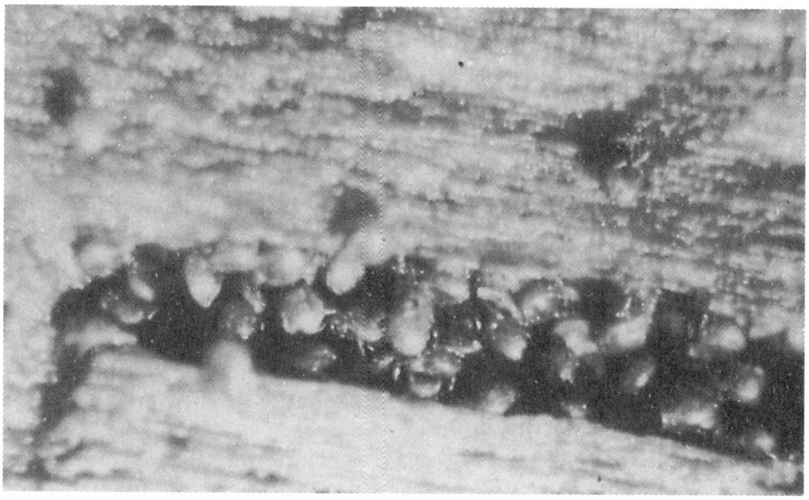

Fig. 15.12. Aggregation of *Hydrozetes* crowded into a crevice in a twig. (Courtesy of J. Bushnell and D. Burford.)

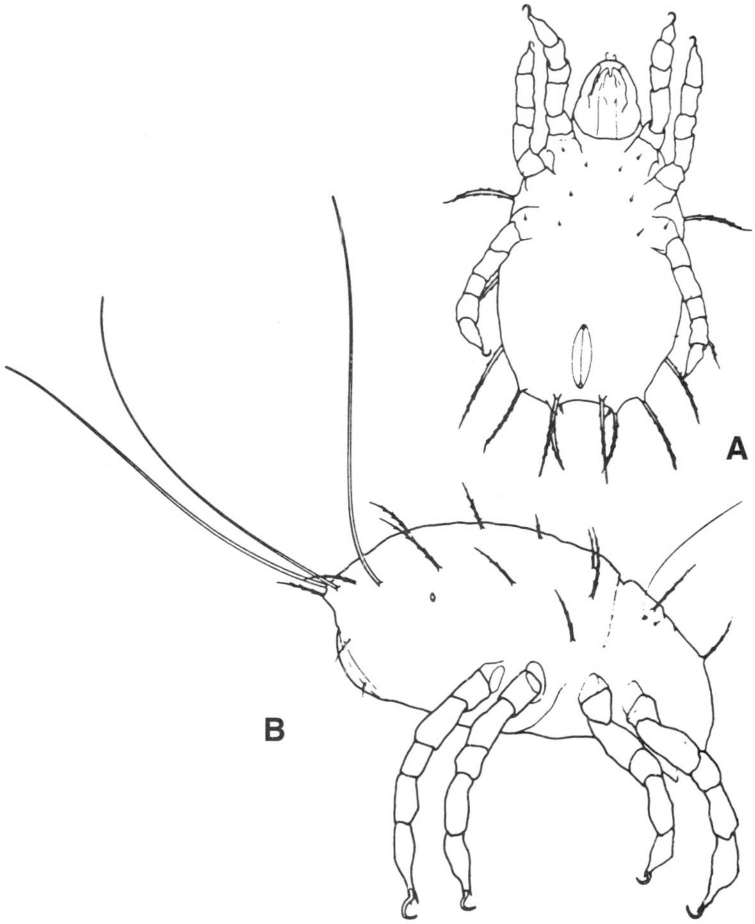

Fig. 15.13. Immature *Hydrozetes.* A, larva, ventral, 250 μm; B, protonymph, lateral, 300 μm. (Courtesy of J. Bushnell and D. Burford.)

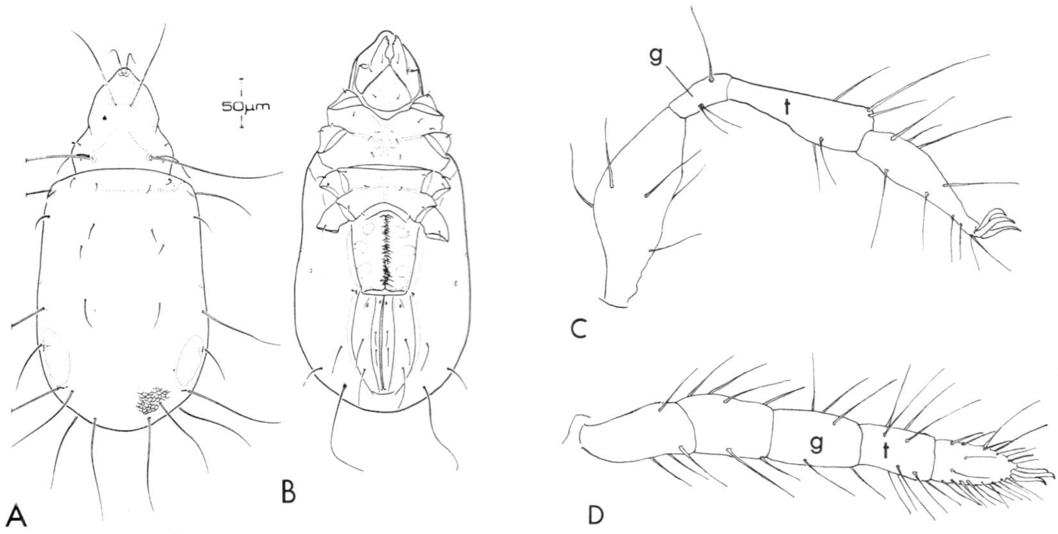

Fig. 15.14. Structure of orbatid mites. A, *Mucronothrus* dorsal view; B, *Mucronothrus* ventral view; C, leg of Hydrozetodae; D, leg of Trhypochthoniidae and Malaconothridae. g, gena; t, tibia. (A, B after Norton et al. 1996; C, D redrawn from Evans 1992).

nothrus is found in streams in the northwestern states, the other two widely distributed.

The American aquatic oribatids are too sketchily known to merit the construction of a key. The three families can be differentiated by features of the leg and cuticle. The Hydrozetodae can be diagnosed by the leg segments; the genu and tibia are quite dissimilar and of unequal size (Fig. 15.14C). In the Trhypochthoniidae and Malaconothridae the genu and tibia are similar of equal size (Fig. 15.14D). The Malaconothridae possess an ornamented cuticle that is birefringent, a feature absent in the Trhypochthoniidae. About 20 freshwater species are known worldwide in these families. Six species of *Hydrozetes* have been reported from the United States, whereas only 2 in *Mucronothrus*. The number of species of Malaconothridae inhabiting the United States is unclear.

SPIDERS

Among the true spiders, or Araneae, there are no American species that are strictly aquatic. A European and Asiatic species, *Argyroneta aquatica*, has become curiously adapted to the aquatic environment. It spins a dense, flat, underwater web in vegetation and brings small air bubbles down from the surface film and releases them beneath the web, thus forming a large bell-shaped bubble held in place by the web. The spider spends most of its time within the bubble. Many American species, however, are semiaquatic. They run about along the beach or among shore vegetation and often skitter out over the surface of ponds and marshes, where they may feed on Hemiptera and other insects supported on the surface film. Some even dive beneath the surface when alarmed. Many build their webs on emergent vegetation. Some of the genera having common semiaquatic species are *Dolomedes*, *Lycosa* (wolf spiders), *Pirata*, *Pachygnatha*, and *Tetragnatha* (orb weavers). *Dolomedes* is a colorful and spectacular genus, often reaching a diameter of 5 cm. It dives with difficulty and remains submerged by hanging on to objects, but may remain below the surface as long as 45 minutes. Oxygen is probably obtained from air bubbles adhering to the body near the respiratory openings. *Dolomedes* feeds chiefly on aquatic insects, but it has also been reported catching small fish.

REFERENCES

Anderson, T. M., and N. H. Anderson. 1998. The life history of *Arrenurus hamrumi*, a water mite from rangeland springs in central Oregon, USA. In: *Studies in Crenobiology—The Biology of Springs and Springbrooks*, pp. 63–74, edited by L. Botosaneanu. Backhuys, Leiden.

Angelier, E., et al. 1985. Recherches sur l'écologie des Hydracariens (Hydrachnellae, Acari) dans les eaux courantes. *Ann. Limnol.* 21:25–64.

Arndt, W., and K. Viets. 1938. Die biologischen (parasitologischen) Beziehungen zwischen Arachnoideen und Spongien. *Z. Parasitenkd.* 10:67–93.

Baker, G. T. 1996. Chemoreception in four species of water mites (Acari: Hydrachnida): behavioural and morphological evidence. *Exp. Appl. Acarol.* 20:445–455.

Balogh, J. 1972. *The Oribatid Genera of the World.* 188 pp. Akademiai Kiado, Budapest.

Barr, D. 1979. Water mites (Acari, Parasitengona) sampled with chemoluminescent bait in under-water traps. *Int. J. Acarol.* 5:187–194.

Bartsch, I. 1982. Halacaridae (Acari) im Süsswasser von Rhode Island, U.S.A., mit einer Diskussion uber Verbreitung und Abstammung der Halacaridae. *Gewässer und Abwässer* 68/69:41–58.

——. 1987. Zur Biologie, Ökologie und Verbreitung der süsswasserbewohnenden Halacaride *Porohalacarus alpinus* (Acari). *Arch. Hydrobiol.* 111:83–93.

——. 1999. *Peregrinacarus reticulatus* gen. nov. spec. nov., a freshwater halacarid mite (Acari, Halacaridae) from Marion Island. *Hydrobiologia* 392:225–232.

Böttger, K. 1970. Die Ernährung der Wassermilben (Hydrachnellae Acari). *Int. Rev. gesamten Hydrobiol.* 55:895–912.

——. 1972. Vergleichend biologisch-ökologische Studien zurn Entwicklungszyklus der Süsswassermilben (Hydrachnellae, Acari), 1: Der Entwicklungszyklus von *Hydrachna globosa* und *Limnochares aquatica. Int. Rev. gesamten Hydrobiol.* 57:109–152.

——. 1977. The general life cycle of fresh water mites. *Acarologia* 18:496–502.

Cook, D. R. 1954a. Preliminary list of the Arrenuri of Michigan, Part I: the subgenus *Arrenurus. Trans. Am. Microsc. Soc.* 73:39–58.

——. 1954b. Preliminary list of the Arrenuri of Michigan, Part II: the subgenus *Megaluracarus. Trans. Am. Microsc. Soc.* 73:367–380.

——. 1955a. Two new genera of Hydracarina from a spring in northern Michigan. *Am. Midl. Nat.* 53:412–418.

——. 1955b. Preliminary list of the Arrenuri of Michigan, Part III: the subgenera *Micruracarus* and *Truncaturus. Trans. Am. Microsc. Soc.* 74:60–67.

——. 1955c. Preliminary studies of the Hydracarina of Michigan: the Subfamily Foerliinae (Acarina: Pionidae). *Ann. Entomol. Soc. Am.* 48:299–307.

——. 1959. Studies on the Thyasinae of North America (Acarina: Hydryphantidae). *Am. Midl. Nat.* 62:402–428.

——. 1961. New species of *Bandakia, Wettina,* and *Athienemannia* from Michigan (Acarina, Hydracarina). *Proc. Entomol. Soc. Wash.* 63:262–268.

——. 1963a. Studies on the phreaticolous water mites of North America: the genera *Neomamersa* Lundblad and *Kawamuracarus* Uchida. *Am. Midl. Nat.* 70:300–308.

——. 1963b. Studies on the phreaticolous water mites of North America: the Family Neoacaridae. *Ann. Entomol. Soc. Am.* 56:481–487.

——. 1963c. Studies on the phreaticolous water mites of North America: the genus *Feltria* (Acarina: Feltriidae). *Ann. Entomol. Soc. Am.* 56:488–500.

——. 1963d. Omartacaridae, a new family of water mites from the ground waters of North America. *Entomol. News* 74:37–43.

——. 1963e. Studies on the phreaticolous water mites of North America: new or unreported genera of Mideopsoidea and Acalyptonotoidea. *Entomol. News* 74:63–70.

——. 1968a. Water mites of the genus *Stygomomonia* in North America (Acarina: Momoniidae). *Proc. Entomol. Soc. Wash.* 70:210–224.

——. 1968b. New species of *Neoacarus* Halbert and *Volsellacarus* Cook from North America (Acanina: Neoacaridae). *Proc. Entomol. Soc. Wash.* 67–74.

——. 1969. New studies on the water mite genera *Neomamersa* Lundblad and *Kawamuracarus* Uchida (Acarina, Limnesiidae) from North America. *Am. Midl. Nat.* 81:29–38.

——. 1974a. North American species of the genus *Axonopsis* (Acarina: Aturidae: Axonopsinae). *Great Lakes Entomol.* 7:55–79.

——. 1974b. Water mite genera and subgenera. *Mem. Am. Entomol. Inst.* 21:1–860.

Crowell, R. M. 1960. The taxonomy, distribution and developmental stages of Ohio water mites. *Bull. Ohio Biol. Surv.* 1:1–77.

——. 1961. Catalogue of the distribution and ecological relationships of North American Hydracarina. *Can. Entomol.* 93:321–359.

Di Sabatino, A., et al. 2000. The biology and ecology of lotic water mites (Hydrachnidia). *Freshwater Biol.* 44:47–62.

Evans, G. O. 1992. *Principles of Acarology.* 563 pp. CAB International, Wallingford, UK.

Fernandez, N., and J. Trave. 1984. La variabilité chaetotaxique et la neotrichie gastronotique des Hydrozetidae (Oribates). *Acarologia* 25: 407–417.

Forbes, M. R. L., and R. L. Baker. 1990. Susceptibility to parasitism: experiments with the damselfly *Enallagma ebrium* (Odonata: Coenagrionidae) and larval water mites, *Arrenurus* spp. (Acari: Arrenuridae). *Oikos* 58:61–66.

Gerecke, R. et al. 1999. Three new species of *Apheviderulix* gen. nov. and proposal of Apheviderulicidae fam. nov. (Acari: Hydrachnidia: Eylaoidea). *Hydrobiologia* 397:133–147.

Gledhill, T. 1985. Water mites: predators and parasites. *Rep. Freshwater Biol. Assoc.* 1985:45–59.

Husmann, S., and D. Teschner. 1970. Okologie, Morphologie und Verbreitungsgeschichte subterraner Wassermilben (Limnohalacaridae) aus Schweden. *Arch. Hydrobiol.* 67:242–267.

Imamura, T. 1957a. Erste Mitteilung über Porohalacaridae aus unterirdischen Gewässern in Japan. *Abh. Naturw. Ver. Bremen* 35:53–62.

——. 1957b. Subterranean water-mites of the Middle and Southern Japan. *Arch. Hydrobiol.* 53:350–391.

——. 1959. Check list of the troglobiontic Troinbidiidae, Porohalacaridae and Hydrachnellae of Japan. *Bull. Biogeogr. Soc. Jpn.* 21:63–66.

——. 1968. Results of the speleological survey in South Korea 1966, IX: Halacaridae (Acari) found in a limestone cave of South Korea. *Bull. Nat. Sci. Mus.* 11:281–284.

——. 1970. Some psammobiotic water mites of Lake Biwa. *Annot. Zool. Jpn.* 43:200–206.

Imamura, T., and R. Mitchell. 1967. The ecology and life cycle of the water mite, *Piersigia limophila* Prom *Annot. Zool. Jpn.* 40:3 7–44.

Jrantz, G. W., and G. T. Baker. 1982. Observations on the plastron mechanism of *Hydrozetes* sp. (Acari: Oribatida: Hydrozetidae). *Acarologia* 23:273–278.

Lavers, C. H., Jr. 1945. The species of *Arrenurus* of the state of Washington. *Trans. Am. Microsc. Soc.* 44:228–264.

Lundblad, O. 1935. Die nordamerikanischen Arten der Gartung Hydrachna. *Ark. Zool.* 28: 1–44.

Marshall, R. 1914. Some new American water mites. *Trans. Wis. Acad. Sci. Arts Lett.* 17:1300–1304.

——. 1927. Hydracarina of the Douglas Lake region. *Trans. Am. Microsc. Soc.* 46:268–285.

——. 1932. Preliminary list of the Hydracarina of Wisconsin, Part II. *Trans. Wis. Acad. Sci. Arts Lett.* 27:339–358.

——. 1933. Preliminary list of the Hydracarina of Wisconsin, Part III. *Trans. Wis. Acad. Sci. Arts Lett.* 28:37–61.

——. 1935. Preliminary list of the Hydracarina of Wisconsin, IV. *Trans. Wis. Acad. Sci. Arts Lett.* 29:273–298.

——. 1937. Preliminary list of the Hydracarina of Wisconsin, V. *Trans. Wis. Acad. Sci. Arts Lett.* 30:225–252.

——. 1940a. Preliminary list of the Hydracarina of Wisconsin, VI. *Trans. Wis. Acad. Sci. Arts Lett.* 32:135–165.

——. 1940b. The water mite genus *Tyrellia*. *Trans. Wis. Acad. Sci. Arts Lett.* 383–389.

——. 1943a. Hydracarina from California, Part I. *Trans. Am. Microsc. Soc.* 62:306–324.

——. 1943b. Hydracarina from California, Part II. *Trans. Am. Microsc. Soc.* 62:404–415.

Martin, P., and S. Speth. 1996. Die substratspezifische Verteilung der Hydrachnidia (Acari) zweier Baeche in jungglazialen Sandergebieten Schleswig-Holsteins (Osterau und Rodenbek). *Limnologica* 26:93–103.

Mitchell, R. D. 1953. A new species of *Lundbladia* and remarks on the Family Hydryphantidae (water mites). *Am. Midl. Nat.* 49:159–170.

——. 1954a. Water-mites of the genus *Aturus* (Family Axonopsidae). *Trans. Am. Microsc. Soc.* 73:350–367.

——. 1954b. A description of the water-mite, *Hydrovolzia gerhardi* new species, with observations on the life history and ecology. *Nat. Hist. Misc.* 134:1–9.

——. 1954c. Check list of North American watermites. *Fieldiana: Zool.* 35:27–70.

——. 1955. Anatomy, life history, and evolution of the mites parasitizing fresh-water mussels. *Misc. Publ. Mus. Zool. Univ. Mich.* 89:1–28.

——. 1957a. Evolutionary lines in water mites. *Syst. Zool.* 6:137–148.

——. 1957b. The mating behavior of pionid watermites. *Am. Midl. Nat.* 58:360–366.

——. 1960. The evolution of thermophilous water mites. *Evolution* 14:361–377.

——. 1963. A new water mite of the Family Thermacaridae from hot springs. *Trans. Am. Microsc. Soc.* 82:230–233.

——. 1964. A study of sympatry in the water mite genus *Arrenurus* (Family Arrenuridae). *Ecology* 45:546–558.

——. 1965. Population regulation of a water mite parasitic on unionid mussels. *J. Parasitol.* 51:990–996.

——. 1967. Host exploitation of two closely related water mites. *Evolution* 21:59–75.

Mitchell, R. D., and D. R. Cook. 1952. The preservation and mounting of water-mites. *Turtox News* 30:169–172.

Modlin, R. F., and J. E. Gannon. 1973. A contribution to the ecology and distribution of aquatic Acari in the St. Lawrence Great Lakes. *Trans. Am. Microsc. Soc.* 92:217–224.

Newell, I. M. 1945. *Hydrozetes* Berlese (Acari, Oribatoidea): the occurrence of the genus in North America and the phenomenon of levitation. *Trans. Conn. Acad. Sci. Arts Lett.* 36:253–268.

——. 1947. A systematic and ecological study of the Halacaridae of eastern North America. *Bull. Bingham Oceanogr. Coll.* 10:1–232.

Nocentini, A. M. 1961. Primiritrovamenti di Porohalacaridae (Acari) nel Lago Maggiore. *Mem. 1st. Ital. Idrobiol.* 13:127–138.

Norton, R. A., et al. 1996. The aquatic oribatid mite genus *Mucronothrus* in Canada and the western U.S.A. (Acari: Trhypochthoniidae). *Can. J. Zool.* 74:926–949.

Petrova, A. 1974. Sur la migration des Halacariens dans les eaux douces et la position systernatique des Halacariens et Limnohalacariens. *Vie et Milieu* 24:8 7–96.

Pieczynski, E. 1960. (Formation of groupings of water mites (Hydracarina) in different environments of Lake Wilkus.) *Ekol. Pol. A,* 8:169–198.

——. 1961. Numbers, sex ratio, and fecundity of several species of water mites (Hydracarina) of Mikolajskie Lake. *Ekol. Pol. A.* 9:219–228.

——. 1963. Some regularities in the occurrence of water mites (Hydracarina) in the littoral of 41 lakes in the River Krutynia basin and the Mikolajki District. *Ekol. Pol. A* 11:141–157.

Proctor, H. C., and G. Pritchard. 1989. Neglected predators: water mites (Acari: Parasitengona: Hydrachnellae) in freshwater communities. *J. North Am. Benthol. Soc.* 8:100–111.

——. 1990. Variability in the life history of *Unionicola crassipes*, a sponge-associated water mite (Acari: Unionicolidae). *Can. J. Zool.* 68:1227–1232.

Ramazzotti, G., and A. M. Nocentini. 1960. Porohalacaridae (Hydracarina) del Lago di Mergozzo. *Mem. 1st. Ital. Idrobiol.* 12:185–200.

Riessen, H. P. 1982. Predatory behavior and prey selectivity of the pelagic water mite *Piona constricta. Can. J. Fish. Aquat. Sci.* 39:1569–1579.

Schmidt, U. 1935. Beitrage zur Anatomie und Histologie der Hydracarinen, besonders von *Diplodontus despiciens* O. F. Müller. *Z. Morphol. Oekol. Tiere* 30:99–176.

Schwoerbel, J. 1955. Über einige Porohalacariden (Acari) aus dem südlichen Schwarzwald. *Zool. Anz.* 155:146–150.

——. 1959. Ökologische und tiergeographische Untersuchungen über die Milben (Acari, Hydrachnellae) der Quellen und Bäche des südlichen Schwarzwaldes und seiner Randgebiete. *Arch. Hydrobiol. Suppl.* 24:385–546.

Simmons, T., and I. M. Smith. 1984. Morphology of the larvae, deutonymphs, and adults of the water mite *Najadicola ingens* (Prostigmata: Parasitengona: Hygrobatoidea) with remarks on phylogenetic relationships and revision of taxonomic placement of Najadicolinae. *Can. Entomol.* 116:691–701.

Smith, B. P. 1990. Hydrachnidia. In: *Freshwater Macroinvertebrates of Northeastern North America*, pp. 290–334, edited by B. L. Peckarsky et al. Cornell Univ. Press, Ithaca, NY.

Smith, I. M. 1976. A study of the systematics of the water mite family Pionidae (Prostigmata: Parasitengona). *Mem. Ent. Soc. Can.* 98:1–299.

Smith, I. M., and D. R. Cook. 1994. North American species of Neomamersinae Lundblad (Acari: Hydrachnidia: Limnesiidae). *Can. Entomol.* 126:1131–1184.

——. 1999a. North American species of *Tartarothyas* Viets (Acari: Hydrachnidia: Hydrophantidae). *Int. J. Acarol.* 25:37–42.

——. 1999b. North American species of *Parathyasella* Viets stat. nov. and *Amerothyasella* gen.

nov. (Acari: Hydrachnidia: Hydraphanitidae). *Int. J. Acarol.* 25:43–50.

——. 1999c. Proposal of *Thyopsoides* gen. nov. and reassignment of *Thyopsis (Thyopsella) plana* Habeeb. (Acari: Hydrachnidia: Hydraphanitidae). *Int. J. Acarol.* 25:51–53.

Smith, I. M., and D. R. Oliver. 1976. The parasitic associations of larval water mites with imaginal aquatic insects, especially Chironomidae. *Can. Entomol.* 108:1427–1442.

——. 1986. Review of parasitic associations of larval water mites (Acari: Parasitengona: Hydrachnida) with insect hosts. *Can. Entomol.* 118:407–472.

Smith, J. 1979. A review of the water mites of the family Anisitsiellidae (Prostigmata, Lebertioidea) from North America. *Can. Entomol.* 111: 529–550.

Soar, C. D., and W. Williamson. 1925–1929. *British Hydracarina.* 3 vols. 612 pp. Ray Society, London.

Sokolow, I. 1924. Untersuchungen fiber die Eiablage und den Laich der Hydracarina, I. *Arch Hydrobiol.* 15:383–405.

——. 1925. Untersuchungen fiber die Eiablage und den Laich der Hydracarina, II. *Z. Morphol. Oekol. Tiere* 4:301–332.

Strayer, D. 1988. Crustaceans and mites (Acari) from hyporheic and other underground waters in southeastern New York. *Stygologia* 4:192–207.

Szalay, L. 1949. Ueber die Hydracarinen der unterirdischen Gewässer. *Hydrobiologia* 2:141–179.

Uchida, T. 1932. Some ecological observations on water mites. *J. Fac. Sci. Hokkaido Imp. Univ. Ser. VI. Zool.* 1:143–165.

Vidrine, M. F. 1986. Revision of the Unionicolinae (Acari: Unionicolidae). *Int. J. Acarol.* 12:233–243.

——. 1987. Comments on the taxonomy of *Unionicola* (Unionicolidae). *Int. J. Acarol.* 113:157–158.

——. 1996. *North American Najadicola and Unionicola.* (In five parts, separate pagination.) Gail Q. Vidrine Collectibles, Eunice, LA.

Vidrine, M. F., and D. R. Clark. 1993. Freshwater mussel-mite associations in the upper Ouachita River, Montgomery County, Arkansas. *Proc. Louisiana Acad. Sci.* 56:8–12.

Viets, K. H. 1936. Wassermilben oder Hydracarina. Die Tierwelt Deutschl. 31 and 32:1–574.

——. 1938. Über die verscheiden Biotope der Wassermilben, besonders fiber solche mit anormalen Lebensbedingungen und fiber einige neue Wassermilben aus Thermalgewassern. *Verh. Int Ver. Limnol.* 8:209–224.

——. 1950. Porohalacaridae (Acari) aus der Grundwasserfauna des Maingebietes. *Arch. Hydrobiol.* 43:247–257.

——. 1955a. In subterranen Gewdssern Demschlands lebende Wassermilben (Hydrachnellae, Porohalacaridae und Stygothrombiidae). *Arch. Hydrobiol.* 50:33–63.

——. 1955b. Wassermilben aus Nordbayern (Hydrachnellae und Porohalacariclae Acari). *Abh. Bayer. Akad. Wiss. N. S.* 73:1–106.

——. 1955c. *Die Milben des Süsswassers und des Meeres,* Teil 1: *Bibliographie.* 476 pp. Gustav Fischer, Jena, Germany.

——. 1956. *Die Milben des Süsswassers und des Meeres. Hydrachnellae et Halacaridae (Acari),* Teil II und III: *Katalogund Nomenklatur.* 870 pp. Gustav Fischer, Jena, Germany.

Viets, K. O. 1982. *Die Milben des Süsswassers. Hydrachnellae und Halacaridae* (Part), Teil 1: *Bibliographie.* 116 pp. Paul Parey, Hamburg, Germany.

——. 1987. *Die Milben des Süsswassers. Hydrachnellae und Halacaridae* (Part), Teil 2: *Katalog.* 1012 pp. Paul Parey, Hamburg, Germany.

Wesenberg-Lund, C. 1919. Contributions to the knowledge of the postembryonal development of the Hydracarina. *Vidensk. Medd. Dansk. Naturhist. For.* 70:5–57.

Wiles, P. R. 1984. Watermite respiratory systems. *Acarologia* 25:27–32.

Wolcott, R. H. 1900. New genera and species of North American Hydrachnidae. *Trans. Am. Microsc. Soc.* 21:177–200.

——. 1905. A review of the genera of water mites. *Trans. Am. Microsc. Soc.* 26:161–243.

Wooley, T. A. 1969. Two new species of *Hydrozetes,* extant and fossil (Acari: Cryptostigmata, Hydrozetidae). *J. N.Y. Entomol. Soc.* 77:250–256.

Young, W. C. 1968. New species of *Midea, Piona,* and *Sperchon* (Hydracarina) from Colorado. *Trans. Am. Microsc. Soc.* 87:165–177.

——. 1969. Ecological distribution of Hydracarina in north central Colorado. *Am. Midl. Nat.* 82:367–401.

Young, W. C., and A. C. Rhodes. 1974. The influence of dissolved oxygen concentrations on three species of water mites (Hydracarina). *Am. Midl. Nat.* 92:115–129.

16

INTRODUCTION TO THE CRUSTACEA

As stated by Dahl (1963) the number of higher taxa of Crustacea has grown steadily since the turn of the 20th century. The combination of the discovery of new, disparate forms and the revelation of discrete lineages within established groups has led to the recognition of over 60 taxa at the ordinal level or above (Bowman and Abele 1982). The advent of molecular techniques for assessing phylogeny and computerized methods (e.g., Wilson 1992) that assign a simple "+" or "−" to complex and variable structures typically creates a myriad of combinations of taxa. Investigators then try to arrange their results into a system that often requires the creation of additional categories for lineages or clades that cannot fit into a traditional taxonomy. In short, trying to present a workable classification herein that will satisfy the expectations of all investigators is not possible. We are avoiding a lengthy discussion of this whole problem, generally following the plan of Bowman and Abele (1982) and Brusca and Brusca (1990), and are restricting our systematic array to the plan shown in Table 16.1, which involves only those taxa found in the freshwaters of the United States.

The position of the Crustacea among jointed invertebrates with an exoskeleton has been an actively debated topic. Different approaches to the analysis of traditional morphological or novel biochemical characters have led to varying conclusions, ranging from the monophyletic origin and radiation of the Crustacea to the alignment of certain putative crustacean groups with other arthropodan taxa. Certainly, the issue is not soon to be resolved and only the study of specimens, whether by molecular techniques or dissection, will provide the data necessary to clarify patterns of evolution within the Crustacea.

Four obvious features clearly distinguish the members of the phylum Crustacea: With few exceptions, they are all aquatic "arthropods"; respiration occurs through gills or through the general body surface; they all have a head with five pairs of appendages (two pairs of antennae followed immediately by a pair of mandibles, followed by to pairs of maxillae); and most of the body segments, or at least the more anterior ones, bear paired, jointed appendages that are fundamentally biramous. Compound eyes are not considered true segmental appendages. In the Cladocera and Ostracoda the second maxillae are lacking. More posteriorly the situation is variable, with some groups having an additional complement of thoracic or body appendages modified as mouthparts, and others not. Further, the number of thoracic and abdominal appendages and their modifications vary widely from one group to another, even within a single genus.

Some carcinologists distinguish fine gra-

Table 16.1. Classification of Crustacea Occurring in the Freshwaters of the United States

(Phylum, Subphylum, Superclass, or Class) Crustacea
- Class (or Subclass) Branchiopoda
  - Order Anostraca (fairy shrimps)
  - Order Notostraca (tadpole shrimps)
  - Order Laevicaudata (clam shrimps)
  - Order Spinicaudata (clam shrimps)
  - Order Cladocera (water fleas)
- Class (or Subclass) Ostracoda (seed shrimps)
  - Order Podocopa
- Class (or Subclass) Copepoda (copepods)
  - Order Calanoida
  - Order Harpacticoida
  - Order Cyclopoida
  - Order Poecilostomatoida
  - Order Siphonostomatoida
- Class (or Subclass) Branchiura (fish lice)
- Class (or Subclass) Malacostraca
  - Superorder Peracarida
    - Order Mysidacea (opossum shrimps)
    - Order Isopoda (slaters, aquatic sow bugs)
    - Order Amphipoda (scuds, sideswimmers)
  - Superorder Eucarida
    - Order Decapoda (freshwater shrimps, crayfish)
  - Superorder Pancarida
    - Order Thermosbaenacea
  - Superorder Syncarida
    - Order Bathynellacea

dations of meaning in the usage of "somite," "metamere," and "body segment" to denote the primary body divisions of adult Crustacea. Nevertheless, these expressions are more generally used interchangeably and without confusion; in this manual "body segment," "trunk segment," or simply "segment" are used consistently. In the primitive condition the trunk segments are distinct, but the segments forming the head are completely fused. The trunk is usually subdivided into thorax and abdomen so that there are three body regions, though fusion of some thoracic and abdominal segments is the common condition. One or more thoracic segments, for example, are often fused with the head to form a cephalothorax, which in many crustaceans is completely or partly covered by a shieldlike carapace. In groups such as the Cladocera and Ostracoda, body segmentation has disappeared.

In the strict sense, the most posterior division of the body, which bears the anus, is not considered a true body segment and is usually referred to as the telson. This term is universally used in the orders of Malacostraca, where this structure is comparatively small and flattened and does not bear any processes. In the other orders, however, the telson typically resembles the preceding body segment; furthermore, it bears a pair of terminal caudal rami, "cercopods," or heavy claws. In common usage it is often referred to as a "segment" in the nonmalacostracan orders. We can see no real objection to this practice. However, see details of these terminology arguments in Bowman (1971) and Schminke (1976).

The paired appendages are serially homologous; each is fundamentally biramous and consists of a basal protopod, a segmented endopod, and a segmented exopod. They are usually more or less specialized for a wide variety of functions, many such modifications being figured and considered rather completely in Chapters 17 through 25. The terminology of the various appendages and their component parts has undergone a long history. Many carcinologists now refer to the various segments of an appendage as articles. Antenna 1 and antenna 2 are often referred to as antennule and antenna, respectively. For clarity and to aid nonspecialists I (DGS) have retained antenna 1 and antenna 2 throughout. I have also adopted the simplest spelling for pereopod (also spelled paereopod and pereiopod).

All crustaceans go through a free nauplius stage (Fig. 16.1) or at least show some evidence of an egg–nauplius phase in early de-

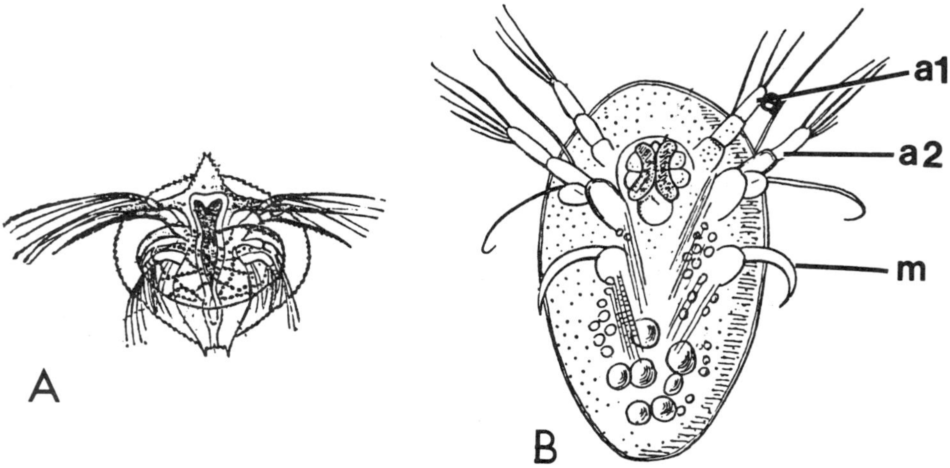

Fig. 16.1. Nauplius larva of freshwater crustaceans, A, a laevicaudate branchiopod; B, a copepod. a1, antenna 1; a2, antenna 2; m, mandibles. (A modified from T. H. Huxley 1885, *A Manual of the Anatomy of the Invertebrated Animals*, Appleton, New York; B modified from Harmer and Shipley (eds.) 1923, *The Cambridge Natural History*, Vol. 4, MacMillan, London.)

velopment. The nauplius is a free-swimming larva that can be best described as a little "head." Typically, it appears with the first three head appendages, antenna 1 and 2 and the mandibles, while adding subsequent appendages as it grows. Although almost universal among marine and estuarine forms, the nauplius is suppressed in most freshwater crustaceans. Among freshwater forms only eubranchiopods, ostracods, and copepods have free-swimming nauplii.

The great majority of the known species of the world are marine, and of the many orders occurring in the United States and surrounding waters, 13 are represented in freshwaters, but only 5 are restricted to our freshwaters. The number of species represented for each of the 13 orders in the United States ranges from one (Thermosbaenacea) to about 350 (Decapoda). Eleven orders are treated in detail in Chapters 15 through 22, so the present chapter is merely a summary of their chief distinguishing characteristics and affinities. Table 16.2 shows the environmental affinities of orders of Crustacea in the freshwaters of the continental United States.

On a worldwide basis, however, the groups listed above are not the sole representatives of the Crustacea in freshwater. The Spelaeogriphacea is a very small group known only from freshwater and to date only a few species have been reported. One occurs in a stream in South Africa and the other in Brazil. The Stygocaridacea, with the single family Stygocaridae and usually considered a group within the Bathynellacea, has just four species occurring in interstitial freshwaters of South America and New Zealand. Exceptional rarities occur among otherwise strictly marine orders. For example, several species of Cumacea are known from freshwater habitats in Kerala; and several freshwater species of Tanaidacea have been found in South America and the Kurile Islands.

To facilitate the identification of a particular crustacean group covered in subsequent chapters a key is provided below to the orders and classes treated individually. The characters have been kept as simple as possible so as to not require discussing the salient features of each group at this point. Specific features of each group are provided in some detail within each chapter.

Table 16.2. Environmental Affinities of Groups of Crustacea Found in Freshwaters of the United States

Chiefly marine; few freshwater species	Well Represented in both marine and freshwaters	Chiefly in freshwaters; few marine species	Exclusively freshwater	Approximate number of species living in U.S. freshwaters
Mysidacea				4
	Amphipoda			150
	Copepoda			230
	Decapoda			350
	Isopoda			130
	Ostracoda			300
		Cladocera		150
			Anostraca	30
			Bathynellacea	5(?)
			Laevicaudata	5
			Notostraca	6
			Spinicaudata	18
			Thermosbaenacea (fresh to brackish)	1

KEY TO THE MAJOR GROUPS OF FRESHWATER CRUSTACEA OCCURRING IN THE UNITED STATES

1. Trunk appendages foliaceous, with mutiple, paddlelike lobes . . . . **BRANCHIOPODA, 2**
Trunk appendages leglike, ambulatory **3**

2. Trunk appendages 10 or more; 5 to 100 mm in length . . . . **Phyllopodous Branchiopoda**
Only 6 or less trunk appendages, usually very small, less than 5 mm in length . . . . . **Cladocera**

3. Animal enclosed in a bivalved carapace . . . . **OSTRACODA**
Animal not enclosed in such a carapace **4**

4. Number of thoracic appendages (feeding and locomoter) 6 or less; abdomen without appendages; free-living or parasitic **5**
Number of thoracic appendages (feeding and locomotor) 8; abdomen with paired appendages; free-living **6**

5. Body flattened and broad; 2 maxillae modified as suckers, parasitic on fish **BRANCHIURA**
Body cylindrical and elongate; maxillae not modified as suckers; parasitic on fish or free-living **COPEPODA***

6. A distinct carapace covering the head and most or all of the thoracic appendages **MALACOSTRACA (in part), 7**
Carapace absent or indistinct . . . . **MALACOSTRACA (in part), 8**

*Some parasitic copepods are highly modified and typical copepod characteristics are obscured. Typically all body segments become fused in more highly specialized forms.

7. Most thoracic segments free; animals rarely over 20 mm in length . . . .**Mysidacea**
All thoracic segments fused; animals usually over 20 mm in length, up to 200 mm **Decapoda**
8. Animals minute, 0.5 to 3.5 mm in length, interstitial in habits; 3 or less abdominal appendages **9**
Animals between 3.5 and 20 mm in length; common in surface and subterranean habitats, rarely interstitial; 6 abdominal appendages **10**
9. First pair of thoracic appendages modified to assist in feeding (maxilliped) **Thermosbaenacea**
First pair of thoracic appendages not so modified . . . .**Bathynellacea**
10. Animals dorsoventrally compressed, abdominal segments fused with telson . . . .**Isopoda**
Animals laterally compressed, abdominal segments not fused with telson . . . . **Amphipoda**

REFERENCES

Abele, L. G., et al. 1992. Phylogeny of selected maxillopodan and other crustacean taxa based on 18S ribosomal nucleotide sequences: a preliminary analysis. *Acta Zool.* 73:373–382.

Bowman, T. E. 1971. The case of the nonubiquitous telson and the fraudulent furca. *Crustaceana* 21:165–175.

Bowman, T. E., and L. G. Abele. 1982. Classification of the recent Crustacea. In: *The Biology of the Crustacea*, Vol. 1: *Systematics, the Fossil Record, and Biogeography*, pp. 1–27, edited by L. G. Abele. Academic, New York.

Brusca, R. C., and G. J. Brusca. 1990. *Invertebrates.* 922 pp. Sinauer Associates, Sunderland, MA.

Crampton, G. C. 1928. The evolution of the head region in lower arthropods and its bearing upon the origin and relationships of the arthropodan groups. *Can. Entomol.* 40:284–302.

Dahl, E. 1963. Main evolutionary lines among recent Crustacea. In: *Phylogeny and Evolution of Crustacea*, edited by H. B. Whittington and W. D. I. Rolfe, pp. 1–15. Museum Comparative Zoology, Harvard University, Cambridge, MA.

Fitzpatrick, J. F. 1983. *How to Know the Freshwater Crustacea.* 227 pp. W. C. Brown, Dubuque, IA.

Gordon, I. 1957. On *Spelaeogriphus*, a new cavernicolous crustacean from South Africa. *Bull. Br. Mus. Nat. Hist. Zool.* 5:29–47.

Kurian, C. V. 1962. Three species of Cumacea from the lakes of Kerala. *Bull. Cent. Res. Inst. Trivandrum* 8C, 1961:55–61.

McLaughlin, P. A. 1980. *Comparative Morphology of Recent Crustacea.* 177 pp. W. H. Freeman, San Francisco.

Mane-Garzon, F. 1943. Tres especies de Tanais de las aquas dulces de Sud America. *Com. Zool. Mus. Hist. Nat. Montevideo* 1(4):1–15.

Moura, G., and M. L. Christoffersen. 1996. The system of the mandibulate arthropods: Tracheata and Remipedia as sister groups, "Crustacea" non-monophyletic. *J. Comp. Biol.* 1:95–113.

Noodt, W. 1964. Naturliches System und Biogeographie der Syncarida (Crustacea Malacostraca). *Gewasser Abwasser* 37:77–186.

——. 1967. Biogeographie der Bathynellacea (Syncarida). *Proc. Symp. Crustacea* 1:411–417.

Schminke, H. K. 1976. The ubiquitous telson and the deceptive furca. *Crustaceana* 30:292–299.

Schram, F. R. 1986. *Crustacea.* 606 pp. Oxford University Press, New York.

Schram, F. R., and M. J. Emerson. 1991. Arthropod pattern theory: a new approach to arthropod phylogeny. *Mem. Queensland Mus.* 31:1–18.

Siewing, R. 1957. Anatomie und Histologie von *Thermosbaena mirabilis. Akad. Wiss. Lit. Abh. Math.Naturwiss. Kl.* 7:195–270.

Wilson, G. D. F. 1992. Computerized analysis of crustacean relationships. *Acta Zool.* 73:383–389.

Zilch, R. 1972. Beitrag zur Verbreitung und Entwicklungsbiologie der Thermosbaenacea. *Int. Rev. ges. Hydrobiol.* 57:75–107.

17

PHYLLOPODOUS BRANCHIOPODA

(Fairy, tadpole, and clam shrimps)

The phyllopodous Branchiopoda are among the most characteristic inhabitants of temporary ponds and pools, especially during spring and early summer. They move along the bottom or swim and glide about gracefully, often with the ventral side uppermost, and frequently in dense concentrations. They are absent from running waters, and there are no true marine representatives, although one genus, *Artemia*, is restricted to highly saline lakes and evaporation basins used for the commercial production of salt.

Except for the immature stages in some families of insects, this is the only major taxonomic category found exclusively in freshwaters in the United States. All others have marine representatives, terrestrial representatives, or both.

General characteristics. All members of this primitive group are distinctly segmented and have 10 to 71 pairs of delicate, flat, lobate swimming and respiratory appendages. Because of the leaflike appearance of these appendages, the members of this division are collectively known as phyllopods. The last body segment bears a pair of short to long cercopods. Body length, exclusive of cercopods, ranges between 2 and 100 mm. Females are usually more abundant than males in natural populations, and sometimes males are rare.

The phyllopodous Branchiopoda consists of four distinct orders. The Anostraca (fairy shrimps) have stalked compound eyes, 20 trunk segments, 11 pairs of swimming legs (in American species) (Fig. 17.1A, B). There are two genital segments, and seven other posterior segments, the last of which is the telson. There is no carapace. The Notostraca (tadpole shrimps) (Fig. 17.1D) have sessile compound eyes, a large, shieldlike carapace covering most of the body, and 35 to 71 pairs of legs. Neither the number of legs nor the number of segments is constant within a species. The clam shrimps, comprising the orders Laevicaudata and Spinicaudata, are laterally compressed and are partially or completely enclosed in a carapace consisting of two lateral valves (Fig. 17.4). The Laevicaudata are partially enclosed by a carapace without growth lines (Fig. 17.1C). They possess a distinct head shield (Fig. 17.1C) and only 10 (male) or 12 (female) thoracic appendages. The telson is very small and simple. The Spinicaudata are completely enclosed in the carapace that shows several growth rings and lack a head shield. There are up to 32 pairs of appendages and the telson is an elaborate, spinose structure.

Coloration ranges from translucent or whitish through gray, blue, green, orange, and reddish. Coloration of some species is highly variable and probably governed to a large extent by the type of food being ingested. All the individuals in a particular pond usually have the same general colora-

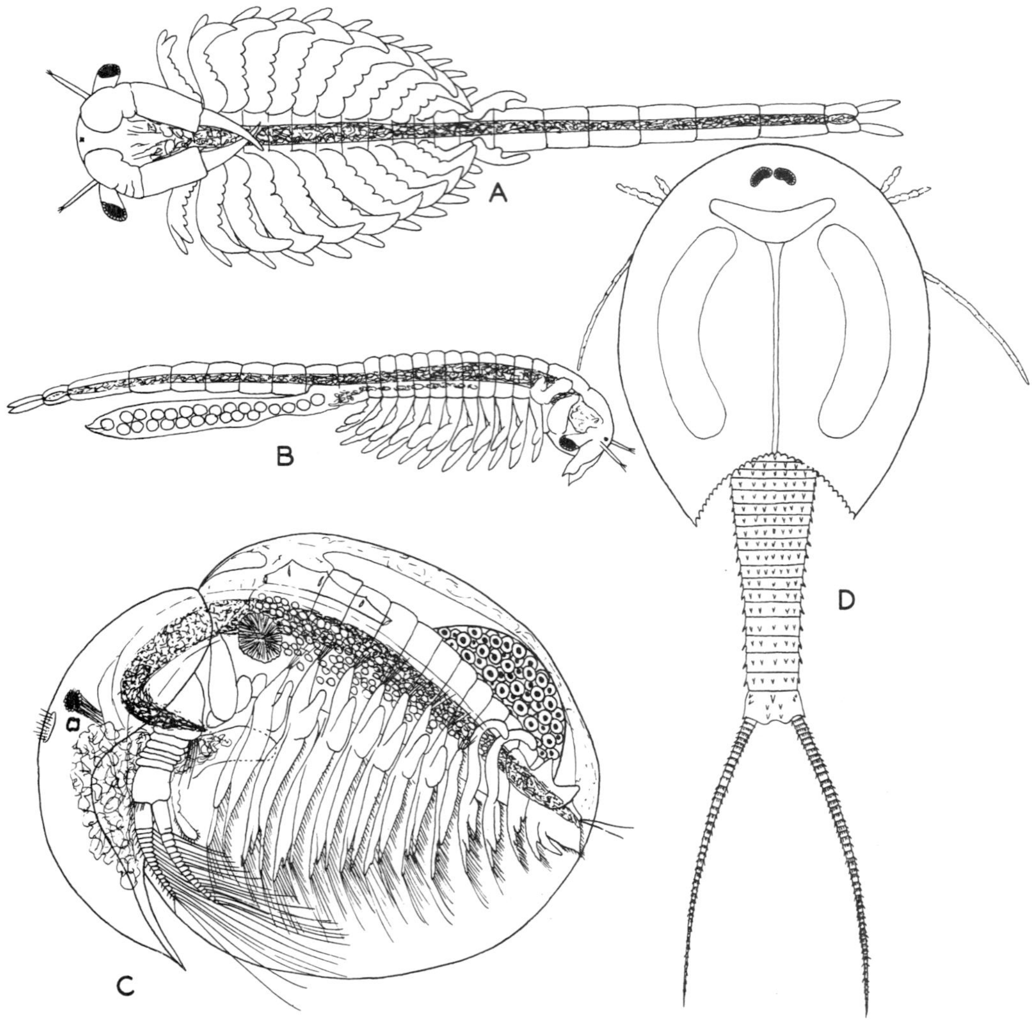

Fig. 17.1. Typical phyllopodous Branchiopoda. A, ventral view of male *Branchinecta paludosa*, ×8 (setation omitted); B, lateral view of female *B. paludosa*, ×6 (setation omitted) showing brood sac; C, female *Lynceus brachyurus*, ×25, with left valve removed (diagrammatic); D, dorsal view of *Triops*, ×3. (A–C modified from Sars 1896.)

tion at any one time, although some species exhibit progressive color changes during the adult instars.

In ventral view the mouthparts are largely concealed by the labrum. The mandibles have undivided molar surfaces and no palp, the first maxillae are triangular plates with setose biting edges, and the second maxillae are small, simple plates that are of little importance in feeding (Fig. 17.2).

The first antennae are small, uniramous, and often unsegmented (Figs. 17.1; 17.3). In female Anostraca the second antennae are cylindrical and elongated, but in the males they are greatly enlarged and specialized for clasping the females during copulation (Fig. 17.3). The males of some species of Anostraca have a median frontal appendage on the head (Fig. 17.3) or an antennal appendage arising near the base of each second antenna (Fig. 17.21). In the Notostraca the second antennae are minute or absent in

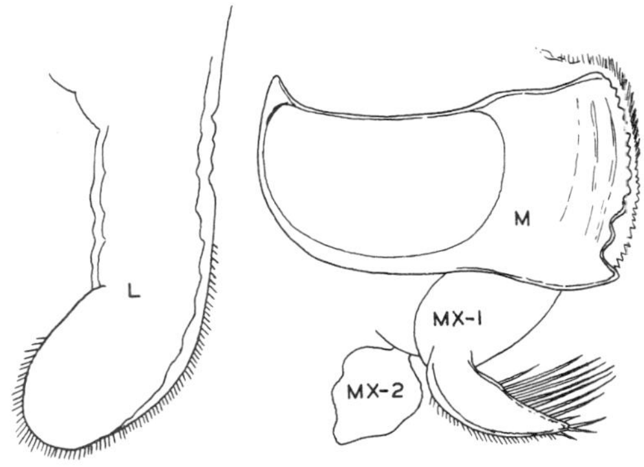

Fig. 17.2. Mouthparts of *Lynceus brachyurus.* L, lateral view of labrum; M, MX-1, and MX-2, right mandible, first maxilla, and second maxilla as seen from below. (Modified from Sars 1896.)

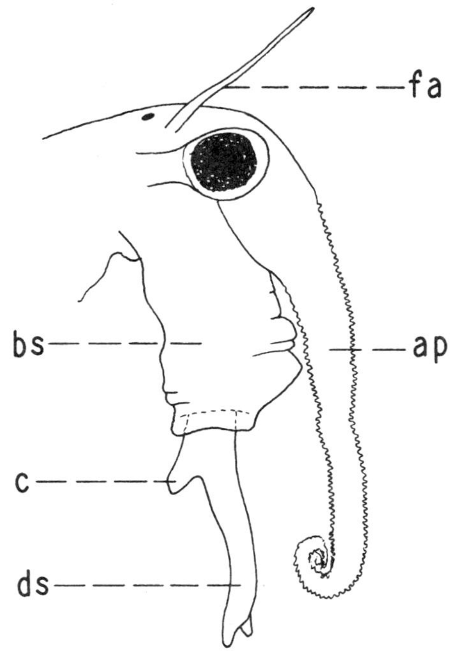

Fig. 17.3. Diagrammatic lateral view of head of male *Eubranchipus bundyi.* ap, antennal appendage; bs, basal segment of second antenna; c, calcar of distal segment of second antenna; ds, distal segment of second antenna; fa, first antenna.

both sexes, but in the five families of clam shrimps they are long, biramous, and setose (Fig. 17.4A–E).

According to prevailing usage, the boundary between head and trunk of phyllopods lies immediately behind the segment that bears the second maxillae, the first trunk segment being the one that bears the first pair of legs. Although the head forms a well-defined body region, the trunk is not clearly divisible into thorax and abdomen as it is in most microcrustaceans. The trunk segments are all similar morphologically and vary widely in number from one genus to another. Furthermore, the number of pairs of legs and the proportion of trunk segments bearing them also vary widely.

The location of the genital pores, however, is sometimes used as a basis for loosely designating a "thorax" (pregenital segments) and an "abdomen" (postgenital segments) in some families. In the Anostraca the first two abdominal segments (collectively the genital segment) are partially fused and bear the external uterine chamber and base of the ovisac in the female, and penes in the male.

The body divisions, segmentation, and leg arrangement of the Notostraca represent a peculiar situation among freshwater crustaceans. Posterior to the second maxillae there are 25 to 44 so-called segments, not counting the terminal telson. Unlike typical segments, however, many of them have two or more pairs of legs, and especially posteriorly some of them bear 5, 10, or more pairs of legs per segment (Fig. 17.1D). The total number of pairs of legs ranges from 35 to 71 among the American species of Notostraca. Following the suggestions of Linder (1952), it therefore seems preferable to refer to "body rings" in the Notostraca rather than "segments." The first 11 rings are sometimes considered the thorax, the remainder being the abdomen. In general, all body rings are of comparable length. A variable number of posterior abdominal rings are without legs. A single species usually shows considerable variation with respect to number of body rings, total number of legs, number of legs on a particular body ring, number of legs bearing abdominal rings, and number of legless abdominal rings (Fig. 17.5). In *Triops longicaudatus,* for example,

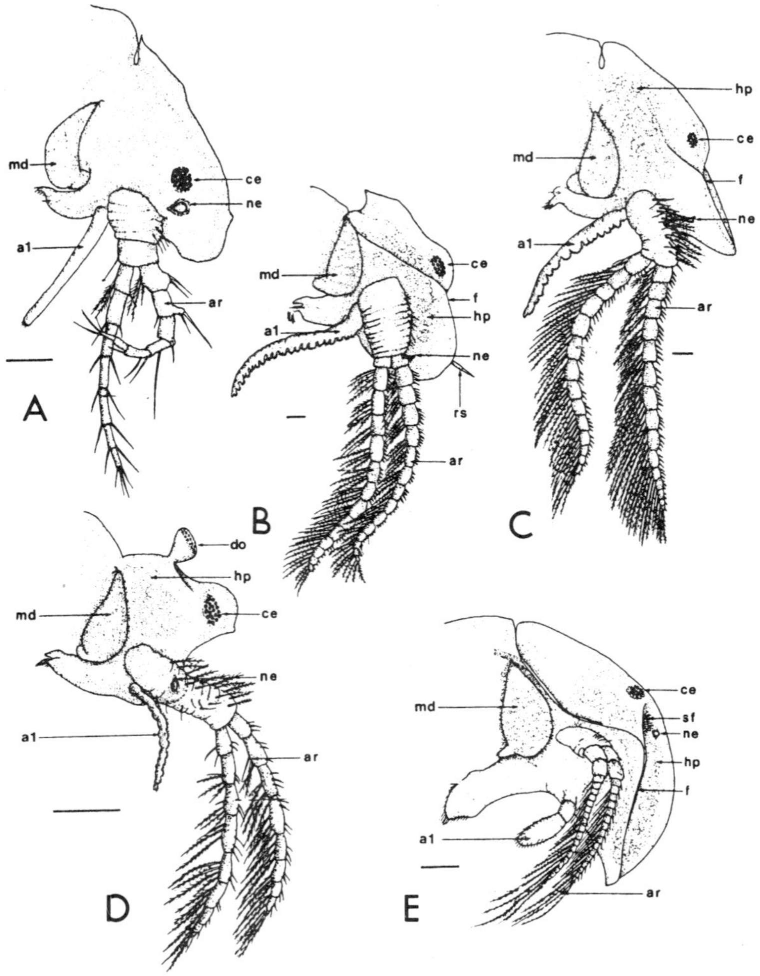

Fig. 17.4. Structure of clam shrimp families. A, Cyclestheriidae; B, Leptestheriidae; C, Cyzicidae; D, Limnadiidae; E, Lynceidae. Scale, 0.25 mm. a1, antenna 1; ar, anterior ramus of antenna 2; ce, compound eye; do, dorsal organ; f, fornix; hp, hepatopancreas; md, mandible; ne, naupliar eye. (A–E from Olesen et al. 1996.)

the number of legless abdominal rings varies from 4 to 16, The number of body rings is usually more constant in a parthenogenetic population than in one containing males.

Tadpole shrimps sometimes have an incomplete body ring just anterior to the telson. Only the dorsal or ventral part of the ring may be present. About 1% of the specimens in many collections show "spiral rings" (Fig. 17.6), usually near the telson. These consist of one to four spirals, running either to the left or to the right.

The telson of the fairy shrimps bears two terminal platelike plumose cercopods (caudal rami, furcal rami). In the tadpole shrimps the cercopods are very long, filamentous, segmented, and spiny. In the clam shrimps the telson is laterally compressed and bears two parallel rows of spines dorsally and two rami (Fig. 17.7).

Most phyllopod trunk appendages, or swimming legs, are basically similar (Fig. 17.8). They are biramous, flat, translucent, lobed, and setose. The median margin bears

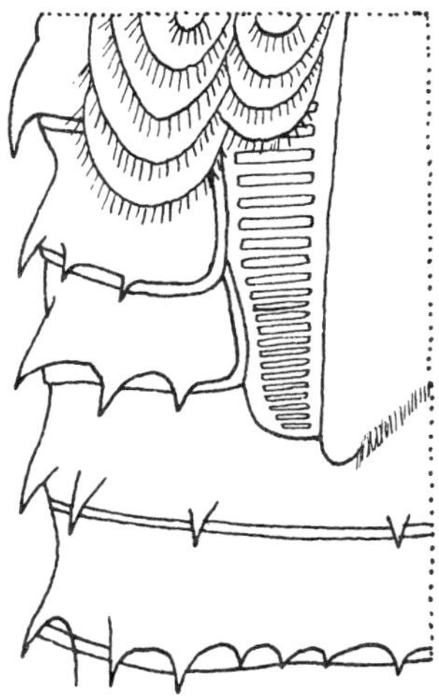

Fig. 17.5. Ventral view of a portion of the right side of a female *Lepidurus lemmoni* from which the last 25 legs have been removed. The small rectangles represent the attachment areas of the removed legs. (Redrawn from Linder 1952.)

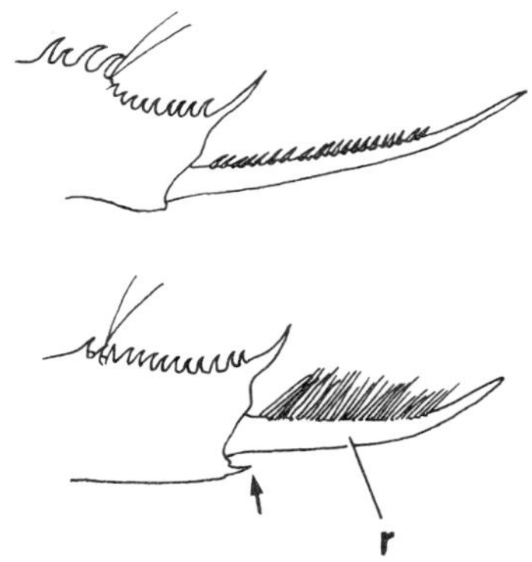

Fig. 17.7. Telson of Limnadiidae; upper, *Limnadia lenticularis;* lower, *Eulimnadia agassizii.* r, ramus; arrow denotes spine.

a series of small lobes, or endites, the basal endite of the Notostraca and clam shrimps being a more robust gnathobase. Near the base of the lateral margin are one or two large lobes, the proepipodites, and more distally is the thin, oval or elongated, nonsetose branchia (epipodite). The two large distal lobes are the exopodite (flabellum) and endopodite.

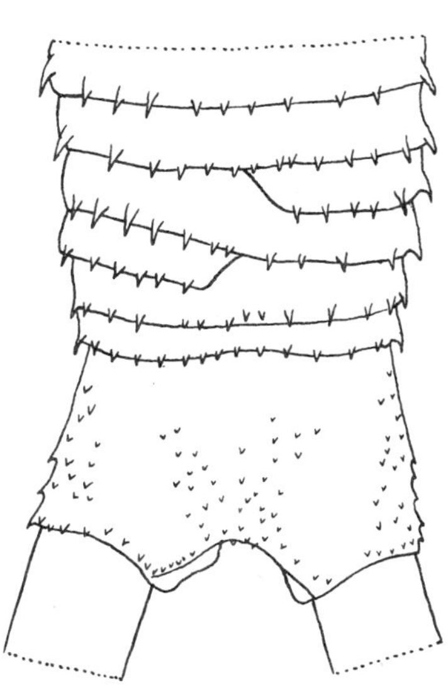

Fig. 17.6. Ventral view of telson and a portion of the abdomen of *Triops longicaudatus* showing a spiral of two rounds located between normal body rings and beginning and ending in the mid-ventral line. (Modified from Linder 1952.)

A few of the legs of all phyllopods are modified from this basic plan. In the Anostraca the more posterior pairs are definitely smaller and less lobate than preceding pairs. In the Notostraca the first one or two pairs of legs have elongated filamentous rami that probably serve as tactile structures (Fig. 17.9A). The flabellum of the eleventh pair of legs of the female is modified into a cup-like brood pouch that carries the eggs after they leave the reproductive tract (Fig. 17.9C). The more posterior legs of the Notostraca taper progressively to a very small size. The first one or two pairs of legs of male clam shrimps are hooked, prehensile (Fig. 17.9D), and sometimes asymmetrical. In the female clam shrimp elongated flabella of two or three pairs of posterior legs aid in keeping the developing eggs in place between body and carapace (Fig. 17.1C).

Locomotion. Fairy shrimps and tadpole shrimps glide or swim gracefully by means of complex beating movements of the legs that pass in a wavelike anterior–posterior di-

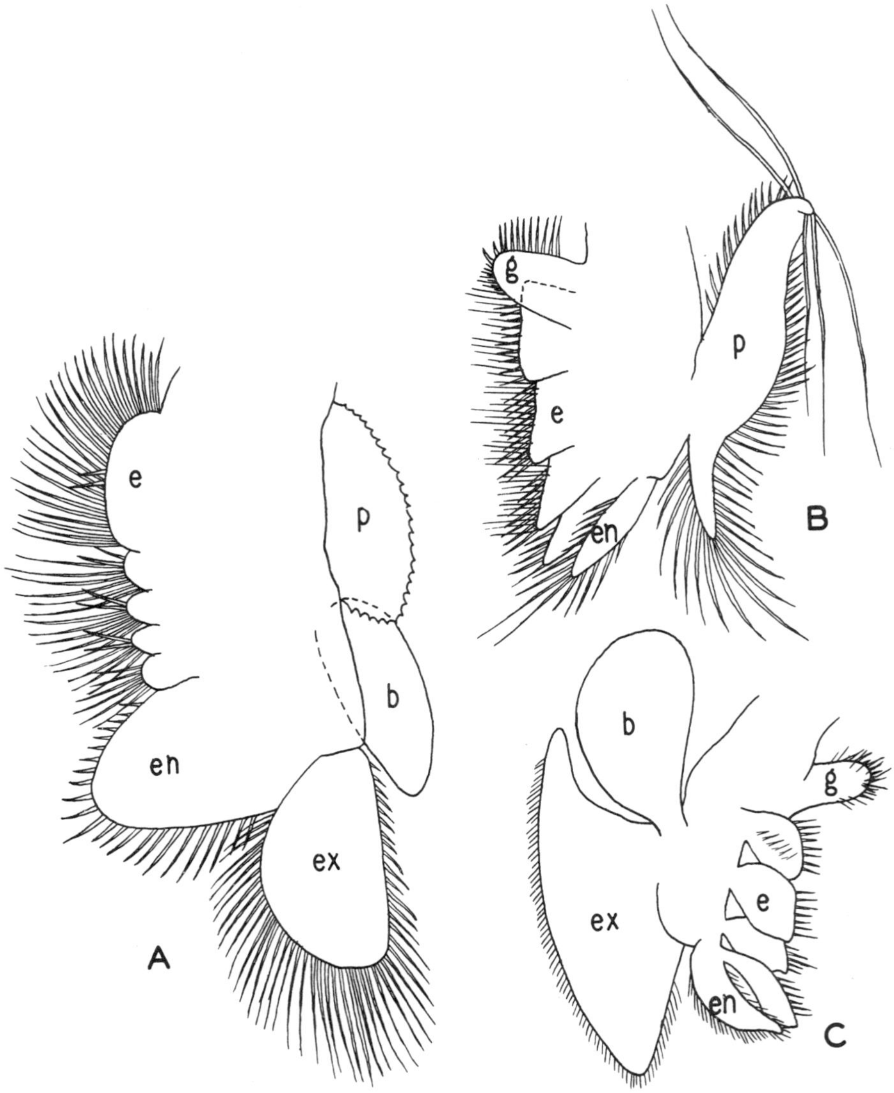

Fig. 17.8. Trunk appendages of typical phyllopodous Branchiopoda. A, *Branchinecta paludosa;* B, *Lynceus brachyurus; C, Lepidurus.* b, branchia; e, endite; en, endopodite; ex, exopodite; g, gnathobase; p, proepipodite. (A and B modified from Sars 1896; C modified from Packard 1883.)

rection (Fig. 17.10). Sometimes they drift along slowly; other times they dart rapidly or come to rest on the bottom. Lowndes (1933) presents evidence to show that locomotion in the Anostraca is not produced by beating of the legs as a whole, but rather by propellerlike movements of the exopodites, the speed of the animal being determined by the angle at which the exopodites are held. In the clam shrimps locomotion is accomplished mainly by "rowing" movements of the large biramous second antennae, the legs being of little importance. Fairy shrimps habitually swim with the ventral side upward, but members of the other two orders are usually seen swimming ventral side downward. Tadpole shrimps creep or burrow superficially in soft substrates much of the

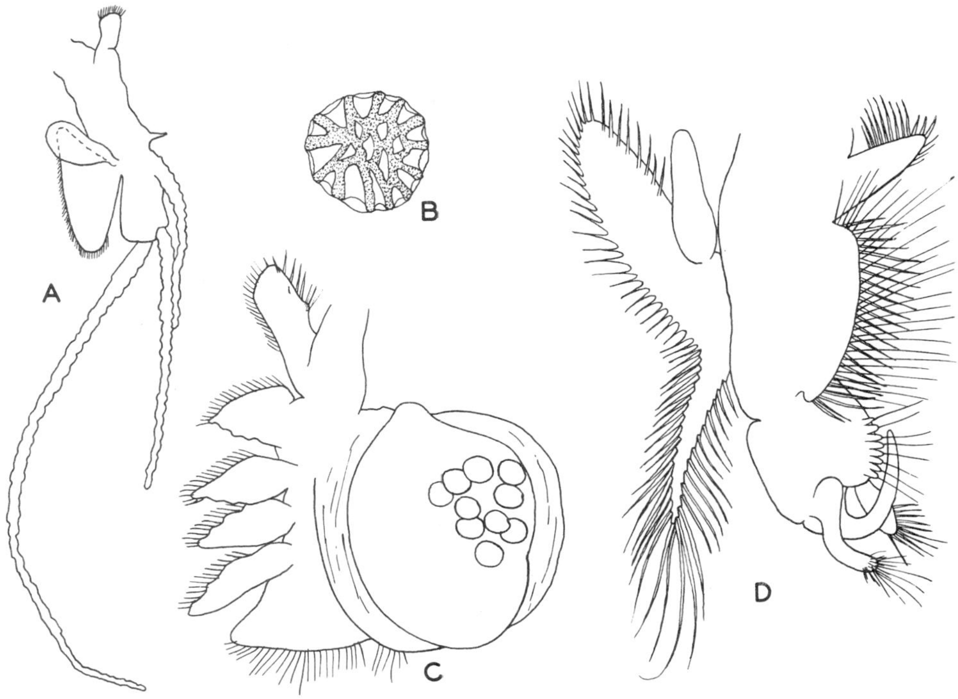

Fig. 17.9. Specialized reproductive structures of the phyllopodous Branchiopoda. A, first leg of male *Triops;* B, resting egg of *Eulimnadia;* C, eleventh leg of female *Triops;* D, first leg of male *Lynceus brachyurus.* (A and C modified from Packard 1883; B redrawn from Mattox 1939; D modified from Sars 1896.)

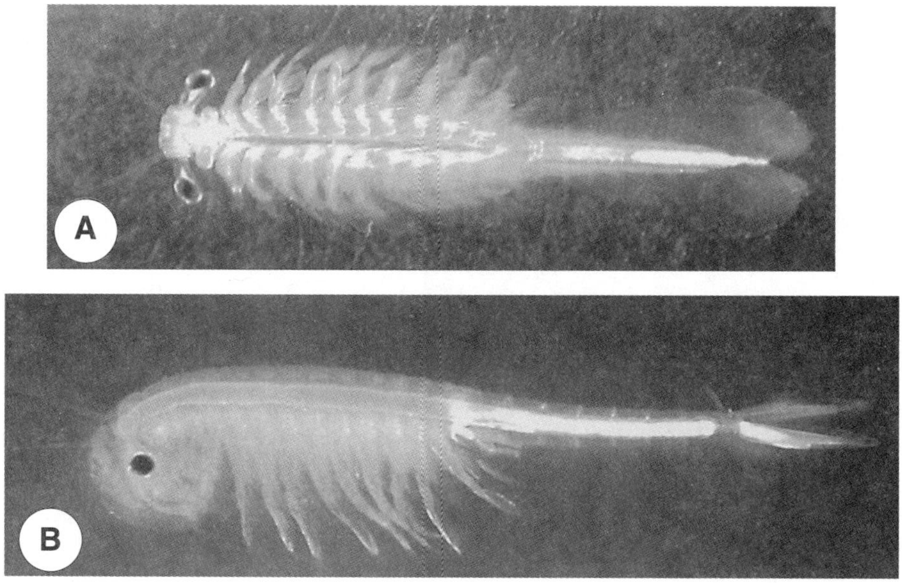

Fig. 17.10. Representative anostracans swimming. A, *Thamnocephalus;* B, *Streptocephalus.* (Photos by Naida Zucker.)

time. Clam shrimps also burrow or move along the surface of the substrate in a clumsy manner. When the antennae stop beating they fall on their side.

Food, feeding. Food consists mostly of algae, bacteria, Protoctista, rotifers, and bits of detritus. Movements of the legs serve as a food-getting mechanism, plankton and materials from the substrate being strained nonselectively from the water or scraped from the substrate by the setose appendages (especially by the basal endites) and concentrated and agglutinated in a median ventral groove running most of the length of the body. This stream of food moves forward, activated chiefly by the gnathobases, and is further agglutinated at the anterior end by a sticky secretion produced by labral glands. Mastication occurs outside the digestive tract in an atrium formed by the overhanging labrum. It is probable that phyllopods feed continuously, but all the food reaching the mouth is not necessarily ingested, the excess being sloughed off. Although microscopic organisms and detritus constitute the sole food of Anostraca and clam shrimp, the Notostraca also utilize larger particles and have even been observed gnawing on dead tadpoles (Fig. 17.11), earthworms, mollusks, and frog eggs. They are best characterized as omnivores. *Branchinecta gigas* is unique among American Anostraca. It is not a filter-feeder and subsists on large food masses as a carnivore, chiefly other crustaceans.

Internal anatomy, physiology. The digestive tract consists of a vertical esophagus, a small lobate stomach lying in the head, a long intestine, a short rectum, and an anus. Two digestive glands in the head empty into the stomach.

A dorsal tubular heart extends throughout most or all of the body segments in the Anostraca, but in the Notostraca and clam shrimp it is much shorter, sometimes extending through only four segments. A variable number of paired, lateral, slitlike ostia bring blood from the hemocoel into the heart, and blood is forced anteriorly and out of the opening at the anterior end of the heart by peristaltic contractions. Colorless amoebocytes occur in phyllopod blood, and a hemoglobin is known to be dissolved in the plasma of some species.

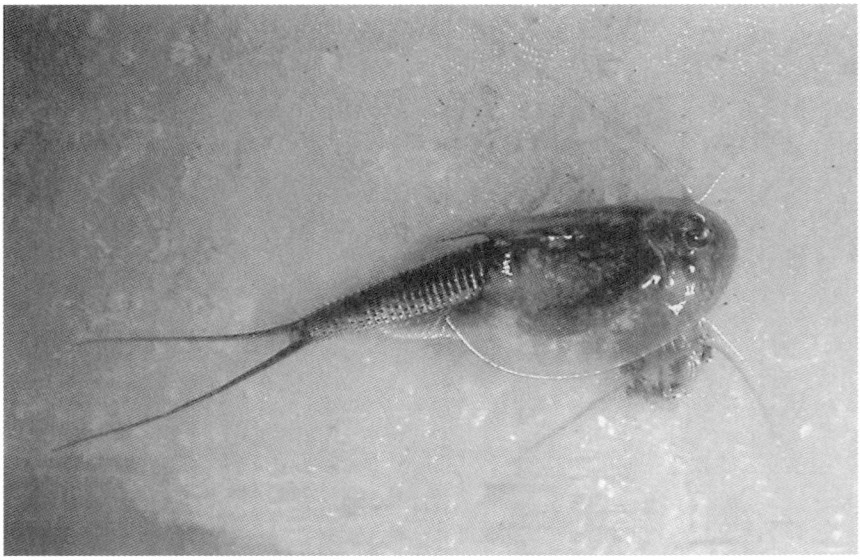

Fig. 17.11. *Triops* feeding on a tadpole. (Photo by Naida Zucker.)

The exchange of oxygen and carbon dioxide probably takes place through most exposed surfaces of the body but especially through the surfaces of the legs and their branchiae.

Excretory organs are located at the base of the second maxillae and are called maxillary glands. The two coiled glands empty their contents, typically urea, to the environment. The majority of metabolic waste is ammonia, which is presumably passed to the aquatic medium through the same surfaces that take up oxygen.

Osmotic regulation presumably occurs chiefly through the wall of the digestive tract and through the branchiae of the first 10 pairs of legs. In some species there is an active uptake of sodium by the gills. The nauplius of many anostracans possesses a circular dorsal gland that has recently been linked to osmoregulatory functions and is now believed to be a salt-regulating organ. It disappears at maturity. *Artemia* continually swallows water, regardless of its salt concentration.

The ladderlike nervous system is diagrammatic and primitive (Fig. 17.12). The two widely separated nerve cords run the full length of the body, and except for the more posterior segments, there are two ganglia and two transverse commissures per segment.

In the Anostraca the compound eyes are stalked and far apart, but in the other two orders they are quite close together, sometimes touching or partially fused. A small median ocellus of the nauplius persists in the adult. Experiments have shown that the upside-down swimming position of some species is a phototropic response governed by stimuli received by both the ocellus and compound eyes, and if the upper side of the container is covered and weak illumination is supplied from below, many of the animals will turn over and swim with the ventral side downward. Swimming on the back is normal in darkness. Immature individuals usually react more strongly to a light source than do the mature animals. Taste areas appear to be concentrated in the region of the mouthparts, and touch is distributed over the whole body. The "dorsal organ" is a small patch of modified cephalic epithelium of unknown function, but it is probably sensory.

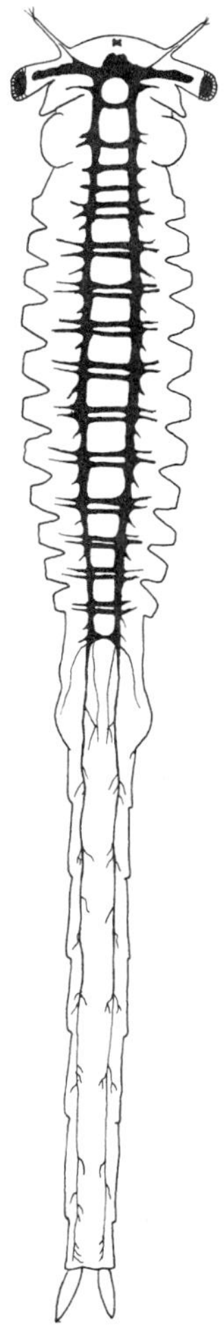

Fig. 17.12. Nervous system of *Branchinecta*. (Modified from Sars 1896.)

Reproduction. The phyllopods demonstrate a wide range of reproductive strategies. Species exhibit sexual reproduction, mixed reproduction involving males and hermaphrodites, or parthenogenesis. Among sexual species or populations of Notostraca and clam shrimps, sperm leave the body through two simple genital pores, through penes in the Anostraca (Fig. 17.13).

Many of the reproductive phenomena in the phyllopods are only beginning to be understood. In many species of clam shrimps, *Artemia* (Anostraca), and Notostraca, the frequency of males is highly variable. In some populations of species in these groups

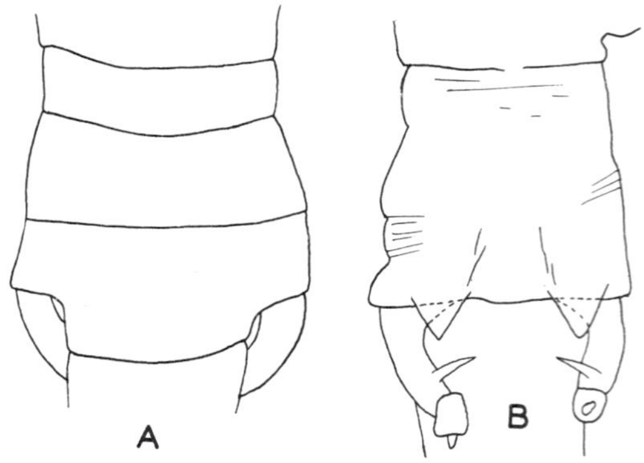

Fig. 17.13. Genital segments of male *Branchinecta paludosa.* A, dorsal view; B, ventral view showing penes. (Redrawn from Linder 1941.)

males are absent altogether. Parthenogenesis is therefore assumed to occur, and, in a few species or populations of Notostraca and clam shrimps, it appears to be the rule. In mixed populations, involving hermaphrodites and males, both selfing and syngamic reproduction may occur. The role and significance of rare males in otherwise parthenogenetic or hermaphroditic populations remains unclear.

Among the Anostraca, sexual reproduction is the rule except in the genus *Artemia*, which can form parthenogenetic populations or, possibly, species. In sexual reproduction, clasping, paired males and females may swim about firmly united or amplexed for varying lengths of time. In *Streptocephalus*, amplexus lasts for only a few moments, while in *Artemia* it can last for up to a few days. In the Anostraca the male approaches the female from a ventral position. The second antennae is then clasped around the female in the region of the genital segment. In the fairy shrimps the male turns the body at an angle to that of the female and recurves the posterior end so that the genital segment is brought into contact with the external uterine chamber of the female. The process is somewhat similar in the tadpole shrimps, and in the clam shrimps the body of the male is placed at a right angle and vertical to that of the female before sperm can be transferred. Females generally live longer than males.

The Laevicaudata appear to engage in sexual reproduction only and males exist in near equivalent numbers as females.

The Spinicaudata, however, display several modes of reproduction, ranging from sexual to mixed breeding to putative parthenogenesis. Members of the Cyzicidae and Leptestheriidae are generally sexual. The single species of the Cyclestheriidae, *Cyclestheria hispida*, consists of females mostly. Males generally appear late in the season when drier conditions prevail, and sexual reproduction apparently occurs. In the Limnadiidae reproduction varies both among and within species. In *Limnadia lenticularis*, males are extremely rare and in northern populations have yet to be observed, suggesting that parthenogenesis is the normal mode of reproduction. Species of *Eulimnadia* exhibit sexual, mixed reproduction including out crossing and selfing hermaphroditism (androdioecy) and parthenogenesis. Mating in spinicaudates is as elaborate as that reported for anostracans. Courting behavior and amplexus (Fig. 17.14) lasts about 30 minutes.

Both fertilized and parthenogenetic eggs are retained externally for one to several days on the body of the female before being released. The eggs of Anostraca are carried in an oval or elongated ventral brood sac. In the Notostraca they are carried in a curious receptable formed by modified and overlapping flabellum and branchia on each eleventh trunk appendage. In clam shrimp the eggs are held together in a mucous mass between body and shell (Fig. 17.1C) except *Cyclestheria*, in which an ephippium like that formed in cladocerans is produced.

The number of eggs produced by a single female varies widely. They are released in a series of fertilized or unfertilized clutches at intervals of 2 to 6 or more days. Usually there are 1 to 6 clutches per female and 10 to 250 eggs/clutch, depending on the species. In syngamic reproduction, typical of anostracans, each clutch of eggs is fertilized by a separate copulation. The female carries the eggs for one to several days, during which time they undergo early development. Thereafter they may be dropped to the bottom or remain attached until the female dies and sinks to the bottom. Hatch-

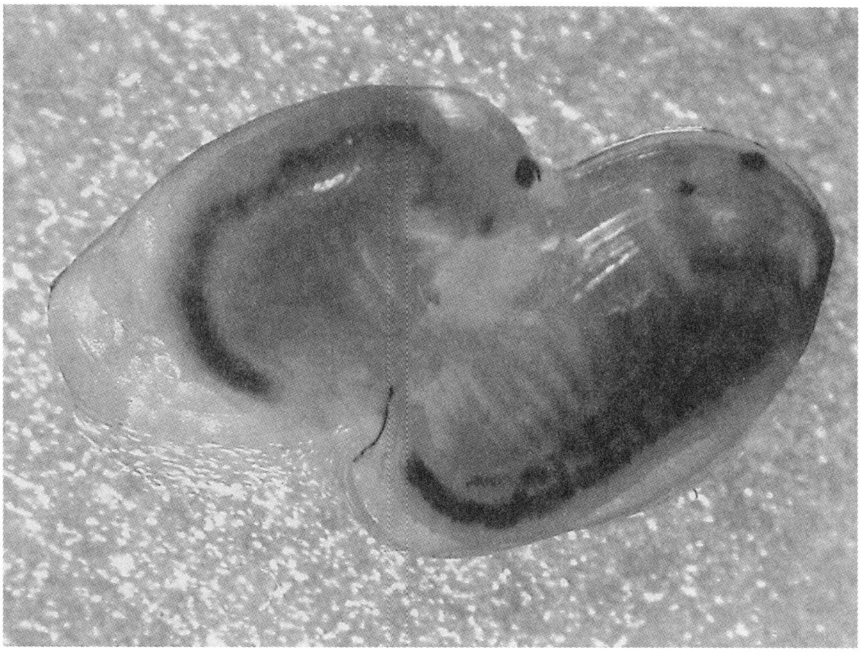

Fig. 17.14. Copulating *Eulimnadia texana.* (Photo by Naida Zucker.)

ing appears to be governed chiefly by environmental queues. In *Artemia*, under certain and as yet unknown circumstances, the eggs hatch within the ovisacs, and the nauplii are released directly into the environment.

Traditionally, it was believed that some phyllopods produced two distinct types of eggs: (1) thin-shelled "summer" eggs, which hatched almost immediately, and (2) thick-shelled "resting" or "winter" eggs, which were capable of withstanding unusual heat, cold, and prolonged desiccation. It now seems that morphology and physiology in eggs vary within and among species, although there is evidence that physiologically distinct eggs occur within some species. The thick shell or cyst characteristic of most species is often ornamented and variously shaped (Figs. 17.9B; 17.15; 17.16). External elaboration affords protection from mechanical injury and sunlight. Ornamented eggs are produced in either the presence or absence of males in the population. In many species eggs hatch within several days after deposition especially in clam shrimp species. Often, development is so rapid that the larva may hatch before the egg is released from the body of the female.

Resting eggs. Phyllopods are chiefly inhabitants of vernal pools and ponds, which dry up completely in the dry, warm months. Resting eggs, therefore, constitute the sole device for tiding a population over from one season to another. It is probable that resting eggs of all species are capable of withstanding desiccation. During the summer the resting eggs become dried in the bottom mud, and during the winter they are frozen for varying periods.

Until rather recently it was commonly believed that both drying and freezing were necessary prerequisites for further development and hatching of these eggs, but experimental evidence has been accumulated to show that this is sometimes not the case. Resting eggs have been hatched in the laboratory after drying without freezing and freezing without drying as well as after both freezing and drying. Groups of eggs subjected to rapid or slow freezing and thawing all showed hatching. There is also some ev-

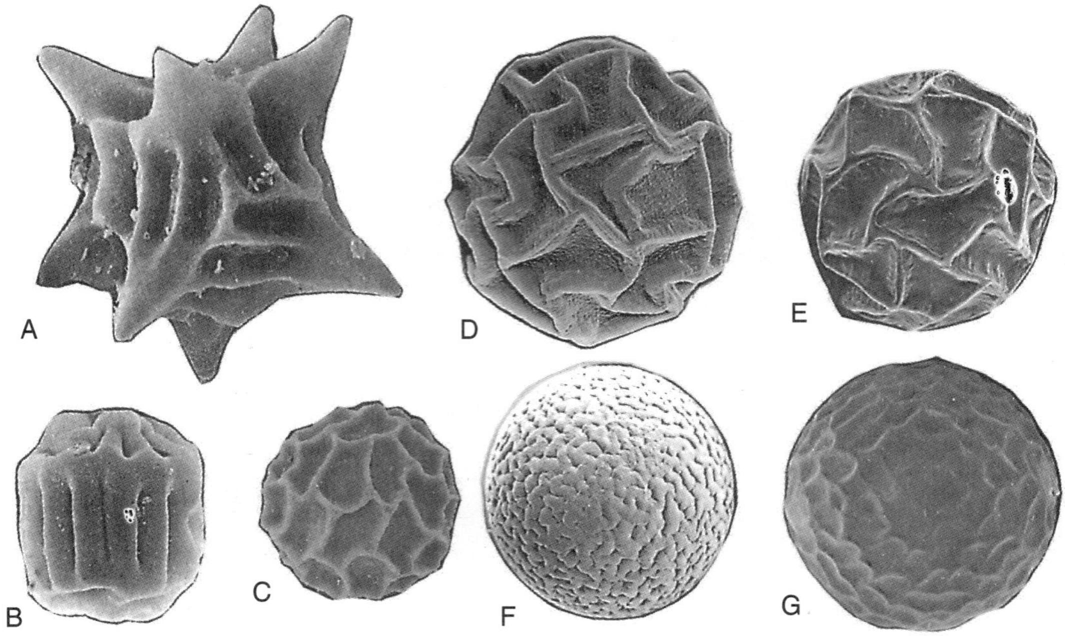

Fig. 17.15. Egg cysts of Limnadiidae and Anostraca. A, *Eulimnadia astrova;* B, *E. texana;* C, *E. agassizii;* D, *Streptocephalus texanus;* E, *S. seali;* F, *Eubranchipus neglectus;* G, *Branchinecta paludosa.* (A, B from Belk 1989; C from Smith 1992; D–G courtesy G. Mura.)

idence to show that resting eggs of a few species will hatch without either drying or freezing.

The observation is sometimes made that phyllopods occur in ponds that never dry up completely during the summer months. This is unquestionably true, but bear in mind that the active phyllopod population disappears with the onset of warm or cool weather and that the animals appearing in the following spring have undoubtedly hatched from resting eggs that overwintered on the bottom or in the dried mud along the pond margin above the lowered water level.

Although freezing and drying usually occur naturally, neither is indispensable. The resting period usually lasts from 6 to 10 months in temperate latitudes, but such a long interval is neither necessary nor inhibitory; dry eggs of certain species may be placed in water and hatched at any time of the year in the laboratory. The viability of resting eggs is very striking. Some species have been described from adults reared from eggs in dried mud sent from distant parts of the world. Viable eggs have been kept in dried pond mud on the laboratory shelf for as long as 15 years. Viable eggs of *Artemia* have withstood a temperature of 81°C for 1 hour, −190°C for 24 hours, and an air pressure of 0.000001 mm of mercury for 6 months. Eggs are more viable when stored in the refrigerator, as compared with room temperature. Eggs of other species can withstand experimental temperatures as high as 99°C. Field study of an anostracan species has revealed that resting eggs can withstand brief range fires.

Development, life cycle. Depending on the species, the eggs hatch into a typical nauplius or a more advanced metanauplius larvae with three pairs of appendages representing the first antennae, second antennae, and mandibles of the adults (Fig. 17.17). A metanauplius usually shows the faint beginning of two or three trunk segments. In the Cyclestheriidae, development is direct and a fully formed juvenile emerges from the egg.

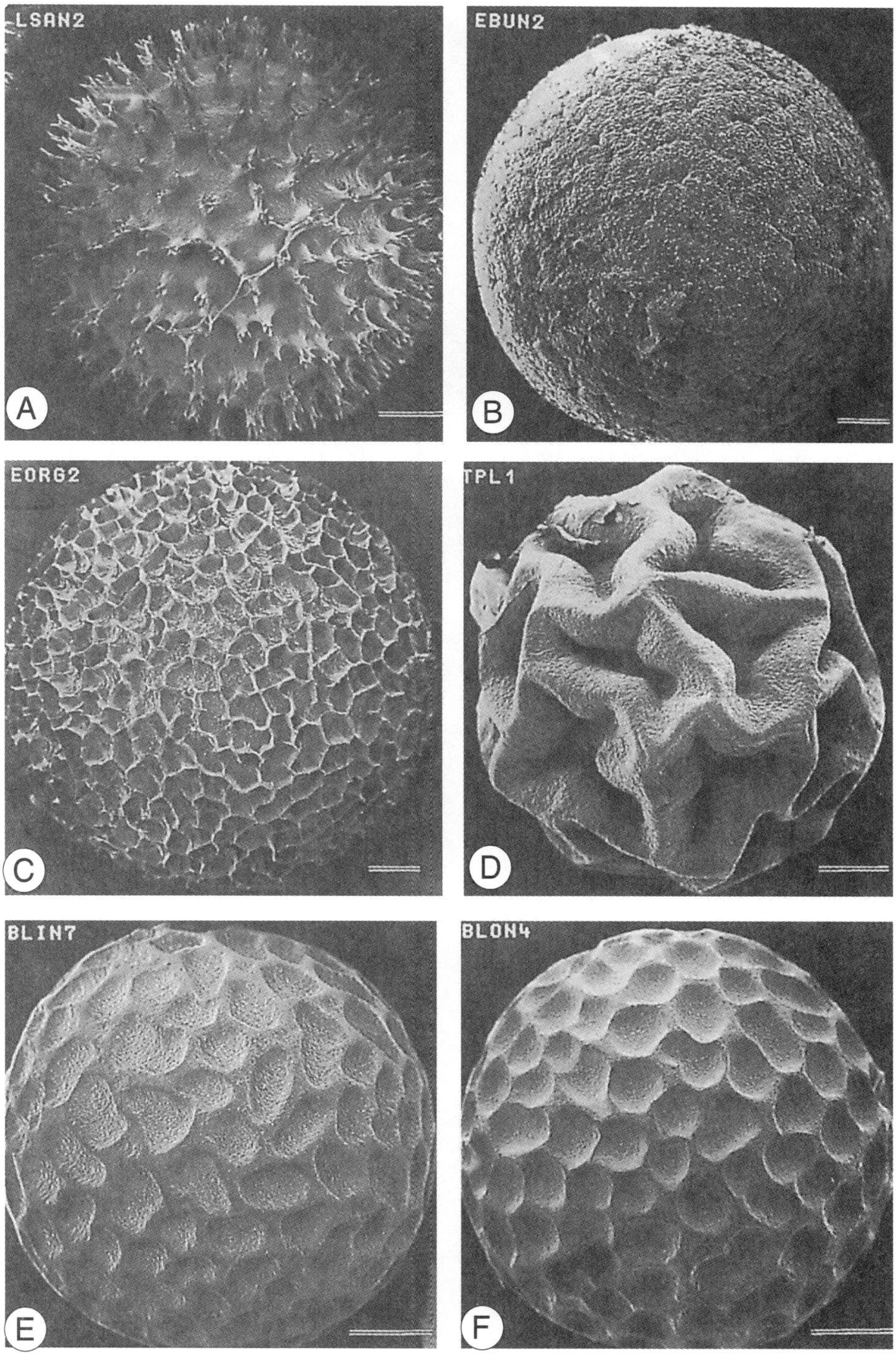

Fig. 17.16. Egg cysts of Anostraca. A, *Linderiella santarosae;* B, *Eubranchipus bundyi;* C, *E. oregonus;* D, *Thamnocephalus platyurus;* E, *Branchinecta lindahli;* F, *B. longiantenna.* Scale, 50 μm. (A–F from Hill and Shepard 1997, with kind permission from Kluwer Academic Publishers.)

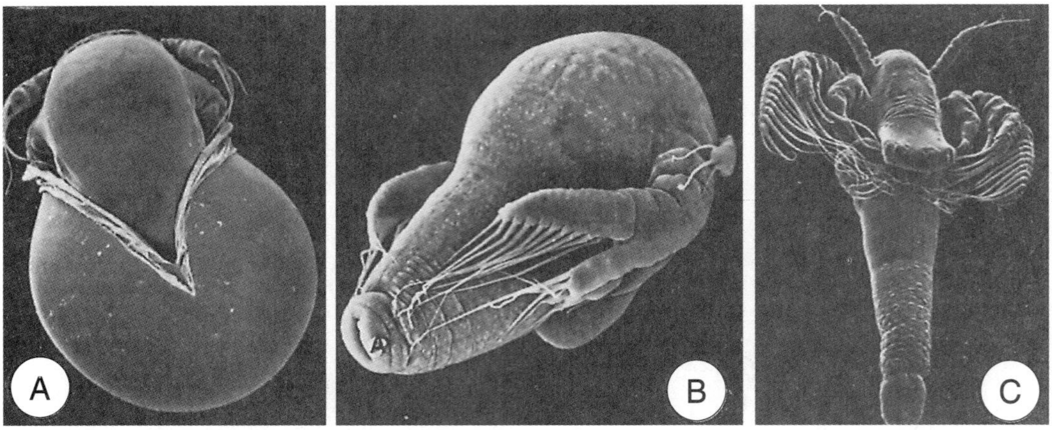

Fig. 17.17. Hatching of *Artemia*. A, emerging nauplius; B, freshly emerged nauplius; C, metanauplius, ×55. (A–C from Rosowski et al. 1997.)

Although "summer" eggs hatch while they are still held on the body of the female or on the bottom shortly after their release, development of resting eggs is much slower. It is thought that early segmentation begins soon after the eggs reach the bottom of the drying pond and before the water temperature gets too low. Development ceases or is greatly retarded by desiccation and low temperatures, however, and is not resumed until early the following spring, when water is available and temperatures become favorable. Hatching of resting eggs in temperate latitudes, therefore, normally occurs sometime between late January and early May. In *Artemia*, hatching is a complex process involving the emergence of a pre-swimming stage (prenauplius) encased in a membranous sac. This is followed by exit of the larva (nauplius 1) from the sac and beginning of a free-swimming lifestyle (Fig. 17.17).

Development proceeds comparatively rapidly in the early spring (or rehydration of the egg), and phyllopods may often be collected soon thereafter. They have even been seen swimming about under ice, especially after an early thaw and subsequent freeze, or in the winter months in deep pools.

Beginning with the newly hatched nauplius, there is a long series of instars, each following a complete ecdysis, or shedding of the exoskeleton. Changes in size from one instar to the next are gradual, and there is progressive appearance of more segments, more appendages, and increasing complexity of appendages. *Linderiella occidentalis*, for example, has 17 instars in the life history. By the third instar the length averages 1.1 mm and all appendages through the fifth or sixth trunk segments are represented to some degree. By the sixth instar the average length is 2.1 mm and all appendages are present, though most of the most posterior ones are completely developed. Complete development, sexual maturity, and copulation are attained in the 16th instar. *Artemia franciscana* probably has 14 instars, sexual maturity being attained in the 12th instar. *Streptocephalus seali* has 18 preadult instars. Some investigators have found that the number of instars is variable for each species, depending on temperatures and food conditions.

The active portion of the life cycle may be completed in a surprisingly short time. The life history of several species of *Eulimnadia* from hatching until death is completed in 2 to 3 weeks. *Lynceus brachyurus* lives about 26 days and *Cyzicus mexicanus* about 2 months. *Artemia franciscana*, on the other hand, reaches sexual maturity in 18 to 21 days after hatching and has a normal life span of 4 months, although specimens have been kept alive for 9 months in the laboratory. *Streptocephalus sealii* can live up to 12 weeks and *Eubranchipus oregonus* has an exceptionally long natural life span, sometimes up to 25 weeks. *Cyclestheria hislopi* dies during the "ovipositing" molt.

In small ponds that contain water for only a few weeks during the spring and early summer, phyllopods usually have one generation per year. The resting eggs hatch early in the spring, and the animals mature rapidly and produce resting eggs that fall to the bottom and do not hatch until the following spring. Some "summer" eggs are usually produced during the short period of activity and, although they may hatch, there is not sufficient time for the second generation to mature. Such immature individuals and any remaining "summer" eggs are presumably destroyed by the adverse conditions accompanying the disappearance of water from the pond. In Great Salt Lake, in permanent ponds, and in ponds that persist for several months, however, there are usually two or more complete generations per year, the summer generations being completed in a short time because the "summer" eggs hatch almost immediately.

Ecology. It seems evident that the development of most phyllopod species in the spring and their sudden disappearance in summer or early autumn are governed largely by temperature conditions. Most species do not appear until water temperatures exceed 4°C; and 13 to 30°C, depending on the species, is the upper limit beyond which individuals die quickly. In Great Salt Lake, *Artemia franciscana* disappears in the autumn when the water temperature drops below 6°C; *Eubranchipus bundyi*, *Eubranchipus serratus*, and *Lynceus brachyurus* are seldom successful in water warmer than 15°C. At the other extreme, phyllopod populations have been recorded from desert ponds where the midday temperatures exceeded 42°C.

Many species exhibit little in the way of specific habitat preferences. The widely distributed *Eubranchipus vernalis* for example, occurs in roadside ditches, grassy vernal ponds, cattail marshes, and woodland pools. Some species, however, appear to be generally restricted to clear or muddy waters. *Eubranchipus bundyi; Eubranchipus oregonus*, and *Branchinecta coloradensis* occur only in clear ponds and pools, while many clam shrimps, *Triops*, *Branchinecta gigas*, and *Thamnocephalus platyurus* are almost invariably found in muddy, alkaline waters. I (RWP) have often collected phyllopods in mountain locations in puddles in pothole depressions of a boulder, sometimes where the entire water volume was less than 20 L.

A compelling feature about phyllopod ecology is their almost complete absence from permanent water bodies, either lentic or lotic. Seldom are they collected from bodies of water having areas exceeding one hectare. Phyllopods are almost defenseless, and it is also notable that they are not often abundant in ponds containing carnivorous insects, and are rarely present along with carnivorous fishes. In fact, they attain greatest densities in vernal ponds and prairie pools that are generally lacking in other macrometazoans. A notable exception to this paradigm is *Cyclestheria hislopi* (Fig. 17.18), which is known from permanent, warm, and nutrient-rich ponds and lakes. This species' preference for thick algal mats probably discourages predators.

In keeping with widely fluctuating water levels and the consequent rapid changes in the dissolved salt content of pond waters, phyllopods have highly efficient but poorly known means of making physiological ad-

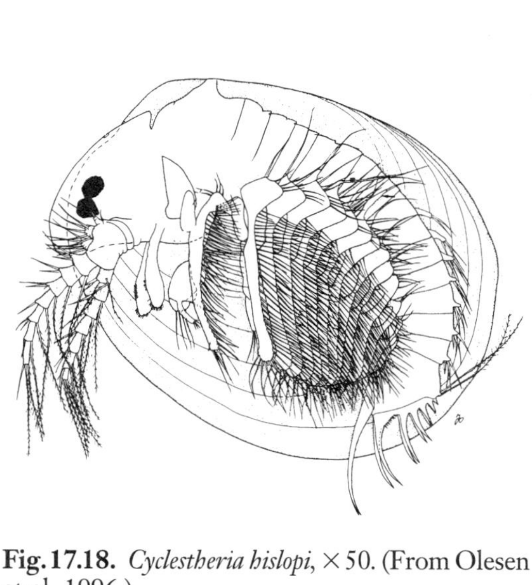

Fig. 17.18. *Cyclestheria hislopi*, × 50. (From Olesen et al. 1996.)

justments to varying osmotic pressures. In prairie pools, for example, the total salt or alkali content may vary between 0.05 and 1.00% within several weeks, yet their phyllopod populations maintain themselves successfully. Similarly, prairie phyllopods are usually capable of tolerating wide ranges in hydrogen ion concentration. *Eubranchipus serratus* has been found in a pH range of 6.6 to 9.5, *E. vernalis* from 5.2 to 7.4, and *Streptocephalus seali* in 5.2 to 10.0. As a whole, however, phyllopods have alkaline affinities; pH 6.8 seems to be about the usual lower limit for most species.

One part per million of dissolved oxygen seems to be a typical limiting factor for many species.

Excellent cultures of *Artemia franciscana* can be maintained in the laboratory, using natural or synthetic seawater, and much experimental work has been done on this species. Mature specimens will tolerate salinities ranging from the saturation point of sodium chloride down to about 3% seawater, though they will not reproduce in salinities below 3% sodium chloride. Adults reared in high salt concentrations reach a larger size than those reared in lower salt concentrations, and sexual maturity is attained more rapidly in populations reared at low salt concentrations. Europeans have found that the pregenital:postgenital body-length ratio and relative development of the cercopods and their setation are correlated with salinity, but these results have not been duplicated with American strains. *Artemia* populations have variously been found to be parthenogenetic, bisexual, diploid, triploid, tetraploid, pentaploid, or octoploid.

Year-to-year field notes on the occurrence of phyllopods in particular ponds present a puzzling problem. Sometimes a species is abundant for several successive years, and then one year it is unaccountably absent. Other ponds may have populations present or absent from year to year in a completely sporadic manner. Furthermore, where there is a group of small ponds and pools in a restricted area, phyllopods are usually present only in certain scattered pools and absent from most or all of the bodies of water immediately surrounding each inhabited one. Often, a pond contains no more than one or two genera at a time, but it is not uncommon for a water body to support several species of different genera or even orders. Several multiple higher taxon combinations have been reported, including *Streptocephalus*, *Thamnocephalus*, and *Triops* in a flooded soybean field in Kansas, *Limnadia*, *Eulimnadia*, and *Eubranchipus* in a small pasture pond in Massachusetts, and *Thamnocephalus*, *Triops*, and *Leptestheria* in playa pools in Arizona. There are many circumstances in which several species of the same genus coexist in the same pool and swim together. In the rich anostracan fauna of California, up to 7 species of *Branchinecta* have been reported in the same pond.

The bodies and appendages of phyllopods sometimes become so completely covered with sessile ciliates and green algae that they swim sluggishly and with difficulty.

An excellent account of the ecology of fairy shrimp in varying habitats can be found in Erikson and Belk (1999).

Geographical distribution. A few genera and species of phyllopodous Branchiopoda are truly cosmopolitan. *Artemia* is known from North and South America, the West Indies, throughout Europe, Asia, Africa, the Middle East, and Australia. *Lepidurus* occurs everywhere except in South America. The species of *Streptocephalus* are found in warm latitudes on both sides of the Atlantic, with most living in North America and Africa. Among the Chirocephalidae, *Eubranchipus* and *Dexteria* are both restricted to North America, and included species occasionally demonstrate extreme examples of endemism. Temporary ponds, alkaline pools, and salt lakes of the Great Plains and westward constitute a rich collecting ground, the majority of the species reported from the United States occurring in this area. California contains 25 described and undescribed species of anostracans.

Phyllopods, represented by *Polyartemiella* and *Artemiopsis*, are found as far north as extreme northern Alaska, Canada, and Greenland. Neither genus is found in the contiguous United States. However, phyllopods have never been taken from the transient freshwater and brackish ponds on the periphery of parts of Antarctica.

Most species are probably effectively distributed through their resting eggs, which may be blown about as dust or transported by birds and insects. The presence of populations of *Branchinecta coloradensis* above timberline in rock pools having a volume of only a few liters may be easily attributed to both of these agencies.

Enemies. Amphibia, dytiscid larvae, caddis larvae, and perhaps a few other insects are the chief predators of phyllopods. In the western United States, the giant fairy shrimp *Branchinecta gigas* feeds on other fairy shrimp species and the notostracan genus *Lepidurus* apparently regularly eats anostracans. Phyllopods are of no great importance in the diet of fish, since they typically occur in bodies of water too alkaline, too small, or too temporary to support fish. However, in floodplain pools, which are sometimes connected to adjacent permanent water, fish can enter and prey on fairy or clam shrimp. Larval cestode worms have been reported to infest *Artemia*.

Economic importance. Many years ago, the Indians in the vicinity of Great Salt Lake are said to have used dried *Artemia* extensively as food, and *Triops* is said to be used as food today by a few natives of the Federal District in Mexico. *Artemia* currently supports a large aquaculture industry and is used as food for fishes and is actively utilized in laboratories and classrooms for research and instruction. Occasionally *Triops* becomes a pest in rice fields, notably in California. There are several reports of clam shrimps becoming a nuisance in fish-rearing ponds, especially where the fish do not feed on the shrimp (goldfish and pike). Shrimp become so numerous as to clog outlet screens.

Collection, preparation. A coarse dipnet is the only necessary implement for collecting adults, but a plankton net is required for the early immature stages. It is suggested, based on observations by Beladial and Mertens (1999), that specimens of anostracans be fixed in 4% formalin prior to preservation in 70 to 80% alcohol, especially in morphologically similar species. These authors have shown that preservation in alcohol alone can distort antennal characters used in taxonomy.

With few exceptions, phyllopods are usually difficult to maintain and rear in the laboratory. *Artemia*, however, is easy to culture: Dry brine shrimp eggs are simply added to seawater or a 0.1 to 6.0% brine solution. The salt should not be iodized. Best results are obtained at 20 to 30°C. See Galtsoff et al. (1937), Cole and Brown (1967), and Dempster (1953) for more details about salt tolerances. *Artemia franciscana* is one of the very few species of metazoans that can be raised in axenic cultures (Provasoli and Shiraishi 1959).

Dissection is not often necessary for identification, but specimens being studied for any length of time should be manipulated in glycerin. Small clam shrimp and isolated appendages of other species may be permanently mounted in glycerin jelly.

Taxonomy. The Anostraca, Notostraca, Laevicaudata, Spinicaudata, and Cladocera constitute the Class Branchiopoda and have the following characters in common: leafy trunk appendages, compound eyes, reduced mouthparts, parthenogenesis as well as gonochorism, and a carapace.

The name "Phyllopoda" as currently interpreted represents an unnatural group, and usually comprises the Anostraca, Notosctraca, and the clam shrimps. It is occasionally still used by some investigators to distinguish certain clusters of branchiopods but its use is not consistent. On the other hand, there is no such serious objection to using "phyllopod" as a common, descriptive, inclusive term for Anostraca, Notostraca, Laevicaudata, and Spinicaudata. The bodies of phyllopods are elongated, distinctly segmented, and have 10 or more pairs of trunk appendages, whereas the Cladocera are short and compact, with no sign of segmentation, and have only five or six pairs of trunk appendages. The abdomen of Cladocera is greatly reduced, but in pyllopods the abdomen (postgenital region) is prominent. It is also significant that the phyllopods have nauplius stages, whereas the Cladocera have direct development. Other less important differences are the occurrence of ephippia in the Cladocera (also present in Cyclestheriidae) and a single compound eye in Cladocera as compared with two in phyllopods (again with the possible exception of the Cyclestheriidae).

Recent study has helped clear up many historical problems in phyllopod taxonomy. Nevertheless, many uncertainties persist and a number of species remain known only from their original, sometimes brief, description. Several species of the clam shrimp family Limnadiidae have been recently reduced to synonymy, but the number of new species of certain anostracan genera continues to increase. Many of the characters used to distinguish anostracans and clam shrimp species are subtle and exist only in the adult. To make matters more complicated, it has been demonstrated that alcohol preservation of anostracans can distort taxonomically important characters. In cryptic species, however, biochemical and biological data can help resolve otherwise highly variable morphological characters. *Lepidurus*, a genus of the Notostraca, has been divided into several poorly defined species in the past, which were difficult to diagnose, thus casting doubt on their validity. Recent biochemical work by King and Hanner (1998) has revealed, however, that several "lineages" exist, each probably consisting of at least one reproductively isolated species. Also, the notostracan *Triops longicaudatus* has at times been considered to be a group of species, but more lately it has been treated as a single, highly variable species. Subsequent biochemical investigations by Sassman et al. (1997) have revealed that at least two morphologically similar but biologically discrete species exist, *Triops longicaudatus* and *T. newberryi*.

Intraspecific variations are further complicated by the occurrence of hybrids. Wiman (1979), for example, cites evidence for "natural" hybrids in the genus *Streptocephalus*, notably *S. mackini* × *S. dorothae* and *S. mackini* × *S. texanus*.

Unfortunately, the general outline of the "conchostracan" shell is somewhat variable and does not have the taxonomic importance formerly attributed to it. Furthermore, the original descriptions of many American "conchostracans" are too generalized to be used for critical work.

The taxonomic status of *Artemia* has long been controversial, especially in Europe, where *Artemia* shows much greater variability than it does in the United States. The reproductive habits of different populations vary markedly. In parts of Europe parthenogenesis is the rule, males being rare or absent; some such populations are diploid, but others are tetraploid and octoploid. *Artemia* in Great Salt Lake is diploid and males are common.

Thus, the *Artemia* complex is thought to contain at least three sibling species, including *A. franciscana* from North and Central America and the Caribbean region. The other two species, *A. monica* and *Artemia* sp. are restricted to Mono Lake, California, and ponds in Nebraska, respectively. Up to 15 strains from North American waters of differing ionic compositions have been identified (Bowen et al. 1985) but most are interfertile (Browne and Bowen 1991). Although these strains differ by only subtle morphological details, they are more clearly reproductively isolated by the specific ionic concentrations of the habitat. They may be designated as cross-fertile population clusters, which are either incipient species or sibling species. Currently, perhaps it is best to use *Artemia franciscana* for North American *Artemia* existing outside of Mono Lake, California. *Artemia monica* from Mono Lake is the only other North American species of *Artemia* currently recognized, the Nebraska populations remaining undescribed. The distribution of *Artemia* is confined to western parts of the country with the exception of water casks in New Haven, Connecticut. Specimens from these barrels were described as *Artemia gracilis* by A. E. Verrill. *Artemia* has not since been collected east of the Mississippi River and the origin of Verrill's specimens has remained a mystery. *Artemia* is known to have occurred in salterns around San Francisco. In that water casks were common on ships that were constructed in and sailed from San Francisco to New England during the 19th century and that these same casks were often converted to fire barrels along railroad tracks near the coast provides a means for *Artemia* to appear in New Haven. It is probable that the occurrence of *Artemia* in New Haven was the result of transport of eggs in wood of water casks brought from San Francisco.

To use the following key effectively, mature males and females are required. The scanty and spotty distribution records are usually the result of inadequate collecting; undoubtedly such ranges will eventually be greatly extended.

Negrea et al. (1999) have proposed separating the family Cyclestheriidae from the Spinicaudata as a separate order, the Cyclestherida. It remains to be seen whether this suggestion is accepted or rejected.

KEY TO FAMILIES AND GENERA OF PHYLLOPODOUS BRANCHIOPODA OF THE UNITED STATES

1. Eyes stalked; no carapace; body elongated; 11 pairs of swimming legs; second antennae uniramous, greatly enlarged and used as clasping organs in male; some genera with one frontal appendage or two antennal appendages in male (Figs. 17.1A, B; 17.10A, B) (most of the Anostraca couplets are based on detailed morphology of the male second antennae) **Order ANOSTRACA (fairy shrimps), 11**

   Eyes sessile; with a large carapace **2**
2. With a low, arched, univalve carapace covering head and much of trunk; 35 to 78 pairs of trunk appendages; second antennae minute or absent; two long cercopods (Figs. 17.1D; 17.11); living animals often brightly colored, blue-green, red, brown, or marbled; total body length 11 to 58 mm, not including supraanal plate; not known east of the Mississippi River **Order NOTOSTRACA (tadpole shrimps), 18**

   Body compressed laterally and completely enclosed in a carapace composed of two lateral, shell-like valves; 10 to 32 pairs of legs; second antennae large and biramous (Figs. 17.1C; 17.4A–E) . . . . **(clam shrimps), 3**
3. Head enclosed in a shield (Figs. 17.1C; 17.4E), not entirely covered by carapace; telson diminutive; carapace without growth lines . . . . . . **Order LAEVICAUDATA, LYNCEIDAE, 4**

   Head naked, entirely covered by carapace; telson well developed; carapace with growth lines **Order SPINICAUDATA, 5**
4. Thoracopods not obviously modified (Fig. 17.19A); rostral carina not bifurcate, as seen in frontal view; 4 species of which 1 is widespread and common . . . . ***Lynceus***

   Second male right or left thoracopod obviously modified (Fig. 17.19B); rostral carina bifurcate near top of head region; 1 species known only from two ponds near Llano, TX ***Paralimnetus***
5. Head with a frontal appendage (dorsal organ) (Fig. 17.4D) . . . . **LIMNADIIDAE, 7**

   Head without a frontal appendage **6**
6. Apex of rostrum armed with a heavy spine (Fig. 17.4B); about 11 mm long; 1 species; CA, CO, KS, OK, TX, UT . . . . **LEPTESTHERIIDAE, *Leptestheria***

   Apex of rostrum without a heavy spine **8**
7. Telson without a spine at posteroventral corner; ramus with short setae (Fig. 17.7); shell with about 7 to 18 growth lines, broadly oval, and 10 to 17 mm long; 1 species, FL, MA ***Limnadia***

   Telson with a spine at posterolateral corner (Fig. 17.7); ramus with long setae; shell with 2 to 11 growth lines; narrow, oval, and 3 to 11 mm long, 6 species with scattered distributions ***Eulimnadia***
8. Anntenule (antenna 1) with sensillae along anterior margin, appearing as minute lobes (Fig. 17.4C) . . . . **CYZICIDAE, 9**

   Anntenule (antenna 1) with sensillae restricted to tip, anterior margin smooth (Figs. 17.4A; 17.18), 1 rare species, TX . . . . **CYCLESTHERIIDAE, *Cyclestheria***
9. Occipital notch shallow and inconspicuous (Fig. 17.19C); 2 rare species . . . . ***Eocyzicus***

   Occipital notch acute and pronounced (Fig. 17.19D) **10**

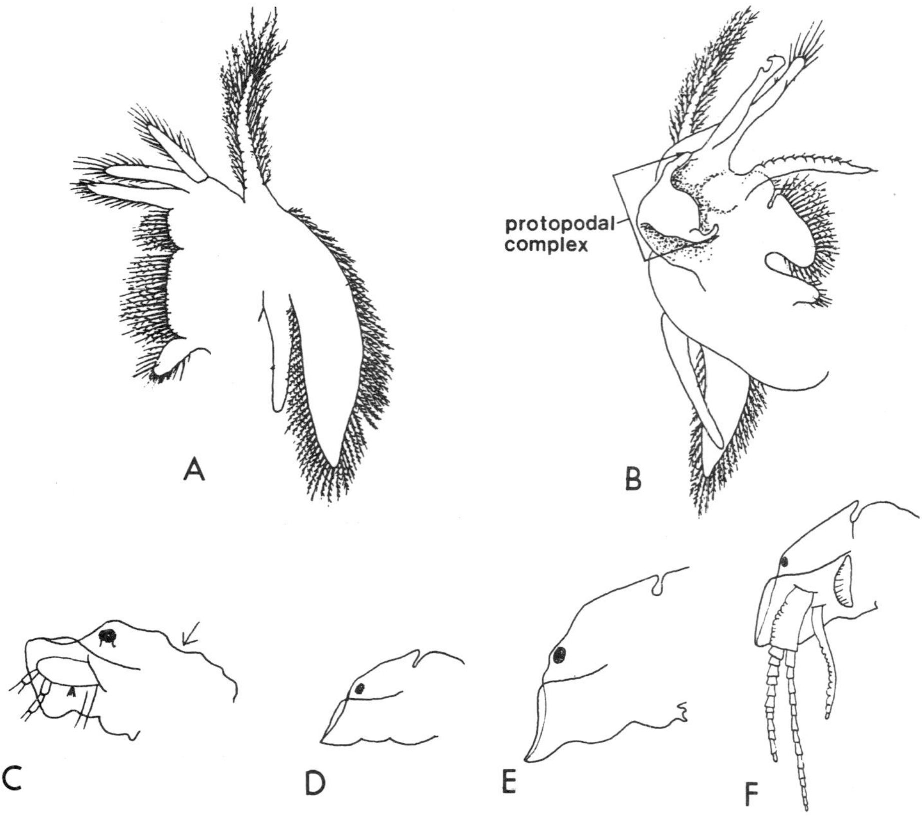

Fig. 17.19. Structure of Laevicaudata and Spinicaudata. A, thoracopod of *Lynceus;* B, modified male thoracopod of *Paralimnetus;* C, anterior end of *Eocyzicus concavus* showing shallow occipital notch (arrow); D, anterior end of *Caenestheriella* showing acute rostrum; E, head of female *Cyzicus mexicanus;* F, head of male *Cyzicus mexicanus.* (A, B from Martin and Belk 1988; E, F redrawn from Mattox 1939.)

10. Rostrum of both male and female acute (Fig. 17.19D), or males absent; second antennae 14 or 15 segmented; 3 species in the central and western U.S. . . . . ***Caenestheriella***
Rostrum of male spatulate (Fig. 17.19F); rostrum of female acute (Fig. 17.19E); second antennae 16 to 22 segmented; 4 species, mostly in the western U.S. . . . . ***Cyzicus***

11. Frontal appendages present and fleshy (Fig. 17.20D–G) **12**
Frontal appendages absent (Fig. 17.20A, B) or very small (Fig. 17.20C) **17**

12. Second antennae of male long and with a long, folded cheliform outgrowth overlapping rest of appendage (Fig. 17.20E); with a small median frontal appendage between the bases of the second antennae; 8 species mostly restricted to the southwestern U.S., a few widely distributed . . . . **STREPTOCEPHALIDAE, *Streptocephalus***
Second antennae of male shorter and more simply constructed **13**

13. Abdomen flattened and possessing paddle-shaped cercopods (Fig. 17.10A); up to 50 mm long; central and southwestern states, 2 species
THAMNOCEPHALIDAE (in part), *Thamnocephalus*
Abdomen not flattened, cercopods extending beyond posterior margin (Fig. 17.1A) **14**

14. Frontal appendage fleshy and complex, with papillae, lobes, or feathery outgrowths (Fig. 17.21A, B); 4 species, FL, GA, TX . . . **THAMNOCEPHALIDAE (in part), *Branchinella***
Frontal appendage otherwise **15**

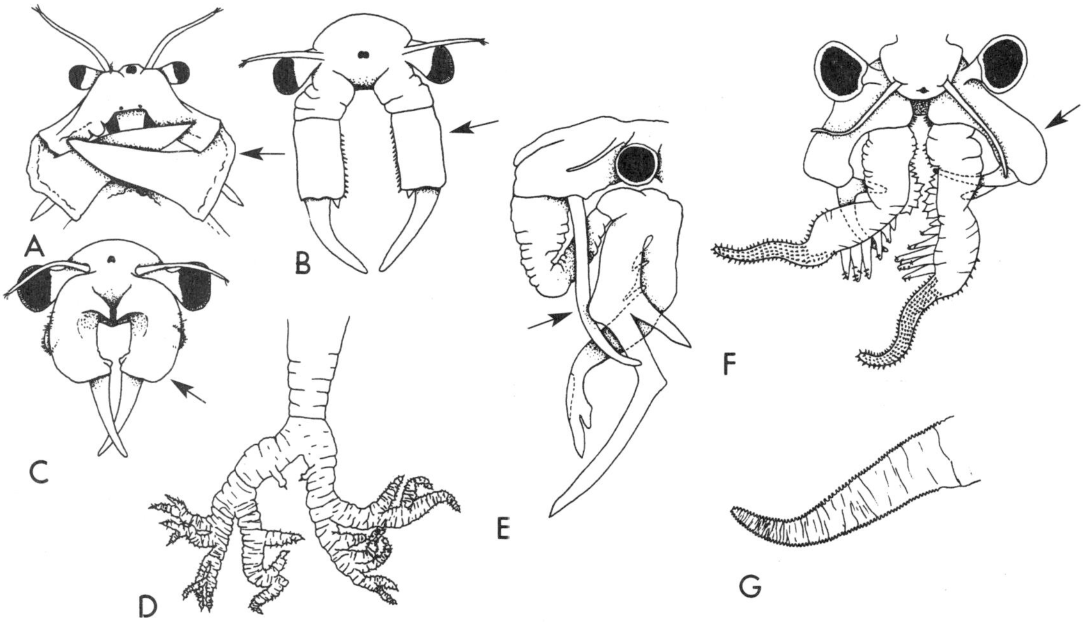

Fig. 17.20. Structure of Anostraca. A, head of male *Artemia franciscana;* B, head of male *Branchinecta paludosa;* C, head of male *Linderiella occidentalis;* D, frontal appendage of male *Triops platyurus;* E, lateral view of anterior part of head and appendages of male *Streptocephalus texanus;* F, anterior view of head of male *Eubranchipus holmani;* G, antennal appendage of *Eubranchipus bundyi.* Arrow indicates antenna 2. (A modified from Heath 1924; B modified from Sars 1896; C modified from Dodds 1923; E modified from Mackin 1942; F modified from Mattox 1936.)

15. Antennal appendages short, conical, and slightly bent medially, with short spines along mesial margin (Fig. 17.20C); 2 species, CA .... **LINDERIELLIDAE, *Linderiella***
Antennal appendages lamelliform, fleshy, with conical projections or serration (teeth) along at least part of one or both edges of distal part (Fig. 17.20F, G) **CHIROCEPHALIDAE, 16**

16. Distal portion of antenna 2 with a basal process (Fig. 17.22A, B, F), antennal appendage with marginal serration (Fig. 17.22C–E); 8 species, several widespread in the U.S. ***Eubranchipus***
Distal portion of antenna 2 without a basal process, antennal appendage with spinous papillae; 1 species, FL .... ***Dexteria***

17. Male second antennae somewhat fused basally and with a laminate or bladelike terminal segment (Fig. 17.20A); average length 10 mm; 2 species, primarily western states (see text) **ARTEMIIDAE, *Artemia***
Male second antennae not fused basally and terminal segment not laminate (Fig. 17.20B); about 14 species in the western U.S. .... **BRANCHINECTIDAE, *Branchinecta***

18. Telson extended as a flat, paddle-shaped protuberance (supra-anal plate) (Fig. 17.23A–D); 25 to 34 body rings, of which two to eight are legless; second maxillae well developed; about five species, generally in midnorthern and western states .... ***Lepidurus***
Telson not extended as a supra-anal plate (Figs. 17.1D; 17.11); 35 to 44 body rings, of which 5 to 15 are legless; second maxillae reduced in adult; a single, possibly two, highly variable species generally distributed west of meridian 99 or 100 in the U.S.; temporary waters and rice fields. .... ***Triops***

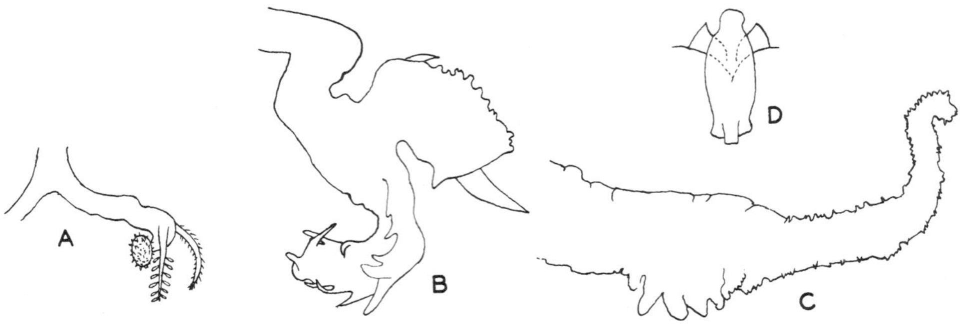

Fig. 17.21. Structure of *Branchinella* and *Eubranchipus.* A, half of frontal appendage of *Branchinella alachua;* B, half of frontal appendage of *B. lithaca;* C, antennal appendage of *Eubranchipus moorei;* D, ventral view of labrum of *E. bundyi.* (A modified from Dexter 1953; C modified from Brtek 1967; D modified from Brtek 1966.)

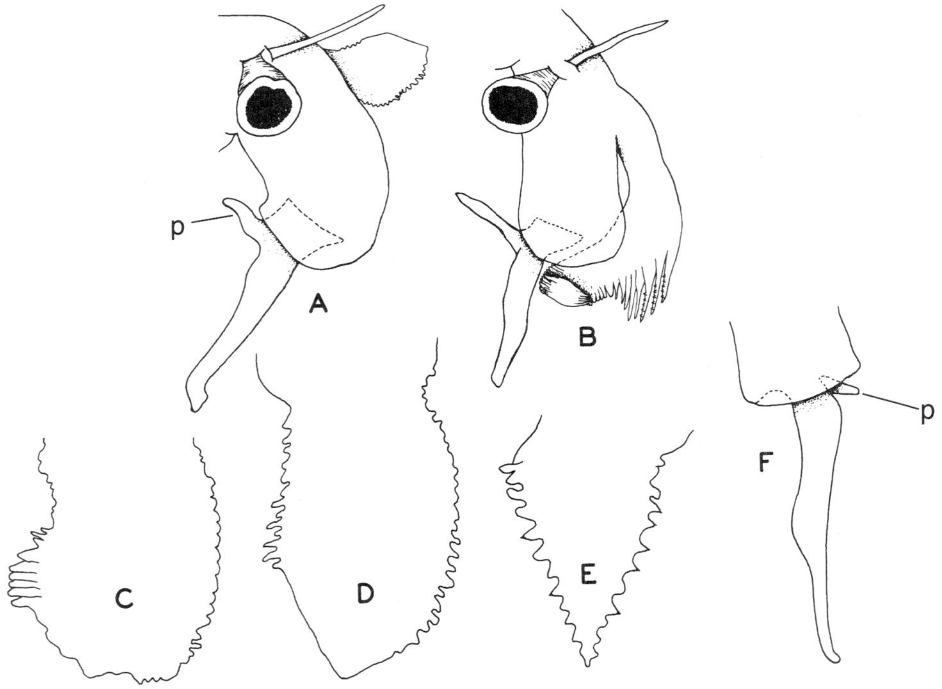

Fig. 17.22. Structure of *Eubranchipus.* A, lateral view of head of *Eubranchipus vernalis;* B, lateral view of head of *E. serratus* (antennal appendage partly coiled); C, antennal appendage of *E. neglectus;* D, antennal appendage of *E. vernalis;* E, antennal appendage of *E. oregonus;* F, male second antenna of *E. ornatus.* p, basal process. (A and B modified from Mattox 1939; C–F redrawn from Creaser 1930a.)

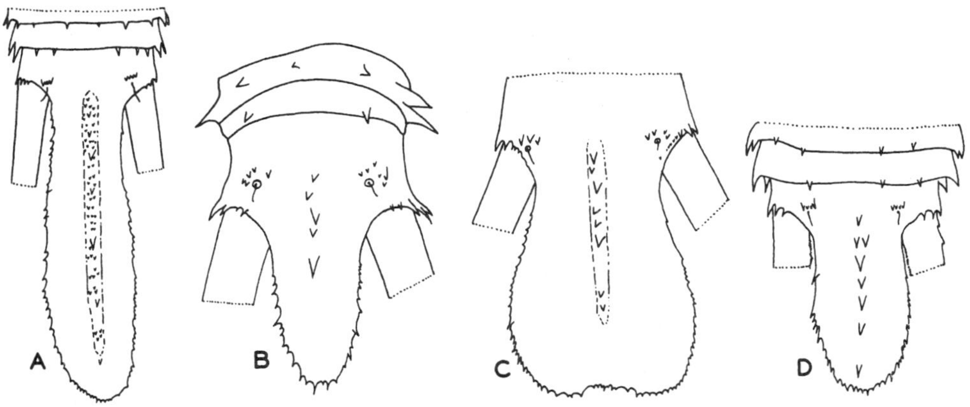

Fig. 17.23. Structure of Notostraca. A, dorsal view of telson and last two body rings of *Lepidurus couesi;* B, dorsal view of telson and last two body rings of *L. lemmoni;* C, dorsal view of telson of *L. bilobatus;* D, dorsal view of telson and last two body rings of *L. packardi.* (A–D modified from Linder 1952.)

REFERENCES

Amat, F., et al. 1991. Some aspects of *Artemia* biology affected by cestode parasitism. *Hydrobiologia* 212:39–44.

Baqui, I. U. 1963. Studies on the postembryonic development of the fairy shrimp *Streptocephalus seali* Ryder. *Tulane Stud. Zool.* 10:91–120.

Beladial, L., and J. Mertens. 1999. Direct preservation in alcohol causes deformation of taxonomic key-characters in Anostraca (Crustacea). *Int. Rev. Hydrobiol.* 84:17–22.

Belk, D. 1975. Key to the Anostraca (fairy shrimps) of North America. *Southwest. Nat.* 20:91–103.

——. 1977. Zoogeography of the Arizona fairy shrimps (Crustacea: Anostraca). *J. Ariz. Acad. Sci.* 12:70–78.

——. 1989. Identification of species in the conchostracan genus *Eulimnadia* by egg shell morphology. *J. Crust. Biol.* 9:115–125.

——. 1991. Anostracan mating behavior: a case of scramble-competiton polygyny. In: *Crustacean Sexual Biology,* pp.111–125, edited by R. T. Bauer and J. W. Martin. Columbia Univ. Press, New York.

——. 1992. Observations on the clam shrimps of Arizona. *J. Arizona–Nevada Acad. Sci.* 26:132–138

Belk, D., and J. Brtek. 1995. Checklist of the Anostraca. *Hydrobiologia* 298:315–353.

——. 1997. Supplement to 'Checklist of the Anostraca.' *Hydrobiologia* 359:243–245.

Bowen, S. T., et al. 1985. Ecological isolation in *Artemia*: population differences in tolerance of anion concentrations. *J. Crust. Biol.* 5:106–129.

Brown, R. A., and S. T. Bowen. 1991. Taxonomy and population genetics of *Artemia.* In: *Artemia Biology,* pp. 221–235, edited by R. A. Browne et al. CRC Press, Boca Raton, FL.

Brtek, J. 1966. Einige notizenzur taxonomie der familie Chirocepohalidae Daday 1910. *Annotat. Zool. Bot.* 33:1–42.

Brtek, J. 1967. *Eubranchipus (Creaseria) moorei* n. sp. *Annotat. Zool. Botan.* 36:1–7.

Cannon, H. G., and F. M. C. Leak. 1933. On the feeding mechanism of the Branchiopoda. *Philos. Trans. R. Soc. London Ser. B,* 222:267–352.

Cole, G. A., and R. J. Brown. 1967. The chemistry of *Artemia* habitats. *Ecology* 48:858–861.

Creaser, E. P. 1930a. Revision of the phyllopod genus *Eubranchipus* with description of a new species. *Occas. Pap. Mus. Zool. Univ. Mich.* 208: 1–13.

——. 1930b. The North American phyllopods

of the genus *Streptocephalus. Occas. Pap. Mus. Zool. Univ. Mich.* 217:1–10.

Criel, G. R. J. 1991. Ontogeny of *Artemia*. In: *Artemia Biology*, pp. 155–186, edited by R. A. Browne et al. CRC Press, Boca Raton, FL.

Daborn, G. R. 1975. Life history and energy relations of the giant fairy shrimp *Branchinecta gigas* Lynch 1937 (Crustacea: Anostraca). *Ecology* 56:1025–1039.

Dempster, R. P. 1953. The use of larval and adult brine shrimp in aquarium fish culture. *Calif. Fish Game* 39:355–364.

Dexter, R. W. 1946. Further studies on the life history and distribution of *Eubranchipus vernalis* (Verrill). *Ohio J. Sci.* 46:31–44.

——. 1953. Studies on North American fairy shrimps with the description of two new species. *Am. Midl. Nat.* 49:751–771.

——. 1956. A new fairy shrimp from Western United States, with notes on other North American species. *J. Wash. Acad. Sci.* 46:159–165.

Dexter, R. W., and D. B. McCarraher. 1967. Clam shrimps as pests in fish rearing ponds. *Prog. Fish Cult.* 29:105–107.

De Walsche, C., et al. 1991. Structural differences between the cyst walls of *Streptocephalus dichotomus* (Baird), *S. torvicornis* (Waga), and *Thamnocephalus platyurus* (Packard) (Crustacea: Anostraca), and a comparison with other genera and species. *Hydrobiologia* 212:195–202.

Dodds, G. S. 1923. A new species of phyllopod. *Occas. Pap. Mus. Zool., Univ. Mich.* 141:1–3.

Eriksen, C., and D. Belk. 1999. *Fairy Shrimps of California's Puddles, Pools, and Playas.* 196 pp. Mad River Press, Eureka, CA.

Eriksen, C. H., and R. J. Brown. 1980. Comparative respiratory physiology and ecology of phyllopod Crustacea, III: Notostraca. *Crustaceana* 39:22–32.

Fryer, G. 1966. *Branchinecta gigas* Lynch, a non-filterfeeding raptatory anostracan, with notes on the feeding habits of certain other anostracans. *Proc. Linn. Soc. London* 177:19–34.

——. 1987. A new classification of the branchiopod Crustacea. *Zool. J. Linn. Soc.* 91:357–383.

Galtsoff, P. S., et al. 1937. *Culture Methods for Invertebrate Animals.* 590 pp. Comstock Publishing, Ithaca, NY.

Graham, T. B. 1995. Sympatric occurrence of Eubranchiopoda in ephemeral pools: a comment. *Am. Midl. Nat.* 133:371–372.

Heath, H. 1924. The external development of certain phyllopods. *J. Morphol.* 38:453–483.

Hill, R. E., and W. D. Shepard. 1997. Observations on the identification of California anostracan cysts. *Hydrobiologia* 359:113–123.

Huggins, D. G. 1976. The sympatric occurrence of three species of Eubranchiopoda in Douglas County, Kansas. *Southwest. Nat.* 20:577–578.

Jennings, R. H., and D. M. Whitaker. 1941. The effect of salinity upon the rate of excystment of *Artemia. Biol. Bull.* 80:194–201.

Jensen, A. C. 1918. Some observations on *Artemia gracilis*, the brine shrimp of Great Salt Lake. *Biol. Bull.* 34:18–32.

King, J. L., and R. Hanner. 1998. Cryptic species in a "living fossil" lineage: taxonomic and phylogenetic relationships within the genus *Lepidurus* (Crustacea: Notostraca) in North America. *Molecular Phylogenies and Evol.* 10: 23–36.

Knoll, L. 1995. Mating behavior and time budget of an androdioecious crustacean, *Eulimnadia texana* (Crustacea: Conchostraca). *Hydrobiologia* 298:73–81.

Linder, F. 1941. Contributions to the morphology and the taxonomy of the Branchiopoda Anostraca. *Zool. Bidr. Uppsala* 20:101–302.

——. 1952. Contributions to the morphology and taxonomy of the Branchiopoda Notostraca, with special reference to the North American species. *Proc. U. S. Nat. Mus.* 102:1–69.

Longhurst, A. R. 1955. A review of the Notostraca. *Bull. Br. Mus. Nat. Hist. Zool.* 3:1–57.

Lowndes, A. G. 1933. The feeding mechanism of *Chirocephalus diaphanus* Prevost, the fairy shrimp. *Proc. Zool. Soc. London* (B) (1933): 1093–1118.

Lynch, J. E. 1937. A giant new species of fairy shrimp of the genus *Branchinecta* from the State of Washington. *Proc U. S. Nat. Mus.* 84:555–562.

——. 1958. *Branchinecta conigera*, a new species of anostracan phyllopod from the state of Washington. *Proc. U.S. Nat. Mus.* 108:25–37.

——. 1966. *Lepidurus lemmoni* Holmes: a redescription with notes on variation and distribution. *Trans. Am. Microsc. Soc.* 85:181–192.

——. 1972a. *Lepidurus couesii* Packard (Notostraca) redescribed with a discussion of specific characters in the genus. *Crustaceana* 23:43–49.

——. 1972b. *Branchinecta dissimilis* n. sp. a new species of fairy shrimp, with a discussion of specific characters in the genus. *Trans. Am. Microsc. Soc.* 91:240–243.

Mackin, J. G. 1940. A new species of conchostracan phyllopod, *Eulimnadia antlei*, from Oklahoma. *Am. Midl. Nat.* 23:219–221.

——. 1942. A new species of phyllopod crustacean from the southwestern short-grass prairies. *Proc U. S. Nat. Mus.* 92:33–39.

——. 1952. On the correct specific names of several North American species of *Branchinecta* Verrill. *Am. Midl. Nat.* 47:61–65.

Maeda-Martínez, A. M., et al. 1997. Large branchiopod assemblages common to Mexico and the United States. *Hydrobiologia* 359:45–62.

Martin, J. W., and D. Belk. 1988. A review of the clam shrimp family Lynceidae Stebbing, 1902 (Branchiopoda, Conchostraca) in the Americas. *J. Crust. Biol.* 8:451–482.

Martin, J. W., et al. 1986. Redescription of the clam shrimp *Lynceus gracilicornis* (Packard) (Branchiopoda, Conchostraca, Lynceidae) from Florida, with notes on its biology. *Zool. Scripta* 15:221–232.

Mattox, N. T. 1936. A new fairy shrimp belonging to the genus *Pristicephalus*. *Am. Midl. Nat.* 17:950–953.

——. 1937. Studies on the life history of a new species of fairy shrimp, *Eulimnadia diversa*. *Trans. Am. Microsc. Soc.* 56:249–255.

——. 1939. Descriptions of two new species of the genus *Eulimnadia* and notes on the other Phyllopoda of Illinois. *Am. Midl. Nat.* 22:642–653.

——. 1954. Description of *Eocyzicus concavus* (Mackin) with a review of other North American species of the genus (Crustacea: Conchostraca). *J. Wash. Acad. Sci.* 44:46–49.

——. 1957. A new estheriid conchostracan with a review of the other North American forms. *Am. Midl. Nat.* 58:367–377.

McCarraher, D. B. 1970. Some ecological relations of fairy shrimps in alkaline habitats of Nebraska. *Am. Midl. Nat.* 84:59–68.

Moore, W. G. 1951. Observations on the biology of *Streptocephalus seali*. *Proc. La. Acad. Sci.* 14:57–65.

——. 1957. Studies on the laboratory culture of Anostraca. *Trans. Am. Microsc. Soc.* 76:159–173.

——. 1966. New World fairy shrimps of the genus *Streptocephalus* (Branchiopoda, Anostraca). *Southwest. Nat.* 1:24–48.

Moore, W. G., and J. B. Young. 1964. Fairy shrimps of the genus *Thamnocephalus* (Branchiopoda, Anostraca) in the United States and Mexico. *Southwest. Nat.* 9:68–77.

Mura, G. 1992. Pattern of egg shell morphology in thamnocephalids and streptocephalids of the new World. *Crustaceana* 62:300–311.

Negrea, S., et al. 1999. Phylogeny, evolution and classification of the Branchiopoda (Crustacea). *Hydrobiologia* 412:191–212.

Olesen, J., et al. 1996. External morphology of the male of *Cyclestheria hislopi* (Baird, 1859) (Crustacea, Branchiopoda, Spinicaudata), with a comparison of male claspers among the Conchostraca and Cladocera and its bearing on phylogeny of the 'bivalved' Branchiopoda. *Zool. Scripta* 25:291–316.

Packard, A. 1883. A monograph of the phyllopod Crustacea of North America, with remarks on the Order Phyllocarida. *Annu. Rep. U.S. Geog. Surv. Terr.* 12:295–592.

Provasoli, L., and K. Shiraishi. 1959. Axenic cultivation of the brine shrimp *Artemia salina*. *Biol. Bull.* 117:347–355.

Relyea, G. M. 1937. The brine shrimp of Great Salt Lake. *Am. Nat.* 71:612–616.

Roessler, E. W. 1995a. Review of Columbian Conchostraca (Crustacea)—ecological aspects and life cycles—family Cyclestheriidae. *Hydrobiologia* 298:113–124.

——. 1995b. Review of Columbian Conchostraca (Crustacea)—ecological aspects and life cycles—families Lynceidae, Limnadiidae, Leptestheriidae and Metalimnadiidae. *Hydrobiologia* 298:125–132.

Rosenberg, L. E. 1946. Fairy shrimps in California rice fields. *Science* 104:111–112.

Rosowski, J. R., et al. 1997. Ultrastructure of the cyst shell and underlying membranes of the brine shrimp *Artemia franciscana* Kellogg (Anostraca) during postencystic development, emergence, and hatching. *J. Shellfish Res.* 16:233–249.

Sars, G. O. 1896. Phyllocarida og Phyllopoda. *Fauna Noro.* 1:1–140.

Sassaman, C. 1995. Sex determination and evolution of unisexuality in the Conchostraca. *Hydrobiologia* 298:45–65.

Sassaman, C., et al. 1997. Reproductive isolation and genetic differentiation in North American species of *Triops* (Crustacea: Branchiopoda: Notostraca). *Hydrobiologia* 359:125–147.

Saunders, J. F. 1980. A redescription of *Lepidu-*

rus bilobatus Packard (Crustacea: Notostraca). *Trans. Am. Microsc. Soc.* 99:179–186.

Schantz, H. L. 1905. Notes on North American species of *Branchinecta* and their habits. *Biol. Bull.* 9:249–264.

Sissom, S. L. 1976. Studies on a new fairy shrimp from the playa lakes of West Texas (Branchiopoda, Anostraca, Thamnocephalidae). *Crustaceana* 30:39–42.

Smith, D. G. 1992. A redescription of types of the clam shrimp *Eulimnadia agassizii* (Spinicaudata: Limnadiidae). *Trans. Am. Microsc. Soc.* 111:223–228.

———. 1995. Notes on the status and natural history of limnadiid clam shrimp in southern New England. *Anostracan News* 3(2):3–4.

Warren, H, S. 1938. The segmental excretory glands of *Artemia salina* Linn. var. principalis Simon. (The brine shrimp.) *J. Morphol.* 62: 263–298.

Weaver, C. R. 1943. Observations on the life cycle of the fairy shrimp *Eubranchipus vernalis*. *Ecology* 24:500–502.

Wells, M. L., et al. 1997. Resilience of anostracan cysts to fire. *Hydrobiologia* 359:199–202.

Whitaker, D. M. 1940. The tolerance of *Artemia* for cold and high vacuum. *J. Exp. Zool.* 83:391–399.

Wiman, F. H. 1979. Hybridization and the detection of hybrids in the fairy shrimp genus *Streptocephalus*. *Am. Midl. Nat.* 102:149–156.

Zucker, N., et al. 1997. Anatomical evidence for androdioecy in the clam shrimp *Eulimnadia texana*. *Hydrobiologia* 359:171–175.

18

CLADOCERAN BRANCHIOPODA

(Water fleas)*

Because of their interesting habits, their availability in nearly all types of freshwater habitats, and their complex but easily studied anatomy, the water fleas have been favorite subjects of observation by both amateur and professional biologists ever since the invention of the microscope.

General characteristics. Most members of the Cladocera (Fig. 18.1) are between 0.2 and 3.0 mm long. The body is not clearly segmented, and in the great majority of species the thoracic and abdominal regions are covered by a secreted shell or carapace that has a general bivalved appearance but is actually a single folded piece that gapes ventrally. In lateral view the shell is variously shaped; it may be oval, circular, elongated, or angular. There are often surface reticulations, striations, or other types of markings. In many species the posterior end has a spinule or spine, and the ventral edges of the valves usually bear setae. The inner surface of the valves is lined with the delicate body wall.

The head is a compact structure and does not open ventrally as the valves do. It is bent downward, and dorsally it is sometimes indistinctly set off from the body by a cervical sinus or notch. The most conspicuous internal structure of the head is the large compound eye. It consists of a few to many small hyaline lenses surrounding a mass of pigment granules. It is rotated constantly and jerkily by three pairs of small muscles. A single small ocellus often lies posterior or ventral to the compound eye.

The first antennae (antennules) are inserted on the ventral side of the head near its posterior margin. They are usually inconspicuous, unsegmented, and bear olfactory setae. The second antennae, however, are very large and are inserted laterally near the posterior margin of the head. Each consists of a stout basal segment, a segmented dorsal ramus, and a segmented ventral ramus. The two rami bear a variable number of plumose setae. Setation formulas are often used as a means of identifying genera and species. Thus, the setation formula for *Daphnia* is 0-0-1-3/1-1-3, which indicates that the dorsal ramus of the antenna is four-segmented and that the basal, second, third, and fourth segments bear 0, 0, 1, and 3 setae, respectively; it also indicates a three-segmented ventral ramus of which the basal, second, and third segments bear 1, 1, and 3 setae, respectively. The antennae are activated by a set of powerful muscles that originate dorsally in the "neck" region, Externally, there is a strengthening ridge, or fornix, above the base of each antenna. The beak, or rostrum, is a more or less well-defined projection of the head between or in front of the first an-

*Prepared with the assistance of Kirsten Work.

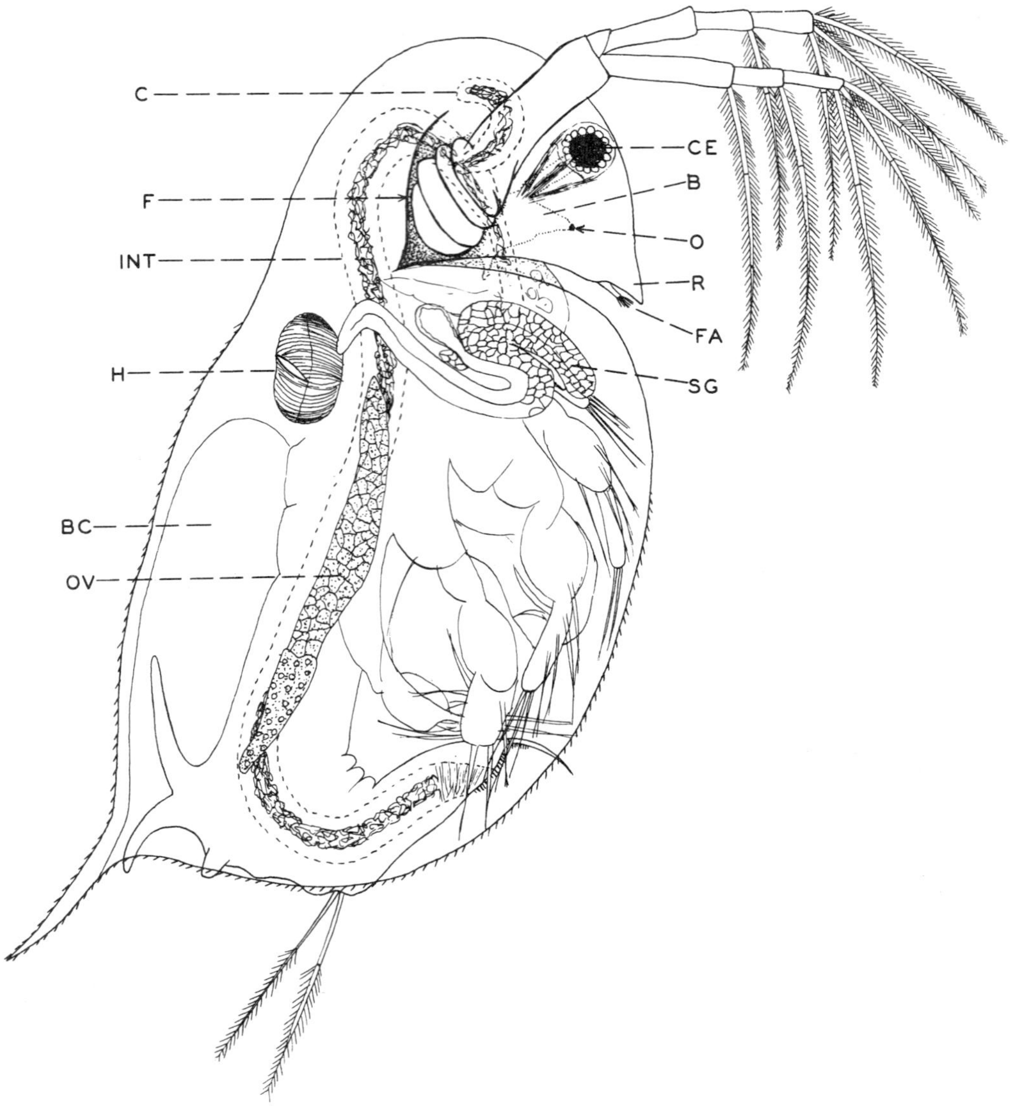

Fig. 18.1. Anatomy of female *Daphnia pulex*, ×70, diagrammatic, muscles not shown. B, brain; BC, brood chamber; C, digestive cecum; CE, compound eye; F, fornix; FA, first antenna (antennule); H, heart; INT, intestine; O, ocellus; OV, ovary; R, rostrum or beak; SG, shell gland.

tennae. The vertex of the head is that part that lies anterior to the compound eye.

The small mouthparts are situated near the junction of head and body (Fig. 18.2). Beginning at the anterior end, they consist of (1) a median labrum; (2) a pair of stout, sclerotized, toothed, or ridged grinding mandibles; (3) a pair of small, pointed maxillae, used in pushing food between the mandibles; and (4) a single median labium. In some of the Macrothricidae and Chydoridae the labium bears a median keel that is of systematic value.

There are five or six pairs of lobed, leaflike thoracic legs, bearing numerous hairs and setae (Figs. 18.3, 18.5). Although the legs are fundamentally biramous, this condition is not clearly evident because of their considerable modification. In the Sididae and Holopedidae all legs are similar, but in the

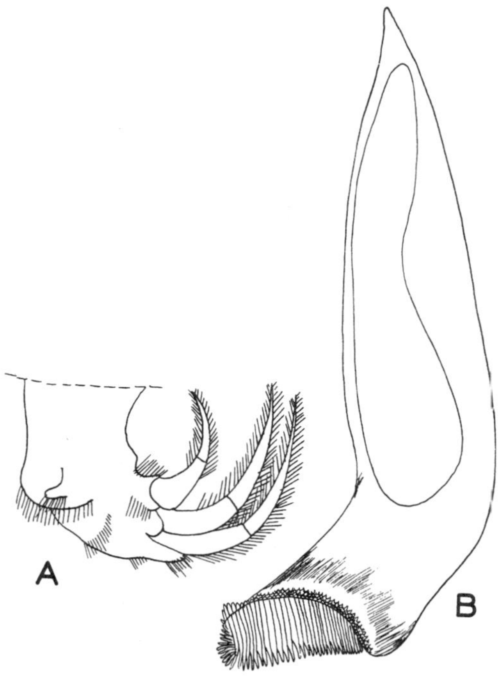

Fig. 18.2. Mouthparts of *Daphnia pulex*. A, maxilla; B, mandible. (Redrawn from Lilljeborg 1900.)

other families the first two pairs are more or less prehensile and may aid in clinging to a substrate.

The true abdomen is suppressed, but there is a large postabdomen at the posterior end of the body. It is usually bent forward so that the dorsal side is downward. It bears two long abdominal setae, two terminal claws, and usually a series of marginal or lateral teeth or denticles. The postabdomen seems to be used chiefly for cleaning debris from the thoracic legs, although it may aid in locomotion.

Although often rare or absent, males may be distinguished from females by their smaller size, larger first antennae, and modified first thoracic legs and postabdomen (Fig. 18.6E). The first thoracic legs of males are armed with a stout hook used for clasping females and the postabdomen may possess a pair of penes.

Limnetic species are usually light-colored and translucent. Pond, littoral, and bottom species are darker in color, ranging from light yellowish brown to reddish brown, grayish, or almost black. Pigmentation occurs in both the carapace and general body tissues. *Holopedium* has chromatophores in the carapace.

Locomotion. For the most part, Cladocera keep intermittently in motion, the antennae being the chief organs of locomotion. In some genera, such as *Latona*, single vigorous strokes of the antennae produce powerful leaps. *Daphnia* and its relatives move by a series of "hops," produced by more rapid and less vigorous strokes. In some other genera the swimming movements have been characterized as "uncertain" and "tottering." A few genera, including *Holopedium* and *Scapholeberis*, habitually swim upside down. Species occurring on a substrate have less frequent periods of active swimming and often use the antennae incidentally for obtaining a purchase on vegetation and bottom debris. The postabdomen also may be used for pushing off surfaces in bottom-inhabiting species. A cladoceran may literally kick itself along, sometimes by quite respectable hops.

A few species have become remarkably adapted for living in damp moss and debris of tropical forest floors; they are unable to swim and only creep about (Frey 1980a). A few species (chiefly Chydoridae) inhabit the interstices of stream gravel deposits; they show no obvious adaptations to this habitat.

Feeding. Limnetic species use complex movements of the highly setose thoracic legs (Figs. 18.3, 18.4) to produce a constant current of water between the valves. These movements further serve to filter or electrochemically attract food particles from the water and collect them in a median ventral groove at the base of the legs. This stream of food is fed forward to the mouthparts where the particles may be ground between the surfaces of the mandibles before being taken into the mouth. In contrast, most littoral-dwelling chydorid species either use the thoracic legs to scrape food off surfaces or use the carapace as a suction pump.

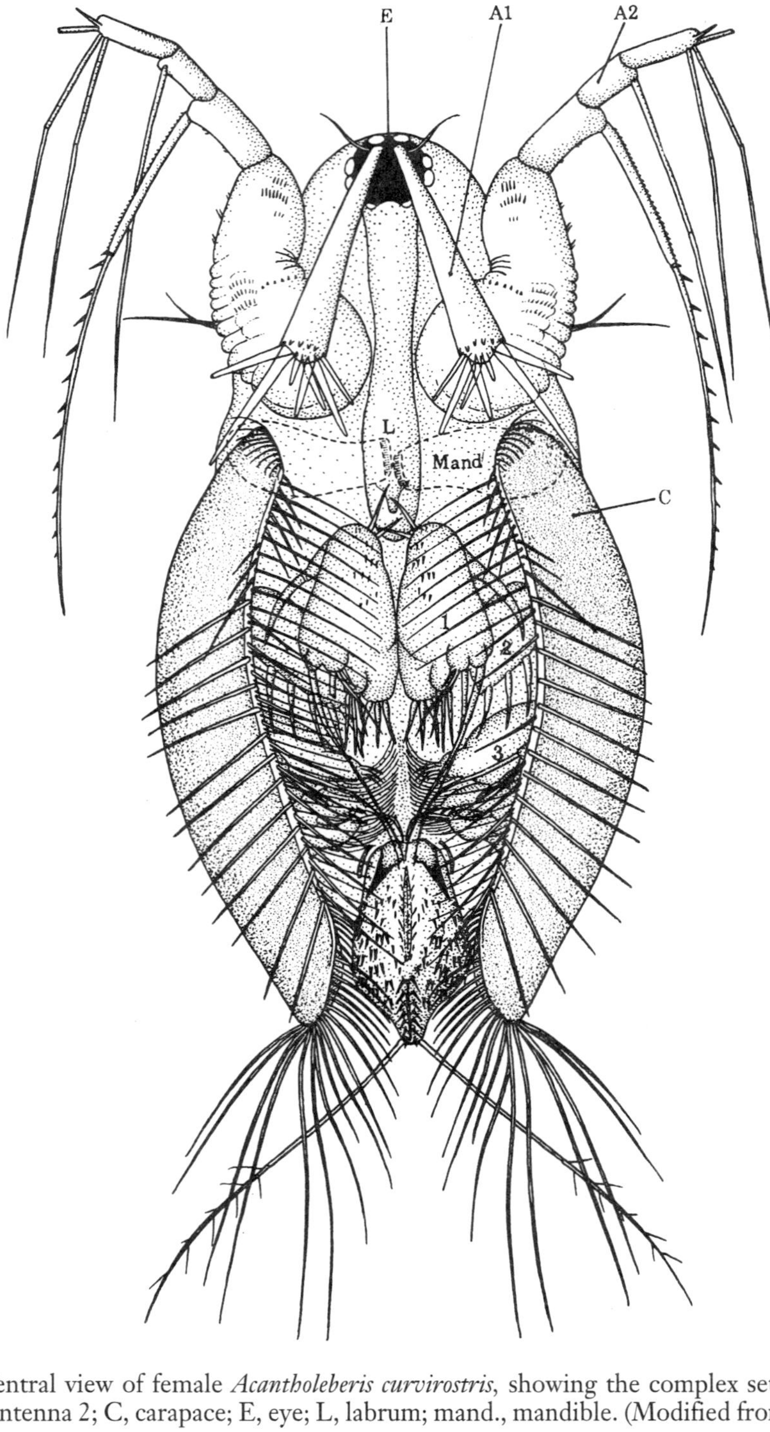

Fig. 18.3. Ventral view of female *Acantholeberis curvirostris*, showing the complex setation. A1, antenna 1; A2, antenna 2; C, carapace; E, eye; L, labrum; mand., mandible. (Modified from Fryer 1974.)

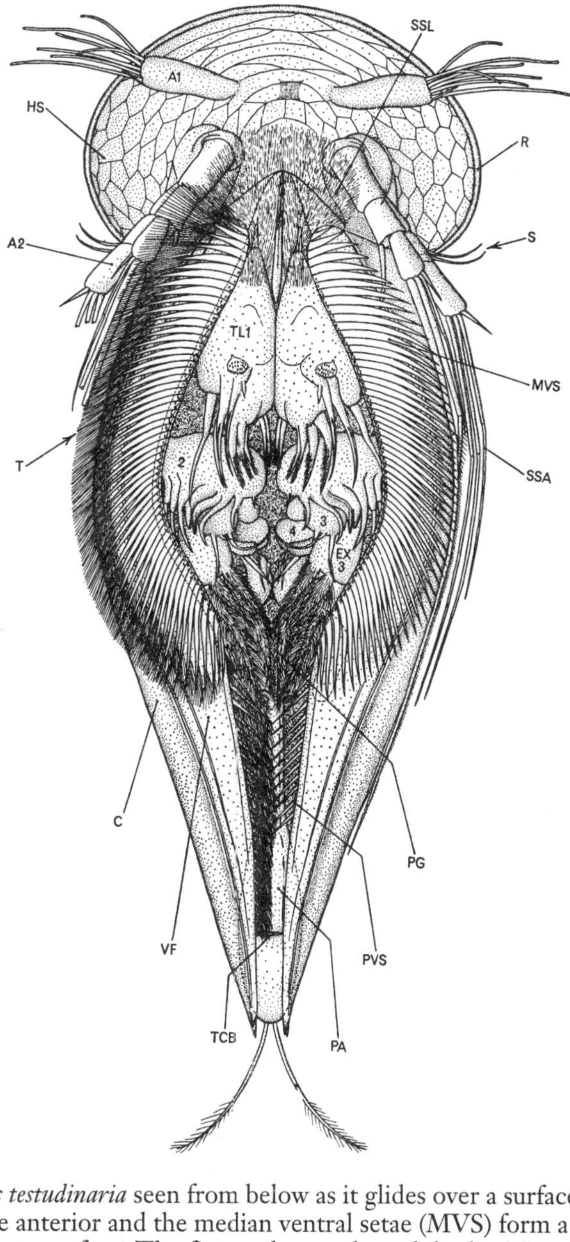

Fig. 18.4. *Graptoleberis testudinaria* seen from below as it glides over a surface. The head shield (HS) and rostrum (R) seal the anterior and the median ventral setae (MVS) form a tight chamber between the body and the substrate surface. The first and second trunk limbs (TL1, 2) scrape food particles from the surface; there are no significant currents within the carapace (C) chamber. Other morphological embellishments (setae, claws) assist the animal's locomotion and food sorting capabilities. (From Fryer 1968).

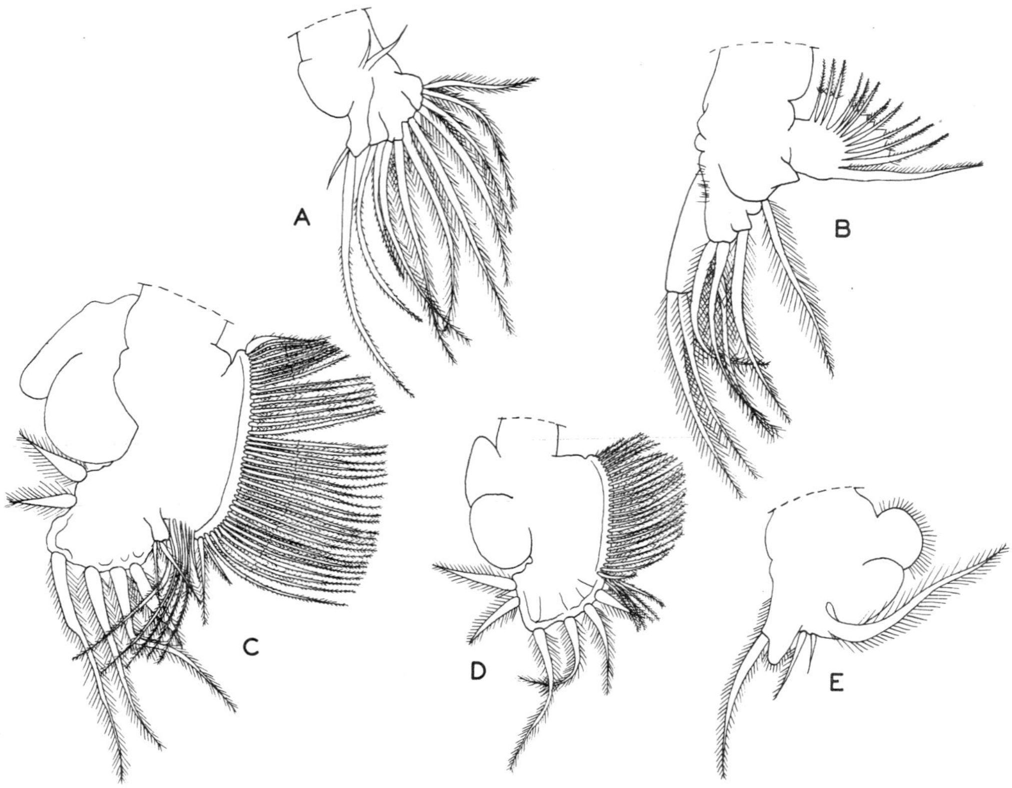

Fig. 18.5. Legs of adult female *Daphnia pulex.* A, first leg; B, second leg; C, third leg; D, fourth leg; E, fifth leg. (Modified from Ueno 1927.)

Algae and Protoctista often have been assumed to be the chief foods, to the exclusion of other materials, but it is now well known that organic detritus of all kinds, as well as bacteria, are very important and commonly form the great bulk of material ingested. It also has been successfully demonstrated that the feeding movements of the legs of plankton species are so efficient that they separate some colloidal organic particles from the water. Although initially thought to be nonselective filter feeders, cladocerans may select food based on size, shape, or taste. Small green algae or diatoms (1–25 mm) generally are preferred, with larger filamentous, spined, or colonial forms rejected. Blue-green algae often are considered unsuitable food for cladocerans due their large or awkward shapes, the gelatinous sheath that surrounds most colonial forms, or the toxicity of some species. Despite these obstacles, some cladocerans can consume blue-green algae; however, de Bernardi and Giussani (1990) suggest that the survival of the animal often is reduced after consuming blue-green algae and depends upon the species or strain of algae. The diet of many cladocerans also is supplemented by consumption of protoctistans.

When undesirable material or large, tangled masses are introduced between the mandibles, they may be removed by spines at the base of the first legs and then kicked out of the shell by the postabdomen. In *Daphnia,* the higher the concentration of algal food, the more narrow is the gap between the valves. This may be a mechanism for surviving algal blooms. Although the food stream in the median groove is continuous, ingestion may cease for varying periods. Minute amounts of dissolved organic

Fig. 18.6. Sex characters of Cladocera. A, rostrum region of female *Daphnia pulex;* B, rostrum region of male *D. pulex;* C, postabdomen of male *Simocephalus exspinosus;* D, postabdomen of female *S. exspinosus;* E, first leg of male *D. pulex.* (C and D redrawn from Banta 1939; E modified from Ueno 1927.)

materials may be absorbed by the general body surface, but they are of no apparent metabolic significance.

A few genera, such as *Polyphemus*, *Leptodora*, and *Bythotrephes*, are predaceous and have the legs modified for seizing. Their prey consists mostly of other microcrustacea and rotifers.

The finely divided food is passed from the mouthparts to the esophagus in small masses of definite size. The time required to fill the digestive tract varies considerably, depending on the species, temperature, concentration of food, and other factors. Laboratory observations show a range of 10 to 240 minutes. Hasler (1937) demonstrated the presence of fat, carbohydrate, and protein digesting enzymes in *Daphnia* and *Polyphemus*. He found the pH of the digestive tract of *Daphnia* to range between 6.8 and 7.2. The indigestible residue leaves the body through the anus, which usually is situated on the lower (dorsal) border of the postabdomen.

Internal anatomy. Although the complex muscular system obscures some of the smaller anatomical features, the essential parts of most of the organ systems can be easily distinguished. The digestive system is relatively unspecialized. In the head region the narrow esophagus opens into a stomach that is often indistinguishable from the remainder of the tubular digestive tract, the intestine. In some genera there are one or two digestive ceca that open into the anterior part of the intestine. The course of the intestine may be relatively straight or convoluted. It can be distinguished in both living and preserved specimens because of its contained dark food mass. In some of the Chydoridae there is a posterior ventral cecum.

The posterior part of the digestive tract is somewhat specialized as a rectum. It is now known that many algae cells are still viable after traversing the cladoceran digestive tract. Many chydorids and macrothricid cladocerans exhibit "anal drinking," which would appear to prevent loss of digestive products and to increase the efficiency of food utilization.

The simple oval or football-shaped heart lies behind the head on the dorsal side. Blood enters the heart through two lateral ostia and leaves through an anterior opening. At room temperature the heart beats about as rapidly as one can open and close his fist. There are no blood vessels, the blood being roughly guided about in the hemocoel by a series of thin mesenteries. The blood plasma usually is colorless or has a faint yellowish cast; it contains numerous colorless corpuscles. However, the blood plasma can appear pinkish or reddish in pond species or occasionally in lake species, owing to dissolved hemoglobin that may be formed when dissolved oxygen concentrations are low. This hemoglobin production appears to be associated with an increased life span, more energetic swimming, and faster embryonic development. Presumably the hemoglobin functions in the respiration of embryos that are retained in a more or less closed brood pouch with little access to the outside water.

The exchange of oxygen and carbon dioxide occurs through the general body surface. Irregular or looped shell glands, situated near the anterior end of the valves, are thought to function as excretory organs. In some of the Chydoridae there is a long tubular gland that arises from the rectum and lies freely in the hemocoel. Its function is unknown.

The nervous system consists of the usual ventral double nerve cord, relatively few ganglia, paired nerves, and a brain just anterior (dorsal) to the esophagus. Both the compound eye and the ocellus appear to be important in orientation to light sources and light intensity. Olfactory stimuli are thought to be received by the setae at the edges of the valves, by the first antennae, and by areas around the mouth. The sense of touch is centered particularly in the abdominal setae and sensory hairs on the basal segment of the antennae.

All Cladocera have small structures on or near the middorsal line known as "head pores." These are borne on the head shield, an unpaired plate covering the frontal and lateral surfaces of the head. In adult Chydoridae there are one to five minute head pores on the midline of the posterior portion of the head shield (Frey 1959). The number and arrangement of these structures may be used as taxonomic characters in the Chydoridae. All the other families of Cladocera have (larger) head pores, at least in some stage of their ontogeny, but here they are associated with certain underlying tissues. Such structures have been called by various names, such as neck organs, dorsal organs, and cervical glands, and they help to maintain osmotic balance in early instars by taking up cations.

Gonads are easily distinguished only in sexually mature specimens. The two elongated ovaries lie lateral or somewhat ventral to the intestine in the thoracic region of the body (Fig. 18.7B). Depending on the stage in reproduction, an ovary is densely granular, with large oogonial nuclei, or it contains a mass of large, pigmented, granular, yolky eggs. Each oviduct is located dorsally at the posterior end; it is so delicate that it cannot be distinguished except when an egg is passing through it. The testes are smaller and contain small, semitransparent, granule-like sperm (Fig. 18.7A). The sperm ducts are posterior continuations of the testes, which pass laterally to the intestine and rectum and open on the postabdomen near the anus or claws. A part of the postabdomen is sometimes specialized to form a copulatory organ.

Reproduction in general. Reproduction is parthenogenetic during the greater part of the year in most habitats, and only female young are produced. The eggs undergo a single maturation division in the ovary and a number are released at a time via the oviducts into the brood chamber. The latter is a cavity dorsal to the body proper and between the valves; it is effectively closed posteriorly by the abdominal processes. Under optimal conditions, one clutch of eggs is re-

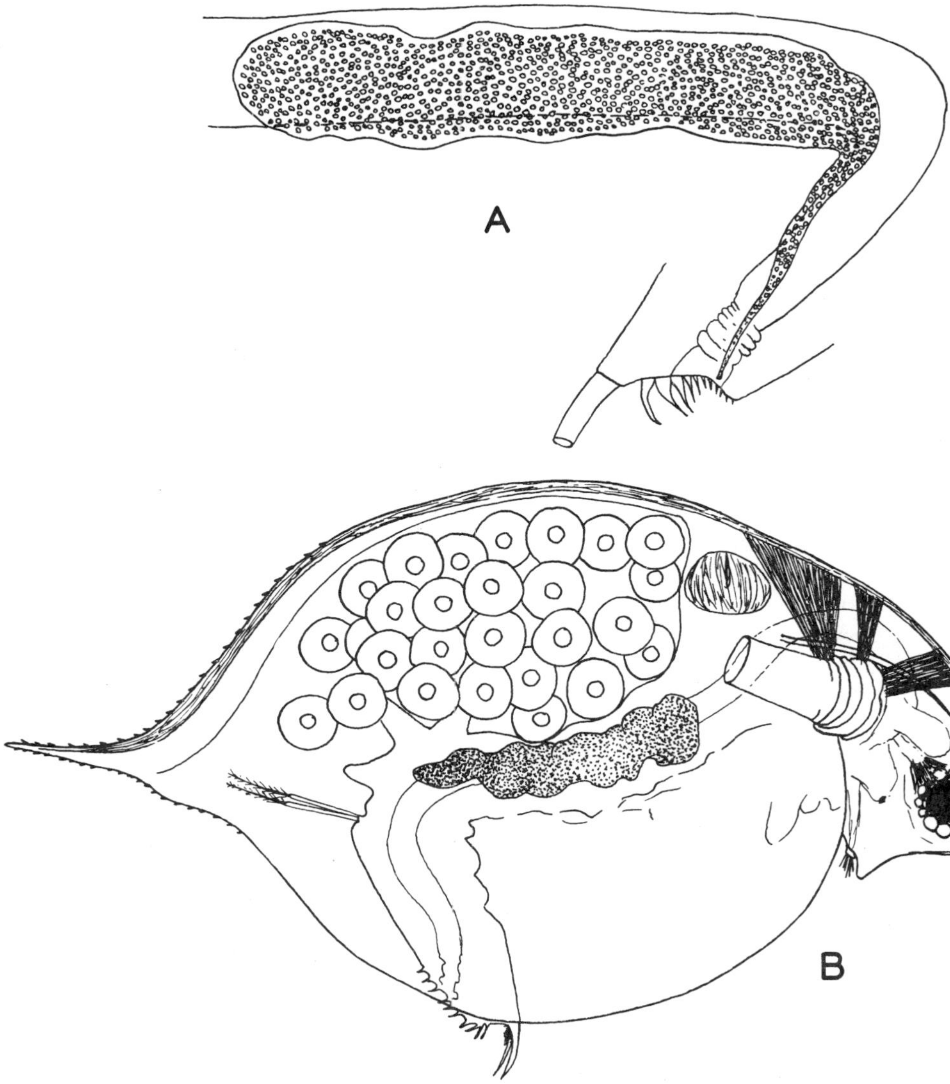

Fig. 18.7. Cladoceran structures. A, postabdomen region of male *Simocephalus exspinosus* showing testis, sperm duct, intestine, and rectum; B, mature female *Daphnia pulex* (×70) showing ovary, parthenogenetic eggs in brood chamber, heart, antennal muscles, and compound eye. (A modified from Banta 1930; B modified from Ueno 1927.)

leased into the brood chamber during each adult instar. Depending on the species and environmental conditions, the number of eggs per clutch varies considerably; usually there are between 2 and 40, and most frequently between 10 and 20. The parthenogenetic eggs undergo further development in the brood chamber and hatch into young similar in form to the adult. Just before molting, the parent liberates them to the outside by moving the postabdomen downward.

Seasonal abundance. In early spring relatively few cladocerans are to be found in temperate lakes and ponds. Such populations consist of females that survived over the winter or recently hatched from winter (resting) eggs. As the water reaches a temperature of 6 to 12°C, active reproduction begins and subsequently speeds up tremendously so that large populations result (in exceptional cases as high as 200 to 500 individuals per liter of water). Populations in ponds then soon begin to wane, so that few individuals can be found during the summer months. In the autumn there may or may not be a second population pulse, but during the winter the population is invariably low, with little or no reproduction. Thus, in a pond where the common species is *Daphnia pulex* or *D. magna* the population may be monocyclic or dicyclic, that is, having one or two population maxima during the year.

In larger or warmer bodies of water, however, seasonal variations in the abundance of Cladocera are not usually so pronounced. There is commonly a spring maximum and sometimes a less well-defined fall maximum, but summer and winter populations are large compared with those of ponds. In some lakes *D. pulex* is monocyclic, in others dicyclic. *Daphnia rosea* may be monocyclic, dicyclic, and in some cold lakes, acyclic, that is, without any pronounced population maxima during the entire year. Other common forms exhibiting monocycly or dicycly are *Simocephalus* and *Ceriodaphnia*. Occasionally, as in *Sida crystallina* in some lakes, there may be a single autumn pulse. There are also many Cladocera that exhibit single, long population pulses during the warmer months in certain lakes. Examples are *Diaphanosoma brachyurum*, *Chydorus sphaericus*, *Bosmina longirostris*, and *Moina*.

Prediction of the seasonal abundance of Cladocera in a particular lake or pond is difficult. Species differ greatly from one another in their seasonal abundance; a single species may have quite different population curves in two adjacent bodies of water. Furthermore, relative abundance and the specific time of maximum and minimum populations may vary considerably within a single species in the same lake from one year to the next. The abundance of any given population depends on many physical, chemical, and biological factors, particularly temperature, food quantity and quality, and the abundance of predators.

Male production, resting eggs. Numbers of males begin to appear in ponds in the spring or summer, late in the intensive reproductive cycle. Sometimes only 5% of the population may consist of males; sometimes more than half are males. The factors responsible for the appearance of males in cladoceran populations have been the object of long and intensive studies by many investigators. We do not as yet have a detailed explanation of the phenomenon. By way of partial explanation, however, it is probable that a complex interplay of factors is responsible. Production of male eggs seems to be induced mostly by (1) crowding of the females and the subsequent accumulation of excretory products, (2) a decrease in available food, (3) an increase in water temperature, (4) high light intensity, and (5) long photoperiod. These conditions (and probably others also), by altering the metabolism, appear to affect the chromosome mechanism in such a way that parthenogenetic male eggs rather than parthenogenetic female eggs are released into the brood chamber.

The same conditions responsible for male production, if continued for a longer time, appear to induce the appearance of sexual eggs. Females producing such eggs are morphologically similar to parthenogenetic females, but they produce only one, two, or sometimes several large opaque "resting" eggs. The fertilized eggs pass into the brood chamber, the walls of which then thicken and darken to form an ephippium (Fig. 18.8), which usually contains only one egg, sometimes two. In the Daphnidae the ephippium separates from the rest of the shell at the subsequent molt. In the Chydoridae it remains attached to the shell at the molt. Sometimes as many as a third or half of the females in a population may bear ephippia.

On being released, the ephippia sink to the bottom, become attached to the substrate (certain Macrothricidae), or float on the surface. In the last event, they may be blown ashore and accumulated in diminutive windrows. Ephippia and their contained eggs are capable of withstanding drying and freezing, and their production is clearly an adaptation to adverse environmental conditions. It has been shown that ephippia are

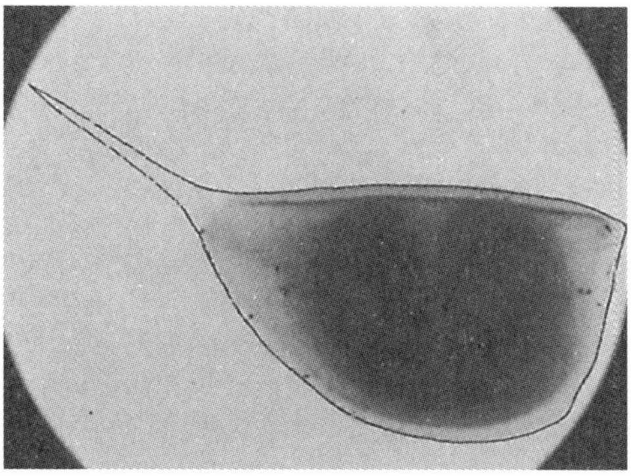

Fig. 18.8. Ephippium of *Daphnia pulex.*

sometimes viable even after passing through the digestive tract of fish. Ephippia are of special value in small ponds that dry up during the summer months. When such basins become filled with water in the early autumn, a large percentage of the ephippial eggs hatch into parthenogenetic females. Then there may be an autumn pulse followed by the appearance of males, sexual females, and ephippia that winter over and give rise to at least a part of the small seed population in the following spring. There is considerable evidence to show that ephippial females are exceptionally susceptible to predation.

Thus, it is seen that during most of the year the population consists entirely of parthenogenetic females, but for short periods the population may be complex, consisting of parthenogenetic females, males, unfertilized sexual females, and fertilized sexual females. In bodies of water larger than ponds, ephippia are produced by smaller percentages of the populations, and in limnetic populations in large lakes it is thought that reproduction may be entirely parthenogenetic the year round, especially in *Daphnia rosea.*

The expressions "monocyclic" and "dicyclic" are really used in two different ways in the literature. They may refer to an annual population curve having one or two pronounced maxima, without reference to types of reproduction, or they may be used to designate populations having one or two periods of syngamic reproduction during the year. The older idea of internal genetic or physiological "rhythms" being responsible for the periodic occurrence of syngamic reproduction has been discarded. The experimental work of Banta and others has demonstrated conclusively that the mode of reproduction is influenced by environmental factors. By changing the culture solutions frequently, Banta reared several common Cladocera for 800 to 1600 successive parthenogenetic generations (more than 27 years). Males, sexual females, and subsequent ephippia may be easily produced by crowding the mothers in a small amount of culture solution or by chilling the culture.

Laboratory observations have shown that under appropriate conditions a mature parthenogenetic female may begin to produce sexual eggs, or a sexual female may change over and produce parthenogenetic eggs. Males usually appear in both natural and controlled populations before females begin the formation of sexual eggs in the ovary. However, parthenogenetic females can produce ephippia that resemble a sexual ephippium without copulation, although the unfertilized egg in these ephippia may not develop and hatch.

Wood (1938) has contributed much to our understanding of the induction of hatching of ephippial eggs. Working with *Daphnia,* she found that (1) the considerable majority of eggs died when stored either wet or dry for one to many weeks; (2) the longer the period of storage, the greater the mortality; (3) a few eggs stored wet for 7 years were viable; (4) exposure to low temperatures decreased the percentage of hatch; (5) placing eggs in fresh culture medium or allowing them to dry out only a day or two gave the highest percentage of hatch, usually 35 to 45%; and (6) aeration and changing the culture solution frequently induced hatching of a comparable percentage of eggs that had not been allowed to dry out.

Development, life cycle. Length of life, from release of the egg into the brood chamber until the death of the adult, is highly variable, depending on the species and environmental conditions. *Daphnia rosea* usually lives for 28 to 33 days in laboratory cultures. MacArthur and Baillie (1929) found that *D. magna* lived an average of 26, 42, and 108 days at 28, 18, and 8°C, respectively. Limnetic individuals may live through the entire winter at low temperatures, although high-latitude and high-altitude populations overwinter by means of ephippial eggs.

Four distinct periods may be recognized in the life history of a cladoceran: egg, juvenile, adolescent, and adult. When a clutch of eggs is released into the brood chamber, segmentation begins promptly; the young, in the first juvenile instar and similar in form to the adult, are released from the brood chamber within a few days. The exact duration of egg development depends on temperature. There are but few juvenile instars, although greatest growth occurs during these stages. *Moina macrocopa* has two juvenile instars, *Daphnia rosea* has three, *D. pulex* has three or four, and *D. magna* three to five. The adolescent period is a single instar between the last juvenile instar and the first adult instar; during this instar the first clutch of eggs reaches full development in the ovary. As soon as the animal molts at the end of the adolescent instar and enters the first adult instar, the first clutch of parthenogenetic eggs is released into the brood chamber. During the first adult instar the second clutch of eggs is developing in the ovary. Successive adult instars and new clutches of young are produced in a similar manner. However, there is often a sterile period during the last few instars of life.

As in all other Crustacea, growth, in terms of increase in size, becomes apparent only immediately after each molt. During juvenile instars there may be almost a doubling of size after each molt, the increase in volume occurring within a few seconds or minutes and before the new exoskeleton hardens and loses its elasticity.

The number of adult instars is much more variable than the number of juvenile instars. *Daphnia pulex* usually has 18 to 25 adult instars; *D. rosea*, 10 to 19; and *D. magna*, 6 to 22. The duration of a single adult instar is highly variable, from a day to several weeks, depending on temperature. At the close of each adult instar four events follow one another in rapid succession, usually in a few minutes to a few hours. These are the release of young from the brood chamber to the outside, molting, increase in size, and the release of a new clutch of eggs to the brood chamber.

Overall size of adults varies considerably and appears to be controlled, in *Daphnia* at least, by temperature. The largest individuals thus occur in middle latitudes in which the mean annual temperature of surface water is between 6 and 8°C. Laboratory work also has shown that growth and reproduction of *Daphnia* may be biochemically slowed or inhibited by dense populations of blue-green algae (e.g., Infante and Abella 1985).

Cyclomorphosis. One of the most intriguing and puzzling of all cladoceran problems is the matter of cyclomorphosis, or seasonal changes in morphology, especially in females of limnetic species, such as *Daphnia pulex* and *D. rosea*. A population of a typical cyclomorphic species has a homogeneous "normal," or round headed, form during the late fall, winter, and early spring. As the water becomes warmer and the population develops, however, there is commonly a progressive increase in the longitudinal axis produced by a general elongation of the head and the appearance of a "helmet" (Fig. 18.9E, P, Q). Characteristically, the helmets become fully developed by midsummer, when they may be quite bizarre (Fig. 18.9G, H, Q). Beginning in the late summer or early autumn, the morphology of the head progressively reverts so that the "normal" head condition prevails by late autumn. However, a few species, such as the recent invader of North America, *D. lumholtzi*, possess a helmet year-round and cyclomorphosis only affects the length of that helmet. In North America, *D. lumholtzi* have been observed helmetless only in culture. In addition to changes in the shape of the head, cyclomorphosis may also involve changes in size of eye and length of the posterior spine.

The problem is greatly complicated by the fact that the degree of summer helmet development differs widely in the same species, even in two neighboring lakes. Cyclomorphosis is less pronounced in ponds and

Fig. 18.9. Some variations in head form in *Daphnia*. A to C, J, and K, *D. pulex;* D to H. *D. retrocurva;* L, M, *D. galeata mendotae;* N, O, *D. ambigua;* P, Q, *D. dubia.*

shallow lakes, and the degree of helmet development is relatively consistent from one individual to another at any one time. In larger and deeper lakes cyclomorphosis and the degree of helmet development in a population is much more variable, and both strongly developed and poorly developed helmets, as well as intergrades, may be found in the same townet sample.

Explanations of these seasonal changes in form naturally have long been sought for in the changing seasonal ecological conditions in bodies of water. Between 1900 and 1910 Wesenberg-Lund and Ostwald elaborated their "Buoyance Theory," which attempted to explain the occurrence of helmets as a flotation adaptation to decreased viscosities of water at summer temperatures. Subsequent observations and experimental work have discredited this viewpoint. Many other proposals have centered around photoperiod, light intensity, turbulence, temperature, and predation.

Temperature and predation have received the most support as causes for cyclomorphosis. The effect of temperature appears to be strongest during the early stages of development. Coker and Addleston (1938) found that the last one-third of the period of development of the first juvenile instar in the brood chamber was the critical period. *Daphnia rosea* raised at a temperature at 10°C or below during this period all had round heads; 15% of those raised at 12°C had pointed heads; 33% of those raised at 13°C had pointed heads; 87% at 14°C had pointed heads; and all of those raised at 16°C or more had pointed heads. In general, the higher temperatures produced more prominent helmets. Brooks (1946) found comparable temperature relationships for helmet formation in *Daphnia retrocurva* and head spine formation in *D. galeata*. Kiang (1942) demonstrated a relation between mean temperature of a lake and relative head length in *D. rosea*.

Nevertheless, the early work of Banta (1939) indicates that temperature is by no means the whole story and that genetic factors probably play an important role. He collected 10 distinct morphological types of *Daphnia pulex* in the field and cultured them in the laboratory. Under such circumstances the pronounced helmets were lost or minimized, but each of the 10 types remained recognizable and peculiar to itself. Recent genetic studies of cyclomorphosis have used electrophoresis to quantify genetic variability. Lampert and Wolf (1986) determined that although some level of genetic variation exists in any population, cyclomorphosis is not the result of a succession of morphologically different genetic clones.

Most recent studies of cyclomorphosis have found that the presence of predators has a strong effect on spine production and elongation. Dodson (1988) and others have shown that predators can "induce" longer helmets and sometimes tailspines in *Daphnia*. Both invertebrate and vertebrate predators can elicit cyclomorphosis through the release of an as yet unidentified low molecular weight molecule, called a kairomone. Chemical induction by invertebrate predators usually affects younger instars, whereas induction by vertebrate predators usually affects larger animals. Hanazato (1991a) showed that chemical induction of cyclomorphosis also may occur in *Daphnia* after exposure to pesticides and herbicides.

In temperate lakes, we may broadly summarize cyclomorphosis phenomena as environmental–genetic–predation pressure interrelationships. In the anticipatory words of Coker (1939), "The changes in form are not simple functions of external conditions or of any inherent cycle, but rather of a combination of internal and external conditions in a way that becomes exceedingly baffling the more we know about it."

Vertical migrations. Most limnetic cladocerans undergo vertical migratory movements during each 24-hour period. These "drifting" movements of the general population are commonly upward with the onset of darkness and downward with the coming of the light of dawn. In some species, however, and in some habitats, there is a rapid upward and downward cycle near dusk and a similar upward and downward cycle around dawn. In rare instances some species show "reverse" migratory movements. The average amplitude ranges from 2 to over 30 m, depending on the species and many environmental factors.

Most of the conclusions about vertical migrations have been based on light sensitivity plus a cluster of environmental variables. Cladocerans exhibit phototaxis, which may be either positive or negative. Individual variation in phototaxis is present in most populations, such that the majority of the population usually exhibits one type of phototaxis, with a few individuals exhibiting the opposite type. The discovery that experimentally blinded *Daphnia* still exhibit vertical movements suggests that some part of the body other than the eyes also may be light sensitive. Furthermore, some investigators have revived the "endogenous rhythm" explanation because they have found that *Daphnia* migrates even when kept in continuous darkness.

The vertical movements of cladocerans appear to be variously affected by species, age, size, temperature, food supply, water chemistry, water color, season of year, light intensity, and turbulence. These factors may mediate vertical migration, but more recent studies (e.g., Gliwicz 1986) have proposed that vertical migration acts as a mechanism to reduce predation pressure. The strategy of migrating down in the water column during daylight reduces the exposure of zooplankton to visual fish predators which require light to feed. Furthermore, Ohman et al. (1983) observed the "reverse" migration pattern in populations exposed to large numbers of invertebrate predators that migrate to the depths during daylight. As with cyclomorphosis, it appears that vertical migration is chemically mediated; Ringelberg (1991) showed that vertical migration can be induced with exposure to water in which predators had been cultured.

General ecology. Over fairly large areas, including various types of connecting standing water habitats (i.e. marshes, ponds, impoundments), as many as 32 species among 21 genera of cladocerans can be found. Aside from rapid streams, brooks, and grossly polluted waters, they are abundant everywhere. Common open-water and limnetic forms are *Daphnia*, *Bosmina*, *Diaphanosoma*, *Chydorus*, *Ceriodaphnia*, and *Holopedium*. Such forms are seldom found in vegetated areas, and it has been demonstrated experimentally that rooted aquatics have a repellent effect on limnetic species, especially *Daphnia* (Pennak 1973). However, the greatest abundance of species may be collected in the vegetation at margins of lakes and rivers, including *D. pulex*, *Sida crystallina*, *Ophryoxus gracilis*, most Chydoridae, and most Macrothricidae. The two most common inhabitants of ponds, permanent pools, and temporary pools are *D. pulex* and *D. magna*, *Alona quadrangularis*, *Drepanothrix*, *Ilyocryptus*, and *Monospilus* are among the most common forms to be found on or near the bottom in weedy littoral areas. Affinities are indicated for most genera in the key at the end of this chapter.

Over fairly large areas, including various types of connecting standing water habitats (i.e., marshes, ponds, impoundments), as many as 25 species of cladocerans can be found. However, most natural populations of *Daphnia* and other limnetic genera rarely consist of more than one numerically dominant species. In limnetic habitats, if two species in the same genus do occur together, one is usually 20 or more times as abundant as the other. At the other habitat extreme, Ranta (1979), working on very small rock pools, found three species of *Daphnia: D. magna* (common), *D. longispina*, and *D. pulex* (rare). No pool contained all three species, and two of these species were found together in only 10% of the pools. In littoral and substrate environments, co-occurrence of two or more species within a genus is common.

Many Cladocera can tolerate a wide range of temperatures, although the distributions of a few species are limited by temperature. *Latona*, *Holopedium gibberum*, and *Daphnia longiremis* are cold stenotherms, whereas *Pseudosida bidentata*, *Ceriodaphnia rigaudi*, and *Euryalona occidentalis* are restricted to warm waters of southern states. Nevertheless, temperatures seem to affect abundance and in some cladocerans, numbers decline at temperatures in excess of 20°C. Other species, such as *Diaphanosoma brachyrum* and the exotic *Daphnia lumholtzi*, become most abundant when temperatures exceed 20°C and other species decline (e.g., Work and Gophen 1999). Many species that occur in north temperate water bodies do not occur in warm southern lakes, perhaps due to temperature exclusion, although differences in lake morphometry, water chemistry, phytoplankton assemblage composition, and predator abundance between north and south temperate lakes undoubtedly affect zooplankton assemblages as well.

Information on chemical limiting factors is somewhat scarce. Dissolved oxygen is sufficiently abundant in most water bodies to sustain Cladocera. Low concentrations occur during summer and winter periods of complete oxygen exhaustion in the hypolimnion of lakes and in water bodies with high oxygen demand from pollutants or dense decaying vegetation. However, even low oxygen concentrations may not exclude Cladocera, since many species can withstand oxygen concentrations of less than 1 ppm by producing hemoglobin. Most Cladocera occur in waters containing a wide range of concentrations of calcium, although some species appear to be sensitive to large differences in water hardness. *Holopedium*, in particular, is confined to calcium-poor waters. Hutchinson (1932) found that magnesium may act as a limiting factor by inhibiting reproduction; critical concentrations for *Daphnia magna*, *D. pulex*, and *D. rosea* were found to be 240, 120, and 30 to 60 mg/L. Although a few species are restricted to acid and bog waters, most Cladocera occur over a wide range in pH, nearly all species being found at 6.5 to 8.5.

Cladocera are very sensitive to manmade chemicals, such as herbicides and pesticides. Therefore, many laboratories use *Daphnia*, especially *D. magna*, as an assay organism for the detection of toxic substances in water. The broad use of this organism has resulted in standardization of methods for comparative purposes.

The species makeup in lakes and ponds also may be significantly altered by invertebrate and vertebrate predation (e.g., Black and Hairston 1988). Sight-feeding planktivorous fishes often select larger and more obvious individuals in preference to smaller individuals and as a result, the zooplankton assemblage may become skewed toward smaller species and individuals. Invertebrate predators, such as the phantom midge, *Chaoborus*, exert a contrasting pressure on cladoceran populations by ingesting smaller instars and species. Therefore, the type of zooplankton species that occur in a given lake or pond depends upon (1) the physical and chemical tolerances of the species and (2) the predator type and density.

Probably no other major freshwater taxon is currently the object of such investigative activity as the Cladocera. The literature is vast and controversial, with variable and debatable information. Many aspects of cladoceran biology are thus deserving of critical published reviews, but few investigators (e.g., Hebert 1978) have had the courage to undertake such chores. Unfortunately, experimental studies based on laboratory cultures are often inconsistent and give results that cannot be transposed to natural, open-water situations.

Most recent interests are centered around cyclomorphosis, clone genetics, species flocks, competition, selective food habits, microanatomy as shown by ultramicroscopy, effects of nutrients and trace elements, population fluctuations, and paleolimnological history. Any detailed discussion of these topics is beyond the scope of this volume.

Geographical distribution. Because of their resistant ephippial eggs, Cladocera may be transported overland to new habitats. As a result, many species have been considered widespread and some cosmopolitan. The general occurrence of a few common species is indicated in Table 18.1. Less than 20 of these species are known only from the United States, and most of these have been collected in only one or two localities. However, our knowledge of the geographical limits of many species in North America is incomplete. Frey (1982a) and others have shown that many species formerly considered widespread or cosmopolitan actually are clusters of two or more species or morphs, each of which has a more restricted distribution.

In recent years several aquatic exotic species have made their way to North America either in company with other deliberately introduced species or in ballast water of oceanic ships. *Daphnia lumholtzi*, originating in Africa, Asia, and Australia, probably invaded North America in the late 1980s and quickly expanded its new range to cover most of the southern United States within 5 years. *Bythotrephes* (or *Cercopagis*) (Fig. 18.10) is a large, bizarre European carnivorous cladoceran that was introduced into Lakes Erie and Huron in the early 1980s most likely via discharged ballast water. It is now a common plankton species in all of the Great Lakes.

Culturing. Several species of *Daphnia* are among the most easily cultured of all freshwater invertebrates. Fish hatcheries and tropical fish fanciers raise them in tanks or

Table 18.1. General Geographic Distribution of Some Common Species of Cladocera

Cosmopolitan
Alona guttata, A. rectangula
Ceriodaphnia laticaudata
Chydorus
Daphnia magna, D. pulex
Eurycercus
North America, Europe, Asia, and South America
Alona affinis
Bosmina longirostris
Ceriodaphnia quadrangula
Macrothrix laticornis
Simocephalus vetulus
North America, Europe, and Asia
Acroperus harpae
Bosmina coregoni
Camptocercus rectirostris
Daphnia rosea
Diaphanosoma brachyurum
Eurycercus lamellatus
Holopedium gibberum
Leptodora kindti
Moina macrocopa
Pleuroxus strictus
Polyphemus pediculus
Scapholeberis mucronata
Sida crystallina

in concrete or plastic outdoor wading pools. Cottonseed meal, manure, some kinds of commercial agricultural fertilizers, dried yeast, dried milk, and dried and chopped hay may be suitable foods, especially for the suspended organic particles and the bacteria that become abundant. However, some species may require a mixture of algae and bacteria for consistent survival and reproduction. It is important to avoid adding too much food to the cultures; otherwise, the medium becomes foul. The writer (RWP) has kept 20 cultures of *Daphnia rosea* going continuously in 1- to 10-gallon aquariums for 8 years. Once per week each aquarium receives dried yeast suspended in 20 to 200 mL of water. Each aquarium is then thoroughly stirred for a few seconds. If *Daphnia* are fed algae, the culture water should be green, but the animals should be clearly visible. If animals are overfed, algal respiration can reduce oxygen concentrations to lethal levels at night. However, Dodson and Frey (1991) indicate that underfeeding animals can selectively kill clones with lower starvation thresholds.

Economic significance. The great importance of Cladocera in the aquatic food chain as food for both young and adult fish was emphasized first by Forbes in 1883, and since then by innumerable investigators. Various studies of the stomach contents of young fish show from 1 to 95% Cladocera by volume, and very few studies show less than 10%. Although of less direct economic importance, Hydra and immature and mature insects also feed on Cladocera. Goulden (1971), for example, found that midsummer populations of littoral Chydoridae were greatly reduced by Diptera larvae predation.

Cladocerans also can play an important role in the restoration of water clarity to overly productive lakes (e.g., Carpenter et al. 1985). Many lakes in North America have become eutrophic as result of agricultural, industrial, or residential nutrient release or runoff. The increase in nutrients often results in blooms of algae, particularly blue-green species, that may create surface scums and unpleasant odors. Although the grazing efficiency of cladocerans for blue-green algae is low, increases of cladocerans have been shown to reduce algal populations and increase water clarity in many lakes.

Collecting, preserving. An ordinary plankton townet is suitable for limnetic species but is of little use for collecting in littoral vegetation. For such habitats the Birge cone net does an excellent job. Bottom forms may be taken by bringing the top centimeter of mud and debris into the laboratory and allowing it to settle. By looking horizontally along the surface toward a light

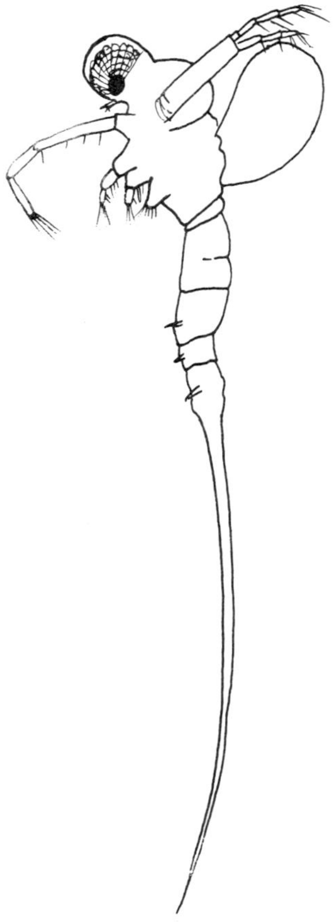

Fig. 18.10. *Bythotrephes* (or *Cercopagis*), a bizarre cladoceran that has recently made its way to the United States from Europe; ×18.

source, the Cladocera may be located and picked up with a long pipette. In small ponds and pools, and to some extent in vegetation, a small dipnet is convenient.

The best general killing agent is 95% ethyl alcohol. Because of their delicate body and strong muscles, however, *Moina* and most of the Sididae are badly distorted when placed in alcohol. These forms should first be narcotized with chloral hydrate, club soda, or a similar deadening agent. Seventy percent alcohol is a standard preservative, although 4% sucrose–formalin also is commonly used. The writer (RWP) always adds 5% glycerin to his storage vials so that specimens will not be destroyed through evaporation if the corks dry out or the vials break.

Fortunately, little or no dissection is necessary for identification purposes, but permanent slide mounts in glycerin or glycerin jelly are highly desirable. Coverslips should be supported by bits of cardboard of appropriate thickness so that specimens will not be crushed. Sometimes it is necessary to disarticulate and clear the exoskeletons of cladocerans to make out minute anatomical details. Judicious use of cold dilute hydrochloric acid for 24 hours or hot acid for a few minutes usually will give the desired results.

Taxonomy. Brooks (1957) has given the genus *Daphnia* its most thorough treatment and Dodson and Frey (1991) present detailed descriptions of chydorids; many of their suggestions have been followed in this chapter. Nevertheless, much work still remains to be done on cladoceran taxonomy, particularly for *Daphnia*. Most of *Daphnia* species feature irksome seasonal and geographic variations in carapace and head morphology, and details of setation and the postabdomen. The *retrocurva*, *carinata*, *rosea-longispina*, and *pulex-pulicaria* problems, as well as several other questions, deserve further consideration. The situation is confused by what seems to be "hybridization" and species (or subspecies) "flocks" in some geographical areas. Our major difficulty is that definitive species descriptions often are based on specimens collected from only a limited number of field sites. We need widely distributed massive collecting and careful analyses of large numbers of specimens to elucidate taxonomic relationships.

Wolf (1987) has demonstrated naturally coexisting hybrids between *Daphnia hyalina*, *D. galeata*, and *D. cucullata* by means of gel electrophoresis. Even more puzzling is the report of Hebert and Crease (1980) who found, by electrophoresis, 22 clones of *D. pulex* in 11 separate populations, with as many as 7 clones being present in a single population. Dodson (1981) concludes that *D. pulex* is either one highly variable species or a cluster of many species differing from each other only slightly.

Some species found in the United States have long been considered Holarctic and identical with those found in Europe. More careful scrutiny of our American forms is needed, however, because some forms may be sufficiently different from their European counterparts as to require different species designations.

Aside from the *Daphnia* problem, certain other genera will be clarified as more specimens are collected from many additional field sites, notably *Ceriodaphnia*, *Diaphanosoma*, *Moina*, *Alonella*, *Alona*, *Pleuroxus*, and *Chydorus*. The papers of Frey, especially, have pointed out the minute anatomical and taxonomic distinctions of species in the chydorids. Edwards (1980) has developed a new means of studying *Daphnia* that involves examining the details of the mandibular surfaces of newly molted daphnids.

Fryer (1987) has suggested that "Cladocera" as an Order designation should be abandoned, and that this taxon should be split into four orders: Ctenopoda, Anomopoda, Onychopoda, and Haplopoda. This system has been adopted, although the term "Cladocera" remains commonly used.

Many genera in the following key currently contain only a single species in the United States. Future work, however, will undoubtedly reveal additional species in several of these genera. In the key that follows, lengths refer only to females and do not include spines or antennae. Additional keys giving more specific information include Balcer et al. (1984) who have produced a key to the common species of the Great Lakes region.

KEY TO GENERA OF CLADOCERA OF THE UNITED STATES

1. Body and legs covered with a bivalve carapace; legs foliaceous, not clearly segmented **2**
   Body and legs not covered with a carapace; legs subcylindrical or somewhat flattened, clearly segmented, and prehensile **51**
2. With six pairs of foliaceous legs; first and second pairs not prehensile (Figs. 18.11; 18.12) ... **3**
   With five or six pairs of legs; first and second pairs more or less prehensile **8**
3. Carapace of the usual type; antennae of female biramous and flattened (Figs. 18.11F; 18.12A, C) **SIDIDAE, 4**
   Animal enclosed in a large gelatinous mantle, open ventrally and forming two valves (Fig. 18.11A); the mantle is shed during ecdysis but regenerated within 2 hours; antennae of female simple and cylindrical; swims ventral side up; found in low-calcium waters; 1 to 2 mm long ...... **HOLOPEDIDAE, *Holopedium***
4. Dorsal ramus of antenna three segmented (Fig. 18.11E); 3 to 4 mm long; common among aquatic plants in lakes and ponds ...... ***Sida***
   Dorsal ramus of antenna two segmented **5**
5. With a lateral expansion on basal segment of dorsal ramus of antenna; with a tonguelike projection on ventral surface of head (Fig. 18.11F, G); 2 to 3 mm long; in littoral vegetation; chiefly in northern states and mountainous areas ...... ***Latona***
   Without lateral expansion on antenna **6**
6. Without spines on postabdomen; ocellus absent; 0.8 to 1.2 mm long; common and widely distributed (Fig. 18.12A) ...... ***Diaphanosoma***
   With spines on postabdomen (Fig. 18.12B); ocellus present; about 2 mm long **7**
7. Rostrum present (Fig. 18.12B), southern states ...... ***Pseudosida***
   Rostrum absent (Fig. 18.12C); usually in vegetation ...... ***Latonopsis***
8. Antenna 1 attached to ventral side of head, not covered by fornices (Figs. 18.13–18.23) **9**
   Fornices extended so as to cover first antennae more or less, and united with rostrum into a beak that projects ventrally in front of first antennae (Fig. 18.20) .... **CHYDORIDAE, 30**
9. Antennae 1 of female small, often rudimentary; if large, then never inserted at anterior end of ventral edge of head (Figs. 18.13–18.15) **11**
   Antennae 1 of female large, inserted at anterior end of ventral edge of head (Figs. 18.16–18.18 **10**

Fig. 18.11. Structure of Holopedidae and Sididae. A, *Holopedium gibberum*, ×40; B, postabdomen of *H. gibberum* C, first leg of *H. gibberum;* D, postabdomen of *H. amazonicum;* E, *Sida crystallina*, ×14; F, antenna of *Latona setifera;* G, anterior end of *L. parviremis.* (B and D modified from Ward and Whipple 1918; E and F modified from Lilljeborg 1900.)

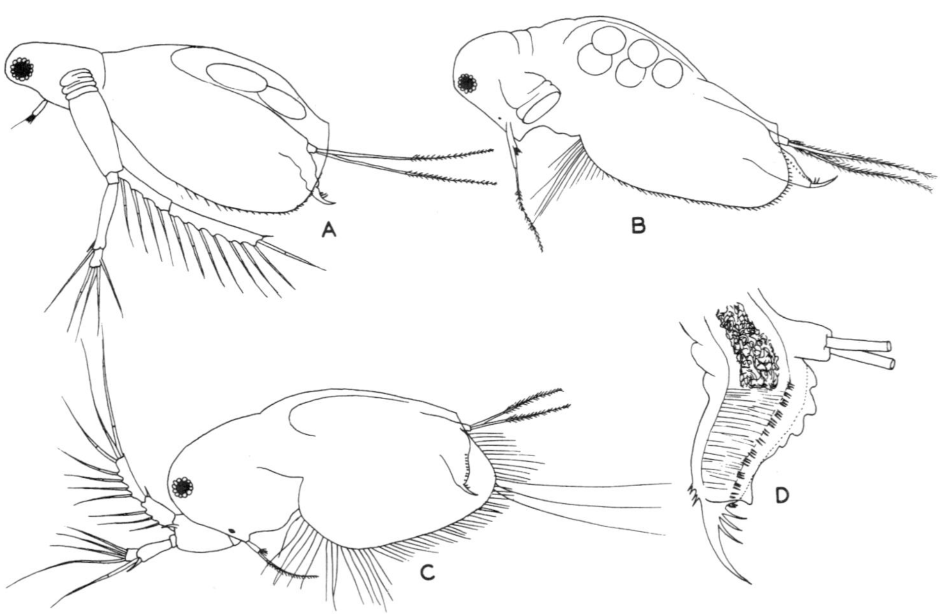

Fig. 18.12. Structure of Sididae. A, *Diaphanosoma brachyurum*, ×45; B, *Pseudosida bidentata*, ×27; C, *Latonopsis occidentalis*, ×22; D, postabdomen of *L. fasciculata.* (A modified from Lilljeborg 1900; B modified from Birge 1910; C and D modified from Ward and Whipple 1918.)

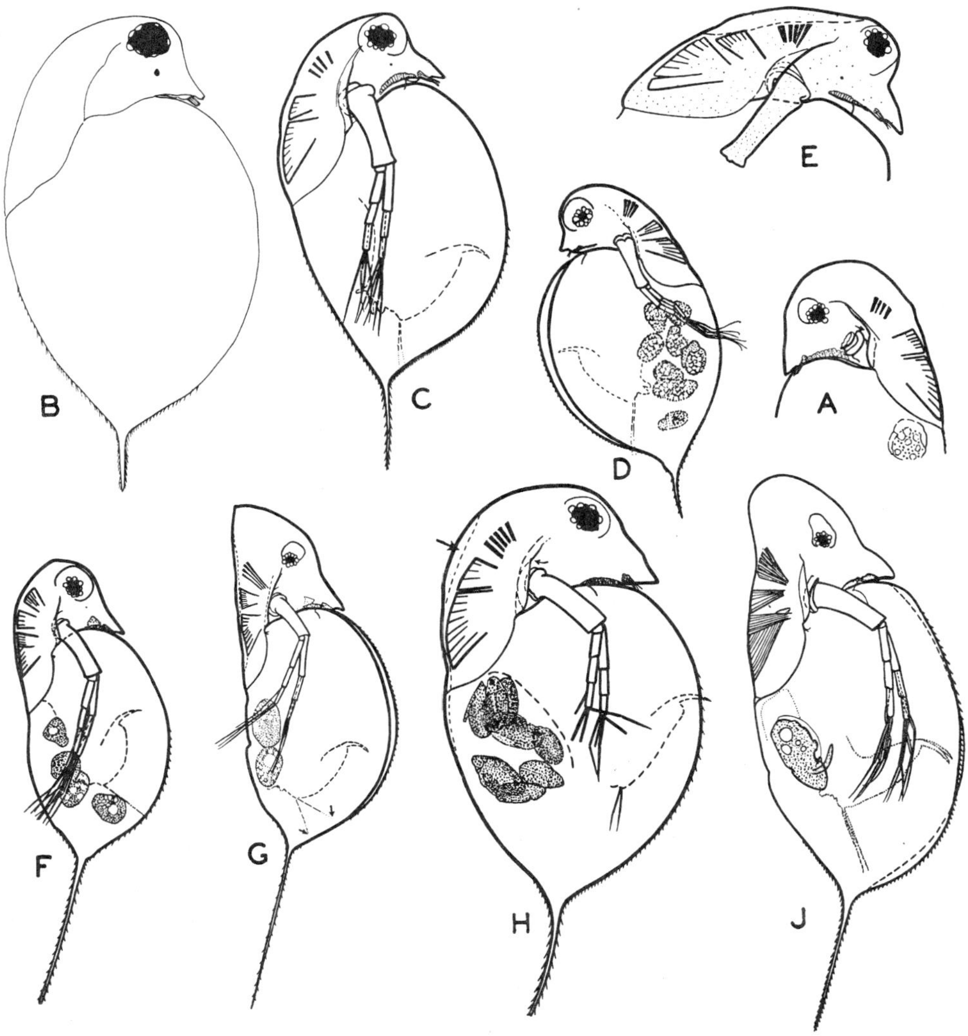

Fig. 18.13. Structure of *Daphnia*. A, head of *D. schødleri;* B and C, *D. pulex;* D, *D. parvula;* E, head of *D. middendorffiana;* F, *D. laevis;* G, *D. dubia;* H, *D. rosea;* J, *D. thorata*. (A, C–J from Brooks 1957; B from Brandlova et al. 1972.)

10. First antennae of female fixed (Fig. 18.16); intestine simple; 0.3 to 0.5 mm long **BOSMINIDAE, 18**
First antennae of female freely movable (Fig. 18.17); intestine simple or convoluted .... **MACROTHRICIDAE, 20**

11. Rostrum present .... **DAPHNIDAE (in part), 12**
Rostrum absent **16**

12. Dorsal and ventral margins lacking spines; pubescence on the middle part of the ventral carapace margin; no dorsal carina; no posterior spine; rare, known only from small ponds in CT, IL .... ***Daphniopsis***
Dorsal and ventral margins more or less spinous; pubescence absent; dorsal carina present; posterior spine usually present **13**

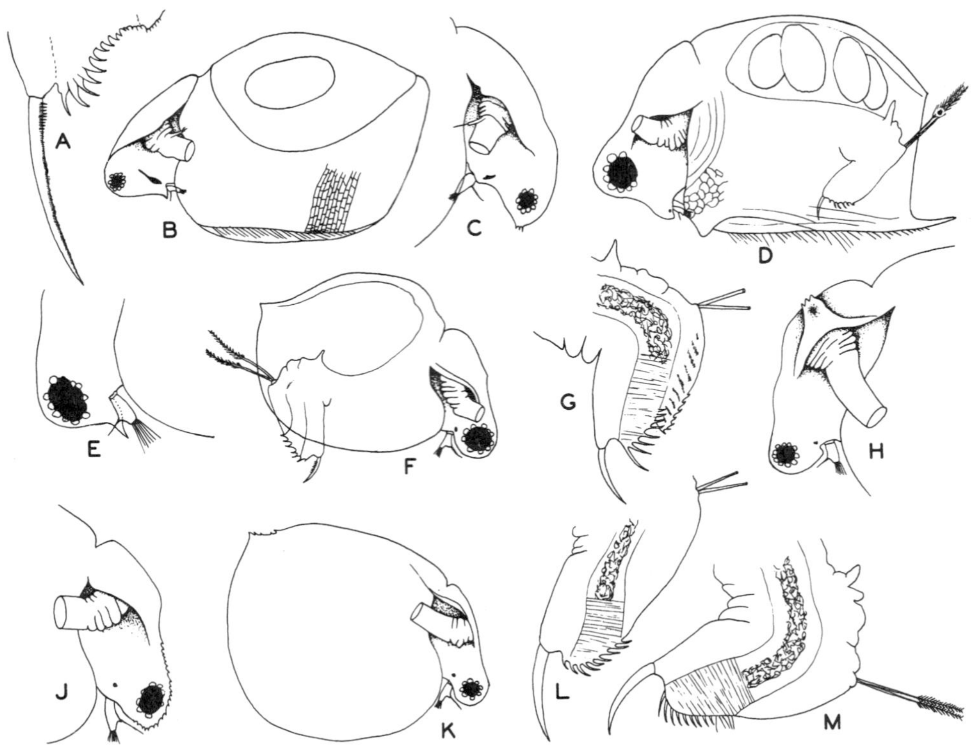

Fig. 18.14. Structure of Daphnidae. A, claw of *Simocephalus exspinosus;* B, *S. vetulus,* showing a portion of transverse striations, ×15; C, head of *S. serrulatus;* D, *Scapholeberis mucronata,* ×50; E, head of *Ceriodaphnia rigaudi;* F, *C. reticulata,* ×25; G, postabdomen of *C. megalops;* H, head of *C. lacustris;* J, head of *C. rotunda;* K, *C. laticaudata,* ×50; L, postabdomen of *C. quadrangula;* M, postabdomen of *C. pulchella.* (B–H, and L modified from Ward and Whipple 1918; J and M modified from Lilljeborg 1900.)

13. Without a cervical sinus . . . . ***Daphnia***
With a cervical sinus (Fig. 18.14B–D) **14**

14. Valves transversely striated (Fig. 18.14B); up to 3 mm long; the species in this genus are known to hybridize . . . . ***Simocephalus***
Valves not transversely striated; up to 1 mm long; usually in vegetation **15**

15. Rostrum of female pointed; color whitish or greenish, translucent or opaque; uncommon in northern states . . . . ***Megafenestra***
Rostrum more pointed (Fig. 18.14D) . . . . ***Scapholeberis***

16. First antennae small; head small and depressed (Fig. 18.14E–K); up to 1.4 mm long **DAPHNIDAE (in part), *Ceriodaphnia***
First antennae large; head large and extended (Fig. 18.15A, B, D) . . . . **MOINIDAE, 17**

17. Body laterally compressed; valves completely covering body (Fig. 18.15A); up to 1 mm long; on bottom and in vegetation; reported only from LA . . . . ***Moinodaphnia***
Body thick and heavy; valves not completely covering body (Fig. 18.15B, D); usually in pools and ponds . . . . ***Moina***

18. First antennae united at base and diverging at apex (Fig. 18.16A, B); widely distributed in southern states; uncommon . . . . ***Bosminopsis***
First antennae not united at base, parallel; highly variable; littoral and limnetic; widely distributed and common **19**

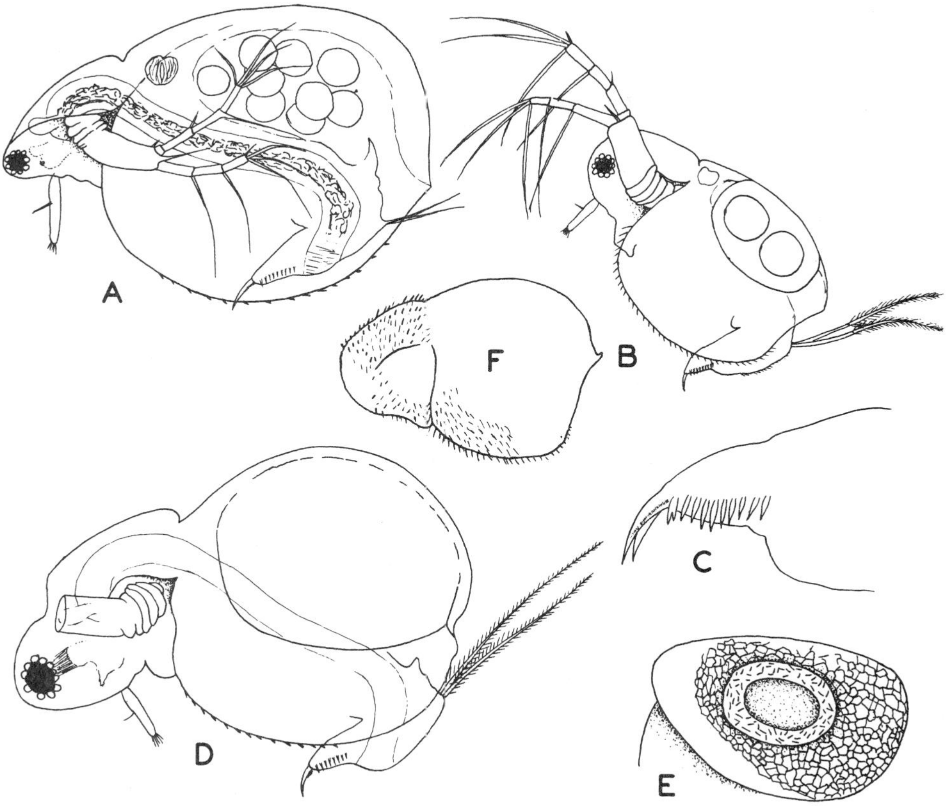

Fig. 18.15. Structure of Moinidae. A, *Moinodaphnia macleayi* with parthenogenetic eggs, ×60; B, *Moina macrocopa* with ephippium, ×24; C, postabdomen of same; D, *M. micrura*, ×40; E, ephippium of *M. micrura;* F, head and carapace of *Moina affinis* showing distribution of hairs. (A, D, and E modified from Ward and Whipple 1918; B and C modified from Ueno 1927.)

19. Female postabdominal claw with a very fine distal pecten (Fig. 18.16D); frontal sensory bristle near midpoint between eye and tip of rostrum (Fig. 18.16B); widely distributed and common ***Bosmina***

Female postabdominal claw with proximal pecten only, emarginate near tip (Fig. 18.16E); frontal sensory bristle near tip of rostrum, near or at base of antenna 1 (Fig. 18.16C); sometimes considered a subgenus of *Bosmina* .... ***Eubosmina***

20. Intestine convoluted (Fig. 18.17A, B) **21**

Intestine simple **25**

21. Valves with spine at posterior dorsal angle (Fig. 18.17A); up to 2 min long; in vegetation; widely distributed .... ***Ophryoxus***

Valves without such a spine **22**

22. Antennal setae 0–0-1–3/1–1-3; 0.2 to 0.5 min long (Fig. 18.18A) .... ***Streblocerus***

With another antennal setae formula; more than 0.5 mm long **23**

23. Valves narrowed posteriorly and prolonged into a short tube (Fig. 18.17C); up to 1.2 mm long; in vegetation; New England states and rare farther west .... ***Parophyroxus***

Valves not narrowed posteriorly; widely distributed **24**

24. Dorsal and posterior margin arched; with a conspicuous dorsal tooth (Fig. 18.17D); up to 0.7 mm long; in littoral zones; uncommon in northern states .... ***Drepanothrix***

Without a dorsal tooth (Fig. 18.17E); up to 1.8 mm long; in littoral, especially in sphagnum bogs .... ***Acantholeberis***

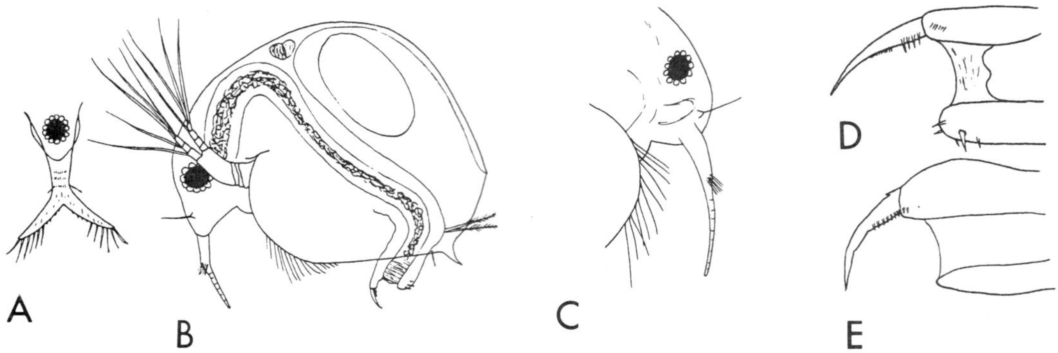

Fig. 18.16. Structure of Bosminidae. A, dorsal view of *Bosminopsis dietersi;* B, *Bosmina longirostris,* ×140; C, anterior end of *Eubosmina coregoni;* D, postabdomen of *B. longirostris;* E, postabdomen of *E. longispina.* (E modified from Deevey and Deevey 1971.)

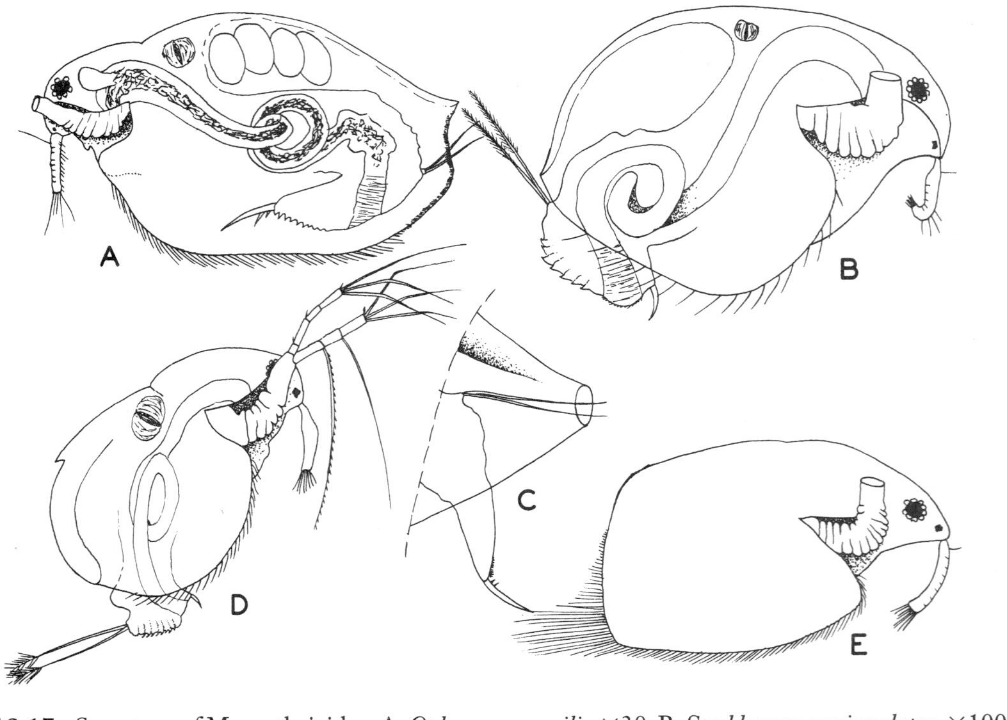

Fig. 18.17. Structure of Macrothricidae. A, *Ophryoxus gracilis,* ×30; B, *Streblocerus serricaudatus,* ×100; C, posterior end of *Parophryoxus tubulatus;* D, *Drepanothrix dentata,* ×60; E, *Acantholeberis curvirostris,* ×26. (Modified from Ward and Whipple 1918.)

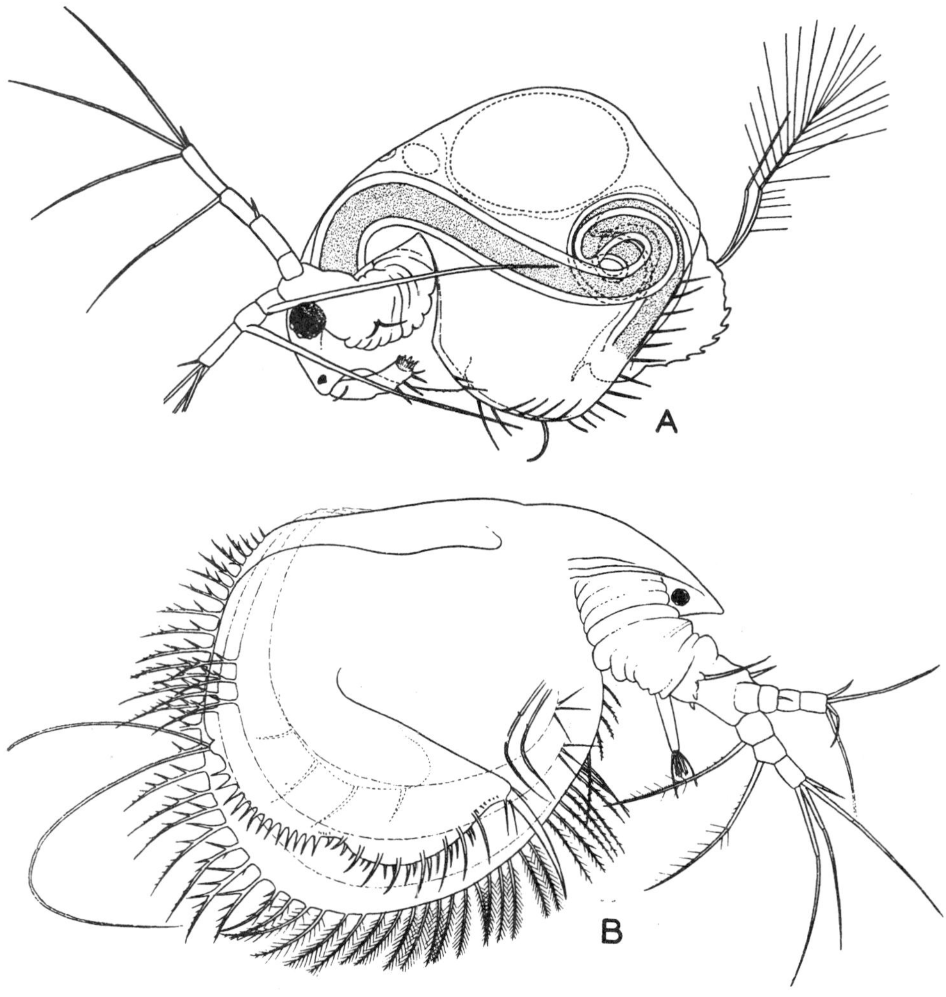

Fig. 18.18. Structure of Macrothricidae. A, *Streblocerus serricaudatus;* B, *Ilyocryptus sordidus.* (From Fryer 1974.)

25. With a wide crest on dorsal margin of valves (Fig. 18.19B); up to 1 mm long; reported from northern states . . . . *Bunops*
Without such a crest **26**
26. Postabdomen with numerous long spines (Fig. 18.19C–E); antennal setae 0-0-0-3/1-1-3; often more or less covered with detritus; usually found creeping about on the substrate, especially in vegetated areas (Fig. 18.18B) . . . . *Ilyocryptus*
Postabdomen without numerous long spines; with another setae formula **27**
27. Hepatic ceca present: 0.9 mm long; in vegetation in shallow water; LA . . . . *Grimaldina*
Hepatic ceca absent **28**
28. Antennal setae 0-0-1-3/1-1-3; setae of basal segment of lower ramus stout (Fig. 18.19G) . . **29**
Antennal setae 0-1-1-3/1-1-3; all setae slender; 1 mm long (Fig. 18.19F); widely distributed but uncommon in vegetation of littoral . . . . *Lathonura*
29. Dorsal margin of head evenly rounded (Fig. 18.19H) . . . . *Macrothrix*
Dorsal margin of head not evenly rounded (Fig. 18.19G); 0.7 mm long; in vegetation; widely distributed . . . . *Echinisca*

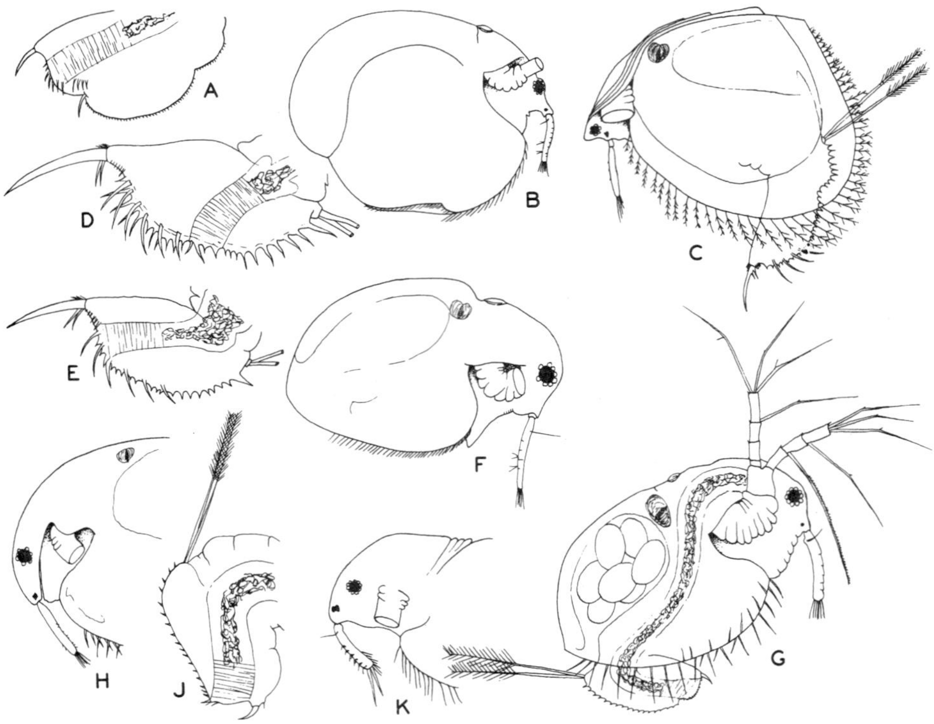

Fig. 18.19. Structure of Macrothricidae. A, postabdomen of *Grimaldina brazzai;* B, *Bunops serricaudata*, ×35; C, *Ilyocryptus spinifer*, ×40; D, postabdomen of *I. sordidus;* E, postabdomen of *I. acutifrons;* F, *Lathonura rectirostris*, ×38; G, *Echinisca rosea*, ×55; H, anterior end of *Macrothrix borysthenica;* J, postabdomen of *M. laticornis;* K, anterior end of *M. montana.* (B–H and K modified from Ward and Whipple 1918; J modified from Lilljeborg 1900.)

30. Anus terminal (Fig. 18.20A); up to 5 mm long; often with many ephippial eggs; common everywhere in weed beds; a complex of closely related species, designations uncertain ***Eurycercus***

Anus on dorsal side of abdomen (Fig. 18.20B, C) **31**

31. Compound eye present **32**

Compound eye absent but ocellus large; 0.5 mm long; rare; northern states . . . . . . ***Monospilus***

32. Eye and ocellus of usual size **33**

Compound eye and ocellus extremely large and of similar size; rare in southern states ***Dadaya***

33. Posterior margin of valves not greatly less than maximum height (Figs. 18.20; 18.21), mandible articulates at meeting point of headshield and shell, usually on basal spine on claw and lateral denticles on postabdomen **34**

Posterior margin of valves considerably less than maximum height (Figs. 18.22; 18.23), mandible articulates in pocket of headshield, usually two basal spines on claw and no lateral denticles on postabdomen **44**

34. Claws with secondary tooth in the middle (Fig. 18.20B, C); secondary tooth sometimes very small **35**

Claws without secondary tooth in the middle **39**

35. Postabdomen with both marginal and lateral denticles (Fig. 18.20C) **36**

Postabdomen with lateral denticles only, sometimes inconspicuous (Fig. 18.20H); 0.8 to 0.9 mm long (Fig. 18.20J); common everywhere in vegetation . . . . ***Acroperus***

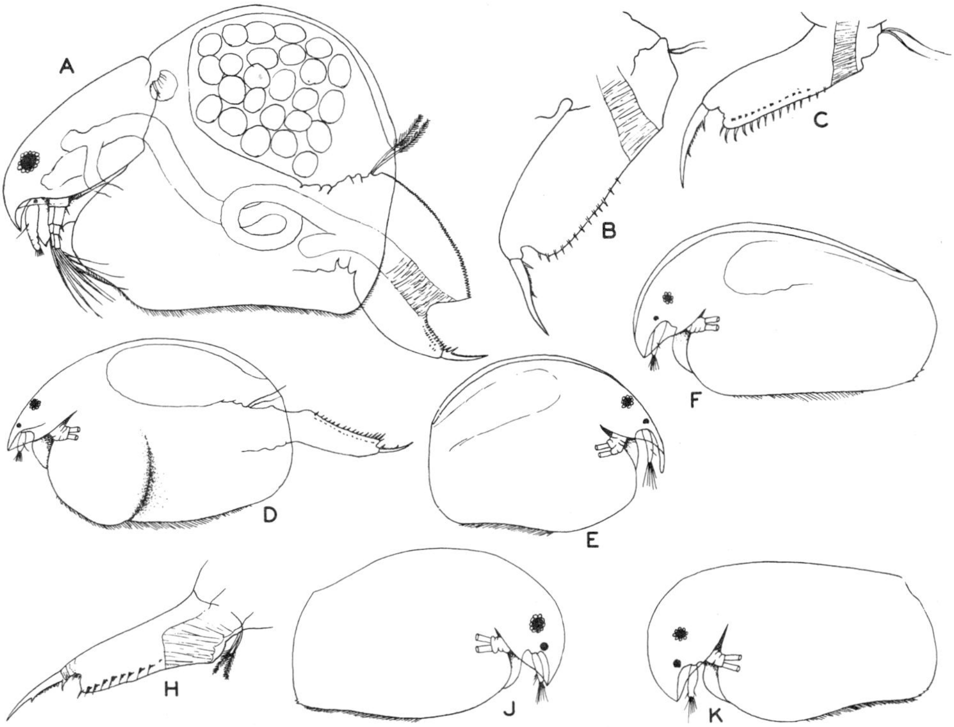

Fig. 18.20. Structure of Eurycercinae and Chydorinae. A, *Eurycercus lamellatus*, ×17; B, postabdomen of *Alonopsis aureola;* C, postabdomen of *A. americana;* D, *Euryalona occidentalis*, ×38; E, *Kurzia latissima*, ×48; F, *Camptocercus rectirostris*, ×40; H, postabdomen of *Acroperus harpae;* J, *A. harpae*, ×42; K, *A. angustatus*, ×46. (A modified from Lilljeborg 1900; B and C modified from Doolittle 1911; D–F, J, and K modified from Ward and Whipple 1918; H modified from Ueno 1927.)

36. Postabdomen relatively broad, terminal claw may have two medium-sized teeth (Fig. 18.20C); up to 1.9 mm long, in littoral vegetation reported from ME, NH . . . . ***Alonopsis***
Postabdomen relatively narrow (as in Fig. 18.20H) **37**

37. Anterior portion of valves swollen (Fig. 18.20D); 1 mm long; in vegetation; southern states ***Euryalona***
Anterior portion of valves not swollen; usually in vegetation in shallows **38**

38. Crest on head and valves (Fig. 18.20F); about 1 mm long . . . . ***Camptocercus***
Crest on valves only (Fig. 18.20E); 0.6 mm long; common in temperate America . . . . . ***Kurzia***

39. Rostrum not greatly exceeding antenna 1 (Fig. 18.21A, G, H) **40**
Rostrum much longer than antenna 1, postabdomen may have numerous lateral denticles and only two to four marginal denticles (Fig. 18.22A); 0.5 mm long; northern states ***Rhynchotalona***

40. Rostrum abruptly narrowed and pointed near tip (Fig. 18.21G, H) **41**
Rostrum not abruptly narrowed near tip, ventroposterior angle with teeth (Fig. 18.21A); 0.5 to 0.7 mm long; usually on bottom in shallows; common and widely distributed .***Graptoleberis***

41. Marginal denticles markedly longer distally **42**
Marginal denticles not markedly longer distally **43**

42. Postabdomen with clusters of large spines (Fig. 18.21B); about 1 mm long; uncommon ***Leydigia***
Postabdomen without clusters of large spines (Fig. 18.21D); about 0.5 mm long . . . ***Oxyurella***

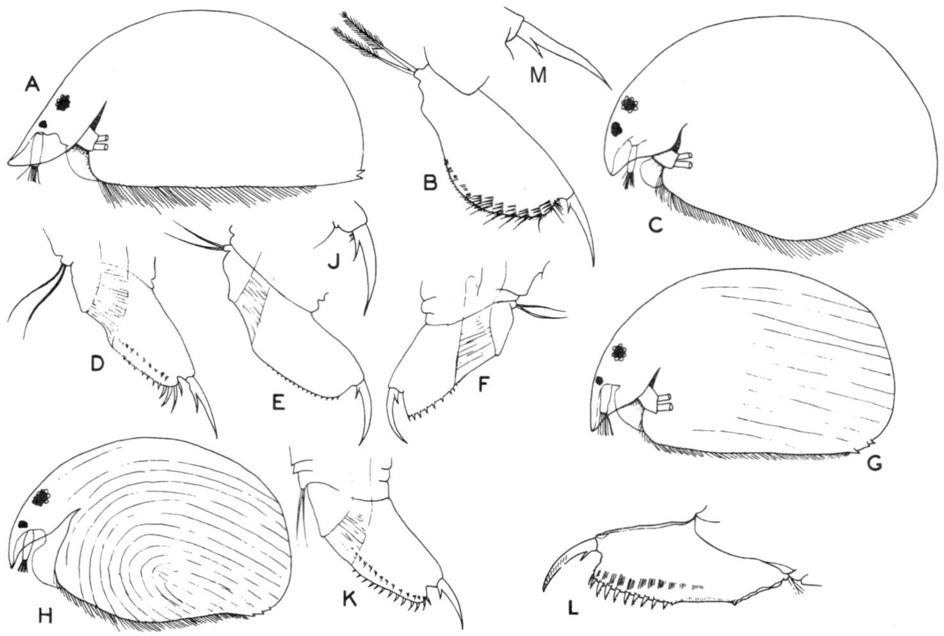

Fig. 18.21. Structure of Chydorinae. A, *Graptoleberis testudinaria*, ×80; B, postabdomen of *Leydigia leydigi;* C, *L. leydigi*, ×45; D, postabdomen of *Oxyurella brevicaudis;* E, postabdomen of *Alonella diaphana;* F, postabdomen of *Alona guttata;* G, *A. monacantha*, ×100; H, *Biapertura karua*, ×75; J, claw of *Biapertura affinis;* K and L, postabdomen of *A. costata;* M, claw of *A. quadrangularis* (A, E, G, and H modified from Ward and Whipple 1918; B modified from Volterra D'Ancona 1933; C, F, and K modified from Lilljeborg 1900; D modified from Birge 1910.)

43. Valves only rarely with striae, fringe of setae on ventroposterior margin, postabdomen usually with many marginal denticles, ventroposterior angle without teeth, claw without spinules at base (Fig. 18.21F, G, K, L, M); 0.9 mm long, common everywhere . . . . ***Alona***

Valves with oblique striae, may have a cluster of fine spinules at base of claw (Fig. 18.21H, J); about 0.5 to 1 mm long, in vegetation of littoral areas . . . . ***Biapertura***

44. Body elongated **45**

Body spherical or broadly ellipsoidal **49**

45. Postabdomen bent abruptly behind anus; one or two teeth at inferoposterior angle (Fig. 18.22F) ***Dunhevedia***

Postabdomen not bent abruptly behind anus **46**

46. One median head pore in first instar, none in later instars, rostrum blunt, ventral edge of keel of labrum may be toothed, claw may have one major basal spine (Fig. 18.24A–C) ***Ephemeroporus***

Two median head pores **47**

47. Head shield very long, extending to middle of dorsal side, rostrum long and may be bent sharply into a hook, claw may have two basal spines (Fig. 18.22B, D, G, H, L, M), 0.5 mm long, common in northern states . . . . ***Pleuroxus***

Head shield shorter **48**

48. Postabdomen expanded beyond anus, with many marginal denticles, sometimes lateral; rostrum of variable length, valves may be striated with one to three large teeth at ventroposterior angle; claw with one large tooth (Figs. 18.21E; 18.22C; 18.23A; 18.24E, F, J, K); up to 0.4 mm long; southern states . . . . ***Alonella***

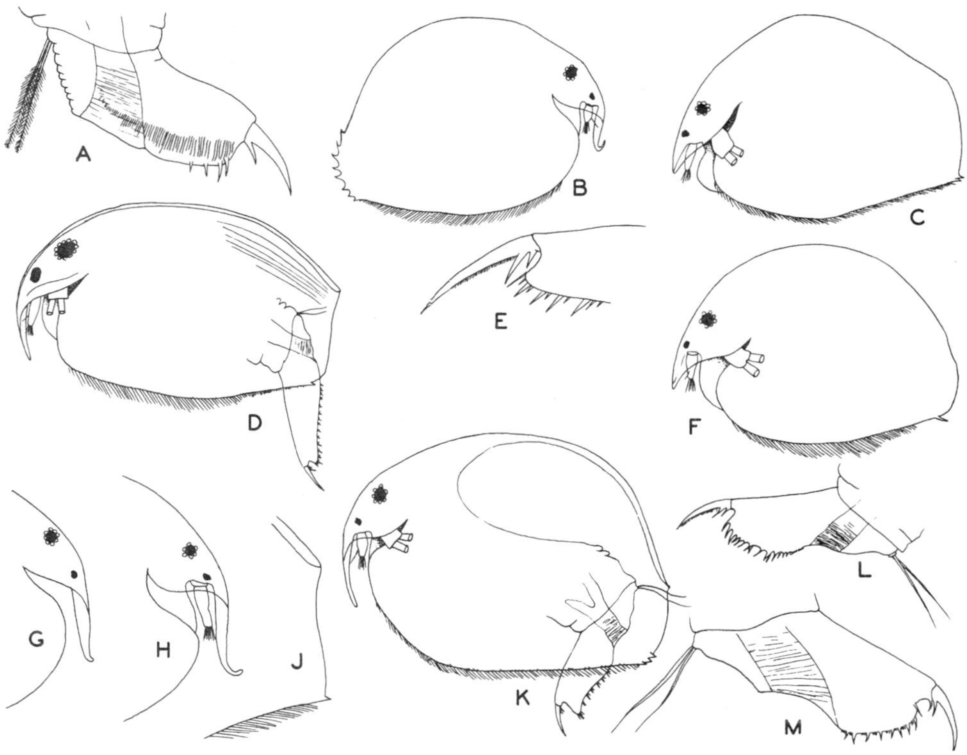

Fig. 18.22. Structure of Chydorinae. A, postabdomen of *Rhynchotalona falcata;* B, *Pleuroxus procurvatus,* ×70; C, *Alonella dentifera,* ×95; D, *Pleuroxus striatus,* ×55; E, claw of *P. striatus;* F, *Dunhevedia crassa,* ×75; G and H, anterior end of *Pleuroxus;* J, posterior end of *P. laevis;* K, *P. denticulatus,* ×85; L, postabdomen of *P. trigonellus;* M, postabdomen of *P. aduncus.* (A modified from Lilljeborg 1900; B, C, and F–M modified from Ward and Whipple 1918; D and E redrawn from Ueno 1927.)

Postabdomen not expanded beyond anus, with many marginal denticles; rostrum of variable length (Fig. 18.24D); up to 0.5 mm long; widely distributed; some workers feel that *Alonella* and *Disparalona* are synonymous . . . . ***Disparalona***

49. Valves with conspicuous projection at anteroventral margin (Fig. 18.23B); 0.35 mm long; widely distributed . . . . ***Anchistropus***

Valves without such a projection **50**

50. Postabdomen long, narrow (Fig. 18.23C); widely distributed; in vegetation . . . ***Pseudochydorus***

Postabdomen short, broad, rostrum pointed (Fig. 18.23F); total length 0.3 to 0.5 mm; the most common of all Cladocera; widely distributed; undoubtedly this is a group of closely related species and morphs, only slightly differing from each other (Fig. 18.23D–F) . . . . . ***Chydorus***

51. With four pairs of stout legs bearing branchial appendages; eye borne on a short, broad lobe **52**

With six pairs of legs; branchial appendages absent; eye borne on an elongate lobe; up to 17 mm long (Fig. 18.25); limnetic; common in northern states but found as far south as TX **LEPTODORIDAE, *Leptodora***

52. Abdomen with short posterior extension and two short spines (Fig. 18.24H); up to 1.5 mm long; common in northern lakes, ponds, and marshes . . . **POLYPHEMIDAE, *Polyphemus***

Abdomen with an elongate spine (Fig. 18.10), up to 7 mm long; introduced into the Great Lakes . . . . **CERCOPAGIDAE, *Bythotrephes***

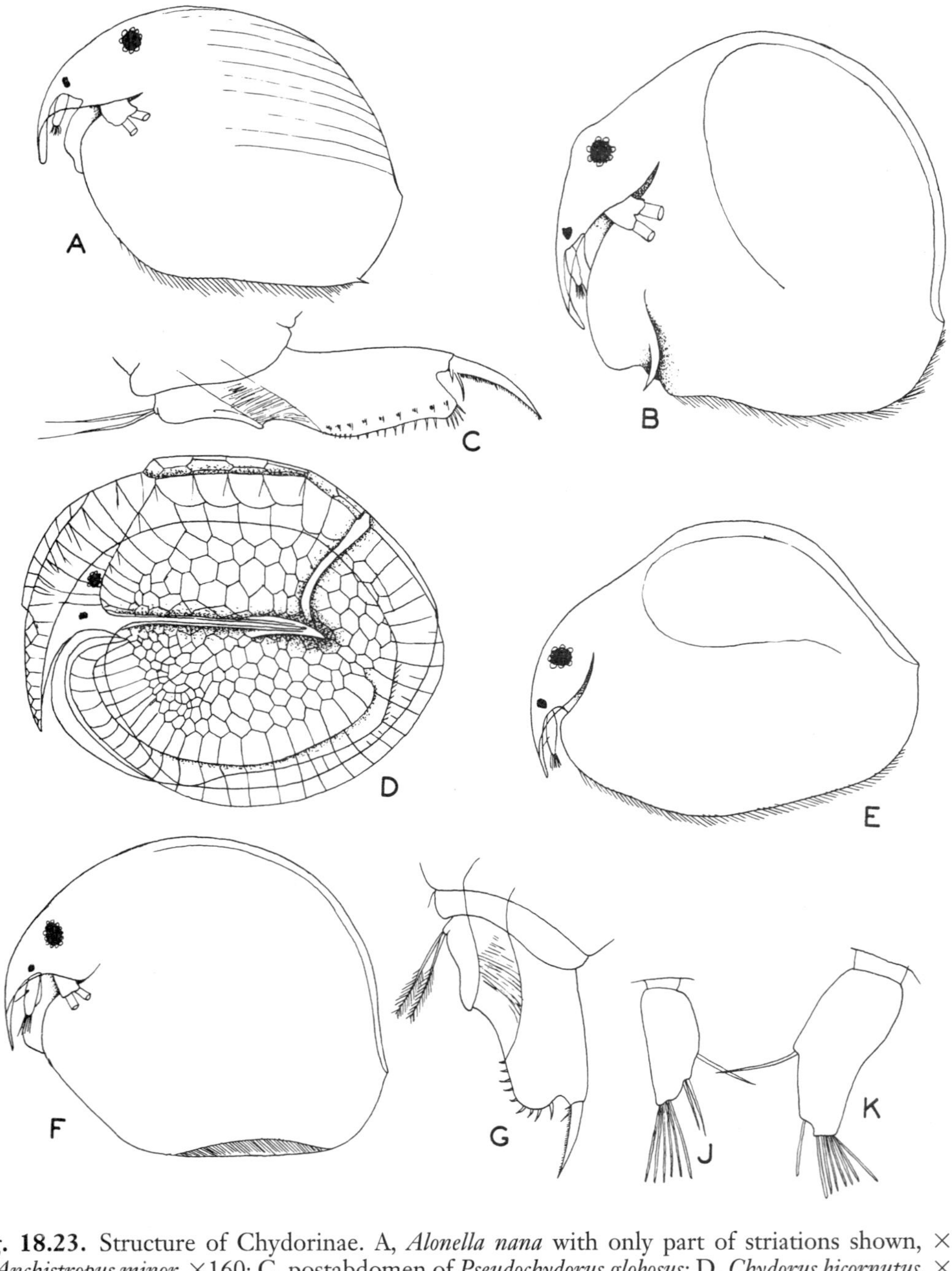

Fig. 18.23. Structure of Chydorinae. A, *Alonella nana* with only part of striations shown, ×200; B, *Anchistropus minor,* ×160; C, postabdomen of *Pseudochydorus globosus;* D, *Chydorus bicornutus,* ×120; E, *C. gibbus,* ×110; F, *C. sphaericus,* ×130; G, postabdomen of *C. sphaericus;* J, antenna 1 of *C. piger;* K, antenna 1 of *C. ovalis.* (A–E modified from Ward and Whipple 1918; G–K modified from Lilljeborg 1900.)

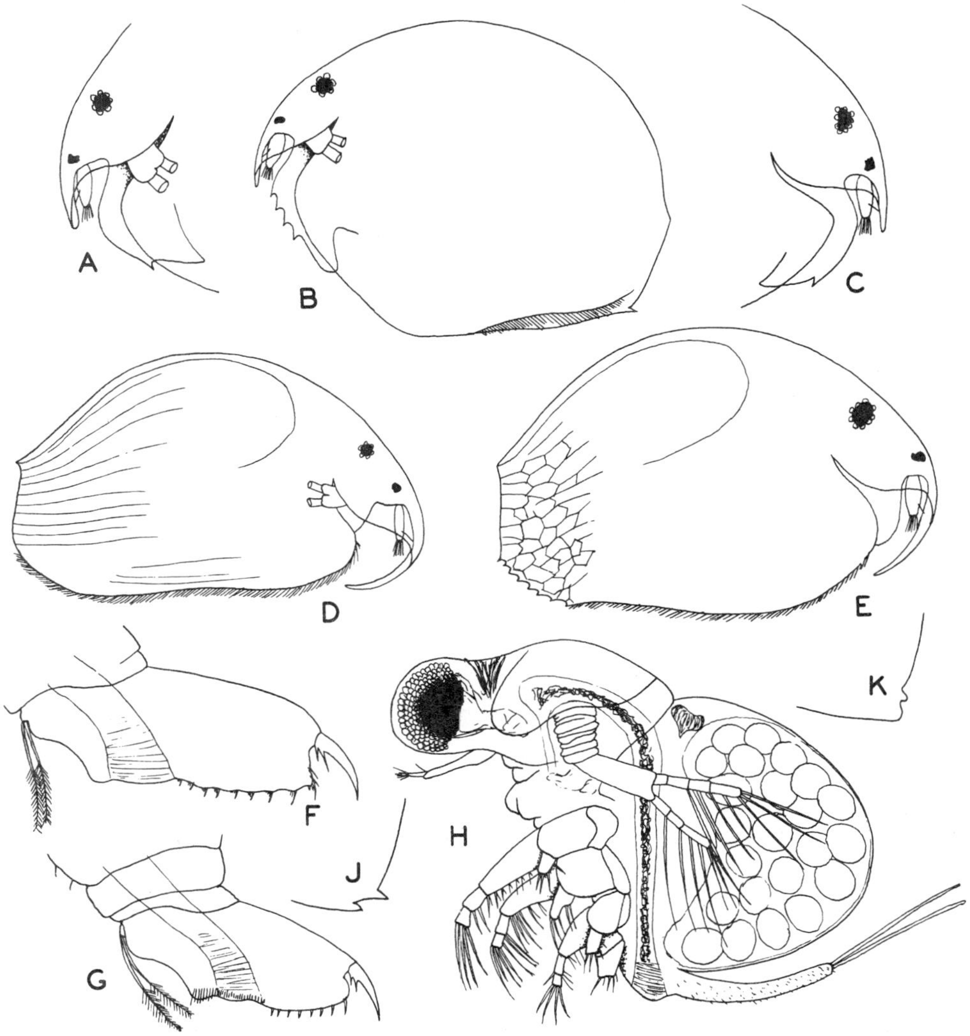

Fig. 18.24. Structure of Chydorinae and Polyphemidae. A, anterior end of *Ephemeroporus poppei;* B, *E. barroisi,* ×140; C, anterior end of *E. hybridus;* D, *Disparalona rostrata* with only a portion of valve reticulations shown, ×110; E, *Alonella dadayi* with only a portion of valve reticulations shown, ×200; F, postabdomen of *Alonella excisa;* G, postabdomen of *A. exigua;* H, *Polyphemus pediculus,* ×50; J and K, ventroposterior angle of *Alonella excisa.* (A–E modified from Ward and Whipple 1918; F–H modified from Lilljeborg 1900.)

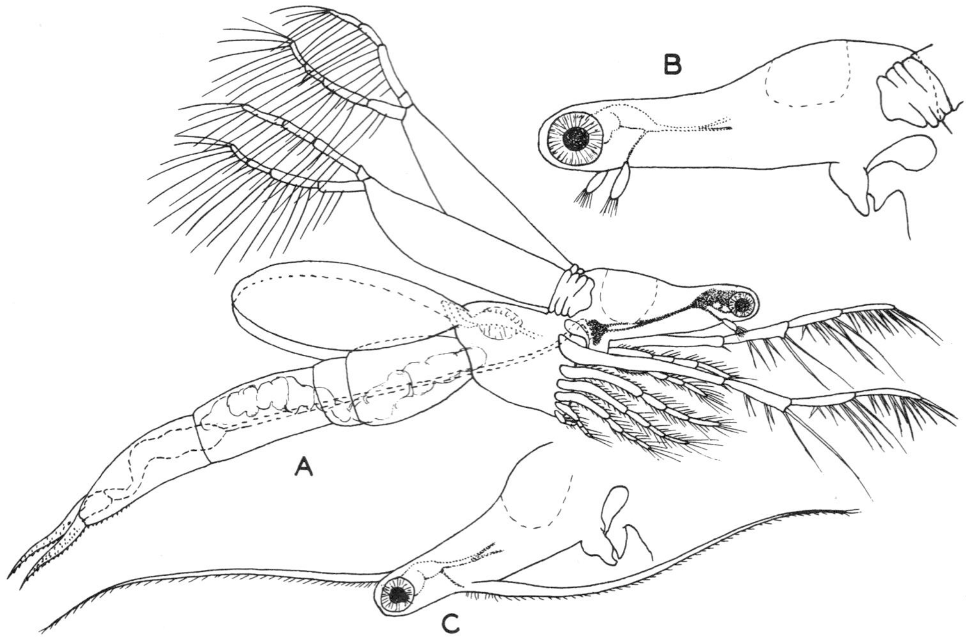

Fig. 18.25. Structure of *Leptodora kindti.* A, adult female with winter eggs, ×85; B, anterior end of female; C, anterior end of male. (Modified from Sebestyen 1931.)

REFERENCES

Anderson, B. G., and J. C. Jenkins. 1942. A time study of events in the life span of *Daphnia magna*. *Biol. Bull.* 83:260–272.

Angino, E. E., et. al. 1973. Population dynamics of pond zooplankton, II: *Daphnia ambigua* Scourfield. *Hydrobiologia* 42:491–507.

Applegate, E. R., and J. W. Mullan. 1970. Ecology of *Daphnia* in Bull Shoals Reservoir. *Res. Rep. U. S. Bur. Sport Fish. Wildlife* 74:1–23.

Balcer, M. D., et al. 1984. *Zooplankton of the Great Lakes.* 174 pp. Univ. of Wisconsin Press, Madison.

Banta, A. M. 1939. Studies on the physiology, genetics, and evolution of some Cladocera. *Publ. Carnegie Inst. Wash.* 513:1–285.

Berg, K. 1931. Studies on the genus *Daphnia* O. F. Müller with especial reference to the mode of reproduction. *Vidensk. Medd. Dansk. Naturhist. Foren.* 92:1–222.

——. 1934. Cyclic reproduction, sex determination and depression in the Cladocera. *Biol. Rev.* 9:139–174.

de Bernardi, R., and G. Giussani. 1990. Are blue-green algae a suitable food for zooplankton? An overview. *Hydrobiologia* 200:29–41.

Biernbaum, C. K. 1989. Distribution and seasonality of branchiopod and malacostracan crustaceans of the Santee National Wildlife Refuge, South Carolina. *Brimleyana* 15:7–30.

Birge, E. A. 1910. Notes on Cladocera, IV. *Trans. Wis. Acad. Sci. Arts Lett.* 16:1017–1066.

Black, R. W., II, and N. G. Hairston, Jr. 1988. Predator driven changes in community structure. *Oecologia* 77:468–479.

Brandlova, J., et al. 1972. The Cladocera of Ontario with remarks on some species and distribution. *Can. J. Zool.* 50:1373–1403.

Brooks, J. L. 1946. Cyclomorphosis in *Daphnia*, 1: an analysis of *D. retrocurva* and *D. galeata*. *Ecol. Monogr.* 16:409–447.

——. 1953a. A re-description of typical *Daphnia clathrata* Forbes and *Daphnia arcuata* Forbes. *Am. Midl. Nat.* 49:193–209.

——. 1953b. Re-description of *Daphnia pulex* var. *pulicaria* Forbes, *D. thorata* F. and *D. dentifera* F. *Am. Midl. Nat.* 49:772–800.

——. 1957. The systematics of North American *Daphnia*. *Mem. Conn. Acad. Arts and Sci.* 13:1–180.

Cannon, H. G., and F. M. C. Leak. 1933. On the feeding mechanism of the Branchiopoda. *Philos. Trans. B. Soc. London Ser. B* 222:267–352.

Carpenter, S. R., et al. 1985. Cascading trophic interactions and lake productivity. *Bioscience* 35:634–639.

Coker, R. E. 1939. The problem of cyclomorphosis in *Daphnia*. *Q. Rev. Biol.* 14:137–148.

Coker, R. E., and H. H. Addlestone. 1938. Influence of temperature on cyclomorphosis of *D. longispina*. *J. Elisha Mitchell Sci. Soc.* 54:45–75.

Davison, J. 1969. Activation of the ephippial egg of *Daphnia pulex*. *J. Gen. Physiol.* 53:562–575.

Deevey, E. S., and G. B. Deevey. 1971. The American species of *Eubosmina* Seligo (Crustacea, Cladocera). *Limnol. Oceanogr.* 16:201–218.

Dodson, S. I. 1974. Zooplankton competition and predation: an experimental test of the size-efficiency hypothesis. *Ecology* 55:605–613.

——. 1981. Morphological variation of *Daphnia pulex* Leydig (Crustacea: Cladocera) and related species m North America. *Hydrobiologia* 83:101–114.

——. 1988. Cyclomorphosis in *Daphnia galeata mendotae* Birge and *D. retrocurva* Forbes as a predator induced response. *Freshwater Biol.* 19:109–114.

Dodson, S. I., and D. G. Frey. 1991. Cladocera and other Branchiopoda. In: *Ecology and Classification of North American Freshwater Invertebrates*, edited by J. H. Thorp and A. P. Covich. 911pp. Academic, New York.

Doolittle, A. A. 1911. Descriptions of recently discovered Cladocera from New England. *Proc. U.S. Nat. Mus.* 41:161–170.

Duigan, C. A., and D. A. Murray. 1987. A contribution to the taxonomy of *C. sphaericus* sen. lat. (Cladocera, Chydoridae). *Hydrobiologia* 145: 113–124.

Dumont, H. J., and J. Pensaert. 1983. A revision of the Scapholeberinae (Crustacea: Cladocera). *Hydrobiologia* 100:3–45.

East, T. L., et al. 1999. *Daphnia lumholtzi* and *Daphnia ambigua:* population comparisons of an exotic and a native cladoceran in Lake Okeechobee, Florida. *J. Plankt. Res.* 21:1537–1551.

Edmondson, W. T. 1955. The seasonal life history of *Daphnia* in an arctic lake. *Ecology* 36:439–451.

Edmondson, W. T., and A. H. Litt. 1982. *Daphnia* in Lake Washington. *Limnol. Oceanogr.* 27: 292–293.

Edwards, C. 1980. The anatomy of *Daphnia* mandibles. *Trans. Am. Microsc. Soc.* 99:2–24.

Epp, G. T. 1996. Grazing on filamentous cyanobacteria by *Daphnia pulicaria*. *Limnol Oceanogr.* 41:560–567.

Eriksson, S. 1936. Studien uber die Fangapparate der Branchiopoden nebst einigen phylogenetischen Bernerkungen. *Zool. Bid. Upps.* 15: 23–287.

Evans, M. S. 1985. The morphology of *Daphnia pulicaria*, a species newly dominating the offshore southeastern Lake Michigan summer daphnia community. *Trans. Am. Microsc. Soc.* 104:223–231.

Frey, D. G. 1959. The taxonomic and phylogenetic significance of the head pores of the Chydoridae (Cladocera). *Int. Rev. gesamten Hydrobiol.* 44:27–50.

——. 1962. Supplement to: The taxonomic and phylogenetic significance of the head pores of the Chydoridae (Cladocera). *Int. Rev. gesamten Hydrobiol.* 47:603–609.

——. 1965. Differentiation of *Alona costata* Sars from two related species (Cladocera, Chydoridae). *Crustaceana* 8:159–173.

——. 1973. Comparative morphology and biology of three species of *Eurycercus* (Chydoridae, Cladocera) with a description of *Eurycercus macrocanthis* sp. nov. *Int. Rev. gesamten Hydrobiol.* 58:221–267.

——. 1974. Reassignment of *Alonella fitzpatricki* and *A. leei* Chien, 1970 (Cladocera, Chydoridae). *Trans. Am. Microsc. Soc.* 93:162–170.

——. 1980a. The non-swimming chydorid Cladocera of wet forests, with descriptions of a new genus and two new species. *Int. Rev. gesamten Hydrobiol.* 65:613–641.

——. 1980b. On the plurality of *Chydorus sphaericus* (O. F. Müller) (Cladocera, Chydoridae) and designation of a neotype from Sjaelso, Denmark. *Hydrobiologia* 69:83–123.

——. 1982a. Relocation of *Chydorus barroisi* and related species (Cladocera, Chydoridae) to a new genus and description of two new species. *Hydrobiologia* 86:231–269.

——. 1982b. The honeycombed species of *Chydorus* (Cladocera, Chydoridae): comparison of *C. bicornutus* and *C. bicollaris* n. sp. with some preliminary comments on faviformis. *Can. J. Zool.* 60:1892–1916.

——. 1982c. Questions concerning cosmopolitanism in Cladocera. *Arch. Hydrobiol.* 93:484–502.

——. 1987a. The taxonomy and biogeography of the Cladocera. *Hydrobiologia* 145:5–17.

——.1987b. The North American *Chydorus faviformis* (Cladocera, Chydoridae) and the honeycombed taxa of other continents. *Philos. Trans. R. Soc. London* 315:353–402.

Fryer, G. 1968a. Evolution and adaptive radiation in the Chydoridae (Crustacea: Cladocera): a study in comparative functional morphology and ecology. *Philos. Trans. R. Soc. London, Ser. B, Biol. Sci.* 254:221–385.

——. 1968b. Tubular and glandular organs in the Cladocera, Chydoridae. *Zool. J. Linn. Soc.* 48: 1–8.

——. 1970. Defaecation in some macrothricid and chydorid cladocerans, and some problems of water intake and digestion in the Anomopoda. *Zool. J. Linn. Soc.* 49:225–269.

——. 1972. Observations on the ephippia of certain macrothricid cladocerans. *Zool. J. Linn. Soc.* 51:79–96.

——. 1974. Evolution and adaptive radiation in the Macrothricidae (Crustacea: Cladocera): a study in comparative functional morphology and ecology. *Philos. Trans. R. Soc. London, Ser. B, Biol. Sci.* 269:137–274.

——. 1987. A new classification of the branchiopod Crustacea. *Zool. J. Linn. Soc.* 91:357–383.

Gellis, S. S., and G. L. Clarke. 1935. Organic matter in dissolved and in colloidal form as food for *Daphnia magna*. *Physiol. Zool.* 8:127–137.

Gillooly, J. F., and S. I. Dodson. 2000. Latitudinal patterns in the size distribution and seasonal dynamics of new world, freshwater cladocerans. *Limnol. Oceanogr.* 45:22–30.

Gliwicz, M. Z. 1986. Predation and the evolution of vertical migration in zooplankton. *Nature* 320:746–748.

Goulden, C. E. 1968. The systematics and evolution of the Moinidae. *Trans. Am. Philos. Soc.* 58(6):1–101.

——. 1971. Environmental control of the abundance and distribution of the chydorid Cladocera. *Limnol. Oceanogr.* 16:320–331.

Goulden, C. E., and D. G. Frey. 1963. The occurrence and significance of lateral head pores in the genus *Bosmina* (Cladocera). *Int. Rev. gesamten Hydrobiol.* 48:513–522.

Grigorovich, I. A., et al. 2000. Comparative biology of the predatory cladocera *Cercopagis pengoi* from Lake Ontario, Baltic Sea, and Caspian Sea. *Arch. Hydrobiol.* 149:23–50.

Hanazato, T. 1991a. Pesticides as chemical agents inducing helmet formation in *Daphnia ambigua*. *Limnol. Oceanogr.* 26:419–424.

——. 1991b. Induction of development of high helmets by a *Chaoborus*-released chemical in *Daphnia galeata*. *Arch. Hydrobiol.* 122:167–175.

Harris, J. E., and P. Mason. 1956. Vertical migration in eyeless *Daphnia*. *Proc. R. Soc. London.* 145B:280–290.

Hasler, A. D. 1937. The physiology of digestion in plankton Crustacea, II: further studies on the digestive enzymes of (A) *Daphnia* and *Polyphemus;* (B) *Diaptomus* and *Calanus*. *Biol. Bull.* 72:290–298.

Hebert, P. D. N. 1978. The population biology of *Daphnia* (Crustacea, Daphnidae). *Biol. Bull.* 53:387–426.

Hebert, P. D. N., and T. J. Crease. 1980. Clonal coexistence in *Daphnia pulex* (Leydig): another planktonic paradox. *Science* 207:1363–1365.

Hellors, W. K. 1975. Selective predation of ephippial *Daphnia* and the resistance of ephippial eggs to digestion. *Ecology* 56:974–980.

Hrbacek, I. 1959. Circulation of water as a main factor influencing the development of helmets in *Daphnia cucullata* Sars. *Hydrobiologia* 13: 170–185.

Hutchinson, G. E. 1932. Experimental studies in ecology, 1: the Mg tolerance of Daphniidae and its ecological significance. *Int. Rev.* 28:90–108.

Idris, B. A. G. 1983. *Freshwater Zooplankton of Malaysia (Crustacea: Cladocera)*. 153 pp. Penerbit Universiti, Pertanian, Malaysia.

Infante, A., and S. E. B. Abella. 1985. Inhibition of *Daphnia* by Oscillatoria in Lake Washington. *Limnol. Oceanogr.* 30:1046–1052.

Ingle, L., et al. 1937. A study of longevity, growth, reproduction, and heart rate in *Daphnia longispina* as influenced by limitations in quantity of food. *J. Exp. Zool.* 76:325–352.

Jacobs, J. 1961. Cyclomorphosis in *Daphnia galeata mendotae* Birge, a case of environmentally controlled allometry. *Arch. Hydrobiol.* 58: 7–71.

——. 1962. Light and turbulence as co-determinant of relative growth rates in cyclomorphic *Daphnia*. *Int Rev. gesamten Hydrobiol.* 47:146–156.

——. 1987. Cyclomorphosis in *Daphnia*. *Mem. Ist. Ital. Idrobiol.* 45:325–352.

Johannsson, O. E., and R. O'Gorman. 1991. Roles of predation, food, and temperature in structuring the epilimnetic zooplankton populations in Lake Ontario, 1981–1986. *Trans. Am. Fish. Soc.* 120:193–208.

Kerfoot, W. C. 1975a. The divergence of adjacent populations. *Ecology* 56:1298–1313.

——. 1975b. Seasonal changes of *Bosmina* (Crustacea Cladocera) in Frains Lake, Michigan; laboratory observations of phenotypic changes induced by in organic factors. *Freshwater Biol.* 5:227–243.

Kiang, H. M. 1942. Über die Cyclomorphose de Daphnien einiger Voralpenseen. *Int. Rev.* 41: 345–408.

Kubersky E. S. 1977. Worldwide distribution and ecology of *Alonopsis* (Cladocera: Chydoridae) with a description of *Alonopsis americana* sp. nov. *Int. Rev. gesamten Hydrobiol.* 52:649–685.

Lampert, W., and H. G. Wolf. 1986. Cyclomorphosis in *Daphnia cucullata:* morphometric and population genetic analyses. *J. Plank. Res.* 8: 289–303.

Lilljeborg, W. 1900. *Cladocera succiae. Nova Acta Regiae Soc. Sci. Ups. Ser.* 3, 19:1–701.

MacArthur, J. W., and W. H. T. Baillie. 1929. Metabolic activity and duration of life, 1: influence of temperature on longevity in *Daphnia magna*. *J. Exp. Zool.* 53:221–242.

Mazumder, A., et al. 1990. A comparison of lakes and lake enclosures with contrasting abundances of planktivorous fish. *J. Plank. Res.* 12: 109–124.

McMahon, J. W., and F. H. Rigler. 1963. Mechanism regulating the feeding rate of *Daphnia magna* Straus. *Can. J. Zool.* 41:321–332.

McNaught, D, C., and A. D. Hasler. 1963. Rate of movement of populations of *Daphnia* in relation to changes in light intensity. *J. Fish. Res. B. Can.* 21:291–318,

Megard, R. O. 1965. A chemical technique for disarticulating the exoskeletons of chydorid Cladocera. *Crustaceana* 9:208–210.

——. 1967. Three new species of *Alona* (Cladocera Chydoridae) from the United States. *Int. Rev. gesamten Hydrobiol.* 52:37–50.

Meijering, M. P. D. 1975. Notes on the systematics and ecology of *Daphnia pulex* Leydig in northern Canada. *Int. Rev. gesamten Hydrobiol.* 60:691–703.

Negra, S. 1983. Cladocera. *Fauna Rep. Soc. Romania, Crustacea* IV (12):1–399.

O'Brien, W. J., et al. 1979. Helmets and invisible armor: structures reducing predation from tactile and visual planktivores. *Ecology* 60: 287–294.

Ohman, M. D., et al. 1983. Reverse diel verticalmigration: an escape from invertebrate predators. *Science* 220:1404–1407.

Pacaud, A. 1939. Contribution a l'ecologie des Cladoceres. *Bull. Biol. Fr. Belg. Suppl.* 25:1–260.

Pejler, B. 1973. On the taxonomy of limnoplanktic *Daphnia* species in northern Sweden. *Zoon* 1:23–27

Pennak, R. W. 1957. Species composition of limnetic zooplankton communities. *Limnol. Oceanogr.* 2:222–232.

——. 1973. Some evidence for aquatic macrophytes as repellants for a limnetic species of *Daphnia*. *Int. Rev. gesamten Hydrobiol.* 58:569–576.

Pratt, D. M. 1943. Analysis of population development in *Daphnia* at different temperatures. *Biol. Bull.* 85:116–140.

Ranta, E. 1979, Niche of *Daphnia* species in rock pools. *Arch. Hydrobiol.* 87:205–223.

Ringelberg, J. 1964. The positively phototactic reaction of *Daphnia magna* Straus: a contribution to the understanding of diurnal vertical migration. *Neth. J. Sea Res.* 2:319–406.

——. 1991. Enhancement of the phototactic reaction in *Daphnia hyalina* by a chemical mediated by juvenile perch (*Perca fluviatilis*). *J. Plankt. Res.* 13:17–25.

Sanders, R. W., et al. 1996. Reproductive success of herbivorous zooplankton algal and nonalgal food resources. *Limnol. Oceanogr.* 41:1295–1305.

Schindler, D. W. 1968. Feeding, assimilation and respiration rates of *Daphnia magna* under various environmental conditions and their relation to production estimates. *J. Anim. Ecol.* 37:369–385.

Schwartz, S. S., et al. 1985. Morphological separation of *Daphnia pulex* and *Daphnia obtusa* in North America. *Limnol. Oceanogr.* 30:189–197.

Scourfield, D. J. 1942. The "pulex" forms of *Daphnia* and their separation into two distinct series represented by *D. pulex* (De Geer) and *D. obtusa* Kurz. *Annu. Mag. Nat. Hist.* 9:202–219.

Scourfield, D. J., and J. P. Harding. 1966. A key to the British species of freshwater Cladocera. *Sci. Publ. Freshwater Biol. Assoc.* 5:1–55.

Sebestyen, O. 1931. Contribution to the biology and morphology of *Leptodora kindtii* (Focke) (Crustacea, Cladocera). *Arb. Ung. Biol. Inst.* 4:151–170.

Smirnov, N. N. 1966a. *Alonopsis* (Chydoridae, Cladocera): morphology and taxonomic position. *Hydrobiologia* 27:113–136.

——. 1966b. *Pleuroxus* (Chydoridae): morphology and taxonomy. *Hydrobiologia.* 28:161–194.

——. 1968. On comparative functional morphology of limbs of Chydoridae (Cladocera). *Crustaceana* 14:76–96.

——. 1972. Detailed morphology of trunk limbs of some Aloninae. *Hydrobiologia* 40:393–422.

——. 1974. Chydoridae of the world's fauna. *Fauna of the U.S.S.R. Crustacea* 1(2):1–644.

Smirnov, N. N., and B. V. Timms. 1983. A revision of the Australian Cladocera (Crustacea). *Rec. Aust. Mus. Suppl.* 1:1–132.

Tappa, D. W. 1965. The dynamics of the association of six limnetic species of *Daphnia* in Aziscoos Lake, Maine. *Ecol. Monogr.* 35:395–423.

Thomas, I. F. 1961. Review of the genera *Pseudosida* Herrick, 1884 and *Latonopsis* Sars, 1888 (Cladocera). *Crustaceana* 3:1–8.

Tollrian, R., and E. von Elert. 1994. Enrichment and purification of *Chaoborus kairomone* for water: further steps toward its chemical characterization. *Limnol. Oceanogr.* 39:788–796.

Ueno, M. 1927. The freshwater Branchiopoda of Japan, 1. *Mem. Coll. Sci. Kyoto Imp. Univ. Ser.* B, 2:259–311.

Volterra D'Ancoma, L. 1933. Contributo allo studio dei Cladoceri dell' Italia centrale. I Cladoceri di un Laghetto dei dintorni di Siena. *Int. Revue ges. Hydrobiol. Hydrogr.* 29:33–83.

Ward, H. B., and G. C. Whipple. 1918. *Freshwater Biology.* 1111 pp. Wiley, New York.

Wesenberg-Lund, C. 1926. Contributions to the biology and morphology of the genus *Daphnia* with some remarks on heredity. *Mem. Acad. B. Sci. Lett. Dan., Sect. Sci., Ser.* 8, 11:89–251.

Wolf, H. G. 1987. Interspecific hybridization between *Daphnia hyalina*, *D. galeata*, and *D. cucullata* and seasonal abundance of these species and their hybrids. *Hydrobiologia* 145:213–217.

Woltereck, R. 1932. Races, associations and stratification of pelagic daphnids in some lakes of Wisconsin and other regions of the United States and Canada. *Trans. Wis. Acad. Sci. Arts Lett.* 27:487–522,

Wood, T. R. 1938. *Activation of the Dormant Form of Freshwater Animals, with Special Reference to Cladocera.* 87 pp. Thesis, Brown University, Providence, RI.

Wood, T. R., and A. M. Banta. 1937. Hatchability of *Daphnia* and *Moina* sexual eggs without drying. *Int. Rev. gesamten Hydrobiol.* 35:229–242.

Work, K. A., and M. Gophen. 1999. Environmental variability and the population dynamics of the exotic *Daphnia lumholtzi* and native zooplankton in Lake Texoma, U.S.A. *Hydrobiologia* 405:11–23.

Zaret, T. M., and W. C. Kerfoot. 1975. Fish predation on *Bosmina longirostris:* body-size selection versus visibility selection. *Ecology* 56:232–237.

Zimmerman, B. D., and J. H. Hubscham. 1990. Open water Cladocera of the little Miami Drainage Basin. *Ohio Biol. Surv. Bull.* 8:1–58.

19

COPEPODA

Like the Cladocera, the Copepoda are almost universally distributed in the plankton, benthic, and littoral regions of freshwaters. They are either absent or present in small numbers in rapid headwater brooks and streams. Including marine and freshwater species, there are probably 8000 to 10,000 described species worldwide.

General characteristics. The body length of the American species ranges from 0.3 to 4.5 mm, but the great majority are less than 2.0 mm long. The smallest species are apparently found in interstitial environments and are as small as 250 μm. Most species are drab grayish or brownish, but others, especially littoral species in the spring of the year and at high altitudes, are brilliant orange, purple, or red. Such colorations are usually produced by a wide variety of carotenoids, and are thought to be a means of photoprotection in some species. Conversely, pigmented individuals are more prone to fish predation. As a group, the members of the Class Copepoda are much more homogeneous in their general structure than the Cladocera, and accurate identification is based largely on anatomical details of the appendages.

The American freshwater copepods comprise five orders: Calanoida (Figs. 19.1, 19.2), Cyclopoida (Fig. 19.1), Harpacticoida, Poecilostomatoida, and Siphonostomatoida. The first three orders are free-living, except Cyclopoida, which contains a parasitic family. The latter two orders are exclusively parasitic, and the morphology is greatly modified and specialized. In the three free-living orders, however, the body is clearly segmented, more or less elongated, cylindrical, and is divided into a head, thorax, and abdomen. The thoracic region is here considered to be composed of seven segments, but the first one or two are fused with the head to form a cephalothorax, or cephalosome, which is covered with a carapace. In addition, the fourth and fifth or fifth and sixth thoracic segments are also occasionally fused. The abdomen consists of three to five segments, typically four. Commonly, the last thoracic (genital) segment and the first abdominal segment are fused. Conversely, there is subdivision of some segments in a few species.

Each segment is a rigid, sclerotized cylinder that is attached to adjacent segments by short, flexible, annular portions. One of the articulations is particularly movable, and, for convenience, the copepod body is often divided into metasome and urosome, the former being that part anterior to this articulation and the latter posterior (Fig. 19.1). The urosome includes all the abdomen, the seventh thoracic (genital) segment, and sometimes the sixth thoracic segment.

Appendages. The true head has five pairs of appendages: first antennae (antennules),

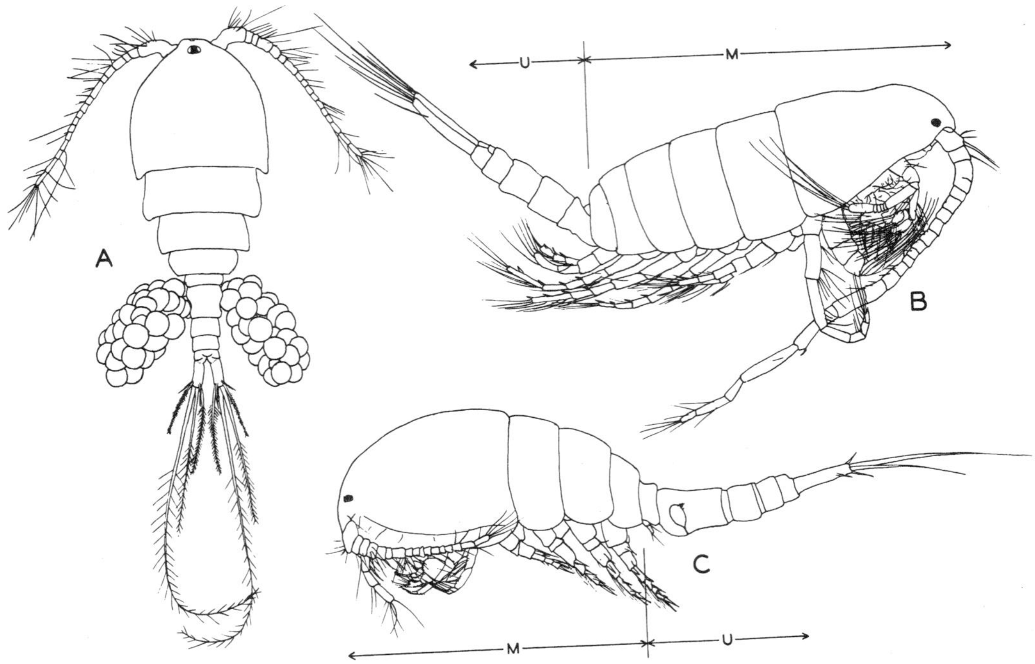

Fig. 19.1. Typical copepods. A, female *Macrocyclops ater* with egg masses, ×18; B, male *Limnocalanus macrurus*, ×33; C, female *Cyclops*, ×42. M, metasome; U, urosome. (A modified from Coker 1938; B modified from Gurney 1931–1933.)

second antennae, mandibles, first maxillae, and second maxillae. The first thoracic segment (fused with the head) bears a pair of maxillipeds, and each of the five subsequent thoracic segments bears one pair of swimming legs (Fig. 19.11B, C). In a few species the seventh thoracic (genital) segment bears a pair of vestigial swimming legs. There are no abdominal appendages.

In Fig. 19.1A and C the first two thoracic segments are fused with the head; the metasome therefore consists of cephalothorax and free thoracic segments III to V; the urosome consists of thoracic segment VI, thoracic segment VII (fused with abdominal segment 1), and free abdominal segments II to IV. In Fig. 19.1B only the first thoracic segment is fused with the head; the metasome therefore consists of the cephalothorax and free thoracic segments II to VI; the urosome consists of thoracic segment VII (genital segment) and abdominal segments I to IV.

Some investigators consider the maxillipeds to be a part of the true head, so that there are only six thoracic segments. Others consider the genital segment to be a part of the abdomen.

In studying species with confusing fusion or subdivision of the thoracic segments, it is easy to orient oneself by locating the first and subsequent swimming legs. They correspond to thoracic segments II to VI. Fundamentally, all head and thoracic appendages are biramous, and a primitive copepod swimming leg consists of a three-segmented basipod having attached at its distal end a median three-segmented endopod and a lateral three-segmented exopod. A typical and relatively unmodified swimming leg of a cyclopoid is shown in Fig. 19.4.

The appendages of the head are highly modified for various special functions. The first antennae are uniramous and long, consisting of as many as 25 segments; they are sensory appendages but are used also for locomotion. In male cyclopoids and harpacticoids both first antennae are geniculate and

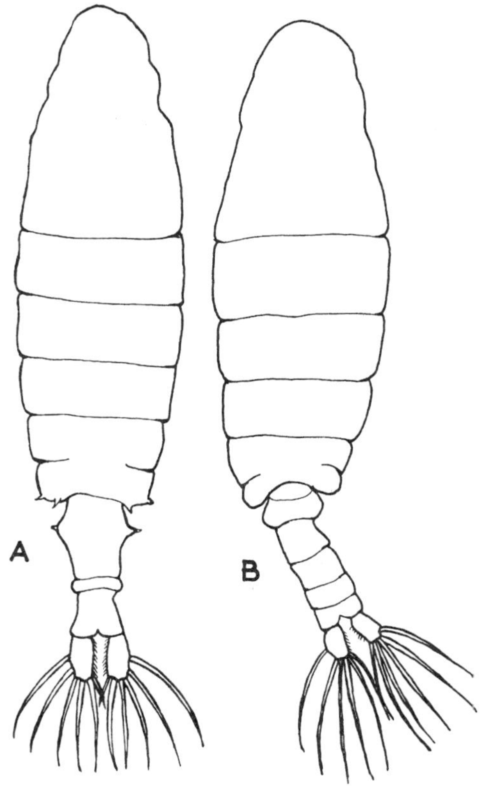

Fig. 19.2. Dorsal view of *Diaptomus* (*Leptodiaptomus*) *siciloides* (appendages and most setation omitted). A, female; B, male. (From Comita and Tommerdahl 1960.)

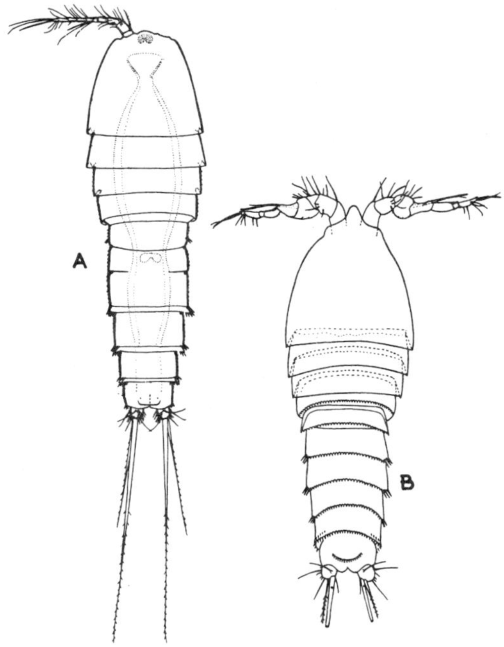

Fig. 19.3. Harpacticoid copepods. A, female *Attheyella;* B, male *Attheyella.* (From Coker 1934.)

modified for copulation (Figs. 19.3; 19.5A). In male calanoids only the right is geniculate (Fig. 19.1).

The second antennae (Fig. 19.5C) are much shorter, either biramous or uniramous, and are probably most important as sensory structures, although they are prehensile in male harpacticoids. The mandibles, first maxillae, second maxillae, and maxillipeds (occurring in that order) are highly modified for feeding (Fig. 19.5D–G).

The segmentation of the legs varies considerably from the primitive condition, and the modifications are most pronounced in the last two pairs, in the free-living species the sixth legs are always lacking in the female and are either rudimentary or lacking in the male. The fifth legs are reduced or vestigial in both sexes in the Cyclopoida

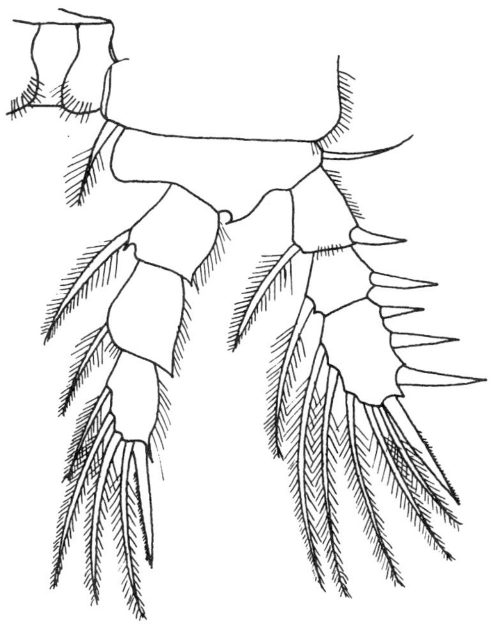

Fig. 19.4. Typical swimming leg (right, second) of a cyclopoid, showing two-segmented basipod, three-segmented exopod (right), and three segmented endopod (left). (Redrawn from Gurney 1931–1933.)

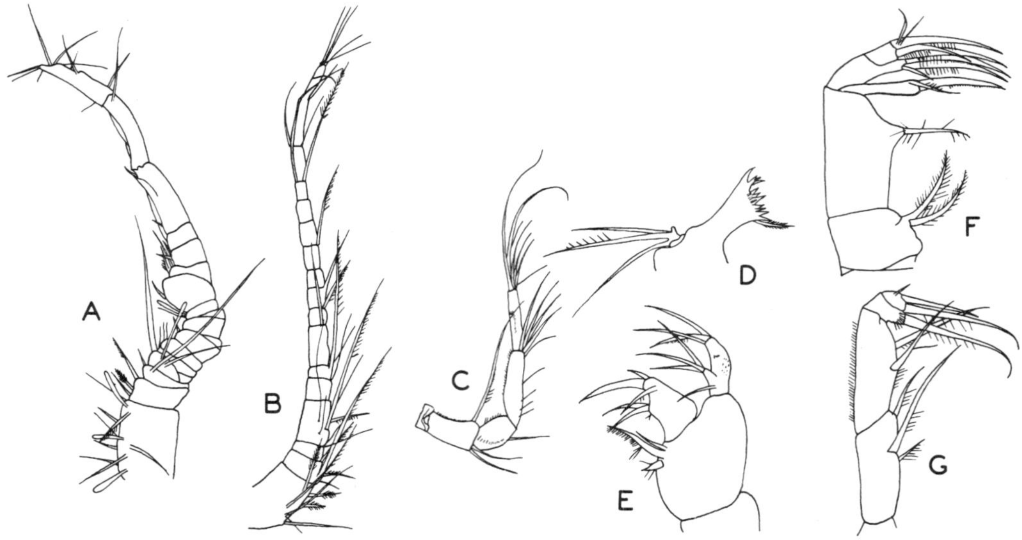

Fig. 19.5. Copepod head appendages. A, first antenna of male *Mesocyclops leuckarti;* B, first antenna of female *Macrocyclops fuscus;* C, second antenna of same; D, mandible of same; E, first maxilla of same; F, second maxilla of same; G, maxilliped of same. (Redrawn from Gurney 1931–1933.)

(Fig. 19.11E, F) and Harpacticoida, but in the Calanoida they are well developed and symmetrical in the female and asymmetrical and modified for clasping in the male (Fig. 19.9A, B).

The last abdominal (anal) segment bears two posteriorly directed caudal rami. These are simple, more or less cylindrical structures that are neither biramous nor serially homologous with the head and thoracic appendages.

In general, there are six types of slender outgrowths from the exoskeleton. One type, the aesthetasks, presumably function as sensory receptors. They are delicate, elongated, blunt-tipped, and occur on the first and second antennae. Aesthetasks are shown in Fig. 19.5A. The other five types of outgrowths occur on all copepod appendages and less commonly on the body segments; they form a morphological series: pegs–spines–spinules–setae–hairs. Spines and spinules are generally short, stout, strong, and inflexible. Setae, on the other hand, are relatively long, flexible, and thin. Hairs are the very finest outgrowths and they usually occur as fringes or rows on setae and spines. None of the categories spines, spinules, setae, and hairs can be precisely defined to the exclusion of the others, and there are wide variations in size and all possible and imperceptible intergrades. All of them, however, may have one or several functions, depending on their location. Generally, they aid in swimming, crawling, and feeding, but some are sensory, and it is possible that the long setae of the caudal rami may act as balancers or stabilizers during locomotion. Pegs are a type of sensory organelle that can be distinguished only with the electron microscope. The same is true for many fine hairs. Sensillae, another type of organelle, are very short, triangular sensory spines found especially on the trunk and legs.

Feeding. Harpacticoid mouthparts are adapted for raking, seizing, and scraping food from the bottom. Calanoids feed chiefly by filtration of plankton. The first antennae function as screws and the second antennae beat continuously to produce a current, from which the particles are filtered by the mouthparts, chiefly the maxillae. There is considerable evidence to show that calanoids exhibit some selectivity in the size and kind of algae ingested, and species differ markedly in the sizes of particles that they remove from the water. Differential feeding has been called the "leaky sieve" concept. There is also some evidence to show that calanoids combine raptorial feeding with filter feeding. Cyclopoids have the mouthparts modified for seizing and biting. The food consists mostly of unicellular plants and animals, small metazoans, and especially other crustaceans, as well as organic debris; and it has now been well established that debris may, under some circumstances, form the majority of material ingested. Cannibalism on immature stages is common. The course of the food through the digestive tract can easily be seen in the living animal under the microscope. The anus is on the last abdominal segment between the bases of the caudal rami. The accumulation of reserve food in the form of oil globules is usually chiefly responsible for the brilliant red coloration of some diaptomids.

Locomotion. Harpacticoids crawl or run on, and to some extent in, the substrate, and although many cyclopoids and a few calanoids also occur in close association with the substrate, they are characteristically swimmers. The swimming mechanisms of *Diaptomus* have been carefully analyzed by Storch (1929) and Kerfoot et al. (1980). A constant, smooth, and relatively slow type of locomotion is produced by the feeding movements of the mouthparts and second antennae. This is punctuated at frequent intervals by a jerky type of locomotion caused by rapid backward movements of the legs. Between beats of the legs the first antennae are thrust out stiffly and laterally; they are thus thought to act chiefly as parachutes that greatly retard sinking. A fraction of a second before each beat of the legs, however, the first antennae are bent medially into the general longitudinal plane of the body, the beat of the legs thereby being much more effective. Immediately following each beat of the legs, the antennae are again extended laterally. According to Storch, a single beat of the legs and the accompanying antennal movements are completed in less than one-twelfth of a second. Cyclopoid mouthparts are of little importance, locomotion being brought about by relatively frequent beats of the legs. The accompanying antennal movements are similar to those of diaptomids. In both cyclopoids and calanoids the abdomen is thought to function as a rudder.

Internal anatomy. Internal details are generally difficult to study because they are largely obscured by the very complicated system of muscles. A saccular heart occurs only in the Calanoida, and in the other suborders the blood simply circulates about in the hemocoel as the result of the movements of muscles, appendages, and the digestive tract. The oxygen and carbon dioxide exchange occurs through the general body surface and probably to some extent through the posterior part of the digestive tract where water is drawn in and out by extrinsic muscles. Excretion is thought to occur through the maxillary glands near the anterior end, but it is possible that the general body surface and the posterior part of the digestive tract may function in excretion also. The nervous system exhibits considerable fusion and cephalization. Aside from tactile structures, the only recognizable sensory area is the eyespot. In free-living species this is a small, median dorsal, clear or pigmented area at the anterior end of the cephalothorax. With the aid of a scanning electron microscope, however, Strickler (1975) reports an abundance of "sensory pegs" and "sensory hairs" scattered over the copepod body.

Reproduction. Although reproductive habits are similar throughout the free-living copepods, the various species differ in their detailed behavior and breeding periods. Calanoids, especially, have complicated mating behaviors. Some species, such as *Acanthocyclops vernalis* and *Eucyclops agilis*, reproduce throughout the year; *Tropocyclops prasinus* breeds between July and October. Other species are monocyclic and breed for short periods: *Macrocyclops fuscus* and *M. ater*, for example, reproduce only during the summer months, and some species of diaptomids and *Limnocalanus macruras* have only one generation per year. Little is known about the breeding habits of harpacticoids. Some have only one generation per year, but others undoubtedly have three or more. Three species of harpacticoids are known to reproduce by parthenogenesis.

The antennae and, in some genera, the modified fifth legs of the male are used in clasping the female. The period of clasping may last only a few minutes or it may persist for several days. Sometimes the male clasps the female before she has had her final molt, and such clasping pairs have been observed for as long as 10 days.

Male calanoids have a single pore located asymmetrically on the genital segment, but female calanoids and both sexes in the other free-living suborders have paired genital pores. During clasping, the male transfers the sperm to the female in small, packetlike spermatophores, usually with the aid of the legs. The sperm are stored in a special ventral area of the female genital segment that serves as a seminal receptacle.

Actual fertilization occurs sometime after the two sexes have separated and as the eggs leave the female reproductive tract. This process may be completed within a few minutes or as long as 2 months after copulation. Fertilized eggs are carried by the female in one or two ovisacs (Fig. 19.1A). Ovisacs usually contain from 5 to 40 eggs each and are attached to the genital segment ventrally, laterally, or subdorsally. In some species the clutch size varies seasonally, with the largest numbers of eggs being produced in the spring months. Possibly this variation is correlated with temperature or food conditions.

In cyclopoids the approximate incubation period ranges from 12 hours to 5 days, and when the newly hatched larvae leave the disintegrating ovisacs the female releases another group of eggs from the reproductive tract; these, in turn, are fertilized and retained in a new pair of ovisacs. In this way from 7 to 13 pairs of ovisacs full of eggs may be produced successively at intervals of 1 to 6 days, all resulting from the sperm retained from a single copulation, Sometimes unfertilized females release sterile, nonhatching eggs, and a single European species of *Attheyella* is known to reproduce by parthenogenesis.

Although it is believed that ordinary fertilized eggs may occasionally overwinter in an extended incubation period, adverse environmental conditions are usually withstood by special thick-walled "resting" eggs. Such eggs do not occur in the cyclopoids, but they are known to be produced by several species of diaptomids and some harpacticoids, and it is the only type of egg produced by *Limnocalanus macrurus.*

Metamorphosis. The copepod egg hatches into a small, compact, active larva called a nauplius, which has three pairs of abbreviated appendages representing the first antennae, second antennae, and mandibles (Fig. 19.6A). After a period of feeding, it molts and becomes the nauplius II stage, which usually has the first maxillae in addition. In a similar manner there are four additional nauplius stages and five to seven copepodid stages before the last molt, which results in the sexually mature adult. As metamorphosis proceeds, the larvae become progressively larger and more elongated and acquire additional appendages (Fig. 19.6B–D). Nauplius VI, for example, usually has all appendages through the second pair of legs, and copepodid I has four distinct thoracic segments and all appendages through the fourth legs.

In summary, the typical life cycle consists of the egg, six nauplius stages, five to seven copepodid stages, and the adult. However, in all parasitic species and some free-living species this series is abbreviated. Thus, there are five nauplius stages in some Cyclopoida, and only four or five in certain Harpacticoida.

The time necessary for the complete life cycle from egg to egg is highly variable, depending on the species and environmental conditions. In cyclopoids it lasts from 7 to about 180 days, whereas some species of diaptomids and *Limnocalanus macrurus* have a 1-year cycle.

Some cyclopoids are known to reproduce year-round, even in the winter months, although maturation and growth proceed very slowly in cold water. Thus, a copepod may have four or more generations during the eight warmest months of the year but only

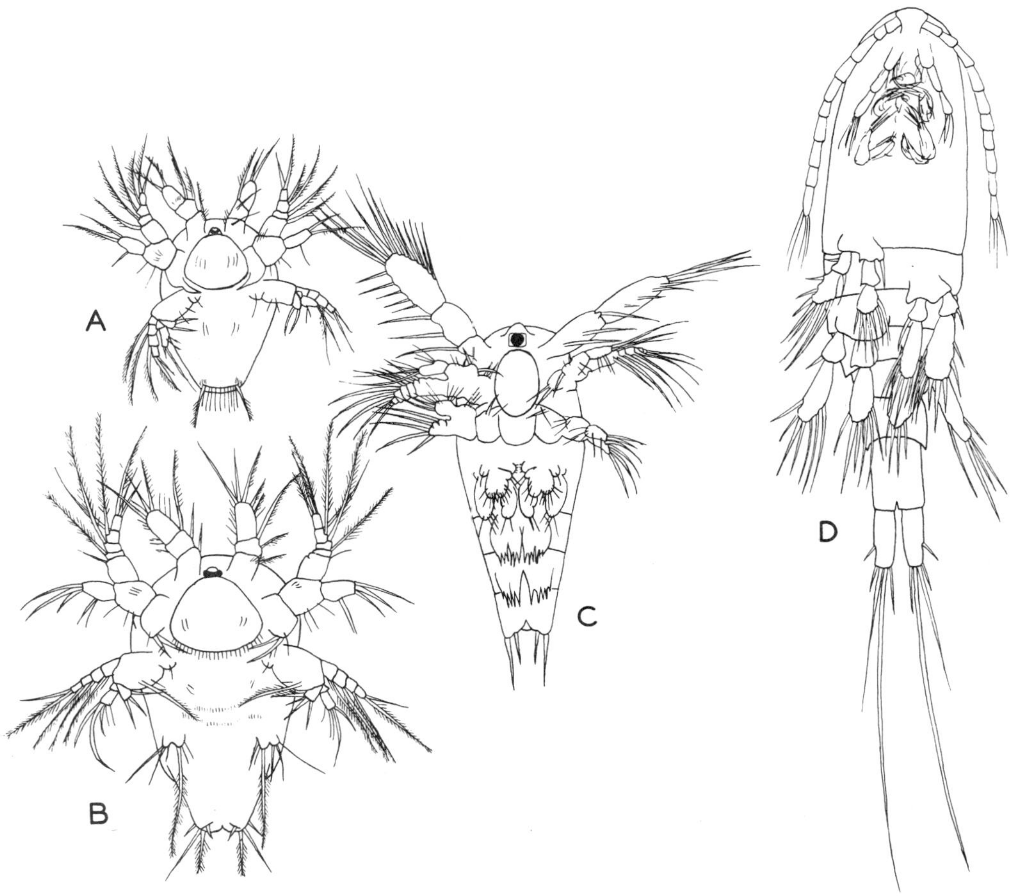

Fig. 19.6. Immature copepods. A, cyclopoid nauplius 1, ×100; B, cyclopoid nauplius IV, ×100; C, diaptomid nauplius VI, ×80; D, diagram of cyclopoid copepodid, ×90 (only alternate legs are shown on each side). (A and B redrawn from Gurney 1931–1933; C modified from Gurney 1931–1933.)

one generation during the four coldest months. Some cyclopoid copepods regularly go into an inactive diapause on lake and pond bottoms during the cold months. This is simply a nonencysted dormant period found in the copepodid II to the copepodid V stages, especially after the fall overturn. Most commonly diapause occurs in copepodid IV. In the spring these stages resume their active existence when the water becomes warmer. Concentrations of more than a million diapausing copepods per square meter of bottom have been reported.

Evidence is accumulating to show that some cyclopoids may have a 2- or 3-year life cycle. Considerably less is known about the life cycles of harpacticoids. Littoral populations of *Canthocamptus staphylinus*, however, have the greatest reproductive interval in January. The new generation becomes adult in May or June, and the adults then estivate as cysts, to emerge again in the autumn. Apparently encystment is induced by high temperatures and long days.

Parasitic copepods. Copepod parasites occur on most common species of American fish and they are among the most highly modified and bizarre of all freshwater animals. Adult stages of some genera do not even resemble copepods, and, indeed, some investigators insist that they are not copepods. They are found on the general body surface, fins, and gills, and obtain nourishment from the tissues of the host.

Among the Poecilostomatioda, the genera *Ergasilus* and *Thersitina* are parasitic on freshwater fishes. These parasites are the least specialized for their parasitic existence. Only the adult female is parasitic, the immature females and all stages of males being free living. Superficially, the female looks like a *Cyclops* (Cyclopoida) but may be easily distinguished by the greatly enlarged, three-segmented, clawlike second antennae (Fig. 19.8) used for clinging to the gills of fish where it feeds on blood and gill tissues.

The Siphonostomatioda is a relatively large order composed entirely of parasites of marine fishes. Only four genera are found on freshwater fish. The Caligidae is represented by one species, *Lepeophtheirus salmonis* (the sea louse) in freshwater. It is scarcely recognizable as a copepod in the adult stage (Fig. 19.7B); there is a large, broad, flat, suckerlike cephalothorax, and most of the segments are fused and the legs reduced. This species is only an incidental one in freshwaters; it is found attached to the surface of the body near the anus of salmon that have recently migrated from the sea, and may live only a week or two in freshwater. Coloration is blackish and they are thus easily seen. Males are very rare. *Lepeophtheirus* feeds by rasping the surface tissues.

The Lerneopodidae contains three genera in freshwater. The genus *Cauloxenus* is known only from its original description and occurs on cave fish (*Amblyopsis*). *Achtheres* and *Salmincola* are found on the gills and fins of many fishes and are perhaps the ultimate among Copepoda in modification for the parasitic mode of life. The soft-bodied adult females (Fig. 19.7C–F) have no legs, little evidence of segmentation, and greatly reduced mouthparts. The second maxillae,

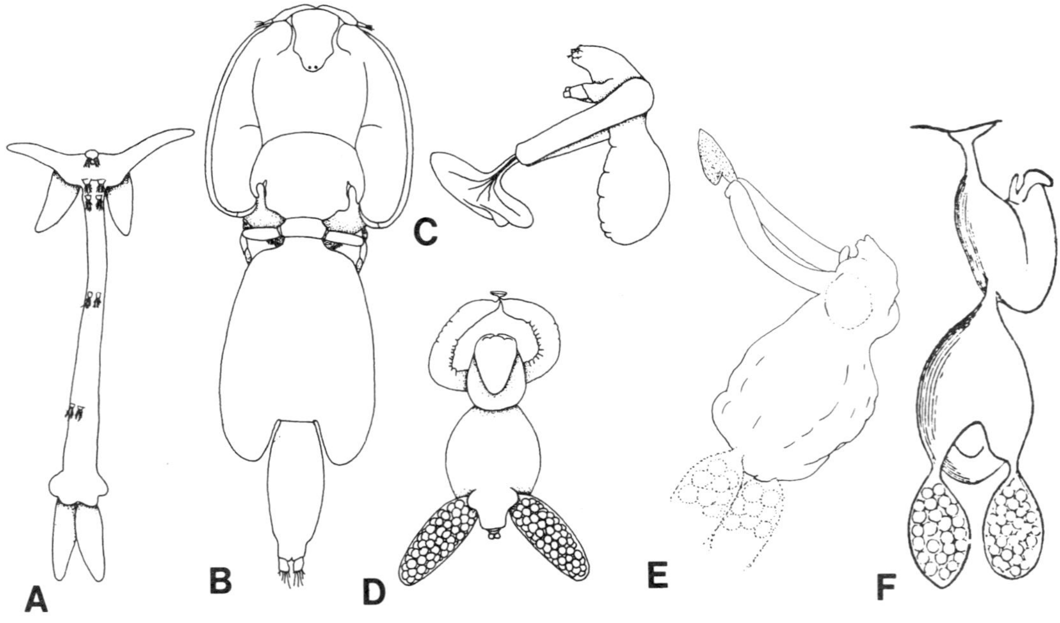

Fig. 19.7. Parasitic copepods. A, ventral view of female *Lernea cruciata*, ×6.5; B, dorsal view of female *Lepeophtheirus salmonis*, ×6; C, lateral view of female *Salmincola inermis*, ×6; D, dorsal view of female *Achtheres microptera*, ×8; E, lateral view of a female *Salmincola salmoneus*, ×5; F, *Cauloxenus stygius*, ×7. (A modified from Wilson 1944; B modified from Gurney 1931–1933; C and D modified from Wilson 1915; E redrawn from Kabata 1988; F from Cope 1872.)

however, are large cylindrical structures that have a single saucer- or umbrella-shaped organ at their distal end. The latter is imbedded in the host tissues and functions in absorbing nutriment. Little is known about the males; they are dwarfed, attached to the host fish during the immature stages, but cling to the body of the female before and after copulation. The nauplii stages are passed within the egg, but the first copepodid is a free-swimming plankter for a brief period before becoming attached to a fish.

The Order Cyclopoida includes one genus occurring in freshwater of the United States *Lernaea*, which comprises several species. The nauplii are free-living plankters, but the copepodids seek out a temporary host fish, cling to the gills, and copulate in this immature stage. Soon thereafter the male copepodid dies, but the female leaves the host and is free-living for a short time. She then attaches to the general body surface of any one of a great variety of freshwater fishes and becomes completely altered morphologically into a long, wormlike creature. The anterior half of the body is buried in the superficial host tissues and produces several large anchoring processes (Fig. 19.7A), which cause inflammation and suppuration.

Economic significance. Under natural conditions parasitic copepods are rarely present in sufficient numbers to cause serious injury to the host. In hatchery ponds, however, where fish are crowded in a limited area, there is much greater opportunity for the free-swimming immature stages to find a host, and, as a result, heavy and serious infections may break out. When firmly established, copepod parasites are difficult to control, chiefly because the sclerotized exoskeleton of the adult is resistant to chemical solutions. Newly attached larval parasites are relatively delicate, however, and sometimes it is feasible to combat infections by giving the fish frequent salt or acid baths. In extreme cases there is no alternative but to get rid of the parasitized fish, clean the hatchery ponds, and start anew.

Free-living copepods constitute an essential link in the aquatic food chain. They are on an intermediate trophic level between bacteria, algae, and protoctists, on the one hand, and small and large plankton predators (chiefly fish), on the other. In general, however, freshwater copepods are not as important an element in the fish diet as Cladocera.

Copepods are of further importance as intermediate hosts of parasites of higher animals, particularly some tapeworms of fish, waterfowl, and carnivores; some flukes of fish, amphibians, and birds; and a few nematodes of fish and birds. In Africa, southern Asia, the West Indies, and parts of the Guianas and Brazil, *Cyclops* is of great importance in the transmission of the nematode parasite of humans, *Dracunculus medinensis* (the Guinea worm), and some species of both cyclopoids and diaptomids are intermediate hosts of the broad or fish tapeworm of humans and numerous carnivores.

Recently, it has been substantiated that nonparasitic cyclopoid copepods may occasionally be a nuisance in hatcheries. They may merely nibble at the fins of fingerling fish, or they may be so abundant as to feed on and kill the fingerlings.

Ecology. Of the free-living copepods, the limnetic cyclopoid and calanoid species are by far the best known and most frequently collected. Sometimes they have been found in almost unbelievable numbers—up to 1000 individuals per liter has been recorded. In most lakes the limnetic copepod plankton is monotonous; usually it is composed of one dominant cyclopoid, such as *Diacyclops bicuspidatus*, *Acanthocyclops vernalis*, or *Mesocyclops leuckarti*. Often there is a dominant calanoid present also. In exceptional instances a limnetic population may consist of two or more cyclopoid species and two or more calanoid species all at the same time. Under such circumstances one cyclopoid is the numerical dominant and one calanoid is the numerical dominant. Coexistence of two species in the same genus is usually explained by vertical, seasonal, and food particle size preferences.

Perhaps because of the greater variety of ecological niches, the shallow littoral, especially in aquatic vegetation, is much richer in cyclopoid and calanoid species, but the population density is not usually high. Some copepods demonstrate extremely wide physiological niches. *Eurytemora affinis* has apparently moved from saltwater to freshwater on several occasions and rarely saltwater forms can successfully reproduce in freshwater.

In rivers and lakes the Harpacticoida are restricted to bottom debris of both littoral and benthic regions. In addition, however, many Cyclopoida and Calanoida have become adapted to this habitat. Common bottom forms are *Paracyclops fimbriatus*, *Eucyclops agilis*, *Macrocyclops fusus*, *M. albidus*, and species of *Canthocamptus*, *Bryocamptus*, and *Attheyella*. Very few copepods are found in rapid streams.

As ecological groups, the deep-water and limnetic species are somewhat distinct from those occurring in the shallow littoral. The former are generally smaller, slender, and more translucent, whereas the latter are larger, stout, and more deeply pigmented, and have more robust appendages.

Harpacticoid copepods have been reported from some unusual habitats. Large numbers (up to 277/10 cm^3 of sand) have been found in the interstitial waters of sandy lake beaches, with greatest concentrations at depths of 1 to 6 cm, and at distances of 1.0 to 2.5 m from the water's edge (Pennak 1940). European workers have described unusual endemic species from springs, underground streams, and cave systems. Study is underway in such habitats in this country and a very high diversity among the three free-living orders is being revealed. Harpacticoids have also been reported from damp moss in forests far from any body of water.

Small cyclopoids and harpacticoids form important elements of the meiobenthic communities of streams and rivers, especially in hyporheic and shoreline gravels. Study on meiobenthic copepods in the United States has recently intensified. Several new taxa, including a new genus, have been discovered in interstitial sands along with the observation that copepods make up a large measure of the meiobenthic fauna. Over 30 taxa representing all three orders of free-living copepods have been reported from hyporheic habitats (Strayer 1988a).

It is thought that temperature plays an important role in copepod distribution and activity. Some species, such as *Limnocalanus macrurus*, *Senecella calanoides*, *Epischura lacustris*, *Diaptomus* (*Leptodiaptomus minutus*), *D.* (*L.*) *sicilis*, and *D.* (*Skistodiaptomus*) *oregonensis*, are definitely cold-water forms occurring only in the northern states and in deep lakes. Others, such as *D.* (*S.*) *mississippiensis* and *D.* (*Arctodiaptomus*) *dorsalis*, are restricted to the southern states. The rare *Acanthocyclops crassicaudis* occurs only in temporary, warm pools.

In some interesting experiments, Coker (1933) and Aycock (1942) reared *Acanthocyclops vernalis* and *Eucyclops agilis* under controlled conditions and found that at low temperatures (7.7 to 8.0°C) adults were as much as 50% longer than adults that had been reared at high temperatures (28.0 to 30.0°C). These workers believe that such size variations with temperature are internally controlled and are not dependent on the available food supply.

In general, copepods are much more tolerant of a deficiency of oxygen than are Cladocera. Cyclopoids and *Canthocamptus* have been collected many times from the surface of bottom muds in stratified lakes during summer and winter periods of stagnation and oxygen depletion. Although it is possible for cyclopoids to swim up into the upper oxygenated layers intermittently, *Canthocamptus*, not being a swimmer, is necessarily restricted to the substrate. The whole problem of the metabolism of copepods and other organisms in anaerobic benthic regions deserves much more attention.

Some adult or advanced copepodids may form cysts or cocoons during unfavorable environmental conditions, and the fact that cysts have been found in abundance on lake bottoms in midsummer is an indication that they may be formed as a response to anaerobic conditions. In cyclopoids, cysts also constitute an estivation mechanism and are resistant to drying. The cyst wall is composed of a secreted organic membrane and an enveloping mass of detritus.

Like the Cladocera, some common species of plankton copepods show a daily rhythmic cycle of vertical migrations in lakes, with a greater concentration of individuals in the upper waters during the hours of darkness and correspondingly large numbers in the bottom waters during the hours of daylight. Individuals of some species may move upward at dusk and downward at dawn a vertical distance of as much as 10 or 20 m. It is believed that the primary stimulus for vertical migrations is the daily cycle of subsurface illumination.

Geographical distribution. The Copepoda undoubtedly originated in the ocean, and in this environment they have undergone much greater evolutionary specialization and speciation than in freshwater. The accompanying physiological divergence has been quite complete, and only a very few American species found in fresh water are known also to inhabit brackish or salt water. Some of these are *Limnocalanus macrurus*, *Eurytemora affinis*, *Harpacticus gracilis*, *Nitokra spinipes*, and *Cletocamptus*.

Limnocalanus macrurus is generally considered a relict species, that is, a saltwater form that has gradually become acclimated to freshwater. It is found only in the deep waters of larger takes such as Green Lake (Wisconsin), the Finger Lakes, and the Great Lakes. It is also widely distributed in similar European lakes and in saltwater. *Eurytemora* has close relatives in saltwater and is probably a recent migrant. One species, *E. affinis*, is spreading rapidly in the interior United States. *Senecella* is thought to be of postglacial origin; it has been reported from several localities in Siberia, including some saline lakes. *Epischura* is confined to North America except for two Asiatic species. *Cyclops* and its relatives have a worldwide distribution, some species, such as *Diacyclops bicuspidatus*, being essentially cosmopolitan. The closest marine and brackish relative is probably *Halicyclops*. As research continues, however, the actual ratio of intercontinental species to total known taxa is declining (Reid 1997/1998). Europe has about 110 species of cyclopoids, of which about half are known only from subterranean and interstitial waters.

The Diaptomidae are confined to freshwaters, and there are no close marine relatives. Unlike the cyclopoids, they show much evidence of recent speciation, and many of them seem to be confined to small geographic areas; numerous examples are given in the keys. Europe has about 70 species of calanoid copepods. Only 10 of these occur also in the United States.

Most freshwater harpacticoids belong to the family Canthocamptidae, a group that has a worldwide distribution in salt, brackish, and fresh waters. Many species occur in both North American and European freshwaters, but so little systematic collecting has been done on the other continents that there are insufficient records to warrant the use of the term "cosmopolitan" for them. The great majority of species in the other harpacticoid families covered by this manual are typical of coastal, brackish, and saline waters.

Unquestionably, active migration is responsible for the geographic dispersal of species throughout a single drainage system, but overland transport to adjacent systems presents some difficult questions. Calanoida and Harpacticoida are probably dispersed as resting eggs, and although cysts are produced by some of the latter, they are not resistant to drying. The Cyclopoida, on the other hand, are not known to produce resting eggs, but resistant copepodid cysts are presumably of common occurrence, and it is likely that passive dispersal is brought about in this stage. It is significant that the first cyclopoids to appear in dried mud cultures are in the copepodid stages. Many copepods are found year after year in vernal ponds that are completely dried up during the greater part of the year, and we may safely assume that resting eggs and cysts in the dry mud and debris tide the species over. Although passive transfer by waterfowl, aquatic insects, and in windblown dust explains the extensive distribution of some species, it is difficult to account for the fact that others are restricted to very small geographic areas, sometimes a single pond.

Collecting. Littoral and plankton copepods are easily collected with a townet or dipnet. An abundance of species may often be obtained by drawing nets through rooted aquatic vegetation and by lightly skimming the bottom. Collecting harpacticoids is a more tedious procedure. The top centimeter or two of mud and debris should be scooped up, brought into the laboratory, and allowed to settle. If the material is examined by looking toward a light source along the surface of the settled mud, the harpacticoids can be seen moving about at the mud–water interface and can be removed with a long pipette. Sometimes washings of aquatic mosses produce surprising numbers of specimens.

Phreatic and other interstitial species are easily collected from lakeside or streamside situations by digging a hole up to a meter deep and straining the accumulating water through a fine net. Hyporheic species may be taken with a Bou-Rouch sampler in shallow water (see figure in Appendix). The Bou-Rouch sampler is also useful as a semiquantitative device for lakeside and streamside sampling.

Preservation. The best specimens are obtained by killing them in 95% alcohol. Seventy percent alcohol is a suitable preservative and much superior to formalin solutions.

Culturing. Copepods are much more difficult than Cladocera to maintain as laboratory cultures, and they apparently never do well in dense numbers. Cyclopoids should be fed large protoctists or small metazoans, preferably crustaceans. Calanoids will sometimes thrive if a bit of algal culture suspension is added to the culture twice a week. Most copepods do best in cultures kept below 15°C.

Identification. It is unfortunate that the identification of species is usually based on minute anatomical details. Some characters, such as the structure of the antennae, first legs, and caudal rami, may be studied from the whole animal, but it is often necessary to dissect the animal to study the mouthparts and legs, and it is an imperative procedure in formulating descriptions of new species. Harpacticoids must be dissected more frequently because of their general opacity and smaller appendages. Specimens to be dissected should be placed in a solution of 20% glycerin in 70% alcohol in a partially covered watch glass. After one to several days, most of the alcohol and water will have evaporated, leaving the undistorted specimens in glycerin. Then they may be easily picked up and transferred with an Irwin loop. The animals should be dissected, one at a time, in a drop of glycerin on a slide on the stage of a high-power binocular microscope. The necessary appendages are removed with a pair of sharp, mounted "minuten nadeln." As a fair warning, it should be said that requisite skill and patience in dissection come only with considerable practice! Appendages and whole animals are conveniently and permanently mounted for examination in glycerin jelly under a coverslip ringed with Murrayite. Mounting in fluid glycerin is more of a nuisance and less permanent. Immature forms of the great majority of species cannot be identified.

If an abundance of specimens is available, the writers have found that all the necessary anatomical details can usually be seen, even in difficult species, if 5 to 10 whole specimens are mounted, ventral side up, on the same slide. The appendages will become spread out sufficiently in a variety of ways so that all structures can be observed by studying several specimens under the high powers of the compound microscope.

Sometimes it is helpful if the animals are stained before dissection and mounting. From 70% alcohol they should be transferred to alum cochineal overnight, washed in 70% alcohol for 5 minutes, and then transferred to the glycerin–alcohol mixture.

Body length extends from the anterior margin of the cephalothorax to the posterior end of the caudal rami. Accurate measurements of body length are of limited use as a taxonomic character because the trunk segments may be more or less retracted into each other.

Taxonomy. Much work remains to be done on species taxonomy and the relationships of many described genera or subgenera. While many genera are known from only a single species in the United States, it is very likely that additional species will be discovered and described. After several years of little systematic activity on the freshwater copepods of the United States, a recent resurgence of study, led mostly by J. W. Reid, has revealed the existence of several previously unknown genera and species. Intensified study of interstitial habitats has revealed many new species and even new genera (e.g., Strayer 1988a; Reid et al. 1999).

Concerning cyclopoids, the key to genera has been modified from Hudson et al. (1998). The key is drawn up specifically for identifying mature females, rather than males, although males can sometimes be successfully run through the couplets because the fifth legs and caudal rami are similar in both sexes. Male cyclopoids often have been misidentified, confused, or not identified at all because the investigator looks at the sixth pair of legs and mistakes them for the fifth legs. The sixth legs are not present in females. Both of the first antennae are geniculate in the male, whereas neither are geniculate in the female. Mature females often carry egg sacs. The number of species listed with each genus is for the most part approximate as revisions of certain genera continue and species names are revised. Figure 19.10 is semidiagrammatic and does not show minute details such as plumose ornamentation on setae and spinules on certain of the spines. Keys and additional information on species of cyclopoids can be found in Balcer et al. (1984) and Hudson et al. (1998).

Immature (copepodid) cyclopoids are difficult to identify, even to the genus level. In plankton populations, however, they may often be identified by association with the adults. In copepodid stages the last body segment is about twice as long as the preceding segment; in adults the last body segment is shorter than the preceding segment.

Sexual dimorphism in freshwater copepods is most pronounced in diaptomid calanoids in which the fifth leg of the male is strongly assymmetrical. In other calanoids the differences between sexes are not as sharply marked. It is generally considered important to have males in a collection to insure accurate identification. Taxonomic characters of the females are less definitive, and in keys they too often amount to unclear judgment decisions. Fortunately, however, when field collections are made there are almost invariably adequate numbers of both males and females in the populations, so that the general facies of associated males and females clearly show that they are in the same species.

The family Diaptomidae is very diverse and the large genus *Diaptomous* has been split into several species groups originally proposed as subgenera (Light 1938, 1939), but now generally considered to represent distinct and separate genera (listed parenthetically in Table 19.1 under *Diaptomus*). This philosophy was reluctantly accepted in the previous edition but no attempt was made to discriminate different genera. The characters used to establish these genera do not lend themselves well to a key format. Thus, the following key includes only the speciose *Diaptomus* and *Actinodiaptomus*, a Eurasian genus with a single species in the United States, *A. denticornis*. Many calanoid species are described and discussed by Balcer et al. (1984).

The harpacticoids of the United States are poorly known. The writer (RWP), for example, has seen more than 30 undescribed species during his field work since the publication of the first edition of this book. A good key to many species in the north central United States can be found in Hudson et al. (1998).

Most of the key characters apply to both sexes of harpacticoids. When they apply to only one sex, the exception is indicated in the couplet. Body length of harpacticoids ranges from 0.3 to 1.4 mm, but the great majority of species are less than 0.8 mm long. Their small size may be the chief reason they have been neglected. An exceptionally large number of new species of this group and other copepods are awaiting discovery in the interstitial habitat of sandy beaches and in the phreatic zone of streams.

TABLE 19.1. Orders, Families, and Genera of Copepods in Fresh Waters of the United States

- Order Cyclopoida
  - Family Cyclopoidae
    - *Acanthocyclops, Crypytocylops, Cyclops, Diacyclops, Ectocyclops, Eucyclops, Halicyclops, Homocyclops, Macrocyclops, Megacyclops, Mesocyclops, Microcyclops, Orthocyclops, Paracyclops, Rheocyclops, Stolonicyclops, Thermocyclops, Tropocyclops*
  - Family Lernaeidae
    - *Lernea*
- Order Calanoida
  - Family Centropagidae
    - *Limnocalanus, Osphranticum*
  - Family Diaptomidae
    - *Actinodiaptomus, Diaptomus (Aglaodiaptomus, Arctodiaptomus, Diaptomus, Eudiaptomus, Hesperodiaptomus, Leptodiaptomus, Mastigodiaptomus, Microdiaptomus, Mixodiaptomus, Nordodiaptomus, Onychodiaptomus, Prionodiaptomus, Sinodiaptomus, Skistodiaptomus)*
  - Family Temoridae
    - *Epischura, Eurytemura*
  - Family Senecellidae
    - *Senecella*
- Order Harpacticoida
  - Family Ameiridae
    - *Nitocrella, Nitokra*
  - Family Canthocamptidae
    - *Attheyella, Bryocamptus, Camthocamptus, Elaphoidella, Epactophanes, Gulcamptus, Maraenobiotus, Moraria*
  - Family Cletodidae
    - *Cletocampus, Huntemannia*
  - Family Cylindropsyllidae
    - *Stenocaris*
  - Family Harpacticidae
    - *Harpacticus, Tigriopus*
  - Family Laophontidae
    - *Onychocamptus*
  - Family Metidae
    - *Metis*
  - Family Parastenocaridae
    - *Parastenocaris*
  - Family Phyllognathopidae
    - *Phyllognathopus*
  - Family Tachidiidae
    - *Microarthridion, Tachidus*
  - Family Thalestridae
    - *Paradactylopodia*
- Order Poecilostomatoida
  - Family Ergasilidae
    - *Ergasilus, Thersitina*
- Order Siphonostomatoida
  - Family Caligidae
    - *Lepeophtheirus*
  - Family Lernaeopodidae
    - *Achtheres, Cauloxenus, Salmincola*

KEY TO THE ORDERS, FAMILIES, AND GENERA OF FRESHWATER COPEPODA OF THE UNITED STATES

1. Adult free-living; body with several pairs of appendages; antennae multisegmented and elongate; segmentation well defined **8**

Adult usually attached to a fish (parasitic), rarely free-living; appendages typically reduced and highly modified; antennae modified, greatly reduced, or absent; body segmentation reduced to absent **2**

2. Adults long and wormlike (Fig. 19.7A); on gills and general external surface of many common freshwater fishes; copepodid clings to gills of temporary host fish, then leaves and is free-swimming for a short time before becoming attached to final host; adult females 7 to 15 mm long; about 6 species, several maybe introduced **Order CYCLOPOIDA, LERNAEIDAE, *Lernaea***

Adults not long and wormlike; males rare **3**

3. Cephalothorax forming a broad, flat disk (Fig. 19.7B); first to fourth legs well developed; males not reduced in size, rare; attached to surface of body near anus of salmon that have recently migrated from the sea; may live for a week or more in freshwater; 6 to 16 mm long; 1 species **Order SIPHONOSTOMATOIDA, CALIGIDAE, *Lepeophtheirus***

Cephalothorax not forming a broad, flat disk; less than 6 mm in length **4**

4. Legs and segmentation present, 2nd antennae modified as claws **Order POECILOSTOMATOIDA, ERGASILIDAE, 7**

Legs absent; males small to minute pygmies clinging to females; early immature stages passed in egg; first copepodid free living for a brief period before seeking host; attached to gills and fins of host by greatly enlarged second maxillae (Fig. 19.7C–E); adult females 3 to 8 mm long without egg sacs ....**Order SIPHONOSTOMATOIDA, LERNAEOPODIDAE, 5**

5. Posterior end of trunk rounded (Fig. 19.7C); not extending beyond insertion of egg sacs **6**

Posterior end of trunk conical or pointed; segmentation present or absent (Fig. 19.7D); about 6 species parasitizing many fish species .... ***Achtheres***

6. Anterior cephalothorax and trunk separated by a groove, bulla mushroom-shaped (Fig. 19.7C, E); about 10 species attacking a wide range of freshwater fishes .... ***Salmincola***

Anterior cephalothorax separated from trunk by a narrow "waist"; bulla disklike; 1 poorly known species on cavefish .... ***Cauloxenus***

7. Cephalothorax not exceptionally large (Fig. 19.8A) or rounded, antenna 2 prominent; over 20 species; common parasites of freshwater fish .... ***Ergasilus***

Cephalothorax greatly expanded and rounded (Fig. 19.8B); antenna 2 not prominent; 1 euryhaline species commonly entering freshwater .... ***Thersitina***

8. Urosome about as wide as metasome and both more or less cylindrical; basal segment of fifth legs with inner expansion; first antennae with five to nine segments (Fig. 19.3); very small, 0.3 to 1.0 mm long .... **Order HARPACTICOIDA, 37**

Urosome much narrower than metasome, cylindrical or flattened; size ranges from 0.3 to 4.5 mm **9**

9. With a marked constriction between the segments bearing the 4th and 5th legs (Figs. 19.1A, C; 19.9); first antennae of female with 5 to 18 segments **Order CYCLOPOIDA, CYCLOPIDAE, 10**

With a marked constriction between segments bearing 5th legs and genital segment (Fig. 19.1B); first antennae of female with 22 to 25 segments ....**Order CALANOIDA, 31**

10. First antenna six segmented; last segment of fifth leg with four or five setae and/or spines; several species in brackish coastal waters .... ***Halicyclops***

First antenna with 8 or more segments; last segment of fifth leg with three or less setae and/or spines **11**

11. Fifth leg of three distinct segments; caudal ramus with distinct dorsal groove running obliquely backwards from insertion of lateral seta; antenna 1 of 16 segments; 1 common species ***Orthocyclops***

Fifth leg of one or two distinct segments; caudal ramus otherwise **12**

12. Fifth leg of one distinct (separate) segment or else completely fused to urosomal segment (Fig. 19.11H) **13**

Fifth leg of two distinct (separate) segments **20**

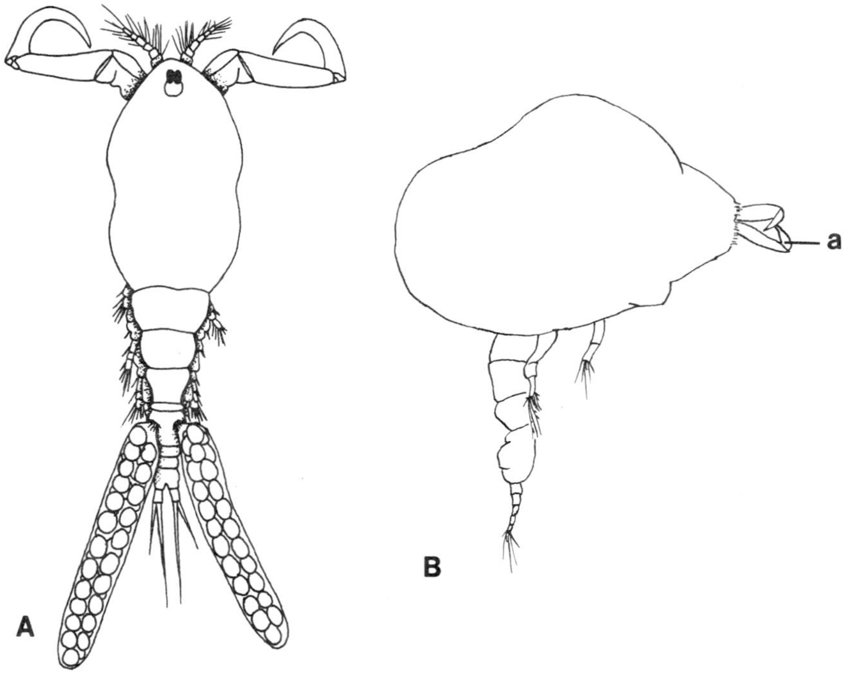

Fig. 19.8. Parasitic copepods, Ergasilidae. A, dorsal view of female *Ergasilus versicolor*, ×70; B, lateral view of *Thersitina*, ×50. a2, antenna 2. (A modified from Wilson 1911a; B redrawn from Kabata 1988.)

13. Caudal ramus at least two times longer than wide (Fig. 19.10C); spines, if present on caudal ramus, limited to a single row **14**

Caudal ramus less than twice as long as wide (Fig. 19.10A), its inner and dorsal surface with several transverse rows of spines; fifth leg not distinct from first urosomal segment and located laterally on the segment (Fig. 19.11B), armed with 3 setae and spines; 1 widespread species *Ectocyclops*

14. Small species (<1.6 mm); antenna 1 of 8–12 segments **15**

Large species (>1.6 mm); antenna 1 of 17 segments; 1 widely distributed species *Homocyclops*

15. Leg 5 with single terminal seta (Fig. 19.11E); first urosomal segment (body) with 1 stout seta protruding laterally on each side **16**

Leg 5 with 3 spines or setae (Fig. 19.11F–G); first urosomal segment (body) with lateral row of fine or coarse hairs **17**

16. Caudal ramus 3 to 4 times as long as wide; first antenna with 10–11 segments; 1 scattered and rare species . . . . *Cryptocyclops*

Caudal ramus less than 2 to 3 times as long as wide, first antenna with 11–12 segments; a single widely distributed variable species . . . . *Microcyclops*

17. Antenna 1 of 12 segments; dorsal surface of caudal ramus plain (ramus may have row of tiny spines along outer margin) **19**

Antenna 1 of 8–11 (rarely 12) segments; caudal ramus plain or with small dorsal spines **18**

18. Caudal ramus usually with row of small spines on dorsal surface; leg 5 with one segment (Fig. 19.11F); about 5 species . . . . *Paracyclops*

Caudal ramus without row of dorsal spines; leg 5 fused to body (Fig. 19.9C); 1 species, GA *Stolonicyclops*

19. Caudal ramus with row of small spines along outer margin (females only) and usually more than four times as long as wide (Figs. 19.10E; 19.11D); total length usually less than 0.70 mm; about 4 species . . . . *Eucyclops*

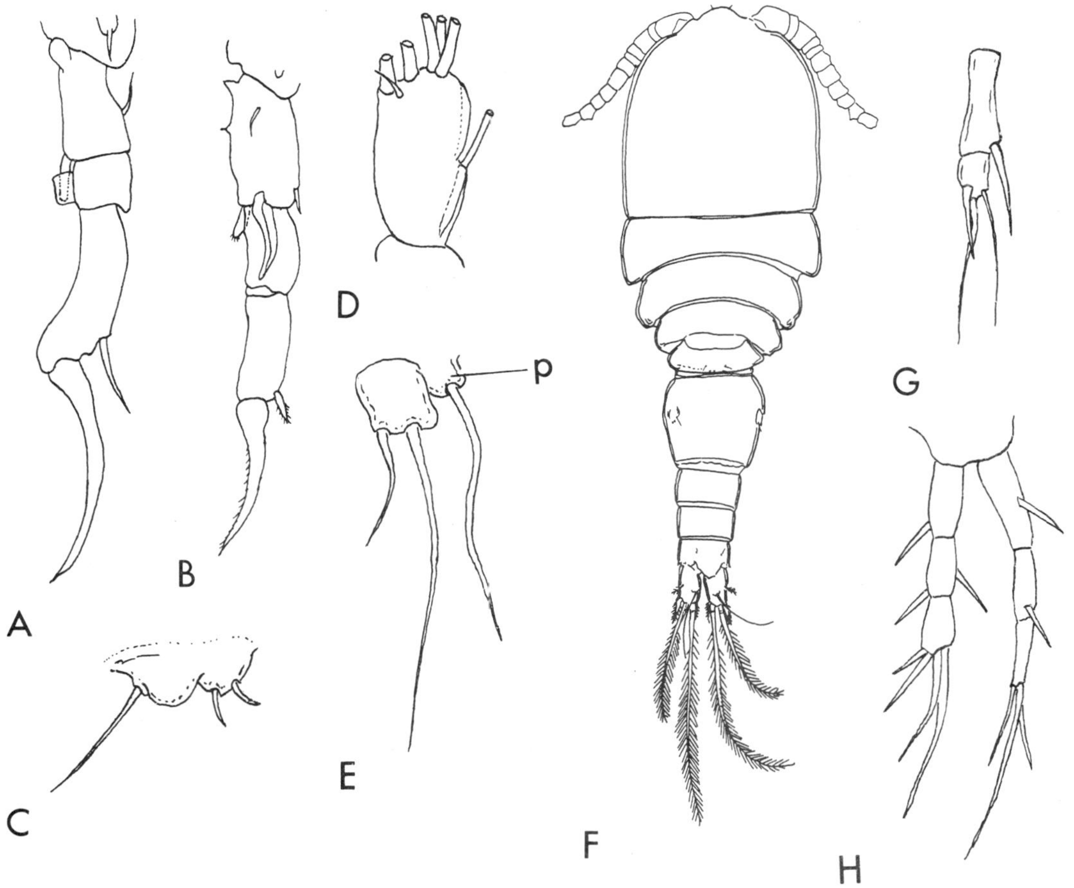

Fig. 19.9. Structure of copepods. A, leg 5 of male *Diaptomus* (*Leptodiaptomus*); B, leg 5 of male *Diaptomus* (*Aglaodiaptomus*); C, leg 5 of *Stolonicyclops;* D, terminal segment of antenna 1 of *Thermocyclops;* E, leg 5 of *Rheocyclops carolinianus;* F, dorsal view of *R. carolinianus,* ×100; G, exopodite of antenna 2 of *Gulcamptus huronensis;* H, leg 1 of *Elaphoidella shawangunkensis.* p, protrusion of fused segment. (A modified from Wilson 1954; B modified from Edmondson 1959; C redrawn from Reid and Spooner 1998; D modified after Duchovnay et al. 1992; E, F redrawn from Reid et al. 1999; G redrawn from Reid and Ishida 1996; H redrawn from Strayer 1988b.)

Caudal ramus without row of small spines along lateral margin and less than four times as long as wide (Fig. 19.10B); total length usually less than 0.60 mm; a common and widespread species . . . .*Tropocyclops*

20. Distal segment of leg 5 with total of 3 spines and/or setae (Figs. 19.10F; 19.11K); about 3 species . . . .*Macrocyclops*

Distal segment of leg 5 with total of 1 or 2 spines or setae (occasionally a few species of *Acanthocyclops* have 3 spines or setae, but the inner and/or outermost spines are always shorter than the distal segment) **21**

21. Distal segment of leg 5 with 2 spines or setae located subterminally to terminally **22**

Distal segment of leg 5 with long apical seta and a scalelike projection about midlength of inner side; innermost terminal caudal seta much longer than ramus; a single introduced species *Megacyclops*

22. Both apical seta and terminal or subterminal inner spine or seta of distal segment of leg 5 quite long, at least twice length of distal segment **23**

Terminal or subterminal inner spine or seta of distal segment of leg 5 shorter than apical seta, less than twice as long as distal segment **25**

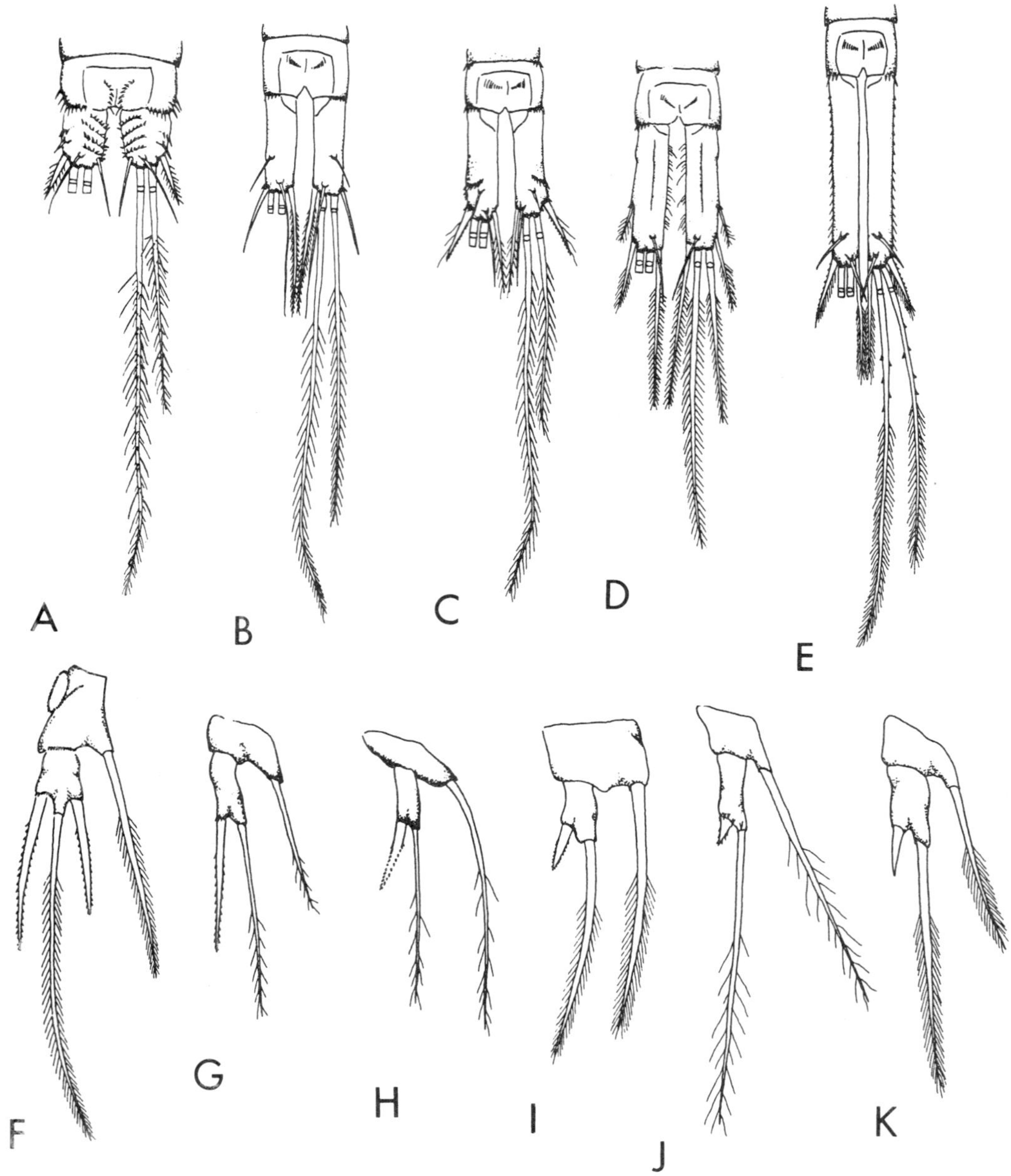

Fig. 19.10. Structure of cyclopoid copepods. A, caudal rami of *Ectocyclops phalerutus;* B, *Tropocyclops prasinus;* C, *Paracyclops chiltoni;* D, *Cyclops scutifer;* E, *Eucyclops elegans;* F, fifth leg of *Macrocyclops albidus;* G, fifth leg of *Diacyclops navus;* H, fifth leg of *D. nearcticus;* I, fifth leg of *Acanthocyclops brevispinosus;* J, fifth leg of *A. carolinianus;* K, fifth leg of *Acanthocyclops robustus.* (A–J after Hudson et al. 1998).

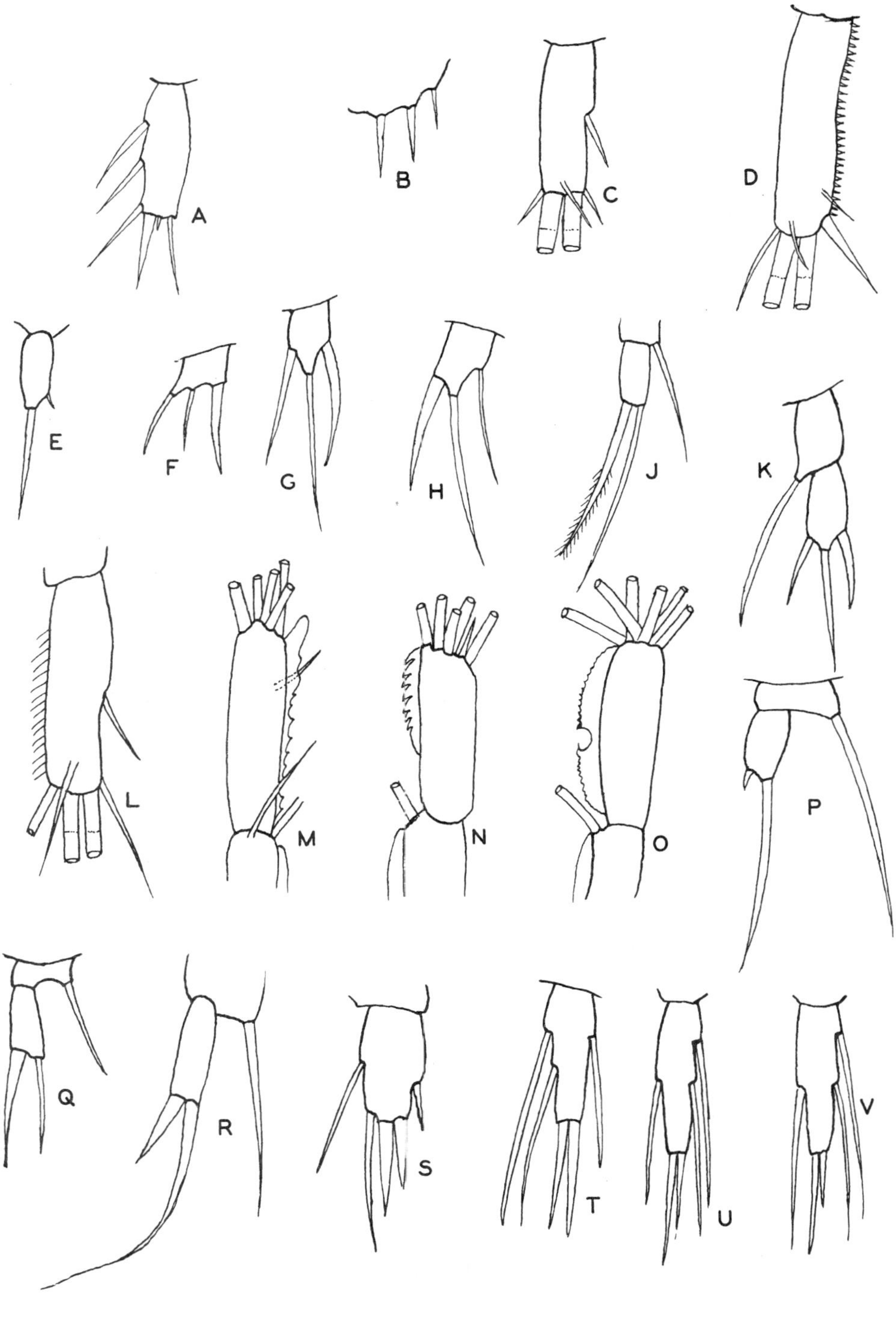

Fig. 19.11. Structure of cyclopoid copepods (fine setation and ornamentation usually omitted). A, terminal endopod segment of fourth leg of *Cryptocyclops bicolor;* B, fifth leg of female *Ectocyclops phaleratus;* C, caudal ramus of *Diacyclops nanus;* D, caudal ramus of *Eucyclops agilis;* E, fifth leg of *Microcyclops varicans;* F, fifth leg of *Paracyclops canadensis;* G, fifth leg of *Tropocyclops prasinus;* H, fifth leg of *Homocyclops ater;* J, fifth leg of *Mesocyclops tenuis;* K, fifth leg of *Macrocyclops albidus;* L, caudal ramus of *Mesocyclops edax;* M, last segment of first antenna of *Macrocyclops fuscus;* N, terminal segment of first antenna of *Mesocyclops edax;* O, terminal segment of first antenna of *Mesocyclops leuckarti;* P, fifth leg of *Acanthocyclops vernalis;* Q, fifth leg of *Diacyclops navus;* R, fifth leg of *Mesocyclops tenuis;* S, terminal endopod segment of fourth leg of *Diacyclops nearcticus;* T, terminal endopod segment of fourth leg of *Diacyclops navus;* U, terminal endopod segment of fourth leg of *Diacyclops bicuspidatus;* V, terminal endopod segment of fourth leg of *Diacyclops thomasi.* (R modified from Yeatmen 1964.)

23. Inner spine on distal segment of leg 5 distinctly subtermineal (Fig. 19.11J), or located in middle of margin, lateral seta of caudal ramus located in middle third of ramus (Fig. 19.11L); free margin of coupler of leg 4 with obvious rounded or triangular protrusions **24**

Seta and spine of distal segment of leg 5 both located near tip of segment (Fig. 19.10G); lateral seta on caudal ramus located in distal third of ramus (Fig. 19.11C); free margin of coupler of leg 4 with small, slightly rounded protrusions or no protrusions; 1 widely distributed species ***Diacyclops* (in part)**

24. Last segment of antenna 1 with or without hyaline process, if process present, with notched margin (Fig. 19.11N, O); about 4 species . . . . ***Mesocyclops***

Last segment of antenna 1 with hyaline process, but without notches (Fig. 19.9D); 4 rare species . . . . ***Thermocyclops***

25. Antenna 1 with 12 or fewer segments **26**

Antenna 1 with 17 or more segments **28**

26. Antenna 1 with 12 segments (two rare species have 11 segments, but are equal to or less than 0.8 mm in length); distal segment of leg 5 less than 2 times as long as wide (Fig. 19.10I); large species (greater than 0.8 mm); 4 species . . . . ***Acanthocyclops* (in part)**

Antenna 1 with 11 segments; distal segment of legs short to long, less than twice to three times as long as wide; smaller species (0.3–0.9 mm) **27**

27. Proximal segment of leg 5 distinct, distal segment about 3 times as long as wide; several species ***Diacyclops***

Proximal segment of leg 5 evidenced by a small knob (Fig. 19.9E); distal segment less than twice as long as wide; about 5 species (Fig. 19.9F) . . . . ***Rheocyclops***

28. Caudal ramus with inner surface haired, with longitudinal dorsal ridge in some species **29**

Caudal ramus with inner surface bare, never with longitudinal dorsal ridge **30**

29. Distal segment of leg 5 short, less than twice as long as wide, subterminal spine long (almost as long as segment) and located medially on segment; longitudinal dorsal ridge on caudal ramus might be present (Fig. 19.10D); 10 or more species . . . . ***Cyclops***

Distal segment of leg 5 long, more than twice as long as wide, subterminal spine short (less than half length of segment) and located subterminally on segment (Figs. 19.10J; 19.11P); 2 species . . . . ***Acanthocyclops* (in part)**

30. Distal segment of leg 5 with long, inner, subapical spine about 7 to 12 times as long as wide and always longer (1.2–1.5) than distal segment (Fig. 19.10H); 2 species . . . ***Diacyclops* (in part)**

Distal segment of leg 5 with smaller subapical medial spine, about 3 to 6 times as long as wide and usually less (0.3–1.2) than length of terminal segment (Fig. 19.10K), about 2 species ***Acanthocyclops* (in part)**

31. Endopod of first leg two-segmented; 0.9 to 4.5 mm long but usually 1.2 to 2.5 mm long **DIAPTOMIDAE, 36**

Endopod of first leg with one or three segments; limnetic **32**

32. Endopod of first leg one-segmented **33**

Endopod of first leg three-segmented . . . . **CENTROPAGIDAE, 35**

33. Endopods of second to fourth legs one-segmented; abdomen of male asymmetrical (Fig. 19.12D, F); 1.1 to 3.6 min long; about 4 species in the northeastern and western United States . . . . **TEMORIDAE, *Epischura***

Endopods of second to fourth legs two- or three-segmented; abdomen of male symmetrical **34**

34. Endopods of second to fourth legs two-segmented; male right antenna geniculate; 1.0 to 1.5 min long; 1 species in salt and brackish waters, as well as freshwater ponds and lakes along all coasts, recently in Mississippi River system and Great Lakes **TEMORIDAE, *Eurytemora***

Endopod of second leg two-segmented, those of third and fourth legs three-segmented; male left antenna not geniculate; 2.4 to 2.9 mm long (Fig. 19.12A); deep, cold lakes of Great Lakes region and MT . . . . **SENECELLIDAE, *Senecella***

35. Caudal rami short (Fig. 19.12B); 1.5 to 2.5 mm long; uncommon and erratically distributed over most of the United States . . . . ***Osphranticum***

Caudal rami long (Fig. 19.12C); 2.0 to 3.2 min long; occurs in deep, cold lakes in northeast, such as Great Lakes, Finger Lakes, and Green Lake (WI), but also in Cascade Range lakes in WA; usually considered a relict marine species . . . . ***Linmocalanus***

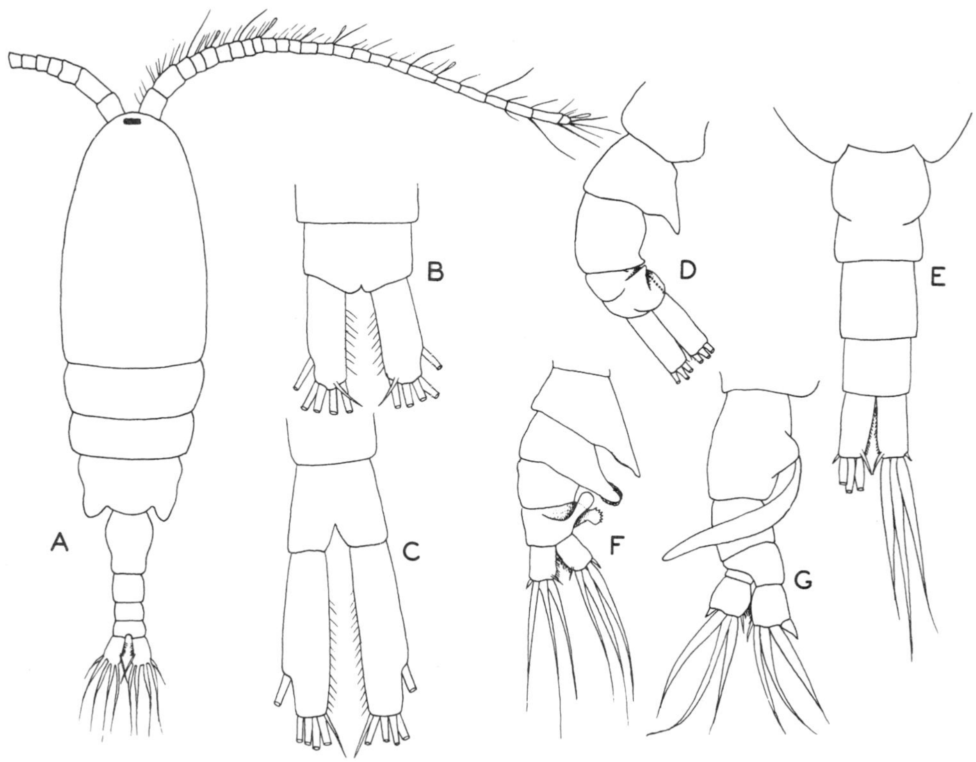

Fig. 19.12. Structure of calanoid copepods. A, female *Senecella calanoides*, ×26; B, caudal rami of female *Osphranticum labronectum;* C, caudal rami of male *Limnocalanus;* D, abdomen of male *Epischura nordenskioldi;* E, female abdomen of same; F, abdomen of male *Epischura lacustris;* G, female abdomen of same. (A modified from Juday 1925; D–G modified from Marsh 1933.)

36. Tip of terminal segment of right antenna 1 produced laterally (Fig.13C, arrow); 1 species in higher elevations of Rocky Mountains . . . . ***Acanthodiaptomus***
Tip of terminal segment of right antenna 1 not produced laterally (Fig. 19.13A, B); over 55 species (usually distributed among 15 groups, see Table 19.1) . . . . ***Diaptomus***

37. Male body composed of 11 segments; female, 10 segments (Fig. 19.3) **38**
Male body composed of 10 segments; female, 9 segments (Fig. 19.14G) **39**

38. First antenna eight-segmented; a small genus with one cosmopolitan species found in many kinds of habitats . . . . ***Phyllognathopus* (in part)**
First antenna seven-segmented; several uncommon and widely distributed species ***Chappuisius***

39. Distal exopod segment of first leg with three to five short, clawlike spines; fresh and brackish water **47**
Distal exopod segment of first leg with the usual long spines and setae **40**

40. Exopod of first leg three-segmented **41**
Exopod of first leg two-segmented **48**

41. Terminal segment of exopod of first leg with five or six spines and setae (Fig. 19.15M) **57**
Terminal segment of exopod of first leg with four spines and setae **42**

42. Second exopod segment of first leg with an inner seta (Fig. 19.15C) **46**
Second exopod segment of first leg without an inner seta **43**

43. Second exopod segment of first leg with an outer spine **44**
Second exopod segment of first leg without an outer spine; vermiform (Fig. 19.14A); a large, worldwide genus; usually found in interstitial and subterranean waters . . . . . ***Parastenocaris***

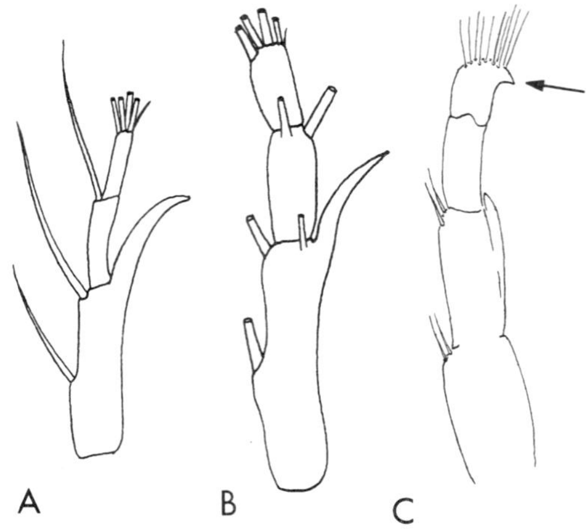

Fig. 19.13. Structure of diaptomids. A, terminal antennal segments of *Diaptomus* (*Hesperodiaptomus*) *franciscanus;* B, terminal antennal segments of *D.* (*Aglaodiaptomus*) *lintoni;* C, terminal segments of *Acanthodiaptomus denticornis.* (A modified from De Guerne and Richard 1889; B modified from Marsh 1907; C redrawn from Edmondson 1959.)

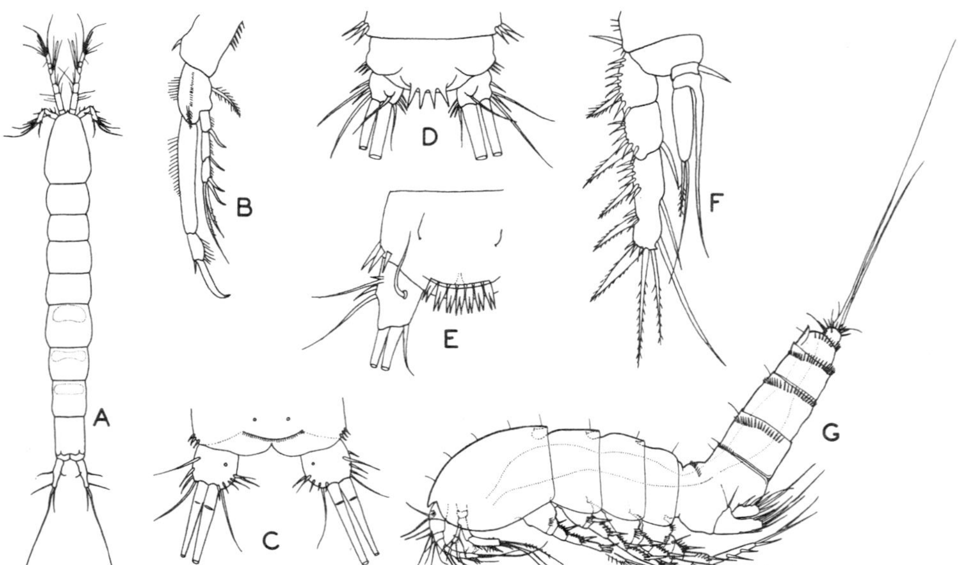

Fig. 19.14. Structure of Harpacticoida. A, dorsal view of *Parastenocaris,* ×100; B, female first leg of *Pseudonychocamptus;* C, D, and E, posterior end of *Bryocamptus* spp.; F, male third leg of *Bryocamptus;* G, female *Bryocamptus,* ×85. (B modified from Wilson 1932; C, D, F, and G modified from Coker 1934; E redrawn from Gurney 1931–1933.)

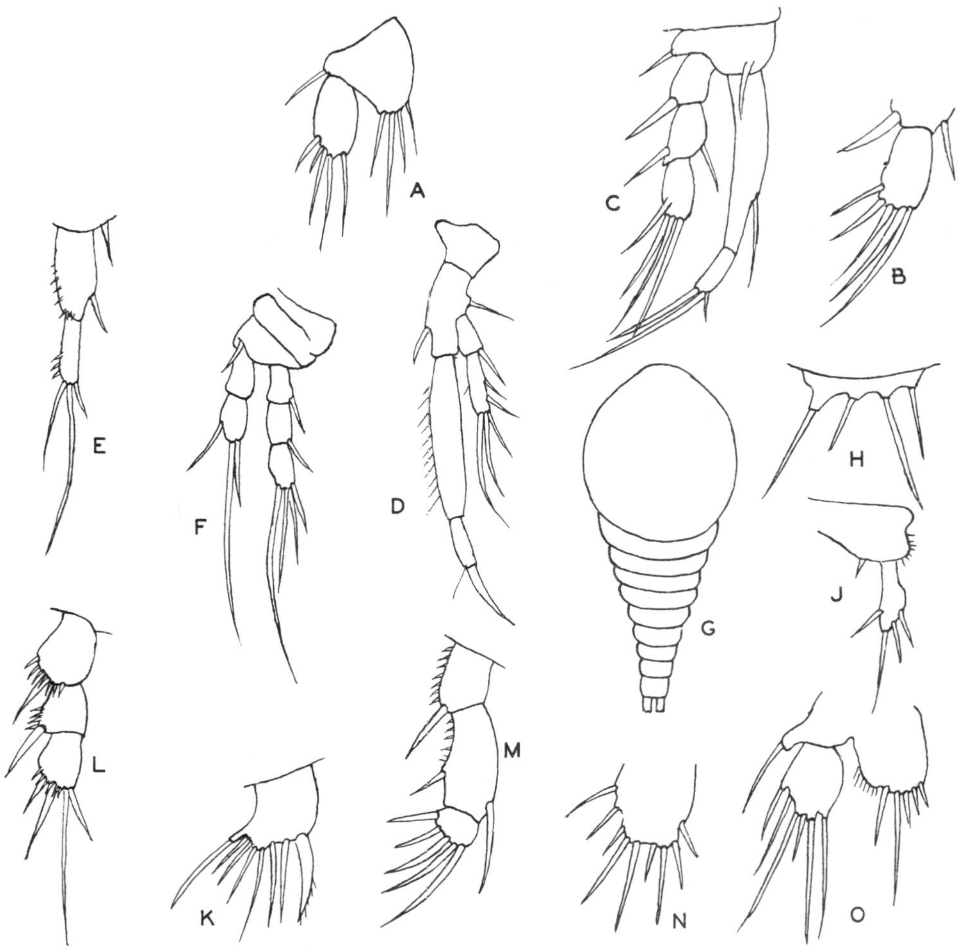

Fig. 19.15. Structure of Harpacticoida (fine hairs and other ornamentation not shown). A, fifth leg of *Harpacticus;* B, terminal segment of exopod of first leg of *Nitokra;* C, first leg of *Mesochra;* D, fifth leg of *Onychocamptus;* E, endopod of first leg of *Maraenobiotus;* F, first leg of *Epactophanes;* G, dorsal view of body of *Metis;* H, fifth leg of *Heterolaophonte;* J, fifth leg of male *Pseudonychocamptus;* K, fifth leg of female *Stenocaris;* L, exopod of second leg of *Moraria;* M, exopod of first leg of *Paradactylopodia;* N, fifth leg of *Tachidius;* O, fifth leg of *Nitokra.* (O redrawn and modified from Edmondson 1959.)

44. Endopod of first leg one-segmented **49**
Endopod of first leg with more than one segment **45**

45. Endopod of first leg two-segmented **50**
Endopod of first leg three-segmented **61**

46. Endopod of first leg two-segmented (Fig. 19.15C) **62**
Endopod of first leg three-segmented **63**

47. Basal expansion of female fifth leg with three or four setae (Fig. 19.15A); New England coast **_Harpacticus_**
Basal expansion of female fifth leg with five or six setae; Pacific coast . . . . **_Tigriopus_**

48. Endopod of first leg about twice or more as long as exopod (Fig. 19.15D); several species reported from fresh and inland saline waters . . . . **_Onychocamptus_**
Endopod of first leg no longer than exopod **49**

49. Endopod of first leg two-segmented (Fig. 19.15E); in vegetation at margins of lotic and lentic waters; uncommon but widely distributed . . . . ***Maraenobiotus***
Endopod of first leg one-segmented and small; a marine genus reported from Bear Lake, Utah; undoubtedly occurs elsewhere . . . . ***Huntemannia***
50. Endopod of first leg not extending beyond second exopod segment (Fig. 19.14B) **51**
Endopod of first leg longer **54**
51. Anterior end of body expanded, tapered posteriorly (Fig. 19.15G); brackish and fresh waters of Atlantic Coast . . . . ***Metis***
Body slender and vermiform **52**
52. First segment of first exopod more than three times as long as wide (Fig. 19.14B) **53**
First segment of first exopod no more than twice as long as wide (Fig. 19.15F); one cosmopolitan species in aquatic mosses and other vegetation; 0.3 to 0.5 mm long . . . . . ***Epactophanes***
53. Fifth leg of male unsegmented (Fig. 19.15H); brackish ponds on northeast coast ***Heterolaophonte***
Fifth leg of male segmented (Fig. 19.15J); fresh and brackish ponds on northeast coast ***Pseudonychocamptus***
54. Fifth leg unsegmented (Fig. 19.15K); vermiform; brackish and fresh waters of northeastern coast . . . . ***Stenocaris***
Fifth leg segmented **55**
55. Second exopod segment of second to fourth legs without a median seta (Fig. 19.15L) **56**
Second exopod segment of second to fourth legs with a median seta (Fig. 19.14F); many species, widely distributed, very common on all kinds of substrates . . . . ***Bryocamptus* (in part)**
56. Antenna 2 exopodite with 1 segment; third segment of exopodite of leg 3 with 1 medial seta; many species, widely distributed . . . . ***Moraria***
Antenna 2 exopodite with 2 segments (Fig. 19.9G); third segment of exopodite of leg 3 with 2 median setae; 1 species in lake Huron . . . . ***Gulcamptus***
57. Third exopod segment of first leg about one-third as long as the second exopod segment (Fig. 19.15M); uncommon, euryhaline species of Atlantic coast . . . . ***Paradactylopodia***
Third exopod segment of first leg about as long as the second exopod segment **58**
58. Fifth leg unsegmented **59**
Fifth leg segmented (Fig. 19.15O) **60**
59. Third exopod segment of first leg with six spines and setae; several euryhaline species of Atlantic coast . . . . ***Microarthridion***
Third exopod segment of first leg with five spines and setae (Fig. 19.16A); several freshwater and brackish coastal species . . . . ***Tachidius***
60. Third exopod segment of second and third legs with three lateral marginal spines (Fig. 19.16B); mostly coastal brackish and freshwater species . . . . ***Nitokra***
Third exopod segment of second and third legs with two lateral marginal spines (Fig. 19.16C); expect to find this genus in phreatic and subterranean waters . . . . ***Nitocrella***
61. Endopod of second leg two-segmented (Fig. 19.16D); many species, widely distributed, and very common on all kinds of substrates . . . . ***Bryocamptus* (in part)**
Endopod of second leg three-segmented; one common cosmopolitan species in many habitats ***Phyllognathopus***
62. Exopod of fifth leg separated from base by only a notch or gap (Fig. 19.16E); several species in brackish and saline lakes and ponds . . . . ***Cletocamptus***
Exopod of fifth leg clearly separated from base (Fig. 19.16K) **63**
63. Rostrum greatly enlarged (Fig. 19.16G); endopod of first leg with two or three segments (Fig. 19.15C); coastal brackish and freshwater ponds, and inland ponds; several species ***Mesochra***
Rostrum small to moderate size (Figs. 19.14G; 19.16H) **64**
64. Rostrum small (Fig. 19.14G); endopod of first leg two-segmented (Fig. 19.16F); many widely distributed species in a variety of freshwater habitats . . . . ***Bryocamptus* (in part)**
Rostrum small to moderate size (Fig. 19.16H); endopod of first leg three-segmented (Fig. 19.16J) **65**
65. Second seta of basal expansion of female fifth leg greatly reduced (Fig. 19.16K); several species in a wide variety of freshwater habitats . . . . ***Canthocamptus***
Second seta of basal expansion of female fifth leg not reduced **66**

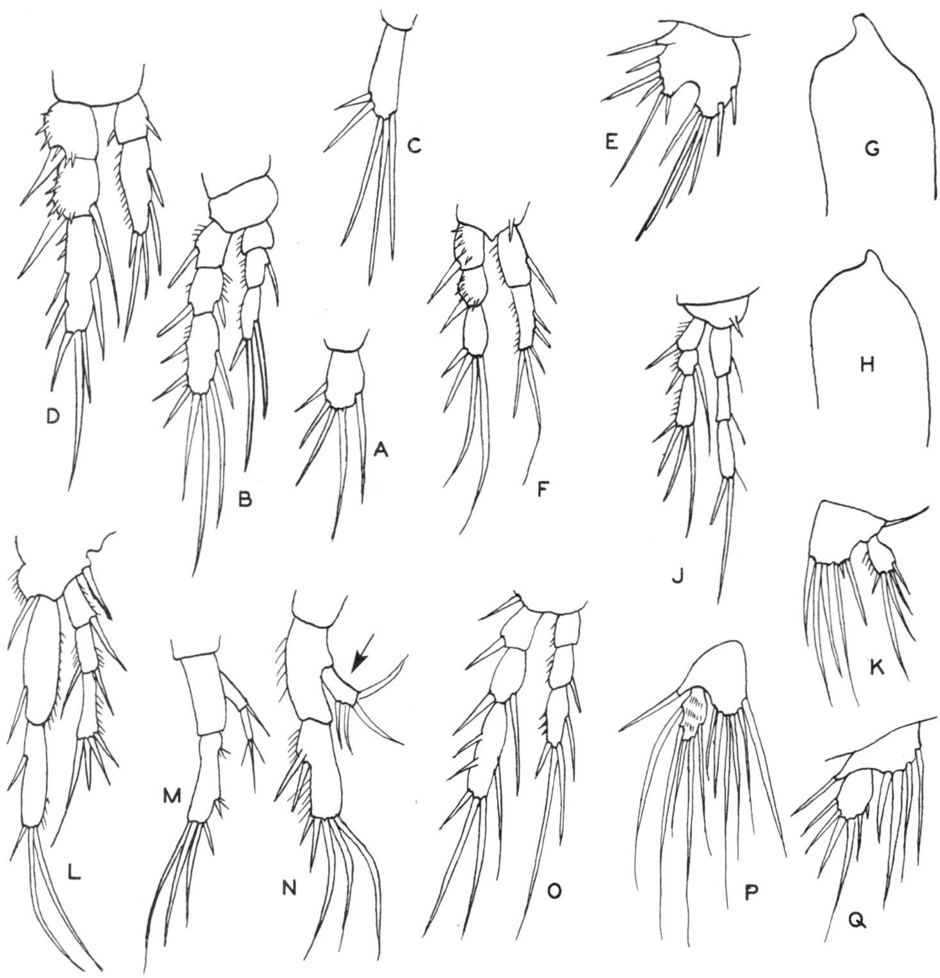

Fig. 19.16. Structure of Harpacticoida (fine hairs and other ornamentation not shown). A, third exopod segment of first leg of *Tachidius;* B, second leg of *Nitokra;* C, third exopod segment of third leg of *Nitocrella;* D, second leg of *Bryocamptus;* E, fifth leg of *Cletocamptus;* F, first leg of *Bryocamptus;* G, lateral view of *Mesochra* cephalothorax outline; H, lateral view of *Bryocamptus* cephalothorax outline; J, first leg of *Bryocamptus;* K, fifth leg of *Canthocamptus;* L, first leg of *Attheyella;* M, second antenna of *Bryocamptus;* N, second antenna of *Attheyella;* O, second leg of *Bryocamptus;* P, fifth leg of *Attheyella;* Q, fifth leg of *Elaphoidella.* (K, M, N, and P modified from Coker 1934.)

66. First endopod segment of first leg reaching or exceeding middle of third exopod segment (Fig. 19.16L); many species in a wide variety of habitats .... ***Attheyella* (in part)**
First endopod segment of first leg not extending beyond second exopod segment (Fig. 19.9H) **67**

67. Palp of second antenna one-segmented (Fig. 19.16N, arrow) **68**
Palp of second antenna two-segmented (Fig. 19.16M); endopods of female second and third legs three segmented (Fig. 19.16O); many widely distributed species in a variety of freshwater habitats .... ***Bryocamptus* (in part)**

68. Fifth leg of female with rows of fine setae on the ventral surface, and basal expansion with six setae; several species .... ***Attheyella* (in part)**
Fifth leg of female without rows of fine setae on the ventral surface, and basal expansion with four setae (Fig. 19.16Q); several uncommon and widely distributed species ... ***Elaphoidella***

REFERENCES

Aycock, D. 1942. Influence of temperature on size and form of *Cyclops vernalis* Fischer. *J. Elisha Mitchell Sci. Soc.* 58:84–93.

Balcer, M. D., et al. 1984. *Zooplankton of the Great Lakes.* 174 pp. Univ. Wisconsin Press, Madison.

Bowman, T. E. 1991. *Epischura fluviatilis* Herrick, 1883 (Copepoda: Calanoida: Temoridae), first report from South Carolina, with notes on sexual dimorphism in the caudal rami and its possible role in mating. *J. Crust. Biol.* 11:460–462.

Bowman, T. E., and J. J. Lewis. 1989. Occurrence of the calanoid copepod *Eurytemora affinis* (Poppe) in the Ohio river at Louisville, Kentucky. *J. Crust. Biol.* 9:83–84.

Brandl, Z. 1973. Laboratory culture of cyclopoid copepods on a definite food. *Vestn. Cesk. Spol. Zool.* 37:81–88.

Brandl, Z., and C. H. Fernando. 1975. Food consumption and utilization in two freshwater cyclopoid copepods (*Mesocyclops edax* and *Cyclops vicious*). *Int. Rev. gesamten Hydrobiol.* 60:471–494.

Byron, E. R. 1982. The adaptive significance of calanoid copepod pigmentation: a comparative and experimental analysis. *Ecology* 63: 1871–1886.

Chappuis, P. A. 1927. Freilebende Süsswasser-Copepoden aus Nordamerika, 2: Harpacticiden. *Zool. Anz.* 74:302–313.

——. 1929a. Die Unterfamilie der Canthocamptinae. *Arch. Hydrobiol.* 20:471–516.

——. 1929b. Copépodes cavernicoles de I'Amérique du Nord. *Bull. Soc. Sci. Cluj* 4:51–58.

——. 1957. Le genre *Parasterrocaris* Kessler. *Vie et Milieu* 8:423–432.

Coker, R. E. 1933. Influence of temperature on size of freshwater copepods (*Cyclops*). *Int. Rev.* 29:406–436.

——. 1934. Contribution to knowledge of North American freshwater harpacticoid copepod Crustacea. *J. Elisha Mitchell Sci. Soc.* 50:75–141.

——. 1938. Anomalies of crustacean distribution of the general region of Chapel Hill, N.C. *J. Elisha Mitchell Sci. Soc.* 54:76–87.

Comita, G. W., and S. J. McNett. 1976. The postembryonic developmental instars of *Diaptomus oregonensis* Lilljeborg, 1889 (Copepoda). *Crustaceana* 30:123–163.

Comita, G. W., and D. M. Tommerdahl. 1960. The postembryonic developmental instars of *Diaptomus siciloides* Lilljeborg. *J. Morphol.* 107: 297–355.

Cope, E. D. 1872. On the Wyandotte Cave and its fauna. *Am. Nat.* 6:406–422.

Davis, C. C. 1959. Osmotic hatching in the eggs of some fresh-water copepods. *Biol. Bull.* 116: 15–29.

Deevey, E. S. 1941. Notes on the encystment of the harpacticoid copepod *Canthocamptus staphylinoides* Pearse. *Ecology* 22:197–200.

De Guerne, J., and J. Richard. 1889. Révision des Calanides d'eau douce. *Mint. Soc. Zool. Fr.* 2: 53–144.

Dodds, G. S. 1915. Descriptions of two new species of Entomostraca from Colorado, with notes on other species. *Proc. U.S. Nat. Mus.* 49: 97–102.

Duchovnay, A., et al. 1992. *Thermocyclops crassus* (Crustacea: Copepods) present in north America: a new record from Lake Champlain. *J. Great Lakes Res.* 18:415–419.

Dussart, B., and D. Defaye. 1983. *Repértoire Mondial des Crustacés Copepodes des Eaux Intérieures.* 224 pp. Editions du CNRS, Paris.

Dussart, B. H., and C. H. Fernando. 1990. A review of the taxonomy of five Ontario genera of freshwater Copepoda (Crustacea). *Can. J. Zool.* 68:2594–2604.

Edmondson, W. T. (ed.). 1959. *Ward and Whipple's Fresh-water Biology*, 2d ed. 1248 pp. Wiley, New York.

Elgmork, K. 1967a. Ecological aspects of diapause in copepods. *Proc. Symp. Crustacea* 3: 947–954.

——. 1967b. On the distribution and ecology of *Cyclops scutifer* Sars in New England (Copepoda, Crustacea). *Ecology* 48:967–971.

Ewers, L. A. 1930. The larval development of freshwater Copepoda. *Ohio State Univ, Contrib. Franz Theodore Stone Lab.* 3:1–43.

——. 1936. Propagation and rate of reproduction of some freshwater Copepoda. *Trans. Am. Microsc. Soc.* 55:230–238.

Forbes, S. A. 1897. A contribution to a knowledge of North American fresh-water Cyclopidae. *Bull. Ill. State Lab.* 5:27–83.

Frenzel, P. 1980. Die Populationsclynamik von

Canthocamptus staphylinus (Jurine) (Copepoda, Harpacticoida) im Littoral des Bodensees. *Crustaceana* 39:282–286.

Fryer, G. 1982. The parasitic Copepoda and Branchiura of British freshwater fishes. *Freshwater Biol. Assoc. Publ.* 46:1–87.

——. 1985. An ecological validation of a taxonomic distinction: the ecology of *Acanthocyclops vernalis* and *A. robustus* (Crustacea: Copepoda). *Zool. J. Linn. Soc.* 84:165–180.

Gurney, R. 1931–1933. *British Fresh-water Copepoda.* 3 vols. 958 pp. Ray Society, London.

Harding, J. P., and W. A. Smith. 1974. A key to the British fresh-water cyclopoid and calanoid copepods. *Sci. Publ. Freshwater Biol. Assoc.* 18: 1–56.

Herrick, C. L., and C. H. Turner. 1895. A synopsis of the Entomostraca of Minnesota. *2d Rept. State Zool., Geol. Nat. Hist. Surv. Minn.* 1–337.

Herzig, A. 1983. The ecological significance of the relationship between temperature and duration of embryonic development in planktonic freshwater copepods. *Hydrobiologia* 100: 65–91.

Hill, L. I., and R. E. Coker. 1930. Observations on mating habits of *Cyclops. J. Elisha Mitchell Sci. Soc.* 45:206–220.

Hoffman, G. L. 1998. *Parasites of North American Freshwater Fishes*, 2d ed. 539 pp. Comstock, Ithaca, NY.

Hudson, P. L., et al. 1998. Cyclopoid and harpacticoid copepods of the Laurentian Great Lakes. *Ohio Biol Surv. Bull.* 12:1–50.

Juday, C. 1925. *Senecella calanoides*, a recently described fresh-water copepod. *Proc. U.S. Nat. Mus.* 66:1–6.

Kabata, Z. 1969. Revision of the genus *Salmincola* Wilson 1915 (Copepoda: Lernaeopodidae). *J. Fish. Res. B. Can.* 26:2987–3041.

——. 1988. Copepoda and Branchiura. In: *Guide to the Parasites of Fishes of Canada*, Part II: Crustacea. pp. 3–127, edited by L. Margolis and Z. Kabata. *Can. Spec. Publ. Fish. Aquat. Sci.* 101: 1–184.

Kerfoot, W. C., et al. 1980. Visual observations of live zooplankters: evasion, escape, and chemical defenses. In: *Evolution and Ecology of Zooplankton Communities*, pp. 10–27, edited by W. C. Kerfoot. University Press of New England, Hanover, NH.

Kiefer, F. 1967. Copepoda. *Limnofauna Europaea* 173–185.

Lang, K. 1948. *Monographie der Harpacticiden.* Vols. 1 and 2, 1662 pp. Nordiska Bokhandeln, Stockholm, Sweden.

Lee, C. E. 1999. Rapid and repeated invasions of fresh water by the copepod *Eurytemora affinis. Evolution* 53:1423–1434.

Lescher-Moutoué, F. 1973. Sur le biologie et l'écologie des copépodes cyclopides hypogés (Crustaces). *Ann. Speleol.* 28:429–502, 581–674.

Light, S. F. 1938. New subgenera and species of diaptomid copepods from the inland waters of California and Nevada. *Univ. Calif. Publ. Zool.* 43:67–78.

——. 1939. New American subgenera of *Diaptomus* Westwood (Copepoda, Calanoida). *Trans. Am. Microsc. Soc.* 58:473–484.

Marsh, C. D. 1907. A revision of the North American species of *Diaptomus. Trans. Wis. Acad. Sci. Arts Lett.* 15:381–516.

——. 1926. On a collection of copepods from Florida with a description of *Diaptomus floridanus*, new species. *Proc. U.S. Nat. Mus.* 70: 1–4.

——. 1929. Distribution and key to the North American copepods of the genus *Diaptomus*, with the description of a new species. *Proc. U.S. Nat. Mus.* 75:1–27.

——. 1933. Synopsis of the calanoid crustaceans, exclusive of the Diaptomidae, found in fresh and brackish waters, chiefly of North America. *Proc. U.S. Nat. Mus.* 82:1–58.

Papinska, K. 1985. Carnivores and detritivores feeding of *Mesocyclops leuckarti* Claus (Cyclopoida, Copepoda). *Hydrobiologia* 120:249–257.

Pennak, R. W. 1940. Ecology of the microscopic metazoa inhabiting the sandy beaches of some Wisconsin lakes. *Ecol. Monogr.* 10:537–615.

Pennak, R. W., and J. V. Ward. 1985. New cyclopoid copepods from interstitial habitats of a Colorado mountain stream. *Trans. Am. Microsc. Soc.* 104:216–222.

Reed, E. B. 1986. Esteval phenology of an *Acanthocyclops* (Crustacea, Copepoda) in a Colorado tarn with remarks on the vernalis–robustus complex. *Hydrobiologia* 139:127–144.

Reid, J. W. 1992a. *Diacyclops albus* n. sp. and *Parastenocaris palmerae* n. sp. (Crustacea: Copepoda) from the meiofauna of a stream bed in Viginia, U.S.A. *Can. J. Zool.* 69:2893–2902.

——. 1992b. Redescription of *Diacyclops nearcti-*

cus (Kiefer, 1934) and description of four similar new congeners from North America, with comments on *D. crassicaudis* (G. O. Sars, 1863) and *D. crassicaudis* var. brachycercus (Kiefer, 1927) (Crustacea: Copepoda). *Can. J. Zool.* 70: 1445–1469.

——. 1993. New records and redescriptions of American species of *Mesocyclops* and of *Diacyclops bernardi* (Petkovski, 1986) (Copepoda: Cyclopoida). *Bijd. Dierk.* 63:173–191.

——. 1997/1998. How "cosmopolitan" are the Continental cyclopoid copepods? Comparison of the North American and Eurasian faunas, with description of *Acanthocyclops parasensitivus* sp.n. (Copepoda: Cyclopoida) from the U.S.A. *Zool. Anz.* 236:109–118.

Reid, J. W., and T. Ishida. 1996. Two new species of *Gulcamptus* (Crustacea: Copepoda: Harpacticoida) from North America. *Jpn. J. Limnol.* 57:133–144.

Reid, J. W., and J. D. Spooner. 1998. *Stolonicyclops heggiensis*, new genus, new species, from Georgia, U.S.A. *J. Crust. Biol.* 18:405–411.

Reid, J. W., and D. L. Strayer. 1994. *Diacyclops dimorphus*, a new species of copepod from Florida, with comments on morphology of interstitial cyclopine cyclopoids. *J. North Am. Benthol. Soc.* 13:250–265.

Reid, J. W., et al. 1989. *Diacyclops navus* (Crustacea: Copepoda) redescribed from Louisiana, U.S.A. *Trans. Am. Microsc. Soc.* 108:332–344.

——. 1991. *Diacyclops languidoides* (Lilljeborg, 1901) s.l. and *Acanthocyclops montana*, new species (Copepoda, Cyclopoida), from groundwater in Montana, USA. *Hydrobiologia* 218: 133–149.

——. 1999. *Rheocyclops*, a new genus of copepods from the southeastern and central U.S.A. (Copepoda: Cyclopoida: Cyclopidae). *J. Crust. Biol.* 19:384–396.

Roberts, L. S. 1970. *Ergasilus* (Copepoda: Cyclopoida): revision and key to species in North America. *Trans. Am. Microsc. Soc.* 89:134–161.

Roberts, L. S., and J. Janovy, Jr. 2000. *Foundations of Parasitology*, 6th ed. 670 pp. McGraw-Hill Higher Education, Boston.

Ruber, E. 1968. Descriptions of a salt marsh copepod *Cyclops* (*Apocyclops*) *spartinus* n. sp. and a comparison with closely related species. *Trans. Am. Microsc. Soc.* 87:368–375.

Rylov, V. M. 1948. Freshwater Cyclopoida. Crustacea. *Fauna USSR* 3(3):1–314.

Sandercock, G. A. 1967. A study of selected mechanisms for the coexistence of *Diaptomus* spp. in Clarke Lake, Ontario. *Limnol. Oceanogr.* 12:97–112.

Sarvala, J. 1979a. Effect of temperature on the duration of egg, nauplius and copepodite development of some freshwater benthic Copepoda. *Freshwater Biol.* 9:515–534.

——. 1979b. A parthenogenetic life cycle in a population of *Canthocamptus staphylinus* (Copepoda, Harpacticoida). *Hydrobiologia* 62:113–129.

Siefken, M., and K. B. Armitage. 1968. Seasonal variations in metabolism and organic nutrients in three *Diaptomus* (Crustacea: Copepoda). *Comp. Biochem. Physiol.* 24:591–609.

Smyly, W. J. P. 1957. Observations on the life-history of the harpacticoid copepod *Canthocamptus staphylinus* (Jurine). *Ann. Mag. Nat. Hist.* 10(115):509–512.

Storch, O. 1929. Die Schwimmbewegung der Copepoden, auf Grund von Mikro-Zeitlupenaufnahmen analysiert. *Zool. Anz., Suppl. Bd.* 4:118–129.

Strayer, D. 1988a. Crustaceans and mites (Acari) from hyporheic and other underground waters in southeastern New York. *Stygologia* 4:192–207.

——. 1988b. New and rare copepods (Cyclopoida and Harpacticoida) from freshwater interstitial habitats in southeastern New York. *Stygologia* 4:279–290.

Strickler, J. R. 1975. Intra- and interspecific information flow among planktonic copepods: receptors. *Proc. Int. Assoc. Theor. Appl. Limnol.* 19:2951–2958.

Walter, E. 1922. Ober die Lebensdauer der freilebenden Süsswasser-Cyclopiden und andere Fragen ihrer Biologie. *Zool. Jahrb. Abt. Syst.* 44:375–420.

Ward, H. B., and G. C. Whipple. 1918. *Freshwater Biology.* 1111 pp. Wiley, New York.

Wilson, C. B. 1911a. North American parasitic copepods, Part 9: the Lernaeopodidae. *Proc. U.S. Nat. Mus.* 39:189–226.

——. 1911b. North American parasitic copepods belonging to the family Ergasilidae. *Proc. U.S. Nat. Mus.* 39:263–400.

——. 1915. North American parasitic copepods belonging to the Lernaeopodidae, with a revision of the entire family. *Proc. U.S. Nat. Mus.* 47:565–729.

——. 1916. Copepod parasites of fresh-water

fishes and their economic relations to mussel glochidia. *Bull. U.S. Bur. Fish.* 34:333–374.

——. 1917. North American parasitic copepods belonging to the Lernaeidae with a revision of the entire family. *Proc. U.S. Nat. Mus.* 53:1–150.

——. 1918. The economic relations, anatomy, and life history of the genus *Lernaea. Bull. U.S. Bur. Fish.* 35:16–198.

——. 1932. The copepods of the Woods Hole region, Massachusetts. *Bull. U.S. Nat. Mus.* 158:1–635.

——. 1944. Parasitic copepods in the United States National Museum. *Proc. U.S. Nat. Mus.* 94:529–582.

Wilson, M. S. 1941. New species and distribution records of diaptomid copepods from the Marsh collection in the United States National Museum. *J. Wash. Acad. Sci.* 31:509–515.

——. 1953. New and inadequately known North American species of the copepod genus *Diaptomus. Smithson. Misc. Coll.* 122:1–30.

——. 1954. A new species of *Diaptomus* from Louisiana and Texas with notes on the subgenus *Leptodiaptomus* (Copepoda, Calanoida). *Tulane Stud. Zool.* 2:51–60.

——. 1955. A new Louisiana copepod related to *Diaptomus* (*Aglaodiaptomus*) *clavipes* Schacht. (Copepoda, Calanoida). *Tulane Stud. Zool.* 3:37–47.

——. 1956. North American harpacticoid copepods, 1: comments on the known fresh water species of the Canthocamptidae; 2: *Canthocamptus oregonensis* n. sp. from Oregon and California. *Trans. Am. Microsc. Soc.* 75:290–306.

——. 1958a. New records and species of calanoid copepods from Saskatchewan and Louisiana. *Can. J. Zool.* 36:489–497.

——. 1958b. The copepod genus *Halicyclops* in North America, with description of new species from Lake Pontchartrain, Louisiana, and the Texas coast. *Tulane Stud. Zool.* 6:176–189.

——. 1958c. North American harpacticoid copepods, 4: diagnoses of new species of freshwater Canthocamptidae and Cletodidae (genus *Huntemannia*). *Proc. Biol. Soc. Wash.* 71: 43–48.

——. 1972. Copepods of marine affinities from mountain lakes of western North America. *Limnol. Oceanogr.* 17:762–763.

——. 1975. North American harpacticoid copepods, II: new records and species of *Elaphoidella* (Canthocamptidae) from the United States and Canada. *Crustaceana* 28:125–138.

Wilson, M. S., and S. F. Light. 1951. Description of anew species of diaptomid copepod from Oregon. *Trans. Am. Microsc. Soc.* 70: 25–30.

Wilson, M. S., and W. G. Moore. 1953a. Diagnosis of a new species of diaptomid copepod from Louisiana. *Trans. Am. Microsc. Soc.* 72: 292–295.

——. 1953b. New records of *Diaptomus sanguineus* and allied species from Louisiana, with the description of a new species (Crustacea: Copepoda). *J. Wash. Acad. Sci.* 43:121–127.

Yeatman, H. C. 1944. American cyclopoid copepods of the viridis–vernalis group (including a description of *Cyclops carolinianus* n. sp.). *Am. Midl. Nat.* 32:1–90.

——. 1959. Some effects of temperature and turbulence on the external morphology of *Cyclops carolinianus. J. Elisha Mitchell Sci. Soc.* 75:154–167.

——. 1964. A new cavernicolous cyclopoid copepod from Tennessee and Illinois. *Tenn. Acad. Sci.* 39:95–98.

20

BRANCHIURA

(Fish Lice)

Adult branchiurans, familiarly known as a fish lice, are parasites of fish, both freshwater and marine. Although superficially similar to caligoid copepods, branchiurans are very distinct crustaceans.

General characteristics. Branchiurans range from 5 to 25 mm in length, depending on species and are usually found attached in the branchial chamber and general body surface of the fish host. Fish lice (Fig. 20.1) are greatly flattened dorsoventrally, and the head and thorax are covered with an expanded carapace whose lateral areas have a respiratory function. Dorsally, there are two large movable compound eyes. The first and second antenna (Fig. 20.2) are highly modified for grasping the host. The mouthparts are greatly modified or reduced, with the mouth itself borne on a long tube or piercing organ used in obtaining food; the second maxillae are completely transformed into curious suction cups. The first four pairs of legs are well developed and used for swimming, but the fifth and sixth legs are absent. The abdominal segments are fused.

Ecology and distribution. Most species of *Argulus* are not host specific. The most common species, *Argulus stizostethi* and *A. catastomi*, infest between 10 and 25 fish species, respectively. Food includes blood and body fluids. The genus *Argulus* is found throughout the United States in fresh and brackish water. Many species can withstand both fresh and brackish water and will remain on a fish from the estuary to well upstream. Presumably, when finished feeding, they detach from the host and drift back downstream. Generally, infestations are light with only a few specimens being found on a fish at a given time. However, extreme infestations can occur, often resulting in lesions on the fish and secondary infection. I (DGS) have observed between 25 and 100 specimens of *A. stizostethi* on individual Atlantic salmon (*Salmo salar*).

Reproduction. Reproduction takes place while *Argulus* is attached or free swimming. Females must detach from the host to lay eggs, which are deposited on some form of substrate, usually sticks, stones, or plants. Since females spend more time away from the host, males make up the majority of the parasites at a given time. Development involves several stages and is poorly known among most species. Juvenile stages in *Argulus* may include a nauplius larva or not. Juveniles pass through several molts before maturing.

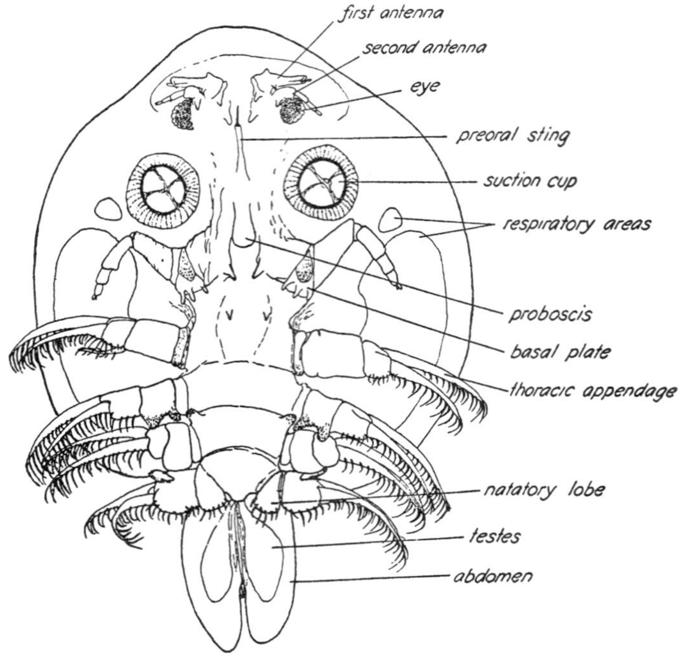

Fig. 20.1. Ventral view of *Argulus*, ×23 (Modified from Meehean 1940.)

Taxonomy. Some workers place the branchiurans with the Copepoda, but the majority consider the Branchiura to be a separate class (or subclass) within the Crustacea. The Branchiura contains a single family, the Argulidae, with about four genera worldwide. Only a single genus occurs in the United States, *Argulus*, with about 27 species. All are typically found parasitizing fish and occasionally amphibians, although individuals can be free-living.

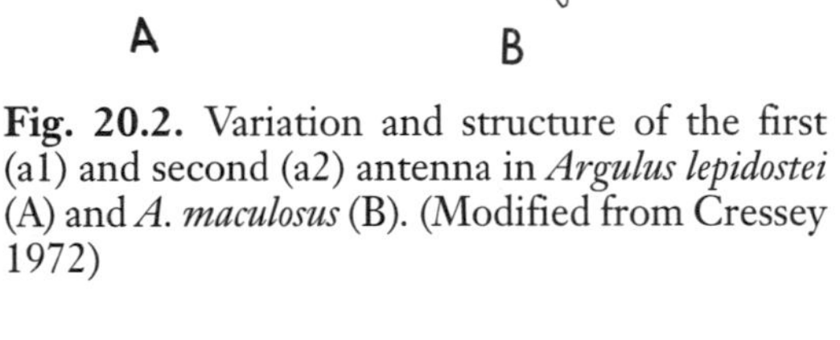

Fig. 20.2. Variation and structure of the first (a1) and second (a2) antenna in *Argulus lepidostei* (A) and *A. maculosus* (B). (Modified from Cressey 1972)

REFERENCES

Cressey, R. F. 1972. The genus *Argulus* (Crustacea: Branchiura) of the United States. *Biota Freshwater Ecosystems Identification Manual* 2: 1–14.

Fryer, G. 1982. Branchiura. In: *Synopsis and Classification of Living Organisms*, pp. 218–220, edited by S. P. Parker, McGraw-Hill, New York.

Hoffman, G. L. 1999. *Parasites of North American Freshwater Fishes*, 2d ed. 539 pp. Comstock, Ithaca, NY.

Meehean, O. L. 1940. A review of the parasitic Crustacea of the genus *Argulus* in the collections of the United States National Museum. *Proc. U. S. Nat. Mus.* 88:459–522.

Poly, W. J. 1997. Host and locality records of the fish ectoparasite, *Argulus* (Branchiura), from Ohio (U.S.A.). *Crustaceana* 70:867–874.

——. 1998. New state, host, and distribution records of the fish ectoparasite, *Argulus* (Branchiura), from Illinois (U.S.A.). *Crustaceana* 71:1–8.

Poulin, R., and G. J. FitzGerald. 1989. Male-biased sex ratio in *Argulus canadensis* Wilson, 1916 (Crustacea: Branchiura) ectoparasitic on sticklebacks. *Can. J. Zool.* 67:2078–2080.

21

OSTRACODA

(Seed shrimps)

Although ostracods are abundant and widely distributed, they have received much less attention than the Cladocera and Copepoda. They inhabit all types of substrates in both standing and running waters, including rooted vegetation, algal mats, debris, mud, sand, and rubble. In Europe, many species have been found in hyporheic and hypogean waters. Similar assemblages are beginning to be discovered in the United States, but such studies are still few in number. A few species swim about actively above the substrate, especially *Cypridopsis*, *Physocypria*, *Heterocypris*, and *Hemicypris*. Representatives of one family occur only as commensals (or parasites) on the gills of malacostracan crustaceans. Ostracods are comparatively difficult to study, chiefly because of their somewhat opaque bivalve shell (more accurately, a carapace). Identification usually involves dissection. Ostracods are richly represented as living marine groups and as fossils; indeed the fossil species record is enormous.

Superficially, the members of the Class Ostracoda resemble miniature mussels, and "mussel shrimps" is an old European vernacular name. However, this name is so easily confused with "clam shrimps," which is used for certain groups within the Branchiopoda, that it does not seem advisable to adopt it. "Seed shrimps" has been suggested as an appropriate alternative, for, indeed, without a lens ostracods do look much like small seeds.

General characteristics. The American freshwater species are seldom more than 3 mm long and usually less than 1 mm long. A South African freshwater species reaches a length of nearly 8 mm and the largest marine species is 21 mm long.

Coloration ranges from chalky white through yellow, green, gray, red, brown, and blackish. Light-colored valves are often blotched with darker colors, and species occurring among algae and rooted aquatics are usually gray, green, or brown.

Each sclerotized lime-impregnated valve consists of an inner and an outer lamella that are fused along the anterior, posterior, and ventral margins. The space between the two lamella is occupied by a thin skin fold that secretes the valve material. Dorsally, the skin folds are continuous with the main body of the animal. The valves are connected on the dorsal margin by an elastic band and may be tightly closed by a group of adductor muscle fibers passing transversely through the body of the ostracod and attached to the inner surfaces of the valves (Figs. 21.10, 21.12, 21.13). When the animal is active, however, the valves gape and the locomotor appendages protrude. The outer surface of the valves may be smooth, pitted, papillate, or setose, and the margins are often tuberculate, crenulate, or with liplike edges. There are no concentric growth lines as in some clam shrimp (Branchiopoda).

Nearly all traces of body segmentation are

lost (Figs. 21.1, 21.2), but the region corresponding to the head bears four paired appendages: first antennae (antennules), second antennae (antennae), mandibles, and maxillae. The five to seven-segmented first antennae are uniramous, an exopod being absent; they bear short, stiff, clawlike bristles for digging and climbing or long setae for swimming. The four- to six-segmented second antennae are biramous, the exopod in the freshwater Podocopida being reduced to a scale or seta; these appendages are used in locomotion and feeding, and in the male they are modified for clasping the female during copulation. The mouth is surrounded by an upper and a lower lip and two lateral mandibles. Each mandible consists of a strongly toothed and sclerotized base, a branchial plate, and a three-segmented palp. Each maxilla has a large branchial plate and four basal processes, the outermost of which is largest and palplike.

The thoracic region bears three pairs of legs. In the Limnocytheridae and Entocytheridae all pairs are morphologically similar; in the Darwinulidae the second and third are similar; and in the Cyprididae all legs are different. The first legs may be modified for mastication, prehension, copulation, or respiration (respiratory plate). The second legs usually have a long terminal claw. The third legs of the Cyprididae are bent dorsally and used in keeping the body and inner surface of the shell free of foreign material.

The abdomen is represented only by two long caudal rami (furcal rami, or furca), which are articulated to the body. Each ramus usually has two terminal claws and two terminal setae. In some species the rami are reduced or lacking.

Locomotion. The great majority of individuals are found moving about on the substrate by means of beating movements of the first and second antennae, and to some extent by kicking of the caudal rami. Such locomotion ranges from creeping and uncertain, weak, tottering movements to rapid bouncing or scurrying. Some of the Candonidae burrow superficially in soft substrates. Those species that leave the bottom and swim about actively are characterized by long, plumose antennae 1 and 2.

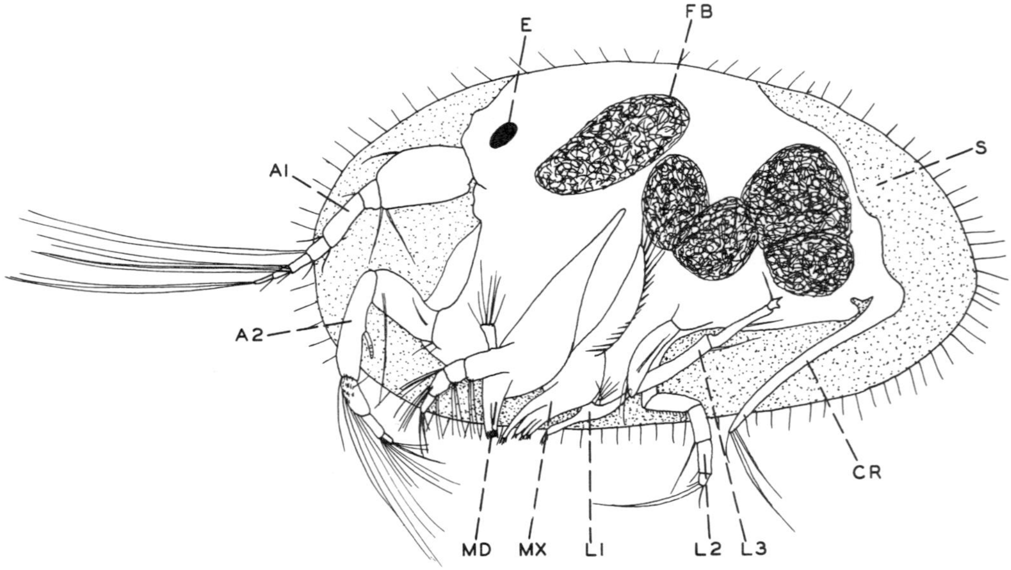

Fig. 21.1. Female *Cypricercus reticulatus*, ×80, with left valve removed to show appendages, diagrammatic. Al, first antenna; A2, second antenna; CR, caudal ramus; E, eye; FB, food ball in digestive tract; L1, first leg; L2, second leg; L3, third leg; MD, mandible; MX, maxilla; S, right valve of shell. (Modified from Hoff 1942a.)

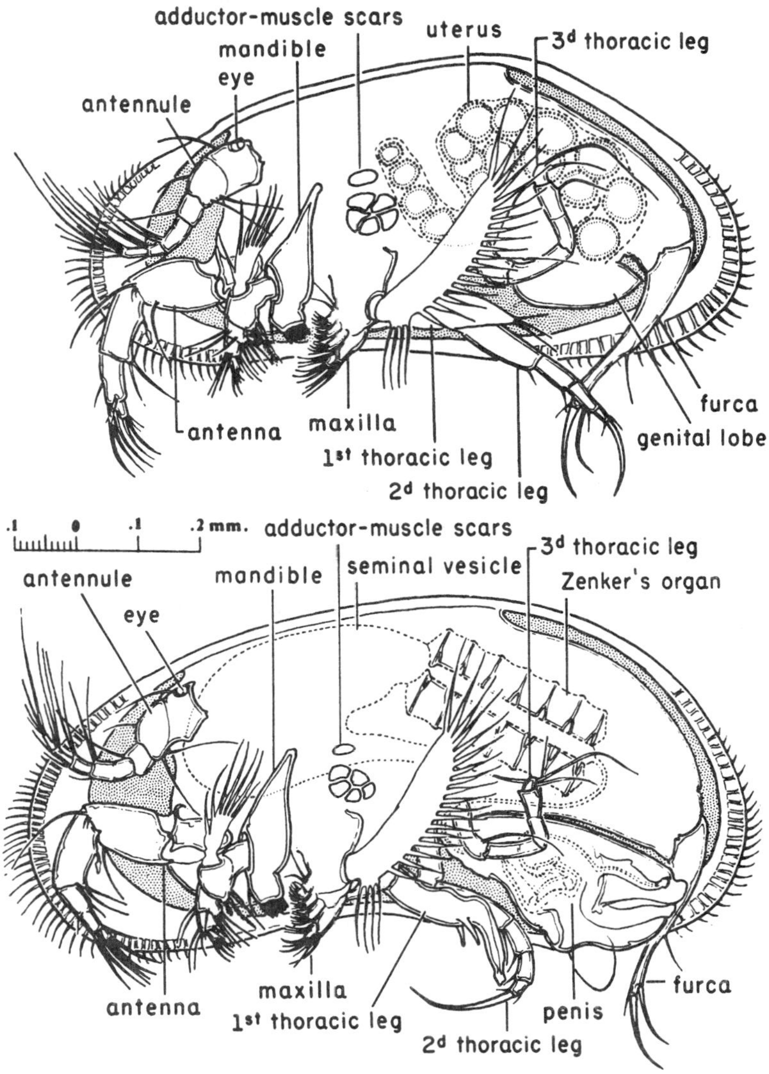

Fig. 21.2. General anatomy of *Candona suburbana*. Upper, female with left valve removed; lower, male with left valve removed. (From Kesling 1956, courtesy Turtox/Cambosco, Macmillan Science Co., Chicago, IL.)

Food, feeding. Food consists mostly of bacteria, molds, algae, and fine detritus, but some of the larger Cyprididae have been observed feeding on living and dead animals. Ecologically, ostracods are omnivorous scavengers. Beating movements of the setose mandibular palps, maxillary processes (Fig. 21.3), and the branchial plates of the first legs create a current of water between the valves. The current picks up fine particles from the substrate that are strained out by the setae and brought to the mouth. Inedible particles are mostly rejected and removed by the respiratory plates of the maxillae. Large edible bits of food are pulled or pushed toward the mouth by the antennae, mandibles, and first legs. The mandibles are used to rasp and break up the larger pieces. Small balls of finely divided food are swept into the mouth by the mandibular palps.

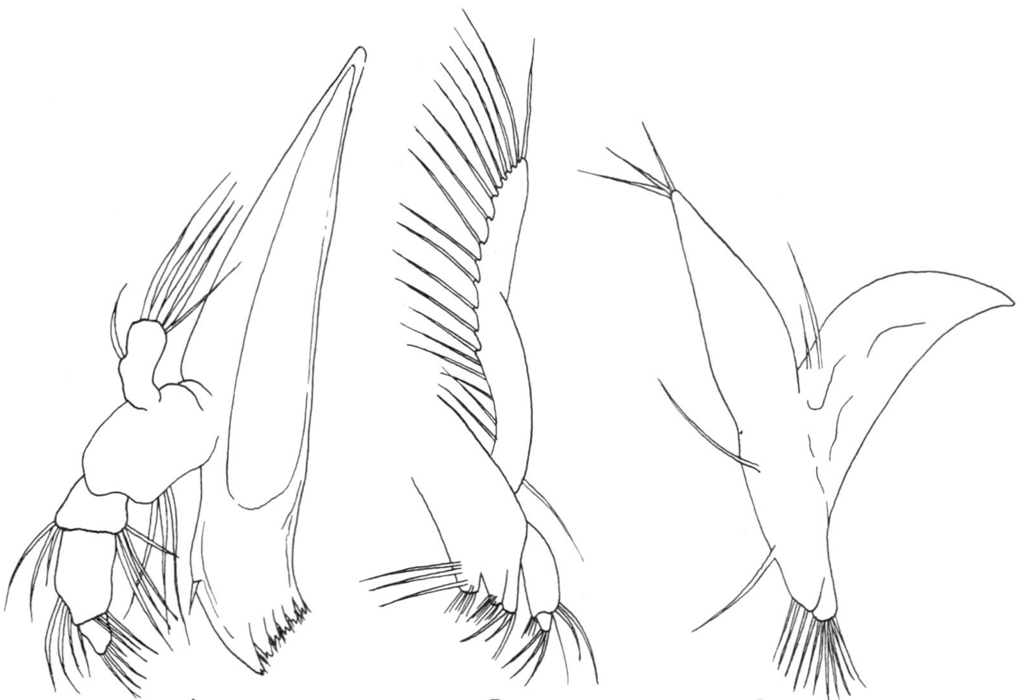

Fig. 21.3. Some ostracod appendages. A, mandible of *Cypridopsis vidua*, showing the long, sclerotized, toothed basal portion, the single-segmented branchial plate, and the larger segmented palp; B, maxilla of *Candona acuta*, showing the four basal processes and the long branchial plate; C, first leg of *C. fluviatilis*. (A modified from Sars 1928; B and C redrawn from Hoff 1942a.)

Internal anatomy. The digestive tract consists of a short esophagus, a secretory and absorptive midgut, and a hindgut (Fig. 21.4). One digestive gland on each side between the lamellae of the valves empties into the midgut. The anus lies at the base of the caudal rami.

Respiration occurs through the general body surface, and a constant supply of oxygenated water is ensured by the movements of the branchial plates. The hemocoel is filled with a circulating fluid, but no heart has ever been described for freshwater species.

Three distinct, paired glands are found in ostracods, but for no one of them has a definite function been assigned. The large convoluted shell glands lie between the shell lamellae and in the hemocoel; they open near the base of the second antennae and have an unknown function. The small antennal glands open near the base of the first antennae; presumably they have an excretory function, especially during immature stages. The small maxillary glands are imbedded in the tissues just posterior to the lower lip; some investigators believe that they have an excretory function.

The nervous system consists of a large supraesophageal ganglion (brain), two esophageal connectives, a subesophageal ganglion, and a ventral chain of two paired ganglia. Two or three prominent eyespots, each with a small lens, and fused in varying degrees, are mounted on a protuberance near the bases of the first antennae. In a few species the eyespots are absent. Many setae of the appendages undoubtedly have a tactile function. In the Cyprididae the second antenna bears a small clublike sensory structure similar to the aesthetasks of copepods (Figs. 21.1, 21.9M).

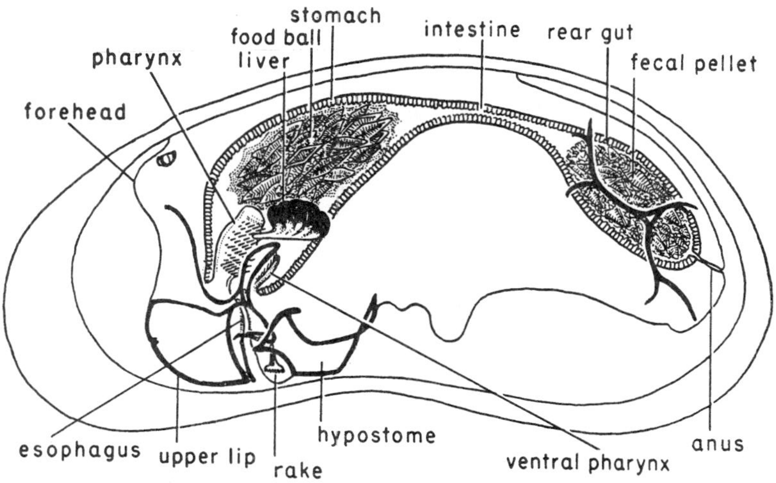

Fig. 21.4. Digestive system of *Candona suburbana*, left valve removed. Sclerotized body framework of left side is shown in heavy lines. (From Kesling 1956, courtesy Turtox/Cambosco, Macmillan Science Co., Chicago, IL.)

Reproduction. A superficial examination of any field collection of ostracods usually reveals a great majority of females. The various species differ widely in their reproductive habits, depending on the relative occurrence of males, and Martens (1998) has recognized three forms of reproduction in freshwater ostracods: fully sexual reproduction; mixed reproduction in which males occur and breed with females which are also capable of parthenogenesis; and asexual reproduction. However, the divisions separating these categories are not always absolute. Asexual reproduction is certainly very important in freshwater ostracods as, among European species, over 50% of the Cyprididae and over 25% of the Cytheridae have yet to reveal males. The Darwinulidae, especially *Darwinula,* is apparently exclusively parthenogenetic and is believed to have been so since the end of the Mesozoic Period!

The female reproductive system (Fig. 21.5A) is simple, consisting of paired tubular ovaries, oviducts, seminal receptacles, and genital openings. In the Cyprididae the ovaries lie between the valve lamellae. The genital openings are located between the bases of the third legs and the caudal rami.

The male reproductive system is more complicated, especially in the Cyprididae (Figs. 21.5B; 21.6) and Entocytheridae (Fig. 21.15). In the Cyprididae, each testis consists of four long, tubular branches lying between the valve lamellae. In the body proper these unite to form a vas deferens that enters the dorsal and posterior ejaculatory duct, a spiny, sclerotized cylinder (Zenker's organ) that forces sperm into the penis during copulation. The two hemipenes are large, irregular, lobed, complex, sclerotized structures located near the posterior end of the animal. The entocytherids have a complex copulatory apparatus in the male, involving a shieldlike peniferum containing the penial organ and associated claspers and processes (Fig. 21.15B). The detailed anatomy of the penis is of importance in separating species taxonomically. The spermatazoa are among the largest in the animal kingdom.

Reproduction is usually restricted to the summer months, but a few species produce more than one generation per year. In some northern species, eggs are laid in spring. Copulation in many other sexual species involves the attachment of the male to the dorsal and posterior margin of the female cara-

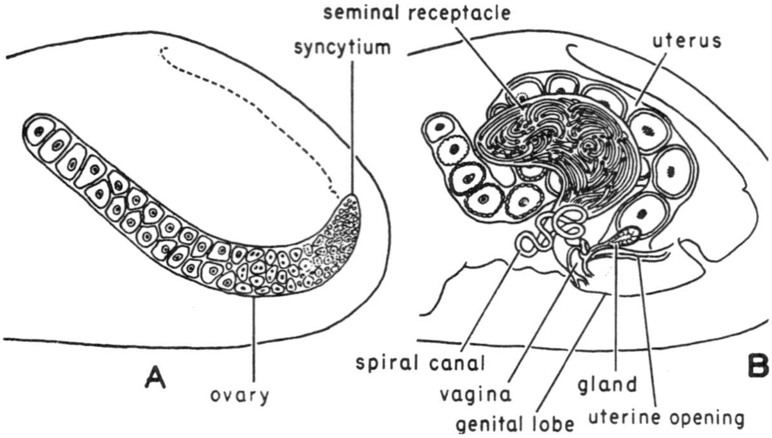

Fig. 21.5. Female ostracod reproductive system. A, ovary in position in hypodermis; B, right half of female reproductive system. (From Kesling 1956, courtesy Turtox/Cambosco, Macmillan Science, Chicago, IL.)

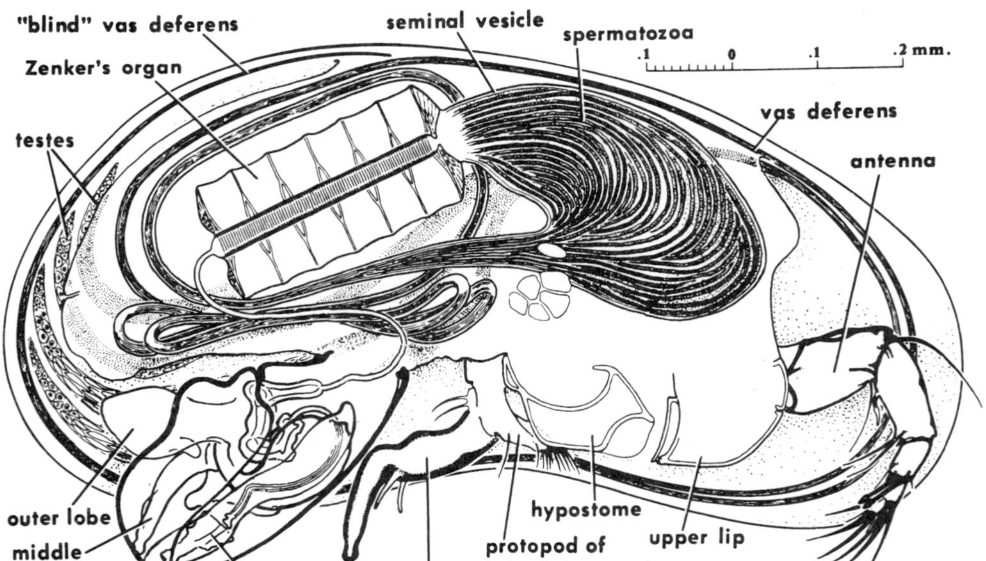

Fig. 21.6. Male reproductive system of *Candona suburbana.* Right valve and most appendages removed. Hemipenis slightly lowered from normal noncopulatory position. (From McGregor and Kesling 1969a.)

pace. The second antennae of the male aid in this attachment, and the hemipenes are inserted between the valves and into the female genital pores. Fertilization occurs in the seminal receptacles. Hatching occurs in May and growth gradually intensifies until August when adults emerge. The gravid adults decline over winter; those surviving winter lay eggs the following spring.

Darwinula stevensoni retains eggs in the dorsal part of the shell cavity until embryonic development is completed, but nearly all other American freshwater species are thought to be oviparous. The spherical eggs have a delicate, double-walled, limy shell and are usually white, yellow, orange, red, or green. They are deposited singly or in clumps or rows on rocks, twigs, other bits of debris, or aquatic vegetation.

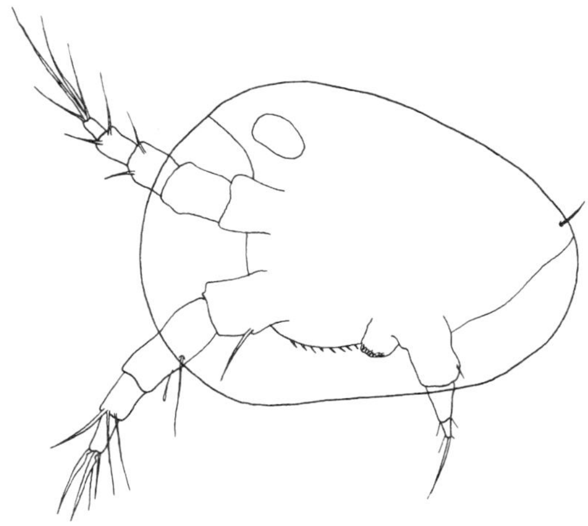

Fig. 21.7. Nauplius larva of *Cyprinotus incongruens*, ×260. (Redrawn from Schreiber 1922.)

Development, life history. Egg development is usually suspended during the cold months and unfavorable moisture conditions, and there are numerous records of ostracods hatching from eggs kept in dried pond mud in the laboratory for many weeks and months. One report indicates that viable eggs may be kept for more than 20 years in the desiccated condition.

In normal habitats, however, the eggs usually hatch in a few days to several months. The liberated larva (Fig. 21.7) is a shelled nauplius with three pairs of appendages, representing the first antennae, second antennae, and mandibles of the adult at intervals of 1 to 3 days. Eight molts and eight additional instars occur during the life history, the adult being the ninth instar. Other appendages and increased structural complexity develop with each successive molt. The caudal rami, for example, first appear as two small bristles in the fourth instar. The copulatory organs appear in the eighth instar, but sexual maturity and copulation do not occur until the ninth and last instar.

Many species exhibit a definite seasonal periodicity. Ferguson (1944), working in Missouri, found that *Cypridopsis vidua* was active between February and December, *Potamocypris smaragdina* between March and October, and *Physotypria pustulosa* between March and September. McGregor (1969a) found that *Darwinula stevensoni* had its reproductive period between May and October, while Strayer (1988b) reported that *Cypris turneri* in New Hampshire had a single annual cycle with reproduction occurring in the spring. Many species of *Candona* may be collected only in the spring; a few species occur only in the autumn. Aside from eggs and some adults that winter over, the usual life span is completed in several weeks to 7 or 8 months, depending on the species and environmental conditions. An extreme example is *Darwinula stevensoni*, which will live up to four years. Sometimes the adult instar may last as long as 6 months. A single generation per year is the common condition in vernal ponds; a few species have two or three generations per year.

The Entocytheridae have a unique life history among freshwater ostracods because they spend their entire life associated with other crustaceans. The sexes are always separate and following copulation eggs are deposited on setae of the host. After about a week or so, each egg hatches out the first instar. Seven subsequent molts (instars) occur, the final molt producing an adult. The distribution on crayfish of ostracod life stages shows no specific pattern and generally, the larger the crayfish, the more the ostracods.

Ecology. The nature of the substrate and the general type of environment seem to have little influence on the distribution of most seed shrimps. In many cases the same species may be found on algae, decaying vegetation, rooted aquatics, mud, and gravel. *Cyprinotus incongruens* is exceptional, however, and may be found in swift streams, ponds, and puddles. Only a minority of species are restricted to such places as vernal ponds and temporary streams. Out of 31 species collected in Illinois by Hoff (1942a), 4 were restricted to temporary still waters, 6 were restricted to permanent still waters, 2 were found only in temporary running waters, and 1 in permanent running waters. The other 18 species were collected in two or more of these habitats. The largest numbers of species may be collected on mud bottoms where there is little or no current, often in aquatic vegetation; few species occur on bare rocks. Habitats having wave action are usually without ostracods. Core samples taken in soft substrates show few ostracods deeper than 5 cm. Many forms, especially clamberers and active swimmers, such as *Cheterocypris* and *Bradleystrandesia*, are more active on lighted than on shaded bottoms. Most species occur in water less than 1 m deep, but some are found regularly as deep as 15 m. *Cypridopsis vidua* has been found in depths ranging from 1 cm to 300 m. *Notodromas monacha* is unusual in that it clings to the surface film with the dorsal surface downward.

On several occasions, I (RWP) have found dense clusters of ostracods, amounting to several thousand individuals, covering 100 to 200 cm^2 on the substrates of slow streams. These did not appear to be feeding aggregations, but may have been a temporary reproductive phenomenon. As a group, ostracods occur in waters having pH readings of 4.0 to 8.0, but most species are definitely restricted to alkaline conditions. A pH of less than 7.0 interferes with calcium deposition in the valves.

In general, ostracods tolerate wide ranges of ecological factors usually regarded as limiting for microcrustaceans, however, some species exhibit very narrow ranges of temperature and chemistry and their remains are important in evaluating past ecological conditions. Though they do not occur in grossly polluted waters, such genera as *Candona* and *Cypris* survive long periods of stagnation and oxygen exhaustion on lake bottoms. It is now known that advanced instars may tide over dry and cold periods by being buried in a torpid state in the mud of dried ponds. Aquaria and old protoctist cultures sometimes become thick with them.

A few hypogean species occur in the waters of underground caves and in interstitial gravels, but in the few studies that have been carried out, they seem to be far less represented than copepods and cladocerans. Several species of entocytherids are found on troglobitic crayfish and a few are adapted to cave environments. The genus *Sagittocythere* is apparently restricted to cave-dwelling crayfish.

Terrestrial species are present throughout much of the world, but have not been discovered in North America. In tropical America, the genus *Elpidium* lives in the captured water of bromeliads.

In North America, the Entocytheridae are common commensals on the gills and articular membranes of cambarid crayfishes and, to a lessor extent, troglobitic isopods. On crayfish they exist at times in great densities. Two subfamilies occur in the United States, the largest and most diverse being the Entocytherinae, which are commensals on crayfish. Although most species occur on epigean crayfishes, troglobitic species have been discovered. They may be removed from their hosts and will live under such artificial conditions for at least several months. However, they will not reproduce unless they are attached to the crayfish gills. Some host specificity occurs among entocytherids, but, most species show little preference for particular crayfish species. As many as 10 species distributed among several genera of entocytherids can be found on a single crayfish. The Sphaeromicolinae is represented in the United States by a single genus that is commensal on cirolanid isopods found in caves.

Geographical distribution. Active and passive distribution are brought about much as in the Cladocera and Copepoda. Various means of transport even include the intestine of snails. Most genera and many species are cosmopolitan or Holarctic, *Chylamydotheca* and the Entocytheridae being the only important major North American taxa not found in the Palaearctic region. Even in New Zealand most of the ostracod fauna consists of cosmopolitan species. Several species, for example *Darwinula stevensoni*, *Herpetocypris incongruens*, and *Ilyocypris gibba*, occur throughout most of the globe.

Because little intensive collecting has been done in the United States, about half of the reported species are known from only a single state or collecting locality and the known distribution of most species is very incomplete. With increasing study, however, genera once thought to be limited to other continents are being discovered in the United States, especially in hyporheic habitats. The case of the Entocytheridae is quite different, however. The subfamily Entocytherinae has received considerable study and the geographical patterns of many genera have been worked out. The entocytherids are concentrated in the Appalachian Mountain region of the United States. From here the group spreads southward and westward into the Mississippi River valley and, to a lessor extent, the Atlantic coast. Only three genera are apparently widespread (*Donaldsoncythere*, *Entocythere*, *Uncinocythere*) in the country. The subfamily Sphaeromicolinae is known only from Texas in the United States.

Economic significance. With occasional exceptions, ostracods are only a minor element in the diet of young and adult fish. A few (*Cypria* spp., *Cyclocypris* sp.) are the intermediate hosts of Acanthocephala (*Neoechinorhynchus* and *Octospinifer*) parasitizing fish. Hoff (1942) estimates that 1.0 to 1.5% of the ostracods in permanent waters are infected. Ostracods are also the intermediate hosts of some tapeworms of waterfowl.

Collection, preparation. Perhaps the chief reason for the general neglect of the American Ostracoda lies in the fact that they are nearly always necessarily collected in masses of debris from which they must be laboriously separated. Surface debris of the bottom and its contained ostracods may be scooped up with a dipnet, but the Birge cone net towed along the bottom, or better, the Birge cone net mounted on a tobogganlike piece of sheet metal is more useful. Ekman dredge samples are equally useful. By washing the debris thoroughly through a fine net or sieve, the smaller particles may be removed, and if the remaining mass is allowed to stand quietly in jars of water, the ostracods soon become active at the debris–water interface, where, with a little patience, they may be picked up under the dissecting binoculars with a long pipette or Irwin loop. Also, when the contents are stirred well, some of the ostracods get a small bubble of air caught between the valves and rise to the surface, where they may be easily removed.

Specimens should be killed in 50% alcohol and preserved in 85% alcohol. Levinson (1951) gives a good method for anesthetizing and killing with the shell gaping. A few drops of 10% chloral hydrate or 0.1% chloretone are good relaxers. Formaldehyde makes the animals brittle and may decalcify the valves.

Stained or unstained whole animals may be mounted in balsam or glycerin jelly, but cover-glass supports should be used to avoid crushing. Dissections of adult specimens are usually necessary for accurate identification, and the operation is best carried out in glycerin with fine needles. The valves should be removed first, leaving the body intact, and then the appendages should be carefully dissected off. Valves and appendages are conveniently mounted on separate slides.

Culturing. Little is known about culturing ostracods in the laboratory, but some investigators report success by feeding dried yeast and bits of lettuce, celery, or vegetable debris.

Taxonomy. The Class Ostracoda is divided into three major living taxa, the orders Myodocopida, Platycopida, and Podocopida. The first two of these are confined to marine waters and are not considered here. The Podocopida are found in both marine and freshwater environments: three super-

Table 21.1. Families and Genera of the Freshwater Ostracoda of the United States

DARWINULIDAE	CANDONIDAE (Continued)
Alicenula	*Fabaeformiscandona*
Darwinula	*Nannocandona*
Microdarwinula	*Paracandona*
Penthesilenula	*Physocypria*
CYPRIDIDAE	*Pseudocandona*
Bradleycypris	ILYOCYPRIDIDAE
Bradleystrandesia	*Ilyocypris*
Candocypria	*Pelocypris*
Candonocypris	CYTHERIDEIDAE
Chlamydotheca	*Cytherissa*
Cyclocypria	NOTODROMADIDAE
Cypretta	*Cyprois*
Cypriconcha	*Notodromus*
Cypris	LIMNOCYTHERIDAE
Dolerocypris	*Elpidium*
Eucypris	*Limnocythere*
Hemicypris	ENTOCYTHERIDAE
Herpetocypris	Sphaeromicolinae
Heterocypris	*Hobbsiella*
Ilyodromus	Entocytherinae
Isocypris	*Ascetocythere*
Prionocypris	*Cymocythere*
Scottia	*Dactylocythere*
Stenocypris	*Harpagocythere*
Strandesia	*Litocythere*
CYPRIDOPSIDAE	*Lordocythere*
Cavernocypris	*Okriocythere*
Cypridopsis	*Ornithocythere*
Potamocypris	*Phymocythere*
CANDONIDAE	*Plectocythere*
Candona	*Psittocythere*
Candocyprinotus	*Sagittocythere*
Cyclocypris	*Saurocythere*
Cypria	*Waltoncythere*

families, the Cytheroidea, Cypridoidea, and Darwinuliidea contain freshwater families.

The size of the fauna in the United States is probably a great deal higher than published surveys indicate. Ward and Whipple (1918) listed 54 species from the entire United States, but in 1933 Furtos listed 57 species from Ohio alone. Dobbin (1941) collected 33 species in Washington, of which 14 were new, and Hoff (1942a) collected 39 in Illinois, of which 10 were new. Goins (1971) reported 20 nonentocytherid species from Indiana of which two were new. There are now more than 300 species known from the United States. Delorme (1967) lists 74 Canadian species.

The larger genera include *Candona*, which contains about 40 species in the United States, and *Dactylocythere* with 46 species. Most of the other genera have less than 20 United States species and several, particularly those within the Entocytheridae, contain only 1 species in the United States. Actually, the number of genera in the United States is growing as fast as the number of species, a function largely the result of increased investigation of narrowly defined habitats. Complicating matters is the fact that there seems to be little consensus on the status of many genera and certain genera considered valid by some workers have been synonymized by others. For example, *Pseudocandona*, recently reported in the United States (Marmonier and Ward 1990), is believed to represent an artificial assemblage (Martens et al. 1992), and *Candona* is a large genus undergoing necessary revision (K. Martens, pers. comm.). Thus, certain confusing genera have been omitted from the key because they cannot be determined satisfactorily.

Many of the characters used to distinguish genera are subtle and often new species are discovered that require a revised definition of a particular genus. With this in mind, generic characters used in the key often apply only to U.S. species, particularly if only one species is represented. Consequently, the key is not going to satisfy all authorities. It should be borne in mind that the key can be used successfully only for specimens in the late instars, preferably the last. The higher classification of freshwater ostracods is equally inconsistent. The classification of the non-entocytherids used in this manual largely follows Martens and Behen (1994), Martens et al. (1998), and K. Martens (pers. comm.). Aspects of the key have been modified from Henderson (1990) and Meisch (1996). The family Notodromadidae is recognized in the key. The cypridid *Dolerocypris fasciata* is known from Canada but not the contiguous United States, thus the genus is omitted from the key. The key to darwinulids is modified from Rossetti and Martens (1996). The account of Meisch (2000) should be a significant reference for American workers.

Most of the ostracod research during the past 20 years in the United States has been restricted to the Entocytheridae (Entocytherinae) and the systematics of the family appears to be relatively stable. Currently, there are 22 described genera and 179 species of entocytherids in North America. The key to genera requires adult males and is after Hart and Hart (1974), kindly modified by D. J. Peters.

KEY TO GENERA OF FRESHWATER OSTRACODA OF THE UNITED STATES

1. Exopod of second antenna in the form of a long, hollow seta carrying the secretion from a gland located at the base of the antenna (Fig. 21.13E); all legs morphologically similar **39**
   Exopod otherwise; all legs not morphologically similar **2**
2. Second and third legs similar (Fig. 21.9A, B); caudal rami lacking, body terminating in a single process (Fig. 21.9C); 0.70 to 0.80 mm long; on mud bottoms of larger bodies of water; rare **DARWINULIDAE, 3**
   Second and third legs not similar (Fig. 21.9D, E); caudal rami present but sometimes reduced **6**
3. Carapace rounded in lateral view (Fig. 21.8A), no externally visible brood pouch; 1 species, Great Lakes region .... ***Microdarwinula***
   Carapace elongate or sub-squarish in lateral view (Fig. 21.8B), with externally visible brood pouch **4**
4. Left valve without internal teeth; last segment of mandible palp with less than five claws (3 or 4), penultimate segment with 3 or 4 terminal setae (Fig. 21.8C), if 4 than only one long **5**
   Left valve with internal teeth; last segment of mandible palp with 5 claws, penultimate segment with four long setae (Fig. 21.8D); a single undetermined species, FL ...... ***Penthesilenula***

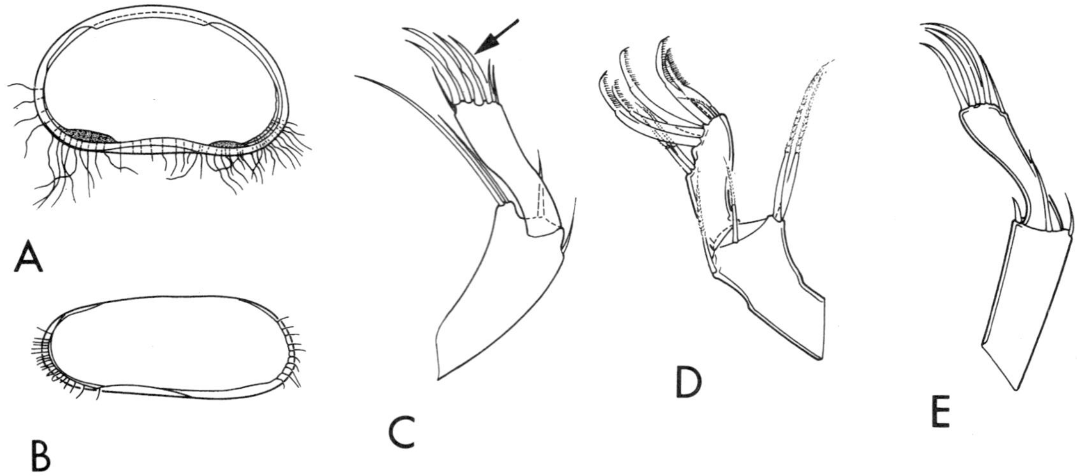

Fig. 21.8. Features of Darwinulidae. A, shell of *Microdarwinula;* B, shell of *Alicenula;* C, mandibular palp of *Darwinula*, arrow denotes claws; D, mandibular palp of *Penthesilenula;* E, mandibular palp of *Alicenula.* (A–E modified from Rossetti and Martens 1998.)

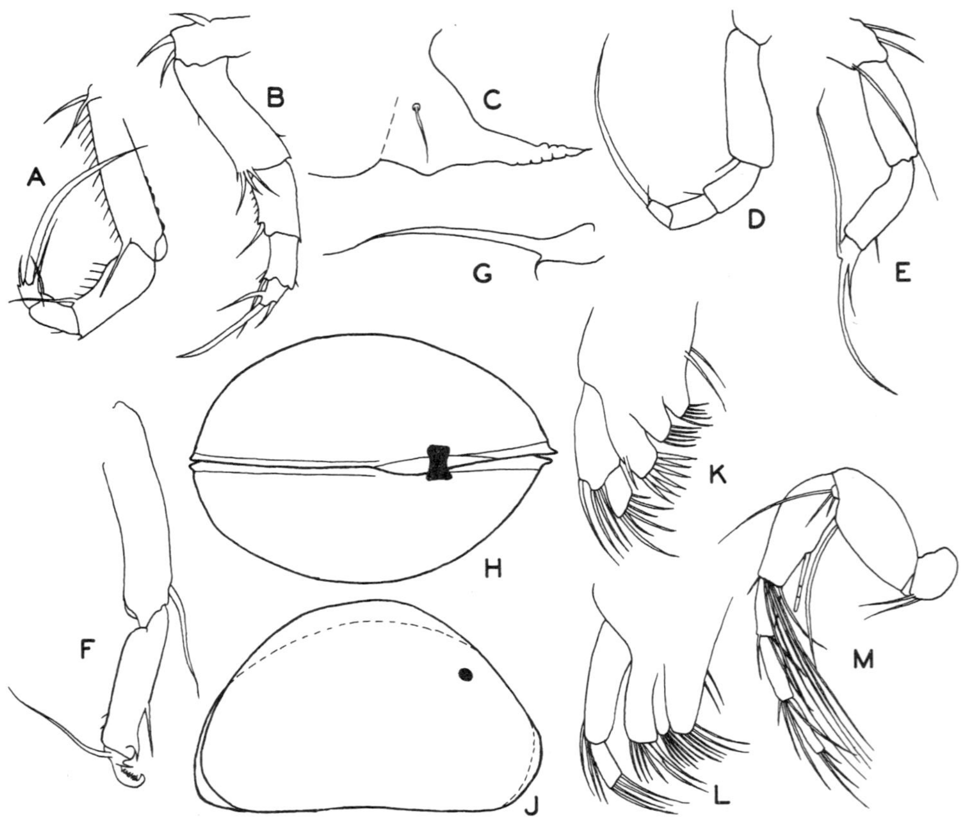

Fig. 21.9. Structure of Ostracoda. A, second leg of female *Darwinula stevensoni;* B, third leg of same; C, end of thorax of same; D, second leg of female *Candona punctata;* E, third leg of same; F, third leg of female *Cypridopsis vidua;* G, caudal ramus of female *Potamocypris smaragdina*; H, dorsal view of female *Cypridopsis vidua*, ×70; J, lateral view of female *Potamocypris hyboforma*, ×40; K, maxilla of *Cyprois marginata;* L, maxilla of female *Heterocypris incongruens;* M, second antenna of female *Notodromus monacha.* (A, B, F, and G redrawn from Furtos 1933; C–E redrawn from Hoff 1942a; H modified from Furtos 1933; J and L modified from Dobbin 1941.)

5. Right valve overlapping left valve; mandible palp penultimate segment with 4 setae (Fig. 21.8C); length greater than 0.65 mm; 1 widely distributed but rare species . . . . . *Darwinula*
Either valve overlapping other; mandible palp penultimate segment with 3 setae (Fig. 21.8E); length less than or equal to 0.65 mm; 1 rare species, FL . . . . *Alicenula*
6. Valves are thick and pitted; lateral surface of valve with vertical folds, dorsal hinge straight; natatory setae of second antenna used for swimming but extending only slightly beyond tips of terminal claws . . . . **ILYOCYPRIDIDAE, 7**
Valves are not thickened, surface not pitted, dorsal hinge arched **8**
7. Margins of shell without mammillary pustules (Fig. 21.12A); several rare species in streams *Ilyocypris*
Margins of shell with numerous mammillary pustules (Fig. 21.11B); rare . . . . *Pelocypris*
8. Furca reduced to setae or missing . . . . **CYPRIDOPSIDAE, 9**
Furca well developed with terminal claws **11**
9. Right valve broadly overlapping left ventrally (Fig. 21.9J), common in standing and running waters and springs . . . . *Potamocypris*
Left valve slightly overlapping right (Fig. 21.9H) **10**
10. Pore canals not prominent along margins (Fig. 21.9H), widely distributed and very common in many types of habitats . . . . *Cypridopsis*
Pore canals prominent along margin, 1 or 2 interstitial species, CO, MT . . . . . . *Cavernocypris*
11. Segment 4 of leg 3 (cleaning leg) very small and obscure, without setae **CYPRIDIDAE, 22**
Segment 4 of leg 3 well developed and with 3 setae **12**
12. Furca with 3 terminal claws but missing either apical or dorsal setae (Fig. 21.9K) **NOTODROMADIDAE, 13**
Furca with 2 terminal claws and apical and dorsal setae . . . . **CANDONIDAE, 14**
13. Second antenna six segmented (Fig. 21.9M); 1.10 to 1.20 mm long; active swimmers; rare; usually in aquatic vegetation . . . . *Notodromus*
Second antenna five segmented (Fig. 21.10A); 1.10 to 1.70 mm long; widely distributed in creeks and vernal ponds . . . . *Cyprois*
14. Natatory setae of second antenna lacking (Fig. 21.10M); shell white when dry; creeping and burrowing forms **15**
Natatory setae of second antenna present; shell not white when dry **20**
15. Shell ornamented with polygonal areas and tubercles (Fig. 21.12D); 0.50 to 0.60 mm long; rare, scattered distribution . . . . *Paracandona*
Shell not ornamented (Figs. 21.12E; 21.14D) **16**
16. Furca short and stout, 1 subterranean species, CO . . . . *Nannocandona*
Furca elongate and slender, surface and subterranean species **17**
17. Leg 3 (cleaning leg) basal segment with 3 setae, status in U.S. unclear, possibly 4 species, CO *Pseudocandona*
Leg 3 basal segment with 2 setae (Fig. 21.11C) **18**
18. Carapace broad in dorsal view **19**
Carapace laterally compressed in dorsal view; 2 species in interstitial habitats, CO (Fig. 21.14D) . . . . *Fabaeformiscandona*
19. Terminal segment of third thoracic leg with three unequal setae (Fig. 21.11C); widely distributed and very common; about 40 species . . . . *Candona*
Terminal segment of third thoracic leg with two setae; rare, FL . . . . *Candocyprinotus*
20. Shell tumid (Fig. 21.12B); last segment of third leg at least twice as long as wide (Fig. 21.12H); 0.80 to 1.40 mm long; widely distributed . . . . *Cyclocypris*
Shell compressed; last segment of third leg less than twice as long as wide (Fig. 21.12J) **21**
21. Valves nearly equal in size; margins of both valves smooth (Fig. 21.13A); 0.40 to 1.00 mm long; many widely distributed species . . . . *Cypria*
Valves unequal in height or length or both; margin of either valve, usually right valve, more or less tuberculate (Fig. 21.13B, C); 0.40 to 0.70 mm long; widely distributed . . . . *Physocypria*
22. Third leg with a cylindrical terminal segment without a claw and armed with three setae, one of which is reflexed (Fig. 21.13D) **23**
Third leg with a short terminal segment, a short beaklike claw, and with or without a reflexed seta (Fig. 21.11D) **24**

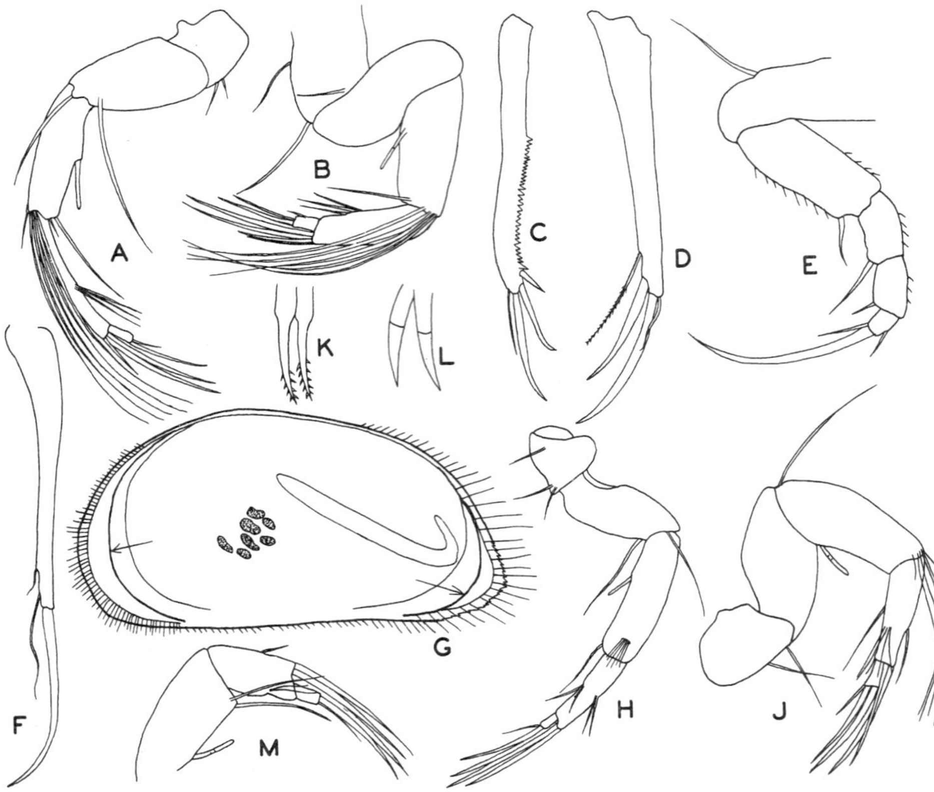

Fig. 21.10. Structure of Ostracoda. A, second antenna of female *Cyprois marginata;* B, second antenna of female *Heterocypris incongruens;* C, caudal ramus of *Ilyodromus pectinatus;* D, caudal ramus of female *H. incongruens;* E, second leg of female *Eucypris virens;* F, caudal ramus of *Cypretta turgida;* G, lateral view of female *Candonocypris pugionis,* ×15 (anterior and posterior ridges indicated by arrows); H, second antenna of female *Herpetocypris reptans;* J, second antenna of female *Prionocypris longiforma;* K, spines of third maxillary process of *H. reptans;* L, spines of third maxillary process of *Cypriconcha gigantea;* M, end of female second antenna of *Candona caudata.* (B, H, J, and M modified from Dobbin 1941; C redrawn from Ward and Whipple 1918; D redrawn from Furtos 1933; E modified from Sars 1928; F redrawn from Furtos 1933, 1936; G modified from Furtos 1936; K and L redrawn from Dobbin 1941.)

23. Natatory setae of second antenna rudimentary (Fig. 21.12F); 0.80 to 1.00 mm long; a burrower and creeper; reported from brooks in OH . . . . ***Candocypria***

Natatory setae of second antenna well developed; with two large setae on the penultimate segment of third leg (Fig. 21.13D); 0.45 mm long; WA . . . . ***Cyclocypria***

24. Shell with reticulate patterns anteriorly and posteriorly; caudal ramus with three claws and one seta (Fig. 21.10C); 1.10 to 1.20 mm long; rare in ponds and slow streams . . . . . ***Ilyodromus***

Shell smooth (margins tuberculate in some species) **25**

25. Caudal rami slightly different in shape (Fig. 21.11A); similar to *Isocypris;* one species, OH ***Stenocypris***

Caudal rami similar **26**

26. Natatory setae of second antenna well developed, extending beyond midlength of claws (Fig. 21.10B) **27**

Natatory setae of second antenna greatly reduced, not attaining midlength of claws (Fig. 21.10H, J) **35**

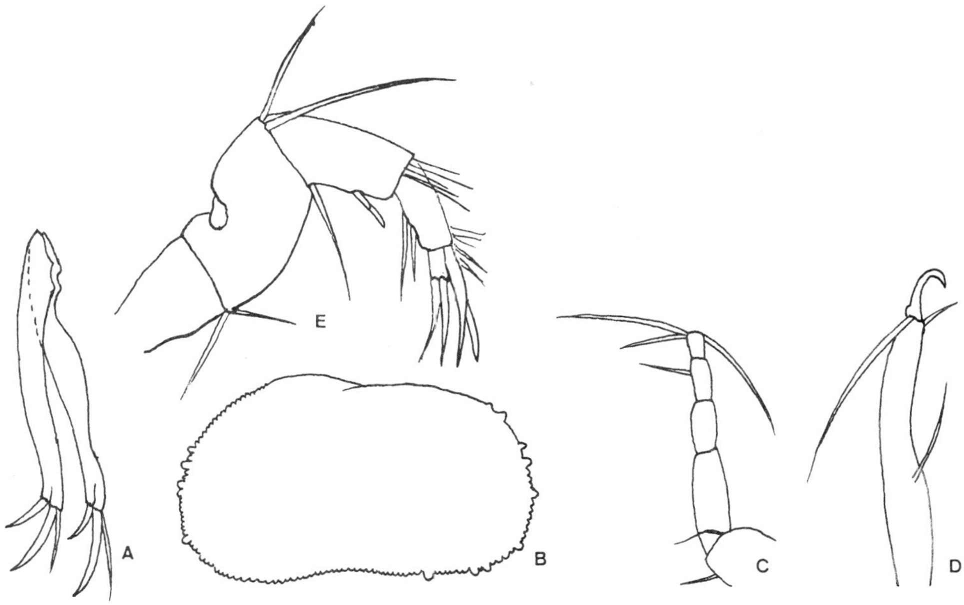

Fig. 21.11. Structure of Ostracoda. A, caudal rami of *Stenocypris;* B, outline of *Pelocypris* shell; C, third leg of *Candona;* D, tip of third leg of *Herpetocypris;* E, second antenna of female *Scottia.*

27. Second segment of second leg with two large setae; shell compressed anteriorly, somewhat flared (Fig. 21.14A, B), 2.60 to 3.30 mm long; several uncommon species in southern half of the United States . . . . **_Chlamydotheca_**
Second segment of second leg with a single large seta (Fig. 21.10E); 0.50 to 4.20 mm long **28**

28. Caudal ramus usually less than one half as long as valves; males usually absent; common and widely distributed **29**
Caudal ramus usually at least one-half as long as valves; males present; 0.80 to 4.20 mm long **33**

29. Subterminal claw of caudal ramus stout, usually nearly as long as terminal claw (Fig. 21.10D); 1.00 to 2.30 mm long; very common; many species **30**
Subterminal claw of caudal ramus setalike, usually much shorter than terminal claw (Fig. 21.10F); 0.50 to 0.95 mm long; several uncommon U.S. species . . . . **_Cypretta_**

30. Second leg four segmented, third and fourth segments fused . . . . **_Cypris_**
Second leg five segmented (Fig. 21.10E) **31**

31. Margins of both valves smooth; about 10 species . . . . **_Eucypris_**
At least the anterior edge of one valve tuberculate **32**

32. Right valve tuberculated, nearly 20 species . . . . **_Heterocypris_**
Left valve tuberculated, few species . . . . **_Hemicypris_**

33. Shell without two long posterior setae **34**
Shell with two long posterior setae (Fig. 21.14C); similar to *Stenocypris;* 1 species in U.S. **_Isocypris_**

34. Anterior and posterior extremity of each valve with distinct ridge between inner duplicature and submarginal line Fig. 21.10G); several scattered and rare southern species **_Candonocypris_**
Anterior and posterior extremity of each valve without such a ridge; about 15 species; widely distributed and common (includes *Bradleycypris*) . . . . **_Bradleystrandesia_**

35. Natatory setae of second antenna usually very small (Fig. 21.10H) **38**
Natatory setae of second antenna longer, reaching nearly halfway to tip of terminal claws (Fig. 21.10J); less than 3 mm long; several widely distributed but uncommon species **36**

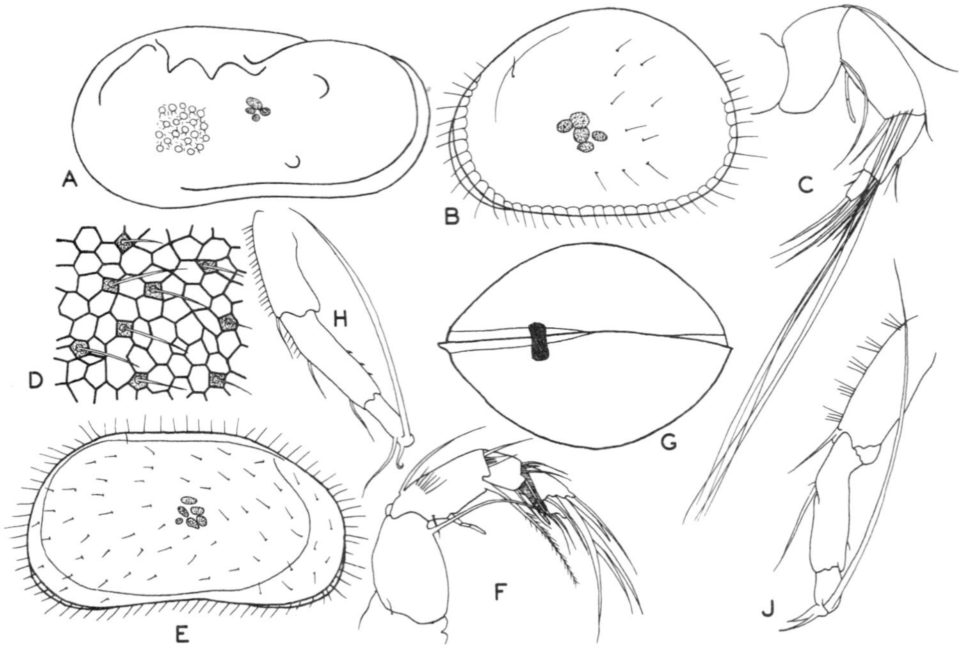

Fig. 21.12. Structure of Ostracoda. A, left valve of *Ilyocypris gibba*, ×47, showing a small portion of sculpturing; B, left valve of *Cyclocypris sharpei*, ×75; C, second antenna of female *Cypria elegantula;* D, valve sculpturing of *Paracandona euplectella;* E, right valve of female *Candona parallela*, ×55; F, second antenna of male *Candocypria osburni;* G, dorsal view of female *Cyclocypris sharpei*, ×75; H, third leg of female *C. sharpei;* J, third leg of female *Cypria elegantula.* (A, B, D, E, and G modified from Furtos 1933; F, H, and J redrawn from Furtos 1933; C modified from Dobbin 1941.)

36. Second antenna robust (Fig. 21.11E); two genera with high diversity on other continents, rare in the U.S. **37**

Second antennae not robust (Fig. 21.10H); valves 1.25 mm long; known only from WA ***Prionocypris***

37. Dorsal surface of shell without dorsal spine; known only from a TN spring . . . . ***Scottia***

Dorsal surface of shell with a prominent spine (Fig. 21.14E); 1 species, FL . . . . . . ***Strandesia***

38. Spines of third maxillary process distinctly denticulate (Fig. 21.10K); less than 2.5 mm long ***Herpetocypris***

Spines of third maxillary process smooth (Fig. 21.10L); 2.5 to 5.0 mm long . . . . . ***Cypriconcha***

39. Dorsal margin ascending towards anterior, ventral margin not distinctly concave; one species in deep, cold lakes . . . . **CYTHERIDEIDAE, *Cytherissa***

Dorsal margin horizontal or rounded, ventral margin distinctly concave or not **40**

40. Margins or ventral surface of valves generally flattened, with long pore canals (Fig. 21.13F); valves often with protuberances or sculpture; free-living in a variety of habitats; 0.50 to 0.90 mm long **LIMNOCYTHERIDAE, 41**

Free margins of valves not flattened, without many long pore canals; valve usually without strong protuberances; commensals on gills and articular membranes of crayfish, occasionally free-living; 0.30 to 0.70 mm long . . . . **ENTOCYTHERIDAE, 42**

41. Shell somewhat flattened, typically ornamented, dorsal margin of shell without teeth; possibly several widespread species (Fig. 21.13F) . . . . ***Limnocythere***

Shell broad (Fig. 21.14F, G) not strongly ornamented, dorsal margin of one valve with teeth; 1 species living in bromeliads; FL . . . . ***Elpidium***

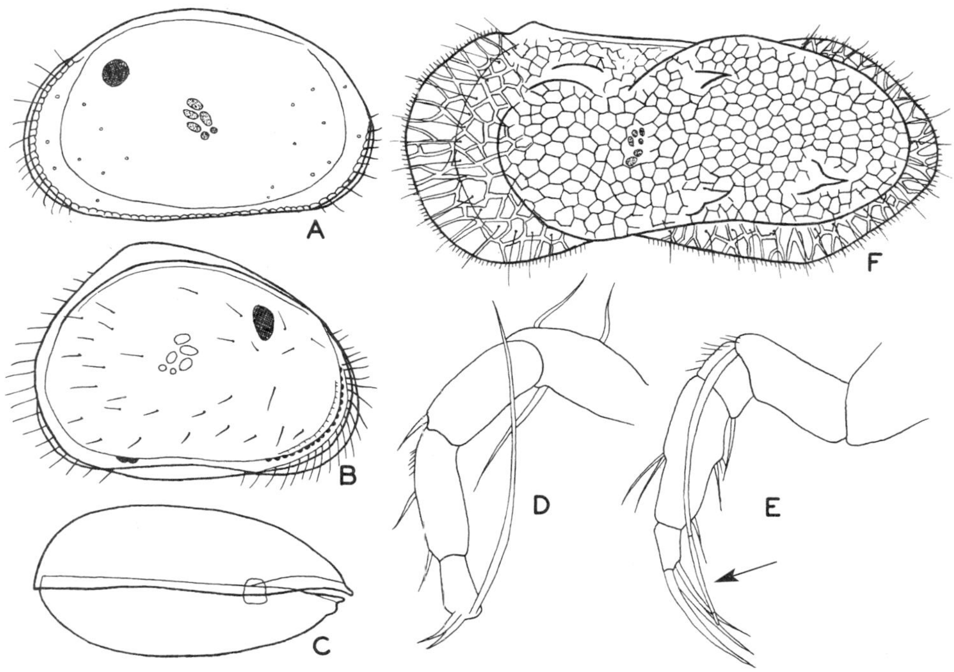

Fig. 21.13. Structure of Ostracoda. A, left view of male *Cypria obesa*, ×60; B, right view of female *Physocypria pustulosa*, ×70; C, dorsal view of same, ×70; D, third leg of female *Cyclocypria kincaidia;* E, second antenna of female *Limnocythere verrucosa*; F, left valve of male *Limnocythere ornata*, ×80. Arrow, exopod. (A, C, and F redrawn from Furtos 1933; B modified from Furtos 1933; D redrawn from Dobbin 1941; E redrawn from Hoff 1942a.)

42. Male copulatory apparatus elongate, somewhat narrow (Fig. 21.15B, C); commensals of crayfish (Cambaridae) and shrimp (Palaemonidae) .... **Entocytherinae, 43**
Male copulatory apparatus compressed, broad (Fig. 21.15D); commensal of troglobitic ispods (Cirolanidae); 1 species, TX .... **Sphaeromicolinae, *Hobbsiella***
43. Penis with prostatic and spermatic elements widely separated along much of their lengths **44**
Penis simple; or if two elements recognizable, contiguous along their entire lengths **46**
44. Ventral portion of peniferum tapering with tip of penis reaching or almost reaching apex; 4 species .... ***Plectocythere***
Ventral portion of peniferum usually rounded or with one or more prominences, seldom tapering; if tapering, tip of penis never approaching apex **45**
45. Ventral portion of peniferum rounded, without prominences; 2 species ...... ***Phymocythere***
Ventral portion of peniferum with one or more prominences ventrally and/or anteriorly (Fig. 21.15C); 19 species .... ***Ascetocythere***
46. Penis directed posteroventrally from base (Fig. 21.16A); 1 species .... ***Lordocythere***
Penis directed anteroventrally from base **47**
47. Finger guard present **48**
Finger guard absent **53**
48. Peniferum with accessory groove, except in *Dactylocythere leptophylax* where finger guard is always slender and trifid **49**
Peniferum without accessory groove; finger guard never slender and trifid **50**

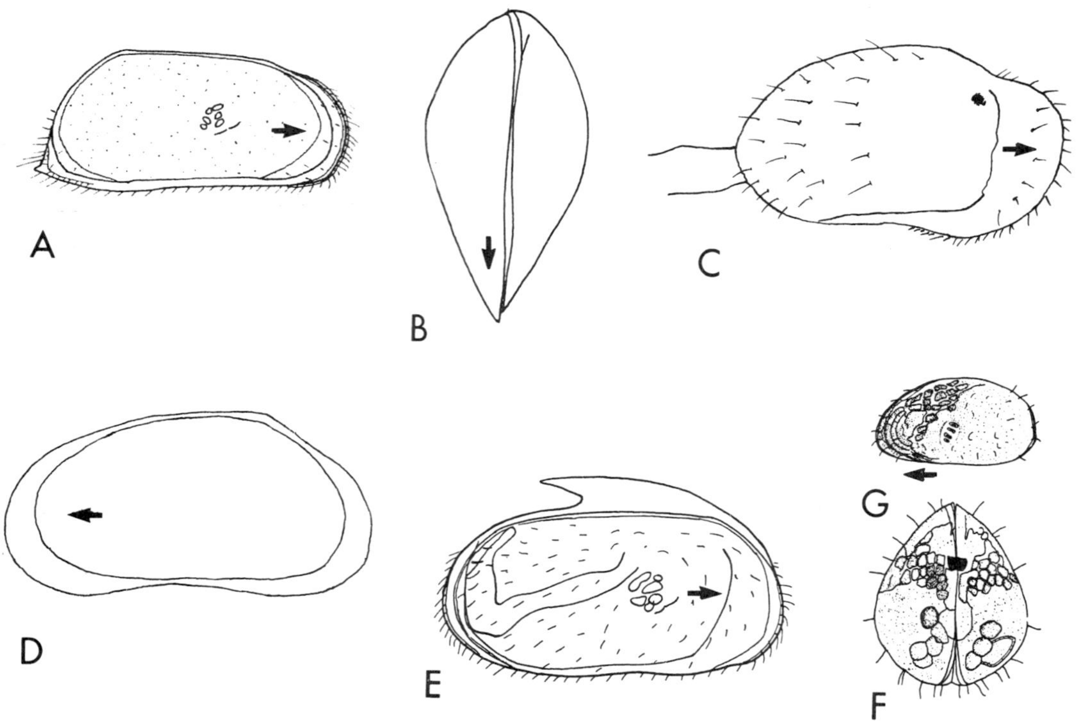

Fig. 21.14. Shells of ostracods. A, lateral view of *Chlamydotheca;* B, dorsal view same; C, lateral view *Isocypris beauchampi;* D, lateral view of *Fabaeformiscandona wegelini;* E, lateral view of *Strandesia bicuspis;* F, dorsal view of *Elpidium maricaoensis;* G, lateral view of same. Arrows point anteriorly. (A, B, E–G from Tressler 1959; C redrawn from Mastrantuono 1989; D modified from Rogulj et al. 1994.)

49. Posteroventral portion of peniferum terminating in barbed point; 2 species from caves **_Sagittocythere_**
Posteroventral portion of peniferum variable, but never ending in barbed point (Fig. 21.16B); 46 species . . . . **_Dactylocythere_**

50. Anteroventral portion of peniferum with slender anteroventral projection (Fig. 21.16D); 1 species . . . . **_Waltoncythere_**
Anteroventral portion of peniferum without slender anteroventral projection **51**

51. Ventral portion of peniferum bulbiform, clasping apparatus never extending so far ventrally as does peniferum (Fig. 21.16C); 3 species . . . . **_Cymocythere_**
Ventral portion of peniferum slender or strongly flattened; clasping apparatus extending ventrally to or beyond ventral extremity of peniferum **52**

52. Ventral portion of peniferum slender, terminating in small recurved projection (Fig. 21.16E); 3 species . . . . **_Harpagocythere_**
Ventral portion of peniferum flattened and with concave ventral border (Fig. 21.16F); 1 species . . . . **_Litocythere_**

53. Peniferum bearing single, acute beak-like projection directed anterodorsally **54**
Anteroventral portion of peniferum never with beak-like projection directed anterodorsally **55**

54. Anteroventral portion of peniferum bearing single, acute beaklike projection (Fig. 21.16G); 8 species . . . . **_Ornithocythere_**
Posteroventral portion of peniferum bearing single, acute hooklike projection; 1 species **_Psittocythere_**

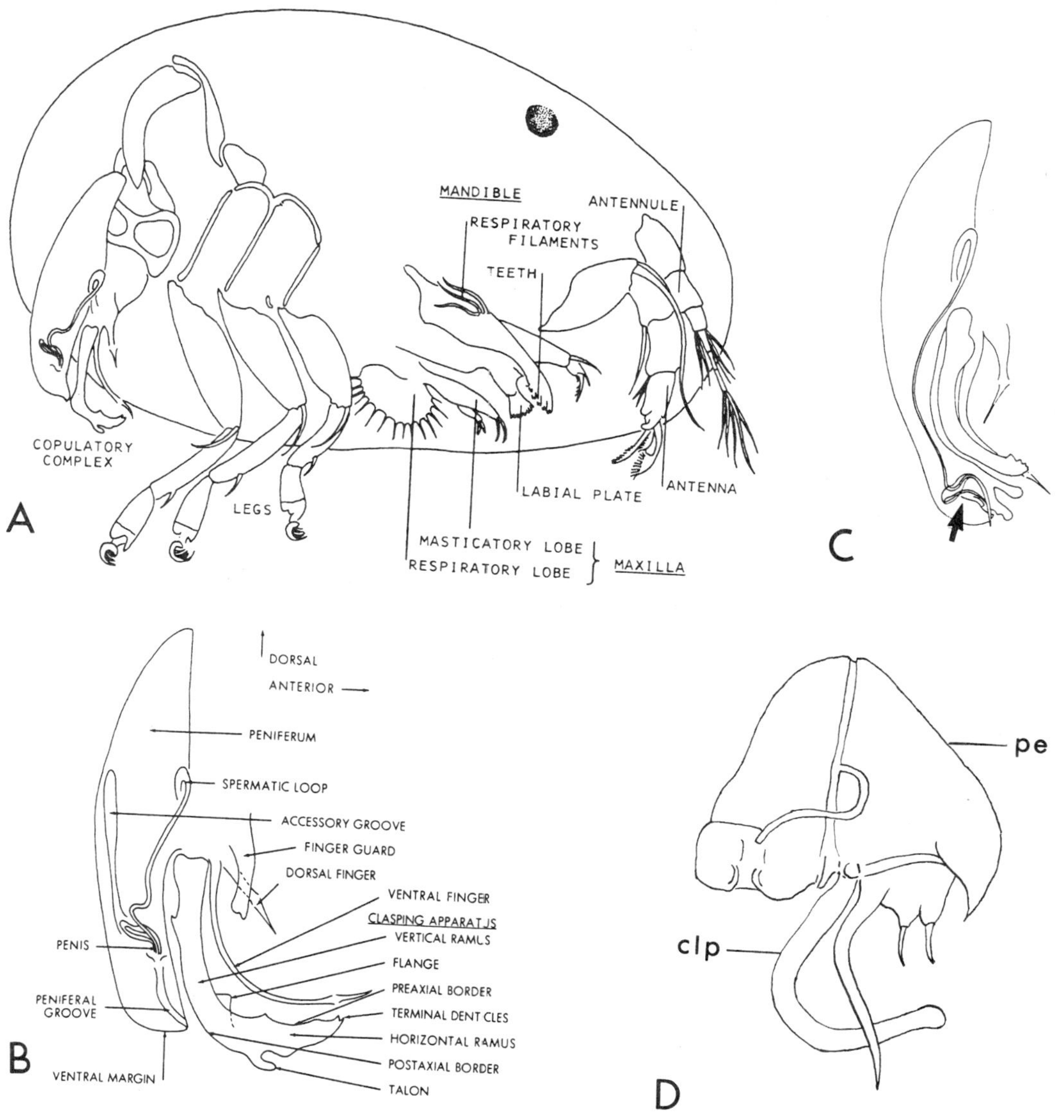

Fig. 21.15. Structure of entocytherid ostracods. A, adult male, ×50; B, copulatory apparatus of a male entocytherid; C, copulatory apparatus of *Ascetocythere*, arrow, penial and prostatic canals; D, copulatory apparatus of *Hobbsiella*. clp, clasping apparatus; pe, peniferum. (A–C after Hobbs and Peters 1977; D redrawn from Hart 1978.)

55. External border of horizontal ramus of clasping apparatus with one or more excrescences . . . **56**
External border of horizontal ramus of clasping apparatus entire or with a few shallow subapical grooves . . . **59**

56. Anteroventral portion of peniferum produced ventrally in rounded lobe . . . **57**
Anteroventral portion of peniferum never produced ventrally in rounded lobe; if produced, apex acute or truncate . . . **58**

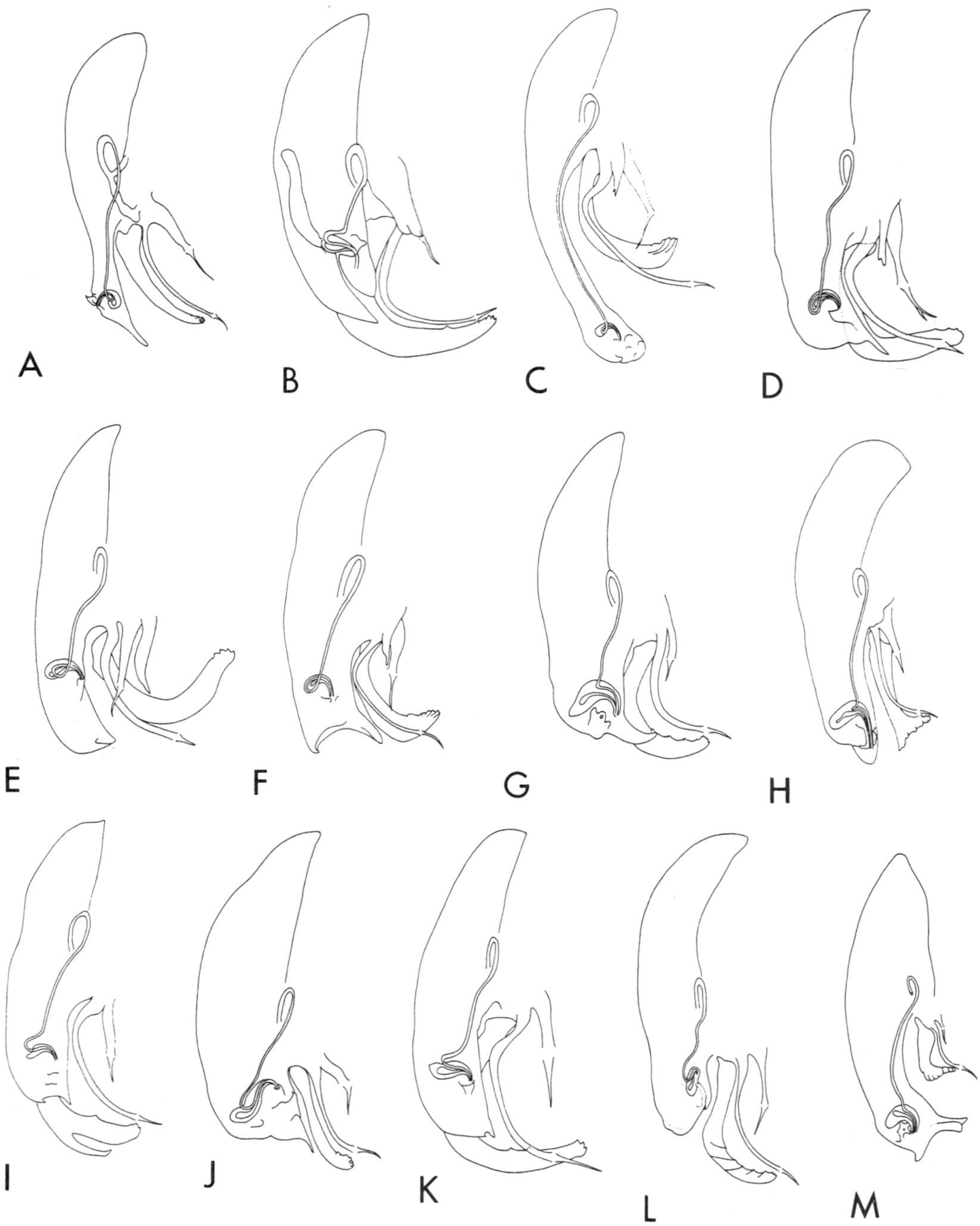

Fig. 21.16. Copulatory apparatus of entocytherids. A, *Lordocythere;* B. *Dactylocythere;* C, *Cymocythere;* D, *Waltoncythere;* E, *Harpagocythere;* F, *Litocythere;* G, *Ornithocythere;* H, *Okricythere;* I, *Ankylocythere;* J, *Donnaldsoncythere;* K, *Uncinocythere;* L, *Entocythere;* M, *Geocythere.* (A–L after Hobbs and Peters 1977; M after Hobbs and Hobbs 1970.)

57. Spermatic loop horizontal; peniferum distal to dorsal margin of spermatic loop at least twice as long as portion dorsal to loop; clasping apparatus with external border bearing single tubercle and terminating in fan-like cluster of serrations; 1 species . . . . ***Saurocythere***
Spermatic loop vertical; peniferum distal to dorsal margin of spermatic loop much less than twice as long as portion dorsal to loop; clasping apparatus with external border broadly serrate and terminating in annulations (Fig. 21.16H); 8 species . . . . ***Okriocythere***

58. Anteroventral portion of peniferum with conspicuous anterodorsally directed projection **60**
Anteroventral portion of peniferum never with conspicuous anterodorsally directed projection (Fig. 21.16I); 27 species . . . . ***Ankylocythere***

59. Internal border of clasping apparatus usually with no more than three teeth; if more than three, with only two apical denticles or vertical ramus strongly convex posteriorly **61**
Internal border of clasping apparatus with more than three teeth, apical cluster with more than two denticles, vertical ramus straight **62**

60. Clasping apparatus not clearly divisible into vertical and horizontal rami, extremities directed at angle of at least 100 degrees (Fig. 21.16J); 14 species . . . . ***Donnaldsoncythere***
Clasping apparatus usually divisible into vertical and horizontal rami, extremities directed at angle of no more than 90 degrees **63**

61. Penis large, S-shaped or sinuous, and with curved posteroventral thickening of peniferum giving forcipate appearance to ventral portion of peniferum; 1 species . . . . ***Thermastrocythere***
Penis of moderate size, L-shaped, and never so disposed as to contribute forcipate appearance to ventral portion of peniferum (Fig. 21.16K); 24 species . . . . ***Uncinocythere***

62. Ventral portion of peniferum tapering to slender tip; 1 species . . . . ***Rhadinocythere***
Ventral portion of peniferum never slender nor tapering (Fig. 21.16L); 16 species ***Entocythere***

63. Anteroventral portion of peniferum entire (Fig. 21.16M); 4 species . . . . ***Geocythere***
Anteroventral portion of peniferum bifid; 1 species . . . . ***Hartocythere***

REFERENCES

Andolshek, M. D., and H. H. Hobbs, Jr. 1986. The entocytherid ostracod fauna of southeastern Georgia. *Smithson. Contrib. Zool.* 424: 1–43.

Butlin, R., et al. 1998. Asexual reproduction in nonmarine ostracods. *Heredity* 81:473–480.

Danielopol, D. L. 1978. Über Herkunft und Morphologie der Süsswasser-hypogäischen Candoninae (Crustacea, Ostracoda). *Sitzungsber. Öster. Akad. Wiss. Math. Naturw. Kl. Abt.* 1:1–160.

Danielopol, D. L., et al. 1994. World subterranean ostracod biogeography: dispersal or vicariance. *Hydrobiologia* 287:119–129.

Delorme, L. D. 1967. Field key and methods of collecting freshwater ostracodes in Canada. *Can. J. Zool.* 45:1275–1281.

——. 1969. On the identity of the ostracode genera *Cypriconcha* and *Megalocypris*. *Can. J. Zool.* 47:271–281.

——. 1970a. Freshwater ostracodes of Canada, Part I: Subfamily Cypridinae. *Can. J. Zool.* 48:153–168.

——. 1970b. Freshwater ostracodes of Canada, Part II: Subfamily Cypriclopsinae and Herpetocypridinae, and Family Cyclocyprididae. *Can. J. Zool.* 253–266.

——. 1970c. Freshwater ostracodes of Canada, Part III: Family Candonidae. *Can. J. Zool.* 1099–1127.

——. 1970d. Freshwater ostracodes of Canada, Part IV: Families Ilyocyprididae, Notodromadidae, Darwinulidae, Cytherideidae, and Entocytheridae. *Can. J. Zool.* 1251–1259.

——. 1971. Freshwater ostracodes of Canada, Part V: Families Limnocytheridae, Ooxochonchidae. *Can. J. Zool.* 49:49–64.

Delorme, L. D., and D. Donald. 1969. Torpidity of fresh-water ostracodes. *Can. J. Zool.* 47: 997–999.

Dobbin, C. N. 1941. Fresh-water Ostracoda from Washington and other western localities. *Univ. Wash. Publ. Biol.* 4:175–245.

Ferguson, E. 1944. Studies of the seasonal life history of three species of freshwater Ostracoda. *Am. Midl. Nat.* 32:713–727.

——. 1957. Ostracoda (Crustacea) from the northern lower Peninsula of Michigan. *Trans. Am. Microsc. Soc.* 76:212–218.

——. 1958. A supplementary list of species and records of distribution for North American freshwater Ostracoda. *Proc. Biol. Soc. Wash.* 71:197–202.

——. 1959a. A synopsis of the ostracod (Crustacea) genus *Cypridopsis* with the description of a new species. *Proc. Biol. Soc. Wash.* 72:59–68.

——. 1959b. The ostracod genus *Potamocypris* with the description of a new species. *Proc. Biol. Soc. Wash.* 72:133–138.

——. 1964a. Stenocyprinae, a new subfamily of freshwater cyprid ostracods (Crustacea) with description of a new species from California. *Proc. Biol. Soc. Wash.* 77:17–24.

——. 1964b. The ostracod (Crustacea) genus *Cypridopsis* in North America and a description of *Cypridopsis howei*, sp. nov. *Trans. Am. Microsc. Soc.* 83:380–384.

——. 1967a. *Cyprinotus newmexicoensis*, a new cyprid ostracod. *Am. Midl. Nat.* 78:248–250.

——. 1967b. New ostracods from the playa lakes of eastern New Mexico and western Texas. *Trans. Am. Microsc. Soc.* 86:244–249.

——. 1968. Recently described species and distributional records for North American freshwater Ostracoda. *Am. Midl. Nat.* 79:499–506.

Furtos, N. C. 1933. The Ostracoda of Ohio. *Bull. Ohio Biol. Surv.* 29:413–524,

——. 1935. Fresh-water Ostracoda from Massachusetts. *J. Wash. Acad. Sci.* 25:530–544.

——. 1936. Freshwater Ostracoda from Florida and North Carolina. *Am. Midl. Nat.* 17:491–522.

Goins, D. R. 1971. A check list of freshwater ostracods of Indiana-including sixteen new recordings and a description of two new species. *Proc. Indiana Acad. Sci.* 81:355–358.

Gonzalez Mozo, M. E., et al. 1996. A taxonomic revision of European *Herpetocypris* Brady and Norman, 1889 (Crustacea, Ostracoda). *Bull. L'Inst. R. Sci. Nat. Belgique* 66:93–132.

Hart, C. W. 1962. A revision of the ostracods of the Family Entocytheridae. *Proc. Acad. Nat. Sci.* 114:121–147.

——. 1978. A new species of the genus *Sphaeromicola* (Ostracoda: Entocytheridae: Sphaeromicolinae) from Texas, with notes on relationships between European and North American species. *Proc. Biol. Soc. Wash.* 91:724–730.

Hart, C. W., and D. G. Hart. 1969. Evolutionary trends in the ostracod Family Entocytheridae, with notes on the distributional patterns in the southern Appalachians. *Va. Polytech. Inst. Res. Div. Monogr.* 1: 179–190.

Hart, C. W., Jr., et al. 1985. The life history and ecology of the entocytherid ostracod *Uncinocythere occidentalis* (Kozloff and Whitman) in Idaho. *Smithson. Contrib. Zool.* 419:1–22.

Hart, D. G., and C. W. Hart. 1974. The ostracod family Entocytheridae. *Monogr. Acad. Nat. Sci.* 18:1–239.

Henderson, P. A. 1990. Freshwater ostracods. *Synopses B. Fauna (New Ser.)* 42:1–228.

Hobbs, H. H., Jr., and H. H. Hobbs, III. 1970. New entocytherid ostracods with a key to the genera of the subfamily Entocytherinae. *Synopses Br. Fauna* (*New Ser.*) 47:1–19.

Hobbs, H. H., Jr., and A. C. McClure. 1983. On a small collection of entocytherid ostracods with the descriptions of three new species. *Proc. Biol. Soc. Wash.* 96:770–779.

Hobbs, H. H., Jr., and D. J. Peters. 1977. The entocytherid ostracods of North Carolina. *Smithson. Contrib. Zool.* 247:1–73.

——. 1982. The entocytherid ostracod fauna of northern Georgia. *Proc. Biol. Soc. Wash.* 95: 297–318.

——. 1989. New records of entocytherid ostracods infesting burrowing crayfishes, with the description of a new species, *Ascetocythere stockeri*. *Proc. Biol. Soc. Wash.* 102:324–330.

——. 1991. Additional records of entocytherid ostracods infesting burrowing crayfishes, with descriptions of five new species. *Proc. Bio. Soc. Wash.* 104:64–75.

Hobbs, H. H., III. 1975. Distribution of Indiana cavernicolous crayfishes and their ectocommensal ostracods. *Int. J. Speleol.* 7:273–302.

Hoff, C. C. 1942a. The ostracods of Illinois. *Univ. Ill. Biol. Monogr.* 19:1–196.

——. 1942b. The subfamily Entocytherinae, a new subfamily of fresh-water cytherid Ostracoda, with descriptions of two new species of the genus *Entocythere*. *Am. Midl. Nat.* 27:63–73.

——. 1943. Seasonal changes in the ostracod fauna of temporary ponds. *Ecology* 24:116–118.

——. 1944. The origin of Nearctic fresh-water ostracods. *Ecology*. 25:369–372.

Howe, H. V. 1962. *Handbook of Ostracod Taxonomy*. 386 pp. Louisiana State University Press, Baton Rouge.

Kesling, R. V. 1951. The morphology of ostracod molt stages. *Ill. Biol. Monogr.* 21:1–324.

——. 1956. The ostracod: a neglected little crustacean. *Turtox News* 34:82–86, 90–94, 114–115.

——. 1957. Notes on Zenker's organs in the ostracod *Candona*. *Am. Midl. Nat.* 57:175–182.

Kosmal, A. 1968. On the distribution of Ostracoda in the littoral of Lake Kisajno (Masurian Lake District). *Pot. Arch. Hydrobiol.* 15:87–102.

Levinson, S. A. 1951. A technique for anesthetizing fresh water ostracodes. *Am. Midl. Nat.* 46:254–255.

Liang, Y-S., et al. 1973 Transmission of ostracods in snails. *Malacol. Rev.* 6:66.

Lowndes, A. G. 1935. The sperms of freshwater ostracods. *Proc. Zool. Soc. London* (1935) 1:35–48.

Marmonier, P., and J. V. Ward. 1990. Surficial and interstitial Ostracoda of the South Platte River (Colorado, U.S.A.): systematics and biogeography. *Stygologia* 5:225–239.

Martens, K. 1998. Sex and ostracods: a new synthesis. In: *Sex and Parthenogenesis: Evolutionary Ecology of Reproductive Modes in Non-Marine Ostracods*, pp. 295–321, edited by K. Martens. Backhuys, Leiden.

Martens, K., and F. Behen. 1994. A checklist of the Recent non-marine ostracods (Crustacea, Ostracoda) from the inland waters of South America and adjacent islands. *Minist. Affaires Clut, Travaux Sci. Mus. Nat. D'Hist. Nat. Luxembourg* 22:1–84.

Martens, K., et al. 1992. On *Pseudocandona gajewskajae* Bronstein. *Stereo-Atlas of Ostracod Shells* 19:49–56.

——. 1998. Maxillopoda. Non-marine Ostracoda. In: *Catalogue of Crustacea of Brazil*, pp. 45–65, edited by P. S. Young, Mus. Nac., Rio de Janeiro (Ser. Livros 6).

Mastrantuono, L. 1989. On *Isocypris beauchampi* (Paris) (Ostracoda) and some suggestions on the taxonomy of the genus. *Crustaceana* 57: 140–144.

McGregor, D. L. 1969. The reproductive potential, life history and parasitism of the freshwater ostracod *Darwinula stevensoni* (Brady and Robertson). In: *Taxonomy, Morphology and Ecology of Recent Ostracoda*, pp. 194–221, edited by J. W. Neale. Oliver & Boyd, Edinburgh.

McGregor, D. L., and R. V. Kesling. 1969a. Copulatory adaptations in ostracods, Part I: Hemipenes of *Candona*. *Univ. Mich. Contrib. Paleontol.* 22:169–191.

——. 1969b. Copulatory adaptations in ostracods, Part II: adaptations in living ostracods. *Univ. Mich. Contrib. Paleontol.* 22:221–239.

Meisch, C. 1996. Contribution to the taxonomy of *Pseudocandona* and four related genera, with the description of *Schellencandona* nov. gen., a list of the Candoninae genera, and a key to the European genera of the subfamily (Crustacea, Ostracoda). *Bull. Soc. Nat. Luxemb.* 97:211–237.

——. 2000. *Freshwater Ostracoda of Western and Central Europe*. 522 pp. Spektrum - Verlag/ Gustav Fischer, Luxembourg.

Peters, D. J., and J. E. Pugh. 1999. On the entocytherid ostracods of the Brazos River basin and adjacent coastal region of Texas. *Proc. Biol. Soc. Wash.* 112:338–351.

Rogulj, B., et al. 1994. Fine-scale distribution of hypogean Ostracoda in the interstitial habitats of the Rivers Sava and Rhône. *Hydrobiologia* 287:19–28.

Rossetti, G., and K. Martens. 1996. Redescription and morphological variability of *Darwinula stevensoni* (Brady & Robertson, 1870) (Crustacea, Ostracoda). *Bull. L'Inst. R. Sci. Nat. Belgique* 66:73–92.

——. 1998. Taxonomic revision of the Recent and Holocene respresentatives of the Family Darwinulidae (Crustacea, Ostracoda), with a description of three new genera. *Bull. L'Inst. R. Sci. Nat. Belgique.* 68:55–110.

Rossetti, G., et al. 1998. On *Darwinula stevensoni* (Brady & Robertson). *Stereo-Atlas of Ostracod Shells* 25:17–22.

Sars, G. O. 1928. Ostracoda. In: *An Account of the Crustacea of Norway* 9:1–277. Bergen Museum, Bergen, Norway.

Scheerer-Ostermeyer, E. 1940. Beitrag zur Entwicklungsgeschichte der Süsswasserostrakoden. *Zool. Jahrb. Abt. Anat. Ontog. Tiere* 66: 349–370.

Schon, I., et al. 2000. Persistence of asexuality through mixed reproduction in *Eucypris virens* (Crustacea, Ostracoda). *Heredity* 84:161–169.

Schreiber, E. 1922. Beiträdge zur Kenntnis der Morphologie, Entwicklung und Lebensweise der Süsswasser-Ostrakoden. *Zool. Jahrb. Abt. Anat. Ontog. Tiere* 43:485–538.

Sohn, I. G., and L. S. Kornicker. 1973. Morphology of *Cypretta kawarai* Sohn and Kornicker, 1972 (Crustacea, Ostracoda), with a discussion of the genus. *Smithson. Contrib. Zool.* 141:1–28.

Strayer, D. 1988a. Crustaceans and mites (Acari) from hyporheic and other underground waters in southeastern New York. *Stygologia* 4:192–207.

——. 1988b. Life history of a lacustrine ostracod. *Hydrobiologia* 160:189–191.

Tressler, W. L. 1947. A check list of the known species of North American freshwater Ostracoda. *Am. Midl. Nat.* 38:698–707.

——. 1959. Ostracoda. In: *Fresh-water Biology*, pp. 657–734, edited by W. T. Edmondson. Wiley, New York.

Turpen, J. B., and R. W. Angell. 1971. Aspects of molting and calcification in the ostracod *Heterocypris. Biol. Bull.* 140:331–338.

Ward, H. B., and G. G. Whipple. 1918. *Fresh-water Biology*. 1111 pp. Wiley, New York.

Ward, J. V., and N. J. Voelz. 1994. Goundwater fauna of the South Platte River system, Colorado. In: *Groundwater Ecology*, pp. 391–423, edited by J. Gilbert et al. Academic, San Diego.

Young, W. 1971. Ecological studies of the Entocytheridae (Ostracoda). *Am. Midl. Nat.* 85: 399–409.

22

MINOR MALACOSTRACA

The Class Malacostraca contains several small to large orders of crustaceans distributed in marine, freshwater, and terrestrial environments. All members of the class have in common an arrangement of appendages distributed among body regions or tagmata composed of a specific number of segments. The most familiar freshwater crustaceans are the large decapods, and include the crayfish and crabs. The less noticeable but nevertheless common and widespread Isopoda and Amphipoda are also important groups within the Malacostraca and are treated in subsequent chapters. Several minor groups of malacostracans are found in freshwaters of the United States as well. These orders are either largely marine with only a small number of freshwater genera and species (Mysidacea, Fig. 22.1) or are very poorly known stygobionts no more than 3.5 mm in length (Bathynellacea, Thermosbaenacea, Figs. 22.2; 22.3).

Order Bathynellacea. This group is distributed almost worldwide in temperate and tropical latitudes, chiefly in wells, subterranean waters, artesian systems, and interstitial waters of streams. The animals are very small, usually less than 1 mm in length. They lack a carapace and have only a few pairs of abdominal appendages. They presumably feed on bacteria and fungi occupying the surfaces of sand grains. Bathynellids are blind, vermiform, and well adapted to interstitial situations. Reproduction is poorly known but it is believed that eggs are not brooded but are simply scattered following fertilization. The bathynellid literature is large and chiefly European. Only since 1974 have bathynellids been reported from the United States, and they are now known from many Rocky and Appalachian Mountain states plus Texas and Ohio. Four species distributed among two families are now known from the United States and others await description (Ward and Voelz 1994). Among the Bathynellidae, *Bathynella riparia* (and possibly other species) (Fig. 22.3B, C) appears to be the most widely distributed, known from many western and eastern states, while *Pacificabathynella sequoiae* (Fig. 22.2) is found in northern California and Wyoming. The Parabathynellidae includes *Texanobathynella bowmani*, known from wells in Texas, and *Californibathynella californica*, which occurs in southern California and Wyoming. Undoubtedly some of these species are more widely distributed in this country.

Order Thermosbaenacea. Another crustacean rarity in freshwaters is the Order Thermosbaenacea (Fig. 22.3A). These are likewise small crustaceans, normally found in interstitial and subterranean waters and in springs. Superficially, they are similar to bathynellids, but they have a shieldlike carapace that roofs the anterior part of the thorax (Fig 22.3A). The adult female uses this

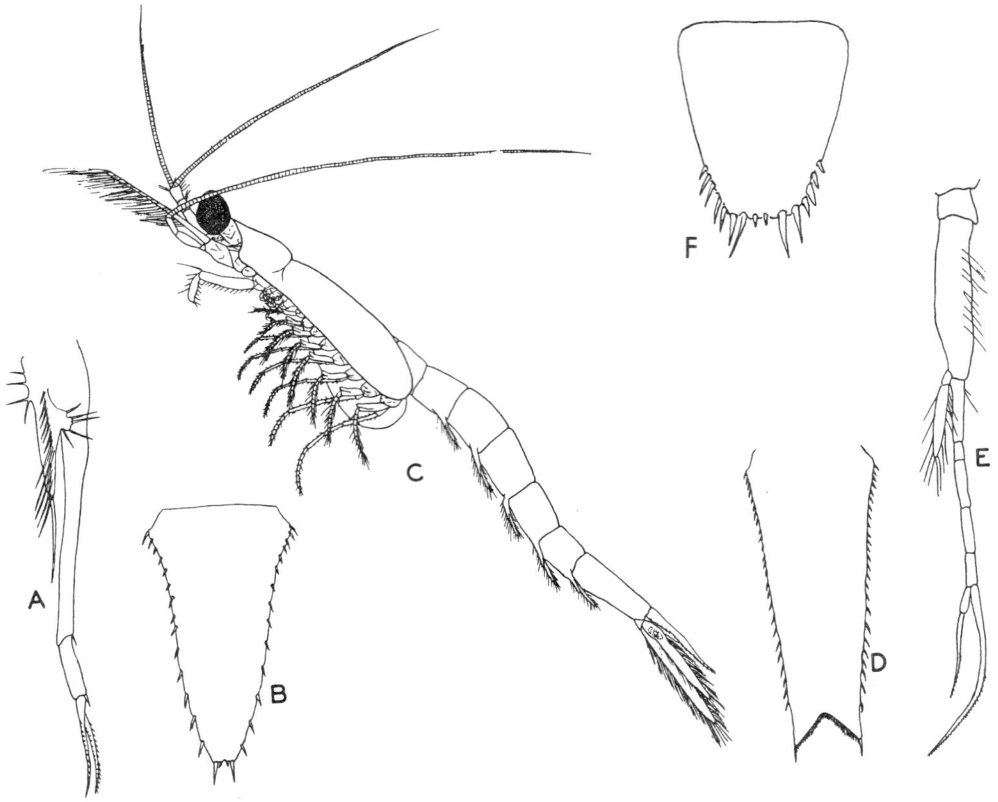

Fig. 22.1. Mysidacea. A, male fourth pleopod of *Neomysis mercedis;* B, telson of *N. mercedis;* C, female *Mysis relicta,* ×8; D, telson of same; E, male fourth pleopod of same; F, telson of *Deltamysis holmquistae.* (A and B modified from Tattersall 1951; F redrawn from Bowman and Orsi 1992.)

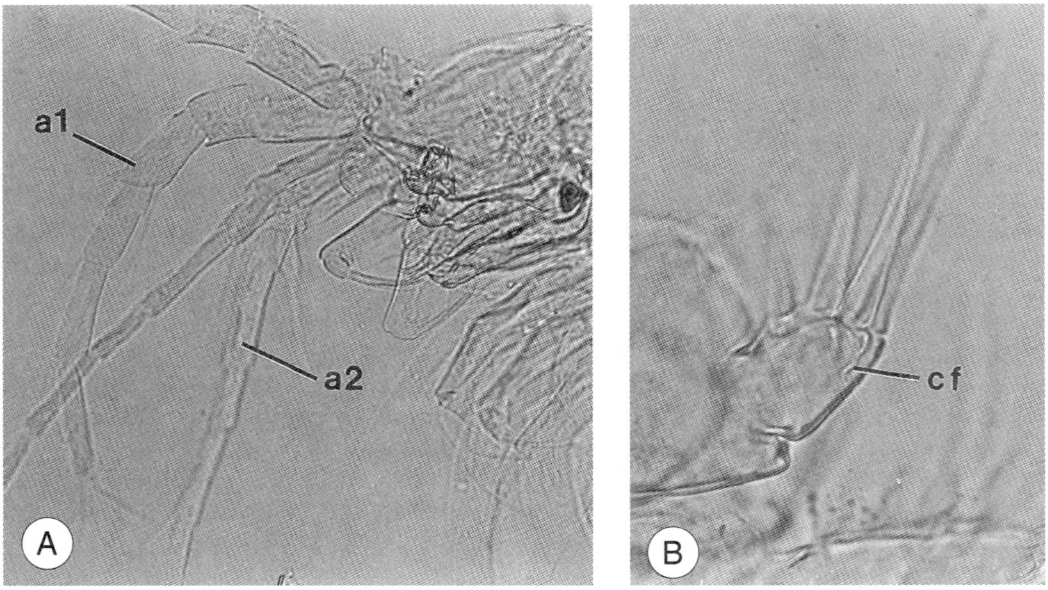

Fig. 22.2. *Pacificabathynella sequoiae,* × 400. A, cephalothorax with appendages; B, uronites. a1, antenna 1; a2, antenna 2; cf, caudal furca. (from a specimen provided by M. Marshall.)

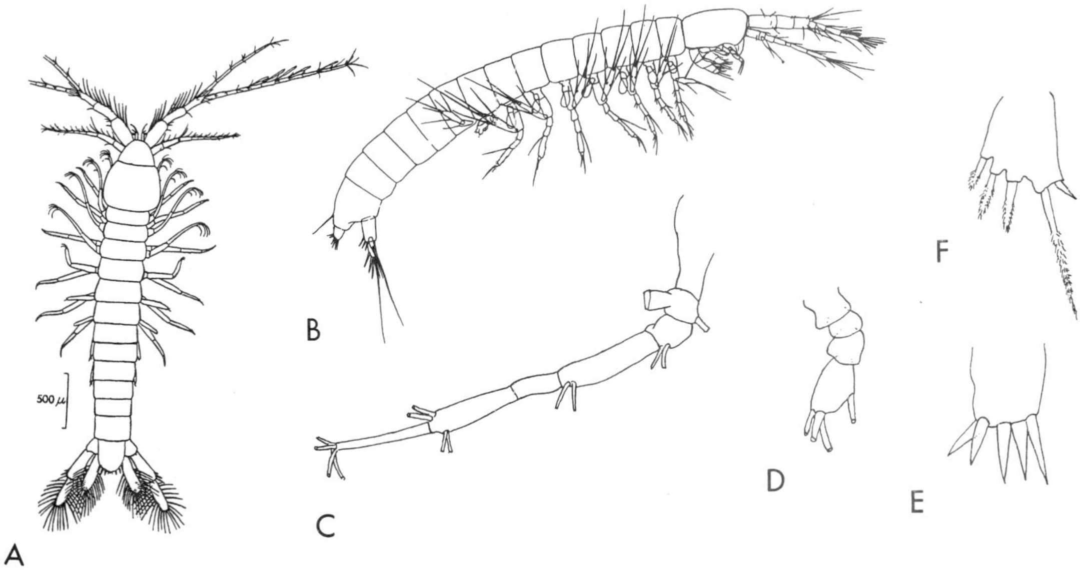

Fig. 22.3. Structure of Thermosbaenacea and Bathynellacea. A, Dorsal view of *Monodella argentarii,* a typical member of the Thermosbaenacea; B, typical *Bathynella;* C, antenna 2 of *B. riparia;* D, antenna 2 of *Californibathynella californica;* E, caudal furca of *B. riparia*; F, caudal furca of *Pacificabathynella sequoiae.* (A from Fryer 1964; B from Noodt 1964; C, E redrawn from Pennak and Ward 1985; D, F redrawn from Schminke and Noodt 1988.)

device as a marsupium. The appendages of Thermosbaenaceans are reduced in development and number; the abdomen contains only a few pairs of minute appendages. As indicated above, eggs are brooded in the carapace of the parent until some point when they become self sufficient. Other species have more recently been described from Italy, Yugoslavia, Canary Islands, West Indies, Somali, Mexico, and Israel. All species discovered prior to 1964 occurred in salty or brackish waters relatively close to the sea, but Fryer (1964) reports keeping thermosbaenaceans in freshwater for long periods. In 1965 Maguire described *Monodella texana* from fresh, cool water in a cave near San Marcos, Texas, about 200 km from the sea. Careful searching should reveal Thermosbaenacea in other subterranean habitats.

Order Mysidacea. The opossum shrimps form a large order that is almost exclusively marine, only three species having become adapted to fresh and mildly brackish waters in the United States. *Mysis relicta* is a native plankter in a few deep, cold, oligotrophic lakes of the northern states east of the Great Plains, in the Great Lakes, and in many Canadian lakes. *Neomysis mercedis* occurs in lakes, rivers, and brackish estuaries of the Washington, Oregon, and California coasts. *Taphromysis louisianae* is now known from the Gulf Coast, large reservoirs in the Mississippi River system, and the Ohio River. *Taphromysis bowmani* is known only from mildly brackish waters in Florida and Alabama. The recently described *Deltamysis holmquistae* is known only from estuaries and tidal freshwater of the Sacramento River delta of California. In contrast, Europe has about 20 freshwater, troglobitic, and mildly brackish species, and recently several troglobitic species have been discovered in Mexico and the West Indies. It seems logical that future study will reveal additional species in the southern United States.

Superficially, mysids resemble miniature crayfish, but a closer examination reveals some striking differences. The carapace for example, is thin and does not completely cover the thorax, the last two segments being exposed dorsally and, to some degree, laterally. In place of the five pairs of walking legs of decapods, mysids have long, thin, setose, biramous, many-segmented appendages. The first thoracic segment bears a pair of maxillipeds, and is fused with the head, but the other seven thoracic segments are distinct. The second thoracic segment bears the second maxillipeds, and each of the remaining six segments has a pair of swimming legs. The stalked compound eyes are extremely large, the scale of the second antenna is large, but the pleopods are proportionately much more reduced than those of decapods, *Mysis relicta* reaches a length of 30 mm, *Neomysis mercedis* reaches 15 mm, *Taphromysis* is 8 mm, and *Deltamysis* reaches only 3 or 5 mm.

Mature male mysids may be distinguished from females by their long and specialized fourth pleopods, which may extend as far as the posterior end of the telson, whereas females are characterized by a marsupium consisting of four ventral oostegites originating at the bases of the last two pairs of legs (hence the name "opossum shrimps"). Females are generally larger and more abundant than males.

The exopods of the six pairs of legs project laterally somewhat and their beating produces smooth, rapid swimming. The endopods of these appendages extend more medially, and their movements create a current of water that passes anteriorly over the mouthparts. The two pairs of maxillipeds are particularly efficient in straining zooplankton (cladocerans, rotifers), phytoplankton, and particulate debris from the current. Such food is then passed on toward the second maxillae, first maxillae, mandibles, and mouth. *Mysis* has been characterized as an "opportunistic" feeder. If zooplankton is abundant, *Mysis* uses it as a primary food source. If zooplankton is sparse, then *Mysis* appears to obtain food from suspended organic detritus or from the surface of bottom deposits. Thus, the feeding habits are both raptorial and filter feeding.

Internal anatomy and physiology are generally comparable to conditions found in the crayfish. There are no gills, however; most respiration occurs through the thin lining of the carapace. A current of water is drawn under the carapace by the action of the epipodites of the first maxillipeds. A unique feature is the occurrence of a statocyst in the basal portion of the endopod of each uropod.

Reproduction occurs during the summer months. *Neomysis mercedis* commences egg deposition in the spring when year-old females lay eggs. Females born in spring reproduce in late summer, over winter, and produce eggs the following spring. A female may produce up to 40 eggs per clutch, depending on her size. Developing eggs and young are carried in the marsupium for 1 to 3 months, the young leaving the mother when 3 or 4 mm long. Development is direct and the life cycle is thought to extend through 2–4 years in northern latitudes or high altitudes but only 1 year farther south. Undoubtedly, food availability is also involved in this range of life cycles. Little is known about the number of clutches of eggs that may be produced by a female during her life cycle.

Freshwater mysids, and especially *Mysis relicta*, are cold-water forms. During the warm months this species is almost restricted to the hypolimnion, where temperatures range as low as 4°C. It is thought that 14°C is the maximum temperature that can be tolerated for any length of time, and certainly *Mysis* has never been reported from lakes where the hypolimnial maximum exceeds this figure. Newly released juveniles migrate in spring to warm shallow waters. Later they move back to deep waters.

Of all the freshwater Crustacea known to exhibit daily vertical migrations, *Mysis* has the most extensive and rapid movements. Populations are confined to the meter of water just above the bottom during the middle of the day and no amount of sampling above that depth will take specimens. At dusk, however, most of the individuals, especially the older ones, migrate into the surface waters where temperatures as high as 20°C may be tolerated for a few hours. Reverse downward movements begin at dawn.

Mysis relicta has been recorded a few times from water that was only about 15% saturated with oxygen or 1 ppm, but the great majority of investigations show that 30 or 40% saturation is the usual lower limit. In exceptional years the oxygen content of the water just above the bottom of oligotrophic lakes falls below this limit, and populations are then confined to strata above the deepest waters during the daytime.

Original native populations of *M. relicta* have been reported from the Great Lakes (as deep as 270 m in Lake Superior); Geneva, Trout, and Green lakes in Wisconsin; some of the Finger Lakes in New York; a few other lakes in the northern states; and many deep Canadian lakes. *M. relicta* is especially abundant in Waterton Lake on the border between Montana and Canada. Undoubtedly, it is much more common and widely distributed in the United States than the published literature indicates. This form is circumpolar and occurs also in the British Isles, the Scandinavian countries, northern Germany, and northern Russia. Recent study in northern Europe has determined that several biochemically and ecologically discrete populations of *Mysis relicta* exist.

Neomysis mercedis is a species that has apparently evolved from marine ancestry very recently. It occurs in brackish bays and estuaries on the west coast where the salinity is usually less than 20‰ and in strictly freshwaters of rivers and lakes near the coast.

Freshwater mysids are an extremely important food source for a group of commercially important fishes inhabiting deep, cold waters, especially lake trout and coregonids. Food studies show that often 80 to 100% of the stomach contents of these fishes may be *Mysis*. *Taphromysis louisianae* can make up to 50% of the diet in young largemouth bass (*Micropterus salmoides*).

Because of its value in the trout diet, *Mysis* is a logical artificial transplant, especially because the "*Mysis*-niche" is unoccupied in most lakes. The genus has already been successfully established in several western Canadian lakes, a few Colorado mountain lakes, and some lakes in California. Experimental plants have also been made in Wisconsin, Minnesota, New York, Nevada, Idaho, and undoubtedly elsewhere in the western states. Similar transplants have been tried in Canada, Sweden, and Russia.

Often the introduction of *Mysis* has resulted in drastic decreases in densities of *Daphnia*, *Bosmina*, and rotifers, as well as shifts in density ratios of cladoceran zooplankters, much to the detriment of trout and salmon populations. Such interactions seem to be more common in shallow (rather than very deep) oligotrophic lakes. Less is known about interactions between *Mysis* and copepod zooplankters. One group of investigators established the following order of prey preferences: cladocerans > copepod copepodites > copepod nauplii > adult copepods. Because of the effect of *Mysis* on plankton populations, a moratorium has been established on future introductions (Nesler and Bergersen 1991).

The American species belong to the Subfamily Mysinae in the Family Mysidae. Aside from their distinctive geographic ranges and size differences, they are easily distinguished morphologically. The Bathynellacea is represented by two families, the Bathynellidae and Parabathynellidae, and Thermosbaenacea is represented by the Thermosbaenidae.

KEY TO THE GENERA OF FRESHWATER MINOR MALACOSTRACA OF THE UNITED STATES

1. Animal shrimplike (Fig. 22.1C); 3 to 30 mm in length . . . . **MYSIDACEA, 2**
   Animal wormlike and minute, interstitial or phreatic; 0.5 to 3.5 mm in length 5
2. Telson with a narrow truncated tip (Fig. 22.1B, F); pleopod of male with 2 or less segments (Fig. 22.1A) **3**
   Telson with a bifurcate tip (Fig. 22.1D); pleopod of male with over 3 segments **4**
3. Entire margin of telson spined; pleopods segmented; 1 species along the west coast ***Neomysis***
   Only distal margin of telson spined; pleopods unsegmented; 1 species, CA . . . . ***Deltamysis***

4. Exopod of the 4th pleopod of male with 6 segments (Fig. 22.1E); 1 species, northern states . . . ***Mysis***
Exopod of the 4th pleopod of male with 7 segments; 2 species . . . .***Taphromysis***

5. Head and anterior thoracic segments covered by a carapace (Fig. 22.3A); 1 species, TX . . . **THERMOSBAENACEA, *Monodella***
Body without a carapace (Figs. 22.2; 22.3B) . . . .**BATHYNELLACEA, 6**

6. Antenna 2 with 7 segments (Fig. 22.3C) **7**
Antenna 2 with 3 segments (Fig. 22.3D) **8**

7. Caudal furca with 5 stout spines of slightly unequal length (Fig. 22.3E); 1 species, widely distributed . . . .***Bathynella***
Caudal furca with 5 slender spines, terminal spine considerably longer than others (Figs. 22.2B, 22.3F); 1 species in CA, WY . . . .***Pacificabathynella***

8. Leg (thoracopod) 6 exopod with 3 segments; 1 species from wells in TX . . .***Texanobathynella***
Leg (thoracopod) 6 exopod with 2 segments; 1 species from CA . . . .***Californibathynella***

REFERENCES

Absolon, K. 1935. O zene fossilii *Thermosbaena mirabilis* z korkych vod Sahary. *Priorda* 38: 1–11.

Banner, A. H. 1953. On a new genus and species of mysid from southern Louisiana. *Tulane Stud. Zool.* 1:1–8.

Barker, D. 1959. The distribution and systematic position of the Thermosbaenacea. *Hydrobiologia* 13:209–237.

——. 1962. A study of *Thermosbaena mirabilis* (Malacostraca, Peracarida) and its reproduction. *Q. J. Microsc. Sci.* 103:261–286.

Bowman, T. E., and J. J. Orsi. 1992. *Deltamysis holmquistae*, a new genus and species of Mysidacea from the Sacramento–San Joaquin estuary of California (Mysidae: Mysinae: Heteromysini). *Proc. Biol. Soc. Wash.* 105:733–742.

Brooks, C. P., et al. 1998. New records of *Taphromysis louisiana* Banner, 1953 (Mysidae) with notes on its ecology. *Crustaceana* 71:955–960.

Bruun, A. F. 1939. Observations on *Thermosbaena mirabilis* Monod from hot springs of El-Hamma, Tunisia. *Vid. Medd. Dansk Nat. For.* 103:493–501.

Camancho, A. L., et al. 2000. Phylogenetic review and biogeographic remarks on the interstitial and subterranean freshwater iberobathynellids (Crustacea, Syncarida, Parabathynellidae). *J. Nat. Hist.* 34:563–585.

Delamare Deboutteville, C., et al. 1975. Decouverte de la famille des Parabathynellidae (Bathynellacea) en Amerique du Nord: *Texanobathynella bowmani* n. g. n. sp. *C. R. Hebd. Seances Acad. Sci. Paris, Ser.* D, 280:2223–2226.

Fryer, G. 1964. Studies on the functional morphology and feeding mechanism of *Monodella argentarii* Stella (Crustacea, Thermosbaenacea). *Trans. R. Soc. Edinburgh* 66:51–90.

García-Garza, M. E., et al. 1992. Distribution of the possum shrimp *Taphromysis louisianae* (Mysidacea), including new far inland records. *J. Crust. Biol.* 12:101–103.

Giere, O. 1993. *Meiobenthology—The Microscopic Fauna in Aquatic Sediments.* 328 pp. Springer, Berlin.

Husmann, S. 1967. Okologie, Systematik und Verbreitung zweier in *Nordcleutschland sympatrisch lebender* Bathynella-Arten (Crustacea, Syncarida). *Int. J. Speleol.* 3:111–114.

Juday, C., and E. A. Birge. 1927. *Pontoporeia* and *Mysis* in Wisconsin lakes. *Ecology* 7:445–452.

Kinsten, B., and P. Olsen. 1981. Impact of *Mysis relicta* Lovén introduction on the plankton of two mountain lakes, Sweden. *Rep. Inst. Freshwater Res. Drottningholm* 59:64–74.

Kost, A. L. B., and A. W. Knight. 1975. The food of *Neomysis mercedis* Holmes in the Sacramento–San Joaquin estuary. *Calif. Fish Game* 61:35–46.

Maguire, B. 1965. *Monodella texana* n. sp., an extension of the range of the crustacean order Thermosbaenacea to the Western Hemisphere. *Crustaceana* 9:149–154.

Morgan, M. D. 1981. Impact of introduced populations of *Mysis relicta* on zooplankton in oligotrophic subalpine lakes. *Verh. Int. Ver. Theor. Angew. Limnol.* 21:339–345.

——. (ed.) 1982. Ecology of Mysidacea. *Hydrobiologia* 93:1–222.

Murtagh, P. A. 1981. Selective predation by *Neomysis mercedis* in Lake Washington. *Limnol. Oceanogr.* 26:445–453.

——. 1989. Fecundity of *Neomysis mercedis* Holmes in Lake Washington (Mysidacea). *Crustaceana* 57:194–200.

Nesler, T. P., and E. P. Bergersen. 1991. Mysids and their impacts on fisheries: an introduction to the 1988 mysid–fisheries symposium. *Am. Fish. Soc. Symp.* 9:1–4.

Noodt, W. 1964. Naturliches System und Biogeographie der Syncarida (Crustacea Malacostraca). *Gewasser Abwasser* 37:77–186.

——. 1967. Biogeographie der Bathynellacea (Syncarida). *Proc. Symp. Crustacea* 1:411–417.

Pennak, R. W., and J. V. Ward. 1985. Bathynellacea (Crustacea: Syncarida) in the United States, and a new species from the phreatic zone of a Colorado mountain stream. *Trans. Am. Microsc. Soc.* 104:209–215.

Pesce, G. L. 1976. Stato attuale delle conoscenze sui misidacei cavernicoli e freatici (Crustacea). *Notiz. Circolo Speleol. Romano* 1:47–57.

Rieman, B. E., and C. M. Falter. 1981. Effects of the establishment of *Mysis relicta* on the macrozooplankton of a large lake. *Trans. Am. Fish. Soc.* 110:613–620.

Schminke, H. K., and W. Noodt. 1988. Groundwater Crustacea of the Order Bathynellacea (Malacostraca) from North America. *J. Crust. Biol.* 8:290–299.

Strayer, D. L., et al. 1995. An endemic groundwater fauna in unglaciated eastern North America. *Can. J. Zool.* 73:502–508.

Tattersall, W. M. 1951. A review of the Mysidacea of the United States National Museum. *Bull. U.S. Nat. Mus.* 201:1–292.

Thienemann, A. 1925. *Mysis relicta. Z. Morphol. Ökol. Tiere* 3:389–440.

Väinölä, R. 1986. Sibling species and phylogenetic relationships of *Mysis relicta* (Crustacea: Mysidacea). *Ann. Zool. Fenn.* 23:207–221.

Ward, J. V., and N. J. Voelz. 1994. Goundwater fauna of the South Platte River system, Colorado. In: *Groundwater Ecology*, pp.391–423, edited by J. Gilbert et al. Academic, San Diego.

Zilch, R. 1972. Beitrag zur Verbreitung und Entwicklungsbiologie der Thermosbaenacea. *Int. Rev. ges. Hydrobiol.* 57:75–107.

23

ISOPODA

(Aquatic sow bugs)

Pill bugs and sow bugs are chiefly marine and terrestrial, only about 5% of the total known North American species are found in freshwaters. Approximately 130 freshwater species occur in the United States distributed among suborders Asellota, Epicaridea, and Flabellifera (Figs. 23.1; 23.4). For the most part freshwater isopods are restricted to springs, spring brooks, streams, and hyporheic, interstitial, and subterranean waters. A decided minority are found in ponds and the shallows of lakes. The biology of European species has been thoroughly investigated, but less is known about our American species.

General characteristics. Unlike amphipods, the members of the Order Isopoda, with few exceptions, are strongly flattened dorsoventrally. The "head" of an isopod is actually a cephalothorax, since it represents the fusion of the true head and the first thoracic segment. The seven remaining thoracic segments are all similar and are expanded laterally in the form of eavelike lamellae over the basal portions of the thoracic appendages. In most American freshwater species the last four abdominal segments and true telson are completely fused into a large shieldlike region, often, but possibly incorrectly, referred to as a pleotelson. The free abdominal segments are greatly reduced and scarcely visible at the anterior end of the abdomen. In the Microparasellidae the abdomen has only one free segment, in Sphaeromatidae, Stenasellidae, and Asellidae the abdomen consists of two segments, and in the Cirolanidae and Cymothidae there are five or six distinct segments.

Total length, from the anterior margin of the cephalothorax to the posterior margin of the abdomen, usually ranges between 5 and 20 mm. Coloration may be blackish, brown, gray, dusky, reddish, or yellowish. Some species are variously marked or mottled. Subterranean species are generally whitish or creamy.

The first antennae arise near the median line and anterior margin of the cephalothorax. Each consists of a three-segmented peduncle and a many-segmented flagellum. The second antennae are much longer than the first; they arise ventral and lateral to the first antennae and have five-segmented peduncles.

The eyes are dorsal, unstalked, compound, and usually reduced or absent in subterranean species.

In typical, common species the mouthparts form a compact buccal mass that is covered anteriorly by an upper lip. The sclerotized mandibles are strong, toothed, and may or may not have a palp. Each first maxilla has a basal portion and two elongated, toothed and setose palps. The second maxillae are smaller, weaker, and more platelike and have three short, terminal divisions. The maxillipeds, representing the appendages of the

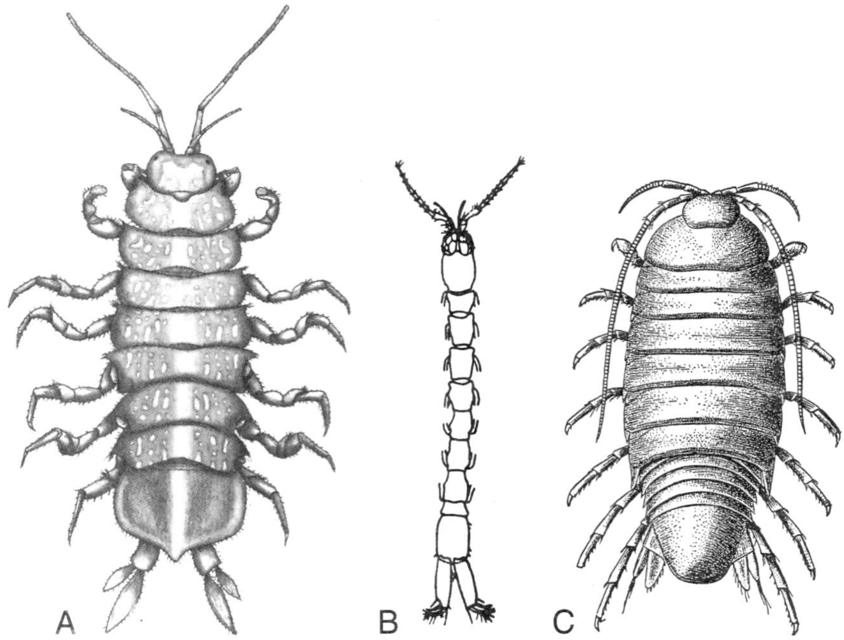

Fig. 23.1. Isopod form. A, *Caecidotea* (Asellidae), ×6.5; B, *Microcharon* (Microparasellidae), ×100; C, *Cirolanides* (Cirolanidae), ×4.5. (A © Ethan Nedeau, with permission; B modified from Birstein 1964; C from Richardson 1905.)

fused thoracic segment, are flattened and palped; they fit closely together at the median line, thus effectively forming the lower and posterior surfaces of the buccal mass.

Each of the seven free thoracic segments bears a pair of long walking legs or pereopods (Figs. 23.1; 23.3), all of which are similar except the first pair (gnathopods), which are usually subchelate and used for grasping. In general, the more posterior legs are the largest. The coxae are completely fused with the body, but the basis, ischium, merus, carpus, propodus, and dactylus are distinct in all the legs. Mature females have large, platelike oostegites attached at the inner bases of several of the anterior pairs of legs. These oostegites extend inward and collectively form a shallow chamber, or marsupium, on the ventral side of the thorax in which incubating eggs and young are carried. A similar device occurs in the amphipods.

The first five pairs of abdominal appendages (pleopods) are considerably modified and hidden beneath the abdomen. The sixth pair of appendages (uropods), however, are unmodified and project well beyond the posterior end of the body. In the female the first pleopods are absent, and the second pleopods are reduced and one-segmented. The third, fourth, and fifth pleopods, however, are large, flattened, and biramous. The exopod of the third pleopod forms a sclerotized protective operculum, but the third endopod and both rami of the fourth and fifth pleopods are delicate and respiratory. The male first pleopods are small, elongated, uniramous, and two-segmented. The second are biramous and greatly specialized for copulation and the transfer of sperm to the female. The third to sixth abdominal appendages are similar to those of the female. Males may be distinguished from females by their larger body size, larger and more specialized gnathopods, absence of oostegites, and the structure of the first and second pleopods.

Locomotion is restricted to slow crawling in freshwater species, except for the subterranean *Remasellus*, which apparently can swim. Only the Sphaeromatidae are capable of

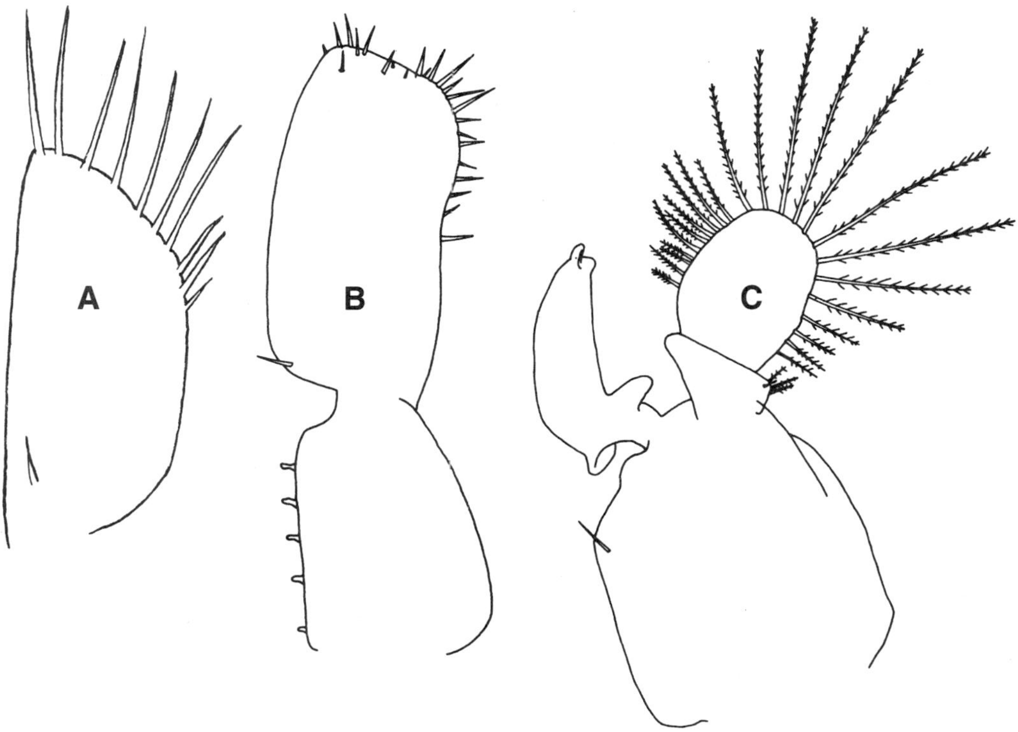

Fig. 23.2. *Caecidotea* appendages. A, "first" pleopod of female *Caecidotea* (actually this is the second pleopod, the true first pleopod being absent in all females.); B, first pleopod of male *C. communis;* C, second pleopod of male *C. communis.* (A and B from Williams 1972.)

rolling up into a ball as do some of the land isopods.

Species inhabiting subterranean waters are notable for their more elongated body, long, weak legs, and long antennae, as well as for their abundance of tactile hairs.

Food, feeding. Isopods are perhaps best characterized as scavengers, since they have been observed eating dead and injured aquatic animals of all kinds and both green and decaying leaves, grass, and aquatic vegetation. A detailed study on stomachs of *Caecidotea r. racovitzai* revealed the following identifiable food: *Tetraedron* (desmid), *Pediastrum* (Chlorophycean), and pieces of cladocerans (Kerr 1978). Minute interstitial forms presumably feed on the film of fungi and other microorganisms that coat pebbles and grains of sand.

Internal anatomy, physiology. The digestive system consists of a short esophagus, a stomach containing a gastric mill, a long intestine, and a rectum. Four long caeca arise at the junction of the stomach and intestine.

The large heart is located in the posterior part of the thorax. It receives blood from the pericardial chamber via one to three pairs of ostia and sends it out to the hemocoel through a surprisingly large number of arteries (up to 11 in some species). In at least one species, however, the heart has no arteries and no ostia. The head contains an auxillary or "frontal" heart that is controlled by muscles primarily associated with food processing. Blood is moved from this organ to the antennae and other anterior structures. Most of the carbon dioxide–oxygen exchange occurs through the thin-walled third to fifth pleopods, although it is also

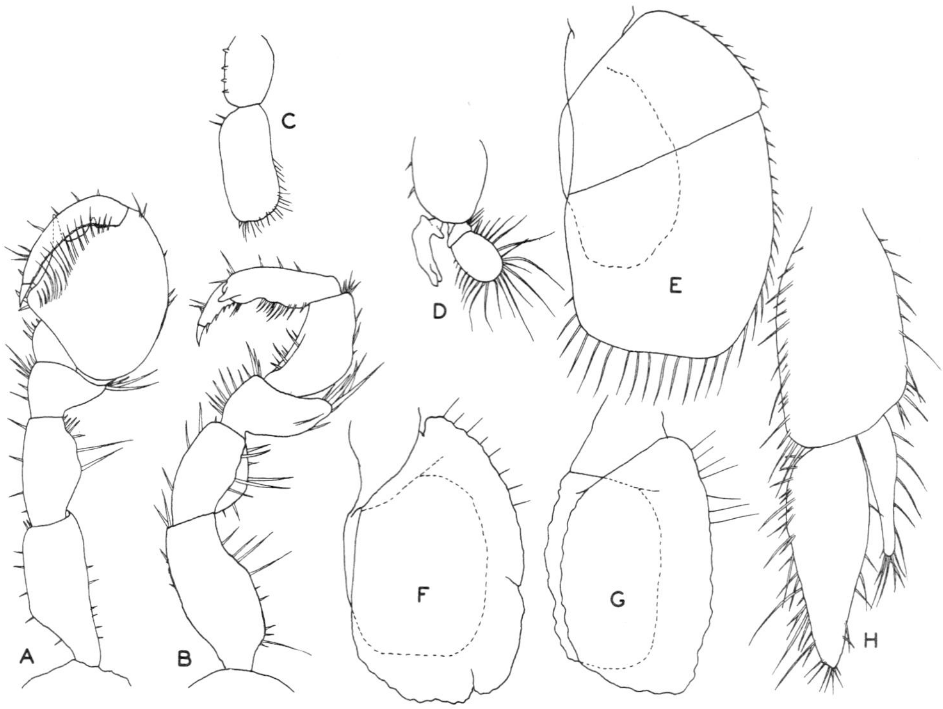

Fig. 23.3. Appendages of a male *Caecidotea;* all in ventral view, from left side, and all to same magnification. A, first leg or pereopod (gnathopod); B, fourth leg or pereopod; C, first pleopod; D, second pleopod; E, third pleopod; F, fourth pleopod; G, fifth pleopod; H, uropod.

possible that the general body surface may also be of some respiratory significance.

Two coiled maxillary glands, probably excretory, open at the base of the second maxillae. In addition, other small paired glands in the cephalothorax have sometimes been interpreted as excretory devices.

The nervous system is quite simple in structure. It consists of a brain, circumesophageal connectives, subesophageal ganglia, a double ventral nerve cord with double segmental ganglia in the thorax, and a group of fused ganglia in the anterior portion of the abdomen.

Reproduction. Males have three pairs of ovoid testes in the middle of the thorax and dorsal to the intestine. The vasa deferentia open at the tips of two fingerlike projections on the ventral surface of the seventh thoracic segment. (These structures should not be mistaken for the first pleopods, which immediately follow.) The two ovaries are dorsal, long, and baglike, and the oviducts open through two slitlike genital pores on the ventral surface of the fifth thoracic segment. Just beneath the surface of the body each oviduct is swollen to form a seminal receptacle.

Reproductive habits have been observed for a number of species. In general, breeding may occur throughout the year and does not seem to be governed by temperature. Among epigean species, reproduction appears to be seasonal. In northern climates, females are brooding eggs or young during the spring. There can be one or two broods of young per year, depending on species. Typically, spring-born females can mature and breed in late summer. Few gravid females can be found between November and March.

When ready for copulation, a mature male seizes a female and carries her, ventral side down, beneath his body with his second to fourth legs. Such females either may be in the preadult instar or may already have shed the posterior half of their exoskeleton in the preadult molt. If a male seizes an immature female or one that has already been impregnated, he releases her almost immediately.

Little is known about the duration of the clasping and copulatory period, but in *Lirceus*, the American genus whose reproductive habits have been most carefully studied, it lasts up to 24 hours.

The female genital pores are not exposed until the preadult exoskeleton has been shed, and if the female has not already lost the posterior half of her exoskeleton, the clasping male usually aids in pulling it off. The male then assumes a copulatory position by sliding sideways so that his ventral surface is pressed against the side of the body of the female. This places the male genital pores and the first and second pleopods in the vicinity of the female genital pore on that side. As sperm are released, the male first and second pleopods undergo rapid vibratory movements that probably aid in transmitting the sperm into the female genital pore and into a seminal receptacle. The process is then repeated on the other side of the female's body, the entire bilateral copulation being completed in less than 2 hours.

The two individuals then separate, and the female promptly sheds the anterior half of her preadult exoskeleton, the oostegites thereby being transformed from small buds to large functional plates.

After an interval of as long as several days, the mature eggs pass down the oviducts, through the seminal receptacles, where they are fertilized, out the genital pores, and into the marsupium. Depending on the species and size of the female, the number of eggs per brood ranges between 20 and 250. Incubating eggs and newly hatched young are retained in the marsupium for 20 to 30 days. They are aerated by slow up-and-down movements of the oostegites and maxilliped movements that produce an anterior–posterior current of water through the marsupium. The young eventually find their way out of the posterior end of the marsupium by trial and error.

Whereas epigean species are active while brooding young, evidence suggests that troglobitic species become inactive and secretive.

Development, life cycle. First instar young have the general characteristics of adults. Little is known about the total number of instars during the isopod life cycle. Presumably there may be at least 15, of which the first 5 to 8 are probably preadult. Casting of the exoskeleton is completed in a few minutes to as much as three days. The process is initiated by a transverse split on the fourth thoracic segment. The posterior half of the exoskeleton is lost first, then, after a variable interval, the anterior half. The life span is thought to be about one year or less. Little specific information is available concerning the number of broods that may be produced during the female life cycle.

Ecology and distribution. Freshwater isopods seldom come into open waters but remain secreted under rocks, vegetation, and debris. They are primarily inhabitants of the unpolluted shallows, rarely being found in water more than a meter deep. *Lirceus lineatus* is a notable exception in having been collected at depths up to 55 m, and *Caecidotea racovitzai* has been found in the Great Lakes as deep as 42 m. Some species of asellids are indicators of organic pollution, inasmuch as they may be abundant in the "recovery" zone of streams polluted by domestic sewage.

Several species of marine and estuarine isopods regularly enter freshwater. Species of the Sphaeromatidae are often found in tidal freshwater but must return to saltwater to breed. A few parasitic species of the Cymothidae will remain attached to host fish well into freshwater before detaching themselves.

Most aquatic habitats usually contain only one species, and it is very unusual to find two or more species in the same habitat. In Massachusetts, Smith (1983) reported that *Caecidotea communis* and *C. racovitzai*, though sympatric, rarely occur together, suggesting that competition maintains separation of congeneric species. Competition between similar species of troglobitic *Caeci-*

dotea for space in cave streams is resolved somewhat by size differences (Culver and Ehlinger 1980). A few species, such as *Caecidotea forbesi*, *C. obtusus*, and *C. scrupulosus*, may occur in temporary ponds, where they burrow into the substrate during periods of drought, but adaptations for withstanding adverse environmenial conditions are not generally developed in this order.

Striking masses, or aggregations, of *Caecidotea*, consisting of scores to thousands of individuals have been observed in small streams. Such aggregations are thought to be formed chiefly as the result of reactions to current velocity. The migration of some individuals upstream as far as possible or, if the current is too swift, being washed downstream until they are able to maintain a footing has the net result of concentrating the isopods in a segment of the stream where the current is neither too fast nor too slow.

Except for *Sphaeroma terebrans*, all of our species of freshwater isopods are restricted to North America. Furthermore, the epigean genera *Caecidotea* and *Lirceus* are not known outside of North America. Conversely, several subterranean or interstitial genera, which have their nearest relatives in shallow marine environments, are found in several continents. *Microcharon* is a large genus of interstitial isopods barely over 1 mm in length. Most species are found in shallow marine habitats in the Mediterranean Sea and Atlantic Ocean. Freshwater interstitial species have been known from areas bordering the Mediterranean Sea for some time. Recently, the genus has been discovered in groundwater in the United States and reported inland as far as Montana. Another remarkable inland record is *Antrolina lira* (Plate 13), known only from cave waters of Virginia some 100 km from known marine deposits. This genus is a member of the largely marine family Cirolanidae, which is distributed throughout most of the globe. Other freshwater genera are known, but their ranges overlap documented marine incursions.

The genus *Caecidotea* is the most common and widespread genus of freshwater isopods in the United States. Most species, either epigean or hypogean, are found in the eastern half of the country, especially in the unglaciated interior plateaus and Appalachian Mountains. The number of species decline west of the Mississippi river and only a few are known from the entire west coast. However, in terms of genera, Texas has the most, with about 8 genera in 4 families occurring within the state.

Though the loss of sight is generally characteristic of species confined to subterranean waters, and functional eyes are characteristic of surface water species, there are some interesting exceptions and variations. *Caecidotea oculata* is found in springs and streams, yet the eyes are reduced. *Lirceus hoppinae*, with functional eyes, occurs in both caves and surface waters. *Caecidotea hobbsi* is subterranean and has eyes present or absent. *Caecidotea spatulata* is a secretive surface form, but the eyes may be present or absent. Finally, *Stenasellus* lives in caves, is pigmented but lacks eyes.

Economic significance. Most species of isopods are restricted to small lakes and streams; consequently, they are probably of little importance in the diet of fishes, except in hatcheries and ponds with large populations isopods. Economically, *Sphaeroma terebrans* is perhaps the most important species. It is a borer in salt and brackish waters, as well as in freshwater estuaries of the Gulf Coast, and causes extensive damage to wharves and piling. This species has numerous important relatives restricted to saltwater. *Lirceus brachyurus* sometimes becomes a pest in commercial beds of watercress. Several species of isopods serve as intermediate hosts for parasitic nematodes and Acanthocephala of birds, fishes, and amphibians.

Members of the Cymothidae are parasites of marine and anadromous fish. Some damage the gill tissue of several important food and game fish. A particularly problematic genus, *Vanamea* from South America, has been imported on aquarium fish to the southern United States where it has probably become established. It attaches to the tongue of small freshwater fish and does considerable damage.

Collection, preparation. Isopods are easily collected by hand picking, washing out aquatic vegetation, or with a small net. Seventy percent alcohol is a suitable killing agent and preservative. Appendages may be dissected off under a dissecting microscope and semipermanently mounted in glycerin jelly or Hoyers medium.

Taxonomy. Most emphasis has been placed on the detailed structure of the mature male second pleopods, pereopods, and shape of the carapace. These features are constant within each species and there are considerable differences from one genus and species to another.

Most studies in the United States have been scattered and uncorrelated. In 1918 only 21 species were known; now there are about 130, and the several active American investigators in this group have specimens of many more undescribed species in their collections. There are still differences of opinion as to which characters are of generic and specific rank. The status of *Caecidotea* and *Asellus* has been the object of a long controversy; some workers contend that the two should be united under *Caecidotea.* Most workers, however, maintain the two as separate, geographically disjunct genera and this policy has been followed in this manual.

The families and genera of freshwater isopods are rather well understood. Unfortunately, identification of most genera and species of the Asellidae is dependent on anatomical differences in the sclerotized tip of the endopod of the second pleopod of the male. However, familiarization with the appendage is a simple process and preparation for examination is relatively easy.

Several imported cymothid isopod parasites, thus far known from aquarium fish only, are not included in the key. These are *Artystone*, *Ichthyoxenus*, and *Vanamea* (Fig. 23.8C–E). Any or all of these could or have become established in southern states. See Hoffman (1999) for additional information and keys.

KEY TO FAMILIES AND GENERA OF FRESHWATER ISOPODA OF THE UNITED STATES

1. Free living or parasitic; morphologically unspecialized; body symmetrical 2
Parasitic in branchial chambers of shrimp (Palaemonidae, *Macrobrachium* and *Palaemonetes*), greatly modified for parasitic existence, appendages reduced, female asymmetrical (Fig. 23.4); 1 species . . . .**Suborder EPICARIDEA, BOPYRIDAE, *Probopyrus***

2. Parasitic of anadromous or euryhaline fish; all pereopods subchelate (prehensile) **Suborder FLABELLIFERA, CYMOTHIDAE, 4**
Free-living; only perieropods 1 or 1 and 2 subchelate (prehensile) 3

3. Uropods terminal (Fig. 23.1); epigean and hypogean . . . .**Suborder ASELLOTA, 11**
Uropods broad and inserted laterally on abdomen, forming with it a large, horizontally expanded, fanlike structure (Fig. 23.5B) . . . .**Suborder FLABELLIFERA, 6**

4. Junction of cephalothorax and first thoracic segment sinuous, producing a median sinus (Fig. 23.8E); a few species entering freshwater on east and west coasts . . . . ***Nerocila***
Junction of cephalothorax and first thoracic segment straight 5

5. Anterior segments of abdomen narrow, enclosed by lateral projections of last thoracic segment (Fig. 23.8A); darkly pigmented; 1 species on surface of coastal freshwater fish, southeast coast ***Anilocra***
Abdomen broader anteriorly, width uniform with thorax; lightly pigmented; in gill cavity of estuarine fishes; 1 common species, entire east and Gulf coasts . . . . ***Livoneca***

6. Abdomen with two distinct segments (Fig. 23.5B); eyes present, pigmented; up to 10 mm long **SPHAEROMATIDAE, 9**
Abdomen with six distinct segments (Fig. 23.5A); eyes absent and without pigment **CIROLANIDAE, 7**

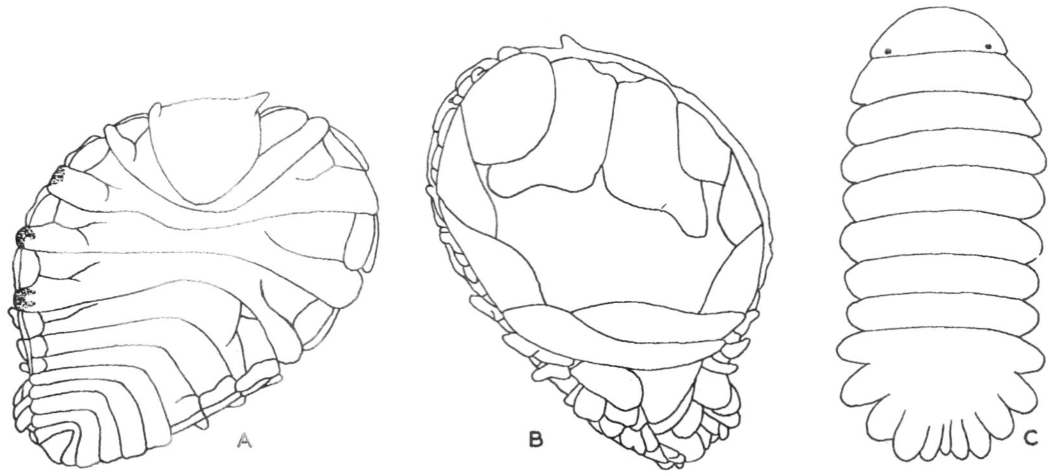

Fig. 23.4. *Probopyrus bithynis,* parasitic on *Palaemonetes ohionis.* A, dorsal view of female, ×16; B, ventral view of female, ×16; C, dorsal view of male, ×40. (Modified from Richardson 1904.)

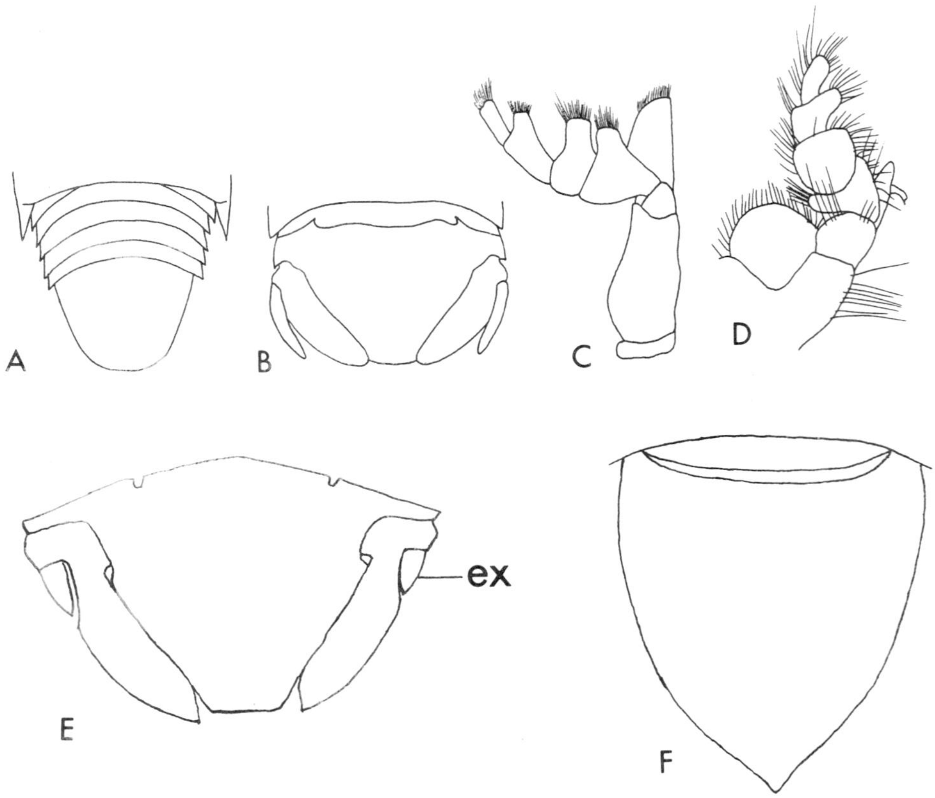

Fig. 23.5. Structure of Flabellifera. A, abdomen of *Cirolanides texensis;* B, posterior end of *Thermosphaeroma;* C. palp of maxilliped of same; D, palp of maxilliped of *Sphaeroma terebrans;* E, pleotelson of *Cassidinidea;* F pletotelson of *Speocirolana hardeni.* ex, exopod. (A–C modified from Richardson 1905, courtesy of the American Museum of Natural History; D modified from Van Name 1936; F redrawn from Bowman 1992.)

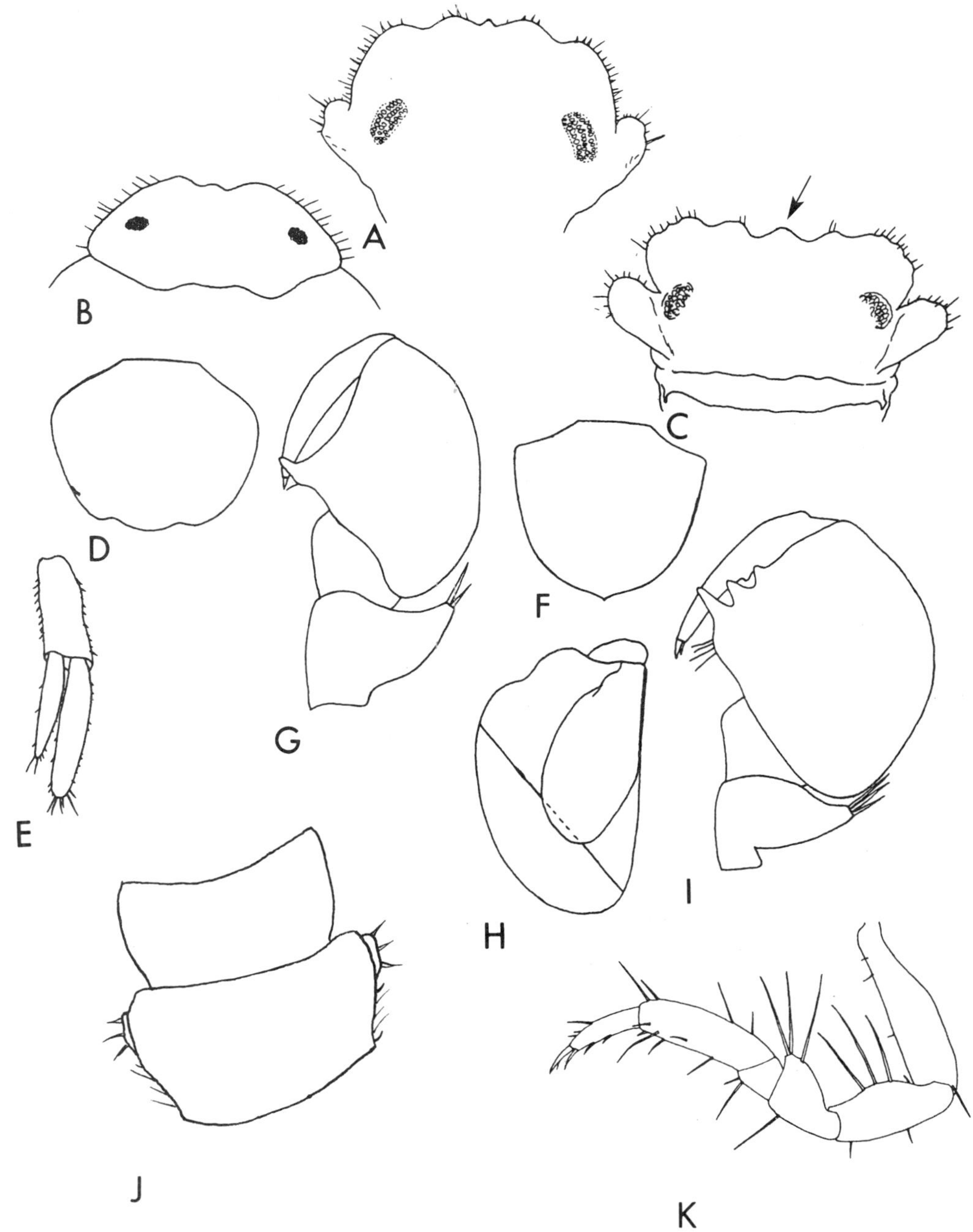

Fig. 23.6. Structure of asellids. A, cephalothorax of *Lirceus fontinalis;* B, cephalothorax of *L. brachyurus;* C, cephalothorax of *L. lineatus;* D, abdomen of *L. hoppinae;* E, uropod of *L. hoppinae;* F, abdomen of *L. fontinalis;* G, gnathopod of *L. brachyurus;* H, third pleopod of *L. brachyurus;* I, gnathopod of *L. garmani;* J, cephalothorax and first thoracic segment of *Remasellus;* K, pereopod of *Remasellus.* Arrow, carina. (A and C modified from Racovitza 1920a; D, E, G, and I modified from Hubricht and Mackin 1949; J, K redrawn from Bowman and Sket 1985).

7. Pereopods 1–3 subchelate; pleotelson somewhat acute (Fig. 23.5F); 1 species in southern TX . . . ***Speocirolana***

Only pereopod 1 subchelate; pleotelson rounded or broadly acute . . . **8**

8. Exopod of second pleopod and endopods of third to fifth pleopods two-segmented; pleotelson rounded (Fig. 23.5A); 1 species in cave systems of south central TX . . . ***Cirolanides***

Exopod of second pleopod and endopods of third to fifth pleopods one-segmented; pleotelson broadly acute; 1 species, caves of VA . . . ***Antrolana***

9. Exopods of uropods very small, set within border of endopod (Fig. 23.5E); 1 species entering freshwater along northeast coast . . . ***Cassidinidea***

Exopods of uropods larger, extending beyond border of endopod . . . **10**

10. Second, third, and fourth segments of palp of maxilliped not produced into lobes (Fig. 23.5D); outer margins of uropods serrate; a borer in timbers, roots, and mangroves in fresh, brackish, and salt waters; FL to TX . . . ***Sphaeroma***

Second, third, and fourth segments of palp of maxilliped produced into lobes (Fig. 23.5C); outer margins of uropods smooth; one species in hot springs near Socorro, NM, and another in hot springs in western TX . . . ***Thermosphaeroma***

11. Body vermiform, minute (Fig. 23.1B), less than 2 mm in length; abdomen with 1 free segment and pleotelson; eyeless; 1 or 2 species in groundwater, GA, MT . . . **MICROPARASELLIDAE,** ***Microcharon***

Body dorsoventrally flattened; 3–15 mm in length; abdomen with 2 free segments and pleotelson; with or without eyes and pigment . . . **12**

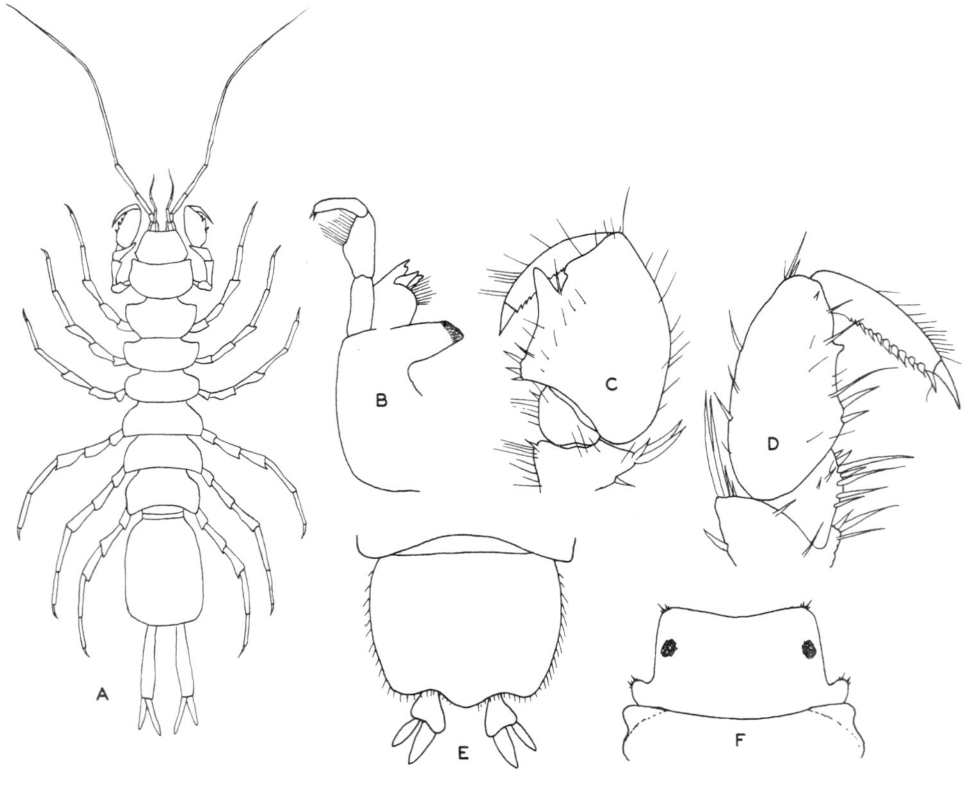

Fig. 23.7. Structure of *Caecidotea.* A, *C. stygius,* ×7 (setation omitted), a subterranean species; B, mandible of *Caecidotea;* C, gnathopod of male *C. communis;* D, gnathopod of female *C. communis;* E, posterior end of *C. brevicaudus;* F, anterior end of *C. intermedius.* (A modified from Hay 1903; B modified from Huntsman 1915; C and D modified from Racovitza 1920a.)

12. Terminal segment of exopod of male pleopod 2 acute with single terminal seta (Fig. 23.9A), endopod U-shaped with a free lobe encircling a style; color bright red in life but without eyes; 1 species, TX . . . . **STENASELLIDAE, *Mexistenasellus***

Terminal segment of exopod of male pleopod 2 broad and rounded, with 0–20 setae, endopod L-shaped or straight; with or without eyes and pigment, never red . . . . . . **ASELLIDAE, 13**

13. Lateral margins of cephalothorax parallel, anterior margin smoothly concave (Fig. 23.6J); pereopod 1 not subchelate (Fig. 23.6K); able to swim weakly; eyeless and unpigmented; 1 species in FL cave systems . . . . ***Remasellus***

Lateral margins of cephalothorax convergent, anterior margin of head not concave, or if so with uneven margin; pereopod 1 subchelate, raptorial, used for grasping (Fig. 23.6G, I); with or without eyes and pigment **14**

14. Distal segment of male pleopod 2 exopod with 0–4 setae (Fig. 23.9B); very small (2–4 mm), blind, unpigmented; 4 hypogean species, TX . . . . ***Lirceolus***

Distal segment of male pleopod 2 exopod setose, with 10 or more setae; with or without eyes and pigment **15**

15. Anterior margin of cephalothorax with a low but distinct carina (Figs. 23.6A–C) **16**

Anterior margin of cephalothorax without a carina (Figs. 23.1A; 23.7A, F; 23.9F) **17**

16. Endopod of pleopod 2 of male with long setae entrained within a groove in the distal portion of endopod (Fig. 23.9D); eyeless and without pigment; 1 species, MT . . . . ***Salmasellus***

Endopod of pleopod 2 of male otherwise; up to 25 mm long; eyes present; only one hypogean species; eastern states as far west as the Great Plains; at least 15 epigean species . . . . ***Lirceus***

17. Endopod of pleopod 2 of the male with a distinct, curved spur or projection at the base (Fig. 23.9E, G) **18**

Endopod of pleopod 2 of male with short, blunt spurs or spurs absent; about 80 epigean and subterranean species in the United States . . . . ***Caecidotea***

18. Endopod of pleopod 2 of male 2 or more times longer than the exopod, with a long, filiform cannula and an associated long lobe (Fig. 23.9E); eyeless and without pigment; 2 species in CA . . . . ***Calasellus***

Endopod of pleopod 2 of male less then twice length of exopod (Fig. 23.9G), cannula not long and filiform; with eyes; 1 introduced species, CA . . . . ***Asellus***

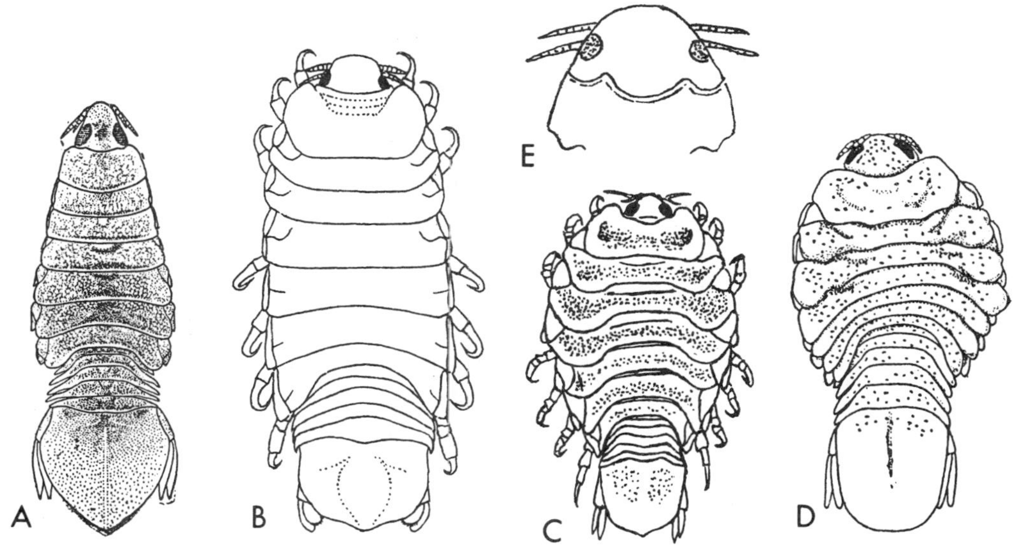

Fig. 23.8. Parasitic cymothid isopods. A, *Anilocra*, ×2; B, *Vanamea*, ×4; C, *Artystone*, ×4; D, *Ichthyoxenus*, ×5; E, *Nerocila*, ×5. (A from Richardson 1910; B, C from Van Name 1936, courtesy American Museum of Natural History; D from Richardson 1913; E modified from Van Name 1936.)

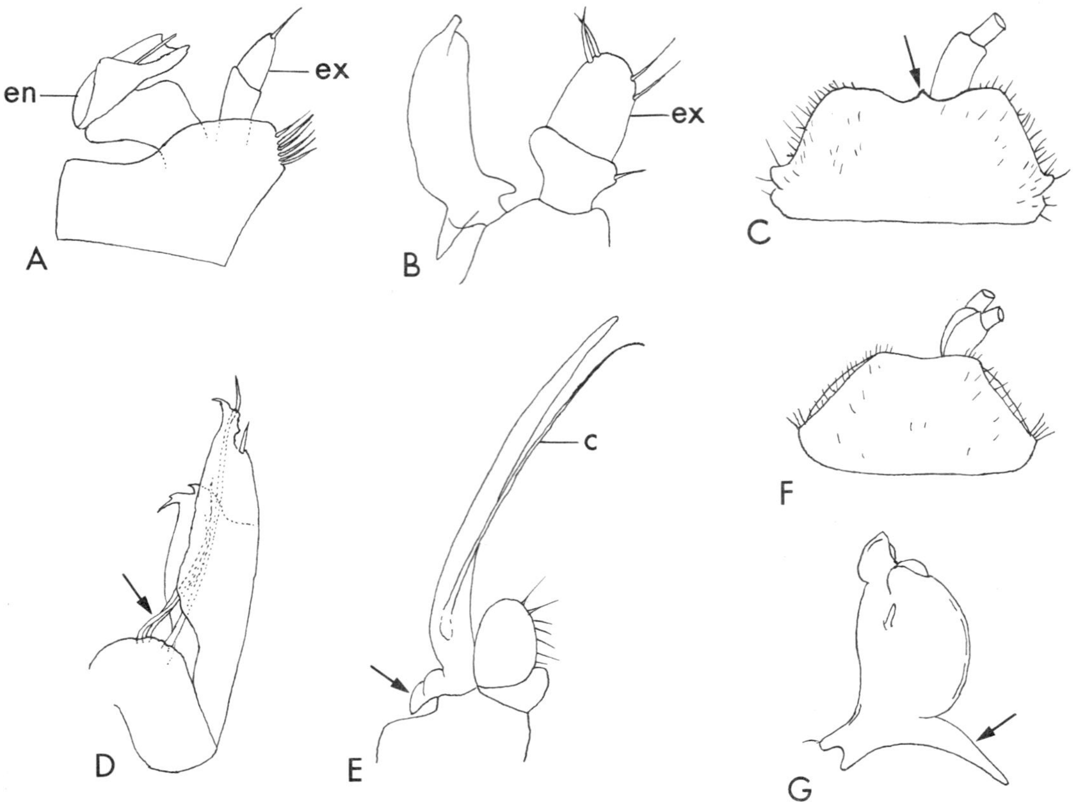

Fig. 23.9. Structure of Stenasellidae and Asellidae. A, male pleopod 2 of *Mexistenasellus;* B, male pleopod 2 of *Lirceolus;* C, cephalothorax of *Salmasellus*, arrow denotes carina; D, endopod of male pleopod 2 of *Salmasellus*, arrow denotes setae; E, male pleopod 2 of *Calasellus*, arrow denotes basal spur; F, cephalothorax of *Calasellus;* G, endopod of male pleopod 2 of *Asellus*, arrow denotes spur. c, cannula or style; en, endopod; ex, exopod. (A redrawn from Cole and Minckley 1972; B redrawn from Lewis and Bowman 1996; C, D redrawn from Bowman 1975; E, F redrawn from Bowman 1981a; G redrawn from Birstein 1964.)

REFERENCES

Allee, W. C. 1929. Studies in animal aggregations: natural aggregations of the isopod, *Asellus communis*. *Ecology* 10:14–36.

Birstein, Y. A. 1964. Crustacea Freshwater Isopods (Asellota). Fauna of U.S.S.R. 7(5):1–148. (Isreal Program Scientific Translations, Jerusalem.)

Bowman, T. E. 1964. *Antrolana lira*, a new genus and species of troglobitic cirolanid isopod from Madison cave, Virginia. *Int. J. Speleol.* 1:229–236.

——. 1967. *Asellus kenki*, a new isopod crustacean from springs in the eastern United States. *Proc. Biol. Soc. Wash.* 80:131–140.

——. 1974. The California freshwater isopod, *Asellus tomalensis*, rediscovered and compared with *Asellus occidentalis*. *Hydrobiologia* 44:431–441.

——. 1975. Three new troglobitic asellids from western North America (Crustacea: Isopoda: Asellidae). *Int. J. Speleol.* 7:339–356.

——. 1981a. *Calasellus longus*, a new genus and species of troglobitic asellid from Shaver Lake, California (Crustacea: Isopoda: Asellidae). *Proc Biol. Soc. Wash.* 94:866–872.

——. 1981b. *Thermosphaeroma milleri* and *T. smithi*, new sphaeromatid isopod crustaceans from hot springs in Chihuahua, Mexico, with a review of the genus. *J. Crust. Biol.* 1:105–122.

——. 1992. Two subterranean aquatic isopod crustaceans new to Texas: *Mexistenasellus coahuila* (Cole and Minckley, 1972) (Asellota: Stenasellidae) and *Speocirolana hardeni*, new species (Flabellifera: Cirolaniidae). *Texas Mem. Mus. Speleol. Monogr.* 3:23–30.

Bowman, T. E., and G. Longley. 1976. Redescription and assignment to the new genus *Lirceolus* of the Texas troglobitic water slater, *Asellus smithii* (Ulrich) (Crustacea: isopoda: Asellidae). *Proc. Biol. Soc. Wash.* 88:489–496.

Bowman, T. E., and B. Sket. 1985. *Remasellus*, a new genus for the troglobitic asellid isopod, *Asellus parvus* Steeves. *Proc. Biol. Soc. Wash.* 98:554–560.

Coineau, N. 1994. Evolutionary biogeography of the microparasellid *Microcharon* (Crustacea) in the Mediterranean basin. *Hydrobiologia* 287:77–93.

Cole, G. A., and C. A. Bane, 1978. *Thermosphaeroma subequalum*, n. gen., n. sp. (Crustacea: Isopoda) from Big Bend National Park, Texas. *Hydrobiologia* 59:223–228.

Cole, G. A., and W. L. Minckley. 1972. Stenasellid isopod crustaceans in the Western Hemisphere—a new genus and species from Mexico: with a review of other North American freshwater isopod genera. *Proc. Biol. Soc. Wash.* 84:313–326.

Collins, T. L., and J. R. Holsinger. 1981. Population ecology of the troglobitic isopod *Antrolina lira* Bowman (Cirolanidae). *Proc. 8th Internat. Cong. Speleol. (Bowling Green, KY)* 1:129–132.

Culver, D. C., and T. J. Ehlinger. 1980. Effects of microhabitat size and competitor size on two cave isopods. *Brimleyana* 4:103–113.

Ellis, R. J, 1961. A life history study of *Asellus intermedius* Forbes. *Trans. Am. Microsc. Soc.* 80: 80–102.

——. 1971. Notes on the biology of the isopod *Asellus tomalensis* Harford in an intermittent pond. *Trans. Am. Microsc. Soc.* 90:51–61.

Fleming, L. E. 1973. The evolution of North American isopods of the genus *Asellus* (Crustacea: Asellidae). *Int. J. Speleol.* 5:283–310.

Giere, O. 1993. *Meiobenthology—The Microscopic Fauna in Aquatic Sediments.* 328 pp. Springer, Berlin.

Hatchett, S. P. 1947. Biology of the Isopoda of Michigan. *Ecol. Monogr.* 17:47–79.

Hay, W. P. 1903. Observations on the crustacean fauna of Nickajack Cave, Tennessee, and vicinity. *Proc. U.S. Nat. Mus.* 25:417–439.

Henry, J. P., and G. Magniez. 1970. Contribution a la systematique des amenities (Crustacea Isopoda). *Ann. Speleol.* 25:335–367.

Hoffman, G. L. 1999. *Parasites of North American Freshwater Fishes.* 539 pp. Cornell University Press, Ithaca, NY.

Holsinger, J. R., and T. E. Bowman. 1973. A new troglobitic isopod of the genus *Lirceus* (Asellidae) from southwestern Virginia, with notes on its ecology and additional cave records for the genus in the Appalachians. *Int. J. Speleol.* 5:261–271.

Holsinger, J. R., and H. R. Steeves. 1971. A new species of subterranean isopod crustacean (Asellidae) from the central Appalachians, with remarks on the distribution of other isopods of the region. *Proc. Biol. Soc. Wash.* 84:189–199.

Holsinger, J. R., et al. 1994. Biogeographic and ecological implications of newly discovered populations of the stygobiont isopod crustacean *Antrolina lira* Bowman (Cirolanidae). *J. Nat. Hist.* 28:1047–1058.

Huber, B, A. 1992. Frontal heart and arterial system in the head of Isopoda. *Crustaceana* 63:57–69.

Hubricht, L., and J. G. Mackin. 1949. The freshwater isopods of the genus *Lirceus* (Asellota, Asellidae). *Am. Midl. Nat.* 42:334–349.

Huntsman, A. G. 1915. The fresh-water Malacostraca of Ontario. *Contrib. Canada Biol. II (Suppl.)* 47:145–163.

Jass, J., and B. Klausmeier. 1997. Wisconsin freshwater isopods (Asellidae). *Univ. Wisc. Field Sta. Bull.* 30:10–18.

Kerr, J. R. 1978. Some aspects of life history and ecology of the isopod *Asellus r. racovitzai in* western and central Lake Erie. *Ohio J. Sci.* 78:298–300.

Lewis, J. J. 1983. The assignment of the Texas troglobitic water stater *Caecidotea pilus* to the genus *Lirceolus*, with an amended diagnosis of the genus (Crustacea: Isopoda: Asellidae). *Proc. Biol. Soc. Wash.* 96:145–148.

Lewis. J. J., and T. E. Bowman. 1981. The subterranean asellids (*Caecidotea*) of Illinois (Crustacea: Isopoda: Asellidae). *Smithson. Contrib. Zool.* 335:1–66.

——. 1996. The subterranean asellids of Texas (Crustacea: Isopoda: Asellidae). *Proc. Biol. Soc. Wash.* 109:482–500.

Mackin, J. G., and L. Hubricht. 1938. Records of the distribution of species of isopods in Central and Southern United States, with descriptions of four new species of *Mancasellus* and *Asellus* (Asellota, Asellidae). *Am. Midl. Nat.* 19:628–637.

——. 1940. Descriptions of seven new species of *Caecidotea* (Isopoda, Asellidae) from central United States. *Trans. Am. Microsc. Soc.* 59:383–397.

Magniez, G. J., and J. D. Toft. 2000. Presence d'un Asellide epige originaire d'Extreme-Orient en Californie. *Bull. Mensuel Soc. Linn. Lyon* 69:127–132.

Markus, H. C. 1930. Studies on the morphology and life history of the isopod, *Mancasellus macrourus*. *Trans. Am. Microsc. Soc.* 49:220–237.

Menzies, R. J. 1954. A review of the systematics and ecology of the genus "*Exosphaeroma*," with the description of a new genus, a new species, and a new subspecies (Crustacea, Isopoda, Sphaeromidae). *Am. Mus. Novit.* 1683:1–24.

Miller, M. A. 1933. A new blind isopod, *Asellus californicus*, and a revision of the subterranean asellids. *Univ. Calif Publ. Zool.* 39:97–110.

Minckley W. L. 1961. Occurrence of subterranean isopods in epigean waters. *Am. Midl. Nat.* 66:452–455.

Muzzall, P. M. 1981. Parasites of the isopod, *Caecidotea communis*, and amphipod, *Hyalella azteca*, in New Hampshire. *Proc. Helminthol. Soc. Wash.* 48:91–92.

Needham, A. E. 1941. Abdominal appendages of *Asellus*, II. *Q. J. Microsc Sci.* 83:61–89.

Racovitza, E. G. 1920a. Notes sur les isopodes, 6: *Asellus communis* Say; 7: Les pléopodes I and II dés asellides; morphologie et developpement. *Arch. Zool. Exp. Gen.* 58:79–115.

——. 1920b. Notes sur les isopodes, 8: *Mancasellus tenax* (Smith); 9: *Mancasellus macrourus* (Garman). *Arch. Zool. Exp. Gen.* 59:28–66.

Richardson, H. 1904. Contributions to the natural history of the Isopoda. *Proc. U.S. Nat. Mus.* 27:1–89.

——. 1905. A monograph of the isopods of North America. *Bull. U.S. Nat. Mus.* 54:1–727.

——. 1910. Description of a new species of *Anilocra* from the Atlantic coast of North America. *Proc. U.S. Nat. Mus.* 39:137–138.

——. 1913. The isopod genus *Ichthyoxenus* Herklots, with a description of a new species, from Japan. *Proc. U.S. Nat. Mus.* 45:559–562.

Schultz, G. A. 1969. *How to Know the Marine Isopod Crustaceans.* 359 pp. Wm. C. Brown, Dubuque, IA.

Schultz, T. W. 1973a. Digestive anatomy of *Lirceus fontinalis* (Crustacea: Isopoda). *Trans. Am. Microsc. Soc.* 92:13–25.

——. 1973b. Functional morphology of the oral appendages and foregut of *Lirceus garmani* (Crustacea: Isopoda). *Trans. Am. Microsc. Soc.* 92:349–364.

Shuster, S. M. 1981. Life history characteristics of *Thermosphaeroma thermophilum*, the Socorro isopod (Crustacea: Peracarida). *Biol. Bull.* 161:291–302.

Smith, D. G. 1983. A comparison of the distribution of two isopod species (*Caecidotea*) in Massachusetts. *Ohio J. Sci.* 83:122–125.

——. 1997. An annotated checklist of malacostracans (Crustacea) inhabiting southern New England fresh waters. *J. Freshwater Ecol.* 12: 217–223.

Stanford, J. A., et al. 1994. Ecology of the alluvial aquifers of the Flathead River, Montana. In: *Groundwater Ecology*, pp. 367–390, edited by J. Gibert et al. Academic, San Diego.

Steeves, H. R., III. 1963. The troglobitic asellids of the United States: the Stygius group. *Am. Midl. Nat.* 69:470–481.

——. 1964. The troglobitic asellids of the United States; the Hobbsi group. *Am. Midl. Nat.* 71:445–451.

——. 1966. Evolutionary aspects of the troglobitic asellids of the United States: the hobbsi, stygius and cannulus groups. *Am. Midl. Nat.* 75:392–403.

——. 1968. Three new species of troglobitic asellids from Texas. *Am. Midl. Nat.* 79:183–188.

——. 1969. The origin and affinities of the troglobitic asellids of the southern Appalachians. *Va. Polytech. Inst. Res. Div. Monogr.* 1:51–65.

Steeves, H. R., III, and J. R. Holsinger. 1968. Biology of three new species of troglobitic asellids from Tennessee. *Am. Midl. Nat.* 80:75–83.

Strayer, D. L., et al. 1994. An endemic groundwater fauna in unglaciated eastern North America. *Can. J. Zool.* 73:502–508.

Styron, C. E. 1969. Taxonomy of two populations of an aquatic isopod, *Lirceus fontinalis* Raf. *Am. Midl. Nat.* 82:402–416.

Unwin, E. E. 1921. Note upon the reproduction of *Asellus aquaticus*. *J. Linn. Soc.* 34:335–343.

Van Name, W. G. 1936. The American land and freshwater isopod Crustacea. *Bull. Am. Mus. Nat. Hist.* 71:1–535.

——. 1940. A supplement to the American land and fresh-water Isopoda. *Bull. Am. Mus. Nat. Hist.* 77:109–142.

——. 1942. A second supplement to the American land and fresh-water isopod Crustacea. *Bull. Am. Mus. Nat. Hist.* 80:299–329.

Williams, W. D. 1970. A revision of North American epigean species of *Asellus* (Crustacea: Isopoda). *Smithson. Contrib. Zool.* 49:1–80.

——. 1972. Freshwater isopods (Asellidae) of North America. *Biota of Freshwater Ecosystems, Identification Manual* 7:1–45.

24

AMPHIPODA

(Scuds, sideswimmers)

Like the Decapoda, the Amphipoda are chiefly marine, only about 150 described American species being confined to freshwaters. Worldwide there are about 900 freshwater species. They occur in a wide variety of unpolluted lakes, ponds, streams, brooks, springs, and subterranean waters. With the exception of one species that may swim about as a plankter in lakes, they are all more or less confined to the substrate. The layman often calls amphipods "freshwater shrimp," but of course they are only distantly related to (marine) shrimp.

General characteristics. The great majority of species are 5 to 20 mm long. The body is laterally compressed and consists of a cephalothorax (first thoracic segment fused with the head), seven free thoracic segments, a six-segmented abdomen, and a small, terminal telson (Figs. 24.1; 24.2). The pleura of the first four abdominal segments are prolonged ventrally so that these segments are deeper than any of the other body segments. In some species the last three abdominal segments are more or less fused.

Eyes are usually well developed in species living in surface waters, but subterranean species show varying degrees of degeneration or absence of eyes. They are unstalked, compound, and circular, oval, reniform, or elongated.

Both pairs of antennae range from short to long. Each first antenna consists of a three-segmented peduncle and a long flagellum; sometimes there is a very small accessory flagellum (Fig. 24.2A, B). Each second antenna has a flagellum and a five-segmented peduncle, but the basal segment of the peduncle is fused with the head, and the second and third segments are very short; only the last two segments are well developed.

Like that of many crustaceans, the exoskeleton of Amphipoda is covered by a variety of minute structures involved with sensory functions. Many of these structures are evident under ordinary magnification and have taxonomic value.

The mouthparts are relatively small, compactly arranged, and hidden by the basal segments of the appendages of the first thoracic segment. They include an upper lip; a pair of tearing and cutting mandibles; a pair of laminar and spinous first maxillae; a pair of small, flexible second maxillae; a lower lip; and a pair of maxillipeds. The maxillipeds are homologs of the first maxillipeds of decapods; each consists of a small inner plate, a small outer plate, and a long palp.

The seven pairs of thoracic legs all have seven segments corresponding to the segments of crayfish legs. Beginning proximally, these are as follows: coxa, basis, ischium, merus, carpus, propodus, and dactylus. The coxae are usually called coxal plates; they are greatly enlarged extensions of the coxae, flattened, relatively immovable, and project ventrally at the edges of the body.

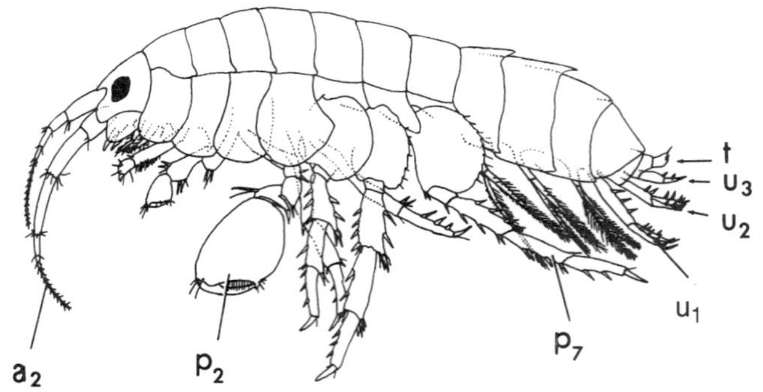

Fig. 24.1. *Hyalella azteca,* ×14, from Massachusetts. a2, antenna 2; p2, pereopod 2; p7, pereopod 7; t, telson; u1, uropod 1; u2, uropod 2; u3, uropod 3.

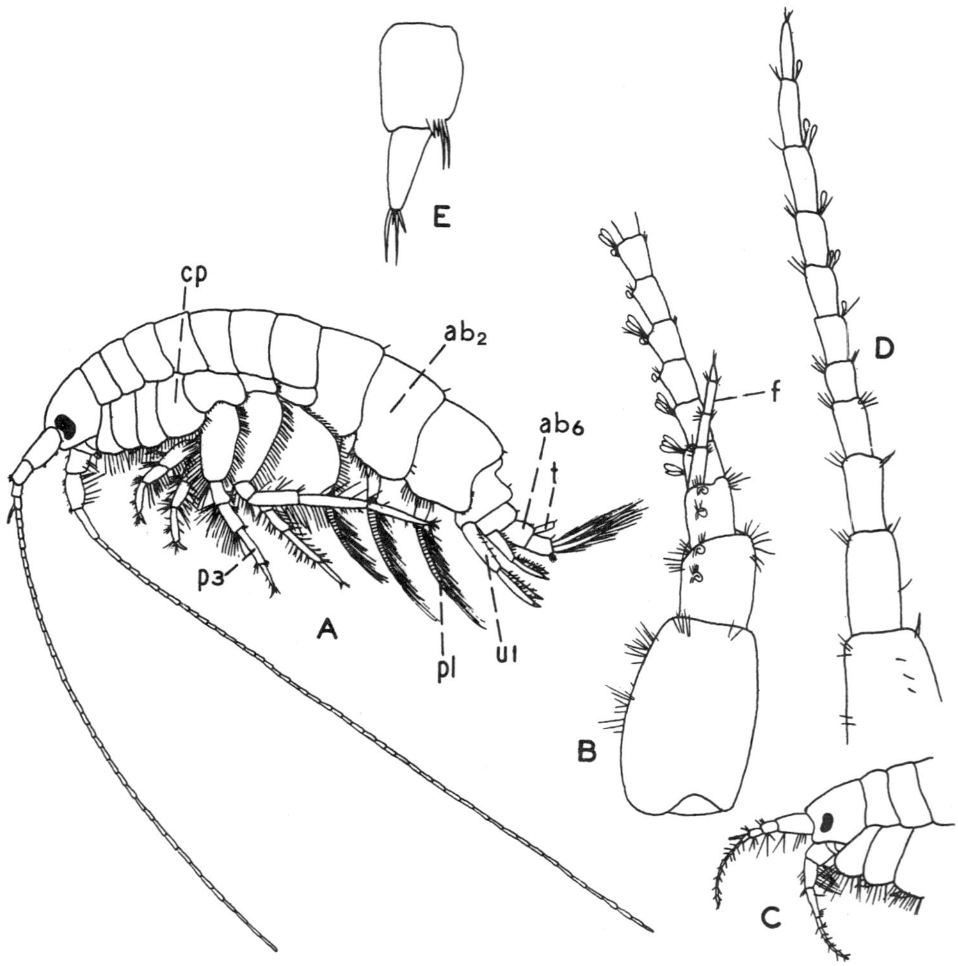

Fig. 24.2. Structure of Amphipoda. A, male *Diporeia hoyi,* ×9; B, basal portion of first antenna of male *D. hoyi;* C, anterior end of female *D. hoyi;* D, first antenna of female *Hyalella azteca;* E, third uropod of *H. azteca;* ab2, second abdominal segment; ab6, sixth abdominal segment; cp, coxal plate; f, flagellum; p3, pereopod 3; pl, pleopod 3; t, telson; ul, uropod 1. (A–C modified from Segerstråle 1937; D redrawn from Geisler 1944; E redrawn from Weckel 1907.)

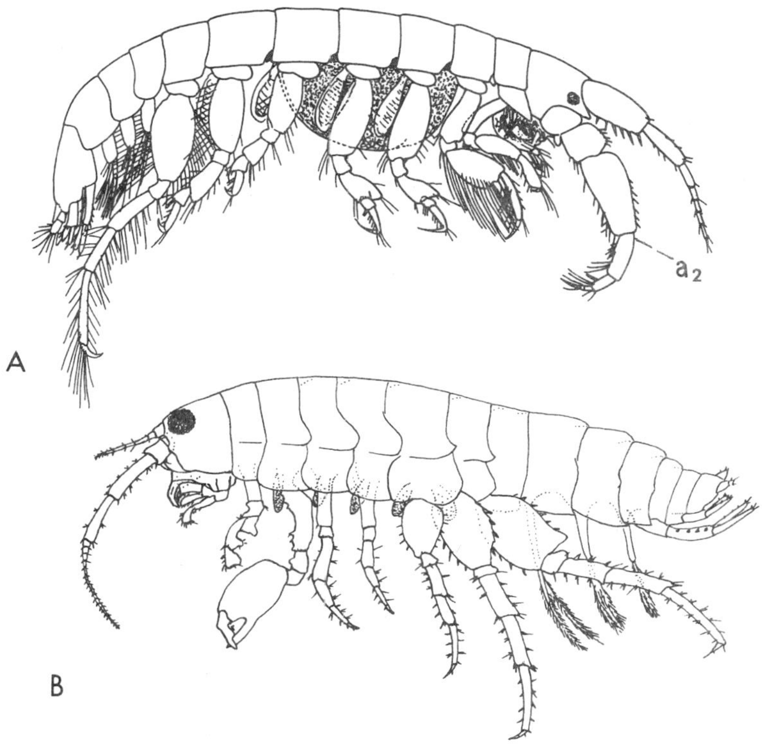

Fig. 24.3. Coastal Amphipoda. A, *Corophium*, ×10; B, *Chelorchestia*, ×10. a2, antenna 2. (A from Barnard 1971, B © D.G. Smith.)

The first two pairs of legs, known as gnathopods, are subchelate and adapted for grasping (Fig. 24.4I), but the third to seventh legs are relatively unspecialized pereopods. The first two pereopods are usually flexed forward, the last three backward.

The coxal gills are flattened oval sacs extending downward from the inner surface of the upper posterior corner of the coxal plates. They occur on the second to seventh or second to sixth legs. In addition, there are sometimes very small lateral sternal gills on some of the thoracic segments (Fig. 24.7B).

The first three abdominal segments bear paired pleopods. Each consists of a basal peduncle and two flexible, many-segmented rami. The last three abdominal segments bear paired uropods. These appendages are never very movable, always directed posteriorly, and closely approximated. The peduncles are unsegmented, but the rami are sometimes composed of two segments, the terminal one being small.

Coloration in living amphipods is occasionally brilliant, but preserved specimens bleach out so that they are whitish, gray, or cream-colored. Some common species show a wide range in color from one place or time to another. *Hyalella azteca*, for example, is usually light brown to greenish, but bluish, purple, dark brown, or reddish populations are often found. *Synurella dentata* is usually gray blue but sometimes lavender, brown, or greenish. *Gammarus fasciatus* is whitish, with the body and proximal segments of the appendages banded with green or brown. Very little is known about the causes of color variations within single species, but diet, temperature, and age are presumably important factors. Subterranean amphipods are almost invariably whitish, cream-colored, gray, or translucent.

The sexes are most readily separated on the basis of large marsupial plates, or oostegites, on the inner surface of the coxal plates of some of the legs (usually second to fifth) of the female. Males often have the larger gnathopods.

Locomotion. In general, amphipods are much more active at night than during daylight. The pereopods are used primarily for holding or stabilizing position, less so for crawling and walking. Swimming is the essential locomotor function and is aided by pushing or thrusting with the relatively inflexible uropods. The last two pereopods are flexed outward somewhat and are important in a clumsy, skittering sort of locomotion when the animal is on its side, a form best developed in *Gammarus*. The pleopods are the chief locomotor appendages for rapid undulatory swimming just above the substrate, the body being kept well straightened and the pereopods directed backward. As an amphipod swims it often rolls over on its side or back (hence the name "sideswimmer"). The few tube building forms that enter freshwater rely on the expanded antennae (Fig. 24.3A) for movement up and down the tube. Especially when disturbed, many amphipods will wriggle into the superficial layers of soft substrates.

Food, feeding. Amphipods are voracious feeders, all kinds of animal and plant matter being consumed. Only rarely do they attack and feed on living animals, but freshly killed animals are consumed readily. Species occurring in aquatic vegetation may often be seen browsing on the film of microscopic plants, animals, and organic debris covering leaves, stems, and other substrates (aufwuchs). Like the decapods, then, scuds are omnivorous, general scavengers or detritus feeders. *Diporeia* is a notable detritus feeder.

The food mass is held by the gnathopods and anterior pereopods and chewed directly without being first torn into smaller pieces. A few species, however, are filter feeders, at least in part.

Internal anatomy, physiology. Food passes through a short esophagus and into a small stomach containing a gastric mill. The long, undifferentiated intestine opens via the anus just below the telson. Six or nine caeca are associated with the digestive tract. The tubular heart is in the dorsal portion of the thorax. It has three pairs of ostia and anterior and posterior aortas. Although it is possible that some respiration occurs through the general body surface, the thin-walled coxal gills are undoubtedly the chief organs for the oxygen–carbon dioxide exchange. The beating of the pleopods creates a current of water over the gills.

Antennal glands, probably osmoregulatory and excretory, open at the base of the second antennae. The typical nervous system consists of two ventral nerve cords united at intervals by large segmental ganglia, two circumesophageal connectives, and a supraesophageal mass, or "brain." Both pairs of antennae are olfactory and highly sensitive to touch. They sometimes bear minute, stalked, clublike, or wine-glass-shaped sensory appendages (Fig. 24.2B, D), presumably comparable to the aesthetasks of copepods.

Reproduction. Testes and ovaries are paired, elongated strands ventral to the heart. The oviducts open at the base of the fifth coxal plates, and the male genital pores open on papillae on the ventral side of the last thoracic segment.

With the exception of *Hyalella azteca* and two or three other species, our knowledge of the details of reproductive behavior is rather sketchy. Syngamic reproduction is the rule, but males are sometimes rare. Most common species breed sometime between February and October, depending largely on water temperatures. Mature males pair for the first time when they are in their ninth instar, females in their eighth (nuptial) instar. When two males meet both are active and fall to pair and soon separate, but a female is passive when seized by a male.

In *Gammarus*, *Hyalella azteca*, and one species of *Synurella*, pairing consists of the males carrying the females on their backs and keeping them in place with their gnathopods. Paired individuals feed and swim about for 1 to as much as 7 days, or until the female is ready to molt to the first adult instar. The two animals then separate for a few minutes to several hours while the female loses her old exoskeleton. The male then returns, the two individuals pair again, and copulation usually occurs within the subsequent 24 hours.

Copulation and transfer of sperm are completed in less than a minute. The male extends the posterior part of his body around and to the ventral side of the female until the tips of his uropods touch the marsupium, which is formed by the marsupial plates, or oostegites. Ejected sperm are then swept into the marsupium by movements of the female pleopods. This process of sperm transfer may be repeated several times at intervals of a few minutes, the male holding on to the female all the time.

The animals finally separate, and the female promptly releases her eggs from the oviducts into the marsupium where they are fertilized. The number of eggs released by the female depends on her size and age as well as the particular species. In general, the larger and older females release the larger number of eggs. One species of *Stygobromus* releases only one or two eggs at a time, but the more common species usually average from 15 to 50 eggs per brood. *Hyalella azteca* averages about 18 eggs per brood.

This same species has been shown to have "strains" in different lakes; these do not interbreed.

Development, life cycle. The oostegites are bordered with stiff hairs that are hooked distally and entangled somewhat, and the developing eggs are therefore retained securely within the marsupium. In the few species that have been carefully studied, the incubation period ranges from 1 to 3 weeks. Newly hatched young are retained in the marsupium an additional 1 to 8 days and are released to the outside when the mother has her first molt following copulation.

Judging from the many observations that have been recorded, the females of most species produce only a single brood during the life cycle. A good example of a demonstrative annual life cycle is *Crangonyx aberrans*, restricted to wetlands of southeastern New England. The writer (DGS) has observed this species for 2 decades and determined aspects of its biology. In spring, males and females mature and breed. Females deposit eggs, usually between early April and mid-May, into a marsupium formed by elongate oostegites. The eggs hatch and larvae remain with the parent through a few molts. Following dispersal of the juveniles, from late May until mid-June, the females become senescent. Adult females are gone by late June–early July. The small males live a bit longer than the females. During summer when water level is low, the small juveniles sequester themselves in the wet, cool root systems of skunk cabbage (*Symplocarpus*), tussocks of sedges and grasses (e.g., *Scirpus*), and cattails (*Typha*), and growth is arrested. In the fall, when wet conditions return, growth resumes, although very slowly. Growth is arrested during winter and does not resume until spring, at which time it becomes rapid.

In a few species, each female customarily produces a series of broods during the breeding months. *Hyalella azteca*, for example averages 15 broods in 152 days (Embody 1912). Copulation is necessary for the production of each brood, so that females pair for the second or subsequent time while they are still carrying their previous brood of eggs or young in the marsupium. These events therefore closely follow each other within a few hours: (1) male leaves female, (2) female molts and releases young, (3) male and female pair again, (4) release of eggs to marsupium, (5) repeated copulation and fertilization, and (6) male leaves female.

Hyalella azteca has a minimum of nine instars in its life history (Geisler 1944). As in all other species, development is direct, and the newly hatched young have all the adult appendages. The first five instars form the juvenile period, during which the sexes are indistinguishable. In the sixth and seventh instars, or adolescent period, the sexes can be differentiated; the female has small oostegites and a few small eggs in the ovaries, and the male has slightly enlarged gnathopods. Pairing occurs for the first time in the eighth, or nuptial, instar, and the ninth and subsequent instars form the adult period. The number of molts that may occur during the adult period is variable but may be as high as 15 or 20. In northern areas, *Hyalella azteca* is univoltine, with major reproduction occurring during early summer, and a life span of about one year.

When the old exoskeleton is ready to be cast, a dorsal, transverse split appears halfway around the body between the first and second thoracic segments or occasionally between the cephalothorax and first thoracic segment. Sometimes there are accessory longitudinal splits along the upper margins of the coxal plates. The cephalothorax is withdrawn first and then the rest of the body. Molting is completed in less than an hour. The interval between molts ranges from a minimum of 3 to a maximum of 40 days, depending on food conditions, temperature, and the species. Immature amphipods molt at much shorter intervals than adults.

Most species complete the life cycle in a year or less. *Diporeia hoyi* is unusual in that it is thought to have a life cycle of 30 months or more. This species breeds between December and April, and the young are released from the marsupium in the spring.

Ecology. *Hyalella azteca*, *Gammarus pseudolimnaeus*, *G. fasciatus*, and *Crangonyx* spp. constitute the great majority of specimens taken by casual collectors. These species are widely distributed and common in unpolluted clear waters, including springs, spring brooks, streams, pools, ponds, and lakes. Less common species are more restricted to certain types of environments. *Diporeia hoyi*, for example, occurs only in deep, cold, oligotrophic, northern lakes. *Gammarus minus* occurs only in caves, springs, and small streams from Pennsylvania, Maryland, Virginia, and West Virginia west-southwest to Illinois, Missouri, and Arkansas. *Hyalella azteca* is an extremely common species and probably represents several sibling species; the other known *Hyalella* species are highly restricted to specific habitats. Recently described species, including *H. longicornis* known only from a spring in Utah, and *H. muerta* and *H. sandra* from Death Valley, California, are known from a few localities. Many amphipod species are restricted to seeps, springs, and subterranean waters in one or a few states, or, indeed, to a single cave or cave system.

As a group, amphipods are cold stenotherms, strongly thigmotactic, and react negatively to light. Consequently, during the daytime they are in vegetation or hidden under and between debris and stones. They usually congregate in the corners of culture jars in the laboratory. The true burrowing habit has not evolved in freshwater species. Amphipods are sometimes unbelievably abundant. The writer (RWP) has collected *Gammarus* from spring brooks rich in rooted vegetation where populations have exceeded 10,000/m^2. Juday and Birge (1927) reported an average population of 4553 *Diporeia hoyi* per square meter at a depth of 50 to 60 m on the bottom of Green Lake, Wisconsin.

Except for *D. hoyi*, which is both a benthic and plankton organism and which has been found as deep as 300 m in Lake Superior, amphipods are restricted to shallow waters. *Hyalella azteca* is about the only other species ever found at depths exceeding 1 m. Large rivers contain few amphipods, except in the shallow backwater and overflow pond areas.

The majority of species appear to be restricted to waters of low or medium carbonate content, but there are a few notable exceptions. *Gammarus lacustris* is common in cool hard waters, and *Hyalella azteca* is often found in warm alkaline waters and can tolerate brackish waters. An abundance of dissolved oxygen and calcium appear to be environmental necessities. *Diporeia hoyi* has been collected from the bottom waters of lakes where the concentration of dissolved oxygen was less than 7% saturation, which is an exception. Acidity (pH) is an essential limiting factor. Several amphipods will tolerate pH from 6.0 to 8.0 but drop out beyond either value; *Hyalella azteca* and *Gammarus minus* have difficulty surviving pH values below 6.0. Because of their wide distribution and broad tolerance for a number of environmental variables, *Hyalella* and

Gammarus have proved to be useful bioassay organisms for studies on pesticides and other toxic chemicals.

The restriction of many species to subterranean waters and springs strongly suggests that they are cold stenotherms. About 100 species are terrestrial or semiterrestrial on a worldwide basis. Some occur in the debris of the high-tide line on marine shores. Others may be found in commercial greenhouses or in terrestrial habitats in the tropics. A few species burrow in the substrate during times of drought and high temperatures; with the onset of normal conditions they resume activities in the water.

Geographic distribution, dispersal. Most of the scuds of the United States are restricted to this continent. *Hyalella azteca* occurs also in South America. *Gammarus fasciatus* and *Crangonyx gracilis* have recently been reported from England; undoubtedly they have been introduced from the United States. In the United States, native species have been transplanted and have apparently affected the distributions of previously existing species. The European genus *Echinogammarus* (Fig. 1.11E) has been introduced into the Great Lakes but its affects have yet to be determined.

Most genera of freshwater amphipods occurring in the United States are found east of the Rocky Mountains. Only a small number of truly freshwater adapted species are found along the west coast. The genera *Eogammarus* and *Ramellogamarus* of the Pacific northwest have close ties with marine forms or are essentially marine forms that are tolerant of freshwater. The recently described *Stygonyx* is restricted to groundwaters of Oregon.

Very little is known about the passive transport of amphipods from one drainage system to another. Unlike many microcrustacea and phyllopod branchiopods, amphipods are not generally adapted for withstanding drought and other adverse environmental conditions. *Crangonyx gracilis*, *C. shoemakeri*, and *Synurella bifurca*, however, are inhabitants of temporary as well as permanent ponds and streams, and it has been suggested that these species tide over unfavorable conditions by burrowing into the substrate.

Many amphipods undergo active movements, either seasonally or gradually over long periods of time. *Gammarus fasciatus* has recently moved through the altered Ohio River to become the dominant amphipod (Beckett et al. 1998). *Crangonyx forbesi* undergoes interesting seasonal migrations. In the spring, populations move upstream into spring brooks and springs. There the populations often become so dense that cannibalism results. In the autumn the remnants of the population migrate back downstream. Breeding occurs during the winter months.

Subterranean amphipods. The central and south central states, from eastern Kansas and Oklahoma to Indiana, Kentucky, and Tennessee, comprise a region abundant in large caves and cave systems, most of which contain brooks, streams, and pools. It is in such bodies of water that interesting subterranean amphipods are found, but usually as sparse populations. Texas has a rich fauna of subterranean species, the Appalachian Mountains have long been known to support a rich fauna, as do the Ozarks, and recently many species have been reported from interstitial habitats in the Rocky Mountain region.

Each major geographical region appears to have an endemic group or groups that demonstrate rather high taxonomic disparity. The most obvious group is that represented by the Hadziidae, Bogidiellidae, Artesidae, and Sebidae, all restricted to aquifers of Texas and adjacent to marine sediments. The Hadziidae almost certainly represents stranding of ancestors that lived in receding shallow marine waters. The Appalachian region is dominated by Crangonyctidae, with *Crangonyx* and *Stygobromus* as the principal subterranean genera. In this region it is more likely that the more cave-adapted *Stygobromus* evolved from a freshwater *Crangonyx* ancestor. The Ozarkian region also is dominated by crangonyctids with *Bactrurus* an important subterranean genus along with *Allocrangonyx* (Allocrangonyctidae). Here one group (*Bactrurus*) seems to have had a freshwater ancestor, while the other (*Allocrangonyx*) may reach back to a marine stranding.

Undoubtedly, interstitial and subterranean species are generally distributed in the United States. Although the cave habitat has evolved in many genera, the species, as a group, are similarly modified. They are generally whitish, creamy, or straw-colored, the antennae and tactile hairs are well developed, the body is often relatively fragile, and the eyes are usually reduced, vestigial, or absent. Their food consists of bits of dead vegetation washed into the caves and the thin bacterial scum covering submerged surfaces. Some subterranean species are generally distributed throughout large areas, but others are rare and have been reported only from single or a few localities.

Subterranean amphipods are often collected just outside of caves and underground streams where the water forms surface seeps, wells, and springs. Some species are represented by two distinct varieties, one subterranean and the other a surface form. The cave variety of *Crangonyx gracilis*, for example, has degenerate eyes and no pigmentation, whereas its pond and stream counterpart has functional eyes and good pigmentation. Comparable differences occur in the varieties of *C. forbesi* and *Gammarus troglophilus*. *Gammarus minus* is unusual in having three varieties. The one occurring in springs has large eyes, short antennae, and brown coloration. A cave variety has slightly reduced eyes, long antennae, and bluish coloration. Another rare cave variety has a fragile body, greatly reduced eyes, long antennae, and bluish coloration.

Enemies, commensals, parasites. Fishes are the chief predators, although only a few species of amphipods occur in streams and ponds large enough to support natural fish populations. The planting of amphipods in small western lakes has, in a few instances, been at least temporarily successful in augmenting the natural food supply of trout. Birds, predaceous aquatic insects, and amphibians probably take an appreciable toll.

Like crayfish, amphipods support an amazing population of algae and sessile Protoctista on the external body surfaces. Scuds serve as intermediate hosts for a wide variety of parasites, including tapeworms of waterfowl and fishes, and a few nematodes, trematodes, and Acanthocephala of birds, fishes, and amphibians.

Collection, preparation. Where specimens are abundant they may be easily taken with a dipnet and by rinsing out masses of aquatic vegetation and bottom debris. Stony bottoms require hand picking with forceps or a small aquarium net. Seventy percent alcohol is a satisfactory killing agent and preservative.

Live specimens may be maintained in aquaria, especially if they are well supplied with aquatic vegetation.

Taxonomy. Of the three suborders of the Order Amphipoda, only one, the Gammaroidea, is represented in American freshwaters. The list of species described has grown from 16 in 1907 (Weckel) to about 150 at the present time. Our American specialists have representatives of almost as many undescribed freshwater species in their possession. For accurate identification it is important that mature males and females be used, since the keys below are based largely on structures that are best differentiated in the later instars.

Diporeia has been a particularly puzzling genus, and several species have been described from the United States. The historical work of Segerstrale proposed that there is only one valid species, *D. hoyi*, that is closely related to *D. affinis* occurring in Europe. Recently, Vaniola (1990) has argued that, on the basis of electrophoretic evidence, several freshwater species exist, with at least two in North America. *Hyalella*, a large genus in North and South America, is another genus that has been troublesome because many populations appear to be distinct, albeit slightly. Much study is being carried out currently on the relationships of these populations and factors contributing to their differences.

We use reflected light and watch glasses containing black wax for our dissections under binoculars. Watchmaker's forceps are mandatory. The following key is greatly modified from that of Holsinger (1972) and Holsinger and Longley (1980).

KEY TO THE FAMILIES AND GENERA OF FRESHWATER AMPHIPODA OF THE UNITED STATES

1. First antenna with an accessory flagellum (even if very rudimentary) (Figs. 24.2B; 24.6C), if flagellum absent, then without eyes or pigment (rare) 5
First antenna without an accessory flagellum, with eyes and pigment 2

2. With eyes and pigment 3
Without eyes or pigment 8

3. Antenna 2 expanded, especially in male (Fig. 24.3A), penduncle segments broad, 1 species; inhabits tubes in tidal freshwater, CA, OR, WA . . . . **COROPHIIDAE, *Corophium***
Antenna 2 not as above, non-tube-building species 4

4. First antenna very short, not extending beyond peduncle of second antenna (Fig. 24.3B); maxilliped palp 3 segmented and without a terminal dactyl; known from fresh and tidal marshes along Gulf Coast and FL, 1 or more species . . . . **TALITRIDAE, *Chelorchestia***
First antenna extending beyond peduncle of second antenna (Fig. 24.1); maxilliped palp 4 segmented and with a terminal dactyl; 5 species, 1 widespread and very common **HYALELLIDAE, *Hyalella***

5. Gnathopods 1 and 2 dimunitive; antennae strongly sexually dimorphic (Fig. 24.2A, C); 1 or 2 species restricted to large lakes in the northern U.S. . . . . . . **HAUSTORIIDAE, *Diporeia***
Gnathopods 1 and 2 well developed, conspicuous; antennae not strongly sexually dimorphic **6**

6. Accessory flagellum of antenna 1 typically 1 or 2 segmented, one segment prominent 7
Accessory flagellum of antenna 1 typically 3 or more segments, at least two segments prominent (Fig. 24.7A) . . . . **GAMMARIDAE, 21**

7. Uropod 3 biramous, even if one ramus only a scale 8
Uropod 3 uniramous (Fig. 24.4A, B) **18**

8. Without eyes or pigment; rami of uropod 3 developed and of equal or subequal length (Fig. 24.4C, D); restricted to aquifers in southcentral TX, NM 9
With or without eyes or pigment; uropod 3 rami significantly unequal in length or only one ramus present (Fig. 24.4B, E, F) **16**

9. Antenna 1 longer than body; telson without prominent plumose setae on lateral margins **HADZIIDAE, 10**
Antenna 1 shorter than body; telson with prominent plumose setae on margins (Fig. 24.4G) 15

10. Feeding appendages (maxillae, mandibles) extending anteriorly beyond face, to peduncular segment 2 of antenna 1 . . . . **11**
Feeding appendages not extending beyond face (Fig. 24.4H) **12**

11. Gnathopods 1 and 2 dimorphic, propods in both sexes mittenform; 1 species ***Texiweckeliopsis***
Gnathopods 1 and 2 dimorphic, male gnathopod 2 stronger, propod subchelate (as in Fig. 24.4I); 2 species . . . . ***.Holsingererius***

12. Telson barely cleft, with only a slight apical notch; 1 species . . . . ***.Mexiweckelia***
Telson with a distinct cleft (Fig. 24.4J) **13**

13. Antenna 1 with a distinct single segmented flagellum (Fig. 24.4K); 1 species ***Paramexiweckelia***
Antenna 1 with flagellum absent or represented by a tiny lobe **14**

14. Antenna 2 penduncle with spines (Fig. 24.4L); gnathopods not sexually dimorphic; 3 species ***Allotexiweckelia***
Antenna 2 peduncle not spinose (Fig. 24.4M); ganthopods sexually dimorphic; 1 species ***Texiweckelia***

15. Coxal plates uniformly broader than long (Fig. 24.5B); telson with shallow cleft, with few apical and subapical spines; 1 species . . . . **BOGIDIELLIDAE, *Parabogidiella***
Coxal plates 3 and 4 as long or longer than broad (Fig. 24.5A); telson deeply cleft, with several apical spines; 1 species . . . . **ARTESIIDAE, *Artesia***

16. Without eyes or pigment; one ramus of uropod 3 greatly elongate (Fig. 24.6I); propod of pereopods 6 and 7 with pairs of ventral spines (Fig. 24.6F); 2 species; hypogean habitats in MO, OK . . . . **ALLOCRANGONYCTIDAE, *Allocrangonyx***

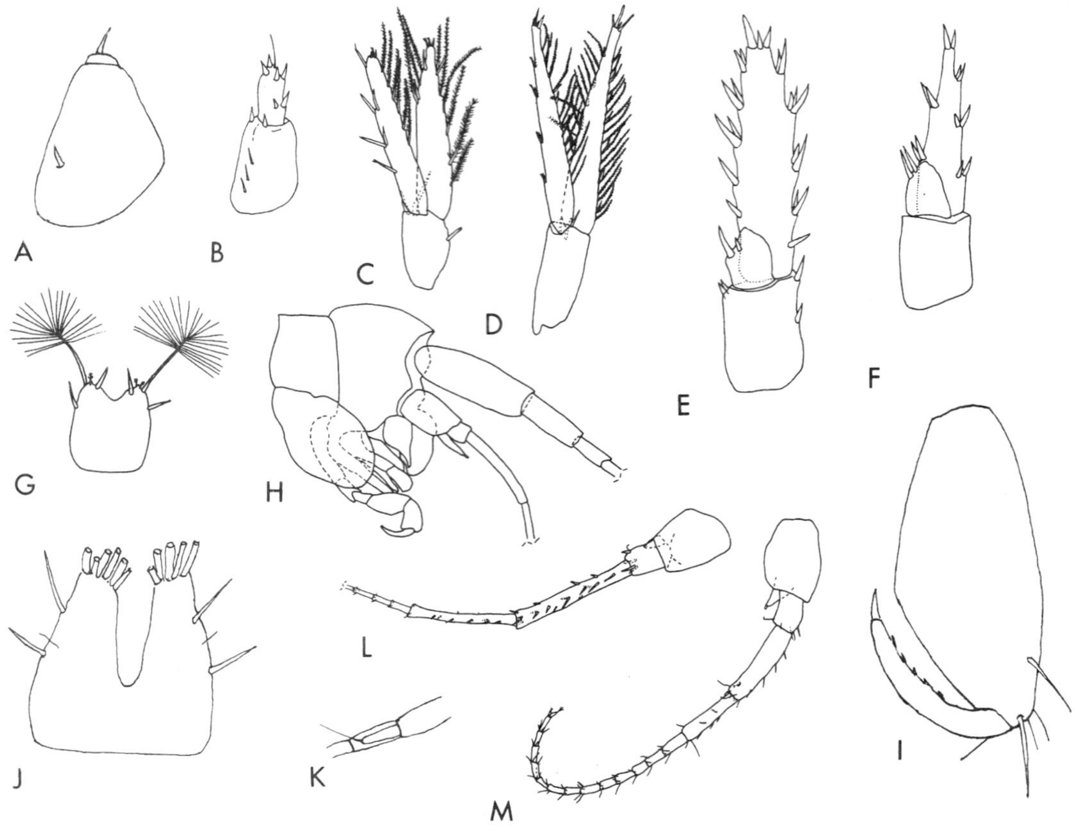

Fig. 24.4. Structure of Amphipoda. A, uropod 3 of *Stygobromus araeus;* B, uropod 3 of *Synurella chamberlaini;* C, uropod 3 of *Artesia subterranea;* D, uropod 3 of *Texiweckelia texensis;* E, uropod 3 of *Crangonyx richmondensis;* F, uropod 3 of *C. pseudogracilis;* G, telson of *Parabogidiella americana;* H, 'head' of *Texiweckelia texensis;* I, propod of pereopod 2 of *Mexiweckelia hardeni* (most spination omitted); J, telson of *Paramexiweckelia ruffoi;* K, antenna 2 and flagellum of *P. ruffoi;* L, antenna 2 of *Allotexiweckelia hirsuta;* M, antenna 2 of *Texiweckelia insolata.* (A redrawn from Barnard and Barnard 1983; C, D, G, H, L, M after Holsinger and Longley 1980; I redrawn from Holsinger 1992; J, K redrawn from Holsinger 1996.)

With or without eyes or pigment; rami of uropod 3 disproportionate but neither greatly elongate (Fig. 24.6A); propod of pereopods 6 and 7 without ventral spines **CRANGONYCTIDAE (in part), 17**

17. Longer ramus of uropod 3 with lateral spine clusters (Fig. 24.4E, F); 5 eyeless and unpigmented hypogean species in eastern and central states, about 15 eyed and pigmented epigean species throughout the U.S. . . . .*Crangonyx*

Longer ramus of uropod 3 without lateral spine groups (Fig. 24.6G), eyeless and without pigment; 3 species; hypogean; IL, KA, MO, OH, OK . . . .*Bactrurus*

18. Without eyes or pigment (of hypogean facies); telson without cleft (Fig. 24.6B), or cleft very shallow **19**

With eyes and pigment; telson with deep cleft (Fig. 24.6D); 3 disjunct species in the eastern U.S. . . . .**CRANGONYCTIDAE (in part), *Synurella***

19. Telson and uropod 3 without spines (Fig. 24.6E, H); adults less than 3 mm total length; known only from wells in TX; 2 species . . . .**SEBIDAE, *Seborgia***

Telson with spines, uropod 3 usually with spines (Fig. 24.4A); adults from 3 to 10 mm in length; about 100 species; throughout the U.S. . . . .**CRANGONYCTIDAE (in part), 20**

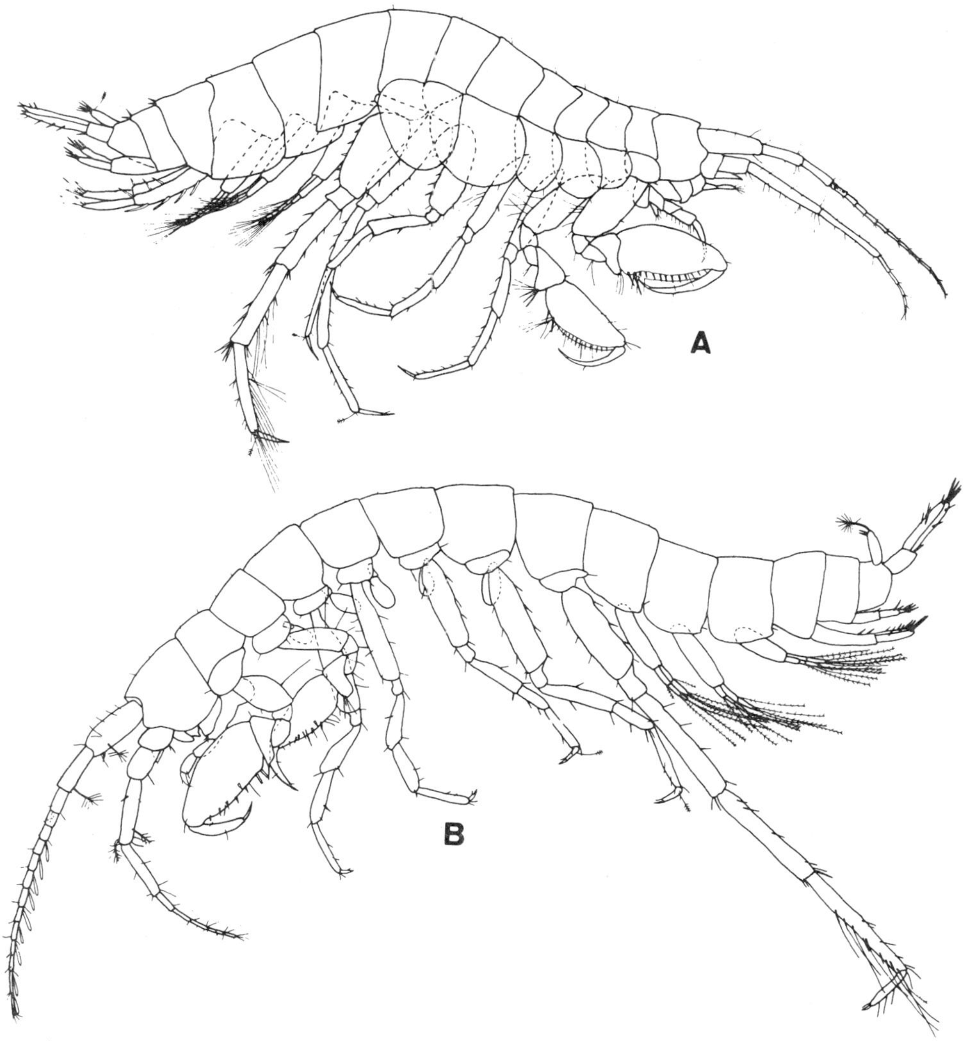

Fig. 24.5. A, *Artesia*, ×22; B, *Parabogidiella*, ×47. (From Holsinger and Longley 1980.)

20. Uropod 3 very short, equal to or less than the length of the telson; many species in the U.S. ***Stygobromus***

Uropod 3 twice the length of the telson; 1 species, OR .... ***Stygonyx***

21. Coxal gills without accessory lobes; gnathopod 1 smaller than gnathopod 2; most species restricted to the interior and eastern U.S. **22**

Coxal gills with accessory lobes (Fig. 24.7B, C); gnathopod 1 slightly larger than gnathopod 2; restricted to the northwest coast of the U.S. **23**

22. Inner ramus of uropod 3 longer, 0.40 or more as long as outer ramus; 15 species; eastern half of the U.S. (Fig. 24.7A) .... ***Gammarus***

Inner ramus of uropod 3 shorter, 0.33 or less as long as outer ramus; 1 introduced species in the Great Lakes (Fig. 1.11E) .... ***Echinogammarus***

23. Pleosome (anterior abdominal segments) without dorsal spines; 2 species; tidal freshwater of CA, OR, WA .... ***Eogammarus***

Pleosome with dorsal spines (Fig. 24.7D); 3 species in inland and tidal freshwater, CA, OR, WA ***Ramellogammarus***

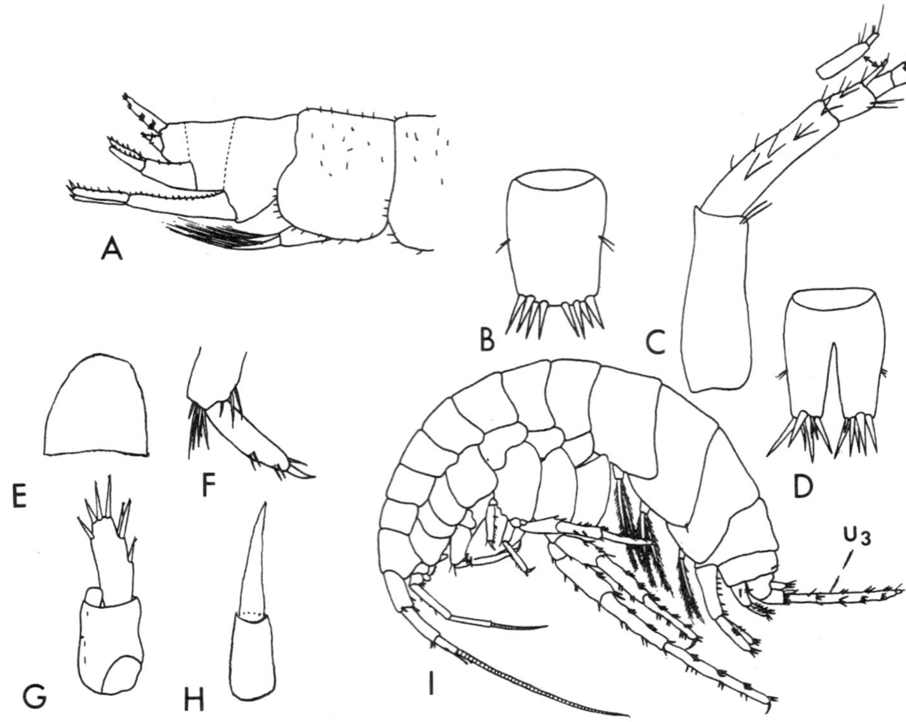

Fig. 24.6. Structure of subterranean amphipods. A, posterior end of *Stygobromus pizzini;* B, telson of *S. balconis;* C, basal portion of first antenna of *S. flagellatus*, note flagellum; D, telson of *Synurella dentata;* E, telson of *Seborgia relicta;* F, end of seventh pereopod of *Allocrangonyx pellucidus;* G, uropod 3 of *Bactrurus;* H, uropod 3 of *Seborgia relicta;* I, *Stygobromus pellucidus*, a blind species, ×7; u3, uropod 3.(A modified from Shoemaker 1942; B and D modified from Hubricht 1943; C redrawn from Weckel 1907; E, H redrawn from Holsonger and Longley 1980; G from Hubricht and Mackin 1940.)

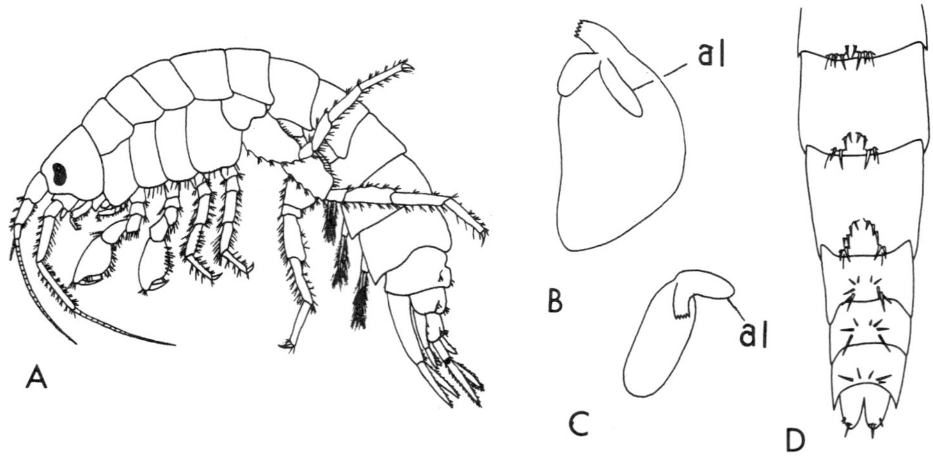

Fig. 24.7. Structure of Gammaridae. A, male *Gammarus fasciatus*, ×5; B and C, coxal gills of *Anisogammarus oregonensis;* D, dorsal view of abdomen of *Ramellogammarus;* al, accessory lobe; arrow, accessory flagellum. (A modified from Kunkel; B and C modified from Shoemaker 1944; D from Barnard and Barnard 1983.)

REFERENCES

Adamstone, F. B. 1928. Relict amphipods of the genus *Pontoporeia*. *Trans. Am. Microsc. Soc.* 47: 366–371.

Baldinger, A. J., et al. 2000. Two new species of *Hyalella* (Crustacea: Amphipoda: Hyalellidae) from Death Valley National Park, California, U.S.A. *Proc. Biol. Soc. Wash.* 113:443–457.

Barnard, J. L. 1958. Index to the families, genera, and species of the gammaridean Amphipoda (Crustacea). *Occas. Pap. Allan Hancock Found. Publ.* 19:1–145.

——. 1971. Keys to the Hawaian marine Gammaridea, 0–30 meters. *Smithson. Contrib. Zool.* 58:1–35.

Barnard, J. L., and C. M. Barnard. 1983. *Freshwater Amphipoda of the World*, 2: *Handbook and Bibliography*, pp. 359–830, Hayfield Associates, Mt. Vernon, VA.

Barnard, J. L., and W. S. Gray. 1968. Introduction of an amphipod crustacean into the Salton Sea, California. *Bull. S. Calif. Acad. Sci.* 67: 219–232.

Beckett, D. C., et al. 1998. Where have all the *Crangonyx* gone? The disappearance of the amphipod *Crangonyx pseudogracilis*, the subsequent appearance of *Gammarus* nr. *fasciatus*, in the Ohio River. *Am. Midl. Nat.* 139:201–209.

Bousfield, E. L. 1958. Fresh-water amphipod crustaceans of glaciated North America. *Can. Field Nat.* 72:55–113.

——. 1977. A new look at the systematics of gammaroidean amphipods of the world. *Crustaceana (Suppl.)* 4:282–316.

——. 1984. Recent advances in the systematics and biogeography of landhoppers (Amphipoda: Talitridae) of the Indo-Pacific Region. In: *Biogeography of the Tropical Pacific*. Bishop Museum Special Publication 72, pp. 171–210, edited by F. J. Radovsky, P. H. Raven, and S. H. Sohmer.

——. 1993. A contribution to the reclassification of Neotropical freshwater hyalellid amphipods (Crustacea: Gammaridea, Talitroidea). *Boll. Mus. Civ. Nat. Verona* 20:175–224.

Bousfield, E. L., and J. R. Holsinger. 1989. A new crangonyctid amphipod crustacean from hypogean fresh waters of Oregon. *Can. J. Zool.* 67:963–968.

Bowman, T. E., and F. Phillips. 1984. Bioluminescence in the freshwater amphipod, *Hyalella azteca*, caused by pathogenic bacteria. *Proc. Biol. Soc. Wash.* 97:526–528.

Clampitt, P. T, 1965. Dispersal related to density in the amphipods *Hyalella azteca* and *Gammarus pseudolimnaeus*. *Proc. Iowa Acad. Sci.* 71: 474–484,

Cole, G. A. 1970. The epimera. of North American freshwater species of *Gammarus* (Crustacea: Amphipoda). *Proc. Biol. Soc. Wash.* 83: 333–348.

——. 1985. Analysis of the *Gammarus pecos* complex (Crustacea: Amphipoda) in Texas and New Mexico, USA. *J. Ariz. Nev. Acad. Sci.* 20: 93–103.

Cole, G. A., and R. L. Watkins. 1977. *Hyalella montezuma*, a new species (Crustacea: Amphipoda) from Montezuma Well, Arizona. *Hydrobiologia* 52:175–184.

Cooper, W. E. 1965. Dynamics and production of a natural population of a fresh-water amphipod, *Hyalella azteca*. *Ecol. Monogr.* 35:377–394.

Creaser, E. P. 1934. A new genus and species of blind amphipod with notes on parallel evolution in certain amphipod genera. *Mus. Zool. Univ. Mich. Occas. Pap.* 282:1–6.

Delong, M. D. 1992. A new species of *Gammarus* (Crustacea: Amphipoda: Gammaridae) from the lower Mississippi River, Louisiana. *Am. Midl. Nat.* 127:241–247.

Duan, Y., et al. 2000. Genetic structure and relationships among populations of *Hyalella azteca* and *H. montezuma* (Crustacea: Amphipoda). *J. North Am. Benthol. Soc.* 19:308–320.

Embody, G. C. 1912. A preliminary study of the distribution, food and reproductive capacity of some fresh-water amphipods. *Int. Rev. (Suppl.)* 4:1–33.

Gaylor, D. 1922. A study of the life history and productivity of *Hyalella knickerbockeri* Bate. *Proc. Ind. Acad. Sci.* (1921):239–250.

Geisler, Sister F. S. 1944. Studies on the postembryonic development of *Hyallela azteca* (Saussure). *Biol. Bull.* 86:6–22.

Glazier, D. S., et al. 1992. Abundance, body composition and reproductive output of *Gammarus minus* (Crustacea: Amphipoda) in cold springs differing in pH and ionic content. *Freshwater Biol.* 28:149–163.

Gledhill, T. 1977. Numerical fluctuations of four species of subterranean amphipods during a five year period. *Crustaceana (Suppl.)* 4:44–152.

Grapentine, L. C., and D. M. Rosenberg. 1992. Responses of the freshwater amphipod *Hyalella azteca* to environmental acidification. *Can. J. Fish. Aquat. Sci.* 49:52–64.

Grossa, L. E., and M. Peralta. 1999. Anfípodos de agua dulce sudamericanos: revisión del género *Hyalella* Smith, I. *Acta Zool. Lilloana* 45:79–98.

Hargrave, B. T. 1970. Distribution, growth, and seasonal abundance of *Hyalella azteca* (Amphipoda) in relation to sediment microflora. *J. Fish. Res. B. Can.* 27:685–699.

Hogg, I. D., et al. 1997. Genetic evidence for a *Hyalella* species complex within the Great Lakes–St. Lawrence River drainage basin: implications for ecotoxicology and conservation biology. *Can. J. Zool.* 76:1134–1140.

Holsinger, J. R. 1969a. The systematics of the North American subterranean amphipod genus *Apocrangonyx* (Gammaridae), with remarks on ecology and zoogeography. *Am. Midl. Nat.* 81:1–28.

——. 1969b. Biogeography of the freshwater amphipod crustaceans (Gammaridae) of the Central and Southern Appalachians. *Va. Polytech. Inst. Res. Div. Monogr.* 1:19–50.

——. 1971. A new species of the subterranean amphipod genus *Allocrangonyx* (Gammaridae) with a redescription of the genus and remarks on its zoogeography. *Int. J. Speleol.* 3:317–331.

——. 1972. The freshwater amphipod crustaceans (Gammaridae) of North America. *Biota of Freshwater Ecosystems, Identification Manual* 5:1–89.

——. 1974. Systematics of the subterranean amphipod genus *Stygobromus* (Gammaridae), Part 1: species of the western United States. *Smithson. Contrib. Zool.* 160:1–63.

——. 1977. A review of the systematics of the Holarctic amphipod Family Crangonyctidae. *Crustaceana (Suppl.)* 4:244–281.

——. 1978. Systematics of the subterranean amphipod genus *Stygobromus* (Crangonyctidae), Part II: species of the eastern United States. *Smithson. Contrib. Zool.* 266:1–144.

——. 1981. *Stygobromus canadensis*, a troglobitic crustacean from Castleguard Cave, with remarks on the concept of cave glacial refugia. *Proc. 8th Int. Cong. Speleol.* 1:93–95.

——. 1986. Zoogeographic patterns of North American subterranean amphipod crustaceans. In: *Crustacean Biogeography*, pp. 85–106. A. A. Balkema, Rotterdam.

——. 1989. Allocrangonyctidae and Pseudocrangonyctidae, two new families of Holarctic subterranean amphipod crustaceans (Gammaridea), with comments on their phylogenetic and zoogeographic relationships. *Proc. Biol. Soc. Wash.* 102:947–959.

——. 1992. Four new species of subterranean amphipod crustaceans (Artesiidae, Hadziidae, Sebidae) from Texas, with comments on their phylogenetic and biogeographic relationships. *Texas Mem. Mus. Speleol. Monogr.* 3:1–22.

——. 1994. Pattern and process in the biogeography of subterranean amphipods. *Hydrobiologia* 287:131–145.

——. 1996. *Paramexiweckelia ruffoi*, a new species of subterranean amphipod crustacean (Hadziidae) from south-central Texas, with observations on phylogenetic and biogeographic relationships. *Boll. Mus. Civ. St. Nat. Verona* 20:89–103.

Holsinger, J. R., and G. W. Dickson. 1977. Burrowing as a means of survival in the troglobitic amphipod crustacean *Crangonyx antennatus* Packard (Crangonyctidae). *Hydrobiologia* 54: 195–199.

Holsinger, J. R., and G. Longley. 1980. The subterranean amphipod crustacean fauna of an artesian well in Texas. *Smithson. Contrib. Zool.* 308:1–62.

Holsinger, J. R., et al. 1983. The subterranean crustacean fauna of Casdeguard Cave, Columbia icefields, Alberta, Canada, and its zoogeographic significance. *Arctic Alpine Res.* 15: 543–549.

Hubricht, L. 1943. Studies in the Nearctic freshwater Amphipoda, III: notes on the freshwater Amphipoda of eastern United States, with descriptions of 10 new species. *Am.. Midl. Nat.* 29:683–712.

Hubricht, L., and J. G. Mackin. 1940. Descriptions of nine new species of fresh-water amphipod Crustaceans with notes and new localities for other species. *Am. Midl. Nat.* 23: 187–218.

Hultin, L. 1971. Upstream movements of *Gam-*

marus pulex pulex (Amphipoda) in a South Swedish stream. *Oikos* 22:329–347.

Juday, C., and E. A. Birge. 1927. *Pontoporeia* and *Mysis* in Wisconsin lakes. *Ecology* 7:445–452.

Karaman, G. S. 1974. Contribution to the knowledge of the Amphipoda: revision of the genus *Stygobromus* Cope 1872 (Fam. Gammaridae) from North America. *Glas. Repub. Zavoda Zast. Prir.-Prir. Muz. Titograd* 7:9 7–125.

Kunkel, B. W. 1918. The Arthrostraca of Connecticut. *Bull. Conn. Geol. Nat. Hist. Surv.* 26: 1–261.

Marzolf, G. R. 1965. Substrate relations of the burrowing amphipod *Pontoporeia affinis* in Lake Michigan. *Ecology* 46:579–591.

Mathias, J. A. 1971. Energy flow and secondary production of the amphipods *Hyallela azteca* and *Crangonyx richmondensis occidentalis* in Marion Lake, British Columbia. *J. Fish. Res. B. Can.* 28:711–726.

Minckley, W. L. 1964. Upstream movements of *Gammarus* (Amphipoda) Doe Run, Meade County, Kentucky. *Ecology* 45:195–197.

Minckley, W. L., and G. A. Cole. 1963. Ecological and morphological studies on gammarid amphipods (*Gammarus* spp.) in spring-fed streams of northern Kentucky. *Occ. Pap. C. C. Adams Center Ecol. Stud.* 10:1–35.

Read, A. T., and D. D. Williams. 1991. The distribution, external morphology, and presumptive function of the surface microstructures of *Gammarus pseudolimnaeus* (Crustacea: Amphipoda), with emphasis on the calceolus. *Can. J. Zool.* 69:853–865.

Rees, C. P. 1972. The distribution of the amphipod *Gammarus pseudolimnaeus* Bousfield as influenced by oxygen concentration, substratum, and current velocity. *Trans. Am. Microsc. Soc.* 91:514–528.

Sanders, H. O. 1969. Toxicity of pesticides to the crustacean *Gammarus lacustris*. *Tech. Pap. Bur. Sport Fish. Wildl.* 25:1–18.

Schmitz, E. H. 1967. Visceral anatomy of *Gammarus lacustris lacustris* Sars (Crustacea: Amphipoda). *Am. Midl. Nat.* 78:10–54.

Schmitz, E. H., and P. M. Scherrey. 1983. Digestive anatomy of *Hyalella azteca* (Crustacea, Amphipoda). *J. Morphol.* 175:91–100.

Segerstråle, S. G. 1937. Studien über die Bodentierwelt in südfinnländischen Küstengewässern, III: Zur Morphologie und Biologie des Amphipoden *Pontoporeia affinis*, nebst einer Revision der *Pontoporeia*-Systematic. *Soc. Sci. Fenn. Comment. Biol.* 7:1–183.

——. 1954. The freshwater amphipods *Gammarus pulex* (L.) and *Gammarus lacustris* G. O. Sars, in Denmark and Fennoscandia: a contribution to the late- and postglacial immigration history of the aquatic fauna of northern Europe. *Soc. Sci. Fenn. Comment. Biol.* 15:1–91.

——. 1971. On summer breeding in populations of *Pontoporia affinis* (Crustacea, Amphipoda) living in lakes of North America. *Soc. Sci. Fenn. Comment. Biol.* 44:1–18.

Shoemaker, C. R. 1938. A new species of freshwater amphipod of the genus *Synpleonia*, with remarks on related genera. *Proc. Biol. Soc. Wash.* 51:137–142.

——. 1942. Notes on some American freshwater amphipod crustaceans and descriptions of a new genus and two new species. *Smithson. Misc. Coll.* 101:1–31.

——. 1944. Description of a new species of Amphipoda of the genus *Anisogammarus* from Oregon. *J. Wash. Acad. Sci.* 34:89–93.

Smith, D. G. 1977. The genus *Crangonyx* in (Amphipoda: Gammaridae) in the Connecticut River system. *Can. Field Nat.* 91:256–261.

——. 1983. A new species of fresh-water Gammaroidean amphipod (Crangonyctidae) from southeastern New England. *Trans. Am. Micros. Soc.* 102:355–365.

——. 1997. An annotated checklist of malacostracans (Crustacea) inhabiting southern New England fresh waters. *J. Freshwater Ecol.* 12: 217–223.

Smith, W. E. 1972. Culture, reproduction, and temperature tolerance of *Pontoporeia affinis* in the laboratory. *Trans. Am. Fish. Soc.* 101:253–256.

Steele, D. H. 1988. What is the amphipod life style? *Crustaceana, Suppl.* 13:134–142.

Vainola, R. 1990. Molecular time scales for evolution in *Mysis* and *Pontoporeia*. *Ann. Zool. Fennici* 27:211–214.

Wagner, V. T., and D. W. Blinn. 1987. A comparative study of the maxillary setae for two coexisting species of *Hyalella* (Amphipoda), a filter feeder and a detritus feeder. *Arch. Hydrobiol.* 109:409–419.

Ward, J. V. 1977. First records of subterranean amphipods from Colorado (Crangonyctidae). *Trans. Am. Microsc. Soc.* 86:452–466.

Weckel, A. L. 1907. The fresh-water Amphipoda of North America. *Proc. U.S. Nat. Mus.* 32:25–58.

Wellborn, G. A. 1994. Size-biased predation and prey life histories: a comparative study of freshwater amphipod populations. *Ecology* 75: 2104–2117.

Wen, Y-H. 1992. Life history and production of *Hyalella azteca* (Crustacea: Amphipoda) in a hypereutrophic prairie pond in southern Alberta. *Can. J. Zool.* 70:1417–1424.

Wilder, J. 1940. The effects of population density upon growth, reproduction, and survival of *Hyallela azteca. Physiol. Zool.* 13:439–462.

25

DECAPODA

(Crayfishes, shrimps)

Of the great array of species constituting the Order Decapoda or Reptantia, the vast majority are marine. In the United States only the Astacidae and Cambaridae (crayfishes), about 12 species of Palaemonidae (freshwater prawns and river shrimps), and four species of Atyidae are found in freshwaters. These representatives, totaling about 350 species, are characteristic and common inhabitants of a wide variety of environments, including most types of running waters, shallows of lakes, ponds, sloughs, swamps, underground waters, and even wet meadows where there is no open water. They are absent from the greater portion of the Rocky Mountain region, although one species occurring in Pacific slope drainages has migrated as far eastward as the Great Basin and the headwaters of the Snake, Yellowstone, and Missouri rivers. Compared with the United States, other parts of the globe have a meager crayfish fauna.

No native, strictly freshwater adapted crabs inhabit the United States. A few species of marine crabs have become increasingly tolerant of freshwater conditions. *Callinectes sapidus*, the well-known edible blue crab, is a frequent visitor to freshwater and in several rivers along the east coast, the species has moved far upstream. At least one grapsoid crab, *Platychirograpsus spectabilis*, native to Mexico, has become well established in several rivers in Florida (see Fig. 1.11F and Plate 16).

General characteristics. Crayfish (Fig. 25.1), crawfish, crawdads, mudbugs, or crabs, as they are locally and variously known, are all more or less cylindrical, and the body and appendages are strongly sclerotized. The compound eyes are large, stalked, and movable. Trunk length, not including the antennae, ranges from 10 to 150 mm. The six abdominal segments are all distinct, but the head and thoracic segments are fused to form a large cephalothorax. A carapace covers the cephalothorax, and there is a large gill chamber on each side between the body and the downward extensions of the carapace. The cervical groove of the carapace roughly delimits the head and thoracic regions, and the areola is a narrow more or less hourglass-shaped median area of the carapace just posterior to the cervical groove. The anterior end of the carapace forms an elongated rostrum that bears a short spine (acumen) at its extremity.

The crayfish has been one of the most important invertebrate models in the study and teaching of zoology and a few species are commercially grown for food. Since Huxley's (1880) little book on the anatomy of the crayfish, most introductory biology and invertebrate text books have used the crayfish as the representative crustacean. Interest in the biology and management of crayfish has produced an enormous literature over the years, approximately 12,500 papers being published through 1987 (Hart and Clark 1989).

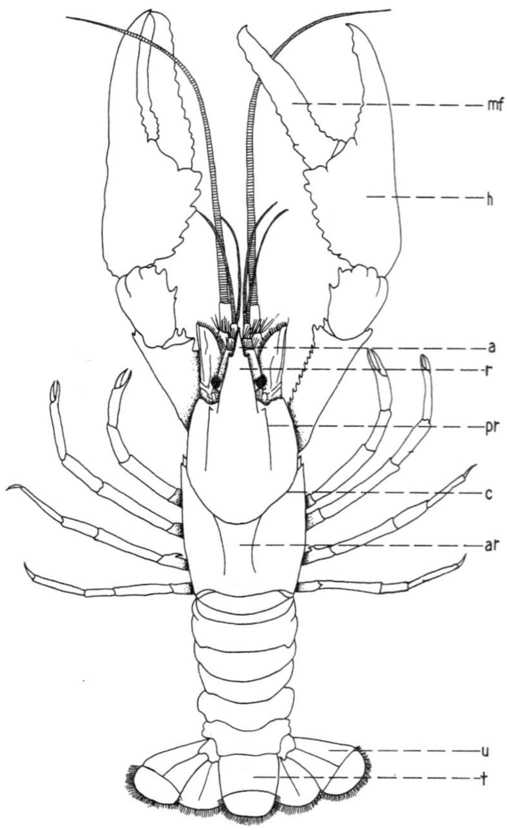

Fig. 25.1. Structure of a typical crayfish. a, acumen; ar, areola; c, cervical groove; h, hand; mf, movable finger (dactyl) of chela; pr, postorbital ridge; r, rostrum; t, telson; u, uropod. (Modified from Hobbs 1942b).

The body of the crayfish is organized with different regions (tagmata) specialized for specific duties. The 19 pairs of serially homologous appendages are basically biramous but modified for a wide variety of functions. Beginning at the anterior end, each of the first antennae (antennules) has two slender, segmented, sensory flagella. The dorsal surface of the basal segment of the first antenna bears the opening of the statocyst, an organ of balance. The second antenna has a flattened basal scale and a single, long flagellum.

The next five pairs of appendages greatly overlap one another and are used chiefly for handling and mincing the food. The heavy, crushing mandibles lie on either side of the mouth. The first maxillae, which follow, are small, flattened, and delicate. Each of the second maxillae has a long, flat extension projecting laterally beyond the point of attachment of this appendage. This is the "bailer"; its rapid beating creates a posterior–anterior current of water through the gill chamber, thereby ensuring a continuous supply of oxygenated water for the gills in the gill chamber. The first, second, and third maxillipeds are progressively more robust and clearly biramous. Generally, the third maxillipeds are used for cleaning the antennae.

The last five pairs of appendages of the cephalothorax are large walking legs, or pereopods (Fig. 25.2) (hence the name "Decapoda"). Each consists of seven segments. Beginning at the base, these are as follows: coxa, basis, ischium, merus, carpus, propodus, and dactylus. Sometimes the suffix "podite" is used for each segment; thus, coxopodite, basiopodite, and so on. The coxa and basis are always short and compact. The first three pereopods are clawed, or chelate, the dactylus being the movable "finger" of the claw, and the propodus constituting the "hand" and immovable finger. The first leg, or chela, is greatly enlarged in the crayfish and is used for crushing the food and as an offensive and defensive weapon. It is of little use in locomotion, walking being chiefly a function of the last four pairs of legs. The second and third legs are also used in handling and mincing the food.

Near the base of the ischium of each walking leg and marked by a double crease is a "breaking joint." When a crayfish leg is seized or irritated, reflex action causes strong contraction of special muscles at the breaking joint, and the whole leg breaks off at this point. Consequently, collectors must exercise care if they are to secure suitable specimens. Perfect specimens are uncommon since appendages are often lost as the result of crayfish fighting among themselves. Missing legs are regenerated in miniature at the subsequent molt but later become larger.

The first five abdominal segments bear pleopods, or swimmerets. In the female they are all more or less biramous and are used

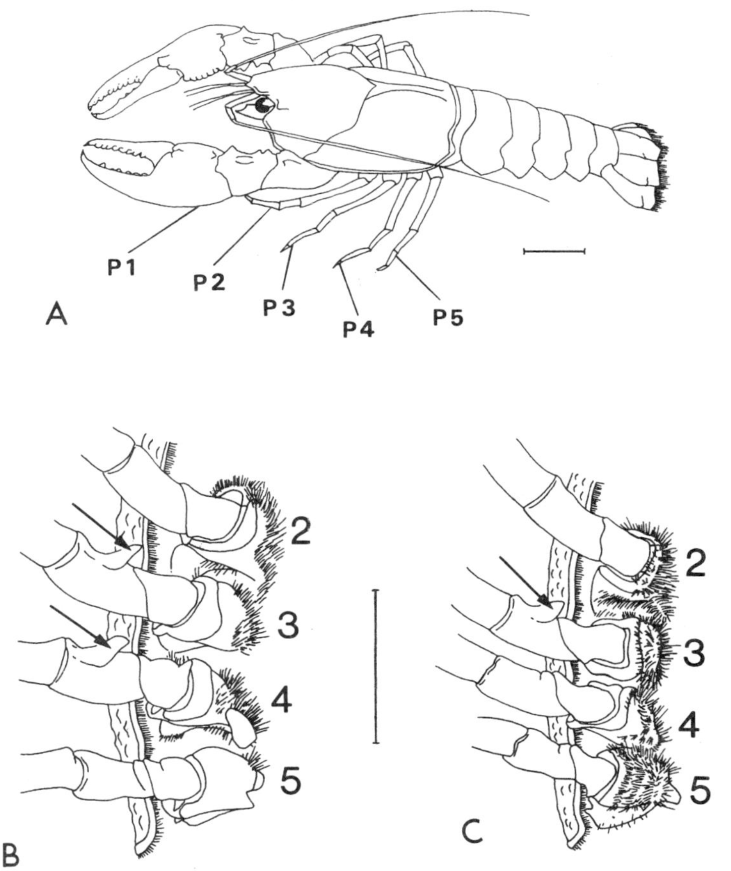

Fig. 25.2. Structure of Cambaridae. A, *Cambarus bartonii;* B, pereopods 2–5 of *Procambarus acutus;* C, pereopods 2–5 of *Orconectes limosus.* p1–5, pereopods 1–5. Arrows denote hooks. Scale, 5 mm.

as places of attachment for the incubating eggs, but in the male the first two pairs are modified and used during copulation (Fig. 25.3). The last abdominal segment bears a flat terminal telson and a pair of flat biramous uropods. The telson and two uropods collectively form the broad tail fan.

This pattern is repeated, with slight variation, in other freshwater representatives of the Decapoda. Shrimp of the families Palaemonidae (Fig. 25.4) and Atyidae are very similar to crayfish except that the appendages are more delicate and elongate, appropriate adaptations for swimming and hovering just above the substrate. Crabs depart somewhat in that the abdomen is folded under the cephalothorax and somewhat reduced (Fig. 25.5).

Mature crayfish exhibit such sexual dimorphism that there is no difficulty in distinguishing the two sexes. Males usually have the larger chelae and narrower abdomen. In the Cambaridae the third segment of some of the male pereopods bears a small hook used during copulation (Fig. 25.2B, C). In the female the genital pores are on the basal segments of the third pereopods but in the male they are on the fifth pereopods The annulus ventralis, or seminal receptacle, is a grooved, elliptical, calcified area in the midline between the bases of the fourth and fifth pereopods of the fe-

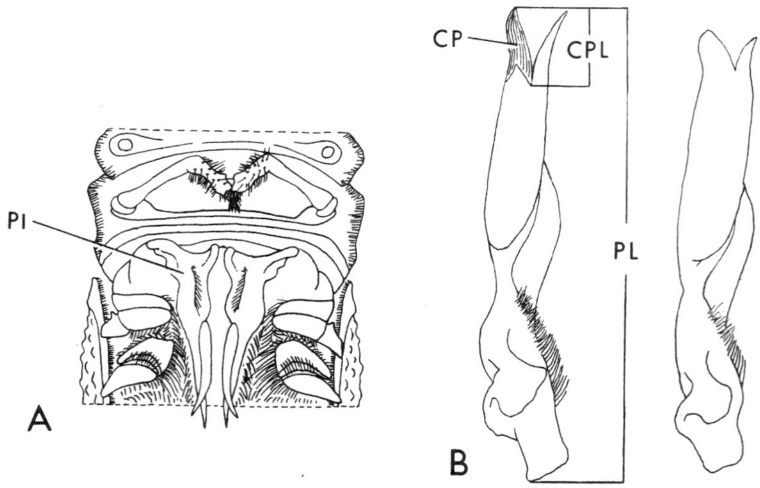

Fig. 25.3. Sructure of Cambaridae. A, location of pleopod 1, anterior down; B, first (left) and second (right) form pleopod 1 of male *Orconectes limosus* and method of measurement. CP, central projection; CPL, central projection length; PL, pleopod length; p1, pleopod 1.

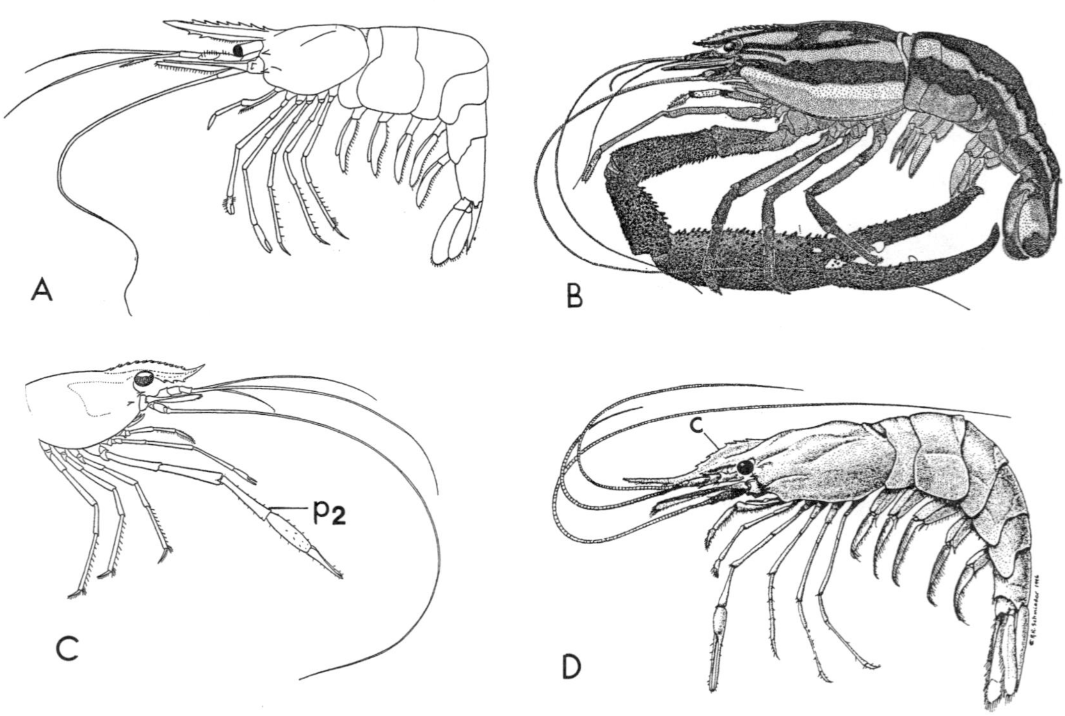

Fig. 25.4. Representative Palaemonidae. A, *Palaemonetes kadiakensis;* B, *Macrobrachium carcinus;* C, *M. ohione;* D, *Exopalaemon.* c, crest; p2, pereopod 2. (A from Creaser 1931; B from Chase and Hobbs 1969; C, modified from Holthuis 1952; D, from Wicksten 1997.)

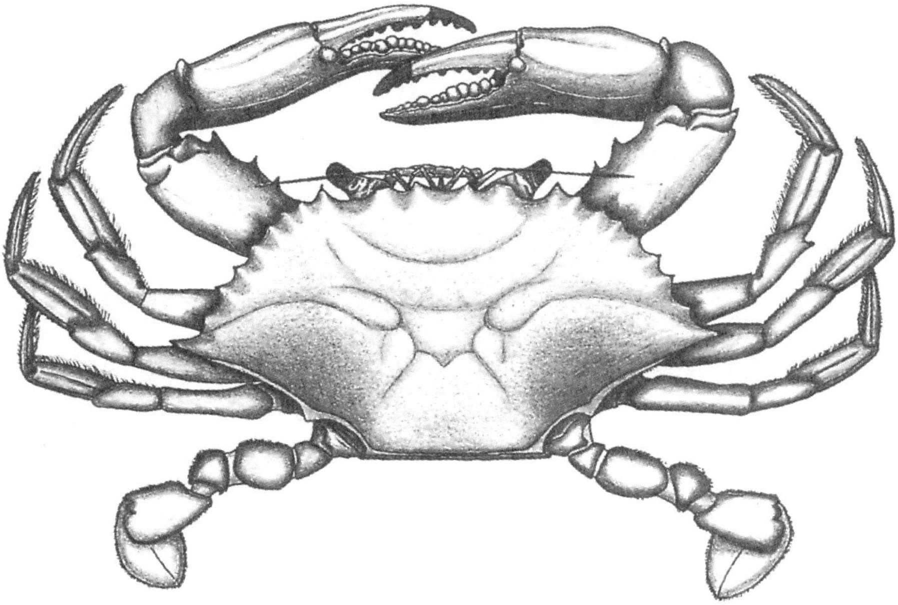

Fig. 25.5. *Callinectes sapidus*, the common blue crab. (© Ethan Nedeau with permission.)

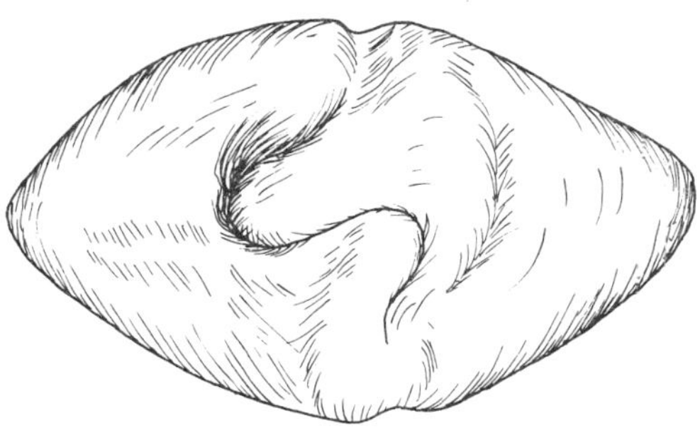

Fig. 25.6. Annulus ventralis of female *Procambarus acutus*. (Redrawn from Hobbs 1942b.)

male (Fig. 25.6). The first pleopods of the female are usually similar to the others, but in the male they have lost their biramous character and are heavily sclerotized, grooved, and specialized at the tip for sperm transfer. The male second pleopods are also somewhat specialized, and both pairs are bent forward so that they lie against the median ventral surface of the body between the bases of the pereopods.

General body coloration ranges from blackish through brown, red, orange, green, and occasionally blue, with innumerable intermediate shades (Plates 14, 15). Mottling is common, and the pereopods are often tipped or banded with brighter colors than the rest of the animal. Many species show spectacular color, which unfortunately disappears soon after death. In some cases, color is affected by diet. Genetic mutation that suppresses or obfuscates normal pigmentation often results in a blue color, a phenomenon that has been often reported. In pigment-suppressed animals, the blue color is due to the presence of hemocyanin in the blood. Other specimens have been collected in which the flesh pigment is intact but the exoskeleton is a deep blue. Many species show a wide range of color. *Orconectes immunis*, for example, may be bluish, blackish, brown, greenish, or red. In general, newly molted specimens are more brightly colored than older specimens, especially since the latter usually have an accumulated coat of dirt, debris, and algae.

The coloration of a species can vary with the coloration of the substrate. In most crayfish such color adaptations are decidedly limited and occur only over a period of weeks or months. In *Palaemonetes paludosus*, however, variable coloration and background resemblance have become developed to a striking degree. This species has four types of chromatophores containing white, red, yellow, and blue pigment granules, respectively, and general body coloration is determined by the relative concentration or dispersion of these granules in various combinations. Color adaptation is completed in less than 24 hours after being placed on a new substrate.

Locomotion. Undisturbed crayfish walk or climb about slowly with the pereopods; they may move forward, backward, or occasionally sideways. When alarmed, however, they quickly dart backward by ventrally doubling up the tail fan and the posterior end of the abdomen. They may also leave the substrate and move through the water for short distances by such abdominal contractions. In addition to these two modes of locomotion, the Palaemonidae are distinct swimmers and may move continuously forward through the water by movements of the pleopods. Crabs move over the substrate laterally, their nervous movements appearing more as a "scurry" than a walk.

Food, feeding. In general, crayfish are omnivorous and are thus "polytrophs." Their feeding systems emulate those of insects and thus crayfish are involved with the transformation and breaking down of both coarse and fine material. Juveniles are able to filter feed and consume phytoplankton. Vegetation is an important food for crayfish and it is well documented that crayfish can control the aquatic flora of a particular lake. Under certain conditions most species will probably resort to predation. The fast-spreading *Orconectes rusticus* preys heavily on snails. There is some evidence that crayfish will at times feed on living aquatic insects and sponges. They eat all kinds of succulent aquatic vegetation, and animal food is usually a minor part of the diet when there is abundant vegetation. They also prefer fresh to stale meat, and in the laboratory they have been fed raw and cooked meats of all kinds, prepared fish foods, hay, whole seeds, cottonseed meal, and soybean meal. Ecologically, they are usually considered scavengers. Species that live in upland headwater streams (e.g., *Cambarus bartonii*) are possibly ordinarily predaceous.

The chelate appendages are used for crushing, picking up the food, and tearing it into pieces. The pieces are then passed forward where they are further cut and masticated by the maxillipeds. The maxillae strain out larger particles and mince the smaller ones further. The mandibles are of limited use in grinding and chewing; mostly they hold food while the maxillipeds tear off fragments.

Internal anatomy, physiology. These topics are treated more completely in most zoology texts and will be only outlined here.

A short esophagus enters a large cardiac chamber of the stomach, which, in turn, enters the smaller pyloric chamber. These chambers contain the gastric mill, which is composed of heavy ossicles for grinding the food and rows of stiff setae for straining out larger particles. The long intestine extends to the anus on the ventral side of the telson. Two large glands are associated with the digestive tract in the cephalothorax; they function in digestion, absorption, and food storage.

The pentagonal heart lies in the dorsal portion of the cephalothorax. It receives blood from the surrounding pericardial sinus through three pairs of ostia and sends it out anteriorly, posteriorly, and ventrally through arteries. The blood passes out of the fine terminal endings of the blood vessels and into the various organs and hemocoel. It then circulates back to the sternal sinus, thence to the gills, and back to the pericardial sinus. The blood is almost colorless and contains amoebocytes and a dissolved hemocyanin.

Seventeen or 18 featherlike gills lie within the branchial chamber on each side. They are attached on or near the bases of the maxillipeds and walking legs and extend dorsally. Blood circulates along definite channels through the central stalk and minute lateral processes of each gill.

Excretion and osmotic control are maintained by two complicated, compact green glands, which are situated in the anterior end of the cephalothorax and open ventrally at the base of the second antennae. A crayfish does not drink, but quantities of water are continuously diffusing the blood through the gill surfaces. Blood (minus its proteins) is absorbed by the distal coelomic sac and labyrinth of the green-gland apparatus. As it moves on through the nephridial canal, the useful materials, such as salts, carbohydrates, and some water, are reabsorbed back into the surrounding blood. The material remaining in the canal is a very dilute urine; it passes into the urinary bladder, from which it is periodically voided to the outside through a small pore. There is also some evidence to show that the gills regulate the internal concentration of certain ions, such as sodium.

The nervous system is of the usual crustacean plan, with a ventral nerve cord and chain of ganglia, two circumesophageal connectives, and dorsal brain. The basal segment of each first antenna contains a statocyst, or organ of balance. Taste receptors are especially abundant on the mouthparts and antennae. The antennae are also sensitive tactile organs. The compound eyes are large and consist of hundreds of ommatidia.

In addition to the optic nerve and ganglia, each eyestalk contains a minute sinus gland, an endocrine gland whose secretions influence many important aspects of crayfish behavior and physiology. Among other things, the sinus gland, to a greater or lesser extent, governs chromatophore contraction, frequency of molting, metabolic rate, growth, viability, and light adaptation of the eyes.

Reproduction. With reference to the structure of the first pleopods, there are two distinct types of adult male Cambaridae: first form and second form. Each form results from a molt. The first pleopods of a first-form male are corneous, hard, and distinctly sculptured at the tip. In a second-form male the pleopod tips are soft and generally lacking in sculpturing. These differences are shown in Fig. 25.3B. Only a first-form male is capable of copulating and transferring sperm to the body of a female. Though the instar that precedes the adult first-form male is a juvenile instar, the first pleopods are almost indistinguishable from those of a true second-form male and for all practical purposes this stage is considered second form. The true second-form male, however, is the instar immediately following the first form. Depending on the species and ecological conditions, a true second-form male may undergo a further molt to become a first-form male again before death. *Orconectes immunis*, a typical example, is commonly first form during the spring, when copulation occurs, second form during the summer, and first form again beginning in the autumn; copulation may also occur in the autumn. No other crayfish, including *Pacifastacus*, demonstrate this phenomenon.

Careful work has been done on more than a dozen decapod species in regard to breeding habits, but the available data indicate that time of copulation is variable and occurs between early spring and autumn. Among noncrayfish decapods, shrimp in the Genus *Palaemonetes* generally breed throughout the summer. Species of *Macrobrachium*, a few of which spend their adult life in freshwater, apparently require some salt to complete their life cycle. Breeding occurs throughout the spring and early summer.

Many studies have been carried out on the reproductive cycle of cambarid crayfish. Generally, in northern species, especially in the genus *Orconectes*, there is a distinct seasonal effect in which copulation occurs in the fall and spring, with oviposition taking place in spring and early summer. Seasonality is not so evident in southern species, particularly within the genus *Procambarus*. During active periods of copulation males seem to locate females as a matter of chance, although sex pheromones have been demonstrated in at least one species. Interestingly, among southern *Procambarus*, seasonality occurs in reproduction accompanied by a specific behavior. Male first-form *Procambarus clarkii* spawn in the autumn and then migrate overland by the thousands. Such exhausted males are weak and soon die.

The male has no power of sex discrimination, and during the mating season he seizes and turns over every crayfish coming his way. Another male will always resist strongly, but a female will either resist or remain passive and receptive.

The actual process of copulation, sperm transfer, and egg deposition has been described most carefully for *Orconectes limosus* and *Pacifastacus trowbridgi*. The male seizes the female and turns her over on her back. He mounts on her ventral side and holds all her clawed appendages securely with his two chelae. The tips of the first two pleopods are then inserted into the chink of the annulus ventralis, and the tips of the two vasa deferentia are extruded into the bases of the grooves that extend along the first pleopods. Sperm move along these grooves in macaroni-like cords and are deposited in the chink. The animals then separate, and either the male or female may copulate later with additional individuals. A single copulation may take from a few minutes to as much as 10 hours. The second pleopods are closely applied to the first during sperm transfer, but their exact function is unknown. After copulation a white waxy plug projects from the chink of the annulus ventralis. In *Pacifastacus* there is no annulus, and the spermatophores are deposited at random on the ventral surface of the female.

The female lays her eggs several weeks to several months after copulation, depending on the season of the year, but before laying them she cleans the ventral side of the abdomen thoroughly with the tips of the pereopods. Immediately preceding egg extrusion she secretes a sticky, cement-like substance (glair) from ventral glands. The glair covers the ventral surface of the abdomen, tail fan, and pleopods. The sperm plug breaks and sperm are released into the glair. She then lies on her back, curls the abdomen, and releases the eggs from the genital pores. By curious turning movements of the entire body, the eggs become dispersed through the glair, fertilized, and securely attached to the pleopods by a short stalk and surrounding capsule. The female is then said to be "in berry." Most females of nearly all species are seen carrying eggs between March and June (Fig. 25.7). A female may carry as few as 10 eggs and as many as 700. In general, within a species, the larger the female, the greater the number of eggs. Rhythmic movements of the pleopods effectively aerate the developing eggs. Few eggs die.

Subterranean decapods depart somewhat from reproductive patterns of epigean species. Whereas epigean species of *Palaemonetes* demonstrate seasonal reproductive patterns, the troglobitic shrimp *Palaemonetes antrorum* apparently undergoes nonseasonal recruitment. Copulation possibly is controlled more by water levels and nutrient availability than other factors.

Development, life cycle. The incubation period usually ranges from 2 to 20 weeks, depending largely on temperature, and during this interval the eggs change from dark and opaque to pale and translucent. Hatching involves the rupture and shedding of the embryonic skin and egg membrane, both of which, however, remain attached to the parent pleopod. A newly hatched first instar young clings to the egg stalk with the chelae, and the posterior end remains attached to the cast exoskeleton by a membrane (Fig. 25.7). Although clearly recognizable as a crayfish, the first instar differs markedly from advanced instars and adults. The cephalothorax is proportionately very large; the rostrum is large, pointed, and bent downward; the eyes are large; the chelae are slender; and there are no first and sixth abdominal appendages. The first instar usually persists for 2 to 7 days, and when it molts the posterior connection with the remains of the egg is lost. The second instar lasts for about 4 to 12 days and remains attached to the pleopods of the mother with its chelae and pereopods. Third instar crayfish leave the parent pleopods intermittently and then permanently. The young of burrowing species leave the burrow of the mother when they have a body length of 8 to 20 mm. Individuals hatching in the spring usually have a total of 6 to 10 molts by autumn, when growth and molting cease. In the freshwater Palaemonidae there are five to eight larval molts. During these early instars the body and its appendages gradually assume adult proportions and anatomical details.

Great differences in growth rate may be found within a single pond or stream, the

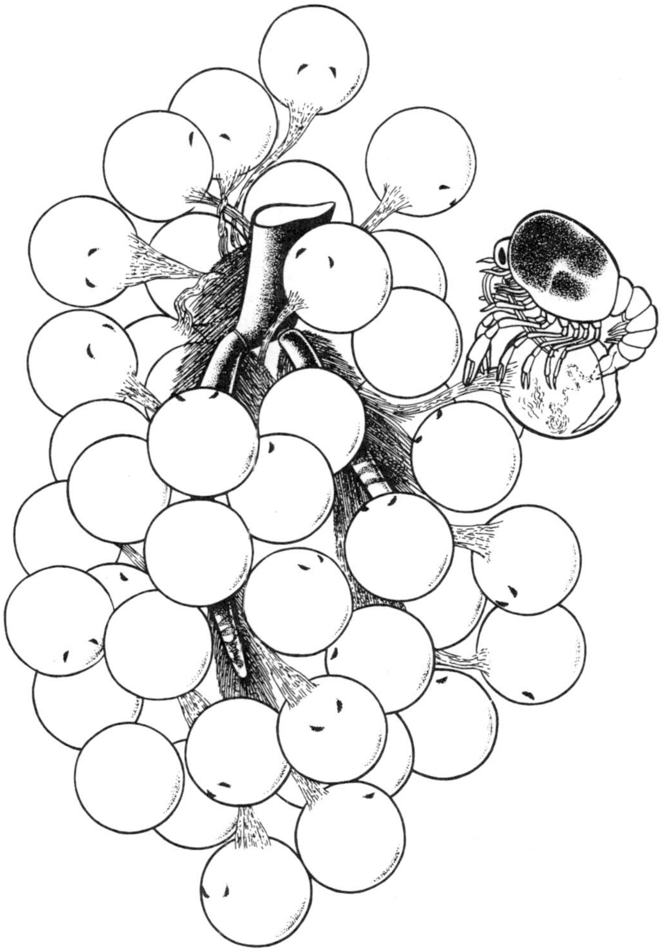

Fig. 25.7. Diagram of one pleopod of a female crayfish with eggs attached. One egg has hatched and the stage I juvenile is grasping the egg stalk with its hooked chelipeds. The pigmented eye regions of juveniles just ready to hatch show through the shells of other eggs. (From Crocker and Barr 1968, modified from Andrews 1907.)

larger specimens being twice or more as long as the smaller ones by autumn. Such differential growth rates are probably due mainly to varying activity and amounts of food consumed.

By autumn many females are sexually mature and many males are first form, so that copulation commonly occurs in the first autumn of life. Some females, however, do not become sexually mature until after a molt early the following spring; similarly, some males do not become first form until after one or two spring molts. Such individuals copulate in the spring, and the female in berry after a very short interval. Some females that copulate in the autumn are not in berry until the following spring. Other species, especially burrowers, copulate and are in berry in the autumn. The young leave the burrow some time during the following spring or summer.

Subsequent to the first mating season, most crayfish have only two to four molts before they die. Following copulation, a female does not molt until after her third instar young have left her. If a male is first form in its first autumn, it usually remains first form during the winter, molts to second form early in the following spring, and after a very short interval molts back again to first form. Such males may therefore copulate again during their second summer or autumn. Among the few species on which life history studies have been made, the normal life span of both male and female crayfish is usually less than 24 months (see Payne 1972), and a few survive their second winter. Females that pass through a second winter may be in berry for the second time the following spring. Lyle (1938) reports an exceptional longevity for *Procambarus hagenianus* in Mississippi and Alabama, the species attaining an age of 6 or 7 years. The molting processes are a complex physiological cycle that is modified and controlled chiefly by temperature, light, and hormonal interrelationships.

Most freshwater shrimp in the genus *Palaemonetes* occurring in the United States have a somewhat reduced larval series than their marine counterparts. Only three zoeal stages are evident in freshwater species. The exception is *P. kadiakensis* in which the larval stages are somewhat more similar to marine forms.

Ecology. Adults remain hidden in their burrows, under stones or debris, or half buried in small depressions in the substrate during the daytime, but between dusk and dawn they feed and move about, even to the extent of coming out on banks of streams and ponds. In shaded streams, and especially on cloudy days, adults may venture into open water in daylight. Immature specimens of most species, on the other hand, regularly crawl about actively on the bottom and in vegetation during the daytime.

There is some evidence to show that the 24-hour cycle of activity is governed by an endogenous mechanism. Solar and lunar rhythms have also been suggested. For example, a few studies have shown that crayfish have a remarkable daily cycle in their oxygen consumption. Fingerman (1955) showed that *Cambarellus shufeldti* had its highest oxygen consumption around 6 A.M., with a secondary maximum at 3 to 6 P.M. Minima occurred from 9 A.M. to noon and from 9 P.M. to midnight. Certainly, with increased activity there is an increased demand for oxygen and crayfish have evolved mechanisms to accommodate fluctuations. Many species live in low-oxygen environments, such as burrows and ditches. Crayfish have apparently evolved physiological responses to levels of oxygen by "reengineering" their circulatory system to direct blood to critical tissues such as nervous and respiratory.

Many crayfish have specific habitat requirements. Some are burrowers in wet meadows and do not enter bodies of water; some occur only in muddy ponds and ditches; some are found only in lakes or slow rivers; others are restricted to swift, stony streams. Beyond physical characteristics, chemical features of the environment are important in determining presence or absence of crayfish. A growing factor is declining pH in especially northern waters. *Orconectes virilis* was experimentally eliminated from a habitat in which the pH was lowered to 5.6. Egg viability and larval development were the most seriously affected (Davies 1989). *Cambarus bartonii* and *C. robustus* can apparently withstand lower pH, populations are known to live in lakes with pH of 5.0. Experimentally, *C. bartonii* can withstand pH levels of less than 3.0. Even variations within a "normal" range of pH can cause differences in production, growth rate, and larval abundance.

Most species tolerate normal but wide ranges in temperature, hydrogen-ion concentration, and free and bound carbon dioxide, though stream species are usually less tolerant than lake and pond species. During winter months crayfish are found in deeper waters of lakes. Similar migrations during the warm months are thought to be associated with gonad maturation. In streams, however, some investigators have found that most individuals have a home range of less than 30 in.

Ordinarilly, crayfish are inhabitants of shallow waters, seldom being found deeper than a meter. An exceptional record is the occurrence of *Orconectes virilis* at a depth of 30 m in Lake Michigan.

Closely related species sometimes have similar ranges but are ecologically isolated within these ranges. For example, *Orconectes virilis* and *O. immunis* are often found in the same area but the former is restricted to streams and lake margins, and the latter to ponds and sloughs.

Population densities vary greatly, depending on the species and habitat. Pond populations usually amount to less than 100 lb/acre, but in exceptional cases may attain 500, 1000, or even 1500 lb/acre. In general, stream populations are less dense, although there is one record of 1176 lb of crayfish per acre. Abrahamson and Goldman (1970) calculated an average of 0.9 adult *Pacifastacus leniusculus* per square meter in Lake Tahoe. Momot et al. (1978) found as high as 15 crayfish per square meter.

Several laboratory studies have shown that in groups of four to eight crayfish, dominance–subdominance relationships soon become established in a linear fashion, and that the dominance order is based on size and learned recognition of individuals. Such work has been done on several species of *Orconectes*, *Procambarus*, and *Cambarellus*.

Burrows. Some species, such as *Procambarus gracilis* and *Cambarus carolinus*, habitually burrow in wet pastures and marshy areas where there is no open water. Species in a second group live in bodies of water during the greater part of the year, but when streams or ponds dry up in late summer or when the temperature drops in the autumn they construct burrows along the margins and live in them until the water level rises and the weather becomes warmer. *Procambarus simulans*, *P. acutus*, and *Orconectes immunis* are included in this group. Species of a third group live continuously in permanent waters and do not make burrows.

Burrows differ widely in construction, depending on the species, the soil, and the depth of the water table (Fig. 25.8). Usually there is only one entrance, though there may be as many as three. The tunnel leading from the entrance may proceed vertically, at an angle, or almost laterally in a sloping bank. Sometimes the tunnels are branched or irregular, but there is always a chamber at the lower end where the crayfish remains during the hours of daylight. Occasionally a burrow has a lateral chamber. The depth of a burrow ranges from 5 cm to as much as 3 m and is partially determined by the level of the water table, since the chamber must contain water to keep the gills wet. Burrows close to the edge of a pond or stream are shallow, those farther away are deeper.

Except during the breeding season, each burrow houses a single crayfish. Burrows are constructed only at night, and the crayfish brings up pellets of mud and deposits them at the entrance to form a chimney. Such chimneys are usually about 15 cm high, but a few as high as 45 cm have been reported. They do not serve any particular purpose, but are simply the result of the safest and most convenient method of disposing of the mud pellets.

Geographic distribution, dispersal. The Cambaridae are restricted to North America. Species of Cambaridae are widely distributed throughout central and eastern North America. In the Appalachian Mountain region and southward the family reaches its greatest diversity. However, in more northern areas affected by Pleistocene glaciations crayfish diversity is low. For purposes of comparison, Hobbs (1981) reported 66 species from Georgia, while Crocker (1979) listed 10 for all of New England, of which 6 are introduced. The distribution of cambarids is highly controlled by drainage evolution. With the isolation of stream systems by geological processes, captured populations of crayfishes evolved into new species. As has been shown for other invertebrate groups, this process has resulted in very high endemism. The ability of crayfish to adapt to semiterrestrial and subterranean habitats has increased their diversity.

The five American species of *Pacifastacus* are restricted to Pacific slope drainages, although two have migrated over to the Great Basin and headwaters of the Missouri River. The genus probably originated in Asia and northern Europe, where a total of only about 11 species occur.

Dry land forms an effective barrier to the

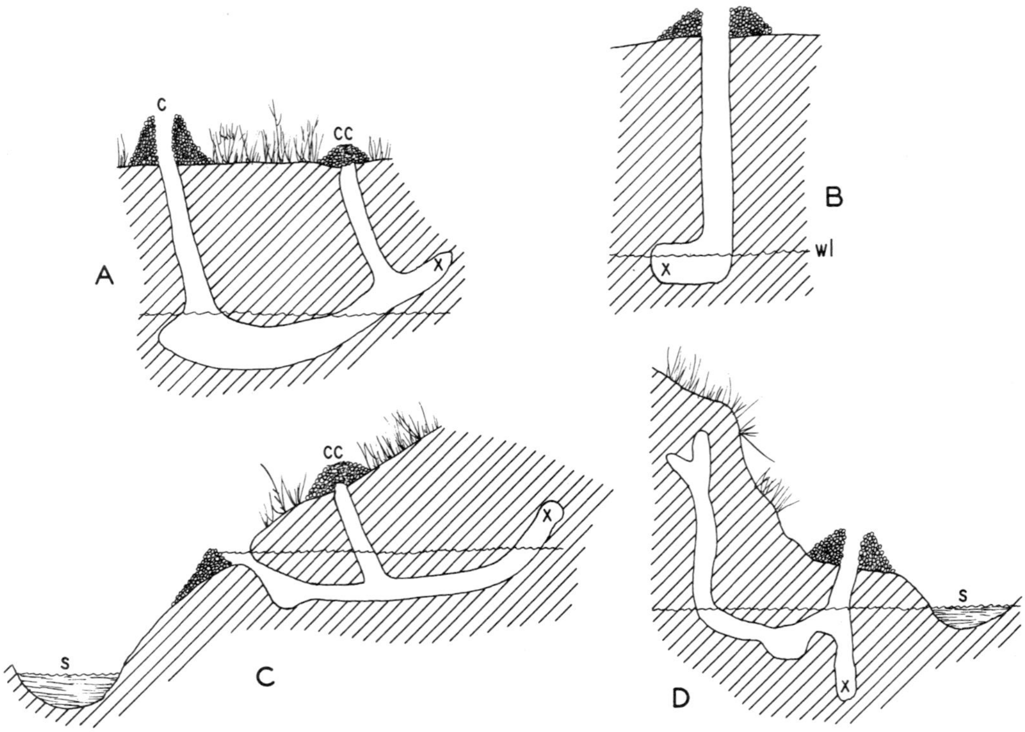

Fig. 25.8. Typical crayfish burrows (A–D). c, chimney and opening of burrow; cc, closed chimney; s, stream; wl, ground water level; x, place where crayfish were taken. (A, C, and D modified from Ortmann 1906.)

migration and geographical spread of lake and stream species, but burrowers such as *Cambarus diogenes* have migrated rapidly and are now found commonly in wet meadows east of the Rocky Mountains.

Orconectes virilis and *O. immunis* form an interesting example of ecological isolation. The two species are very closely related and are often in the same general geographic areas. Yet the former occurs only in clear, running waters with stony bottoms, whereas the latter is restricted to stagnant waters and mud bottoms. Very little is known about interspecific competition among crayfishes. Many species, especially in the southern states, are known only from a single state or single collecting locality.

Perhaps some of the most interesting examples of ecological isolation are to be found among those subterranean crayfish and shrimp that are restricted to the waters of single caves, cave systems, large springs, and sinks. Cave species are strikingly modified for that environment. They are mostly small, the body length ranging from 10 to 80 mm. The body is slender, the chelae are not well developed, and the antennae and other appendages are very long and highly specialized as tactile organs. Pigmentation is absent, and the body is translucent or whitish. All are blind, the eyes being atrophied and the eyestalks reduced or rudimentary. About 25 North American crayfish species and subspecies are true troglobites.

In addition to true troglobitic species, there are some that are normally found above ground but have migrated into caves where they have successfully persisted. With the exception of one or two species that have become differentiated into subspecies, these forms are identical with individuals occurring above ground.

Shrimp in the families Palaemonidae and Atyidae also have become adapted to cave environments, though not to the extent shown by crayfish. A few troglobitic species of *Palaemonetes* are found in Texas and the atyid genus *Palaemonias* is found in caves of the Appalachian Mountains.

A disturbing and unfortunate aspect of the popularity of crayfish in aquatic management, recreation, aquaculture, biological research, and consumption is the transport and introduction of crayfish species into new habitats. Several species of *Orconectes*, particularly *O. rusticus* and *O. virilis*, and *Procambarus clarkii* have been introduced widely outside of their natural range in North America. *Orconectes rusticus*, in particular, has disrupted vegetation patterns and eliminated native congeners and other invertebrates in lakes in which it has become established. The species is aggressive and competes and hybridizes with other crayfish, in many cases eventually eliminating them. *Orconectes rusticus* seems to have direct effects on native *O. propinquus* populations. Most troubling is the introduction of aggressive species into the habitats of environmentally and geographically restricted species, the result tending to reduce or eliminate the endemic form. The effects of *O. rusticus* and other species have been documented in many states and now some of these states ban the import and release of crayfish species. Several American crayfishes have been intentionally introduced into Europe, dating back to the transportation of *Orconectes limosus* into France as early as 1890; it is now widely distributed in western Europe. *Pacifastacus leniusculus* was brought to Sweden and *Procambarus clarkii* into both Hawaii and Japan. Along the west coast, species of the Asian shrimp genus *Exopalaemon* are beginning to appear and at least one freshwater species, *Exopalaemon modestus*, is present in the Columbia River system of Oregon and Washington.

Parasites, commensals, enemies. The external surfaces of crayfish form a convenient and favorable habitat for microorganisms, and when the exoskeleton has not been shed for a long time it is often covered with debris and a dense growth of algae and sessile Protoctista, especially at winter's end.

The Branchiobdellida (see Chapter 13) are small, leechlike, parasitic or commensal clitellates that live on the gills, in the gill chambers, on the general body surface, and at the bases of the legs, antennae, and mouthparts. The ostracod family Entocytheridae is commensal on the gills (see Chapter 20). *Macrobrachium* and *Palaemonetes* often carry parasitic isopods (*Probopyrus*) in their gill chambers (see Chapter 23).

Many species of crayfish east of the Mississippi River may act as the second intermediate host of *Paragonimus westemani*, a lung fluke parasitic in humans and many carnivores. The metacercariae of this parasite become imbedded in the muscles and viscera of the crayfish from which they may be transmitted to the definitive host when the raw or poorly cooked crayfish are used as food.

Crayfish and shrimp are important elements in the diet of an enormous array of invertebrates and vertebrates: wading birds, frogs, turtles, snakes, raccoons, otter, and mink consume appreciable numbers. Crayfish and shrimp are in the diet of 22 families of fish.

Economic importance. In poorly drained clay agricultural lands of West Virginia, Mississippi, Alabama, and a few other southern states, burrowing crayfish become so abundant (as many as 30,000 burrows per acre) that crops cannot be raised because the crayfish denude the land of the young grain, sugarcane, and cotton seedlings. Also, their chimneys clog farm machinery. If the land cannot be drained, the best control is the application of some poison to the burrows. Pyrethrum, creosote, cyanide, carbon disulfide, and unslaked lime are some of the more useful poisons. They are most effective when the water table is high. In California, *Procambarus clarkii* is notorious because of its depredations on rice seedlings. Occasionally crayfish become a nuisance in small reservoirs and irrigation ditches when their lateral burrows through earthen dams and dikes drain the reservoirs.

Europeans use crayfish for the table to a much greater extent than do Americans. Most of the edible crayfish in this country come from the Great Lakes, certain parts of the Mississippi River system, Louisiana, and Oregon. The most important species are *Pacifastacus leniusculus; Orconectes rusticus*, and *Procambarus clarkii*. Probably the most important is *P. clarkii*, occurring in southern states. The biggest farms are in Louisiana and occupy about 46,000 ha. The total production from farm-raised and wild caught animals from 1991 to 1995 averaged about 48,000 metric tons.

In addition to the harvest of "wild crayfish," there is an increasing commercial production, especially in Louisiana, where there is an estimated total of 7000 hectares of crayfish "farms." Only two species are cultured, *Procambarus clarkii* (the red crayfish) and *P. acutus* (the white crayfish). They are annual species, which reach maturity in 6 months. Artificial impoundments produce 80% of cultured Louisiana crayfish, and rice fields produce the remainder. Both species are euryokous under natural conditions and variables. New impoundments are originally stocked with immature crayfish, but thereafter the population is usually self-sustaining. The food source is mainly a wide variety of hydrophytes. Harvesting occurs in November to June and is done chiefly with chicken wire traps baited with chunks of rough fish, canned dog food, and other high-protein materials.

The giant freshwater prawn, *Macrobrachium*, is raised commercially in southeastern Asia, but in the United States attempts to culture it commercially in the Gulf Coast states have thus far not been financially successful. Small species of crayfish have been raised experimentally at hatcheries for use as fish food, but the practice has not attained any importance. Crayfish often become a nuisance in fish-hatchery rearing ponds, where they may be excessively abundant and feed on small fish.

Collection, preparation. Stream and pond species may be collected with long-handled dipnets, minnow seines, and minnow traps. When crayfish are not abundant, however, it is often necessary to collect by hand and examine overhanging banks, masses of vegetation, and the undersides of submerged logs and stones.

A skillful collector with a headlamp may find it profitable to collect burrowers (as well as open-water species) at night, when they usually leave the burrow and roam about, but often there is no alternative but to dig up the burrow carefully with a spade or trowel until the specimen can be taken in its chamber. Surface traps baited with meat or peanut butter will sometimes take burrowing crayfish.

Crayfish should be killed by freezing and then immersed in 5–10% buffered formalin. Simply thrusting live animals into fixative often results in involuntary ablation of limbs. Large specimens should also be hypodermically injected with formalin to insure penetration of formalin to deepest organs. Following fixation, 1 or 2 weeks, specimens should be washed and transferred to 80% alcohol and stored in a change to 70% alcohol. Retain the original fixative to sieve for commensal branchiobdellids and ostracods. This preservative keeps the joints pliable. Some workers remove the male first pleopods and the female annulus ventralis, dehydrate in absolute alcohol, clear in xylol, and mount on insect pins.

Taxonomy. In 1902 only 66 species of North American Cambaridae were known, but now there are 55 in Florida alone and about 350 from the United States, the southern states by far producing the greatest number of species. In addition, some of the common and widely distributed species are represented by several well-defined subspecies. Europe has only 15 native species of crayfish.

Characters that are useful for identifying both males and females of certain genera and species include the structure of the rostrum, shape of the areola, structure of the chelae, spines on the cephalothorax, and general body proportions. Unfortunately, however, most species can be identified only from first-form males, the copulatory hooks on some of the legs, and the detailed morphology of the first pleopods, including accurate measurement of various components (Fig. 25.3B). In making collections in the field it is therefore essential to get first-form males whenever possible. Characters used in the key to the Cambaridae are based on the first-form male. Several genera of crayfish are restricted geographically and locality data can be very useful in distinguishing groups.

Beginning with the publication of Hagen's monograph in 1870, various classification schemes for the Cambaridae have been suggested. Both Faxon and Ortmann made major contributions to the North American crayfish systematics, but the most significant work on these animals has been done by H. H. Hobbs, Jr. Hobbs stabilized the classification of the Camabridae and provided many significant publications on the taxonomy of the various genera and species. With the increasing use of molecular methods (allozyme and DNA sequencing), relationships among living species will hopefully become more sharply defined. Aspects of the key below are adapted or modified from his work. The key to the genera of the Palaemonidae is modified from Chace and Bruce (1993).

KEY TO THE FAMILIES AND GENERA OF FRESHWATER DECAPODS OF THE UNITED STATES

1. Abdomen reduced and flexed under broad cephalothorax; locomotion lateral **BRACHYURA, 2**

Abdomen well developed and extended beyond narrow cephalothorax; locomotion forward or backward . . . . **MACRURA, 3**

2. One claw of adult male enlarged and expanded beyond articulation with carpus (Fig. 1.11F; Plate 16); dorsal surface of carapace tuberculate towards lateral margins; 1 introduced species, west coast of FL . . . . **GRAPSIDAE, *Platychirograpsus***

Claws not dimorphic and neither expanded (Fig. 25.5); surface of carapace without tubercles; 1 species ranging along Gulf and east coasts . . . . **PORTUNIDAE, *Callinectes***

3. First three pairs of legs chelate; cephalothorax subcylindrical; abdomen more or less flattened dorsoventrally (Figs. 25.1; 25.2) **4**

First two pairs of legs chelate; cephalothorax and abdomen laterally compressed (Fig. 25.4) **15**

4. Ischia of pereopods of male without hooks; male first pleopods simply rolled at tips (Fig. 25.9); female without first pleopods; lakes and rivers west of the Continental Divide, also upper Missouri River; 5 species, a few presumed extinct . . . . **ASTACIDAE, *Pacifastacus***

Ischia of one or more pairs of pereopods with hooks (Fig. 25.2B, C); male first pleopods bifid or toothed at tips (Figs. 25.3B; 25.11); females with first pleopods; east of the Continental Divide (some species introduced to western states) . . . . **CAMBARIDAE, 5**

5. Ischia of second pereopods without hooks **6**

Ischia of second and third pereopods with hooks; pleopod 1 of male with three distinct tips (Fig. 25.10A, B); 8 small species; southern states through Mississippi River system, epigean ***Cambarellus***

6. Third maxilliped greatly enlarged (Fig. 25.10C), inner margin of ischium without teeth; without eyes or pigment; 1 species, FL caves . . . . ***Troglocambarus***

Third maxilliped not enlarged; inner margin of ischium with teeth (Fig. 25.10D); with or without eyes and pigment **7**

7. Second antennae conspicuously fringed (Fig. 25.10E); 1 epigean lotic species, Barren and Green River systems, KY, TN . . . . ***Barbicambarus***

Second antennae not conspicuously fringed **8**

8. First pleopod of male ending in two clearly defined terminal elements, which may be long and broadly separated or short and closely set (certain species of *Fallicambarus* have a third, small process but both major tips are sharply bent); hooks on ischium of pereopod 3 only (Fig. 25.2C) **9**

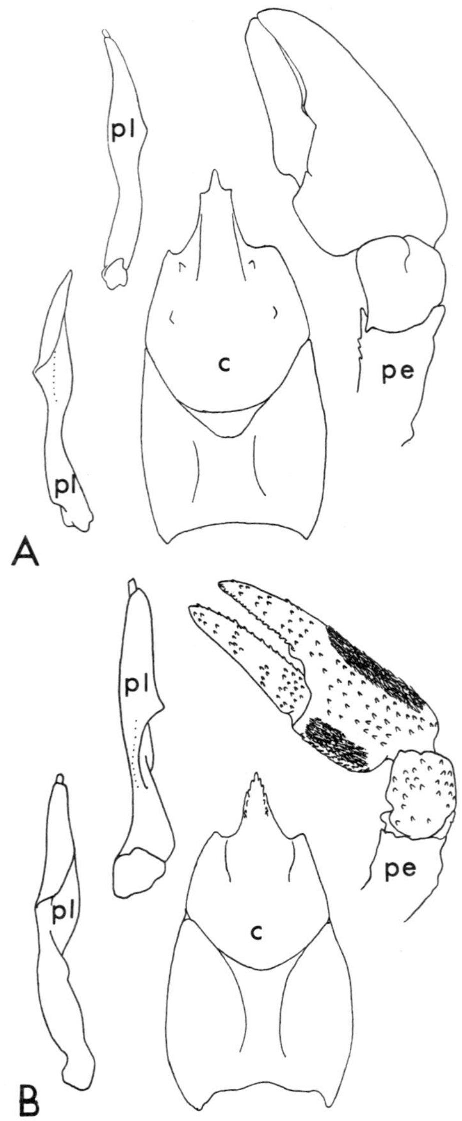

Fig. 25.9. Representative *Pacifastacus.* A, *P. leniusculus;* B, *P. gambeli.* c, carapace; pe, pereopod 1; pl, pleopod 1. (A, B modified from Riegel 1959.)

First pleopod of male ending in more than two terminal elements, often as many as four, which are usually short and highly sculptured, setiferous, or bladelike, and occasionally accompanied by a terminal prominence (Fig. 25.11); hooks on ischia of pereopods 3, 4, or both (Fig. 25.2B); very large genus with over 110 epigean and hypogean species; generally in southern states . . . .*Procambarus*

9. One tip (central projection) of male pleopod 1 extremely long relative to the other, slightly bent or gently curved, triangular in cross section, and tapering to a point (Fig. 25.12A, B); rostrum broad and tapering to acumen without interruption; 4 epigean species in sluggish, lentic environments in southcentral states . . . .*Faxonella*

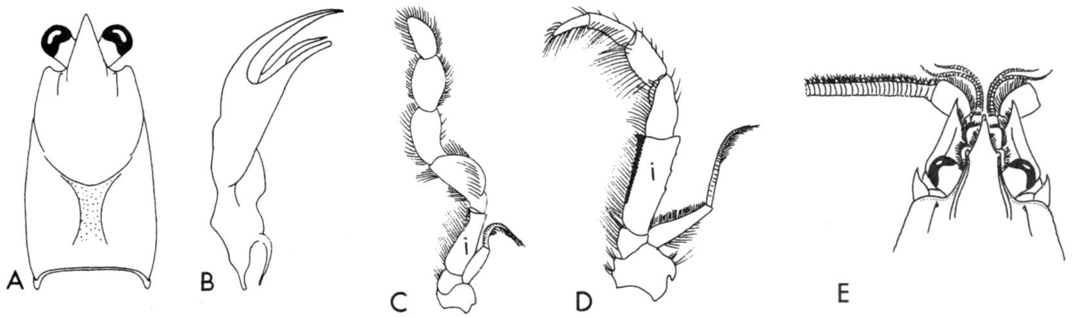

Fig. 25.10. Features of Cambaridae. A, carapace of *Cambarellus texanus;* B, pleopod 1 of first-form male, *C. texanus;* C, maxilliped of *Troglocambarus maclanei;* D, maxilliped of *Procambarus;* E, head and antenna 2 of *Barbicambarus cornutus.* i, ischium. (A, B from Hobbs 1989; C–E from Hobbs 1972a.)

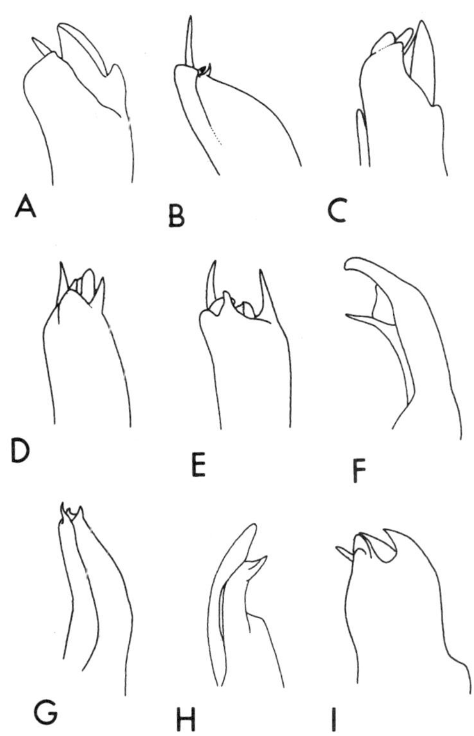

Fig. 25.11. Variation in tips of pleopod 1, first form male, of representative *Procambarus* subgenera. A, *Hagenides;* B, *Leconticambarus;* C, *Lonnbergius;* D, *Pennides;* E, *Villalobosus;* F, *Tenuicambarus;* G, *Capillicambarus;* H, *Remoticambarus;* I, *Scapulicambarus.* (A–I from Hobbs 1972b.)

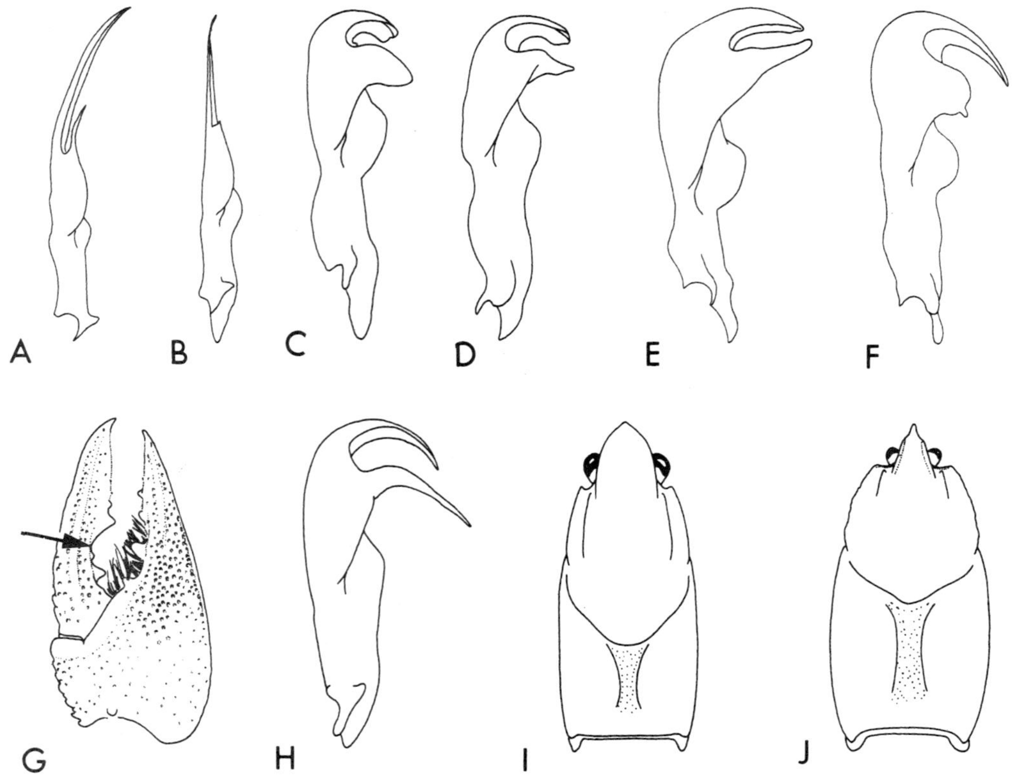

Fig. 25.12. Structure of Cambaridae. A, male pleopod 1 of *Faxonella creaseri;* B, male pleopod 1 of *F. clypeata;* C, male pleopod 1 of *Fallicambarus oryktes;* D, male pleopod 1 of *Cambarus tenebrosus;* E, male pleopod 1 of *C. hamulatus;* F, male pleopod 1 of *C. causeyi;* G, chela of *F. oryktes,* arrow denotes excision; H, male pleopod 1 of *Hobbseus valleculus;* I, carapace of *H. orconectoides;* J, carapace of *C. rusticiformis.* (A–J from Hobbs 1989.)

Male with one tip of pleopod 1 not much longer than other; if considerable difference in lengths exist then longer tip spiniform and rostrum with marginal spines or interruption (Fig. 25.13A) **10**

10. Tips of pleopod 1 of male sharply bent at right angle to shaft (Fig. 25.12C–F, H), central projection bladelike and abruptly acuminate **11**

Tips of pleopod 1 of male not sharply bent, or if so central projection not bladelike **13**

11. Mesial process of pleopod 1 longer than central projection (Fig. 25.12H), narrow, and pointed; rostrum broad, lateral margins uninterrupted (Fig. 25.12I); 7 epigean species in lowland streams and stagnant water, AL, MS . . . . ***Hobbseus***

Mesial process of pleopod 1 otherwise; rostrum margins interrupted (Figs. 25.12J; 25.13A) **12**

12. Opposable margin of dactyl of chela with abrupt excision marked by a large tubercle (Fig. 25.12G); 13 epigean species in burrows and shallow waters; widespread in southern and central states . . . . ***Fallicambarus***

Opposable margin of dactyl of chela without excision or large tubercle; 80 or more species in nearly all aquatic habitats and distributed throughout the United States east of the Continental Divide . . . . ***Cambarus***

13. Longer tip (central projection) of pleopod 1 narrow to spiniform (Fig. 25.13B, D, F, H); rostrum margins typically interrupted by spines or tubercles (Fig. 25.13A, C, E, G); about 75

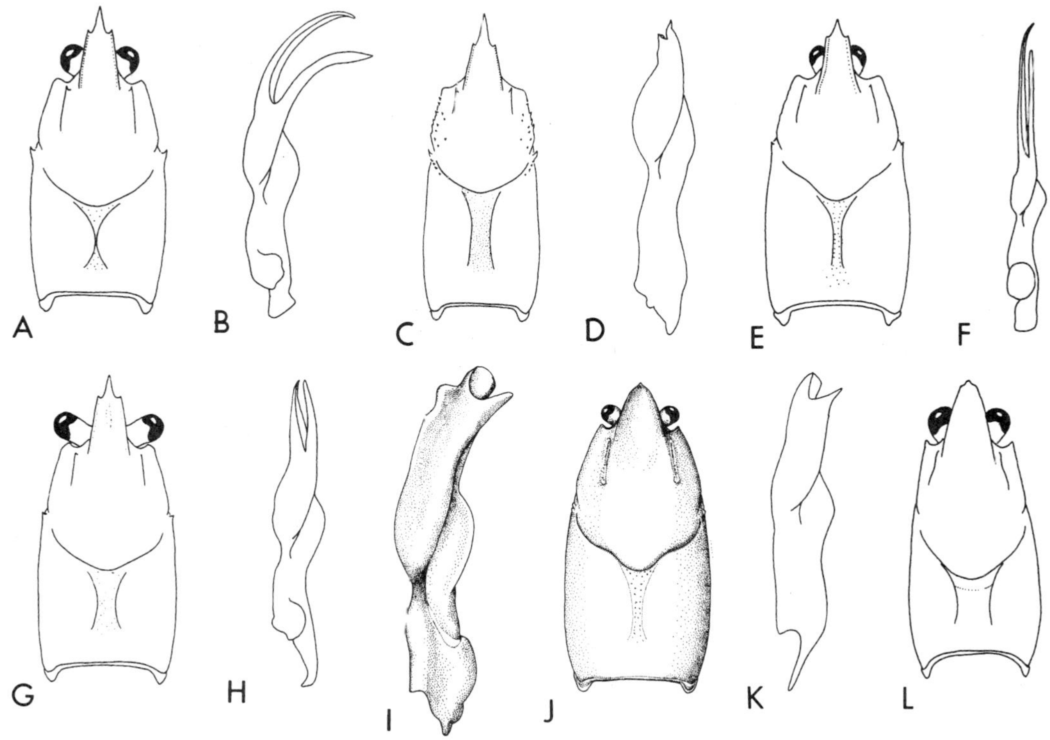

Fig. 25.13. Structure of Cambaridae. A, carapace of *Orconectes hobbsi;* B. male pleopod 1 of *O. hobbsi;* C, carapace of *O. pellucidus;* D, male pleopod 1 of *O. pellucidus;* E, carapace of *O. ozarkae;* F, male pleopod 1 of *O. ozarkae;* G, carapace of *O. propinquus;* H, male pleopod 1 of *O. propinquus;* I, male pleopod 1 of *Distocambarus devexus;* J, carapace of *D. devexus;* K, male pleopod 1 of *Bouchardina robinsoni;* L, carapace of *B. robisoni.* (A–H, K, L from Hobbs 1989; I, J from Hobbs 1981.)

species distributed throughout the central and eastern U.S., includes several hypogean species . . . . ***Orconectes***

Longer tip of pleopod 1 broad and truncated (Fig.13I, K) **14**

14. Areola narrow, more than 5 times as long as wide (Fig. 25.13J); 5 species, burrowers in GA, SC ***Distocambarus***

Areola broad, 2 to 2.5 times as long as wide (Fig. 25.13L); 1 species in lentic waters, AR ***Bouchardina***

15. Chelae of pereopods 1 and 2 partially obscured by hairlike setae (Fig. 25.14A) **ATYIDAE, 16**

Chelae not obscured by setiferous patches . . . . **PALAEMONIDAE, 17**

16. Eyestalks rudimentary, no trace of facets or pigments; rostrum with many dorsal spines (Fig. 25.14B); 2 species known only from caves in AL, KY . . . . ***Palaemonias***

Eyestalks well developed; rostrum with none to one or two dorsal spines (Fig. 25.14C); 2 species in coastal streams of California . . . . ***Syncaris***

17. Carapace with a branchiostegal spine (Fig. 25.14D) **18**

Carapace without a branchiostegal spine; pereopod 2 strongly dimorphic in male, one leg greatly enlarged (Fig. 25.4B, C); 5 species, 2 exotic, chiefly in southern states, 1 species ranging into Mississippi River system . . . . ***Macrobrachium***

18. Elevated crest of spines at base of rostrum (Fig. 25.4D); 1 introduced species, OR, WA ***Exopalaemon***

No elevated crest at base of rostrum (Figs. 25.4A; 25.14D, E); 6 epigean or hypogean species, 2 widely distributed . . . . ***Palaemonetes***

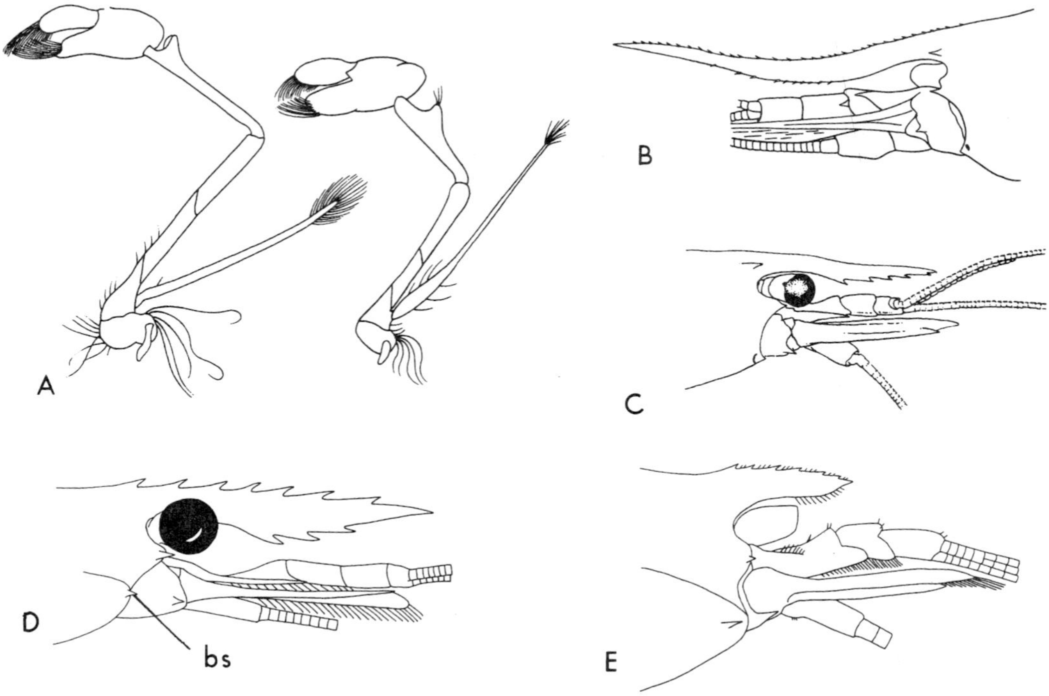

Fig. 25.14. Structure of Atyidae and Palaemonidae. A, pereopods 1 and 2 of *Palaemonias;* B, head of *Palaemonias;* C, head of *Syncaris;* D, head of epigean *Palaemonetes;* E, head of hypogean *Palaemonetes.* bs, branchiostegal spine. (A, B from Hobbs et al. 1977; C from Chace et al. 1959; D, E from Strenth 1976.)

REFERENCES

Abrahamson, S. A. A., and C. R. Goldman. 1970. Distribution, density and production of the crayfish *Pacifastacus leniusculus* Dana in Lake Tahoe California–Nevada. *Oikos* 21:83–91.

Ameyaw-Akumfi, C., and B. A. Hazlett. 1975. Sex recognition in the crayfish *Procambarus clarkii. Science* 190:1225–1226.

Andrews, E. A. 1904. Breeding habits of crayfish. *Am. Nat.* 38:165–206.

——. 1907. The young of the crayfishes *Astacus* and *Cambarus. Smithson. Contrib. Knowl.* 35:5–79.

Black, J. B. 1963. Observations on the home range of stream-dwelling crawfishes. *Ecology* 44:592–595.

Bovbjerg, R. V. 1952. Comparative ecology and physiology of the crayfish *Orconectes propinquus* and *Cambarus fodiens. Physiol. Zool.* 25:34–56.

——. 1953. Dominance order in the crayfish *Orconectes virilis* (Hagen). *Physiol. Zool.* 26:173–178.

——. 1956. Some factors affecting aggressive behavior in crayfish. *Physiol. Zool.* 29:127–136.

——. 1970. Ecological isolation and competitive exclusion in two crayfish (*Orconectes virilis* and *Orconectes immunis*). *Ecology* 51:225–236.

Broad, A. C., and J. H. Hubschman. 1963. The larval development of *Palaemonetes kadiakensis* M. J. Rathbun in the laboratory. *Trans. Am. Microsc. Soc.* 82:185–197.

Caine, E. A. 1974. Zoogeography of the Florida troglobitic crayfishes, genus *Procambarus. Am. Midl. Nat.* 92:487–492.

Capelli, G. M. 1982. Displacement of northern Wisconsin crayfish by *Orconectes rusticus* (Girard). *Limnol. Oceanogr.* 27:741–744.

Capelli, G. M., and J. J. Magnuson. 1983. Morphoedaphic and biogeographic analysis of crayfish distribution in northern Wisconsin. *J. Crust. Biol.* 3:548–564.

Chace, F. A., Jr., and A. J. Bruce. 1993. The Caridean shrimps (Crustacea: Decapoda) of the Albatross Philippine expedition 1907–1910, Part 6: Superfamily Palaemonoidea. *Smithson. Contrib. Zool.* 543:1–152.

Chace, F. A., Jr., and H. H. Hobbs, Jr. 1969. The freshwater and terrestrial decapod crustaceans of the West Indies with special reference to Dominica. *Bull. U. S. Nat. Mus.* 292:1–258.

Chace, F. A., Jr., et al. 1959. Malacostraca. In: *Fresh-Water Biology*, 2nd ed., pp. 869–901, edited by W. T. Edmondson. Wiley, New York.

Cooper, J. E., and A. L. Braswell. 1995. Observations on North Carolina crayfishes (Decapoda:Cambaridae). *Brimleyana* 22:87–132.

Corey, S. 1988a. Comparative life histories of four populations of *Orconectes propinquus* (Girard, 1852) in southwestern Ontario, Canada (Deacpoda, Astacidea). *Crustaceana* 54:129–138.

——. 1988b. Comparative life histories of two populations of the introduced crayfish *Orconectes rusticus* (Girard, 1852) in Ontario. *Crustaceana.* 55:29–38.

Crandall, K. A., and J. F. Fitzpatrick, Jr. 1996. Crayfish molecular systematics: using a combination of procedures to estimate phylogeny. *Syst. Biol.* 45:1–26.

Creaser, E. P. 1931. The Michigan decapod crustaceans. *Pap. Mich. Acad. Sci Arts Lett.* 13:257–276.

Creaser, E. P., and A. I. Ortenburger. 1933. The decapod crustaceans of Oklahoma. *Publ. Univ. Okla. Biol. Surv.* 5:14–80.

Crocker, D. W. 1957. The crayfishes of New York State (Decapoda, Astacidac). *Bull. N.Y. State Mus.* 355(1): 197.

——. 1979. The crayfishes of New England. *Proc. Biol. Soc. Wash.* 92:225–252.

Crocker, D. W., and D. W. Barr. 1968. Handbook of the crayfishes of Ontario. *Life. Sci. Misc. Publ. R. Ont. Mus.*, pp. 1–158.

Daniels, R. A. 1998. Changes in the distribution of stream-dwelling crayfishes in the Schoharie Creek system, eastern New York State. *Northeast. Nat.* 5:231–248.

Davies, I. J. 1989. Population collapse of the crayfish *Orconectes virilis* in response to experimental whole-lake acidification. *Can. J. Fish. Aquat. Sci.* 46:910–922.

DiStefano, R. J., et al. 1991. Response of the crayfish *Cambarus bartonii bartonii* to acid exposure in southern Appalachian streams. *Can. J. Zool.* 69:1585–1591.

Faxon, W. 1885. Revision of the Astacidae, Part I: the genera *Cambarus* and *Astacus*. *Mem. Mus. Comp. Zool. Harvard Coll.* 10:1–186.

——. 1914. Notes on the crayfishes in the United States National Museum and the Museum of Comparative Zoology with descriptions of new species and subspecies to which is appended a catalogue of the known species and subspecies. *Mem. Mus. Comp. Zool. Harvard Coll.* 40:347–427.

Fingerman, M. 1955. Factors influencing the rate of oxygen consumption of the dwarf crawfish, *Cambarellus shufeldtii* (Decapoda, Astacidae). *Tulane Stud. Zool.* 3:103–116.

Fingerman, M., and A. D. Lago. 1957. Endogenous twenty-four hour rhythms of locomotor activity and oxygen consumption in the crawfish *Orconectes clypeatus*. *Am. Midl. Nat.* 58: 383–393.

Fitzpatrick, J. F. 1967. The propinquus group of the crawfish genus *Orconectes* (Decapoda: Astacidae). *Ohio J. Sci.* 67:129–177.

——. 1987. The subgenera of the crawfish genus *Orconectes* (Decapoda; Cambafidae). *Proc. Biol. Soc. Wash.* 100:44–74.

Fitzpatrick, J. F., and J. F. Payne. 1968. A new genus and species of crawfish from the southeastern United States (Decapoda, Astacidae). *Proc. Biol. Soc. Wash.* 81:11–22.

Franz, R., and D. S. Lee. 1982. Distribution and evolution of Florida's troglobitic crayfishes. *Bull. Fla. State Mus. Biol. Sci.* 28:53–78.

Gunter, G. 1937. Observations on the river shrimp *Macrobrachium ohionis* (Smith). *Am. Midl. Nat.* 18:1038–1042.

Guyselman, J. B. 1957. Solar and lunar rhythms of locomotor activity in the crayfish *Cambarus virilis*. *Physiol. Zool.* 30:70–87.

Hagen, H. A. 1870. Monograph of the North American Astacidae. *Illus. Cat. Mus. Comp. Zool. Harvard Coll.* 3: 1109.

Harris, J. A. 1903. An ecological catalogue of the crayfishes belonging to the genus *Cambarus*. *Kans. Univ. Sci. Bull.* 2:51–187.

Hart, C. W., Jr., and J. Clark. 1989. *An Interdisciplinary Bibliography of Freshwater Crayfishes*. 498 pp. Smithsonian Institution Press, Washington, DC.

Hay, W. P. 1903. Observations on the crustacean

fauna of the region about the Mammoth Cave, Kentucky. *Proc. U.S. Nat. Mus.* 25:223–226.

Hazlett, B., et al. 1974. Behavioral biology of the crayfish *Orconectes virilis*, I: home range. *Am. Midl. Nat.* 92:301–319.

Hedgpeth, J. W. 1947. River shrimps. *Prog. Fish Cult.* 9:181–184.

——. 1949. The North American species of *Macrobrachium* (river shrimp). *Tex. J. Sci.* 1:28–38.

——. 1968. The atyid shrimp of the genus *Syncaris* in California. *Int. Rev. gesamten Hydrobiol.* 53:511–524.

Hobbs, H. H., Jr. 1942a. A generic revision of the crayfishes of the Subfamily Carribarinae (Decapoda, Astacidae) with the description of a new genus and species. *Am. Midl. Nat.* 28:334–357.

——. 1942b. The crayfishes of Florida. *Univ. Fla. Publ. Biol. Sci. Ser.* 3:1–179.

——. 1945. Two new species of crayfishes of the genus *Cambarellus* from the Gulf Coast states, with a key to the species of the genus (Decapoda, Astacidae). *Am. Midl. Nat.* 34:466–474.

——. 1967. A new crayfish from Alabama caves with notes on the origin of the genera *Orconectes* and *Cambarus* (Decapoda: Astacidae). *Proc. U.S. Nat. Mus.* 123:1–17.

——. 1969. On the distribution and phylogeny of the crayfish genus *Cambarus. Va. Polytech. Inst. Res. Div. Monogr.* 1:93–178.

——. 1972a. Crayfishes (Astacidae) of North and Middle America. *Biota of Freshwater Ecosystems, Identification Manual* 9:1–173.

——. 1972b. The subgenera of the crayfish genus *Procambarus* (Decapoda: Astacidae). *Smithson. Contrib. Zool.* 117:1–22.

——. 1981. The crayfishes of Georgia. *Smithson. Contrib. Zool.* 318:1–549.

——. 1989. An illustrated checklist of the American crayfishes (Decapoda: Astacidae, Cambaridae, and Parastacidae). *Smithson. Contrib. Zool.* 480:1–236.

Hobbs, H. H., Jr., and H. H. Hobbs III. 1991. An illustrated key to the crayfishes of Florida (based on first form males). *Fla. Sci.* 54:13–24.

Hobbs, H. H., Jr., et al. 1977. A review of the troglobitic decapod crustaceans of the Americas. *Fla. Sci.* 244:1–183.

Hobbs, H. H., III. 1993. Trophic relationships of North American freshwater crayfishes and shrimps. *Contrib. Milwaukee Publ. Mus.* 85:1–110.

Hobbs, H. H., III, and J. P. Jass. 1988. *The Crayfishes and Shrimp of Wisconsin.* 177 pp. Milwaukee Public Museum, Milwaukee, WI.

Holthuis, L. B. 1949. Note on the species of *Palaemonetes* (Crustacea Decapoda) found in the United States of America. *Proc. Konink. Ned. Akad. Wet.* 52:87–95.

——. 1952. A general revision of the Palaemonidae (Crustacea, Decapoda, Natantia) of the Americas, II: the subfamily Palaemoninae. *Occas. Pap. Allan Hancock Found.* 12:1–396.

Huner, J. V. 1997. The crayfish industry in North America. *Fisheries* 22:28–31.

Huntsman, A. G. 1915. The fresh-water Malacostraca of Ontario. *Contrib. Can. Biol. II (Suppl.)* 47:145–163.

Huxley, T. H. 1906. *The Crayfish*, 2d ed., 371 pp. Kegan Paul, London.

Jezerinac, R. F., et al. 1995.The crayfishes (Decapoda: Cambaridae) of West Virginia. *Bull Ohio Biol. Surv.* 10:1–193.

Koppelman, J. B., and D. E. Figg. 1995. Genetic estimates of variability and relatedness for conservation of an Ozark cave crayfish species complex. *Conserv. Biol.* 9:1288–1294.

Lagler, K. F., and M. J. Lagler. 1944. Natural enemies of crayfishes in Michigan. *Pap. Mich. Acad. Sci. Arts Lett.* 29:293–303.

Lowe, M. E. 1956. Dominance–subordinance relationships in the crawfish *Cambarellus shufeldtii. Tulane Stud. Zool.* 4:139–170.

Lyle, C. 1938. The crawfishes of Mississippi, with special reference to the biology and control of destructive species. *Iowa State Coll. J. Sci.* 13: 75–77.

Maluf, N. S. R. 1939. On the anatomy of the kidney of the crayfish and on the absorption of chloride from freshwater by this animal. *Zool. Jahrb. Abt. allg. Zool. Physiol. Tiere* 59:515–534.

Mason, J. C. 1970a. Egg-laying in the western North American crayfish, *Pacifastacus trowbridgii* (Stimpson) (Decapoda, Astacidae). *Crustaceana* 19:37–44.

——. 1970b. Copulatory behavior of the crayfish *Pacifastacus trowbridgii* (Stimpson). *Can. J. Zool.* 48:969–976.

——. 1970c. Maternal-offspring behavior of the crayfish, *Pacifastacus trowbridgii* (Stimpson). *Am. Midl. Nat.* 84:463–473.

Meehean, O. L. 1936. Notes on the freshwater shrimp *Palaemonetes paludosa* (Gibbes). *Trans. Am. Microsc. Soc.* 55:433–441.

Merkel, E. L. 1969. Home range of crayfish *Orconectes juvenalis*. *Am. Midl. Nat.* 81:228–235.

Mobberly, W. C. 1963. Hormonal and environmental regulation of the molting cycle in the crayfish *Faxonella clypeata*. *Tulane Stud. Zool.* 11:79–96.

Momot, W. T. 1966. Upstream movements of crayfish in an intermittent Oklahoma stream. *Am. Midl. Nat.* 75:150–159.

Momot, W. T., and H. Gowing. 1972. Differential seasonal migration of the crayfish, *Orconectes virilis* (Hagen) in marl lakes. *Ecology* 53:479–483.

Momot, W. T. et al. 1978. The dynamics of crayfish and their role in ecosystems. *Am. Midl. Nat.* 99:10–35.

Olsen, T. M., et al. 1991. Mechanisms of impact of an introduced crayfish (*Orconectes rusticus*) on littoral congeners, snails, and macrophytes. *Can. J. Fish. Aquat. Sci.* 48:1853–1861.

Ortmann, A. E. 1906. The crawfishes of the state of Pennsylvania. *Mem. Carnegie Mus.* 2:343–521.

——. 1931. Crawfishes of the southern Appalachians and the Cumberland Plateau. *Ann. Carnegie Mus.* 20:61–160.

Payne, J. F. 1972. The life history of *Procambarus hayi*. *Am. Midl. Nat.* 87:25–35.

Pearse, A. S. 1909. Observations on copulation among crayfishes. *Am. Nat.* 43:746–753.

Penn, G. H. 1942. Observations on the biology of the dwarf crawfish, *Cambarellus shufeldtii* (Faxon). *Am. Midl. Nat.* 28:644–647.

——. 1943. A study of the life history of the Louisiana red-crawfish, *Cambarus clarkii* Girard. *Ecology* 24:118.

——. 1957. Variation and subspecies of the crawfish *Orconectes palmeri* (Faxon). *Tulane Stud. Zool.* 5:229–262.

——. 1959. An illustrated key to the crayfishes of Louisiana with a summary of their distribution within the state (Decapoda, Astacidae). *Tulane Stud. Zool.* 7:3–20.

Penn, G. H., and J. F. Fitzpatrick. 1963. Interspecific competition between two sympatric species of dwarf crawfishes. *Ecology* 44:793–797.

Penn, G. H., and H. H. Hobbs. 1958. A contribution toward a knowledge of the crawfishes of Texas (Decapoda, Astacidae). *Tex. J. Sci.* 10:452–483.

Reiber, C. L. 1995. Physiological adaptations of crayfish to the hypoxic environment. *Am. Zool.* 35:1–11.

Rhoades, R. 1944. The crayfishes of Kentucky, with notes on variation, distribution and descriptions of new species and subspecies. *Am. Midl. Nat.* 31:111–149.

Riegel, J. A. 1959. The systematics and distribution of crayfishes in California. *Calif. Fish Game* 45:29–50.

Riggert, C. M., et al. 1999. Distributions and selected aspects of the life histories and habitat associations of the crayfishes *Orconectes peruncus* (Creaser, 1931) and *O. quadruncus* (Creaser, 1933) in Missouri. *Am. Midl. Nat.* 142:348–362.

Roberts, T. W. 1944. Light, eyestalk chemical, and certain other factors as regulators of community activity for the crayfish, *Cambarus virilis* Hagen. *Ecol. Monogr.* 14:359–392.

Schmitt, W. L. 1933. Notes on shrimps of the genus *Macrobrachium* found in the United States. *J. Wash. Acad. Sci.* 23:312–317.

Scudamore, H. H. 1947. The influence of the sinus glands upon molting and associated changes in the crayfish. *Physiol. Zool.* 20:187–208.

——. 1948. Factors influencing molting and the sexual cycles in the crayfish. *Biol. Bull.* 95:229–237.

Smith, D. G. 1981. Evidence for hybridization between two crayfish species (Decapoda: Cambaridae: *Orconectes*) with a comment on the phenomenon in cambarid crayfish. *Am. Midl. Nat.* 105:405–407.

Smith, E. W. 1953. The life history of the crawfish, *Orconectes* (Faxonella) *clypeatus* (Hay). *Tulane Stud Zool.* 1:79–96.

Steele, M. 1902. The crayfish of Missouri. *Univ. Cincin. Bull.* 10:1–53.

Stetzer, R. S., and T. M. Burton. 1993. Growth and abundance of the crayfish *Orconectes propinquus* in a hard water and a soft water stream. *J. Freshwater Ecol.* 8:329–340.

Strenth, N. E. 1976. A review of the systematics and zoogeography of the freshwater species of *Palaemonetes* Heller of North America (Crustacea: Decapoda). *Smithson. Contrib. Zool.* 228:1–27.

——. 1990. Reproductive patterns of the subterranean shrimp *Palaemonetes antrorum* Benedict (Decapoda, Palaemonidae) from central Texas. *Stygologia* 5:221–224.

——. 1991. The significance of the larval development of *Palaemonetes antrorum* Benedict (Crustacea, Decapoda) on the origin of subterranean palemonids. *Stygologia* 6:149–153.

Strenth, N. E., and G. Longley. 1990. Reproductive patterns of the subterranean shrimp *Palaemonetes antrorum* Benedict (Decapoda, Palaemonidae) from central Texas. *Stygologia* 5:221–224.

Tack, P. I. 1941. The life history and ecology of the crayfish *Cambarus immunis* Hagen. *Am. Midl. Nat.* 25:420–446.

Turner, C. L. 1926. The crayfish of Ohio. *Bull. Ohio Biol. Surv.* 13:145–195.

Van Deventer, W. C. 1937. Studies on the biology of the crayfish, *Cambarus propinquus* Girard. *Ill. Biol. Monogr.* 15:1–67.

Wicksten, M. 1997. Introduction of the ridgetail prawn, *Exopalaemon carinicauda*, into San Fransisco Bay, California. *Calif. Fish Game* 83: 43–44.

Williams, A. B. 1954. Speciation and distribution of the crayfishes of the Ozark plateaus and Ouachita provinces. *Univ. Kans. Sci. Bull.* 36: 803–918.

——. 1965. Marine decapod crustaceans of the Carolinas. *Fish. Bull.* 65:1–298.

Williams, A. B., and A. B. Leonard. 1952. The crayfishes of Kansas. *Fish. Bull.* 34:961–1012.

Young, L. B., and H. H. Harvey. 1990. Metal concentrations in crayfish tissues in relation to lake pH and metal concentrations in water and sediments. *Can. J. Zool.* 69:1076–1082.

APPENDIX

REAGENTS, SOLUTIONS, AND LABORATORY ITEMS

For the most part, this brief appendix lists only those reagents and solutions that are mentioned more than once in the text portion of this volume. Information concerning additional reagents and procedures, including fixation, embedding, sectioning, staining, and so on, can be obtained by consulting any of the standard microtechnique handbooks.

In the laboratory all transfers of small specimens from one solution to another can be performed in the same vial. When solutions are added they are squirted into the vial with a pipette so that there is thorough mixing; when solutions are removed they are sucked up with a pipette after the specimens have all settled to the bottom of the vial. Although two or three such portions are usually necessary to effect a complete change of solutions in a vial, the small additional time is well compensated for by the convenience and compactness of the method. Furthermore, there is little danger of drying, spilling, or loss of specimens. One-piece disposable plastic pipettes are preferred, and, despite their name, are reusable.

The following list includes a variety of anesthetics. In general, such solutions should be slowly and gradually added to the water containing the specimens. Experience will determine proper amounts. For more information on stains and staining technique see Presnell and Schreibman (1997).

Acetic acid. Usually supplied as glacial acetic acid, which is 99.5% CH_3COOH. Two to 10 parts of glacial acetic acid are often added to 100 parts saturated mercuric chloride as a fixative.

Alcohol. Wherever "alcohol" is indicated in this manual, ethyl alcohol, or C_2H_5OH, is meant. It is used in varying concentrations as a fixative and preservative. Alcohol is usually supplied commercially in about 95% concentration (190 proof). All alcohol dilutions are made by measuring out the number of rnilliliters of 95% alcohol equivalent to the percentage of alcohol desired and diluting to 95 mL with water. For example, to make a 60% solution, add 35 mL of water to 60 mL of 95% alcohol. Absolute (essentially 100%) alcohol is expensive to buy but may be easily made in the laboratory by dehydrating 95% alcohol with anhydrous copper sulfate. Crystals of copper (cupric) sulfate should first be heated in an evaporating dish until the water of crystallization is driven off and a white powder remains. Place some of this powder in a bottle and add 95% alcohol. The water in the alcohol immediately unites with the copper sulfate, turning it blue. Add more anhydrous copper sulfate until it no longer turns blue. Then quickly filter the alcohol into a dry bottle and stopper with a tight cork or ground glass stopper.

Alum cochineal. An excellent whole-mount stain for many small invertebrates. Add 6 g of potassium alum and 6 g of powdered cochineal to 90 mL of distilled water and boil for 30 minutes. Allow the fluid to settle, decant the supernatant, add a little water, and boil until about 90 mL of solution remains. Filter when cool and add 0.5 g thymol to inhibit molds. Depending on size and density of specimens, stain for 4 to 30 hours. After staining, wash specimens in several changes of water for 15 to 30 minutes to extract the alum. The material is then dehydrated and mounted in balsam, permount, or some similar nonaqueous material.

Balsam. (Canada balsam). A general permanent mounting medium for dehydrated whole specimens or microtome sections. It is easiest to buy a thick solution and dilute with xylol (xylene) to the proper consistency. Add a marble chip to keep the solution neutral.

Benzamine hydrochloride, benzamine lactate. Anesthetics for certain small metazoans; usually used in 2% solutions.

Borax carmine. This is a most useful stain for whole mounts of small metazoans. Boil 2 g of carmine in 100 mL of 4% borax until dissolved. Then add 100 mL of 70% alcohol; filter after 24 hours. Stain objects for 10 hours to several days, depending on size and density. Transfer to 70% acid alcohol and keep there until the stain no longer forms clouds. Wash the specimens in neutral alcohol.

Bottom fauna samplers. Many different devices are described in the literature. Boulton (1985) lists some of these references.

Bouin's fluid. A general fixative consisting of 75 mL of saturated aqueous picric acid, 25 mL of formalin, and 5 mL of glacial acetic acid. The normal fixing time is 2 to 16 hours, but tissues are not harmed if left in the fluid up to several weeks.

Butyn, butyn sulfate. Anesthetics for certain small metazoans; usually used in a 0.1 to 2% solution.

Carmine, Gower's carmine stain. This stain is especially useful for staining small freshwater invertebrates to be mounted in glycerin or glycerin jelly. It will not "bleed" into the mounting medium. Add 10 g of carmine to 100 mL of 45% acetic acid. Heat to a boil, cool, filter, dry, and save the residue. Then to 1 g of dry residue add 10 g of alum and 200 mL of distilled water. Heat to dissolve, cool, filter, and add one crystal of thymol. (See Staining with Gower's carmine.)

Carnoy's fluid. Especially useful for killing and fixing fragile immature or small crustaceans. Consists of 10% acetic acid, 60% alcohol, and 30% chloroform. Fix for 12 to 24 hours; wash in 70% alcohol; store in 70% alcohol.

Chloral hydrate. A general anesthetic for small metazoans. Make up as a 10% solution.

Chloretone. A common anesthetic (chlorbutanol) usually used in 0.1 to 0.6% solution.

Chlorine removal. Tapwater is generally not suitable for freshwater invertebrates because of its chlorine content. If allowed to stand for a day or two, with frequent stirring, it will usually be usable. Otherwise, add 0.5 mg/L of sodium bisulfite or 2 to 3 mg/L of sodium thiosulfate.

Chloroform. Occasionally used as a saturated aqueous solution (1.0%) for anesthetizing small metazoans. Chloroform vapors are sometimes useful if the organisms are concentrated in a small amount of water under a bell jar beside a small beaker of chloroform.

Chloroplatinic acid. This compound, $H_2PtCl_6 \cdot 6H_2O$, is especially valuable for preserving and keeping osmic acid solutions stable. A gram should be added to every 100 mL of osmic acid solution.

Cleaning solution. Dissolve 60 to 65 g of sodium or potassium bichromate by heating in 30 to 65 mL of water. Cool to room temperature and then slowly add concentrated sulfuric acid to make 1 L of solution. Handle this solution with great care. After using it for cleaning glassware, rinse the glassware thoroughly.

Clorox. A commercial caustic useful for isolating small sclerotized structures from musculature and connective tissues, and especially for isolating and cleaning trophi of rotifers and radula of gastropod mollusks. Usually diluted 1 to 10 with water.

CMCP (nonresinous mounting medium). This is a permanent mounting material useful for whole mounts of aquatic insects, crustaceans, and other small metazoans. It is available in a thick, syrupy solution in which specimens may be mounted directly and satisfactorily from water, formalin preservative solutions, alcohols, and glycerin. It dries hard (but slowly) and has the additional advantage of being a clearing agent. For real permanency, however, coverslips should be ringed with Murrayite or some similar cement after the mounts have dried for several days. CMCP may be thinned with water. It is just as permanent as balsam, and is indeed an important contribution to biological technique and slide making. Some researchers tint CMCP with acid fuchsin or anilin blue. The product is marketed by Polysciences (400 Valley Road, Warrington, PA 18976).

Cocaine, Novocain. Excellent anesthetics for small metazoans but difficult to obtain legally, because of narcotics regulations. Best used as a 1% solution in 95% alcohol or water.

Corn syrup mounting medium. Microzoans, and especially plankton material, may be satisfactorily mounted on glass slides in a solution consisting of 70% aqueous 4% formalin, and 30% commercial "light" corn syrup obtainable in grocery stores everywhere. Slides should be stored horizontally for several days as the medium forms a seal at the edge of the coverslip. If the mountant draws away from the edge, simply add another drop or two. After the seal is completely hardened, slides may be stored on edge. The writer (RWP) has had such plankton slides in satisfactory condition for 20 years.

Culturing. See the appropriate sections in the chapters in this volume. Also see Boulton (1985), Galtsoff et al. (1937), and Sudia (1951).

Eosin. General purpose cytoplasmic stain for small metazoans. Usually used as a 0.5% solution in 95% alcohol or water. Putt's variation adds picric acid (5% of solution) and sometimes produces a sharper stain.

Euparal. A permanent slide mounting medium. Specimens may be transferred directly from 95% alcohol into euparal.

FAA (formalin-aceto-alcohol). A general-purpose fixative in which material may be kept for several days without harm. Numerous modifications of this fixative are used; a typical one consists of 40 mL of distilled water, 2 mL of glacial acetic acid, 50 mL of 95% alcohol, and 10 mL of formalin.

Formalin. A common and cheap preservative consisting of a 37 to 40% aqueous solution of formaldehyde, H_2CO. Depending on the nature of the preserved material, 2 to 6% solutions of formalin (not 2–6% formaldehyde) are usually used. Commercial formalin is slightly acid and may be neutralized by maintaining a slight deposit of magnesium carbonate, household borax, or marble chips in the bottom of the reagent bottle. When formalin dries it forms a whitish irreversible polymer.

Fuchsin (basic). A nuclear stain sometimes useful in temporary whole mounts of micrometazoans. Make up a 0.5% aqueous solution.

Glycerin. An excellent medium for dissecting and for temporary and permanent mounts for many kinds of micrometazoans. Glycerin is miscible with water and alcohol at all concentrations and has the advantage of being a slight clearing agent. Because glycerin is nondrying, one should add about 5%

of it to stored vials of alcoholic and formalin specimens. If such vials should break, or the water and alcohol should dry, the specimens are still available in good condition. Specimens to be used in concentrated glycerin should first be placed in a 5 to 20% solution of glycerin in 50 to 95% alcohol. If the container is uncovered (but protected from dust), the alcohol and water evaporate in a few days, leaving the specimens in pure glycerin. However, beware that glycerin is a clearing agent.

Glycerin jelly, glycerin jelly slide mounts. A valuable permanent mounting medium that is solid at ordinary temperatures. It is sold by most biological supply firms, but may be easily made in the laboratory. Soak 1.5 g of clean gelatin overnight in 100 mL of water, then dissolve with gentle heat. Add half its volume of clean glycerin and 0.5 g of carbolic acid. Warm gently and stir for 10 minutes until homogeneous. Store in a small wide-mouth bottle with a screw cap. Specimens to be mounted in glycerin jelly should first be in at least 50% glycerin. Melt the glycerin jelly gently in warm water or on a hot plate. (If it is heated too much it polymerizes and is useless.) Place the specimen on a slide in a droplet of glycerin, drop on one to several drops of liquid glycerin jelly with a thin glass rod, and quickly orient the specimen with a warm needle. Then pick up a clean, circular coverslip with a forceps, breathe on the under surface, and gently put it in place on top of the fluid. Set the mount aside to congeal for at least 30 minutes. Later the excess glycerin jelly may be removed from the edge of the coverslip with a razor blade and clean cloth. For real permanence the coverslip should be ringed on a turntable with several coats of Murrayite.

Heat-absorbing fluid. To be used in a glass container between a light source and heat-sensitive living animals being studied on the stage of a binocular dissecting microscope. Use 2% aqueous calcium chloride; or dissolve 200 g of ferrous ammonium sulfate in water, filter, and add 1.7 mL of concentrated sulfuric acid.

Helly's fluid. A fixative for soft-bodied animals with a reputation for reducing shrinkage. Mix 2.5 g of potassium dichromate, 5 g of mercuric chloride, 1 g of sodium sulfate, 100 mL of water, and 5 mL of formalin.

Hematoxylin. Delafield's hematoxylin is a general purpose nuclear stain. Dissolve 1 g of hematoxylin crystals in 10 mL of absolute alcohol and add it drop by drop to 100 mL of a saturated aqueous solution of aluminum ammonium sulfate. Expose this mixture to air and light for several weeks to "ripen." Filter, then add 25 mL of glycerin and 25 mL of methyl alcohol. After staining, differentiate blue with Scotts solution. Good when used alone for soft-bodied, slide-mounted invertebrates.

Hoyer's mounting medium. Dissolve 30 g of pulverized gum arabic in 50 mL of water. Add 20 mL of glycerin. Add and dissolve 125 g of chloral hydrate. Add 2 g of iodine crystals. Add 1 g of potassium iodide. Filter through cotton wool.

Hydroxylamine hydrochloride. Anesthetic for certain small metazoans; usually used in a 2% aqueous solution.

Irwin loop. This item is indispensable for sorting, picking up, and transferring microscopic species. It replaces needles, toothpicks, bristles, and fine pipettes.

Isopropyl alcohol. Occasionally useful as an anesthetic, especially for certain protoctistans. Sometimes used in histological work as a substitute for ethyl alcohol.

Jeweler's forceps. These are extremely fine and well-made forceps that are indispensable for dissections. Also called watchmaker's forceps, they are obtainable from jewelers or medical supply houses.

Kahle's fluid. Useful for killing and fixing fragile, small crustaceans. Consists of 11% formalin, 28% alcohol, 2% glacial acetic acid, and 59% water. Drain off after a week, rinse in 70% alcohol, and store the specimens in 70% alcohol.

Lactophenol. A special mounting medium with index of refraction especially favorable for cuticular details. Made up of 20 g of phenol crystals, 40 g of glycerin, 20 g of lactic acid, and 20 mL of distilled water.

Lugol's solution. An excellent fixative and preservative for microzoans, especially for cellular details. Consists of 5 g of iodine, 10 g of potassium iodide, 10 g of glacial acetic acid, and 100 mL of water. Add 1 mL of this solution to each 100 mL of water containing the specimens.

Magnesium sulfate. Occasionally used as an anesthetic for freshwater invertebrates, but more useful for marine forms.

Menthol. An anesthetic occasionally used for freshwater invertebrates. Sprinkle fine crystals on the surface of the water containing the specimens. Depending on the specimen, relaxation takes from a few hours to several days.

Menthol–chloral hydrate narcotic. General-purpose narcotic for small invertebrates. Grind 48 g menthol and 52 g of chloral hydrate with a mortar and pestle until an oily fluid results. Place a few drops of this liquid on the surface of the water containing the animals.

Methyl cellulose. A fluffy white substance, easily soluble in water, and especially useful for slowing the movements of proctistans and platyhelminth worms. Usually sold as 1 or 10% syrupy water solutions. Add a drop or two to water containing organisms on a slide, mix with a needle, and add a coverslip. Keep it away from the end of the microscope objective!

Methylene blue. An excellent intravitam stain for translucent forms. Make up a stock of 0.01 g of methylene blue in 100 mL of absolute alcohol. Add enough of this solution to the water containing the organisms to tinge it a light blue.

Monk's fluid. An aqueous mounting medium that is useful for making permanent slides of copepods and other small crustaceans. It consists of 5 mL of white corn syrup, 5 mL of Certo (fruit pectin), and 3 mL of water. Add a crystal of thymol as a preservative. The mixture sets in a few minutes. Dry to hardness over gentle heat, preferably a hot plate.

Murrayite. A synthetic cement that is dissolved in benzol to a thin, syrupy consistency. Since it is permanently waterproof, alcohol proof, and glycerin proof, it is an excellent secaling cement for museum jars and coverslips. If circular coverslips are ringed on a turntable, several coats should be applied with a camel's hair brush, allowing at least 10 hours between coats.

Neosynephrin hydrochloride. Anesthetic for small metazoans; usually used in a 1% water solution.

Newcomer's fixative. A good general fixative for small invertebrates. Six parts isopropyl alcohol, 3 parts propionic acid, 1 part petroleum ether, 1 part acetone, and 1 part dioxane. Specimens may be stored in this fixative.

Nicotine sulfate. This salt is occasionally used as an anesthetic for small freshwater invertebrates in about 2% solution.

Nitric acid. Concentrated nitric acid is used in cleaning and isolating sponge spicules and gemmules. Two percent nitric acid is useful for fixing turbellarians, and a 1% solution is used for dissociating tissues of hydras.

Novolac. One of several related alcohol-soluble resins suitable for high-quality permanent mounts of a wide variety of small aquatic invertebrates. See Crumpton and Wetzel (1980).

Osmic acid. Really osmium tetroxide, OsO_4, this compound is poisonous and extremely volatile. Used in a 1% solution in distilled water, it is an excellent and rapid fixative for a variety of small metazoans. Prepare the solution by crushing the cleaned, hermetically sealed glass container in an appropriate amount of water and allowing the released osmic acid crystals to dissolve. This compound is easily oxidized by light and dust and should therefore be kept in a blackened dropping bottle. Solutions have much greater permanence if they contain 1% chloroplatinic acid. If specimens are in a drop of water on a slide, fixation is accomplished in a few seconds if the slide is inverted over the mouth of the osmic acid bottle. A few drops of osmic acid are sufficient for fixing specimens in 5 to 10 mL of water. All fixed material should be washed in several changes of water. If the material is colored dark brown or black by the action of the osmic acid, place in commercial hydrogen peroxide until sufficiently bleached. Osmic acid is, unfortunately, very expensive.

Perforated pipe technique. Several types of perforated pipe samplers are in use for taking interstitial hyporheic and phreatic water samples. If the pipe is sturdy and has a pointed end it may be driven into the substrate with a sledgehammer. Otherwise, it is placed in an excavation dug with a shovel, and the hole is back-filled (Fig. A1). If the substrate is saturated, water may be pumped out immediately and passed through a fine net to recover the contained micrometazoans. The pipe may also be left in place for a variable time before any pumping is done. Sampling pipes up to 1.5 in or more long are effective in suitable substrates. The BouRouche sampler is an elaborate version.

Permount. A synthetic mounting medium widely used in histology, useful for permanent preparations of cleared (in xylene) specimens.

Phenol. Melt 400 g of solid phenol by immersing container in hot water; then add 10 mL of water.

Potassium hydroxide. A 2 to 10% solution of potassium hydroxide is often used for clearing opaque chitinous invertebrates. Specimens may be heated gently in the solution for a few minutes or left in it cold for 8 to 48 hours. Specimens should then be neutralized with 10% acetic acid and washed with several changes of water.

Propylene phenoxetol. Same as hydroxyethyl phenyl ether. A useful narcotic for highly contractile aquatic invertebrates. Use only 0.2 mL/L of water. Requires 15 minutes to 2 hours to attain complete relaxation.

Schaudinn's fixative. A general fixative consisting of 2 parts saturated mercuric chloride and 1 part 95% alcohol. Fix for 5 minutes to 3 hours, depending on size and density of the objects. Remove all the salt by washing in several changes of 70% alcohol for a total of 3 to 8 days. Salt removal may be greatly speeded by adding a small amount of tincture of iodine to the alcohol. The action of this fixative is enhanced by the addi-

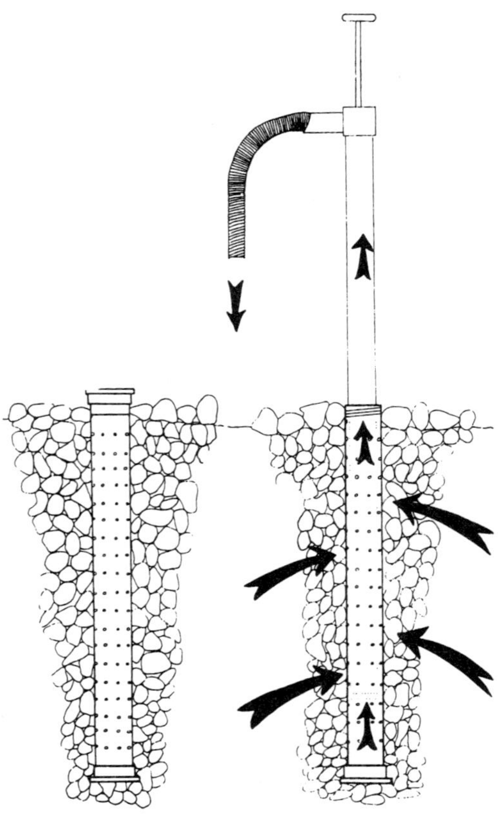

Fig. A.1. Perforated pipe sampler, similar to BouRouch sampler. (From Gledhill 1977.)

tion of sufficient glacial acetic acid to make a 1% solution immediately before using.

Seawater salt. Generally available in quantities at aquarium shops and biological supply houses. Not worth the effort involved in making synthetically from component salts.

Spoon microcompressor. A rotary compressor that permits delicately controlled compression of living microzoans between coverslip and slide. When mounted on a microscope stage, it is designed to gently slow and stop swimming organisms. See construction details in Spoon (1978).

Staining with Gower's carmine. Transfer specimens to water. Add Gower's carmine solution to make 20 to 90% of the total volume. Let stand a half-hour to 24 hours. Wash twice in water. If necessary, destain in 70% alcohol. Transfer to dilute glycerin in readiness for further processing.

Storage jars. Glass is preferred over plastic, because alcohol eventually causes plastic to crack; however, glass jars are expensive. Excellent storage capacity and longevity can be had cost-free with commercial screw-cap glass jars used by the food industry for mayonnaise and related products. These jars come in several sizes but use fairly standardized caps. Plastic inserts cut from coffee can lids provide very good sealing and specimens can be stored with little maintenance for many years (Smith 1998).

Terpineol (synthetic oil of lilac). A good clearing agent, readily miscible with 90% alcohol. Does not make specimens brittle.

Urethane. A good general narcotic; used as a 2% aqueous solution.

Waxed dish. A small, high-sided glass dish (e.g., Stender dish) $\frac{1}{3}$ to $\frac{1}{2}$ filled with wax. Very useful for dissection of worms and other soft-bodied invertebrates. Heat dish gently and add molten wax, allow to cool slowly. Immerse specimen in water and pin out with insect pins.

Zut. Trade name of a cement that is especially useful for tacking or ringing temporary or permanent slides, especially glycerin jelly mounts. Thin with butyl acetate. Apply with a turntable and a small camel's hair brush.

REFERENCES

Boulton, A. 1985. A sampling device that quantitatively collects benthos in flowing or standing waters. *Hydrobiologia* 127:31–40.

Crumpton, W. G., and R. G. Wetzel. 1980. Novalacs: synthetic resins suitable for mounting biological materials. *Trans. Am. Microsc. Soc.* 99:347–348.

Galtsoff, P. W. et al. 1937. *Culture Methods for Invertebrate Animals.* 590 pp. Dover, New York.

Gledhill, T. 1977. Numerical fluctuations of four species of subterranean amphipods during a five-year period. *Crustaceana Suppl.* 4:144–152.

Higgins, R. P., and H. Thiel (eds.). 1988. *Introduction to the Study of Meiofauna.* 487 pp. Smithsonian Institution Press, Washington, DC.

Lawrence, S. G. 1981. Manual for the culture of selected freshwater invertebrates. *Can. Spec. Publ. Fish. Aquatic Sci.* 54:1–169.

Lind, O. T. 1979. *Handbook of common methods in limnology.* 199 pp. Mosby, St. Louis.

Presnell, J. K., and M. P. Schreibman. 1997. *Humason's Animal Tissue Techniques*, 5th ed. 572 pp. Johns Hopkins University Press, Baltimore.

Saur, J. F. T., et al. 1964. Sources of limnological and oceanographic apparatus and supplies, 3rd rev. *Limnol. Oceanogr. Spec. Publ.* 1:i–xxxi.

Schwoerbel, J. 1966. *Methoden der Hydrobiologie.* 207 pp. Franckh'sche Verlagshandlung, Keller, Stuttgart.

Smith, D. G. 1998. A free and effective storage method for biology specimens. *Am. Biol. Teacher* 60:606–607.

Spoon, D. M. 1978. A new rotary microcompressor. *Trans. Am. Microsc. Soc.* 97:412–416.

Sudia, W. D. 1951. A device for rearing animals requiring a flowing water environment. *Ohio J. Sci.* 51:197–202.

Welch, P. S. 1948. *Limnological Methods.* 381 pp. Blakiston, Philadelphia.

INDEX

For subjects such as coloration, ecology, predation, respiration, reproduction—which are characteristic of almost all animals—the reader is advised to see appropriate sections in the introductory sections of each chapter.

Page numbers of figures are italicized; bold numbers indicate taxa in keys; page numbers of tables are followed by the letter t. The plates are inserted following page 326.